ENGINEERING
MATHEMATICS
EXAM PREP

ENGINEERING MATHEMATICS EXAM PREP

Problems and Solutions

A. SAHA, PhD

D. DUTTA, PhD

S. KAR, PhD

P. MAJUMDER, PhD

A. PAUL, PhD

MERCURY LEARNING AND INFORMATION
Boston, Massachusetts

Publisher: David Pallai
MERCURY LEARNING AND INFORMATION
121 High Street, 3rd Floor
Boston, MA 02210
info@merclearning.com
www.merclearning.com
800-232-0223

A.Saha et al. *Engineering Mathematics Exam Prep: Problems and Solutions.*
ISBN: 9781683929109

The publisher recognizes and respects all marks used by companies, manufacturers, and developers as a means to distinguish their products. All brand names and product names mentioned in this book are trademarks or service marks of their respective companies. Any omission or misuse (of any kind) of service marks or trademarks, etc. is not an attempt to infringe on the property of others.

Library of Congress Control Number: 2023939598

212223 321   Printed on acid-free paper in the United States of America.

Our titles are available for adoption, license, or bulk purchase by institutions, corporations, etc. For additional information, please contact the Customer Service Dept. at 800-232-0223(toll free).

All of our titles are available in digital format at *academiccourseware.com* and other digital vendors. The sole obligation of MERCURY LEARNING AND INFORMATION to the purchaser is to replace the book, based on defective materials or faulty workmanship, but not based on the operation or functionality of the product.

CONTENTS

CHAPTER 1: LINEAR ALGEBRA   1

  1.1 MATRICES AND THEIR TYPES   1

    1.1.1 Definition of a Matrix   1

    1.1.2 Types of Matrices   1

  1.2 ALGEBRA OF MATRICES   2

    1.2.1 Negative, Sum, and Differences of Matrices   2

    1.2.2 Multiplication of a Matrix by a Scalar   3

    1.2.3 Transpose of a Matrix   3

    1.2.4 Multiplication of Matrices (Product of Matrices)   3

  1.3 DETERMINANT OF A SQUARE MATRIX   4

    1.3.1 Definition of Determinant   4

    1.3.2 Properties of a Determinant   4

    1.3.3 Minors and Cofactors   6

  1.4 ADJOINT AND INVERSE OF A MATRIX   6

    1.4.1 Adjoint of a Matrix   6

    1.4.2 Inverse of a Matrix   7

  1.5 VARIOUS TYPES OF REAL SQUARE MATRICES   7

    1.5.1 Symmetric Matrix   7

    1.5.2 Skew-Symmetric Matrix   7

    1.5.3 Orthogonal Matrix   8

    1.5.4 Idempotent Matrix   8

    1.5.5 Involutary Matrix   8

    1.5.6 Nilpotent Matrix   8

  1.6 COMPLEX MATRICES AND THEIR TYPES   8

    1.6.1 Complex Conjugate of a Matrix   8

    1.6.2 Transposed Conjugate of a Matrix   8

    1.6.3 Unitary Matrix   9

    1.6.4 Hermitian Matrix   9

    1.6.5 Skew-Hermitian Matrix   9

1.7 RANK OF A MATRIX	9
1.7.1 Elementary Transformations	9
1.7.2 Equivalent Matrices	10
1.7.3 Rank of a Matrix	10
1.7.4 Determination of the Rank of a Matrix	10
1.8 SYSTEM OF LINEAR EQUATIONS AND THEIR SOLUTIONS	11
1.8.1 Introduction	11
1.8.2 Methods for Solving Non-Homogeneous System of Linear Equations	11
1.8.2.1 Cramer's Rule	11
1.8.2.2 Matrix Method	13
1.8.2.3 Rank Method	13
1.8.3 Homogeneous System of Linear Equations	15
1.9 EIGENVALUES AND EIGENVECTORS	15
1.9.1 Characteristic Roots (Eigenvalues) of a Matrix	15
1.9.2. Trace of a Matrix	16
1.9.3. Eigenvectors or Characteristic Vectors	16
1.10 Vectors	17
1.10.1 Introduction	17
1.10.2 Linear Dependence and Linear Independence	17
1.10.3 Inner Product and Norm of Vectors	18
1.10.4 Orthogonal and Orthonormal Vectors	19
1.10.5 Basis and Dimension	19
Fully Solved MCQs (Level-I)	19
Answer Key	22
Explanation	22
Fully Solved MCQs (Level-II)	28
Answer Key	34
Explanation	34
Previous Years Solved Papers (2000-2018)	52
Answer Key	71
Answer Key	71
Explanation	71
Questions for Practice	102
Answer Key	108
Hints	108
CHAPTER 2: CALCULUS	**113**
2.1 FUNCTIONS AND LIMITS	113
2.1.1 Definition of a Function	113
2.1.2 Some Special Functions	113
2.1.3 Introduction to Limits	114

2.1.4 Definition of Limit 114

2.1.5 Fundamental Theorems on Limits 114

2.1.6 Fundamental Formulas on Limits 115

2.1.7 The Sandwich Theorem 115

2.1.8 Infinite Limits 115

2.1.9 Limits at Infinity 116

2.1.10 Infinite Limits at Infinity 116

2.2 CONTINUITY AND DIFFERENTIABILITY 116

2.2.1 Continuity 116

2.2.2 Discontinuity 117

2.2.3 Derivative 117

2.2.4 Computation of Derivatives 118

2.3 INDETERMINATE FORMS 118

2.3.1 Introduction 118

2.3.2 The L'Hospital Rule 118

2.4 MEAN VALUE THEOREMS 119

2.4.1 Rolle's Theorem 119

2.4.2 Lagrange's Mean Value Theorem 119

2.4.3 Cauchy's Mean Value Theorem 119

2.5 INCREASING AND DECREASING FUNCTIONS 119

2.6 MAXIMA AND MINIMA OF FUNCTIONS OF A SINGLE VARIABLE 120

2.6.1 First Derivative Test 120

2.6.2 Second Derivative Test 120

2.6.3 Higher Order Derivative Test 120

2.7 INFINITE SERIES AND EXPANSION OF FUNCTIONS 121

2.7.1 Infinite Series 121

2.7.2 Test for Convergence of Infinite Series 121

2.7.3 Taylor's Theorem With Lagrange's Form of Remainder 123

2.7.4 The Taylor Series 123

2.7.5 Maclaurin's Series 123

2.8 INDEFINITE AND DEFINITE INTEGRALS 123

2.8.1 Indefinite Integral 123

2.8.2 Fundamental Formulas of Indefinite Integral 123

2.8.3 Advanced Formulas of Indefinite Integrals 124

2.8.4 Definite Integral 124

2.8.5 Properties of Definite Integral 125

2.8.6 Definite Integral as a Limit of Sum 125

2.8.7 Differentiation Under the Sign of Integration 125

2.9 IMPROPER INTEGRALS, BETA, AND GAMMA FUNCTIONS 126

2.9.1 Improper Integral 126

2.9.2 Evaluation of Improper Integrals 126

2.9.3   Beta Function   127

2.9.4   Gamma Function   127

2.10   FUNCTIONS OF SEVERAL VARIABLES AND PARTIAL DERIVATIVES   128

2.10.1   Functions of Two Variables   128

2.10.2   Limit of Functions of Two Variables   128

2.10.3   Continuity of Functions of Two Variables   128

2.10.4   Partial Derivatives   129

2.10.5   Homogeneous Function   129

2.10.6   Euler's Theorem   129

2.10.7   Total Differential and Total Derivative   129

2.10.8   Jacobian   130

2.11   MAXIMA AND MINIMA OF FUNCTIONS OF TWO VARIABLES   131

2.11.1   Introduction   131

2.11.2   Working Rule to Find the Maximum and Minimum Values of $f(x, y)$ 131

2.11.3   Lagrange's Method for Undetermined Multipliers   131

2.12   CHANGE OF ORDER OF INTEGRATION   132

2.13   DOUBLE AND TRIPLE INTEGRALS   133

2.13.1   Double Integrals   133

2.13.2   Triple Integrals   133

2.14   ARC LENGTH OF A CURVE   134

2.15   VOLUMES OF SOLIDS OF REVOLUTION   134

2.15.1   Working Formulas   134

2.16   SURFACE AREAS OF SOLIDS OF REVOLUTION   135

Fully Solved MCQs (Level-I)   135

Answer Key   141

Explanation   141

Fully Solved MCQs (Level-II)   157

Answer Key   164

Explanation   164

Previous Years Solved Papers (2000-2018)   191

Answer Key   208

Explanation   209

Questions for Practice   241

Answer Key   246

Explanation   246

CHAPTER 3: VECTORS   **249**

3.1   BASIC CONCEPTS   249

3.1.1   Scalars and Vectors   249

3.1.2   Position Vector   249

3.1.3   Equal Vectors   249

3.1.4   Negative of a Vector   249

3.1.5   Unit Vectors   249
3.1.6   Sum and Difference of Two Vectors   249
3.1.7   Triangle Law of Addition   250
3.1.8   Product of a Vector with a Scalar   250
3.1.9   Collinear Vectors   250
3.1.10 Coplanar Vectors   250
3.1.11 Section Formula   250
3.2 PRODUCT OF VECTORS   250
3.2.1   Scalar Product (Dot Product)   250
3.2.2   Vector Product (Cross Product)   250
3.2.3   Scalar Triple Product   251
3.2.4   Vector Triple Product   251
3.3 VECTOR DIFFERENTIATION AND INTEGRATION   251
3.3.1   Derivative of a Vector Function   251
3.3.2   General Rules for Vector Differentiation   251
3.3.3   Velocity and Acceleration   252
3.3.4   Vector Integration   252
3.4 GRADIENT, DIVERGENCE AND CURL   252
3.4.1   Del Operator   252
3.4.2   Gradient of a Scalar Point Function   252
3.4.3   Divergence of a Vector Point Function   252
3.4.4   Curl of a Vector Point Function   252
3.4.5   Vector Identities   253
3.4.6   Directional Derivative   253
3.5 LINE, SURFACE, AND VOLUME INTEGRALS   253
3.5.1   Line Integral   253
3.5.2   Surface Integral   254
3.5.3   Volume Integral   255
3.6 GREEN'S, STOKES', AND GAUSS DIVERGENCE THEOREM   255
3.6.1   Greens Theorem (in a Plane)   255
3.6.2   Stokes' Theorem   256
3.6.3   Gauss Divergence Theorem   256
Fully Solved MCQs   257
Answer Key   260
Explanation   260
Previous Years Solved Papers (2000-2018)   269
Answer Key   276
Explanation   277
Questions for Practice   290
Answer Key   292

CHAPTER 4: ORDINARY DIFFERENTIAL EQUATIONS   **293**

  4.1 BASIC CONCEPTS   293

    4.1.1 Definition of a Differential Equation   293

    4.1.2 Classification of Differential Equations   293

    4.1.3 Order of a Differential Equation   293

    4.1.4 Degree of a Differential Equation   294

    4.1.5 Formation of a Differential Equation   294

    4.1.6 Solution of a Differential Equation   294

  4.2 LINEARLY DEPENDENT AND LINEARLY INDEPENDENT SOLUTIONS   295

    4.2.1 Wronskian   295

    4.2.2 Linearly Dependent Solutions   295

    4.2.3 Linearly Independent Solutions   295

  4.3 DIFFERENTIAL EQUATIONS OF 1ST ORDER AND 1ST DEGREE   296

    4.3.1 General Form   296

    4.3.2 Solution by Separation of Variables   296

    4.3.3 Homogeneous Differential Equation   296

    4.3.4 Exact Differential Equations   297

    4.3.5 Linear Differential Equations   298

  4.4 LINEAR DIFFERENTIAL EQUATIONS OF 2ND ORDER   299

    4.4.1 General Form   299

    4.4.2 Complementary Function (C.F)   299

    4.4.3 Particular Integral (P.I)   300

    4.4.4 Complete (General) Solution   301

    4.4.5 Homogeneous Linear Differential Equations of Order Two   301

  Fully Solved MCQs   302

  Answer Key   304

  Explanations   304

  Fully Solved MCQs   310

  Answer Key   311

  Explanations   311

  Previous Years Questions (2000-18)   318

  Answer Key   327

  Explanations   327

  Questions for Practice   344

  Answer Key   346

  Hints   347

CHAPTER 5: PARTIAL DIFFERENTIAL EQUATIONS   **349**

  5.1 BASIC CONCEPTS   349

    5.1.1 Introduction   349

    5.1.2 Order and Degree   349

5.1.3 Linear and No-Linear Partial Differential Equations   349

5.1.4 Formation of Partial Differential Equations   350

5.2 CLASSIFICATION OF 2ND ORDER PARTIAL DIFFERENTIAL EQUATION   350

5.3 HEAT, WAVE, AND LAPLACE EQUATIONS   350

5.3.1 Solution by Separation of Variables   350

5.3.2 One-Dimensional Heat (Diffusion) Equation and Its Solution   350

5.3.3 One-Dimensional Wave Equation and Its Solution   351

5.3.4 The Laplace Equation and Its Solution   351

Fully Solved MCQs   351

Answer Key   352

Explanation   352

Fully Solved MCQs   353

Answer Key   354

Explanation   354

Previous Years Solved Papers (2000-2018)   357

Answer Key   358

Explanation   358

Questions for Practice   359

Answer Key   360

CHAPTER 6: LAPLACE TRANSFORMS   **361**

6.1 BASICS OF LAPLACE TRANSFORMS   361

6.1.1 Definition of the Laplace Transform   361

6.1.2 Linear Property of the Laplace Transform   361

6.1.3 Fundamental Formulas of the Laplace Transform   361

6.1.4 First Shifting Theorem   362

6.1.5 Some Advanced Formulas of the Laplace Transform   362

6.1.6 Change of Scale Property   362

6.2 LAPLACE TRANSFORM ON DERIVATIVES   362

6.3 LAPLACE TRANSFORM ON INTEGRALS   363

6.4 LAPLACE TRANSFORM ON PERIODIC FUNCTIONS   363

6.5 EVALUATION OF INTEGRALS USING LAPLACE TRANSFORMS   363

6.6 INITIAL AND FINAL VALUE THEOREMS   363

6.6.1 Initial Value Theorem   363

6.6.2 Final Value Theorem   363

6.7 FUNDAMENTALS OF INVERSE LAPLACE TRANSFORM   363

6.7.1 Definition of Inverse Laplace Transform   363

6.7.2 Useful Formulas on Inverse Laplace Transforms   363

6.8 IMPORTANT THEOREMS ON INVERSE LAPLACE TRANSFORMS   364

6.9 UNIT STEP FUNCTION AND UNIT IMPULSE FUNCTION   364

     6.9.1 Unit Step Function   364

     6.9.2 Second Shifting Theorem   365

     6.9.3 Unit Impulse Function   365

6.10 SOLVING ORDINARY DIFFERENTIAL EQUATIONS   365

Fully Solved MCQs (Level-I)   365

Answer Key   368

Explanation   368

Fully Solved MCQs (Level-II)   374

Answers Key   377

Explanation   377

Previous Years Questions (2000-2018)   385

Answers Key   389

Explanations   390

Questions for Practice   395

Answers Key   398

Explanation   398

CHAPTER 7: NUMERICAL ANALYSIS   **401**

7.1 ERRORS AND APPROXIMATIONS   401

     7.1.1 Rounding Off   401

     7.1.2 Errors and their Computation   401

7.2 CALCULUS OF FINITE DIFFERENCES   401

     7.2.1 Forward Difference Operator   401

     7.2.2 Backward Difference Operator   402

     7.2.3 Shift Operator   402

7.3 INTERPOLATION   402

     7.3.1 Newton's Forward Difference Interpolation Formula   402

     7.3.2 Newton's Backward Difference Interpolation Formula   402

     7.3.3 Lagrange's Interpolation Formula   403

     7.3.4 Error in Interpolation   403

7.4 NUMERICAL DIFFERENTIATION   403

     7.4.1 Differentiation Formula Based on Newton's Forward Difference Formula   403

     7.4.2 Differentiation Formula Based on Newton's Backward Difference Formula   403

7.5 NUMERICAL INTEGRATION   403

     7.5.1 Trapezoidal Rule   403

     7.5.2 Simpson's 1/3rd Rule   403

     7.5.3 Weddle's Rule   404

     7.5.4 Simpson's 3/8th's Rule   404

7.6 SYSTEM OF LINEAR ALGEBRAIC EQUATIONS   404

     7.6.1 Gauss Elimination Method   404

     7.6.2 LU Decomposition Method   404

     7.6.3 Gauss–Seidel Iteration Method   405

7.7 SOLUTION OF ALGEBRAIC AND TRANSCENDAL EQUATIONS — 405

    7.7.1 Method of Bisection — 405

    7.7.2 Regula Falsi Method — 405

    7.7.3 Newton–Raphson Method — 405

7.8 NUMERICAL SOLUTION OF ORDINARY DIFFERENTIAL EQUATIONS — 405

    7.8.1 Euler's Method — 405

    7.8.2 Modified Euler's Method — 405

    7.8.3 Runge–Kutta Method — 406

        I. Second-Order Runge–Kutta Method — 406

        II. Fourth-Order Runge–Kutta Method — 406

    7.8.4 Predictor-Corrector Method — 406

Fully Solved MCQs — 406

Answer Key — 408

Explanation — 408

Previous Years Solved Papers (2000-2018) — 412

Answer Key — 419

Explanation — 419

Questions for Practice — 430

Answer Key — 432

CHAPTER 8: COMPLEX ANALYSIS — **433**

8.1 BASICS OF COMPLEX ANALYSIS — 433

    8.1.1 Complex Number — 433

    8.1.2 Modulus and Amplitude of a Complex Number — 433

    8.1.3 Conjugate of a Complex Number — 435

    8.1.4 Properties of Modulus, Argument, and Conjugate — 435

    8.1.5 Sum, Difference, and Product of Two Complex Numbers — 435

    8.1.6 Cube Roots of Unity — 435

    8.1.7 De Moivre's Theorem — 435

    8.1.8 Hyperbolic Functions — 436

    8.1.9 Logarithm of a Complex Number — 436

8.2 CALCULUS OF COMPLEX VALUED FUNCTIONS — 436

    8.2.1 Function of a Complex Variable — 436

    8.2.2 Limit of a Complex Valued Function — 436

    8.2.3 Continuity of a Complex Valued Function — 437

    8.2.4 Derivative of a Complex Valued Function — 437

    8.2.5 Analytic Function — 437

    8.2.6 Cauchy Riemann Equations — 437

    8.2.7 Conjugate Function — 437

    8.2.8 Harmonic Function — 437

    8.2.9 Construction of an Analytic Function (by Milne Thomson's method) — 438

    8.2.10 Construction of Harmonic Conjugate — 438

8.3 COMPLEX INTEGRATION	438
8.3.1 Curves	438
8.3.2 Complex Line Integral	439
8.3.3 Cauchy-Goursat Theorem	439
8.3.4 Cauchy's Integral Formula	439
8.3.5 Cauchy's Integral Formula on Higher Order Derivatives	440
8.4 TAYLOR AND LAURENT SERIES	440
8.4.1 The Taylor Series	440
8.4.2 The Laurent Series	440
8.5 SINGULARITIES	440
8.5.1 Singular Point	440
8.5.2 Types of Singularities	441
8.5.2.1 Isolated singularity	441
8.5.2.2 Removable singularity	441
8.5.2.3 Essential singularity	441
8.5.3 Zeros and Poles	441
8.6 RESIDUES	442
8.6.1 Residue at a Simple Pole	442
8.6.2 Residue at a Pole of Order "n"	442
8.6.3 Residue at Infinity	443
8.6.4 Cauchy's Residue Theorem	443
Fully Solved MCQ's	443
Answer Key	446
Explanation	446
Fully Solved MCQ's (Level-II)	452
Answer Key	455
Explanation	456
Previous Years Solved Papers (2000-2018)	466
Answer Key	474
Explanation	474
Questions for Practice	490
Answer Key	493
Explanation	493
CHAPTER 9: PROBABILITY AND STATISTICS	**495**
9.1 BASICS OF PROBABILITY	495
9.1.1 Experiment	495
9.1.2 Random Experiment	495
9.1.3 Sample Space (Event Space)	495
9.1.4 Event	496

9.1.5 Equally Likely Events 496

9.1.6 Mutually Exclusive Events 496

9.1.7 Mutually Exhaustive Events 496

9.1.8 Classical Definition Of Probability 496

9.1.9 Independent Events 497

9.2 CONDITIONAL PROBABILITY AND BAYES' THEOREM 497

9.2.1 Conditional Probability 497

9.2.2 Theorem on Total Probability 498

9.2.3 Bayes' Theorem 498

9.3 RANDOM VARIABLE AND PROBABILITY DISTRIBUTION 498

9.3.1 Random Variable 498

9.3.2 Types of Random Variable 498

9.3.3 Probability Mass Function (P.M.F) 498

9.3.4 Probability Distribution Function 499

9.3.5 Expectation or Mean 500

9.3.6 Variance and Standard Deviation 500

9.4 SPECIAL TYPES OF PROBABILITY DISTRIBUTIONS 501

9.4.1 Binomial Distribution 501

9.4.2 Poisson Distribution 501

9.4.3 Normal Distribution 501

9.4.4 Geometric Distribution 502

9.4.5 Uniform (Rectangular) Distribution 502

9.4.6 Gamma Distribution 502

9.4.7 Exponential Distribution 502

9.5 INTRODUCTION OF STATISTICS 502

9.5.1 Statistics 502

9.5.2 Scopes and limitations of Statistics 502

9.5.3 Frequency Distribution 503

9.5.4 Mean (Arithmetic Mean) 503

9.5.5 Median 503

9.5.6 Mode 504

9.5.7 Standard Deviation (S.D) 504

9.5.8 Correlation 504

9.5.9 Regression 504

Fully Solved MCQs 505

Answers Key 506

Explanations 506

Fully Solved MCQs 510

Answers Key 512

Explanations 512

Previous Years Solved Papers (2000-2018) 517

Answer Key   533
Explanation   533
Questions For Practice   558
Answer Key   562
Explanations   562

CHAPTER 10: FOURIER SERIES   **565**
  10.1 BASICS OF THE FOURIER SERIES   565
    10.1.1 Definition of the Fourier Series   565
    10.1.2 Dirichlet's Condition   565
  10.2 FOURIER SERIES OF EVEN AND ODD FUNCTIONS   566
    10.2.1 Fourier Series of Even Function   566
    10.2.2 Fourier Series for Odd Function   566
  10.3 HALF RANGE FOURIER SERIES   566
    10.3.1 Half Range Sine Series   566
    10.3.2 Fourier Cosine Series   566
  Fully Solved MCQs   566
  Answer Key   567
  Explanation   567
  Previous Years Solved Papers (2000-2018)   569
  Answer Key   570
  Explanation   570
  Questions for Practice   571
  Answer Key   572
  Explanation   572

CHAPTER 11: GRAPH THEORY   **573**
  11.1 GRAPHS   573
    11.1.1 Definition of a Graph   573
    11.1.2 Incidence   573
    11.1.3 Loops   574
    11.1.4 Parallel Edges   574
    11.1.5 Degree of a Vertex   574
    11.1.6 Directed Graph   575
    11.1.7 In Degree and Out Degree   575
    11.1.8 Minimum Degree and Maximum Degree   575
  11.2 DIFFERENT TYPES OF GRAPHS   576
    11.2.1 Mixed Graph   576
    11.2.2 Multi Graph   576
    11.2.3 Simple Graph   576
    11.2.4 Trivial Graph   576
    11.2.5 Null Graph   576
    11.2.6 K-regular Graph   576

11.2.7 Complete Graph   577

11.2.8 Bipartite Graph   577

11.2.9 Complete Bipartite Graph   577

11.3 WALK AND PATH   577

11.3.1 Walk   577

11.3.2 Path and Circuit   577

11.4 MATRIX REPRESENTATION OF GRAPHS   578

11.4.1 Adjacent Matrix   578

11.4.2 Incidence Matrix   579

11.4.3 Path Matrix   579

11.5 PLANAR GRAPHS AND EULER'S FORMULA   579

11.5.1 Planar Graph   579

11.5.2 Euler's Formula   579

11.6 SUB GRAPHS AND ISOMORPHIC GRAPHS   580

11.6.1 Subgraphs   580

11.6.2 Isomorphic Graph   580

11.7 CONNECTEDNESS   581

11.7.1 Connected Graph   581

11.7.2 Strongly Connected Graph   581

11.7.3 Weakly Connected Graph   581

11.7.4 Component   581

11.7.5 Eulerian Graph   581

11.7.6 Hamiltonian Graph   582

11.8 VERTEX AND EDGE CONNECTIVITY   582

11.8.1 Cut Vertex   582

11.8.2 Cut Edge (Bridge)   582

11.8.3 Cut Set   583

11.8.4 Edge Connectivity   583

11.8.5 Vertex Connectivity   583

11.9 GRAPH COLORING, MATCHING, AND COVERING   583

11.9.1 Vertex Coloring   583

11.9.2 Chromatic Number   584

11.9.3 Matching   584

11.9.4 Covering   584

11.10 TREE   584

11.10.1 Definition   584

11.10.2 Spanning Tree   585

Construction of Spanning Trees   585

(I) BFS (Breath First Search) Algorithm   585

(II) DFS (Depth First Search) Algorithm   586

11.10.3 Minimal Spanning Tree   586

    (I) Prim's Algorithm   586

    (II) Kruskal's Algorithm   587

11.10.4 Binary Tree   588

11.10.5 Rooted Tree   588

11.10.6 Traversal of a Tree   588

Fully Solved MCQs   589

Answer Key   592

Explanation   592

Previous Years Solved Paper (2000-2018)   595

Answer Key   599

Explanation   599

Questions for Practice   603

Answer Key   606

Hints   606

APPENDIX A: GATE 2019 SOLVED PAPERS   **607**

APPENDIX B: GATE 2020 SOLVED PAPERS   **623**

CHAPTER 1

LINEAR ALGEBRA

1.1 MATRICES AND THEIR TYPES

1.1.1 Definition of a Matrix

A set of mn numbers (real or complex) arranged in "m" rows and "n" columns is called a rectangular matrix or simply a matrix of order "m by n" (denoted by $m \times n$). Thus,

$$A = \begin{pmatrix} a_{11} & a_{12} & a_{13} & .... & a_{1n} \\ a_{21} & a_{22} & a_{23} & ... & a_{2n} \\ ... & ... & ... & ... & ... \\ ... & ... & ... & ... & ... \\ a_{m1} & a_{m2} & a_{m3} & ... & a_{mn} \end{pmatrix} \qquad ...(i)$$

is a matrix of order $m \times n$.

The numbers $a_{11}, a_{12},...a_{mn}$ are called elements of the matrix, and a_{ij} is the element of the matrix lying in the ith row and j^{th} column. The matrix given in (i) can also be represented by

$$A = [a_{ij}]_{m \times n} \quad \text{or} \quad \text{by } A = (a_{ij})_{m \times n}$$

Examples:

$$A = \begin{pmatrix} 0 & 3 & 5 \\ 1 & 4 & 1 \end{pmatrix} \text{ is a matrix of order } 2 \times 3,$$

$$A = \begin{pmatrix} 4 & 5 \\ 3 & 6 \\ 2 & 1 \end{pmatrix} \text{ is a matrix of order } 3 \times 2.$$

The rows of a matrix are denoted by $R1$ (first row), $R2$ (second row) and so on. The columns of a matrix are denoted by $C1$ (first column), C_2 (second column) and so on.

Remember:

Two matrices $A = [a_{ij}]_{m \times n}$ and $B = [b_{ij}]_{p \times q}$ are said to be equal if $m = p$, $n = q$ (i.e., they have same number of rows and columns) and the elements lying in the corresponding places of the two matrices are the same.

Thus, for
$$A = \begin{pmatrix} x & y \\ 0 & 2 \end{pmatrix}, B = \begin{pmatrix} 3 & 5 \\ a & b \end{pmatrix};$$

$$A = B \quad \Leftrightarrow \quad x = 3, y = 5, a = 0, b = 2.$$

1.1.2 Types of Matrices

(a) Row matrix

A matrix with only one row is called a row matrix.

Example:

$[2\ 5\ 1]$ is a row matrix (order of this matrix is 1×3).

(b) Column matrix

A matrix with only one column is called a column matrix.

Example:
$$\begin{bmatrix} 1 \\ 2 \\ 4 \\ 5 \end{bmatrix} \text{ is a column matrix}$$

(order of this matrix is 4×1)

(c) Square matrix

A matrix that has equal number of rows and columns is called a square matrix.

The elements a_{ij} of the square matrix $A = [a_{ij}]_{n \times n}$ for which $i = j$ are called diagonal elements and the line formed by the diagonal elements is called the principal diagonal of the matrix.

If matrix A has "n" rows and "n" columns, then we say that A is a square matrix of order n. Then A can be represented by

$$\begin{pmatrix} a_{11} & a_{12} & a_{13} & .... & a_{1n} \\ a_{21} & a_{22} & a_{23} & .... & a_{2n} \\ .... & .... & .... & .... & .... \\ .... & .... & .... & .... & .... \\ a_{n1} & a_{n2} & a_{n3} & .... & a_{nn} \end{pmatrix}$$

where $a_{11}, a_{22}, ..., a_{nn}$ are diagonal elements.

Example:

1, 2, 4 are diagonal elements of the matrix
$$\begin{pmatrix} 1 & 0 & 5 \\ 3 & 2 & 1 \\ -1 & 9 & 4 \end{pmatrix}$$

(d) Diagonal matrix

A square matrix whose non-diagonal elements are all zero is called a diagonal matrix.

Example:

$$\begin{pmatrix} 4 & 0 & 0 \\ 0 & 7 & 0 \\ 0 & 0 & -1 \end{pmatrix}$$ is a diagonal matrix of order 3.

The above diagonal matrix can also be written as diag $(4, 7, -1)$ or diag $[-1, 7, 4]$.

(e) Scalar matrix

If all the diagonal elements of a diagonal matrix are equal, then the matrix is called a scalar matrix.

Example:

$$\begin{pmatrix} 3 & 0 & 0 \\ 0 & 3 & 0 \\ 0 & 0 & 3 \end{pmatrix}$$ is a scalar matrix of order 3.

(f) Identity matrix

A diagonal matrix whose diagonal elements are all "1," is called an identity matrix. The identity matrix of order n is denoted by I_n or $I_{n \times n}$ or by I.

Example:

$$I_1 = [1]_{1 \times 1}, I_2 = \begin{pmatrix} 1 & 0 \\ 0 & 1 \end{pmatrix}_{2 \times 2}, \quad I_3 = \begin{pmatrix} 1 & 0 & 0 \\ 0 & 1 & 0 \\ 0 & 0 & 1 \end{pmatrix}_{3 \times 3}, \text{ etc.}$$

(g) Null matrix

If all the elements of a matrix are zero, then it is called a null matrix or zero matrix. The null matrix of order $m \times n$ is denoted by $O_{m \times n}$.

Example:

$$O_{2 \times 2} = \begin{bmatrix} 0 & 0 \\ 0 & 0 \end{bmatrix}, O_{3 \times 2} = \begin{bmatrix} 0 & 0 \\ 0 & 0 \\ 0 & 0 \end{bmatrix},$$

etc. are null matrices.

For the sake of simplicity sometimes, we denote the null matrix by "O."

(h) Upper triangular matrix

A square matrix $A = [a_{ij}]_{n \times n}$ is called an upper triangular matrix if $a_{ij} = 0 \ \forall \ i > j$.

Thus, for any upper triangular matrix, all the elements lying below the principal diagonal are zero.

Example:

$$\begin{pmatrix} 1 & -1 \\ 0 & 2 \end{pmatrix}, \begin{pmatrix} 3 & 5 & 6 \\ 0 & -1 & 1 \\ 0 & 0 & 2 \end{pmatrix} \begin{pmatrix} 0 & 2 & 3 \\ 0 & -1 & 1 \\ 0 & 0 & 0 \end{pmatrix},$$

etc. are all upper triangular matrices.

(i) Lower triangular matrix:

A square matrix $A = [a_{ij}]_{n \times n}$ is called an lower triangular matrix if $a_{ij} = 0 \ \forall \ i < j$.

Thus, for any lower triangular matrix, all the elements lying above the principal diagonal are zero.

Example:

$$\begin{pmatrix} 3 & 0 \\ 1 & 7 \end{pmatrix}, \begin{pmatrix} 4 & 0 & 0 \\ 2 & 6 & 0 \\ 1 & 0 & 0 \end{pmatrix}, \begin{pmatrix} 4 & 0 & 0 \\ 5 & 0 & 0 \\ 1 & 0 & 0 \end{pmatrix},$$

etc. are all lower triangular matrices.

1.2 ALGEBRA OF MATRICES

1.2.1 Negative, Sum, and Differences of Matrices

(i) The negative of a matrix $A = [a_{ij}]_{m \times n}$ is the matrix $[-a_{ij}]_{m \times n}$ and is denoted by '$-A$'.

Example:

If $A = \begin{pmatrix} 1 & 2 & 5 \\ 6 & 3 & 4 \end{pmatrix}$, then $-A = \begin{pmatrix} -1 & -2 & -5 \\ -6 & -3 & -4 \end{pmatrix}$

(ii) The sum of two matrices $A = [a_{ij}]_{m \times n}$ and $B = [b_{ij}]_{m \times n}$ is denoted by $A + B$ and defined by $A + B = [a_{ij} + b_{ij}]_{m \times n}$.

Thus, the sum of two matrices A and B of same order is the matrix obtained by adding the corresponding elements of the two matrices A and B.

Example:

If $A = \begin{pmatrix} 4 & 5 & 6 \\ 2 & -1 & 0 \end{pmatrix}$ and $B = \begin{pmatrix} 7 & 0 & 1 \\ 2 & 1 & 9 \end{pmatrix}$

Then $A + B = \begin{pmatrix} 11 & 5 & 7 \\ 4 & 0 & 9 \end{pmatrix}$

(iii) The difference of two matrices $A = [a_{ij}]_{m \times n}$ and $B = [b_{ij}]_{m \times n}$ is denoted by $A - B$ and is defined to be the sum $A + (-B)$.

Thus, $A - B = A + (-B) = [a_{ij} - b_{ij}]_{m \times n}$.

Example:

For $A = \begin{pmatrix} 4 & 5 \\ 2 & 1 \end{pmatrix}$, $B = \begin{pmatrix} -1 & 0 \\ 0 & 3 \end{pmatrix}$;

$A - B = \begin{pmatrix} 5 & 5 \\ 2 & -2 \end{pmatrix}$ and $B - A = \begin{pmatrix} -5 & -5 \\ -2 & 2 \end{pmatrix}$

Remember: $A - B \neq B - A$ (in general)

Properties of addition of matrices

If $A = [a_{ij}]_{m \times n}$, $B = [b_{ij}]_{m \times n}$, $C = [c_{ij}]_{m \times n}$, are three matrices of same order, then

(i) $A + B = B + A$ (Commutative property)

(ii) $A + (B + C) = (A + B) + C$ (Associative property)

(iii) $A + O = O + A$, where O is the null matrix of order $m \times n$.

(iv) $A + (-A) = (-A) + A = O$

where "$-A$" is called the additive inverse of A w.r.t. the matrix addition.

1.2.2 Multiplication of a Matrix by a Scalar

Let $A = [a_{ij}]_{m \times n}$ be any matrix and k be any scalar quantity. Then the product kA is the matrix of order $m \times n$ defined by $kA = [ka_{ij}]_{m \times n}$.

Example:

If $= 2$, $A = \begin{pmatrix} 4 & 5 \\ 6 & 7 \end{pmatrix}$, then $kA = \begin{pmatrix} 8 & 10 \\ 12 & 14 \end{pmatrix}$.

Properties:

If $A = [a_{ij}]_{m \times n}$ and $B = [b_{ij}]_{m \times n}$ be two matrices and k_1, k_2 be scalars, then

(i) $k_1 (A \pm B) = k_1 A \pm k_1 B$

(ii) $(k_1 \pm k_2)A = k_1 A \pm k_2 A$

(iii) $k_1(k_2 A) = (k_1 k_2)A$

(iv) $-k_1 A = k_1(-A)$

1.2.3 Transpose of a Matrix

The transpose of a matrix A is the matrix obtained from A by changing its rows into columns and columns into rows. The transpose of a matrix A is denoted by A' or A^t or A^T. Thus, if A is matrix of order $m \times n$, then A^T is of order $n \times m$.

Example:

If $A = \begin{pmatrix} 1 & 2 & 3 \\ 4 & 5 & 6 \end{pmatrix}$, then $A^T = \begin{pmatrix} 1 & 4 \\ 2 & 5 \\ 3 & 6 \end{pmatrix}$

Properties of transpose of a matrix

If $A = [a_{ij}]_{m \times n}$ and $B = [b_{ij}]_{m \times n}$ are two matrices of same order, then

(i) $(A^T)^T = A$

(ii) $(A \pm B)^T = A^T \pm B^T$

(iii) $(\lambda A)^T = \lambda A^T$, where λ is any number (constant).

(iv) $(\lambda A \pm \mu B)^T = \lambda A^T \pm \mu B^T$, where λ, and μ are scalars.

(v) $(AB)^T = B^T A^T$ provided AB is defined.

Remember:

If A is a square matrix, then

$$A = \frac{1}{2}(A + A^T) + \frac{1}{2}(A - A^T), \text{ where } \frac{1}{2}(A + A^T)$$

is symmetric and $\frac{1}{2}(A - A^T)$ is skew-symmetric. Thus, every square matrix can be expressed as a sum of a symmetric and skew-symmetric matrix.

1.2.4 Multiplication of Matrices (Product of Matrices)

If the number of columns of a matrix A is equal to the number of rows of another matrix B, then we say that the product AB is defined.

If $A = [a_{ij}]_{m \times n}$, $B = [b_{jk}]_{n \times p}$, then $AB = [c_{ik}]_{m \times p}$

where $c_{ik} = \sum_{j=1}^{n} a_{ij} b_{jk}$, where the suffixes i, j, k ranges

from 1 to m, 1 to n, 1 to p, respectively.

In other words, if A be a matrix of order $m{\times}n$ and B be a matrix of order $n \times p$, then the product AB is a matrix of order $m{\times}p$ and the element lying in the ith row and k^{th} column in AB is the sum of the products of the elements of the ith row of A and the corresponding elements of the kth column of B. Sometimes we denote the matrix product by $A.B$ or $A \times B$ also.

Example:

For $A = \begin{pmatrix} a_1 & a_2 & a_3 \\ b_1 & b_2 & b_3 \end{pmatrix}_{2\times 3}$, $B = \begin{pmatrix} c_1 & c_2 \\ d_1 & d_2 \\ f_1 & f_2 \end{pmatrix}_{3\times 2}$

$AB = \begin{pmatrix} a_1 c_1 + a_2 d_1 + a_3 f_1 & a_1 c_2 + a_2 d_2 + a_3 f_2 \\ b_1 c_1 + b_2 d_1 + b_3 f_1 & b_1 c_2 + b_2 d_2 + b_3 f_2 \end{pmatrix}$

Remember:

(a) In the product AB, A is called pre-factor and B is called the post factor. In general, $AB \neq BA$.

(b) $A(BC) = (AB)C$, provided the products AB and BC are defined (Associative property)

(c) $A(B + C) = AB + AC$, provided the products AB, AC are defined.

(d) $A^2 = AA$, $A^3 = AAA$ and so on (provided A is a square matrix)

(e) $(A \pm B)^2 \neq A^2 \pm 2AB + B^2$ (in general but equality holds if $AB = BA$)

(f) $(A + B)(A - B) \neq A^2 - B^2$ (in general but equality holds if $AB = BA$)

(g) $IA = AI = A$, where A is a square matrix of order n and I is the identity matrix of order n.

1.3 DETERMINANT OF A SQUARE MATRIX

1.3.1 Definition of Determinant

Let $A = [a_{ij}]_{n\times n}$ be a square matrix of order n. Then the determinant of A is denoted by $|A|$ or det (A) and is defined by

$$|A| = \begin{vmatrix} a_{11} & a_{12} & a_{13} & ...... & a_{1n} \\ a_{21} & a_{22} & a_{23} & ...... & a_{2n} \\ a_{31} & a_{32} & a_{33} & ...... & a_{3n} \\ ...... & ...... & ...... & ...... & ...... \\ a_{n1} & a_{n2} & a_{n3} & ...... & a_{nn} \end{vmatrix}$$

Here $|A|$ is called the determinant of order "n." One can use the symbols $\Delta, \Delta_1, \Delta_2$, etc. to denote a determinant. We can find the value of determinant in the following manner:

(i) If $A = \begin{vmatrix} a_{11} & a_{12} \\ a_{21} & a_{22} \end{vmatrix}$, then $|A| = a_{11}a_{22} - a_{12}a_{21}$

(ii) If $A = \begin{vmatrix} a_{11} & a_{12} & a_{13} \\ a_{21} & a_{22} & a_{23} \\ a_{31} & a_{32} & a_{33} \end{vmatrix}$, then

$|A| = a_{11}(a_{22}a_{33} - a_{32}a_{23}) - a_{12}(a_{21}a_{33} - a_{31}a_{23})$
$\qquad + a_{13}(a_{21}a_{32} - a_{31}a_{22})$

The above process of finding the value of a determinant is called *expansion of a determinant.*

1.3.2 Properties of a Determinant

(i) A determinant remains unchanged by changing its rows into columns and columns into rows.

Thus, the determinants $\begin{vmatrix} a_1 & a_2 \\ b_1 & b_2 \end{vmatrix}$ and $\begin{vmatrix} a_1 & b_1 \\ a_2 & b_2 \end{vmatrix}$ have the same value.

(ii) The interchange of two rows (or columns) of determinant changes the sign of the determinant without changing its numerical value.

Thus, if $\Delta_1 = \begin{vmatrix} a_1 & b_1 \\ a_2 & b_2 \end{vmatrix}$ and $\Delta_2 = \begin{vmatrix} a_2 & b_2 \\ a_1 & b_1 \end{vmatrix}$

(Δ_2 is obtained by interchanging the first and second rows), then

$\Delta_2 = a_2 b_1 - a_1 b_2 = -(a_1 b_2 - a_2 b_1) = -\Delta_1.$

The operation "interchange of R_1 and R_2" is denoted by $R_1 \leftrightarrow R_2$. Similar notation can be used for interchange of two columns.

(iii) If two rows (or columns) of a determinant are identical, then the value of the determinant is zero.

For example, in $\Delta = \begin{vmatrix} 1 & 0 & 3 \\ 5 & -1 & 2 \\ 1 & 0 & 3 \end{vmatrix}$,

R_1(first row) and R_3(third row) are identical, and so $\Delta = 0$.

(iv) If AB is defined, then $|AB| = |A||B|$.

(v) If A be a square matrix of order "n," then

$|kA| = k^n|A|.$

Example:

Let $k = 4$ and $A = \begin{bmatrix} 1 & 3 \\ 0 & 5 \end{bmatrix}$. So A is a square matrix of order "2" and $|A| = 5$.

$$\text{Then } |4A| = \begin{vmatrix} 4 \times 1 & 4 \times 3 \\ 4 \times 0 & 4 \times 5 \end{vmatrix} = \begin{vmatrix} 4 & 12 \\ 0 & 20 \end{vmatrix}$$

$$= 80 = 4^2 \times 5$$

$$= 4^{\text{order of } A} \times |A|.$$

(vi) $|A^n| = |A|^n$, n being a positive integer.

(vii) If every element of any row (or column) of a determinant is multiplied by a constant "k," then the determinant is multiplied by the same constant "k."

Example:

If $\Delta = \begin{vmatrix} 1 & 2 \\ 5 & 7 \end{vmatrix}$, then

$$2\Delta = 2\begin{vmatrix} 1 & 2 \\ 5 & 7 \end{vmatrix} = \begin{vmatrix} 2 & 4 \\ 5 & 7 \end{vmatrix} \text{ (by multiplying } R_1 \text{ by 2)}$$

$(viii)$ The value of a determinant corresponding to a lower (or upper) triangular matrix is obtained by taking the product of all the diagonal elements.

Example:

The value of the determinant $\begin{vmatrix} 1 & 2 & 3 \\ 0 & 5 & -4 \\ 0 & 0 & 6 \end{vmatrix}$ is the product of 1, 5, and 6, $i.e$, 30.

(ix) The determinant value of the null and identity matrices are, respectively, 0 and 1.

(x) If all elements of any row (or column) are zero, then the determinant value becomes zero.

Example:

The determinant $\begin{vmatrix} 1 & 2 & 3 \\ 0 & 5 & -4 \\ 0 & 0 & 0 \end{vmatrix}$ has value "0."

(xi) The value of a determinant remains unaltered by adding (or subtracting) "k" times the elements of any row (or column) to (or from) the corresponding elements of any other row (column), where "k" is any constant. In this case, we use the following notations:

$$R_i \to R_i + kR_j, \ R_i \to R_i - kR_j,$$
$$C_i \to C_i + kC_j, \ C_i \to C_i - kC_j.$$

Example:

Let, $\Delta = \begin{vmatrix} 1 & 0 & -1 \\ 2 & 3 & 4 \\ 0 & 1 & 5 \end{vmatrix}$.

Then, $\Delta = (15 - 4) - 0 - 2 = 9$.

Again, $\Delta = \begin{vmatrix} 1 & 0 & -1 \\ 2 & 3 & 4 \\ 0 & 1 & 5 \end{vmatrix} = \begin{vmatrix} 1 + 3 \times (-1) & 0 & -1 \\ 2 + 3 \times 4 & 3 & 4 \\ 0 + 3 \times 5 & 1 & 5 \end{vmatrix}$

$$\text{(by } C_1 \to C_1 + 3C_3 \text{)}$$

$$= \begin{vmatrix} -2 & 0 & -1 \\ 14 & 3 & 4 \\ 15 & 1 & 5 \end{vmatrix} = -2(15 - 4) - 0 - (14 - 45) = 9.$$

Also,

$$\Delta = \begin{vmatrix} 1 & 0 & -1 \\ 2 & 3 & 4 \\ 0 & 1 & 5 \end{vmatrix} = \begin{vmatrix} 1 & 0 & -1 \\ 2 - 2 \times 1 & 3 - 2 \times 0 & 4 - 2 \times (-1) \\ 0 & 1 & 5 \end{vmatrix}$$

$$\text{(by } R_2 \to R_2 - 2R_1 \text{)}$$

$$= \begin{vmatrix} 1 & 0 & -1 \\ 0 & 3 & 6 \\ 0 & 1 & 5 \end{vmatrix} = 15 - 6 = 9.$$

Remember:

If the given determinant is of order 4 (or more), then use row (or column) operations to get only one non-zero element in the first row (or first column). Then the value of the determinant = non-zero element obtained in first row (first column) × its cofactor.

Example:

If $\Delta = \begin{vmatrix} 1 & 2 & 3 & -1 \\ 0 & 2 & 1 & 2 \\ -1 & 5 & 1 & 0 \\ 1 & 0 & -1 & 3 \end{vmatrix}$. Then

$$\Delta = \begin{vmatrix} 1 & 2 & 3 & -1 \\ 0 & 2 & 1 & 2 \\ -1 & 5 & 1 & 0 \\ 1 & 0 & -1 & 3 \end{vmatrix} = \begin{vmatrix} 1 & 2 & 3 & -1 \\ 0 & 2 & 1 & 2 \\ 0 & 7 & 4 & -1 \\ 0 & -2 & -4 & 4 \end{vmatrix}$$

$[\text{by } R_3 \to R_3 + R_1, R_4 \to R_4 - R_1]$

$= 1 \times \text{cofactor of "1"}$

[since "1" is the only non-zero element in first column]

$$= 1 \times (-1)^{1+1} \begin{vmatrix} 2 & 1 & 2 \\ 7 & 4 & -1 \\ -2 & -4 & 4 \end{vmatrix}$$

$$= 2(16 - 4) - 1(28 - 2) + 2(-28 + 8) = -42.$$

1.3.3 Minors and Cofactors

The minor of an element in a determinant Δ is obtained by eliminations from Δ the row and the column containing the element. Thus, if a_{ij} be in i^{th} row and j^{th} column, then the minor of a_{ij} is denoted by M_{ij} and the cofactor of a_{ij} denoted by C_{ij}, is defined by $C_{ij} = (-1)^{i+j} M_{ij}$

Example:

Consider a determinant $\Delta = \begin{vmatrix} 1 & 2 & 3 \\ 4 & 7 & 5 \\ 6 & 9 & 8 \end{vmatrix}$

Then

(i) the minor of "1" is $\begin{vmatrix} 7 & 5 \\ 9 & 8 \end{vmatrix}$ (which is obtained by omitting first row and first column, where the element 1 lies),

(ii) the minor of "5" is $\begin{vmatrix} 1 & 2 \\ 6 & 9 \end{vmatrix}$ (which is obtained by omitting second row and third column, where the element 5 lies), etc.

(iii) the cofactor of "1" is

$$(-1)^{1+1} \begin{vmatrix} 7 & 5 \\ 9 & 8 \end{vmatrix}, i.e., \begin{vmatrix} 7 & 5 \\ 9 & 8 \end{vmatrix}$$

(iv) the cofactor of "5" is

$$(-1)^{2+3} \begin{vmatrix} 1 & 2 \\ 6 & 9 \end{vmatrix}, i.e., -\begin{vmatrix} 1 & 2 \\ 6 & 9 \end{vmatrix}$$

Remember:

The sum of products of all elements of any row (or column) with their respective cofactors give the value of the determinant.

Thus, if $\Delta = \begin{vmatrix} a_1 & a_2 & a_3 \\ b_1 & b_2 & b_3 \\ c_1 & c_2 & c_3 \end{vmatrix}$

then considering first row, we can write

$\Delta = (a_1 \times \text{cofactor of } a_1) + (a_2 \times \text{cofactor of } a_2) + (a_3 \times \text{cofactor of } a_3)$

Again, considering second column, we can write

$\Delta = (a_2 \times \text{cofactor of } a_2) + (b_2 \times \text{cofactor of } b_2) + (c_2 \times \text{cofactor of } c_2)$

1.4 ADJOINT AND INVERSE OF A MATRIX

1.4.1 Adjoint of a Matrix

If $A = [a_{ij}]_{n \times n}$ be a square matrix of order "n" and C_{ij} be the cofactor of a_{ij} in $|A|$. Then the matrix $[C_{ij}]^T_{n \times n}$ is called the adjoint or adjugate of the matrix A and is denoted by adj(A). Thus adj(A) = $[C_{ij}]^T_{n \times n}$.

Example:

Let $A = \begin{bmatrix} 1 & 2 & 3 \\ 4 & 5 & 6 \\ 7 & 8 & 9 \end{bmatrix}$. Then

(i) cofactor of "1" is $(-1)^{1+1} \begin{vmatrix} 5 & 6 \\ 8 & 9 \end{vmatrix}$, i.e., -3

(since "1" lies in first row and first column)

(ii) cofactor of "2" is $(-1)^{1+2} \begin{vmatrix} 4 & 6 \\ 7 & 9 \end{vmatrix}$, i.e., $+6$

(since "2" lies in first row and second column)

(iii) cofactor of "3" is $(-1)^{1+3} \begin{vmatrix} 4 & 5 \\ 7 & 8 \end{vmatrix}$, i.e., -7

(since "3" lies in first row and third column)

(iv) cofactor of "4" is $(-1)^{2+1} \begin{vmatrix} 2 & 3 \\ 8 & 9 \end{vmatrix}$, i.e., $+6$

(since "4" lies in second row and first column)

(v) cofactor of "5" is $(-1)^{2+2} \begin{vmatrix} 1 & 3 \\ 7 & 9 \end{vmatrix}$, i.e., -12

(since "5" lies in second row and second column)

(vi) cofactor of "6" is $(-1)^{2+3} \begin{vmatrix} 1 & 2 \\ 7 & 8 \end{vmatrix}$, i.e., $+6$

(since "6" lies in second row and third column)

(vii) cofactor of "7" is $(-1)^{3+1} \begin{vmatrix} 2 & 3 \\ 5 & 6 \end{vmatrix}$, i.e., -3

(since "7" lies in third row and first column)

(viii) cofactor of "8" is $(-1)^{3+2} \begin{vmatrix} 1 & 3 \\ 4 & 6 \end{vmatrix}$, i.e., $+6$

(since "8" lies in third row and second column)

(ix) cofactor of "9" is $(-1)^{3+3} \begin{vmatrix} 1 & 2 \\ 4 & 5 \end{vmatrix}$, i.e., -3

(since "9" lies in third row and third column)

Hence, adj(A)

$$= \begin{bmatrix} \text{cofactor of 1} & \text{cofactor of 2} & \text{cofactor of 3} \\ \text{cofactor of 4} & \text{cofactor of 5} & \text{cofactor of 6} \\ \text{cofactor of 7} & \text{cofactor of 8} & \text{cofactor of 9} \end{bmatrix}^T$$

$$= \begin{pmatrix} -3 & 6 & -7 \\ 6 & -12 & 6 \\ -3 & 6 & -3 \end{pmatrix}^T = \begin{pmatrix} -3 & 6 & -3 \\ 6 & -12 & 6 \\ -7 & 6 & -3 \end{pmatrix}$$

Properties :

If A be a square matrix of order "n," then

   (i) $A \times (\text{adj } A) = (\text{adj } A) \times A = |A|I$

      I being the identity matrix of order "n."

  (ii) $|\text{adj } A| = |A|^{n-1}$,

 (iii) $|\text{adj (adj } A)| = |\text{adj } (A)|^{n-1} = |A|^{(n-1)^2}$

 (iv) $\text{adj(adj } A) = |A|^{n-2} \times A$

  (v) $\text{adj } (kA) = (\text{adj } A) \times k^{n-1}$, where "$k$" is a real number

 (vi) $\text{adj } (A^T) = (\text{adj } A)^T$.

1.4.2 Inverse of a Matrix

Suppose A be a square matrix of order "n" and there exist an another square matrix B of the same order such that $AB = BA = I$, where I is the identity matrix of order "n," then B is called the inverse of A, denoted by A^{-1} and is defined by

$$A^{-1} = \frac{1}{|A|} \text{adj } (A) \text{ provided } |A| \neq 0$$

Example:

Let $\quad A = \begin{pmatrix} 1 & 2 & 3 \\ 4 & 5 & 6 \\ 7 & 8 & 9 \end{pmatrix}$, then

$\det (A) = 1(45 - 48) - 2(36 - 42) + 3(32 - 35) = 0$

So A^{-1} does not exist.

Let us consider another matrix, $B = \begin{pmatrix} 1 & 0 & 3 \\ 4 & 5 & 6 \\ 0 & 1 & 1 \end{pmatrix}$

Then $\det (B) = 1(5 - 6) - 0 + 3(4 - 0) = 11$.

Therefore B^{-1} exist.

$\therefore \text{adj}(B) = \begin{pmatrix} -1 & -4 & 4 \\ 3 & 1 & -1 \\ -15 & 6 & 5 \end{pmatrix}^T = \begin{pmatrix} -1 & 3 & -15 \\ -4 & 1 & 6 \\ 4 & -1 & 5 \end{pmatrix}.$

Hence, $B^{-1} = \dfrac{1}{|B|} \text{adj}(B) = \dfrac{1}{11} \begin{pmatrix} -1 & 3 & -15 \\ -4 & 1 & 6 \\ 4 & -1 & 5 \end{pmatrix}.$

Properties:

If A be a square matrix of order "n," then

  (a) $A \times A^{-1} = A^{-1} \times A = I$

  (b) $(AB)^{-1} = B^{-1}A^{-1}$

  (c) $(A^{-1})^{-1} = A$

  (d) $(A^T)^{-1} = (A^{-1})^T$

  (e) For $\quad A = \begin{pmatrix} a & b \\ c & d \end{pmatrix}, A^{-1} = \dfrac{1}{|A|}\begin{pmatrix} d & -b \\ -c & a \end{pmatrix}$ provided inverse exist.

  (f) If $|A| \neq 0$, then A is called non-singular (invertible) matrix; otherwise A is called singular.

1.5 VARIOUS TYPES OF REAL SQUARE MATRICES

1.5.1 Symmetric Matrix

A square matrix A is called symmetric if $A^t = A$.

Example:

$\begin{pmatrix} 1 & 3 & 4 \\ 3 & 0 & -7 \\ 4 & -7 & 4 \end{pmatrix}$ is symmetric.

Properties:

   (i) If A, B are symmetric matrices of same order, then kA, $aA + bB$, $AB + BA$, A^n ($n \in N$), are symmetric (where "k," "a," and "b" are real numbers).

  (ii) If A, B are symmetric matrices of same order, then AB is symmetric if and only if $AB = BA$.

 (iii) For any square matrix A, the matrix $A + A^T$ is always symmetric.

1.5.2 Skew-Symmetric Matrix

A square matrix A is called skew-symmetric if $A^T = -A$.

Example:

$\begin{pmatrix} 0 & -3 & 4 \\ 3 & 0 & -7 \\ -4 & 7 & 0 \end{pmatrix}$ is skew-symmetric.

Properties:

   (i) If A, B are skew-symmetric matrices of same order, then kA, $aA \pm bB$, $AB - BA$, A^n ($n \in N$) are all skew-symmetric.

  (ii) If A and B are skew-symmetric matrices of same order, then AB is skew-symmetric if and only if $AB + BA = 0$.

 (iii) Diagonal elements of skew symmetric matrix are all zero.

 (iv) For any square matrix A, the matrix $A - A^T$ is always skew-symmetric.

1.5.3 Orthogonal Matrix

A square matrix A is said to be orthogonal if and only if $AA^T = A^TA = I$

Example:

$A = \begin{pmatrix} \cos\alpha & \sin\alpha \\ -\sin\alpha & \cos\alpha \end{pmatrix}$ is orthogonal.

Properties:

(i) If A is orthogonal, then $|A| = \pm 1$ and hence A is non-singular.

(ii) If A, B are orthogonal matrices of same order, then A^T, AB, BA, and A^{-1} are all orthogonal matrices.

(iii) If A is orthogonal, then $A^T = A^{-1}$.

1.5.4 Idempotent Matrix

A square matrix "A" is called idempotent if $A^2 = A$.

Example:

The matrix $\begin{pmatrix} 2 & -2 & -4 \\ -1 & 3 & 4 \\ 1 & -2 & -3 \end{pmatrix}$ is idempotent.

Properties:

(a) If A, B are idempotent matrices of same order, then

(i) AB is idempotent if and only if $AB = BA$

(ii) $A + B$ is idempotent if and only if $AB = BA = O$

(iii) $I - A$ is idempotent

(b) If $AB = A$, $BA = B$, then A and B are both idempotent matrices.

1.5.5 Involutary Matrix

A square matrix "A" is called involutary if $A^2 = 1$.

Example:

The matrix $\begin{pmatrix} -5 & -8 & 0 \\ 3 & 5 & 0 \\ 1 & 2 & -0 \end{pmatrix}$ is involutary.

1.5.6 Nilpotent Matrix

A square matrix "A" is said to be nilpotent of index "k" if $A^k = O$ for some positive number k "k" should be the least positive integer).

Example:

(i) The matrix $A = \begin{bmatrix} 1 & -1 & 1 \\ -3 & 3 & -3 \\ -4 & 4 & -4 \end{bmatrix}$ is nilpotent of

index "2" (since $A^2 = O$)

(ii) $\begin{pmatrix} 0 & 0 \\ 1 & 0 \end{pmatrix}$ is a nilpotent matrix of index "2"

(since $A^2 = O$).

1.6 COMPLEX MATRICES AND THEIR TYPES

1.6.1 Complex Conjugate of a Matrix

If $A = [a_{ij}]_{m \times n}$ be a complex matrix of order $m \times n$, then the complex conjugate of $A = [a_{ij}]_{m \times n}$, is denoted by $\bar{A}$ and is defined by $\bar{A} = [\bar{a}_{ij}]_{m \times n}$. Thus, the complex conjugate matrix of a given matrix is obtained by replacing all the elements in the given matrix by their respective complex conjugates.

Example:

If $A = \begin{pmatrix} 5 & -8+i & i \\ 3 & 5 & 0 \\ 1-i & 2 & -i \end{pmatrix}$, then

$\bar{A} = \begin{pmatrix} 5 & -8-i & -i \\ 3 & 5 & 0 \\ 1+i & 2 & i \end{pmatrix}$.

Properties:

If A and B are complex matrices, then

(i) $\bar{\bar{A}} = A$

(ii) $\overline{A + B} = \bar{A} + \bar{B}$

(iii) $\overline{kA} = \bar{k} \times \bar{A}$, where "$k$" is any number (real or complex)

(iv) $\overline{AB} = \bar{A} \times \bar{B}$, provided AB is defined.

(v) $\overline{A^n} = (\bar{A})^n$

(vi) $\bar{A} = A$ if and only if A is real.

(vii) $\bar{A} = -A$ if and only if A purely imaginary

1.6.2 Transposed Conjugate of a Matrix

The transposed conjugate of a complex matrix A is denoted by A^θ and is defined by $A^\theta = (\bar{A})^T = \overline{A^T}$.

Example:

If $A = \begin{pmatrix} 5 & -8+i & i \\ 3 & 5 & 0 \\ 1-i & 2 & -i \end{pmatrix}$, then

$$A^\theta = \begin{pmatrix} 5 & 3 & i+1 \\ -8-i & 5 & 2 \\ -i & 0 & i \end{pmatrix}.$$

Properties:

If A and B are complex matrices of same order, then

(i) $(A^\theta)^\theta = A$.

(ii) $(A + B)^\theta = A^\theta + B^\theta$.

(iii) $(kA)^\theta = \bar{k}(A)^\theta$, where "$k$" is any number (real or complex).

(iv) $(AB)^\theta = B^\theta A^\theta$.

(v) $(A^n)^\theta = (A^\theta)^n$.

1.6.3 Unitary Matrix

A square complex matrix A is said to be unitary matrix if $AA^\theta = A^\theta A = I$.

Example:

$A = \dfrac{1}{\sqrt{2}}\begin{pmatrix} 1 & i \\ -i & -1 \end{pmatrix}$ is unitary.

Properties:

(i) If A is unitary then A^T, A^{-1} are both unitary matrix.

(ii) If A, B are unitary matrices of same order, then AB and BA are also unitary matrices.

1.6.4 Hermitian Matrix

A square matrix "A" is called Hermitian if $A^\theta = A$.

Example:

$A = \begin{pmatrix} 1 & 2-i & 3-i \\ 2+i & 2 & 1+2i \\ 3+i & 1-2i & 4 \end{pmatrix}$ is Hermitian.

Properties:

(i) If A, B are Hermitian matrices of same order, then

kA, $aA + bB$, AA^θ, $A^\theta A$, $AB + BA$, $\bar{A}$, A^n, $A + A^\theta$

are also Hermitian matrices.

(ii) If A, B are Hermitian matrices of same order, then AB is Hermitian if and only if $AB = BA$.

1.6.5 Skew-Hermitian Matrix

A square matrix "A" is called skew-Hermitian if $A^\theta = -A$

Example:

$\begin{pmatrix} 3i & 3+4i & 4-5i \\ -3+4i & -4i & 5+6i \\ -4-5i & -5+6i & 0 \end{pmatrix}$ is skew-Hermitian.

Properties:

(i) $A = [a_{ij}]_{n \times n}$ is skew-Hermitian if and only if $\overline{a_{ij}} = -a_{ji}$

(ii) In a skew-Hermitian matrix, all the elements lying in the principal diagonal are purely imaginary or zero.

(iii) If A, B are skew-Hermitian matrices of same order, then kA, $aA + bB$, $AB - BA$, $A - A^\theta$, $\bar{A}$ are also skew-Hermitian matrices.

(iv) If A is Hermitian, then iA is skew-Hermitian.

(v) If A is skew-Hermitian, then iA is Hermitian.

(vi) For any square matrix A, $A = \dfrac{1}{2}(A + A^\theta) + \dfrac{1}{2}(A - A^\theta)$. Thus, every square matrix can be expressed as a sum of a Hermitian matrix $\dfrac{1}{2}(A + A^\theta)$ and a skew-Hermitian matrix $\dfrac{1}{2}(A - A^\theta)$.

1.7 RANK OF A MATRIX

1.7.1 Elementary Transformations

The following transformations are known as elementary transformations:

(i) **Interchange of two rows (or columns).**

Suppose we interchange first row and second row in a matrix, then this operation will be denoted by $R_1 \leftrightarrow R_2$. Again, if we interchange third column and fourth column in a matrix, then this operation will be denoted by $C_3 \leftrightarrow C_4$.

(ii) **The multiplication of the elements of a row (or column) by a non-zero number.**

Suppose we multiply each element of second row by "k," then the operation is denoted by $R_2 \rightarrow kR_2$. In a similar way, if we multiply each element of third column by "k," then the operation is denoted by $C_3 \rightarrow kC_3$.

(iii) **The addition to the elements of a row (or column), the corresponding elements of a row (or column) multiplied by any number.**

Suppose we first multiply the second row by "k" and add it with the first row, then the corresponding operation is denoted by $R_1 \rightarrow R_1 + kR_2$. Similarly, if we first multiply the second column by "k" and add it with the first column, then the corresponding operation is denoted by $C_1 \rightarrow C_1 + kC_2$.

1.7.2 Equivalent Matrices

Two matrices A and B are said to be equivalent if one can be obtained from the other by a sequence of elementary operations. The symbol ~ is used for equivalence. Thus, if A and B are equivalent matrices, then we write A~B. The relation "~" forms an equivalence relation.

Example:

If $A = \begin{pmatrix} 3 & 4 & 5 \\ -3 & -4 & 6 \\ -5 & 6 & 0 \end{pmatrix}$, then

$$A = \begin{pmatrix} 3 & 4 & 5 \\ -3 & -4 & 6 \\ -5 & 6 & 0 \end{pmatrix}$$

$$\sim \begin{pmatrix} 3 & 4 & 5 \\ 0 & 0 & 11 \\ -5 & 6 & 0 \end{pmatrix} \text{ (by } R_2 \rightarrow R_2 + R_1)$$

$$\sim \begin{pmatrix} 3 & 4 & 11 \\ 0 & 0 & 11 \\ -5 & 6 & -10 \end{pmatrix} \text{ (by } C_3 \rightarrow C_3 + 2C_1)$$

$$= B \text{ (say), where } B = \begin{pmatrix} 3 & 4 & 11 \\ 0 & 0 & 11 \\ -5 & 6 & -10 \end{pmatrix}$$

Then A and B are equivalent.

1.7.3 Rank of a Matrix

A positive real number "r" is said to be the rank of a given matrix A of order $m \times n$ if

(i) there exist a square sub-matrix of order "r" whose determinant is non-zero

(ii) the determinant value of every square sub-matrix of order higher than "r" is zero.

The symbols $r(A)$, $\rho(A)$, rank(A) are used to denote the rank of a matrix A.

Example:
Consider a matrix $A = \begin{pmatrix} 3 & 4 & 5 \\ 1 & 1 & 2 \\ 6 & 8 & 10 \end{pmatrix}$. Then det (A) $= 0$. So rank of A cannot be "3."

Let us consider a square sub-matrix $B = \begin{pmatrix} 3 & 4 \\ 1 & 1 \end{pmatrix}$. Then det$(B) = -1 \neq 0$. Thus, there exist a square sub-matrix B of order "2" whose determinant value is non-zero. Hence, rank $(A) = 2$.

Properties:

(i) If A is a matrix A of order $m \times n$, then rank $(A) \leq \min \{m, n\}$.

(ii) rank $(O) = 0$

(iii) rank $(I_n) = n$, where I_n is the identity matrix of order "n."

(iv) rank $(A) = n$ if A is non-singular square matrix of order "n" and vice versa.

(v) if A is singular, then rank $(A) < n$ and vice versa, provided A is a square matrix of order "n."

(vi) If A and B are matrices of same order, then $r(A + B) \leq r(A) + r(B)$.

(vii) $r(AB) \leq \min \{\text{rank } (A), \text{rank } (B)\}$.

(viii) rank (A) = rank (A^T) and rank (A) = rank (AA^T)

(ix) rank (A) = rank (A^θ) and rank (A) = rank (AA^θ)

(x) Two equivalent matrices have same rank.

(xi) Rank of a matrix remains the same when we multiply any non-singular matrix with it.

(xii) The rank of a skew-symmetric matrix can't be "1."

(xiii) Let A be a square matrix of order "n." Then

$$\text{rank(adj}A) = \begin{cases} n, & \text{if rank}(A) = n \\ 1, & \text{if rank}(A) = n - 1. \\ 0, & \text{if rank}(A) < n - 1 \end{cases}$$

1.7.4 Determination of the Rank of a Matrix

There are many forms namely "normal form," "echelon form," and "row reduced echelon form," which are very useful in determining the rank of a matrix. But here we will use a secondary method (quickest method) to find the rank of a matrix discussed below:

"Let A be a matrix of order $m \times n$. Then apply elementary row transformations to get maximum number of zero rows. In this case, the number of non-zero rows appearing in the final equivalent matrix will be equal to the rank of the given matrix."

Example:

$$A = \begin{pmatrix} 1 & 0 & 3 \\ 3 & 5 & 1 \\ 2 & 0 & 6 \end{pmatrix}$$

$$\sim \begin{pmatrix} 1 & 0 & 3 \\ 3 & 5 & 1 \\ 0 & 0 & 0 \end{pmatrix} \left(\text{by } R_2 \to R_2 - 2R_1 \right)$$

which has only two non-zero rows. Therefore, rank$(A) = 2$.

1.8 SYSTEM OF LINEAR EQUATIONS AND THEIR SOLUTIONS

1.8.1 Introduction

A linear equation in "n" unknowns $x_1, x_2, x_3, \dots x_n$ is an equation of the form

$$a_1 x_1 + a_2 x_2 + a_3 x_3 + \dots + a_n x_n = b.$$

If $b = 0$, then the equation is called a homogeneous equation; otherwise it is called a non-homogeneous equation.

A system of "m" non-homogeneous linear equations in "n" unknowns is given by

$$\left. \begin{array}{l} a_{11}x_1 + a_{12}x_2 + a_{13}x_3 + \dots + a_{1n}x_n = b_1 \\ a_{21}x_1 + a_{22}x_2 + a_{23}x_3 + \dots + a_{2n}x_n = b_2 \\ \dots \\ a_{m1}x_1 + a_{m2}x_2 + a_{m3}x_3 + \dots + a_{mn}x_n = b_m \end{array} \right\}$$
$$\dots\dots\dots(i)$$

If there exist at least one set of values of x_1, $x_2, x_3, \dots, x_n$ satisfying all the equations in (i), then the system of equations given by (i), is said to be consistent; otherwise inconsistent. In other words, the system of equations given by (i) is consistent if it has a solution; otherwise, it is inconsistent.

Let us consider the following system of equations:

$$\left. \begin{array}{l} x - y = 2 \\ 2x + y = 4 \end{array} \right\} \qquad \dots(ii)$$

$$\left. \begin{array}{l} x + y = 2 \\ 2x + 2y = 3 \end{array} \right\} \qquad \dots(iii)$$

The system of equations given by (ii) has a solution $x = 2$, $y = 0$ and so it is consistent.

On the other hand, no pair of values of x and y simultaneously satisfy both the equations in (iii). So it is inconsistent.

Remember:

The non-homogeneous system of linear equations given in (i) can also be rewritten as $AX = B$, where

$$A = \begin{pmatrix} a_{11} & a_{12} & a_{13} & \dots & a_{1n} \\ a_{21} & a_{22} & a_{23} & \dots & a_{2n} \\ \dots & \dots & \dots & \dots & \dots \\ a_{m1} & a_{m2} & a_{m3} & \dots & a_{mn} \end{pmatrix},$$

$$X = \begin{bmatrix} x_1 \\ x_2 \\ \dots \\ \dots \\ x_n \end{bmatrix}, B = \begin{bmatrix} b_1 \\ b_2 \\ \dots \\ \dots \\ b_m \end{bmatrix}$$

Here A is called the coefficient matrix. The augmented matrix, denoted by $[A: B]$, is defined by

$$[A:B] = \begin{bmatrix} a_{11} & a_{12} & \dots & a_{1n} & b_1 \\ a_{21} & a_{22} & \dots & a_{2n} & b_2 \\ a_{31} & a_{32} & \dots & a_{3n} & b_3 \\ \dots & \dots & \dots & \dots & \dots \\ a_{m1} & a_{m2} & \dots & a_{mn} & b_m \end{bmatrix}.$$

1.8.2 Methods for Solving Non-Homogeneous System of Linear Equations

1.8.2.1 Cramer's Rule

Let us consider a system of equations consisting 3 non-homogeneous linear equations in 3 unknowns given by:

$$\left. \begin{array}{l} a_{11}x + a_{12}y + a_{13}z = b_1 \\ a_{21}x + a_{22}y + a_{23}z = b_2 \\ a_{31}x + a_{32}y + a_{33}z = b_3 \end{array} \right\} \qquad \dots(i)$$

Let

$$\Delta = \begin{vmatrix} a_{11} & a_{12} & a_{13} \\ a_{21} & a_{22} & a_{23} \\ a_{31} & a_{32} & a_{33} \end{vmatrix}, \Delta_1 = \begin{vmatrix} b_1 & a_{12} & a_{13} \\ b_2 & a_{22} & a_{23} \\ b_3 & a_{32} & a_{33} \end{vmatrix},$$

$$\Delta_2 = \begin{vmatrix} a_{11} & b_1 & a_{13} \\ a_{21} & b_2 & a_{23} \\ a_{31} & b_3 & a_{33} \end{vmatrix}, \Delta_3 = \begin{vmatrix} a_{11} & a_{12} & b_1 \\ a_{21} & a_{22} & b_2 \\ a_{31} & a_{32} & b_3 \end{vmatrix}.$$

Then,

Case-I:

If $\Delta \neq 0$, then (i) has an unique solution and so consistent. In this case, the solution is given by $x = \dfrac{\Delta_1}{\Delta}$, $y = \dfrac{\Delta_2}{\Delta}$, $z = \dfrac{\Delta_3}{\Delta}$.

Example:

Let us consider the system of equations:

$$2x - z = 1, \ 2x + 4y - z = 1, \ x - 8y - 3z = -2.$$

Then $\Delta = \begin{vmatrix} 2 & 0 & -1 \\ 2 & 4 & -1 \\ 1 & -8 & -3 \end{vmatrix} = -20 \neq 0.$

So the given system of equations has a unique solution.

$$\Delta_1 = \begin{vmatrix} 1 & 0 & -1 \\ 1 & 4 & -1 \\ -2 & -8 & -3 \end{vmatrix} = -20,$$

$$\Delta_2 = \begin{vmatrix} 2 & 1 & -1 \\ 2 & 1 & -1 \\ 1 & -2 & -3 \end{vmatrix} = 0,$$

$$\Delta_3 = \begin{vmatrix} 2 & 0 & 1 \\ 2 & 4 & 1 \\ 1 & -8 & -2 \end{vmatrix} = -20.$$

Hence, by Cramer's rule, the solution is

$$x = \frac{\Delta_1}{\Delta} = 1, y = \frac{\Delta_2}{\Delta} = 0, z = \frac{\Delta_3}{\Delta} = 1.$$

Case-II:

If $\Delta = 0$ and at least one of $\Delta_1, \Delta_2, \Delta_3$ be zero, then (i) has no solution and so inconsistent. Thus, the following sub-cases arise:

(i) If $\Delta = 0$, $\Delta_1 \neq 0$, then (i) has no solution

(ii) If $\Delta = 0$, $\Delta_2 \neq 0$, then (i) has no solution

(iii) If $\Delta = 0$, $\Delta_3 \neq 0$, then (i) has no solution

Case-III:

If $\Delta = \Delta_1 = \Delta_2 = \Delta_3 = 0$, then (i) has an infinite number of solutions and so consistent.

Example:

Let us consider the system of equations:

$$x + y + z = 6, \ x + 2y + 3z = 10, \ 2x + 2y + 2z = 12.$$

Then

$$\Delta = \begin{vmatrix} 1 & 1 & 1 \\ 1 & 2 & 3 \\ 2 & 2 & 2 \end{vmatrix} = 0,$$

$$\Delta_1 = \begin{vmatrix} 6 & 1 & 1 \\ 10 & 2 & 3 \\ 12 & 2 & 2 \end{vmatrix} = 0,$$

$$\Delta_2 = \begin{vmatrix} 1 & 6 & 1 \\ 1 & 10 & 3 \\ 2 & 12 & 2 \end{vmatrix} = 0,$$

$$\Delta_3 = \begin{vmatrix} 1 & 1 & 6 \\ 1 & 2 & 10 \\ 2 & 2 & 12 \end{vmatrix} = 0.$$

Since $\Delta = \Delta_1 = \Delta_2 = \Delta_3 = 0$, hence, the system has an infinite number of solutions and so is consistent.

Remark:

If we consider a system of equations consisting two non-homogeneous linear equations in two unknowns given by

$$\left. \begin{array}{l} a_{11}x + a_{12}y = b_1 \\ a_{21}x + a_{22}y = b_2 \end{array} \right\}$$
....................(i)

Then

$$\Delta = \begin{vmatrix} a_{11} & a_{12} \\ a_{21} & a_{22} \end{vmatrix},$$

$$\Delta_1 = \begin{vmatrix} b_1 & a_{12} \\ b_2 & a_{22} \end{vmatrix}, \ \Delta_2 = \begin{vmatrix} a_{11} & b_1 \\ a_{21} & b_2 \end{vmatrix}$$

Then

- If $\Delta \neq 0$, then (i) has an unique solution and so consistent. In this case, the solution is given by $x = \dfrac{\Delta_1}{\Delta}$, $y = \dfrac{\Delta_2}{\Delta}$.
- If $\Delta = 0$ and at least one of Δ_1, Δ_2 be zero, then (1) has no solution and so inconsistent. Thus, the following sub-cases arise:

(i) If $\Delta = 0$, $\Delta_1 \neq 0$, then (i) has no solution

(ii) If $\Delta = 0$, $\Delta_2 \neq 0$, then (i) has no solution

Example:

Let us consider the system of equations:

$$2x + y = 1, \ 4x + 2y = 5$$

Then

$$\Delta = \begin{vmatrix} 2 & 1 \\ 4 & 2 \end{vmatrix} = 0, \text{ but } \Delta_2 = \begin{vmatrix} 2 & 1 \\ 4 & 5 \end{vmatrix} = 6 \neq 0.$$

So the system has no solution.

⌐ If $\Delta = \Delta_1 = \Delta_2 = 0$, then (i) has an infinite number of solutions and so consistent.

1.8.2.2 Matrix Method

Let us consider a system of equations consisting 3 non- homogeneous linear equations in 3 unknowns given by

$$\left.\begin{array}{l} a_{11}x + a_{12}y + a_{13}z = b_1 \\ a_{21}x + a_{22}y + a_{23}z = b_2 \\ a_{31}x + a_{32}y + a_{33}z = b_3 \end{array}\right\}$$

The above system of equations can be rewritten as: $AX = B$, where

$$A = \begin{pmatrix} a_{11} & a_{12} & a_{13} \\ a_{21} & a_{22} & a_{23} \\ a_{31} & a_{32} & a_{33} \end{pmatrix}$$

$$X = \begin{bmatrix} x \\ y \\ z \end{bmatrix}, B = \begin{bmatrix} b_1 \\ b_2 \\ b_3 \end{bmatrix}$$

If $|A| = \begin{vmatrix} a_{11} & a_{12} & a_{13} \\ a_{21} & a_{22} & a_{23} \\ a_{31} & a_{32} & a_{33} \end{vmatrix} \neq 0,$

then we say that A–1 exist and the solution is given by: X = A–1B.

Example:

Let us consider the system of equations:

$$x - y + z = 0, \quad 2x + 3y - 5z = 7, \quad 3x - 4y - 2z = -1.$$

The above system of equations can be re-written as $AX = B$, where

Here

$$A = \begin{pmatrix} 1 & -1 & 1 \\ 2 & 3 & -5 \\ 3 & -4 & -2 \end{pmatrix}, X = \begin{pmatrix} x \\ y \\ z \end{pmatrix}, B = \begin{pmatrix} 0 \\ 7 \\ -1 \end{pmatrix}.$$

Here $|A| = \begin{vmatrix} 1 & -1 & 1 \\ 2 & 3 & -5 \\ 3 & -4 & -2 \end{vmatrix} = -32 \neq 0.$

so A^{-1} exist and the solution is given by: $X = A^{-1}B$.

Now, $X = A^{-1}B$

$$= \frac{1}{|A|} adj(A) \times B$$

$$= \frac{1}{(-32)} \begin{pmatrix} -26 & -6 & 2 \\ -11 & -5 & 7 \\ -17 & 1 & 5 \end{pmatrix} \times \begin{bmatrix} 0 \\ 7 \\ -1 \end{bmatrix}$$

$$= \frac{-1}{32} \begin{bmatrix} 0 - 42 - 2 \\ 0 - 35 - 7 \\ 0 + 7 - 5 \end{bmatrix}$$

$$\Rightarrow \begin{bmatrix} x \\ y \\ z \end{bmatrix} = \begin{bmatrix} \dfrac{-44}{-32} \\ \dfrac{-42}{-32} \\ \dfrac{2}{-32} \end{bmatrix} = \begin{bmatrix} \dfrac{11}{8} \\ \dfrac{21}{16} \\ \dfrac{-1}{16} \end{bmatrix}$$

$$\therefore x = \frac{11}{8}, \ y = \frac{21}{16}, \ z = \frac{-1}{16}.$$

1.8.2.3 Rank Method

Let us consider a system of equations consisting "m" non- homogeneous linear equations in "n" unknowns given by

$$\left.\begin{array}{l} a_{11}x_1 + a_{12}x_2 + a_{13}x_3 + \dots + a_{1n}x_n = b_1 \\ a_{21}x_1 + a_{22}x_2 + a_{23}x_3 + \dots + a_{2n}x_n = b_2 \\ \dots \dots \dots \dots \dots \dots \dots \dots \dots \dots \dots \dots \dots \dots \\ a_{m1}x_1 + a_{m2}x_2 + a_{m3}x_3 + \dots + a_{mn}x_n = b_m \end{array}\right\}$$

$$\dots \dots \dots \dots (2)$$

Then the above system of equations can be re-written as $AX = B$ (discussed earlier).

Case-I:

If rank (A) = rank $([A : B]) = n$, then the system given by (2) has an unique solution and so the system is consistent.

Example:

Let us consider the system of equations:

$$2x - z = 1, \ 2x + 4y - z = 1, \ x - 8y - 3z = -2.$$

$[A : B]$

$$= \begin{pmatrix} 2 & 0 & -1 & 1 \\ 2 & 4 & -1 & 1 \\ 1 & -8 & -3 & -2 \end{pmatrix}$$

$$\sim \begin{pmatrix} 1 & -8 & -3 & -2 \\ 2 & 4 & -1 & 1 \\ 2 & 0 & -1 & 1 \end{pmatrix} \text{ (by } R_1 \leftrightarrow R_3)$$

$$\sim \begin{pmatrix} 1 & -8 & -3 & -2 \\ 0 & 20 & 5 & 5 \\ 0 & 16 & 5 & 5 \end{pmatrix}$$

(by $R_2 \to R_2 - 2R_1, R_3 \to R_3 - 2R_1$)

$$\sim \begin{pmatrix} 1 & -8 & -3 & -2 \\ 0 & 20 & 5 & 5 \\ 0 & -4 & 0 & 0 \end{pmatrix}$$

(by $R_3 \to R_3 - R_2$)

which has 3 non-zero rows.

So rank ($[A{:}B]$) = 3.

Ignoring the last column of the final equivalent matrix of $[A{:}B]$, we see that there are three non-zero rows. So rank(A) = 3.

Thus, rank(A) = rank ($[A{:}B]$) = 3 (=number of unknowns). Hence, the system is consistent and has the unique solution.

Case-II:

If rank (A) = rank $([A{:}B]) < n$, then the system given by (2) has an infinite number of solutions and so the system is consistent.

Example:

Let us consider the system of equations:

$$x + y + z = 6, x + 2y + 3z = 1$$

Here $A = \begin{bmatrix} 1 & 1 & 1 \\ 1 & 2 & 3 \end{bmatrix}$, $[A : B] = \begin{bmatrix} 1 & 1 & 1 & 6 \\ 1 & 2 & 3 & 1 \end{bmatrix}$.

Then $A = \begin{bmatrix} 1 & 1 & 1 \\ 1 & 2 & 3 \end{bmatrix}$

$\sim \begin{bmatrix} 1 & 1 & 1 \\ 0 & 1 & 2 \end{bmatrix}$ (by $R_2 \to R_2 - R_1$)

which have 2 non-zero rows.

Hence, rank(A) = 2.

$[A : B] = \begin{bmatrix} 1 & 1 & 1 & 6 \\ 1 & 2 & 3 & 1 \end{bmatrix}$

$\sim \begin{bmatrix} 1 & 1 & 1 & 6 \\ 0 & 1 & 2 & -5 \end{bmatrix}$ (by $R_2 \to R_2 - R_1$)

which has two non-zero rows.

So rank ($[A{:}B]$) = 2.

Thus, rank(A) = rank ($[A{:}B]$) = 2 < 3 (= number of unknowns). Hence, the system has an infinite number of solutions and so the system is consistent.

Case-III:

If rank $(A) \neq$ rank $([A{:}B])$, then the system given by (2) has no solution and so the system is inconsistent.

Example:

Let us consider the system of equations:

$$2x + y - z = 12, x - y - 2z = -3, 3y + 3z = 10.$$

$$[A : B] = \begin{pmatrix} 2 & 1 & -1 & 12 \\ 1 & -1 & -2 & -3 \\ 0 & 3 & 3 & 10 \end{pmatrix}$$

$$\sim \begin{pmatrix} 0 & 3 & 3 & 18 \\ 1 & -1 & -2 & -3 \\ 0 & 0 & 3 & 10 \end{pmatrix}$$

(by $R_1 \to R_1 - 2R_2$)

$$\sim \begin{pmatrix} 0 & 0 & 0 & 8 \\ 1 & -1 & -2 & -3 \\ 0 & 3 & 3 & 10 \end{pmatrix}$$

(by $R_1 \to R_1 - R_3$)

$$\sim \begin{pmatrix} 0 & 3 & 3 & 10 \\ 1 & -1 & -2 & -3 \\ 0 & 0 & 0 & 8 \end{pmatrix}$$

(by $R_1 \leftrightarrow R_3$)

which has 3 non-zero rows.

So rank ($[A : B]$) = 3.

Ignoring the last column of the final equivalent matrix of $[A : B]$, we see that there are two non-zero rows. So rank (A) = 2.

Thus, rank $(A) \neq$ rank $([A{:}B])$. Hence, the system is inconsistent.

1.8.3 Homogeneous System of Linear Equations

A homogeneous system of m linear equations in n unknowns is given by

$$\left.\begin{array}{l} a_{11}x_1 + a_{12}x_2 + a_{13}x_3 + \ldots\ldots + a_{1n}x_n = 0 \\ a_{21}x_1 + a_{22}x_2 + a_{23}x_3 + \ldots\ldots + a_{2n}x_n = 0 \\ \text{--} \\ a_{m1}x_1 + a_{m2}x_2 + a_{m3}x_3 + \ldots\ldots + a_{mn}x_n = 0 \end{array}\right\} \ldots(i)$$

The above homogeneous system can also be rewritten as $AX = O$, where

$$A = \begin{bmatrix} a_{11} & a_{12} & \ldots & a_{1n} \\ a_{21} & a_{22} & \ldots & a_{2n} \\ \ldots & \ldots & \ldots & \ldots \\ a_{m1} & a_{m2} & \ldots & a_{mn} \end{bmatrix}, X = \begin{pmatrix} x_1 \\ x_2 \\ \ldots \\ x_n \end{pmatrix}.$$

Here, A is called the coefficient matrix. It is easy to verify that $x_1 = x_2 = x_3 = \ldots = x_n = 0$ is a solution of (i) [called the trivial solution]. Hence, the system (i) is consistent.

Further if rank $(A) < n (=$ number of unknowns), then the system (i) has an infinite number of solutions.

Remember:

If a system has "n" homogeneous equations in "n" unknowns, then

System has an infinite number of solutions $\Leftrightarrow$ determinant of the coefficient matrix is zero.

Example:

Let us consider the system of equations:

$x + 2y + 3z = 0, 2x + 3y + 4z = 0, 3x + 4y + 5z = 0.$

$$\text{Here } A = \begin{pmatrix} 1 & 2 & 3 \\ 2 & 3 & 4 \\ 3 & 4 & 5 \end{pmatrix}$$

$$\sim \begin{pmatrix} 1 & 2 & 3 \\ 0 & -1 & -2 \\ 0 & -2 & -4 \end{pmatrix}$$

$$\left(\text{using } R_2 \to R_2 - 2R_1, R_3 \to R_3 - 3R_1\right)$$

$$\sim \begin{pmatrix} 1 & 2 & 3 \\ 0 & -1 & -2 \\ 0 & 0 & 0 \end{pmatrix} \left(\text{using } R_3 \to R_3 - 2R_2\right)$$

which has "2" non-zero rows.

Hence, rank $(A) = 2 < 3$ (= number of unknowns). Thus, the given system of equations has an infinite number of solutions.

Alternative method:

Here $\det(A) = (15 - 16) - 2(10 - 12) + 3(8 - 9) = 0.$

Hence, the system has an infinite number of solutions.

1.9 EIGENVALUES AND EIGENVECTORS

1.9.1 Characteristic Roots (Eigenvalues) of a Matrix

If A be a square matrix of order n and I be an identity matrix of order n, then $A - \lambda I$ is called the characteristic matrix of A, λ being a scalar.

The determinant $|A - \lambda I|$ is called the characteristic polynomial of A, which is basically a polynomial in λ of n th degree.

The equation $|A - \lambda I| = 0$, *i.e.*, $\det (A - \pi I) = 0$ is called characteristic equation of the matrix A and the roots of the characteristic equation are called the eigenvalues or characteristic roots of the matrix A.

Example:

Let $A = \begin{bmatrix} 1 & 3 \\ 1 & -1 \end{bmatrix}.$

Then $A - \lambda I = \begin{bmatrix} 1 & 3 \\ 1 & -1 \end{bmatrix} - \lambda \begin{bmatrix} 1 & 0 \\ 0 & 1 \end{bmatrix}$

$= \begin{bmatrix} 1-\lambda & 3 \\ 1 & -1-\lambda \end{bmatrix}.$

Then the characteristic equation is given by

$|A - \lambda I| = 0$

or, $\begin{vmatrix} 1-\lambda & 3 \\ 1 & -1-\lambda \end{vmatrix} = 0$

or, $(1-\lambda)(-1-\lambda) - 3 = 0$

or, $-(1-\lambda^2) = 3$

or, $\lambda^2 = 4$

or, $\lambda = \pm 2$

Therefore, eigenvalues of A are 2 and -2.

Properties:

(i) If A be a square matrix, then A and A^T both have the same eigenvalues.

(ii) If A be a square matrix of order n, then $\det(A)$ = product of all eigenvalues of A.

(iii) If the matrix A is invertible and λ is an eigenvalues of A, then λ^{-1} will be an eigenvalue of A^{-1}

(iv) If a square matrix A have eigenvalues $\lambda_1, \lambda_2, ..., \lambda_m$ then the matrix kA have eigenvalues $k\lambda_1, k\lambda_2, ..., k\lambda_m$.

(v) If $\lambda_1, \lambda_2, ..., \lambda_k$ are eigenvalues of a square matrix A, then $\lambda_1{}^m, \lambda_2{}^m, ..., \lambda_k{}^m$ will be the eigenvalues of A^m, m being a positive integer.

(vi) If both A and P are square matrices and P is non-singular, then both the matrices A and $P^{-1} AP$ have the same eigenvalues.

(vii) If both A and B are square invertible matrices of same order, then AB and BA will have the same eigenvalues.

(viii) "0" is an eigenvalue of a square matrix A if and only if A is singular.

(ix) All the eigenvalues of an identity matrix is "1."

(x) The eigenvalues of a lower (or upper) triangular matrix are the diagonal elements of the matrix.

(xi) The eigenvalues of a diagonal matrix are the diagonal elements of the matrix.

(xii) The eigenvalues of a unitary matrix are of unit modulus, i.e., if λ be an eigenvalue of an unitary matrix A, then $|\lambda| = 1$.

(xiii) The eigenvalues of a Hermitian matrix are all real.

(xiv) The eigenvalues of a real symmetric matrix are real.

(xv) The eigenvalues of a skew-symmetric matrix are either purely imaginary or zero.

(xvi) The eigenvalues of an orthogonal matrix are of unit modulus.

(xvii) The eigenvalues of an idempotent matrix are 0 and 1.

(xviii) (*Cayley-Hamilton theorem*). Every square matrix satisfies its own characteristic equation.

Application:

If $\lambda^2 + \lambda + 3 = 0$ be the characteristic equation of a square matrix A of order 3, then we can write $A^2 + A + 3I = O$.

1.9.2. Trace of a Matrix

Let $A = [a_{ij}]_{n \times n}$ be a square matrix of order n. Then the trace of the matrix A is denoted by $\text{tr}(A)$ and is defined by

$$tr(A) = \sum_{r=1}^{n} a_{rr}$$

Thus, the trace of a given square matrix is equal to the sum of the diagonal elements of the matrix.

Example:

Let $A = \begin{pmatrix} 2 & 3 & 5 \\ 3 & 7 & 1 \\ 0 & -1 & 4 \end{pmatrix}$

Then $\text{tr}(A)$

= Sum of the diagonal elements = $2 + 7 + 4 = 13$.

Properties:

If A and B be two square matrices of same order. Then

(i) $\text{tr}(kA) = k\,\text{tr}(A)$, k being a scalar

(ii) $\text{tr}(A + B) = \text{tr}(A) + \text{tr}(B)$

(iii) $\text{tr}(AB) = \text{tr}(BA)$

(iv) $\text{tr}(A)$ = sum of all eigenvalues of A.

Remember:

(i) If $\lambda^2 + a\lambda + b$ be the characteristic polynomial of a 2×2 matrix A, then $\text{tr}(A) = -a$ and $\det(A) = b$.

(ii) If $\lambda^3 + a\lambda^2 + b\lambda + c$ be the characteristic polynomial of a 3×3 matrix A, then $\text{tr}(A) = -a$ and $\det(A) = -c$.

1.9.3. Eigenvectors or Characteristic Vectors

Let A be a square matrix of order n and be λ an eigenvalue of A. If there exists a column matrix X ($\neq 0$) of order $n \times 1$ such that $AX = \lambda X$ holds, then X is called an eigenvector or characteristic vector corresponding to the eigenvalue λ.

Remember:

If A is a square matrix of order 2, then we assume X in the form $X = \begin{pmatrix} x \\ y \end{pmatrix}$ and if A is a square matrix of order 3, then we assume X in the form

$$X = \begin{pmatrix} x \\ y \\ z \end{pmatrix}$$

Example:

Let $A = \begin{pmatrix} 1 & 3 \\ 1 & -1 \end{pmatrix}$.

Then as discussed earlier, the eigenvalue of A are 2 and -2 (see Section 1.9.1).

For $\lambda = 2$:

$AX = \lambda X$

$\Rightarrow \begin{pmatrix} 1 & 3 \\ 1 & -1 \end{pmatrix}\begin{pmatrix} x \\ y \end{pmatrix} = 2\begin{pmatrix} x \\ y \end{pmatrix} \left[\text{where } X = \begin{pmatrix} x \\ y \end{pmatrix} \right]$

$\Rightarrow \begin{pmatrix} x+3y \\ x-y \end{pmatrix} = \begin{pmatrix} 2x \\ 2y \end{pmatrix}$

Equating the corresponding elements we get, $x + 3y = 2x$ and $x - y = 2y$ i.e., $x - 3y = 0$

$\therefore X = \begin{pmatrix} x \\ y \end{pmatrix} = \begin{pmatrix} 3y \\ y \end{pmatrix} = y\begin{pmatrix} 3 \\ 1 \end{pmatrix}$

Which is the eigenvector for $\lambda = 2$

For $\lambda = -2$.

$AX = \lambda X$

$\Rightarrow \begin{pmatrix} 1 & 3 \\ 1 & -1 \end{pmatrix}\begin{pmatrix} x \\ y \end{pmatrix} = -2\begin{pmatrix} x \\ y \end{pmatrix} \left[\text{where } X = \begin{pmatrix} x \\ y \end{pmatrix} \right]$

$\Rightarrow \begin{pmatrix} x+3y \\ x-y \end{pmatrix} = \begin{pmatrix} -2x \\ -2y \end{pmatrix}$

Equating the corresponding elements we get, $x + 3y = -2x$ i.e., $x + y = 0$ i.e., $y = -x$

and $x - y = -2y$ i.e., $x + y = 0$ i.e., $y = -x$.

$\therefore X = \begin{pmatrix} x \\ y \end{pmatrix} = \begin{pmatrix} x \\ -x \end{pmatrix} = x\begin{pmatrix} 1 \\ -1 \end{pmatrix}$

which is the eigenvector for $\lambda = -2$.

1.10 Vectors

1.10.1 Introduction

An ordered n-tuple $X = (x_1, x_2, x_3, ..., x_n)$ is called a n-vector. $x_1, x_2, x_3, ..., x_n$ are called components of the vector X.

Remember:

(i) $(x_1, x_2, x_3, ..., x_n)$ is also termed as n-tuple vector.

(ii) Every row matrix as well as every column matrix is also termed as a vector. Thus,

$[2\ 5\ 1]$ and $\begin{bmatrix} 1 \\ 5 \\ 6 \end{bmatrix}$ are vectors.

(iii) A vector whose components belong to a field F is said to be a vector over the field F. The set of all n-tuple vectors over a field F (denoted by $V_n(F)$ is called a vector space over F. The elements of F are called scalars.

(iv) A subset S of a vector space $V_n(F)$ is called a vector subspace if $aX + bY \in S$ $\forall X, Y \in S$ and $a, b \in F$.

Example:

Let $S = \left\{ X = (x,y,z) \in \mathbb{R}^3 : x + y + z = 0 \right\}$

(where $\mathbb{R}^3$ is the set of all points in three dimensional space)

Then for $X = (x, y, z), Y = (x', y', z') \in S$ and $a, b \in F$,

we have

$aX + bY = a(x,y,z) + b(x',y',z')$
$= (ax,ay,az) + (bx',by',bz')$
$= (ax + bx', ay + by', az + bz')$

Now $(ax + bx') + (ay + by') + (az + bz')$
$= a(x + y + z) + b(x' + y' + z')$
$= a \times 0 + b \times 0$
$= 0$

$\begin{pmatrix} \because X, Y \in S \\ \Rightarrow x + y + z = 0, \ x' + y' + z' = 0 \end{pmatrix}$

So $aX + bY \in S$.

Hence, S is a vector sub-space.

1.10.2 Linear Dependence and Linear Independence

(i) The vectors $X_1, X_2, X_3, ..., X_k$ are said to be linearly dependent if there exist scalars $a_1, a_2, a_3, ..., a_k$ (not all zero) such that

$a_1X_1 + a_2X_2 + a_3X_3 + ... + a_kX_k = O$

Example:

Let $X_1 = (1, 0, -2)$ and $X_2 = (-2, 0, 4)$. Then,

$$2X_1 + X_2 = 2(1, 0, -2) + (-2, 0, 4)$$
$$= (2 - 2, 0 + 0, -4 + 4)$$
$$= (0, 0, 0) = O$$

Therefore, the vectors X_1 and X_2 are linearly dependent.

(ii) The vectors $X_1, X_2, X_3, ..., X_k$ are said to be linearly independent if there exist scalars $a_1, a_2, a_3, ..., a_k$ such that

$$a_1 X_1 + a_2 X_2 + a_3 X_3 + ... + a_k X_k = O$$
$$\Rightarrow a_1 = a_2 = a_3 = ... = a_k = 0$$

Example:

Let $X_1 = (1, 0 - 2)$ and $X_2 = (0, 3, 0)$.

Then

$$a_1 X_1 + a_2 X_2 = O$$
$$\Rightarrow a_1(1, 0, -2) + a_2(0, 3, 0) = O$$
$$\Rightarrow (a_1, 0, -2a_1) + (0, 3a_2, 0) = O$$
$$\Rightarrow (a_1 + 0, 0 + 3a_2, -2a_1 + 0) = (0, 0, 0)$$
$$\Rightarrow a_1 = 0, 3a_2 = 0, -2a_1 = 0$$
$$\Rightarrow a_1 = a_2 = 0$$

Therefore, the vectors X_1 and X_2 are linearly independent.

Remember:

(i) Rank of a matrix = number of linearly independent eigenvectors of the matrix.

(ii) Rank of a matrix = number of linearly independent columns (or rows) of the matrix.

(iii) If A be a matrix of order $n \times n$ such that A has "n" distinct eigenvalues, then A has "n" linearly independent eigenvectors.

(iv) Two functions $f(x)$ and $g(x)$ are said to be linearly independent if $\begin{vmatrix} f(x) & g(x) \\ f'(x) & g'(x) \end{vmatrix} \neq 0$; otherwise they called linearly dependent.

(v) Three functions $f(x)$, $g(x)$ and $h(x)$ are said to be linearly independent if

$$\begin{vmatrix} f(x) & g(x) & h(x) \\ f'(x) & g'(x) & h'(x) \\ f''(x) & g''(x) & h''(x) \end{vmatrix} \neq 0; \text{ otherwise they are}$$

called linearly dependent.

(vi) Three vectors $X_1 = (x_1, x_2, x_3)$, $X_2 = (y_1, y_2, y_3)$ and $X_3 = (z_1, z_2, z_3)$ are said to be linearly independent if $\begin{vmatrix} x_1 & x_2 & x_3 \\ y_1 & y_2 & y_3 \\ z_1 & z_2 & z_3 \end{vmatrix} \neq 0$; otherwise they are said to be linearly dependent.

Example:

$$\because \begin{vmatrix} 1 & -1 & 2 \\ 3 & 0 & 5 \\ 2 & -2 & 4 \end{vmatrix} = 0,$$

the vectors $(1, -1, 2), (3, 0, 5)$ and $(2, -2, 4)$ are linearly dependent.

1.10.3 Inner Product and Norm of Vectors

(i) The inner product of two vectors $X = (x_1, x_2, x_3, ..., x_n)$ and $Y = (y_1, y_2, y_3, ..., y_n)$ over the field R of real numbers is denoted by $<X, Y>$ and is defined by

$$<X, Y> = x_1 y_1 + x_2 y_2 + x_3 y_3 + ... + x_n y_n$$

Example:

Let $X_1 = (1, 0, -2)$ and $X_2 = (3, 3, 4)$.

Then $\langle X, Y \rangle = 1 \times 3 + 0 \times 3 + (-2) \times 4 = -5$.

(ii) The norm of a vector $X = (x_1, x_2, x_3, ... x_n)$ over the field R of real numbers is denoted by $\|X\|$ and is defined by

$$\|X\| = \sqrt{x_1^2 + x_2^2 + x_3^2 + ... + x_n^2}$$

Example:

Let $X = (1, 3, -2)$. Then $\|X\| = \sqrt{1^2 + 3^2 + (-2)^2} = \sqrt{14}$.

Remember:

If θ be the angle between the two vectors X and Y, then $\cos\theta = \dfrac{\langle X, Y \rangle}{\|X\| \|Y\|}$.

Example:

Let $X_1 = (2, 1, -2)$ and $X_2 = (3, 0, 4)$.

Then $\langle X, Y \rangle = 2 \times 3 + 0 \times 1 + (-2) \times 4 = -2$. Also

$$\|X\| = \sqrt{2^2 + 1^2 + (-2)^2} = 3,$$
$$\|Y\| = \sqrt{3^2 + 0^2 + 4^2} = 5.$$

$$\therefore \cos\theta = \frac{\langle X, Y\rangle}{\|X\|\|Y\|} = \frac{-2}{3\times 5} = -\frac{2}{15}.$$

Now, So $\theta = \cos^{-1}\left(-\frac{2}{15}\right)$.

Remember : $\|x\|^2 = \langle x, x\rangle$.

1.10.4 Orthogonal and Orthonormal Vectors

(i) Two vectors X and Y are said to be orthogonal if $\langle X, Y\rangle = 0$.

Example:

Let $X_1 = (2, 1, -2)$ and $X_2 = (-3, 0, -3)$.

Then $\langle X, Y\rangle = 2\times(-3) + 1\times 0 + (-2)\times(-3) = 0$.

Therefore, the vectors X and Y are orthogonal.

(ii) A set of n-tuple vectors $X_1, X_2, X_3, ..., X_k$ is called an orthonormal set of vectors if

(i) $\|X_i\| = 1$ for $i = 1, 2, 3, ..., k$

(ii) $\langle X_i, Y_j\rangle = 0$ for $i \neq j;\ i, j = 1, 2, 3, ..., k$

Example:

Let $X_1 = \frac{1}{\sqrt{3}}(1,1,1)$ and $X_2 = \frac{1}{\sqrt{2}}(-1,0,1)$.

Then

$$X_1 = \frac{1}{\sqrt{3}}(1,1,1) = \left(\frac{1}{\sqrt{3}},\frac{1}{\sqrt{3}},\frac{1}{\sqrt{3}}\right) \text{ and}$$

$$X_2 = \frac{1}{\sqrt{2}}(-1,0,1) = \left(\frac{-1}{\sqrt{2}},0,\frac{1}{\sqrt{2}}\right).$$

$$\therefore \langle X_1, X_2\rangle = \frac{1}{\sqrt{3}}\times\left(\frac{-1}{\sqrt{2}}\right) + 0 + \frac{1}{\sqrt{3}}\times\frac{1}{\sqrt{2}}$$
$$= 0$$

Therefore, the vectors X_1 and X_2 are orthogonal.

Also

$$\|X_1\| = \sqrt{\left(\frac{1}{\sqrt{3}}\right)^2 + \left(\frac{1}{\sqrt{3}}\right)^2 + \left(\frac{1}{\sqrt{3}}\right)^2} = 1,$$

$$\|X_2\| = \sqrt{\left(-\frac{1}{\sqrt{2}}\right)^2 + 0^2 + \left(\frac{1}{\sqrt{2}}\right)^2} = 1$$

Hence, the vectors X_1 and X_2 form an orthonormal set of vectors.

1.10.5 Basis and Dimension

A collection of n-tuple vectors $X_1, X_2, X_3, ... X_k$ is said to form a basis of the vector space $V_n(F)$ if

(i) $X_1, X_2, X_3, ..., X_k$ are linearly independent

(ii) any arbitrary n-tuple vector X can be expressed as

$$X = b_1 X_1 + b_2 X_2 + b_3 X_3 + ... + b_k X_k$$

where $b_1, b_2, b_3, ..., b_k \in F$.

If the n-tuple vectors $X_1, X_2, X_3, ..., X_k$ forms a basis of $V_n(F)$, then we say that dimension of the vector space $V_n(F)$ is "k."

Example:

Let $X_1 = (1,0,0), X_2 = (0,1,0)$ and $X_3 = (0,0,1)$.

Then $\begin{vmatrix} 1 & 0 & 0 \\ 0 & 1 & 0 \\ 0 & 0 & 1 \end{vmatrix} = 1 \neq 0.$

Therefore, the vectors X_1, X_2, and X_3 are linearly independent.

Now let $X = (a,b,c)$ be any arbitrary 3-tuple vector.

Then we can write

$$X = (a,b,c) = a(1,0,0) + b(0,1,0) + c(0,0,1)$$
$$= aX_1 + bX_2 + cX_3$$

Hence, the vectors X_1, X_2, and X_3, form a basis of $V_3(R)$ and dimension of $V_3(R)$ is "3."

Fully Solved MCQs (Level-I)

1. For the matrices $A_{3\times 1}, B_{1\times 3}, C_{3\times 5}, D_{5\times 3}$ which of the followings is possible?
(a) AB only   (b) CD only
(c) AB and CD only
(d) AB, BA, BC, CD, DA, DC only.

2. A square matrix "A" of order 2 which commutes with every 2×2 matrix is of the form

(a) $\begin{pmatrix} 0 & a \\ a & 0 \end{pmatrix}$ (b) $\begin{pmatrix} a & 0 \\ 0 & a \end{pmatrix}$

(c) $\begin{pmatrix} a & 0 \\ 0 & b \end{pmatrix}$ (d) $\begin{pmatrix} 0 & a \\ b & 0 \end{pmatrix}$

3. If $A = \begin{pmatrix} 0 & \alpha \\ \beta & 0 \end{pmatrix}$, then $A^3 + A = 0$, whenever
(a) $\alpha\beta = 2$ (b) $\alpha\beta = 1$
(c) $\alpha\beta \neq 0$ (d) $\alpha\beta = -1$

4. A square matrix becomes a diagonal matrix if and only if
    (a) it is upper triangular
    (b) it is lower triangular
    (c) both lower and upper triangular
    (d) none of these

5. If a matrix A is symmetric as well as skew-symmetric, then A is
    (a) Diagonal matrix   (b) Null matrix
    (c) Unit matrix   (d) None of these

6. Let $A = \begin{pmatrix} 6 & x \\ y & 0 \end{pmatrix}$ and $A = A'$. Then
    (a) $x = 0, y = 6$ (b) $x + y = 6$
    (c) $x = y$ (d) None of these

7. If $A = \begin{pmatrix} 1 & -2 & 3 \\ -4 & 2 & 7 \end{pmatrix}$, $B = \begin{pmatrix} 2 & 3 \\ 4 & 5 \\ 2 & 0 \end{pmatrix}$, then
    (a) AB, BA exist but not equal
    (b) AB exists but BA does not exist
    (c) AB does not exist but BA exist
    (d) AB, BA exist and both are equal

8. If A and B are square matrices of order 3 such that $\det(A) = -1$, $\det(B) = 3$; then $\det(3AB) = ?$
    (a) 9   (b) 81
    (c) 27   (d) −81

9. If $A = \begin{pmatrix} 4 & 5 \\ 1 & 2 \end{pmatrix}$, then $A.$ adj $A = ?$
    (a) $\begin{pmatrix} 3 & 0 \\ 0 & 3 \end{pmatrix}$ (b) $\begin{pmatrix} -3 & 0 \\ 0 & -3 \end{pmatrix}$
    (c) $\begin{pmatrix} 10 & 0 \\ 0 & 10 \end{pmatrix}$ (d) $\begin{pmatrix} 1 & 3 \\ -3 & 0 \end{pmatrix}$

10. If A, B, C are square matrix of the same order, then $AB = AC \Rightarrow B = C$ if
    (a) $|A| \neq 0$ (b) $|A| = 0$ (c) $A = I$ (d) $A = 0$

11. If ω is a complex cube root of unity, then the matrix $A = \begin{pmatrix} 1 & \omega & \omega^2 \\ \omega^2 & \omega & 1 \\ \omega & 1 & \omega^2 \end{pmatrix}$ is
    (a) Singular   (b) Non-singular
    (c) Symmetric   (d) None of these

12. If $A = \begin{pmatrix} 2x & 0 \\ x & x \end{pmatrix}$ and $A^{-1} = \begin{pmatrix} 1 & 0 \\ -1 & 2 \end{pmatrix}$, then $x = ?$
    (a) 1   (b) 2
    (c) 1/2   (d) −1

13. If A is a square matrix, then adj(A') − (adj A)′is equal to
    (a) $2|A|$ (b) $2|A|I$
    (c) null matrix   (d) I

14. If A is an idempotent matrix and $A + B = I$, then B is
    (a) idempotent   (b) involutary
    (c) null matrix   (d) none of these

15. If A is a real skew-symmetric matrix such that $A^2 + I = 0$, then A is
    (a) idempotent   (b) involutary
    (c) null matrix   (d) orthogonal

16. The matrix $A = \begin{pmatrix} ab & b^2 \\ -a^2 & -ab \end{pmatrix}$ is
    (a) Idempotent   (b) Orthogonal
    (c) Nilpotent   (d) None of these

17. The matrix $\begin{pmatrix} \frac{1}{\sqrt{3}} & \frac{1}{\sqrt{6}} & \frac{-1}{\sqrt{2}} \\ \frac{1}{\sqrt{3}} & \frac{-2}{\sqrt{6}} & 0 \\ \frac{1}{\sqrt{3}} & \frac{1}{\sqrt{6}} & \frac{1}{\sqrt{2}} \end{pmatrix}$ is
    (a) Orthogonal   (b) Idempotent
    (c) Symmetric   (d) None of these

18. The matrix $A = \frac{1}{\sqrt{2}}\begin{pmatrix} 1 & i \\ -i & -1 \end{pmatrix}$ is
    (a) Unitary   (b) Idempotent
    (c) Symmetric   (d) None of these

19. For the matrix $M = \begin{bmatrix} 7 & 1+i & 2 \\ 1-i & 4 & 3i \\ 2 & -3i & 0 \end{bmatrix}$ which of the following is correct?
    (a) M is skew-Hermitian and iM is Hermitian
    (b) M is Hermitian and iM is Skew-Hermitian
    (c) M and iM both are Hermitian
    (d) M and iM both are Skew-Hermitian

20. If the rank of the matrix $A = \begin{pmatrix} 2 & 4 & 2 \\ 3 & 1 & 2 \\ 1 & 0 & x \end{pmatrix}$ is "3," then the value of x is?
    (a) $\neq \frac{3}{5}$ (b) $\neq \frac{4}{5}$
    (c) $\neq \frac{2}{5}$ (d) none of these

21. The rank of the unit matrix I of order n is

(a) $n - 1$ (b) n
(c) n^2 (d) $n+1$

22. If A is a non-singular matrix of order n, then the rank of A is

(a) $n - 1$ (b) n
(c) 2 (d) $n+1$

23. The rank of the matrix $\begin{pmatrix} 1 & 1 & 1 \\ 2 & 3 & 4 \\ 4 & 6 & 8 \end{pmatrix}$ is

(a) 2 (b) 0
(c) 1 (d) 3

24. If $A = \begin{pmatrix} 0 & 2 & 0 & 0 \\ 0 & 0 & 3 & 0 \\ 0 & 0 & 0 & 4 \\ 0 & 0 & 0 & 0 \end{pmatrix}$, then rank $(A) =$?

(a) 2 (b) 0
(c) 1 (d) 3

25. The rank of the matrix $\begin{pmatrix} 1 & 2 & 3 & 4 \\ 0 & 4 & 1 & 2 \\ 0 & 0 & 5 & 0 \end{pmatrix}$ is

(a) 1 (b) 2
(c) 3 (d) 4

26. The system of given equations $x + y + z = 6$, $x - y + z = 2$, $2x + y - z = 1$ has

(a) unique solution (b) two solutions
(c) an infinite number of solutions
(d) no solution

27. A system of "m" homogeneous linear equations $AX = 0$ in "n" unknowns has only trivial solution if

(a) $m \neq n$ (b) $m = n$
(c) rank $(A) = m$ (d) rank $(A) = n$

28. The system of equations $4x + 6y = 5$, $6x + 9y = 7$ has

(a) a unique solution
(b) an infinite number of solutions
(c) no solution
(d) finite number of solutions

29. The equations $2x - 3y + 6z = 4$, $5x + 7y - 14z = 1$, $3x + 2y - 4z = 0$ has

(a) a unique solution (b) no solution
(c) infinitely many solutions
(d) none of these

30. The values of "a" for which the system of equations $ax + y + z = 0$, $x + ay + z = 0$, $x + y + z = 0$ posses non-zero solutions are given by

(a) 1, 2 (b) 1, 1
(c) 1, − 1 (d) − 1, − 2

31. The system of equations $x + 2y + 3z = 1$, $2x + y + 3z = 2$, $5x + 5y + 9z = 4$ has

(a) a unique solution
(b) no solution
(c) infinitely many solutions
(d) none of these

32. The system of equations $x + y + z = 0$, $2x + y - z = 0$, $3x + 2y = 0$ has

(a) a unique solution
(b) no solution
(c) infinitely many solutions
(d) none of these

33. The system of equations $x + 2y + 3z = 0$, $2x + 3y + 4z = 0$, $3x + 4y + 5z = 0$ has

(a) unique solution
(b) an infinite number of solutions
(c) trivial solution
(d) none of these

34. The system of equations $x - 2y + z = 0$, $x - 2y - z = 0$, $2x - 4y - 5z = 0$ has

(a) unique solution
(b) an infinite number of solutions
(c) trivial solution
(d) a none of these

35. Consider the system of equations $5x + 2y - z = 1$, $2x + 3y + 4z = 7$, $4x - 5y + \lambda z = \lambda - 5$

It will have a unique solution if

(a) $\lambda = 14$ (b) $\lambda \neq 14$
(c) $\lambda \neq -14$ (d) $\lambda = -14$

36. The system of equations

$x_1 + x_2 + x_3 + x_4 = 0$,
$x_1 + 3x_2 + 2x_3 + 4x_4 = 0$,
$2x_1 + x_3 - x_4 = 0$.

has

(a) unique solution
(b) an infinite number of solutions
(c) trivial solution
(d) none of these

37. The system of equations $x + 2y + 3z = 0$, $2x + 3y + z = 0$, $3x + y + 2z = 0$ has

(a) two solutions
(b) an infinite number of solutions
(c) trivial solution
(d) none of these

38. If $A = \begin{pmatrix} 1 & 2 \\ -1 & 3 \end{pmatrix}$, , then $A^4 = ?$

(a) $24 A - 55I$ (b) $-4A + 55I$
(c) $A + I$ (d) $A - I$

39. The eigenvalues of $A = \begin{pmatrix} 2 & 1 & 0 \\ 1 & 2 & 0 \\ 1 & 1 & 1 \end{pmatrix}$ are

(a) 1, 2, 3   (b) 1, 3, 3
(c) 1, 1, 3   (d) 1, 1, 1

40. The eigenvalue of $A = \begin{pmatrix} 2 & 0 & 1 \\ 0 & 1 & 0 \\ 1 & 0 & 2 \end{pmatrix}$ are

(a) 1, 3, 1   (b) 1, 2, 3
(c) 0, 2, 3   (d) 1, 1, 1

41. The eigenvalue of $A = \begin{pmatrix} 2 & -2 & 0 \\ -2 & 1 & -2 \\ 0 & -2 & 0 \end{pmatrix}$ are

(a) 1, 2, 4   (b) 1, -2, 4
(c) 1, -2, -4   (d) -1, 2, 4

42. If $A = \begin{pmatrix} 1 & 0 & 0 \\ 1 & 0 & 1 \\ 0 & 1 & 0 \end{pmatrix}$, then $A^{-1} = ?$

(a) $A^2 + A - I$ (b) $A^2 - A - I$
(c) $A^2 + A + I$ (d) None of these

43. The eigenvalue and eigenvector of $A = \begin{pmatrix} 0 & 1 \\ 0 & 0 \end{pmatrix}$ are, respectively

(a) $0, \begin{pmatrix} k \\ 0 \end{pmatrix}$ (b) $1, \begin{pmatrix} 0 \\ k \end{pmatrix}$

(c) $0, \begin{pmatrix} k \\ k \end{pmatrix}$ (d) $1, \begin{pmatrix} k \\ k \end{pmatrix}$

44. If $\begin{bmatrix} 1 \\ -1 \end{bmatrix}$ is an eigenvector of $\begin{bmatrix} 1 & -n \\ -3 & 2n \end{bmatrix}$, then "$n$" = ?

(a) 2   (b) -2
(c) 3   (d) 1

45. Let P be a 3×3 matrix with real entries such that det $(P) = 6$ and trace $(P) = 0$. If

det $(P + I) = 0$, where I is the identity matrix, then the eigenvalues of P are?

(a) -1, -2, 3   (b) 4, -4, 0
(c) 1, 2, 3   (d) 1, 1, 6

46. If the vectors $(0, 1, a)$, $(1, a, 1)$ and $(a, 1, 0)$ are linearly dependent, then the value of "a" will be

(a) 0, 1   (b) 1, -1

(c) $0, \pm\sqrt{2}$ (d) 0

47. If the vectors (a, b) and (c, d) are linearly dependent, then which of the following is true?
(a) $ab = ac$ (b) $a + c = b + d$
(c) $a = d = 0$ (d) $ad - bc = 0$

48. If the vectors $(1, 0, 1)$, $(1, 1, 0)$ and (p, q, r) are linearly dependent, then which of the following is true?
(a) $p + q + r = 0$ (b) $p = q = r$
(c) $p = q + r$ (d) $r = p + q$

Answer key				
1. (d)	**2.** (b)	**3.** (d)	**4.** (c)	**5.** (b)
6. (c)	**7.** (a)	**8.** (d)	**9.** (a)	**10.** (a)
11. (a)	**12.** (c)	**13.** (c)	**14.** (a)	**15.** (d)
16. (c)	**17.** (a)	**18.** (a)	**19.** (b)	**20.** (a)
21. (b)	**22.** (b)	**23.** (a)	**24.** (d)	**25.** (c)
26. (a)	**27.** (b)	**28.** (c)	**29.** (b)	**30.** (b)
31. (a)	**32.** (c)	**33.** (b)	**34.** (b)	**35.** (c)
36. (b)	**37.** (c)	**38.** (a)	**39.** (c)	**40.** (a)
41. (b)	**42.** (d)	**43.** (a)	**44.** (b)	**45.** (a)
46. (c)	**47.** (d)	**48.** (c)		

Explanation

1. (d) The number of column of A is 1 and the number of rows of B is 1. So AB is possible.

The number of column of B is 3 and the number of rows of both A and C are 3. So BA and BC both are defined.

The number of column of C is 5 and the number of rows of D is 5. So CD is possible.

The number of column of D is 3 and the number of rows of both A and C are 3. So DA and DC are possible.

2. (b) Let, $B = \begin{pmatrix} x & y \\ z & t \end{pmatrix}$ be any arbitrary 2×2

matrix and $A = \begin{pmatrix} a & 0 \\ 0 & a \end{pmatrix}$. Then

$$AB = \begin{pmatrix} a & 0 \\ 0 & a \end{pmatrix}\begin{pmatrix} x & y \\ z & t \end{pmatrix} = \begin{pmatrix} ax & ay \\ az & at \end{pmatrix},$$

$$BA = \begin{pmatrix} x & y \\ z & t \end{pmatrix}\begin{pmatrix} a & 0 \\ 0 & a \end{pmatrix} = \begin{pmatrix} ax & ay \\ az & at \end{pmatrix}$$

Hence, $AB = BA$ which means A commutes with B.

3. (d) $A^2 = AA = \begin{pmatrix} 0 & \alpha \\ \beta & 0 \end{pmatrix}\begin{pmatrix} 0 & \alpha \\ \beta & 0 \end{pmatrix} = \begin{pmatrix} \alpha\beta & 0 \\ 0 & \alpha\beta \end{pmatrix}$

$A^3 = A^2 A = \begin{pmatrix} \alpha\beta & 0 \\ 0 & \alpha\beta \end{pmatrix}\begin{pmatrix} 0 & \alpha \\ \beta & 0 \end{pmatrix} = \begin{pmatrix} 0 & \alpha^2\beta \\ \alpha\beta^2 & 0 \end{pmatrix}$

Then,

$A^3 + A = 0 \Rightarrow \begin{pmatrix} 0 & \alpha^2\beta \\ \alpha\beta^2 & 0 \end{pmatrix} + \begin{pmatrix} 0 & \alpha \\ \beta & 0 \end{pmatrix} = \begin{pmatrix} 0 & 0 \\ 0 & 0 \end{pmatrix}$

$\Rightarrow \begin{pmatrix} 0 & \alpha^2\beta+\alpha \\ \alpha\beta^2+\beta & 0 \end{pmatrix} = \begin{pmatrix} 0 & 0 \\ 0 & 0 \end{pmatrix}$

$\Rightarrow \alpha^2\beta + \alpha = 0, \alpha\beta^2 + \beta = 0$

$\Rightarrow \alpha(\alpha\beta + 1) = 0, \beta(\alpha\beta + 1) = 0$

$\Rightarrow \alpha\beta + 1 = 0$ (assuming $\alpha \neq 0, \beta \neq 0$)

$\Rightarrow \alpha\beta = -1$

4. (c) Consider $A = \begin{pmatrix} a & b & c \\ d & e & f \\ g & h & k \end{pmatrix}$

If A is upper as well as lower triangular, then the elements lying below and above of the principal diagonal will be all zero. So $b = c = d = f = g = h = 0$.

Hence, A becomes $\begin{pmatrix} a & 0 & 0 \\ 0 & e & 0 \\ 0 & 0 & k \end{pmatrix}$, which is

clearly a diagonal matrix.

5. (b) A is symmetric $\Rightarrow A^T = A$ ...(i)

A is skew-symmetric $\Rightarrow A^T = -A$ ...(ii)

Adding (i) and (ii) we get $2A^T = O$

$\Rightarrow (A^T)^T = O^T \Rightarrow A = O$.

6. (c) $A = A' \Rightarrow \begin{pmatrix} 6 & x \\ y & 0 \end{pmatrix} = \begin{pmatrix} 6 & y \\ x & 0 \end{pmatrix}$

Then equating the corresponding elements we get $x = y$.

7. (a) A is a matrix of order 2×3 and B is a matrix of order 3×2. Therefore, AB and BA are defined.

Also AB is a matrix of order 2×2 and BA is a matrix of order 3×3. So obviously they are not equal.

8. (d) Det $(3AB)$

$= 3^3 \det(A) \det(B) = 27 \times (-1) \times 3$

$= -81$

(since AB is a square matrix of order "3").

9. (a) $|A| = \begin{vmatrix} 4 & 5 \\ 1 & 2 \end{vmatrix} = 4 \times 2 - 5 \times 1 = 3$

Then

$A \cdot \text{adj } A = |A|I = 3\begin{pmatrix} 1 & 0 \\ 0 & 1 \end{pmatrix} = \begin{pmatrix} 3 & 0 \\ 0 & 3 \end{pmatrix}$

10. (a) $|A| \neq 0 \Rightarrow A^{-1}$ exist.

$\therefore \quad AB = AC \Rightarrow A^{-1}(AB) = A^{-1}(AC)$

$\Rightarrow \quad (A^{-1}A)B = (A^{-1}A)C$

$\Rightarrow \quad IB = IC$

$\Rightarrow \quad B = C$

11. (a)

$|A| = \begin{vmatrix} 1 & \omega & \omega^2 \\ \omega^2 & \omega & 1 \\ \omega & 1 & \omega^2 \end{vmatrix} = \begin{vmatrix} 1+\omega+\omega^2 & \omega & \omega^2 \\ 1+\omega+\omega^2 & \omega & 1 \\ 1+\omega+\omega^2 & 1 & \omega^2 \end{vmatrix}$

$\left(\text{by } C_1 \to C_1 + C_2 + C_3\right)$

$= \begin{vmatrix} 0 & \omega & \omega^2 \\ 0 & \omega & 1 \\ 0 & 1 & \omega^2 \end{vmatrix} = 0$

$\left(\because 1+\omega+\omega^2 = 0\right)$

Hence, A is singular.

12. (c)

$$|A| = \begin{vmatrix} 2x & 0 \\ x & x \end{vmatrix} = 2x^2$$

$$\therefore A^{-1} = \frac{1}{|A|} adj(A)$$

$$= \frac{1}{2x^2} \begin{pmatrix} x & -x \\ 0 & 2x \end{pmatrix}^T$$

$$= \frac{1}{2x^2} \begin{pmatrix} x & 0 \\ -x & 2x \end{pmatrix}$$

$$= \begin{pmatrix} \dfrac{1}{2x} & 0 \\ -\dfrac{1}{2x} & \dfrac{1}{x} \end{pmatrix} = \begin{pmatrix} 1 & 0 \\ -1 & 2 \end{pmatrix} \text{(given)}$$

$$\Rightarrow \frac{1}{2x} = 1 \Rightarrow x = \frac{1}{2}.$$

13. (c) $adj(A') - (adj\, A') = adj\,(A') - adj(A')$
  $= O =$ null matrix.

14. (a) $A + B = I \Rightarrow B = I - A$
  $\Rightarrow B^2 = (I - A)^2 = I - 2IA + A^2$
  $\Rightarrow B^2 = I - 2A + A$ (since A is idempotent,
        so $A^2 = A$)
  $\Rightarrow B^2 = I - A = B$
  Hence, B is idempotent.

15. (d) A is skew-symmetric $\Rightarrow A^T = -A$
  $\Rightarrow AA^T = -AA = -A^2 = I$ (using $A^2 + I = 0$).
  Hence, A is orthogonal.

16. (c) $A^2 = A \times A = \begin{pmatrix} ab & b^2 \\ -a^2 & -ab \end{pmatrix} \times \begin{pmatrix} ab & b^2 \\ -a^2 & -ab \end{pmatrix}$

$$= \begin{pmatrix} a^2b^2 - a^2b^2 & ab^3 - ab^3 \\ -a^2b + ab & -a^2b^2 + a^2b^2 \end{pmatrix} = O$$

Hence, A is a nilpotent matrix of index 2.

17. (a)

$$AA^T = \begin{pmatrix} \dfrac{1}{\sqrt{3}} & \dfrac{1}{\sqrt{6}} & \dfrac{-1}{\sqrt{2}} \\ \dfrac{1}{\sqrt{3}} & \dfrac{-2}{\sqrt{6}} & 0 \\ \dfrac{1}{\sqrt{3}} & \dfrac{1}{\sqrt{6}} & \dfrac{1}{\sqrt{2}} \end{pmatrix} \begin{pmatrix} \dfrac{1}{\sqrt{3}} & \dfrac{1}{\sqrt{3}} & \dfrac{1}{\sqrt{3}} \\ \dfrac{1}{\sqrt{6}} & \dfrac{-2}{\sqrt{6}} & \dfrac{1}{\sqrt{6}} \\ \dfrac{-1}{\sqrt{2}} & 0 & \dfrac{1}{\sqrt{2}} \end{pmatrix}$$

$$= \begin{vmatrix} \dfrac{1}{3}+\dfrac{1}{6}+\dfrac{1}{2} & \dfrac{1}{3}-\dfrac{2}{6}+0 & \dfrac{1}{3}+\dfrac{1}{6}-\dfrac{1}{2} \\ \dfrac{1}{3}-\dfrac{2}{6}+0 & \dfrac{1}{3}+\dfrac{4}{6}+0 & \dfrac{1}{3}-\dfrac{2}{6}+0 \\ \dfrac{1}{3}+\dfrac{1}{6}-\dfrac{1}{2} & \dfrac{1}{3}-\dfrac{2}{6}+0 & \dfrac{1}{3}+\dfrac{1}{6}+\dfrac{1}{2} \end{vmatrix}$$

$$= \begin{pmatrix} 1 & 0 & 0 \\ 0 & 1 & 0 \\ 0 & 0 & 1 \end{pmatrix} = I$$

So A is orthogonal.

18. (a)

$$A = \frac{1}{\sqrt{2}} \begin{pmatrix} 1 & i \\ -i & -1 \end{pmatrix}$$

$$\Rightarrow A^\theta = \frac{1}{\sqrt{2}} \begin{pmatrix} 1 & -i \\ i & -1 \end{pmatrix}^T = \frac{1}{\sqrt{2}} \begin{pmatrix} 1 & i \\ -i & -1 \end{pmatrix}$$

$$\therefore AA^\theta = \frac{1}{\sqrt{2}} \begin{pmatrix} 1 & i \\ -i & -1 \end{pmatrix} \times \frac{1}{\sqrt{2}} \begin{pmatrix} 1 & i \\ -i & -1 \end{pmatrix}$$

$$= \frac{1}{2} \begin{pmatrix} 1-i^2 & i-i \\ -i+i & -i^2+1 \end{pmatrix}$$

$$= \frac{1}{2} \begin{pmatrix} 2 & 0 \\ 0 & 2 \end{pmatrix} = \begin{pmatrix} 1 & 0 \\ 0 & 1 \end{pmatrix} = I$$

Similarly $AA^\theta = I$. Hence, $A^\theta A = AA^\theta = I$. So A is unitary.

19. (b)

$$\bar{M} = \begin{bmatrix} 7 & 1-i & 2 \\ 1+i & 4 & -3i \\ 2 & 3i & 0 \end{bmatrix} \text{ and so}$$

$$M^\theta = (\bar{M})^T = \begin{bmatrix} 7 & 1+i & 2 \\ 1-i & 4 & 3i \\ 2 & -3i & 0 \end{bmatrix}^T$$

$$= \begin{bmatrix} 7 & 1-i & 2 \\ 1+i & 4 & -3i \\ 2 & 3i & 0 \end{bmatrix} = M$$

Hence, M is Hermitian matrix and so iM is skew-Hermitian.

20. (a) Rank $(A) = 3$
  $\Rightarrow A$ has non-zero determinant
  $\Rightarrow \det(A) \neq 0.$

$$\Rightarrow \begin{vmatrix} 2 & 4 & 2 \\ 3 & 1 & 2 \\ 1 & 0 & x \end{vmatrix} \neq 0$$

$$\Rightarrow 2(x-0) - 4(3x-2) + 2(0-1) \neq 0$$

$$\Rightarrow -10x + 6 \neq 0$$

$$\Rightarrow x \neq \frac{3}{5}.$$

21. (b) Since I is an unit matrix of order "n" and det $(I) = 1 \neq 0$. So rank (I) = order $(I) = n$.

22. (b) Since A is non-singular, so det $(A) \neq 0$.
Hence, rank (A) = order of $A = n$.

23. (a)

$$\begin{pmatrix} 1 & 1 & 1 \\ 2 & 3 & 4 \\ 4 & 6 & 8 \end{pmatrix} \sim \begin{pmatrix} 1 & 1 & 1 \\ 2 & 3 & 4 \\ 0 & 0 & 0 \end{pmatrix} (by \ R_3 \rightarrow R_3 - 2R_2)$$

which has two non-zero rows.
Hence, rank = 2.

24. (d) A has only three non-zero rows. Hence, rank $(A) = 3$.

25. (c)

Let, $A = \begin{pmatrix} 1 & 2 & 3 & 4 \\ 0 & 4 & 1 & 2 \\ 0 & 0 & 5 & 0 \end{pmatrix}$

Then A is a matrix of order 3×4. Therefore, rank $(A) \leq \min \{3, 4\} = 3$.
Now consider a sub matrix B of order 3, where

$B = \begin{pmatrix} 1 & 2 & 3 \\ 0 & 4 & 1 \\ 0 & 0 & 5 \end{pmatrix}$. Then det $(B) = 1 \times 4 \times 5 = 20 \neq 0$

Therefore, rank (A) = order of the square sub-matrix $B = 3$.

Alternative method:

The given matrix has three non-zero rows. Hence, its rank = 3.

26. (a)

Here $\Delta = \begin{vmatrix} 1 & 1 & 1 \\ 1 & -1 & 1 \\ 2 & 1 & -1 \end{vmatrix} = 6 \neq 0$.

Therefore by Cramer's rule, the given system of equations has a unique solution.

27. (b) Any system of "n" homogeneous linear equations in "n" unknowns has a trivial solution.

28. (c)

$$\Delta = \begin{vmatrix} 4 & 6 \\ 6 & 9 \end{vmatrix} = 36 - 36 = 0$$

$$\Delta_1 = \begin{vmatrix} 5 & 6 \\ 7 & 9 \end{vmatrix} = 45 - 42 = 3 \neq 0$$

Therefore, by Cramer's rule, the system has no solution.

29. (b)

$$\text{Here } \Delta = \begin{vmatrix} 2 & -3 & 6 \\ 5 & 7 & -14 \\ 3 & 2 & -4 \end{vmatrix} = 0,$$

$$\Delta_1 = \begin{vmatrix} 4 & -3 & 6 \\ 1 & 7 & -14 \\ 0 & 2 & -4 \end{vmatrix} = 0,$$

$$\Delta_2 = \begin{vmatrix} 2 & 4 & 6 \\ 5 & 1 & -14 \\ 3 & 0 & -4 \end{vmatrix} \neq 0$$

Therefore, by Cramer's rule, the system has no solution.

30. (b) The homogeneous system has a non-zero solution
$\Rightarrow$ rank of the co-efficient matrix < 3
(here number of unknowns = 3)

$$\Rightarrow |A| = 0$$

$$\Rightarrow \begin{vmatrix} a & 1 & 1 \\ 1 & a & 1 \\ 1 & 1 & 1 \end{vmatrix} = 0$$

$$\Rightarrow a^2 - 2a + 1 = 0$$

$$\Rightarrow a = 1, 1$$

31. (a)

Here $\Delta = \begin{vmatrix} 1 & 2 & 3 \\ 2 & 1 & 3 \\ 5 & 5 & 9 \end{vmatrix} = 3 \neq 0$.

Therefore, by Cramer's rule, the system has a unique solution.

32. (c)

$$\text{Here } |A| = \begin{vmatrix} 1 & 1 & 1 \\ 2 & 1 & -1 \\ 3 & 2 & 0 \end{vmatrix} = 0.$$

Therefore, rank of the coefficient matrix < 3 (here number of variables = 3).

Hence, the system has an infinite number of solutions.

33. (b)

$$A = \begin{pmatrix} 1 & 2 & 3 \\ 2 & 3 & 4 \\ 3 & 4 & 5 \end{pmatrix}$$

$$\sim \begin{pmatrix} 1 & 2 & 3 \\ 0 & -1 & -2 \\ 0 & -2 & -4 \end{pmatrix}$$

(by $R_2 \to R_2 - 2R_1, R_3 \to R_3 - 3R_1$)

$$\sim \begin{pmatrix} 1 & 2 & 3 \\ 0 & -1 & -2 \\ 0 & 0 & 0 \end{pmatrix}$$

(by $R_3 \to R_3 - 2R_2$)

which has two non-zero rows.

Therefore, rank $(A) = 2 < 3$ (= number of unknowns)

Hence, the system has an infinite number of solutions.

Alternative method:

Here, co-efficient determinant

$$= \begin{vmatrix} 1 & 2 & 3 \\ 2 & 3 & 4 \\ 3 & 4 & 5 \end{vmatrix}$$

$$= (15 - 16) - 2(10 - 12) + 3(8 - 9)$$

$$= 0.$$

Therefore, the rank of the coefficient matrix < 3 (here number of variables = 3).

Hence, the system has an infinite number of solutions.

34. (b)

$$A = \begin{pmatrix} 1 & -2 & 1 \\ 1 & -2 & -1 \\ 2 & -4 & -5 \end{pmatrix}$$

$$\sim \begin{pmatrix} 1 & -2 & 1 \\ 0 & 0 & -2 \\ 0 & 0 & 7 \end{pmatrix}$$

(by $R_2 \to R_2 - R_1, R_3 \to R_3 - 2R_1$)

$$\sim \begin{pmatrix} 1 & -2 & 1 \\ 0 & 0 & -2 \\ 0 & 0 & 0 \end{pmatrix}$$

(by $R_3 \to R_3 - \dfrac{7}{2}R_2$)

which has two non-zero rows.

Therefore rank $(A) = 2 < 3$ (= number of unknowns)

Hence, the system has an infinite number of solutions.

Alternative method:

Here, co-efficient determinant

$$= \begin{vmatrix} 1 & -2 & 1 \\ 1 & -2 & -1 \\ 2 & -4 & -5 \end{vmatrix}$$

$$= (10 - 4) + 2(-5 + 2) + 1(-4 + 4)$$

$$= 0.$$

Therefore, rank of the co-efficient matrix < 3 (here number of variables = 3).

Hence, the system has an infinite number of solutions.

35. (c)

$$|A| = \begin{vmatrix} 5 & 2 & -1 \\ 2 & 3 & 4 \\ 4 & -5 & \lambda \end{vmatrix}$$

$$= 5(3\lambda + 20) - 2(2\lambda - 16) - 1(-10 - 12)$$

$$= 11\lambda + 154$$

System has a unique solution

$$\Rightarrow |A| \neq 0 \Rightarrow 11\lambda + 154 \neq 0 \Rightarrow \lambda \neq -14.$$

36. (b)

$$A = \begin{pmatrix} 1 & 1 & 1 & 1 \\ 1 & 3 & 2 & 4 \\ 2 & 0 & 1 & -1 \end{pmatrix}$$

$$\sim \begin{pmatrix} 1 & 1 & 1 & 1 \\ 0 & 2 & 1 & 3 \\ 0 & -2 & -1 & -3 \end{pmatrix}$$

(by $R_2 \to R_2 - R_1$, $R_3 \to R_3 - 2R_1$)

$$\sim \begin{pmatrix} 1 & 1 & 1 & 1 \\ 0 & 2 & 1 & 3 \\ 0 & 0 & 0 & 0 \end{pmatrix}$$

(by $R_3 \to R_3 + R_2$)

which has two non-zero rows

Therefore rank $(A) = 2 < 4$.

(here number of unknowns = 4)

Hence, the system has an infinite number of solutions.

37. (c)

$$A = \begin{pmatrix} 1 & 2 & 3 \\ 2 & 3 & 1 \\ 3 & 1 & 2 \end{pmatrix}$$

$$\sim \begin{pmatrix} 1 & 2 & 3 \\ 0 & -1 & -5 \\ 0 & -5 & -7 \end{pmatrix}$$

(by $R_2 \to R_2 - 2R_1$, $R_3 \to R_3 - 3R_1$)

which has three non-zero rows

Therefore, rank $(A) = 3$ = number of unknowns

Hence, the system has a unique solution. So, the only solution is $x = y = z = 0$ (trivial solution).

Alternative method:

Here, co-efficient determinat

$$= \begin{vmatrix} 1 & 2 & 3 \\ 2 & 3 & 1 \\ 3 & 1 & 2 \end{vmatrix}$$

$$= (6 - 1) - 2(4 - 3) + 3(2 - 9)$$

$$= -18 \neq 0.$$

Hence, the system has a unique solution which is the trivial solution.

38. (a)

$$|A - \lambda I| = 0 \Rightarrow \begin{vmatrix} 1 - \lambda & 2 \\ -1 & 3 - \lambda \end{vmatrix} = 0$$

$$\Rightarrow \quad (1 - \lambda)(3 - \lambda) + 2 = 0$$

$$\Rightarrow \quad \lambda^2 - 4\lambda + 5 = 0$$

Therefore, by the Cayley Hamilton theorem,

$$A^2 - 4A + 5I = O$$

$$\Rightarrow A^2 = 4A - 5I$$

$$\Rightarrow \left(A^2\right)^2 = \left(4A - 5I\right)^2 = 16A^2 - 40A + 25I$$

$$\Rightarrow A^4 = 16\left(4A - 5I\right) - 40A + 25I = 24A - 55I.$$

39. (c)

$$|A| = \begin{vmatrix} 2 & 1 & 0 \\ 1 & 2 & 0 \\ 1 & 1 & 1 \end{vmatrix} = 2(2 - 0) + 1(0 - 1) = 3 = 1 \times 1 \times 3$$

= product of the eigenvalues.

40. (a)

$$|A| = \begin{vmatrix} 2 & 0 & 1 \\ 0 & 1 & 0 \\ 1 & 0 & 2 \end{vmatrix} = 2(2 - 0) + 1(0 - 1) = 3 = 1 \times 1 \times 3$$

= product of the eigenvalues.

41. (b) Trace of the matrix = $2 + 1 + 0 = 3$ = sum of eigenvalues = $1 + 4 + (-2)$.

42. (d)

$$|A - \lambda I| = 0 \Rightarrow \begin{vmatrix} 1 - \lambda & 0 & 0 \\ 1 & 0 - \lambda & 1 \\ 0 & 1 & 0 - \lambda \end{vmatrix} = 0$$

$$\Rightarrow \quad (1 - \lambda) \times (\lambda^2 - 1) = 0$$

$$\Rightarrow \quad \lambda^2 - 1 - \lambda^3 + \lambda = 0$$

$$\Rightarrow \quad \lambda^3 - \lambda^2 - \lambda + 1 = 0$$

Therefore, by the Cayley Hamilton theorem,

$$A^3 - A^2 - A + I = 0$$

$$\Rightarrow \quad A^{-1}(A^3 - A^2 - A + I) = A^{-1}0$$

$$\Rightarrow \quad A^{-1}A^3 - A^{-1}A^2 - A^{-1}A + A^{-1}I = 0$$

$$\Rightarrow \quad A^2 - A - I + A^{-1} = 0$$

$$\Rightarrow \quad A^{-1} = A + I - A^2$$

43. (a)

$$A - \lambda I = \begin{pmatrix} 0 & 1 \\ 0 & 0 \end{pmatrix} - \lambda \begin{pmatrix} 1 & 0 \\ 0 & 1 \end{pmatrix} = \begin{pmatrix} -\lambda & 1 \\ 0 & -\lambda \end{pmatrix}$$

$$\therefore \quad |A - \lambda I| = 0 \Rightarrow \begin{vmatrix} -\lambda & 1 \\ 0 & -\lambda \end{vmatrix} = 0$$

$$\Rightarrow \quad \lambda^2 = 0 \Rightarrow \lambda = 0, 0$$

Therefore, 0 is the only eigenvalue.

$$\therefore AX = \lambda X$$

$$\Rightarrow \begin{pmatrix} 0 & 1 \\ 0 & 0 \end{pmatrix} \begin{pmatrix} x \\ y \end{pmatrix} = \pi \begin{pmatrix} x \\ y \end{pmatrix}$$

$$\Rightarrow \begin{pmatrix} y \\ 0 \end{pmatrix} = \begin{pmatrix} 0 \\ 0 \end{pmatrix}$$

$$\Rightarrow y = 0$$

Hence, $X = \begin{pmatrix} x \\ 0 \end{pmatrix} = \begin{pmatrix} k \\ 0 \end{pmatrix}$

(by replacing x by k)

which is the eigenvector for eigenvalue "0."

44. (b)

$$AX = \lambda X$$

$$\Rightarrow \begin{bmatrix} 1 & -n \\ -3 & 2n \end{bmatrix} \begin{bmatrix} 1 \\ -1 \end{bmatrix} = \lambda \begin{bmatrix} 1 \\ -1 \end{bmatrix}$$

$$\Rightarrow \begin{bmatrix} 1+n \\ -3-2n \end{bmatrix} = \begin{bmatrix} \lambda \\ -\lambda \end{bmatrix}$$

$$\Rightarrow 1 + n = \lambda, \; -3 - 2n = -\lambda$$

$$\Rightarrow -3 - 2n = -(1 + n)$$

$$\Rightarrow n = -2.$$

45. (a) $\text{Det}(P) = 6 = $ product of the eigenvalues
$$= (-1) \times (-2) \times 3$$

$\text{Trace}(P) = 0 = $ sum of the eigenvalues $= (-1) + (-2) + 3$.

46. (c) The vectors $(0, 1, a)$, $(1, a, 1)$ and $(a, 1, 0)$ are linearly dependent

$$\Rightarrow \begin{vmatrix} 0 & 1 & a \\ 1 & a & 1 \\ a & 1 & 0 \end{vmatrix} = 0$$

$$\Rightarrow 0 - (0 - a) + a(1 - a^2) = 0$$

$$\Rightarrow 2a - a^3 = 0$$

$$\Rightarrow a(2 - a^2) = 0$$

$$\Rightarrow a = 0, \pm\sqrt{2}$$

47. (d) The vectors (a, b) and (c, d) are linearly dependent

$$\Rightarrow \begin{vmatrix} a & b \\ c & d \end{vmatrix} = 0$$

$$\Rightarrow ad - bc = 0$$

48. (c) The vectors $(1, 0, 1)$, $(1, 1, 0)$ and (p, q, r) are linearly dependent

$$\Rightarrow \begin{vmatrix} 1 & 0 & 1 \\ 1 & 1 & 0 \\ p & q & r \end{vmatrix} = 0$$

$$\Rightarrow (r - 0) - 0 + (q - p) = 0$$

$$\Rightarrow p = r + q$$

Fully Solved MCQs (Level-II)

1. If A and B are square matrices of the same order such that $(A + B)^2 = A^2 + B^2 + 2AB$, then
(a) $AB = BA$ (b) $A = B$
(c) $A + B = 0$ (d) $A = -B^T$

2. Let A be a 3×5 matrix and B be a matrix such that $A^T B$ and BA^T are both defined. Then B is of the type:
(a) 3×5 (b) 5×3
(c) 3×3 (d) 5×5

3. If $A_\alpha = \begin{pmatrix} \cos\alpha & \sin\alpha \\ -\sin\alpha & \cos\alpha \end{pmatrix}$, then
(a) $A_\alpha^n = A_\alpha$ (b) $A_\alpha^n = nA_\alpha$
(c) $A_\alpha^n = A_{n\alpha}$ (d) $A_\alpha^n = 0$

4. If A and B are square matrices of same order so that $AB = A$, $BA = B$, then
(a) Both A and B are singular.
(b) Both A and B are non-singular.
(c) Both A and B are unit matrix.
(d) $A^2 = A$, $B^2 = B$

5. If the matrices A and B commute, then
(a) $(AB)^n = A^n B^n$ (b) $(AB)^n = AB$
(c) $(AB)^n = B^n$ (d) none of these

6. If $A = \begin{pmatrix} 1 & 0 \\ -1 & 1 \end{pmatrix}$, then $A^n = ?$

(a) $\begin{pmatrix} n & 0 \\ -n & n \end{pmatrix}$ (b) $\begin{pmatrix} n & 0 \\ -1 & n \end{pmatrix}$

(c) $\begin{pmatrix} 1 & 0 \\ -n & 0 \end{pmatrix}$ (d) $\begin{pmatrix} 1 & 0 \\ -n & 1 \end{pmatrix}$

7. If $A = \begin{pmatrix} i & -i \\ -i & i \end{pmatrix}$ and $B = \begin{pmatrix} 1 & -1 \\ -1 & 1 \end{pmatrix}$ and then $A^8 = ?$

(a) $128B$ (b) $130B$
(c) $116B$ (d) $8B$

8. Two matrices $\begin{bmatrix} \cos\theta & -\sin\theta \\ \sin\theta & \cos\theta \end{bmatrix}$ and $\begin{bmatrix} a & 0 \\ 0 & b \end{bmatrix}$ commute under multiplication. Then
(a) $a = b$ or $\theta = n\varpi$ (n is an integer)
(b) always
(c) $a = 4b$ (d) $\theta = \varpi/3$

9. Find P^{50} if P is given by

$$P = \begin{bmatrix} 1 & 1 & 1 \\ 0 & 1 & 1 \\ 0 & 0 & 1 \end{bmatrix}$$

(a) $\begin{bmatrix} 1 & 100 & 500 \\ 0 & 1 & 100 \\ 0 & 0 & 1 \end{bmatrix}$ (b) $\begin{bmatrix} 1 & 50 & 100 \\ 0 & 1 & 50 \\ 0 & 0 & 1 \end{bmatrix}$

(c) $\begin{bmatrix} 50 & 100 & 150 \\ 0 & 50 & 100 \\ 0 & 0 & 50 \end{bmatrix}$ (d) $\begin{bmatrix} 1 & 50 & 1275 \\ 0 & 1 & 50 \\ 0 & 0 & 1 \end{bmatrix}$

10. Let $P = \begin{bmatrix} 0 & \omega \\ \omega & 0 \end{bmatrix}$, where ω is the complex cube root of unity . Then P^{24} is equal to
(a) P^2 (b) P
(c) Identity matrix   (d) 0

11. Let $A = (a_{ij})_{3\times3}$ be a matrix with $a_{ij} \in R$. Let B be a matrix obtained by interchanging two columns of A. Then det $(A + B) = ?$
(a) $2\det(A) + \det(B)$ (b) 0
(c) $2\det(A)$ (d) $\det(A) - \det(B)$

12. What is the value of the following determinant of order "n"?

$$\begin{vmatrix} 1 & 0 & 0 & \dots & \dots & 0 \\ 1 & \frac{1}{2} & 0 & \dots & \dots & 0 \\ 1 & \frac{1}{2} & \frac{1}{3} & \dots & \dots & 0 \\ .. & .. & .. & .. & .. & .. \\ .. & .. & .. & .. & .. & .. \\ 1 & \frac{1}{2} & \frac{1}{3} & .. & .. & \frac{1}{n} \end{vmatrix}$$

(a) $\dfrac{1}{n!}$ (b) $\dfrac{1}{n}$
(c) 0 (d) $\dfrac{1}{n(n+1)}$

13. If $A = \begin{pmatrix} \cos x & \sin x & 0 \\ -\sin x & \cos x & 0 \\ 0 & 0 & 1 \end{pmatrix} = f(x)$, then $A^{-1} = ?$
(a) $f(x)$ (b) $f(-x)$
(c) $-f(x)$ (d) $-f(-x)$

14. If $A^k = O$, then $I + A + A^2 + \dots + A^{k-1}$ is equal to
(a) Null matrix   (b) $(I + A)^k$
(c) I (d) $(I - A)^{-1}$

15. If the matrix $A, B, A + B$ are non-singular, then $[A(A + B)^{-1}B]^{-1}$?
(a) $A + B$ (b) $A^{-1} + B^{-1}$
(c) $(A + B)^{-1}$ (d) AB

16. Let A, B be two 3×3 invertible matrices with $A + B = AB$. Then
(a) $A^{-1} + B^{-1} = 0$ (b) $A^{-1} + B^{-1} = B^{-1}A^{-1}$
(c) $I - A^{-1}$ is invertible
(d) $I + B^{-1}$ is invertible

17. Let A and B be two non-zero 2×2 matrices such that $AB = 0$. Then
(a) both A and B are non-singular
(b) exactly one of A and B is singular
(c) both A and B are singular
(d) $A + B$ is singular

18. If $A = \begin{pmatrix} 0 & 1 & 2 \\ 1 & 2 & 3 \\ 3 & x & 1 \end{pmatrix}$, $A^{-1} = \begin{pmatrix} \frac{1}{2} & -\frac{1}{2} & \frac{1}{2} \\ -4 & 3 & y \\ \frac{5}{2} & -\frac{3}{2} & \frac{1}{2} \end{pmatrix}$, then
(a) $x = 1, y = -1$ (b) $x = -1, y = 1$
(c) $x = y = \dfrac{1}{2}$ (d) $x = y = -\dfrac{1}{2}$

19. Let A and B be symmetric and skew-symmetric matrices, respectively, of same order. Then $A^m B^n A^m$ is
(a) Skew-symmetric for all m, n
(b) Symmetric for all m, n
(c) Skew-symmetric if n is odd and symmetric if n is even.
(d) Symmetric if m is even and skew-symmetric if m is odd.

20. If A and B are symmetric and commute, then
(a) $A^{-1}B$ is symmetric only
(b) AB^{-1} is symmetric only
(c) $A^{-1}B^{-1}$ is symmetric only
(d) $A^{-1}B$, AB^{-1}, $A^{-1}B^{-1}$ are all symmetric

21. Let A (a square matrix of order n) be a nilpotent matrix of index "p." Then
(a) $I_n - A$ is invertible
(b) $I_n + A$ is a zero matrix
(c) $I_n + A$ is nilpotent
(d) $I_n - A$ is a zero matrix

22. If A, a square matrix of order "n," be nilpotent, then
(a) $I_n - A$ is singular and $I_n + A$ is non-singular
(b) $I_n + A$ is singular and $I_n - A$ is non-singular
(c) Both $I_n - A$ and $I_n + A$ are singular.
(d) Both $I_n + A$ and $I_n - A$ are non-singular.

23. The matrix $A = \begin{pmatrix} 1 & 1 & 3 \\ 5 & 2 & 6 \\ -2 & -1 & -1 \end{pmatrix}$ is
(a) idempotent
(b) Nilpotent of index 3
(c) skew-symmetric (d) none of these

24. The values of α, β, and γ so that the matrix
$A = \begin{pmatrix} 0 & 2\beta & \gamma \\ \alpha & \beta & -\gamma \\ \alpha & -\beta & \gamma \end{pmatrix}$ is orthogonal are
(a) $\alpha = \pm\dfrac{1}{\sqrt{2}}, \beta = \pm\dfrac{1}{\sqrt{6}}, \gamma = \pm\dfrac{1}{\sqrt{3}}$
(b) $\alpha = \pm\dfrac{1}{\sqrt{2}}, \beta = \pm\dfrac{1}{\sqrt{2}}, \gamma = \pm\dfrac{1}{\sqrt{3}}$
(c) $\alpha = \pm\dfrac{1}{\sqrt{2}}, \beta = \pm\dfrac{1}{\sqrt{6}}, \gamma = \pm\dfrac{1}{\sqrt{2}}$
(d) none of these

25. If A is a square matrix and $A - \dfrac{I}{2}$ and $A + \dfrac{I}{2}$ are both orthogonal, then
(i) A is
(a) idempotent   (b) involutary
(c) skew-symmetric (d) symmetric
(ii) $A^2 =$
(a) $-\dfrac{3}{4}I$ (b) $-\dfrac{3}{2}I$
(c) $-\dfrac{3}{4}I$ (d) I

26. Let A be a $n \times n$ matrix with integral entries and $B = A + (1/2)I$, where I is the $n \times n$ identity matrix. Then B is
(a) idempotent   (b) nilpotent
(c) invertible   (d) none of these

27. Suppose A and B are two orthogonal matrices such that det (A) + det $(B) = 0$, then
(a) $A + B = -I$ (b) $A + B = I$
(c) det $(A + B) = 0$ (d) $A + B =$ null matrix

28. If A and B are 3×3 matrices such that rank $(AB) = 1$, then rank (BA) cannot be
(a) 0   (b) 1
(c) 2   (d) 3

29. The rank of the matrix $A = \begin{pmatrix} 1 & 2 & 3 & 1 \\ 2 & 4 & 6 & 4 \\ 1 & 2 & 3 & 2 \end{pmatrix}$ is
(a) 0   (b) 1   (c) 2   (d) 3

30. The rank of the matrix $A = \begin{pmatrix} 6 & 1 & 3 & 8 \\ 4 & 2 & 6 & -1 \\ 10 & 3 & 9 & 7 \\ 16 & 4 & 12 & 15 \end{pmatrix}$ is
(a) 0   (b) 1   (c) 2   (d) 3

31. The rank of the matrix $A = \begin{pmatrix} 1 & 1 & -1 & 1 \\ 1 & -1 & 2 & -1 \\ 3 & 1 & 0 & 1 \end{pmatrix}$ is
(a) 0   (b) 1   (c) 2   (d) 3

32. The rank of the matrix $A = \begin{pmatrix} 2 & 3 & -1 & -1 \\ 1 & -1 & -2 & -4 \\ 3 & 1 & 3 & -2 \\ 6 & 3 & 0 & -7 \end{pmatrix}$ is
(a) 4   (b) 1   (c) 2   (d) 3

33. The rank of the matrix $A = \begin{pmatrix} 0 & 1 & -3 & -1 \\ 1 & 0 & 1 & 1 \\ 3 & 1 & 0 & 2 \\ 1 & 1 & -2 & 0 \end{pmatrix}$ is
(a) 4   (b) 1   (c) 2   (d) 3

34. The rank of the matrix $A = \begin{pmatrix} 0 & 1 & 2 & 1 \\ 1 & 2 & 3 & 2 \\ 3 & 1 & 1 & 3 \end{pmatrix}$ is
(a) 4   (b) 1   (c) 2   (d) 3

35. The rank of the matrix

$$A = \begin{pmatrix} 1 & 0 & 1 & 0 & 0 \\ 1 & 1 & 0 & 0 & 0 \\ 0 & 1 & 1 & 0 & 0 \\ 0 & 0 & 1 & 1 & 0 \\ 0 & 1 & 0 & 1 & 0 \end{pmatrix} \text{ is}$$

(a) 4   (b) 1   (c) 2   (d) 3

36. The rank of the matrix $A = \begin{pmatrix} 1 & 1 & 1 \\ -1 & 1 & -1 \\ 1 & 1 & 5 \\ -1 & 0 & -3 \end{pmatrix}$ is

(a) 4   (b) 1   (c) 2   (d) 3

37. The rank of the matrix

$$A = \begin{pmatrix} -2 & -1 & -3 & -1 \\ 1 & 2 & 3 & -1 \\ 1 & 0 & 1 & 1 \\ 0 & 1 & 1 & -1 \end{pmatrix} \text{ is}$$

(a) 4   (b) 1   (c) 2   (d) 3

38. The rank of the matrix

$$A = \begin{pmatrix} 2 & 0 & 2 & 0 & 2 \\ 0 & 1 & 0 & 1 & 0 \\ 2 & 1 & 0 & 2 & 1 \\ 0 & 1 & 0 & 0 & 0 \end{pmatrix} \text{ is}$$

(a) 4   (b) 1   (c) 2   (d) 3

39. The rank of the matrix $\begin{pmatrix} 1 & 3 & 5 & 9 \\ 0 & 7 & 2 & -1 \\ 0 & 0 & 1 & 1 \\ 0 & 0 & 0 & 2 \end{pmatrix}$ is

(a) 3   (b) 4   (c) 1   (d) 2

40. The rank $A = \begin{pmatrix} \lambda & -1 & 0 & 0 \\ 0 & \lambda & -1 & 0 \\ 0 & 0 & \lambda & -1 \\ -6 & 11 & -6 & 1 \end{pmatrix}$ is 3. Then the value of λ is

(a) 1   (b) – 1   (c) 0   (d) 4

41. Let $A = \begin{pmatrix} 1 & 1 & 0 \\ -1 & 1 & 2 \\ 2 & 2 & 0 \\ -1 & 0 & 1 \end{pmatrix}$, then rank (A) = ?

(a) 1   (b) 2   (c) 3   (d) 4

42. If a, b, c are in A.P. with common difference "d" and the rank of the matrix $\begin{bmatrix} 4 & 5 & a \\ 5 & 6 & b \\ 6 & \lambda & c \end{bmatrix}$ is "2." Then the value of d and λ are given by

(a) $d = 0$, λ = any arbitrary number
(b) d = any arbitrary number, $\lambda = 7$
(c) $d = \lambda = 7$
(d) none of these

43. The system of given equations $x + 2y + z = 8$, $2x + y + 3z = 13$, $3x + 4y - \lambda z = \mu$

(*i*) has an unique solution if

(a) $\lambda \neq -\dfrac{11}{3}$ (b) $\lambda = -\dfrac{11}{3}$

(c) $\lambda \neq -1$ (d) $\lambda = 0$

(*ii*) No solution for

(a) $\lambda = -\dfrac{11}{3}, \mu = 22$ (b) $\lambda = -\dfrac{11}{3}, \mu \neq 22$

(c) $\lambda \neq -\dfrac{11}{3}, \mu \neq 22$ (d) $\lambda \neq -\dfrac{11}{3}, \mu \neq 22$

(*iii*) Infinite solution

(a) $\lambda = -\dfrac{11}{3}, \mu = 22$ (b) $\lambda \neq -\dfrac{11}{3}, \mu \neq 22$

(c) $\lambda \neq -\dfrac{11}{3}, \mu = 22$ (d) $\lambda \neq -\dfrac{11}{3}, \mu \neq 22$

44. Consider the system of equations $x + y + z = 6$, $x + 2y + 3z = 10$, $x + 2y + \lambda z = \mu$

Then the system has
(*i*) An unique solution if
(a) $\lambda = 3$ (b) $\lambda \neq 3$
(c) $\lambda \neq 1$ (d) $\lambda = 1$

(*ii*) No solution for
(a) $\lambda = 3, \mu \neq 10$ (b) $\lambda \neq 3, \mu \neq 10$
(c) $\lambda = 3, \mu = 10$ (d) $\lambda \neq 3, \mu = 10$

(*iii*) Infinite solution
(a) $\lambda = 3, \mu \neq 10$ (b) $\lambda \neq 3, \mu \neq 10$
(c) $\lambda = 3, \mu = 10$ (d) $\lambda \neq 3, \mu = 10$

45. Consider the system of equations
$x + 4y + 2z = 1$,
$2x + 7y + 5z = 2\mu$,   It has
$4x + \lambda y + 10z = 2\mu + 1$.
(*i*) no solution for

(a) $\lambda = 14, \mu \neq \dfrac{1}{2}$ (b) $\lambda = 14, \mu = \dfrac{1}{2}$

(c) $\mu \neq 14$ (d) None of these

(*ii*) a unique solution if

(a) $\lambda = 14$ (b) $\lambda \neq 14$

(c) $\lambda \neq \dfrac{1}{2}$ (d) $\lambda = \dfrac{1}{2}$

(*iii*) infinite number of solutions for

(a) $\lambda = 14, \mu = \dfrac{1}{2}$ (b) $\lambda \neq 14, \mu = \dfrac{1}{2}$

(c) $\lambda = \mu = 14$ (d) None of these

46. Consider the system of equations

$x_1 + x_2 + x_3 = 1,$

$x_1 + 2x_2 - x_3 = \mu,$ It has

$5x_1 + 7x_2 + \lambda x_3 = \mu^2$

(*i*) no solution for

(a) $\lambda = 1, \mu \neq -1, 3$

(b) $\lambda \neq 1, \mu = -1, 3$

(c) $\lambda \neq 1, \mu = -1, 3$

(d) none of these

(*ii*) a unique solution if

(a) $\lambda \neq -1$ (b) $\lambda \neq 3$

(c) $\lambda \neq 1$ (d) $\lambda = 1$

(*iii*) infinite number of solutions

(a) $\lambda = 1, \mu = -1$ or 3

(b) $\lambda \neq 1, \mu \neq -1$ or 3

(c) $\lambda = 1, \mu \neq -1$ or 3

(d) None of these

47. The value of λ for which the system

$x + 2y + 3z = \lambda x,$

$3x + y + 2z = \lambda y,$

$2x + 3y + z = \lambda z;$

has a non-trivial soluation, is

(a) $\lambda = 4$ (b) $\lambda = 5$

(c) $\lambda = 6$ (d) $\lambda = 1$

48. The system of equations given below is

$x_1 + 2x_2 + 3x_3 + x_4 = 1,$

$x_1 - x_2 + 2x_3 - x_4 = -3,$

$3x_1 + 3x_2 + 8x_3 + x_4 = -3$

(a) consistent

(b) inconsistent

(c) non-trivial solution

(d) none of these

49. The system of equations given below is

$2x_1 + x_2 + 4x_3 = 4,$

$x_1 - 3x_2 - x_3 = -5,$

$-3x_1 + 2x_2 - 2x_3 = 1,$

$8x_1 - 3x_2 + 8x_3 = 2$

(a) consistent

(b) inconsistent

(c) non-trivial solution

(d) none of these

50. The characteristic values of $\begin{pmatrix} 2 & -1 & 1 \\ -1 & 2 & -1 \\ 1 & -1 & 2 \end{pmatrix}$ are

(a) 1, 1, 4   (b) 1, 4, 4

(c) 4, 4, 4   (d) 1, 1, 1

51. The eigenvectors of the matrix

$A = \begin{pmatrix} 6 & -2 & 2 \\ -2 & 3 & -1 \\ 2 & -1 & 3 \end{pmatrix}$ are

(a) $\begin{pmatrix} 1 \\ 2 \\ 0 \end{pmatrix}, \begin{pmatrix} 0 \\ 1 \\ 1 \end{pmatrix}, \begin{pmatrix} 2 \\ -1 \\ 1 \end{pmatrix},$ (b) $\begin{pmatrix} 1 \\ 0 \\ 1 \end{pmatrix}, \begin{pmatrix} 0 \\ 1 \\ 0 \end{pmatrix}, \begin{pmatrix} 2 \\ 1 \\ 1 \end{pmatrix}$

(c) $\begin{pmatrix} 1 \\ 0 \\ 1 \end{pmatrix}, \begin{pmatrix} 1 \\ 1 \\ 1 \end{pmatrix}, \begin{pmatrix} 2 \\ 0 \\ 1 \end{pmatrix}$ (d) None of these

52. If $A = \begin{pmatrix} 1 & 0 & 2 \\ 0 & -1 & 1 \\ 0 & 1 & 0 \end{pmatrix}$, then $A^9 = ?$

(a) $12 A^2 - 22A - 9I$ (b) $-12A^2 + 22A - 9I$

(c) $12A^2 + 22A + 9I$ (d) None of these

53. Which of the following can be an eigenvector

of $A = \begin{pmatrix} 1 & 1 & 1 \\ 1 & 1 & 1 \\ 1 & 1 & 1 \end{pmatrix}$?

(a) $\begin{pmatrix} 2 \\ 2 \\ 9 \end{pmatrix}$ (b) $\begin{pmatrix} 1 \\ 1 \\ 2 \end{pmatrix}$

(c) $\begin{pmatrix} 1 \\ 1 \\ 1 \end{pmatrix}$ (d) $\begin{pmatrix} 0 \\ 0 \\ 1 \end{pmatrix}$

54. The possible set of eigenvalues of a 4×4 skew symmetric orthogonal real matrix is

(a) $\pm i$ (b) $\pm 1, \pm i$

(c) ± 1 (d) $0, \pm i$

55. Let A be a 2×2 complex matrix such that $\text{tr}(A) = 1$ and $\det(A) = -6$. Then $\text{tr}(A^4 - A^3) = ?$

(a) 55   (b) 78   (c) 70   (d) 88

56. Let $\alpha = e^{\frac{2\pi i}{5}}$ and $M = \begin{bmatrix} 1 & \alpha & \alpha^2 & \alpha^3 & \alpha^4 \\ 0 & \alpha & \alpha^2 & \alpha^3 & \alpha^4 \\ 0 & 0 & \alpha^2 & \alpha^3 & \alpha^4 \\ 0 & 0 & 0 & \alpha^3 & \alpha^4 \\ 0 & 0 & 0 & 0 & \alpha^4 \end{bmatrix}$

then trace of $I + M + M^2$ is

(a) 5   (b) -5
(c) 1   (d) 0

57. If a 3×3 real skew symmetric matrix has an eigenvalue $3i$, then one of the remaining eigenvalues is

(a) 0   (b) $1/3i$
(c) $-1/3i$ (d) 1

58. Let H be a 3×3 complex Hermitian matrix which is unitary. Then the distinct eigenvalues of H are

(a) ± 1 (b) $\pm i, \pm i$
(c) $1 \pm i$ (d) $(1 \pm i)/2$

59. If $A = \begin{bmatrix} 1 & 0 & 0 \\ i & \dfrac{-1+i\sqrt{3}}{2} & 0 \\ 0 & 1+2i & \dfrac{-1-i\sqrt{3}}{2} \end{bmatrix}$, then

$\operatorname{tr}(A^{102}) = ?$

(a) 0   (b) 1   (c) 2   (d) 3

60. Suppose that the matrix

$A = \begin{bmatrix} 40 & -29 & -11 \\ -18 & 30 & -12 \\ 26 & 24 & -50 \end{bmatrix}$ has a certain complex

number $\lambda \neq 0$ as an eigenvalue. Then which of the following must also be an eigenvalue of A?

(a) $\lambda + 20$ (b) $\lambda - 20$
(c) $20 - \lambda$ (d) None of these

61. Consider the matrix $M = \begin{bmatrix} 0 & 3 & 2 & 0 \\ 3 & 0 & 1 & 0 \\ 2 & 1 & 0 & 2 \\ 0 & 0 & 2 & 0 \end{bmatrix}$. Then

(a) M has no real eigenvalues
(b) all eigenvalues of M are positive
(c) all eigenvalues of M are negative
(d) M has both are positive and negative real eigenvalues

62. The linear operation $L(X)$ is defined by the cross product $L(X) = b \times X$, where $b = [0\ 1\ 0]^T$ and $X = [x_1\ x_2\ x_3]^T$ are three dimensional vectors. Also, given that $L(X) = A \begin{bmatrix} x_1 \\ x_2 \\ x_3 \end{bmatrix}$. Then the eigenvalues of A are

(a) $0, \pm 1$ (b) $0, \pm i$
(c) $1, \pm i$ (d) $-1, \pm i$

63. Let A be a 3×3 matrix with eigenvalues 0, 1 and -1. Then $\det (I + A^{100}) = ?$

(a) 3   (b) 4   (c) 5   (d) 6

64. If ω is a non-real cube roots of unity, then the eigenvalues of the following matrix are

$\begin{bmatrix} 1 & 1 & 1 \\ 1 & \omega & \omega^2 \\ 1 & \omega^2 & \omega^4 \end{bmatrix} \times \begin{bmatrix} 1 & 0 & 0 \\ 0 & -1 & 0 \\ 0 & 0 & 0 \end{bmatrix} \times \begin{bmatrix} 1 & 1 & 1 \\ 1 & \dfrac{1}{\omega} & \dfrac{1}{\omega^2} \\ 1 & \dfrac{1}{\omega^2} & \dfrac{1}{\omega^4} \end{bmatrix}$

(a) 1, 1, -1 (b) 1/3, 1/3, 0
(c) $1, -\omega, -\omega^2$ (d) 3, -3, 0

65. Let P, M, N be square matrices of order "n" such that the matrices M and N are non-singular. If X be an eigenvector of P corresponding to the eigenvalue λ, then an eigenvector of $N^{-1}MPM^{-1}N$ corresponding to the eigenvalue λ is

(a) $MN^{-1}X$ (b) $M^{-1}NX$
(c) $M^{-1}N^{-1}X$ (d) $N^{-1}MX$

66. Let $A = \begin{bmatrix} a & -1 & 4 \\ 0 & b & 7 \\ 0 & 0 & 3 \end{bmatrix}$ be a matrix with real entries. If the sum and product of all the eigenvalues of A are 10 and 30, respectively, then $a^2 + b^2$ equals

(a) 29   (b) 45   (c) 58   (d) 60

67. If A be a 3×3 non-zero matrix such that $A^2 = O$, then the number of non-zero eigenvalues of A is

(a) 0   (b) 1   (c) 2   (d) 4

68. Let A be a 3×3 matrix with eigenvalues 1, -1 and 3. Then

(a) $A^2 + A$ is non-singular
(b) $A^2 - A$ is non-singular
(c) $A^2 + 3A$ is non-singular
(d) $A^2 - 3A$ is non-singular

69. If $a_{ij} = 1$ for $1 \le i, j \le m$, then the characteristic equation of the matrix $A = \{a_{ij}\}_{m \times m}$ is
(a) $\lambda^m - 2\lambda^{m-1} + 2 = 0$
(b) $\lambda^m - m = 0$
(c) $\lambda^m - m\lambda^{m-1} = 0$
(d) $\lambda^m + m = 0$

70. Consider the 2×2 matrix $\begin{pmatrix} 4 & 0 \\ 0 & 4 \end{pmatrix}$. Then which of the following vectors is not a valid eigenvector:
(a) $\begin{bmatrix} 1 \\ 0 \end{bmatrix}$ (b) $\begin{bmatrix} -2 \\ 1 \end{bmatrix}$ (c) $\begin{bmatrix} 4 \\ -3 \end{bmatrix}$ (d) $\begin{bmatrix} 0 \\ 0 \end{bmatrix}$

71. One of the eigenvectors of the matrix $A = \begin{bmatrix} 1 & -1 & 0 \\ 0 & 1 & -1 \\ -1 & 0 & 1 \end{bmatrix}$ is $V = \begin{pmatrix} 1 \\ 1 \\ 1 \end{pmatrix}$. Then the corresponding eigenvalue is?
(a) −1 (b) 2 (c) 0 (d) 4

72. Let $f(x) = x - 1$, $g(x) = x + 1$, $h(x) = x^2 - 1$ and $q(x) = x^2 + 1$.
Then these functions are
(a) linearly independent
(b) linearly dependent because $f(x) \times g(x) = h(x)$
(c) linearly dependent because $f(x) - g(x) - h(x) = q(x)$
(d) none of the above

73. If $V_n(R)$ be the vector space over the field of real numbers and $W = \left\{ (a_1, a_2, ...) : \lim_{n \to \infty} a_n = \beta \right\}$ be a subspace of $V_n(R)$. Then
(a) β is real (b) $0 < \beta < 1$
(c) $\beta = 0$ (d) $\beta = 1$

74. The set $\{1, x, x(1-x)\}$ is
(a) linearly independent for all x
(b) linearly independent for $x = 0$ only
(c) linearly independent for $x = 1$ only
(d) linearly dependent

Answer key								
1. (a)	2. (a)	3. (c)	4. (d)	5. (a)	6. (d)	7. (a)	8. (a)	9. (d)
10. (c)	11. (b)	12. (a)	13. (b)	14. (d)	15. (b)	16. (c)	17. (c)	18. (a)
19. (c)	20. (d)	21. (a)	22. (d)	23. (b)	24. (a)	25. (i)-(c), (ii)-(a)	26. (c)	27. (c)
28. (d)	29. (c)	30. (c)	31. (c)	32. (d)	33. (c)	34. (d)	35. (a)	36. (d)
37. (c)	38. (a)	39. (b)	40. (a)	41. (b)	42. (b)	43. (i) (a), (ii) (b), (iii) **(a)**		
44. (i) (b), (ii) (a), (iii) **(c)**			45. (i) (a), (ii) (b), (iii) **(a)**			46. (i) (a), (ii) (c), (iii) **(a)**		
47. (c)	48. (b)	49. (a)	50. (a)	51. (a)	52. (b)	53. (c)	54. (a)	55. (b)
56. (a)	57. (a)	58. (a)	59. (d)	60. (c)	61. (d)	62. (b)	63. (b)	64. (d)
65. (d)	66. (a)	67. (a)	68. (c)	69. (c)	70. (d)	71. (c)	72. (c)	73. (c)
74. (a)								

Explanation

1. (a) $(A + B)^2 = A^2 + 2AB + B^2$
$\Rightarrow (A + B)(A + B) = A^2 + 2AB + B^2$
$\Rightarrow A^2 + AB + BA + B^2 = A^2 + AB + AB + B^2$
$\Rightarrow AB = BA$

2. (a) Order of A is $3 \times 5 \Rightarrow$ Order of A^T is 5×3
$A^T B$ is defined if number of columns of $A^T = $ Number of rows of B.
Therefore number of rows of $B = 3$.
BA^T is defined if number of columns of $B = $ number of rows of A^T.

Therefore, the number of columns of $B = 5$.
Hence, B is a matrix of order 3×5.

3. (c) A^2_α
$= A_\alpha \times A_\alpha$
$= \begin{pmatrix} \cos\alpha & \sin\alpha \\ -\sin\alpha & \cos\alpha \end{pmatrix} \times \begin{pmatrix} \cos\alpha & \sin\alpha \\ -\sin\alpha & \cos\alpha \end{pmatrix}$
$= \begin{pmatrix} \cos^2\alpha - \sin^2\alpha & \sin\alpha\cos\alpha + \sin\alpha\cos\alpha \\ -\sin\alpha\cos\alpha - \sin\alpha\cos\alpha & -\sin^2\alpha + \cos^2\alpha \end{pmatrix}$
$= \begin{pmatrix} \cos 2\alpha & \sin 2\alpha \\ -\sin 2\alpha & \cos 2\alpha \end{pmatrix}$
$\therefore A^n_\alpha = A_{n\alpha}$ is satisfied for $n = 2$.

4. (d) $AB = A$

$\Rightarrow \quad A(BA) = A$ $\qquad [\because \quad BA = B]$

$\Rightarrow \quad (AB)A = A$ $\qquad$ [by associative law]

$\Rightarrow \quad AA = A$ $\qquad [\because \quad AB = A]$

$\Rightarrow \quad A^2 = A$

Similarly considering $BA = B$ it can be shown that $B^2 = B$.

5. (a) $(AB)^2 = (AB)(AB)$

$= A(BA)B$

$= A(AB)B$

$[\because \quad A$ and B commutative, so $AB = BA]$

$= A^2B^2,$

which satisfies $(AB)^n = A^nB^n$ for $n = 2$.

6. (d) $A^2 = \begin{pmatrix} 1 & 0 \\ -1 & 1 \end{pmatrix} \times \begin{pmatrix} 1 & 0 \\ -1 & 1 \end{pmatrix} = \begin{pmatrix} 1-0 & 0 \\ -1-1 & 1 \end{pmatrix}$

$= \begin{pmatrix} 1 & 0 \\ -2 & 1 \end{pmatrix}$

$\therefore \qquad A^n = \begin{pmatrix} 1 & 0 \\ -n & 1 \end{pmatrix}$ is true for $n = 2$.

7. (a)

$A^2 = \begin{pmatrix} i & -i \\ -i & i \end{pmatrix} \times \begin{pmatrix} i & -i \\ -i & i \end{pmatrix}$

$= \begin{pmatrix} i^2 + i^2 & -i^2 - i^2 \\ -i^2 - i^2 & i^2 + i^2 \end{pmatrix}$

$= \begin{pmatrix} -2 & 2 \\ 2 & -2 \end{pmatrix}$

$= -2\begin{pmatrix} 1 & -1 \\ -1 & 1 \end{pmatrix} = -2B$

$A^8 = (-2B)^4 = 16B^4$ $\qquad \ldots(i)$

$B = \begin{pmatrix} 1 & -1 \\ -1 & 1 \end{pmatrix}$

$\Rightarrow B^2 = \begin{pmatrix} 1 & -1 \\ -1 & 1 \end{pmatrix} \times \begin{pmatrix} 1 & -1 \\ -1 & 1 \end{pmatrix}$

$= \begin{pmatrix} 1+1 & -1-1 \\ -1-1 & 1+1 \end{pmatrix}$

$= \begin{pmatrix} 2 & -2 \\ -2 & 2 \end{pmatrix}$

$= 2\begin{pmatrix} 1 & -1 \\ -1 & 1 \end{pmatrix} = 2B$

$\Rightarrow B^4 = (2B)^2 = 4B^2 = 4 \times 2B = 8B$

$\therefore (i) \Rightarrow A^8 = 16 \times 8B = 128B.$

8. (a) Let $A = \begin{bmatrix} \cos\theta & -\sin\theta \\ \sin\theta & \cos\theta \end{bmatrix}$ and $B = \begin{bmatrix} a & 0 \\ 0 & b \end{bmatrix}$

Then, $BA = AB$

$\Rightarrow \begin{bmatrix} a & 0 \\ 0 & b \end{bmatrix}\begin{bmatrix} \cos\theta & -\sin\theta \\ \sin\theta & \cos\theta \end{bmatrix}$

$= \begin{bmatrix} \cos\theta & -\sin\theta \\ \sin\theta & \cos\theta \end{bmatrix}\begin{bmatrix} a & 0 \\ 0 & b \end{bmatrix}$

$\Rightarrow \begin{bmatrix} a\cos\theta & -a\sin\theta \\ b\sin\theta & b\cos\theta \end{bmatrix} = \begin{bmatrix} a\cos\theta & -b\sin\theta \\ a\sin\theta & b\cos\theta \end{bmatrix}$

$\Rightarrow -a\sin\theta = -b\sin\theta$

$\Rightarrow (a-b)\sin\theta = 0$

$\Rightarrow a = b$ or $\sin\theta = 0$

$\Rightarrow a = b$ or $\theta = n\pi$

$[\because \sin\theta = 0 = \sin 0 \Rightarrow \theta = n\pi + (-1)^n 0]$

9. (d) $P = \begin{bmatrix} 1 & 1 & 1 \\ 0 & 1 & 1 \\ 0 & 0 & 1 \end{bmatrix}$

$P^2 = P \times P = \begin{bmatrix} 1 & 1 & 1 \\ 0 & 1 & 1 \\ 0 & 0 & 1 \end{bmatrix}\begin{bmatrix} 1 & 1 & 1 \\ 0 & 1 & 1 \\ 0 & 0 & 1 \end{bmatrix}$

$= \begin{bmatrix} 1 & 2 & 3 \\ 0 & 1 & 2 \\ 0 & 0 & 1 \end{bmatrix}$

$= \begin{bmatrix} 1 & 2 & 1+2 \\ 0 & 1 & 2 \\ 0 & 0 & 1 \end{bmatrix}$

$P^3 = P^2 \times P = \begin{bmatrix} 1 & 2 & 3 \\ 0 & 1 & 2 \\ 0 & 0 & 1 \end{bmatrix}\begin{bmatrix} 1 & 1 & 1 \\ 0 & 1 & 1 \\ 0 & 0 & 1 \end{bmatrix}$

$= \begin{bmatrix} 1 & 3 & 6 \\ 0 & 1 & 3 \\ 0 & 0 & 1 \end{bmatrix}$

$= \begin{bmatrix} 1 & 3 & 1+2+3 \\ 0 & 1 & 3 \\ 0 & 0 & 1 \end{bmatrix}$

$$P^4 = P^3 \times P = \begin{bmatrix} 1 & 3 & 6 \\ 0 & 1 & 3 \\ 0 & 0 & 1 \end{bmatrix}\begin{bmatrix} 1 & 1 & 1 \\ 0 & 1 & 1 \\ 0 & 0 & 1 \end{bmatrix}$$

$$= \begin{bmatrix} 1 & 4 & 10 \\ 0 & 1 & 4 \\ 0 & 0 & 1 \end{bmatrix}$$

$$= \begin{bmatrix} 1 & 4 & 1+2+3+4 \\ 0 & 1 & 4 \\ 0 & 0 & 1 \end{bmatrix}$$

Continuing like this we get,

$$= \begin{bmatrix} 1 & 50 & 1+2+3+\ldots\ldots+50 \\ 0 & 1 & 50 \\ 0 & 0 & 1 \end{bmatrix}$$

$$P^{50} = \begin{bmatrix} 1 & 50 & \dfrac{50(50+1)}{2} \\ 0 & 1 & 50 \\ 0 & 0 & 1 \end{bmatrix}$$

$$= \begin{bmatrix} 1 & 50 & 1275 \\ 0 & 1 & 50 \\ 0 & 0 & 1 \end{bmatrix}$$

10. (c) $P = \begin{bmatrix} 0 & \omega \\ \omega & 0 \end{bmatrix}$

$$P^2 = P \times P = \begin{bmatrix} 0 & \omega \\ \omega & 0 \end{bmatrix}\begin{bmatrix} 0 & \omega \\ \omega & 0 \end{bmatrix} = \begin{bmatrix} \omega^2 & 0 \\ 0 & \omega^2 \end{bmatrix}$$

$$P^4 = P^2 \times P^2 = \begin{bmatrix} \omega^2 & 0 \\ 0 & \omega^2 \end{bmatrix}\begin{bmatrix} \omega^2 & 0 \\ 0 & \omega^2 \end{bmatrix}$$

$$\begin{bmatrix} \omega^4 & 0 \\ 0 & \omega^4 \end{bmatrix}$$

$$P^6 = P^4 \times P^2 = \begin{bmatrix} \omega^4 & 0 \\ 0 & \omega^4 \end{bmatrix}\begin{bmatrix} \omega^2 & 0 \\ 0 & \omega^2 \end{bmatrix} = \begin{bmatrix} \omega^6 & 0 \\ 0 & \omega^6 \end{bmatrix}$$

Continuing like this we get, P^{24}

$$= \begin{bmatrix} \omega^{24} & 0 \\ 0 & \omega^{24} \end{bmatrix} = \begin{bmatrix} 1 & 0 \\ 0 & 1 \end{bmatrix} = I \text{ (since } \omega^{24} = (\omega^3)^8 = 1).$$

11. (b) Let $A = \begin{pmatrix} 1 & 2 & 3 \\ 4 & 1 & 2 \\ 6 & 7 & 5 \end{pmatrix}$

Let B be a matrix obtained by interchanging two columns C_2 and C_3 of A.

Then, $B = \begin{pmatrix} 1 & 3 & 2 \\ 4 & 2 & 1 \\ 6 & 5 & 7 \end{pmatrix}$

and $A + B = \begin{pmatrix} 2 & 5 & 5 \\ 8 & 3 & 3 \\ 12 & 12 & 12 \end{pmatrix}$

$\therefore$ det $(A + B)$

$= 2(36 - 36) - 5(96 - 36) + 5(96 - 36) = 0$

12. (a) Since the associated matrix is lower triangular, so value of the determinant = product of the elements lying in the principal diagonal

$$= 1 \times \frac{1}{2} \times \frac{1}{3} \times \ldots \times \frac{1}{n} = \frac{1}{n!}.$$

13. (b) $|A| = \begin{vmatrix} \cos x & \sin x & 0 \\ -\sin x & \cos x & 0 \\ 0 & 0 & 1 \end{vmatrix}$

$$= \cos^2 x + \sin^2 x = 1$$

$$A^{-1} = \frac{1}{|A|} \text{adj}(A)$$

$$= \begin{pmatrix} \cos x & \sin x & 0 \\ -\sin x & \cos x & 0 \\ 0 & 0 & 1 \end{pmatrix}^T$$

$$= \begin{pmatrix} \cos x & -\sin x & 0 \\ \sin x & \cos x & 0 \\ 0 & 0 & 1 \end{pmatrix}$$

$$= \begin{pmatrix} \cos(-x) & \sin(-x) & 0 \\ -\sin(-x) & \cos(-x) & 0 \\ 0 & 0 & 1 \end{pmatrix}$$

$$= f(-x).$$

14. (d) $I - A^k = (I - A)(I + A + A^2 + \ldots + A^{k-1})$

$\Rightarrow I = (I - A)(I + A + A^2 + \ldots + A^{k-1})$

$[\because A^k = O]\ldots\ldots\ldots(1)$

$\Rightarrow |(I - A)(I + A + A^2 + \ldots + A^{k-1})| = |I|$

$\Rightarrow |I - A||I + A + A^2 + \ldots + A^{k-1}| = 1$

(by $|AB| = |A||B|$)

$\Rightarrow |I - A| \neq 0$

$\Rightarrow I - A$ is invertible

$\Rightarrow (I - A)^{-1}$ exist

Then multiplying both sides of (1) by $(I - A)^{-1}$, we get

$(I-A)^{-1}I = (I-A)^{-1}(I-A)(I+A+A^2+...+A^{k-1})$

$\Rightarrow (I-A)^{-1} = I(I+A+A^2+...+A^{k-1})$

$= I+A+A^2+...+A^{k-1}$

15. (b) $\left[A(A+B)^{-1}B\right]^{-1}$

$= B^{-1}\left[(A+B)^{-1}\right]^{-1}A^{-1}$

$= B^{-1}(A+B)A^{-1}$

$= B^{-1}\left(AA^{-1} + BA^{-1}\right)$

$= B^{-1}\left(I + BA^{-1}\right)$

$= B^{-1}I + \left(B^{-1}B\right)A^{-1}$

$= B^{-1} + IA^{-1}$

$= B^{-1} + A^{-1}$

16. (c) $A+B=AB \Rightarrow A^{-1}(A+B) = A^{-1}(AB)$

$\Rightarrow A^{-1}A + A^{-1}B = (A^{-1}A)B$

$\Rightarrow A^{-1}A + A^{-1}B = (A^{-1}A)B$

$\Rightarrow I + A^{-1}B = IB = B$

$\Rightarrow I = B - A^{-1}B = (I - A^{-1})B$

$\Rightarrow |I| = |(I - A^{-1})B|$

$\Rightarrow |I - A^{-1}||B| = 1$

$\Rightarrow |I - A^{-1}| \neq 0$

$\Rightarrow I - A^{-1}$ is invertible.

17. (c) Let $A = \begin{pmatrix} 1 & 0 \\ 0 & 0 \end{pmatrix}, B = \begin{pmatrix} 0 & 0 \\ 0 & 1 \end{pmatrix}.$

Then $AB = \begin{pmatrix} 1 & 0 \\ 0 & 0 \end{pmatrix}\begin{pmatrix} 0 & 0 \\ 0 & 1 \end{pmatrix} = \begin{pmatrix} 0 & 0 \\ 0 & 0 \end{pmatrix}.$

Here $|A| = 0, |B| = 0$ and so both are singular.

Now $A + B = \begin{pmatrix} 1 & 0 \\ 0 & 0 \end{pmatrix} + \begin{pmatrix} 0 & 0 \\ 0 & 1 \end{pmatrix}$

$= \begin{pmatrix} 1 & 0 \\ 0 & 1 \end{pmatrix}.$

Hence, $|A+B| = \begin{vmatrix} 1 & 0 \\ 0 & 1 \end{vmatrix} = 1 \neq 0.$

So $A + B$ is non-singular.

Here, $|A| = 0, |B| = 0$ and so both are singular

NOTE *If A and B be two non-zero n × n matrices such that AB = 0. Then both of them are singular.*

18. (a) $|A| = \begin{vmatrix} 0 & 1 & 2 \\ 1 & 2 & 3 \\ 3 & x & 1 \end{vmatrix} = 0 - (1-9) + 2(x-6)$

$= 2x - 4.$

$\text{adj}(A) = \begin{bmatrix} \begin{vmatrix} 2 & 3 \\ x & 1 \end{vmatrix} & -\begin{vmatrix} 1 & 3 \\ 3 & 1 \end{vmatrix} & \begin{vmatrix} 1 & 2 \\ 3 & x \end{vmatrix} \\ -\begin{vmatrix} 1 & 2 \\ x & 1 \end{vmatrix} & \begin{vmatrix} 0 & 2 \\ 3 & 1 \end{vmatrix} & -\begin{vmatrix} 0 & 1 \\ 3 & x \end{vmatrix} \\ \begin{vmatrix} 1 & 2 \\ 2 & 3 \end{vmatrix} & -\begin{vmatrix} 0 & 2 \\ 1 & 3 \end{vmatrix} & \begin{vmatrix} 0 & 1 \\ 1 & 2 \end{vmatrix} \end{bmatrix}^T$

$= \begin{bmatrix} 2-3x & 8 & x-6 \\ 2x-1 & -6 & 3 \\ -1 & 2 & -1 \end{bmatrix}^T$

$= \begin{bmatrix} 2-3x & 2x-1 & -1 \\ 8 & -6 & 2 \\ x-6 & 3 & -1 \end{bmatrix}$

Now, $A^{-1} = \dfrac{1}{|A|}\text{adj}(A)$

$\Rightarrow \begin{pmatrix} \dfrac{1}{2} & -\dfrac{1}{2} & \dfrac{1}{2} \\ -4 & 3 & y \\ \dfrac{5}{2} & -\dfrac{3}{2} & \dfrac{1}{2} \end{pmatrix} = \dfrac{1}{2x-4}\begin{pmatrix} 2-3x & 2x-1 & -1 \\ 8 & -6 & 2 \\ x-6 & 3 & -1 \end{pmatrix}$

$\Rightarrow \begin{pmatrix} \dfrac{1}{2} & -\dfrac{1}{2} & \dfrac{1}{2} \\ -4 & 3 & y \\ \dfrac{5}{2} & -\dfrac{3}{2} & \dfrac{1}{2} \end{pmatrix} = \begin{pmatrix} \dfrac{2-3x}{2x-4} & \dfrac{2x-1}{2x-4} & \dfrac{-1}{2x-4} \\ \dfrac{8}{2x-4} & \dfrac{-6}{2x-4} & \dfrac{2}{2x-4} \\ \dfrac{x-6}{2x-4} & \dfrac{3}{2x-4} & \dfrac{-1}{2x-4} \end{pmatrix}$

$\Rightarrow \dfrac{1}{2} = \dfrac{2-3x}{2x-4}, \quad y = \dfrac{2}{2x-4}$

$\Rightarrow 2x - 4 = 4 - 6x, \quad y = \dfrac{2}{2x-4}$

$\Rightarrow 8x = 8, \quad y = \dfrac{2}{2x-4}$

$\Rightarrow x = 1, \quad y = \dfrac{2}{2-4} = -1.$

19. (c) A is symmetric $\Rightarrow A^T = A$. Also, B is skew-symmetric $\Rightarrow B^T = -B$.

$(A^m B^n A^m)^T$

$= (A^m)^T (B^n)^T (A^m)^T$

$= (A^T)^m (B^T)^n (A^T)^m$

$= A^m (-B)^n A^m$

$= (-1)^n A^m B^n A^m$

$= \begin{cases} A^m B^n A^m & \text{if } n \text{ is even} \\ -A^m B^n A^m & \text{if } n \text{ is odd} \end{cases}$

Thus $(A^m B^n A^m)^T = A^m B^n A^m$ if n is even

$\Rightarrow A^m B^n A^m$ is symmetric if n is even

Also $(A^m B^n A^m)^T = -A^m B^n A^m$ if n is odd

$\Rightarrow A^m B^n A^m$ is skew-symmetric if n is odd

20. (d) As A and B are symmetric, so $A^T = A, B^T = B$. Also A and B commute, so $AB = BA$. Then

$\left(A^{-1}B\right)^T = B^T \left(A^{-1}\right)^T = B\left(A^T\right)^{-1} = BA^{-1}$ ..(1)

Now $AB = BA \Rightarrow A^{-1}AB = A^{-1}BA$

$\Rightarrow IB = \left(A^{-1}B\right)A$

$\Rightarrow BA^{-1} = \left(A^{-1}B\right)AA^{-1}$

$\Rightarrow BA^{-1} = \left(A^{-1}B\right)I = A^{-1}B$

$\therefore (1) \Rightarrow \left(A^{-1}B\right)^T = A^{-1}B$

So $A^{-1}B$ is symmetric.

Similarly, it can be shown that AB^{-1} is symmetric.

Now $\left(A^{-1}B^{-1}\right)^T = \left(B^{-1}\right)^T \left(A^{-1}\right)^T$

$= \left(B^T\right)^{-1} \left(A^T\right)^{-1}$

$= B^{-1}A^{-1}$

$= \left(AB\right)^{-1}$

$= \left(BA\right)^{-1} = A^{-1}B^{-1}$

Hence, $A^{-1}B^{-1}$ is symmetric.

21. (a) Since A is a nilpotent matrix of index p, so $A^p = 0$.

Then,

$I_n - A^p = (I_n - A)(I_n + A + A^2 + ... + A^{p-1})$

$\Rightarrow I_n = (I_n - A)(I_n + A + A^2 + ... + A^{p-1})$

$\Rightarrow |I_n| = \left|(I_n - A)(I_n + A + A^2 + ... + A^{p-1})\right|$

$\Rightarrow 1 = |I_n - A|\left|I_n + A + A^2 + ... + A^{p-1}\right|$

$(\because |AB| = |A||B|)$

$\Rightarrow |I_n - A| \neq 0$

$\Rightarrow I_n - A$ is invertible

22. (d) Let A be a nilpotent matrix of index p. So, $A^p = 0$.

Then,

$I_n + A^n = (I_n + A)(I_n - A + A^2 - ... - A^{p-1})$

$\Rightarrow I_n = (I_n + A)(I_n - A + A^2 - ... - A^{p-1})$

$\Rightarrow |I_n| = \left|(I_n + A)(I_n - A + A^2 - ... - A^{p-1})\right|$

$\Rightarrow 1 = |I_n + A|\left|I_n - A + A^2 - ... - A^{p-1}\right|$

$(by |AB| = |A||B|)$

$\Rightarrow |I_n + A| \neq 0$

$\Rightarrow I_n + A$ is non-singular

Again,

$I_n - A^n = (I_n - A)(I_n + A + A^2 + ... + A^{p-1})$

$\Rightarrow I_n = (I_n - A)(I_n + A + A^2 + ... + A^{p-1})$

$\Rightarrow |I_n| = \left|(I_n - A)(I_n + A + A^2 + ... + A^{p-1})\right|$

$\Rightarrow 1 = |I_n - A|\left|I_n + A + A^2 + ... + A^{p-1}\right|$

$(by |AB| = |A||B|)$

$\Rightarrow |I_n - A| \neq 0$

$\Rightarrow I_n - A$ is non-singular.

23. (b) $A^2 = A \times A$

$= \begin{pmatrix} 1 & 1 & 3 \\ 5 & 2 & 6 \\ -2 & -1 & -3 \end{pmatrix} \times \begin{pmatrix} 1 & 1 & 3 \\ 5 & 2 & 6 \\ -2 & -1 & -3 \end{pmatrix}$

$= \begin{pmatrix} 1+5-6 & 1+2-3 & 3+6-9 \\ 5+10-12 & 5+4-6 & 15+12-18 \\ -2-5+6 & -2-2+3 & -6-6+9 \end{pmatrix}$

$= \begin{pmatrix} 0 & 0 & 0 \\ 3 & 3 & 9 \\ -1 & -1 & -3 \end{pmatrix}$,

$A^3 = A^2 \times A$

$$= \begin{pmatrix} 0 & 0 & 0 \\ 3 & 3 & 9 \\ -1 & -1 & -3 \end{pmatrix} \times \begin{pmatrix} 1 & 1 & 3 \\ 5 & 2 & 6 \\ -2 & -1 & -3 \end{pmatrix}$$

$$= \begin{pmatrix} 0+0+0 & 0+0+0 & 0+0+0 \\ 3+15-18 & 3+6-9 & 9+18-27 \\ -3-5+6 & -1-2+3 & -3-6+9 \end{pmatrix}$$

$$= \begin{pmatrix} 0 & 0 & 0 \\ 0 & 0 & 0 \\ 0 & 0 & 0 \end{pmatrix} = 0.$$

Hence, A is a nilpotent matrix of index 3.

24. (*a*) A is orthogonal

$$\Rightarrow \quad A \times A^T = I$$

$$\Rightarrow \begin{pmatrix} 0 & 2\beta & \gamma \\ \alpha & \beta & -\gamma \\ \alpha & -\beta & \gamma \end{pmatrix} \times \begin{pmatrix} 0 & \alpha & \alpha \\ 2\beta & \beta & -\beta \\ \gamma & -\gamma & \gamma \end{pmatrix} = I$$

$$\Rightarrow \begin{pmatrix} 4\beta^2+\gamma^2 & 2\beta^2-\gamma^2 & -2\beta^2+\gamma^2 \\ 2\beta^2-\gamma^2 & \alpha^2+\beta^2+\gamma^2 & \alpha^2-\beta^2-\gamma^2 \\ -2\beta^2+\gamma^2 & \alpha^2-\beta^2-\gamma^2 & \alpha^2+\beta^2+\gamma^2 \end{pmatrix}$$

$$= \begin{pmatrix} 1 & 0 & 0 \\ 0 & 1 & 0 \\ 0 & 0 & 1 \end{pmatrix}$$

$$\Rightarrow \quad 4\beta^2 + \gamma^2 = 1 \qquad \dots(i)$$
$$\quad 2\beta^2 - \gamma^2 = 0 \qquad \dots(ii)$$
$$\quad \alpha^2 + \beta^2 + \gamma^2 = 1 \qquad \dots(iii)$$

Solving the above three equations, we get

$$\alpha = \pm \frac{1}{\sqrt{2}}, \beta = \pm \frac{1}{\sqrt{6}}, \gamma = \pm \frac{1}{\sqrt{3}}.$$

25. (*i*)-(*c*); (*ii*)-(*a*)

$A - \dfrac{I}{2}$ is orthogonal

$$\Rightarrow \left(A - \frac{I}{2}\right)\left(A - \frac{I}{2}\right)^T = I \Rightarrow \left(A - \frac{I}{2}\right)\left(A^T - \frac{I}{2}\right) = I$$

$$\Rightarrow AA^T - \frac{AI}{2} - \frac{IA^T}{2} + \frac{I}{4} = I$$

$$\Rightarrow AA^T - \frac{A}{2} - \frac{A^T}{2} + \frac{I}{4} = I \qquad \dots(i)$$

Similarly, $A + \dfrac{I}{2}$ is orthogonal

$$\Rightarrow AA^T + \frac{A}{2} + \frac{A^T}{2} + \frac{I}{4} = I \qquad \dots(ii)$$

Subtracting (*i*) from (*ii*), we get, $A + A^T = O$ which gives $A^T = -A$. Thus, A is skew-symmetric.

Adding (*i*) and (*ii*), we get, $2AA^T + \dfrac{I}{2} = 2I$

which implies $AA^T = \dfrac{3I}{4}$ and so $A(-A) = \dfrac{3I}{4}$

i.e., $A^2 = -\dfrac{3}{4}I.$

26. (*c*) Let $A = \begin{pmatrix} 1 & -1 \\ 2 & 3 \end{pmatrix}$

Then $B = \begin{pmatrix} 1 & -1 \\ 2 & 3 \end{pmatrix} + \dfrac{1}{2}\begin{pmatrix} 1 & 0 \\ 0 & 1 \end{pmatrix}$

$$= \begin{pmatrix} 1 & -1 \\ 2 & 3 \end{pmatrix} + \begin{pmatrix} \frac{1}{2} & 0 \\ 0 & \frac{1}{2} \end{pmatrix} = \begin{pmatrix} \frac{3}{2} & -1 \\ 2 & \frac{7}{2} \end{pmatrix}$$

$\therefore \det(B) = \dfrac{21}{4} + 2 \neq 0.$

So, B is invertible.

27. (*c*) Since A and B are two orthogonal matrices, so
$A^T A = A A^T = I$ and $B^T B = B B^T = I$. Then

$$\left| A^T(A+B)B^T \right| = \left| (A^T A + A^T B)B^T \right|$$

$$\Rightarrow \left| A^T \right|\left| A+B \right|\left| B^T \right| = \left| (I + A^T B)B^T \right|$$

$$\Rightarrow |A||A+B||B| = \left| B^T + A^T BB^T \right|$$

$$\Rightarrow |A||A+B||B| = \left| B^T + A^T I \right| = \left| B^T + A^T \right|$$

$$\Rightarrow |A||A+B||B| = \left| (B+A)^T \right| = |A+B|$$

$$\Rightarrow |A+B|(|A||B|-1) = 0$$

$$\Rightarrow |A+B| = 0$$

$$\begin{pmatrix} \because |A||B| - 1 = 0 \\ \Rightarrow |A| = |B| = -1 \ or \ |A| = |B| = -1 \\ \Rightarrow |A| + |B| \neq 0, \text{ a contradiction} \end{pmatrix}$$

28. (*d*) $\text{rank}(AB) = 1 \neq 3 = \text{order of } AB$
$$\Rightarrow |AB| = 0 \Rightarrow |A||B| = 0.$$

So, at least either $|A|$ or $|B|$ should be zero
Hence, $|BA| = |B||A| = 0.$ Therefore, BA is singular.
Hence, rank (BA) cannot be 3.

29. (c) $A = \begin{pmatrix} 1 & 2 & 3 & 1 \\ 2 & 4 & 6 & 4 \\ 1 & 2 & 3 & 2 \end{pmatrix}$

$\sim \begin{pmatrix} 1 & 2 & 3 & 1 \\ 0 & 0 & 0 & 0 \\ 1 & 2 & 3 & 2 \end{pmatrix}$ (by $R_2 \rightarrow R_2 - 2R_3$)

which has two non-zero rows.

Hence, rank $(A) = 2$.

30. (c) $A = \begin{pmatrix} 6 & 1 & 3 & 8 \\ 4 & 2 & 6 & -1 \\ 10 & 3 & 9 & 7 \\ 16 & 4 & 12 & 15 \end{pmatrix}$

$\sim \begin{pmatrix} 6 & 1 & 3 & 8 \\ 4 & 2 & 6 & -1 \\ 10 & 3 & 9 & 7 \\ 0 & 0 & 0 & 0 \end{pmatrix}$ $\left[\text{by } R_4 \rightarrow R_4 - (R_1 + R_3)\right]$

$\sim \begin{pmatrix} 6 & 1 & 3 & 8 \\ 4 & 2 & 6 & -1 \\ 0 & 0 & 0 & 0 \\ 0 & 0 & 0 & 0 \end{pmatrix}$ $\left[\text{by } R_3 \rightarrow R_3 - (R_1 + R_2)\right]$

which has two non-zero rows.

Hence, rank $(A) = 2$.

31. (c) $A = \begin{pmatrix} 1 & 1 & -1 & 1 \\ 1 & -1 & 2 & -1 \\ 3 & 1 & 0 & 1 \end{pmatrix}$

$\sim \begin{pmatrix} 1 & 1 & -1 & 1 \\ 1 & -1 & 2 & -1 \\ 0 & 0 & 0 & 0 \end{pmatrix}$

$\left(\text{by } R_3 \rightarrow R_3 - (2R_1 + R_2)\right)$

Which has two non-zero rows.

Hence, rank $(A) = 2$.

32. (d) $A = \begin{bmatrix} 2 & 3 & -1 & -1 \\ 1 & -1 & -2 & -4 \\ 3 & 1 & 3 & -2 \\ 6 & 3 & 0 & -7 \end{bmatrix}$

$\sim \begin{bmatrix} 2 & 3 & -1 & -1 \\ 1 & -1 & -2 & -4 \\ 3 & 1 & 3 & -2 \\ 0 & 0 & 0 & 0 \end{bmatrix}$

$\left(\text{by } R_4 \rightarrow R_4 - (R_1 + R_2 + R_3)\right)$

which has three non-zero rows.

Hence, rank $(A) = 3$.

33. (c) $A = \begin{pmatrix} 0 & 1 & -3 & -1 \\ 1 & 0 & 1 & 1 \\ 3 & 1 & 0 & 2 \\ 1 & 1 & -2 & 0 \end{pmatrix}$

$\sim \begin{pmatrix} 1 & 0 & 1 & 1 \\ 0 & 1 & -3 & -1 \\ 3 & 1 & 0 & 2 \\ 1 & 1 & -2 & 0 \end{pmatrix}$

(by $R_1 \leftrightarrow R_2$)

$\sim \begin{pmatrix} 1 & 0 & 1 & 1 \\ 0 & 1 & -3 & -1 \\ 0 & 1 & -3 & -1 \\ 0 & 1 & -3 & -1 \end{pmatrix}$

(by $R_3 \rightarrow R_3 - 3R_1, R_4 \rightarrow R_4 - R_1$)

$\sim \begin{pmatrix} 1 & 0 & 1 & 1 \\ 0 & 1 & -3 & -1 \\ 0 & 0 & 0 & 0 \\ 0 & 0 & 0 & 0 \end{pmatrix}$

(by $R_3 \rightarrow R_3 - R_2, R_4 \rightarrow R_4 - R_2$)

which has two non-zero rows.

Hence, rank $(A) = 2$.

34. (d) Since order of A is 3×4, so rank $(A) \leq \min$ $\{3, 4\} = 3$. Consider a sub-matrix B, where

$B = \begin{pmatrix} 0 & 1 & 2 \\ 1 & 2 & 3 \\ 3 & 1 & 1 \end{pmatrix}$.

Then $|B| = 0 - (1 - 9) + 2(1 - 6)$

$= -2 \neq 0$.

Thus, there exist a square sub-matrix B of order "3" whose determinant value is non-zero.

Hence, rank $(A) = 3$.

35. (a) $A = \begin{pmatrix} 1 & 0 & 1 & 0 & 0 \\ 1 & 1 & 0 & 0 & 0 \\ 0 & 1 & 1 & 0 & 0 \\ 0 & 0 & 1 & 1 & 0 \\ 0 & 1 & 0 & 1 & 0 \end{pmatrix}$

$$\sim \begin{pmatrix} 0 & -2 & 0 & 0 & 0 \\ 1 & 1 & 0 & 0 & 0 \\ 0 & 1 & 1 & 0 & 0 \\ 0 & 0 & 1 & 1 & 0 \\ 0 & 1 & 0 & 1 & 0 \end{pmatrix} [\text{by } R_1 \to R_1 - (R_2 + R_3)]$$

$$\sim \begin{pmatrix} 0 & 1 & 0 & 0 & 0 \\ 1 & 1 & 0 & 0 & 0 \\ 0 & 1 & 1 & 0 & 0 \\ 0 & 0 & 1 & 1 & 0 \\ 0 & 1 & 0 & 1 & 0 \end{pmatrix} \left[\text{by } R_1 \to -\frac{1}{2} R_1\right]$$

$$\sim \begin{pmatrix} 0 & 1 & 0 & 0 & 0 \\ 1 & 1 & 0 & 0 & 0 \\ 0 & 1 & 1 & 0 & 0 \\ 0 & 0 & 1 & 1 & 0 \\ 0 & 0 & 0 & 1 & 0 \end{pmatrix} (\text{by } R_5 \to R_5 - R_1)$$

$$\sim \begin{pmatrix} 0 & 1 & 0 & 0 & 0 \\ 1 & 0 & 0 & 0 & 0 \\ 0 & 1 & 1 & 0 & 0 \\ 0 & 0 & 1 & 0 & 0 \\ 0 & 0 & 0 & 1 & 0 \end{pmatrix}$$
$$(\text{by } R_2 \to R_2 - R_1, \ R_4 \to R_4 - R_5)$$

$$\sim \begin{pmatrix} 0 & 1 & 0 & 0 & 0 \\ 1 & 0 & 0 & 0 & 0 \\ 0 & 0 & 1 & 0 & 0 \\ 0 & 0 & 1 & 0 & 0 \\ 0 & 0 & 0 & 1 & 0 \end{pmatrix} (\text{by } R_3 \to R_3 - R_1)$$

$$\sim \begin{pmatrix} 0 & 1 & 0 & 0 & 0 \\ 1 & 0 & 0 & 0 & 0 \\ 0 & 0 & 1 & 0 & 0 \\ 0 & 0 & 0 & 0 & 0 \\ 0 & 0 & 0 & 1 & 0 \end{pmatrix} (\text{by } R_4 \to R_4 - R_3)$$

which has four non-zero rows.

Hence, rank $(A) = 4$.

36. (d) $A = \begin{pmatrix} 1 & 1 & 1 \\ -1 & 1 & -1 \\ 1 & 1 & 5 \\ -1 & 0 & -3 \end{pmatrix}$

$$\sim \begin{pmatrix} 1 & 1 & 1 \\ 0 & 2 & 0 \\ 1 & 1 & 5 \\ -1 & 0 & -3 \end{pmatrix} (\text{by } R_2 \to R_2 + R_1)$$

$$\sim \begin{pmatrix} 1 & 1 & 1 \\ 0 & 2 & 0 \\ 0 & 0 & 4 \\ -1 & 0 & -3 \end{pmatrix} (\text{by } R_3 \to R_3 - R_1)$$

$$\sim \begin{pmatrix} 0 & 1 & -2 \\ 0 & 1 & 0 \\ 0 & 0 & 1 \\ -1 & 0 & -3 \end{pmatrix}$$
$$\left[\begin{array}{c} \text{by } R_1 \to R_1 + R_4, \ R_2 \to \dfrac{1}{2} R_2, \\ R_3 \to \dfrac{1}{4} R_3 \end{array}\right]$$

$$\sim \begin{pmatrix} 0 & 1 & 0 \\ 0 & 1 & 0 \\ 0 & 0 & 1 \\ -1 & 0 & -3 \end{pmatrix} (\text{by } R_1 \to R_1 + 2R_3)$$

$$\sim \begin{pmatrix} 0 & 0 & 0 \\ 0 & 1 & 0 \\ 0 & 0 & 1 \\ -1 & 0 & -3 \end{pmatrix} (\text{by } R_1 \to R_1 - R_2)$$

which has three non-zero rows.

Hence, rank $(A) = 3$.

Alternative method:

Rank$(A) \le$ min {number of rows, number of colums} = min {4, 3}. So rank $(A) \le 3$.

Let us consider a third order square sub-matrix

$$B = \begin{pmatrix} 1 & 1 & 1 \\ -1 & 1 & -1 \\ 1 & 1 & 5 \end{pmatrix}$$

Then $\det(B) = 1(5+1) - (-5+1) + 1(-1-1) \ne 0$.

Thus, there exist a square sub-matrix B of order 3 whose determinant value is non-zero. Hence, rank $(A) = 3$.

37. (c) $A = \begin{pmatrix} -2 & -1 & -3 & -1 \\ 1 & 2 & 3 & -1 \\ 1 & 0 & 1 & 1 \\ 0 & 1 & 1 & -1 \end{pmatrix}$

$$\sim \begin{pmatrix} 0 & -1 & -1 & 1 \\ 0 & 2 & 2 & -2 \\ 1 & 0 & 1 & 1 \\ 0 & 1 & 1 & -1 \end{pmatrix}$$

(by $R_1 \to R_1 + 2R_3$, $R_2 \to R_2 - R_3$)

$$\sim \begin{pmatrix} 0 & -1 & -1 & 1 \\ 0 & 0 & 0 & 0 \\ 1 & 0 & 1 & 1 \\ 0 & 0 & 0 & 0 \end{pmatrix}$$

(by $R_2 \to R_2 + 2R_1$, $R_4 \to R_4 + R_1$)

which has two non-zero rows.

Hence, rank $(A) = 2$.

38. (a) rank$(A) \le$ min {number of rows of A, number of columns of A} = min {4, 5}. So rank$(A) \le 4$.

Let us consider a 4th order square sub-matrix

$$B = \begin{pmatrix} 0 & 2 & 0 & 2 \\ 1 & 0 & 1 & 0 \\ 1 & 0 & 2 & 1 \\ 1 & 0 & 0 & 0 \end{pmatrix}$$

$$\det(B) = \begin{vmatrix} 0 & 2 & 0 & 2 \\ 1 & 0 & 1 & 0 \\ 1 & 0 & 2 & 1 \\ 1 & 0 & 0 & 0 \end{vmatrix}$$

$$= \begin{vmatrix} 1 & 0 & 0 & 0 \\ 1 & 0 & 1 & 0 \\ 1 & 0 & 2 & 1 \\ 0 & 2 & 0 & 2 \end{vmatrix} \quad (\text{by } R_1 \leftrightarrow R_4)$$

$$= 1 \times \text{cofactor of } 1$$

(since 1 is the only non-zero element in first row)

$$= \begin{vmatrix} 0 & 1 & 0 \\ 0 & 2 & 1 \\ 2 & 0 & 2 \end{vmatrix} = 2. \ne 0$$

Thus, there exist a square sub-matrix B of order 4 whose determinant value is non-zero. Hence, rank $(A) = 4$.

39. (b) Let A be the given matrix. Since A is upper triangular, so $\det(A) = 1 \times 7 \times 1 \times 2 = 14$ $\ne 0$.

So rank (A) = order of the square matrix $A = 4$.

40. (a) rank$(A) = 3$

$\Rightarrow$ rank$(A) < 4$ (= order of the square matrix)

$\Rightarrow |A| = 0$

$$\Rightarrow \begin{vmatrix} \lambda & -1 & 0 & 0 \\ 0 & \lambda & -1 & 0 \\ 0 & 0 & \lambda & -1 \\ -6 & 11 & -6 & 1 \end{vmatrix} = 0$$

$$\Rightarrow \begin{vmatrix} \lambda & -1 & 0 & 0 \\ 0 & \lambda & -1 & 0 \\ -6 & 11 & \lambda-6 & 0 \\ -6 & 11 & -6 & 1 \end{vmatrix} = 0$$

(by $R_3 \to R_3 + R_4$)

$$\Rightarrow \begin{vmatrix} \lambda & -1 & 0 \\ 0 & \lambda & -1 \\ -6 & 11 & \lambda-6 \end{vmatrix} = 0$$

$\Rightarrow \lambda[\lambda(\lambda - 6) + 11] - (-1)(0 - 6) = 0$

$\Rightarrow \lambda^3 - 6\lambda^2 + 11\lambda - 6 = 0$

$\Rightarrow \lambda^2(\lambda - 1) - 5\lambda(\lambda - 1) + 6(\lambda - 1) = 0$

$\Rightarrow (\lambda - 1)(\lambda^2 - 5\lambda + 6) = 0$

$\Rightarrow (\lambda - 1)(\lambda - 2)(\lambda - 3) = 0$

So, $\lambda = 1, 2, 3$.

41. (b)

$$A = \begin{bmatrix} 1 & 1 & 0 \\ -1 & 1 & 2 \\ 2 & 2 & 0 \\ -1 & 0 & 1 \end{bmatrix}$$

$$\sim \begin{bmatrix} 1 & 1 & 0 \\ 0 & 2 & 2 \\ 0 & 0 & 0 \\ 0 & 1 & 1 \end{bmatrix}$$

$$\begin{pmatrix} \text{by } R_2 \to R_2 + R_1, R_3 \to R_3 - 2R_1, \\ R_4 \to R_4 + R_1 \end{pmatrix}$$

$$\sim \begin{bmatrix} 1 & 1 & 0 \\ 0 & 0 & 0 \\ 0 & 0 & 0 \\ 0 & 1 & 1 \end{bmatrix} \ (\text{by } R_2 \to R_2 - 2R_4)$$

which has two non-zero rows.

Therefore, rank $(A) = 2$.

42. (*b*)
$$\begin{bmatrix} 4 & 5 & a \\ 5 & 6 & b \\ 6 & \lambda & c \end{bmatrix}$$

$$\sim \begin{bmatrix} 4 & 5 & a \\ 10 & 12 & 2b \\ 6 & \lambda & c \end{bmatrix} \quad \text{(by } R_2 \to 2R_2\text{)}$$

$$\sim \begin{bmatrix} 4 & 5 & a \\ 0 & 7-\lambda & 2b-(a+c) \\ 6 & \lambda & c \end{bmatrix}$$
$$[\text{by } R_2 \to R_2 - (R_1 + R_3)]$$

$$\sim \begin{bmatrix} 4 & 5 & a \\ 0 & 7-\lambda & 0 \\ 6 & \lambda & c \end{bmatrix}$$
$$[\because a, b, c \text{ are in A.P, so } 2b = a + c]$$

$$\sim \begin{bmatrix} 4 & 5 & a \\ 0 & 0 & 0 \\ 6 & \lambda & c \end{bmatrix} \quad \text{(if } 7-\lambda=0\text{)}$$

Thus, for $\lambda = 7$, we get two non-zero rows in the final equivalent matrix and so rank becomes "2."

But since the rank is independent of "*d*," so we have d = any arbitrary number.

43. $(i) - (a), (ii) - (b), (iii) - (a)$

(i) By Cramer's rule the system has a unique solution

$$\Leftrightarrow \Delta \neq 0$$

$$\Leftrightarrow \begin{vmatrix} 1 & 2 & 1 \\ 2 & 1 & 3 \\ 3 & 4 & -\lambda \end{vmatrix} \neq 0$$

$$\Leftrightarrow 3\lambda + 11 \neq 0$$

$$\Leftrightarrow \lambda \neq -\frac{11}{3}.$$

(ii) Now if the system has no solution, then $\Delta = 0$.

$$\Delta = 0 \Rightarrow = -\frac{11}{3}$$

$$\Delta_3 = \begin{vmatrix} 1 & 2 & 8 \\ 2 & 1 & 13 \\ 3 & 4 & \mu \end{vmatrix} = -3\mu + 66$$

If $\Delta_3 \neq 0$, then $-3\mu + 66 \neq 0$ *i.e*; $\mu \neq 22$ and vice versa.

But $\Delta = 0$, $\Delta_3 \neq 0 \Rightarrow$ system has no solutions (by Cramer's rule).

Hence, the system has no solution for $\lambda = -\frac{11}{3}$ and $\mu \neq 22$

(*iii*) The system has an infinite number of solutions

$$\Delta = \Delta_1 = \Delta_2 = \Delta_3 = 0 \text{ (by Cramer's rule)}$$
$$\Rightarrow \Delta = \Delta_3 = 0$$
$$\Rightarrow \lambda = -\frac{11}{3}, \mu = 22$$

Alternative method:

$$\begin{bmatrix} A : B \end{bmatrix}$$

$$= \begin{bmatrix} 1 & 2 & 1 & 8 \\ 2 & 1 & 3 & 13 \\ 3 & 4 & -\lambda & \mu \end{bmatrix}$$

$$\sim \begin{bmatrix} 1 & 2 & 1 & 8 \\ 0 & -3 & 1 & -3 \\ 0 & -2 & -\lambda-3 & \mu-24 \end{bmatrix}$$
$$\text{(by } R_2 \to R_2 - 2R_1, R_3 \to R_3 - 3R_1\text{)}$$

$$\sim \begin{bmatrix} 1 & 2 & 1 & 8 \\ 0 & -3 & 1 & -3 \\ 0 & 0 & -\lambda-\frac{11}{3} & \mu-22 \end{bmatrix}$$
$$\text{(by } R_3 \to R_3 - \frac{2}{3}R_2\text{)}$$

Case-I: $-\lambda - \frac{11}{3} \neq 0$ i.e; $\lambda \neq -\frac{11}{3}$.

In this case, rank $(A) = 3$ = number of variables. So the system has a unique solution.

Case-II:

$-\lambda - \frac{11}{3} = 0$ and $\mu - 22 \neq 0$ i.e; $\lambda = -\frac{11}{3}$ and $\mu \neq 22$.

In this case, rank $(A) = 2 \neq 3 = $ rank $([A:B])$. So the system has no solution.

Case-III:

$-\lambda - \frac{11}{3} = 0$ and $\mu - 22 = 0$, i.e; $\lambda = -\frac{11}{3}$ and $\mu = 22$.

In this case, rank $(A) = $ rank$([A:B]) = 2 < 3 = $ number of variables. Hence, the system has an infinite number of solutions.

44. $(i) - (b), (ii) - (a), (iii) - (c)$

(*i*) The system has a unique solution

$\Rightarrow \Delta \neq 0$

$\Rightarrow \begin{vmatrix} 1 & 1 & 1 \\ 1 & 2 & 3 \\ 1 & 2 & \lambda \end{vmatrix} \neq 0$

$\Rightarrow \lambda - 3 \neq 0$

$\Rightarrow \lambda \neq 3.$

(*ii*) Now if the system has no solution, then $\Delta = 0$.

But $\Delta = 0 \Rightarrow \lambda = 3$

$\Delta_3 = \begin{vmatrix} 1 & 1 & 6 \\ 1 & 2 & 10 \\ 1 & 2 & \mu \end{vmatrix} = \mu - 10$

If $\Delta_3 \neq 0$, then $\mu \neq 10$ and vice versa.

But $\Delta = 0, \Delta_3 \neq 0 \Rightarrow$ System has no solution.

Hence, the system has no solution for $\lambda = 3$ and $\mu \neq 10$

(*iii*) The system has an infinite number of solutions

$\Leftrightarrow \quad \Delta = \Delta_1 = \Delta_2 = \Delta_3 = 0 \Leftrightarrow \Delta = \Delta_3 = 0 \Leftrightarrow \pi = 3, \mu = 10$

(by Cramer's rule).

Alternative method

$[A:B]$

$= \begin{bmatrix} 1 & 1 & 1 & 6 \\ 1 & 2 & 3 & 10 \\ 1 & 2 & \lambda & \mu \end{bmatrix}$

$\sim \begin{bmatrix} 1 & 1 & 1 & 6 \\ 0 & 1 & 2 & 4 \\ 0 & 1 & \lambda - 1 & \mu - 6 \end{bmatrix}$

(by $R_2 \to R_2 - R_1, \; R_3 \to R_3 - R_1$)

$\sim \begin{bmatrix} 1 & 1 & 1 & 6 \\ 0 & 1 & 2 & 4 \\ 0 & 0 & \lambda - 3 & \mu - 10 \end{bmatrix}$

(by $R_3 \to R_3 - R_2$)

Case-I: $\lambda - 3 \neq 0$ *i.e.*, $\lambda \neq 3$.

In this case, rank $(A) = 3 =$ number of variables. So the system has a unique solution.

Case-II: $\lambda - 3 = 0$ and $\mu - 10 \neq 0$ *i.e*; $\lambda = 3$ and $\mu \neq 10$.

In this case, rank $(A) = 2 \neq 3 =$ rank$([A:B])$. So the system has no solution.

Case-III:

$\lambda - 3 = 0$ and $\mu - 10 = 0$ *i.e.*, $\lambda = 3$ and $\mu = 10$.

In this case, rank $(A) =$ rank$([A:B]) = 2 < 3 =$ number of variables. Hence, the system has an infinite number of solutions.

45. $(i) - (a), \; (ii) - (b), \; (iii) - (a)$

(*i*) $\Delta = \begin{vmatrix} 1 & 4 & 2 \\ 2 & 7 & 5 \\ 4 & \lambda & 10 \end{vmatrix}$

$= \begin{vmatrix} 1 & 4 & 2 \\ 0 & -1 & 1 \\ 0 & \lambda - 16 & 2 \end{vmatrix}$

(by $R_2 \to R_2 - 2R_1, R_3 \to R_3 - 4R_1$)

$= 1(-2 - \lambda + 16)$

$= -\lambda + 14$

$\Delta_2 = \begin{vmatrix} 1 & 1 & 2 \\ 2 & 2\mu & 5 \\ 4 & 2\mu + 1 & 10 \end{vmatrix}$

$= \begin{vmatrix} 1 & 1 & 2 \\ 0 & 2\mu - 2 & 1 \\ 0 & 2\mu - 3 & 2 \end{vmatrix}$

(by $R_2 \to R_2 - 2R_1, R_3 \to R_3 - 4R_1$)

$= 1(4\mu - 4 - 2\mu + 3)$

$= 2\mu - 1$

Then $\Delta = 0 \Rightarrow -\lambda + 14 = 0$, *i.e.*, $\lambda = 14$ and,

$\Delta_2 \neq 0 \Rightarrow 2\mu - 1 \neq 0$ *i.e*; $\mu \neq \dfrac{1}{2}$.

But $\Delta = 0, \Delta_2 \neq 0 \Rightarrow$ System has no solution (by Cramer's rule).

Hence, the system has no solution for $\lambda = 14$ and $\mu \neq \dfrac{1}{2}$

(*ii*) Now if the system has unique solution, then $\Delta \neq 0$. $\Delta \neq 0 \Leftrightarrow \lambda \neq 14$. Thus, system has a unique solution for $\pi \neq 14$

(*iii*) The system has an infinite number of solutions

$\Leftrightarrow \Delta = \Delta_1 = \Delta_2 = \Delta_3$

$\Leftrightarrow \Delta = \Delta_2 = 0 \Leftrightarrow \Delta = 0, \Delta_2 = 0 \Leftrightarrow \pi = 14, \mu = \dfrac{1}{2}$

Alternative method:

$$[A : B]$$

$$= \begin{bmatrix} 1 & 4 & 2 & 1 \\ 2 & 7 & 5 & 2\mu \\ 4 & \lambda & 10 & 2\mu+1 \end{bmatrix}$$

$$\sim \begin{bmatrix} 1 & 4 & 2 & 1 \\ 0 & -1 & 1 & 2\mu-2 \\ 0 & \lambda-16 & 2 & 2\mu-3 \end{bmatrix}$$

(by $R_2 \to R_2 - 2R_1$, $R_3 \to R_3 - 3R_1$)

$$\sim \begin{bmatrix} 1 & 4 & 2 & 1 \\ 0 & -1 & 1 & 2\mu-2 \\ 0 & \lambda-14 & 0 & 1-2\mu \end{bmatrix}$$

(by $R_3 \to R_3 - 2R_2$)

Case-I: $\lambda - 14 \neq 0$ *i.e.*, $\lambda \neq 14$

In this case, rank $(A) = 3 =$ number of variables. So the system has a unique solution.

Case-II:

$\lambda - 14 = 0$ and $1 - 2\mu \neq 0$ *i.e.*, $\lambda = 14$ and $\mu \neq \dfrac{1}{2}$.

In this case, rank $(A) = 2 \neq 3 = $ rank($[A{:}B]$). So the system has no solution.

Case-III:

$\lambda - 14 = 0$ and $1 - 2\mu = 0$ *i.e.*, $\lambda = 14$ and $\mu = \dfrac{1}{2}$.

In this case, rank $(A) = $ rank $([A{:}B]) = 2 < 3$ = number of variables. Hence, the system has an infinite number of solutions.

46. $(i) - (a)$, $(ii) - (c)$, $(iii) - (a)$

(i) $\Delta = 0$

$$\Rightarrow \begin{vmatrix} 1 & 1 & 1 \\ 1 & 2 & -1 \\ 5 & 7 & \lambda \end{vmatrix} = 0$$

$$\Rightarrow \begin{vmatrix} 1 & 1 & 1 \\ 0 & 1 & -2 \\ 0 & 2 & \lambda-5 \end{vmatrix} = 0$$

(by $R_2 \to R_2 - R_1, R_3 \to R_3 - 5R_1$)

$$\Rightarrow 1(\lambda - 5 + 4) = 0$$

$$\Rightarrow \lambda = 1$$

$\Delta_3 \neq 0$

$$\Rightarrow \begin{vmatrix} 1 & 1 & 1 \\ 1 & 2 & \mu \\ 5 & 7 & \mu^2 \end{vmatrix} \neq 0$$

$$\Rightarrow \begin{vmatrix} 1 & 1 & 1 \\ 0 & 1 & \mu-1 \\ 0 & 2 & \mu^2-5 \end{vmatrix} \neq 0$$

(by $R_2 \to R_2 - R_1, R_3 \to R_3 - 5R_1$)

$$\Rightarrow 1(\mu^2 - 5 - 2\mu + 2) \neq 0$$

$$\Rightarrow (\mu+1)(\mu-3) \neq 0$$

$$\Rightarrow \mu \neq -1 \text{ and } \mu \neq 3.$$

But $\Delta = 0$, $\Delta_3 \neq 0 \Rightarrow$ system has no solution (by Cramer's rule)

Hence, the system has no solution for $\lambda = 1$ and $\mu \neq -1$ or 3.

(ii) If the system has a unique solution, then $\Delta \neq 0$ (by Cramer's rule) and vice versa.

But $\Delta \neq 0 \Rightarrow \lambda \neq 1$. Hence, the system has a unique solution for $\lambda \neq 1$.

(iii) The system has an infinite number of solutions

$$\Leftrightarrow \Delta = \Delta_1 = \Delta_2 = \Delta_3 = 0 \text{ (by Cramer's rule)}$$

$$\Leftrightarrow \Delta = \Delta_3 = 0$$

$$\Leftrightarrow \lambda = 1, \mu = -1 \text{ or } 3$$

Alternative method:

$$[A : B]$$

$$= \begin{bmatrix} 1 & 1 & 1 & 1 \\ 1 & 2 & -1 & \mu \\ 5 & 7 & \lambda & \mu^2 \end{bmatrix}$$

$$\Delta = \begin{vmatrix} 1 & 4 & 2 \\ 2 & 7 & 5 \\ 4 & \lambda & 10 \end{vmatrix}$$

$$= \begin{vmatrix} 1 & 4 & 2 \\ 0 & -1 & 1 \\ 0 & \lambda-16 & 2 \end{vmatrix}$$

(by $R_2 \to R_2 - 2R_1, R_3 \to R_3 - 4R_1$)

$$= 1(-2 - \lambda + 16)$$

$$= -\lambda + 14$$

Case-I: $\lambda - 1 \neq 0$ *i.e.*, $\lambda \neq 1$.

In this case, rank $(A) = 3 =$ number of variables. So the system has a unique solution.

Case-II:

$\lambda - 1 = 0$ and $\mu^2 - 2\mu - 3 \neq 0$ *i.e.*, $\lambda = 1$ and $\mu \neq -1, 3$

In this case, rank $(A) = 2 \neq 3 = $ rank$([A:B])$. So the system has no solution.

Case-III:

$\lambda - 1 = 0$ and $\mu^2 - 2\mu - 3 = 0$ *i.e.*, $\lambda = 1$ and $\mu = -1$ or 3.

In this case, rank $(A) = $ rank $([A:B]) = 2 < 3 = $ number of variables. Hence, the system has an infinite number of solutions.

47. (c) The given system of equations can be written as:

$$(1 - \lambda)x + 2y + 3z = 0$$
$$3x + (1 - \lambda)y + 2z = 0$$
$$2x + 3y + (1 - \lambda)z = 0$$

The system of equation has a non-trivial solution

$\Rightarrow$ Rank of the coefficient matrix < 3 (here Number of variables = 3)

$\Rightarrow$ co-efficient determinant = 0

$$\Rightarrow \begin{vmatrix} 1-\lambda & 2 & 3 \\ 3 & 1-\lambda & 2 \\ 2 & 3 & 1-\lambda \end{vmatrix} = 0$$

$$\Rightarrow \begin{vmatrix} 6-\lambda & 2 & 3 \\ 6-\lambda & 1-\lambda & 2 \\ 6-\lambda & 3 & 1-\lambda \end{vmatrix} = 0 \text{ (by } C_1 \to C_1 + C_2 + C_3)$$

$$\Rightarrow (6-\lambda)\begin{vmatrix} 1 & 2 & 3 \\ 1 & 1-\lambda & 2 \\ 1 & 3 & 1-\lambda \end{vmatrix} = 0$$

$$\Rightarrow 6-\lambda = 0$$
$$\Rightarrow \lambda = 6.$$

48. (b)

$$[A:B] = \begin{bmatrix} 1 & 2 & 3 & 1 & 1 \\ 1 & -1 & 2 & -1 & -3 \\ 3 & 3 & 8 & 1 & -3 \end{bmatrix}$$

$$\sim \begin{bmatrix} 1 & 2 & 3 & 1 & 1 \\ 0 & -3 & -1 & -2 & -4 \\ 0 & -3 & -1 & -2 & -6 \end{bmatrix}$$

$$(by \ R_2 \to R_2 - R_1, R_3 \to R_3 - 3R_1)$$

$$\sim \begin{bmatrix} 1 & 2 & 3 & 1 & 1 \\ 0 & -3 & -1 & -2 & -4 \\ 0 & 0 & 0 & 0 & -2 \end{bmatrix}$$

$$(by \ R_3 \to R_3 - R_2)$$

which has three non-zero rows.

Therefore, rank $([A:B]) = 3$. Now ignoring the last column in the final equivalent matrix, we see that there are only two non-zero rows. Hence, rank $(A) = 2$.

Thus, rank $(A) \neq$ rank $([A:B])$ and so the system is inconsistent. (*i.e.*, no solution exist)

49. (a)

$$[A:B] = \begin{bmatrix} 2 & 1 & 4 & 4 \\ 1 & -3 & -1 & -5 \\ -3 & 2 & -2 & 1 \\ 8 & -3 & 8 & 2 \end{bmatrix}$$

$$\sim \begin{bmatrix} 1 & -3 & -1 & -5 \\ 2 & 1 & 4 & 4 \\ -3 & 2 & -2 & 1 \\ 8 & -3 & 8 & 2 \end{bmatrix}$$

$$(by \ R_1 \leftrightarrow R_2)$$

$$\sim \begin{bmatrix} 1 & -3 & -1 & -5 \\ 0 & 7 & 6 & 14 \\ 0 & -7 & -5 & -14 \\ 0 & 21 & 16 & 42 \end{bmatrix}$$

$$[by \ R_2 \to R_2 - 2R_1,$$
$$R_3 \to R_3 + 3R_1,$$
$$R_4 \to R_4 - 8R_1]$$

$$\sim \begin{bmatrix} 1 & -3 & -1 & -5 \\ 0 & 7 & 6 & 14 \\ 0 & 0 & 1 & 0 \\ 0 & 0 & -2 & 0 \end{bmatrix}$$

$$[by \ R_3 \to R_3 + R_2$$
$$R_4 \to R_4 - 3R_2]$$

$$\sim \begin{bmatrix} 1 & -3 & -1 & -5 \\ 0 & 7 & 6 & 14 \\ 0 & 0 & 1 & 0 \\ 0 & 0 & 0 & 0 \end{bmatrix} \ [by \ R_4 \to R_4 + 2R_3]$$

which has three non-zero rows.

Therefore, rank ([A: B]) = 3. Now ignoring the last column in the final equivalent matrix, we see that there are only three non-zero rows. Hence, rank(A) = 3.

Thus, rank (A) = rank ([A: B]) and so the system is consistent.

50. (a)

$$|A| = \begin{vmatrix} 2 & -1 & 1 \\ -1 & 2 & -1 \\ 1 & -1 & 2 \end{vmatrix} = 2(4-1)+(-2+1)+(1-2) = 4$$

Using |A|= product of eigenvalue, we get $\lambda = 1$, 1, 4.

51. (a) $A - \lambda I$

$$= \begin{pmatrix} 6 & -2 & 2 \\ -2 & 3 & -1 \\ 2 & -1 & 3 \end{pmatrix} - \lambda \begin{pmatrix} 1 & 0 & 0 \\ 0 & 1 & 0 \\ 0 & 0 & 1 \end{pmatrix}$$

$$= \begin{pmatrix} 6-\lambda & -2 & 2 \\ -2 & 3-\lambda & -1 \\ 2 & -1 & 3-\lambda \end{pmatrix}$$

$$\therefore |A - \lambda I| = 0$$

$$\Rightarrow \begin{vmatrix} 6-\lambda & -2 & 2 \\ -2 & 3-\lambda & -1 \\ 2 & -1 & 3-\lambda \end{vmatrix} = 0$$

$$\Rightarrow (6-\lambda)\{(3-\lambda)^2 - (-1)\times(-1)\}$$
$$-(-2)\{-2(3-\lambda)-(-1)\times 2\}$$
$$+2\{(-1)\times(-2)-2\times(3-\lambda)\} = 0$$

$$\Rightarrow (6-\lambda)\{9-6\lambda+\lambda^2-1\}-(12-4\lambda-4)$$
$$+(-8+4\lambda) = 0$$

$$\Rightarrow (6-\lambda)(\lambda^2-6\lambda+8)+8\lambda-16 = 0$$

$$\Rightarrow 6\lambda^2-36\lambda+48-\lambda^3+6\lambda^2-8\lambda+8\lambda-16 = 0$$

$$\Rightarrow \lambda^3-12\lambda^2+36\lambda-32 = 0$$

$$\Rightarrow \lambda^2(\lambda-2)-10\lambda(\lambda-2)+16(\lambda-2) = 0$$

$$\Rightarrow (\lambda-2)(\lambda^2-10\lambda+16) = 0$$

$$\Rightarrow (\lambda-2)(\lambda-2)(\lambda-8) = 0$$

$$\Rightarrow \lambda = 2, 2, 8$$

For $\lambda = 2$:

$AX = \lambda X$

$$\Rightarrow \begin{pmatrix} 6 & -2 & 2 \\ -2 & 3 & -1 \\ 2 & -1 & 3 \end{pmatrix} \begin{pmatrix} x \\ y \\ z \end{pmatrix} = 2 \begin{pmatrix} x \\ y \\ z \end{pmatrix}$$

$$\Rightarrow \begin{pmatrix} 6x-2y+2z \\ -2x+3y-z \\ 2x-y+3z \end{pmatrix} = \begin{pmatrix} 2x \\ 2y \\ 2z \end{pmatrix}$$

$$\Rightarrow 6x-2y+2z = 2x, -2x+3y-z = 2y,$$
$$2x-y+3z = 2z$$

$$\Rightarrow 4x-2y+2z = 0, -2x+y-z = 0,$$
$$2x-y+z = 0$$

$$\Rightarrow 2x-y+z = 0, 2x-y+z = 0,$$
$$2x-y+z = 0$$

$$\Rightarrow y = 2x+z$$

$$\therefore X = \begin{pmatrix} x \\ y \\ z \end{pmatrix} = \begin{pmatrix} x \\ 2x+z \\ z \end{pmatrix} = x\begin{pmatrix} 1 \\ 2 \\ 0 \end{pmatrix} + z\begin{pmatrix} 0 \\ 1 \\ 1 \end{pmatrix}$$

Therefore, x = 1, z = 0 gives eigenvector = $\begin{pmatrix} 1 \\ 2 \\ 0 \end{pmatrix}$ and

x = 0, z = 1 gives eigenvector = $\begin{pmatrix} 0 \\ 1 \\ 1 \end{pmatrix}$

For $\lambda = 8$:

$AX = \lambda X$

$$\Rightarrow \begin{pmatrix} 6 & -2 & 2 \\ -2 & 3 & -1 \\ 2 & -1 & 3 \end{pmatrix} \begin{pmatrix} x \\ y \\ z \end{pmatrix} = 8 \begin{pmatrix} x \\ y \\ z \end{pmatrix}$$

$$\Rightarrow \begin{pmatrix} 6x-2y+2z \\ -2x+3y-z \\ 2x-y+3z \end{pmatrix} = \begin{pmatrix} 8x \\ 8y \\ 8z \end{pmatrix}$$

$$\Rightarrow 6x-2y+2z = 8x, -2x+3y-z = 8y,$$
$$2x-y+3z = 8z$$

$$\Rightarrow -2x-2y+2z = 0, -2x-5y-z = 0,$$
$$2x-y-5z = 0$$

$$\Rightarrow x+y-z = 0 \ i.e., z = x+y............(1)$$
$$2x+5y+z = 0..........(2)$$
$$2x-y-5z = 0..........(3)$$

(3) & (1) $\Rightarrow 2x-y-5(x+y) = 0$
$$\Rightarrow x = -2y........(4)$$

(4) & (1) $\Rightarrow z = -2y+y = -y$

$$\therefore X = \begin{pmatrix} x \\ y \\ z \end{pmatrix} = \begin{pmatrix} -2y \\ y \\ -y \end{pmatrix} = \begin{pmatrix} 2 \\ -1 \\ 1 \end{pmatrix} \text{(for } y = -1)$$

Therefore, y = −1 gives eigenvector = $\begin{pmatrix} 2 \\ -1 \\ 1 \end{pmatrix}$

52. (*b*)

$\therefore |A - \lambda I| = 0$

$$\Rightarrow \begin{vmatrix} 1-\lambda & 0 & 2 \\ 0 & -1-\lambda & 1 \\ 0 & 1 & -\lambda \end{vmatrix} = 0$$

$\Rightarrow (1-\lambda)\{(-\lambda)(-1-\lambda) - 1\} - 0 + 0 = 0$

$\Rightarrow (1-\lambda)(\lambda^2 + \lambda - 1) = 0$

$\Rightarrow \lambda^2 + \lambda - 1 - \lambda^3 - \lambda^2 + \lambda = 0$

$\Rightarrow \lambda^3 - 2\lambda + 1 = 0$

Therefore, by the Cayley Hamilton theorem,

$A^3 - 2A + I = 0$

$or, A^3 = 2A - I$ ....(1)

$\therefore A^9 = (A^3)^3 = (2A - I)^3$

$= 8A^3 - 12A^2 + 6A - I$

$= 8(2A - I) - 12A^2 + 6A - I$ [Using (1)]

$= -12A^2 + 22A - 9I$

53. (*c*)

$|A - \lambda I| = 0$

$$\Rightarrow \begin{vmatrix} 1-\lambda & 1 & 1 \\ 1 & 1-\lambda & 1 \\ 1 & 1 & 1-\lambda \end{vmatrix} = 0$$

$\Rightarrow (1-\lambda) \times \{(1-\lambda)^2 - 1\} - 1 \times (1-\lambda-1)$

$\quad + 1 \times (1 - 1 + \lambda) = 0$

$\Rightarrow (1-\lambda)(\lambda^2 - 2\lambda) + 2\lambda = 0$

$\Rightarrow \lambda^2 - 2\lambda - \lambda^3 + 2\lambda^2 + 2\lambda = 0$

$\Rightarrow \lambda^3 - 3\lambda^2 = 0$

$\Rightarrow \lambda^2(\lambda - 3) = 0$

$\Rightarrow \lambda = 0, 0, 3$

Therefore, the eigenvalues are 0, 0, 3

For $\lambda = 0$

$AX = \lambda X$

$$\Rightarrow \begin{pmatrix} 1 & 1 & 1 \\ 1 & 1 & 1 \\ 1 & 1 & 1 \end{pmatrix} \begin{pmatrix} x \\ y \\ z \end{pmatrix} = 0 \begin{pmatrix} x \\ y \\ z \end{pmatrix}$$

$$\Rightarrow \begin{pmatrix} x+y+z \\ x+y+z \\ x+y+z \end{pmatrix} = \begin{pmatrix} 0 \\ 0 \\ 0 \end{pmatrix}$$

$\Rightarrow x + y + z = 0$

$\Rightarrow x = -y - z$

$$\therefore X = \begin{pmatrix} -y-z \\ y \\ z \end{pmatrix} = y \begin{pmatrix} -1 \\ 1 \\ 0 \end{pmatrix} + z \begin{pmatrix} -1 \\ 0 \\ 1 \end{pmatrix}$$

Therefore, $y = 1$, $z = 0$ gives eigenvector

$$= \begin{pmatrix} -1 \\ 1 \\ 0 \end{pmatrix}$$

and $y = 0$, $z = 1$ gives eigenvector $= \begin{pmatrix} -1 \\ 0 \\ 1 \end{pmatrix}$

For $\lambda = 3$:

$AX = \lambda X$

$$\Rightarrow \begin{pmatrix} 1 & 1 & 1 \\ 1 & 1 & 1 \\ 1 & 1 & 1 \end{pmatrix} \begin{pmatrix} x \\ y \\ z \end{pmatrix} = 3 \begin{pmatrix} x \\ y \\ z \end{pmatrix}$$

$$\Rightarrow \begin{pmatrix} x+y+z \\ x+y+z \\ x+y+z \end{pmatrix} = \begin{pmatrix} 3x \\ 3y \\ 3z \end{pmatrix}$$

$\Rightarrow x + y + z = 3x, x + y + z = 3y, x + y + z = 3y$

$$\Rightarrow \begin{cases} -2x + y + z = 0 & ....(1) \\ x - 2y + z = 0 & ....(2) \\ x + y - 2z = 0 & ....(3) \end{cases}$$

$(1) - (2) \Rightarrow -3x + 3y = 0 \Rightarrow x = y$

$\therefore (3) \Rightarrow x + x - 2z = 0 \Rightarrow z = x$

$$\therefore X = \begin{pmatrix} -y-z \\ y \\ z \end{pmatrix} = \begin{pmatrix} x \\ x \\ x \end{pmatrix} = x \begin{pmatrix} 1 \\ 1 \\ 1 \end{pmatrix}$$

Therefore, $x = 1$ gives eigenvector $= \begin{pmatrix} 1 \\ 1 \\ 1 \end{pmatrix}$.

54. (*a*) We know that the eigenvalues of a skew symmetric matrix are either imaginary number or zero. But since the eigenvalues of an orthogonal matrix are of unit modulus, so the possible eigenvalues are ± i.

55. (*b*)

Let $A = \begin{pmatrix} 0 & 3 \\ 2 & 1 \end{pmatrix}$

Then, $tr(A) = 0 + 1 = 1$; $\det(A) = 0 - 6 = -6$

Now, $A^2 = \begin{pmatrix} 0 & 3 \\ 2 & 1 \end{pmatrix}\begin{pmatrix} 0 & 3 \\ 2 & 1 \end{pmatrix} = \begin{pmatrix} 6 & 3 \\ 2 & 7 \end{pmatrix}$

$\therefore A^3 = A^2 \times A = \begin{pmatrix} 6 & 3 \\ 2 & 7 \end{pmatrix}\begin{pmatrix} 0 & 3 \\ 2 & 1 \end{pmatrix}$

$= \begin{pmatrix} 6 & 21 \\ 14 & 13 \end{pmatrix}$

$A^4 = A^3 \times A = \begin{pmatrix} 6 & 21 \\ 14 & 13 \end{pmatrix}\begin{pmatrix} 0 & 3 \\ 2 & 1 \end{pmatrix}$

$= \begin{pmatrix} 42 & 39 \\ 26 & 55 \end{pmatrix}$

Hence,

$A^4 - A^3 = \begin{pmatrix} 42 & 39 \\ 26 & 55 \end{pmatrix} - \begin{pmatrix} 6 & 21 \\ 14 & 13 \end{pmatrix}$

$= \begin{pmatrix} 36 & 18 \\ 12 & 42 \end{pmatrix}$

So $tr(A^4 - A^3) = 36 + 42 = 78$.

56. (*a*)

$\alpha = e^{\frac{2\pi i}{5}} = \left(e^{\pi i}\right)^{\frac{2}{5}}$

$= (\cos \pi + i \sin \pi)^{\frac{2}{5}}$

$= (-1)^{\frac{2}{5}} = \left[(-1)^2\right]^{\frac{1}{5}} = 1^{\frac{1}{5}}$

$\therefore \alpha$ is a fifth root of 1.

So, $1 + \alpha + \alpha^2 + \alpha^3 + \alpha^4 = 0$ and

$1.\alpha.\alpha^2.\alpha^3.\alpha^4 = 1$ *i.e.*, $\alpha^{10} = 1$

$$\text{Here } I = \begin{bmatrix} 1 & 0 & 0 & 0 & 0 \\ 0 & 1 & 0 & 0 & 0 \\ 0 & 0 & 1 & 0 & 0 \\ 0 & 0 & 0 & 1 & 0 \\ 0 & 0 & 0 & 0 & 1 \end{bmatrix}$$

$\therefore tr(I) = 1 + 1 + 1 + 1 + 1 = 5$

Also $tr(M) = 1 + \alpha + \alpha^2 + \alpha^3 + \alpha^4 = 0$

Now $M^2 = MM$

$$= \begin{bmatrix} 1 & \alpha & \alpha^2 & \alpha^3 & \alpha^4 \\ 0 & \alpha & \alpha^2 & \alpha^3 & \alpha^4 \\ 0 & 0 & \alpha^2 & \alpha^3 & \alpha^4 \\ 0 & 0 & 0 & \alpha^3 & \alpha^4 \\ 0 & 0 & 0 & 0 & \alpha^4 \end{bmatrix}\begin{bmatrix} 1 & \alpha & \alpha^2 & \alpha^3 & \alpha^4 \\ 0 & \alpha & \alpha^2 & \alpha^3 & \alpha^4 \\ 0 & 0 & \alpha^2 & \alpha^3 & \alpha^4 \\ 0 & 0 & 0 & \alpha^3 & \alpha^4 \\ 0 & 0 & 0 & 0 & \alpha^4 \end{bmatrix}$$

$$= \begin{bmatrix} 1 & - & - & - & - \\ - & \alpha^2 & - & - & - \\ - & - & \alpha^4 & - & - \\ - & - & - & \alpha^6 & - \\ - & - & - & - & \alpha^8 \end{bmatrix}$$

$tr(M^2) = 1 + \alpha^2 + \alpha^4 + \alpha^6 + \alpha^8$

$= \dfrac{1 \times \left[1 - \left(\alpha^2\right)^5\right]}{1 - \alpha^2}$

$= \dfrac{1 - \alpha^{10}}{1 - \alpha^2} = \dfrac{1 - 1}{1 - \alpha^2} = 0$

Hence, $tr(I + M + M^2) = tr(I) + tr(M) + tr(M^2)$

$= 5 + 0 + 0 = 5.$

57. (*a*) We know that the eigenvalue of a 3 × 3 real skew symmetric matrix are either purely imaginary number or zero and since complex eigenvalues occurs in conjugate pairs, so the other eigenvalues are 0 and $-3i$.

58. (*a*) Since the eigenvalue of Hermitian matrix are all real and the eigenvalues of an unitary matrix are of unit modulus. So ± 1 are the only possible eigenvalues for H.

59. (*d*) We know that 1, ω and ω^2 are cube root of 1, where $\omega = \dfrac{-1 + i\sqrt{3}}{2}, \omega^2 = \dfrac{-1 - i\sqrt{3}}{2}$

Also, $\omega^3 = 1$ and $1 + \omega + \omega^2 = 0$

$$A = \begin{bmatrix} 1 & 0 & 0 \\ i & \omega & 0 \\ 0 & 1+2i & \omega^2 \end{bmatrix} \text{ which is the lower triangu-}$$

lar matrix with 1, ω, ω^2 as the diagonal elements.

Therefore, eigenvalue of A are 1, ω, ω^2.

So $tr(A) = $ Sum of eigenvalues $= 1 + \omega + \omega^2 = 0$

Now 1, ω, ω^2 are eigenvalues of $A \Rightarrow 1^{102}$, ω^{102}, $(\omega^2)^{102}$ are eigenvalues of A^{102}.

Therefore, $tr(A^{102}) = 1^{102} + \omega^{102} + (\omega^2)^{102} = 1 + (\omega^3)^{34} + (\omega^3)^{68} = 1 + (1)^{34} + (1)^{68} = 1 + 1 + 1 = 3.$

60. (*c*) $\det(A) = \begin{vmatrix} 40 & -29 & -11 \\ -18 & 30 & -12 \\ 26 & 24 & -50 \end{vmatrix} = \begin{vmatrix} 0 & -29 & -11 \\ 0 & 30 & -12 \\ 0 & 24 & -50 \end{vmatrix}$

$[By \ C_1 \to C_1 + C_2 + C_3]$

$= 0$

Therefore, A is singular and so "0" is an eigenvalue of A. Let λ_1 and λ_2 be other two eigenvalue.

Now the sum of eigenvalues = trace of the matrix

$$\Rightarrow \quad 0 + \lambda_1 + \lambda_2 = 40 + 30 - 50 \Rightarrow \lambda_2 = 20 - \lambda_1.$$

Thus, if $\lambda(\ne 0)$ be the eigenvalue of A, then $20 - \lambda$ will be another eigenvalue.

61. (d) Since $M^T = M$, so M is a real symmetric matrix with real eigenvalues.

But since $\text{tr}(M) = 0$, so M has some positive and some negative eigenvalues.

62. (b) $L(X) = b \times X = $ cross product of b and X

$$= \begin{vmatrix} \hat{i} & \hat{j} & \hat{k} \\ 0 & 1 & 0 \\ x_1 & x_2 & x_3 \end{vmatrix} = x_3 \hat{i} - x_1 \hat{k} = [x_3 \ \ 0 \ \ -x_1]^T$$

Let $A = \begin{pmatrix} a_1 & b_1 & c_1 \\ a_2 & b_2 & c_2 \\ a_3 & b_3 & c_3 \end{pmatrix}$

Then, $L(X) = A \begin{bmatrix} x_1 \\ x_2 \\ x_3 \end{bmatrix}$

$$\Rightarrow \begin{bmatrix} x_3 \\ 0 \\ -x_1 \end{bmatrix} = \begin{pmatrix} a_1 & b_1 & c_1 \\ a_2 & b_2 & c_2 \\ a_3 & b_3 & c_3 \end{pmatrix} \begin{bmatrix} x_1 \\ x_2 \\ x_3 \end{bmatrix}$$

$$\Rightarrow \begin{bmatrix} a_1 x_1 + b_1 x_2 + c_1 x_3 \\ a_2 x_1 + b_2 x_2 + c_2 x_3 \\ a_3 x_1 + b_3 x_2 + c_3 x_3 \end{bmatrix} = \begin{bmatrix} x_3 \\ 0 \\ -x_1 \end{bmatrix}$$

Comparing both sides we get,

$$a_1 x_1 + b_1 x_2 + c_1 x_3 = x_3 = 0x_1 + 0x_2 + 1x_3$$

$$a_2 x_1 + b_2 x_2 + c_2 x_3 = 0 = 0x_1 + 0x_2 + 0x_3,$$

$$a_3 x_1 + b_3 x_2 + c_3 x_3 = -x_1 = (-1)x_1 + 0x_2 + 0x_3.$$

Therefore, we have,

$a_1 = b_1 = 0, c_1 = 1, a_2 = b_2 = c_2 = 0,$
$a_3 = -1, b_3 = c_3 = 0.$

Therefore, $A = \begin{bmatrix} 0 & 0 & 1 \\ 0 & 0 & 0 \\ -1 & 0 & 0 \end{bmatrix}$

Then, $|A - \lambda I| = 0$

$$\Rightarrow \begin{vmatrix} 0-\lambda & 0 & 1 \\ 0 & 0-\lambda & 0 \\ -1 & 0 & 0-\lambda \end{vmatrix} = 0$$

$$\Rightarrow \lambda^3 + \lambda = 0$$

$$\Rightarrow \lambda(\lambda^2 + 1) = 0$$

$$\Rightarrow \lambda = 0, \pm i \quad (\because \lambda^2 + 1 = 0 \Rightarrow \lambda^2 = -1 = i^2)$$

63. (b) Let $A = \begin{bmatrix} 1 & 0 & 0 \\ 0 & -1 & 0 \\ 0 & 0 & 0 \end{bmatrix}$. Then, A has eigen-

values 0, 1 and −1 (since A is a diagonal matrix). Then,

$$A^2 = A \times A$$

$$= \begin{bmatrix} 1 & 0 & 0 \\ 0 & -1 & 0 \\ 0 & 0 & 0 \end{bmatrix} \times \begin{bmatrix} 1 & 0 & 0 \\ 0 & -1 & 0 \\ 0 & 0 & 0 \end{bmatrix}$$

$$= \begin{bmatrix} 1 & 0 & 0 \\ 0 & 1 & 0 \\ 0 & 0 & 0 \end{bmatrix},$$

$$A^4 = A^2 \times A^2$$

$$= \begin{bmatrix} 1 & 0 & 0 \\ 0 & 1 & 0 \\ 0 & 0 & 0 \end{bmatrix} \times \begin{bmatrix} 1 & 0 & 0 \\ 0 & 1 & 0 \\ 0 & 0 & 0 \end{bmatrix}$$

$$= \begin{bmatrix} 1 & 0 & 0 \\ 0 & 1 & 0 \\ 0 & 0 & 0 \end{bmatrix} \text{ and so on.}$$

$$\therefore A^{100} = \begin{bmatrix} 1 & 0 & 0 \\ 0 & 1 & 0 \\ 0 & 0 & 0 \end{bmatrix} \text{ and so}$$

$$I + A^{100} = \begin{bmatrix} 1 & 0 & 0 \\ 0 & 1 & 0 \\ 0 & 0 & 1 \end{bmatrix} + \begin{bmatrix} 1 & 0 & 0 \\ 0 & 1 & 0 \\ 0 & 0 & 0 \end{bmatrix}$$

$$= \begin{bmatrix} 2 & 0 & 0 \\ 0 & 2 & 0 \\ 0 & 0 & 1 \end{bmatrix}.$$

Hence, $\det(I + A^{100}) = 2 \times 2 \times 1 = 4$.

64. (d)

$$A = \begin{bmatrix} 1 & 1 & 1 \\ 1 & \omega & \omega^2 \\ 1 & \omega^2 & \omega^4 \end{bmatrix} \times \begin{bmatrix} 1 & 0 & 0 \\ 0 & -1 & 0 \\ 0 & 0 & 0 \end{bmatrix} \times \begin{bmatrix} 1 & 1 & 1 \\ 1 & \dfrac{1}{\omega} & \dfrac{1}{\omega^2} \\ 1 & \dfrac{1}{\omega^2} & \dfrac{1}{\omega^4} \end{bmatrix}$$

$$= \begin{bmatrix} 1 & -1 & 0 \\ 1 & -\omega & 0 \\ 1 & -\omega^2 & 0 \end{bmatrix} \times \begin{bmatrix} 1 & 1 & 1 \\ 1 & \omega^2 & \omega \\ 1 & \omega & \omega^2 \end{bmatrix} \quad \begin{bmatrix} \because \dfrac{1}{\omega} = \dfrac{\omega^3}{\omega} = \omega^2, \\ \dfrac{1}{\omega^2} = \dfrac{\omega^3}{\omega^2} = \omega \\ \dfrac{1}{\omega^4} = \dfrac{\omega^6}{\omega^4} = \omega^2 \end{bmatrix}$$

$$= \begin{bmatrix} 1-1 & 1-\omega^2 & 1-\omega \\ 1-\omega & 1-\omega^3 & 1-\omega^2 \\ 1-\omega^2 & 1-\omega^4 & 1-\omega^3 \end{bmatrix}$$

$$= \begin{bmatrix} 0 & 1-\omega^2 & 1-\omega \\ 1-\omega & 0 & 1-\omega^2 \\ 1-\omega^2 & 1-\omega & 0 \end{bmatrix}$$

$$[\because \omega^3 = 1 \text{ and } \omega^4 = (\omega^3)\omega = \omega]$$

Therefore, $\text{tr}(A) = 0 = $ sum of eigenvalues $= 3 + (-3) + 0$.

65. (d) Since X is the eigenvector of P corresponding to the eigenvalue λ, so $PX = \lambda X$.

Now $(N^{-1}MPM^{-1}N)(N^{-1}MX)$

$= N^{-1}MPM^{-1}(NN^{-1})MX$

$= N^{-1}MP(M^{-1}M)X$

$= (N^{-1}M)PX$

$= (N^{-1}M)\lambda X$

$= \lambda(N^{-1}MX)$ [since $NN^{-1} = I = M^{-1}M$]

This shows that $N^{-1}MX$ is an eigenvector of $N^{-1}MPM^{-1}N$ corresponding to the eigenvalue λ.

66. (a) Trace of a matrix = sum of eigenvalues

$a + b + 3 = 10 \Rightarrow a + b = 7$ ...(i)

determinant of a matrix = product of eigenvalues

$3ab = 30 \Rightarrow ab = 10$ ...(ii)

By (i) and (ii)

$a^2 + b^2 = (a+b)^2 - 2ab = 49 - 20 = 29$

67. (a) The eigenvalues of A^2 are square of eigenvalues of A. Since A^2 is a zero matrix. So, all its eigenvalues will be zero. Hence, all the eigenvalues of A will also be zero.

68. (c) We knot that the characteristic equation for A is

$|A - \lambda I| = 0$ ...(i)

Since "1" is an eigenvalue, so (i) gives

$|A - I| = 0 \Rightarrow |A^2 - A| = 0.$

Hence, $A^2 - A$ is singular.

"-1" is eigenvalue

$\Rightarrow \quad |A + I| = 0.$

$\Rightarrow \quad |A^2 + A| = 0.$ Hence, $A^2 + A$ is singular

Since "3" is an eigenvalue, so $|A - 3I| = 0$(by (i))

$\therefore \quad |A^2 - 3A| = 0.$ Hence, $A^2 - 3A$ is singular

Since "0" and "-3" are not eigenvalues,

so, $|A| \neq 0$ and $|A + 3I| \neq 0.$

Hence, $\left|A^2 + 3A\right| = |A| \times |A + 3I| \neq 0.$

Therefore, $A^2 + 3A$ is non–singular.

69. (c)

Let $m = 2$. Then $A = \begin{bmatrix} 1 & 1 \\ 1 & 1 \end{bmatrix}$

The characteristic equation of A is given by

$\det(A - \lambda I) = 0$

or, $\begin{vmatrix} 1-\lambda & 1 \\ 1 & 1-\lambda \end{vmatrix} = (1-\lambda)^2 - 1 = 0$

or, $\lambda^2 - 2\lambda = 0$

This satisfies the equation $\lambda^m - m\lambda^{m-1} = 0$ for $m = 2$.

70. (d) Let A be the given matrix. Then $A = 4I$ and the eigenvalues of A are 4 and 4.

Then $AX = \lambda X \Rightarrow 4IX = 4I$, which is true for any $X \neq O = \begin{bmatrix} 0 \\ 0 \end{bmatrix}.$

71. (c)

$$AV = \lambda V \Rightarrow \begin{bmatrix} 1 & -1 & 0 \\ 0 & 1 & -1 \\ -1 & 0 & 1 \end{bmatrix} \begin{pmatrix} 1 \\ 1 \\ 1 \end{pmatrix} = \lambda \begin{pmatrix} 1 \\ 1 \\ 1 \end{pmatrix}$$

$$\Rightarrow \begin{pmatrix} 1-1 \\ 1-1 \\ -1+1 \end{pmatrix} = \begin{pmatrix} \lambda \\ \lambda \\ \lambda \end{pmatrix}$$

$$\Rightarrow \lambda = 0$$

72. (c) $f(x) - g(x) - h(x) = (x-1) - (x+1) - (x^2 - 1)$

$= x^2 + 1 = q(x)$

$\Rightarrow \quad f(x) - g(x) - h(x) - q(x) = 0$

$\Rightarrow \quad$ the functions are linearly dependent.

73. (c)

Let $X = (a_1, a_2, ...)$ and $Y = (b_1, b_2, ...) \in W$.

Then $\lim\limits_{n \to \infty} a_n = \beta$ and $\lim\limits_{n \to \infty} b_n = \beta$.

Now,

$aX + bY$

$= a(a_1, a_2, ...) + b(b_1, b_2, ...)$

$= (aa_1 + bb_1, aa_2 + bb_2, ..., aa_n + bb_n, ...)$

$\therefore \lim\limits_{n \to \infty} (aa_n + bb_n) = a \lim\limits_{n \to \infty} a_n + \lim\limits_{n \to \infty} b_n$

$= a\beta + b\beta$

$= \beta$ (if $\beta = 0$)

Thus, W is a subspace of $V_n(R)$ if $\beta = 0$.

74. (a)

Let $f(x) = 1, g(x) = x, h(x) = x(1-x) = x - x^2$.

Then

$$\begin{vmatrix} f(x) & g(x) & h(x) \\ f'(x) & g'(x) & h'(x) \\ f''(x) & g''(x) & h''(x) \end{vmatrix}$$

$$= \begin{vmatrix} 1 & x & x - x^2 \\ 0 & 1 & 1 - 2x \\ 0 & 0 & -2 \end{vmatrix}$$

$$= -2(1 - 0) = -2 \neq 0$$

Therefore, the given set of vectors is linearly independent for all x.

**PREVIOUS YEARS SOLVED PAPERS
(2000-2018)**

1. The eigenvalues of the matrix

$$\begin{bmatrix} 2 & -1 & 0 & 0 \\ 0 & 3 & 0 & 0 \\ 0 & 0 & -2 & 0 \\ 0 & 0 & -1 & 4 \end{bmatrix} \text{ are}$$

(a) 2, −2, 1, −1   (b) 2,3,−2,4
(c) 1, 2, 3, 4   (d) none of these
                      (EC GATE 2000)

2. An $n \times n$ array V is defined as follows:
$V[i,j] = i - j$ for all i, j, where $1 \le i, j \le n$. Then the sum of the elements of the array V is

(a) 0   (b) $n-1$
(c) $n^2 - 3n + 2$ (d) $n(n+1)$
                   (CS GATE 2000)

3. The determinant of the matrix $\begin{bmatrix} 2 & 0 & 0 & 0 \\ 8 & 1 & 7 & 2 \\ 2 & 0 & 2 & 0 \\ 9 & 0 & 6 & 1 \end{bmatrix}$ is

(a) 4   (b) 0   (c) 15   (d) 20
                             (CS GATE 2000)

4. If A, B, C are square matrices of the same order, then $(ABC)^{-1}$ is equal to

(a) $C^{-1}A^{-1}B^{-1}$ (b) $C^{-1}B^{-1}A^{-1}$
(c) $A^{-1}B^{-1}C^{-1}$ (d) $A^{-1}C^{-1}B^{-1}$
                            (CE GATE 2000)

5. The rank of the matrix $A = \begin{bmatrix} 1 & 2 & 3 \\ 3 & 4 & 5 \\ 4 & 6 & 8 \end{bmatrix}$ is

(a) 0   (b) 1   (c) 2   (d) 3
                            (IN GATE 2000)

6. The product $[P][Q]^T$ of the following two matrices $[P]$ and $[Q]$ is

$$[P] = \begin{bmatrix} 2 & 3 \\ 4 & 5 \end{bmatrix}, [Q] = \begin{bmatrix} 4 & 8 \\ 9 & 2 \end{bmatrix}$$

(a) $\begin{bmatrix} 32 & 24 \\ 56 & 46 \end{bmatrix}$ (b) $\begin{bmatrix} 46 & 56 \\ 24 & 32 \end{bmatrix}$

(c) $\begin{bmatrix} 35 & 22 \\ 61 & 42 \end{bmatrix}$ (d) $\begin{bmatrix} 32 & 56 \\ 24 & 46 \end{bmatrix}$

                                      (CE GATE 2001)

7. The rank of a 3×3 matrix $C = BA$, found by multiplying a non-zero column matrix A of size 3×1 and a non-zero row matrix B of size 1×3 is

(a) 0   b) 1   (c) 2   (d) 3

(GATE 2001)

8. The rank of the matrix $\begin{pmatrix} 1 & 1 \\ 0 & 0 \end{pmatrix}$ is

(a) 4   (b) 1   (c) 2   (d) 0

(CE GATE 2001)

9. The eigenvalues of the matrix $\begin{pmatrix} 5 & 3 \\ 2 & 9 \end{pmatrix}$ are

(a) (5.13, 9.42)   (b) (3.85, 2.93)
(c) (9, 5)   (d) (10.16, 3.84)

(CE GATE 2001)

10. Consider the system of equations given below:

$x + y = 2$

$2x + 2y = 5$

This system has

(a) one solution   (*b*) no solution
(c) infinite solution   (d) four solutions

(GATE 2001)

11. The necessary condition to diagonalize a matrix is that

(a) its all eigenvalues should be distinct
(b) its eigenvalues should be independent
(c) its eigenvalues should be real
(d) the matrix is non-singular

(IN GATE 2001)

12. The following set of equations has

$3x + 2y + z = x - y + z = 2, -2x + 2z = 5$

(a) no solution   (b) a unique solution
(c) multiple solutions   (d) an inconsistency

(GATE 2001)

13. Consider the following statements:

S_1 : The sum of two singular matrices may be singular

S_2 : The sum of two non-singular matrices may be non-singular

Which of the following statements is true?

(a) S_1 and S_2 are both true
(b) S_1 and S_2 are both false
(c) S_1 is true and S_2 is false
(d) S_1 is false and S_2 is true

(CS GATE 2001)

14. Obtain the eigenvalues of the matrix

$$\begin{bmatrix} 1 & 2 & 34 & 49 \\ 0 & 2 & 43 & 94 \\ 0 & 0 & -2 & 104 \\ 0 & 0 & 0 & -1 \end{bmatrix}$$

(a) 1, 2, –2, –1   (b) –1, –2, –1, –2
(c) 1, 2, 2, 1   (d) none

(CS GATE 2002)

15. The determinant of the matrix

$$\begin{bmatrix} 1 & 0 & 0 & 0 \\ 100 & 1 & 0 & 0 \\ 100 & 200 & 1 & 0 \\ 100 & 200 & 300 & 1 \end{bmatrix}$$ is

(a) 100   (b) 200
(c) 1   (d) 300

(EE GATE 2002)

16. If $X = \begin{bmatrix} a & 1 \\ -a^2 + a - 1 & 1 - a \end{bmatrix}$ and $X^2 - X + I = 0$, then the inverse of $X =$?

(a) $\begin{bmatrix} 1-a & -1 \\ a^2 & a \end{bmatrix}$ (b) $\begin{bmatrix} 1-a & -1 \\ a^2 - a + 1 & a \end{bmatrix}$

(c) $\begin{bmatrix} -a & 1 \\ -a^2 + a - 1 & a - 1 \end{bmatrix}$

(d) $\begin{bmatrix} a^2 - a + 1 & a \\ 1 & 1 - a \end{bmatrix}$

(GATE 2002)

17. Consider the system of simultaneous equations $x + 2y + z = 6, 2x + y + 2z = 6, x + y + z = 5$

This system has

(a) a unique solution
(b) an infinite number of solutions
(c) no solution
(d) exactly two solutions

(ME GATE 2003)

18. Consider the following system of linear equations:

$$\begin{pmatrix} 2 & 1 & -4 \\ 4 & 3 & -12 \\ 1 & 2 & -8 \end{pmatrix} \begin{bmatrix} x \\ y \\ z \end{bmatrix} = \begin{bmatrix} \alpha \\ 5 \\ 7 \end{bmatrix}$$

Notice that the second and third columns of the coefficient matrix are linearly dependent. For how many values of α, does this system of equations have infinitely many solutions?

(a) 0   (b) 1

(c) 2   (d) infinitely many

(CS GATE 2003)

19. A system of equation represented by $AX = 0$, when X is a column vector of unknowns and A is a matrix containing coefficients has a non-trivial solution when A is

(a) non-singular   (b) singular

(c) symmetric   (d) Hermitian

(GATE 2003)

20. Given matrix $A = \begin{pmatrix} 4 & 2 & 1 & 3 \\ 6 & 3 & 4 & 7 \\ 2 & 1 & 0 & 1 \end{pmatrix}$. The rank of the matrix is

(a) 4   (b) 3   (c) 2   (d) 1

(CE GATE 2003)

21. For the matrix $\begin{pmatrix} 4 & 1 \\ 1 & 4 \end{pmatrix}$, the eigenvalues are

(a) 3 and –3   (b) –3 and –5

(c) 3 and 5   (d) 5 and 0

(ME GATE 2003)

22. The eigenvalues of the matrix $A = \begin{pmatrix} 4 & -2 \\ -2 & 1 \end{pmatrix}$ are

(a) 1, 4   (b) –1, 2   (c) 0, 5

(d) cannot be determined

(CE GATE 2004)

23. The sum of the eigenvalues of the matrix given below is

$$\begin{bmatrix} 1 & 2 & 3 \\ 1 & 5 & 1 \\ 3 & 1 & 1 \end{bmatrix}$$

(a) 5   (b) 7   (c) 9   (d) 18

(ME GATE 2004)

24. The number of $n \times n$ symmetric matrices with each elements being either 0 or 1 is

(a) 2^n (b) $2n^2$ (c) $2^{\frac{n^2+n}{2}}$ (d) $2^{\frac{n^2-n}{2}}$

(CS GATE 2004)

25. For which value of "x" will the matrix $\begin{pmatrix} 8 & x & 0 \\ 4 & 0 & 2 \\ 12 & 6 & 0 \end{pmatrix}$ become singular?

(a) 4   (b) 6   (c) 12   (d) 8

(ME GATE 2004)

26. Let A, B, C, and D be $n \times n$ matrices, each with non-zero determinant. If $ABCD = I$, then $B^{-1} = ?$

(a) $D^{-1}C^{-1}A^{-1}$ (b) CDA (c) ADC

(d) does not necessarily exist.

(CS GATE 2004)

27. Real matrices $A_{3\times1}$, $B_{3\times3}$, $C_{3\times5}$, $D_{5\times3}$, $E_{5\times5}$, $F_{5\times1}$ are given, where the matrices B and E are symmetric.

Following statements are made with respect to these matrices:

(i) The matrix product $F^T C^T BCF$ is a scalar

(ii) The matrix product $D^T FD$ is always symmetric

With reference to above statements, which of the following is true?

(a) Statement (i) is true but (ii) is false.

(b) Statement (i) false but (ii) is true.

(c) Both the statements are true.

(d) Both the statements are false.

(CE GATE 2004)

28. How many solutions does the following system of linear equations have?

$-x + 5y = -1$, $x - y = 2$, $x + 3y = 3$

(a) infinitely many

(b) two distinct solutions

(c) unique

(d) none

(CS GATE 2004)

29. What values of x, y, z satisfy the following system of linear equations

$$\begin{bmatrix} 1 & 2 & 3 \\ 1 & 3 & 4 \\ 2 & 2 & 3 \end{bmatrix}\begin{bmatrix} x \\ y \\ z \end{bmatrix} = \begin{bmatrix} 6 \\ 8 \\ 12 \end{bmatrix}$$

(a) $x = 6, y = 3, z = 2$

(b) $x = 12, y = 3, z = -4$

(c) $x = 6, y = 6, z = -4$

(d) $x = 12, y = -3, z = 4$

(GATE 2005)

30. Which of the following is an eigenvector of

the matrix $\begin{bmatrix} 5 & 0 & 0 & 0 \\ 0 & 5 & 5 & 5 \\ 0 & 0 & 2 & 1 \\ 0 & 0 & 3 & 1 \end{bmatrix}$?

(a) $\begin{bmatrix} 1 \\ -2 \\ 0 \\ 0 \end{bmatrix}$ (b) $\begin{bmatrix} 0 \\ 0 \\ 1 \\ 0 \end{bmatrix}$ (c) $\begin{bmatrix} 1 \\ 0 \\ 0 \\ -2 \end{bmatrix}$ (d) $\begin{bmatrix} 1 \\ -1 \\ 2 \\ 1 \end{bmatrix}$

(ME GATE 2005)

31. The eigenvalues of the matrix $A = \begin{pmatrix} 1 & 4 \\ a & 2 \end{pmatrix}$ are real and non-negative for the condition

(a) $-1/16 \leq a \leq 1/16$ (b) $-1/2 \leq a \leq 1/2$
(c) $-1/2 \leq a \leq 1/16$ (d) $-1/16 \leq a \leq 1/2$

(CS GATE 2005)

32. For the matrix $P = \begin{pmatrix} 3 & -2 & 0 \\ 0 & -1 & 1 \\ 0 & 0 & 1 \end{pmatrix}$, one of the eigenvalue is equal to −2. Then which of the following is an eigenvector?

(a) $\begin{bmatrix} 3 \\ -2 \\ 1 \end{bmatrix}$ (b) $\begin{bmatrix} -3 \\ 2 \\ -1 \end{bmatrix}$ (c) $\begin{bmatrix} 1 \\ -2 \\ 3 \end{bmatrix}$ (d) $\begin{bmatrix} 2 \\ 5 \\ 0 \end{bmatrix}$

(EE GATE 2005)

33. What are the eigenvalues of the following 2 × 2 matrix:

$\begin{bmatrix} 2 & -1 \\ -4 & 5 \end{bmatrix}$

(a) −1 and 1 (b) 1 and 6
(c) 2 and 5 (d) 4 and −1

(CS GATE 2005)

34. Given the matrix $\begin{bmatrix} -4 & 2 \\ 4 & 3 \end{bmatrix}$. The eigenvector is

(a) $\begin{bmatrix} 3 \\ 2 \end{bmatrix}$ (b) $\begin{bmatrix} 4 \\ 3 \end{bmatrix}$ (c) $\begin{bmatrix} 2 \\ -1 \end{bmatrix}$ (d) $\begin{bmatrix} -1 \\ 2 \end{bmatrix}$

(EC GATE 2005)

35. Consider the system of equations $A_{n\times n}X_{n\times 1} = \lambda X_{n\times 1}$, where λ is a scalar. Let (λ_1, X_i) be an eigenpair of an eigenvalue and its corresponding eigenvector for real matrix A.

Let I be a $n\times n$ unit matrix. Then which of the following statements is not correct?

(a) For a homogeneous $n\times n$ system of linear equations $(A - \lambda I)X = 0$, having a non-trivial solution, the rank of $A - \lambda I$ is less than n.

(b) For matrix A^m, m be a positive integer, (λ_i^m, X_i^m) will be eigenpair for all i.

(c) If $A^T = A^{-1}$, then $|\lambda_i| = 1$ for all i.

(d) If $A^T = A$, then λ_i is real for all i.

(CE GATE 2005)

36. The eigenvalue of the matrix M given below are 15, 3, 0.

$M = \begin{bmatrix} 8 & -6 & 2 \\ -6 & 7 & -4 \\ 2 & -4 & 3 \end{bmatrix}$

Then the value of the determinant of the matrix is

(a) 20 (b) 10 (c) 0 (d) −10

(PI GATE 2005)

37. Identify which one of the followings is are eigenvector of the matrix $A = \begin{bmatrix} 1 & 0 \\ -1 & -2 \end{bmatrix}$

(a) $\begin{bmatrix} -1 \\ 1 \end{bmatrix}$ (b) $\begin{bmatrix} 3 \\ -1 \end{bmatrix}$ (c) $\begin{bmatrix} 1 \\ -1 \end{bmatrix}$ (d) $\begin{bmatrix} -2 \\ 1 \end{bmatrix}$

(ME GATE 2005)

38. A is a 3 × 4 real matrix and $AX = B$ is an inconsistent system of equations. The highest possible rank of A is

(a) 1 (b) 2 (c) 3 (d) 4

(ME GATE 2005)

39. Consider the following system of equations in three real variables x_1, x_2, and x_3:

$2x_1 - x_2 + 3x_3 = 1$,
$3x_1 - 2x_2 + 5x_3 = 2$,
$-x_1 - 4x_2 + x_3 = 3$.

This system of equations has
(a) no solution
(b) a unique solution
(c) more than one but finite number of solution
(d) an infinite number of solutions

(CS GATE 2005)

40. Consider a non-homogeneous system of linear equations representing mathematically an over determined system. Such a system will be

(a) consistent having a unique solution
(b) consistent having many solutions
(c) inconsistent having a unique solution
(d) inconsistent having no solution

(CE GATE 2005)

41. Let A be a 3×3 matrix with rank 2. Then $AX = 0$ has

(a) only the trivial solution $X = 0$
(b) one independent solution
(c) two independent solutions
(d) three independent solutions

(GATE 2005)

42. In the matrix equation $PX = Q$, which of the following is a necessary condition for the existence of at least one solution for the unknown vector X:

(a) The augmented matrix $[P{:}Q]$ must have the same rank as matrix P
(b) Vector Q must have only one-zero elements
(c) Matrix P must be singular
(d) Matrix P must be square

(EE GATE 2005)

43. The determinant of the matrix given below is

$$\begin{vmatrix} 0 & 1 & 0 & 2 \\ -1 & 1 & 1 & 3 \\ 0 & 0 & 0 & 1 \\ 1 & -2 & 0 & 1 \end{vmatrix}$$

(a) -1 (b) 0   (c) 1   (d) 2

(GATE 2005)

44. For an orthogonal matrix A of order 4×4, $(AA')^{-1}$ is

(a) $\begin{pmatrix} \frac{1}{4} & 0 & 0 & 0 \\ 0 & \frac{1}{4} & 0 & 0 \\ 0 & 0 & \frac{1}{2} & 0 \\ 0 & 0 & 0 & \frac{1}{2} \end{pmatrix}$ (b) $\begin{pmatrix} \frac{1}{2} & 0 & 0 & 0 \\ 0 & \frac{1}{2} & 0 & 0 \\ 0 & 0 & \frac{1}{2} & 0 \\ 0 & 0 & 0 & \frac{1}{2} \end{pmatrix}$

(c) $\begin{pmatrix} 1 & 0 & 0 & 0 \\ 0 & 1 & 0 & 0 \\ 0 & 0 & 1 & 0 \\ 0 & 0 & 0 & 1 \end{pmatrix}$ (d) $\begin{pmatrix} \frac{1}{4} & 0 & 0 & 0 \\ 0 & \frac{1}{4} & 0 & 0 \\ 0 & 0 & \frac{1}{4} & 0 \\ 0 & 0 & 0 & \frac{1}{4} \end{pmatrix}$

(EC GATE 2005)

45. Let $A = \begin{pmatrix} 2 & -0.1 \\ 0 & 3 \end{pmatrix}$, $A^{-1} = \begin{pmatrix} \frac{1}{2} & a \\ 0 & b \end{pmatrix}$, then $a + b = ?$

(a) 7/20   (b) 3/20   (c) 19/60   (d) 11/20

(EC GATE 2005)

46. Consider the matrices $X_{(4\times3)}$, $Y_{(4\times3)}$, and $P_{(2\times3)}$. Then the order of $[P(X^TY)^{-1}P^T]^T$ will be

(a) 2×2 (b) 3×3 (c) 4×3 (d) 3×4

(CE GATE 2005)

47. If $R = \begin{pmatrix} 1 & 0 & -1 \\ 2 & 1 & -1 \\ 2 & 3 & 2 \end{pmatrix}$ then top row R^{-1} is

(a) $[5\ 6\ 4]$ (b) $[5\ -3\ 1]$

(c) $[2\ 0\ -1]$ (d) $\left[2\ \ -1\ \ \frac{1}{2}\right]$

(EE GATE 2005)

48. Multiplication of matrices E and F is G. Matrices E and G are as follows

$$E = \begin{pmatrix} \cos\theta & \sin\theta & 0 \\ -\sin\theta & \cos\theta & 0 \\ 0 & 0 & 1 \end{pmatrix}, G = \begin{pmatrix} 1 & 0 & 0 \\ 0 & 1 & 0 \\ 0 & 0 & 1 \end{pmatrix}.$$

What is the matrix F ?

(a) $\begin{pmatrix} \cos\theta & -\sin\theta & 0 \\ \sin\theta & \cos\theta & 0 \\ 0 & 0 & 1 \end{pmatrix}$

(b) $\begin{pmatrix} \cos\theta & \cos\theta & 0 \\ -\cos\theta & \sin\theta & 0 \\ 0 & 0 & 1 \end{pmatrix}$

(c) $\begin{pmatrix} \cos\theta & \sin\theta & 0 \\ -\sin\theta & \cos\theta & 0 \\ 0 & 0 & 1 \end{pmatrix}$

(d) $\begin{pmatrix} -\cos\theta & \sin\theta & 0 \\ \sin\theta & \cos\theta & 0 \\ 0 & 0 & 1 \end{pmatrix}$

(ME GATE 2006)

49. A system of linear simultaneous equations is given as $AX = B$, where

$$A = \begin{pmatrix} 1 & 0 & 1 & 0 \\ 0 & 1 & 0 & 1 \\ 1 & 1 & 0 & 0 \\ 0 & 0 & 0 & 1 \end{pmatrix}, \; B = \begin{pmatrix} 0 \\ 0 \\ 0 \\ 1 \end{pmatrix}$$

Then the rank of the matrix A is

(a) 1   (b) 2   (c) 3   (d) 4

(IN GATE 2006)

50. The rank of the matrix $\begin{pmatrix} 1 & 1 & 1 \\ 1 & -1 & 0 \\ 1 & 1 & 1 \end{pmatrix}$ is

(a) 0   (b) 1   (c) 2   (d) 3

(EC GATE 2006)

51. For the matrix $\begin{bmatrix} 4 & 2 \\ 2 & 4 \end{bmatrix}$, the eigenvalue corresponding to the eigenvector $\begin{bmatrix} 101 \\ 101 \end{bmatrix}$ is

(a) 2   (b) 4   (c) 6   (d) 8

(EC GATE 2006)

52. Match the items in Columns I and II

Column-I	Column-II
P. Singular matrix	1. Determinant is not defined
Q. Non square matrix	2. Determinant is always one.
R. Real symmetric	3. Determinant is zero
S. Orthogonal matrix	4. Eigenvalues are always real
	5. Eigenvalues are not defined

(a) $P \to 3, Q \to 1, R \to 4, S \to 2$
(b) $P \to 2, Q \to 3, R \to 4, S \to 1$
(c) $P \to 3, Q \to 2, R \to 5, S \to 4$
(d) $P \to 3, Q \to 4, R \to 2, S \to 1$

(ME GATE 2006)

53. For a given matrix, $A = \begin{pmatrix} 2 & -2 & 3 \\ -2 & -1 & 6 \\ 1 & 2 & 0 \end{pmatrix}$ one of

the eigenvalue is 3. The other eigenvalues are

(a) 2, –5   (b) 3, –5   (c) 2, 5   (d) 3, 5

(CE GATE 2006)

54. The eigenvalues of a matrix $S = \begin{pmatrix} 3 & 2 \\ 2 & 3 \end{pmatrix}$ are 5 and 1. What are the eigenvalues of S^2 ?

(a) 1, 25   (b) 6, 4
(c) 1, 5   (d) 2, 10

(ME GATE 2006)

55. The eigenvalues and the corresponding eigenvectors of a 2×2 matrix are given by

Eigenvalue	Eigenvector
$\lambda_1 = 8$	$v_1 = \begin{bmatrix} 1 \\ 1 \end{bmatrix}$
$\lambda_2 = 4$	$v_2 = \begin{bmatrix} 1 \\ -1 \end{bmatrix}$

Then the matrix is

(a) $\begin{pmatrix} 6 & 2 \\ 2 & 6 \end{pmatrix}$ (b) $\begin{pmatrix} 4 & 6 \\ 6 & 4 \end{pmatrix}$

(c) $\begin{pmatrix} 2 & 4 \\ 4 & 2 \end{pmatrix}$ (d) $\begin{pmatrix} 4 & 8 \\ 8 & 4 \end{pmatrix}$

(EC GATE 2006)

56. For a given 2×2 matrix A, it is observed that

$$A\begin{bmatrix} 1 \\ -1 \end{bmatrix} = -\begin{bmatrix} 1 \\ -1 \end{bmatrix}, A\begin{bmatrix} 1 \\ -2 \end{bmatrix} = -2\begin{bmatrix} 1 \\ -2 \end{bmatrix}.$$

Then the matrix A is

(a) $\begin{bmatrix} 2 & 1 \\ -1 & -1 \end{bmatrix} \times \begin{bmatrix} -1 & 0 \\ 0 & -2 \end{bmatrix} \times \begin{bmatrix} 1 & 1 \\ -1 & -2 \end{bmatrix}$

(b) $\begin{bmatrix} 1 & 1 \\ -1 & -2 \end{bmatrix} \times \begin{bmatrix} 1 & 0 \\ 0 & 2 \end{bmatrix} \times \begin{bmatrix} 2 & 1 \\ -1 & -1 \end{bmatrix}$

(c) $\begin{bmatrix} 1 & 1 \\ -1 & -2 \end{bmatrix} \times \begin{bmatrix} -1 & 0 \\ 0 & -2 \end{bmatrix} \times \begin{bmatrix} 2 & 1 \\ -1 & -1 \end{bmatrix}$

(d) $\begin{bmatrix} 1 & -2 \\ 1 & -3 \end{bmatrix}$

(IN GATE 2006)

57. If a square matrix A is real and symmetric, then the eigenvalues

(a) are always real
(b) are always real and positive
(c) are always real and non-positive
(d) occur in complex conjugate pairs

(ME GATE 2007)

58. The minimum and maximum eigenvalues of the matrix $\begin{bmatrix} 1 & 1 & 3 \\ 1 & 5 & 1 \\ 3 & 1 & 1 \end{bmatrix}$ are -2 and 6, respectively. Then what is the other eigenvalue?

(a) 5   (b) 3   (c) 1   (d) -1

(CE GATE 2007)

59. Consider the set of column vectors defined by
$$X = \left\{ x \in R^3 : x_1 + x_2 + x_3 = 0, \text{where } x^T = [x_1, x_2, x_3]^T. \right\}$$
Then which of the following is true?

(a) $\{[1, -1, 0]^T, [1, 0, -1]^T\}$ is the basis for the subspace X

(b) $\{[1, -1, 0]^T, [1, 0, -1]^T\}$ is a linearly independent set, but it does not span X and therefore not a basis of X

(c) X is not a subspace of R^3

(d) none of the above

(CS GATE 2007)

60. (common data linked question)

The Cayley Hamilton theorem states that a square matrix satisfies its own characteristic equation. Consider a matrix $A = \begin{bmatrix} -3 & 2 \\ -1 & 0 \end{bmatrix}$

(i) A satisfies the relation

(a) $A + 3I + 2A^{-1} = O$

(b) $A^2 + 2A + 2I = O$

(c) $(A + I)(A + 2I) = I$

(d) $\exp(A) = O$

(ii) A^9 Equal to

(a) $511A + 510I$ (b) $309A + 104I$

(c) $154A + 155I$ (d) $\exp(9A)$

(EE GATE 2007)

61. The determinant $\begin{vmatrix} 1+b & b & 1 \\ b & b+1 & 1 \\ 1 & 2b & 1 \end{vmatrix}$ equals to

(a) 0   (b) $2b(b-1)$

(c) $2(1-b)(2b+1)$ (d) $3b(1+b)$

(PI GATE 2007)

62. $X = \begin{bmatrix} x_1 & x_2 & \cdots & x_n \end{bmatrix}^T$ is a n-tuple non-zero vector. Then the $n \times n$ matrix $V = XX'$

(a) has rank zero   (b) has rank 1

(c) is orthogonal   (d) has rank n.

(EE GATE 2007)

63. The inverse of the 2×2 matrix $A = \begin{pmatrix} 1 & 2 \\ 5 & 7 \end{pmatrix}$ is

(a) $\dfrac{1}{3}\begin{pmatrix} -7 & 2 \\ 5 & -1 \end{pmatrix}$ (b) $\dfrac{1}{3}\begin{pmatrix} 7 & 2 \\ 5 & 1 \end{pmatrix}$

(c) $\dfrac{1}{3}\begin{pmatrix} 7 & -2 \\ -5 & 1 \end{pmatrix}$ (d) $\dfrac{1}{3}\begin{pmatrix} -7 & 2 \\ -5 & 1 \end{pmatrix}$

(CE GATE 2007)

64. The solution for the system defined by the set of equations:
$$4y + 3z = 8, \ 2x - z = 2 \text{ and } 3x + 2y = 5 \text{ is}$$

(a) $x = 0, y = 1, z = \dfrac{4}{3}$,   (b) $x = 0, y = \dfrac{1}{2}, z = 2$

(c) $x = 1, y = \dfrac{1}{2}, z = 2$ (d) non-existent

(GATE 2007)

65. For what value of α and β, the following simultaneous equations have an infinite number of solutions?
$$x + y + z = 5, \ x + 3y + 3z = 9, \ x + 2y + \alpha z = \beta$$

(a) 2, 7   (b) 3, 8   (c) 8, 3   (d) 7, 2

(CE GATE 2007)

66. Let A be an $n \times n$ real matrix such that $A^2 = I$ and Y be an n-dimensional vector. Then, the linear system of equations $AX = Y$ has

(a) no solution

(b) unique solution

(c) more than one but finitely many dependent solutions

(d) infinitely many dependent solution.

(IN GATE 2007)

67. The number of linearly independent eigenvectors of $\begin{pmatrix} 2 & 1 \\ 0 & 2 \end{pmatrix}$ is

(a) 0   (b) 1

(c) 2   (d) infinite

(ME GATE 2007)

68. It is given that $X_1, X_2, ..., X_M$ are M non-zero orthogonal vectors. Then the dimension of the vector space spanned by the $2M$ vectors $X_1, X_2, ..., X_M, -X_1, -X_2, ... -X_M$ is

(a) $2M$ (b) $M + 1$ (c) M

(d) dependent on the choices of $X_1, X_2, ..., X_M$

(EC GATE 2007)

69. The following simultaneous equations $x + y + z = 3$, $x + 2y + 3z = 4$, $x + 4y + kz = 6$ will not have a unique solution for k is equal to

(a) 0   (b) 5   (c) 6   (d) 7

(CE GATE 2008)

70. For what value of "a" if any, the following system of equations is x, y, z have a solution?

$2x + 3y = 4$, $x + y + z = 4$, $x + 2y - z = a$

(a) 0   (b) any real number
(c) 1   (d) there is no such value

(ME GATE 2008)

71. The following system of equations

$x_1 + x_2 + 2x_3 = 1$,

$x_1 + 2x_2 + 3x_3 = 2$,

$x_1 + 4x_2 + \alpha x_3 = 4$.

has a unique solution. Then the only possible values of "a" is / are ?

(a) 0

(b) either 0 or 1

(c) one of 0, 1 and − 1

(d) any real number other than 5

(CS GATE 2008)

72. The system of linear equations $4x + 2y = 7$, $2x + y = 6$ has

(a) a unique solution

(b) an infinite number of solutions

(c) no solution

(d) exactly two distinct solutions

(EC GATE 2008)

73. The product of matrices $(PQ)^{-1}P$ is

(a) P^{-1} (b) Q^{-1}
(c) $P^{-1}Q^{-1}P$ (d) PQP^{-1}

(CE GATE 2008)

74. A is $m{\times}n$ full rank matrix with $m > n$ and I is an identity matrix. Let $A^1 = (A^TA)^{-1}A^T$. Then which of the following statement is true?

(a) $AA^1A = A$ (b) $(AA^1)^2 = A$
(c) $AA^1A = I$ (d) $AA^1A = A^1$

(EE GATE 2008)

75. All the four entries of the 2×2 matrix

$P = \begin{pmatrix} P_{11} & P_{12} \\ P_{21} & P_{22} \end{pmatrix}$ are non-zero and one of its eigenvalues is zero. Then which of the following is true?

(a) $P_{11}P_{22} - P_{12}P_{21} = 1$
(b) $P_{11}P_{22} - P_{12}P_{21} = -1$
(c) $P_{11}P_{22} - P_{12}P_{21} = 0$
(d) $P_{11}P_{22} + P_{12}P_{21} = 0$

(EC GATE 2008)

76. The matrix $\begin{bmatrix} 1 & 2 & 4 \\ 3 & 0 & 6 \\ 1 & 1 & P \end{bmatrix}$ has eigenvalue equal to 3. Then the sum of other two eigenvalues is

(a) p (b) $p-1$ (c) $p - 2$ (d) $p - 3$

(ME GATE 2008)

77. How many of the following matrices have eigenvalue "1"?

$\begin{pmatrix} 1 & 0 \\ 0 & 0 \end{pmatrix}, \begin{pmatrix} 0 & 1 \\ 0 & 0 \end{pmatrix}, \begin{pmatrix} 1 & -1 \\ 1 & 1 \end{pmatrix}, \begin{pmatrix} -1 & 0 \\ 0 & -1 \end{pmatrix}$

(a) one   (b) two   (c) three   (d) four

(CS GATE 2008)

78. The eigenvalues of the matrix $P = \begin{pmatrix} 4 & 5 \\ 2 & -5 \end{pmatrix}$ are

(a) −7, 8   (b) −6, 5   (c) 3, 4   (d) 1, 2

(CE GATE 2008)

79. The eigenvectors of the matrix $\begin{pmatrix} 1 & 2 \\ 0 & 2 \end{pmatrix}$ are written in the form $\begin{bmatrix} 1 \\ a \end{bmatrix}$ and $\begin{bmatrix} 1 \\ b \end{bmatrix}$. Then $a + b = ?$

(a) 0   (b) 1/2   (c) 1   (d) 2

(ME GATE 2008)

80. The characteristic equation of a 3×3 matrix P is defined as

$P(\lambda) = |P - \lambda I| = p^3 + p^2 + 2p + 1 = 0$.

If I denotes identify matrix, then the inverse of the matrix P will be

(a) $P^2 + P + 2I$ (b) $P^2 + P + I$
(c) $-(P^2 - P + 2I)$ (d) $-(P^2 + P + 2I)$

(EE GATE 2008)

81. The trace and determinant of a 2×2 a matrix are known to be −2 and −35, respectively. Its eigenvalues are?

(a) −30, −5   (b) −37, −1
(c) −7, 5   (d) 17.5, −2

(EE GATE 2009)

82. The eigenvalues of the following matrix
$$\begin{pmatrix} -1 & 3 & 5 \\ -3 & -1 & 6 \\ 0 & 0 & 3 \end{pmatrix} \text{ are}$$
    (a) $3, 3 + 5j, 6 - j$
    (b) $-6 + 5j, 3 + j, 3 - j$
    (c) $3 + j, 3 - j, 5 + j$
    (d) $3, -1 + 3j, -1 - 3j$
    (EC GATE 2009)

83. The eigenvalues of a 2×2 matrix X are -2 and -3. Then, the eigenvalues of the matrix $(X + I)^{-1} (X + 5I)$ are
    (a) $-3, -4$ (b) $-1, -2$
    (c) $-1, -3$ (d) $-2, -4$
    (IN GATE 2009)

84. For a matrix $M = \begin{pmatrix} \dfrac{3}{5} & \dfrac{4}{5} \\ x & \dfrac{3}{5} \end{pmatrix}$, the transpose of the matrix is equal to the inverse of the matrix. Then, the value of $x = ?$
    (a) $-4/5$ (b) $-3/5$ (c) $3/5$ (d) $4/5$
    (ME GATE 2009)

85. A square matrix B is skew-symmetric if
    (a) $B^T = -B$ (b) $B^T = B$
    (c) $B^{-1} = B$ (d) $B^{-1} = B^T$
    (CE GATE 2009)

86. The value of x_3 obtained by solving the followings systems of linear equation is
$$x_1 + 2x_2 - 2x_3 = 4$$
$$2x_1 + x_2 + x_3 = -2,$$
$$-x_1 + x_2 - x_3 = 2$$
    (a) -12 (b) -2 (c) 0 (d) 12
    (GATE 2009)

87. The value of "q" for which the following set of linear equations $2x + 3y = 0$, $6x + qy = 0$ can have non-trivial solution is
    (a) 2 (b) 7 (c) 9 (d) 11
    (GATE 2009)

88. The inverse of the matrix $\begin{pmatrix} 3 + 2i & i \\ -i & 3 - 2i \end{pmatrix}$ is
    (a) $\dfrac{1}{12}\begin{pmatrix} 3 + 2i & -i \\ i & 3 - 2i \end{pmatrix}$ (b) $\dfrac{1}{12}\begin{pmatrix} 3 - 2i & -i \\ i & 3 + 2i \end{pmatrix}$

    (c) $\dfrac{1}{14}\begin{pmatrix} 3 + 2i & -i \\ i & 3 - 2i \end{pmatrix}$ (d) $\dfrac{1}{14}\begin{pmatrix} 3 + 2i & -i \\ i & 3 + 2i \end{pmatrix}$
    (CE GATE 2010)

89. For the set of equations $x_1 + 2x_2 + x_3 + 4x_4 = 2$, $3x_1 + 6x_2 + 3x_3 + 12x_4 = 6$ which of the following statement is true?
    (a) Only the trivial solution $x_1 = x_2 = x_3 = x_4 = 0$ exist
    (b) There are no solutions
    (c) A unique non-trivial solution exists
    (d) Multiple non-trivial solution exists
    (EE GATE 2010)

90. A real $n \times n$ matrix $A = [a_{ij}]_{n \times n}$ is defined as follows:
$$a_{ij} = \begin{cases} i, & \text{for } i = j \\ 0, & \text{otherwise} \end{cases}$$
    Then, the sum of all n eigenvalues of A is
    (a) $\dfrac{n(n+1)}{2}$ (b) $n(n-1)$

    (c) $\dfrac{n(n+1)(2n+1)}{2}$ (d) n^2
    (IN GATE 2010)

91. If $(1, 0, -1)^T$ is an eigenvector of the following matrix
$$\begin{bmatrix} 1 & -1 & 0 \\ -1 & 2 & -1 \\ 0 & -1 & 1 \end{bmatrix}, \text{ then the corresponding eigenvalue is}$$
    (a) 1 (b) 2 (c) 3 (d) 5
    (PI GATE 2010)

92. Consider the matrix $A = \begin{pmatrix} 2 & 3 \\ x & y \end{pmatrix}$. If the eigenvalues are 4 and 8, then
    (a) $x = 4, y = 10$ (b) $x = 5, y = 8$
    (c) $x = -3, y = 9$ (d) $x = -4, y = 10$
    (CS GATE 2010)

93. One of the eigenvector of the matrix $A = \begin{pmatrix} 2 & 2 \\ 1 & 3 \end{pmatrix}$ is
    (a) $\begin{bmatrix} 2 \\ -1 \end{bmatrix}$ (b) $\begin{bmatrix} 2 \\ 1 \end{bmatrix}$ (c) $\begin{bmatrix} 4 \\ 1 \end{bmatrix}$ (d) $\begin{bmatrix} 1 \\ -1 \end{bmatrix}$
    (ME GATE 2010)

94. An eigenvector of $P = \begin{bmatrix} 1 & 1 & 0 \\ 0 & 2 & 2 \\ 0 & 0 & 3 \end{bmatrix}$ is

    (a) $[-1\ 1\ 1]^T$ (b) $[1\ 2\ 1]^T$
    (c) $[1\ -1\ 2]^T$ (d) $[2\ 1\ -1]^T$

    (EE GATE 2010)

95. The matrix $M = \begin{bmatrix} -2 & 2 & -3 \\ 2 & 1 & 6 \\ -1 & -2 & 0 \end{bmatrix}$ has eigenval-

    ues 3, −3, and 5. An eigenvector correspond-
    ing to the eigenvalue 5 is $[1\ 2\ -1]^T$. One of the
    eigenvector of the matrix M^3 is
    (a) $[1\ 8\ -1]^T$ (b) $[1\ 2\ -1]^T$

    (c) $\left[1\ \sqrt[3]{2}\ -1\right]^T$ (d) $[1\ 1\ -1]^T$

    (IN GATE 2011)

96. Consider the matrix as given below:

    $\begin{bmatrix} 1 & 2 & 3 \\ 0 & 4 & 7 \\ 0 & 0 & 3 \end{bmatrix}$

    Which one of the following options provides the
    correct values of the eigenvalues of the matrix?

    (a) 1,4, 3   (b) 3, 7, 3   (c) 7, 3, 2   (d) 1, 2, 3

    (CS GATE 2011)

97. The system of equations $x + y + z = 6$, $x + 4y$
    $+ 6z = 20$, $x + 4y + \lambda z = \mu$ has no solution for
    values of λ and μ and given by
    (a) $\lambda = 6, \mu = 20$ (b) $\lambda = 6, \mu \neq 20$
    (c) $\lambda \neq 6, \mu = 20$ (d) $\lambda = 6, \mu \neq 20$

    (EC GATE 2011)

98. $[A]$ is a square matrix which is neither sym-
    metric nor skew symmetric and $[A]^T$ is its
    transpose. The sum and difference of these
    matrices are defined as
    $[S] = [A] + [A]^T$ and $[D] = [A] - [A]^T$, respec-
    tively. Then which of the following statement
    is true?
    (a) both $[S]$ and $[D]$ are symmetric.
    (b) both $[S]$ and $[D]$ are skew-symmetric.
    (c) $[S]$ is skew-symmetric and $[D]$ is symmetric
    (d) $[S]$ is symmetric and $[D]$ is skew-symmetric

    (CS GATE 2011)

99. The matrix $[A] = \begin{bmatrix} 2 & 1 \\ 4 & -1 \end{bmatrix}$ is decomposed
    into a product of lower triangular matrix $[L]$
    and an upper triangular matrix $[U]$. The prop-
    erly decomposed $[L]$ and $[U]$ matrices, re-
    spectively, are

    (a) $\begin{bmatrix} 1 & 0 \\ 4 & -1 \end{bmatrix}$ and $\begin{bmatrix} 1 & 1 \\ 0 & -2 \end{bmatrix}$

    (b) $\begin{bmatrix} 1 & 0 \\ 2 & 1 \end{bmatrix}$ and $\begin{bmatrix} 2 & 1 \\ 0 & -3 \end{bmatrix}$

    (c) $\begin{bmatrix} 1 & 0 \\ 4 & 1 \end{bmatrix}$ and $\begin{bmatrix} 2 & 1 \\ 0 & -1 \end{bmatrix}$

    (d) $\begin{bmatrix} 2 & 0 \\ 4 & -3 \end{bmatrix}$ and $\begin{bmatrix} 1 & 0.5 \\ 0 & 1 \end{bmatrix}$

    (EE GATE 2011)

100. If $A = \begin{pmatrix} -5 & -3 \\ 2 & 0 \end{pmatrix}, I = \begin{pmatrix} 1 & 0 \\ 0 & 1 \end{pmatrix}$, then $A^3 = ?$

    (a) $15A + 12I$ (b) $19A + 30I$

    (c) $17A + 15I$ (d) $17A + 21I$

    (EC GATE 2012)

101. Consider the following system of equations:
    $2x_1 + x_2 + x_3 = 0$, $x_2 - x_3 = 0$, $x_1 + x_2 = 0$
    This system has
    (a) a unique solution
    (b) no solution
    (c) an infinite number of solutions
    (d) five solutions

    (GATE 2012)

102. Consider
    $x + 2y + z = 4$, $2x + y + 2z = 5$, $x - y + z = 1$.
    Then the system of algebraic equations given
    above has
    (a) a unique solution: $x = 1, y = 1, z = 1$
    (b) only two solutions: $x = 1, y = 1, z = 1$ and x
        $= 2, y = 1, z = 0$
    (c) an infinite number of solutions
    (d) no feasible solution

    (ME GATE 2012)

103. The eigenvalues of the matrix $\begin{bmatrix} 9 & 5 \\ 5 & 8 \end{bmatrix}$ are

(a) −2.42 and 6.86   (b) 3.48 and 13.52

(c) 4.70 and 6.86   (d) 6.86 and 9.50

(CE GATE 2012)

104. Let A be the 2×2 matrix with elements $a_{11} = a_{12} = a_{21} = 1$ and $a_{22} = -1$. Then, the eigenvalues of the matrix A^{19} are

(a) $1024, -1024$

(b) $1024\sqrt{2}, -1024\sqrt{2}$

(c) $4\sqrt{2}, -4\sqrt{2}$

(d) $512\sqrt{2}, -512\sqrt{2}$

(CS GATE 2012)

105. One pair of eigenvectors corresponding to the eigenvalues of the matrix $\begin{bmatrix} 0 & -1 \\ 1 & 0 \end{bmatrix}$ is

(a) $\begin{bmatrix} 1 \\ -j \end{bmatrix}, \begin{bmatrix} j \\ -1 \end{bmatrix}$ (b) $\begin{bmatrix} 0 \\ 1 \end{bmatrix}, \begin{bmatrix} -1 \\ 0 \end{bmatrix}$

(c) $\begin{bmatrix} 1 \\ j \end{bmatrix}, \begin{bmatrix} 0 \\ 1 \end{bmatrix}$ (d) $\begin{bmatrix} 1 \\ j \end{bmatrix}, \begin{bmatrix} j \\ 1 \end{bmatrix}$

(IN GATE 2013)

106. A matrix has eigenvalues −1 and −2. The corresponding eigenvectors are $\begin{bmatrix} 1 \\ -1 \end{bmatrix}$ and $\begin{bmatrix} 1 \\ -2 \end{bmatrix}$, respectively. The matrix is

(a) $\begin{bmatrix} 1 & 1 \\ -1 & -2 \end{bmatrix}$ (b) $\begin{bmatrix} 1 & 2 \\ -2 & -4 \end{bmatrix}$

(c) $\begin{bmatrix} -1 & 0 \\ 0 & -2 \end{bmatrix}$ (d) $\begin{bmatrix} 0 & 1 \\ -2 & -3 \end{bmatrix}$

(EE GATE 2013)

107. The minimum eigenvalue of the following matrix $\begin{bmatrix} 3 & 5 & 2 \\ 5 & 12 & 7 \\ 2 & 7 & 5 \end{bmatrix}$ is

(a) 0   (b) 1   (c) 2   (d) 3

(EC GATE 2013)

108. Choose the correct set of functions that are linearly dependent

(a) $\sin x, \sin^2 x$, and $\cos^2 x$

(b) $\cos x, \sin x$, and $\tan x$

(c) $\cos 2x, \sin^2 x$, and $\cos^2 x$

(d) $\cos 2x, \sin x$, and $\cos x$

(ME GATE 2013)

109. The equation $\begin{bmatrix} 2 & -2 \\ 1 & -1 \end{bmatrix}\begin{bmatrix} x_1 \\ x_2 \end{bmatrix} = \begin{bmatrix} 0 \\ 0 \end{bmatrix}$ has

(a) no solution

(b) only one solution $\begin{bmatrix} x_1 \\ x_2 \end{bmatrix} = \begin{bmatrix} 0 \\ 0 \end{bmatrix}$

(c) non zero unique solution

(d) multiple solutions

(GATE 2013)

110. There are three matrices $P_{4 \times 2}$, $Q_{2 \times 4}$, and $R_{4 \times 1}$. Then, the minimum number of multiplication required to compute the matrix PQR is _____

(CE GATE 2013)

111. Which of the followings does not equal to

$$\begin{vmatrix} 1 & x & x^2 \\ 1 & y & y^2 \\ 1 & z & z^2 \end{vmatrix}?$$

(a) $\begin{vmatrix} 1 & x(x+1) & x+1 \\ 1 & y(y+1) & y+1 \\ 1 & z(z+1) & z+1 \end{vmatrix}$

(b) $\begin{vmatrix} 1 & x+1 & x^2+1 \\ 1 & y+1 & y^2+1 \\ 1 & z+1 & z^2+1 \end{vmatrix}$

(c) $\begin{vmatrix} 0 & x-y & x^2-y^2 \\ 0 & y-z & y^2-z^2 \\ 1 & z & z^2 \end{vmatrix}$

(d) $\begin{vmatrix} 2 & x+y & x^2+y^2 \\ 2 & y+z & y^2+z^2 \\ 1 & z & z^2 \end{vmatrix}$

(CS GATE 2013)

112. Let A be a $m \times n$ matrix and B be a $n \times m$ matrix. It is given that $\det(I_m + AB) = \det(I_n + BA)$, where I_k denote the identity matrix of order k. Use the above property, the determinant of the matrix given below is

$$\begin{bmatrix} 2 & 1 & 1 & 1 \\ 1 & 2 & 1 & 1 \\ 1 & 1 & 2 & 1 \\ 1 & 1 & 1 & 2 \end{bmatrix}$$

(a) 2   (b) 5   (c) 8   (d) 16

(CE GATE 2013)

113. Consider the matrix

$$J_6 = \begin{bmatrix} 0 & 0 & 0 & 0 & 0 & 1 \\ 0 & 0 & 0 & 0 & 1 & 0 \\ 0 & 0 & 0 & 1 & 0 & 0 \\ 0 & 0 & 1 & 0 & 0 & 0 \\ 1 & 0 & 0 & 0 & 0 & 0 \end{bmatrix}$$

which is obtained by reversing the order of the columns of the identity matrix I_6. Let $P = I_6 + \alpha J_6$, where a is non-negative real number. Then the value of α for which $\det(P) = 0$ is_____

(ECE GATE 2014)

114. Which one of the following equations is a correct identity for arbitrary 3×3 real matrices P, Q, and R?
(a) $P(Q + R) = PQ + RP$
(b) $(P - Q)^2 = P^2 - 2PQ + Q^2$
(c) $\det(P + Q) = \det(P) + \det(Q)$
(d) $(P + Q)^2 = P^2 + PQ + QP + Q^2$

(ME GATE 2014)

115. If the matrix A is such that $A = \begin{bmatrix} 2 \\ -4 \\ 7 \end{bmatrix} \begin{bmatrix} 1 & 9 & 5 \end{bmatrix}$,

then the det(A) =?   **(CS GATE 2014)**

116. The rank of the matrix $\begin{bmatrix} 6 & 0 & 4 & 4 \\ -2 & 14 & 8 & 18 \\ 14 & -14 & 0 & -10 \end{bmatrix}$ is

_____.

(CE GATE 2014)

117. Two matrices A and B are given below:

$$A = \begin{pmatrix} p & q \\ r & s \end{pmatrix}, B = \begin{pmatrix} p^2 + q^2 & pr + qs \\ pr + qs & r^2 + s^2 \end{pmatrix}.$$

If the rank of A is "N," then the rank of matrix B is

(a) $N/2$ (b) $N - 1$ (c) N (d) $2N$

(EE GATE 2014)

118. With reference to the conventional Cartesian (x, y) co-ordinate system, the vertices of a triangle have the following co-ordinates:
$(x_1, y_1) = (1, 0), (x_2, y_2) = (2, 2)$ and $(x_3, y_3) = (4, 3)$.
The area of the triangle is equal to

(a) 3/2   (b) 3/4   (c) 4/5   (d) 5/2

(CE GATE 2014)

119. The matrix form of the linear system

$$\frac{dx}{dt} = 3x - 5y \text{ and } \frac{dy}{dt} = 4x + 8y \text{ is}$$

(a) $\dfrac{d}{dt}\begin{bmatrix} x \\ y \end{bmatrix} = \begin{bmatrix} 3 & -5 \\ 4 & 8 \end{bmatrix}\begin{bmatrix} x \\ y \end{bmatrix}$

(b) $\dfrac{d}{dt}\begin{bmatrix} x \\ y \end{bmatrix} = \begin{bmatrix} 3 & 8 \\ 4 & -5 \end{bmatrix}\begin{bmatrix} x \\ y \end{bmatrix}$

(c) $\dfrac{d}{dt}\begin{bmatrix} x \\ y \end{bmatrix} = \begin{bmatrix} 4 & -5 \\ 3 & 8 \end{bmatrix}\begin{bmatrix} x \\ y \end{bmatrix}$

(d) $\dfrac{d}{dt}\begin{bmatrix} x \\ y \end{bmatrix} = \begin{bmatrix} 4 & 8 \\ 3 & -5 \end{bmatrix}\begin{bmatrix} x \\ y \end{bmatrix}$

(GATE 2014)

120. Consider the following system of equations:
$3x + 2y = 1, 4x + 7z = 1, x + y + z = 3, x - 2y + 7z = 0$.

The number of solutions for this system is_____.

(CS GATE 2014)

121. The system of equations given below has
$x + 2y + 4z = 2, 4x + 3y + z = 5, 3x + 2y + 3z = 1$
(a) a unique solution
(b) two solutions
(c) no solution
(d) more than two solutions

(GATE 2014)

122. The system of linear equations has
$$\begin{pmatrix} 2 & 1 & 3 \\ 3 & 0 & 1 \\ 1 & 2 & 5 \end{pmatrix}\begin{pmatrix} a \\ b \\ c \end{pmatrix} = \begin{pmatrix} 5 \\ -4 \\ 14 \end{pmatrix}$$

(a) a unique solution
(b) infinitely many solutions
(c) no solution
(d) exactly two solutions

(EC GATE 2014)

123. Given a system of equations: $x + 2y + 2z = b_1$, $5x + y + 3z = b_2$

which of the following is true regarding it's solution?

(a) The system has a unique solution for any given b_1 and b_2

(b) The system will have infinitely many solutions for any given b_1 and b_2

(c) Whether or not a solution exists depends on the given b_1 and b_2

(d) The system would have no solution for any values of b_1 and b_2

(EE GATE 2014)

124. The determinant of matrix A is 5 and the determinant of matrix B is 40. Then determinant of matrix AB is____?

(EC GATE 2014)

125. For the matrix A satisfying the equation given below, the eigenvalues are

$$A\begin{bmatrix} 1 & 2 & 3 \\ 7 & 8 & 9 \\ 4 & 5 & 6 \end{bmatrix} = \begin{bmatrix} 1 & 2 & 3 \\ 4 & 5 & 6 \\ 7 & 8 & 9 \end{bmatrix} \text{ is}$$

(a) (1, -1, 1)   (b) (1, 1, 0)
(c) (1, 1, −1)   (d) (1, 0, 0)

(IN GATE 2014)

126. Which of the following statements is true for all symmetric matrices?

(a) all eigenvalues are real
(b) all eigenvalues are positive
(c) all eigenvalues are distinct
(d) sum of the eigenvalues is zero

(EE GATE 2014)

127. Which of the following statements is not true for a square matrix A?

(a) If A is upper triangular, the eigenvalues of A are the diagonal elements of it

(b) If A is real symmetric, the eigenvalues of A are always real and positive

(c) If A is real, the eigenvalues of A and A^T are always the same

(d) If all the principal minors of A are positive, all the eigenvalues are all positive.

(EC GATE 2014)

128. The sum of eigenvalues of matrix M is, where

$$M = \begin{bmatrix} 215 & 650 & 795 \\ 655 & 150 & 835 \\ 485 & 355 & 550 \end{bmatrix}$$

(a) 915   (b) 1355
(c) 1640   (d) 2180

(CE GATE 2014)

129. The maximum value of the determinant among all 2×2 real symmetric matrices with the trace of 14 is_____

(EC GATE 2014)

130. One of the eigenvectors of the matrix $\begin{bmatrix} -5 & 2 \\ -9 & 6 \end{bmatrix}$ is

(a) $\begin{bmatrix} -1 \\ 1 \end{bmatrix}$ (b) $\begin{bmatrix} -2 \\ 9 \end{bmatrix}$

(c) $\begin{bmatrix} 2 \\ -1 \end{bmatrix}$ (d) $\begin{bmatrix} 1 \\ 1 \end{bmatrix}$

(ME GATE 2014)

131. Which one of the following statements is true about every $n \times n$ matrix with only real eigenvalues?

(a) If the trace of the matrix is positive and the determinant of the matrix is negative, at least one of its eigenvalues is negative

(b) If the trace of the matrix is positive, all its eigenvalues are positive

(c) If the determinant of the matrix is positive, all its eigenvalues are positive

(d) If the product of the trace and determinant of the matrix is positive, all its eigenvalues are positive

(CS GATE 2014)

132. Consider a 3×3 real symmetric matrix A such that two of its eigenvalues are $a \neq 0$, $b \neq 0$ with respective eigenvectors $\begin{bmatrix} x_1 \\ x_2 \\ x_3 \end{bmatrix}$ and $\begin{bmatrix} y_1 \\ y_2 \\ y_3 \end{bmatrix}$. If $a \neq b$, then $x_1y_1 + x_2y_2 + x_3y_3$ equals

(a) a   (b) b   (c) ab (d) 0

(ME GATE 2014)

133. The product of the non-zero eigenvalues of

the matrix $\begin{bmatrix} 1 & 0 & 0 & 0 & 1 \\ 0 & 1 & 1 & 1 & 0 \\ 0 & 1 & 1 & 1 & 0 \\ 0 & 1 & 1 & 1 & 0 \\ 1 & 0 & 0 & 0 & 1 \end{bmatrix}$ is_____.

(CS GATE 2014)

134. A real 4×4 matrix A satisfies the equation $A^2 = I$, where I is the identity matrix of order 4. Then, the positive eigenvalue of A is_____.

(EC GATE 2014)

135. A system matrix is given as follows:

$$A = \begin{pmatrix} 0 & 1 & -1 \\ -6 & -11 & 6 \\ -6 & -11 & 5 \end{pmatrix}$$

The absolute value of the ratio of the maximum eigenvalue to the minimum eigenvalue is_____.

(EE GATE 2014)

136. The smallest and the largest eigenvalues of the following matrix are

$$\begin{bmatrix} 3 & -2 & 2 \\ 4 & -4 & 6 \\ 2 & -3 & 5 \end{bmatrix}:$$

(a) 1.5 and 2.5   (b) 0.5 and 2.5
(c) 1 and 3   (d) 1 and 2

(CE GATE 2015)

137. The value of "p" such that the vector $\begin{bmatrix} 1 \\ 2 \\ 3 \end{bmatrix}$ is an eigenvector of the matrix

$\begin{bmatrix} 4 & 1 & 2 \\ p & 2 & 1 \\ 14 & -4 & 10 \end{bmatrix}$ is _____

(EC GATE 2015)

138. In the given matrix $\begin{bmatrix} 1 & -1 & 2 \\ 0 & 1 & 0 \\ 1 & 2 & 1 \end{bmatrix}$, one of the eigenvalues is "1." The eigenvectors corresponding to the eigenvalue "1" are

(a) $\{\alpha(4, 2, 1): \alpha \neq 0, \alpha \in R\}$
(b) $\{\alpha(-4, 2, 1): \alpha \neq 0, \alpha \in R\}$

(c) $\left\{\alpha(\sqrt{2}, 0, 1): \alpha \neq 0, \alpha \in R\right\}$

(d) $\left\{\alpha(-\sqrt{2}, 0, 1): \alpha \neq 0, \alpha \in R\right\}$

(CS GATE 2015)

139. The two eigenvalues of the matrix $\begin{pmatrix} 2 & 1 \\ 1 & p \end{pmatrix}$ have a ratio $3:1$ for $p = 2$. What is the another value of "p" for which the eigenvalues have the same ratio of 3:1 ?

(a) –2   (b) 1   (c) 7/3   (d) 14/3

(CE GATE 2015)

140. At least one eigenvalue of a singular matrix is

(a) positive   (b) zero
(c) negative   (d) imaginary

(ME GATE 2015)

141. The maximum value of "a" such that the matrix $\begin{bmatrix} -3 & 0 & -2 \\ 1 & -1 & 0 \\ 0 & a & -2 \end{bmatrix}$ has three linearly independent real eigenvectors is

(a) $\dfrac{2}{3\sqrt{3}}$ (b) $\dfrac{1}{3\sqrt{3}}$

(c) $\dfrac{1 + 2\sqrt{3}}{3\sqrt{3}}$ (d) $\dfrac{1 + \sqrt{3}}{3\sqrt{3}}$

(EE GATE 2015)

142. Consider the following 2×2 matrix, A where two elements are unknown and are marked by "a" and "b." The eigenvalues of this matrix are -1 and 7. What are the values of "a" and "b"?

$$\begin{bmatrix} 1 & 4 \\ b & a \end{bmatrix}$$

(a) $a = 6, b = 4$ (b) $a = 4, b = 6$
(c) $a = 3, b = 5$ (d) $a = 5, b = 3$

(CS GATE 2015)

143. The value of "x" for which all the eigenvalues of the matrix given below are real is

$$\begin{bmatrix} 10 & 5+j & 4 \\ x & 20 & 2 \\ 4 & 2 & -10 \end{bmatrix}$$

(a) $5 + j$ (b) $5 - j$ (c) $1 - 5j$ (d) $1 + 5j$

(EC GATE 2015)

144. For $A = \begin{bmatrix} 1 & \tan x \\ -\tan x & 1 \end{bmatrix}$, the determinant of $A^T A^{-1}$ is

(a) $\sec^2 x$ (b) $\cos^4 x$ (c) 1   (d) 0

(EC GATE 2015)

145. If any two columns of a determinant $\begin{bmatrix} 4 & 7 & 8 \\ 3 & 1 & 5 \\ 9 & 6 & 2 \end{bmatrix}$ are interchanged, which one of the statement is correct?

(a) absolute value remains unchanged but sign will change.
(b) both value and sign will change.
(c) absolute value will change but sign will not change.
(d) both absolute value and sign will remain unchanged.

(ME GATE 2015)

146. Let $A = [a_{ij}]$, $1 \leq i, j \leq n$ with $n \geq 3$ and $a_{ij} = i \times j$. Then the rank of A is

(a) 0   (b) 1   (c) $n-1$ (d) n

(CE GATE 2015)

147. Consider a system of linear equations:
$x - 2y + 3z = -1$, $x - 3y + 4z = 1$, $-2x + 4y - 6z = k$

The value of "k" for which the system has infinitely many solutions is_____

(EC GATE 2015)

148. For what value of "p," the following set of equations will have no solution?
$2x + 3y = 5$, $3x + py = 10$

(CE GATE 2015)

149. If the following system has non-trivial solution:

$px + qy + rz = 0$, $qx + ry + pz = 0$, $rx + py + qz = 0$;

then which of the following options is true?
(a) $p - q + r = 0$ or $p = q = -r$
(b) $p + q - r = 0$ or $p = -q = r$
(c) $p + q + r = 0$ or $p = q = r$
(d) $p - q + r = 0$ or $p = -q = -r$

(CS GATE 2015)

150. The matrix $A = \begin{bmatrix} a & 0 & 3 & 7 \\ 2 & 5 & 1 & 3 \\ 0 & 0 & 2 & 4 \\ 0 & 0 & 0 & b \end{bmatrix}$ has det(A) = 100 and trace(A) = 14. The value of |a-b| is_____

(EC GATE 2016)

151. Let $P = \begin{bmatrix} 3 & 1 \\ 1 & 3 \end{bmatrix}$. Consider the set S of all vectors $\begin{bmatrix} x \\ y \end{bmatrix}$ such that $a^2 + b^2 = 1$ where $\begin{bmatrix} a \\ b \end{bmatrix} = P \begin{bmatrix} x \\ y \end{bmatrix}$. Then S is

(a) a circle of radius $\sqrt{10}$
(b) a circle of radius $\dfrac{1}{\sqrt{10}}$
(c) an ellipse with major axis along $\begin{bmatrix} 1 \\ 1 \end{bmatrix}$
(d) an ellipse with minor axis along $\begin{bmatrix} 1 \\ 1 \end{bmatrix}$

[EE GATE 2016]

152. If the entries in each column of a square matrix M add up to "1," then an eigenvalue of M is

(a) 4   (b) 3   (c) 2   (d) 1

[CE GATE 2016]

153. Consider a 3 × 3 matrix with every element being equal to "1." Its only non-zero eigenvalue is _____.

(EE GATE 2016]

154. Let $M^4 = I$ (where I denote the identity matrix) and $M \neq I$, $M^2 \neq I$, $M^3 \neq I$. Then for any natural number k, M^{-1} equals

(a) M^{4k+1} (b) M^{4k+2} (c) M^{4k+3} (d) M^{4k}

(EC GATE 2016)

155. The condition for which the eigenvalues of the matrix $\begin{bmatrix} 2 & 1 \\ 1 & k \end{bmatrix}$ are positive is

(a) $k > 1/2$ (b) $k > -2$
(c) $k > 0$ (d) $k < -1/2$

(ME GATE 2016)

156. Two eigenvalues of a 3×3 matrix real matrix P are $2 + \sqrt{-1}$ and 3. Then, the determinant of P is_____

(CS GATE 2016)

157. Consider the following linear system:

$x + 2y - 3z = a$, $2x + 3y + 3z = b$, $5x + 9y - 6z = c$.

The system is consistent if a, b, and c satisfy the equation

(a) $7a - b - c = 0$ (b) $3a + b - c = 0$

(c) $3a - b + c = 0$ (d) $7a - b + c = 0$

(CE GATE 2016)

158. The solution to the system of equations $\begin{bmatrix} 2 & 5 \\ -4 & 3 \end{bmatrix}\begin{bmatrix} x \\ y \end{bmatrix} = \begin{bmatrix} 2 \\ -30 \end{bmatrix}$ is

(a) 6, 2   (b) –6, 2   (c) –6, –2   (d) 6, –2

(ME GATE 2016)

159. The number of solutions of the simultaneous algebraic equations $y = 3x + 3$ and $y = 3x + 5$ is?

(a) zero   (b) 1

(c) 2   (d) infinite

(PI GATE 2016)

160. Consider the systems, each consisting of "m" linear equations in "n" variables.

  I. If $m < n$, then all such systems have a solution

 II. If $m > n$, then none of the system has a solution

III. If $m = n$, there exist a system which has a solution.

Then which of the following is correct?

(a) I, II, and III are true

(b) only II and III are true

(c) only III is true

(d) none of them is true

(CSE GATE 2016)

161. Consider a 2×2 square matrix $P = \begin{bmatrix} \sigma & x \\ \omega & \sigma \end{bmatrix}$ where "x," is unknown. If the eigenvalues of the matrix A are $(\sigma + j\omega)$ and $(\sigma - j\omega)$, then "x" is equal to

(a) $j\omega$ (b) $-j\omega$ (c) w (d) $-\omega$

(EC GATE 2016)

162. The eigenvalue of the matrix $A = \begin{pmatrix} 0 & 1 \\ -1 & 0 \end{pmatrix}$ are

(a) $i, -i$ (b) $1, -1$ (c) $0, 1$ (d) $0, -1$

(PI GATE 2016)

163. Consider the matrix $A = \begin{pmatrix} 2 & 1 & 1 \\ 2 & 3 & 4 \\ -1 & -1 & -2 \end{pmatrix}$ whose eigenvalues are 1, – 1, and 3. Then the trace of $A^3 - A^2$ is_____

(IN GATE 2016)

164. The value of "x" for which the matrix $A = \begin{bmatrix} 3 & 2 & 4 \\ 9 & 7 & 13 \\ -6 & -4 & -9+x \end{bmatrix}$ has zero as an eigen-value, is _____

(EC GATE 2016)

165. Suppose that the eigenvalues of a matrix A are 1,2,4. Then the determinant of $(A^{-1})^T$ is

_____

(CS GATE 2016)

166. A real squire matrix A is called skew-symmetric if

(a) $A^T = A$ (b) $A^T = A^{-1}$

(c) $A^T = -A$ (d) $A^T = A + A^{-1}$

(ME GATE 2016)

167. A 3×3 matrix P is such that $P^3 = P$. Then the eigenvalues of P are

(a) 1, 1, –1

(b) 1, 0.5 + 0.866j, 0.5–0.866j

(c) 1, –0.5 + 0.866j, –0.5 – 0.866j

(d) 0, 1, –1

(EE GATE 2016)

168. A sequence $x[n]$ is defined as

$\begin{bmatrix} x[n] \\ x[n-1] \end{bmatrix}\begin{pmatrix} 1 & 1 \\ 1 & 0 \end{pmatrix}^n = \begin{bmatrix} 1 \\ 0 \end{bmatrix}$ for $n \geq 2$.

The initial conditions are : $x[0] = 1$, $x[1] = 1$ and $x[n] = 0$ for $n < 0$. The the value of $x[12]$ is_____

(EC GATE 2016)

169. Let A be a 4×3 real matrix with rank 2. Then which of the following statements is true?
(a) Rank of $A^T A$ is less than 2
(b) Rank of $A^T A$ is equal to 2
(c) Rank of $A^T A$ is greater than 2
(d) Rank of $A^T A$ can be any number between 1 and 3

(EE GATE 2016)

170. If the vectors $e_1 = (1, 0, 2)$, $e_2 = (0, 1, 0)$ and $e_3 = (-2, 0, 1)$ form an orthogonal basis of the three-dimensional real space R^3, then the vector u $= (4, 3, -3) \in R^3$ can be expressed as

(a) $u = -\dfrac{2}{5}e_1 - 3e_2 - \dfrac{11}{5}e_3$

(b) $u = -\dfrac{2}{5}e_1 - 3e_2 + \dfrac{11}{5}e_3$

(c) $u = -\dfrac{2}{5}e_1 + 3e_2 + \dfrac{11}{5}e_3$

(d) $u = -\dfrac{2}{5}e_1 + 3e_2 - \dfrac{11}{5}e_3$

(EC GATE 2016)

171. The rank of the matrix $M = \begin{bmatrix} 5 & 10 & 10 \\ 1 & 0 & 2 \\ 3 & 6 & 6 \end{bmatrix}$ is

(a) 0   (b) 1   (c) 2   (d) 3

(EC GATE 2017)

172. If the determinant of a 2×2 matrix is 50 and one eigenvalue is 10, then the other eigenvalue is_____

(ME GATE 2017)

173. The product of the eigenvalues of the matrix P is

$$\begin{pmatrix} 2 & 0 & 1 \\ 4 & -3 & 3 \\ 0 & 2 & -1 \end{pmatrix}$$

(a) −6   (b) 2   (c) 6   (d) −2

(ME GATE 2017)

174. Let A be a $n \times n$ real valued square symmetric matrix of rank 2 with $\sum\limits_{i=1}^{n}\sum\limits_{j=1}^{n} A_{ij}^{2} = 50$. Consider the following statements:
I. one eigenvalue must be in $[-5, 5]$
II. The eigenvalue with the largest magnitude must be strictly greater than "5."

Which of the above statements about eigenvalues of A is/are necessarily correct?
(a) both I and II
(b) I only
(c) II only
(d) neither I nor II

(CS GATE 2017)

175. Consider the matrix $A = \begin{pmatrix} 50 & 70 \\ 70 & 80 \end{pmatrix}$ whose eigenvectors corresponding to eigenvalues λ_1 and λ_2 are $X_1 = \begin{pmatrix} 70 \\ \lambda_1 - 50 \end{pmatrix}$ and $X_2 = \begin{pmatrix} \lambda_2 - 80 \\ 70 \end{pmatrix}$, respectively.
The value of $X_1{}^T X_2$ is_____

(ME GATE 2017)

176. The eigenvalues of the matrix
$$A = \begin{bmatrix} 1 & -1 & 5 \\ 0 & 5 & 6 \\ 0 & -6 & 5 \end{bmatrix} \text{ are}$$
(a) 1, 5, 6   (b) 1, −5 ± 6j
(c) 1, 5 ± 6j (d) 1, 5, 5

(IN GATE 2017)

177. Consider the matrix $\begin{bmatrix} 5 & -1 \\ 4 & 1 \end{bmatrix}$. Then which one of the following statements is true for the eigenvalues and eigenvectors of this matrix?
(a) Eigenvalue 3 has a multiplicity of 2 and only one independent eigenvector exists
(b) Eigenvalue 3 has a multiplicity of 2 and two independent eigenvector exist
(c) Eigenvalue 3 has a multiplicity of 2 and no independent eigenvector exists
(d) Eigenvalues are 3 and -3 and two independent eigenvector exist

(CE GATE 2017)

178. Consider the 5×5 matrix
$$A = \begin{bmatrix} 1 & 2 & 3 & 4 & 5 \\ 5 & 1 & 2 & 3 & 4 \\ 4 & 5 & 1 & 2 & 3 \\ 3 & 4 & 5 & 1 & 2 \\ 2 & 3 & 4 & 5 & 1 \end{bmatrix}$$

It is given that A has only one real eigenvalue. Then, the real eigenvalue of A is
(a) −25   (b) 0
(c) 15   (d) 25

(EC GATE 2017)

179. Consider the matrix $P = \begin{bmatrix} \dfrac{1}{\sqrt{2}} & 0 & \dfrac{1}{\sqrt{2}} \\ 0 & 1 & 0 \\ -\dfrac{1}{\sqrt{2}} & 0 & \dfrac{1}{\sqrt{2}} \end{bmatrix}$.

Which one of the following statements about P is incorrect?

(a) $Det(P) = 1$ (b) P is orthogonal
(c) $P^{-1} = P^{T}$
(d) All eigenvalues of P are real numbers
(ME GATE 2017)

180. Consider the following simultaneous equations (c_1 and c_2 being constants):
$$3x_1 + 2x_2 = c_1$$
$$4x_1 + x_2 = c_2$$

The characteristic equation for these simultaneous equations is
(a) $\lambda^2 - 4\lambda - 5 = 0$ (b) $\lambda^2 - 4\lambda + 5 = 0$
(c) $\lambda^2 + 4\lambda - 5 = 0$ (d) $\lambda^2 + 4\lambda + 5 = 0$
(CE GATE 2017)

181. The matrix $A = \begin{bmatrix} \dfrac{3}{2} & 0 & \dfrac{1}{2} \\ 0 & -1 & 0 \\ \dfrac{1}{2} & 0 & \dfrac{3}{2} \end{bmatrix}$ has three distinct eigenvalues and one of the eigenvector

is $\begin{bmatrix} 1 \\ 0 \\ 1 \end{bmatrix}$. Which of the following can be another eigenvector of A?

(a) $\begin{bmatrix} 0 \\ 0 \\ -1 \end{bmatrix}$ (b) $\begin{bmatrix} -1 \\ 0 \\ 0 \end{bmatrix}$ (c) $\begin{bmatrix} 1 \\ 0 \\ -1 \end{bmatrix}$ (d) $\begin{bmatrix} 1 \\ -1 \\ 1 \end{bmatrix}$
(EE GATE 2017)

182. If the characteristic polynomial of a 3×3 matrix M over R (where R is the set of real numbers) is $\lambda^3 - 4\lambda^2 + a\lambda + 30$ for $a \in R$ and one eigenvalue of M is 2, then the largest among the absolute values of the eigenvalue of M is
_____.
(CS GATE 2017)

183. The rank of the matrix $\begin{bmatrix} 1 & -1 & 0 & 0 & 0 \\ 0 & 0 & 1 & -1 & 0 \\ 0 & 1 & -1 & 0 & 0 \\ -1 & 0 & 0 & 0 & 1 \\ 0 & 0 & 0 & 1 & -1 \end{bmatrix}$

is_____
(EC GATE 2017)

184. Let $P = \begin{bmatrix} 1 & 1 & -1 \\ 2 & -3 & 4 \\ 3 & -2 & 3 \end{bmatrix}$ and $Q = \begin{bmatrix} -1 & -2 & -1 \\ 6 & 12 & 6 \\ 5 & 10 & 5 \end{bmatrix}$

be two martices, then the rank of $P + Q$ is_____
(CS GATE 2017)

185. The matrix P is the inverse of a matrix Q. If I denotes the identity matrix, which one of the following option is correct?
(a) $PQ = I$ but $QP \neq I$
(b) $QP = I$ but $PQ \neq I$
(c) $PQ = I$ and $QP = I$
(d) $PQ - QP = I$
(CE GATE 2017)

186. If V is a non-zero vector of dimension 3×1, then the matrix $A = VV^T$ has a rank
_____.
(IN GATE 2017)

187. If $A = \begin{bmatrix} 1 & 5 \\ 6 & 2 \end{bmatrix}$ and $B = \begin{bmatrix} 3 & 7 \\ 8 & 4 \end{bmatrix}$, then AB^T is equal to
(a) $\begin{bmatrix} 38 & 28 \\ 32 & 56 \end{bmatrix}$ (b) $\begin{bmatrix} 3 & 40 \\ 42 & 8 \end{bmatrix}$
(c) $\begin{bmatrix} 43 & 27 \\ 34 & 50 \end{bmatrix}$ (d) $\begin{bmatrix} 38 & 32 \\ 28 & 56 \end{bmatrix}$
(CE GATE 2017)

188. Which of the following matrix is singular?
(a) $\begin{pmatrix} 2 & 5 \\ 1 & 3 \end{pmatrix}$ (b) $\begin{pmatrix} 3 & 2 \\ 2 & 3 \end{pmatrix}$
(c) $\begin{pmatrix} 2 & 4 \\ 3 & 6 \end{pmatrix}$ (d) $\begin{pmatrix} 4 & 3 \\ 6 & 2 \end{pmatrix}$
(CE GATE 2018)

189. For the given orthogonal matrix Q,

$$Q = \begin{pmatrix} \dfrac{3}{7} & \dfrac{2}{7} & \dfrac{6}{7} \\[2mm] -\dfrac{6}{7} & \dfrac{3}{7} & \dfrac{2}{7} \\[2mm] \dfrac{2}{7} & \dfrac{6}{7} & -\dfrac{3}{7} \end{pmatrix}$$

The inverse is

(a) $Q = \begin{vmatrix} \dfrac{3}{7} & \dfrac{2}{7} & \dfrac{6}{7} \\[2mm] \dfrac{-6}{7} & \dfrac{3}{7} & \dfrac{2}{7} \\[2mm] \dfrac{2}{7} & \dfrac{6}{7} & \dfrac{-3}{7} \end{vmatrix}$ (b) $Q = \begin{vmatrix} \dfrac{-3}{7} & \dfrac{-2}{7} & \dfrac{6}{7} \\[2mm] \dfrac{6}{7} & \dfrac{-3}{7} & \dfrac{-2}{7} \\[2mm] \dfrac{-2}{7} & \dfrac{-6}{7} & \dfrac{3}{7} \end{vmatrix}$

(c) $Q = \begin{vmatrix} \dfrac{3}{7} & \dfrac{-6}{7} & \dfrac{2}{7} \\[2mm] \dfrac{2}{7} & \dfrac{3}{7} & \dfrac{6}{7} \\[2mm] \dfrac{6}{7} & \dfrac{2}{7} & \dfrac{-3}{7} \end{vmatrix}$ (d) $Q = \begin{vmatrix} \dfrac{-3}{7} & \dfrac{6}{7} & \dfrac{-2}{7} \\[2mm] \dfrac{-2}{7} & \dfrac{-3}{7} & \dfrac{-6}{7} \\[2mm] \dfrac{-6}{7} & \dfrac{-2}{7} & \dfrac{3}{7} \end{vmatrix}$

(CE GATE 2018)

190. The rank of the following matrix is

$$\begin{pmatrix} 1 & 1 & 0 & -2 \\ 2 & 0 & 2 & 2 \\ 4 & 1 & 3 & 1 \end{pmatrix}$$

(a) 1   (b) 2   (c) 3   (d) 4

(CE GATE 2018)

191. The rank of the matrix $\begin{pmatrix} -4 & 1 & -1 \\ -1 & -1 & -1 \\ 7 & -3 & 1 \end{pmatrix}$ is

(a) 1   (b) 2   (c) 3   (d) 4

[ME GATE 2018]

192. The matrix $\begin{pmatrix} 2 & -4 \\ 4 & -2 \end{pmatrix}$ has

(a) Real eigenvalues and eigenvectors.
(b) Real eigenvalues but complex eigenvectors.
(c) Complex eigenvalues but real eigenvectors.
(d) Complex eigenvalues and eigenvectors.

(CS/IT-GATE-2018)

193. Let M be a 4×4 matrix. Consider the following statements:

S1: M has 4 linearly independent eigenvectors.

S2: M has distinct eigenvalues.

S3: M is non-singular (invertible).

Which one among the following is TRUE?

(a) S1 implies S2   (b) S1 implies S3
(c) S2 implies S1   (d) S3 implies S2

[EC GATE 2018]

194. Let $A = \begin{pmatrix} 1 & 0 & -1 \\ -1 & 2 & 0 \\ 0 & 0 & -2 \end{pmatrix}$ and $B = A^3 - A^2 - 4A$ $+ 5I$, where I is the 3×3 identity matrix. The determinant of B is _____? (Up to 1 decimal places).

[EC GATE 2018]

195. Consider the following system of linear equation

$3x + 2ky = -2$, $kx + 6y = 2$

Here x and y are the unknowns and k is a real constant. The value of k for which there are an infinite number of solutions is

(a) 3   (b) 1   (c) -3 (d) -6

[IN GATE 2018]

196. Consider matrix $A = \begin{pmatrix} k & 2k \\ k^2 - k & k^2 \end{pmatrix}$ and vector $X = \begin{bmatrix} x_1 \\ x_2 \end{bmatrix}$. The number of distinct real values of k for which the equation $AX = O$ has an infinitely many solutions is _____?

[EC GATE 2018]

197. The diagonal elements of a 3×3 matrix are $-10, 5$, and 0, respectively. If two of its eigenvalues are -15 each, the third eigenvalues is _____?

[PI GATE 2018]

198. Consider a matrix $A = UV^T$, where $U = \begin{bmatrix} 1 \\ 2 \end{bmatrix}$, $V = \begin{bmatrix} 1 \\ 1 \end{bmatrix}$ Note that V^T denotes the transpose of V. The largest eigenvalue of A is _____?

[CS/IT GATE 2018]

199. Consider a non-singular 2×2 square matrix A. If trace $(A) = 4$ and trace $(A^2) = 5$, the determinant of the matrix A is _____ ?

[EC GATE 2018]

200. Let N be a 3×3 matrix with real numbers entries. The matrix N is such that $N^2 = 0$. The eigenvalues of N are

(a) 0, 0, 0   (b) 0, 0, 1

(c) 0, 1, 1   (d) 1, 1, 1

Answer key				
177. (a)	178. (c)	179. (d)	180. (a)	181. (c)
182. 5.	183. 4	184. 2.	185. (c)	186. 1
187. (a)	188. (c)	189. (c)	190. (b)	191. (b)
192. (d)	193. (c)	194. 1.	195. (c)	196. 2.
197. 25.	198. 3.	199. $\frac{11}{2}$.		200. (a)

Answer key				
1. (b)	**2.** (a)	**3.** (a)	**4.** (b)	**5.** (c)
6. (a)	**7.** (b)	**8.** (b)	**9.** (d)	**10.** (b)
11. (d)	**12.** (b)	**13.** (a)	**14.** (a)	**15.** (c)
16. (b)	**17.** (c)	**18.** (b)	**19.** (b)	**20.** (c)
21. (c)	**22.** (c)	**24.** (c)	**25.** (a)	**26.** (b)
27. (d)	**28.** (c)	**29.** (c)	**30.** (a)	**31.** (d)
32. (d)	**33.** (b)	**34.** (c)	**35.** (b)	**36.** (c)
37. (b)	**38.** (b)	**39.** (b)	**40.** (a), (b), (d)	**41.** (b)
41. (b)	**42.** (a)	**43.** (a)	**44.** (c)	**45.** (a)
46. (a)	**47.** (b)	**48.** (a)	**49.** (d)	**50.** (c)
51. (c)	**52.** (a)	**53.** (b)	**54.** (a)	**55.** (a)
56. (c)	**57.** (a)	**58.** (b)	**59.** (a)	
60. (i) – (a); (ii) – (a)			**61.** (a)	**62.** (b)
63. (a)	**64.** (d)	**65.** (a)	**66.** (b)	**67.** (b)
68. (c)	**69.** (d)	**70.** (a)	**71.** (d)	**72.** (c)
73. (b)	**74.** (a)	**75.** (c)	**76.** (c)	**77.** (a)
78. (b)	**79.** (b)	**80.** (d)	**81.** (c)	**82.** (d)
83. (c)	**84.** (a)	**85.** (a)	**86.** (b)	**87.** (c)
88. (b)	**89.** (d)	**90.** (a)	**91.** (a)	**92.** (d)
93. (a)	**94.** (b)	**95.** (b)	**96.** (a)	**97.** (b)
98. (d)	**99.** (b)	**100.** (b)	**101.** (c)	**102.** (c)
103. (b)	**104.** (d)	**105.** (a),(d)		**106.** (d)
107. (a)	**108.** (c)	**109.** (d)	**110.** 16	**111.** (a)
112. (b)	**113.** 1,–1	**114.** (d)	**115.** 0.	**116.** 2.
117. (c)	**118.** (a)	**119.** (a)	**120.** one	**121.** (a)
122. (b)	**123.** (b)	**124.** 200	**125.** (c)	**126.** (a)
127. (b)	**128.** (a)	**129.** 49	**130.** (d)	**131.** (a)
132. (d)	**133.** 6.	**134.** 1.	**135.** 3.	**136.** (d)
137. 17	**138.** (b)	**139.** (d)	**140.** (b)	**141.** (b)
142. (d)	**143.** (a)	**144.** (c)	**145.** (a)	**146.** (b)
147. 2	**148.** 4.5	**149.** (c)	**150.** 3	**151.** (d)
152. (d)	**153.** 3	**154.** (c)	**155.** (a)	**156.** 15.
157. (b)	**158.** (d)	**159.** (a)	**160.** (c)	**161.** (d)
162. (a)	**163.** -6.	**164.** 1.	**165.** 1/8.	**166.** (c)
167. (d)	**168.** 233	**169.** (b)	**170.** (d)	**171.** (c)
172. 5.	**173.** (b)	**174.** (b)	**175.** [0]	**176.** (c)

Explanation

1. (b) The trace of the matrix $= 2 + 3 + (-2) + 4 = 7 = $ sum of eigenvalues $= 2 + 3 + (-2) + 4$.

2. (a) Here $V[i, j]$ denotes the element lying in i^{th} row and j^{th} column.

$$\text{For } n = 3, \ V = \begin{pmatrix} V[1,1] & V[1,2] & V[1,3] \\ V[2,1] & V[2,2] & V[2,3] \\ V[3,1] & V[3,2] & V[3,3] \end{pmatrix}$$

$$= \begin{pmatrix} 1-1 & 1-2 & 1-3 \\ 2-1 & 2-2 & 2-3 \\ 3-1 & 3-2 & 3-3 \end{pmatrix}$$

$$= \begin{pmatrix} 0 & -1 & -2 \\ 1 & 0 & -1 \\ 2 & 1 & 0 \end{pmatrix} (\because V[i,j] = i - j)$$

Therefore, the sum of the elements in the array V is zero.

3. (a)

$$\begin{vmatrix} 2 & 0 & 0 & 0 \\ 8 & 1 & 7 & 2 \\ 2 & 0 & 2 & 0 \\ 9 & 0 & 6 & 1 \end{vmatrix}$$

$= 2 \times$ cofactor of "2"

(since "2" is the only non-zero element in first row)

$$= 2 \begin{vmatrix} 1 & 7 & 2 \\ 0 & 2 & 0 \\ 0 & 6 & 1 \end{vmatrix}$$

$= 2 \times 2$

$= 4.$

4. (b) $(ABC)^{-1} = [(A(BC)]^{-1} = (BC)^{-1} A^{-1} = C^{-1}B^{-1}A^{-1}$.

5. (c)

$$A = \begin{bmatrix} 1 & 2 & 3 \\ 3 & 4 & 5 \\ 4 & 6 & 8 \end{bmatrix}$$

$$\sim \begin{bmatrix} 1 & 2 & 3 \\ 3 & 4 & 5 \\ 0 & 0 & 0 \end{bmatrix} [\text{by } R_3 \to R_3 - (R_1 + R_2)]$$

which has 2 non-zero rows
Therefore, rank $(A) = 2$.

6. (a)

$$[P][Q]^T =$$

$$\begin{bmatrix} 2 & 3 \\ 4 & 5 \end{bmatrix} \times \begin{bmatrix} 4 & 9 \\ 8 & 2 \end{bmatrix} = \begin{bmatrix} 2 \times 4 + 3 \times 8 & 2 \times 9 + 3 \times 2 \\ 4 \times 4 + 5 \times 8 & 4 \times 9 + 5 \times 2 \end{bmatrix}$$

$$= \begin{bmatrix} 32 & 24 \\ 56 & 46 \end{bmatrix}.$$

7. (b) $C = BA = B_{1 \times 3} A_{3 \times 1}$. Then, C is a matrix of order 1×1. Therefore, rank$(A) \leq \min \{1, 1\} = 1$.

So the possible cases are
rank$(C) = 0$ or rank$(C) = 1$.
But since both A and B are non-zero matrices, so the product AB ($=C$) contains at least one non-zero element. Hence, rank$(C) = 1$.

8. (b) Since the given matrix has only one non-zero row, so rank $= 1$.

9. (d) The trace of the matrix $= 5 + 9 = 14 = $ sum of eigenvalues $= 10.16 + 3.84$.

10. (b)

$$\Delta = \begin{vmatrix} 1 & 1 \\ 2 & 2 \end{vmatrix} = 0, \qquad \Delta_1 = \begin{vmatrix} 2 & 1 \\ 5 & 2 \end{vmatrix} \neq 0$$

Therefore, the system has no solution (by Cramer's rule).

11. (d) The necessary condition to diagonalize a matrix is that the matrix is non-singular, whereas the sufficient condition is that the matrix has n linearly independent eigenvectors (where "n" is the order of the square matrix).

12. (b)

$$\Delta = \begin{vmatrix} 3 & 2 & 1 \\ 1 & -1 & 1 \\ -2 & 0 & 2 \end{vmatrix}$$

$$= 3(-2 - 0) - 2(2 + 2) + 1(0 - 2) \neq 0$$

Thus, the system has a unique solution.

13. (a)

Case-I: Let $A = \begin{bmatrix} 0 & 1 \\ 0 & 0 \end{bmatrix}, B = \begin{bmatrix} 0 & 0 \\ 0 & 1 \end{bmatrix}$. Then det$(A)$

$= 0$, det$(B) = 0$ and det$(A + B) = \begin{vmatrix} 0 & 1 \\ 0 & 1 \end{vmatrix} = 0.$

In this case, each of A and B is a singular matrix and A+B is also singular. Thus, the statement S_1 is true.

Case-II: Let $A = \begin{bmatrix} -1 & 0 \\ 0 & 2 \end{bmatrix}, B = \begin{bmatrix} 3 & 0 \\ 0 & 2 \end{bmatrix}$. Then,

det$(A) \neq 0$, det$(B) \neq 0$ and det$(A + B) = \begin{vmatrix} 2 & 0 \\ 0 & 4 \end{vmatrix} \neq 0.$

In this case, each of A and B is non-singular matrix and $A + B$ is also non-singular. Thus, the statement $S2$ is true.

14. (a) The trace of the matrix $= 1 + 2 + (-2) + (-1) = 0 = $ sum of eigenvalues $= 1 + 2 + (-2) + (-1)$.

15. (c) Since the determinant value of an upper triangular matrix is the product of the diagonal elements, so the determinant value of the given matrix $= 1 \times 1 \times 1 \times 1 = 1$.

16. (b)

$$X^2 - X + I = O$$

$$\Rightarrow I = X - X^2$$

$$\Rightarrow IX^{-1} = (X - X^2) X^{-1}$$

$$\Rightarrow X^{-1} = XX^{-1} - X^2 X^{-1} = I - X$$

$$= \begin{bmatrix} 1 & 0 \\ 0 & 1 \end{bmatrix} - \begin{bmatrix} a & 1 \\ -a^2 + a - 1 & 1 - a \end{bmatrix}$$

$$= \begin{bmatrix} 1 - a & -1 \\ a^2 - a + 1 & a \end{bmatrix}$$

17. (c)

$$\Delta = \begin{vmatrix} 1 & 2 & 1 \\ 2 & 1 & 2 \\ 1 & 1 & 1 \end{vmatrix} = 0$$

$$\Delta_1 = \begin{vmatrix} 6 & 2 & 1 \\ 6 & 1 & 2 \\ 5 & 1 & 1 \end{vmatrix} = 3 \neq 0$$

$\because \quad \Delta = 0, \Delta_1 \neq 0$, so the system has no solutions (by Cramer's rule).

Alternative method:

$$[A:B] = \begin{bmatrix} 1 & 2 & 1 & 6 \\ 2 & 1 & 2 & 6 \\ 1 & 1 & 1 & 5 \end{bmatrix}$$

$$\sim \begin{bmatrix} 1 & 2 & 1 & 6 \\ 0 & -3 & 0 & -6 \\ 0 & -1 & 0 & -1 \end{bmatrix}$$

(by $R_2 \to R_2 - 2R_1, R_3 \to R_3 - R_1$)

$$\sim \begin{bmatrix} 1 & 2 & 1 & 6 \\ 0 & 0 & 0 & -3 \\ 0 & -1 & 0 & -1 \end{bmatrix} \text{ by } (R_2 \to R_2 - 3R_3)$$

$$\sim \begin{bmatrix} 1 & 2 & 1 & 6 \\ 0 & -1 & 0 & -1 \\ 0 & 0 & 0 & -3 \end{bmatrix}$$

(by $R_2 \leftrightarrow R_3$)

Hence, rank ($[A:B]$) = 3 and rank(A) = 2. Since they are unequal, the system has no solution.

18. (b) The given system of equations can be written as

$2x + y - 4z = \alpha, 4x + 3y - 12z = 5, x + 2y - 8z = 7.$

The system has infinitely many solutions

$\Rightarrow \Delta_3 = 0$

$$\Rightarrow \begin{vmatrix} 2 & 1 & \alpha \\ 4 & 3 & 5 \\ 1 & 2 & 7 \end{vmatrix} = 0$$

$\Rightarrow 2(21 - 10) - (28 - 5) + \alpha(8 - 3) = 0$

$\Rightarrow \alpha = \dfrac{1}{5}.$

So α has only one value.

19. (b) A homogeneous system of equations has a non-trivial solution if the coefficient matrix is singular.

20. (c)

$$A = \begin{pmatrix} 4 & 2 & 1 & 3 \\ 6 & 3 & 4 & 7 \\ 2 & 1 & 0 & 1 \end{pmatrix} \sim \begin{pmatrix} 4 & 2 & 1 & 3 \\ 0 & 0 & \dfrac{5}{2} & \dfrac{5}{2} \\ 2 & 1 & 0 & 1 \end{pmatrix}$$

(by $R_2 \to R_2 - \dfrac{3}{2}R_1$)

$$\sim \begin{pmatrix} 4 & 2 & 1 & 3 \\ 0 & 0 & 1 & 1 \\ 2 & 1 & 0 & 1 \end{pmatrix} \text{(by } R_2 \to \dfrac{2}{5}R_2\text{)}$$

$$\sim \begin{pmatrix} 0 & 0 & 1 & 1 \\ 0 & 0 & 1 & 1 \\ 2 & 1 & 0 & 1 \end{pmatrix} \text{(by } R_1 \to R_1 - 2R_3\text{)}$$

$$\sim \begin{pmatrix} 0 & 0 & 0 & 0 \\ 0 & 0 & 1 & 1 \\ 2 & 1 & 0 & 1 \end{pmatrix} \text{(by } R_1 \to R_1 - R_2\text{)}$$

Which has two non-zero rows

Therefore, rank (A) = 2.

21. (c) The trace of the matrix = 4 + 4 = sum of eigenvalues = 3 + 5.

22. (c)

$$|A - \lambda I| = O \Rightarrow \begin{vmatrix} 4 - \lambda & -2 \\ -2 & 1 - \lambda \end{vmatrix} = 0$$

$\Rightarrow (4 - \lambda)(1 - 1) - 4 = 0$

$\Rightarrow \lambda^2 - 5\lambda = 0$

$\Rightarrow \lambda = 0, 5$

Alternative Method

If we take the eigenvalues 0 and 5, then product of the eigenvalues = 0, which is equal to det(A).

23. (b) The sum of eigenvalues = trace of the matrix = 1 + 5 + 1 = 7.

24. (c) For $n = 2$, the symmetric matrices with each element equal to either 0 or 1 are

$$\begin{bmatrix} 0 & 0 \\ 0 & 0 \end{bmatrix}, \begin{bmatrix} 1 & 0 \\ 0 & 1 \end{bmatrix}, \begin{bmatrix} 0 & 1 \\ 1 & 0 \end{bmatrix}, \begin{bmatrix} 1 & 1 \\ 1 & 1 \end{bmatrix},$$

$$\begin{bmatrix} 1 & 0 \\ 0 & 0 \end{bmatrix}, \begin{bmatrix} 0 & 0 \\ 0 & 1 \end{bmatrix}, \begin{bmatrix} 1 & 1 \\ 1 & 0 \end{bmatrix}, \begin{bmatrix} 0 & 1 \\ 1 & 1 \end{bmatrix}.$$

Thus, for $n = 2$, there exist eight such symmetric matrices.

Now, $8 = 2^{\frac{2^2 + 2}{2}}$

Hence, option (c) is correct.

25. (a)

$$A = \begin{pmatrix} 8 & x & 0 \\ 4 & 0 & 2 \\ 12 & 6 & 0 \end{pmatrix}$$

Then, A is singular $\Rightarrow$ det(A) = 0 $\Rightarrow x = 4.$

26. (b) A, B, C, D has non-zero determinants A^{-1}, B^{-1}, C^{-1}, D^{-1} exist

Then,
$$ABCD = I \Rightarrow A^{-1}(ABCD) = A^{-1}I = A^{-1}$$
$$\Rightarrow (A^{-1}A)(BCD)D^{-1} = A^{-1}D^{-1}$$
$$\Rightarrow IBC(DD^{-1}) = A^{-1}D^{-1}$$
$$\Rightarrow BCI = A^{-1}D^{-1}$$
$$\Rightarrow BC = A^{-1}D^{-1}$$
$$\Rightarrow (BC)C^{-1} = A^{-1}D^{-1}C^{-1}$$
$$\Rightarrow B(CC^{-1}) = A^{-1}D^{-1}C^{-1}$$
$$\Rightarrow BI = A^{-1}D^{-1}C^{-1}$$
$$\Rightarrow B = A^{-1}D^{-1}C^{-1} = (CDA)^{-1}$$
$$\Rightarrow B^{-1} = ((CDA)^{-1})^{-1} = CDA$$

27. (d) Since product of matrices (if defined) is again matrix. Hence, the product $F^T C^T\text{-}BCF$ cannot be a scalar. Hence, the statement (i) is false.

Again, F is a matrix of order 5×1 and D is a matrix of order 5×3 $\Rightarrow FD$ is not defined (since number of columns of $F \neq$ number of rows of D and so $D^T FD$ is not defined. Therefore, the statement (ii) is false.

28. (c) Solving the first two equations we get, $x = 9/4$ and $y = 1/4$; which satisfies the third equation. So the system has a unique solution given by $x = 9/4$ and $y = 1/4$.

29. (c) Using Cramer's rule we have

$$\Delta = \begin{vmatrix} 1 & 2 & 3 \\ 1 & 3 & 4 \\ 2 & 2 & 3 \end{vmatrix} = -1$$

$$\Delta_1 = \begin{vmatrix} 6 & 2 & 3 \\ 8 & 3 & 4 \\ 12 & 2 & 3 \end{vmatrix} = \begin{vmatrix} 6 & 2 & 3 \\ 2 & 1 & 1 \\ 6 & 0 & 0 \end{vmatrix} = -6$$
$$(by\ R_2 \to R_2 - R_1,\ \ R_3 \to R_3 - R_1)$$

$$\Delta_2 = \begin{vmatrix} 1 & 6 & 3 \\ 1 & 8 & 4 \\ 2 & 12 & 3 \end{vmatrix}$$

$$= \begin{vmatrix} 1 & 6 & 3 \\ 0 & 2 & 1 \\ 0 & 0 & -3 \end{vmatrix}$$
$$(by\ R_2 \to R_2 - R_1,\ \ R_3 \to R_3 - 2R_1)$$

$$= 1(-6 - 0) = -6$$

$$\Delta_3 = \begin{vmatrix} 1 & 2 & 6 \\ 1 & 3 & 8 \\ 2 & 2 & 12 \end{vmatrix}$$

$$= \begin{vmatrix} 1 & 2 & 6 \\ 0 & 1 & 2 \\ 0 & -2 & 0 \end{vmatrix}$$
$$(by\ R_2 \to R_2 - R_1,\ \ R_3 \to R_3 - 2R_1)$$

$$= 4$$

Thus,
$$x = \frac{\Delta_1}{\Delta} = 6, y = \frac{\Delta_2}{\Delta} = 6, z = \frac{\Delta_3}{\Delta} = -4.$$

Alternative Method

The given system of equations can be written as
$$x + 2y + 3z = 6, x + 3y + 4z = 8, 2x + 2y + 3z = 12.$$
These equations are satisfied for $x = y = 6$ and $z = -4$.

30. (a)
$$|A - \lambda I| = 0$$

$$\Rightarrow \begin{vmatrix} 5-\lambda & 0 & 0 & 0 \\ 0 & 5-\lambda & 5 & 0 \\ 0 & 0 & 2-\lambda & 1 \\ 0 & 0 & 3 & 1-\lambda \end{vmatrix} = 0$$

$$\Rightarrow (5-\lambda) \begin{vmatrix} 5-\lambda & 5 & 0 \\ 0 & 2-\lambda & 1 \\ 0 & 3 & 1-\lambda \end{vmatrix} = 0$$

$$\Rightarrow \lambda = 5 \text{ is an eigenvalue.}$$

Then $AX = \lambda X$

$$\Rightarrow \begin{bmatrix} 5 & 0 & 0 & 0 \\ 0 & 5 & 5 & 0 \\ 0 & 0 & 2 & 1 \\ 0 & 0 & 3 & 1 \end{bmatrix} \begin{bmatrix} x \\ y \\ z \\ t \end{bmatrix} = 5 \begin{bmatrix} x \\ y \\ z \\ t \end{bmatrix}$$

$$\Rightarrow \begin{bmatrix} 5x \\ 5y+5z \\ 2z+t \\ 3z+t \end{bmatrix} = \begin{bmatrix} 5x \\ 5y \\ 5z \\ 5t \end{bmatrix}$$

$$\Rightarrow 5x = 5x, 5y + 5z = 5y, 2z + t = 5z, 3z + t = 5t$$

$$\Rightarrow z = 0, t = 0$$

Then eigenvector,

$$X = \begin{bmatrix} x \\ y \\ z \\ t \end{bmatrix} = \begin{bmatrix} x \\ y \\ 0 \\ 0 \end{bmatrix} = \begin{bmatrix} 1 \\ -2 \\ 0 \\ 0 \end{bmatrix} \text{ (taking } x = 1, y = -2)$$

31. (d)

$|A - \lambda I| = 0$

$\Rightarrow \begin{vmatrix} 1-\lambda & 4 \\ a & 2-\lambda \end{vmatrix} = 0$

$\Rightarrow (1-\lambda)(2-\lambda) - 4a = 0$

$\Rightarrow \lambda^2 - 3\lambda + (2 - 4a) = 0$

$\Rightarrow \lambda = \dfrac{3 \pm \sqrt{(-3)^2 - 4(2-4a)}}{2}$

$\Rightarrow \lambda = \dfrac{3 \pm \sqrt{16a+1}}{2}$,

which is real if

$16a + 1 \geq 0$, i.e; $a \geq -\dfrac{1}{16}$

Also, $\lambda \geq 0$

$\Rightarrow \dfrac{3 \pm \sqrt{16a+1}}{2} \geq 0$

$\Rightarrow \dfrac{3}{2} \geq \mp \dfrac{\sqrt{16a+1}}{2}$

$\Rightarrow \dfrac{9}{4} \geq \dfrac{16a+1}{4}$

$\Rightarrow a \leq \dfrac{1}{2}$

$\therefore -\dfrac{1}{16} \leq a \leq \dfrac{1}{2}$.

32. (d)

$PX = \lambda X$

$\Rightarrow \begin{pmatrix} 3 & -2 & 2 \\ 0 & -2 & 1 \\ 0 & 0 & 1 \end{pmatrix} \begin{bmatrix} x \\ y \\ z \end{bmatrix} = -2 \begin{bmatrix} x \\ y \\ z \end{bmatrix}$

$\Rightarrow \begin{bmatrix} 3x - 2y + 2z \\ -2y + z \\ z \end{bmatrix} = \begin{bmatrix} -2x \\ -2y \\ -2z \end{bmatrix}$

$\Rightarrow 3x - 2y + 2z = -2x, -2y + z = -2y, z = -2z$

$\Rightarrow 5x - 2y + 2z = 0, z = 0 \Rightarrow x = \dfrac{2}{5}y, z = 0$

$\therefore$ eigenvector, $X = \begin{bmatrix} \dfrac{2}{5}y \\ y \\ z \end{bmatrix} = \begin{bmatrix} 2 \\ 5 \\ 0 \end{bmatrix}$ (for $y = 5$)

33. (b) Determinant of the matrix $= 10 - 4 = 6 =$ product of eigenvalues $= 1 \times 6$.

34. (c)

$|A - \lambda I| = 0$

$\Rightarrow \begin{vmatrix} -4-\lambda & 2 \\ 4 & 3-\lambda \end{vmatrix} = 0$

$\Rightarrow (\lambda - 3)(4 + \lambda) - 8 = 0$

$\Rightarrow \lambda^2 + \lambda - 20 = 0$

$\Rightarrow \lambda = -5, 4$

For $\lambda = -5$,

$\underline{AX = \lambda X}$

$\Rightarrow \begin{bmatrix} -4 & 2 \\ 4 & 3 \end{bmatrix} \begin{bmatrix} x \\ y \end{bmatrix} = -5 \begin{bmatrix} x \\ y \end{bmatrix}$

$\Rightarrow \begin{bmatrix} -4x + 2y \\ 4x + 3y \end{bmatrix} = \begin{bmatrix} -5x \\ -5y \end{bmatrix}$

$\Rightarrow -4x + 2y = -5x, \ 4x + 3y = -5y$

$\Rightarrow x = -2y$

$\therefore$ Eigenvector, $X = \begin{bmatrix} x \\ y \end{bmatrix} = \begin{bmatrix} -2y \\ y \end{bmatrix} = \begin{bmatrix} 2 \\ -1 \end{bmatrix}$

(taking $y = -1$)

35. (b) System has a non-trivial solution

$\Rightarrow$ rank $(A - \lambda I) < n$.

Thus, (a) is correct.

$A^T = A^{-1} \Rightarrow A$ is orthogonal

$\Rightarrow |\lambda_i| = 1$, (c) is correct

$A^T = A \Rightarrow A$ is symmetric

$\Rightarrow$ eigenvalues λ_i are real

Thus, (d) is correct.

Therefore, (b) is not a correct statement.

36. (c) Value of the determinant

= product of eigenvalues

$= 15 \times 3 \times 0 = 0$

37. (b) Since A is a lower triangular, so the eigenvalues are the diagonal elements, *i.e.* 1 and -2.

For $\lambda = 1$

$AX = \lambda X$

$\Rightarrow \begin{bmatrix} 1 & 0 \\ -1 & -2 \end{bmatrix} \begin{bmatrix} x \\ y \end{bmatrix} = 1 \begin{bmatrix} x \\ y \end{bmatrix}$

$\Rightarrow \begin{bmatrix} x \\ -x - 2y \end{bmatrix} = \begin{bmatrix} x \\ y \end{bmatrix}$

$\Rightarrow x = x, -x - 2y = y$

$\Rightarrow x = -3y$

Thus, eigenvector,

$$X = \begin{bmatrix} x \\ y \end{bmatrix} = \begin{bmatrix} -3y \\ y \end{bmatrix} = \begin{bmatrix} 3 \\ -1 \end{bmatrix}$$

(for $y = -1$)

38. (b) A has order 3×4

$\Rightarrow$ rank (A) $\leq$ min{3, 4}

$\Rightarrow$ rank (A) ≤ 3

Also rank ([A : B]) $\leq$ min {3, 5} = 3

If rank (A) = rank ([A : B]) = 3, then the system will be consistent. Hence, for the system to be inconsistent, the highest possible rank of A will be 2.

39. (b)

$$\Delta = \begin{vmatrix} 2 & -1 & 3 \\ 3 & -2 & 5 \\ -1 & -4 & 1 \end{vmatrix} = 2 \neq 0$$

Therefore, the system has a unique solution (by Cramer's rule).

40. (a), (b), (d) In case of non-homogeneous system of equations, a consistent system will have either unique solution or infinitely many solutions, where as an inconsistent system will have no solution.

41. (b) No. of independent solution

= order of the square matrix − rank of the co-efficient matrix

$= 3 - 2 = 1$.

42. (a) Rank $(P) = $ rank $([P : Q]) \Leftrightarrow$ at least one solution exists.

43. (a)

$$\begin{vmatrix} 0 & 1 & 0 & 2 \\ -1 & 1 & 1 & 3 \\ 0 & 0 & 0 & 1 \\ 1 & -2 & 0 & 1 \end{vmatrix}$$

$$= \begin{vmatrix} 0 & 1 & 0 & 2 \\ 0 & -1 & 1 & 4 \\ 0 & 0 & 0 & 1 \\ 1 & -2 & 0 & 1 \end{vmatrix} \quad \text{[by } R_2 \to R_2 + R_4]$$

$= 1 \times \text{cofactor of "1"} \quad [\because \text{"1" lies in 1st column}]$

$$= \begin{vmatrix} 1 & 0 & 2 \\ -1 & 1 & 4 \\ 0 & 0 & 1 \end{vmatrix} = \begin{vmatrix} 1 & 0 \\ -1 & 1 \end{vmatrix} = -1$$

44. (c) A is orthogonal $\Rightarrow AA' = I$

$$\Rightarrow (AA')^{-1} = I^{-1} = I = \begin{pmatrix} 1 & 0 & 0 & 0 \\ 0 & 1 & 0 & 0 \\ 0 & 0 & 1 & 0 \\ 0 & 0 & 0 & 1 \end{pmatrix}$$

45. (a)

$$|A| = \begin{vmatrix} 2 & -0.1 \\ 0 & 3 \end{vmatrix} = 6$$

$$A^{-1} = \frac{1}{|A|} adj(A)$$

$$= \frac{1}{6} \begin{bmatrix} 3 & 0.1 \\ 0 & 2 \end{bmatrix}$$

$$= \begin{bmatrix} \frac{1}{2} & \frac{0.1}{6} \\ 0 & \frac{1}{3} \end{bmatrix} = \begin{bmatrix} \frac{1}{2} & a \\ 0 & b \end{bmatrix} \text{(given)}$$

$$\therefore a = \frac{0.1}{6}, b = \frac{1}{3}. \text{ So } a + b = \frac{1}{60} + \frac{1}{3} = \frac{7}{20}.$$

46. (a)

$[P(X^TY)^{-1}P^T]^T$

$= (P^T)^T [(X^TY)^{-1}]^T P^T = P[(X^TY)^{-1}]^T P^T$

$= P_{2\times3}[\{(X^T)_{3\times4} Y_{4\times3}\}^{-1}]^T (P^T)_{3\times2}$

$= P_{2\times3}[\{(X^TY)_{3\times3}\}^{-1}]^T (P^T)_{3\times2}$

$= P_{2\times3}[\{(X^TY)^{-1}\}_{3\times3}]^T (P^T)_{3\times2}$

$= P_{2\times3}[\{(X^TY)^{-1}\}^T]_{3\times3} (P^T)_{3\times2}$

$= P_{2\times3}[\{(X^TY)^{-1}\}^T P^T]_{3\times2}$

$= [P\{(X^TY)^{-1}\}^T P^T]_{2\times2}$

$= P_{2\times3}[\{(X^TY)^{-1}\}^T]_{3\times3}(P^T)_{3\times2}$

$= P_{2\times3}[\{(X^TY)^{-1}\}^T P^T]_{3\times2}$

$= [P\{(X^TY)^{-1}\}^T P^T]_{2\times2}$

47. (b)

$$|R| = \begin{vmatrix} 1 & 0 & -1 \\ 2 & 1 & -1 \\ 2 & 3 & 2 \end{vmatrix} = 1(2+3) - 0 - 1(6-2) = 5 - 4 = 1$$

$$adj(R) = \begin{bmatrix} \begin{vmatrix} 1 & -1 \\ 3 & 2 \end{vmatrix} & -\begin{vmatrix} 2 & -1 \\ 2 & 2 \end{vmatrix} & \begin{vmatrix} 2 & 1 \\ 2 & 3 \end{vmatrix} \\ -\begin{vmatrix} 0 & -1 \\ 3 & 2 \end{vmatrix} & \begin{vmatrix} 1 & -1 \\ 2 & 2 \end{vmatrix} & -\begin{vmatrix} 1 & 0 \\ 2 & 3 \end{vmatrix} \\ \begin{vmatrix} 0 & -1 \\ 1 & -1 \end{vmatrix} & -\begin{vmatrix} 1 & -1 \\ 2 & -1 \end{vmatrix} & \begin{vmatrix} 1 & 0 \\ 2 & 1 \end{vmatrix} \end{bmatrix}^T$$

$$= \begin{bmatrix} 5 & -6 & 4 \\ -3 & 4 & -3 \\ 1 & -1 & 1 \end{bmatrix}^T = \begin{bmatrix} 5 & -3 & 1 \\ -6 & 4 & -1 \\ 4 & -3 & 1 \end{bmatrix}$$

$$R^{-1} = \frac{1}{|R|} adj(R) = \frac{1}{1} \begin{bmatrix} 5 & -3 & 1 \\ -6 & 4 & -1 \\ 4 & -3 & 1 \end{bmatrix}$$

$$\Rightarrow R^{-1} = \begin{bmatrix} 5 & -3 & 1 \\ -6 & 4 & -1 \\ 4 & -3 & 1 \end{bmatrix}$$

Therefore, top row of $R^{-1} = [5\ -3\ 1]$

48. (a)

$$|E| = \begin{vmatrix} \cos\theta & \sin\theta & 0 \\ -\sin\theta & \cos\theta & 0 \\ 0 & 0 & 1 \end{vmatrix}$$

$$= \cos^2\theta + \sin^2\theta = 1 \neq 0.$$

$\therefore E^{-1}$ exist.

adj(E)

$$= \begin{bmatrix} \begin{vmatrix} \cos\theta & 0 \\ 0 & 1 \end{vmatrix} & -\begin{vmatrix} -\sin\theta & 0 \\ 0 & 1 \end{vmatrix} & \begin{vmatrix} -\sin\theta & \cos\theta \\ 0 & 0 \end{vmatrix} \\ -\begin{vmatrix} \sin\theta & 0 \\ 0 & 1 \end{vmatrix} & \begin{vmatrix} \cos\theta & 0 \\ 0 & 1 \end{vmatrix} & -\begin{vmatrix} \cos\theta & \sin\theta \\ 0 & 0 \end{vmatrix} \\ \begin{vmatrix} \sin\theta & 0 \\ \cos\theta & 0 \end{vmatrix} & -\begin{vmatrix} \cos\theta & 0 \\ -\sin\theta & 0 \end{vmatrix} & \begin{vmatrix} \cos\theta & \sin\theta \\ -\sin\theta & \cos\theta \end{vmatrix} \end{bmatrix}^T$$

$$= \begin{bmatrix} \cos\theta & \sin\theta & 0 \\ -\sin\theta & \cos\theta & 0 \\ 0 & 0 & 1 \end{bmatrix}^T = \begin{bmatrix} \cos\theta & -\sin\theta & 0 \\ \sin\theta & \cos\theta & 0 \\ 0 & 0 & 1 \end{bmatrix}$$

$$E^{-1} = \frac{1}{|R|} adj(E) = \begin{bmatrix} \cos\theta & -\sin\theta & 0 \\ \sin\theta & \cos\theta & 0 \\ 0 & 0 & 1 \end{bmatrix}$$

Now $EF = G$

$\Rightarrow E^{-1}(EF) = E^{-1}G$

$\Rightarrow (E^{-1}E)F = E^{-1}I$ ($\because G = I$, the identity matrix)

$\Rightarrow IF = E^{-1}$

$$\Rightarrow F = E^{-1} = \begin{bmatrix} \cos\theta & -\sin\theta & 0 \\ \sin\theta & \cos\theta & 0 \\ 0 & 0 & 1 \end{bmatrix}.$$

49. (d)

$$|A| = \begin{vmatrix} 1 & 0 & 1 & 0 \\ 0 & 1 & 0 & 1 \\ 1 & 1 & 0 & 0 \\ 0 & 0 & 0 & 1 \end{vmatrix}$$

$$= \begin{vmatrix} 1 & 0 & 1 & 0 \\ 0 & 1 & 0 & 1 \\ 0 & 1 & -1 & 0 \\ 0 & 0 & 0 & 1 \end{vmatrix} \text{ (by } R_3 \to R_3 - R_1)$$

$= 1 \times$ cofactor of "1"

(since "1" is the only non-zero element of first column)

$$= \begin{vmatrix} 1 & 0 & 1 \\ 1 & -1 & 0 \\ 0 & 0 & 1 \end{vmatrix} = -1 \neq 0.$$

Thus, A is a non-singular matrix of order 4. Therefore, rank $(A) = 4$

50. (c)

$$\begin{pmatrix} 1 & 1 & 1 \\ 1 & -1 & 0 \\ 1 & 1 & 1 \end{pmatrix} \sim \begin{pmatrix} 1 & 1 & 1 \\ 1 & -1 & 0 \\ 0 & 0 & 0 \end{pmatrix} \text{(by } R_3 \to R_3 - R_1)$$

which has 2 non-zero rows

Therefore, rank = 2.

51. (c)

$AX = \lambda X$

$$\Rightarrow \begin{bmatrix} 4 & 2 \\ 2 & 4 \end{bmatrix}\begin{bmatrix} 101 \\ 101 \end{bmatrix} = \lambda\begin{bmatrix} 101 \\ 101 \end{bmatrix}$$

$$\Rightarrow \begin{bmatrix} 4\times101 + 2\times101 \\ 2\times101 + 4\times101 \end{bmatrix} = \begin{bmatrix} 101\lambda \\ 101\lambda \end{bmatrix}$$

$\Rightarrow 101\lambda = 4\times101 + 2\times101$

$\Rightarrow \lambda = 4 + 2 = 6.$

52. (a)

53. (b) Let λ_1 and λ_2 be the other two eigenvalues.

Then, the sum of eigenvalues = trace of the matrix

$\Rightarrow 3 + \lambda_1 + \lambda_2 = 2 - 1 + 0 = 1$

$\Rightarrow \lambda_2 = -2 - \lambda_1$

If $\lambda_1 = 3$, then $\lambda_2 = -2 - 3 = -5$.

If $\lambda_1 = 2$, then $\lambda_2 = -2 - 2 = -4$.

54. (a) The eigenvalues of S^2 = square of the eigenvalues of $S = 1, 25$.

55. (a)

Let $A = \begin{pmatrix} a & b \\ c & d \end{pmatrix}$

Then,

$Av_1 = 8v_1$

$$\Rightarrow \begin{pmatrix} a & b \\ c & d \end{pmatrix}\begin{bmatrix} 1 \\ 1 \end{bmatrix} = 8\begin{bmatrix} 1 \\ 1 \end{bmatrix}$$

$$\Rightarrow \begin{bmatrix} a+b \\ c+d \end{bmatrix} = \begin{bmatrix} 8 \\ 8 \end{bmatrix}$$

$$\Rightarrow \begin{cases} a+b = 8 \dots\dots(1) \\ c+d = 8 \dots\dots(2) \end{cases}$$

Also,

$Av_2 = 4v_2$

$$\Rightarrow \begin{pmatrix} a & b \\ c & d \end{pmatrix} \begin{bmatrix} 1 \\ -1 \end{bmatrix} = 4 \begin{bmatrix} 1 \\ -1 \end{bmatrix}$$

$$\Rightarrow \begin{bmatrix} a-b \\ c-d \end{bmatrix} = \begin{bmatrix} 4 \\ -4 \end{bmatrix}$$

$$\Rightarrow \begin{cases} a-b = 4 \dots\dots(3) \\ c-d = -4 \dots\dots(4) \end{cases}$$

Solving (1) and (3) we get, $a = 6, b = 2$.

Solving (2) and (4) we get, $c = 2, d = 6$.

Hence, $A = \begin{pmatrix} 6 & 2 \\ 2 & 6 \end{pmatrix}$.

56. (c)

$$A = \begin{bmatrix} a & b \\ c & d \end{bmatrix}$$

Then

$$A \begin{bmatrix} 1 \\ -1 \end{bmatrix} = - \begin{bmatrix} 1 \\ -1 \end{bmatrix}$$

$$\Rightarrow \begin{bmatrix} a & b \\ c & d \end{bmatrix} \begin{bmatrix} 1 \\ -1 \end{bmatrix} = - \begin{bmatrix} 1 \\ -1 \end{bmatrix}$$

$$\Rightarrow \begin{bmatrix} a-b \\ c-d \end{bmatrix} = \begin{bmatrix} -1 \\ 1 \end{bmatrix}$$

$\Rightarrow a-b = -1 \dots\dots(1), c-d = 1\dots\dots(2)$

Again, $A \begin{bmatrix} 1 \\ -2 \end{bmatrix} = -2 \begin{bmatrix} 1 \\ -2 \end{bmatrix}$

$$\Rightarrow \begin{bmatrix} a & b \\ c & d \end{bmatrix} \begin{bmatrix} 1 \\ -2 \end{bmatrix} = -2 \begin{bmatrix} 1 \\ -2 \end{bmatrix}$$

$$\Rightarrow \begin{bmatrix} a-2b \\ c-2d \end{bmatrix} = \begin{bmatrix} -2 \\ 4 \end{bmatrix}$$

$\Rightarrow a-2b = -2\dots\dots(3), c-2d = 4\dots\dots(4)$

Solving (1) and (3) we get $b = 1, a = 0$

Solving (2) and (4) we get $d = -3$, $c = -2$.

Thus, $A = \begin{bmatrix} 0 & 1 \\ -2 & -3 \end{bmatrix}$. Now,

$$\begin{bmatrix} 1 & 1 \\ -1 & -2 \end{bmatrix} \times \begin{bmatrix} -1 & 0 \\ 0 & -2 \end{bmatrix} \times \begin{bmatrix} 2 & 1 \\ -1 & -1 \end{bmatrix}$$

$$= \begin{bmatrix} 1 & 1 \\ -1 & -2 \end{bmatrix} \times \begin{bmatrix} -2 & -1 \\ 2 & 2 \end{bmatrix}$$

$$= \begin{bmatrix} -2+2 & -1+2 \\ 2-4 & 1-4 \end{bmatrix} = \begin{bmatrix} 0 & 1 \\ -2 & -3 \end{bmatrix} = A$$

57. (a) A is symmetric $\Rightarrow$ eigenvalues are all real.

58. (b) Let λ be the other eigenvalue. Then,
the sum of eigenvalues = trace of the matrix $\Rightarrow$
$\lambda - 2 + 6 = 5 + 1 + 1 \Rightarrow \lambda = 3$.

59. (a) Clearly the set X forms a subspace of R^3.

$a[1,-1,0]^T + b[1,0,-1]^T = O$

$\Rightarrow [a,-a,0]^T + [b,0,-b]^T = O$

$\Rightarrow [a+b,-a+0,0-b]^T = [0,0,0]^T$

$\Rightarrow a+b = 0,-a = 0,-b = 0$

$\Rightarrow a = b = 0$

Hence, the vectors $[1,-1,0]^T$ and $[1,0,-1]^T$ are linearly independent.

Now $x^T = [x_1,x_2,x_3]^T = a[1,-1,0]^T + b[1,0,-1]^T$

$\Rightarrow [x_1,x_2,x_3]^T = [a,-a,0]^T + [b,0,-b]^T$

$\Rightarrow [x_1,x_2,x_3]^T = [a+b,-a+0,0-b]^T$

$\Rightarrow x_1 = a+b, x_2 = -a = 0, x_3 = -b$

$\therefore x_1 + x_2 + x_3 = a+b-a-b = 0, a = x_2, b = -x_3$.

so $x^T = [x_1, x_2, x_3]^T$

$= (-x_2)[1, -1, 0]^T + (-x_3)[1, 0, -1]^T$

Thus, the vector "x" can be expressed as a linear combination of the vectors $[1,-1,0]^T$ and $[1,0,-1]^T$

Consequently, $\left\{ [1,-1,0]^T, [1,0,-1]^T \right\}$ is the basis for the subspace X.

60. (i) – (a) ; (ii) – (a)

$|A - \lambda I| = 0$

$$\Rightarrow \begin{vmatrix} -3-\lambda & 2 \\ -1 & 0-\lambda \end{vmatrix} = 0$$

$\Rightarrow \lambda(\lambda+3) + 2 = 0$

$\Rightarrow \lambda^2 + 3\lambda + 2 = 0 \qquad \dots(*)$

$\Rightarrow \lambda = -1, -2$

Using the Cayley Hamilton theorem, we get from (*),

$$A^2 + 3A + 2I = O........(**)$$
$$\Rightarrow A^{-1}(A^2 + 3A + 2I) = A^{-1}O$$
$$\Rightarrow A + 3A^{-1}A + 2A^{-1} = O$$
$$\quad (since, A^{-1}A^2 = (A^{-1}A)A = IA = A)$$
$$\Rightarrow A + 3I + 2A^{-1} = O$$

Again $(**) \Rightarrow A^2 = -3A - 2I$

$$\therefore \quad A^4 = A^2.A^2 = (3A + 2I)^2$$
$$= 9A^2 + 12A + 4I$$
$$= 9(-3A - 2I) + 12A + 4I$$
$$= -15A - 14I$$

$$A^8 = A^4.A^4 = (15A + 14I)^2$$
$$= 225A^2 + 420A + 196I$$
$$= 225(-3A - 2I) + 420A + 196I$$
$$= -255A - 254I$$

Hence, $A^9 = A^8 \times A = (-255A - 254I)A$
$$= -255A^2 - 254A$$
$$= -255(-3A - 2I) - 254A$$
$$= 511A + 510I.$$

61. (a)
$$\begin{vmatrix} 1+b & b & 1 \\ b & b+1 & 1 \\ 1 & 2b & 1 \end{vmatrix}$$
$$= \begin{vmatrix} 1+b & b & 1 \\ -1 & 1 & 0 \\ -b & b & 0 \end{vmatrix}$$
$$[by \ R_2 \to R_2 - R_1, R_3 \to R_3 - R_1]$$
$$= -b + b = 0$$

62. (b)
$$V = XX' = \begin{bmatrix} x_1 \\ x_2 \\ . \\ . \\ . \\ . \\ x_n \end{bmatrix} \times \begin{bmatrix} x_1 & x_2 & . & . & . & . & x_n \end{bmatrix}$$

$$= \begin{bmatrix} x_1^2 & x_1x_2 & .. & .. & .. & .. & x_1x_n \\ x_2x_1 & x_2^2 & .. & .. & .. & .. & x_2x_n \\ .. & .. & .. & .. & .. & .. & .. \\ x_nx_1 & x_nx_2 & .. & .. & .. & .. & x_n^2 \end{bmatrix}$$

$$\sim \begin{bmatrix} x_1 & x_2 & .. & .. & .. & .. & x_n \\ x_1 & x_2 & .. & .. & .. & .. & x_n \\ .. & .. & .. & .. & .. & .. & .. \\ x_1 & x_2 & .. & .. & .. & .. & x_n \end{bmatrix}$$

$$\left(by \ R_1 \to \frac{1}{x_1}R_1, R_2 \to \frac{1}{x_2}R_2, ..., R_n \to \frac{1}{x_n}R_n \right)$$

$$\sim \begin{bmatrix} x_1 & x_2 & .. & .. & .. & .. & x_n \\ 0 & 0 & .. & .. & .. & .. & 0 \\ .. & .. & .. & .. & .. & .. & .. \\ 0 & 0 & .. & .. & .. & .. & 0 \end{bmatrix}$$

$$\left(by \ R_2 \to R_2 - R_1, R_3 \to R_3 - R_1, ..., R_n \to R_n - R_1 \right)$$

which has one non-zero row

Therefore, rank $(A) = 1$.

63. (a)

Let, $A = \begin{pmatrix} 1 & 2 \\ 5 & 7 \end{pmatrix}$. Then, det $(A) = 7 - 10 = -3$

$$adj \ (A) = \begin{pmatrix} 7 & -5 \\ -2 & 1 \end{pmatrix}^T = \begin{pmatrix} 7 & -2 \\ -5 & 1 \end{pmatrix}$$

$$\therefore \quad A^{-1} =$$
$$\frac{1}{|A|}adj(A) = \frac{1}{(-3)}\begin{pmatrix} 7 & -2 \\ -5 & 1 \end{pmatrix} = \frac{1}{3}\begin{pmatrix} -7 & 2 \\ 5 & -1 \end{pmatrix}$$

64. (d)

Here,
$$\Delta = \begin{vmatrix} 0 & 4 & 3 \\ 2 & 0 & -1 \\ 3 & 2 & 0 \end{vmatrix} = 0 - 4(0+3) + 3(4-0) = 0,$$

$$\Delta_1 = \begin{vmatrix} 8 & 4 & 3 \\ 2 & 0 & -1 \\ 5 & 2 & 0 \end{vmatrix} = 8(0+2) - 4(0+5) + 3(4-0)$$
$$\neq 0$$

Since $\Delta = 0$ and $\Delta_1 \neq 0$, so the system is inconsistent (by Cramer's rule) and hence the solution in non-existent.

65. (a) By Cramer's rule, the system has an infinite number of solutions.

$\Leftrightarrow \Delta = 0, \Delta_3 = 0$

$$\Leftrightarrow \begin{vmatrix} 1 & 1 & 1 \\ 1 & 3 & 3 \\ 1 & 2 & \alpha \end{vmatrix} = 0, \begin{vmatrix} 1 & 1 & 5 \\ 1 & 3 & 9 \\ 1 & 2 & \beta \end{vmatrix} = 0$$

$\Leftrightarrow 2\alpha - 4 = 0, 2\beta - 14 = 0$

$\Rightarrow \alpha = 2, \beta = 7$

Alternative Method:

[A : B]

$$= \begin{bmatrix} 1 & 1 & 1 & 5 \\ 1 & 3 & 3 & 9 \\ 1 & 2 & \alpha & \beta \end{bmatrix}$$

$$\sim \begin{bmatrix} 1 & 1 & 1 & 5 \\ 0 & 2 & 2 & 4 \\ 0 & 1 & \alpha-1 & \beta-5 \end{bmatrix}$$

(by $R_2 \rightarrow R_2 - R_1, R_3 \rightarrow R_3 - R_1$)

$$\sim \begin{bmatrix} 1 & 1 & 1 & 5 \\ 0 & 2 & 2 & 4 \\ 0 & 0 & \alpha-2 & \beta-7 \end{bmatrix}$$

(by $R_3 \rightarrow R_3 - \frac{1}{2}R_2$)

Hence, if $\alpha - 2 = 0$ and $\beta - 7 = 0$ *i.e*; if $\alpha = 2$ and $\beta = 7$, then rank([A:B]) = rank(A) = 2 < no. of variables. This means the system has an infinite number of solutions for $\alpha = 2$ and $\beta = 7$.

66. (b) $A^2 = I \Rightarrow |A^2| = |I| = 1$

$\Rightarrow |A|^2 = 1$

$\Rightarrow |A| = 0 \pm 1 \Rightarrow |A| \neq 0$

$\Rightarrow$ The system has a unique solution (by Cramer's rule).

67. (b) Let $A = \begin{pmatrix} 2 & 1 \\ 0 & 2 \end{pmatrix}$. Since A is upper triangular, so its eigenvalues are diagonal elements, *i.e*, 2, 2.

Then $AX = \lambda X$

$\Rightarrow \begin{pmatrix} 2 & 1 \\ 0 & 2 \end{pmatrix} \begin{bmatrix} x \\ y \end{bmatrix} = 2 \begin{bmatrix} x \\ y \end{bmatrix}$

$\Rightarrow \begin{bmatrix} 2x + y \\ 2y \end{bmatrix} = \begin{bmatrix} 2x \\ 2y \end{bmatrix}$

$\Rightarrow 2x + y = 2x, 2y = 2y$

$\Rightarrow y = 0$

$\therefore X = \begin{bmatrix} x \\ y \end{bmatrix} = \begin{bmatrix} x \\ 0 \end{bmatrix} = x \begin{bmatrix} 1 \\ 0 \end{bmatrix}$.

Hence, A has only one linearly independent eigenvector, which is $\begin{bmatrix} 1 \\ 0 \end{bmatrix}$.

68. (c) $X_1, X_2, ..., X_M$ are M non-zero orthogonal vectors $\Rightarrow$ the vectors are linearly independent.

Let X be any arbitrary vector. Then,

$$X = a_1 X_1 + a_2 X_2 + a_3 X_3 + ... + a_M X_M$$
$$+ b_1(-X_1) + b_2(-X_2) + b_3(-X_3) + ... + b_M(-X_M)$$
$$= (a_1 - b_1)X_1 + (a_2 - b_2)X_2 + ... + (a_M - b_M)X_M$$

Thus, X is expressed as a linear combination of M vectors $X_1, X_2, ..., X_M$ instead of 2M vectors.

$$X_1, X_2, ..., X_M, -X_1, -X_2, ..., -X_M.$$

Hence, dimension of the vector space spanned by the 2M vectors $X_1, X_2, ..., X_M, -X_1, -X_2, ..., -X_M$ is M.

69. (d) The system will not have a unique solution

$\Rightarrow$ system will have an infinite number of solutions (assuming that the system is consistent)

$\Rightarrow \Delta = 0$

$$\Rightarrow \begin{vmatrix} 1 & 1 & 1 \\ 1 & 2 & 3 \\ 1 & 4 & k \end{vmatrix} = 0$$

$\Rightarrow (2k - 12) - (k - 3) + (4 - 2) = 0$

$\Rightarrow k = 7$.

70. (a)

$$\Delta = \begin{vmatrix} 2 & 3 & 0 \\ 1 & 1 & 1 \\ 1 & 2 & -1 \end{vmatrix} = 2(-1-2) - 3(-1-1) = 0$$

Therefore, the system cannot have a unique solution and so it must have an infinite number of solutions (since the system is consistent).

Now, the systems has an infinite number of solutions

$\Rightarrow \Delta_1 = 0$

$$\Rightarrow \begin{vmatrix} 4 & 3 & 0 \\ 4 & 1 & 1 \\ a & 2 & -1 \end{vmatrix} = 0$$

$\Rightarrow 4(-1-2) - 3(-4-a) = 0$

$\Rightarrow a = 0$.

Alternative Method

[A : B]

$$= \begin{bmatrix} 2 & 3 & 0 & 4 \\ 1 & 1 & 1 & 4 \\ 1 & 2 & -1 & a \end{bmatrix}$$

$$\sim \begin{bmatrix} 1 & 1 & 1 & 4 \\ 2 & 3 & 0 & 4 \\ 1 & 2 & -1 & a \end{bmatrix} \quad (\text{by } R_2 \leftrightarrow R_1)$$

$$\sim \begin{bmatrix} 1 & 1 & 1 & 4 \\ 0 & 1 & -2 & -4 \\ 0 & 1 & -2 & a-4 \end{bmatrix}$$

$(\text{by } R_2 \rightarrow R_2 - 2R_1, R_3 \rightarrow R_3 - R_1)$

$$\sim \begin{bmatrix} 1 & 1 & 1 & 4 \\ 0 & 1 & -2 & -4 \\ 0 & 0 & 0 & a \end{bmatrix} \quad (\text{by } R_3 \rightarrow R_3 - R_2)$$

Hence, if $a = 0$, then rank([A:B]) = rank(A) = 2 < no. of variables. This means the system has an infinite number of solutions for $a = 0$.

71. (d) System has a unique solution

$\Rightarrow \Delta \neq 0$ (by Cramer's Rule)

$$\Rightarrow \begin{vmatrix} 1 & 1 & 2 \\ 1 & 2 & 3 \\ 1 & 4 & a \end{vmatrix} \neq 0$$

$\Rightarrow (2a - 12) - (a - 3) + 2(4 - 2) \neq 0$

$\Rightarrow a - 5 \neq 0$

$\Rightarrow a \neq 5$.

72. (c)

$$\Delta = \begin{vmatrix} 4 & 2 \\ 2 & 1 \end{vmatrix} = 0$$

$$\Delta_1 = \begin{vmatrix} 7 & 2 \\ 6 & 1 \end{vmatrix} = 7 - 12 = 5 \neq 0$$

∵ $\Delta = 0$, $\Delta_1 \neq 0$, so the system has no solutions (by Cramer's rule).

73. (b) $(PQ)^{-1}P = Q^{-1}P^{-1}P = Q^{-1}I = Q^{-1}$

74. (a) Here A^I does not mean A^T.

$$AA'A = A\left\{\left(A^T A\right)^{-1} A^T\right\} A$$

$$= A\left\{A^{-1}\left(A^T\right)^{-1}\right\} A^T A$$

$$= AA^{-1}\left\{\left(A^T\right)^{-1} A^T\right\} A$$

$$= I(IA)$$

$$= A.$$

75. (c) One of the eigenvalue is zero

$\Rightarrow \det (P) = 0$

$$\Rightarrow \begin{vmatrix} P_{11} & P_{12} \\ P_{21} & P_{22} \end{vmatrix} = 0$$

$\Rightarrow P_{11}P_{22} - P_{12}P_{21} = 0$

76. (c) Let λ_1 and λ_2 be the other two eigenvalues. Then, the sum of eigenvalues = trace of the matrix

$\Rightarrow 3 + \lambda_1 + \lambda_2 = 1 + 0 + p$

$\Rightarrow \lambda_1 + \lambda_2 = p - 2$.

77. (a) If $A = \begin{pmatrix} 1 & 0 \\ 0 & 0 \end{pmatrix}$, then the eigenvalues of A are 0 and 1 (since A is an upper triangular).

If $A = \begin{pmatrix} 0 & 1 \\ 0 & 0 \end{pmatrix}$, then the eigenvalues of A are 0 and 0 (since A is an upper triangular).

If $A = \begin{pmatrix} 1 & -1 \\ 1 & 1 \end{pmatrix}$, then

$|A - \lambda I| = 0$

$$\Rightarrow \begin{vmatrix} 1-\lambda & -1 \\ 1 & 1-\lambda \end{vmatrix} = 0$$

$\Rightarrow (1 - \lambda)^2 + 1 = 0$

$\Rightarrow \lambda \neq 1$.

If $A = \begin{pmatrix} -1 & 0 \\ 0 & -1 \end{pmatrix}$, the eigenvalues of A are -1 and -1 (since A is diagonal).

Thus, only one matrix is there with eigenvalue 1.

78. (b) The trace of the matrix = 4 + (−5) = −1 = sum of eigenvalues = (−6) + 5.

79. (b) Let A be the square matrix. Then the eigenvalues of A are 1 and 2 (since A is upper triangular).

For $\lambda = 1$:

$AX = \lambda X$

$$\Rightarrow \begin{pmatrix} 1 & 2 \\ 0 & 2 \end{pmatrix}\begin{bmatrix} x \\ y \end{bmatrix} = 1\begin{bmatrix} x \\ y \end{bmatrix}$$

$$\Rightarrow \begin{bmatrix} x + 2y \\ 2y \end{bmatrix} = \begin{bmatrix} x \\ y \end{bmatrix}$$

$\Rightarrow x + 2y = x, \ 2y = y$

$\Rightarrow y = 0$

$$\therefore X = \begin{bmatrix} x \\ y \end{bmatrix} = \begin{bmatrix} x \\ 0 \end{bmatrix} = \begin{bmatrix} 1 \\ 0 \end{bmatrix} \text{ (for } x = 1)$$

Comparing $\begin{bmatrix} 1 \\ 0 \end{bmatrix}$ with $\begin{bmatrix} 1 \\ a \end{bmatrix}$ we get $a = 0$.

For $\lambda = 2$:

$AX = \lambda X$

$$\Rightarrow \begin{pmatrix} 1 & 2 \\ 0 & 2 \end{pmatrix}\begin{bmatrix} x \\ y \end{bmatrix} = 2\begin{bmatrix} x \\ y \end{bmatrix}$$

$$\Rightarrow \begin{bmatrix} x + 2y \\ 2y \end{bmatrix} = \begin{bmatrix} 2x \\ 2y \end{bmatrix}$$

$\Rightarrow x + 2y = 2x, \ 2y = 2y$

$\Rightarrow 2y = x$

$\Rightarrow y = \dfrac{x}{2}$

$$\therefore X = \begin{bmatrix} x \\ y \end{bmatrix} = \begin{bmatrix} x \\ \frac{x}{2} \end{bmatrix} = \begin{bmatrix} 1 \\ \frac{1}{2} \end{bmatrix} \text{ (for } x = 1)$$

Comparing $\begin{bmatrix} 1 \\ \frac{1}{2} \end{bmatrix}$ with $\begin{bmatrix} 1 \\ b \end{bmatrix}$ we get $b = \dfrac{1}{2}$.

Hence, a + b = 1/2.

80. (d) The characteristic equation is

$p^3 + p^2 + 2p + 1 = 0$

By the Cayley Hamilton theorem,

$P^3 + P^2 + 2P + I = O$

$\Rightarrow P^{-1}(P^3 + P^2 + 2P + I) = P^{-1}O$

$\Rightarrow P^2 + P + 2I + P^{-1} = O$

$\Rightarrow P^{-1} = -(P^2 + P + 2I)$

81. (c) Let λ_1 and λ_2 be the two eigenvalues. Then, the sum of eigenvalues = trace of the matrix

$\Rightarrow \lambda_1 + \lambda_2 = -2$

$\Rightarrow \lambda_2 = -2 - \lambda_1 \qquad \qquad \ldots(1)$

The product of the eigenvalues

= determinant of the matrix

$\Rightarrow \lambda_1\lambda_2 = -35$

$\Rightarrow \lambda_1(-2 - \lambda_1) = -35$ [using (1)]

$\Rightarrow -2\lambda_1 + \lambda_1^2 + 35 = 0$

$\Rightarrow (\lambda_1 + 7)(\lambda_1 - 5) = 0$

$\Rightarrow \lambda_1 = -7, 5.$

82. (d) The trace of the matrix $= (-1) + (-1) + 3$ $= 1 = $ sum of eigenvalues $= 3 + (-1 + 3i) + (-1 -3i)$.

83. (c) If λ be an eigenvalue of X, then $(\lambda + 1)^{-1}(\lambda + 5)$ will be an eigenvalue of $(X + I)^{-1}(X + 5I)$

Now

$\lambda = -2 \Rightarrow (\lambda + 1)^{-1}(\lambda + 5) = -3$

$\lambda = -3 \Rightarrow (\lambda + 1)^{-1}(\lambda + 5) = -1$

84. (a) $M^T = M^{-1} \Rightarrow M$ is orthogonal

$\Rightarrow \det(M) = \pm 1$

$$\Rightarrow \begin{vmatrix} \frac{3}{5} & \frac{4}{5} \\ x & \frac{3}{5} \end{vmatrix} = \pm 1$$

$\Rightarrow \dfrac{9}{25} - \dfrac{4x}{5} = \pm 1$

$\Rightarrow 9 - 20x = \pm 25$

$\Rightarrow 20x = 9 \mp 25 = -16, 34$

$\Rightarrow x = -\dfrac{4}{5}, \dfrac{17}{10}$

85. (a) By the definition of the skew-symmetric matrix.

86. (b)

$$\Delta = \begin{vmatrix} 1 & 2 & -2 \\ 2 & 1 & 1 \\ -1 & 1 & -1 \end{vmatrix} = -6,$$

$$\Delta_3 = \begin{vmatrix} 1 & 2 & 4 \\ 2 & 1 & -2 \\ -1 & 1 & 2 \end{vmatrix} = 12$$

Therefore by Cramer's rule,

$$x_3 = \frac{\Delta_3}{\Delta} = \frac{12}{-6} = -2.$$

87. (c) The homogeneous system of equations

have a non-trivial solution

⇒ co-efficient determinant = 0

$$\Rightarrow \begin{vmatrix} 2 & 3 \\ 6 & q \end{vmatrix} = 0$$

$$\Rightarrow 2q - 18 = 0$$

$$\Rightarrow q = 9.$$

88. (b)

Let $A = \begin{pmatrix} 3 + 2i & i \\ -i & 3 - 2i \end{pmatrix}$

Then $|A| = (3 + 2i)(3 - 2i) - (-i)i$

$$= 3^2 + 2^2 + i^2 = 9 + 4 - 1 = 12$$

$$adj(A) = \begin{pmatrix} 3 - 2i & i \\ -i & 3 + 2i \end{pmatrix}^T$$

$$= \begin{pmatrix} 3 - 2i & -i \\ i & 3 + 2i \end{pmatrix}$$

$$\therefore A^{-1} = \frac{1}{|A|} adj(A) = \frac{1}{12} \begin{pmatrix} 3 - 2i & -i \\ i & 3 + 2i \end{pmatrix}.$$

89. (d)

$$[A:B]$$

$$= \begin{pmatrix} 1 & 2 & 1 & 4 & 2 \\ 3 & 6 & 3 & 12 & 6 \end{pmatrix}$$

$$\sim \begin{pmatrix} 1 & 2 & 1 & 4 & 2 \\ 0 & 0 & 0 & 0 & 0 \end{pmatrix} (by\ R_2 \rightarrow R_2 - 3R_1)$$

Which has one non-zero row.
Therefore, rank ([A:B]) = 1. Now ignoring the last column in the final equivalent matrix, we see that there are only one non-zero row and so rank (A) = 1. Thus, rank (A) = rank ([A:B]) = 1 < 4 = number of unknowns.

Hence, the system has an infinite number of solutions and also multiple non-trivial solutions exist.

90. (a)

$$A = \begin{bmatrix} a_{11} & a_{12} & \cdots\cdots & a_{1n} \\ a_{21} & a_{22} & \cdots\cdots & a_{2n} \\ a_{31} & a_{32} & \cdots\cdots & a_{3n} \\ \cdots & \cdots & \cdots & \cdots \\ a_{n1} & a_{n2} & \cdots\cdots & a_{nn} \end{bmatrix}_{n \times n}$$

$$= \begin{bmatrix} 1 & 0 & 0 & \cdots & 0 \\ 0 & 2 & 0 & \cdots & 0 \\ 0 & 0 & 3 & \cdots & 0 \\ \cdots & \cdots & \cdots & \cdots & \cdots \\ 0 & 0 & \cdots & \cdots & n \end{bmatrix}_{n \times n}$$

Thus, the sum of n eigenvalues
= trace of A
= sum of the diagonal elements
$$= 1 + 2 + 3 + \cdots\cdots + n = \frac{n(n + 1)}{2}.$$

91. (a)

$$AX = \lambda X$$

$$\Rightarrow \begin{bmatrix} 1 & -1 & 0 \\ -1 & 2 & -1 \\ 0 & -1 & 1 \end{bmatrix} \begin{bmatrix} 1 \\ 0 \\ -1 \end{bmatrix} = \lambda \begin{bmatrix} 1 \\ 0 \\ -1 \end{bmatrix}$$

$$\Rightarrow \begin{bmatrix} 1 + 0 + 0 \\ -1 + 0 + 1 \\ 0 + 0 - 1 \end{bmatrix} = \begin{bmatrix} \lambda \\ 0 \\ -\lambda \end{bmatrix}$$

$$\Rightarrow \lambda = 1$$

92. (d) trace of A = sum of eigenvalues of A

$$\Rightarrow 2 + y = 4 + 8$$

$$\Rightarrow y = 10$$

The product of eigenvalues of A = det(A)

$$\Rightarrow 4 \times 8 = \begin{vmatrix} 2 & 3 \\ x & y \end{vmatrix} = \begin{vmatrix} 2 & 3 \\ x & 10 \end{vmatrix}$$

$$\Rightarrow 32 = 20 - 3x$$

$$\Rightarrow x = -4$$

93. (a)

$$|A - \lambda I| = 0$$

$$\Rightarrow \begin{vmatrix} 2-\lambda & 2 \\ 1 & 3-\lambda \end{vmatrix} = 0$$

$$\Rightarrow (2-\lambda)(3-\lambda) - 2 = 0$$

$$\Rightarrow \lambda^2 - 5\lambda + 4 = 0$$

$$\Rightarrow \lambda = 1, 4$$

Then $AX = \lambda X$

$$\Rightarrow \begin{bmatrix} 2 & 2 \\ 1 & 3 \end{bmatrix}\begin{bmatrix} x \\ y \end{bmatrix} = 1\begin{bmatrix} x \\ y \end{bmatrix} \text{ (for } \lambda = 1)$$

$$\Rightarrow \begin{bmatrix} 2x+2y \\ x+3y \end{bmatrix} = \begin{bmatrix} x \\ y \end{bmatrix}$$

$$\Rightarrow 2x+2y = x, \ x+3y = y$$

$$\Rightarrow x = -2y$$

$$\therefore X = \begin{bmatrix} x \\ y \end{bmatrix} = \begin{bmatrix} -2y \\ y \end{bmatrix} = \begin{bmatrix} 2 \\ -1 \end{bmatrix} \text{ (for } y = -1)$$

94. (b) Since P is upper triangular, so eigenvalues of P are 1, 2, 3 (since 1, 2, and 3 are the diagonal elements of the upper triangular matrix P).

Let us consider $\lambda = 3$. Then,

$$PX = \lambda X$$

$$\Rightarrow \begin{bmatrix} 1 & 1 & 0 \\ 0 & 2 & 2 \\ 0 & 0 & 3 \end{bmatrix}\begin{bmatrix} x \\ y \\ z \end{bmatrix} = 3\begin{bmatrix} x \\ y \\ z \end{bmatrix}$$

$$\Rightarrow \begin{bmatrix} x+y \\ 2y+2z \\ 3z \end{bmatrix} = \begin{bmatrix} 3x \\ 3y \\ 3z \end{bmatrix}$$

$$\Rightarrow x+y = 3x, \ 2y+2z = 3y, \ 3z = 3z$$

$$\Rightarrow y = 2x, \ y = 2z$$

$$\Rightarrow x = z, \ y = 2z$$

$$\therefore X = \begin{bmatrix} x \\ y \\ z \end{bmatrix} = \begin{bmatrix} z \\ 2z \\ z \end{bmatrix} = z\begin{bmatrix} 1 \\ 2 \\ 1 \end{bmatrix} = \begin{bmatrix} 1 \\ 2 \\ 1 \end{bmatrix}$$

$$= \begin{bmatrix} 1 & 2 & 1 \end{bmatrix}^T \text{ (for } z = 1)$$

95. (b) If λ and X are, respectively, the eigenvalue and the eigenvector (corresponding to λ) of the matrix M, then λ^K and X will be the eigenvalue and the eigenvector (corresponding to λ^K) of the matrix M^K.

96. (a) The given matrix is an upper triangular, so its eigenvalues are the diagonal elements, *i.e.*, 1, 4, 3.

97. (b)

$$\Delta = 0$$

$$\Rightarrow \begin{vmatrix} 1 & 1 & 1 \\ 1 & 4 & 6 \\ 1 & 4 & \lambda \end{vmatrix} = 0$$

$$\Rightarrow 1(4\lambda - 24) - 1(\lambda - 6) + 6(4 - 4) = 0$$

$$\Rightarrow \lambda = 6.$$

$$\Delta_3 \neq 0$$

$$\Rightarrow \begin{vmatrix} 1 & 1 & 6 \\ 1 & 4 & 20 \\ 1 & 4 & \mu \end{vmatrix} \neq 0$$

$$\Rightarrow 1(4\mu - 80) - 1(\mu - 20) + 6(4 - 4) \neq 0$$

$$\Rightarrow \mu \neq 20.$$

$\because \Delta = 0, \Delta_3 \neq 0 \Leftrightarrow$ the system has no solution, so we must have $\lambda = 6, \mu \neq 20$.

Alternative Method:

$[A : B]$

$$= \begin{bmatrix} 1 & 1 & 1 & 6 \\ 1 & 4 & 6 & 20 \\ 1 & 4 & \lambda & \mu \end{bmatrix}$$

$$\sim \begin{bmatrix} 1 & 1 & 1 & 6 \\ 0 & 3 & 5 & 14 \\ 0 & 3 & \lambda-1 & \mu-6 \end{bmatrix}$$

(by $R_2 \rightarrow R_2 - R_1; R_3 \rightarrow R_3 - R_1$)

$$\sim \begin{bmatrix} 1 & 1 & 1 & 6 \\ 0 & 3 & 5 & 14 \\ 0 & 0 & \lambda-6 & \mu-20 \end{bmatrix}$$

(by $R_3 \rightarrow R_3 - R_2$)

Hence, if $\lambda = 6$ and $\mu \neq 20$, then rank $([A:B])$ $= 3 \neq 2 = \text{rank}(A)$. This means the system has no solution for $\lambda = 6$ and $\mu \neq 20$.

98. (d) We know that for any square matrix A, $A + A^T$ is symmetric and $A - A^T$ is skew-symmetric.

99. (b)

Let $[A] = \begin{bmatrix} 2 & 1 \\ 4 & -1 \end{bmatrix} = [L][U] = \begin{bmatrix} a & 0 \\ b & c \end{bmatrix}\begin{bmatrix} d & g \\ 0 & f \end{bmatrix}$

Then $\begin{bmatrix} 2 & 1 \\ 4 & -1 \end{bmatrix} = \begin{bmatrix} ad & ag \\ bd & bg+cf \end{bmatrix}$

Equating the corresponding elements we get

$ad = 2$ ...(i)

$ag = 1$ ...(ii)

$bd = 4$ ...(iii)

$bg + cf = -1$ ...(iv)

All these equations are satisfied if

$[L] = \begin{bmatrix} a & 0 \\ b & c \end{bmatrix} = \begin{bmatrix} 1 & 0 \\ 2 & 1 \end{bmatrix}$ and

$[U] = \begin{bmatrix} d & g \\ 0 & f \end{bmatrix} = \begin{bmatrix} 2 & 1 \\ 0 & -3 \end{bmatrix}$

100. (b)

$A^2 = \begin{pmatrix} -5 & -3 \\ 2 & 0 \end{pmatrix} \times \begin{pmatrix} -5 & -3 \\ 2 & 0 \end{pmatrix}$

$\quad = \begin{pmatrix} 25-6 & 15-0 \\ -10+0 & -6+0 \end{pmatrix} = \begin{pmatrix} 19 & 15 \\ -10 & -6 \end{pmatrix}$

$\therefore A^3 = A^2 \times A$

$\quad = \begin{pmatrix} 19 & 15 \\ -10 & -6 \end{pmatrix} \times \begin{pmatrix} -5 & -3 \\ 2 & 0 \end{pmatrix}$

$\quad = \begin{pmatrix} -95+30 & -57+0 \\ 50-12 & 30+0 \end{pmatrix}$

$\quad = \begin{pmatrix} 19\times(-5)+30\times1 & 19\times(-3)+30\times0 \\ 19\times2+30\times0 & 19\times0+30\times1 \end{pmatrix}$

$\quad = 19A + 30I.$

101. (c)

$|A| = \begin{vmatrix} 2 & 1 & 1 \\ 0 & 1 & -1 \\ 1 & 1 & 0 \end{vmatrix}$

$\quad = 2(0+1) - 1(0+1) + 1(0-1) = 0.$

So rank (A) < 3 = number of variables.
Therefore, the system has an infinite number of solutions.

102. (c)

Here $\Delta = \begin{vmatrix} 1 & 2 & 1 \\ 2 & 1 & 2 \\ 1 & -1 & 1 \end{vmatrix} = 0,$

$\Delta_1 = \begin{vmatrix} 4 & 2 & 1 \\ 5 & 1 & 2 \\ 1 & -1 & 1 \end{vmatrix} = 0,$

$\Delta_2 = \begin{vmatrix} 1 & 4 & 1 \\ 2 & 5 & 2 \\ 1 & 1 & 1 \end{vmatrix} = 0,$

$\Delta_3 = \begin{vmatrix} 1 & 2 & 4 \\ 2 & 1 & 5 \\ 1 & -1 & 1 \end{vmatrix} = 0.$

$\Delta = \Delta_1 = \Delta_2 = \Delta_3 = 0$, so the system has an infinite number of solutions (by Cramer's rule).

Alternative Method:

[A : B]

$= \begin{bmatrix} 1 & 2 & 1 & 4 \\ 2 & 1 & 2 & 5 \\ 1 & -1 & 1 & 1 \end{bmatrix}$

$\sim \begin{bmatrix} 1 & 1 & 1 & 6 \\ 0 & -3 & 0 & -3 \\ 0 & -3 & 0 & -3 \end{bmatrix}$

(by $R_2 \rightarrow R_2 - 2R_1; R_3 \rightarrow R_3 - R_1$)

$\sim \begin{bmatrix} 1 & 1 & 1 & 6 \\ 0 & -3 & 0 & -3 \\ 0 & 0 & 0 & 0 \end{bmatrix}$

(by $R_3 \rightarrow R_3 - R_2$)

Hence, rank ([A:B]) = 2 = rank(A) < 3 (=number of variables). This means the system has an infinite number of solutions.

103. (b) The sum of eigenvalues = trace of the matrix = 9 + 8 = 17, which is satisfied for the eigenvalues 3.48 and 13.52 only.

104. (d) Here $A = \begin{pmatrix} 1 & 1 \\ 1 & -1 \end{pmatrix}$. Then,

$|A - \lambda I| = 0$

$\Rightarrow \begin{vmatrix} 1-\lambda & 1 \\ 1 & -1-\lambda \end{vmatrix} = 0$

$\Rightarrow -(1-\lambda)(1+\lambda) - 1 = 0$

$\Rightarrow \lambda^2 - 1 - 1 = 0$

$\Rightarrow \lambda = \pm\sqrt{2}$

A has an eigenvalue $\pi \Rightarrow A^{19}$ has an eigenvalue λ^{19}. Hence, the eigenvalues of A^{19} are $\left(\pm\sqrt{2}\right)^{19}$, i.e; $512\sqrt{2}$, $-512\sqrt{2}$.

105. (a), (d)

$|A - \lambda I| = 0$

$\Rightarrow \begin{vmatrix} 0-\lambda & -1 \\ 1 & 0-\lambda \end{vmatrix} = 0$

$\Rightarrow \lambda^2 + 1 = 0$

$\Rightarrow \lambda^2 = -1 = j^2$

$\Rightarrow \lambda = \pm j \quad (j = \sqrt{-1})$

For $\lambda = j$

$AX = \lambda X \Rightarrow \begin{bmatrix} 0 & -1 \\ 1 & 0 \end{bmatrix}\begin{bmatrix} x \\ y \end{bmatrix} = j\begin{bmatrix} x \\ y \end{bmatrix}$

$\Rightarrow \begin{bmatrix} -y \\ x \end{bmatrix} = \begin{bmatrix} jx \\ jy \end{bmatrix}$

$\Rightarrow -y = jx, \quad x = jy \dots\dots(*)$

$\Rightarrow -jy = j^2x, \quad x = jy$

$\Rightarrow x = jy \ (\text{since } j^2 = -1)$

Thus, $X = \begin{bmatrix} x \\ y \end{bmatrix} = \begin{bmatrix} jy \\ y \end{bmatrix} = \begin{bmatrix} j \\ 1 \end{bmatrix}$ (for $y = 1$)

Again, from $(*)$ we can write,

$y = -jx, x = jy$

$\Rightarrow y = -j\,x, xj = j^2 y$

$\Rightarrow y = -jx \ (\text{since } j^2 = -1)$

Thus, $X = \begin{bmatrix} x \\ y \end{bmatrix} = \begin{bmatrix} x \\ -jx \end{bmatrix} = \begin{bmatrix} 1 \\ -j \end{bmatrix}$ (for $x = 1$)

For $\lambda = -j$

$AX = \lambda X \Rightarrow \begin{bmatrix} 0 & -1 \\ 1 & 0 \end{bmatrix}\begin{bmatrix} x \\ y \end{bmatrix} = -j\begin{bmatrix} x \\ y \end{bmatrix}$

$\Rightarrow \begin{bmatrix} -y \\ x \end{bmatrix} = \begin{bmatrix} -jx \\ -jy \end{bmatrix}$

$\Rightarrow -y = -jx, \quad x = -jy$

$\Rightarrow y = jx, \quad x = -jy \ \ \dots(**)$

$\Rightarrow y = jx, jx = -j^2 y = y \ (\text{since } j^2 = -1)$

$\Rightarrow y = jx$

Thus, $X = \begin{bmatrix} x \\ y \end{bmatrix} = \begin{bmatrix} x \\ jx \end{bmatrix} = \begin{bmatrix} 1 \\ j \end{bmatrix}$ (for $x = 1$)

Again, $(**) \Rightarrow x = -jy$,

Thus, $X = \begin{bmatrix} x \\ y \end{bmatrix} = \begin{bmatrix} -jy \\ y \end{bmatrix} = \begin{bmatrix} j \\ -1 \end{bmatrix}$ (for $y = -1$)

Therefore, two possible pair of eigenvectors

are $\begin{bmatrix} 1 \\ -j \end{bmatrix}$ and $\begin{bmatrix} j \\ -1 \end{bmatrix}$; $\begin{bmatrix} 1 \\ j \end{bmatrix}$ and $\begin{bmatrix} j \\ 1 \end{bmatrix}$.

106. (d)

Let $A = \begin{bmatrix} x & y \\ z & t \end{bmatrix}$

For $\lambda = -1$,

$AX = \lambda X$

$\Rightarrow \begin{bmatrix} x & y \\ z & t \end{bmatrix}\begin{bmatrix} 1 \\ -1 \end{bmatrix} = -1\begin{bmatrix} 1 \\ -1 \end{bmatrix}$

$\Rightarrow \begin{bmatrix} x-y \\ z-t \end{bmatrix} = \begin{bmatrix} -1 \\ 1 \end{bmatrix}$

$\Rightarrow \begin{cases} x - y = -1 \dots\dots(1) \\ z - t = 1 \dots\dots\dots(2) \end{cases}$

For $\lambda = -2$,

$AX = \lambda X \Rightarrow \begin{bmatrix} x & y \\ z & t \end{bmatrix}\begin{bmatrix} 1 \\ -2 \end{bmatrix} = -2\begin{bmatrix} 1 \\ -2 \end{bmatrix}$

$\Rightarrow \begin{bmatrix} x-2y \\ z-2t \end{bmatrix} = \begin{bmatrix} -2 \\ 4 \end{bmatrix}$

$\Rightarrow \begin{cases} x - 2y = -2 \dots\dots(3) \\ z - 2t = 4 \dots\dots\dots(4) \end{cases}$

Solving (1) and (3) we get, $x = 0$, $y = 1$ and solving (2) and (4) we get, $z = -2$, $t = -3$

Therefore, $A = \begin{bmatrix} 0 & 1 \\ -2 & -3 \end{bmatrix}$

107. (a)

$|A - \lambda I| = 0$

$\Rightarrow \begin{vmatrix} 3-\lambda & 5 & 2 \\ 5 & 12-\lambda & 7 \\ 2 & 7 & 5-\lambda \end{vmatrix} = 0$

$\Rightarrow \begin{vmatrix} -\lambda & \lambda & -\lambda \\ 5 & 12-\lambda & 7 \\ 2 & 7 & 5-\lambda \end{vmatrix} = 0$

$[by \ R_1 \rightarrow R_1 + (R_3 - R_2)]$

$\Rightarrow (-\lambda)\begin{vmatrix} 1 & -1 & 1 \\ 5 & 12-\lambda & 7 \\ 2 & 7 & 5-\lambda \end{vmatrix} = 0$

$$\Rightarrow \lambda \begin{vmatrix} 1 & 0 & 0 \\ 5 & 17-\lambda & 2 \\ 2 & 9 & 3-\lambda \end{vmatrix} = 0$$

$$[by \ C_2 \to C_2 + C_1, C_3 \to C_3 - C_1]$$

$$\Rightarrow \lambda\{(17-\lambda)(3-\lambda)-18\} = 0$$

$$\Rightarrow \lambda\{\lambda^2 - 20\lambda + 33\} = 0$$

$$\Rightarrow \lambda = 0, \frac{20 \pm \sqrt{400 - 4\times 33}}{2}$$

$$\Rightarrow \lambda = 0, \frac{20 \pm \sqrt{268}}{2}$$

Therefore, the minimum eigenvalue $= 0$.

108. (c)

$$\cos 2x = \cos^2 x - \sin^2 x$$

$$\Rightarrow \cos 2x + (-1)\cos^2 x + \sin^2 x = 0$$

$$\Rightarrow 1 \times \cos 2x + (-1)\cos^2 x + 1 \times \sin^2 x = 0$$

Therefore, the functions $\cos 2x$, $\sin^2 x$, and $\cos^2 x$ are linearly dependent.

109. (d) The given system represents a homogeneous system and the coefficient determinant

$$= \begin{vmatrix} 2 & -2 \\ 1 & -1 \end{vmatrix} = -2 + 2 = 0.$$

Therefore, the system has an infinite number of solutions (multiple solutions).

110. 16

We know that number of multiplication required to multiply $A_{m\times n}$ with $B_{n\times k}$ is mnk.

Now $PQR = (PQ)R = (P_{4\times 2} \ Q_{2\times 4}) \ R_{4\times 1} = (PQ)_{4\times 4} R_{4\times 1}$

So the number of multiplication required to compute $(PQ)_{4\times 4} R_{4\times 1}$ is $(4\times 2\times 4) + (4\times 4\times 1)$ i.e., 48

Again $PQR = P(QR) = P_{4\times 2}(Q_{2\times 4}R_{4\times 1}) = P_{4\times 2}(QR)_{2\times 1}$

So the number of multiplication required to compute $P_{4\times 2}(QR)_{2\times 1}$ is $(2\times 4\times 1) + (4\times 2\times 1)$, i.e., 16

Hence, minimum number of multiplication required to compute PQR is 16.

111. (a)

$$\begin{vmatrix} 1 & x(x+1) & x+1 \\ 1 & y(y+1) & y+1 \\ 1 & z(z+1) & z+1 \end{vmatrix}$$

$$= \begin{vmatrix} 1 & x(x+1) & x \\ 1 & y(y+1) & y \\ 1 & z(z+1) & z \end{vmatrix} \ (by \ C_3 \to C_3 - C_1)$$

$$= \begin{vmatrix} 1 & x^2 & x \\ 1 & y^2 & y \\ 1 & z^2 & z \end{vmatrix} \ (by \ C_2 \to C_2 - C_3)$$

$$= -\begin{vmatrix} 1 & x & x^2 \\ 1 & y & y^2 \\ 1 & z & z^2 \end{vmatrix} \ (by \ C_2 \Leftrightarrow C_3)$$

Therefore, (a) is correct.

$$\begin{vmatrix} 1 & x+1 & x^2+1 \\ 1 & y+1 & y^2+1 \\ 1 & z+1 & z^2+1 \end{vmatrix} = \begin{vmatrix} 1 & x & x^2 \\ 1 & y & y^2 \\ 1 & z & z^2 \end{vmatrix}$$

$$(by \ C_3 \to C_3 - C_1, C_2 \to C_2 - C_1)$$

$$\begin{vmatrix} 1 & x & x^2 \\ 1 & y & y^2 \\ 1 & z & z^2 \end{vmatrix} = \begin{vmatrix} 0 & x-y & x^2-y^2 \\ 1 & y & y^2 \\ 1 & z & z^2 \end{vmatrix} \ \begin{pmatrix} by \ R_1 \to \\ R_1 - R_2 \end{pmatrix}$$

$$= \begin{vmatrix} 0 & x-y & x^2-y^2 \\ 0 & y-z & y^2-z^2 \\ 1 & z & z^2 \end{vmatrix} \ (by \ R_2 \to R_2 - R_3)$$

$\therefore$ (c) is incorrect.

$$\begin{vmatrix} 1 & x & x^2 \\ 1 & y & y^2 \\ 1 & z & z^2 \end{vmatrix} = \begin{vmatrix} 2 & x+y & x^2+y^2 \\ 1 & y & y^2 \\ 1 & z & z^2 \end{vmatrix} \ \begin{pmatrix} by \ R_1 \to \\ R_1 + R_2 \end{pmatrix}$$

$$= \begin{vmatrix} 2 & x+y & x^2+y^2 \\ 2 & y+z & y^2+z^2 \\ 1 & z & z^2 \end{vmatrix} \ (by \ R_2 \to R_2 + R_3)$$

$\therefore$ (d) is incorrect.

112. (b)

$$\begin{vmatrix} 2 & 1 & 1 & 1 \\ 1 & 2 & 1 & 1 \\ 1 & 1 & 2 & 1 \\ 1 & 1 & 1 & 2 \end{vmatrix}$$

$$= \begin{vmatrix} 0 & -1 & -1 & -3 \\ 0 & 1 & 0 & -1 \\ 0 & 0 & 1 & -1 \\ 1 & 1 & 1 & 2 \end{vmatrix}$$

$[by\ R_1 \to R_1 - 2R_4, R_2 \to R_2 - R_4,$
$R_3 \to R_3 - R_4]$

$= 1 \times$ cofactor of "1"

$$= (-1)^{4+1} \begin{vmatrix} -1 & -1 & -3 \\ 1 & 0 & -1 \\ 0 & 1 & -1 \end{vmatrix}$$

$= -\{-1(0+1) + 1(-1+0) - 3(1-0)\} = 5.$

113. 1.

$P = I_6 + \alpha J_6$

$$= \begin{bmatrix} 1 & 0 & 0 & 0 & 0 & 0 \\ 0 & 1 & 0 & 0 & 0 & 0 \\ 0 & 0 & 1 & 0 & 0 & 0 \\ 0 & 0 & 0 & 1 & 0 & 0 \\ 0 & 0 & 0 & 0 & 1 & 0 \\ 0 & 0 & 0 & 0 & 0 & 1 \end{bmatrix} + \alpha \begin{bmatrix} 0 & 0 & 0 & 0 & 0 & 1 \\ 0 & 0 & 0 & 0 & 1 & 0 \\ 0 & 0 & 0 & 1 & 0 & 0 \\ 0 & 0 & 1 & 0 & 0 & 0 \\ 0 & 1 & 0 & 0 & 0 & 0 \\ 1 & 0 & 0 & 0 & 0 & 0 \end{bmatrix}$$

$$= \begin{bmatrix} 1 & 0 & 0 & 0 & 0 & \alpha \\ 0 & 1 & 0 & 0 & \alpha & 0 \\ 0 & 0 & 1 & \alpha & 0 & 0 \\ 0 & 0 & \alpha & 1 & 0 & 0 \\ 0 & \alpha & 0 & 0 & 1 & 0 \\ \alpha & 0 & 0 & 0 & 0 & 1 \end{bmatrix}$$

$$\therefore \det(P) = \begin{vmatrix} 1 & 0 & 0 & 0 & 0 & \alpha \\ 0 & 1 & 0 & 0 & \alpha & 0 \\ 0 & 0 & 1 & \alpha & 0 & 0 \\ 0 & 0 & \alpha & 1 & 0 & 0 \\ 0 & \alpha & 0 & 0 & 1 & 0 \\ \alpha & 0 & 0 & 0 & 0 & 1 \end{vmatrix}$$

$$= \begin{vmatrix} 1-\alpha^2 & 0 & 0 & 0 & 0 & 0 \\ 0 & 1 & 0 & 0 & \alpha & 0 \\ 0 & 0 & 1 & \alpha & 0 & 0 \\ 0 & 0 & \alpha & 1 & 0 & 0 \\ 0 & \alpha & 0 & 0 & 1 & 0 \\ \alpha & 0 & 0 & 0 & 0 & 1 \end{vmatrix}$$

$(by\ R_1 \to R_1 - \alpha R_6)$

$= (1-\alpha^2) \times$ cofactor of $(1-\alpha^2)$

$$= (1-\alpha^2) \begin{vmatrix} 1 & 0 & 0 & \alpha & 0 \\ 0 & 1 & \alpha & 0 & 0 \\ 0 & \alpha & 1 & 0 & 0 \\ \alpha & 0 & 0 & 1 & 0 \\ 0 & 0 & 0 & 0 & 1 \end{vmatrix}$$

$$= (1-\alpha^2) \begin{vmatrix} 1-\alpha^2 & 0 & 0 & \alpha & 0 \\ 0 & 1 & \alpha & 0 & 0 \\ 0 & \alpha & 1 & 0 & 0 \\ 0 & 0 & 0 & 1 & 0 \\ 0 & 0 & 0 & 0 & 1 \end{vmatrix}$$

$(by\ C_1 \to C_1 - \alpha C_4)$

$= (1-\alpha^2) \times \{(1-\alpha^2) \times$ cofactor of $(1-\alpha^2)\}$

$$= (1-\alpha^2)^2 \begin{vmatrix} 1 & \alpha & 0 & 0 \\ \alpha & 1 & 0 & 0 \\ 0 & 0 & 1 & 0 \\ 0 & 0 & 0 & 1 \end{vmatrix}$$

$$= (1-\alpha^2)^2 \begin{vmatrix} 1-\alpha^2 & 0 & 0 & 0 \\ \alpha & 1 & 0 & 0 \\ 0 & 0 & 1 & 0 \\ 0 & 0 & 0 & 1 \end{vmatrix}$$

$(by\ R_1 \to R_1 - \alpha R_2)$

$$= (1-\alpha^2)^3 \begin{vmatrix} 1 & 0 & 0 \\ 0 & 1 & 0 \\ 0 & 0 & 1 \end{vmatrix}$$

(by taking the element $(1-\alpha^2)$ and its cofactor)

$= (1-\alpha^2)^3.$

Now, $\det(P) = 0$ gives $(1-\alpha^2)^3 = 0$, which implies $\alpha = -1, 1.$

114. (d) Since matrix product is not commutative, so options (a) and (b) are not correct.

Let

$P = \begin{bmatrix} 2 & 3 \\ 1 & 1 \end{bmatrix}, Q = \begin{bmatrix} -2 & 0 \\ 6 & 4 \end{bmatrix}.$ So $P + Q = \begin{pmatrix} 0 & 3 \\ 7 & 5 \end{pmatrix}.$

Then $\det(P) = -1$, $\det(Q) = -8$ and so $\det(P) + \det(Q) = -9$. But $\det(P + Q) = -21.$ Hence, option (c) is not correct.

115. 0.

$$A = \begin{bmatrix} 2 \\ -4 \\ 7 \end{bmatrix}_{3 \times 1} \begin{bmatrix} 1 & 9 & 5 \end{bmatrix}_{1 \times 3} = \begin{bmatrix} 2 & 18 & 10 \\ -4 & -36 & -20 \\ 7 & 63 & 35 \end{bmatrix}$$

$$\therefore \det(A) = \begin{vmatrix} 2 & 18 & 10 \\ -4 & -36 & -20 \\ 7 & 63 & 35 \end{vmatrix}$$

$$= \begin{vmatrix} 2 & 18 & 10 \\ 0 & 0 & 0 \\ 7 & 63 & 35 \end{vmatrix} \quad [\text{by } R_2 \to R_2 + 2R_1]$$

$$= 0.$$

116. 2.

$$\begin{bmatrix} 6 & 0 & 4 & 4 \\ -2 & 14 & 8 & 18 \\ 14 & -14 & 0 & -10 \end{bmatrix} \sim \begin{bmatrix} 6 & 0 & 4 & 4 \\ -2 & 14 & 8 & 18 \\ 0 & 0 & 0 & 0 \end{bmatrix}$$
$$[\text{by } R_3 \to R_3 - (2R_1 - R_2)]$$

which has two non-zero rows
Therefore, rank = 2.

117. (c)

$$|A| = \begin{vmatrix} p & q \\ r & s \end{vmatrix} = ps - qr$$

$$|B|$$

$$= \begin{vmatrix} p^2 + q^2 & pr + qs \\ pr + qs & r^2 + s^2 \end{vmatrix}$$

$$= (p^2 + q^2)(r^2 + s^2) - (pr + qs)^2$$

$$= p^2 r^2 + p^2 s^2 + q^2 r^2 + q^2 s^2 - p^2 r^2$$

$$\quad - q^2 s^2 - 2pqrs$$

$$= p^2 s^2 - 2ps \times qr + q^2 r^2$$

$$= (ps - qs)^2 = |A|^2$$

Thus, the matrices A and B become singular or non-singular together. So, rank(A) = Rank(B) = N.

118. (a) The area of the triangle with vertices (x_1, y_1), (x_2, y_2), and (x_3, y_3) is given by

$$\Delta = \frac{1}{2} \begin{vmatrix} x_1 & y_1 & 1 \\ x_2 & y_2 & 1 \\ x_3 & y_3 & 1 \end{vmatrix} = \frac{1}{2} \begin{vmatrix} 1 & 0 & 1 \\ 2 & 2 & 1 \\ 4 & 3 & 1 \end{vmatrix}$$

$$= \frac{1}{2} \{1(2-3) + 1(6-8)\} = -\frac{3}{2}.$$

Therefore, the magnitude of the area = 3/2

119. (a)

$$\frac{d}{dt}\begin{bmatrix} x \\ y \end{bmatrix} = \begin{bmatrix} \dfrac{dx}{dt} \\ \dfrac{dy}{dt} \end{bmatrix} = \begin{bmatrix} 3x - 5y \\ 4x + 8y \end{bmatrix} = \begin{bmatrix} 3 & -5 \\ 4 & 8 \end{bmatrix}\begin{bmatrix} x \\ y \end{bmatrix}$$

120. one

$$[A : B]$$

$$= \begin{pmatrix} 3 & 2 & 0 & 1 \\ 4 & 0 & 7 & 1 \\ 1 & 1 & 1 & 3 \\ 1 & -2 & 7 & 0 \end{pmatrix}$$

$$\sim \begin{pmatrix} 1 & 1 & 1 & 3 \\ 4 & 0 & 7 & 1 \\ 3 & 2 & 0 & 1 \\ 1 & -2 & 7 & 0 \end{pmatrix} \quad [\text{by } R_1 \leftrightarrow R_3]$$

$$\sim \begin{pmatrix} 1 & 1 & 1 & 3 \\ 0 & -4 & 3 & -11 \\ 0 & -1 & -3 & -8 \\ 0 & -3 & 6 & -3 \end{pmatrix}$$

$$[\text{by } R_2 \to R_2 - 4R_1, R_3 \to R_3 - 3R_1,$$
$$R_4 \to R_4 - R_1]$$

$$\sim \begin{pmatrix} 1 & 1 & 1 & 3 \\ 0 & -4 & 3 & -11 \\ 0 & -1 & -3 & -8 \\ 0 & 0 & 0 & 0 \end{pmatrix}$$

$$[\text{by } R_4 \to R_4 - (R_2 - R_3)]$$

which has three non-zero rows.
So $r([A:B]) = 3$. Now ignoring the last column in the final equivalent matrix, we see that there are only three non-zero rows. Therefore, $r(A) = 3$. Thus, $r([A:B]) = r(A) = 3$ = number of unknowns. Hence, the system has a unique solution. Thus, the number of solutions is one.

121. (a)

$$\Delta = \begin{vmatrix} 1 & 2 & 4 \\ 4 & 3 & 1 \\ 3 & 2 & 3 \end{vmatrix}$$

$$= 1(9-2) - 2(12-3) + 4(8-9)$$

$$\neq 0$$

So the system has a unique solution.

122. (b)

$$\Delta = \begin{vmatrix} 2 & 1 & 3 \\ 3 & 0 & 1 \\ 1 & 2 & 5 \end{vmatrix}$$

$$= 2(0-2) - 1(15-1) + 3(6-0) = 0,$$

$$\Delta_1 = \begin{vmatrix} 5 & 1 & 3 \\ -4 & 0 & 1 \\ 14 & 2 & 5 \end{vmatrix}$$

$$= 5(0-2) - 1(-20-14) + 3(-8-0) = 0,$$

$$\Delta_2 = \begin{vmatrix} 2 & 5 & 3 \\ 3 & -4 & 1 \\ 1 & 14 & 5 \end{vmatrix}$$

$$= 2(-20-14) - 5(15-1) + 3(42+4) = 0,$$

$$\Delta_3 = \begin{vmatrix} 2 & 1 & 5 \\ 3 & 0 & -4 \\ 1 & 2 & 14 \end{vmatrix}$$

$$= 2(0+8) - 1(42+4) + 5(6-0) = 0.$$

$\because \quad \Delta = \Delta_1 = \Delta_2 = \Delta_3 = 0$ so the system has an infinite number of solution (by Cramer's rule).

Alternative Method

[A : B]

$$= \begin{bmatrix} 2 & 1 & 3 & 5 \\ 3 & 0 & 1 & -4 \\ 1 & 2 & 5 & 14 \end{bmatrix} \sim \begin{bmatrix} 1 & 2 & 5 & 14 \\ 3 & 0 & 1 & -4 \\ 2 & 1 & 3 & 5 \end{bmatrix}$$

$$(\text{by } R_3 \leftrightarrow R_1)$$

$$\sim \begin{bmatrix} 1 & 2 & 5 & 14 \\ 0 & -6 & -14 & -46 \\ 0 & -3 & -7 & -23 \end{bmatrix}$$

$$(\text{by } R_2 \rightarrow R_2 - 3R_1; \ R_3 \rightarrow R_3 - 2R_1)$$

$$\sim \begin{bmatrix} 1 & 2 & 5 & 14 \\ 0 & -6 & -14 & -46 \\ 0 & 0 & 0 & 0 \end{bmatrix}$$

$$(\text{by } R_3 \rightarrow R_3 - \frac{1}{2}R_2)$$

Hence, rank ([A:B]) = rank(A) = 2 < 3 (=number of variables). This means the system has an infinite number of solutions.

123. (b)

$$A = \begin{pmatrix} 1 & 2 & 2 \\ 5 & 1 & 3 \end{pmatrix}$$

$$\sim \begin{pmatrix} 1 & 2 & 2 \\ 0 & -9 & -7 \end{pmatrix} (\text{by } R_2 \rightarrow R_2 - 5R_1)$$

Which has two non-zero rows.

Therefore, rank (A) = 2

$$[A:B] = \begin{pmatrix} 1 & 2 & 2 & b_1 \\ 5 & 1 & 3 & b_2 \end{pmatrix} \sim \begin{pmatrix} 1 & 2 & 2 & b_1 \\ 0 & -9 & -7 & b_2 - 5b_1 \end{pmatrix}$$

$$(\text{by } R_2 \rightarrow R_2 - 5R_1)$$

Which has two non-zero rows.

Therefore, rank ([A:B]) = 2

So rank (A) = rank ([A:B]) = 2 < 3 (= number of unknowns)

Hence, the system has infinitely many solutions for any given b_1 and b_2.

124. 200

$$|AB| = |A||B| = 5 \times 40 = 200.$$

125. (c)

$$\text{Let } B = \begin{bmatrix} 1 & 2 & 3 \\ 7 & 8 & 9 \\ 4 & 5 & 6 \end{bmatrix} \text{ and } C = \begin{bmatrix} 1 & 2 & 3 \\ 4 & 5 & 6 \\ 7 & 8 & 9 \end{bmatrix}.$$

Then, we have $AB = C$.

Then $AB = C$

$\Rightarrow |AB| = |C| \Rightarrow |A||B| = -|B|$

($\because C$ is obtained by interchanging R_1

and R_2 in B)

$\Rightarrow |B|(|A|+1) = 0$

$\Rightarrow |A| = -1$ (since $|B| \neq 0$)

Let us take the eigenvalues 1, 1 and −1. Then, the product of these eigenvalues = −1 = det(A).

126. (a)

127. (b) Let $A = \begin{pmatrix} -1 & 0 \\ 0 & 2 \end{pmatrix}$. Then A is a symmetric matrix with one negative eigenvalue "−1" and one positive eigenvalue "2."

128. (a) The sum of eigenvalues = trace of the matrix = $215 + 150 + 550 = 915$.

129. 49 Let A = $\begin{bmatrix} x & a \\ a & 14-x \end{bmatrix}$. Then A is a real symmetric matrix with trace $x + (14 - x)$ i.e; 14.

Then det(A) = $x(14 - x) - a^2$, which is maximum when $a = 0$. Let $f(x) = x(14 - x)$.

Now det(A) is maximum $\Rightarrow f(x)$ is maximum and vice versa.

$f'(x) = 0 \Rightarrow 14 - 2x = 0 \Rightarrow x = 7.$

Also $f''(x)|_{x=7} = -2 < 0.$ Thus, $x = 7$ is a point of maxima and so $f(x)$ is maximum when $x = 7.$ Consequently, $\det(A)$ is maximum when $x = 7$ and maximum value of $\det(A)$ is $7(14 - 7),$ *i.e*; 49.

130. (d)

$|A - \lambda I| = 0$

$\Rightarrow \begin{vmatrix} -5 - \lambda & 2 \\ -9 & 6 - \lambda \end{vmatrix} = 0$

$\Rightarrow \begin{vmatrix} -3 - \lambda & 2 \\ -3 - \lambda & 6 - \lambda \end{vmatrix} = 0$

$\qquad [by\ C_1 \to C_1 + C_2]$

$\Rightarrow (-3 - \lambda)\begin{vmatrix} 1 & 2 \\ 1 & 6 - \lambda \end{vmatrix} = 0$

$\Rightarrow -3 - \lambda = 0$

Therefore, $\lambda = -3$ is an eigenvalue. Then,

$AX = \lambda X$

$\Rightarrow \begin{bmatrix} -5 & 2 \\ -9 & 6 \end{bmatrix}\begin{bmatrix} x \\ y \end{bmatrix} = -3\begin{bmatrix} x \\ y \end{bmatrix}$

$\Rightarrow \begin{bmatrix} -5x + 2y \\ -9x + 6y \end{bmatrix} = \begin{bmatrix} -3x \\ -3y \end{bmatrix}$

$\Rightarrow -5x + 2y = -3x,\ -9x + 6y = -3y$

$\Rightarrow x = y$

$\therefore$ Eigenvector, $X = \begin{bmatrix} x \\ y \end{bmatrix} = \begin{bmatrix} x \\ x \end{bmatrix} = \begin{bmatrix} 1 \\ 1 \end{bmatrix}$

(putting $x = 1$)

131. (a) Let $A = \begin{bmatrix} 2 & 0 \\ 0 & -1 \end{bmatrix}.$ Then trace$(A) = 2 + (-1)$ $> 0,$ but the eigenvalues are -1 and 2 (since A is diagonal matrix). So option (b) is incorrect.

Let $B = \begin{bmatrix} -1 & 2 \\ 0 & -1 \end{bmatrix}.$ Then, $\det(B) = 1 > 0,$ but the eigenvalues are -1 and -1 (since A is upper triangular). So option (c) is incorrect.

Let $C = \begin{bmatrix} 1 & 0 \\ 0 & -2 \end{bmatrix}.$ Then trace$(C) = 1 + (-2)$ $= -1$ and $\det(C) = -2.$ Therefore, trace$(C) \times \det(C) = 2 > 0,$ but the eigenvalues are 1 and -2 (since A is diagonal matrix). So option (d) is incorrect.

132. (d) Let $A = \begin{bmatrix} 0 & 0 & 0 \\ 0 & 1 & 0 \\ 0 & 0 & 2 \end{bmatrix}.$ Then, clearly A is symmetric with two non-zero eigenvalues 1 and 2 with respective eigenvectors X and $Y.$

For $\lambda = 1$:

$AX = \lambda X$

$\Rightarrow \begin{bmatrix} 0 & 0 & 0 \\ 0 & 1 & 0 \\ 0 & 0 & 2 \end{bmatrix}\begin{bmatrix} x_1 \\ x_2 \\ x_3 \end{bmatrix} = \begin{bmatrix} x_1 \\ x_2 \\ x_3 \end{bmatrix}$

$\Rightarrow \begin{bmatrix} 0 \\ x_2 \\ 2x_3 \end{bmatrix} = \begin{bmatrix} x_1 \\ x_2 \\ x_3 \end{bmatrix}$

$\Rightarrow x_1 = 0,\ x_3 = 0$

For $\lambda = 2$:

$AY = \lambda Y$

$\Rightarrow \begin{bmatrix} 0 & 0 & 0 \\ 0 & 1 & 0 \\ 0 & 0 & 2 \end{bmatrix}\begin{bmatrix} y_1 \\ y_2 \\ y_3 \end{bmatrix} = 2\begin{bmatrix} y_1 \\ y_2 \\ y_3 \end{bmatrix}$

$\Rightarrow \begin{bmatrix} 0 \\ y_2 \\ 2y_3 \end{bmatrix} = \begin{bmatrix} 2y_1 \\ 2y_2 \\ 2y_3 \end{bmatrix}$

$\Rightarrow y_1 = 0,\ y_2 = 0.$

Hence,

$x_1 y_1 + x_2 y_2 + x_3 y_3$

$= 0 \times 0 + x_2 \times 0 + 0 \times y_3$

$= 0.$

133. 6.

$|A - \lambda I| = 0$

$\Rightarrow \begin{vmatrix} 1 - \lambda & 0 & 0 & 0 & 1 \\ 0 & 1 - \lambda & 1 & 1 & 0 \\ 0 & 1 & 1 - \lambda & 1 & 0 \\ 0 & 1 & 1 & 1 - \lambda & 0 \\ 1 & 0 & 0 & 0 & 1 - \lambda \end{vmatrix} = 0$

$\Rightarrow \begin{vmatrix} 0 & 0 & 0 & 0 & 1 \\ 0 & 1 - \lambda & 1 & 1 & 0 \\ 0 & 1 & 1 - \lambda & 1 & 0 \\ 0 & 1 & 1 & 1 - \lambda & 0 \\ 1 - (1 - \lambda)^2 & 0 & 0 & 0 & 1 - \lambda \end{vmatrix} = 0$

$\qquad [by\ C_1 \to C_1 - (1 - \lambda)C_5]$

$$\Rightarrow \begin{vmatrix} 0 & 1-\lambda & 1 & 1 \\ 0 & 1 & 1-\lambda & 1 \\ 0 & 1 & 1 & 1-\lambda \\ 1-(1-\lambda)^2 & 0 & 0 & 0 \end{vmatrix} = 0$$

(by expanding using R_1)

$$\Rightarrow \left\{ 1-(1-\lambda)^2 \right\} \begin{vmatrix} 1-\lambda & 1 & 1 \\ 1 & 1-\lambda & 1 \\ 1 & 1 & 1-\lambda \end{vmatrix} = 0$$

(by expanding using C_1)

$$\Rightarrow \lambda(2-\lambda) \begin{vmatrix} 3-\lambda & 1 & 1 \\ 3-\lambda & 1-\lambda & 1 \\ 3-\lambda & 1 & 1-\lambda \end{vmatrix} = 0$$

$$[\text{by } C_1 \to C_1 + C_2 + C_3]$$

$$\Rightarrow \lambda(2-\lambda)(3-\lambda) \begin{vmatrix} 1 & 1 & 1 \\ 1 & 1-\lambda & 1 \\ 1 & 1 & 1-\lambda \end{vmatrix} = 0$$

$$\Rightarrow \lambda(2-\lambda)(3-\lambda) \begin{vmatrix} 1 & 0 & 0 \\ 1 & -\lambda & 0 \\ 1 & 0 & -\lambda \end{vmatrix} = 0$$

$$[\text{by } C_2 \to C_2 - C_1, C_3 \to C_3 - C_1]$$

$$\Rightarrow \lambda^3(2-\lambda)(3-\lambda) = 0$$

$$\Rightarrow \lambda = 0,0,0,2,3.$$

Hence, the product of the non-zero eigenvalues $= 2 \times 3 = 6$.

134. 1.

$$A^2 = I \Rightarrow \lambda^2 = 1$$

(by the Cayley Hamilton Theorem)

$$\Rightarrow \lambda = \pm 1$$

Hence, the positive eigenvalue of A is "1."

135. 3.

$$|A - \lambda I| = 0$$

$$\Rightarrow \begin{vmatrix} 0-\lambda & 1 & -1 \\ -6 & -11-\lambda & 6 \\ -6 & -11 & 5-\lambda \end{vmatrix} = 0$$

$$\Rightarrow \begin{vmatrix} -1-\lambda & 1 & -1 \\ 0 & -11-\lambda & 6 \\ -1-\lambda & -11 & 5-\lambda \end{vmatrix} = 0$$

$$[\text{by } C_1 \to C_1 + C_3]$$

$$\Rightarrow (-1-\lambda) \begin{vmatrix} 1 & 1 & -1 \\ 0 & -11-\lambda & 6 \\ 1 & -11 & 5-\lambda \end{vmatrix} = 0$$

$$\Rightarrow (1+\lambda) \begin{vmatrix} 1 & 1 & -1 \\ 0 & -11-\lambda & 6 \\ 0 & -12 & 6-\lambda \end{vmatrix} = 0$$

$$[\text{by } R_3 \to R_3 - R_1]$$

$$\Rightarrow (\lambda+1)\left\{ (\lambda-6)(\lambda+11) + 72 \right\} = 0$$

$$\Rightarrow (\lambda+1)(\lambda+2)(\lambda+3) = 0$$

$$\Rightarrow \lambda = -1, -2, -3.$$

Hence, the required absolute ratio $= \left| \dfrac{-3}{-1} \right| = 3$.

136. (d)

$$|A - \lambda I| = 0$$

$$\Rightarrow \begin{vmatrix} 3-\lambda & -2 & 2 \\ 4 & -4-\lambda & 6 \\ 2 & -3 & 5-\lambda \end{vmatrix} = 0$$

$$\Rightarrow \begin{vmatrix} 1-\lambda & -1+\lambda & 1-\lambda \\ 4 & -4-\lambda & 6 \\ 2 & -3 & 5-\lambda \end{vmatrix} = 0$$

$$(\text{by } R_1 \to R_1 + (R_3 - R_2))$$

$$\Rightarrow (1-\lambda) \begin{vmatrix} 1 & -1 & 1 \\ 4 & -4-\lambda & 6 \\ 2 & -3 & 5-\lambda \end{vmatrix} = 0$$

$$\Rightarrow (1-\lambda) \begin{vmatrix} 1 & 0 & 0 \\ 4 & -\lambda & 2 \\ 2 & -1 & 3-\lambda \end{vmatrix} = 0$$

$$(\text{by } C_2 \to C_2 + C_1, C_3 \to C_3 - C_1)$$

$$\Rightarrow (1-\lambda)\{\lambda(\lambda-3) + 2\} = 0$$

$$\Rightarrow \lambda = 1,1,2.$$

137. 17

$$AX = \lambda X$$

$$\Rightarrow \begin{bmatrix} 4 & 1 & 2 \\ p & 2 & 1 \\ 14 & -4 & 10 \end{bmatrix} \begin{bmatrix} 1 \\ 2 \\ 3 \end{bmatrix} = \lambda \begin{bmatrix} 1 \\ 2 \\ 3 \end{bmatrix}$$

$$\Rightarrow \begin{bmatrix} 12 \\ p+7 \\ 36 \end{bmatrix} = \begin{bmatrix} \lambda \\ 2\lambda \\ 3\lambda \end{bmatrix}$$

$$\Rightarrow \lambda = 12, \quad p+7 = 2\lambda$$
$$\Rightarrow \lambda = 12, \quad p = 17$$

138. (b)

$AX = \lambda X$

$$\Rightarrow \begin{bmatrix} 1 & -1 & 2 \\ 0 & 1 & 0 \\ 1 & 2 & 1 \end{bmatrix} \begin{bmatrix} x \\ y \\ z \end{bmatrix} = 1 \begin{bmatrix} x \\ y \\ z \end{bmatrix}$$

$$\Rightarrow \begin{bmatrix} x-y+2z \\ y \\ x+2y+z \end{bmatrix} = \begin{bmatrix} x \\ y \\ z \end{bmatrix}$$

$$\Rightarrow x - y + 2z = x, \quad y = y, \quad x + 2y + z = z$$

$$\Rightarrow -y + 2z = 0, \quad x + 2y = 0$$

$$\Rightarrow x = -2y, \quad z = \frac{y}{2}$$

Therefore, eigenvector,

$$X = \begin{bmatrix} x \\ y \\ z \end{bmatrix} = \begin{bmatrix} -2y \\ y \\ y/2 \end{bmatrix} = \alpha \begin{bmatrix} -4 \\ 2 \\ 1 \end{bmatrix}$$

(replacing y by 2α and then taking α as common)

$= \alpha(-4, 2, 1)$

139. (d)

Let λ_1 and λ_2 be two eigenvalues.

Then, ATQ, $\dfrac{\lambda_1}{\lambda_2} = \dfrac{3}{1}$ and so $\lambda_1 = 3\lambda_2$.

Now, $\lambda_1 + \lambda_2 = $ trace of the matrix

$\qquad = 2 + p$

$$\Rightarrow 3\lambda_2 + \lambda_2 = 2+p \Rightarrow \lambda_2 = \frac{2+p}{4}$$

Again, $\lambda_1 \lambda_2 = $ determinant of the matrix

$$= \begin{vmatrix} 2 & 1 \\ 1 & p \end{vmatrix} = 2p - 1$$

$$\Rightarrow 3\lambda_2^{\,2} = 2p - 1$$

$$\Rightarrow 3\left(\frac{2+p}{4}\right)^2 = 2p - 1$$

$$\Rightarrow 3p^2 - 20p + 28 = 0$$

$$\Rightarrow p = 2, \frac{14}{3}$$

140. (b) A square matrix is singular

$\Leftrightarrow$ "0" is a eigenvalue of the matrix.

141. (b)

$|A - \lambda I| = 0$

$$\Rightarrow \begin{vmatrix} -3-\lambda & 0 & -2 \\ 1 & -1-\lambda & 0 \\ 0 & a & -2-\lambda \end{vmatrix} = 0$$

$$\Rightarrow -(3+\lambda)(1+\lambda)(2+\lambda) - 2a = 0$$

$$\Rightarrow a = -\frac{1}{2}(3+\lambda)(1+\lambda)(2+\lambda)$$

$$= f(\lambda) \text{ (say)}$$

Then, $f'(\lambda)$

$$= -\frac{1}{2}\left[\begin{array}{c} (1+\lambda)(2+\lambda) + (1+\lambda)(3+\lambda) \\ + (2+\lambda)(3+\lambda) \end{array} \right] = 0$$

$$\Rightarrow -\frac{1}{2}\left[3\lambda^2 + 3\lambda + 4\lambda + 5\lambda + 11 \right] = 0$$

$$\Rightarrow 3\lambda^2 + 12\lambda + 11 = 0$$

$$\Rightarrow \lambda = \frac{-12 \pm \sqrt{144 - 12 \times 11}}{2 \times 3}$$

$$\Rightarrow \lambda = \frac{-12 \pm \sqrt{12}}{6} = \frac{-12 \pm 2\sqrt{3}}{6}$$

$$\Rightarrow \lambda = -2 \pm \frac{1}{\sqrt{3}}$$

Now, $f''(\lambda)$

$$= -\frac{1}{2}\left[6\lambda + 12 \right]$$

$$= -\frac{1}{2}\left[6\left(-2 + \frac{1}{\sqrt{3}}\right) + 12 \right] \left(\text{at } \lambda = -2 + \frac{1}{\sqrt{3}} \right)$$

$$= -\frac{3}{\sqrt{3}} < 0$$

$\therefore$ "a" i.e., $f(\lambda)$ is maximum for

$\lambda = -2 + \dfrac{1}{\sqrt{3}}$

$\therefore a_{max}$

$= f(\lambda)\Big|_{\lambda = -2 + \frac{1}{\sqrt{3}}}$

$$= -\frac{1}{2}\left(3 + \left(-2 + \frac{1}{\sqrt{3}}\right)\right) \times \left(1 + \left(-2 + \frac{1}{\sqrt{3}}\right)\right)$$

$$\times \left(2 + \left(-2 + \frac{1}{\sqrt{3}}\right)\right)$$

$$= -\frac{1}{2}\left(-1 + \frac{1}{\sqrt{3}}\right)\left(\frac{1}{\sqrt{3}}\right)\left(1 + \frac{1}{\sqrt{3}}\right)$$

$$= -\frac{1}{2\sqrt{3}}\left\{\left(\frac{1}{\sqrt{3}}\right)^2 - 1^2\right\}$$

$$= \frac{1}{3\sqrt{3}}.$$

142. (d)

$trace(A) = $ sum of eigenvalue

$\Rightarrow 1 + a = (-1) + 7 = 6 \Rightarrow a = 5$

$\det(A) = $ product of eigenvalue

$\Rightarrow a - 4b = (-1) \times 7$

$\Rightarrow 5 - 4b = -7 \Rightarrow b = 3.$

143. (a) The eigenvalue will be real if the matrix is symmetric, *i.e.*, if $5 + j = x$.

144. (c)

$$\left|A^T A^{-1}\right| = \left|A^T\right|\left|A^{-1}\right| = |A| \times \frac{1}{|A|} = 1$$

$$\left(\because AA^{-1} = I \Rightarrow \left|AA^{-1}\right| = |I| \Rightarrow |A|\left|A^{-1}\right| = 1\right)$$

145. (a)

$$\Delta = \begin{vmatrix} 4 & 7 & 8 \\ 3 & 1 & 5 \\ 9 & 6 & 2 \end{vmatrix} = 4(2-30) - 7(6-45) + 8(18-9)$$

$$= -112 + 273 + 72 = 233$$

Now interchanging C_1 and C_2 in Δ, we get

$$\begin{vmatrix} 7 & 4 & 8 \\ 1 & 3 & 5 \\ 6 & 9 & 2 \end{vmatrix} = 7(6-45) - 4(2-30) + 8(9-18)$$

$$= -273 + 112 - 72 = -233 = -\Delta$$

Therefore, (a) is correct.

146. (b) For $n = 3$, we have

$$A = \begin{bmatrix} a_{11} & a_{12} & a_{13} \\ a_{21} & a_{22} & a_{23} \\ a_{31} & a_{32} & a_{33} \end{bmatrix}$$

$$= \begin{bmatrix} 1\times 1 & 1\times 2 & 1\times 3 \\ 2\times 1 & 2\times 2 & 2\times 3 \\ 3\times 1 & 3\times 2 & 3\times 3 \end{bmatrix}$$

$$= \begin{bmatrix} 1 & 2 & 3 \\ 2 & 4 & 6 \\ 3 & 6 & 9 \end{bmatrix} \sim \begin{bmatrix} 1 & 2 & 3 \\ 0 & 0 & 0 \\ 0 & 0 & 0 \end{bmatrix}$$

$[by\ R_2 \to R_2 - 2R_1,\ R_3 \to R_3 - 3R_1]$

which has only one non-zero row

Therefore, rank = 1.

147. 2 The system has infinitely many solutions

$\Leftrightarrow \Delta_3 = 0$ (by Cramer's rule)

$$\Leftrightarrow \begin{vmatrix} 1 & -2 & -1 \\ 1 & -3 & 1 \\ -2 & 4 & k \end{vmatrix} = 0$$

$$\Rightarrow (-3k - 4) + 2(k+2) - (4-6) = 0$$

$$\Rightarrow k = 2.$$

148. 4.5 The system has no solution

$\Rightarrow \Delta = 0$ (by Cramer's rule)

$$\Rightarrow \begin{vmatrix} 2 & 3 \\ 3 & p \end{vmatrix} = 0 \Rightarrow 2p - 9 = 0$$

$$\Rightarrow p = 4.5$$

149. (c) The homogeneous system has a non-trivial solution $\Rightarrow$ coefficient determinant = 0

$$\Rightarrow \begin{vmatrix} p & q & r \\ q & r & p \\ r & p & q \end{vmatrix} = 0$$

$$\Rightarrow \begin{vmatrix} p+q+r & q & r \\ p+q+r & r & p \\ p+q+r & p & q \end{vmatrix} = 0$$

$(by\ C_1 \to C_1 + C_2 + C_3)$

$$\Rightarrow (p+q+r)\begin{vmatrix} 1 & q & r \\ 1 & r & p \\ 1 & p & q \end{vmatrix} = 0$$

(taking $p + q + r$ common from the first column)

$$\Rightarrow (p+q+r)(p^2 + q^2 + r^2 - pq - pr - qr) = 0$$

$$\Rightarrow (p+q+r)(2p^2 + 2q^2 + 2r^2 - 2pq - 2pr - 2qr) = 0$$

$$\Rightarrow (p+q+r)\{(p-q)^2 + (q-r)^2 + (r-p)^2\} = 0$$

$$\Rightarrow p+q+r = 0 \text{ or } p = q = r$$

$(\because a^2 + b^2 + c^2 = 0 \Leftrightarrow a = b = c = 0)$

150. 3

$trace(A) = 14$

$\Rightarrow a + 5 + 2 + b = 14$

$\Rightarrow a + b = 7 \Rightarrow b = 7 - a$

$det(A) = 100$

$\Rightarrow \begin{vmatrix} a & 0 & 3 & 7 \\ 2 & 5 & 1 & 3 \\ 0 & 0 & 2 & 4 \\ 0 & 0 & 0 & b \end{vmatrix} = 100$

$\Rightarrow b \begin{vmatrix} a & 0 & 3 \\ 2 & 5 & 1 \\ 0 & 0 & 2 \end{vmatrix} = 100$

$\Rightarrow 2b \begin{vmatrix} a & 0 \\ 2 & 5 \end{vmatrix} = 100$

$\Rightarrow 2b \times 5a = 100$

$\Rightarrow ab = 10$

$\Rightarrow a(7 - a) = 10$

$\Rightarrow a^2 - 7a + 10 = 0$

$\Rightarrow a = 2, 5$

For $a = 2, b = 5$ and for $a = 5, b = 2$.

So $|a - b| = 3$.

151. (d)

$P \begin{bmatrix} x \\ y \end{bmatrix} = \begin{bmatrix} a \\ b \end{bmatrix}$

$\Rightarrow \begin{bmatrix} 3 & 1 \\ 1 & 3 \end{bmatrix} \begin{bmatrix} x \\ y \end{bmatrix} = \begin{bmatrix} a \\ b \end{bmatrix}$

$\Rightarrow \begin{bmatrix} 3x + y \\ x + 3y \end{bmatrix} = \begin{bmatrix} a \\ b \end{bmatrix}$

$\Rightarrow \begin{cases} 3x + y = a \dots\dots\dots(1) \\ x + 3y = b \dots\dots\dots(2) \end{cases}$

Then $(1)^2 + (2)^2$

$\Rightarrow (3x + y)^2 + (x + 3y)^2 = a^2 + b^2 = 1$

$\Rightarrow 10x^2 + 10y^2 + 12xy = 1 \dots\dots\dots(3)$

Comparing (3) with $Ax^2 + By^2 + 2Hxy = 1$

we get $A = 10, B = 10, H = 6$

Since, $H^2 - AB = 36 - 100 = -64 < 0$,

so (3) represents an ellipse with a major axis along $y = -x$

and a minor axis along $y = x$

Now on $y = x$, $\begin{bmatrix} x \\ y \end{bmatrix} = \begin{bmatrix} x \\ x \end{bmatrix} = \begin{bmatrix} 1 \\ 1 \end{bmatrix}$ (for x=1)

152. (d)

Let $M = \begin{pmatrix} 1 & 0 & 1 \\ 0 & 0 & 0 \\ 0 & 1 & 0 \end{pmatrix}$

Then $|M - \lambda I| = 0$

$\Rightarrow \begin{vmatrix} 1 - \lambda & 0 & 1 \\ 0 & -\lambda & 0 \\ 0 & 1 & -\lambda \end{vmatrix} = 0$

$\Rightarrow (1 - \lambda) \begin{vmatrix} -\lambda & 0 \\ 1 & -\lambda \end{vmatrix} = 0$

$\Rightarrow (1 - \lambda)\lambda^2 = 0$

$\Rightarrow \lambda = 0, 0, 1$

$\therefore \lambda = 1$ is an eigenvalue of M.

153. 3

Let $A = \begin{pmatrix} 1 & 1 & 1 \\ 1 & 1 & 1 \\ 1 & 1 & 1 \end{pmatrix}$

Then $|A - \lambda I| = 0$

$\Rightarrow \begin{vmatrix} 1 - \lambda & 1 & 1 \\ 1 - \lambda & 1 - \lambda & 1 \\ 1 - \lambda & 1 & 1 - \lambda \end{vmatrix} = 0$

$\Rightarrow \begin{vmatrix} 3 - \lambda & 1 & 1 \\ 3 - \lambda & 1 - \lambda & 1 \\ 3 - \lambda & 1 & 1 - \lambda \end{vmatrix} = 0$

$[by\ C_1 \to C_1 + C_2 + C_3]$

$\Rightarrow (3 - \lambda) \begin{vmatrix} 1 & 1 & 1 \\ 1 & 1 - \lambda & 1 \\ 1 & 1 & 1 - \lambda \end{vmatrix} = 0$

$\Rightarrow 3 - \lambda = 0$

$\Rightarrow \lambda = 3$ is a non-zero eigenvalue of A.

154. (c) $M^4 = I \Rightarrow M^{-1} M^4 = M^{-1} I \Rightarrow M^{-1} = M^3$

For $k = 1$, $M^{-1} \neq M^{4 \times 1 + 1}$, $M^{-1} \neq M^{4 \times 1 + 2}$, $M^{-1} \neq M^{4 \times 1}$ but $M^{-1} = M^{4 \times 1 + 3} = M^4 M^3 = M^3$ (Since $M^4 = I$)

Hence, (c) is correct only.

155. (a)

$|A - \lambda I| = 0$

$\Rightarrow \begin{vmatrix} 2 - \lambda & 1 \\ 1 & k - \lambda \end{vmatrix} = 0$

$\Rightarrow (2 - \lambda)(k - \lambda) - 1 = 0$

$$\Rightarrow \lambda^2 - (2+k)\lambda + (-1+2k) = 0$$

$$\Rightarrow \lambda = \frac{(2+k) \pm \sqrt{(2+k)^2 - 4(-1+2k)}}{2}$$

$$= \frac{(2+k) \pm \sqrt{k^2 - 4k + 8}}{2}$$

Now λ is $+ve \Rightarrow \lambda > 0$

$$\Rightarrow (2+k) \pm \sqrt{k^2 - 4k + 8} > 0$$

$$\Rightarrow (2+k) > \pm\sqrt{k^2 - 4k + 8}$$

$$\Rightarrow (2+k)^2 > k^2 - 4k + 8$$

$$\Rightarrow k > 1/2$$

156. 15. Since the complex eigenvalues occurs in conjugate pairs, so $2 + \sqrt{-1}$ *i.e.*, $2 + i$ is an eigenvalue implies $2 - i$ will be another eigenvalue. Then the eigenvalues are $2 + i$, $2 - i$ and 3.

Then $|P|$

= product of the eigenvalues

$= (2+i)(2-i)3 = 15$.

157. (b)

$$\Delta = \begin{vmatrix} 1 & 2 & -3 \\ 2 & 3 & 3 \\ 5 & 9 & -6 \end{vmatrix}$$

$$= (-18-27) - 2(-12-15) - 3(18-15) = 0.$$

Therefore, the system cannot have a unique solution. Since the system is consistent, so it must has an infinite number of solutions.

Now the system has an infinite number of solutions

$\Delta_3 = 0$ (by Cramer's rule)

$$\Rightarrow \begin{vmatrix} 1 & 2 & a \\ 2 & 3 & b \\ 5 & 9 & c \end{vmatrix} = 0$$

$$\Rightarrow (3c - 9b) - 2(2c - 5b) + a(18 - 15) = 0$$

$$\Rightarrow 3a + b - c = 0.$$

158. (d) The given system of equations can be rewritten as: $2x + 5y = 2$ and $-4x + 3y = -30$.

Solving these two equations we get $x = 6$ and $y = -2$.

159. (a) $y = 3x + 3 = 3x + 5 \Rightarrow 3 = 5$, which is absurd. Hence, the system has no solution.

160. (c)

(*i*) Consider the system of equations: $x + y + z = 1$, $2x + 2y + 2z = 3$ (here $m = 2 < 3 = n$).

Clearly this system does not have any solution. So statement I is not correct.`

(*ii*) Consider the system of equations: $x + y = 1$, $2x - y = 2$, $x + 2y = 1$ (here $m = 3 > 2 = n$).

Clearly this system has a solution $x = 1$ and $y = 0$. So statement II is not correct.

(*iii*) Consider the system of equations: $x + y = 1$, $2x - y = 2$ (here $m = 2 = n$).

Clearly this system has a solution $x = 1$ and $y = 0$. So statement III is correct.

161. (d)

$|P| = $ product of the eigenvalues

$$\Rightarrow \sigma^2 - \omega x = (\sigma + j\omega)(\sigma - j\omega)$$

$$= \sigma^2 - j^2 \omega^2$$

$$\Rightarrow -\omega x = -j^2 \omega^2$$

$$\Rightarrow x = -\omega \quad [\because j = \sqrt{-1}]$$

162. (a) Determinant of the matrix $= 0 + 1 = 1 = $ product of the eigenvalue $= i \times (-i)$ [since $i^2 = -1$].

163. -6.

A has eigenvalue λ

$\Rightarrow A^3 - 3A^2$ has eigenvalue $\lambda^3 - 3\lambda^2$

$\therefore \lambda = 1 \Rightarrow \lambda^3 - 3\lambda^2 = 1 - 3 = -2$,

$\lambda = -1 \Rightarrow \lambda^3 - 3\lambda^2 = -1 - 3 = -4$,

$\lambda = 3 \Rightarrow \lambda^3 - 3\lambda^2 = 27 - 27 = 0$.

Therefore, the eigenvalue of $A^3 - 3A^2$ are -2, -4, 0.

Hence,

$tr\left(A^3 - 3A^2\right) = $ sum of the eigenvalues

$$= (-2) + (-4) + 0 = -6$$

164. 1.

$\lambda = 0 \Rightarrow A$ is singular

$\Rightarrow |A| = 0$

$$\Rightarrow \begin{vmatrix} 3 & 2 & 4 \\ 9 & 7 & 13 \\ -6 & -4 & -9+x \end{vmatrix} = 0$$

$$\Rightarrow \begin{vmatrix} 3 & 2 & 4 \\ 9 & 7 & 13 \\ 0 & 0 & x-1 \end{vmatrix} = 0 \; [by \; R_3 \to R_3 + 2R_1]$$

$\Rightarrow (x-1)\,(21-18)=0$

$\Rightarrow x=1.$

165. 1/8.

$|A|$ = product of the eigenvalues=$1 \times 2 \times 4 = 8.$

Then $\left|A^{-1}\right| = \dfrac{1}{|A|} = \dfrac{1}{8}$

$\left(\because AA^{-1} = I \Rightarrow \left|AA^{-1}\right| = |I| = 1 \Rightarrow \left|A^{-1}\right|\left|A\right| = 1\right)$

$\therefore \left|(A^{-1})^T\right| = \left|A^{-1}\right| = \dfrac{1}{8}.$

166. (c)

167. (d).

$P^3 = P$

$\Rightarrow \lambda^3 = \lambda$

  (by the Cayley Hamilton theorem)

$\Rightarrow \lambda^3 - \lambda = 0$

$\Rightarrow \lambda(\lambda^2 - 1) = 0$

$\Rightarrow \lambda = 0, 1, -1$

168. 233

Let $A = \begin{pmatrix} 1 & 1 \\ 1 & 0 \end{pmatrix}.$ Then

$|A - \lambda I| = 0 \Rightarrow \begin{vmatrix} 1-\lambda & 1 \\ 1 & 0-\lambda \end{vmatrix} = 0$

$\Rightarrow \lambda^2 - \lambda - 1 = 0.$

By the Cayley Hamilton theorem, we can write

$A^2 - A - I = 0$ i.e; $A^2 = A + I$

$\therefore A^4 = A^2 \times A^2$

$= (A+I) \times (A+I)$

$= A^2 + 2AI + I^2$

$= (A+I) + 2A + I$

$= 3A + 2I,$

$A^6 = A^4 \times A^2$

$= (3A+2I) \times (A+I)$

$= 3A^2 + 5AI + 2I^2$

$= 3(A+I) + 5A + 2I$

$= 8A + 5I,$

$A^{12} = A^6 \times A^6$

$= (8A+5I) \times (8A+5I)$

$= 64A^2 + 80AI + 25I^2$

$= 64(A+I) + 80A + 25I$

$= 144A + 89I$

$= 144\begin{pmatrix} 1 & 1 \\ 1 & 0 \end{pmatrix} + 89\begin{pmatrix} 1 & 0 \\ 0 & 1 \end{pmatrix}$

$= \begin{pmatrix} 144+89 & 144+0 \\ 144+0 & 0+89 \end{pmatrix}$

$= \begin{pmatrix} 233 & 144 \\ 144 & 89 \end{pmatrix}$

Now $\begin{bmatrix} x[12] \\ x[11] \end{bmatrix} = \begin{pmatrix} 1 & 1 \\ 1 & 0 \end{pmatrix}^{12} \begin{bmatrix} 1 \\ 0 \end{bmatrix}$

$= A^{12} \begin{bmatrix} 1 \\ 0 \end{bmatrix}$

$= \begin{pmatrix} 233 & 144 \\ 144 & 89 \end{pmatrix}\begin{bmatrix} 1 \\ 0 \end{bmatrix}$

$= \begin{bmatrix} 233 \\ 144 \end{bmatrix}$

Comparing both sides we get, $x[12] = 233.$

169. (b) For any matrix A, rank(A^TA) = rank(AA^T) = rank(A).

170. (d)

$u = ae_1 + be_2 + ce_3$

$\Rightarrow (4,3,-3) = a(1,0,2) + b(0,1,0) + c(-2,0,1)$

$= (a,0,2a) + (0,b,0) + (-2c,0,c)$

$\Rightarrow (4,3,-3) = (a-2c, b, 2a+c)$

$\Rightarrow a - 2c = 4, b = 3, 2a + c = -3.$

Solving we get,

$a = -\dfrac{2}{5}, b = 3, c = -\dfrac{11}{5}.$

171. (c)

$M = \begin{bmatrix} 5 & 10 & 10 \\ 1 & 0 & 2 \\ 3 & 6 & 6 \end{bmatrix}$

$\sim \begin{bmatrix} 1 & 2 & 2 \\ 1 & 0 & 2 \\ 3 & 6 & 6 \end{bmatrix} [by\, R_1 \to \frac{1}{5}R_1]$

$\sim \begin{bmatrix} 1 & 2 & 2 \\ 1 & 0 & 2 \\ 0 & 0 & 0 \end{bmatrix} [by\, R_3 \to R_3 - 3R_1]$

which has two non-zero rows.

Therefore, rank(M) = 2.

172. 5. Let λ be the other eigenvalue.

Then determinant of the matrix = product of the eigenvalues $\Rightarrow 50 = 10\lambda$.

So $\lambda = 5$.

173. (b) Product of the eigenvalues = determinant of $P = 2(3\text{-}6) + 0 + 1(8 - 0) = 2$.

174. (b).

Let $A = \begin{pmatrix} 5 & 0 \\ 0 & -5 \end{pmatrix} = \begin{pmatrix} A_{11} & A_{12} \\ A_{21} & A_{22} \end{pmatrix}$.

Then $\text{rank}(A) = 2$

($\because A$ has two non-zero rows)

Also $\displaystyle\sum_{i=1}^{2}\sum_{j=1}^{2} A_{ij}^2 = A_{11}^2 + A_{12}^2 + A_{21}^2 + A_{22}^2$

$$= 5^2 + 0^2 + 0^2 + (-5)^2$$
$$= 50.$$

Thus, the matrix A satisfies the given condition.

Since A is upper triangular, so its eigenvalues are the diagonal elements *i.e*; 5 and -5. So statement–I is correct. But since $|\pm 5| = 5$, which is not greater than 5, so statement–II is not correct.

175. [0] Sum of eigenvalues = trace(A)

$\Rightarrow \lambda_1 + \lambda_2 = 50 + 80 = 130$

$X_1^T X_2$

$= \begin{bmatrix} 70 \\ \lambda_1 - 50 \end{bmatrix}^T \begin{bmatrix} \lambda_2 - 80 \\ 70 \end{bmatrix}$

$= [70 \quad \lambda_1 - 50]\begin{bmatrix} \lambda_2 - 80 \\ 70 \end{bmatrix}$

$= [70(\lambda_2 - 80) + 70\,(\lambda_1 - 50)]$

$= [70(\lambda_1 + \lambda_2) - 9100]$

$= [70 \times 130 - 9100]$

$= [0]$

176. (c).

$|A - \lambda I| = 0$

$\Rightarrow \begin{vmatrix} 1-\lambda & -1 & 5 \\ 0 & 5-\lambda & 6 \\ 0 & -6 & 5-\lambda \end{vmatrix} = 0$

$\Rightarrow (1-\lambda)\begin{vmatrix} 5-\lambda & 6 \\ -6 & 5-\lambda \end{vmatrix} = 0$

$\Rightarrow (1-\lambda)(\lambda^2 - 10\lambda + 61) = 0$

$\Rightarrow \lambda = 1, \dfrac{10 \pm \sqrt{10^2 - 4\times 61}}{2}$

$\Rightarrow \lambda = 1, \dfrac{10 \pm 12j}{2}$ (where $j = \sqrt{-1}$)

$\Rightarrow \lambda = 1.5 \pm 6j$

177. (a) Let $A = \begin{pmatrix} 5 & -1 \\ 4 & 1 \end{pmatrix}$.

Then $|A - \lambda I| = 0$

$\Rightarrow \begin{vmatrix} 5-\lambda & -1 \\ 4 & 1-\lambda \end{vmatrix} = 0$

$\Rightarrow (5-\lambda)(1-\lambda) + 4 = 0$

$\Rightarrow \lambda^2 - 6\lambda + 9 = 0$

$\Rightarrow \lambda = 3, 3$

Eigenvalue 3 has a multiplicity of 2. Then $AX = \lambda X$

$\Rightarrow \begin{pmatrix} 5 & -1 \\ 4 & 1 \end{pmatrix}\begin{bmatrix} x \\ y \end{bmatrix} = 3\begin{bmatrix} x \\ y \end{bmatrix}$

$\Rightarrow \begin{bmatrix} 5x - y \\ 4x + y \end{bmatrix} = \begin{bmatrix} 3x \\ 3y \end{bmatrix}$

$\Rightarrow 5x - y = 3x, 4x + y = 3y$

$\Rightarrow 2x - y = 0, 4x = 2y$

$\Rightarrow y = 2x$

$\therefore X = \begin{bmatrix} x \\ y \end{bmatrix} = \begin{bmatrix} x \\ 2x \end{bmatrix} = x\begin{bmatrix} 1 \\ 2 \end{bmatrix}$.

Hence, A has only one linearly independent eigenvector, which is $\begin{bmatrix} 1 \\ 2 \end{bmatrix}$.

178. (c)

$|A - \lambda I| = 0$

$\Rightarrow \begin{vmatrix} 1-\lambda & 2 & 3 & 4 & 5 \\ 5 & 1-\lambda & 2 & 3 & 4 \\ 4 & 5 & 1-\lambda & 2 & 3 \\ 3 & 4 & 5 & 1-\lambda & 2 \\ 2 & 3 & 4 & 5 & 1-\lambda \end{vmatrix} = 0$

$\Rightarrow \begin{vmatrix} 15-\lambda & 2 & 3 & 4 & 5 \\ 15-\lambda & 1-\lambda & 2 & 3 & 4 \\ 15-\lambda & 5 & 1-\lambda & 2 & 3 \\ 15-\lambda & 4 & 5 & 1-\lambda & 2 \\ 15-\lambda & 3 & 4 & 5 & 1-\lambda \end{vmatrix} = 0$

[by $C_1 \to C_1 + C_2 + C_3 + C_4 + C_5$]

$$\Rightarrow (15-\lambda)\begin{vmatrix} 1 & 2 & 3 & 4 & 5 \\ 1 & 1-\lambda & 2 & 3 & 4 \\ 1 & 5 & 1-\lambda & 2 & 3 \\ 1 & 4 & 5 & 1-\lambda & 2 \\ 1 & 3 & 4 & 5 & 1-\lambda \end{vmatrix} = 0$$

$$\Rightarrow 15-\lambda = 0$$

$\Rightarrow \lambda = 15$ is a real eigenvalue.

179. (d)

$$|P| = \begin{vmatrix} \dfrac{1}{\sqrt{2}} & 0 & \dfrac{1}{\sqrt{2}} \\ 0 & 1 & 0 \\ -\dfrac{1}{\sqrt{2}} & 0 & \dfrac{1}{\sqrt{2}} \end{vmatrix} = \dfrac{1}{2} \times 2 = 1.$$

So statement (a) is correct.

PP^T

$$= \begin{bmatrix} \dfrac{1}{\sqrt{2}} & 0 & \dfrac{1}{\sqrt{2}} \\ 0 & 1 & 0 \\ -\dfrac{1}{\sqrt{2}} & 0 & \dfrac{1}{\sqrt{2}} \end{bmatrix} \times \begin{bmatrix} \dfrac{1}{\sqrt{2}} & 0 & -\dfrac{1}{\sqrt{2}} \\ 0 & 1 & 0 \\ \dfrac{1}{\sqrt{2}} & 0 & \dfrac{1}{\sqrt{2}} \end{bmatrix}$$

$$= \begin{bmatrix} \dfrac{1}{2}+\dfrac{1}{2} & 0 & -\dfrac{1}{2}+\dfrac{1}{2} \\ 0 & 1 & 0 \\ -\dfrac{1}{2}+\dfrac{1}{2} & 0 & \dfrac{1}{2}+\dfrac{1}{2} \end{bmatrix} = I$$

Hence, P is orthogonal and so statement (b) is also correct.

P is orthogonal $\Rightarrow P^T = P^{-1}$. So option (c) is also correct.

Alternative method:

$$|P - \lambda I| = 0$$

$$\Rightarrow \begin{vmatrix} \dfrac{1}{\sqrt{2}}-\lambda & 0 & \dfrac{1}{\sqrt{2}} \\ 0 & 1-\lambda & 0 \\ -\dfrac{1}{\sqrt{2}} & 0 & \dfrac{1}{\sqrt{2}}-\lambda \end{vmatrix} = 0$$

$$\Rightarrow (1-\lambda)\begin{vmatrix} \dfrac{1}{\sqrt{2}}-\lambda & 0 & \dfrac{1}{\sqrt{2}} \\ 0 & 1 & 0 \\ -\dfrac{1}{\sqrt{2}} & 0 & \dfrac{1}{\sqrt{2}}-\lambda \end{vmatrix} = 0$$

$$\Rightarrow (1-\lambda)\left\{ \left(\dfrac{1}{\sqrt{2}}-\lambda\right)^2 + \left(\dfrac{1}{\sqrt{2}}\right)^2 \right\} = 0$$

$$\Rightarrow (1-\lambda)(\lambda^2 - \sqrt{2}\lambda + 1) = 0$$

$$\Rightarrow \lambda = 1, \dfrac{\sqrt{2} \pm \sqrt{2}i}{2}$$

$$\Rightarrow \lambda = 1, \dfrac{1 \pm i}{\sqrt{2}}$$

Hence, all eigenvalues are not real. Thus, option (d) is in correct.

180. (a) Here the co-efficient matrix is, $A = \begin{pmatrix} 3 & 2 \\ 4 & 1 \end{pmatrix}$.

Then the characteristic equation is given by
$$|A - \lambda I| = 0$$

or, $\begin{vmatrix} 3-\lambda & 2 \\ 4 & 1-\lambda \end{vmatrix} = 0$

or, $\lambda^2 - 4\lambda - 5 = 0.$

181. (c) If X and Y are any two eigenvectors of a symmetric matrix, then we must have $X^T Y = O.$

Let $X = \begin{bmatrix} 1 \\ 0 \\ 1 \end{bmatrix}$ (given) and $Y = \begin{bmatrix} 1 \\ 0 \\ -1 \end{bmatrix}$.

Then $X^T Y = \begin{pmatrix} 1 & 0 & 1 \end{pmatrix}\begin{bmatrix} 1 \\ 0 \\ -1 \end{bmatrix}$

$$= \left[1 + 0 + (-1)\right] = \left[0\right]_{1\times 1}$$

$$= O \,(\text{null matrix}).$$

182. 5. We know that for a 3×3 matrix M, the characteristic polynomial is given by
$$\lambda^3 - \text{trace}(M) \times \lambda^2 + a\lambda - \det(M) \qquad \ldots(1)$$

Comparing (1) with $\lambda^3 - 4\lambda^2 + a\lambda + 30$, we get,

$\text{trace}(M) = 4 \qquad \ldots(2)$

and $\det(M) = -30 \qquad \ldots(3)$

Given that "2" is an eigenvalue of M. Let λ_1 and λ_2 be the other two eigenvalues of M.

Then equation (2)

$\Rightarrow 2 + \lambda_1 + \lambda_2 = 4$

$\Rightarrow \lambda_2 = 2 - \lambda_1$

Now equation (3)

$$\Rightarrow 2\lambda_1 \lambda_2 = -30$$
$$\Rightarrow \lambda_1(2 - \lambda_1) = -15$$
$$\Rightarrow \lambda_1^2 - 2\lambda_1 - 15 = 0$$
$$\Rightarrow \lambda_1 = 5, -3.$$

Hence, the eigenvalues of M are 2, –3, and 5.

183. 4

$$\begin{bmatrix} 1 & -1 & 0 & 0 & 0 \\ 0 & 0 & 1 & -1 & 0 \\ 0 & 1 & -1 & 0 & 0 \\ -1 & 0 & 0 & 0 & 1 \\ 0 & 0 & 0 & 1 & -1 \end{bmatrix} \sim \begin{bmatrix} 0 & 0 & 0 & 0 & 0 \\ 0 & 0 & 1 & -1 & 0 \\ 0 & 1 & -1 & 0 & 0 \\ -1 & 0 & 0 & 0 & 1 \\ 0 & 0 & 0 & 1 & -1 \end{bmatrix}$$

$$[by\ R_1 \to R_1 + R_2 + R_3 + R_4 + R_5]$$

which has four non-zero rows.

Therefore, rank = 4.

184. 2.

$$P + Q = \begin{bmatrix} 0 & -1 & -2 \\ 8 & 9 & 10 \\ 8 & 8 & 8 \end{bmatrix} \sim \begin{bmatrix} 0 & -1 & -2 \\ 0 & 0 & 0 \\ 8 & 8 & 8 \end{bmatrix}$$

$$[by\ R_2 \to R_2 - (R_3 - R_1)]$$

which has two non-zero rows.

Therefore, rank$(P + Q) = 2$.

185. (c) P is the inverse of $Q \Rightarrow PQ = PQ = I$.

186. 1

Let $V = \begin{bmatrix} v_1 \\ v_2 \\ v_3 \end{bmatrix}$. Then,

$$A = VV^T$$

$$= \begin{bmatrix} v_1 \\ v_2 \\ v_3 \end{bmatrix} \times \begin{bmatrix} v_1 & v_2 & v_3 \end{bmatrix}$$

$$= \begin{bmatrix} v_1^2 & v_1 v_2 & v_1 v_3 \\ v_2 v_1 & v_2^2 & v_2 v_3 \\ v_3 v_1 & v_3 v_2 & v_3^2 \end{bmatrix}$$

$$\sim \begin{bmatrix} v_1 & v_2 & v_3 \\ v_1 & v_2 & v_3 \\ v_1 & v_2 & v_3 \end{bmatrix}$$

$$\sim \begin{bmatrix} v_1 & v_2 & v_3 \\ v_1 & v_2 & v_3 \\ v_1 & v_2 & v_3 \end{bmatrix}$$

$$[by\ R_1 \to \frac{1}{v_1} R_1,\ R_2 \to \frac{1}{v_2} R_2,\ R_3 \to \frac{1}{v_3} R_3]$$

$$\sim \begin{bmatrix} v_1 & v_2 & v_3 \\ 0 & 0 & 0 \\ 0 & 0 & 0 \end{bmatrix}$$

$$[by\ R_2 \to R_2 - R_1,\ R_3 \to R_3 - R_1]$$

which has one non-zero row.

Therefore, rank = 1.

187. (a)

$$AB^T = \begin{bmatrix} 1 & 5 \\ 6 & 2 \end{bmatrix} \begin{bmatrix} 3 & 8 \\ 7 & 4 \end{bmatrix} = \begin{bmatrix} 3+35 & 8+20 \\ 18+14 & 48+8 \end{bmatrix} = \begin{bmatrix} 38 & 28 \\ 32 & 56 \end{bmatrix}$$

188. (c)

$$A = \begin{pmatrix} 2 & 4 \\ 3 & 6 \end{pmatrix} \Rightarrow |A| = \begin{vmatrix} 2 & 4 \\ 3 & 6 \end{vmatrix} = 12 - 12 = 0.$$

Therefore, the given matrix is singular.

189. (c) We know that if $A_{n \times n}$ is an orthogonal matrix then

$A^{-1} = A^T$. Here given that $Q = \begin{pmatrix} \dfrac{3}{7} & \dfrac{2}{7} & \dfrac{6}{7} \\ \dfrac{-6}{7} & \dfrac{3}{7} & \dfrac{2}{7} \\ \dfrac{2}{7} & \dfrac{6}{7} & \dfrac{-3}{7} \end{pmatrix}$ is an orthogonal matrix.

Therefore, $Q^{-1} = \begin{pmatrix} \dfrac{3}{7} & \dfrac{-6}{7} & \dfrac{2}{7} \\ \dfrac{2}{7} & \dfrac{3}{7} & \dfrac{6}{7} \\ \dfrac{6}{7} & \dfrac{2}{7} & \dfrac{-3}{7} \end{pmatrix}$.

190. (b)

$$A = \begin{pmatrix} 1 & 1 & 0 & -2 \\ 2 & 0 & 2 & 2 \\ 4 & 1 & 3 & 1 \end{pmatrix}$$

$$\sim \begin{pmatrix} 1 & 1 & 0 & -2 \\ 0 & -2 & 2 & 6 \\ 0 & -3 & 3 & 9 \end{pmatrix}$$

$$(by\ R_2 \to R_2 - 2R_1;\ R_3 \to R_3 - 4R_1)$$

$$\sim \begin{pmatrix} 1 & 1 & 0 & -2 \\ 0 & -2 & 2 & 6 \\ 0 & 0 & 0 & 0 \end{pmatrix} \text{ (by } R_3 \to 2R_3 - 3R_2\text{)}$$

which has two non-zero rows.

Hence, the rank of the given matrix = 2.

191. (b)

$$A = \begin{pmatrix} -4 & 1 & -1 \\ -1 & -1 & -1 \\ 7 & -3 & 1 \end{pmatrix}$$

$$\sim \begin{pmatrix} -4 & 1 & -1 \\ -1 & -1 & -1 \\ 0 & 0 & 0 \end{pmatrix}$$

(by R3 $\to R_3 + 2R_1 - R_2$)

which has two non-zero rows. So the rank of the given matrix = 2.

192. (d) $|A - \lambda I| = 0$

$$\Rightarrow \begin{vmatrix} 2-\lambda & -4 \\ 4 & -2-\lambda \end{vmatrix} = 0$$

$$\Rightarrow \lambda^2 - (2-2)\lambda + 12 = 0 \Rightarrow \lambda^2 + 12 = 0$$

$\therefore \lambda = \pm i\sqrt{12}$ are eigenvalues of A.

Case (i) : $\lambda = i\sqrt{12}$

$AX = \lambda X$

$$\Rightarrow \begin{bmatrix} 2 & -4 \\ 4 & -2 \end{bmatrix} \begin{bmatrix} x_1 \\ x_2 \end{bmatrix} = i\sqrt{12} \begin{bmatrix} x_1 \\ x_2 \end{bmatrix}$$

$$\Rightarrow \begin{bmatrix} 2x_1 - 4x_2 \\ 4x_1 - 2x_2 \end{bmatrix} = \begin{bmatrix} i\sqrt{12}x_1 \\ i\sqrt{12}x_2 \end{bmatrix}$$

$$\Rightarrow 2x_1 - 4x_2 = i\sqrt{12}x_1, 4x_1 - 2x_2 = i\sqrt{12}x_2$$

$$\Rightarrow (2-i\sqrt{12})x_1 - 4x_2 = 0, 4x_1 - (2+i\sqrt{12})x_2 = 0$$

Let $x_1 = K$. Then $x_2 = \dfrac{(2-i\sqrt{12})}{4}k$

Therefore,

$$X = \begin{bmatrix} x_1 \\ x_2 \end{bmatrix} = \begin{bmatrix} k \\ \dfrac{(2-i\sqrt{12})}{4}k \end{bmatrix} = k \begin{bmatrix} 1 \\ \dfrac{(2-i\sqrt{12})}{4} \end{bmatrix} \text{ is an}$$

eigenvector corresponding to the eigenvalue $\lambda = i\sqrt{12}$ for $k \neq 0$.

Case (ii) : $\lambda = -i\sqrt{12}$

$AX = \lambda X$

$$\Rightarrow \begin{bmatrix} 2 & -4 \\ 4 & -2 \end{bmatrix} \begin{bmatrix} x_1 \\ x_2 \end{bmatrix} = -i\sqrt{12} \begin{bmatrix} x_1 \\ x_2 \end{bmatrix}$$

$$\Rightarrow \begin{bmatrix} 2x_1 - 4x_2 \\ 4x_1 - 2x_2 \end{bmatrix} = \begin{bmatrix} -i\sqrt{12}x_1 \\ i\sqrt{12}x_2 \end{bmatrix}$$

$$\Rightarrow 2x_1 - 4x_2 = -i\sqrt{12}x_1, 4x_1 - 2x_2 = -i\sqrt{12}x_2$$

$$\Rightarrow (2+i\sqrt{12})x_1 - 4x_2 = 0, 4x_1 - (2-i\sqrt{12})x_2 = 0$$

Let $x_1 = k$. Then $x_2 = \dfrac{(2+i\sqrt{12})}{4}k$.

Therefore,

$$X = \begin{bmatrix} x_1 \\ x_2 \end{bmatrix} = \begin{bmatrix} k \\ \dfrac{(2+i\sqrt{12})}{4}k \end{bmatrix} = k \begin{bmatrix} 1 \\ \dfrac{(2+i\sqrt{12})}{4} \end{bmatrix} \text{ is an}$$

eigenvector corresponding to the eigenvalue $\lambda = -i\sqrt{12}$ for $k \neq 0$.

Hence, $\begin{bmatrix} 1 \\ \dfrac{(2-i\sqrt{12})}{4} \end{bmatrix}$ and $\begin{bmatrix} 1 \\ \dfrac{(2+i\sqrt{12})}{4} \end{bmatrix}$ are the

eigenvectors corresponding to the eigenvectors $i\sqrt{12}$ and $-i\sqrt{12}$, respectively.

Hence, the given matrix has complex eigenvalues and eigenvectors.

193. (c) We know that for a square matrix of order "n," the following results hold:

(*i*) If a matrix $M_{n \times n}$ has n different eigenvalues, then the matrix $M_{n \times n}$ will have n linearly independent eigenvectors.

(*ii*) If a matrix $M_{n \times n}$ has some repeated eigenvalues then the eigenvectors may or may not be linearly independent.

(*iii*) If a matrix $M_{n \times n}$ is non-singular then the eigenvalues may or may not be repeated.

Thus, it is clear that the statement "S2 implies S1" is correct.

194. 1. $|A - \lambda I| = 0$

$$\Rightarrow \begin{vmatrix} 1-\lambda & 0 & -1 \\ -1 & 2-\lambda & 0 \\ 0 & 0 & -2-\lambda \end{vmatrix} = 0$$

$$\Rightarrow (1-\lambda)(2-\lambda)(-2-\lambda) - 0 = 0$$

$$\Rightarrow \lambda = 1, -2, 2.$$

$\therefore$ Eigenvalues of A are 1, 2, -2

If λ is an eigenvalue of A then the eigenvalue of $B = A^3 - A^2 - 4A + 5I$ is $\lambda^3 - \lambda^2 - 4\lambda + 5$.

Thus, eigenvalues of B are $1^3 - 1^2 - 4 \times 1 + 5\,1$, $2^3 - 2^2 - 4 \times 2 + 5$ and $(-2)^3 - (-2)^2 - 4 \times (-2) + 5$, *i.e.*, 1.

∴ $|B|$ = Product of eigenvalues of $B = 1 \times 1 \times 1 = 1$.

195. (c) The given system of equations can be represented as $AX = B$, where $A = \begin{bmatrix} 3 & 2k \\ k & 6 \end{bmatrix}$.

Then, the system has an infinite number of solutions

$$\Rightarrow |A| = 0 \Rightarrow \begin{vmatrix} 3 & 2k \\ k & 6 \end{vmatrix} = 0 \Rightarrow 18 - 2k^2 = 0$$

$$\Rightarrow k^2 = 9 \Rightarrow k = \pm 3.$$

Case-I: $k = 3$,

$$[A:B] = \begin{bmatrix} 3 & 6 & -2 \\ 3 & 6 & 2 \end{bmatrix} \sim \begin{bmatrix} 3 & 6 & -2 \\ 0 & 0 & 4 \end{bmatrix}$$

(by $R_2 \to R_2 - R_1$)

Thus, rank ($[A:B]$) = 2. But det(A) = $\begin{vmatrix} 3 & 6 \\ 3 & 6 \end{vmatrix} = 0$. So rank($A$) = 1.

Hence, rank ($[A:B]$) ≠ rank(A). Consequently, the system has no solution in this case.

Case-II: $k = -3$,

$$[A:B] = \begin{bmatrix} 3 & -6 & -2 \\ -3 & 6 & 2 \end{bmatrix} \sim \begin{bmatrix} 3 & 6 & -2 \\ 0 & 0 & 0 \end{bmatrix}$$

(by $R_2 \to R_2 + R_1$)

Thus, rank ($[A:B]$) = 1. But det(A) = $\begin{vmatrix} 3 & 6 \\ 3 & 6 \end{vmatrix} = 0$ and so rank(A) = 1.

Hence, rank ($[A:B]$) = rank(A) < number of variables. Consequently, the system has an infinite number of solutions in this case.

196. 2. The system has infinitely many solutions
$$\Rightarrow |A| = 0.$$

$$\Rightarrow \begin{vmatrix} k & 2k \\ k^2 - k & k^2 \end{vmatrix} = 0 \Rightarrow k^3 - 2k^3 + 2k^2 = 0$$

$$\Rightarrow 2k^2 - k^3 = 0 \Rightarrow k^2(2 - k) = 0 \Rightarrow 0, 2$$

∴ The system $AX = 0$ will have infinitely many solutions for two distinct (*i.e.*, different) real values namely, $k = 0$ and $k = 2$.

197. 25. Let A be a 3 × 3 matrix with diagonal elements −10, 5 and 0. Given that two eigenvalues of A are −15 and −15.

Let λ be the third eigenvalue of A.

Then sum of the eigenvalues = trace of the matrix

$$\Rightarrow -15 - 15 + \lambda = -10 + 5 + 0 \Rightarrow \lambda = 25.$$

198. 3.

$$A = UV^T = \begin{bmatrix} 1 \\ 2 \end{bmatrix}\begin{bmatrix} 1 \\ 1 \end{bmatrix}^T \Rightarrow A = \begin{bmatrix} 1 \\ 2 \end{bmatrix}\begin{bmatrix} 1 & 1 \end{bmatrix} = \begin{bmatrix} 1 & 1 \\ 2 & 2 \end{bmatrix}.$$

Then, $|A - \lambda I| = 0$

$$\Rightarrow \begin{vmatrix} 1-\lambda & 1 \\ 2 & 2-\lambda \end{vmatrix} = 0 \Rightarrow (1-\lambda)(2-\lambda) - 2 = 0$$

$$\Rightarrow \lambda^2 - 3\lambda = 0 \Rightarrow \lambda(\lambda - 3) = 0 \Rightarrow \lambda = 0, 3.$$

Therefore, the largest eigenvalue of A is 3.

199. $\dfrac{11}{2}$. Given trace (A) = 4 and trace (A^2) = 5

Let λ_1, λ_2 be two eigenvalues of A. Then, λ_1^2, λ_2^2 will be two eigenvalues of A^2

∴ trace (A) = 4 and trace (A^2) = 5

$$\Rightarrow \lambda_1 + \lambda_2 = 4 \text{ and } \lambda_1^2 + \lambda_2^2 = 5$$

(since trace = sum of eigenvalues)

Now $(\lambda_1 + \lambda_2)^2 = \lambda_1^2 + 2\lambda_1\lambda_2 + \lambda_2^2$

$$\Rightarrow 16 = 5 + 2\lambda_1\lambda_2 \Rightarrow \lambda_1\lambda_2 = \frac{11}{2}.$$

∴ $|A|$ = product of eigenvalues = $\lambda_1\lambda_2 = \dfrac{11}{2}$.

200. (a) $N^2 = 0 \Rightarrow N_{3\times3}$ is a Nilpotent matrix. We know that the eigenvalues of a non-zero nilpotent matrix are always zero.

Therefore, the eigenvalues of a given matrix N are 0, 0, 0.

Questions for Practice

1. The matrix $A = \begin{pmatrix} -1/3 & 2/3 & 2/3 \\ 2/3 & -1/3 & 2/3 \\ 2/3 & 2/3 & -1/3 \end{pmatrix}$ is

(a) idempotent   (b) orthogonal
(c) skew symmetric   (d) none of these

2. The matrix $A = \dfrac{1}{6}\begin{pmatrix} 1 & -2 & 1 \\ -2 & 4 & -2 \\ 1 & -2 & 1 \end{pmatrix}$ is

(a) idempotent   (b) orthogonal
(c) skew symmetric   (d) none of these

3. If $\begin{pmatrix} 5 & x+2 \\ x+1 & -2 \end{pmatrix} = \begin{pmatrix} x+3 & 4 \\ 3 & -4 \end{pmatrix}$, then $x = ?$

(a) 0   (b) 2   (c) –2   (d) 1

4. If $A = \begin{pmatrix} 1 & 0 \\ 2 & 0 \end{pmatrix}$, $B = \begin{pmatrix} 0 & 0 \\ 1 & 10 \end{pmatrix}$, then

(a) $AB = O, BA = O$ (b) $AB = O, BA \neq O$
(c) $AB \neq O, BA = O$ (d) $AB \neq O, BA \neq O$

5. If $A = \begin{pmatrix} 0 & -1 \\ 1 & 0 \end{pmatrix}$, $B = \begin{pmatrix} 0 & i \\ i & 0 \end{pmatrix}$, then

(a) $A^2 = B^2 = I$ (b) $A^2 = B^2 = -I$
(c) $A^2 = I, B^2 = -I$ (d) $A^2 = -I, B^2 = I$

6. If A and B are skew-symmetric matrices of same order n, then $A + B$ is

(a) skew-symmetric   (b) null matrix
(c) identity matrix   (d) symmetric

7. A necessary and sufficient condition for a square matrix A to possess inverse is that

(a) $A \neq O$ (b) adj $(A) \neq O$
(c) $\det(A) \neq 0$ (d) $\det(A) = 0$

8. If $A = A = \begin{pmatrix} 1 & 2 \\ 3 & 5 \end{pmatrix}$, then $A^{-1} = ?$

(a) $\begin{pmatrix} 1 & 2 \\ 3 & 5 \end{pmatrix}$ (b) $\begin{pmatrix} 5 & -2 \\ -3 & 1 \end{pmatrix}$

(c) $\begin{pmatrix} 5 & -2 \\ 3 & 1 \end{pmatrix}$ (d) $\begin{pmatrix} -1 & 2 \\ -3 & 5 \end{pmatrix}$

9. If I_3 be the identity matrix of order 3, then $(2I_3) - 1 = ?$

(a) $\dfrac{1}{4} I_3$ (b) $2I_3$
(c) I_3 (d) none of these

10. Non-zero matrices A and B are called divisors of zero if

(a) $AB \neq O$ (b) adj $(A) \neq O$
(c) $\det(A) \neq O$ (d) $AB = O$

11. If A is Hermitian, then iA is

(a) symmetric   (b) unitary
(c) skew-Hermitian   (d) Hermitian

12. If $1, \omega, \omega^2$ are cube roots of unity, then the inverse of which of the following matrices exist?

(a) $\begin{pmatrix} 1 & w \\ w & w^2 \end{pmatrix}$ (b) $\begin{pmatrix} \omega & \omega^2 \\ \omega^2 & 1 \end{pmatrix}$

(c) $\begin{pmatrix} \omega^2 & 1 \\ 1 & \omega \end{pmatrix}$ (d) none of these

13. If A is an orthogonal matrix, then $A^t = ?$

(a) A (b) A^2
(c) A^{-1} (d) Identity matrix

14. If $A = \begin{pmatrix} 3 & 2 \\ 0 & 1 \end{pmatrix}$, then $(A^{-1})^3 = ?$

(a) $\dfrac{1}{27}\begin{pmatrix} 1 & -26 \\ 0 & 27 \end{pmatrix}$ (b) $\dfrac{1}{27}\begin{pmatrix} -1 & 26 \\ 0 & 27 \end{pmatrix}$

(c) $\dfrac{1}{27}\begin{pmatrix} 1 & -26 \\ 0 & -27 \end{pmatrix}$ (d) $\dfrac{1}{27}\begin{pmatrix} -1 & -26 \\ 0 & -27 \end{pmatrix}$

15. The rank of the following $(n + 1) \times (n + 1)$ matrix is

$$\begin{pmatrix} 1 & a & a^2 & .. & .. & .. & a^n \\ 1 & a & a^2 & .. & .. & .. & a^n \\ .. & .. & .. & .. & .. & .. & .. \\ 1 & a & a^2 & .. & .. & .. & a^n \end{pmatrix}$$

(a) 1   (b) 2
(c) n (d) depends on n

16. If A is a square matrix of order n, then adj (adj A) $= ?$

(a) $|A|^{n-2} A$ (b) $|A|^{n-2}$
(c) $|A|^{n-1} A$ (d) identity matrix

17. If $A = \begin{pmatrix} 1+i & 3-5i \\ 2i & 5 \end{pmatrix}$, then $A^{\theta} = ?$

(a) $\begin{pmatrix} 1-i & -2i \\ 3+5i & 5 \end{pmatrix}$ (b) $\begin{pmatrix} 1+i & -2i \\ 3-5i & 5 \end{pmatrix}$

(c) $\begin{pmatrix} 1-i & 2i \\ 3+5i & 5 \end{pmatrix}$ (d) $\begin{pmatrix} 1+i & -2i \\ 3+5i & -5 \end{pmatrix}$

18. If inverse of a matrix A exist, then A is

(a) singular   (b) non-singular
(c) $\det(A) = 0$ (d) none of these

19. If $A = \begin{pmatrix} 0 & 1 & -2 \\ -1 & 0 & 5 \\ 2 & -5 & 0 \end{pmatrix}$ then $A^{-1} = ?$

(a) A (b) A^2
(c) $-A$ (d) doesn't exist

20. If $A = \begin{pmatrix} 0 & 0 & 1 \\ 0 & 1 & 0 \\ 1 & 0 & 0 \end{pmatrix}$ then $A^{-1} = ?$

(a) $\begin{pmatrix} 1 & 0 & 1 \\ 0 & 1 & 0 \\ 0 & 0 & 0 \end{pmatrix}$ (b) $\begin{pmatrix} 0 & 1 & 0 \\ 1 & 0 & 0 \\ 0 & 0 & 1 \end{pmatrix}$

(c) $\begin{pmatrix} 0 & 0 & 1 \\ 0 & 1 & 0 \\ 1 & 0 & 0 \end{pmatrix}$ (d) none of these

21. If $A = \begin{pmatrix} 0 & 2i \\ -2i & 0 \end{pmatrix}$, then A is

(a) Hermitian   (b) skew Hermitian
(c) symmetric   (d) unitary

22. If A and B be two square matrices of same order, then which of the followings is not true in general?
(a) $2A - 3B = -3B + 2A$ (b) $AB = BA$
(c) $4(A + B) = 4A + 4B$ (d) $AIB = AB$

23. If $A = \begin{pmatrix} \cos x & \sin x \\ -\sin x & \cos x \end{pmatrix}$ and $A + A^T = I$, then $x = ?$

(a) $\dfrac{\pi}{3}$ (b) $\dfrac{2\pi}{3}$ (c) $\dfrac{5\pi}{3}$ (d) 0

24. The system of equations: $x + y + z = 2$, $2x + y - z = 3$, $3x + 2y + kz = 4$ has a unique solution for
(a) $k = 0$ (b) $k \neq 0$
(c) $k = 4$ (d) $k = 2$

25. The value of the determinant $\begin{vmatrix} \dfrac{1}{a} & a & bc \\ \dfrac{1}{b} & 1 & ca \\ \dfrac{1}{c} & 1 & cb \end{vmatrix}$ is equals to:

(a) 0   (b) abc
(c) $a + b + c$ (d) 1

26. If $A + B + C = \pi$, then the value of the determinant $\begin{vmatrix} \sin(A+B+C) & \sin B & \cos C \\ -\sin B & 0 & \tan A \\ \cos(A+B) & -\tan A & 0 \end{vmatrix}$ is
(a) 0   (b) 1
(c) $2\sin B \tan A \cos C$
(d) none of these

27. If $\Delta_1 = \begin{vmatrix} x & b & b \\ a & x & b \\ a & a & x \end{vmatrix}$ and $\Delta_2 = \begin{vmatrix} x & b \\ a & x \end{vmatrix}$ are the given determinants, then

(a) $\Delta_1 = 3(\Delta_2)^2$ (b) $\dfrac{d}{dx}\Delta_1 = 3\Delta_2$
(c) $\dfrac{d}{dx}\Delta_2 = 3\Delta_1$ (d) $\Delta_1 = 3\Delta_2$

28. If $f(x) = \begin{vmatrix} \cos x & 1 & 0 \\ 1 & 2\cos x & 1 \\ 0 & 1 & 2\cos x \end{vmatrix}$ then $\displaystyle\int_0^{\frac{\pi}{2}} f(x)\,dx = ? = ?$

(a) $\dfrac{1}{3}$ (b) $\dfrac{1}{2}$ (c) $\dfrac{1}{4}$ (d) $\dfrac{1}{5}$

29. For what value of x, $\begin{vmatrix} x+\omega^2 & \omega & 1 \\ \omega & 1+x & \omega^2 \\ 1 & \omega^2 & x+\omega \end{vmatrix} = 0$

(a) $x = 0$ (b) $x = 1$ (c) $x = 2$ (d) $x = -1$

30. If $a + b + c = 0$, then one root of $\begin{vmatrix} a-x & c & b \\ c & b-x & a \\ b & a & c-x \end{vmatrix} = 0$ is is
(a) $x = 0$ (b) $x = 1$
(c) $x = 2$ (7) $x = a^2 + b^2 + c^2$

31. The value of $\begin{vmatrix} 100 & 101 & 102 \\ 105 & 106 & 107 \\ 110 & 111 & 112 \end{vmatrix} = ?$
(a) 0   (b) 1   (c) 101   (d) -1

32. If $A = \begin{pmatrix} 1 & -1 & 1 \\ 2 & 1 & -3 \\ 1 & 1 & 1 \end{pmatrix}$ and $B = \begin{pmatrix} 4 & 2 & 2 \\ -5 & 1 & \alpha \\ 0 & -2 & 3 \end{pmatrix}$ and $B = A^{-1}$, then $\alpha = ?$

(a) -5 (b) 2   (c) -2 (d) 5

33. Which of the followings is correct?
(a) Skew-symmetric matrix of even order is always singular.
(b) Skew-symmetric matrix of odd order is always non-singular.
(c) Skew-symmetric matrix of odd order is always singular.
(d) None of these.

34. If $A = \begin{pmatrix} 3 & 4 \\ 2 & 4 \end{pmatrix}, B = \begin{pmatrix} 3 & 1 \\ 2 & 0 \end{pmatrix}$, then $(A+B)^{-1}$?

(a) $A^{-1} + B^{-1}$ (b) $A + B$

(c) $(A+B)^{-1}$ does not exist

(d) none of these

35. Given that the rank of the matrix $A = \begin{pmatrix} -1 & 2 & 5 \\ 2 & 4 & k-4 \\ 1 & -2 & k+1 \end{pmatrix}$ is 1. Then $k = ?$

(a) 4   (b) 5   (c) –6   (d) 7

36. If $X = \begin{pmatrix} 0 & 1 & 0 \\ 0 & 0 & 1 \\ 1 & 0 & 0 \end{pmatrix}$, , then the rank of $X^2 + X$ is

(a) 1   (b) 2   (c) 3   (d) 0

37. If the rank of the matrix $A = \begin{bmatrix} \lambda & 1 & 1 & 1 \\ 1 & \lambda & 1 & 1 \\ 1 & 1 & \lambda & 1 \\ 1 & 1 & 1 & \lambda \end{bmatrix}$ is less than 4, then $\lambda = ?$

(a) 1, –3   (b) 1, 3   (c) –1, 3   (d) –1, –3

38. The rank of the matrix $X = \begin{pmatrix} 0 & i & -i \\ -i & 0 & i \\ i & -i & 0 \end{pmatrix}$, is

(a) 0   (b) 1   (c) 2   (d) 4

39. The system of equations: $x + y + z = 6, x - y - z = -4, x + y - z = 0$ is/has

(a) consistent   (b) inconsistent

(c) an unique solution   (d) trivial solution

40. The system of equations: $x + 2y - z = 10, x - y - 2z = -2, 2x + y - 3z = 8$ has

(a) only trivial solution

(b) unique non-trivial solution

(c) no solution

(d) an infinite number of solutions

41. The system of equations: $x - 5y + 3z = -1, 2x - y - z = 5, 5x - 7y + z = 2$ has

(a) no solution

(b) an infinite number of solutions

(c) an unique solution

(d) finite number of solutions

42. The system of equations: $6x + 20y - 6z = -3, 2x + 6y = -11, 6y - 18z = -1$ has

(a) no solution

(b) an infinite number of solutions

(c) an unique solution

(d) finite number of solutions

43. The system of equations: $x + 2y + 3z = 1, 2x + y + 3z = 2, 5x + 5y + 9z = 4$ has

(a) no solution

(b) an infinite number of solutions

(c) an unique solution

(d) finite number of solutions

44. The system of equations:

$x + 2y + 3z = 0, 3x + 4y + 4z = 0, 7x + 10y + 12z = 0$ has

(a) no solution

(b) an infinite number of solutions

(c) an unique solution

(d) only trivial solution

45. If $3x + 2y + z = 0, x + 4y + z = 0, 2x + y + 4z = 0$ be the system of equations, then

(a) it is consistent

(b) it has only the trivial solution

(c) it can be reduced to a single equation

(d) the determinant of the co-efficient matrix is zero

46. The equations

$kx + y + z = 0, -x + ky + z = 0, -x - y + kz = 0$ will have a non-trivial solution if k=?

(a) 0   (b) –1   (c) 1   (d) 2

47. The system of equations

$kx + 2y - z = 1, (k-1)y - 2z = 2, (k+2)z = 3$ will have an unique solution if $k = ?$

(a) 0   (b) –1   (c) 1   (d) –2

48. The sum of eigenvalues of $\begin{pmatrix} 8 & -6 & 0 \\ 6 & 7 & 4 \\ -2 & 4 & 3 \end{pmatrix}$ is

(a) 4   (b) 10   (c) 12   (d) 18

49. The eigenvalues of A^5 if $A = \begin{pmatrix} 3 & 0 & 0 \\ 5 & 4 & 0 \\ 3 & 0 & 1 \end{pmatrix}$ are

(a) 1, 3, 4   (b) 1, 9, 16

(c) 1, 27, 64   (d) 1, 243, 1024

50. The sum and the product of the eigenvalues of $\begin{pmatrix} 2 & 3 & -2 \\ -2 & 1 & 1 \\ 1 & 0 & 2 \end{pmatrix}$ are, respectively

(a) 5, 20   (b) 25, 30   (c) 5, 21   (d) 21, 30

51. Let A be a square matrix of order 3 with trace$(A) = 9$ and det$(A) = 24$ and if one eigenvalue is 2, then the other eigenvalues are

(a) 8, 3   (b) 3,4   (c) 3, 0   (d) 4, 4

52. Let A be a square matrix of order n such that for some scalar λ, the matrix $A-\lambda I$ is singular. Then
(a) λ is a characteristic root of A
(b) $\lambda = 0$
(c) λ is not a characteristic root of A
(d) none of these

53. Which of the followings is true?
(a) a matrix may have many eigenvectors corresponding to a eigenvalue
(b) a matrix have an unique eigenvector corresponding to a eigenvalue
(c) a matrix may not have an eigenvector corresponding to a eigenvalue
(d) none of these

54. If A and B be two non-singular square matrices, then
(a) A and B have the same eigenvalues
(b) A^{-1} and B^{-1} have the same eigenvalues
(c) $A^{-1}B$ and BA^{-1} have the same eigenvalues
(d) none of these

55. If A be a singular matrix, then which of the following can be a characteristic polynomial of A?
(a) $\lambda^2 + 2\lambda$ (b) $\lambda^3 + 2\lambda + 1$
(c) $\lambda + 3$ (d) $\lambda^2 - 4\lambda + 5$

56. If $\lambda^3 - 6\lambda^2 + 9\lambda - 4$ is the characteristic polynomial of a square matrix A, then $A^{-1} = ?$
(a) $A^2 - 6A + 9I$ (b) $\frac{1}{4}A^2 - \frac{3}{2}A + \frac{9}{4}$
(c) $\frac{1}{4}A^2 - \frac{3}{2}A + \frac{9}{4}I$ (d) none of these

57. The eigenvalues of $\begin{pmatrix} 8 & -6 & 2 \\ -6 & 7 & -4 \\ 2 & -4 & 3 \end{pmatrix}$ are
(a) 5, 7, 8 (b) 3, 0, –7
(c) 8, 3, –7 (d) 3, 0, 15

58. The determinant of the matrix
$\begin{pmatrix} 6 & -8 & 1 & 1 \\ 0 & 2 & 4 & 6 \\ 0 & 0 & 4 & 8 \\ 0 & 0 & 0 & -1 \end{pmatrix}$ is
(a) 11 (b) – 48 (c) 0 (d) – 24

59. The minimum eigenvalue of the following matrix is?
$\begin{bmatrix} 3 & 5 & 2 \\ 5 & 12 & 7 \\ 2 & 7 & 5 \end{bmatrix}$
(a) 0 (b) 1 (c) 2 (d) 3

60. The rank of the matrix given below is
$\begin{pmatrix} 1 & 4 & 8 & 7 \\ 0 & 0 & 3 & 0 \\ 4 & 2 & 3 & 1 \\ 3 & 12 & 24 & 21 \end{pmatrix}$
(a) 3 (b) 1 (c) 2 (d) 4

61. The value of the determinant
$\begin{vmatrix} 1 & 0 & 0 & 0 & 0 \\ -2 & 2 & 0 & 0 & 0 \\ 3 & 5 & 3 & 0 & 0 \\ -1 & 4 & 7 & 4 & 0 \\ -5 & 6 & 2 & 1 & 1 \end{vmatrix}$ is
(a) 24 (b) 32 (c) – 112 (d) 0

62. The solution for the system defined by the set of equations: $4y + 3z = 8$, $2x - z = 2$, $3x + 2y - z = 5$
(a) $x = 0, y = 1/2, z = 2$
(b) $x = 0, y = 1, z = 4/3$
(c) $x = 1, y = 1/2, z = 2$
(d) non existent

63. The solution of the following set of equation $x + 2y + 3z = 20$, $7x + 3y + z = 13$, $x + 6y + 2z = 0$ is
(a) $x = -2, y = 2, z = 8$
(b) $x = 2, y = -3, z = 8$
(c) $x = -2, y = 3, z = -8$
(d) $x = 8, y = 2, z = -3$

64. The solution of the following set of equations: $5x + 4y + 10z = 13$, $x + 3y + z = 7$, $4x - 2y + z = 0$ is
(a) $x = 2, y = 1, z = 1$
(b) $x = 1, y = 2, z = 0$
(c) $x = 1, y = 0, z = 2$
(d) $x = 0, y = 1, z = 2$

65. Consider the system of equations $-x + 2y - 3z = 2$, $x + 6y + 12z = 1$, $2x - 4y + 3kz = -4$ What is the value of k for which the system of equations has an infinite number of solution?
(a) 1 (b) 2 (c) 3 (d) 0

66. Consider $2x_1 + x_2 = 3$, $5x_1 + bx_2 = 7.5$

The system of linear combinations in two variables shown above will have an infinite number of solutions, if and only if "b" = ?

(a) 2.5   (b) 3   (c) 3.5   (d) 4

67. If $A = \begin{bmatrix} 1 & 0 & 0 & 1 \\ 0 & -1 & 0 & -1 \\ 0 & 0 & i & i \\ 0 & 0 & 0 & -i \end{bmatrix}$, then the matrix A^4,

calculated by the use of the Cayley Hamilton theorem or otherwise is

(a) $\begin{bmatrix} 1 & 0 & 0 & 0 \\ 0 & 1 & 0 & 0 \\ 0 & 0 & 1 & 0 \\ 0 & 0 & 0 & 1 \end{bmatrix}$ (b) $\begin{bmatrix} -1 & 0 & 0 & -1 \\ 0 & 0 & 1 & 0 \\ 0 & 0 & 0 & 1 \\ 0 & 1 & 0 & 0 \end{bmatrix}$

(c) $\begin{bmatrix} 1 & 0 & 1 & 0 \\ 0 & 1 & 0 & 1 \\ 0 & 0 & 1 & 0 \\ 0 & 0 & 0 & 1 \end{bmatrix}$ (d) none of these

68. For the matrix $\begin{pmatrix} 4 & 1 \\ 1 & 4 \end{pmatrix}$ the eigenvalues are

(a) 3, –3   (b) –3, –5   (c) 3, 5   (d) 5, 0

69. The number of values of "k" for which the system of equations:

$kx + (k + 3)y = 10z$, $(k - 1)x + (k - 2)y = 5z$, $2x + (k + 4)y = kz$ has infinitely many solutions

(a) 2   (b) 3   (c) 5   (d) 0

70. If $A = \begin{bmatrix} 1 & -2 & -1 \\ 2 & 3 & 1 \\ 0 & 5 & -2 \end{bmatrix}$ and $\text{adj}(A) = $

$\begin{bmatrix} -11 & -9 & 1 \\ 4 & -2 & -3 \\ 10 & k & 7 \end{bmatrix}$, then "$k$" = ?

(a) 3   (b) –3   (c) 5   (d) –5

71. Let N be a nilpotent matrix of order 4 with real entries. Then which one of the following statements is true about eigenvalues of N?

(a) All eigenvalues are non-zero real numbers
(b) All eigenvalues are purely imaginary
(c) Zero is the only eigenvalue
(d) At least one eigenvalue is real and at least one eigenvalue has non-zero imaginary part.

72. Let A be a 3×3 matrix with trace $(A) = 3$ and det $(A) = 2$. If "1" is an eigenvalue of A, then the eigenvalues of the matrix $A^2 - 2I$ are

(a) 1, 2 $(i - 1)$, $-2(i + 1)$
(b) -1, 2 $(i - 1)$, $2(i + 1)$
(c) 1, 2 $(i + 1)$, $-2(i + 1)$
(d) -1, 2 $(i - 1)$, $-2(i + 1)$

73. $P = \begin{pmatrix} 1 & 1 \\ 2 & 2 \end{pmatrix}$, $Q = \begin{pmatrix} 2 & 1 \\ 2 & 2 \end{pmatrix}$, and $R = \begin{pmatrix} 3 & 0 \\ 1 & 3 \end{pmatrix}$, then which of the following statements are true?

(a) $PQ = PR$ (b) $QR = RP$
(c) $QP = RP$ (d) $PQ = QR$

74. If A be a square symmetric real valued matrix of dimension $2n$, then the eigenvalues of A are

(a) $2n$ distinct real values
(b) $2n$ real values not necessarily distinct
(c) n distinct pairs of complex conjugate numbers
(d) n pairs of complex conjugate numbers not necessarily distinct

75. Which of the following is an eigenvector of

the matrix $\begin{pmatrix} 5 & 0 & 0 & 0 \\ 0 & 5 & 5 & 0 \\ 0 & 0 & 2 & 1 \\ 0 & 0 & 3 & 1 \end{pmatrix}$?

(a) $\begin{bmatrix} 1 \\ -2 \\ 0 \\ 0 \end{bmatrix}$ (b) $\begin{bmatrix} 0 \\ 0 \\ 1 \\ 0 \end{bmatrix}$ (c) $\begin{bmatrix} 1 \\ 0 \\ 0 \\ -2 \end{bmatrix}$ (d) $\begin{bmatrix} 1 \\ -1 \\ 2 \\ 1 \end{bmatrix}$

76. The eigenvalues of the matrix $A = \begin{pmatrix} 1 & 2 \\ 4 & 3 \end{pmatrix}$ are

5 and -1. Then, the eigenvalues of $-2A + 3I$ are

(a) -7 and 5   (b) -7 and -5
(c) $-7/5$ and 1/5   (d) 1/7 and $-1/2$

77. Let $c_1, c_2, ..., c_n$ be scalars, not all zero, such

that $\sum_{i=1}^{n} c_i a_i = 0$, where a_i are column vectors

in R^n. Consider the set of linear equations Ax

$= b$, where $A = [a_1, a_2, ..., a_n]$ and $b = \sum_{i=1}^{n} a_i$.
The set of equations has

(a) a unique solution at $x = J_n$, where J_n denotes a n-dimensional vector of all 1

(b) no solution

(c) infinitely many solutions

(d) finitely many solutions

Answer key				
1. (b)	**2.** (a)	**3.** (b)	**4.** (b)	**5.** (c)
6. (a)	**7.** (c)	**8.** (b)	**9.** (a)	**10.** (d)
11. (c)	**12.** (d)	**13.** (c)	**14.** (a)	**15.** (a)
16. (a)	**17.** (a)	**18.** (b)	**19.** (d)	**20.** (c)
21. (a)	**22.** (b)	**23.** (a)	**24.** (b)	**25.** (a)
26. (a)	**27.** (b)	**28.** (a)	**29.** (a)	**30.** (a)
31. (a)	**32.** (d)	**33.** (c)	**34.** (d)	**35.** (c)
36. (c)	**37.** (a)	**38.** (c)	**39.** (a)	**40.** (d)
41. (a)	**42.** (a)	**43.** (c)	**44.** (d)	**45.** (b)
46. (a)	**47.** (b)	**48.** (d)	**49.** (d)	**50.** (c)
51. (b)	**52.** (a)	**53.** (a)	**54.** (c)	**55.** (a)
56. (c)	**57.** (d)	**58.** (b)	**59.** (a)	**60.** (a)
61. (a)	**62.** (c)	**63.** (b)	**64.** (b)	**65.** (b)
66. (a)	**67.** (a)	**68.** (c)	**69.** (a)	**70.** (d)
71. (c)	**72.** (d)	**73.** (a)	**74.** (b)	**75.** (a)
76. (a)	**77.** (c)			

Hints

1. (b) Show that $AA^T = I$

2. (a) Show that $A^2 = A$

3. (b) Comparing we get, $x + 3 = 5$ and so $x = 2$

4. (b) Compute AB and BA.

5. (c)

6. (a) See the properties of a skew-symmetric matrix

7. (c)

8. (b) See the properties of the inverse of a matrix.

9. (a)

10. (d)

11. (c) See the properties of a Hermitian matrix.

12. (d)

$$\begin{vmatrix} 1 & \omega \\ \omega & \omega^2 \end{vmatrix} = \begin{vmatrix} \omega & \omega^2 \\ \omega^2 & 1 \end{vmatrix} = \begin{vmatrix} \omega^2 & 1 \\ 1 & \omega \end{vmatrix} = 0 \quad .$$

$$\left(since \ 1 + \omega + \omega^2 = 0, \ \omega^3 = 1, \ \omega^4 = \omega \right)$$

13. (c)

14. (a)

15. (a)

16. (a)

17. (a)

18. (b)

19. (d)

20. (c)

21. (a)

22. (b)

23. (a)

$$A + A^T = I$$

$$\Rightarrow \begin{pmatrix} \cos x & \sin x \\ -\sin x & \cos x \end{pmatrix} + \begin{pmatrix} \cos x & -\sin x \\ \sin x & \cos x \end{pmatrix} = I$$

$$\Rightarrow \begin{pmatrix} 2\cos x & 0 \\ 0 & 2\cos x \end{pmatrix} = \begin{pmatrix} 1 & 0 \\ 0 & 1 \end{pmatrix}$$

$$\Rightarrow 2\cos x = 1$$

$$\Rightarrow \cos x = \frac{1}{2} = \cos\frac{\pi}{3}$$

24. (b)

25. (a)

$$\begin{vmatrix} \frac{1}{a} & 1 & bc \\ \frac{1}{b} & 1 & ca \\ \frac{1}{c} & 1 & ab \end{vmatrix} = \frac{1}{a} \times \frac{1}{b} \times \frac{1}{c} \begin{vmatrix} 1 & a & abc \\ 1 & b & bca \\ 1 & c & abc \end{vmatrix}$$

$$= \frac{1}{abc} \times abc \begin{vmatrix} 1 & a & 1 \\ 1 & b & 1 \\ 1 & c & 1 \end{vmatrix}$$

$$= 0$$

(since C_1 and C_3 are identical)

26. (a)

$$\begin{vmatrix} \sin(A+B+C) & \sin B & \cos C \\ -\sin B & 0 & \tan A \\ \cos(A+B) & -\tan A & 0 \end{vmatrix}$$

$$= \begin{vmatrix} \sin \pi & \sin B & \cos C \\ -\sin B & 0 & \tan A \\ \cos(\pi - C) & -\tan A & 0 \end{vmatrix}$$

$$= \begin{vmatrix} 0 & \sin B & \cos C \\ -\sin B & 0 & \tan A \\ -\cos C & -\tan A & 0 \end{vmatrix}$$

$= -\sin B(0 + \cos C \times \tan A)$

$\quad + \cos C(\tan A \times \sin B - 0)$

$= 0]$

27. (b)

28. (a)

$$f(x) = \begin{vmatrix} \cos x & 1 & 0 \\ 1 & 2\cos x & 1 \\ 0 & 1 & 2\cos x \end{vmatrix}$$

$= \cos x(4\cos^2 x - 1) - 2\cos x$

$= \cos x \left\{ 2(2\cos^2 x - 1) + 1 \right\} - 2\cos x$

$= \cos x \left\{ 2\cos 2x + 1 \right\} - 2\cos x$

$= 2\cos x \cos 2x + \cos x - 2\cos x$

$= \cos(2x + x) + \cos(2x - x) - \cos x$

$= \cos 3x$

29. (a) Expand the determinant and use the results $1 + \omega + \omega^2 = 0$, $\omega^3 = 1$, $\omega^4 = \omega$

30. (a) Use the operation $C_1 \to C_1 + C_2 + C_3$ and then take $a + b + c - x$ common from C_1

31. (a) Use the operation $R_3 \to R_3 + R_1 - 2R_2$.

32. (d)

33. (c)

$$\text{Let, } A = \begin{pmatrix} 0 & -1 & 2 \\ 1 & 0 & -3 \\ -2 & 3 & 0 \end{pmatrix}$$

Then A is Skew-symmetric matrix of order 3 (odd order)

$$\text{Then } |A| = \begin{vmatrix} 0 & -1 & 2 \\ 1 & 0 & -3 \\ -2 & 3 & 0 \end{vmatrix} = 0.$$

Therefore, (c) is satisfied.

$$\text{Let } B = \begin{pmatrix} 0 & 2 \\ -2 & 0 \end{pmatrix}$$

Then B is Skew-symmetric matrix of order 2 (even order)

$|B| = 0 + 4 = 4 \neq 0$

Therefore, B is non-singular.

So (a) is not correct

34. (d)

35. (c) Rank(A) = 1

$\Rightarrow$ rank $(A) < 3(= \text{order of } A)$

$\Rightarrow \det(A) = 0]$

36. (c)

37. (a)

38. (c)

39. (a)

40. (d)

41. (a)

42. (a)

43. (c)

44. (d)

45. (b)

46. (a)

47. (b)

48. (d)

49. (d)

A is a lower triangular matrix and so its eigenvalues are diagonal elements *i.e*; 1, 3, 4.

Hence, the eigenvalues of A^5 are 1^5, 3^5, 4^5 *i.e*; 1, 243, 1024.

50. (c) The sum of eigenvalues

= trace of the matrix and determinant of the matrix

= product of the eigenvalues

51. (b)

Let λ_1 and λ_2 be the other two eigenvalues.

Then, the sum of eigenvalues of $A = \text{trace}(A)$

$\Rightarrow 2 + \lambda_1 + \lambda_2 = 9$

$\Rightarrow \lambda_1 + \lambda_2 = 7$ ...(1)

Again, the product of the eigenvalues of $A = \det(A)$

$\Rightarrow 2 \times \lambda_1 \times \lambda_2 = 24$

$\Rightarrow \lambda_1 \lambda_2 = 12$ ...(2)

Then, solve (1) and (2)

52. (a) The matrix $A - \lambda I$ is singular

$\Rightarrow \det(A - \lambda I) = 0$

$\Rightarrow \lambda$ is a characteristic root of A

53. (a)

54. (c) A and B are non-singular implies both A^{-1} and B^{-1} exist.

Now, $A^{-1}B = A^{-1}BI = A^{-1}BA^{-1}A = A^{-1}(BA^{-1})A$. Since the matrices $A^{-1}(BA^{-1})$ A and BA^{-1} have the same eigenvalues, so $A^{-1}B$ and BA^{-1} have the same eigenvalues.

55. (a) A is singular $\Rightarrow \lambda = 0$ is an eigenvalue of A, which can be obtained by considering the characteristic polynomial $\lambda^2 + 2\lambda$.

56. (c)

$\lambda^3 - 6\lambda^2 + 9\lambda - 4$ is the characteristic polynomial

$\Rightarrow \lambda^3 - 6\lambda^2 + 9\lambda - 4 = 0$ is the characteristic equation

$\Rightarrow A^3 - 6A^2 + 9A - 4I = O$ (using the Cayley Hamilton theorem)

$\Rightarrow A^{-1}(A^3 - 6A^2 + 9A - 4I) = A^{-1}O = O$

$\Rightarrow A^2 - 6A + 9I - 4A^{-1} = O$

$\Rightarrow A^{-1} = \dfrac{1}{4}A^2 - \dfrac{3}{2}A + \dfrac{9}{4}I$

57. (d)

The sum of eigenvalues = trace of the matrix

58. (b) The matrix corresponding to the given determinant is upper triangular. Therefore, given determinant = products of the all elements lying in the principal diagonal $= 6 \times 2 \times 4 \times (-1) = -48$.

59. (a)

Let A be the given matrix. Then, it can be shown that $\det(A) = 0$ and so one eigenvalue is zero, which is the minimum.

60. (a)

61. (a)

The matrix corresponding to the given determinant is a lower triangular.

Therefore, given determinant = products of the all elements lying in the principal diagonal

$= 1 \times 2 \times 3 \times 4 \times 1 = 24$.

62. (c)

$x = 1$, $y = 1/2$, $z = 2$ satisfy all the three equations.

63. (b)

64. (b)

65. (b)

$\Rightarrow \Delta = 0$

$\Rightarrow \begin{vmatrix} -1 & 2 & -3 \\ 1 & 6 & 12 \\ 2 & -4 & 3k \end{vmatrix} = 0$

$\Rightarrow \begin{vmatrix} -1 & 2 & -3 \\ 0 & 8 & 9 \\ 0 & 0 & 3k-6 \end{vmatrix} = 0$

(by $R_1 \to R_1 + R_2, R_3 \to R_3 + 2R_1$)

$\Rightarrow (3k-6)(-8-2) = 0$

$\Rightarrow k = 2$

66. (a) The system of equations has an infinite number of solutions $\Rightarrow$ coefficient determinant $= 0$.

67. (a)

$|A - \lambda I| = 0$

$\Rightarrow \begin{bmatrix} 1-\lambda & 0 & 0 & 1 \\ 0 & -1-\lambda & 0 & -1 \\ 0 & 0 & i-\lambda & i \\ 0 & 0 & 0 & -i-\lambda \end{bmatrix} = 0$

$\Rightarrow (1-\lambda)(-1-\lambda)(i-\lambda)(-i-\lambda) = 0$

$\Rightarrow (\lambda-1)(\lambda+1)(\lambda-i)(\lambda+i) = 0$

$\Rightarrow (\lambda^2 - 1^2)(\lambda^2 - i^2) = 0$

$\Rightarrow (\lambda^2 - 1)(\lambda^2 + 1) = 0$

$\Rightarrow \lambda^4 - 1 = 0$

$\Rightarrow \lambda^4 = 1$

Now use the Cayley Hamilton theorem.

68. (c) The sum of the eigenvalues = trace of the matrix $= 8$]

69. (a) The system has infinitely many solutions $\Rightarrow$ co-efficient determinant $= 0$.

70. (d) Using the matrix A, find adj(A) and then compare it with given adj(A).

71. (c)

Zero is the only eigenvalue of a nilpotent matrix.

72. (d), 73. (a), 74. (b), 75. (a)

73. (a)

If λ is an eigenvalue of **A**, then the eigenvalue of $-2A + 3I$ has the form $-2\lambda + 3$.

74. (c)

$$\sum_{i=1}^{n} c_i \, a_i = 0$$

$\Rightarrow$ the columns a_1, a_2, ...a_n are not linearly independent

$\Rightarrow$ rank$(A) < n$

$\Rightarrow$ the system has an infinite number of solutions

CALCULUS

2.1 FUNCTIONS AND LIMITS

2.1.1 Definition of a Function

Let A and B be two empty sets. Then, a function "*f*" from A to B is a rule (correspondence) that assigns each element $x \in A$ to a unique element of $y \in B$. A is said to be the domain of the function and *B* is said to be the co-domain of the function "*f*," respectively. Symbolically, we write $f : A \to B$ is a function which is defined by $y = f(x)$ for $x \in A$.

Example:

Let A = {–1, 2, 3} and B = {1,4,9,10} be two sets. Let us apply a rule "squaring" on each of the elements of A. If "*f*" denotes the rule "squaring," then

$f(-1)$ = square of –1 = $1 \in B$ for $-1 \in A$,

$f(2)$ = square of 2 = $4 \in B$ for $2 \in A$,

$f(3)$ = square of 3 = $9 \in B$ for $3 \in A$.

Thus, there exist a rule "squaring" (f) which assigns each element of A to a unique element of B. Consequently, *f* is a function from A to B. This function can be defined by $f(x) = x^2$ for $x \in A$.

Remember:

Range of the function $f : A \to B$ is given by Range (f) = {f(x):x ∈ A}. Thus, for the above example, Range (f) = {f(x): x ∈ A} = {f(–1), f(2), f(3)} = {1, 4, 9}.

2.1.2 Some Special Functions

I. Modulus Function:

Let $f : R \to R$ be a function defined by f(x) = |x| for $x \in R$, where

$$|x| = \begin{cases} x, x > 0 \\ -x, x < 0 \\ 0, x = 0 \end{cases}$$ Here, f is called modulus function.

II. Greatest Integer Function:

Let $f : R \to R$ be a function defined by f(x) = [x] for $x \in R$, where [x] denotes the greatest integer not greater than "x." Here, f is called greatest integer function.

Remember:

(*i*) x – 1 < [x] ≤ x for x ∈ R.

(*ii*) [4] = 4, [–7] = –7, [5.01] = 5, [–2.9] = –3 etc. Thus

$$|x| = \begin{cases} x, \text{if } x \text{ is an integer} \\ \text{integer part of x,} \\ \quad \text{if } x = +ve \text{ fraction} \\ (\text{integer part of } x) - 1, \\ \quad \text{if } x = -ve \text{ fraction} \end{cases}$$

III. Even Function:

A function $f : R \to R$ is called an even function if $f(x) = f(-x) \; \forall x \in R$.

Example:

cos x is an even function of "x"

(since cos (–x) = cosx).

IV. Odd Function:

A function $f : R \to R$ is called an odd function if $f(x) = -f(-x) \; \forall x \in R$.

Example:

sin x is an odd function of "x" (since – sin(–x) = sin x).

V. Periodic Function:

A function $f : R \to R$ is called a periodic function of period "T" (T should be least positive integer) if $f(x + T) = f(x) \ \forall \ x \in R$.

Example:

sin x is a periodic function with period 2π.

2.1.3 Introduction to Limits

Suppose x is a real variable which takes the values 2.9, 2.99, 2.999, and so on. Then it is clear that "x" takes the values very close to "3" (but less than "3"), and the values of x are gradually approaches to the number 3 but never becomes equal to "3." This is mathematically expressed as $x \to 3-$ (read as 'x tends to three minus).

Next, consider the values 3.1, 3.01, 3.001, and so on of the variable "x." Then clearly "x" takes the values very close to "3" (but greater than "3"), and the values of x are gradually approaches to the number 3 but never becomes equal to "3." This is mathematically expressed as $x \to 3+$ (read as ' x tends to three plus).

When $x \to 3-$ and $x \to 3+$, we say that "x" tends to "3" and we write x → 3. Sometimes we write lim $x = 3$.

Now let us consider a function "f" defined by

$$f(x) = \frac{x^2 - 9}{x - 3}, x \neq 3$$

x	2.99	2.999	...	3.01	3.001	...
f(x)	5.99	5.999	...	6.01	6.001	...

From the above table, it is clear that when x → 3 –, we have f(x) → 6 – and when x → 3+, we have f(x) → 6 +. Thus, x → 3 ⇒ f(x) → 6. This can be expressed mathematically by writing $\lim_{x \to 3} f(x) = 6$.

Also, we can write $\lim_{x \to 3+} f(x) = 6$ and $\lim_{x \to 3-} f(x) = 6$

2.1.4 Definition of Limit

Let $f : R \to R$ be a function. Then a real number "l" is said to be the limit of f at x = a if for a preassigned number $\varepsilon > 0$, there exist a number $\delta > 0$ such that $|f(x) - l| < \varepsilon$ whenever $|x - a| < \delta$.

Symbolically, we write $\lim_{x \to a} f(x) = l$.

Example:

Let $f(x) = \frac{x^2 - 9}{x - 3}, x \neq 3$. Then

$$|f(x) - 6| = \left| \frac{x^2 - 9}{x - 3} - 6 \right| = |x + 3 - 6| = |x - 3|$$

Let us choose $\varepsilon > 0$. Then there exist a number δ $(= \varepsilon) > 0$ such that $|f(x) - 6| < \varepsilon$ whenever $|x - 3| < \delta$.

Hence, $\lim_{x \to 3} f(x) = 6$.

Remember:

I. Right-hand limit (RHL) at x = a is denoted by $\lim_{x \to a+} f(x)$ and left-hand limit (LHL) at x = a is

denoted by $\lim_{x \to a-} f(x)$ If $\lim_{x \to a+} f(x) = \lim_{x \to a-} f(x)$,

then we say that $\lim_{x \to a} f(x)$ exists.

II. $\lim_{x \to a} f(x)$ may be equal to f(a) or may not be

equal to $f(a)$. This is illustrated by the following examples:

Let us consider a function "f" defined by

$$f(x) = \begin{cases} x + 1, & x < 1 \\ 2, & x = 1 \\ 3x - 1, & x > 1 \end{cases}$$

Then, $\lim_{x \to 1+} f(x) = \lim_{x \to 1+} (3x - 1) = 3 \times 1 - 1 = 2$,

$\lim_{x \to 1-} f(x) = \lim_{x \to 1-} (x + 1) = 1 + 1 = 2$.

Hence, $\lim_{x \to 1} f(x) = 2 = f(1)$.

Let us consider another function "g" defined by

$$g(x) = \begin{cases} x, & x < 1 \\ 3, & x = 1 \\ 2x - 1, & x > 1 \end{cases}$$

Then $\lim_{x \to 1+} g(x) = \lim_{x \to 1+} (2x - 1) = 2 \times 1 - 1 = 1$,

$\lim_{x \to 1-} g(x) = \lim_{x \to 1-} (x) = 1$.

Hence, $\lim_{x \to 1} g(x) = 1 \neq g(1) = 3$.

2.1.5 Fundamental Theorems on Limits

If $\lim_{x \to a} f(x) = l_1$ and $\lim_{x \to a} g(x) = l_2$ (where l_1 and l_2 are finite quantities), then

(i) $\lim_{x \to a} \{ f(x) \pm g(x) \} = l_1 \pm l_2$

(ii) $\lim_{x \to a} \{ f(x) \times g(x) \} = l_1 \times l_2$

(iii) $\lim_{x \to a} \left\{ \dfrac{f(x)}{g(x)} \right\} = \dfrac{l_1}{l_2}$, provided $l_2 \neq 0$

(iv) $\lim_{x \to a} |f(x)| = |l_1|$

(v) $\lim_{x \to a} f(x)^{g(x)} = l_1^{l_2}$

2.1.6 Fundamental Formulas on Limits

(i) $\lim_{x \to a} \dfrac{x^n - a^n}{x - a} = na^{n-1}$

(ii) $\lim_{x \to 0} \dfrac{e^x - 1}{x} = 1$

(iii) $\lim_{x \to 0} \dfrac{\log_e (1 + x)}{x} = 1$

(iv) $\lim_{x \to 0} \dfrac{\sin x}{x} = 1$

(v) $\lim_{x \to 0} \dfrac{\tan x}{x} = 1$

(vi) $\lim_{x \to 0} \cos x = 1$

(vii) $\lim_{x \to 0} \dfrac{(l + x)^n - 1}{x} = n$

(viii) $\lim_{x \to 0} \dfrac{a^x - 1}{x} = \log_e a$

(ix) $\lim_{x \to 0} (1 + x)^{\frac{1}{x}} = e$

(x) $\lim_{x \to \infty} \left(1 + \dfrac{1}{x} \right)^x = e$

(xi) $\lim_{x \to 0} (1 + ax)^{\frac{1}{x}} = e^a$

(xii) $\lim_{x \to \infty} \left(1 + \dfrac{a}{x} \right)^x = e^a$

Remember:

(i) $\lim_{x \to \infty} \dfrac{\log_e x}{x^m} = 0$ (provided m>0)

(ii) $\lim_{x \to \infty} x^n = 0$ (provided "n" is a positive integer and $-1 < x < 1$)

(iii) $\lim_{n \to \infty} \dfrac{x^n}{n} = \begin{cases} 0, \text{ if } |x| \leq 1 \\ \infty, \text{ if } x > 1 \end{cases}$

(iv) $\lim_{x \to 0} [x]$ does not exist

(v) None of the limits $\lim_{x \to 0} \cos \dfrac{1}{x}$, $\lim_{x \to 0} \sin \dfrac{1}{x}$ exist.

(vi) $\lim_{x \to 0} \text{sgn}(x)$ does not exist, where

$$\text{sgn}(x) = \begin{cases} \dfrac{|x|}{x}, & x \neq 0 \\ 0, & x = 0 \end{cases} = \begin{cases} 1, & x > 0 \\ 0, & x = 0 \\ -1, & x < 0 \end{cases}$$

(vii) $\lim_{x \to 0} x^n \sin \dfrac{1}{x} = \lim_{x \to 0} x^n \cos \dfrac{1}{x} = 0$

(provided n > 0)

(viii) $\lim_{x \to a} \dfrac{|x - a|}{x - a}$ does not exist.

2.1.7 The Sandwich Theorem

If $f, g, h : R \to R$ be three functions such that $f(x) \leq g(x) \leq h(x)$ for all $x \in R$ and $\lim_{x \to a} f(x) = \lim_{x \to a} h(x) = l$, then $\lim_{x \to a} g(x) = l$.

Application:

Let us evaluate $\lim_{x \to 0} x \cos \dfrac{1}{x^2}$.

$-1 \leq \cos \dfrac{1}{x^2} \leq 1 \Rightarrow -x \leq x \cos \dfrac{1}{x^2} \leq x$ for all $x(\neq 0)$

$\in R$.

Let $f(x) = -x$, $g(x) = x \cos \dfrac{1}{x^2}$, $h(x) = x$. Then

$\lim_{x \to 0} f(x) = \lim_{x \to 0} h(x) = 0$. Hence by the Sandwich

theorem, $\lim_{x \to 0} g(x) = 0$, i.e, $\lim_{x \to 0} x \cos \dfrac{1}{x^2} = 0$.

2.1.8 Infinite Limits

I. Let $D \subseteq R$ and $f : D \to R$ be a function. If corresponding to a preassigned positive number G, there exist $\delta > 0$ such that f(x) > G whenever $|x - a| < \delta$, then we say that f tends to ∞ as $x \to a$. Symbolically, we write $\lim_{x \to a} f(x) = \infty$.

Example: $\lim_{x \to 0} \dfrac{1}{x} = \infty$.

II. Let $D \subseteq R$ and $f : D \to R$ be a function. If corresponding to a pre-assigned positive number G, there exist $\delta > 0$ such that f(x) < -G whenever

$|x-a| < \delta$, then we say that f tends to $-\infty$ as $x \to a$. Symbolically we write $\lim\limits_{x \to a} f(x) = -\infty$.

Example: $\lim\limits_{x \to -\infty} \left(\dfrac{-4}{x} \right) = -\infty$.

2.1.9 Limits at Infinity

I.   Let $D \subseteq R$ and $f : D \to R$ be a function. Then we say that f tends to "l" as $x \to \infty$ if corresponding to a preassigned $\varepsilon > 0$, there exist a real number $G > 0$ such that $|f(x) - l| < \varepsilon$ for all x > G. Symbolically, we write $\lim\limits_{x \to \infty} f(x) = l$.

Example: $\lim\limits_{x \to \infty} \dfrac{1}{x} = 0$.

II.   Let $D \subseteq R$ and $f : D \to R$ be a function. Then we say that f tends to "l" as $x \to -\infty$ if corresponding to a preassigned $\varepsilon > 0$, there exist a real number $G > 0$ such that $|f(x) - l| < \varepsilon$ for all x < $-$G. Symbolically, we write $\lim\limits_{x \to -\infty} f(x) = l$.

Example: $\lim\limits_{x \to -\infty} \dfrac{-1}{x} = 0$.

2.1.10 Infinite Limits at Infinity

I.   Let $D \subseteq R$ and $f : D \to R$ be a function. If corresponding to a preassigned positive number G, there exist a positive real number K such that $f(x)$ > G for all x > K, then we say that f tends to ∞ as $x \to \infty$. Symbolically, we write $\lim\limits_{x \to \infty} f(x) = \infty$.

Example: $\lim\limits_{x \to \infty} 4x = \infty$.

II.   Let $D \subseteq R$ and $f : D \to R$ be a function. If corresponding to a preassigned positive number G, there exist a positive real number K such that $f(x) < -$G for all x < $-$K, then we say that f tends to $-\infty$ as $x \to -\infty$. Symbolically, we write $\lim\limits_{x \to -\infty} f(x) = -\infty$.

Example: $\lim\limits_{x \to -\infty} x = -\infty$.

III. Let $D \subseteq R$ and $f : D \to R$ be a function. If corresponding to a preassigned positive number G, there exist a positive real number K such that $f(x) < -$G for all x > K, then we say that f tends to $-\infty$ as $x \to \infty$. Symbolically, we write $\lim\limits_{x \to \infty} f(x) = -\infty$.

Example: $\lim\limits_{x \to \infty} (-x) = -\infty$.

IV. Let $D \subseteq R$ and $f : D \to R$ be a function. If corresponding to a preassigned positive number

G, there exists a positive real number K such that $f(x) >$G for all x<$-$K, then we say that f tends to ∞ as $x \to -\infty$. Symbolically, we write $\lim\limits_{x \to -\infty} f(x) = \infty$.

Example: $\lim\limits_{x \to -\infty} (-x) = \infty$.

2.2 CONTINUITY AND DIFFERENTIABILITY

2.2.1 Continuity

A function is said to be continuous at $x = a$ if the followings condition is satisfied

$$\lim_{x \to a+} f(x) = \lim_{x \to a-} f(x) = f(a) \text{ i.e., } \lim_{x \to a} f(x) = f(a).$$

In other words, a function

$f : D \to R$ (where $D \subseteq R$) is said to be continuous at $x = a$ if for a preassigned $\varepsilon > 0$, $\exists \delta > 0$ such that $|f(x) - f(a)| < \varepsilon$ whenever $|x - a| < \delta$.

If f is continuous at each point of D, then we say that f is continuous on D.

Remember:

(i)   A function f defined by $y = f(x)$ is said to be continuous if the graph of the function is a continuous curve i.e., if we can draw the graph at one go.

(ii)   If a function f : R $\to$ R is continuous on R and $f(x + y) = f(x) + f(y)$ $\forall x, y \in R$ and if $f(1) = k \text{ (constant)}$, then $f(x) = kx$ $\forall x \in \mathbb{R}$.

(iii)   If a function $f : \mathbb{R} \to \mathbb{R}$ is continuous on R and $f(x + y) = f(x) f(y)$ $\forall x, y \in \mathbb{R}$ then either $f(x) = 0$ $\forall x \in \mathbb{R}$ or $f(x) = k^x$ $\forall x \in \mathbb{R}$ (where k is a real number).

(iv)   If $f, g : D \to \mathbb{R}$ (where $D \subseteq \mathbb{R}$) are continuous functions, then each of the functions $f + g$, $f - g$, $cf (c = \text{constant})$, fg, $\dfrac{f}{g}$ (assuming $g(x) \neq 0$ $\forall x \in D$) and $|f(x)|$ is continuous.

(v) $f(x) = |x - a|$ is continuous $\forall x \in \mathbb{R}$ but $f(x) = [x]$ is not continuous for any $x \in \mathbb{R}$

(vi) If $f:[a, b] \rightarrow \mathbb{R}$ is continuous and $f(a)f(b) < 0$, then $\exists$ a point $c \in (a, b)$ such that $f(c) = 0$.

2.2.2 Discontinuity

A function f is said to be discontinuous at x = a if at least one of the following condition is satisfied

(i) $\lim_{x \to a} f(x) \neq f(a)$

(ii) $\lim_{x \to a+} f(x) \neq \lim_{x \to a-} f(x)$

(iii) $\lim_{x \to a+} f(x)$ does not exist

(iv) $\lim_{x \to a-} f(x)$ does not exist

(v) $f(x)$ is undefined at x = a.

There are three types of discontinuities:

I. Removable Discontinuity:

If $\lim_{x \to a} f(x) \neq f(a)$ i.e; $\lim_{x \to a+} f(x) = \lim_{x \to a-} f(x) \neq f(a)$, then we say that f has a removable discontinuity at $x = a$.

Example:

Let, $f(x) = \begin{cases} \dfrac{\sin x}{x}, & x \neq 0 \\ 0, & x = 0 \end{cases}$

Here, $\lim_{x \to 0+} f(x) = \lim_{x \to 0+} \dfrac{\sin x}{x} = 1$ and

$\lim_{x \to 0-} f(x) = \lim_{x \to 0-} \dfrac{\sin x}{x} = 1$.

Therefore, $\lim_{x \to 0} f(x) = 1$. But $f(0) = 0$.

Hence, $\lim_{x \to 0} f(x) \neq f(0)$ and so f has a removable discontinuity at $x = 0$.

II. Jump Discontinuity:

If $\lim_{x \to a+} f(x)$ and $\lim_{x \to a-} f(x)$ both exist but $\lim_{x \to a+} f(x) \neq \lim_{x \to a-} f(x)$, then we say that f has a jump discontinuity at $x = a$. In this case,

$\lim_{x \to a+} f(x) - \lim_{x \to a-} f(x)$ (denoted by J_f)

is called the total jump of the function f at

$x = a$. Here the function f is said to have discontinuity of the first kind at $x = a$.

Example:

Let $f(x) = \begin{cases} 0, & 0 < x < 1 \\ 1, & 1 \leq x < 2 \end{cases}$

Here, $\lim_{x \to 1+} f(x) = \lim_{x \to 1+} 1 = 1$ and

$\lim_{x \to 1-} f(x) = \lim_{x \to 1-} 0 = 0$.

Therefore, $\lim_{x \to 1+} f(x)$ and $\lim_{x \to 1-} f(x)$ exist

but $\lim_{x \to 1+} f(x) \neq \lim_{x \to 1-} f(x)$.

So, f has a jump discontinuity at $x = 1$.

Here, total jump of f at $x = 1$ is given by

$J_f = \lim_{x \to 1+} f(x) - \lim_{x \to 1-} f(x) = 1 - 0 = 1$.

III. Discontinuity of the 2nd kind:

A function f is said to have a discontinuity of 2nd kind if $\lim_{x \to a+} f(x)$ or $\lim_{x \to a-} f(x)$ or both does not exist.

Example:

Let $f(x) = \begin{cases} \sin\dfrac{1}{x}, & x \neq 0 \\ 0, & x = 0 \end{cases}$

Here $\lim_{x \to 0+} f(x)$

$= \lim_{x \to 0+} \sin \dfrac{1}{x}$

= an oscillatory value which lies between −1 and 1.

Hence, $\lim_{x \to 0+} f(x)$ does not exist. Therefore, the function f has discontinuity of the 2nd kind at $x = 0$.

2.2.3 Derivative

The right-hand derivative of the function "f" at $x = a$ is denoted by $Rf'(a)$ and is defined by

$$Rf'(a) = \lim_{x \to a+} \frac{f(x) - f(a)}{x - a} = \lim_{h \to 0+} \frac{f(a+h) - f(a)}{h}.$$

The left-hand derivative of the function "f" at $x = a$ is denoted by $Lf'(a)$ and is defined by

$$Lf'(a) = \lim_{x \to a-} \frac{f(x)-f(a)}{x-a} = \lim_{h \to 0-} \frac{f(a+h)-f(a)}{h}$$
$$= \lim_{h \to 0+} \frac{f(a-h)-f(a)}{(-h)}$$

If $Rf'(a)$ and $Lf'(a)$ exist and are equal, then we say that f is differentiable or $f'(x)$ exist and we write $f'(x) = \lim_{h \to 0} \frac{f(a+h)-f(a)}{h}$.

If $y = f(x)$, then we write $\frac{dy}{dn} = 5(x)$.

Remember:

(i) If f is differentiable at $x = a$, then
$$f'(a) = \lim_{h \to 0} \frac{f(a+h)-f(a-h)}{2h}.$$

(ii) If f is differentiable at $x = a$, then f is continuous at $x = a$. But the converse is not true in general.

(iii) $f(x) = |x-a|$ is not differentiable at $x = a$.

(iv) If $f, g : R \to R$ are both differentiable, then $f+g, f-g, kf$ (k = constant), $f.g$, $\frac{f}{g}$ (where $g(x) \neq 0 \; \forall x$ are also differentiable).

(v) $f'(0) = \lim_{x \to 0} \frac{f(x)-f(0)}{x}$.

2.2.4 Computation of Derivatives

(i) $(f+g)'(x) = f'(x)+g'(x)$

(ii) $(f-g)'(x) = f'(x)-g'(x)$

(iii) $(kf)'(x) = kf'(x)$

(iv) $(fg)'(x) = f'(x)\times g(x)+f(x)\times g'(x)$

(v) $\left(\dfrac{f}{g}\right)'(x) = \dfrac{g(x)\times f'(x)-f(x)\times g'(x)}{\{g(x)\}^2}$

(vi) If $x = f(t)$, $y = g(t)$, then $\dfrac{dy}{dx} = \dfrac{\frac{dy}{dt}}{\frac{dx}{dt}}$

(vii) If $f(x) = (g(x))^{h(x)}$, then to compute $f'(x)$ proceed as follows:

$$f(x) = \{g(x)\}^{h(x)}$$
$$\Rightarrow \log f(x) = \log\{g(x)\}^{h(x)} = h(x)\log\{g(x)\}$$

Differentiating both sides w.r.t x, we get

$$\frac{1}{f(x)} f'(x) = h'(x)\times \log\{g(x)\}+h(x)\times \frac{1}{g(x)} g'(x)$$

$$\Rightarrow f'(x) = f(x)[h'(x)\times \log\{g(x)\}+h(x)\times \frac{1}{g(x)} g'(x)]$$

(viii) If $f(x)$ contains the expression a^2-x^2, then put $x = a\sin\theta$ or $x = a\cos\theta$.

(ix) If $f(x)$ contains the expression x^2-a^2, then put $x = a\sec\theta$ or $x = a\cosec\theta$.

(x) If $f(x)$ contains the expression x^2+a^2, then put $x = a\tan\theta$ or $x = a\cot\theta$.

2.3 INDETERMINATE FORMS

2.3.1 Introduction

In general, $\lim_{x \to a} \dfrac{f(x)}{g(x)} = \dfrac{\lim_{x \to a} f(x)}{\lim_{x \to a} g(x)}$. But when $\lim_{x \to a} f(x) = 0 = \lim_{x \to a} g(x)$, then the quotient reduces to the indeterminate form $\dfrac{0}{0}$. The other indeterminate forms are represented by the symbols $\dfrac{\infty}{\infty}, \infty-\infty, 0\times\infty, \infty^0$ etc.

2.3.2 The L'Hospital Rule

If $f(x), g(x), f'(x), g'(x)$ are all continuous at $x = a$ and if $\lim_{x \to a} f(x) = 0 = \lim_{x \to a} g(x)$, then

$$\lim_{x \to a} \frac{f(x)}{g(x)} = \lim_{x \to a} \frac{f'(x)}{g'(x)} = \frac{f'(a)}{g'(a)}.$$

If further $\lim_{x \to a} f'(x) = 0 = \lim_{x \to a} g'(x)$, then

$$\lim_{x \to a} \frac{f(x)}{g(x)} = \lim_{x \to a} \frac{f''(x)}{g''(x)} = \frac{f''(a)}{g''(a)} \text{ and so on.}$$

2.4 MEAN VALUE THEOREMS

2.4.1 Rolle's Theorem

If a function $f : [a,b] \to R$ is

(i) continuous on $[a,b]$,

(ii) differentiable on (a,b)

(iii) $f(a) = f(b)$,

then there exists a point $c \in (a,b)$ such that $f'(c) = 0$.

Geometrically, $f'(c) = 0$ implies that at the point $(c, f(c))$, the tangent to the curve $y = f(x)$ is parallel to the x-axis.

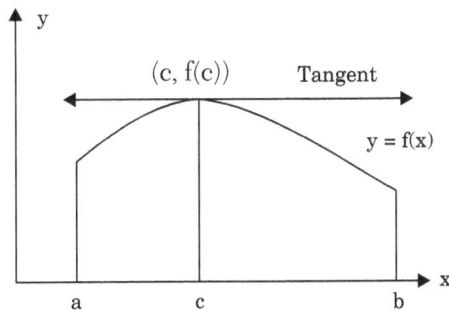

2.4.2 Lagrange's Mean Value Theorem

If a function $f : [a,b] \to R$ is

(i) continuous on $[a,b]$,

(ii) differentiable on (a,b)

then there exists a point $c \in (a,b)$ such that

$$f'(c) = \frac{f(b) - f(a)}{b - a}.$$

Geometrically, $f'(c) = \dfrac{f(b) - f(a)}{b - a}$ implies that at the point $(c, f(c))$, the tangent to the curve $y = f(x)$ is parallel to the chord joining the points $(a, f(a))$ and $(b, f(b))$.

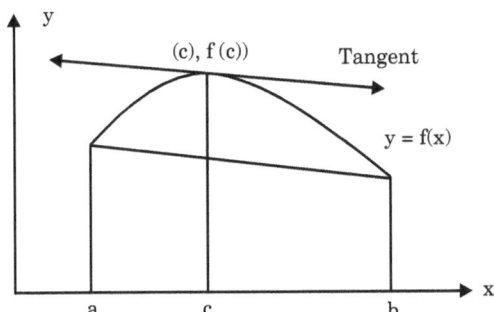

2.4.3 Cauchy's Mean Value Theorem

If the functions $f, g : [a,b] \to R$ are

(i) continuous on $[a,b]$,

(ii) differentiable on (a,b) and $g'(x) \neq 0$

$\forall x \in (a,b)$

then there exists a point $c \in (a,b)$ such that

$$\frac{f'(c)}{g'(c)} = \frac{f(b) - f(a)}{g(b) - g(a)}.$$

2.5 INCREASING AND DECREASING FUNCTIONS

A function $f : D \to R$ (where $D \subseteq R$) is said to be

(i) increasing if $f(x_2) \geq f(x_1)$ for $x_2 > x_1$ and $x_1, x_2 \in D$.

(ii) decreasing if $f(x_2) \leq f(x_1)$ for $x_2 > x_1$ and $x_1, x_2 \in D$.

(iii) strictly increasing if $f(x_2) > f(x_1)$ for $x_2 > x_1$ and $x_1, x_2 \in D$.

(iv) strictly decreasing if $f(x_2) < f(x_1)$ for $x_2 > x_1$ and $x_1, x_2 \in D$.

Remember:

A function $f : D \to R$ (where $D \subseteq R$) is said to be

(i) increasing if $f'(x) \geq 0$ for $x \in D$

(ii) decreasing if $f'(x) \leq 0$ for $x \in D$

(iii) strictly increasing if $f'(x) > 0$ for $x \in D$

(iv) strictly decreasing if $f'(x) < 0$ for $x \in D$

(v) $ab \geq 0 \Rightarrow a \geq 0, b \geq 0$ or $a \leq 0, b \leq 0$

(vi) $ab \leq 0 \Rightarrow a \geq 0, b \leq 0$ or $a \leq 0, b \geq 0$

(vii) $ab > 0 \Rightarrow a > 0, b > 0$ or $a < 0, b < 0$

(viii) $ab < 0 \Rightarrow a > 0, b < 0$ or $a < 0, b > 0$

2.6 MAXIMA AND MINIMA OF FUNCTIONS OF A SINGLE VARIABLE

2.6.1 First Derivative Test

1. $x = a$ is said to be point of local maxima of $f(x)$ if

   (i) $f'(a) = 0$

   (ii) $f'(x) = 0$ changes sign from positive to negative as x increases through a, i.e, $f'(x) > 0$ at every point sufficiently close to a and to the left of a.

2. $x = a$ is said to be point of local minima of $f(x)$ if

   (i) $f'(a) = 0$

   (ii) $f'(x) = 0$ changes sign from negative to positive as x increases through a, i.e, $f'(x) < 0$ at every point sufficiently close to a and to the right of a.

Example:

Let $f(x) = x^3 - 6x^2 + 9x - 8$.

Then $f'(x) = 3x^2 - 12x + 9 = 3(x-1)(x-3)$.

$\therefore f'(x) = 0 \Rightarrow x = 1, 3$.

Now clearly $f'(x)$ changes sign from positive to negative as "x" increases through "1".

$$\left[\because f'(0) > 0, f'(0.5) > 0, f'(1.5) < 0, f'(2) < 0 \text{ etc}\right]$$

$\therefore x = 1$ is a point of local maxima.

Again $f'(x)$ changes sign from negative to positive as "x" increases through "3"

$$\left[\because f'(2) < 0, f'(2.5) < 0, f'(3.5) > 0, f'(4) > 0 \text{ etc}\right]$$

$\therefore x = 3$ is a point of local minima.

2.6.2 Second Derivative Test

$x = a$ is said to be point of local maxima of $f(x)$ if $f'(a) = 0$ and $f''(a) < 0$.

$x = a$ is said to be point of local minima of $f(x)$ if $f'(a) = 0$ and $f''(a) > 0$.

Example:

Let $f(x) = x^3 - 6x^2 + 9x - 18$.

Then, $f'(x) = 3x^2 - 12x + 9 = 3(x-1)(x-3)$.

$\therefore f'(x) = 0 \Rightarrow x = 1, 3$.

Now $f''(x) = 6x - 12$.

$\therefore f''(1) = 6 - 12 < 0$ and $f''(3) = 18 - 12 > 0$.

Hence, $x = 1$ is a point of local maxima and $x = 3$ is a point of local minima.

2.6.3 Higher Order Derivative Test

Let $f(x)$ be a differentiable function such that

(I) $f'(a) = f''(a) = \ldots\ldots = f^{(n-1)}(a) = 0$

(II) $f^{(n)}(a)$ exists and nonzero.

   Then

   (i) $f^{(n)}(a) < 0$ and n is even $\Rightarrow x = a$ is a point of local maxima.

   (ii) $f^{(n)}(a) > 0$ and n is even $\Rightarrow x = a$ is a point of local minima.

   (iii) n = odd $\Rightarrow x = a$ is neither a point of local minima nor a point of local maxima.

Example:

(1) Let $f(x) = (x-1)^3$.

Then $f'(x) = 3(x-1)^2$

$\therefore f'(x) = 0 \Rightarrow x = 1$.

Now $f''(x) = 6(x-1)$, $f'''(x) = 6$.

$\therefore f(1) = f'(1) = f''(1) = 0$ but $f'''(1) = 6 \neq 0$. Since n = 3(odd), so $x = 1$ is neither a point of local minima nor a point of local maxima.

(2) Let $f(x) = (x-2)^2$.

Then $f'(x) = 2(x-2)$

$\therefore f'(x) = 0 \Rightarrow x = 2$.

Now $f''(x) = 2$.

$\therefore f(2) = f'(2) = 0$ but $f''(2) = 2 \neq 0$.

Since n = 2(even) and $f''(2) > 0$, so $x = 2$ is a point of local minima.

2.7 INFINITE SERIES AND EXPANSION OF FUNCTIONS

2.7.1 Infinite Series

An infinite series is denoted by $\sum_{n=1}^{\infty} u_n$ and is

defined by $\sum_{n=1}^{\infty} u_n = u_1 + u_2 + \ldots + u_n + \ldots \infty$, where

u_n is called the n th term of the infinite series.

An infinite series $\sum_{n=1}^{\infty} u_n$ is said to

(i) converge if

$\lim_{n \to \infty}(u_1 + u_2 + \ldots + u_n) = a$ finite real number

(ii) diverge if $\lim_{n \to \infty}(u_1 + u_2 + \ldots + u_n) = \infty$ or $-\infty$

If the series $\sum_{n=1}^{\infty} u_n$ is convergent, then we must

have $\lim_{n \to \infty} u_n = 0$.

Example:

Consider the series $\dfrac{1}{1 \times 2} + \dfrac{1}{2 \times 3} + \dfrac{1}{3 \times 4} + \ldots \infty$

Let $\sum_{n=1}^{\infty} u_n$ be the given series. Then,

$u_1 = \dfrac{1}{1 \times 2}, \ u_2 = \dfrac{1}{2 \times 3}, \ldots, \ u_n = \dfrac{1}{n(n+1)}, \ldots$

$\therefore \lim_{n \to \infty}(u_1 + u_2 + \ldots + u_n)$

$= \lim_{n \to \infty}\left[\dfrac{1}{1 \times 2} + \dfrac{1}{2 \times 3} + \ldots + \dfrac{1}{n(n+1)}\right]$

$= \lim_{n \to \infty}\left[\dfrac{2-1}{1 \times 2} + \dfrac{3-2}{2 \times 3} + \ldots + \dfrac{(n+1)-n}{n(n+1)}\right]$

$= \lim_{n \to \infty}\left[1 - \dfrac{1}{2} + \dfrac{1}{2} - \dfrac{1}{3} + \ldots + \dfrac{1}{n} - \dfrac{1}{n+1}\right]$

$= \lim_{n \to \infty}\left(1 - \dfrac{1}{n+1}\right) = 1 - 0 = 1$, a finite real number.

$= \lim_{x \to \infty}\left(1 - \dfrac{1}{n+1}\right) = 1 - 0 = 1$, a finite real number.

Hence, the given series is convergent.

Remember:

A. $e^x = 1 + \dfrac{x}{1!} + \dfrac{x^2}{2!} + \dfrac{x^3}{3!} + \ldots \infty$

B. $e^{-x} = 1 - \dfrac{x}{1!} + \dfrac{x^2}{2!} - \dfrac{x^3}{3!} + \ldots \infty$

C. $e = 1 + \dfrac{1}{1!} + \dfrac{1}{2!} + \dfrac{1}{3!} + \ldots \infty$

D. $e^{-1} = 1 - \dfrac{1}{1!} + \dfrac{1}{2!} - \dfrac{1}{3!} + \ldots \infty$

E. $\sin x = x - \dfrac{x^3}{3!} + \dfrac{x^5}{5!} - \ldots \infty$

F. $\cos x = 1 - \dfrac{x^2}{2!} + \dfrac{x^4}{4!} \ldots \infty$

G. $\log(1+x) = x - \dfrac{x^2}{2} + \dfrac{x^3}{3} - \dfrac{x^4}{4} + \ldots \infty$

H. $\log(1-x) = -x - \dfrac{x^2}{2} - \dfrac{x^3}{3} - \dfrac{x^4}{4} - \ldots \infty$

I. $a + ar + ar^2 + ar^3 + \ldots \infty$ (infinite G.P.) $= \dfrac{a}{1-r}$

J. $(1+x)^{-1} = 1 - x + x^2 - x^3 + \ldots \infty$;

(provided $|x| < 1$)

K. $(1-x)^{-1} = 1 + x + x^2 + x^3 + \ldots \infty$;

(provided $|x| < 1$)

L. The series $1 + \dfrac{1}{2} + \dfrac{1}{3} + \dfrac{1}{4} + \ldots \infty$ is divergent.

M. The series $1 - \dfrac{1}{2} + \dfrac{1}{3} - \dfrac{1}{4} + \ldots \infty$ is convergent

N. $1 + 2 + 3 + 4 + \ldots + n = \dfrac{n(n+1)}{2}$

O. $1^2 + 2^2 + 3^2 + 4^2 + \ldots + n^2 = \dfrac{n(n+1)(2n+1)}{6}$

P. $1^3 + 2^3 + 3^3 + 4^3 + \ldots + n^3 = \left\{\dfrac{n(n+1)}{2}\right\}^2$

2.7.2 Test for Convergence of Infinite Series

I. The series $\dfrac{1}{1^p} + \dfrac{1}{2^p} + \dfrac{1}{3^p} + \ldots \infty$, i.e, $\sum_{n=1}^{\infty} \dfrac{1}{n^p}$

(known as p-series) is convergent for $p > 1$ and divergent for $p \leqslant 1$

Example: $1 + \dfrac{1}{2^2} + \dfrac{1}{3^2} + \dfrac{1}{4^2} \ldots \infty$ is convergent

(since $p = 2 > 1$)

II. **Comparison test (particular form):** Let $\sum_{n=1}^{\infty} u_n$ and $\sum_{n=1}^{\infty} v_n$ be two series of positive real numbers such that $u_n \leqslant v_n \ \forall n \in N$. Then,

A. If $\sum_{n=1}^{\infty} v_n$ is convergent, then $\sum_{n=1}^{\infty} u_n$ is convergent

B. If $\sum_{n=1}^{\infty} u_n$ is divergent, then $\sum_{n=1}^{\infty} v_n$ is divergent.

Example:

Consider $\sum_{n=1}^{\infty} u_n = \sum_{n=1}^{\infty} \frac{1}{n^2+1}$ and $\sum_{n=1}^{\infty} v_n = \sum_{n=1}^{\infty} \frac{1}{n^2}$

Then, $u_n = \frac{1}{n^2+1}$ and $v_n = \frac{1}{n^2}$.

Now, $n^2+1 > n^2 \Rightarrow \frac{1}{n^2+1} < \frac{1}{n^2} \Rightarrow u_n < v_n \ \forall n \in N$.

Also $\sum_{n=1}^{\infty} \frac{1}{n^2}$ is convergent ($\because$ it is a p-series

with $p = 2$). Hence, by comparison test, $\sum_{n=1}^{\infty} \frac{1}{n^2+1}$ is convergent.

III. **Comparison Test (limit form):**

Let $\sum_{n=1}^{\infty} u_n$ and $\sum_{n=1}^{\infty} v_n$ be two series of positive real numbers such that

$\lim_{x \to \infty} \frac{u_n}{v_n} = a$ nonzero finite real number.

Then the series $\sum_{n=1}^{\infty} u_n$ and $\sum_{n=1}^{\infty} v_n$ converge or diverge together.

Example:

Consider $\frac{1}{1 \times 2^2} + \frac{1}{2 \times 3^2} + \frac{1}{3 \times 4^2} + \ldots \ldots \infty$

Let $\sum_{n=1}^{\infty} u_n$ be the given series. Then,

$u_n = \frac{1}{n(n+1)^2}$.

Let $\sum_{n=1}^{\infty} v_n = \sum_{n=1}^{\infty} \frac{1}{n^3}$. Then, $v_n = \frac{1}{n^3}$.

$\therefore \lim_{n \to \infty} \frac{u_n}{v_n} = \lim_{n \to \infty} \frac{n^3}{n(n+1)^2} = \lim_{n \to \infty} \frac{n^3}{n^3\left(1+\frac{1}{n}\right)^2}$

$= \lim_{n \to \infty} \frac{1}{\left(1+\frac{1}{n}\right)^2} = \frac{1}{1+0} = 1$,

a non-zero finite real number.

Also $\sum_{n=1}^{\infty} v_n = \sum_{n=1}^{\infty} \frac{1}{n^3}$ is convergent ($\because$ it is a p-series with $p=3 > 1$). Hence, by comparison test (limit form), $\sum_{n=1}^{\infty} u_n$, i.e, the given series is convergent.

IV. **D'Alembert's ratio test:**

Let $\sum_{n=1}^{\infty} u_n$ be a series of positive real numbers. Then

A. $\lim_{n \to \infty} \frac{u_{n+1}}{u_n} < 1 \Rightarrow$ the series is convergent.

B. $\lim_{n \to \infty} \frac{u_{n+1}}{u_n} > 1 \Rightarrow$ the series is divergent.

C. $\lim_{n \to \infty} \frac{u_{n+1}}{u_n} = 1 \Rightarrow$ the test fails.

Example:

Consider the series $\sum_{n=1}^{\infty} \frac{n!}{2 \times 4 \times 6 \times \ldots \ldots \times 2n}$.

Comparing it with $\sum_{n=1}^{\infty} u_n$, we get,

$u_n = \frac{n!}{2 \times 4 \times 6 \times \ldots \ldots \times 2n}$.

$\therefore u_{n+1} = \frac{(n+1)!}{2 \times 4 \times 6 \times \ldots \ldots \times \{2(n+1)\}}$

$= \frac{(n+1)n!}{2 \times 4 \times 6 \times \ldots \ldots \times 2n \times (2n+2)}$

Then, $\lim_{n \to \infty} \frac{u_{n+1}}{u_n} = \lim_{n \to \infty} \frac{n+1}{2n+2}$

$= \lim_{n \to \infty} \frac{1+\frac{1}{n}}{2+\frac{2}{n}} = \frac{1+0}{2+0} = \frac{1}{2} < 1$

Hence, by the D'Alembert's ratio test, the given series is convergent.

V. **Cauchy root test:**

Let $\sum_{n=1}^{\infty} u_n$ be a series of positive real numbers.

and $\lim_{n\to\infty}(a_n)^{\frac{1}{n}}=\lambda$. Then,

(i) $\lambda < 1 \Rightarrow$ the series is convergent

(ii) $\lambda > 1 \Rightarrow$ the series is divergent

(iii) $\lambda = 1 \Rightarrow$ the test fails.

Example:

Consider $\dfrac{1^4}{2^1}+\dfrac{2^4}{2^2}+\dfrac{3^4}{2^3}+\dfrac{4^4}{2^4}+......\infty$

Comparing it with $\sum_{n=1}^{\infty} u_n$, we get $u_n=\dfrac{n^4}{2^n}$.

$$\therefore \lim_{n\to\infty}(u_n)^{\frac{1}{n}}=\lim_{n\to\infty}\left(\dfrac{n^4}{2^n}\right)^{\frac{1}{n}}=\lim_{n\to\infty}\dfrac{\left(n^{\frac{1}{n}}\right)^4}{2}$$

$$=\dfrac{1}{2}\left\{\lim_{n\to\infty} n^{\frac{1}{n}}\right\}^4$$

$$=\dfrac{1}{2}\times1^4=\dfrac{1}{2}<1 \qquad \left(\because \lim_{x\to\infty} n^{\frac{1}{n}}=1\right)$$

Hence, by Cauchy's root test, the given series is convergent.

2.7.3 Taylor's Theorem With Lagrange's Form of Remainder

If f be a function of x such that

(i) $f', f'', f''',......, f^{(n-1)}$ are all continuous in $[a, a+h]$ and

(ii) $f^{(n)}(x)$ exists $\forall\, x\in (a,a+h)$,

then $\exists$ at least one number θ in $(0,1)$ such that

$$f(a+h)=f(a)+hf'(a)+\dfrac{h^2}{2!}f''(a)+.....$$

$$+\dfrac{h^n}{n!}f^{(n)}(a+\theta h) \qquad ...(1)$$

Remember:

Putting $a = 0$ and $h = x$ in (1), we get,

$$f(x)=f(0)+xf'(0)+\dfrac{x^2}{2!}f''(0)+......$$

$$+\dfrac{x^n}{n!}f^{(n)}(\theta x)$$

This is known as Maclaurin's theorem with Lagrange's form of remainder.

2.7.4 The Taylor Series

If $f(x)$ possesses derivatives of all orders, then

$$f(x+h)=f(x)+hf'(x)+\dfrac{h^2}{2!}f''(x)+......\infty$$

Remember:

Replacing x by a and h by $x - a$, we get

$$f(x)=f(a)+(x-a)f'(a)+\dfrac{(x-a)^2}{2!}f''(a)+...\infty,$$

which is called Taylor series about $x = a$ (or Taylor series in powers of $x - a$).

2.7.5 Maclaurin's Series

If $f(x)$ possesses derivatives of all orders, then

$$f(x)=f(0)+xf'(0)+\dfrac{x^2}{2!}f''(0)+\dfrac{x^3}{3!}f'''(0)+......\infty$$

2.8 INDEFINITE AND DEFINITE INTEGRALS

2.8.1 Indefinite Integral

If $\dfrac{d}{dx}\{F(x)+C\}= f(x)$, where C is a constant, then we write $\int f(x)\,dx =F(x)+C$

$$\left[\int f(x)\,dx \text{ is read as "integral } f(x) \text{ dx"}\right]$$

Here $f(x)$ is called the integrand and the process of finding the integral of a function is called integration.

Example:

$$\because \dfrac{d}{dx}\{\sin x+C\}= \cos x, \text{ so } \int \cos x\,dx = \sin x+C.$$

2.8.2 Fundamental Formulas of Indefinite Integral

1. $\int x^n\, dx =\dfrac{x^{n+1}}{n+1}+C \qquad (n\neq -1)$

2. $\int \sin x\, dx =-\cos x +C$

3. $\int \cos x\, dx = \sin x +C$

4. $\int \tan x\, dx = \log|\sec x|+C$

5. $\int \cot x \, dx = \log|\sin x| + C$

6. $\int \cos ecx \, dx = \log|\cos ecx - \cot x| + C$

7. $\int \sec x \, dx = \log|\sec x + \tan x| + C$

8. $\int \sec^2 x \, dx = \tan x + C$

9. $\int \cos ec^2 x \, dx = -\cot x + C$

10. $\int \sec x \tan x \, dx = \sec x + C$

11. $\int \cos ecx \cot x \, dx = -\cos ecx + C$

12. $\int \dfrac{1}{x} dx = \log x + C \qquad (x > 0)$

13. $\int a^x \, dx = \dfrac{a^x}{\log_e a} + C$

14. $\int e^x \, dx = e^x + C$

2.8.3 Advanced Formulas of Indefinite Integrals

1. $\int \dfrac{1}{\sqrt{a^2 - x^2}} dx = \sin^{-1}\left(\dfrac{x}{a}\right) + C$

2. $\int \dfrac{1}{a^2 + x^2} dx = \dfrac{1}{a}\tan^{-1}\left(\dfrac{x}{a}\right) + C$

3. $\int \dfrac{1}{x\sqrt{x^2 - a^2}} dx = \sec^{-1}\left(\dfrac{x}{a}\right) + C$

4. $\int \dfrac{1}{\sqrt{a^2 + x^2}} dx = \sinh^{-1}\left(\dfrac{x}{a}\right) + C$

$\qquad = \log\left(x + \sqrt{a^2 + x^2}\right) + C$

5. $\int \dfrac{1}{\sqrt{x^2 - a^2}} dx = \cosh^{-1}\left(\dfrac{x}{a}\right) + C$

$\qquad = \log\left(x + \sqrt{x^2 - a^2}\right) + C$

6. $\int \sqrt{x^2 + a^2} \, dx$

$\qquad = \dfrac{x}{2}\sqrt{x^2 + a^2} + \dfrac{a^2}{2}\log\left(x + \sqrt{x^2 + a^2}\right) + C$

7. $\int \sqrt{x^2 - a^2} \, dx$

8. $\int \sqrt{a^2 - x^2} \, dx$

$\qquad = \dfrac{x}{2}\sqrt{a^2 - x^2} + \dfrac{a^2}{2}\sin^{-1}\left(\dfrac{x}{a}\right) + C$

9. $\int \dfrac{1}{x^2 - a^2} dx = \dfrac{1}{2a}\log\left|\dfrac{x - a}{x + a}\right| + C$

10. $\int e^{ax} \sin bx \, dx$

$\qquad = \dfrac{e^{ax}(a \sin bx - b \cos bx)}{a^2 + b^2} + C$

11. $\int e^{ax} \cos bx \, dx$

$\qquad = \dfrac{e^{ax}(a \cos bx + b \sin bx)}{a^2 + b^2} + C$

12. (BY PARTS RULE) $\int f(x) g(x) \, dx$

$\qquad = f(x) \int g(x) \, dx - \int\left[\dfrac{d}{dx}\{f(x)\} \times \int g(x) \, dx\right] dx$

[**Remember:**

To obtain $\int f(x) g(x) \, dx$ using By Parts rule, use LIATE method to chose the first function, where

"L" stands for logarithmic functions,

"I" stands for inverse trigonometric functions,

"A" stands for algebraic functions,

"T" stands for trigonometric functions,

"E" stands for exponential functions]

13. $\int \dfrac{f'(x)}{f(x)} dx = \log\{f(x)\} + C$

14. $\int [f(x)]^n f'(x) dx = \dfrac{[f(x)]^{n+1}}{n+1} + C$

15. $\int f(x) f'(x) dx = \dfrac{[f(x)]^2}{2} + C$

16. $\int e^x \{f(x) + f'(x)\} dx = e^x f(x) + C$

2.8.4 Definite Integral

The definite integral of the function $f(x)$ within the limits $x = a$ and $x = b$ is denoted by $\int_a^b f(x) \, dx$

and is defined by $\int_a^b f(x)\,dx = \phi(b) - \phi(a),$

where

$$\frac{d}{dx}\phi(x) = f(x).$$

2.8.5 Properties of Definite Integral

1. $\int_a^b f(x)\,dx = \int_a^b f(t)\,dt$

2. $\int_a^b f(x)\,dx = -\int_b^a f(x)\,dx$

3. $\int_a^b f(x)\,dx = \int_a^c f(x)\,dx + \int_c^b f(x)\,dx$

   (where $a < c < b$)

4. $\int_a^b f(x)\,dx = \int_a^b f(a+b-x)\,dx$

5. $\int_0^a f(x)\,dx = \int_0^a f(a-x)\,dx$

6. $\int_{-a}^a f(x)\,dx = \begin{cases} 2\int_0^a f(x)\,dx \text{ if } f(-x) = f(x) \end{cases}$

7. $\int_0^{2a} f(x)\,dx = \begin{cases} 2\int_0^a f(x)\,dx \text{ if } f(2a-x) = f(x) \\ 0 \text{ if } f(2a-x) = -f(x) \end{cases}$

8. $\int_a^b f(x)\,dx = (b-a)\int_0^1 f[(b-a)x+a]\,dx$

9. $\int_0^{2a} f(x)\,dx = \int_0^a [f(a-x)+f(a+x)]\,dx$

Remember:

I. $\int_0^{\frac{\pi}{2}} \sin^m x \times \cos^n x\,dx = \dfrac{1}{2} \times \dfrac{\Gamma\left(\dfrac{m+1}{2}\right)\Gamma\left(\dfrac{n+1}{2}\right)}{\Gamma\left(\dfrac{m+n+2}{2}\right)}$

II. $\int_0^{\frac{\pi}{2}} \log(\sin x)\,dx = \int_0^{\frac{\pi}{2}} \log(\cos x)\,dx = -\dfrac{\pi}{2}\log 2.$

2.8.6 Definite Integral as a Limit of Sum

$$\int_a^b f(x)\,dx = \lim_{h\to 0} h\sum_{r=1}^n f(a+rh), \text{ where } nh = b-a$$

$$= \lim_{n\to\infty} \frac{1}{n}\sum_{r=1}^n f\left(a+\frac{r}{n}\right), \text{ where } nh = b-a$$

In particular,

$$\int_0^1 f(x)\,dx = \lim_{h\to 0} h\sum_{r=1}^n f(rh) = \lim_{n\to\infty} \frac{1}{n}\sum_{r=1}^n f\left(\frac{r}{n}\right),$$

where $nh = 1$

$\int_0^1 f(x)\,dx$ can also be written as

$$\int_0^1 f(x)\,dx = \lim_{h\to 0} h\sum_{r=0}^{n-1} f(rh) = \lim_{h\to 0} \frac{1}{n}\sum_{r=0}^{n-1} f\left(\frac{r}{n}\right)$$

where $nh = 1$

Application:

$$\lim_{n\to\infty} \frac{1^m + 2^m + 3^m + \dots\dots + n^m}{n^{m+1}}$$

$$= \lim_{n\to\infty} \frac{1}{n}\left[\frac{1^m + 2^m + 3^m + \dots\dots + n^m}{n^m}\right]$$

$$= \lim_{n\to\infty} \frac{1}{n}\left[\left(\frac{1}{n}\right)^m + \left(\frac{2}{n}\right)^m + \left(\frac{3}{n}\right)^m + \dots\dots + \left(\frac{n}{n}\right)^m\right]$$

$$= \lim_{n\to\infty} \frac{1}{n}\sum_{r=1}^n \left(\frac{r}{n}\right)^m \quad \left[\text{here } f\left(\frac{r}{n}\right) = \left(\frac{r}{n}\right)^m\right]$$

$$= \int_0^1 x^m\,dx = \left[\frac{x^{m+1}}{m+1}\right]_0^1 = \frac{1}{m+1}.$$

2.8.7 Differentiation Under the Sign of Integration

The Leibnitz's rule for differentiation under the sign of integration is given by

$$\frac{d}{dx}\int_{\psi_1(x)}^{\psi_2(x)} f(x,t)\,dt = \int_{\psi_1(x)}^{\psi_2(x)} \frac{\partial}{\partial x} f(x,t)\,dt$$

$$+ \frac{d\psi_2}{dx} \times f(x, \psi_2(x))$$

$$- \frac{d\psi_1}{dx} \times f(x, \psi_1(x))$$

Remember:

If the integrand is a function of "t" alone, i.e, if $f(x,t) = f(t)$, then

$$\frac{d}{dx} \int_{\psi_1(x)}^{\psi_2(x)} f(t)\, dt$$

$$= \frac{d\psi_2}{dx} \times f(\psi_2(x)) - \frac{d\psi_1}{dx} \times f(\psi_1(x)).$$

Application:

$$\frac{d}{dx} \int_0^{\cos^2 x} \cos^{-1}\sqrt{t}\, dt$$

$$= \left\{ \frac{d}{dx}(\cos^2 x) \right\} \times \cos^{-1}\sqrt{\cos^2 x}$$

$$- \left\{ \frac{d}{dx}(0) \right\} \times \cos^{-1}\sqrt{0}$$

$$= (-2\sin x \cos x) \times x - 0$$

$$= -x \sin 2x.$$

2.9 IMPROPER INTEGRALS, BETA, AND GAMMA FUNCTIONS

2.9.1 Improper Integral

An integral $\int_a^b f(x)\, dx$ is called an improper integral if at least one of the following conditions is satisfied:

(i) either a or b or both are infinite

(ii) "f" is not continuous in $[a,b]$.

An improper integral $\int_a^b f(x)\, dx$ is said to be convergent if it is value is a finite real number; otherwise, it is called divergent.

2.9.2 Evaluation of Improper Integrals

Let us consider the improper integral $\int_a^b f(x)\, dx$.

Case-I: $a = -\infty$ or ∞.

Then, $\int_\infty^b f(x)\, dx = \lim_{X \to \infty} \int_X^b f(x)\, dx$ and

$$\int_{-\infty}^b f(x)\, dx = \lim_{X \to -\infty} \int_X^b f(x)\, dx.$$

Example:

$$\int_{-\infty}^1 \frac{1}{1+x^2}\, dx$$

$$= \lim_{X \to -\infty} \int_X^1 \frac{1}{1+x^2}\, dx = \lim_{X \to -\infty} \left[\tan^{-1} x\right]_X^1$$

$$= \lim_{X \to -\infty}\left[\tan^{-1}1 - \tan^{-1}X\right] = \lim_{X \to -\infty}\left[\frac{\pi}{4} - \tan^{-1}X\right]$$

$$= \frac{\pi}{4} - \tan^{-1}(-\infty) = \frac{\pi}{4} - \left(-\frac{\pi}{2}\right) = \frac{3\pi}{4}.$$

Case-II: $b = -\infty$ or ∞.

Then $\int_a^\infty f(x)\, dx = \lim_{X \to \infty} \int_a^X f(x)\, dx$ and

$$\int_a^{-\infty} f(x)\, dx = \lim_{X \to -\infty} \int_a^X f(x)\, dx.$$

Example:

$$\int_0^\infty \frac{1}{1+x^2}\, dx$$

$$= \lim_{X \to \infty} \int_0^X \frac{1}{1+x^2}\, dx = \lim_{X \to \infty}\left[\tan^{-1} x\right]_0^X$$

$$= \lim_{X \to \infty}\left[\tan^{-1}X - \tan^{-1}0\right] = \lim_{X \to \infty}\left[\tan^{-1}X - 0\right]$$

$$= \tan^{-1}(\infty) = \frac{\pi}{2}.$$

Case-III: $a = -\infty$ and $b = \infty$.

Here $\int_{-\infty}^\infty f(x)\, dx = \lim_{X \to \infty} \int_{-X}^X f(x)\, dx$

Example:

$$\int_{-\infty}^{\infty} \frac{1}{1+x^2}\,dx$$

$$= \lim_{X\to\infty}\int_{-X}^{X}\frac{1}{1+x^2}\,dx = \lim_{X\to\infty}\left[\tan^{-1}x\right]_{-X}^{X}$$

$$= \lim_{X\to\infty}\left[\tan^{-1}X - \tan^{-1}(-X)\right]$$

$$= \tan^{-1}(\infty) - \tan^{-1}(-\infty) = \frac{\pi}{2} - \left(-\frac{\pi}{2}\right) = \pi.$$

Case-IV: f has a discontinuity at $x = a$.

Here $\int_a^b f(x)\,dx = \lim_{\varepsilon\to a+}\int_\varepsilon^b f(x)\,dx.$

Example:

$$\int_0^2 \frac{1}{x}\,dx = \lim_{\varepsilon\to 0+}\int_\varepsilon^2 \frac{1}{x}\,dx = \lim_{\varepsilon\to 0+}\left[\log x\right]_\varepsilon^2$$

$$= \lim_{\varepsilon\to 0+}\left[\log 2 - \log\varepsilon\right]$$

$$= \log 2 - \lim_{\varepsilon\to 0+}\log\varepsilon,$$

which doesn't exist.

Case-V: f has a discontinuity at $x = b$

Here $\int_a^b f(x)\,dx = \lim_{\varepsilon\to b-}\int_a^\varepsilon f(x)\,dx.$

Example:

$$\int_0^2 \frac{1}{(x-2)^2}\,dx$$

$$= \lim_{\varepsilon\to 2-}\int_0^\varepsilon \frac{1}{(x-2)^2}\,dx = \lim_{\varepsilon\to 2-}\left[-\frac{1}{x-2}\right]_0^\varepsilon$$

$$= -\lim_{\varepsilon\to 2-}\left[\frac{1}{\varepsilon-2}-\left(-\frac{1}{2}\right)\right] = -\lim_{\varepsilon\to 2-}\frac{1}{\varepsilon-2}-\frac{1}{2},$$

which does not exist.

Case-VI: f has a discontinuity at $x = c\in(a,b)$.

Here $\int_a^b f(x)\,dx$

$$= \int_a^c f(x)\,dx + \int_c^b f(x)\,dx.$$

$$= \lim_{\varepsilon_1\to 0+}\int_a^{c-\varepsilon_1} f(x)\,dx + \lim_{\varepsilon_2\to 0+}\int_{c+\varepsilon_2}^b f(x)\,dx$$

Example:

$$\int_0^2 \frac{1}{(x-1)^2}\,dx$$

$$= \int_0^1 \frac{1}{(x-1)^2}\,dx + \int_1^2 \frac{1}{(x-1)^2}\,dx$$

$$= \lim_{\varepsilon_1\to 0+}\int_0^{1-\varepsilon_1}\frac{1}{(x-1)^2}\,dx + \lim_{\varepsilon_2\to 0+}\int_{1+\varepsilon_1}^2 \frac{1}{(x-1)^2}\,dx$$

$$= \lim_{\varepsilon_1\to 0+}\left[-\frac{1}{x-1}\right]_0^{1-\varepsilon_1} + \lim_{\varepsilon_2\to 0+}\left[-\frac{1}{x-1}\right]_{1+\varepsilon_2}^2$$

$$= -\lim_{\varepsilon_1\to 0+}\left[\frac{1}{-\varepsilon_1}-(-1)\right] - \lim_{\varepsilon_2\to 0+}\left[1-\frac{1}{\varepsilon_2}\right]$$

$$= \lim_{\varepsilon_1\to 0+}\frac{1}{\varepsilon_1} + \lim_{\varepsilon_2\to 0+}\frac{1}{\varepsilon_2} - 2,\text{ which does not exist.}$$

Remember:

If $\varepsilon_1 = \varepsilon_2$, then the value (if exist) of the improper integral is called the Cauchy Principal value of the integral.

2.9.3 Beta Function
The beta function is denoted by $\beta(m,n)$ and is defined by $\beta(m,n) = \int_0^1 x^{m-1}(1-x)^{n-1}\,dx\ (m,n>0)$

Properties:

(i) $\beta(m,n) = \beta(n,m)$

(ii) $\beta(m,n) = \int_0^1 \frac{x^{m-1}+x^{n-1}}{(1+x)^{m+n}}\,dx$

2.9.4 Gamma Function
The gamma function is denoted by $\Gamma(n)$ and is defined by $\Gamma(n) = \int_0^\infty e^{-x}x^{n-1}\,dx\ (n>0)$

Sometimes $\Gamma(n)$ is denoted also by $\lfloor n$.

Properties:

(I) $\beta(m,n) = \dfrac{\Gamma(m)\,\Gamma(n)}{\Gamma(m+n)}$

(II) $\displaystyle\int_0^{\frac{\pi}{2}} \sin^m\theta\cos^n\theta\,d\theta = \dfrac{\Gamma\left(\dfrac{m+1}{2}\right)\Gamma\left(\dfrac{n+1}{2}\right)}{2\Gamma\left(\dfrac{m+n+2}{2}\right)}$

(III) $\Gamma(n)\Gamma(1-n) = \dfrac{\pi}{\sin n\pi}$

(IV) $\Gamma(n+1) = \begin{cases} n\Gamma(n), & \text{if } n > 0 \\ n!, & \text{if } n \text{ is a positive integer} \end{cases}$

(V) $\Gamma\left(\dfrac{1}{2}\right) = \sqrt{\pi}$

2.10 FUNCTIONS OF SEVERAL VARIABLES AND PARTIAL DERIVATIVES

2.10.1 Functions of Two Variables

Let x, y be two independent variables and z be a variable which takes a value corresponding to a pair of values (x, y). Then we say that z depends on x & y and z is a function of x and y. We write $z = f(x, y)$.

For example, $z = x^2 + y^2 + xy$,

$z = f(x,y) = x^2 + e^y \sin x$ etc.

Remember:

$z = f(x, y)$ represents a surface in three dimensional space.

For example, $z = \sqrt{1 - x^2 - y^2}$ represents a sphere with center $(0, 0, 0)$ and radius "1" unit in three dimensional space.

2.10.2 Limit of Functions of Two Variables

Let $z = f(x, y)$ be a function of two variables. If "l" be a real number and for a given $\varepsilon > 0$, there exist $\delta > 0$ (δ depends on ε) such that $|f(x, y) - l| < \varepsilon$ whenever $0 < |x - a| < \delta$ and $0 < |y - b| < \delta$,

then we say that $\displaystyle\lim_{(x,y)\to(a,b)} f(x,y) = l$.

Remember:

1. If there exist two functions

    $y = \varphi(x)$ and $y = \psi(x)$

such that $\displaystyle\lim_{x\to a}\varphi(x) = b = \lim_{x\to a}\psi(x)$, but

$\displaystyle\lim_{x\to a} f(x,\varphi(x)) \neq \lim_{x\to a} f(x,\psi(x))$, then we say

that $\displaystyle\lim_{(x,y)\to(a,b)} f(x,y)$ does not exist.

2. $\displaystyle\lim_{(x,y)\to(a,b)} f(x,y)$ is called double or simultaneous limit.

3. $\displaystyle\lim_{x\to a}\lim_{y\to b} f(x,y)$ and $\displaystyle\lim_{y\to b}\lim_{x\to a} f(x,y)$ are called repeated or iterated limits. These two limits may or may not exist. If the double limits exists, then the repeated limits will exist but the converse is not true in general.

Consider, $f(x,y) = \dfrac{(x+2)(y-x)}{(y+2)(y+x)}$

Then,

$\displaystyle\lim_{(x,y)\to(0,0)} f(x,y) = \lim_{(x,y)\to(0,0)}\dfrac{(x+2)(y-x)}{(y+2)(y+x)}$

$= \displaystyle\lim_{x\to 0}\dfrac{(x+2)(mx-x)}{(mx+2)(mx+x)}$

(Taking the curve $y = mx$, so that when $y \to 0$, $x \to 0$)

$= \displaystyle\lim_{x\to 0}\dfrac{(x+2)(m-1)}{(mx+2)(m+1)} = \dfrac{(0+2)(m-1)}{(0+2)(m+1)} = \dfrac{m-1}{m+1}$,

which depends on "m" and so is not unique.

Hence, the double limit does not exist.

Now, $\displaystyle\lim_{x\to 0}\lim_{y\to 0} f(x,y) = \lim_{x\to 0}\left\{\lim_{y\to 0}\dfrac{(x+2)(y-x)}{(y+2)(y+x)}\right\}$

$= \displaystyle\lim_{x\to 0}\dfrac{(x+2)(0-x)}{(0+2)(0+x)} = \lim_{x\to 0}\left\{\dfrac{-(x+2)}{2}\right\}$

$= -\dfrac{(0+2)}{2} = -1$

Again, $\displaystyle\lim_{y\to 0}\lim_{x\to 0} f(x,y) = \lim_{y\to 0}\left\{\lim_{x\to 0}\dfrac{(x+2)(y-x)}{(y+2)(y+x)}\right\}$

$= \displaystyle\lim_{y\to 0}\dfrac{(0+2)(y-0)}{(y+2)(y+0)} = \lim_{y\to 0}\dfrac{2}{y+2}$

$= \dfrac{2}{0+2} = 1$

Thus, $\displaystyle\lim_{x\to 0}\lim_{y\to 0} f(x,y) \neq \lim_{y\to 0}\lim_{x\to 0} f(x,y)$.

2.10.3 Continuity of Functions of Two Variables

A function $z = f(x, y)$ is called continuous at (a, b) if

(i) $\displaystyle\lim_{(x,y)\to(a,b)} f(x,y)$ exists and

(ii) $\displaystyle\lim_{(x,y)\to(a,b)} f(x,y)=f(a,b)$, provided $f(a,b)$ is defined.

Alternatively, $f(x, y)$ is said to be continuous at (a, b), if for a a given $\varepsilon > 0$, there exist $\delta > 0$ (δ depends on ε) such that $|f(x, y) - f(a, b)| < \varepsilon$ whenever $0 <|x - a| < \delta$ and $0< |y - b| < \delta$.

2.10.4 Partial Derivatives

Let $z = f(x, y)$ be a function of two variables x and y. Then

(i) the partial derivative of $z = f(x, y)$ w.r.t "x" (keeping y constant) is denoted by $\dfrac{\partial f}{\partial x}$ or f_x or $\dfrac{\partial z}{\partial x}$ and is defined by

$$\frac{\partial f}{\partial x}=\lim_{h\to 0}\frac{f(x+h,y)-f(x,y)}{h}$$

(ii) The partial derivative of $z = f(x, y)$ w.r.t "y" (keeping "x" constant) is denoted by $\dfrac{\partial z}{\partial y}$ or $\dfrac{\partial f}{\partial y}$ or f_y and is defined by

$$\frac{\partial f}{\partial y}=\lim_{k\to 0}\frac{f(x,y+k)-f(x,y)}{k}$$

Remember:

(i) $\left.\dfrac{\partial f}{\partial x}\right|_{(0,0)}=\lim_{x\to 0}\dfrac{f(x,0)-f(0,0)}{x}$

(ii) $\left.\dfrac{\partial f}{\partial y}\right|_{(0,0)}=\lim_{y\to 0}\dfrac{f(0,y)-f(0,0)}{y}$

2.10.5 Homogeneous Function

A function $f(x, y)$ is said to be homogeneous function of degree "n" if it can be expressed as $f(x, y) = x^n\varphi\left(\dfrac{y}{x}\right)$.

Alternatively, a function $f(x, y)$ is said to be homogeneous function of degree "n" if

$$f(tx, ty) = t^n f(x, y).$$

Example:

Consider $f(x, y) = x^2 + xy$.

Then, $f(x, y) = x^2\left(1+\dfrac{y}{x}\right) = x^2\varphi\left(\dfrac{y}{x}\right)$.

$\therefore$ $f(x, y)$ is a homogeneous function of degree "2."

2.10.6 Euler's Theorem

If $u = f(x, y)$ be a homogeneous function of degree "n," then $x\dfrac{\partial u}{\partial x}+y\dfrac{\partial u}{\partial y}=nu$, i.e

$$x\frac{\partial f}{\partial x}+y\frac{\partial f}{\partial y}=nf(x,y).$$

Application:

Consider the function $u(x, y) = x^2 + xy + y^2$

Then $u(x,y)=x^2\left\{1+\dfrac{y}{x}+\left(\dfrac{y}{x}\right)^2\right\}=x^2\varphi\left(\dfrac{y}{x}\right)$,

where $\varphi\left(\dfrac{y}{x}\right)=1+\dfrac{y}{x}+\left(\dfrac{y}{x}\right)^2$.

$\therefore$ $u(x, y)$ is a homogeneous function of degree

"2." So by Euler's theorem,

$$x\frac{\partial u}{\partial x}+y\frac{\partial u}{\partial y}=2u=2\left(x^2+xy+y^2\right)$$

2.10.7 Total Differential and Total Derivative

If $u = f(x, y)$ be a function of two variables and $x = \varphi(t)$, $y = \psi(t)$; then

(i) Total differential of u, denoted by du, is given by $du=\dfrac{\partial u}{\partial x}dx+\dfrac{\partial u}{\partial y}dy$.

(ii) The total derivative of u w.r.t another variable "t" is given by $\dfrac{du}{dt}=\dfrac{\partial u}{\partial x}\dfrac{du}{dt}+\dfrac{\partial u}{\partial y}\dfrac{dy}{dt}$.

Remember:

1. If $u = f(x, y)$, then $\dfrac{du}{dx}=\dfrac{\partial u}{\partial x}+\dfrac{\partial u}{\partial y}\dfrac{dy}{dx}$.

2. If $u = f(x, y, z)$, then
$$\frac{du}{dt}=\frac{\partial u}{\partial x}\frac{dx}{dt}+\frac{\partial u}{\partial y}\frac{dy}{dt}+\frac{\partial u}{\partial z}\frac{dz}{dt}.$$

3. If $f(x, y) = k$ be an implicit relation between the variable x and y, then
$$\frac{dy}{dx}=\frac{-\dfrac{\partial f}{\partial x}}{\dfrac{\partial f}{\partial y}}=\frac{-f_x}{f_y}.$$

4. If $w = f(u, v)$ be a function of u and v; $u = \varphi(x, y)$, $v = \psi(x, y)$; then

$$\frac{\partial f}{\partial x} = \frac{\partial f}{\partial u} \times \frac{\partial u}{\partial x} + \frac{\partial f}{\partial v} \times \frac{\partial v}{\partial x} \text{ and}$$

$$\frac{\partial f}{\partial y} = \frac{\partial f}{\partial u} \times \frac{\partial u}{\partial y} + \frac{\partial f}{\partial v} \times \frac{\partial v}{\partial y}.$$

5. If $v = f(u)$ and $u = g(x, y)$, then

$$\frac{\partial f}{\partial x} = \frac{df}{du} \times \frac{\partial u}{\partial x} \text{ and } \frac{\partial f}{\partial y} = \frac{df}{du} \times \frac{\partial u}{\partial y}$$

Example:

Let $v = \sin u$ and $u = x^2 y$

Then $\dfrac{\partial v}{\partial y} = \dfrac{dv}{du} \times \dfrac{\partial u}{\partial y} = \cos u \times \left(x^2 \times 1\right)$

$$= x^2 \cos u = x^2 \cos\left(x^2 y\right).$$

2.10.8 Jacobian

(i) The Jacobian of $u = f(x, y)$ and $v = g(x, y)$ is denoted by $\dfrac{\partial(u,v)}{\partial(x,y)}$ or J and is defined by

$$J = \begin{vmatrix} \dfrac{\partial u}{\partial x} & \dfrac{\partial u}{\partial y} \\ \dfrac{\partial v}{\partial x} & \dfrac{\partial v}{\partial y} \end{vmatrix}$$

Example:

Let $x = r\cos\theta$ and $y = r\sin\theta$

Then clearly $x = f(r, \theta)$ and $y = g(r, \theta)$.

So, $J = \dfrac{\partial(x,y)}{\partial(r,\theta)} = \begin{vmatrix} \dfrac{\partial x}{\partial r} & \dfrac{\partial x}{\partial \theta} \\ \dfrac{\partial y}{\partial r} & \dfrac{\partial y}{\partial \theta} \end{vmatrix} = \begin{vmatrix} \cos\theta & -r\sin\theta \\ \sin\theta & r\cos\theta \end{vmatrix}$

$$= r\cos^2\theta - \left(-r\sin^2\theta\right) = r.$$

(ii) The Jacobian of $u = f(x, y, z)$, $v = g(x, y, z)$ and $w = h(x, y, z)$ is denoted by $\dfrac{\partial(u,v,w)}{\partial(x,y,z)}$ or J and is defined by

$$J = \begin{vmatrix} \dfrac{\partial u}{\partial x} & \dfrac{\partial u}{\partial y} & \dfrac{\partial u}{\partial z} \\ \dfrac{\partial v}{\partial x} & \dfrac{\partial v}{\partial y} & \dfrac{\partial v}{\partial z} \\ \dfrac{\partial w}{\partial x} & \dfrac{\partial w}{\partial y} & \dfrac{\partial w}{\partial z} \end{vmatrix}$$

Example:

Let, $x = r\sin\theta\cos\varphi$, $y = r\sin\theta\sin\varphi$ and $z = r\cos\theta$.

Then, clearly $x = f(r,\theta,\varphi)$, $y = g(r,\theta,\varphi)$ and $z = h(r,\theta,\varphi)$

$$\therefore J = \frac{\partial(x,y,z)}{\partial(r,\theta,\varphi)} = \begin{vmatrix} \dfrac{\partial x}{\partial r} & \dfrac{\partial x}{\partial \theta} & \dfrac{\partial x}{\partial \varphi} \\ \dfrac{\partial y}{\partial r} & \dfrac{\partial y}{\partial \theta} & \dfrac{\partial y}{\partial \varphi} \\ \dfrac{\partial z}{\partial r} & \dfrac{\partial z}{\partial \theta} & \dfrac{\partial z}{\partial \varphi} \end{vmatrix}.$$

$$= \begin{vmatrix} \sin\theta\cos\varphi & r\cos\theta\cos\varphi & -r\sin\theta\sin\varphi \\ \sin\theta\sin\varphi & r\cos\theta\sin\varphi & r\sin\theta\cos\varphi \\ \cos\theta & -r\sin\theta & 0 \end{vmatrix}$$

$$= \sin\theta\cos\varphi\left\{0 + r^2\sin^2\theta\cos\varphi\right\}$$
$$\quad - r\cos\theta\cos\varphi\left\{0 - r\sin\theta\cos\theta\cos\varphi\right\}$$
$$\quad - r\sin\theta\sin\varphi\left\{-r\sin^2\theta\sin\varphi - r\cos^2\theta\sin\varphi\right\}$$

$$= r^2\sin^3\theta\cos^2\varphi + r^2\sin\theta\cos^2\varphi\cos^2\theta$$
$$\quad + r^2\sin\theta\sin^2\varphi\left(\cos^2\theta + \sin^2\theta\right)$$

$$= r^2\sin\theta\cos^2\varphi\left(\sin^2\theta + \cos^2\theta\right) + r^2\sin\theta\sin^2\varphi$$
$$= r^2\sin\theta\cos^2\varphi + r^2\sin\theta\sin^2\varphi$$
$$= r^2\sin\theta\left(\cos^2\varphi + \sin^2\varphi\right) = r^2\sin\theta.$$

Properties of Jacobian:

1. If $w_1 = f_1(u,v,x,y) = 0$, $w_2 = f_2(u,v,x,y) = 0$ then

$$\frac{\partial(u,v)}{\partial(x,y)} = \frac{\dfrac{\partial(w_1,w_2)}{\partial(x,y)}}{\dfrac{\partial(w_1,w_2)}{\partial(u,v)}}$$

2. If $f_1(u,v,w,x,y,z) = 0$, $f_2(u,v,w,x,y,z) = 0$ and $f_3(u,v,w,x,y,z) = 0$; then

$$\frac{\partial(u,v,w)}{\partial(x,y,z)} = \frac{\dfrac{\partial(f_1,f_2,f_3)}{\partial(x,y,z)}}{\dfrac{\partial(f_1,f_2,f_3)}{\partial(u,v,w)}}$$

2.11 MAXIMA AND MINIMA OF FUNCTIONS OF TWO VARIABLES

2.11.1 Introduction

1. A function $f(x,y)$ is said to have a maximum value at (a,b) if $f(a+h,b+k) < f(a,b)$ for all positive or negative small values of h and k.

2. A function $f(x,y)$ is said to have a minimum value at (a,b) if $f(a+h,b+k) > f(a,b)$ for all positive or negative small values of h and k.

3. A saddle point is a point where the function $f(x,y)$ is neither maximum nor minimum.

2.11.2 Working Rule to Find the Maximum and Minimum Values of $f(x,y)$:

Step-I:

Find $\dfrac{\partial f}{\partial x}, \dfrac{\partial f}{\partial y}, \dfrac{\partial^2 f}{\partial x^2}, \dfrac{\partial^2 f}{\partial x\,\partial y}$ and $\dfrac{\partial^2 f}{\partial y^2}$.

Step-II: Consider $\dfrac{\partial f}{\partial x} = 0$ and $\dfrac{\partial f}{\partial y} = 0$ and then solve these equations for x and y. Let (a,b) be one of the values of (x,y).

Step-III: Calculate the followings:

$$A = \frac{\partial^2 f}{\partial x^2}\bigg|_{(a,b)}, \; B = \frac{\partial^2 f}{\partial x\,\partial y}\bigg|_{(a,b)} \;\&\; C = \frac{\partial^2 f}{\partial y^2}\bigg|_{(a,b)}.$$

Step-IV:

(i) If $AC - B^2 > 0$ and $A < 0$, then $f(x,y)$ has a maximum value at (a,b).

(ii) If $AC - B^2 > 0$ and $A > 0$, then $f(x,y)$ has a minimum value at (a,b).

(iii) If $AC - B^2 < 0$, then $f(x,y)$ has neither a maximum value nor a minimum value at (a,b). In this case, the point (a,b) is called a saddle point.

(iv) If $AC - B^2 = 0$, then no conclusion can be withdrawn and further investigation is needed.

Example:

Let $f(x,y) = x^2 + y^2 - 2x - 2y - 1$

Then $\dfrac{\partial f}{\partial x} = 2x - 2, \; \dfrac{\partial f}{\partial y} = 2y - 2,$

$\dfrac{\partial^2 f}{\partial x^2} = 2, \; \dfrac{\partial^2 f}{\partial x\,\partial y} = 0$ and $\dfrac{\partial^2 f}{\partial y^2} = 2.$

$\therefore \dfrac{\partial f}{\partial x} = 0, \; \dfrac{\partial f}{\partial y} = 0$

$\Rightarrow 2x - 2 = 0, 2y - 2 = 0$

$\Rightarrow x = 1, y = 1.$

So, $A = \dfrac{\partial^2 f}{\partial x^2}\bigg|_{(1,1)} = 2, \; B = \dfrac{\partial^2 f}{\partial x\,\partial y}\bigg|_{(1,1)} = 0$

and $C = \dfrac{\partial^2 f}{\partial y^2}\bigg|_{(1,1)} = 2.$

$\therefore AC - B^2 = 4 > 0$ and $A = 2 > 0$

Hence, $f(x,y)$ has a minimum value at $(1,1)$ and minimum value of $f(x,y)$ is $f(1,1)$, i.e, $1 + 1 - 2 - 2 - 1 = -3$.

2.11.3 Lagrange's Method for Undetermined Multipliers

Let $u = f(x,y,z)$ be a function of three variables x, y, z which are connected by the relation $\phi(x,y,z) = 0$. Suppose we need to find the extreme values (maximum or minimum values) of the function $u = f(x,y,z)$. Then we need to follow the steps given below:

Step-I:

Take $F(x,y,z) = f(x,y,z) + \lambda\phi(x,y,z)$

Step-II:

Consider $\dfrac{\partial F}{\partial x} = 0, \; \dfrac{\partial F}{\partial y} = 0$ and $\dfrac{\partial F}{\partial z} = 0$ and then solve these equations for $x, y,$ and z. Let (a,b,c) be one of the values of (x,y,z). Then, $u = f(x,y,z)$ will have a stationary value (maximum/minimum value) at (a,b,c).

Application:

Given that $x + y + z = 1$. Let us find the maximum value of $x\,y^2\,z^3$. Here, $\phi(x,y,z) = x + y + z - 1$, $f(x,y,z) = xy^2z^3$.

$$\therefore F(x,y,z) = f(x,y,z) + \lambda\phi(x,y,z)$$
$$= xy^2z^3 + \lambda(x + y + z - 1)$$

Then, $\dfrac{\partial F}{\partial x} = 0$, $\dfrac{\partial F}{\partial y} = 0$ and $\dfrac{\partial F}{\partial z} = 0$

$$\Rightarrow y^2z^3 + \lambda = 0,\ 2xyz^3 + \lambda = 0 \text{ and}$$
$$3xy^2z^2 + \lambda = 0$$

$$\Rightarrow -\lambda = y^2z^3 = 2xyz^3 = 3xy^2z^2$$

$$\Rightarrow \frac{1}{y^2z^3} = \frac{1}{2xyz^3} = \frac{1}{3xy^2z^2}$$

$$\Rightarrow \frac{x}{xy^2z^3} = \frac{y}{2xy^2z^3} = \frac{z}{3xy^2z^3}$$

$$= \frac{x + y + z}{xy^2z^3 + 2x\,y^2z^3 + 3xy^2z^3}$$

$$\Rightarrow \frac{x}{xy^2z^3} = \frac{y}{2xy^2z^3} = \frac{z}{3xy^2z^3} = \frac{1}{6xy^2z^3}$$

$$(\because x + y + z = 1)$$

$$\therefore F(x,y,z) = f(x,y,z) + \lambda\phi(x,y,z)$$
$$= xy^2z^3 + \lambda(x + y + z - 1)$$

Then, $\dfrac{\partial F}{\partial x} = 0$, $\dfrac{\partial F}{\partial y} = 0$ and $\dfrac{\partial F}{\partial z} = 0$

$$\Rightarrow y^2z^3 + \lambda = 0,\ 2xyz^3 + \lambda = 0 \text{ and}$$
$$3xy^2z^2 + \lambda = 0$$

$$\Rightarrow -\lambda = y^2z^3 = 2xyz^3 = 3xy^2z^2$$

$$\Rightarrow \frac{1}{y^2z^3} = \frac{1}{2xyz^3} = \frac{1}{3xy^2z^2}$$

$$\Rightarrow \frac{x}{xy^2z^3} = \frac{y}{2xy^2z^3} = \frac{z}{3xy^2z^3}$$

$$= \frac{x + y + z}{xy^2z^3 + 2x\,y^2z^3 + 3xy^2z^3}$$

$$\Rightarrow \frac{x}{xy^2z^3} = \frac{y}{2xy^2z^3} = \frac{z}{3xy^2z^3} = \frac{1}{6xy^2z^3}$$

$$(\because x + y + z = 1)$$

$$\Rightarrow \frac{x}{xy^2z^3} = \frac{1}{6xy^2z^3}, \frac{y}{2xy^2z^3} = \frac{1}{6xy^2z^3},$$

$$\frac{z}{3xy^2z^3} = \frac{1}{6xy^2z^3}$$

$$\Rightarrow x = \frac{1}{6}, y = \frac{1}{3}, z = \frac{1}{2}.$$

Therefore, $f(x,y,z) = xy^2z^3$ has a maximum

value at $\left(\dfrac{1}{6}, \dfrac{1}{3}, \dfrac{1}{2}\right)$. Hence, the maximum value of

$x\,y^2\,z^3$ is $\left(\dfrac{1}{6}\right)\left(\dfrac{1}{3}\right)^2\left(\dfrac{1}{2}\right)^3$.

2.12 CHANGE OF ORDER OF INTEGRATION

Let us consider a double integral $\iint\limits_R f(x,y)dxdy$, where R is the region. Now assume that the region R lies between the lines $x = x_1$ and $x = x_2$; $y = y_1$ and $y = y_2$.

Consider any value of x in $x1 \leqslant x \leqslant x2$ and keep the value constant. y varies between $\varphi1(x)$ and $\varphi2(x)$ where $\varphi1(x)$ and $\varphi2(x)$ are the ordinate of the points at which the boundary of R is intersected by the line through (x,y) and parallel to y-axis. Now first integrate $f(x,y)$ w.r.t y and then integrate the result w.r.t. "x."

Thus, $\iint\limits_R f(x,y)dxdy = \int\limits_{x_1}^{x_2}\left\{\int\limits_{\varphi_1(x)}^{\varphi_2(x)} f(x,y)dy\right\}dx$.

Again consider any y in the interval $y_1 \leq y \leq y_2$ and keep the value constant. x varies between $\psi_1(y)$ and $\psi_2(y)$ where $\psi_1(y)$ and $\psi_2(y)$ are the abscissa of the points at which the boundary of R is intersected by the line through (x,y) and parallel to the x-axis. Now first integrate $f(x,y)$ w.r.t x and then integrate the result w.r.t. "y."

Thus, $\iint\limits_R f(x,y)dxdy = \int\limits_{y_1}^{y_2}\left\{\int\limits_{\psi_1(y)}^{\psi_2(y)} f(x,y)dx\right\}dy$.

Hence, we can conclude that in a region, the order of integration when evaluating a double integral can always be changed.

Remember:

The easiest way to find the limits of integration is to sketch the region of integration and determine the limits from the sketch.

Example:

Let us consider $\int\limits_{0}^{1}\int\limits_{x^2}^{x}\left(x^2+y^2\right)dydx$

Hence, the region of integration is described by $0 \le x \le 1$, $x^2 \le y \le x$.

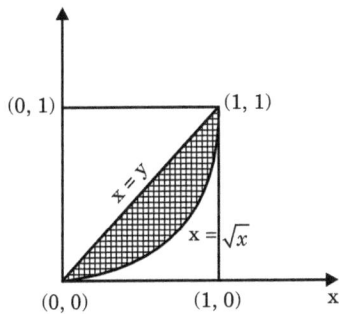

From the figure above, it is clear that the region of integration can be described as $0 \le x \le 1$ and $y \le x \le \sqrt{y}$

Hence, $\int\limits_{0}^{1}\left\{\int\limits_{x^2}^{x}\left(x^2+y^2\right)dy\right\}dx=\int\limits_{0}^{1}\left\{\int\limits_{y}^{\sqrt{y}}\left(x^2+y^2\right)dx\right\}dy$.

2.13 DOUBLE AND TRIPLE INTEGRALS

2.13.1 Double Integrals

Let the region R be divided into rectangular partitions and dx be the length of a subrectangular partition and dy be its width so that $dxdy$ represents an elementary area. Then, the integral $\iint\limits_{R} f(x,y)dxdy$ is called the double integral of the single valued function $f(x,y)$.

Evaluation of double integral:

1. Let the region R be bounded by the continuous curves $y = \varphi(x)$, $y = \psi(x)$ and the ordinates $x = a$ and $x = b$. Then we write

$$\iint\limits_{R} f(x,y)dxdy=\int\limits_{x=a}^{b}\left\{\int\limits_{y=\varphi(x)}^{\psi(x)} f(x,y)dy\right\}dx$$

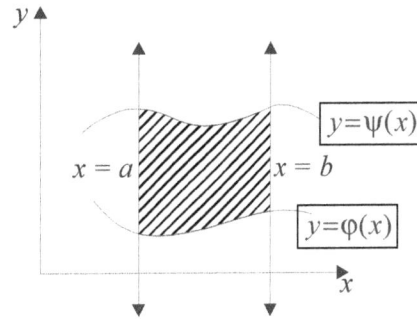

Example:

$$\int\limits_{1}^{3}\int\limits_{0}^{x}\frac{1}{\left(x^2+y^2\right)}dxdy=\int\limits_{x=1}^{3}\left\{\int\limits_{y=0}^{x}\frac{dy}{x^2+y^2}\right\}dx$$

$$=\int\limits_{x=1}^{3}\left[\frac{1}{x}\tan^{-1}\frac{y}{x}\right]_{y=0}^{x}dx$$

$$=\int\limits_{x=1}^{3}\left\{\frac{1}{x}\tan^{-1}\frac{x}{x}-\frac{1}{x}\tan^{-1}\frac{0}{x}\right\}dx$$

$$=\int\limits_{1}^{3}\frac{1}{x}\tan^{-1}1dx=\frac{\pi}{4}\int\limits_{1}^{3}\frac{dx}{x}\quad\left(\because\tan^{-1}1=\frac{\pi}{4}\right)\left(\because\tan^{-1}1=\frac{\pi}{4}\right)$$

$$=\frac{\pi}{4}\left[\log_e x\right]_{1}^{3}=\frac{\pi}{4}\left(\log_e 3-\log_e 1\right)$$

$$=\frac{\pi}{4}\log_e 3\quad\left(\because\log_e 1=0\right)$$

$$=\frac{\pi}{4}\log_e 3\ (\because\log_e 1=0)$$

2. Let the region R be bounded by the continuous curves $x = \varphi(y)$, $x = \psi(y)$ and the abscissas $y = c$ and $y = d$. Then we write

$$\iint\limits_{R} f(x,y)dxdy=\int\limits_{y=c}^{d}\left\{\int\limits_{x=\varphi(y)}^{\psi(y)} f(x,y)dx\right\}dy$$

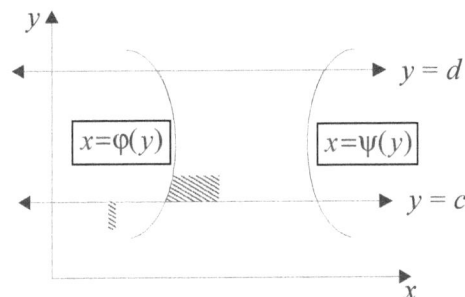

2.13.2 Triple Integrals

If the region of integration R is given by

$$R=\left\{(x,y,x):a\le x\le b,\ c\le y\le d,\ e\le z\le f\right\},$$

then we define

$$\iiint_R f(x,y,z)dxdydz = \int_{x=a}^{b} \int_{y=c}^{d} \int_{z=e}^{f} f(x,y,z)dxdydz$$

where $f(x,y,z)$ is a continuous function in R.

Remember: If the region of integration R is given by

$$R = \{(x,y,z) : a \leqslant x \leqslant b,\ \varphi_1(x) \leqslant y \leqslant \varphi_2(x),$$

$$\psi_1(x,y) \leqslant z \leqslant \psi_2(x,y)\}$$

Then we define $\iiint_R f(x,y,z)dxdydz$

$$= \int_{x=a}^{b} \left[\int_{y=\varphi_1(x)}^{\varphi_2(x)} \left(\left\{ \int_{z=\psi_1(x,y)}^{\psi_2(x,y)} f(x,y,z)dz \right\} \right) dy \right] dx$$

where $f(x,y,z)$ is a continuous function in R.

Example:

Let us evaluate $\int_0^1 \int_0^{\sqrt{1-x^2}} \int_0^{\sqrt{1-x^2-y^2}} xyz\,dxdydz$

Here region of integration,

$$R = \left\{ (x,y,z) : 0 \leqslant x \leqslant 1,\ 0 \leqslant y \leqslant \sqrt{1-x^2}, \right.$$

$$\left. 0 \leqslant z \leqslant \sqrt{1-x^2-y^2} \right\}$$

$$\therefore \int_0^1 \int_0^{\sqrt{1-x^2}} \int_0^{\sqrt{1-x^2-y^2}} xyz\,dxdydz$$

$$= \int_{x=0}^1 \left\{ \int_{y=0}^{\sqrt{1-x^2}} \left(\int_{z=0}^{\sqrt{1-x^2-y^2}} xyz\,dz \right) dy \right\} dx$$

$$= \int_{x=0}^1 \left(\int_{y=0}^{\sqrt{1-x^2}} xy \left[\frac{z^2}{2} \right]_{z=0}^{\sqrt{1-x^2-y^2}} dy \right) dx$$

$$= \int_{x=0}^1 \left(\int_{y=0}^{\sqrt{1-x^2}} xy \left[\frac{1-x^2-y^2}{2} \right] dy \right) dx$$

$$= \int_{x=0}^1 \left(\int_{y=0}^{\sqrt{1-x^2}} \left\{ \frac{x(1-x^2)}{2} y - \frac{x}{2} y^3 \right\} dy \right) dx$$

$$= \int_{x=0}^1 \left(\frac{x(1-x^2)}{4} \left[y^2 \right]_0^{\sqrt{1-x^2}} - \frac{x}{8} \left[y^4 \right]_0^{\sqrt{1-x^2}} \right) dx$$

$$= \int_{x=0}^1 \left(\frac{x(1-x^2)}{4} \left[y^2 \right]_0^{\sqrt{1-x^2}} - \frac{x}{8} \left[y^4 \right]_0^{\sqrt{1-x^2}} \right) dx$$

$$= \int_0^1 \left\{ \frac{x(1-x^2)}{4}(1-x^2) - \frac{x}{8}(1-x^2)^2 \right\} dx$$

$$= \int_0^1 (1-x^2)^2 \times \frac{x}{8} dx$$

$$= \frac{1}{8} \int_0^1 (x - 2x^3 + x^5) dx = \frac{1}{8} \left[\frac{x^2}{2} - \frac{x^4}{2} + \frac{x^6}{6} \right]_0^1$$

$$= \frac{1}{8} \left(\frac{1}{2} - \frac{1}{2} + \frac{1}{6} \right) = \frac{1}{48}.$$

2.14 ARC LENGTH OF A CURVE

(i) The arc length of a curve $y = f(x)$ between the points where $x = a$ and $x = b$ is given by

$$\int_a^b \sqrt{1 + \left(\frac{dy}{dx} \right)^2}\, dx$$

(ii) The arc length of a curve $x = f(y)$ between the points where $y = c$ and $y = d$ is given by

$$\int_c^d \sqrt{1 + \left(\frac{dx}{dy} \right)^2}\, dy$$

(iii) The arc length of the parametric curve $x = f(t)$, $y = g(t)$ between the points where $t = \alpha$ and $t = \beta$ is given by

$$\int_\alpha^\beta \sqrt{\left(\frac{dx}{dt} \right)^2 + \left(\frac{dy}{dt} \right)^2}\, dt$$

(iv) The arc length of the curve $\theta = f(r)$ between the points where $r = a$ and $r = b$ is given by

$$\int_a^b \sqrt{1 + r^2 \left(\frac{d\theta}{dr} \right)^2}\, dr\ .$$

2.15 VOLUMES OF SOLIDS OF REVOLUTION

2.15.1 Working Formulas

(i) The volume V of a solid formed by the revolution of a plane area bounded by the continuous curve $y = f(x)$, the ordinates $x = a$, $x = b$ and the x-axis about the x-axis is given by

$$V = \pi \int_a^b y^2 dx\ .$$

(ii) The volume V of a solid formed by the revolution of a plane area bounded by the curve $x = f(y)$, the abscissas $y = c$, $y = d$ and y-axis about y-axis is given by

$$V = \pi \int_c^d x^2 dy .$$

(iii) The volume V of a solid formed by the revolution of a plane area bounded by the parametric curve $x = f(t)$, $y = g(t)$, x-axis and the ordinates where $t = a$, $t = b$, about x-axis is given by

$$V = \pi \int_a^b y^2 \left(\frac{dx}{dt}\right) dt = \pi \int_a^b \{g(t)\}^2 f'(t) dt$$

(iv) The volume V of a solid formed by the revolution of a plane area bounded by the parametric curve $x = f(t)$, $y = g(t)$, y-axis and the abscissas where $t = c$, $t = d$, about y-axis is given by

$$V = \pi \int_c^d x^2 \left(\frac{dy}{dt}\right) dt = \pi \int_c^d \{f(t)\}^2 g'(t) dt.$$

(v) The volume V of a solid formed by the revolution of a plane area bounded by the polar curve $r = f(\theta)$, the radii vectors $\theta = \theta_1$ and $\theta = \theta_2$, about the initial line i.e; $\theta = 0$ (x axis) is given by

$$V = \frac{2}{3}\pi \int_{\theta_1}^{\theta_2} \{f(\theta)\}^3 \sin\theta d\theta$$

2.16 SURFACE AREAS OF SOLIDS OF REVOLUTION

(i) The surface area S of a solid generated by the revolution of the curve $y = f(x)$ about x-axis between the ordinates $x = a$ and $x = b$

is given by $S = 2\pi \int_a^b y \sqrt{1 + \left(\frac{dy}{dx}\right)^2} dx$

(ii) The surface area S of a solid generated by the revolution of the curve $x = f(y)$ about y-axis between the abscissas $y = a$ and $y = b$

is given by $S = 2\pi \int_a^b x \sqrt{1 + \left(\frac{dx}{dy}\right)^2} dy$

(iii) The surface area S of a solid generated by the revolution of the parametric curve

$x = f(t), y = g(t)$ about x-axis between the limits $t = t_1$ and $t = t_2$ is given by

$$S = 2\pi \int_{t_1}^{t_2} y \sqrt{\left(\frac{dx}{dt}\right)^2 + \left(\frac{dy}{dt}\right)^2} dt$$

(iv) The surface area S of a solid generated by the revolution of the polar curve $r = f(\theta)$ about x-axis between the limits $\theta = \theta_1$ and $\theta = \theta_2$ is given by

$$S = 2\pi \int_{\theta_1}^{\theta_2} r\sin\theta \times \sqrt{r^2 + \left(\frac{dr}{d\theta}\right)^2} d\theta$$

Fully Solved MCQs (Level-I)

1. If $f(x) = \begin{cases} \dfrac{|x-2|}{x-2}, & x \neq 2 \\ 0, & x = 2 \end{cases}$, then

(a) $\lim_{x\to 2} f(x)$ exist

(b) $\lim_{x\to 2} f(x)$ does not exist

(c) $\lim_{x\to 2+} f(x) = -1$

(d) $\lim_{x\to 2-} f(x) = 1$

2. Given that $f(x) = \begin{cases} \dfrac{x-|x|}{x}, & x \neq 0 \\ 1, & x = 0 \end{cases}$, then

(a) $\lim_{x\to 0+} f(x)$ does not exist

(b) $\lim_{x\to 0} f(x)$ does not exist

(c) $\lim_{x\to 0-} f(x)$ does not exist

(d) $\lim_{x\to 0} f(x) = f(1)$

3. $\lim_{x\to 0} \dfrac{\sin x^0}{x} = ?$

(a) π (b) 0 (c) $\pi/90$ (d) $\pi/180$

4. $\lim_{x\to 0} \dfrac{3^x - 1}{\sqrt{1+x} - 1} = ?$

(a) 9 (b) 4 (c) 1
(d) none of these

5. $\lim_{x \to 0} (1+x)^{\cos ecx} = ?$

   (a) 0   (b) 1   (c) e   (d) e^2

6. $\lim_{x \to \infty} \dfrac{\sin^4 x - \sin^2 x + 1}{\cos^4 x - \cos^2 x + 1} = ?$

   (a) 1/2   (b) –1   (c) 0   (d) 1

7. $\lim_{x \to a} \dfrac{\sqrt{3x-a} - \sqrt{x+a}}{x-a} = ?$

   (a) $\sqrt{\dfrac{2}{a}}$ (b) $\dfrac{1}{\sqrt{2a}}$ (c) $\dfrac{1}{a\sqrt{2}}$ (d) 1

8. $\lim_{x \to 0} \dfrac{\tan x - \sin x}{x^3} = ?$

   (a) 0   (b) 1/2   (c) 1   (d) –1

9. $\lim_{x \to 0} \dfrac{\tan 2x - x}{3x - \sin x} = ?$

   (a) 1/2   (b) 1/3   (c) 2/3   (d) 0

10. If $f : R \to R$ be a continuous function, then

   (a) f is bounded
   (b) f is differentiable
   (c) f may be unbounded
   (d) none of these

11. The graph of a function f is given below:

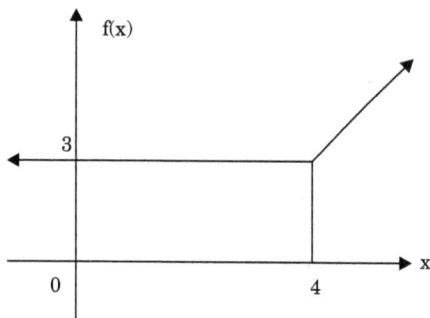

Then, f is differentiable
   (a) everywhere except at x = 4
   (b) nowhere
   (c) everywhere
   (d) only at x = 4

12. If $f(x) = k|\sin x| + 2x$ and $f'(0)$ exist, then the value of "k" is?

   (a) 4   (b) 3   (c) 1   (d) 0

13. Let $f(x) = \begin{cases} x^2 - ax + b, & x < 1 \\ a + b, & x = 1 \\ 4x^2 - bx + a, & x > 1 \end{cases}$.

If f is continuous at x = 1, then the values of "a" and "b" are, respectively

   (a) $\dfrac{1}{2}$, 2   (b) 0, 3   (c) 1, 1   (d) 0, 2

14. $\lim_{x \to 0} \dfrac{\sin\left(\pi \cos^2 x\right)}{x^2} = ?$

   (a) 0   (b) 1   (c) 2π (d) π

15. $\lim_{x \to 0} \dfrac{\log(4+x) - \log(4-x)}{x} = ?$

   (a) $\dfrac{1}{2}$ (b) $-\dfrac{1}{2}$ (c) 0   (d) 1

16. $\lim_{x \to \frac{\pi}{6}} \left\{ \dfrac{2 - \sqrt{3}\cos x - \sin x}{(6x - \pi)^2} \right\} = ?$

   (a) 0   (b) 1   (c) $\dfrac{1}{36}$ (d) $\dfrac{1}{18}$

17. $\lim_{x \to \frac{\pi}{2}} \left\{ \dfrac{1 + \cos 2x}{(\pi - 2x)^2} \right\} = ?$

   (a) $\dfrac{1}{2}$ (b) $-\dfrac{1}{2}$ (c) 0   (d) 1

18. $\lim_{x \to 0} \left\{ \dfrac{4^{3x} - 3^{4x}}{x} \right\} = ?$

   (a) $\log\left(\dfrac{4}{31}\right)$ (b) $\log\left(\dfrac{64}{81}\right)$ (c) 0   (d) 1

19. $\lim_{x \to \pi} \left\{ \dfrac{1 + \cos^3 x}{\sin^2 x} \right\} = ?$

   (a) $\dfrac{3}{2}$ (b) $\dfrac{4}{9}$ (c) 0   (d) 1

20. Which of the following is true?

   (a) $\dfrac{b-a}{b} < \log\left(\dfrac{b}{a}\right) < \dfrac{b-a}{a}$

   (b) $\dfrac{b-a}{a} < \log\left(\dfrac{b}{a}\right) < \dfrac{b-a}{b}$

   (c) $\dfrac{1}{a} < \log\left(\dfrac{b}{a}\right) < \dfrac{1}{b}$

   (d) $\dfrac{1}{b} < \log\left(\dfrac{b}{a}\right) < \dfrac{1}{a}$

21. The points on the curve $y = (x+1)^2$ at which the tangent is parallel to x-axis are given by
(a) 2, 4   (b) –1   (c) 1,1   (d) 1,2

22. The point on the curve $y = (x-1)(x-2)$ at which the tangent is parallel to the chord joining the points $(1, 0)$ and $(2, 0)$ is given by
(a) x = 1.7   (b) x = 1.5
(c) $x = 1.3$ (d) x = 1.2

23. Let f, g be two differentiable functions on R and $g'(x) \neq 0$ for $x \in R$. Also let $f(1) = 6$, $g(1) = 0$, $f(2) = 8$ and $g(2) = 4$. Then ,

(a) $\exists c \in (2,4)$ such that $g'(c) = 2f'(c)$

(b) $\exists c \in (2,4)$ such that $g'(c) = 4f'(c)$

(c) $\exists c \in (1,2)$ such that $g'(c) = 2f'(c)$

(d) $\exists c \in (2,4)$ such that $g'(c) = 3f'(c)$

24. Let f' exist $\forall x \in R$ & $f'(x) \geq 3$ $\forall x \in [1,4]$.

If $f(1) = -3$, then
(a) $f(4) \geq 3$ (b) $f(4) \geq 4$

(c) $f(4) \geq 5$ (d) $f(4) \geq 6$

25. The function $f(x) = \dfrac{x-2}{x+1}$ (where $x \neq -1$) is

(a) strictly increasing
(b) strictly decreasing
(c) increasing
(d) decreasing

26. The function $f(x) = \log\left(\sqrt{x^2+1} - x\right)$ is

(a) strictly increasing for all x
(b) strictly decreasing for all x
(c) increasing for all x
(d) decreasing for all x

27. Which of the following functions are decreasing on $\left(0, \dfrac{\pi}{2}\right)$?

(a) $\cos x$ (b) $3x^3 - 5$ (c) $\tan x$ (d) x^9

28. The interval on which the function $f(x) = x^x \, (x > 0)$ is strictly increasing is given by

(a) $\left(-\infty, \dfrac{1}{e}\right)$ (b) $(-\infty, \infty)$

(c) $\left(\dfrac{1}{e}, \infty\right)$ (d) $(0, \infty)$

29. The intervals on which the function $f(x) = \dfrac{x}{\log x}$ are strictly increasing and strictly decreasing are given by, respectively

(a) $(e, \infty); (0, e)$

(b) $(0, e); (e, \infty)$

(c) $(e, \infty); (0, e) - \{1\}$

(d) $(e, \infty); (0, e) - \{-1\}$

30. The value/values of λ for which the function $f(x) = \lambda x^3 - 9\lambda x^2 + 9x + 5$ is strictly increasing on R is/are given by

(a) $\lambda \in (0, \infty)$ (b) $\lambda \in (0,1)$

(c) $\lambda \in \left(1, \dfrac{1}{3}\right)$ (d) $\lambda \in \left(0, \dfrac{1}{3}\right)$

31. The intervals on which the function $f(x) = 6 - 9x - x^2$ are strictly increasing and strictly decreasing are given by, respectively

(a) $\left(-\infty, -\dfrac{9}{2}\right); \left(-\dfrac{9}{2}, \infty\right)$

(b) $\left(-\infty, \dfrac{9}{2}\right); \left(\dfrac{9}{2}, \infty\right)$

(c) $(-\infty, 0); (0, \infty)$

(d) $(1, \infty); (0, 1)$

32. Which of the following functions has a local minima or local maxima?

(a) e^x (b) $\log x$
(c) $\cos x$ in $(0, \pi)$ (d) none of these

33. The function $f(x) = \dfrac{\log x}{x}$ has a local maximum value at

(a) $x = e$ (b) $x = 1$

(c) x = 0   (d) none of these

34. The function $f(x) = x + \sqrt{1-x}$, $x \le 1$ has a local maximum value

(a) 0   (b) 1   (c) $\dfrac{5}{4}$ (d) $\dfrac{1}{4}$

35. If $f(x) = A\log x + Bx^2 + x$ has extreme values at x = 1 and x = 2, then the values of "A" and "B" are, respectively

(a) $\dfrac{2}{3}, \dfrac{1}{3}$ (b) $-\dfrac{2}{3}, -\dfrac{1}{6}$

(c) $\dfrac{1}{2}, \dfrac{1}{3}$ (d) $-\dfrac{2}{3}, \dfrac{1}{6}$

36. The combined resistance R of two resistors R_1 and R_2 is given as

$$\frac{1}{R} = \frac{1}{R_1} + \frac{1}{R_2}$$

If $R_1 + R_2 = k$ (constant), then R is maximum when

(a) $R_1 = 3R_2$ (b) $2R_1 = R_2$

(c) $R_1 = R_2$ (d) none of these

37. A beam is uniformly loaded and is supported at the two ends. The bending moments M at a distance x from one end is given by

$$M = \frac{Wx}{3} - \frac{Wx^3}{3L^2}$$

The M is maximum at x = ?

(a) $\dfrac{L}{\sqrt{3}}$ (b) $\dfrac{L}{\sqrt{2}}$ (c) $\dfrac{L}{2}$ (d) $2L$

38. The series $1 + \dfrac{3}{1!} + \dfrac{9}{2!} + \dfrac{27}{3!} + \dots \infty$ is

(a) convergent
(b) divergent
(c) p-series
(d) neither convergent nor divergent

39. The value of $\log 3 + \dfrac{1}{2!}(\log 3)^2 + \dfrac{1}{3!}(\log 3)^3 + \dots \infty$ is

(a) -1 (b) 0   (c) 1   (d) 2

40. The expansion of $\sin x$ in powers of $x - \dfrac{\pi}{2}$ is given by

(a) $1 - \dfrac{\left(x - \frac{\pi}{2}\right)^2}{2!} + \dfrac{\left(x - \frac{\pi}{2}\right)^4}{4!} - \dots \infty$

(b) $\left(x - \dfrac{\pi}{2}\right) - \dfrac{1}{3!}\left(x - \dfrac{\pi}{2}\right)^3 + \dfrac{1}{5!}\left(x - \dfrac{\pi}{2}\right)^5 - \dots \infty$

(c) $1 + \dfrac{1}{2!}\left(x - \dfrac{\pi}{2}\right)^2 + \dfrac{1}{4!}\left(x - \dfrac{\pi}{2}\right)^4 + \dots \infty$

(d) none of these

41. Which of the followings is true?

(a) $\log\left(\dfrac{1-x}{1+x}\right) = 2\left(x + \dfrac{x^3}{3} - \dfrac{x^5}{5} + \dots \infty\right)$

(b) $\log\left(\dfrac{1+x}{1-x}\right) = 2\left(x + \dfrac{x^3}{3} + \dfrac{x^5}{5} + \dots \infty\right)$

(c) $\log\left(\dfrac{1+x}{1-x}\right) = 2\left(x - \dfrac{x^3}{3} + \dfrac{x^5}{5} - \dots \infty\right)$

(d) $\log\left(\dfrac{1-x}{1+x}\right) = 2\left(x - \dfrac{x^3}{3} + \dfrac{x^5}{5} - \dots \infty\right)$

42. If $\displaystyle\sum_{n=0}^{\infty} a_n x^n$ be the Taylor series expansion of the function $f(x)$ at $x = 0$, then at $x = 0$, what will be the Taylor series expansion of $x^k f(x)$?

(a) $\displaystyle\sum_{n=0}^{\infty} a_n x^{k+n}$ (b) $\displaystyle\sum_{n=0}^{\infty} a_n x^{2k}$

(c) $\displaystyle\sum_{n=0}^{\infty} a_n x^{2n}$ (d) none of these

43. The Taylors series expansion of $x^3 + 5x^2 + 2x - 5$ in a powers of $x - 1$ is given by

(a) $2(x-1)^3 - 10(x-1)^2 + 8(x-1) + 3$

(b) $(x-1)^3 + \dfrac{3}{2}(x-1)^2 - 5(x-1) + 2$

(c) $(x-1)^3 - 5(x-1)^2 + 8(x-1) + 10$

(d) $(x-1)^3 + 8(x-1)^2 + 15(x-1) + 3$

44. Let $f(x) = \dfrac{\sqrt{\tan x}}{\sin x \cos x}$ and $F(x)$ be its anti-derivative. If $F\left(\dfrac{\pi}{4}\right) = 4$, then $F(x)$ = ?

(a) $2\sqrt{\tan x} + 2$ (b) $2\sqrt{\cot x} + 1$

(c) $2\tan x + 1$ (d) $2\cot x + 2$

45. $\int e^{\tan^{-1}x}\left(\dfrac{1+x+x^2}{1+x^2}\right)dx\ =?$

(a) $(x-1)e^{\tan^{-1}x}+C$

(b) $\tan^{-1}x \times e^{\tan^{-1}x}+C$

(c) $xe^{\tan^{-1}x}+C$

(d) none of these

46. If $\int\dfrac{d\theta}{\sin\theta\cos\theta}=\log|f(\theta)|$, then $f(\theta)=?$

(a) $\cot\theta$ (b) $\tan\theta$

(c) $2\sin\theta$ (d) $2\cos\theta$

47. If $\int f(x)\,dx=\phi(x)$, then $\int x^5 f(x^3)\,dx=?$

(a) $\dfrac{1}{3}x^3\phi(x^3)-\int x^2\phi(x^3)\,dx$

(b) $\dfrac{1}{2}x^2\phi(x^2)-\int x^2\phi(x^3)\,dx$

(c) $\dfrac{1}{3}x^3\phi(x^3)-\int x^2\phi(x^2)\,dx$

(d) $\dfrac{1}{4}x^4\phi(x^3)-\int x^3\phi(x^4)\,dx$

48. The value of $\int_0^1 x(1-x)^{49}\,dx$ is

(a) $\dfrac{1}{15}$ (b) $\dfrac{1}{1550}$ (c) 0 (d) $\dfrac{1}{2550}$

49. The value of $\lim\limits_{x\to0}\dfrac{\int_0^x \cos(t^2)\,dt}{x}$ is

(a) $\dfrac{1}{5}$ (b) $\dfrac{1}{2}$ (c) 0 (d) 1

50. If $f(x)=f(a+b-x)$, then $\int_a^b x f(x)\,dx=?$

(a) $\dfrac{a+b}{2}\int_a^b f(x)\,dx$ (b) $\dfrac{a-b}{2}\int_a^b f(x)\,dx$

(c) $\dfrac{a+b}{2}\int_a^b 3f(x)\,dx$

(d) $\dfrac{3a+4b}{2}\int_a^b f(x)\,dx$

51. If $\int_0^{\frac{\pi}{2}}\dfrac{dx}{1+\sin x+\cos x}=\log 2$, then the

value of $\int_0^{\frac{\pi}{2}}\dfrac{\sin x\,dx}{1+\sin x+\cos x}$ is

(a) $\dfrac{\pi}{4}+\dfrac{1}{2}\log 3$ (b) $\dfrac{\pi}{2}-\log 2$

(c) $\dfrac{\pi}{4}-\dfrac{1}{2}\log 2$ (d) $\dfrac{\pi}{3}-\dfrac{1}{3}\log 2$

52. If $\int_0^1 \cot^{-1}(1-x+x^2)\,dx=k\int_0^1 \tan^{-1}x\,dx$, then the value of k is

(a) 0 (b) 1 (c) 2 (d) 3

53. $\int_0^\infty e^{-x^2}x^2\,dx=?$

(a) $\dfrac{\pi}{4}$ (b) $\dfrac{\sqrt{\pi}}{4}$ (c) $\dfrac{\pi}{2}$ (d) 0

54. $\int_0^1 \dfrac{1}{\sqrt{1-x^6}}\,dx=?$

(a) $\dfrac{\sqrt{\pi}}{2}\dfrac{\Gamma\left(\frac{1}{6}\right)}{\Gamma\left(\frac{1}{3}\right)}$ (b) $\dfrac{\sqrt{\pi}}{6}\dfrac{\Gamma\left(\frac{1}{6}\right)}{\Gamma\left(\frac{2}{3}\right)}$

(c) $\dfrac{1}{6}\dfrac{\Gamma\left(\frac{1}{6}\right)}{\Gamma\left(\frac{1}{3}\right)}$ (d) None of these

55. The integral $\int_1^\infty \dfrac{1}{x^2}\,dx$

(a) exist and converges to "0"
(b) exist and converges to "1"
(c) doesn't exist
(d) converges to "2"

56. The value of the integral $\int_{-1}^{1} \frac{1}{x^{2/3}} dx$ is

(a) 6   (b) 4   (c) 0   (d) 2

57. Which of the following is NOT correct?

(a) $\int_{0}^{\infty} e^{-x^2} dx$ has a finite value

(b) $\int_{0}^{\infty} e^{-2x} dx$ is convergent

(c) $\int_{0}^{\infty} e^{-x^2} dx$ is convergent

(d) $\int_{-\infty}^{0} e^{x} dx$ is divergent

58. The function $f(x,y) = x^3 + y^3 - 6xy$ has maximum/minimum value at

(a) (0,0)   (b) (1, 2)   (c) (2, 2)   (d) (1, 1)

59. The minimum value of the function

$u(x,y) = xy + \frac{1}{x} + \frac{1}{y}$ is

(a) 3   (b) 10   (c) 2   (d) 4

60. The minimum value of $x^2 + y^2 + z^2$, given that $xyz = 8$, is

(a) 12   (b) 10   (c) 15   (d) 8

61. By changing the order of integration

$\int_{0}^{4} \int_{0}^{\sqrt{4x-x^2}} f(x,y) dy dx$ can be expressed as

(a) $\int_{0}^{2} \int_{2-\sqrt{4-y^2}}^{2+\sqrt{4+y^2}} f(x,y) dx dy$

(b) $\int_{0}^{2} \int_{-\sqrt{4-y^2}}^{\sqrt{4+y^2}} f(x,y) dx dy$

(c) $\int_{0}^{4} \int_{0}^{\sqrt{4+y^2}} f(x,y) dx dy$

(d) none of these

62. Consider $I = \int_{0}^{\infty} \int_{x}^{\infty} \frac{e^{-y}}{y} dy dx$. Then which of the following is true?

(a) $I = \int_{0}^{\infty} \int_{0}^{\infty} \frac{e^{-y}}{y} dx dy;\ I = 0$

(b) $I = \int_{0}^{\infty} \int_{0}^{y} \frac{e^{-y}}{y} dx dy;\ I = 1$

(c) $I = \int_{0}^{\infty} \int_{-\infty}^{0} \frac{e^{-y}}{y} dx dy;\ I = \frac{1}{2}$

(d) none of these

63. By changing the order of integration,

$\int_{0}^{a} \int_{0}^{\sqrt{a^2-x^2}} f(x,y) dy dx$ is equal to

(a) $\int_{-a}^{a} \int_{0}^{\sqrt{a^2-y^2}} f(x,y) dx dy$

(b) $\int_{0}^{a} \int_{-\sqrt{a^2-y^2}}^{\sqrt{a^2-y^2}} f(x,y) dx dy$

(c) $\int_{0}^{a} \int_{0}^{\sqrt{a^2-y^2}} f(x,y) dx dy$

(d) none of these

64. By changing the order of integration, the value of $\int_{0}^{4} \int_{\frac{x^2}{4}}^{2\sqrt{x}} dy dx$ is equal to

(a) $\frac{16}{3}$ (b) $\frac{4}{3}$ (c) $\frac{2}{3}$ (d) 0

65. What will be the value of $\iint_R (x^2 + y^2) dx dy$, where R is the region bounded by $x = 0$, $y = 0$ and $x + y = 1$

(a) $\frac{1}{2}$ (b) $\frac{1}{3}$ (c) $\frac{1}{6}$ (d) $\frac{1}{12}$

66. The value of the double integral

$\int_{0}^{\frac{\pi}{2}} \int_{0}^{\pi} \sin(x+y) dy dx$ is

(a) 0   (b) 1   (c) 2   (d) 3

67. Let $I = \iint_R \sqrt{4x^2 - y^2} dx dy$, where R is the triangle formed by the straight lines $y = 0$, $x = 1$ and $y = x$. Then which of the following is true?

(a) $I = \frac{1}{18}(3\sqrt{3} + 2\pi)$ (b) $I = \frac{1}{6}(\sqrt{3} + \pi)$

(c) $I = \frac{1}{3}(6\sqrt{3} + 2\pi)$ (d) $I = \frac{2}{3}(2\sqrt{3} + 4\pi)$

68. The value of $\int\limits_{0}^{2}\int\limits_{0}^{\sqrt{4-y^2}}\left(x^2+y^2\right)dydx$ is

(a) $\dfrac{3}{2}\pi$ (b) 2π (c) 4π (d) $\dfrac{\pi}{2}$

69. Find the value of $\int\limits_{0}^{1}\int\limits_{0}^{1-x}e^{\frac{y}{x+y}}\,dydx$

(a) $\dfrac{1}{2}(e-1)$ (b) $\dfrac{1}{2}\left(e^2+1\right)$

(c) $\dfrac{e}{2}$ (d) $e+\dfrac{1}{2}$

70. Find the value of the double integral $\iint\limits_{R} xy\sqrt{1-x-y}$, where R is the region bounded by $x=0$, $y=0$ and $x+y=1$

(a) $\dfrac{2}{745}$ (b) $\dfrac{1}{235}$ (c) $\dfrac{4}{235}$ (d) $\dfrac{16}{945}$

71. The integral $=\int\limits_{0}^{\infty}\int\limits_{0}^{\infty}e^{-\left(x^2+y^2\right)}dxdy$ is equal to

(a) $\dfrac{\pi}{2}$ (b) $\dfrac{\pi}{4}$ (c) π (d) 2π

72. The value $\iint\limits_{R} x^4 y^3 dxdy$, of where

$R=\left\{(x,y):x\geqslant 0,\,y\geqslant 0,\,x+y\leqslant 1\right\}$ is

(a) $\dfrac{1}{504}$ (b) $\dfrac{3}{208}$ (c) $\dfrac{1}{302}$ (d) $\dfrac{5}{12}$

73. The length of the arc of the parabola $y^2=8x$ measured from the vertex to an extremity of the latus rectum is

(a) 12   (b) $\log\left(\dfrac{4+3\sqrt{2}}{2}\right)$

(c) $10\sqrt{2}$ (d) $2\left\{6\sqrt{2}+\log\left(\dfrac{4+3\sqrt{2}}{\sqrt{2}}\right)\right\}$

74. The arc length of the curve $x=e^t\sin t$, $y=e^t\cos t$ between the points where $t=0$ and $t=\dfrac{\pi}{2}$ is given by

(a) $\sqrt{2}\left(e^{\frac{\pi}{2}}-1\right)$ (b) $\sqrt{2}\left(e^{\frac{\pi}{2}}+1\right)$

(c) $2\sqrt{2}$ (d) $e^{\pi}+1$

75. The arc length of the Cartesian curve $y=\log(\sec x)$ between the points where $x=0$ and $x=\dfrac{\pi}{6}$ is

(a) $\dfrac{1}{2}\log 2$ (b) $\dfrac{1}{2}\log 3$

(c) $\log\left(2+\sqrt{2}\right)$ (d) $\log\left(2+\sqrt{3}\right)$

76. The total length of the astroid $x^{\frac{2}{3}}+y^{\frac{2}{3}}=a^{\frac{2}{3}}$ is given by

(a) $6a$ units   (b) $12a$ units
(c) $8a$ units   (d) $4a$ units

Answer key				
1. (b)	**2.** (b)	**3.** (d)	**4.** (d)	**5.** (c)
6. (d)	**7.** (b)	**8.** (b)	**9.** (a)	**10.** (c)
11. (a)	**12.** (d)	**13.** (a)	**14.** (d)	**15.** (a)
16. (c)	**17.** (a)	**18.** (b)	**19.** (a)	**20.** (a)
21. (b)	**22.** (b)	**23.** (c)	**24.** (d)	**25.** (a)
26. (d)	**27.** (a)	**28.** (c)	**29.** (c)	**30.** (d)
31. (a)	**32.** (d)	**33.** (a)	**34.** (c)	**35.** (b)
36. (c)	**37.** (a)	**38.** (a)	**39.** (d)	**40.** (a)
41. (b)	**42.** (a)	**43.** (d)	**44.** (a)	**45.** (c)
46. (b)	**47.** (a)	**48.** (d)	**49.** (d)	**50.** (a)
51. (c)	**52.** (c)	**53.** (b)	**54.** (b)	**55.** (b)
56. (a)	**57.** (d)	**58.** (c)	**59.** (a)	**60.** (a)
61. (a)	**62.** (b)	**63.** (c)	**64.** (a)	**65.** (c)
66. (c)	**67.** (a)	**68.** (b)	**69.** (a)	**70.** (d)
71. (b)	**72.** (a)	**73.** (d)	**74.** (a)	**75.** (b)
76. (a)				

Explanation

1. (b) We know that $\lim\limits_{x\to a}\dfrac{|x-a|}{x-a}$ does not exist.

2. (b) $\lim\limits_{x\to 0+}f(x)=\lim\limits_{x\to 0+}\dfrac{x-|x|}{x}=\lim\limits_{x\to 0+}\dfrac{x-x}{x}=0$,

$\lim\limits_{x\to 0-}f(x)=\lim\limits_{x\to 0-}\dfrac{x-|x|}{x}=\lim\limits_{x\to 0-}\dfrac{x-(-x)}{x}$

$=\lim\limits_{x\to 0-}\dfrac{2x}{x}=2$

$\because \lim\limits_{x\to 0+}f(x)\neq\lim\limits_{x\to 0-}f(x)$, so $\lim\limits_{x\to 0}f(x)$ does not exist.

3. (d) $\lim\limits_{x\to 0}\dfrac{\sin x^0}{x}$

$=\lim\limits_{x\to 0}\dfrac{\sin\left(\dfrac{\pi x}{180}\right)}{x}\left(\because 1^0=\dfrac{\pi^c}{180}\right)$

$$= \lim_{\frac{\pi x}{180} \to 0} \left[\frac{\sin\left(\frac{\pi x}{180}\right)}{\frac{\pi x}{180}} \times \frac{\pi}{180} \right] = \frac{\pi}{180} \times \lim_{\frac{\pi x}{180} \to 0} \left[\frac{\sin\left(\frac{\pi x}{180}\right)}{\frac{\pi x}{180}} \right]$$

$$= \frac{\pi}{180} \times 1 = \frac{\pi}{180} \quad \left(\because \lim_{x \to 0} \frac{\sin x}{x} = 1 \right)$$

4. (d)

$$\lim_{x \to 0} \frac{3^x - 1}{\sqrt{1+x} - 1}$$

$$= \lim_{x \to 0} \left[\frac{\frac{3^x - 1}{x}}{\frac{(1+x)^{\frac{1}{2}} - 1}{x}} \right] = \lim_{x \to 0} \left[\frac{\frac{3^x - 1}{x}}{\frac{(1+x)^{\frac{1}{2}} - 1^{\frac{1}{2}}}{(1+x) - 1}} \right]$$

$$= \frac{\lim_{x \to 0}\left(\frac{3^x - 1}{x}\right)}{\lim_{x \to 0}\left\{ \frac{(1+x)^{\frac{1}{2}} - 1^{\frac{1}{2}}}{(1+x) - 1} \right\}} = \frac{\log_e 3}{\frac{1}{2} \times 1^{\frac{1}{2} - 1}}$$

$$= 2\log_e 3 = \log_e 9.$$

5. (c)

$$\lim_{x \to 0} (1+x)^{\cos ec x}$$

$$= \lim_{x \to 0} \left\{ (1+x)^{\frac{1}{x} \times \frac{x}{\sin x}} \right\} = \lim_{x \to 0} \left\{ (1+x)^{\frac{1}{x}} \right\}^{\lim_{x \to 0}\left(\frac{x}{\sin x}\right)}$$

$$= e^{\left(\frac{1}{\lim_{x \to 0} \frac{\sin x}{x}}\right)} = e^1 = e.$$

6. (d) $\lim_{x \to \infty} \dfrac{\sin^4 x - \sin^2 x + 1}{\cos^4 x - \cos^2 x + 1}$

$$= \lim_{x \to \infty} \frac{\left(1 - \cos^2 x\right)^2 - \left(1 - \cos^2 x\right) + 1}{\cos^4 x - \cos^2 x + 1}$$

$$= \lim_{x \to \infty} \frac{\cos^4 x - \cos^2 x + 1}{\cos^4 x - \cos^2 x + 1}$$

$$= \lim_{x \to \infty} 1 = 1$$

7. (b)

$$\lim_{x \to a} \frac{\sqrt{3x-a} - \sqrt{x+a}}{x-a}$$

$$= \lim_{x \to a} \frac{\left(\sqrt{3x-a} - \sqrt{x+a}\right)\left(\sqrt{3x-a} + \sqrt{x+a}\right)}{(x-a)\left(\sqrt{3x-a} + \sqrt{x+a}\right)}$$

$$= \lim_{x \to a} \frac{(3x-a) - (x+a)}{(x-a)\left(\sqrt{3x-a} + \sqrt{x+a}\right)}$$

$$= \lim_{x \to a} \frac{2(x-a)}{(x-a)\left(\sqrt{3x-a} + \sqrt{x+a}\right)}$$

$$= \lim_{x \to a} \frac{2}{\left(\sqrt{3x-a} + \sqrt{x+a}\right)}$$

$$= \frac{2}{\sqrt{3a-a} + \sqrt{a+a}} = \frac{2}{2\sqrt{2a}} = \frac{1}{\sqrt{2a}}.$$

8. (b)

$$\lim_{x \to 0} \frac{\tan x - \sin x}{x^3}$$

$$= \lim_{x \to 0} \frac{\frac{\sin x}{\cos x} - \sin x}{x^3}$$

$$= \lim_{x \to 0} \frac{\sin x}{x} \times \lim_{x \to 0} \frac{(1 - \cos x)}{x^2 \cos x}$$

$$= \lim_{x \to 0} \left[\frac{\sin x}{x} \times \lim_{x \to 0} \frac{(1 - \cos x)}{x^2 \cos x} \right]$$

$$= 1 \times \lim_{x \to 0} \frac{2\sin^2 \frac{x}{2}}{x^2 \cos x}$$

$$= \lim_{x \to 0} \left\{ \frac{2\sin^2 \frac{x}{2}}{\left(\frac{x}{2}\right)^2 \cos x} \times \frac{1}{4} \right\}$$

$$= \frac{1}{2} \lim_{x \to 0} \left\{ \left(\frac{\sin \frac{x}{2}}{\frac{x}{2}} \right)^2 \times \frac{1}{\cos x} \right\}$$

$$= \frac{1}{2} \times \left\{ \lim_{\frac{x}{2} \to 0} \left(\frac{\sin \frac{x}{2}}{\frac{x}{2}} \right) \right\}^2 \times \frac{1}{\lim_{x \to 0} \cos x}$$

$$= \frac{1}{2} \times 1^2 \times \frac{1}{1} = \frac{1}{2}.$$

9. (a) $\lim\limits_{x \to 0} \dfrac{\tan 2x - x}{3x - \sin x}$

$$= \lim_{x \to 0} \dfrac{\dfrac{\tan 2x}{x} - 1}{3 - \dfrac{\sin x}{x}} = \dfrac{\lim\limits_{x \to 0} \dfrac{\tan 2x}{x} - 1}{3 - \lim\limits_{x \to 0} \dfrac{\sin x}{x}}$$

$$= \dfrac{2 \times \left(\lim\limits_{2x \to 0} \dfrac{\tan 2x}{2x} \right) - 1}{3 - \lim\limits_{x \to 0} \dfrac{\sin x}{x}} = \dfrac{2 \times 1 - 1}{3 - 1} = \dfrac{1}{2}.$$

10. (c) Let $f(x) = x^2 \ \forall x \in R$. Then clearly f is unbounded.

Remember:

A function $f : R \to R$ is said to be bounded if there exists a positive number "K" such that $|f(x)| \le K \ \forall x \in R$.

11. (a) A function is differentiable at only those points where its graph is smooth, i.e; it has no edges. Here, the function has an edge only at x = 4 and so it is differentiable everywhere except at x = 4.

12. (d) Clearly, $f(0) = k|\sin 0| + 0 = 0$.

$$Rf'(0) = \lim_{x \to 0+} \dfrac{f(x) - f(0)}{x}$$

$$= \lim_{x \to 0+} \dfrac{k|\sin x| + 2x - 0}{x}$$

$$= \lim_{x \to 0+} \dfrac{k \sin x + 2x}{x}$$

$$= k \lim_{x \to 0+} \dfrac{\sin x}{x} + 2 \lim_{x \to 0+} \dfrac{x}{x}$$

$$= k \times 1 + 2 = k + 2$$

$$Lf'(0) = \lim_{x \to 0-} \dfrac{f(x) - f(0)}{x}$$

$$= \lim_{x \to 0-} \dfrac{k|\sin x| + 2x - 0}{x}$$

$$= \lim_{x \to 0-} \dfrac{k(-\sin x) + 2x}{x}$$

$$= -k \lim_{x \to 0-} \dfrac{\sin x}{x} + 2 \lim_{x \to 0-} \dfrac{x}{x}$$

$$= -k \times 1 + 2 = 2 - k$$

Now, $f'(0)$ exist

$$\Rightarrow Rf'(0) = Lf'(0) \Rightarrow k + 2 = 2 - k \Rightarrow k = 0.$$

13. (a) f is continuous at x = 1

$$\Rightarrow \lim_{x \to 1-} f(x) = \lim_{x \to 1+} f(x) = f(1)$$

$$\Rightarrow \lim_{x \to 1-} \left(x^2 - ax + b \right) = \lim_{x \to 1+} \left(4x^2 - bx + a \right) = a + b$$

$$\Rightarrow 1 - a + b = 4 - b + a = a + b$$

$$\Rightarrow 1 - a + b = a + b, \ 4 - b + a = a + b$$

$$\Rightarrow 2a = 1, \ 2b = 4 \Rightarrow a = \dfrac{1}{2}, \ b = 2$$

14. (d)

$$\lim_{x \to 0} \dfrac{\sin\left(\pi \cos^2 x \right)}{x^2} \quad \left(\text{form } \dfrac{0}{0} \right)$$

$$= \lim_{x \to 0} \dfrac{\dfrac{d}{dx}\left\{ \sin\left(\pi \cos^2 x \right) \right\}}{\dfrac{d}{dx}\left\{ x^2 \right\}} \quad \text{(by the L'Hospital Rule)}$$

$$= \lim_{x \to 0} \dfrac{\cos\left(\pi \cos^2 x \right) \times \pi \times 2 \cos x \times (-\sin x)}{2x}$$

$$= -\pi \lim_{x \to 0} \left(\cos\left(\pi \cos^2 x \right) \times \cos x \right) \times \lim_{x \to 0} \left(\dfrac{\sin x}{x} \right)$$

$$= -\pi \left(\cos\left(\pi \cos^2 0 \right) \times \cos 0 \right) \times 1$$

$$= -\pi \cos \pi = -\pi \, (-1) = \pi.$$

15. (a) $\lim\limits_{x \to 0} \left[\dfrac{\log(4+x) - \log(4-x)}{x} \right] \left(\text{form } \dfrac{0}{0} \right)$

$$= \lim_{x \to 0} \dfrac{\dfrac{d}{dx}\left\{ \log(4+x) - \log(4-x) \right\}}{\dfrac{d}{dx}(x)}$$

(by the L'Hospital Rule)

$$= \lim_{x \to 0} \left[\dfrac{\dfrac{1}{(4+x)} - \dfrac{(-1)}{(4-x)}}{1} \right]$$

$$= \lim_{x \to 0} \left[\dfrac{1}{(4+x)} + \dfrac{1}{(4-x)} \right]$$

$$= \dfrac{1}{(4+0)} + \dfrac{1}{(4-0)} = \dfrac{1}{2}.$$

16. (c) $\lim\limits_{x \to \frac{\pi}{6}} \left[\dfrac{2 - \sqrt{3} \cos x - \sin x}{(6x - \pi)^2} \right] \left(\text{form } \dfrac{0}{0} \right)$

$$= \lim_{x \to \frac{\pi}{6}} \dfrac{\dfrac{d}{dx}\left\{ 2 - \sqrt{3} \cos x - \sin x \right\}}{\dfrac{d}{dx}\left\{ (6x - \pi)^2 \right\}}$$

(by the L'Hospital Rule)

$$= \lim_{x \to \frac{\pi}{6}} \left\{ \frac{0 + \sqrt{3}\sin x - \cos x}{12(6x - \pi)} \right\} \quad \left(\text{form } \frac{0}{0}\right)$$

$$= \lim_{x \to \frac{\pi}{6}} \frac{\dfrac{d}{dx}\left\{\sqrt{3}\sin x - \cos x\right\}}{\dfrac{d}{dx}\left\{12(6x - \pi)\right\}} \text{ (by the L'Hospital Rule)}$$

$$= \lim_{x \to \frac{\pi}{6}} \left\{ \frac{\sqrt{3}\cos x + \sin x}{12 \times 6} \right\}$$

$$= \frac{\sqrt{3}\cos\dfrac{\pi}{6} + \sin\dfrac{\pi}{6}}{72} = \frac{\sqrt{3} \times \dfrac{\sqrt{3}}{2} + \dfrac{1}{2}}{72} = \frac{1}{36}.$$

17. (a)

$$\lim_{x \to \frac{\pi}{2}} \left\{ \frac{1 + \cos 2x}{(\pi - 2x)^2} \right\} \quad \left(\text{form } \frac{0}{0}\right)$$

$$= \lim_{x \to \frac{\pi}{2}} \frac{\dfrac{d}{dx}\left\{1 + \cos 2x\right\}}{\dfrac{d}{dx}\left\{(\pi - 2x)^2\right\}} \text{ (by the L'Hospital Rule)}$$

$$= \lim_{x \to \frac{\pi}{2}} \left\{ \frac{-2\sin 2x}{-2(\pi - 2x) \times 2} \right\}$$

$$= \frac{1}{2} \lim_{x \to \frac{\pi}{2}} \left\{ \frac{\sin 2x}{(\pi - 2x)} \right\} \quad \left(\text{form } \frac{0}{0}\right)$$

$$= \frac{1}{2} \lim_{x \to \frac{\pi}{2}} \frac{\dfrac{d}{dx}\left\{\sin 2x\right\}}{\dfrac{d}{dx}\left\{(\pi - 2x)\right\}} \text{ (by the L'Hospital Rule)}$$

$$= \frac{1}{2} \lim_{x \to \frac{\pi}{2}} \left\{ \frac{2\cos 2x}{(-2)} \right\} = \frac{1}{2}\left\{ \frac{\cos \pi}{(-1)} \right\} = \frac{1}{2}.$$

18. (b)

$$\lim_{x \to 0} \left\{ \frac{4^{3x} - 3^{4x}}{x} \right\} \quad \left(\text{form } \frac{0}{0}\right)$$

$$= \lim_{x \to 0} \frac{\dfrac{d}{dx}\left(4^{3x} - 3^{4x}\right)}{\dfrac{d}{dx}(x)} \text{ (by the L'Hospital Rule)}$$

$$= \lim_{x \to 0} \left\{ \frac{3 \times 4^{3x} \times \log 4 - 4 \times 3^{4x} \times \log 3}{1} \right\}$$

$$= \lim_{x \to 0} \left\{ 4^{3x} \log 4^3 - 3^{4x} \times \log 3^4 \right\}$$

$$= 4^0 \log 4^3 - 3^0 \times \log 3^4$$

$$= \log 64 - \log 81 = \log\left(\frac{64}{81}\right).$$

19. (a)

$$\lim_{x \to \pi} \left\{ \frac{1 + \cos^3 x}{\sin^2 x} \right\} \quad \left(\text{form } \frac{0}{0}\right)$$

$$= \lim_{x \to \pi} \frac{\dfrac{d}{dx}\left\{1 + \cos^3 x\right\}}{\dfrac{d}{dx}\sin^2 x} \text{ (by the L'Hospital Rule)}$$

$$= \lim_{x \to \pi} \left\{ \frac{0 + 3\cos^2 x(-\sin x)}{2\sin x \cos x} \right\} = -\frac{3}{2}\lim_{x \to \pi}\left\{ \cos x \right\}$$

$$= -\frac{3}{2} \times \cos \pi = -\frac{3}{2} \times (-1) = \frac{3}{2}.$$

20. (a) Let $f(x) = \log x$. Then clearly f satisfies all the conditions of Lagrange's Mean Value Theorem. Hence, there exist a point $c \in (a, b)$ such that $f'(c) = \dfrac{f(b) - f(a)}{b - a}$.

Here, $f(a) = \log a$, $f(b) = \log b$, $f'(x) = \dfrac{1}{x}$.

$$\therefore f'(c) = \frac{f(b) - f(a)}{b - a}$$

$$\Rightarrow \frac{1}{c} = \frac{\log b - \log a}{b - a} = \frac{\log\left(\dfrac{b}{a}\right)}{b - a}$$

Now, $c \in (a, b)$

$$\Rightarrow a < c < b$$

$$\Rightarrow \frac{1}{b} < \frac{1}{c} < \frac{1}{a}$$

$$\Rightarrow \frac{1}{b} < \frac{\log\left(\dfrac{b}{a}\right)}{b - a} < \frac{1}{a}$$

$$\Rightarrow \frac{b - a}{b} < \log\left(\frac{b}{a}\right) < \frac{b - a}{a}.$$

21. (b) The points on the curve $y = (x+1)^2$ at which the tangent is parallel to x-axis are given by $\dfrac{dy}{dx} = 0$.

Here $\dfrac{dy}{dx} = 0 \Rightarrow 2(x+1) = 0 \Rightarrow x = -1$

22. (b) The points on the curve $y = f(x) = (x-1)(x-2)$ at which the tangent is parallel to the chord joining the points (1, 0) and (2, 0) are given by $f'(c) = \dfrac{f(2) - f(1)}{2 - 1}$.

Here, $\dfrac{dy}{dx} = f'(x) = (x-1) + (x-2) = 2x - 3$.

Now, $f'(c) = \dfrac{f(2) - f(1)}{2 - 1} = \dfrac{0-0}{1} = 0$.

$\therefore f'(c) = 0 \Rightarrow 2c - 3 = 0 \Rightarrow c = 1.5$

23. (c) Clearly, the functions f and g satisfy the Cauchy's Mean Value Theorem on $[1,2]$. Hence, $\exists\, c \in (1,2)$

such that $\dfrac{f'(c)}{g'(c)} = \dfrac{f(2) - f(1)}{g(2) - g(1)}$.

Now $\dfrac{f'(c)}{g'(c)} = \dfrac{f(2) - f(1)}{g(2) - g(1)}$

$\Rightarrow \dfrac{f'(c)}{g'(c)} = \dfrac{8-6}{4-0} = \dfrac{1}{2}$

$\Rightarrow g'(c) = 2f'(c)$

24. (d) Clearly, f satisfies all the conditions of Lagrange's Mean Value Theorem on $[1, 4]$. Hence, $\exists\, c \in (1,4)$ such that

$f'(c) = \dfrac{f(4) - f(1)}{4 - 1}$.

Now $f'(c) = \dfrac{f(4) - f(1)}{4 - 1}$

$\Rightarrow f'(c) = \dfrac{f(4) + 3}{3} \Rightarrow \dfrac{f(4) + 3}{3} = f'(c) \geq 3$

$\left(\because f'(x) \geq 3 \ \forall x \in [1,4] \right)$

$\Rightarrow f(4) \geq 9 - 3 = 6.$

25. (a)

$f(x) = \dfrac{x-2}{x+1}$

$\Rightarrow f'(x) = \dfrac{(x+1) - (x-2)}{(x+1)^2} = \dfrac{3}{(x+1)^2} > 0$

Therefore, the function is strictly increasing.

26. (d)

$f(x) = \log\left(\sqrt{x^2 + 1} - x \right)$

$\Rightarrow f'(x) = \dfrac{\dfrac{2x}{2\sqrt{x^2+1}} - 1}{\left(\sqrt{x^2+1} - x \right)}$

$= \dfrac{-\left(\sqrt{x^2+1} - x \right)}{\sqrt{x^2+1}\left(\sqrt{x^2+1} - x \right)} = \dfrac{-1}{\sqrt{x^2+1}} < 0$

Therefore, the function is strictly decreasing.

27. (a)

If $f(x) = \cos x$, then $f'(x) = -\sin x < 0$

for $x \in \left(0, \dfrac{\pi}{2} \right)$.

Therefore, the function $\cos x$ is strictly decreasing on $\left(0, \dfrac{\pi}{2} \right)$.

If $f(x) = 3x^3 - 5$, then $f'(x) = 9x^2 > 0$

for $x \in \left(0, \dfrac{\pi}{2} \right)$.

Therefore, the function $3x^3 - 5$ is strictly increasing on $\left(0, \dfrac{\pi}{2} \right)$.

If $f(x) = \tan x$, then $f'(x) = \sec^2 x > 0$

for $x \in \left(0, \dfrac{\pi}{2} \right)$.

Therefore, the function $\tan x$ is strictly increasing on $\left(0, \dfrac{\pi}{2} \right)$.

If $f(x) = x^9$, then $f'(x) = 9x^8 > 0$

for $x \in \left(0, \dfrac{\pi}{2}\right)$.

Therefore, the function x^9 is strictly increasing on $\left(0, \dfrac{\pi}{2}\right)$.

28. (c) f is strictly increasing

$\Leftrightarrow f'(x) > 0$

$\Leftrightarrow x^x(1 + \log x) > 0$

$\begin{bmatrix} \because f(x) = x^x \Rightarrow \log f(x) = x \log x \\ \Rightarrow \dfrac{f'(x)}{f(x)} = \log x + \dfrac{1}{x} \times x \\ \Rightarrow f'(x) = f(x)(1 + \log x) \end{bmatrix}$

$\Leftrightarrow (1 + \log x) > 0$

$\Leftrightarrow \log x > -1 = \log \dfrac{1}{e} \Leftrightarrow x > \dfrac{1}{e} \Leftrightarrow x \in \left(\dfrac{1}{e}, \infty\right)$

29. (c) f is strictly increasing

$\Leftrightarrow f'(x) > 0$

$\Leftrightarrow \dfrac{\log x - x \times \dfrac{1}{x}}{(\log x)^2} > 0 \Leftrightarrow \log x - 1 > 0$

$\Leftrightarrow \log x > 1 = \log e \Leftrightarrow x > e \Leftrightarrow x \in (e, \infty)$

f is strictly decreasing

$\Leftrightarrow f'(x) < 0$

$\Leftrightarrow \dfrac{\log x - x \times \dfrac{1}{x}}{(\log x)^2} < 0$

$\Leftrightarrow \log x - 1 < 0 \Leftrightarrow \log x < 1 = \log e$

$\Leftrightarrow x < e \Leftrightarrow x \in (0, e) - \{1\}$

$\begin{bmatrix} \because \log x \text{ is defined for } x > 0 \text{ and } f(x) \\ \text{ is undefined for } x = 1 \end{bmatrix}$

30. (d) f is strictly increasing

$\Leftrightarrow f'(x) > 0$

$\Leftrightarrow 3\lambda x^2 - 18\lambda x + 9 > 0$

$\Leftrightarrow \lambda x^2 - 6\lambda x + 3 > 0$

$\Leftrightarrow \lambda > 0 \text{ and } 36\lambda^2 - 12\lambda < 0$

$\begin{bmatrix} \because ax^2 + bx + c > 0 \\ \Rightarrow a > 0 \text{ and } b^2 - 4ac < 0 \end{bmatrix}$

$\Leftrightarrow \lambda > 0 \text{ and } \lambda(3\lambda - 1) < 0$

$\Leftrightarrow \lambda > 0 \text{ and } 3\lambda - 1 < 0$

$\Leftrightarrow \lambda > 0 \text{ and } \lambda < \dfrac{1}{3} \Leftrightarrow \lambda \in \left(0, \dfrac{1}{3}\right)$

31. (a) f is strictly increasing

$\Leftrightarrow f'(x) > 0$

$\Leftrightarrow -9 - 2x > 0 \Leftrightarrow x < -\dfrac{9}{2}$

$\Leftrightarrow x \in \left(-\infty, -\dfrac{9}{2}\right)$

f is strictly decreasing

$\Leftrightarrow f'(x) < 0$

$\Leftrightarrow -9 - 2x < 0 \Leftrightarrow x > -\dfrac{9}{2}$

$\Leftrightarrow x \in \left(-\dfrac{9}{2}, \infty\right)$

32. (d)

Let $f(x) = e^x$.

Then $f'(x) = e^x \neq 0$ for any value of x.

Therefore, e^x has neither a local minima nor a local maxima.

Now let $g(x) = \cos x$, $x \in (0, \pi)$

Then $g'(x) = -\sin x \neq 0$ for $x \in (0, \pi)$

Therefore, $\cos x$ has neither a local minima nor a local maxima in $(0, \pi)$.

Next, let $h(x) = \log x$.

Then, $h'(x) = \dfrac{1}{x} \neq 0$ for any value of x

Therefore, $\log x$ has neither a local minima nor a local maxima.

33. (a)

$$f(x) = \frac{\log x}{x}$$

$$\Rightarrow f'(x) = \frac{x \times \dfrac{1}{x} - \log x}{x^2} = \frac{1 - \log x}{x^2}$$

$$\therefore f'(x) = 0 \Rightarrow \frac{1 - \log x}{x^2} = 0 \Rightarrow 1 - \log x = 0$$

$$\Rightarrow \log x = 1 = \log e \Rightarrow x = e$$

$$\text{Now } f''(x) = \frac{x^2 \times \left(-\dfrac{1}{x}\right) - (1 - \log x) \times 2x}{x^4}$$

$$= \frac{-3x + 2x \log x}{x^4}$$

$$\therefore f''(e) = \frac{-3e + 2e \log e}{e^4} = \frac{-3e + 2e}{e^4} < 0$$

$$(\because \log e = 1)$$

Hence, the function $f(x) = \dfrac{\log x}{x}$ has a local maximum value at $x = e$.

34. (c)

$$f(x) = x + \sqrt{1-x} \Rightarrow f'(x) = 1 - \frac{1}{2\sqrt{1-x}}.$$

$$\therefore f'(x) = 0 \Rightarrow 1 - \frac{1}{2\sqrt{1-x}} = 0$$

$$\Rightarrow 1 - x = \frac{1}{4} \Rightarrow x = \frac{3}{4}.$$

$$\text{Now, } f''(x) = -\frac{1}{4(1-x)^{\frac{3}{2}}}$$

$$\therefore f''\left(\frac{3}{4}\right) = -\frac{1}{4\left(1 - \dfrac{3}{4}\right)^{\frac{3}{2}}} < 0.$$

Hence, the given function has a local maximum value at $x = \dfrac{3}{4}$.

Therefore, local maximum value of the function $= f\left(\dfrac{3}{4}\right) = \dfrac{3}{4} + \sqrt{1 - \dfrac{3}{4}} = \dfrac{5}{4}.$

35. (b)

$$f(x) = A\log x + Bx^2 + x$$

$$\Rightarrow f'(x) = \frac{A}{x} + 2Bx + 1$$

If $f(x) = A\cos x + Bx^2 + x$ has extreme values (minimum or maximum values) at x = 1 and x = 2, we must have $f'(1) = 0$ and $f'(2) = 0$. Now,

$$f'(1) = 0 \Rightarrow A + 2B + 1 = 0 \ldots\ldots(i)$$

$$f'(2) = 0 \Rightarrow \frac{A}{2} + 4B + 1 = 0$$

$$\Rightarrow A + 8B + 2 = 0 \ldots\ldots(ii)$$

Solving (i) and (ii) we get $A = -\dfrac{2}{3}$, $B = -\dfrac{1}{6}$.

36. (c)

$$\frac{1}{R} = \frac{1}{R_1} + \frac{1}{R_2}$$

$$\Rightarrow \frac{1}{R} = \frac{R_1 + R_2}{R_1 R_2} = \frac{k}{R_1 R_2}$$

$$\Rightarrow R = \frac{R_1 R_2}{k} = \frac{R_1(k - R_1)}{k}$$

$$\Rightarrow R = R_1 - \frac{R_1^2}{k} = f(R_1) \text{ (say)}$$

$$\therefore f'(R_1) = 0 \Rightarrow 1 - \frac{2R_1}{k} = 0 \Rightarrow R_1 = \frac{k}{2}.$$

Now $f''(R_1) = -\dfrac{2}{k}$. So $f''\left(\dfrac{k}{2}\right) = -\dfrac{2}{k} < 0$

Hence, $f(R_1)$ has a local maximum at $R_1 = \dfrac{k}{2}$.

When $R_1 = \dfrac{k}{2}$, $R_2 = k - R_1 = k - \dfrac{k}{2} = \dfrac{k}{2}.$

Thus, R i.e; $f(R_1)$ is maximum when $R_1 = R_2$.

37. (a)

$$M = \frac{Wx}{3} - \frac{Wx^3}{3L^2} \Rightarrow \frac{dM}{dx} = \frac{W}{3} - \frac{Wx^2}{L^2}$$

$$\therefore \frac{dM}{dx} = 0 \Rightarrow \frac{W}{3} - \frac{Wx^2}{L^2} = 0 \Rightarrow x = \frac{L}{\sqrt{3}}.$$

Now $\dfrac{d^2M}{dx^2} = -\dfrac{2Wx}{L^2}$ and so

$$\left.\frac{d^2M}{dx^2}\right|_{x=\frac{L}{\sqrt{3}}} = -\frac{2W \times \dfrac{L}{\sqrt{3}}}{L^2} < 0.$$

Thus, M is maximum when $x = \dfrac{L}{\sqrt{3}}$.

38. (a) Ignoring the first term, let $\displaystyle\sum_{n=1}^{\infty} u_n$ be the new series

i.e; $\displaystyle\sum_{n=1}^{\infty} u_n = \frac{3}{1!} + \frac{9}{2!} + \frac{27}{3!} + \dots \infty$. Then

$$u_1 = \frac{3}{1!}, \ u_2 = \frac{3^2}{2!}, \ u_3 = \frac{3^3}{3!}, \dots, u_n = \frac{3^n}{n!}, \dots$$

$$\therefore u_{n+1} = \frac{3^{n+1}}{(n+1)!} = \frac{3 \times 3^n}{(n+1)n!}.$$

So, $\displaystyle\lim_{n \to \infty} \frac{u_{n+1}}{u_n} = \lim_{n \to \infty} \frac{3}{n+1} = 0 < 1$

Hence, by the D'Alembert's ratio test, the given series is convergent.

39. (d) We know that $e^x = 1 + x + \dfrac{x^2}{2!} + \dfrac{x^3}{3!} + \dots \infty$.

Replacing x by log3, we get

$$e^{\log 3} = 1 + \log 3 + \frac{1}{2!}(\log 3)^2 + \frac{1}{3!}(\log 3)^3 + \dots \infty$$

or, $\log 3 + \dfrac{1}{2!}(\log 3)^2 + \dfrac{1}{3!}(\log 3)^3 + \dots \infty$

$= e^{\log 3} - 1 = 3 - 1 = 2$.

40. (a) Let $f(x) = \sin x$. Then we have

$f'(x) = \cos x, \ f''(x) = -\sin x, \ f'''(x) = -\cos x$

$f^{IV}(x) = \sin x, \dots$

$\therefore f\left(\dfrac{\pi}{2}\right) = 1, \ f'\left(\dfrac{\pi}{2}\right) = 0, \ f''\left(\dfrac{\pi}{2}\right) = -1, \ f'''\left(\dfrac{\pi}{2}\right) = 0,$

$f^{IV}\left(\dfrac{\pi}{2}\right) = 1$ etc.

Then,

$$f(x) = f(a) + (x-a)f'(a) + \frac{1}{2!}(x-a)^2 f''(a) +$$

$$\frac{1}{3!}(x-a)^3 f'''(a) + \frac{1}{4!}(x-a)^4 f^{IV}(a) + \dots \infty$$

$$\Rightarrow \sin x = 1 + \left(x - \frac{\pi}{2}\right)f'\left(\frac{\pi}{2}\right) + \frac{1}{2!}\left(x - \frac{\pi}{2}\right)^2 f''\left(\frac{\pi}{2}\right)$$

$$+ \frac{1}{3!}\left(x - \frac{\pi}{2}\right)^3 f'''\left(\frac{\pi}{2}\right) + \frac{1}{4!}\left(x - \frac{\pi}{2}\right)^4 f^{IV}\left(\frac{\pi}{2}\right) + \dots \infty$$

$$\left(\text{putting } a = \frac{\pi}{2}\right)$$

$$\Rightarrow \sin x = 1 + \left(x - \frac{\pi}{2}\right) \times 0 + \frac{1}{2!}\left(x - \frac{\pi}{2}\right)^2 \times (-1)$$

$$+ \frac{1}{3!}\left(x - \frac{\pi}{2}\right)^3 \times 0 + \frac{1}{4!}\left(x - \frac{\pi}{2}\right)^4 \times 1 + \dots \infty$$

$$\Rightarrow \sin x = 1 - \frac{1}{2!}\left(x - \frac{\pi}{2}\right)^2 + \frac{1}{4!}\left(x - \frac{\pi}{2}\right)^4 - \dots \infty.$$

41. (b) We know that,

$$\log(1+x) = x - \frac{x^2}{2} + \frac{x^3}{3} - \frac{x^4}{4} + \frac{x^5}{5} - \dots \infty \dots (i)$$

$$\log(1-x) = -x - \frac{x^2}{2} - \frac{x^3}{3} - \frac{x^4}{4} - \frac{x^5}{5} - \dots \infty \dots (ii)$$

$(i) - (ii) \Rightarrow$

$$\log(1+x) - \log(1-x) = 2x + \frac{2x^3}{3} + \frac{2x^5}{5} + \dots \infty$$

or, $\log\left(\dfrac{1+x}{1-x}\right) = 2\left(x + \dfrac{x^3}{3} + \dfrac{x^5}{5} + \dots \infty\right)$

42. (a)

$$f(x) = \sum_{n=0}^{\infty} a_n x^n \Rightarrow x^k f(x) = x^k \sum_{n=0}^{\infty} a_n x^n$$

$$= \sum_{n=0}^{\infty} a_n x^{n+k}.$$

43. (d) Let $f(x) = x^3 + 5x^2 + 2x - 5$

Then,

$f'(x) = 3x^2 + 10x + 2, \ f''(x) = 6x + 10,$

$f'''(x) = 6, \ f^{iv}(x) = 0, \dots$

$\therefore f'(1) = 3 + 10 + 2 = 15, \ f''(1) = 6 + 10 = 16,$

$f'''(1) = 6, \ f^{iv}(1) = 0, \dots$

Then,

$$f(x) = f(1) + (x-1)f'(1) + \frac{(x-1)^2}{2!}f''(1)$$

$$+ \frac{(x-1)^3}{3!}f'''(1) + \frac{(x-1)^4}{4!}f^{iv}(1) + \ldots\ldots\infty$$

$$\Rightarrow x^3 + 5x^2 + 2x - 5$$

$$= 3 + 15(x-1) + 8(x-1)^2 + (x-1)^3.$$

44. (a)

$F(x)$ be the anti-derivative of $f(x)$

$$\Rightarrow F(x) = \int f(x)\,dx + C$$

$$= \int \frac{\sqrt{\tan x}}{\sin x \cos x}\,dx$$

$$= \int \frac{\sqrt{\tan x}}{\tan x} \times \sec^2 x\,dx$$

$$= \int \frac{1}{\sqrt{\tan x}} \times \sec^2 x\,dx$$

$$= \int \frac{1}{\sqrt{t}}\,dt$$

$$\left[\text{putting } \tan x = t \text{ so that } \sec^2 x\,dx = dt\right]$$

$$= 2\sqrt{t} + C = 2\sqrt{\tan x} + C$$

Now $F\left(\dfrac{\pi}{4}\right) = 4 \Rightarrow 2\sqrt{\tan\dfrac{\pi}{4}} + C = 4$

$$\Rightarrow 2 + C = 4 \Rightarrow C = 2$$

$$\therefore F(x) = 2\sqrt{\tan x} + 2.$$

45. (c)

$$\int e^{\tan^{-1}x}\left(\frac{1+x+x^2}{1+x^2}\right)dx$$

$$= \int e^t \left(1 + \tan t + \tan^2 t\right)dt$$

$$\begin{bmatrix} \text{putting } \tan^{-1}x = t \text{ so that } x = \tan t \text{ and} \\ \dfrac{1}{1+x^2}\,dx = dt \end{bmatrix}$$

$$= \int e^t \left(\tan t + \sec^2 t\right)dt$$

$$= e^t \tan t + C = xe^{\tan^{-1}x} + C$$

$$\begin{bmatrix} \text{using the formula:} \\ \int e^x \left\{f(x) + f'(x)\right\}dx = e^x f(x) + C \end{bmatrix}$$

46. (b)

$$\int \frac{d\theta}{\sin\theta\cos\theta} = \log|f(\theta)|$$

$$\Rightarrow \int \frac{\left(\sin^2\theta + \cos^2\theta\right)d\theta}{\sin\theta\cos\theta} = \log|f(\theta)|$$

$$\Rightarrow \int \left(\tan\theta + \cot\theta\right)d\theta = \log|f(\theta)|$$

$$\Rightarrow \int \tan\theta\,d\theta + \int \cot\theta\,d\theta = \log|f(\theta)|$$

$$\Rightarrow \log|\sec\theta| + \log|\sin\theta| = \log|f(\theta)|$$

$$\Rightarrow \log|\sec\theta \times \sin\theta| = \log|f(\theta)|$$

$$\Rightarrow \log|\tan\theta| = \log|f(\theta)|$$

$$\Rightarrow f(\theta) = \tan\theta.$$

47. (a)

$$\int x^5 f(x^3)\,dx$$

$$= \int x^3 f(x^3) \times x^2\,dx$$

$$= \frac{1}{3}\int t\,f(t)\,dt$$

$$\left[\text{putting } x^3 = t \text{ so that } 3x^2\,dx = dt\right]$$

$$= \frac{1}{3}\left[t\int f(t)\,dt - \int\left\{\frac{d}{dt}(t)\int f(t)\,dt\right\}dt\right]$$

$$= \frac{1}{3}\left[t\,\phi(t) - \int \phi(t)\,dt\right]$$

$$= \frac{1}{3}\left[x^3\,\phi(x^3) - \int \phi(x^3)3x^2\,dx\right]$$

$$= \frac{1}{3}x^3\phi(x^3) - \int x^2\phi(x^3)\,dx.$$

48. (d)

$$\int_0^1 x(1-x)^{49}\,dx$$

$$= \int_0^1 (1-x)\{1-(1-x)\}^{49}\,dx$$

$$= \int_0^1 (1-x)x^{49}\,dx = \int_0^1 \left(x^{49} - x^{50}\right)dx$$

$$= \left[\frac{x^{50}}{50} - \frac{x^{51}}{51}\right]_0^1 = \frac{1}{50} - \frac{1}{51} = \frac{1}{2550}.$$

49. (d)

$$\lim_{x\to 0}\frac{\displaystyle\int_0^x \cos(t^2)\,dt}{x}\quad\left[\text{form }\frac{0}{0}\right]$$

$$=\lim_{x\to 0}\frac{\dfrac{d}{dx}\displaystyle\int_0^x \cos(t^2)\,dt}{\dfrac{d}{dx}x}\quad(\text{by the L'Hospital Rule})$$

$$=\lim_{x\to 0}\frac{\cos(x^2)\times 1-\cos 0\times 0}{1}$$

$$=\lim_{x\to 0}\cos(x^2)=\cos 0=1.$$

50. (a)

Let $I=\displaystyle\int_a^b x\,f(x)\,dx$. Then,

$$I=\int_a^b x\,f(x)\,dx$$

$$\Rightarrow I=\int_a^b (a+b-x)\,f(a+b-x)\,dx$$

$$\Rightarrow I=\int_a^b (a+b)\,f(a+b-x)\,dx$$

$$-\int_a^b x\,f(a+b-x)\,dx$$

$$\Rightarrow I=\int_a^b (a+b)\,f(x)\,dx-\int_a^b x\,f(x)\,dx$$

$$(\because f(a+b-x)=f(x))$$

$$\Rightarrow I=(a+b)\int_a^b f(x)\,dx-I$$

$$\Rightarrow 2I=(a+b)\int_a^b f(x)\,dx$$

$$\Rightarrow I=\frac{(a+b)}{2}\int_a^b f(x)\,dx$$

51. (c)

Let $I=\displaystyle\int_0^{\frac{\pi}{2}}\frac{\sin x\,dx}{1+\sin x+\cos x}$. Then,

$$I=\int_0^{\frac{\pi}{2}}\frac{\sin x\,dx}{1+\sin x+\cos x}$$

$$\Rightarrow I=\int_0^{\frac{\pi}{2}}\frac{\sin\left(\dfrac{\pi}{2}-x\right)dx}{1+\sin\left(\dfrac{\pi}{2}-x\right)+\cos\left(\dfrac{\pi}{2}-x\right)}$$

$$=\int_0^{\frac{\pi}{2}}\frac{\cos x\,dx}{1+\cos x+\sin x}$$

$$\Rightarrow 2I=\int_0^{\frac{\pi}{2}}\frac{\sin x\,dx}{1+\sin x+\cos x}$$

$$+\int_0^{\frac{\pi}{2}}\frac{\cos x\,dx}{1+\cos x+\sin x}$$

$$\Rightarrow 2I=\int_0^{\frac{\pi}{2}}\frac{(\sin x+\cos x)\,dx}{1+\cos x+\sin x}$$

$$=\int_0^{\frac{\pi}{2}}\frac{(1+\sin x+\cos x)-1}{1+\cos x+\sin x}\,dx$$

$$\Rightarrow 2I=\int_0^{\frac{\pi}{2}} dx-\int_0^{\frac{\pi}{2}}\frac{1}{1+\cos x+\sin x}\,dx$$

$$=\frac{\pi}{2}-\log 2$$

$$\Rightarrow I=\frac{\pi}{4}-\frac{1}{2}\log 2.$$

52. (c)

$$\int_0^1 \cot^{-1}(1-x+x^2)\,dx=k\int_0^1 \tan^{-1}x\,dx$$

$$\Rightarrow k\int_0^1 \tan^{-1}x\,dx=\int_0^1 \tan^{-1}\left(\frac{1}{1-x+x^2}\right)dx$$

$$=\int_0^1 \tan^{-1}\left[\frac{(1-x)+x}{1-x(1-x)}\right]dx$$

$$=\int_0^1 \left[\tan^{-1}(1-x)+\tan^{-1}x\right]dx$$

$$\Rightarrow (k-1)\int_0^1 \tan^{-1}x\,dx=\int_0^1 \tan^{-1}(1-x)\,dx$$

$$\Rightarrow (k-1)\int_0^1 \tan^{-1}x\,dx=\int_0^1 \tan^{-1}(1-(1-x))\,dx$$

$$=\int_0^1 \tan^{-1}x\,dx$$

$$\Rightarrow (k-2)\int_0^1 \tan^{-1}x\,dx=0$$

$$\Rightarrow k-2=0\Rightarrow k=2.$$

53. (b)

$$\int\limits_0^\infty e^{-x^2} x^2 \, dx$$

$$= \int\limits_0^\infty y e^{-y} \frac{dy}{2\sqrt{y}}$$

$$\left[\begin{array}{l} \text{putting } x = \sqrt{y} \text{ so that } dx = \dfrac{dy}{2\sqrt{y}}; \\[2mm] \text{Also } x = 0 \Rightarrow y = 0 \ \& \ x = \infty \Rightarrow y = \infty \end{array} \right]$$

$$= \frac{1}{2} \int\limits_0^\infty y^{\frac{1}{2}} e^{-y} \, dy = \frac{1}{2} \int\limits_0^\infty y^{\frac{3}{2}-1} e^{-y} \, dy$$

$$= \frac{1}{2}\Gamma\left(\frac{3}{2}\right) = \frac{1}{2}\Gamma\left(\frac{1}{2}+1\right) = \frac{1}{2}\times\frac{1}{2}\Gamma\left(\frac{1}{2}\right) = \frac{1}{4}\sqrt{\pi}.$$

54. (b)

$$\int\limits_0^1 \frac{1}{\sqrt{1-x^6}} \, dx$$

$$= \frac{1}{6}\int\limits_0^\infty \frac{y^{-\frac{5}{6}}\, dy}{\sqrt{1-y}}$$

$$\left[\begin{array}{l} \text{putting } x = y^{\frac{1}{6}} \text{ so that } dx = \dfrac{1}{6} y^{-\frac{5}{6}} dy; \\[2mm] \text{Also } x = 0 \Rightarrow y = 0 \ \& \ x = 1 \Rightarrow y = 1 \end{array} \right]$$

$$= \frac{1}{6}\int\limits_0^\infty y^{\frac{1}{6}-1}(1-y)^{\frac{1}{2}-1} \, dy$$

$$= \frac{1}{6}\beta\left(\frac{1}{6},\frac{1}{2}\right)$$

$$= \frac{1}{6}\frac{\Gamma\left(\frac{1}{6}\right)\Gamma\left(\frac{1}{2}\right)}{\Gamma\left(\frac{1}{6}+\frac{1}{2}\right)} = \frac{1}{6}\frac{\Gamma\left(\frac{1}{6}\right)\sqrt{\pi}}{\Gamma\left(\frac{2}{3}\right)} = \frac{\sqrt{\pi}}{6}\frac{\Gamma\left(\frac{1}{6}\right)}{\Gamma\left(\frac{2}{3}\right)}.$$

55. (b)

$$\int\limits_1^\infty \frac{1}{x^2} \, dx$$

$$= \lim_{X\to\infty}\int\limits_1^X \frac{1}{x^2} \, dx = \lim_{X\to\infty}\left[-\frac{1}{x}\right]_1^X$$

$$= \lim_{X\to\infty}\left[-\frac{1}{X}+1\right] = -\lim_{X\to\infty}\frac{1}{X}+1 = 0+1 = 1.$$

56. (a)

$$\int\limits_{-1}^1 \frac{1}{x^{2/3}} \, dx$$

$$= \int\limits_{-1}^0 \frac{1}{x^{2/3}} \, dx + \int\limits_0^1 \frac{1}{x^{2/3}} \, dx$$

$$= \lim_{\varepsilon_1\to 0+}\int\limits_{-1}^{0-\varepsilon_1} \frac{1}{x^{2/3}} \, dx + \lim_{\varepsilon_2\to 0+}\int\limits_{0+\varepsilon_2}^1 \frac{1}{x^{2/3}} \, dx$$

$$= \lim_{\varepsilon_1\to 0+}\left[3x^{1/3}\right]_{-1}^{-\varepsilon_1} + \lim_{\varepsilon_2\to 0+}\left[3x^{1/3}\right]_{\varepsilon_2}^1$$

$$= 3\lim_{\varepsilon_1\to 0+}\left[(-\varepsilon_1)^{1/3}-(-1)^{1/3}\right] + 3\lim_{\varepsilon_2\to 0+}\left[1^{1/3}-\varepsilon_2^{1/3}\right]$$

$$= 3\left[0^{1/3}-(-1)\right] + 3\left[1-0^{1/3}\right]$$

$$= 6.$$

57. (d)

$$\int\limits_0^\infty e^{-x^2} \, dx = \int\limits_0^\infty e^{-y}\frac{1}{2\sqrt{y}} \, dy$$

$$\left(\begin{array}{l} \text{putting } x = \sqrt{y} \text{ so that } dx = \dfrac{dy}{2\sqrt{y}}; \\[2mm] \text{Also } x = 0 \Rightarrow y = 0 \ \& \ x = \infty \Rightarrow y = \infty \end{array} \right)$$

$$= \frac{1}{2}\int\limits_0^\infty e^{-y} y^{\frac{1}{2}-1} \, dy$$

$$= \frac{1}{2}\Gamma\left(\frac{1}{2}\right) = \frac{1}{2}\sqrt{\pi}, \text{a finite value}$$

Therefore, $\int\limits_0^\infty e^{-x^2} \, dx$ is convergent.

$$\int\limits_0^\infty e^{-2x} \, dx$$

$$= \frac{1}{2}\int\limits_0^\infty e^{-y} \, dy = \frac{1}{2}\int\limits_0^\infty e^{-y} y^{1-1} \, dy$$

$$\left(\begin{array}{l} \text{putting } 2x = y \text{ so that } dx = \dfrac{dy}{2}; \\[2mm] \text{Also } x = 0 \Rightarrow y = 0 \ \& \ x = \infty \Rightarrow y = \infty \end{array} \right)$$

$$= \frac{1}{2}\Gamma(1) = \frac{1}{2}\times 0! = \frac{1}{2}, \text{a finite value}$$

Therefore, $\displaystyle\int_{0}^{\infty} e^{-2x}\, dx$ is convergent.

$$\int_{-\infty}^{0} e^{x}\, dx$$

$$= \lim_{X \to -\infty} \int_{X}^{0} e^{x}\, dx = \lim_{X \to -\infty} \left[e^{x} \right]_{X}^{0}$$

$$= \lim_{X \to -\infty} \left[e^{0} - e^{X} \right] = 1 - \lim_{X \to -\infty} e^{X}$$

$$= 1 - 0 = 1, \text{a finite real number}$$

Therefore, $\displaystyle\int_{-\infty}^{0} e^{-x}\, dx$ is not divergent.

58. (c)

Here, $\dfrac{\partial f}{\partial x} = 3x^{2} - 6y, \dfrac{\partial f}{\partial y} = 3y^{2} - 6x,$

$\dfrac{\partial^{2} f}{\partial x^{2}} = 6x, \dfrac{\partial^{2} f}{\partial x\, \partial y} = -6 \text{ and } \dfrac{\partial^{2} f}{\partial y^{2}} = 6y.$

$\therefore \dfrac{\partial f}{\partial x} = 0, \dfrac{\partial f}{\partial y} = 0$

$\Rightarrow 3x^{2} - 6y = 0, 3y^{2} - 6x = 0$

$\Rightarrow x^{2} = 2y, \ y^{2} = 2x$

$\Rightarrow y^{4} = 4x^{2} = 8y$

$\Rightarrow y(y^{3} - 8) = 0 \Rightarrow y = 0, 2$

$\therefore$ the points are $(0,0)$ and $(2,2)$.

So $A = \left.\dfrac{\partial^{2} f}{\partial x^{2}}\right|_{(0,0)} = 0, B = \left.\dfrac{\partial^{2} f}{\partial x\, \partial y}\right|_{(0,0)} = -6$

and $C = \left.\dfrac{\partial^{2} f}{\partial y^{2}}\right|_{(0,0)} = 0.$

$\therefore AC - B^{2} = -36 < 0.$

So $(0, 0)$ is neither a point of maximum nor a point of minimum.

Again,

$A = \left.\dfrac{\partial^{2} f}{\partial x^{2}}\right|_{(2,2)} = 12, B = \left.\dfrac{\partial^{2} f}{\partial x\, \partial y}\right|_{(2,2)} = -6$

and $C = \left.\dfrac{\partial^{2} f}{\partial y^{2}}\right|_{(2,2)} = 12.$

$\therefore AC - B^{2} = 144 - 36 > 0 \text{ and } A = 12 > 0.$

Hence, the given function has a minimum value at $(2, 2)$.

59. (a)

$$u(x, y) = xy + \frac{1}{x} + \frac{1}{y}$$

$$\Rightarrow \frac{\partial u}{\partial x} = y - \frac{1}{x^{2}}, \frac{\partial u}{\partial y} = x - \frac{1}{y^{2}},$$

$$\frac{\partial^{2} u}{\partial x^{2}} = \frac{2}{x^{3}}, \frac{\partial^{2} u}{\partial x\, \partial y} = 1, \frac{\partial^{2} u}{\partial y^{2}} = \frac{2}{y^{3}}.$$

$$\therefore \frac{\partial u}{\partial x} = 0, \frac{\partial u}{\partial y} = 0$$

$$\Rightarrow y - \frac{1}{x^{2}} = 0, \ x - \frac{1}{y^{2}} = 0$$

$$\Rightarrow y = \frac{1}{x^{2}} = \frac{1}{\sqrt{x}} \Rightarrow x^{2} = \sqrt{x}$$

$$\Rightarrow x^{4} = x \Rightarrow x^{4} - x = 0$$

$$\Rightarrow x(x^{3} - 1) = 0 \Rightarrow x = 0, 1.$$

Therefore, the only point is $(1,1)$.

$A = \left.\dfrac{\partial^{2} f}{\partial x^{2}}\right|_{(1,1)} = 2, B = \left.\dfrac{\partial^{2} f}{\partial x\, \partial y}\right|_{(1,1)} = 1$

and $C = \left.\dfrac{\partial^{2} f}{\partial y^{2}}\right|_{(1,1)} = 2.$

$\therefore AC - B^{2} = 4 - 1 > 0 \text{ and } A = 2 > 0$

Thus, the given function has a minimum value at $(1, 1)$. Thus, the minimum value of $u(x, y)$ is $1 + 1 + 1$, i.e, 3.

60. (a)

Here,

$f(x, y, z) = x^{2} + y^{2} + z^{2}, \phi(x, y, z) = xyz - 8.$

$\therefore F(x, y, z) = f(x, y, z) + \lambda\phi(x, y, z)$

$\qquad = x^{2} + y^{2} + z^{2} + \lambda(xyz - 8)$

Then, $\dfrac{\partial F}{\partial x} = 0, \dfrac{\partial F}{\partial y} = 0 \text{ and } \dfrac{\partial F}{\partial z} = 0$

$\Rightarrow 2x + \lambda yz = 0, 2y + \lambda xz = 0, 2z + \lambda xy = 0$

$$\Rightarrow -\lambda = \frac{2x}{yz} = \frac{2y}{xz} = \frac{2z}{xy}$$

$$\Rightarrow \frac{x}{yz} = \frac{y}{xz} = \frac{z}{xy} \Rightarrow \frac{x^2}{xyz} = \frac{y^2}{xyz} = \frac{z^2}{xyz}$$

$$\Rightarrow x^2 = y^2 = z^2 = k \text{ (say)}$$

$$\therefore xyz = 8 \Rightarrow x^2 y^2 z^2 = 64$$

$$\Rightarrow k \times k \times k = 64$$

$$\Rightarrow k = 4$$

Therefore, the minimum value of $x^2 + y^2 + z^2$ is $k + k + k$ i.e; $4 + 4 + 4 = 12$.

61. (a)

Let, I $= \int_0^4 \int_0^{\sqrt{4x-x^2}} f(x,y)dydx$ ...(i)

Here the limits of integration are given by

$x=0, x=4; y=0, y=\sqrt{4x-x^2}$

Now $y=\sqrt{4x-x^2} \Rightarrow x^2+y^2-4x=0$

$\Rightarrow (x-2)^2 +(y-0)^2 =2^2 \Rightarrow x-2=\pm\sqrt{4-y^2}$

$\Rightarrow x=2\pm\sqrt{4-y^2}$

Here the region of integration is bounded by $y = 0$, $x = 0$ and the circle $x2 + y2 – 4x = 0$. In the following figure, OABO is the region of integration.

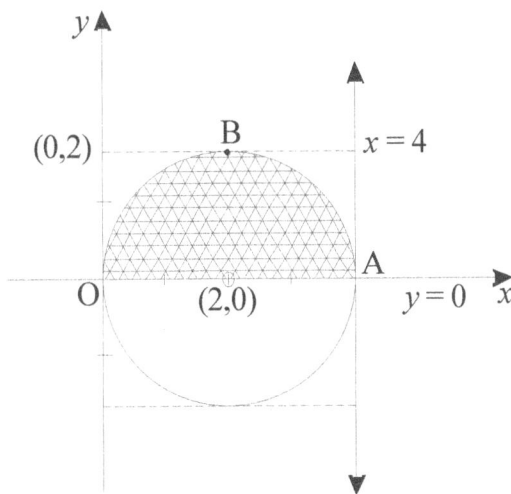

Clearly, y varies from 0 to 2 and x varies from $2-\sqrt{4-y^2}$ to $2+\sqrt{4-y^2}$

Hence,

$$I=\int_0^4 \int_0^{\sqrt{4x-x^2}} f(x,y)dydx=\int_0^2 \int_{2-\sqrt{4-y^2}}^{2+\sqrt{4-y^2}} f(x,y)dxdy.$$

62. (b) Given that $I=\int_0^\infty \int_x^\infty \frac{e^{-y}}{y}dydx$.

Here the limits of integration are

$x = 0, x = \infty; y = x, y = \infty$.

The region of integration is OAB, which is shown in the following figure:

Clearly, y varies from 0 to ∞ and x varies from 0 to y.

Thus, $I=\int_0^\infty \int_x^\infty \frac{e^{-y}}{y}dydx=\int_0^\infty \int_0^y \frac{e^{-y}}{y}dxdy$

Now, $I=\int_0^\infty \int_x^\infty \frac{e^{-y}}{y}dydx=\int_0^\infty \left\{\frac{e^{-y}}{y}\int_0^y dx\right\}dy$

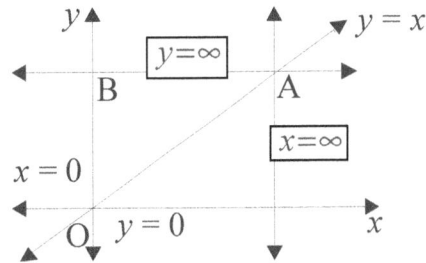

$$=\int_0^\infty \left\{\frac{e^{-y}}{y}\times[x]_0^y\right\}dy$$

$$=\int_0^\infty \frac{e^{-y}}{y}\times ydy=\int_0^\infty e^{-y}dy=\left[-e^{-y}\right]_0^\infty$$

$$=-\left[e^{-\infty}-e^0\right]=1 \quad \left(\because e^{-\infty}=\frac{1}{e^\infty}=0\right)$$

63. (c)

The given integral is $\int_0^a \int_0^{\sqrt{a^2-x^2}} f(x,y)dydx$.

The limits of integration are

$x = 0, x = a; y = 0, y = \sqrt{a^2-x^2}$. Now,

$y = \sqrt{a^2 - x^2} \Rightarrow x^2 + y^2 = a^2 \Rightarrow x = \sqrt{a^2 - y^2}$.

The region of integration is OAB, which is shown in the following figure:

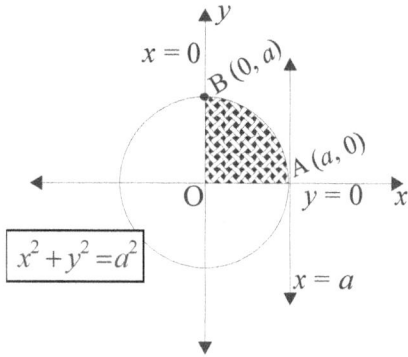

Clearly, y varies from 0 to a and x varies from 0 to $\sqrt{a^2 - y^2}$.

Hence, $\displaystyle\int_0^a \int_0^{\sqrt{a^2-x^2}} f(x,y)\,dy\,dx = \int_0^a \int_0^{\sqrt{a^2-y^2}} f(x,y)\,dx\,dy$.

64. (a) The limits of integration are:

$x = 0, x = 4;\ y = \dfrac{x^2}{4}$ and $y = 2\sqrt{x}$.

Now $y = \dfrac{x^2}{4} \Rightarrow x^2 = 4y$ and $y = 2\sqrt{x} \Rightarrow y^2 = 4x$.

Here the region of integration is OABCO, which is shown in the following figure:

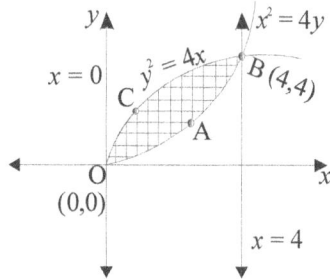

Clearly, y varies from 0 to 4 and x varies from $\dfrac{y^2}{4}$ to $2\sqrt{y}$

$\left(\because y^2 = 4x \Rightarrow x = \dfrac{y^2}{4} \text{ and } x^2 = 4y \Rightarrow x = 2\sqrt{y}\right)$

Hence, by changing order of integration we have,

$\displaystyle\int_0^4 \int_{\frac{x^2}{4}}^{2\sqrt{x}} dy\,dx = \int_0^4 \int_{\frac{y^2}{4}}^{2\sqrt{y}} dx\,dy$

$= \displaystyle\int_0^4 [x]_{\frac{y^2}{4}}^{2\sqrt{y}}\,dy = \int_0^4 \left(2\sqrt{y} - \dfrac{y^2}{4}\right) dy$

$= \left[\dfrac{2y^{\frac{3}{2}}}{\frac{3}{2}} - \dfrac{y^3}{12}\right]_0^4 = \dfrac{4}{3}4^{\frac{3}{2}} - \dfrac{4^3}{12} = \dfrac{4}{3} \times 8 - \dfrac{16}{3} = \dfrac{16}{3}$.

65. (c) Here, $R = \{(x,y): 0 \leqslant x \leqslant 1,\ 0 \leqslant y \leqslant 1 - x\}$.

Thus, x varies from 0 to 1 and y varies from 0 to $(1 - x)$

$\therefore \displaystyle\iint_R (x^2 + y^2)\,dx\,dy = \int_{x=0}^1 \int_{y=0}^{1-x} \{(x^2 + y^2)\,dy\}dx$

$= \displaystyle\int_{x=0}^1 \left[x^2 y + \dfrac{y^3}{3}\right]_{y=0}^{1-x} dx$

$= \displaystyle\int_0^1 \left\{x^2(1-x) + \dfrac{1}{3}(1-x)^3\right\} dx$

$= \displaystyle\int_0^1 \left\{x^2 - x^3 - \dfrac{1}{3}(x-1)^3\right\} dx$

$= \left[\dfrac{x^3}{3} - \dfrac{x^4}{4} - \dfrac{1}{12}(x-1)^4\right]_0^1$

$= \dfrac{1}{3} - \dfrac{1}{4} - \dfrac{1}{12}(1-1) - \left(-\dfrac{1}{12} \times (0-1)^4\right)$

$= \dfrac{1}{12} + \dfrac{1}{12} = \dfrac{1}{6}$.

66. (c) Here, the region

$R = \left\{(x,y): 0 \leqslant x \leqslant \dfrac{\pi}{2},\ 0 \leqslant y \leqslant \pi\right\}$

$\therefore \displaystyle\int_0^{\frac{\pi}{2}} \int_0^{\pi} \sin(x+y)\,dx\,dy = \int_{x=0}^{\frac{\pi}{2}} \left\{\int_{y=0}^{\pi} \sin(x+y)\,dy\right\} dx$

$= \displaystyle\int_{x=0}^{\frac{\pi}{2}} [-\cos(x+y)]_{y=0}^{\pi}\,dx = -\int_0^{\frac{\pi}{2}} \{\cos(\pi + x) - \cos x\}\,dx$

$= -\displaystyle\int_0^{\frac{\pi}{2}} \{-\cos x - \cos x\}\,dx = 2\int_0^{\frac{\pi}{2}} \cos x\,dx$

$= 2[\sin x]_0^{\frac{\pi}{2}} = 2\left(\sin\dfrac{\pi}{2} - \sin 0\right) = 2(1 - 0) = 2$.

67. (a)

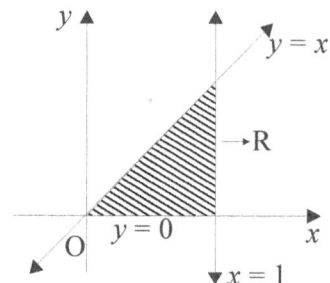

Here, the region of integration,

$R = \left\{(x,y): 0 \leqslant x \leqslant 1, 0 \leqslant y \leqslant x\right\}.$

$\therefore I = \iint\limits_{R} \sqrt{4x^2 - y^2}\, dx\, dy$

$= \int\limits_{x=0}^{1}\left\{\int\limits_{y=0}^{x} \sqrt{4x^2 - y^2}\, dy\right\} dx$

$= \int\limits_{x=0}^{1}\left[\frac{y}{2}\sqrt{4x^2 - y^2} + \frac{4x^2}{2}\sin^{-1}\frac{y}{2x}\right]_{y=0}^{x} dx$

$\left(\text{using } \int \sqrt{4a^2 - y^2}\, dy = \frac{y}{2}\sqrt{4a^2 - y^2} + \frac{a^2}{2}\sin^{-1}\frac{y}{2a}\right)$

$= \int\limits_{x=0}^{1}\left\{\left(\frac{x}{2}\sqrt{4x^2 - x^2} + 2x^2\sin^{-1}\frac{x}{2x}\right) - (0+0)\right\} dx$

$= \int\limits_{0}^{1}\left(\frac{\sqrt{3}}{2}x^2 + 2x^2 \times \frac{\pi}{6}\right) dx \quad \left(\because \sin^{-1}\frac{1}{2} = \frac{\pi}{6}\right)$

$= \left[\frac{\sqrt{3}}{2} \times \frac{x^3}{3} + \frac{\pi}{3} \times \frac{x^3}{3}\right]_{0}^{1} = \frac{\sqrt{3}}{6} + \frac{\pi}{9} = \frac{3\sqrt{3} + 2\pi}{18}.$

68. (b) Here, the region of integration,

$R = \left\{(x,y): 0 \leqslant y \leqslant 2, 0 \leqslant x < \sqrt{4 - y^2}\right\}$

But, $x \leqslant \sqrt{4 - y^2} \Rightarrow x^2 + y^2 \leqslant 2^2$ ....(i)

Let us put $x = r\cos\theta$, $y = r\sin\theta$.

Then Jacobian, $J = r$.

(i) gives, $r^2 \leqslant 2^2$ and so $0 \leqslant r \leqslant 2$.

Since, $x \geqslant 0$, $y \geqslant 0$; so $0 \leqslant \theta \leqslant \frac{\pi}{2}$.

Hence, the new region of integration is

$R' = \left\{(r,\theta): 0 \leqslant r \leqslant 2, 0 \leqslant \theta \leqslant \frac{\pi}{2}\right\}$

Then, $\iint\limits_{R}\left(x^2 + y^2\right) dy\, dx = \iint\limits_{R'}\left(x^2 + y^2\right) dy\, dx$

$= \int\limits_{r=0}^{2}\int\limits_{\theta=0}^{\pi/2} r^2 \times r\, dr\, d\theta \quad (\because dx\, dy = J\, dr\, d\theta)$

$= \left\{\int\limits_{r=0}^{2} r^3\, dr\right\} \times \left\{\int\limits_{\theta=0}^{\pi/2} d\theta\right\} = \left[\frac{r^4}{4}\right]_{0}^{2} \times [\theta]_{0}^{\pi/2}$

$= \frac{16}{4} \times \frac{\pi}{2} = 2\pi.$

69. (a) Here, region of integration,

$R = \left\{(x,y): 0 \leqslant x \leqslant 1, 0 \leqslant y \leqslant 1 - x\right\}.$

Let us put $x + y = u$ and $y = uv$.

So, $y = uv$, $x = u - y = u - uv = u(1 - v)$.

$\therefore x = 0 \Rightarrow u(1-v) = 0 \Rightarrow u = 0, v = 1;$
$\quad y = 1 - x \Rightarrow x + y = 1 \Rightarrow u = 1;$
$\quad y = 0 \Rightarrow uv = 0 \Rightarrow u = 0, v = 0.$

Hence, new region of integration,

$R' = \left\{(u,v): 0 \leqslant u \leqslant 1, 0 \leqslant v \leqslant 1\right\}.$

Jacobian, $J = \begin{vmatrix} \dfrac{\partial x}{\partial u} & \dfrac{\partial x}{\partial v} \\ \dfrac{\partial y}{\partial u} & \dfrac{\partial y}{\partial v} \end{vmatrix} = \begin{vmatrix} 1-v & -u \\ v & u \end{vmatrix}$

$= u(1 - v) - v(-u) = u.$

$\therefore$ Given integral $= \int\limits_{u=0}^{1}\int\limits_{v=0}^{1} e^{\frac{uv}{u}}\, u\, dv\, du$

$(\because \quad dx\, dy = J \times du\, dv)$

$= \int\limits_{u=0}^{1} u\, du \times \int\limits_{v=0}^{1} e^v\, dv$

$= \left[\frac{u^2}{2}\right]_{0}^{1} \times \left[e^v\right]_{0}^{1} = \frac{1}{2}\left(e^1 - e^0\right) = \frac{1}{2}(e - 1).$

70. (d) Let us put $x + y = u$ and $y = uv$
Then, $y = uv$, $x = u - y = u - uv = u(1 - v)$
$\therefore x = 0 \Rightarrow u(1 - v) = 0 \Rightarrow u = 0, v = 1.$

Also, $x + y = 1 \Rightarrow u = 1$ and
$y = 0 \Rightarrow uv = 0 \Rightarrow u = 0, v = 0.$

$\therefore$ New region of integration is
$R' = \left\{(u,v): 0 \leqslant u \leqslant 1, 0 \leqslant v \leqslant 1\right\}$

Jacobian, $J = \begin{vmatrix} \dfrac{\partial x}{\partial u} & \dfrac{\partial x}{\partial v} \\ \dfrac{\partial y}{\partial u} & \dfrac{\partial y}{\partial v} \end{vmatrix} = \begin{vmatrix} 1-v & -u \\ v & u \end{vmatrix}$

$= u(1 - v) - v(-u) = u.$

Hence, given integral

$= \int\limits_{u=0}^{1}\int\limits_{v=0}^{1} u(1-v)uv\sqrt{1-u}\, u\, du\, dv$

$= \int\limits_{u=0}^{1} u^3\sqrt{1-u}\, du \times \int\limits_{v=0}^{1} v(1-v)\, dv$

$= \int\limits_{0}^{1} u^{4-1}(1-u)^{\frac{3}{2}-1}\, du \times \int\limits_{0}^{1} v^{2-1}(1-v)^{2-1}\, dv$

$$=\beta\left(4,\frac{3}{2}\right)\times\beta(2,2)\left(\because\beta(m,n)=\int_0^1 x^{m-1}(1-x)^{n-1}dx\right)$$

$$=\frac{\Gamma(4)\Gamma\left(\frac{3}{2}\right)}{\Gamma\left(4+\frac{3}{2}\right)}\times\frac{\Gamma(2)\Gamma(2)}{\Gamma(2+2)}=\frac{\Gamma\left(\frac{3}{2}\right)\times(1!)^2}{\Gamma\left(\frac{11}{2}\right)}(\because\Gamma(2)=1!)$$

$$=\frac{\sqrt{3/2}}{\frac{9}{2}\times\frac{7}{2}\times\frac{5}{2}\times\frac{3}{2}\sqrt{3/2}}=\frac{16}{945}\cdot\left[\Gamma n=\Gamma(n)\right]$$

71. (b) Here, region of integration,

$$R=\{(x,y):0\leqslant x\leqslant\infty,0\leqslant y\leqslant\infty\}.$$

Let us put $x=r\cos\theta$ and $y=r\sin\theta$

Then Jacobian, $J=r$

$\because \because x\geqslant 0,\ y\geqslant 0$, so we must have $0\leqslant\theta\leqslant\frac{\pi}{2}$.

Also, $0\leq x,y<\infty$

$\Rightarrow 0\leq x^2+y^2<\infty\Rightarrow 0\leq r<\infty$

Hence, new region of integration,

$$R'=\left\{(r,\theta):0\leqslant r\leqslant\infty,0\leqslant\theta\leqslant\frac{\pi}{2}\right\}$$

$\therefore$ given integral

$$=\int_{r=0}^{\infty}\int_{\theta=0}^{\frac{\pi}{2}}e^{-r^2}rd\theta dr\ (\because dxdy=J\times drd\theta)$$

$$=\int_{r=0}^{\infty}re^{-r^2}dr\times\int_{\theta=0}^{\frac{\pi}{2}}d\theta=\frac{\pi}{2}\int_{r=0}^{\infty}re^{-r^2}dr$$

$$=\frac{\pi}{2}\int_{r=0}^{\infty}e^{-t}\frac{dt}{2}$$

[Putting $r^2=t$ so that $2rdr=dt$ i.e; $r\,dr=\frac{dt}{2}$,
Also, $r=0\Rightarrow t=0$ and $r=\infty\Rightarrow t=\infty$]

$$=\frac{\pi}{4}\int_0^{\infty}e^{-t}dt=-\frac{\pi}{4}\left[e^{-t}\right]_0^{\infty}=\frac{\pi}{4}(e^{-\infty}-e^0)$$

$$=\frac{\pi}{4}(\because e^{-\infty}=0,e^0=1)$$

72. (a)

We know that $\iint_R x^{m-1}y^{n-1}dxdy=\frac{\Gamma(m)\Gamma(n)}{\Gamma(m+n+1)}$.

If $R=\{(x,y):x\geqslant 0,y\geqslant 0,x+y\leqslant 1\}$

$\therefore \iint_R x^4y^3dxdy=\iint_R x^{5-1}y^{4-1}dxdy\ (m=5,n=4)$

$$=\frac{\sqrt{5}\sqrt{4}}{\sqrt{(5+4+1)}}=\frac{4!\times 3!}{9!}$$

$$=\frac{4!\times 6}{9\times 8\times 7\times 6\times 5\times 4!}=\frac{1}{504}.$$

73. (d) $y^2=8x=4\times 2\times x$ ...(i)

Comparing (i) with $y^2=4ax$, we get $a=2$. Hence, focus of the parabola is $(a,0)$, i.e. $(2,0)$. When $x=2$, (i) gives $y^2=4\times 2\times 2$ i.e; $y=\pm 4$. So, the extremities of the latus rectum are $(2,4)$ and $(2,-4)$.

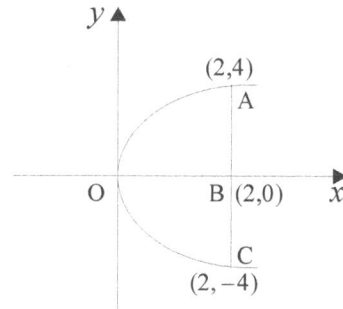

Again, $y^2=8x\Rightarrow y=2\sqrt{2}\sqrt{x}\Rightarrow\frac{dy}{dx}=\frac{\sqrt{2}}{\sqrt{x}}$.

$\therefore$ Required arc length

= length of the arc $\overarc{OA}$

$$=\int_0^2\sqrt{1+\left(\frac{dy}{dx}\right)^2}dx=\int_0^2\sqrt{1+\frac{2}{x}}dx=\int_0^2\sqrt{\frac{x+2}{x}}dx$$

$$=2\int_0^4\sqrt{t^2+\left(\sqrt{2}\right)^2}dt$$

[by putting $x=t^2$ so that $dx=2tdt$;
$x=0\Rightarrow t=0;x=2\Rightarrow t=4$]

$$=2\left\{\left[\frac{t}{2}\sqrt{t^2+2}+\frac{2}{2}\log\left|t+\sqrt{t^2+2}\right|\right]_0^4\right\}$$

$$=2\left\{\frac{4}{2}\sqrt{16+2}+\log\left|4+\sqrt{16+2}\right|-0-\log\sqrt{2}\right\}$$

$$=2\left\{2\times 3\sqrt{2}+\log\left|4+3\sqrt{2}\right|-\log\sqrt{2}\right\}$$

$$=2\left\{6\sqrt{2}+\log\left(\frac{4+3\sqrt{2}}{\sqrt{2}}\right)\right\}.$$

74. (a)

$x=e^t\sin t\Rightarrow\frac{dx}{dt}=e^t\cos t+e^t\sin t=e^t(\cos t+\sin t)$,

$y=e^t\cos t\Rightarrow\frac{dy}{dt}=e^t\cos t-e^t\sin t=e^t(\cos t-\sin t)$.

∴ Required arc length

$$=\int_{t=0}^{\pi/2}\sqrt{\left(\frac{dx}{dt}\right)^2+\left(\frac{dy}{dt}\right)^2}\,dt$$

$$=\int_0^{\pi/2}\sqrt{e^{2t}(\cos t+\sin t)^2+e^{2t}(\cos t-\sin t)^2}\,dt$$

$$=\int_{t=0}^{\pi/2}\sqrt{e^{2t}\times2(\cos^2 t+\sin^2 t)}\,dt$$

$$\left[\text{using }(a+b)^2+(a-b)^2=2(a^2+b^2)\right]$$

$$=\sqrt2\int_{t=0}^{\pi/2}e^t\,dt=\sqrt2\left[e^t\right]_0^{\frac{\pi}{2}}=\sqrt2\left(e^{\frac{\pi}{2}}-1\right).$$

75. (b) $y=\log(\sec x)\Rightarrow\frac{dy}{dx}=\frac{1}{\sec x}\frac{d}{dx}(\sec x)$

$$=\frac{1}{\sec x}(\sec x\tan x)=\tan x$$

∴ Required arc length

$$=\int_0^{\pi/6}\sqrt{1+\left(\frac{dy}{dx}\right)^2}\,dx=\int_0^{\pi/6}\sqrt{1+\tan^2 x}\,dx$$

$$=\int_0^{\pi/6}\sec x\,dx\quad(\because\sec^2 x=1+\tan^2 x)$$

$$=\left[\log(\sec x+\tan x)\right]_0^{\pi/6}$$

$$=\log\left(\sec\frac{\pi}{6}+\tan\frac{\pi}{6}\right)-\log(\sec0+\tan0)$$

$$=\log\left(\frac{2}{\sqrt3}+\frac{1}{\sqrt3}\right)-\log1=\log\left(\frac{3}{\sqrt3}\right)$$

$$=\log\sqrt3=\frac{1}{2}\log3.$$

76. (a) The parametric equation of the astroid $x^{\frac{2}{3}}+y^{\frac{2}{3}}=a^{\frac{2}{3}}$ is $x=a\cos^3 t$, $y=a\sin^3 t$.

Now, $x=a\cos^3 t\Rightarrow\frac{dx}{dt}=-3a\cos^2 t\sin t$,

$y=a\sin^3 t\Rightarrow\frac{dy}{dt}=3a\sin^2 t\cos t$

In first quadrant, "t" varies from 0 to $\frac{\pi}{2}$.

∴ Arc length of the astroid in 1st quadrant

$$=\int_0^{\frac{\pi}{2}}\sqrt{\left(\frac{dx}{dt}\right)^2+\left(\frac{dy}{dt}\right)^2}\,dt$$

$$=\int_0^{\frac{\pi}{2}}\sqrt{9a^2\cos^4 t\sin^2 t+9a^2\sin^4 t\cos^2 t}\,dt$$

$$=3a\int_0^{\frac{\pi}{2}}\sqrt{\cos^2 t\sin^2 t(\cos^2 t+\sin^2 t)}\,dt$$

$$=\frac{3a}{2}\int_0^{\frac{\pi}{2}}2\sin t\cos t\,dt=\frac{3a}{2}\int_0^{\frac{\pi}{2}}\sin2t\,dt$$

$$=\frac{-3a}{4}[\cos2t]_0^{\pi/2}=-\frac{3a}{4}(\cos\pi-\cos0)$$

$$=-\frac{3a}{4}(-1-1)=\frac{3a}{2}.$$

Hence, the total arc length of the astroid

= 4 × length in one quadrant $=4\times\frac{3a}{2}=6a$ unit

| **Fully Solved MCQs (Level-II)** |

1. $\lim\limits_{x\to\infty}\frac{x+3\sin x}{x-5}=?$

(a) –1　　(b) 1　　(c) 0　　(d) ∞

2. $\lim\limits_{x\to\infty}\frac{[x]}{x}=?$

(a) 1　　(b) 0　　(c) –1　　(d) ∞

3. $\lim\limits_{x\to0}\frac{\sin x-x}{x^3}=?$

(a) 0　　(b) 1　　(c) 1/6　　(d) –1/6

4. $\lim\limits_{x\to\infty}\frac{\sum_{k=1}^{50}x^k-50}{x-1}=?$

(a) 1550　　(b) 1275　　(c) 2000　　(d) 2500

5. $\lim\limits_{n\to\infty}\cos\{\pi\sqrt{n^2+n}\}=?$

(a) 0　　(b) 1　　(c) 1/2　　(d) ∞

6. $\lim\limits_{n\to\infty}n\cos\left(\frac{\pi}{8n}\right)\sin\left(\frac{\pi}{8n}\right)=?$

(a) $\frac{\pi}{8}$　　(b) $\frac{\pi}{4}$　　(c) 1/2　　(d) 0

7. If $\lim\limits_{x\to0}\frac{x^m\sin^n x}{\sin x^k}$ has a non-zero value (where m, n, k are non-zero real numbers), then

(a) m + n = k　　　(b) m + k = n
(c) k + n = m　　　(d) m = n

8. $\lim\limits_{n\to\infty}\left(\dfrac{n^2-n+1}{n^2-n-1}\right)^{n(n-1)} = ?$

(a) e (b) e^2 (c) e (d) 1

9. If $f(x)=\lim\limits_{n\to\infty} n\left(x^{\frac{1}{n}}-1\right)$, then for x, y > 0,

$f(xy) = ?$

(a) $f(x)+f(y)$ (b) $f(x)\text{-}f(y)$

(c) $f(x)\times f(y)$ (d) none of these

10. If $f(x)=-\sqrt{25-x^2}$, then $\lim\limits_{x\to 1}\dfrac{f(x)-f(1)}{x-1}=?$

(a) $\dfrac{1}{2\sqrt{6}}$ (b) $\dfrac{1}{\sqrt{6}}$ (c) 1/2   (d) 0

11. For non-negative integers p and q, defined a function f as below:

$$f(p,q)=\begin{cases} q+1, \text{ if } p=0 \\ f(p-1,1), \text{ if } p\neq 0, q=0 \\ f(p-1,f(p,q-1)), \text{ if } p\neq 0, q\neq 0 \end{cases}$$

Then, $f(1,1) = ?$

(a) 1   (b) 2   (c) 3   (d) 4

12. If $f\left(x+\dfrac{1}{x}\right)=x^2+\dfrac{1}{x^2}+3$, then $f(x) = ?$

(a) x (b) x^2 (c) x^2-1 (d) x^2+1

13. The number of solutions of the equation $|x|^2-3|x|+2=0$ is

(a) 1   (b) 2   (c) 3   (d) 4

14. Let $\varphi(x)=\dfrac{1}{1+e^{\frac{1}{x-2}}+e^{\frac{-1}{(x-3)^2}}}$; for $x\neq 2$; and

$\varphi(2)=1$, $\varphi(3)=\dfrac{1}{1+e}$.

Then which of the following is true

(a) is continuous at $x=2,3$

(b) φ is discontinuous at $x=2,3$

(c) φ is discontinuous at $x=2$ but continuous at $x=3$

(d) None of these

15. Let $f(x)=|x\sin x|, x\in(-\pi,\pi)$. Then f is

(a) not differentiable at $x=0$

(b) differentiable except at $x=0,\dfrac{\pi}{2}$

(c) twice continuously differentiable

(d) none of these

16. If $f(x)=q_0\cos|x|+q_1\sin|x|+q_2|x|^3$ is

differentiable at $x=0$, then

(a) $q_1=0$ (b) $q_1=q_2=0$

(c) $q_0=q_1=0$

(d) q_0,q_1,q_2 can take any value.

17. If $f(x+y)=f(x)\times f(y)\;\forall x,y$ and

$f(5)=2, f'(0)=3$ then $f'(5)=?$

(a) 6   (b) 5   (c) 4   (d) 3

18. If $2f(x)+3f(-x)=x^2-x+1$, then find the value of $f'(1)$

(a) $\dfrac{5}{7}$ (b) $\dfrac{7}{5}$ (c) $\dfrac{6}{5}$ (d) $\dfrac{8}{5}$

19. If $f(x)$ is defined by

$$f(x)=\begin{cases} \dfrac{x}{1+e^{\frac{1}{x}}}, & x\neq 0 \\ 0, & x=0 \end{cases}$$

then

(a) f is continuous at $x=0$

(b) f is not continuous at $x=0$

(c) $\lim\limits_{x\to 0+} f(x)$ does not exist

(d) $\lim\limits_{x\to 0+} f(x)$ exist

20. Let $f(x)=\begin{cases} \dfrac{1-\cos kx}{x\sin x}, & x\neq 0 \\ \dfrac{1}{2}, & x=0 \end{cases}$

If f is continuous at $x=0$, find the value of k

(a) ± 1 (b) 1   (c) -1

(d) none of these

21. Let $f(x)=\begin{cases} \dfrac{x^3+x^2-16x+20}{(x-2)^2}, & x\neq 2 \\ k, & x=2 \end{cases}$

If f is continuous at $x=0$, find the value of k

(a) 7   (b) -7 (c) 8

(d) none of these

22. Given that

$$f(x) = \begin{cases} \dfrac{xe^{\frac{1}{x}}}{1+e^{\frac{1}{x}}}, & x \neq 0 \\ 0, & x = 0 \end{cases}$$

(a) f is continuous but $f'(0)$ does not exist

(b) f is not continuous but $f'(0)$ does exist

(c) f is continuous and $f'(0)$ does exist

(d) f is neither continuous nor $f'(0)$ does exist

23. If $f : R \rightarrow R$ satisfies $|f(x)| \leq x^2 \ \forall x \in R$, then

(a) f is differentiable at $x = 0$

(b) f is not differentiable at $x = 0$

(c) f is differentiable except at $x = 0$

(d) none of these

24. $f(x) = \begin{cases} xe^{-\left(\frac{1}{x}+\frac{1}{|x|}\right)}, & x \neq 0 \\ k, & x = 0 \end{cases}$

$f'(0)$ exist for what value of k

(a) no value of k exist

(b) value of k exist

(c) value of k = ±1

(d) none of these

25. If $f(x) = \begin{cases} e^{-\frac{1}{x^2}}, & x > 0 \\ 0, & x \leq 0 \end{cases}$, then f is

(a) discontinuous at x = 0

(b) differentiable at x = 0

(c) continuous but not differentiable at x = 0

(d) none of these

26. If $f(x) = \begin{cases} x^2 \left(\dfrac{e^{\frac{1}{x}} - e^{-\frac{1}{x}}}{e^{\frac{1}{x}} + e^{-\frac{1}{x}}} \right), & x > 0 \\ 0, & x \leq 0 \end{cases}$

Then, $f(x)$ is

(a) discontinuous at $x = 0$

(b) continuous at $x = 0$

(c) $f(x)$ is not defined at $x = 0$

(d) none of these

27. If $\lim\limits_{x \to 0} \dfrac{e^{kx} - e^x - x}{x^2} = \dfrac{3}{2}$, then $k = ?$

(a) $\dfrac{1}{2}$ \qquad\qquad (b) $\dfrac{3}{2}$

(c) 0 \qquad\qquad\quad (d) 1

28. $\lim\limits_{x \to 3} \left\{ \dfrac{x-3}{\log_a (x-2)} \right\} = ?$

(a) $\log_e a$ \qquad\qquad (b) $\log_a e$

(c) 0 \qquad\qquad\quad (d) 1

29. If $\lim\limits_{x \to \infty} \left\{ \dfrac{x^2 + x + 1}{x+1} - ax - b \right\} = 4$, then

(a) a=1, b=4 \qquad (b) a = 4, b = 1

(c) a = 1, b = –4 \qquad (d) a = 4, b = –1

30. Let $f : R \rightarrow R$ be such that $f(1) = 3$ and $f'(1) = 6$. Then, $\lim\limits_{x \to 0} \left\{ \dfrac{f(x+1)}{f(1)} \right\}^{\frac{1}{x}} = ?$

(a) e \quad (b) e^2 \quad (c) 0 \quad (d) 1

31. If $\theta_1, \theta_2,$ and θ satisfies the condition $0 < \theta_1 < \theta < \theta_2 < \dfrac{\pi}{2}$, then which of the following is true?

(a) $\cot \theta = \dfrac{\sin \theta_1 - \sin \theta_2}{\cos \theta_1 - \cos \theta_2}$

(b) $\tan \theta = \dfrac{\sin \theta_1 - \sin \theta_2}{\cos \theta_1 - \cos \theta_2}$

(c) $\cot \theta = \dfrac{\sin \theta_2 - \sin \theta_1}{\cos \theta_1 - \cos \theta_2}$

(d) $\tan \theta = \dfrac{\sin \theta_2 - \sin \theta_1}{\cos \theta_1 - \cos \theta_2}$

32. Let $f, g : [a,b] \rightarrow R$ be continuous on $[a,b]$, differentiable on (a,b) with $g(x) \neq 0$ in $[a,b]$. Then, $\dfrac{g(a)f(b) - f(a)g(b)}{g(c)f'(c) - f(c)g'(c)} = ?$

(a) $\dfrac{g(a)g(b)}{(g(c))^2}$ \qquad (b) $\dfrac{(b-a)g(a)g(b)}{(g(c))^2}$

(c) $\dfrac{(b-a)f(a)f(b)}{(g(c))^2}$ \quad (d) $\dfrac{(b-a)f(a)g(b)}{(f(c))^2}$

33. Let $2p + 3q + 6r = 0$ then at least one root of the equation $px^2 + qx + r = 0$ lies in the interval

(a) $(0, 1)$ (b) $(0, 2)$ (c) $(1, 2)$ (d) $(2, 4)$

34. If $p(x)$ be a polynomial function with real coefficients and a, b be two consecutive roots of the equation $p(x) = 0$ (where a < b), then $\exists\, c \in (a, b)$ such that

(a) $50 p'(c) + p(c) = 0$

(b) $p'(c) + \dfrac{p(c)}{50} = 0$

(c) $p'(c) + 50 p(c) = 0$

(d) none of these

35. If a function $f : [a, b] \to R$ is continuous on $[a, b]$ and differentiable on (a, b), then for $c \in (a, b)$, $\dfrac{bf(a) - af(b)}{b - a} = ?$

(a) $f(c) - cf'(c)$ (b) $f(c) - f'(c)$

(c) $cf(c) - f'(c)$ (d) $cf'(c)$

36. Let f be twice differentiable function for all x and $f(1) = 1, f(2) = 4, f(4) = 16$. Then which of the following is true?

(a) $f''(x) = 2\ \forall x \in (1, 4)$

(b) $f''(x) = f'(x) = 5\ \forall x \in (2, 4)$

(c) $f''(x) = 4\ \forall x \in (2, 4)$

(d) $f''(x) = 2$ for some $x \in (1, 4)$

37. The function $f(x) = \sin x + \cos x$ $(0 \le x \le 2\pi)$ is strictly increasing on

(a) $\left(\dfrac{5\pi}{4}, 2\pi\right)$

(b) $\left(\dfrac{5\pi}{4}, 2\pi\right) \cup \left(0, \dfrac{\pi}{4}\right)$

(c) $\left[0, \dfrac{\pi}{4}\right]$

(d) $\left(\dfrac{5\pi}{4}, 2\pi\right) \cap \left(0, \dfrac{\pi}{4}\right)$

38. The intervals on which the function $f(x) = 2x^3 - 15x^2 + 36x + 1$ are strictly increasing and strictly decreasing are given by, respectively

(a) $(-\infty, 2) \cup (3, \infty); (2, 3)$

(b) $(-\infty, 2); (2, 3)$ (c) $(3, \infty); (2, 3)$

(d) $(-\infty, 2) \cup (2, \infty); (1, 2)$

39. If $f(x)$ be a real valued function defined for all $x \ge 0$ such that $f(0) = 0$ and $f''(x) > 0\ \forall x$, then the function $g(x) = \dfrac{f(x)}{x}$ is

(a) decreasing on $(0, \infty)$

(b) increasing on $(0, \infty)$

(c) increasing on $(0, 1)$ & decreasing on $(1, \infty)$

(d) none of these

40. Which of the following points is the point of a local minima or local maxima of the function $f(x) = \sin^4 x + \cos^4 x$ in $\left(0, \dfrac{\pi}{2}\right)$?

(a) $\dfrac{3\pi}{4}$ (b) $\dfrac{\pi}{2}$ (c) π (d) $\dfrac{\pi}{4}$

41. The maximum value of the function $f(x) = \left(\dfrac{1}{x}\right)^x$ is

(a) $e^{\frac{1}{e}}$ (b) e^e (c) 10 (d) $\dfrac{1}{e} + e$

42. The positive numbers a and b are such that $a + b = 60$ and ab^3 is maximum, are, respectively

(a) 15, 45   (b) 10, 50   (c) 20, 40   (d) 25, 35

43. An open box with a square base is to be made out of a given quantity of card board of area c^2 square units. Then the maximum volume of the box is

(a) $\dfrac{c^3}{\sqrt{3}}$ cubic units   (b) $\dfrac{c^3}{2}$ cubic units

(c) $\dfrac{c^3}{6\sqrt{3}}$ cubic units   (d) $\dfrac{c^3}{2\sqrt{3}}$ cubic units

44. A wire of length 36 m is to be cut into two pieces. One of the pieces is to be made into a square and the other into a circle. What should be the lengths of the two pieces so that the combined area of the circle and square is minimum?

(a) $\dfrac{36\pi}{\pi + 4}; \dfrac{144}{\pi + 4}$ (b) $\dfrac{28}{\pi + 4}; \dfrac{112\pi}{\pi + 4}$

(c) $\dfrac{28}{\pi}; \dfrac{112}{\pi}$ (d) $\dfrac{12\pi}{\pi + 4}; \dfrac{16}{\pi + 4}$

45. The value of $\displaystyle\sum_{n=1}^{\infty} \frac{3n^2-2}{n!}$ is

(a) $4e$ (b) $4e+1$
(c) $4e-1$ (d) $4e+2$

46. Consider the Taylors series expansion of the function $f(x) = e^{\frac{x}{2}}$ about $x = 3$. Then the co-efficient of $(x-3)^5$ in the expansion is

(a) $\dfrac{1}{5!} \times \dfrac{e^{\frac{3}{2}}}{32}$ (b) $\dfrac{1}{3!} \times \dfrac{e^{\frac{5}{2}}}{25}$

(c) -1 (d) 0

47. If $f(h) = f(0) + hf'(0) + \dfrac{h^2}{2!} f''(\theta h)$, where

$\theta \in (0,1)$; then for $h = 1$ and $f(x) = (1-x)^{\frac{5}{2}}$, the value of θ will be?

(a) 0 (b) $\dfrac{1}{2}$ (c) $\dfrac{5}{9}$ (d) $\dfrac{9}{25}$

48. Which of the following is true?

(a) $0 < \dfrac{1}{x} \log\left(\dfrac{e^x-1}{x}\right) < 1$

(b) $\dfrac{x}{1+x} < \log(1+x) < x$ if $x > 0$

(c) $x > \log(1+x) > x - \dfrac{x^2}{2}$ if $x > 0$

(d) all the above

49. If $I_n = \displaystyle\int_0^1 x^n \tan^{-1} x \, dx$, then the value of $(n+1)I_n + (n-1)I_{n-2}$ is

(a) $\dfrac{\pi}{2}+1$ (b) $\dfrac{\pi}{2}-\dfrac{1}{n}$ (c) $\dfrac{\pi}{2}-\dfrac{2}{n}$ (d) $\dfrac{\pi}{2}$

50. Let f be a function such that f', f'', f''' all exist and $u = f'(x)\cos x - f''(x)\sin x$ and $v = f'(x)\sin x + f''(x)\cos x$. Then,

$$\int \sqrt{\left(\frac{du}{dx}\right)^2 + \left(\frac{dv}{dx}\right)^2}\, dx = ?$$

(a) $f(x) + f''(x) + C$
(b) $f'(x) + f'''(x) + C$
(c) $f'(x) + f''(x) + C$
(d) $f''(x) + f'''(x) + C$

51. If $\displaystyle\int_{-1}^{2} |x \sin \pi x|\, dx = \dfrac{k}{\pi^2}$, then the value of "$k$" is?

(a) $\dfrac{3\pi}{2}$ (b) $\dfrac{\pi}{2}$ (c) 4π (d) 0

52. $\displaystyle\int_0^{\pi} x\, f(\sin x)\, dx$ is equal to

(a) $\displaystyle\int_0^{\pi} f(\cos x)\, dx$ (b) $\displaystyle\int_0^{\frac{\pi}{2}} f(\cos x)\, dx$

(c) $\pi \displaystyle\int_0^{\frac{\pi}{2}} f(\cos x)\, dx$ (d) O

53. $\displaystyle\int_{-1}^{2} |x^3 - x|\, dx$ is equal to

(a) $\dfrac{1}{2}$ (b) $\dfrac{2}{21}$ (c) $\dfrac{1}{30}$ (d) $\dfrac{11}{4}$

54. Let $I_1 = \displaystyle\int_0^1 \dfrac{e^x}{1+x}\, dx$ & $I_2 = \displaystyle\int_0^1 \dfrac{x^2}{e^{x^3}(2-x^3)}\, dx$, then $\dfrac{I_1}{I_2} = ?$

(a) $3e$ (b) $2e$ (c) $3e+2$ (d) $2e-1$

55. $\displaystyle\lim_{n\to\infty} \left[\dfrac{1}{\sqrt{2n-1^2}} + \dfrac{1}{\sqrt{4n-2^2}} + \ldots + \dfrac{1}{n}\right] = ?$

(a) $\dfrac{\pi}{2}$ (b) $\dfrac{3\pi}{2}$ (c) $\pi-1$ (d) 0

56. $\displaystyle\lim_{n\to\infty} \left[\dfrac{1}{n} + \dfrac{1}{n+1} + \dfrac{1}{n+2} + \ldots + \dfrac{1}{3n}\right] = ?$

(a) $3+\log 5$ (b) $1-\log 2$
(c) $\log 3$ (d) 0

57. $\displaystyle\int_0^{\frac{\pi}{4}} \log(1+\tan\theta)\, d\theta = ?$

(a) $\dfrac{\pi}{8}\log 2$ (b) $\dfrac{\pi}{2}\log 2$

(c) $\dfrac{\pi}{3}\log 3$ (d) 0

58. If $\displaystyle\int_0^{f(x)} t^2\, dt = x\cos\pi x$, then $f'(4) = ?$

(a) 1 (b) $\dfrac{1}{5^{\frac{2}{3}}}$ (c) $\dfrac{1}{12^{\frac{2}{3}}}$ (d) 0

59. $\displaystyle\lim_{n\to\infty}\left(\dfrac{n!}{n^n}\right)^{\frac{1}{n}} = ?$

(a) 1 (b) $\dfrac{1}{e}$ (c) $2e$ (d) -1

60. If $g(x) = \cos x - \displaystyle\int_0^x (x-t)g(t)\,dt$, then

$g(x) + g''(x) = ?$

(a) 1 (b) $\sin x$ (c) $-\cos x$ (d) 0

61. $\displaystyle\int_0^1 \dfrac{1}{\left(1-x^n\right)^{\frac{1}{n}}}\,dx = ?$

(a) $\dfrac{\pi}{2}\sin\dfrac{\pi}{n}$ (b) $\dfrac{\pi}{n}\sin\dfrac{n\pi}{2}$

(c) $\dfrac{\pi}{n}\cos ec\dfrac{\pi}{n}$ (d) None of these

62. $\displaystyle\int_0^1 \dfrac{x^2}{\sqrt{1-x^4}}\,dx = ?$

(a) $\sqrt{\pi}\,\dfrac{\Gamma\left(\dfrac{3}{2}\right)}{\Gamma\left(\dfrac{1}{2}\right)}$ (b) $\dfrac{\sqrt{\pi}}{4}\,\dfrac{\Gamma\left(\dfrac{3}{4}\right)}{\Gamma\left(\dfrac{1}{4}\right)}$

(c) $\sqrt{\pi}\,\dfrac{\Gamma\left(\dfrac{3}{4}\right)}{\Gamma\left(\dfrac{1}{4}\right)}$ (d) None of these

63. $\displaystyle\int_0^{\frac{\pi}{2}}\sqrt{\sin\theta}\,d\theta \times \int_0^{\frac{\pi}{2}}\dfrac{1}{\sqrt{\sin\theta}}\,d\theta = ?$

(a) π (b) $\dfrac{\pi}{2}$ (c) $\dfrac{\pi}{3}$ (d) 2π

64. $\displaystyle\int_0^\infty e^{-x^4}x^2\,dx \times \int_0^\infty e^{-x^4}\,dx = ?$

(a) $\dfrac{\pi}{4}$ (b) $\dfrac{\pi}{2}$ (c) $\dfrac{\pi}{3}$ (d) $\dfrac{\pi}{8\sqrt{2}}$

65. $\Gamma\left(\dfrac{1}{9}\right)\times\Gamma\left(\dfrac{2}{9}\right)\times\Gamma\left(\dfrac{3}{9}\right)\times\Gamma\left(\dfrac{4}{9}\right)\times.......\times\Gamma\left(\dfrac{8}{9}\right) = ?$

(a) $\dfrac{16}{3}\pi^4$ (b) $3\pi^4$ (c) $\dfrac{8}{3}\pi^4$ (d) $\dfrac{1}{3}\pi^4$

66. If $u = \log\left(x^3 + y^3 + z^3 - 3xyz\right)$, then which of the following is not true?

(a) $u_x + u_y + u_z = \dfrac{3}{x+y+z}$

(b) $\left(\dfrac{\partial}{\partial x}+\dfrac{\partial}{\partial y}+\dfrac{\partial}{\partial z}\right)^2 u = -\dfrac{9}{\left(x+y+z\right)^2}$

(c) $\dfrac{\partial^2 u}{\partial x^2}+\dfrac{\partial^2 u}{\partial y^2}+\dfrac{\partial^2 u}{\partial z^2} = -\dfrac{3}{\left(x+y+z\right)^2}$

(d) none of these

67. If $V = \log(x^2 + y^2 + z^2)$, then which of the following is true?

(a) $\dfrac{\partial^2 V}{\partial x^2}+\dfrac{\partial^2 V}{\partial y^2}+\dfrac{\partial^2 V}{\partial z^2} = \dfrac{-2}{\left(x^2+y^2+z^2\right)^2}$

(b) $\dfrac{\partial^2 V}{\partial x^2}+\dfrac{\partial^2 V}{\partial y^2}+\dfrac{\partial^2 V}{\partial z^2} = \dfrac{2}{\left(x^2+y^2+z^2\right)}$

(c) $\dfrac{\partial^2 V}{\partial x^2}+\dfrac{\partial^2 V}{\partial y^2}+\dfrac{\partial^2 V}{\partial z^2} = -\dfrac{1}{\left(x^2+y^2+z^2\right)^2}$

(d) none of these

68. If $u = f(r)$, $x = r\cos\theta$ and $y = r\sin\theta$, then $u_{xx} + u_{yy} = ?$

(a) $\dfrac{1}{r}f''(r) + f'(r)$ (b) $f''(r) + f'(r)$

(c) $f''(r) + \dfrac{1}{r}f'(r)$ (d) none of these

69. If $u = \sin^{-1}\left(\dfrac{x^2+y^2}{x+y}\right)$, then $x\dfrac{\partial u}{\partial x}+y\dfrac{\partial u}{\partial y} = ?$

(a) $\tan u$ (b) $\cot u$

(c) $\sec u$ (d) $\cos ec\, u$

70. If $u = f(x,y)$ and $x = e^\phi + e^{-\psi}$, $y = e^{-\phi} - e^\psi$, then $\dfrac{\partial u}{\partial \phi}-\dfrac{\partial u}{\partial \psi} = ?$

(a) $x\dfrac{\partial u}{\partial x}+y\dfrac{\partial u}{\partial y}$ (b) $x\dfrac{\partial u}{\partial x}-y\dfrac{\partial u}{\partial y}$

(c) $y\dfrac{\partial u}{\partial x}-x\dfrac{\partial u}{\partial y}$ (d) 0

71. If $u = f(2x - 3y, 3y - 4z, 4z - 2x)$, then,
$$\frac{1}{2}u_x + \frac{1}{3}u_y + \frac{1}{4}u_z = ?$$
  (a) -1 (b) 1   (c) 0   (d) $2u$

72. If $u = \frac{x+y}{1-xy}$ and $v = \tan^{-1}x + \tan^{-1}y$, then which of the following is true?
  (a) u and v are functionally related
  (b) $\frac{\partial(u,v)}{\partial(x,y)} = 0$ (c) $u = \tan v$
  (d) all of these

73. If $u = xf\left(\frac{y}{x}\right) + g\left(\frac{y}{x}\right)$, then $x\frac{\partial u}{\partial x} + y\frac{\partial u}{\partial y} = ?$
  (a) $xf\left(\frac{y}{x}\right)$ (b) $-yf\left(\frac{y}{x}\right)$
  (c) $f\left(\frac{y}{x}\right)$ (d) 0

74. If $x^3 + y^3 + 3xy - 1 = 0$, then $\frac{dy}{dx} = ?$
  (a) $\frac{x^2+y}{x+y^2}$ (b) $-\frac{x^2+y}{x+y^2}$
  (c) $\frac{x^2+y^2}{x+y}$ (d) $-\frac{x^2+y^2}{x+y}$

75. If $u = x\log(xy)$ and $x^2 + y^2 + xy = 1$, then $\frac{du}{dx} = ?$
  (a) $\log xy - \frac{y}{x}$ (b) $\log xy + 1 + \frac{x}{y}$
  (c) $\log xy - 1 + \frac{x}{y}$ (d) $\log xy + 1 - \frac{x}{y}$

76. The function $f(x,y) = xy(1-x-y)$ has
  (a) minimum value at $(0, 1)$
  (b) maximum value at $(1, 0)$
  (c) minimum value at $\left(\frac{1}{3},\frac{1}{3}\right)$
  (d) maximum value at $\left(\frac{1}{3},\frac{1}{3}\right)$

77. The minimum value of $x^2 + y^2 + z^2$, given that $x + 2y + 3z = 4$, is
  (a) 8/7   (b) 1/7   (c) 15   (d) 8

78. The minimum and maximum distance of the point $(1, 2, 3)$ from the sphere $x^2 + y^2 + z^2 = 1$ are, respectively
  (a) $\sqrt{14}-1, \sqrt{14}+1$ (b) $\sqrt{14}, 2\sqrt{14}$
  (c) 15   (d) 8

79. Find the value of $\iiint_V (x^2 + y^2 + z^2)\,dxdydz$, where V is the volume of a cube bounded by the co-ordinate planes $x = 0, y = 0, z = 0$ and the plane $x = y = z = 2$
  (a) 16   (b) 32   (c) 8   (d) $\frac{1}{16}$

80. The value of $\int_0^1 \int_0^x \int_0^y x^3 y^2 z\,dzdydx$ is
  (a) $\frac{1}{10}$ (b) $\frac{1}{25}$ (c) $\frac{1}{90}$ (d) $\frac{4}{21}$

81. The value of $\iiint_V (x+y+z)xyz$, taken over the volume V bounded by $x = 0, y = 0, z = 0$ and $x + y + z = 1$ is
  (a) $\frac{1}{840}$ (b) $\frac{2}{525}$ (c) $\frac{1}{420}$ (d) $\frac{1}{2}$

82. Evaluate $\iiint_V x^3 y^4 z^2\,dxdydz$, where V is the region given by $x\geqslant0, y\geqslant0, z\geqslant0$ and $x+y+z\leqslant1$
  (a) $\frac{1}{(11!)^2}$ (b) $\frac{3!4!5!}{(11)!}$
  (c) $\frac{1!2!3!}{(11)!}$ (d) $\frac{2!3!4!}{(11)!}$

83. If the triple integral bounded by the planes $2x + y + z = 4, x = 0, y = 0$ and $z = 0$ is given by $\int_0^2 \int_0^{\lambda(x)} \int_0^{\mu(x,y)} dzdydx$, then $\lambda(x) - \mu(x,y) = ?$
  (a) $x+y$ (b) $x-y$ (c) x (d) y

84. The value of $\int_{x=0}^1 \int_{y=0}^{x^2} \int_{z=0}^y (y+2z)\,dzdydx$ is
  (a) $\frac{4}{21}$ (b) $\frac{2}{21}$ (c) $\frac{5}{21}$ (d) $\frac{8}{21}$

85. The volume of the solid generated by the revolution of the curve $y=\dfrac{8}{4+x^2}$ about x-axis from $x = -\infty$ to $x = \infty$ is given by

(a) $2\pi^2$ (b) $4\pi^2$ (c) $8\pi^2$ (d) $16\pi^2$

86. The volume of a solid formed by the revolution of the loop of the curve $y^2(2 + x) = x^2(2-x)$ about x-axis is

(a) $8\pi\left(\log 2+\dfrac{2}{3}\right)$ (b) $16\pi\left(\log 2-\dfrac{2}{3}\right)$

(c) $16\pi\left(\log 2+\dfrac{2}{3}\right)$ (d) $8\pi\left(\log 2-\dfrac{2}{3}\right)$

87. The volume of the solid generated by the revolution of the cardiod $r = a(1 + \cos\theta)$ about the initial line is

(a) $\dfrac{8}{3}\pi a^3$ (b) $\dfrac{4}{3}\pi a^3$

(c) $\dfrac{2}{3}\pi a^3$ (d) $4\pi a^3$

88. What will be the surface area of a solid generated by revolving the arc of the parabola $y^2 = 4ax$ about x-axis bounded by its latus rectum?

(a) $\dfrac{4\pi a^2}{3}\left(\sqrt{2}+1\right)$ (b) $\dfrac{16\pi a^2}{3}\left(1+2\sqrt{2}\right)$

(c) $\dfrac{8\pi a^2}{3}\left(2\sqrt{2}-1\right)$ (d) none of these

89. The surface area of the solid generated by the revolution of the astroid $x^{\frac{2}{3}}+y^{\frac{2}{3}}=a^{\frac{2}{3}}$ about x-axis is

(a) $\dfrac{12\pi a^2}{5}$ (b) $\dfrac{3\pi a^2}{5}$ (c) $\dfrac{8\pi a^2}{5}$ (d) $2\pi a^2$

90. What will be the surface area of the solid formed by the revolution of the cardioid $r = a(1 + \cos\theta)$ about the initial line?

(a) $\dfrac{4\pi a^2}{5}$ (b) $\dfrac{8\pi a^2}{5}$ (c) $\dfrac{16\pi a^2}{5}$ (d) $\dfrac{32\pi a^2}{5}$

Answer key				
1. (b)	**2.** (a)	**3.** (d)	**4.** (b)	**5.** (a)
6. (a)	**7.** (a)	**8.** (b)	**9.** (a)	**10.** (a)
11. (c)	**12.** (d)	**13.** (d)	**14.** (c)	**15.** (c)
16. (a)	**17.** (a)	**18.** (b)	**19.** (a)	**20.** (a)
21. (a)	**22.** (a)	**23.** (a)	**24.** (a)	**25.** (b)
26. (b)	**27.** (b)	**28.** (a)	**29.** (c)	**30.** (b)

31. (c)	**32.** (b)	**33.** (a)	**34.** (c)	**35.** (a)
36. (d)	**37.** (b)	**38.** (a)	**39.** (b)	**40.** (d)
41. (a)	**42.** (a)	**43.** (c)	**44.** (a)	**45.** (d)
46. (a)	**47.** (d)	**48.** (d)	**47.** (d)	**48.** (d)
49. (b)	**50.** (b)	**51.** (c)	**52.** (c)	**53.** (d)
54. (a)	**55.** (a)	**56.** (c)	**57.** (a)	**58.** (c)
59. (b)	**60.** (c)	**61.** (c)	**62.** (c)	**63.** (a)
64. (d)	**65.** (a)	**66.** (d)	**67.** (b)	**68.** (c)
69. (a)	**70.** (b)	**71.** (c)	**72.** (d)	**73.** (a)
74. (b)	**75.** (d)	**76.** (d)	**77.** (a)	**78.** (a)
79. (b)	**80.** (c)	**81.** (a)	**82.** (d)	**83.** (d)
84. (b)	**85.** (b)	**85.** (b)	**86.** (b)	**87.** (a)
88. (c)	**89.** (a)	**90.** (d)		

Explanation

1. (b) $-1\le \sin x\le 1$

$\Rightarrow -3\le 3\sin x\le 3$

$\Rightarrow x-3\le x+3\sin x\le x+3$

$\Rightarrow \dfrac{x-3}{x-5}\le \dfrac{x+3\sin x}{x-5}\le \dfrac{x+3}{x-5}$

Now $\lim\limits_{x\to\infty}\dfrac{x-3}{x-5}=\lim\limits_{x\to\infty}\dfrac{1-\dfrac{3}{x}}{1-\dfrac{5}{x}}=\dfrac{1-0}{1-0}=1,$

$\lim\limits_{x\to\infty}\dfrac{x+3}{x-5}=\lim\limits_{x\to\infty}\dfrac{1+\dfrac{3}{x}}{1-\dfrac{5}{x}}=\dfrac{1+0}{1-0}=1.$

Hence, by the Sandwich theorem, we get,

$\lim\limits_{x\to\infty}\dfrac{x+3\sin x}{x-5}=1.$

2. (a) $x-1< [x]\le x$

$\Rightarrow \dfrac{x-1}{x}<\dfrac{[x]}{x}\le 1$

Since, $\lim\limits_{x\to\infty}\dfrac{x-1}{x}=\lim\limits_{x\to\infty}\dfrac{1-\dfrac{1}{x}}{1}=1-0=1$ and

$\lim\limits_{x\to\infty}1=1$, so by the Sandwich theorem we get,

$\lim\limits_{x\to\infty}\dfrac{[x]}{x}=1.$

3. (d) We know that $\sin x = x-\dfrac{x^3}{3!}+\dfrac{x^5}{5!}-........\infty$

Therefore, $\sin x - x = -\dfrac{x^3}{3!}+\dfrac{x^5}{5!}-.........\infty$

$$\therefore \lim_{x\to0}\frac{\sin x-x}{x^3}$$

$$=\lim_{x\to0}\left[\frac{-\dfrac{x^3}{3!}+\dfrac{x^5}{5!}-\dots\dots\infty}{x^3}\right]$$

$$=\lim_{x\to0}\left(-\frac{1}{3!}+\frac{x^2}{5!}-\dots\dots\infty\right)$$

$$=-\frac{1}{3!}+\frac{0}{5!}-\dots\dots\infty=-\frac{1}{3!}=-\frac{1}{6}.$$

4. (b)

$$\lim_{x\to1}\frac{\displaystyle\sum_{k=1}^{50}x^k-50}{x-1}$$

$$=\lim_{x\to1}\left[\frac{\left(x+x^2+x^3+\dots+x^{50}\right)-50}{x-1}\right]$$

$$=\lim_{x\to1}\left[\frac{(x-1)+(x^2-1)+(x^3-1)+\dots(x^{50}-1)}{x-1}\right]$$

$$=\lim_{x\to1}\left[\frac{x-1}{x-1}+\frac{x^2-1}{x-1}+\frac{x^3-1}{x-1}+\dots+\frac{x^{50}-1}{x-1}\right]$$

$$=\lim_{x\to1}\left[\begin{array}{l}1+(x+1)+\left(x^2+x+1\right)+\dots\dots\\ +\left(x^{49}+x^{48}+\dots+1\right)\end{array}\right]$$

$$=1+(1+1)+(1+1+1)+\dots+(1+1+\dots+1)$$

$$=1+2+3+\dots+50=\frac{50(50+1)}{2}=1275$$

$$\left(\because 1+2+3+\dots+n=\frac{n(n+1)}{2}\right)$$

5. (a)

$$\lim_{n\to\infty}\cos\left\{\pi\sqrt{n^2+n}\right\}$$

$$=\lim_{n\to\infty}(-1)^n\cos\left\{n\pi-\pi\sqrt{n^2+n}\right\}$$

$$=\lim_{n\to\infty}(-1)^n\cos\left\{\pi\left(n-\sqrt{n^2+n}\right)\right\}$$

$$=\lim_{n\to\infty}(-1)^n\cos\left\{\pi\frac{\left(n-\sqrt{n^2+n}\right)\left(n+\sqrt{n^2+n}\right)}{\left(n+\sqrt{n^2+n}\right)}\right\}$$

$$=\lim_{n\to\infty}(-1)^n\cos\left\{\pi\frac{\left(n^2-\left(\sqrt{n^2+n}\right)^2\right)}{\left(n+\sqrt{n^2+n}\right)}\right\}$$

$$=\lim_{n\to\infty}(-1)^n\cos\left\{\frac{-n\pi}{\left(n+\sqrt{n^2+n}\right)}\right\}$$

$$=\lim_{n\to\infty}(-1)^n\cos\left\{\frac{-n\pi}{n\left(1+\sqrt{1+\dfrac{1}{n}}\right)}\right\}$$

$$=\lim_{n\to\infty}(-1)^n\times\lim_{n\to\infty}\cos\left\{\frac{-\pi}{\left(1+\sqrt{1+\dfrac{1}{n}}\right)}\right\}$$

$$=\lim_{x\to\infty}(-1)^n\times\lim_{x\to\infty}\left\{\frac{-\pi}{\left(1+\sqrt{1+\dfrac{1}{n}}\right)}\right\}$$

$$=\lim_{n\to\infty}(-1)^n\times\cos\left\{\frac{-\pi}{\left(1+\sqrt{1+0}\right)}\right\}$$

$$=\lim_{n\to\infty}(-1)^n\times\cos\left(-\frac{\pi}{2}\right)=0\left(\because\cos\left(-\frac{\pi}{2}\right)=0\right)$$

6. (a)

$$\lim_{n\to\infty}n\cos\left(\frac{\pi}{8n}\right)\sin\left(\frac{\pi}{8n}\right)$$

$$=\lim_{x\to0}\frac{1}{x}\cos\left(\frac{\pi x}{8}\right)\sin\left(\frac{\pi x}{8}\right)\quad\left(\text{putting }n=\frac{1}{x}\right)$$

$$=\lim_{x\to0}\cos\left(\frac{\pi x}{8}\right)\times\lim_{x\to0}\left[\frac{\sin\left(\dfrac{\pi x}{8}\right)}{x}\right]$$

$$=\cos0\times\lim_{x\to0}\left[\frac{\sin\left(\dfrac{\pi x}{8}\right)}{\dfrac{\pi x}{8}}\times\frac{\pi}{8}\right]$$

$$=1\times\frac{\pi}{8}\times\lim_{\frac{\pi x}{8}\to0}\left[\frac{\sin\left(\dfrac{\pi x}{8}\right)}{\dfrac{\pi x}{8}}\right]=\frac{\pi}{8}\times1=\frac{\pi}{8}.$$

7. (a) $\lim\limits_{x\to 0}\dfrac{x^m \sin^n x}{\sin x^k}$

$$=\lim\limits_{x\to 0}\left\{\left(\dfrac{\sin x}{x}\right)^n \times \dfrac{x^k}{\sin x^k}\times x^{m+n-k}\right\}$$

$$=\left(\lim\limits_{x\to 0}\dfrac{\sin x}{x}\right)^n \times \dfrac{1}{\left(\lim\limits_{x^k\to 0}\dfrac{\sin x^k}{x^k}\right)}\times \lim\limits_{x\to 0}\left(x^{m+n-k}\right)$$

$$\left(\because x\to 0 \Rightarrow x^k \to 0\right)$$

$$=1^n \times \dfrac{1}{1}\times \lim\limits_{x\to 0}x^{m+n-k}=\lim\limits_{x\to 0}x^{m+n-k}$$

$$=\lim\limits_{x\to 0}x^0 = 1, \text{a finite non-zero value}$$

$$(\text{provided } m+n-k=0 \ i.e; \ k=m+n)$$

8. (b)

$$\lim\limits_{n\to\infty}\left(\dfrac{n^2-n+1}{n^2-n-1}\right)^{n(n-1)}$$

$$=\lim\limits_{n\to\infty}\left[\dfrac{n(n-1)+1}{n(n-1)-1}\right]^{n(n-1)}$$

$$=\lim\limits_{n(n-1)\to\infty}\left[\dfrac{1+\dfrac{1}{n(n-1)}}{1-\dfrac{1}{n(n-1)}}\right]^{n(n-1)}$$

$$\left(\because n\to\infty \Rightarrow n(n-1)\to\infty\right)$$

$$=\lim\limits_{x\to\infty}\left[\dfrac{1+\dfrac{1}{x}}{1-\dfrac{1}{x}}\right]^x \quad (\text{putting } n(n-1)=x)$$

$$=\dfrac{\lim\limits_{x\to\infty}\left(1+\dfrac{1}{x}\right)^x}{\lim\limits_{x\to\infty}\left(1-\dfrac{1}{x}\right)^x}=\dfrac{e}{e^{-1}}=e^2.$$

9. (a)

$$f(x)=\lim\limits_{n\to\infty}n\left(x^{\frac{1}{n}}-1\right)=\lim\limits_{n\to\infty}\left(\dfrac{x^{\frac{1}{n}}-1}{\frac{1}{n}}\right)$$

$$=\lim\limits_{m\to 0}\left(\dfrac{x^m-1}{m}\right)\quad \left(\text{putting } m=\dfrac{1}{x}\right)$$

$$=\lim\limits_{m\to 0}\left(\dfrac{x^m-1}{m}\right)\left(\text{putting } m=\dfrac{1}{n}\right)$$

$$=\log_e^x \quad \left(\because \lim\limits_{x\to 0}\left(\dfrac{a^x-1}{x}\right)=\log_e a\right)$$

So, $f(xy) = \log_e(xy) = \log_e(x) + \log_e(y) = f(x)+f(y).$

10. (a) $\lim\limits_{x\to 1}\dfrac{f(x)-f(1)}{x-1}$

$$=\lim\limits_{x\to 1}\dfrac{-\sqrt{25-x^2}-\left(-\sqrt{25-1^2}\right)}{x-1}$$

$$=\lim\limits_{x\to 1}\dfrac{\sqrt{24}-\sqrt{25-x^2}}{x-1}$$

$$=\lim\limits_{x\to 1}\dfrac{\left(\sqrt{24}-\sqrt{25-x^2}\right)\left(\sqrt{24}+\sqrt{25-x^2}\right)}{(x-1)\left(\sqrt{24}+\sqrt{25-x^2}\right)}$$

$$=\lim\limits_{x\to 1}\dfrac{\left(24-25+x^2\right)}{(x-1)\left(\sqrt{24}+\sqrt{25-x^2}\right)}$$

$$=\lim\limits_{x\to 1}\dfrac{\left(x^2-1\right)}{(x-1)\left(\sqrt{24}+\sqrt{25-x^2}\right)}$$

$$=\lim\limits_{x\to 1}\dfrac{(x+1)}{\left(\sqrt{24}+\sqrt{25-x^2}\right)}$$

$$=\dfrac{(1+1)}{\left(\sqrt{24}+\sqrt{25-1^2}\right)}=\dfrac{2}{\left(\sqrt{24}+\sqrt{24}\right)}$$

$$=\dfrac{2}{2\times 2\sqrt{6}}=\dfrac{1}{2\sqrt{6}}.$$

11. (c)

$f(1,1)$

$= f(1-1,\ f(1,1-1)) = f(0,\ f(1,0))$

$= f(0,\ f(1-1,1)) = f(0,\ f(0,1))$

$= f(0,\ 2)=2+1=3.$

12. (d)

$$f\left(x+\dfrac{1}{x}\right)$$

$$=x^2+\dfrac{1}{x^2}+3=\left(x+\dfrac{1}{x}\right)^2-2\times x\times\dfrac{1}{x}+3$$

$$= \left(x + \frac{1}{x}\right)^2 + 1$$

$$\Rightarrow f(x) = x^2 + 1 \text{ (by replacing } x + \frac{1}{x} \text{ by } x)$$

13. (d)

Case-I: $x > 0$

Here $|x| = x$.

$$|x|^2 - 3|x| + 2 = 0$$

$$\Rightarrow x^2 - 3x + 2 = 0 \Rightarrow (x-1)(x-2) = 0$$

$$\Rightarrow x = 1, 2$$

Case-II: $x < 0$

Here $|x| = -x$.

$$|x|^2 - 3|x| + 2 = 0$$

$$\Rightarrow x^2 + 3x + 2 = 0 \Rightarrow (x+1)(x+2) = 0$$

$$\Rightarrow x = -1, -2$$

Thus, four solutions exist.

14. (c)

$$\lim_{x \to 2+} \varphi(x)$$

$$= \lim_{x \to 2+} \frac{1}{1 + e^{\frac{1}{x-2}} + e^{\frac{-1}{(x-3)^2}}}$$

$$= \frac{1}{1 + \lim_{x \to 2+} e^{\frac{1}{x-2}} + \lim_{x \to 2+} e^{\frac{-1}{(x-3)^2}}} = 0$$

$$\left(\because \lim_{x \to 2+} e^{\frac{1}{x-2}} = \infty \right)$$

$$\lim_{x \to 2-} \varphi(x)$$

$$= \lim_{x \to 2-} \frac{1}{1 + e^{\frac{1}{x-2}} + e^{\frac{-1}{(x-3)^2}}}$$

$$= \frac{1}{1 + \lim_{x \to 2-} e^{\frac{1}{x-2}} + \lim_{x \to 2-} e^{\frac{-1}{(x-3)^2}}}$$

$$= \frac{1}{1 + 0 + e^{\frac{-1}{(2-3)^2}}} = \frac{1}{1 + e^{-1}}$$

$$\left(\because \lim_{x \to 2-} e^{\frac{1}{x-2}} = 0 \right)$$

$\because \lim_{x \to 2+} \varphi(x) \neq \lim_{x \to 2-} \varphi(x)$, so φ is discontinuous

at $x = 2$

In a similar manner, it can be shown that $\lim_{x \to 3+} \varphi(x) = \lim_{x \to 3-} \varphi(x) = \varphi(3)$. As a result φ is continuous at $x = 3$.

15. (c)

$$f(x) = |x \sin x| = |x||\sin x|$$

$$= \begin{cases} (-x)(-\sin x), & -\pi < x \leq 0 \\ x \sin x, & 0 \leq x \leq \pi \end{cases}$$

$$= x \sin x, \quad -\pi < x < \pi$$

Clearly, $f'(x)$ exists, i.e, f is differentiable (since each of x and $\sin x$ is differentiable).

Then,

$$f'(x) = x \times \frac{d}{dx}(\sin x) + \sin x \times \frac{d}{dx}(x)$$

$$= x \cos x + \sin x$$

Since each of x, $\sin x$, $\cos x$ is

differentiable, so f' is also differentiable.

$\therefore \quad f''(x)$

$\therefore f''(x)$

$$= \frac{d}{dx}(f'(x)) = \frac{d}{dx}(x \cos x) + \frac{d}{dx}(\sin x)$$

$$= -x \sin x + \cos x + \cos x = -x \sin x + 2 \cos x,$$

which exists $\forall x \in R$.

Thus, we can say that f is twice differentiable.

$\therefore f''$ exist, so it is continuous.

16. (a) We know that

$$|x| = \begin{cases} x, & x \geq 0 \\ -x, & x \leq 0 \end{cases}$$

and $\sin(-x) = -\sin x$, $\cos(-x) = \cos x$.

Now, $Rf'(0)$

$$= \lim_{x \to 0+} \frac{f(x) - f(0)}{x}$$

$$= \lim_{x \to 0+} \frac{q_0 \cos x + q_1 \sin x + q_2 x^3 - q_0}{x} \left(\frac{0}{0} \text{form} \right)$$

$$= \lim_{x \to 0+} \frac{-q_0 \sin x + q_1 \cos x + 3x^2 q_2}{1}$$

(by the L'Hospital rule)

$$= -q_0 \times 0 + q_1 + 0 = q_1.$$

Again, $Lf'(0)$

$$= \lim_{x \to 0-} \frac{f(x) - f(0)}{x}$$

$$= \lim_{x \to 0-} \frac{q_0 \cos x - q_1 \sin x - q_2 x^3 - q_0}{x} \left(\frac{0}{0} \text{ form} \right)$$

$$= \lim_{x \to 0-} \frac{-q_0 \sin x - q_1 \cos x - 3x^2 q_2}{1}$$

(by the L'Hospital rule)

$$= -q_0 \times 0 - q_1 \times 1 - 0 = -q_1$$

$\because f$ is differentiable at $x = 0$,

So $Rf'(0) = Lf'(0)$ i.e; $q_1 = -q_1 \Rightarrow q_1 = 0$.

17. (a)

$f(x+y) = f(x)f(y)$

$\Rightarrow f(x) = f(x) \times f(0)$ (for $y = 0$)

$\Rightarrow f(x) \{f(0) - 1\} = 0$

$\Rightarrow f(0) - 1 = 0$ (assuming $f(x) \neq 0$)

$\Rightarrow f(0) = 0$

$f'5 = \lim_{h \to 0} \frac{f(5+h) - f(5)}{h}$

$\Rightarrow f(x)\{f(0) - 1\} = 0$

$\Rightarrow f(0) - 1 = 0$ (assuming $f(x) \neq 0$)

$\Rightarrow f(0) = 1$

$f'(5) = \lim_{h \to 0} \frac{f(5+h) - f(5)}{h}$

$= \lim_{h \to 0} \frac{f(5)f(h) - f(5)}{h}$

$= \lim_{h \to 0} \frac{2\{f(h) - 1\}}{h}$ ($\because f(5) = 2$)

$= 2 \lim_{h \to 0} \frac{f(h) - f(0)}{h}$ ($\because f(0) = 1$)

$= 2f'(0) = 2 \times 3 = 6$.

18. (b) $2f(x) + 3f(-x) = x^2 - x + 1 \ldots (i)$

Replacing x by $-x$, we get from (i),

$2f(-x) + 3f(x) = x^2 + x + 1 \ldots (ii)$

Then, $3 \times$ Eqn (ii) $- 2 \times$ Eqn (i)

$\Rightarrow 6f(-x) + 9f(x) - 4f(x) - 6f(-x)$

$= 3x^2 + 3x + 3 - 2x^2 + 2x - 2$

$\Rightarrow 5f(x) = x^2 + 5x + 1$

$\Rightarrow 5f'(x) = 2x + 5$

$\Rightarrow f'(x) = \frac{2x+5}{5} \Rightarrow f'(1) = \frac{7}{5}$.

19. (a)

$$\lim_{x \to 0+} f(x) = \lim_{x \to 0+} \frac{x}{1 + e^{\frac{1}{x}}} = \frac{0}{1 + \infty} = 0$$

$$\left(\because x \to 0+ \Rightarrow \frac{1}{x} \to \infty \Rightarrow e^{\frac{1}{x}} \to \infty \right)$$

$$\lim_{x \to 0-} f(x) = \lim_{x \to 0-} \frac{x}{1 + e^{\frac{1}{x}}} = \frac{0}{1 + 0} = 0$$

$$\left(\because x \to 0- \Rightarrow e^{\frac{1}{x}} \to 0 \right)$$

Hence, $\lim_{x \to 0+} f(x) = \lim_{x \to 0-} f(x) = 0 = f(0)$.

Consequently, f is continuous at $x = 0$.

20. (a)

$\lim_{x \to 0} f(x)$

$$= \lim_{x \to 0} \frac{1 - \cos kx}{x \sin x} \left(\text{form } \frac{0}{0} \right)$$

$$= \lim_{x \to 0} \frac{\frac{d}{dx}(1 - \cos kx)}{\frac{d}{dx}(x \sin x)}$$ (by the L'Hospital rule)

$$= \lim_{x \to 0} \frac{k \sin kx}{x \cos x + \sin x} \left(\text{form } \frac{0}{0} \right)$$

$$= \lim_{x \to 0} \frac{\frac{d}{dx}(k \sin kx)}{\frac{d}{dx}(x \cos x + \sin x)} \left(\text{form } \frac{0}{0} \right)$$

$$= \lim_{x \to 0} \frac{k^2 \cos kx}{-x \sin x + \cos x + \cos x}$$

$$= \frac{k^2 \times 1}{0 + 1 + 1} = \frac{k^2}{2}$$

Now f is continuous at $x = 0$

$\Rightarrow \lim_{x \to 0} f(x) = f(0)$

$\Rightarrow \frac{k^2}{2} = \frac{1}{2} \Rightarrow k^2 = 1 \Rightarrow k = \pm 1$.

21. (a)

$$\lim_{x \to 2} f(x)$$

$$= \lim_{x \to 2} \frac{x^3 + x^2 - 16x + 20}{(x-2)^2} \left(\text{form} \frac{0}{0} \right)$$

$$= \lim_{x \to 2} \frac{3x^2 + 2x - 16}{2(x-2)} \text{ (by the L'Hospital rule)}$$

$$= \frac{1}{2} \lim_{x \to 2} \frac{3x^2 + 2x - 16}{(x-2)} \left(\text{form} \frac{0}{0} \right)$$

$$= \frac{1}{2} \lim_{x \to 2} \frac{6x + 2}{1} \text{ (by the L'Hospital rule)}$$

$$= \frac{6 \times 2 + 2}{2} = 7.$$

Now f is continuous at $x = 2$

$$\Rightarrow \lim_{x \to 2} f(x) = f(2) \Rightarrow 7 = k.$$

22. (a) RHL at $x = 0$

$$= \lim_{x \to 0+} f(x) = \lim_{x \to 0+} \frac{xe^{\frac{1}{x}}}{1 + e^{\frac{1}{x}}} = \lim_{x \to 0+} \frac{x}{e^{-\frac{1}{x}} + 1} = \frac{0}{0+1} = 0$$

$$\left(\because x \to 0+ \Rightarrow -\frac{1}{x} \to -\infty \Rightarrow e^{-\frac{1}{x}} \to e^{-\infty} = 0 \right)$$

LHL at $x = 0$

$$= \lim_{x \to 0-} f(x) = \lim_{x \to 0-} \frac{xe^{\frac{1}{x}}}{1 + e^{\frac{1}{x}}} = \lim_{x \to 0-} \frac{x}{e^{-\frac{1}{x}} + 1} = \frac{0}{\infty + 1} = 0$$

$$\left(\because x \to 0- \Rightarrow -\frac{1}{x} \to \infty \Rightarrow e^{-\frac{1}{x}} \to e^{\infty} = \infty \right)$$

Hence, f is continuous at $x = 0$.

Again

23. (a)

$$|f(x)| \le x^2$$

$$\Rightarrow -x^2 \le f(x) \le x^2 \Rightarrow -x \le \frac{f(x)}{x} \le x$$

$$\Rightarrow \lim_{x \to 0}(-x) \le \lim_{x \to \infty} \frac{f(x) - f(0)}{x} \le \lim_{x \to 0} x$$

$$\Rightarrow 0 \le f'(0) \le 0 \ [\because |f(0)| \le 0^2 \Rightarrow f(0) = 0]$$

$$\Rightarrow f'(0) = 0 \text{ [by the Sandwich theorem]}$$

$$Lf'(0) = \lim_{x \to 0-} \frac{f(x) - f(0)}{x}$$

$$= \lim_{x \to 0-} \frac{\frac{xe^{\frac{1}{x}}}{1 + e^{\frac{1}{x}}} - 0}{x} = \lim_{x \to 0-} \frac{1}{e^{-\frac{1}{x}} + 1} = 0,$$

$$Rf'(0) = \lim_{x \to 0+} \frac{f(x) - f(0)}{x}$$

$$= \lim_{x \to 0+} \frac{\frac{xe^{\frac{1}{x}}}{1 + e^{\frac{1}{x}}} - 0}{x} \lim_{x \to 0+} \frac{1}{e^{-\frac{1}{x}} + 1} = 1.$$

Since $Lf'(0) \ne Rf'(0)$, so $f'(0)$ does not exist.

24. (a) Let k = 0

$$Rf'(0) = \lim_{x \to 0+} \frac{f(x) - f(0)}{x}$$

$$= \lim_{x \to 0+} \frac{xe^{-\left(\frac{1}{x} + \frac{1}{|x|} \right)} - 0}{x}$$

$$= \lim_{x \to 0+} \frac{xe^{-\frac{2}{x}} - 0}{x} = \lim_{x \to 0+} e^{-\frac{2}{x}} = 0,$$

$$Lf'(0) = \lim_{x \to 0-} \frac{f(x) - f(0)}{x}$$

$$= \lim_{x \to 0-} \frac{xe^{-\left(\frac{1}{x} + \frac{1}{|x|} \right)} - 0}{x}$$

$$= \lim_{x \to 0-} \frac{xe^{-0} - 0}{x} = \lim_{x \to 0-} 1 = 1.$$

$\because Rf'(0) \ne Lf'(0)$, so f is not differentiable for k = 0. It is easy to verify that $f(x)$ is not continuous for $k \ne 0$ and hence not differentiable for $k \ne 0$.

25. (b) $Rf'(0)$

$$= \lim_{x \to 0+} \frac{f(x) - f(0)}{x}$$

$$= \lim_{x \to 0+} \left[\frac{e^{-\frac{1}{x^2}} - 0}{x} \right] = \lim_{x \to 0+} \left\{ \frac{1}{xe^{\frac{1}{x^2}}} \right\}$$

$$= \lim_{x \to 0+} \left\{ \frac{1}{x\left(1 + \frac{1}{x^2} + \frac{1}{2!} \times \frac{1}{x^4} + \cdots \cdots\right)} \right\}$$

$$= \lim_{x \to 0+} \left\{ \frac{1}{x + \frac{1}{x} + \frac{1}{2x^3} + \cdots \cdots} \right\} = \frac{1}{0 + \infty} = 0.$$

Similarly $Lf'(0) = 0$. $\therefore Rf'(0) = Lf'(0)$.

Hence, f is differentiable at $x = 0$.

26. (b)

$$\lim_{x \to 0+} f(x)$$

$$= \lim_{x \to 0+} \frac{x^2(e^{\frac{1}{x}} - e^{-\frac{1}{x}})}{e^{\frac{1}{x}} + e^{-\frac{1}{x}}} = \lim_{x \to 0+} \frac{x^2(1 - e^{-\frac{2}{x}})}{(1 + e^{-\frac{2}{x}})} = \frac{0(1-0)}{(1+0)} = 0$$

$$\left(\because x \to 0+ \Rightarrow -\frac{2}{x} \to -\infty \Rightarrow e^{-\frac{2}{x}} \to e^{-\infty} \Rightarrow e^{-\frac{2}{x}} \to 0 \right)$$

$$\lim_{x \to 0-} f(x)$$

$$= \lim_{x \to 0-} \frac{x^2(e^{\frac{1}{x}} - e^{-\frac{1}{x}})}{e^{\frac{1}{x}} + e^{-\frac{1}{x}}} = \lim_{x \to 0-} \frac{x^2(e^{\frac{2}{x}} - 1)}{(e^{\frac{2}{x}} + 1)} = \frac{0(0-1)}{(0+1)} = 0$$

$$\left(\because x \to 0- \Rightarrow \frac{2}{x} \to -\infty \Rightarrow e^{\frac{2}{x}} \to e^{-\infty} \Rightarrow e^{\frac{2}{x}} \to 0 \right)$$

Thus, $\lim_{x \to 0+} f(x) = \lim_{x \to 0-} f(x) = 0 = f(0)$, so f is continuous at $x = 0$.

27. (b)

$$\lim_{x \to 0} \frac{\sin\left(\pi \cos^2 x\right)}{x^2} \quad \left(\text{form } \frac{0}{0} \right)$$

$$= \lim_{x \to 0} \frac{\frac{d}{dx}\left\{\sin\left(\pi \cos^2 x\right)\right\}}{\frac{d}{dx}\left\{x^2\right\}} \quad \text{(by the L'Hospital Rule)}$$

$$= \lim_{x \to 0} \frac{\cos\left(\pi \cos^2 x\right) \times \pi \times 2\cos x \times (-\sin x)}{2x}$$

$$= \lim_{x \to 0} \left\{ \frac{2e^{2x} - e^x - 1}{2x} \right\} \quad \left(\text{form } \frac{0}{0} \right)$$

$$= \lim_{x \to 0} \frac{\frac{d}{dx}\left(2e^{2x} - e^x - 1\right)}{\frac{d}{dx}(2x)} \quad \text{(by the L'Hospital rule)}$$

$$= \lim_{x \to 0} \left\{ \frac{4e^{2x} - e^x}{2} \right\} = \frac{4e^0 - e^0}{2} = \frac{4-1}{2} = \frac{3}{2}$$

$\therefore k = 2.$

28. (a)

$$\lim_{x \to 3} \left\{ \frac{x-3}{\log_a(x-2)} \right\}$$

$$= \lim_{x \to 3} \left\{ \frac{x-3}{\log_a e \times \log_e(x-2)} \right\} \quad \left(\text{form } \frac{0}{0} \right)$$

$$= \frac{1}{\log_a e} \times \lim_{x \to 3} \frac{\frac{d}{dx}(x-3)}{\frac{d}{dx}(\log_e(x-2))}$$

(by the L'Hospital rule)

$$= \frac{1}{\log_a e} \times \lim_{x \to 3} \left\{ \frac{1}{\frac{1}{(x-2)}} \right\}$$

$$= \frac{1}{\log_a e} \lim_{x \to 3}(x-2) = \frac{1}{\log_a e} \times 1 = \log_e a.$$

29. (c)

$$\lim_{x \to \infty} \left\{ \frac{x^2 + x + 1}{x + 1} - ax - b \right\}$$

$$= \lim_{x \to \infty} \left\{ \frac{x^2 + x + 1 - ax^2 - ax - bx - b}{x + 1} \right\}$$

30. (b)

$Let y = \left\{ \frac{f(x+1)}{f(1)} \right\}^{\frac{1}{x}}$. Then

$$y = \left\{ \frac{f(x+1)}{f(1)} \right\}^{\frac{1}{x}}$$

Let $y = \left\{ \frac{f(x+1)}{f(1)} \right\}^{\frac{1}{x}}$. Then

$$y = \left\{ \frac{f(x+1)}{f(1)} \right\}^{\frac{1}{x}}$$

$$\Rightarrow \log y = \frac{1}{x} \log \left\{ \frac{f(x+1)}{f(1)} \right\}$$

$$\Rightarrow \log y = \frac{\log[f(x+1)] - \log[f(1)]}{x}$$

$$\Rightarrow \lim_{x \to 0}(\log y)$$

$$= \lim_{x \to 0} \left\{ \frac{\log[f(x+1)] - \log[f(1)]}{x} \right\} \left(\text{form } \frac{0}{0} \right)$$

$$\Rightarrow \log\left(\lim_{x\to 0} y\right) = \lim_{x\to 0}\left[\frac{\dfrac{f'(x+1)}{f(x+1)}-0}{1}\right]$$

(by the L'Hospital rule)

$$= \frac{f'(0+1)}{f(0+1)} = \frac{f'(1)}{f(1)} = \frac{6}{3} = 2$$

$$\Rightarrow \lim_{x\to 0} y = e^2 \Rightarrow \lim_{x\to 0}\left\{\frac{f(x+1)}{f(1)}\right\}^{\frac{1}{x}} = e^2.$$

$$= \lim_{x\to\infty}\left\{\frac{(1-a)x^2+(1-a-b)x+(1-b)}{x+1}\right\}$$

$$= \lim_{x\to\infty}\left\{\frac{(1-a)+\dfrac{(1-a-b)}{x}+\dfrac{(1-b)}{x^2}}{\dfrac{1}{x}+\dfrac{1}{x^2}}\right\}$$

$$\left(\text{form } \frac{0}{0} \text{ if } 1-a=0 \text{ i.e; if } a=1\right)$$

$$= \lim_{x\to\infty}\left\{\frac{\dfrac{(-b)}{x}+\dfrac{(1-b)}{x^2}}{\dfrac{1}{x}+\dfrac{1}{x^2}}\right\}\left(\text{form } \frac{0}{0}\right)$$

$$= \lim_{x\to\infty}\left\{\frac{-b+\dfrac{(1-b)}{x}}{1+\dfrac{1}{x}}\right\}$$

$$= \frac{-b+0}{1+0} = -b = 4 \text{ (given)}$$

$$\therefore b = -4;\ a = 1.$$

31. (c)

Let $f(x) = \sin x$ and $g(x) = \cos x$. Then clearly the functions f and g satisfy the Cauchy's Mean Value Theorem on $[\theta_1, \theta_2]$. Hence, $\exists\ \theta \in (\theta_1, \theta_2)$

such that $\dfrac{f'(\theta)}{g'(\theta)} = \dfrac{f(\theta_2)-f(\theta_1)}{g(\theta_2)-g(\theta_1)}$.

Now, $f'(x) = \cos x$ and $g'(x) = -\sin x$.

$$\therefore \frac{f'(\theta)}{g'(\theta)} = \frac{f(\theta_2)-f(\theta_1)}{g(\theta_2)-g(\theta_1)}$$

$$\Rightarrow \frac{\cos\theta}{-\sin\theta} = \frac{\sin\theta_2 - \sin\theta_1}{\cos\theta_2 - \cos\theta_1}$$

$$\Rightarrow -\cot\theta = \frac{\sin\theta_2 - \sin\theta_1}{\cos\theta_2 - \cos\theta_1}$$

$$\Rightarrow \cot\theta = \frac{\sin\theta_2 - \sin\theta_1}{\cos\theta_1 - \cos\theta_2}.$$

32. (b)

Let $\phi(x) = \dfrac{f(x)}{g(x)}$. Then clearly ϕ satisfies all the conditions of the Lagrange's Mean Value Theorem on [a, b]. Then there exist a point $c \in (a, b)$ such that $\phi'(c) = \dfrac{\phi(b)-\phi(a)}{b-a}$.

Now, $\phi'(c) = \dfrac{\phi(b)-\phi(a)}{b-a}$

$$\Rightarrow \frac{g(c)f'(c)-f(c)g'(c)}{(g(c))^2} = \frac{\dfrac{f(b)}{g(b)}-\dfrac{f(a)}{g(a)}}{b-a}$$

$$\left[\because \phi'(x) = \frac{g(x)f'(x)-f(x)g'(x)}{(g(x))^2}\right]$$

$$\Rightarrow \frac{g(c)f'(c)-f(c)g'(c)}{(g(c))^2} = \frac{f(b)g(a)-f(a)g(b)}{(b-a)g(a)g(b)}$$

$$\Rightarrow \frac{g(a)f(b)-f(a)g(b)}{g(c)f'(c)-f(c)g'(c)} = \frac{(b-a)g(a)g(b)}{(g(c))^2}.$$

33. (a)

Let $f(x) = \dfrac{1}{3}px^3 + \dfrac{1}{2}qx^2 + rx + s.$

Then $f(0) = s$ and

$$f(1) = \frac{1}{3}p + \frac{1}{2}q + r + s = \frac{2p+3q+6r}{6} + s$$

$$= \frac{0}{6} + s = s$$

$$\therefore f(0) = f(1).$$

Now f being a polynomial function, is continuous on [0, 1] and differentiable on (0, 1). Hence, f satisfies all the conditions of Rolle's theorem. Therefore, there exists a point $c \in (0,1)$ such that $f'(c) = 0$. But $f'(x) = px^2 + qx + r$. Consequently, at least one root of the equation $px^2 + qx + r = 0$ lies in the interval (0, 1).

34. (c) Since a, b are two consecutive roots of the equation

$$p(x) = 0, \text{ so } p(a) = 0 = p(b).$$

Let $f(x) = e^{50x} p(x)$ and then

$$f'(x) = 50e^{50x} p(x) + e^{50x} p'(x).$$

Therefore, f is continuous on [a, b] and differentiable on (a, b). Also

$$f(a) = e^{50a} p(a) = 0 \ (\because p(a) = 0),$$
$$f(b) = e^{50b} p(b) = 0 \ (\because p(b) = 0).$$

Thus, f satisfies all the conditions of Rolle's theorem on [a, b]. Hence, there exists a point $c \in (a, b)$ such that $f'(c) = 0$.

Now $f'(c) = 0$

$$\Rightarrow 50e^{50c} p(c) + e^{50c} p'(c) = 0$$
$$\Rightarrow 50p(c) + p'(c) = 0.$$

35. (a) Let $g(x) = \dfrac{f(x)}{x}$ and $h(x) = \dfrac{1}{x}$. Then the functions $g, h : [a, b] \to R$ are continuous on $[a, b]$ and differentiable on (a, b). Hence, by the Cauchy's Mean Value Theorem on [a, b], there exists a point $c \in (a, b)$ such that

$$\frac{g'(c)}{h'(c)} = \frac{g(b) - g(a)}{h(b) - h(a)} \quad \ldots\ldots\ldots(i)$$

Here, $g'(x) = \dfrac{xf'(x) - f(x)}{x^2}$ and $h'(x) = -\dfrac{1}{x^2}$.

Therefore, (i) $\Rightarrow \dfrac{g'(c)}{h'(c)} = \dfrac{g(b) - g(a)}{h(b) - h(a)}$

$$\Rightarrow \frac{\dfrac{cf'(c) - f(c)}{c^2}}{-\dfrac{1}{c^2}} = \frac{\dfrac{f(b)}{b} - \dfrac{f(a)}{a}}{\dfrac{1}{b} - \dfrac{1}{a}}$$

$$\Rightarrow -[cf'(c) - f(c)] = \frac{af(b) - bf(a)}{a - b}$$

$$\Rightarrow \frac{bf(a) - af(b)}{b - a} = f(c) - cf'(c).$$

36. (d) Let $g(x) = f(x) - x^2$. Then

$$g(1) = f(1) - 1^2 = 1 - 1 = 0,$$
$$g(2) = f(2) - 2^2 = 4 - 4 = 0,$$
$$g(4) = f(4) - 4^2 = 16 - 16 = 0.$$

Hence, g satisfies all the conditions of Rolle's theorem on [1, 2] and [2, 4]. Therefore, there exists points $c_1 \in (1, 2)$ and $c_2 \in (2, 4)$ such that $g'(c_1) = 0$ and $g'(c_2) = 0$.

Now consider the function g'. Then clearly g' satisfies all the conditions of the Lagrange's Mean Value Theorem on $[c_1, c_2]$.

Hence, there exists a point $c \in [c_1, c_2]$ such that

$$g''(c) = \frac{g'(c_2) - g'(c_1)}{c_2 - c_1}.$$

But since $g'(c_1) = 0 = g'(c_2)$, so $g''(c) = 0$ for some $c \in [c_1, c_2]$.

Now $g''(c) = 0 \Rightarrow f''(c) - 2 = 0 \Rightarrow f''(c) = 2$.

Hence, we can conclude that $f''(x) = 2$ for some $x \in (1, 4)$.

37. (b) $f(x) = \sin x + \cos x$

$$= \sqrt{2}\left(\frac{1}{\sqrt{2}}\sin x + \frac{1}{\sqrt{2}}\cos x\right)$$

$$= \sqrt{2}\cos\left(x - \frac{\pi}{4}\right)$$

$\therefore f$ is strictly increasing

$$\Leftrightarrow f'(x) > 0$$

$$\Leftrightarrow -\sqrt{2}\sin\left(x - \frac{\pi}{4}\right) > 0$$

$$\Leftrightarrow \sin\left(x - \frac{\pi}{4}\right) < 0 \Leftrightarrow \pi < x - \frac{\pi}{4} < 2\pi$$

$$\Leftrightarrow \pi + \frac{\pi}{4} < x < 2\pi + \frac{\pi}{4} \Leftrightarrow \frac{5\pi}{4} < x < \frac{9\pi}{4}$$

$$\Leftrightarrow \frac{5\pi}{4} < x < 2\pi \text{ or } 2\pi < x < \frac{9\pi}{4}$$

$$\Leftrightarrow x \in \left(\frac{5\pi}{4}, 2\pi\right) \cup \left(0, \frac{\pi}{4}\right)$$

$$\left[\because 2\pi < x < \frac{9\pi}{4} \Rightarrow 0 < x < \frac{\pi}{4}\right]$$

38. (a)

$$f(x) = 2x^3 - 15x^2 + 36x + 1$$
$$\Rightarrow f'(x) = 6x^2 - 30x + 36$$

$\therefore f$ is strictly increasing

$$\Leftrightarrow f'(x) > 0$$

$$\Leftrightarrow 6x^2 - 30x + 36 > 0$$

$$\Leftrightarrow x^2 - 5x + 6 > 0 \Leftrightarrow (x - 2)(x - 3) > 0$$

$$\Leftrightarrow (x - 2) > 0, (x - 3) > 0$$

or

$(x-2)<0, (x-3)<0$

$\Leftrightarrow x>2,\ x>3$ or $x<2,\ x<3$

$\Leftrightarrow x>3$ or $x<2 \Leftrightarrow x\in(-\infty,2)\cup(3,\infty)$

Again, f is strictly decreasing

$\Leftrightarrow f'(x)<0$

$\Leftrightarrow 6x^2-30x+36<0$

$\Leftrightarrow x^2-5x+6<0 \Leftrightarrow (x-2)(x-3)<0$

$\Leftrightarrow (x-2)>0, (x-3)<0$

or

$(x-2)<0, (x-3)>0$

$\Leftrightarrow x>2,\ x<3$ or $x<2,\ x>3$

$\Leftrightarrow x>2,\ x<3 \Leftrightarrow x\in(2,3)$

39. (b)

$f''(x)>0 \Rightarrow f'(x)$ is strictly increasing.

Now, $g(x)=\dfrac{f(x)}{x}$

$\Rightarrow g'(x)=\dfrac{xf'(x)-f(x)}{x^2}$

$=\dfrac{f'(x)-\dfrac{f(x)}{x}}{x}$ ....(i)

By the Lagrange's Mean Value Theorem on $[0,x]$, $\exists$

$c\in(0,x)$ such that $f'(c)=\dfrac{f(x)-f(0)}{x-0}=\dfrac{f(x)}{x}$.

This implies $f'(c)=\dfrac{f(x)}{x}=g(x)<f'(x)$

$\begin{bmatrix} \because f'(x) \text{ is strictly increasing and} \\ 0<c<x \Rightarrow f'(c)<f'(x) \end{bmatrix}$

Now $g(x)<f'(x)$

$\Rightarrow \dfrac{f(x)}{x}<f'(x) \Rightarrow f'(x)-\dfrac{f(x)}{x}>0$

$\Rightarrow g'(x)>0$ (using (i))

Consequently, $g(x)$ is increasing on $(0,\infty)$.

40. (d)

$f(x)=\sin^4 x+\cos^4 x$

$\Rightarrow f'(x)=4\sin^3 x\cos x+4\cos^3 x(-\sin x)$

$=-4\sin x\cos x\,(\cos^2 x-\sin^2 x)$

$=-2\sin 2x\cos 2x=-\sin 4x$

$\therefore f'(x)=0 \Rightarrow -\sin 4x=0 \Rightarrow \sin 4x=\sin\pi$

$\Rightarrow 4x=\pi \Rightarrow x=\dfrac{\pi}{4}\in\left(0,\dfrac{\pi}{2}\right)$

Also, $f''(x)=-4\cos 4x$ and so

$f''\left(\dfrac{\pi}{4}\right)=-4\cos\pi=-4\times(-1)>0$

Therefore, $x=\dfrac{\pi}{4}$ is a point of local minima of the given function.

41. (a)

$f(x)=\left(\dfrac{1}{x}\right)^x$

$\Rightarrow \log f(x)=\log\left(\dfrac{1}{x}\right)^x=x\log\left(\dfrac{1}{x}\right)$

$\Rightarrow \log f(x)=x\{\log 1-\log x\}=-x\log x$

$\Rightarrow \dfrac{d}{dx}[\log f(x)]=-\dfrac{d}{dx}[x\log x]$

$\Rightarrow \dfrac{f'(x)}{f(x)}=-\log x-x\times\dfrac{1}{x}=-\log x-1$

$\Rightarrow f'(x)=-f(x)\times\{\log x+1\}$

$\Rightarrow \dfrac{f'(x)}{f(x)}=-\log x-x\times\dfrac{1}{x}=-\log x-1$

$\Rightarrow f'(x)=-f(x)\times\{\log x+1\}$

$\therefore f'(x)=0$

$\Rightarrow \log x+1=0$

$\Rightarrow \log x=-1=-\log e=\log\left(\dfrac{1}{e}\right)$

$\Rightarrow x=\dfrac{1}{e}$

Now $f''(x)=-f'(x)\times\{\log x+1\}+\dfrac{1}{x}f(x)$.

$\therefore f''\left(\dfrac{1}{e}\right)=-f'\left(\dfrac{1}{e}\right)\times\left\{\log\left(\dfrac{1}{e}\right)+1\right\}-\dfrac{1}{\left(\dfrac{1}{e}\right)}f\left(\dfrac{1}{e}\right)$

$=-e\times f\left(\dfrac{1}{e}\right)=-e\times e^{\frac{1}{e}}<0\left[\because f'\left(\dfrac{1}{e}\right)=0\right]$

Hence, the given function has a local maximum value at $x=\dfrac{1}{e}$. Therefore, the local maximum value of the function $=f\left(\dfrac{1}{e}\right)=e^{\frac{1}{e}}$.

42. (a)

$$ab^3 = (60-b)b^3 \quad (\because a+b=60)$$

$$= 60b^3 - b^4 = f(b) \text{ (say)}$$

$$\therefore f'(b) = 0 \Rightarrow 180b^2 - 4b^3 = 0$$

$$\Rightarrow b = 45$$

$$\therefore a = 60 - 45 = 15.$$

Now $f''(b) = 360b - 12b^2$ and so

$$f''(45) = 360 \times 45 - 12 \times 45^2 < 0$$

Hence, $f(b)$ i.e; ab^3 is maximum when $a = 15$ and $b = 45$.

43. (c) Let the length, breadth, and height of the box be, respectively, x, x and y units. Then ATQ,

$$x^2 + xy + xy + xy + xy = c^2$$

or, $x^2 + 4xy = c^2$ or, $y = \dfrac{c^2 - x^2}{4x}$

The volume of the box, V

$$= x \times x \times y = x^2 y$$

$$= x^2 \left(\frac{c^2 - x^2}{4x} \right) = \frac{c^2 x - x^3}{4}$$

$$\therefore \frac{dV}{dx} = 0 \Rightarrow \frac{c^2 - 3x^2}{4} = 0 \Rightarrow x = \frac{c}{\sqrt{3}}$$

Also, $\dfrac{d^2 V}{dx^2} = \dfrac{0 - 6x}{4}$ and so

$$\left. \frac{d^2 V}{dx^2} \right|_{x=\frac{c}{\sqrt{3}}} = \frac{-6}{4} \times \frac{c}{\sqrt{3}} < 0.$$

Thus, volume V is maximum when $x = \dfrac{c}{\sqrt{3}}$. Hence, maximum volume, V =

$$\frac{c^2 \times \dfrac{c}{\sqrt{3}} - \left(\dfrac{c}{\sqrt{3}} \right)^3}{4} = \frac{c^3}{6\sqrt{3}} \text{ cubic units.}$$

44. (a) Let r be the radius of the circle and x be the length of each side of the square. Then ATQ,

$$2\pi r + 4x = 36 \Rightarrow r = \frac{18 - 2x}{\pi}$$

$\therefore$ Combined area, $A = \pi r^2 + x^2$

$$= \pi \left(\frac{18 - 2x}{\pi} \right)^2 + x^2$$

Then, $\dfrac{dA}{dx} = 2\pi \left(\dfrac{18 - 2x}{\pi^2} \right)(-2) + 2x$

$$= -\left(\frac{72\pi - 8x}{\pi} \right) + 2x$$

$$\therefore \frac{dA}{dx} = 0 \Rightarrow -\left(\frac{72\pi - 8x}{\pi} \right) + 2x = 0$$

$$\Rightarrow x = \frac{36}{\pi + 4}$$

Now, $\dfrac{d^2 A}{dx^2} = \dfrac{8}{\pi} + 2;$ so $\left. \dfrac{d^2 A}{dx^2} \right|_{x=\frac{36}{\pi+4}} > 0.$

Hence, the combined area, A is minimum when $x = \dfrac{36}{\pi + 4}$.

Now $x = \dfrac{36}{\pi + 4}$

$$\Rightarrow r = \frac{18 - 2 \times \left(\dfrac{36}{\pi + 4} \right)}{\pi} = \frac{18}{\pi + 4}.$$

$$\therefore 2\pi r = 2\pi \times \frac{18}{(\pi + 4)} = \frac{36\pi}{\pi + 4} \text{ and }$$

$$4x = 4 \times \frac{36}{\pi + 4} = \frac{144}{\pi + 4}.$$

Hence, the combined area, A is minimum when the lengths of the two pieces are $\dfrac{36\pi}{\pi + 4}$ and $\dfrac{144}{\pi + 4}$.

45. (d)

$$\sum_{n=1}^{\infty}\frac{3n^2-2}{n!}$$

$$=3\sum_{n=1}^{\infty}\frac{n^2}{n!}-2\sum_{n=1}^{\infty}\frac{1}{n!}$$

$$=3\left[\frac{1^2}{1!}+\sum_{n=2}^{\infty}\frac{n^2}{n!}\right]-2\sum_{n=1}^{\infty}\frac{1}{n!}$$

$$=3\left[1+\sum_{n=2}^{\infty}\frac{n}{(n-1)!}\right]-2\left\{\sum_{n=1}^{\infty}\left(\left(1+\frac{1}{n!}\right)-1\right)\right\}$$

$$=3\left[1+\sum_{n=2}^{\infty}\frac{(n-1)+1}{(n-1)!}\right]-2(e-1)$$

$$\left[\because e=1+\frac{1}{1!}+\frac{1}{2!}+\dots\dots\infty=1+\sum_{n=1}^{\infty}\frac{1}{n!}\right]$$

$$=3\left[1+\sum_{n=2}^{\infty}\frac{1}{(n-2)!}+\sum_{n=2}^{\infty}\frac{1}{(n-1)!}\right]-2(e-1)$$

$$=3\left[\begin{array}{c}1+\left(\frac{1}{0!}+\frac{1}{1!}+\frac{1}{2!}+\dots\dots+\infty\right)\\+\left(\frac{1}{1!}+\frac{1}{2!}+\dots\dots+\infty\right)\end{array}\right]$$
$$-2(e-1)$$

$$=3\left[1+e+(e-1)\right]-2e+2=4e+2.$$

46. (a) By the Taylor series expansion about x = 3,

$$f(x)=f(3)+(x-3)f'(3)+\frac{(x-3)^2}{2!}f''(3)$$
$$+\frac{(x-3)^3}{3!}f'''(3)+\frac{(x-3)^4}{4!}f^{iv}(3)$$
$$+\frac{(x-3)^5}{5!}f^{v}(3)$$

Here,

$$f(x)=e^{\frac{x}{2}},\text{ so }f^{v}(x)=\frac{1}{2}\times\frac{1}{2}\times\frac{1}{2}\times\frac{1}{2}\times\frac{1}{2}e^{\frac{x}{2}}=\frac{e^{\frac{x}{2}}}{32}$$

$$\therefore f^{v}(3)=\frac{e^{\frac{3}{2}}}{32}.$$

Hence, **coefficient of** $(x-3)^5$ in the expansion

$$=\frac{1}{5!}\times f^{v}(3)=\frac{1}{5!}\times\frac{e^{\frac{3}{2}}}{32}.$$

47. (d)

$$f(x)=(1-x)^{\frac{5}{2}}\Rightarrow f'(x)=-\frac{5}{2}(1-x)^{\frac{3}{2}},$$

$$f''(x)=\frac{15}{4}(1-x)^{\frac{1}{2}}.$$

$$\therefore f'(0)=-\frac{5}{2}\text{ and }f''(\theta h)=\frac{15}{4}(1-\theta h)^{\frac{1}{2}}.$$

Then $f(h)=f(0)+hf'(0)+\frac{h^2}{2!}f''(\theta h)$

$$\Rightarrow(1-h)^{\frac{5}{2}}=1+h\times\left(-\frac{5}{2}\right)+\frac{h^2}{2}\times\frac{15}{4}(1-\theta h)^{\frac{1}{2}}$$

$$\left(\because f(0)=(1-0)^{\frac{5}{2}}=1\right)$$

$$\Rightarrow(1-1)^{\frac{5}{2}}=1-\frac{5}{2}+\frac{15}{8}(1-\theta)^{\frac{1}{2}}\ \text{(for }h=1)$$

$$\Rightarrow\frac{15}{8}\sqrt{1-\theta}=\frac{5}{2}-1=\frac{3}{2}$$

$$\Rightarrow\sqrt{1-\theta}=\frac{3}{2}\times\frac{8}{15}=\frac{4}{5}$$

$$\Rightarrow\theta=1-\frac{16}{25}=\frac{9}{25}.$$

48. (d)

(a) Let, $f(x)=e^x$. Then by Taylor's theorem,

$$f(x)=f(0)+xf'(\theta x)$$

$$\Rightarrow e^x=e^0+xe^{\theta x}=1+xe^{\theta x}$$

$$\Rightarrow e^{\theta x}=\frac{1}{x}(e^x-1)$$

$$\Rightarrow\log e^{\theta x}=\log\left(\frac{e^x-1}{x}\right)$$

$$\Rightarrow\theta x=\log\left(\frac{e^x-1}{x}\right)\Rightarrow\theta=\frac{1}{x}\log\left(\frac{e^x-1}{x}\right)$$

Now $0<\theta<1\Rightarrow0<\frac{1}{x}\log\left(\frac{e^x-1}{x}\right).$

Hence, (a) is correct.

(b) Let $f(x) = \log(1+x)$. Then $f'(x) = \dfrac{1}{1+x}$.

$\therefore$ By Taylor's theorem,

$f(x) = f(0) + xf'(\theta x)$

$\Rightarrow \log(1+x) = \log 1 + x \times \dfrac{1}{(1+\theta x)} = \dfrac{x}{1+\theta x}$

Now $0 < \theta < 1 \Rightarrow 0 < \theta x < x \Rightarrow 1 < 1+\theta x < 1+x$

$\Rightarrow \dfrac{1}{1+x} < \dfrac{1}{1+\theta x} < 1 \; (\because x > 0)$

$\Rightarrow \dfrac{x}{1+x} < \log(1+x) < x$

$\therefore$ (b) is correct.

(c) Let $f(x) = \log(1+x)$.

Then $f'(x) = \dfrac{1}{1+x}, f'(x) = -\dfrac{1}{(1+x)^2}$

$\therefore$ By Taylor's theorem,

$f(x) = f(0) + xf'(0) + \dfrac{x^2}{2!}f''(\theta x)$

$\Rightarrow \log(1+x) = 0 + x \times 1 + \dfrac{x^2}{2} \times \left\{-\dfrac{1}{(1+\theta x)^2}\right\}$

$\left(\because f(0) = \log 1 = 0, f'(0) = \dfrac{1}{1+0} = 1\right)$

$\qquad = x - \dfrac{x^2}{2(1+\theta x)^2}$

Then,

$\theta > 0 \Rightarrow \theta x > 0 \Rightarrow (1+\theta x)^2 > 1$

$\Rightarrow -\dfrac{1}{(1+\theta x)^2} > -1$

$\Rightarrow -\dfrac{x^2}{2(1+\theta x)^2} > -\dfrac{x^2}{2}$

$\Rightarrow x - \dfrac{x^2}{2(1+\theta x)^2} > x - \dfrac{x^2}{2}$

$\Rightarrow \log(1+x) > x - \dfrac{x^2}{2} \quad ....(i)$

Again, by Taylor's theorem,

$f(x) = f(0) + xf'(\theta x)$

$\Rightarrow \log(1+x) = \log 1 + x \times \dfrac{1}{1+\theta x} = \dfrac{x}{1+\theta x}$.

$\therefore \theta > 0 \Rightarrow \theta x > 0 \Rightarrow 1+\theta x > 1$

$\Rightarrow \dfrac{1}{1+\theta x} < 1 \Rightarrow \dfrac{x}{1+\theta x} < x \; (\because x > 0)$

$\Rightarrow \log(1+x) < x ....(ii)$

Combining (i) and (ii), we get

$x > \log(1+x) > x - \dfrac{x^2}{2}$

$\therefore$ (c) is correct.

49. (b)

$I_n = \displaystyle\int_0^1 x^n \tan^{-1} x \, dx = \int_0^1 x^{n-1}\left(x\tan^{-1}x\right)dx$

$= \left[x^{n-1}\displaystyle\int x\tan^{-1}x\,dx\right]_0^1$

$\quad -\displaystyle\int_0^1\left[\left\{\dfrac{d}{dx}x^{n-1}\right\}\times\int x\tan^{-1}x\,dx\right]dx ....(i)$

But $\displaystyle\int x\tan^{-1}x\,dx$

$= \tan^{-1}x\displaystyle\int x\,dx - \int\left[\left\{\dfrac{d}{dx}\tan^{-1}x\right\}\int x\,dx\right]dx$

$= \dfrac{x^2}{2}\tan^{-1}x - \displaystyle\int\left[\left\{\dfrac{1}{1+x^2}\right\}\dfrac{x^2}{2}\right]dx$

$= \dfrac{x^2}{2}\tan^{-1}x - \dfrac{1}{2}\displaystyle\int\dfrac{(x^2+1)-1}{1+x^2}dx$

$= \dfrac{x^2}{2}\tan^{-1}x - \dfrac{1}{2}\displaystyle\int\left\{1-\dfrac{1}{1+x^2}\right\}dx$

$= \dfrac{x^2}{2}\tan^{-1}x - \dfrac{1}{2}\left\{x-\tan^{-1}x\right\}$

$\therefore$ (i) $\Rightarrow$

$I_n = \left[x^{n-1}\left(\dfrac{x^2}{2}\tan^{-1}x - \dfrac{1}{2}\left\{x-\tan^{-1}x\right\}\right)\right]_0^1$

$\quad -\displaystyle\int_0^1\left[(n-1)x^{n-2}\left(\dfrac{x^2}{2}\tan^{-1}x - \dfrac{x}{2} + \dfrac{\tan^{-1}x}{2}\right)\right]dx$

$$\Rightarrow I_n = \frac{1}{2}\tan^{-1}1 - \frac{1}{2}\{1-\tan^{-1}1\}$$

$$-\frac{(n-1)}{2}\int_0^1 x^n \tan^{-1}x\,dx + \frac{(n-1)}{2}\int_0^1 x^{n-1}dx$$

$$-\frac{(n-1)}{2}\int_0^1 x^{n-2}\tan^{-1}x\,dx$$

$$\Rightarrow I_n = \frac{1}{2}\times\frac{\pi}{4} - \frac{1}{2}\left\{1-\frac{\pi}{4}\right\} - \frac{(n-1)}{2}I_n + \frac{(n-1)}{2}\times\frac{1}{n}$$

$$-\frac{(n-1)}{2}I_{n-2}$$

$$\Rightarrow \left\{1+\frac{(n-1)}{2}\right\}I_n = \frac{\pi}{4} - \frac{1}{2} + \frac{(n-1)}{2n} - \frac{(n-1)}{2}I_{n-2}$$

$$\Rightarrow (n+1)I_n + (n-1)I_{n-2} = \frac{\pi}{2} - 1 + \frac{(n-1)}{n} = \frac{\pi}{2} - \frac{1}{n}.$$

50. (b)

$$u = f'(x)\cos x - f''(x)\sin x$$

$$\Rightarrow \frac{du}{dx} = f''(x)\cos x - f'(x)\sin x - f'''(x)\cos x$$

$$- f'''(x)\sin x$$

$$\Rightarrow \frac{du}{dx} = -\{f'(x)+f'''(x)\}\sin x$$

Again, $v = f'(x)\sin x + f''(x)\cos x$

$$\Rightarrow \frac{dv}{dx} = f''(x)\sin x + f'(x)\cos x - f''(x)\sin x$$

$$+ f'''(x)\cos x$$

$$\Rightarrow \frac{dv}{dx} = \{f'(x)+f'''(x)\}\cos x$$

$$\therefore \left(\frac{du}{dx}\right)^2 + \left(\frac{dv}{dx}\right)^2$$

$$= \{f'(x)+f''(x)\}^2 \sin^2 x + \{f'(x)+f''(x)\}^2 \cos^2 x$$

$$= \{f'(x)+f''(x)\}^2$$

Hence, $\displaystyle\int \sqrt{\left(\frac{du}{dx}\right)^2 + \left(\frac{dv}{dx}\right)^2}\,dx = f'(x)+f''(x)+C$

51. (c)

$$\int_{-1}^{2}|x\sin\pi x|\,dx = \frac{k}{\pi^2}$$

$$|x\sin\pi x| = \begin{cases} x\sin\pi x, & \text{for } -1\le x\le 1 \\ x(-\sin\pi x), & \text{for } 1\le x\le 2 \end{cases}$$

$$\int_{-1}^{2}|x\sin\pi x|\,dx = \frac{k}{\pi^2}$$

$$|x\sin\pi x| = \begin{cases} x\sin\pi x, & \text{for } -1\le x\le 1 \\ x(-\sin\pi x), & \text{for } 1\le x\le 2 \end{cases}$$

$$\therefore \int_{-1}^{2}|x\sin\pi x|\,dx$$

$$= \int_{-1}^{1}|x\sin\pi x|\,dx + \int_{1}^{2}|x\sin\pi x|\,dx$$

$$= \int_{-1}^{1} x\sin\pi x\,dx + \int_{1}^{2} x(-\sin\pi x)\,dx \quad\ldots\ldots(i)$$

Now, $\displaystyle\int x\sin\pi x\,dx$

$$= x\int\sin\pi x\,dx - \int\left[\frac{d}{dx}(x)\times\int\sin\pi x\,dx\right]dx$$

$$= x\left[\frac{-\cos\pi x}{\pi}\right] - \int\left[1\times\left(\frac{-\cos\pi x}{\pi}\right)\right]dx$$

$$= \frac{-x\cos\pi x}{\pi} + \frac{\sin\pi x}{\pi^2}$$

Then (i) $\Rightarrow$

$$\int_{-1}^{2}|x\sin\pi x|\,dx$$

$$= \int_{-1}^{1} x\sin\pi x\,dx - \int_{1}^{2} x\sin\pi x\,dx$$

$$= \left[\frac{-x\cos\pi x}{\pi} + \frac{\sin\pi x}{\pi^2}\right]_{-1}^{1} - \left[\frac{-x\cos\pi x}{\pi} + \frac{\sin\pi x}{\pi^2}\right]_{1}^{2}$$

$$= \left[\frac{-\cos\pi}{\pi} + \frac{-\cos(-\pi)}{\pi}\right] - \left[\frac{-2\cos 2\pi}{\pi} + \frac{\cos\pi}{\pi}\right]$$

$$= \frac{1}{\pi} + \frac{1}{\pi} + \frac{2}{\pi} + \frac{1}{\pi} = \frac{4}{\pi}.$$

$$(\because \sin\pi = \sin(-\pi) = 0, \cos\pi = \cos(-\pi) = -1,$$

$$\cos 2\pi = 1)$$

Hence, $\displaystyle\int_{-1}^{2}|x\sin\pi x|\,dx = \frac{k}{\pi^2} \Rightarrow \frac{4}{\pi} = \frac{k}{\pi^2}$

$$\Rightarrow k = 4\pi$$

52. (c)

Let $I = \int_0^\pi x f(\sin x)\,dx$. Then,

$$I = \int_0^\pi x f(\sin x)\,dx$$

$$\Rightarrow I = \int_0^\pi (\pi - x) f\{\sin(\pi - x)\}\,dx$$

$$= \int_0^\pi (\pi - x) f(\sin x)\,dx$$

$$= \pi \int_0^\pi f(\sin x)\,dx - \int_0^\pi x f(\sin x)\,dx$$

$$= \pi \int_0^\pi f(\sin x)\,dx - I$$

$$\Rightarrow 2I = \pi \int_0^\pi f(\sin x)\,dx = \pi \times 2 \int_0^{\frac{\pi}{2}} f(\sin x)\,dx$$

$$\left(\begin{array}{l} \because f\left\{\sin\left(2\times\frac{\pi}{2}-x\right)\right\} = f\{\sin(\pi-x)\} \\ \qquad\qquad = f(\sin x), \\ \text{so } \int_0^\pi f(\sin x)\,dx = 2\int_0^{\frac{\pi}{2}} f(\sin x)\,dx \end{array} \right)$$

$$\Rightarrow I = \pi \int_0^{\frac{\pi}{2}} f(\sin x)\,dx$$

$$= \pi \int_0^{\frac{\pi}{2}} f\left\{\sin\left(\frac{\pi}{2}-x\right)\right\}\,dx$$

$$= \pi \int_0^{\frac{\pi}{2}} f(\cos x)\,dx$$

53. (d)

$$x^3 - x = x(x^2 - 1) = x(x+1)(x-1)$$
$$\therefore x^3 - x = 0 \Rightarrow x(x+1)(x-1) = 0$$
$$\Rightarrow x = 0, -1, 1$$

Moreover we have,

$$|x^3 - x|$$
$$= |x(x+1)(x-1)| = |x||x+1||x-1|$$

$$= \begin{cases} (-x)(x+1)\{-(x-1)\} & \text{for } -1\le x\le 0 \\ x(x+1)\{-(x-1)\} & \text{for } 0\le x\le 1 \\ x(x+1)(x-1) & \text{for } 1\le x\le 2 \end{cases}$$

$$= \begin{cases} x^3 - x & \text{for } -1\le x\le 0 \\ -(x^3 - x) & \text{for } 0\le x\le 1 \\ x^3 - x & \text{for } 1\le x\le 2 \end{cases}$$

$$\therefore \int_{-1}^2 |x^3 - x|\,dx$$

$$= \int_{-1}^0 |x^3 - x|\,dx + \int_0^1 |x^3 - x|\,dx + \int_1^2 |x^3 - x|\,dx$$

$$= \int_{-1}^0 (x^3 - x)\,dx - \int_0^1 (x^3 - x)\,dx + \int_1^2 (x^3 - x)\,dx$$

$$= \left[\frac{x^4}{4} - \frac{x^2}{2}\right]_{-1}^0 - \left[\frac{x^4}{4} - \frac{x^2}{2}\right]_0^1 + \left[\frac{x^4}{4} - \frac{x^2}{2}\right]_1^2$$

$$= -\left[-\frac{1}{4}+\frac{1}{2}\right] - \left[\frac{1}{4}-\frac{1}{2}\right] + \left[\left(\frac{16}{4}-\frac{4}{2}\right)-\left(\frac{1}{4}-\frac{1}{2}\right)\right]$$

$$= \frac{11}{4}.$$

54. (a)

$$I_2 = \int_0^1 \frac{x^2}{e^{x^3}(2-x^3)}\,dx$$

$$= \frac{1}{3}\int_0^1 \frac{1}{e^t(2-t)}\,dt$$

$$\left(\begin{array}{l}\text{putting } x^3 = t \text{ so that } 3x^2 dx = dt; \\ \text{also } x=0\Rightarrow t=0 \,\&\, x=1\Rightarrow t=1\end{array}\right)$$

$$= \frac{1}{3e}\int_0^1 \frac{e^{1-t}}{\{1+(1-t)\}}\,dt$$

$$= -\frac{1}{3e}\int_1^0 \frac{e^y}{1+y}\,dy$$

$$\left(\begin{array}{l}\text{putting } 1-t=y \text{ so that } -dt = dy; \\ \text{also } t=0\Rightarrow y=1 \,\&\, t=1\Rightarrow y=0\end{array}\right)$$

$$= \frac{1}{3e}\int_0^1 \frac{e^y}{1+y}\,dy = \frac{1}{3e}\int_0^1 \frac{e^x}{1+x}\,dx = I_1$$

(using the properties of definite integral)

$$\therefore \frac{I_1}{I_2} = 3e.$$

55. (a)

$$\lim_{n\to\infty}\left[\frac{1}{\sqrt{2n-1^2}}+\frac{1}{\sqrt{4n-2^2}}+\ldots+\frac{1}{n}\right]$$

$$=\lim_{n\to\infty}\left[\frac{1}{\sqrt{2n-1^2}}+\frac{1}{\sqrt{4n-2^2}}+\ldots+\frac{1}{\sqrt{n^2}}\right]$$

$$=\lim_{n\to\infty}\frac{1}{n}\left[\frac{n}{\sqrt{1\times2n-1^2}}+\frac{n}{\sqrt{2\times2n-2^2}}+\ldots\right.$$

$$\left.\ldots\ldots+\frac{n}{\sqrt{n\times2n-n^2}}\right]$$

$$=\lim_{n\to\infty}\frac{1}{n}\left[\frac{1}{\sqrt{2\times\left(\frac{1}{n}\right)-\left(\frac{1}{n}\right)^2}}+\frac{1}{\sqrt{2\times\left(\frac{2}{n}\right)-\left(\frac{2}{n}\right)^2}}+\right.$$

$$\left.\ldots\ldots+\frac{1}{\sqrt{2\times\left(\frac{n}{n}\right)-\left(\frac{n}{n}\right)^2}}\right]$$

$$=\lim_{n\to\infty}\frac{1}{n}\sum_{r=1}^{n}\frac{1}{\sqrt{2\times\left(\frac{r}{n}\right)-\left(\frac{r}{n}\right)^2}}$$

$$=\int_0^1\frac{1}{\sqrt{2x-x^2}}\left[\because f\left(\frac{r}{n}\right)=\frac{1}{\sqrt{2\times\left(\frac{r}{n}\right)-\left(\frac{r}{n}\right)^2}}\right]$$

$$=\int_0^1\frac{1}{\sqrt{1-(x-1)^2}}$$

$$=\left[\sin^{-1}(x-1)\right]_0^1=0-\sin^{-1}(-1)=\frac{\pi}{2}.$$

56. (c)

$$\lim_{n\to\infty}\left[\frac{1}{n}+\frac{1}{n+1}+\frac{1}{n+2}+\ldots+\frac{1}{3n}\right]$$

$$=\lim_{n\to\infty}\frac{1}{n}+\lim_{n\to\infty}\left[\frac{1}{n+1}+\frac{1}{n+2}+\ldots+\frac{1}{n+n}\right]$$

$$+\lim_{n\to\infty}\left[\frac{1}{2n+1}+\frac{1}{2n+2}+\ldots+\frac{1}{2n+n}\right]$$

$$=0+\lim_{n\to\infty}\frac{1}{n}\left[\frac{n}{n+1}+\frac{n}{n+2}+\ldots+\frac{n}{n+n}\right]$$

$$+\lim_{n\to\infty}\frac{1}{n}\left[\frac{n}{2n+1}+\frac{n}{2n+2}+\ldots+\frac{n}{2n+n}\right]$$

$$=\lim_{n\to\infty}\frac{1}{n}\left[\frac{1}{1+\frac{1}{n}}+\frac{1}{1+\frac{2}{n}}+\ldots+\frac{1}{1+\frac{n}{n}}\right]$$

$$+\lim_{n\to\infty}\frac{1}{n}\left[\frac{1}{2+\frac{1}{n}}+\frac{1}{2+\frac{2}{n}}+\ldots+\frac{1}{2+\frac{n}{n}}\right]$$

$$=\lim_{n\to\infty}\frac{1}{n}\sum_{r=1}^{n}\frac{1}{\left(1+\frac{r}{n}\right)}+\lim_{n\to\infty}\frac{1}{n}\sum_{r=1}^{n}\frac{1}{\left(2+\frac{r}{n}\right)}$$

$$=\int_0^1\frac{1}{1+x}\,dx+\int_0^1\frac{1}{2+x}\,dx$$

$$=\left[\log(1+x)\right]_0^1+\left[\log(2+x)\right]_0^1$$

$$=\log2-\log1+\log3-\log2=\log3.$$

57. (a) Let $I=\int_0^{\frac{\pi}{4}}\log(1+\tan\theta)\,d\theta$.

Then, $I=\int_0^{\frac{\pi}{4}}\log(1+\tan\theta)\,d\theta$

$$\Rightarrow I=\int_0^{\frac{\pi}{4}}\log\left[1+\tan\left(\frac{\pi}{4}-\theta\right)\right]d\theta$$

$$=\int_0^{\frac{\pi}{4}}\log\left[1+\frac{1-\tan\theta}{1+\tan\theta}\right]d\theta$$

$$=\int_0^{\frac{\pi}{4}}\log\left[\frac{2}{1+\tan\theta}\right]d\theta$$

$$=\int_0^{\frac{\pi}{4}}\left[\log2-\log(1+\tan\theta)\right]d\theta$$

$$=\log2\int_0^{\frac{\pi}{4}}d\theta-\int_0^{\frac{\pi}{4}}\log(1+\tan\theta)\,d\theta=\frac{\pi}{4}\log2-I$$

$$\Rightarrow 2I=\frac{\pi}{4}\log2\Rightarrow I=\frac{\pi}{8}\log2.$$

58. (c)

$$\int_0^{f(x)} t^2 \, dt = x \cos \pi x$$

$$\Rightarrow \left[\frac{t^3}{3}\right]_0^{f(x)} = x \cos \pi x$$

$$\Rightarrow \frac{\{f(x)\}^3}{3} = x \cos \pi x$$

$$\Rightarrow f(x) = \left(3x \cos \pi x\right)^{\frac{1}{3}}$$

$$\Rightarrow f'(x)$$

$$= \frac{1}{3}\left(3x \cos \pi x\right)^{-\frac{2}{3}}\left[3 \cos \pi x + 3x\left(-\pi \sin \pi x\right)\right]$$

$$\Rightarrow f'(4)$$

$$= \frac{1}{3}\left(3 \times 4 \cos 4\pi\right)^{-\frac{2}{3}}\left[3 \cos 4\pi + 3 \times 4\left(-\sin 4\pi\right)\right]$$

$$\Rightarrow f'(4) = \frac{1}{3} \times 12^{-\frac{2}{3}}\left[3 + 0\right] = \frac{1}{12^{\frac{2}{3}}}.$$

$$\left(\because \cos 4\pi = 1, \sin 4\pi = 0\right)$$

59. (b)

Let $y = \left(\frac{n!}{n^n}\right)^{\frac{1}{n}}$. Then $y = \left(\frac{n!}{n^n}\right)^{\frac{1}{n}}$

$$\Rightarrow \log y = \frac{1}{n}\log\left(\frac{n!}{n^n}\right)$$

$$= \frac{1}{n}\log\left[\frac{n \times (n-1) \times (n-2) \times \ldots \ldots \times 2 \times 1}{n \times n \times n \times \ldots \ldots \times n \times n}\right]$$

$$\Rightarrow \lim_{x \to \infty} \log y$$

$$= \lim_{x \to \infty} \frac{1}{n}\log\left[\frac{1}{n} \times \frac{2}{n} \times \ldots \ldots \times \frac{n-1}{n} \times \frac{n}{n}\right]$$

$$= \lim_{x \to \infty} \frac{1}{n}\left[\log\frac{1}{n} + \log\frac{2}{n} + \ldots + \log\frac{n}{n}\right]$$

$$= \lim_{x \to \infty} \frac{1}{n}\sum_{r=1}^{n}\log\frac{r}{n}$$

$$= \int_0^1 \log x \, dx = \left[x(\log x - 1)\right]_0^1 = -1$$

$$\left[\because \left[x(\log x - 1)\right]_0^1 = 1(\log 1 - 1) - \lim_{x \to 0+} x \log x\right.$$

$$= -1 - \lim_{x \to 0+} \frac{\log x}{\frac{1}{x}} \left(\text{form } \frac{\infty}{\infty}\right)$$

$$= -1 - \lim_{x \to 0+} \left(\frac{\frac{1}{x}}{-\frac{1}{x^2}}\right) \text{ (by the L'Hospital rule)}$$

$$= -1 + \lim_{x \to 0+} x = -1 + 0 = -1\right]$$

$$\Rightarrow \log\left[\lim_{n \to \infty} y\right] = -1 \Rightarrow \lim_{n \to \infty} y = e^{-1}$$

$$\Rightarrow \lim_{n \to \infty}\left(\frac{n!}{n^n}\right)^{\frac{1}{n}} = \frac{1}{e}.$$

60. (c)

$$g(x) = \cos x - \int_0^x (x-t)g(t)\,dt$$

$$\Rightarrow \frac{d}{dx}g(x) = \frac{d}{dx}\left[\cos x - \int_0^x (x-t)g(t)\,dt\right]$$

$$\Rightarrow g'(x) = -\sin x - \frac{d}{dx}\int_0^x (x-t)g(t)\,dt$$

$$= -\sin x - \left[\frac{d}{dx}\int_0^x h(x,t)\,dt\right]$$

$$\text{(assuming } h(x,t) = (x-t)g(t))$$

$$= -\sin x - \left[\int_0^x \frac{\partial}{\partial x}h(x,t)\,dt + h(x,x) - 0\right]$$

$$= -\sin x - \int_0^x g(t)\,dt$$

$$\Rightarrow \frac{d}{dx}g'(x) = \frac{d}{dx}\left[-\sin x - \int_0^x g(t)\,dt\right]$$

$$\Rightarrow g''(x) = -\cos x - \{0 + g(x)\}$$

$$\Rightarrow g(x) + g''(x) = -\cos x.$$

61. (c)

Let $x^n = \sin^2\theta$.

Then $x^n = \sin^2\theta$

$\Rightarrow x = \sin^{2/n}\theta$ and $dx = \dfrac{2}{n}\sin^{\frac{2}{n}-1}\theta\cos\theta\,d\theta$

Also $x = 0 \Rightarrow \sin\theta = 0 \Rightarrow \theta = 0$ &

$x = 1 \Rightarrow \sin\theta = 1 \Rightarrow \theta = \dfrac{\pi}{2}$

Therefore, given integral

$= \displaystyle\int_0^{\frac{\pi}{2}} \dfrac{1}{\left(1-\sin^2\theta\right)^{\frac{1}{n}}} \times \dfrac{2}{n}\sin^{\frac{2}{n}-1}\theta\cos\theta\,d\theta$

$= \dfrac{2}{n}\displaystyle\int_0^{\frac{\pi}{2}} \dfrac{\sin^{\frac{2}{n}-1}\theta\cos\theta}{\cos^{\frac{2}{n}}\theta}\,d\theta$

$= \dfrac{2}{n}\displaystyle\int_0^{\frac{\pi}{2}} \sin^{\frac{2}{n}-1}\theta\cos^{1-\frac{2}{n}}\theta\,d\theta$

$= \dfrac{2}{n}\dfrac{\Gamma\!\left(\dfrac{\dfrac{2}{n}-1+1}{2}\right)\times\Gamma\!\left(\dfrac{1-\dfrac{2}{n}+1}{2}\right)}{2\,\Gamma\!\left(\dfrac{\dfrac{2}{n}-1+1-\dfrac{2}{n}+2}{2}\right)}$

$= \dfrac{2}{n}\dfrac{\Gamma\!\left(\dfrac{1}{n}\right)\times\Gamma\!\left(1-\dfrac{1}{n}\right)}{2\,\Gamma(1)} = \dfrac{1}{n}\times\dfrac{\pi}{\sin\dfrac{\pi}{n}}\quad (\because \Gamma(1)=0!=1)$

$= \dfrac{\pi}{n}\cos ec\dfrac{\pi}{n}$

62. (c)

Let $x^2 = \sin\theta$.

Then $x^2 = \sin\theta$

$\Rightarrow x = \sqrt{\sin\theta}$ and $dx = \dfrac{1}{2\sqrt{\sin\theta}}\cos\theta\,d\theta$

Also $x = 0 \Rightarrow \sin\theta = 0 \Rightarrow \theta = 0$ &

$x = 1 \Rightarrow \sin\theta = 1 \Rightarrow \theta = \dfrac{\pi}{2}$

Therefore, given integral

$= \displaystyle\int_0^{\frac{\pi}{2}} \dfrac{\sin\theta}{\sqrt{1-\sin^2\theta}} \times \dfrac{1}{2\sqrt{\sin\theta}}\cos\theta\,d\theta$

$= \dfrac{1}{2}\displaystyle\int_0^{\frac{\pi}{2}} \sqrt{\sin\theta}\,d\theta$

$= \dfrac{1}{2}\displaystyle\int_0^{\frac{\pi}{2}} \sin^{\frac{1}{2}}\theta\cos^0\theta\,d\theta$

$= \dfrac{1}{2}\dfrac{\Gamma\!\left(\dfrac{\dfrac{1}{2}+1}{2}\right)\times\Gamma\!\left(\dfrac{0+1}{2}\right)}{2\,\Gamma\!\left(\dfrac{\dfrac{1}{2}+0+2}{2}\right)}$

$= \dfrac{1}{4}\dfrac{\Gamma\!\left(\dfrac{3}{4}\right)\times\sqrt{\pi}}{\Gamma\!\left(\dfrac{5}{4}\right)} = \dfrac{1}{4}\dfrac{\Gamma\!\left(\dfrac{3}{4}\right)\times\sqrt{\pi}}{\Gamma\!\left(\dfrac{1}{4}+1\right)}$

$= \dfrac{\sqrt{\pi}}{4}\dfrac{\Gamma\!\left(\dfrac{3}{4}\right)}{\dfrac{1}{4}\Gamma\!\left(\dfrac{1}{4}\right)} = \sqrt{\pi}\,\dfrac{\Gamma\!\left(\dfrac{3}{4}\right)}{\Gamma\!\left(\dfrac{1}{4}\right)}.$

63. (a)

$\displaystyle\int_0^{\frac{\pi}{2}} \sqrt{\sin\theta}\,d\theta \times \int_0^{\frac{\pi}{2}} \dfrac{1}{\sqrt{\sin\theta}}\,d\theta$

$= \displaystyle\int_0^{\frac{\pi}{2}} \sin^{\frac{1}{2}}\theta\cos^0\theta\,d\theta \times \int_0^{\frac{\pi}{2}} \sin^{-\frac{1}{2}}\theta\cos^0\theta\,d\theta$

$= \dfrac{\Gamma\!\left(\dfrac{\dfrac{1}{2}+1}{2}\right)\times\Gamma\!\left(\dfrac{0+1}{2}\right)}{2\,\Gamma\!\left(\dfrac{\dfrac{1}{2}+0+2}{2}\right)} \times \dfrac{\Gamma\!\left(\dfrac{-\dfrac{1}{2}+1}{2}\right)\times\Gamma\!\left(\dfrac{0+1}{2}\right)}{2\,\Gamma\!\left(\dfrac{-\dfrac{1}{2}+0+2}{2}\right)}$

$= \dfrac{\Gamma\!\left(\dfrac{3}{4}\right)\times\Gamma\!\left(\dfrac{1}{2}\right)}{2\,\Gamma\!\left(\dfrac{5}{4}\right)} \times \dfrac{\Gamma\!\left(\dfrac{1}{4}\right)\times\Gamma\!\left(\dfrac{1}{2}\right)}{2\,\Gamma\!\left(\dfrac{3}{4}\right)}$

$$= \frac{(\sqrt{\pi})^2 \, \Gamma\left(\frac{1}{4}\right)}{4 \times \frac{1}{4}\Gamma\left(\frac{1}{4}\right)} = \pi$$

$$\left(\because \Gamma\left(\frac{1}{2}\right) = \sqrt{\pi} \ \& \ \Gamma\left(\frac{5}{4}\right) = \Gamma\left(\frac{1}{4}+1\right) = \frac{1}{4}\Gamma\left(\frac{1}{4}\right)\right)$$

64. (d)

$$\int_0^\infty e^{-x^4} x^2 \, dx \times \int_0^\infty e^{-x^4} \, dx$$

$$= \int_0^\infty e^{-y} y^{\frac{1}{2}} \frac{1}{4} y^{-\frac{3}{4}} \, dy \times \int_0^\infty e^{-y} \frac{1}{4} y^{-\frac{3}{4}} \, dy$$

$$\left[\begin{array}{l} \text{putting } x = y^{\frac{1}{4}} \text{ so that } dx = \frac{1}{4} y^{-\frac{3}{4}} \, dy; \\ \text{Also } x=0 \Rightarrow y=0 \ \& \ x=\infty \Rightarrow y=\infty \end{array}\right]$$

$$= \frac{1}{16}\int_0^\infty e^{-y} y^{-\frac{1}{4}} \, dy \times \int_0^\infty e^{-y} y^{-\frac{3}{4}} \, dy$$

$$= \frac{1}{16}\int_0^\infty e^{-y} y^{\frac{3}{4}-1} \, dy \times \int_0^\infty e^{-y} y^{\frac{1}{4}-1} \, dy$$

$$= \frac{1}{16}\Gamma\left(\frac{3}{4}\right) \times \Gamma\left(\frac{1}{4}\right) = \frac{1}{16}\Gamma\left(1-\frac{1}{4}\right) \times \Gamma\left(\frac{1}{4}\right)$$

$$= \frac{1}{16} \times \frac{\pi}{\sin\frac{\pi}{4}} = \frac{\pi}{8\sqrt{2}}$$

$$\left(\text{using } \Gamma(n)\,\Gamma(1-n) = \frac{\pi}{\sin n\pi}\right)$$

65. (a)

$$\Gamma\left(\frac{1}{9}\right) \times \Gamma\left(\frac{2}{9}\right) \times \Gamma\left(\frac{3}{9}\right) \times \Gamma\left(\frac{4}{9}\right) \times \dots \times \Gamma\left(\frac{8}{9}\right)$$

$$= \left(\Gamma\left(\frac{1}{9}\right)\Gamma\left(\frac{8}{9}\right)\right) \times \left(\Gamma\left(\frac{2}{9}\right)\Gamma\left(\frac{7}{9}\right)\right) \times \left(\Gamma\left(\frac{3}{9}\right)\Gamma\left(\frac{6}{9}\right)\right)$$

$$\times \left(\Gamma\left(\frac{4}{9}\right)\Gamma\left(\frac{5}{9}\right)\right)$$

$$= \left(\Gamma\left(\frac{1}{9}\right)\Gamma\left(1-\frac{1}{9}\right)\right) \times \left(\Gamma\left(\frac{2}{9}\right)\Gamma\left(1-\frac{2}{9}\right)\right)$$

$$\times \left(\Gamma\left(\frac{3}{9}\right)\Gamma\left(1-\frac{3}{9}\right)\right) \times \left(\Gamma\left(\frac{4}{9}\right)\Gamma\left(1-\frac{4}{9}\right)\right)$$

$$= \frac{\pi}{\sin\frac{\pi}{9}} \times \frac{\pi}{\sin\frac{2\pi}{9}} \times \frac{\pi}{\sin\frac{3\pi}{9}} \times \frac{\pi}{\sin\frac{4\pi}{9}}$$

$$\Gamma\left(\frac{1}{9}\right) \times \Gamma\left(\frac{2}{9}\right) \times \Gamma\left(\frac{3}{9}\right) \times \Gamma\left(\frac{4}{9}\right) \times \dots \times \Gamma\left(\frac{8}{9}\right)$$

$$= \left(\Gamma\left(\frac{1}{9}\right)\Gamma\left(\frac{8}{9}\right)\right) \times \left(\Gamma\left(\frac{2}{9}\right)\Gamma\left(\frac{7}{9}\right)\right) \times \left(\Gamma\left(\frac{3}{9}\right)\Gamma\left(\frac{6}{9}\right)\right)$$

$$\times \left(\Gamma\left(\frac{4}{9}\right)\Gamma\left(\frac{5}{9}\right)\right)$$

$$= \left(\Gamma\left(\frac{1}{9}\right)\Gamma\left(1-\frac{1}{9}\right)\right) \times \left(\Gamma\left(\frac{2}{9}\right)\Gamma\left(1-\frac{2}{9}\right)\right)$$

$$\times \left(\Gamma\left(\frac{3}{9}\right)\Gamma\left(1-\frac{3}{9}\right)\right) \times \left(\Gamma\left(\frac{4}{9}\right)\Gamma\left(1-\frac{4}{9}\right)\right)$$

$$= \frac{\pi}{\sin\frac{\pi}{9}} \times \frac{\pi}{\sin\frac{2\pi}{9}} \times \frac{\pi}{\sin\frac{3\pi}{9}} \times \frac{\pi}{\sin\frac{4\pi}{9}}$$

$$\left(\text{using } \Gamma(n)\,\Gamma(1-n) = \frac{\pi}{\sin n\pi}\right)$$

$$= \frac{\pi^4}{\frac{\sqrt{3}}{2}\sin\frac{\pi}{9}\sin\frac{2\pi}{9}\sin\frac{4\pi}{9}}$$

$$\left(\because \sin\frac{3\pi}{9} = \sin\frac{\pi}{3} = \frac{\sqrt{3}}{2}\right)$$

$$= \frac{4\pi^4}{\sqrt{3}\left(2\sin\frac{\pi}{9}\sin\frac{2\pi}{9}\right)\sin\frac{4\pi}{9}}$$

$$= \frac{4\pi^4}{\sqrt{3}\left(\cos\frac{\pi}{9} - \cos\frac{3\pi}{9}\right)\sin\frac{4\pi}{9}}$$

$$[\because 2\sin A \sin B = \cos(A-B) - \cos(A+B)]$$

$$= \frac{4\pi^4}{\sqrt{3}\left(\cos\frac{\pi}{9} - \frac{1}{2}\right)\sin\frac{4\pi}{9}}$$

$$= \frac{8\pi^4}{\sqrt{3}\left(2\sin\frac{4\pi}{9}\cos\frac{\pi}{9} - \sin\frac{4\pi}{9}\right)}$$

$$= \frac{8\pi^4}{\sqrt{3}\left(\sin\frac{5\pi}{9} + \sin\frac{3\pi}{9} - \sin\frac{4\pi}{9}\right)}$$

$$[\because 2\sin A \cos B = \sin(A+B) + \sin(A-B)]$$

$$= \frac{8\pi^4}{\sqrt{3} \sin \frac{3\pi}{9}} = \frac{8\pi^4}{\sqrt{3} \times \frac{\sqrt{3}}{2}} = \frac{16\pi^4}{3}$$

$$\left[\because \sin \frac{5\pi}{9} = \sin\left(\pi - \frac{4\pi}{9}\right) = \sin \frac{4\pi}{9} \right]$$

66. (d)

$$u = \log\left(x^3 + y^3 + z^3 - 3xyz\right)$$

$$= \log\left\{ (x+y+z)(x+y\omega^2+z\omega)(x+y\omega+z\omega^2) \right\}$$

(where ω is the cube root of unity)

$$= \log(x+y+z) + \log(x+y\omega^2+z\omega)$$
$$+ \log(x+y\omega+z\omega^2)$$

$$\therefore \frac{\partial u}{\partial x} = \frac{1}{x+y+z} + \frac{1}{x+y\omega^2+z\omega} + \frac{1}{x+y\omega+z\omega^2},$$

$$\frac{\partial u}{\partial y} = \frac{1}{x+y+z} + \frac{\omega^2}{x+y\omega^2+z\omega} + \frac{\omega}{x+y\omega+z\omega^2},$$

$$\frac{\partial u}{\partial z} = \frac{1}{x+y+z} + \frac{\omega}{x+y\omega^2+z\omega} + \frac{\omega^2}{x+y\omega+z\omega^2}.$$

Hence, $\dfrac{\partial u}{\partial x} + \dfrac{\partial u}{\partial y} + \dfrac{\partial u}{\partial z}$

$$= \frac{3}{x+y+z} + \frac{1+\omega^2+\omega}{x+y\omega^2+z\omega} + \frac{1+\omega+\omega^2}{x+y\omega+z\omega^2}$$

$$= \frac{3}{x+y+z} \quad \left[\because 1+\omega+\omega^2 = 0 \right]$$

$\therefore$ (a) is true.

Now, $\left(\dfrac{\partial}{\partial x} + \dfrac{\partial}{\partial y} + \dfrac{\partial}{\partial z} \right)^2 u$

$$= \left(\frac{\partial}{\partial x} + \frac{\partial}{\partial y} + \frac{\partial}{\partial z} \right)\left(\frac{\partial u}{\partial x} + \frac{\partial u}{\partial y} + \frac{\partial u}{\partial z} \right)$$

$$= \left(\frac{\partial}{\partial x} + \frac{\partial}{\partial y} + \frac{\partial}{\partial z} \right)\left(\frac{3}{x+y+z} \right)$$

$$= \frac{\partial}{\partial x}\left(\frac{3}{x+y+z} \right) + \frac{\partial}{\partial x}\left(\frac{3}{x+y+z} \right) + \frac{\partial}{\partial z}\left(\frac{3}{x+y+z} \right)$$

$$= -\frac{9}{(x+y+z)^2}$$

So option (b) is true.

Again,

$$\frac{\partial^2 u}{\partial x^2} + \frac{\partial^2 u}{\partial y^2} + \frac{\partial^2 u}{\partial z^2}$$

$$= \frac{\partial}{\partial x}\left(\frac{\partial u}{\partial x} \right) + \frac{\partial}{\partial y}\left(\frac{\partial u}{\partial y} \right) + \frac{\partial}{\partial z}\left(\frac{\partial u}{\partial z} \right)$$

$$= \frac{\partial}{\partial x}\left(\frac{1}{x+y+z} + \frac{1}{x+y\omega^2+z\omega} + \frac{1}{x+y\omega+z\omega^2} \right)$$

$$+ \frac{\partial}{\partial y}\left(\frac{1}{x+y+z} + \frac{\omega^2}{x+y\omega^2+z\omega} + \frac{\omega}{x+y\omega+z\omega^2} \right)$$

$$+ \frac{\partial}{\partial z}\left(\frac{1}{x+y+z} + \frac{\omega}{x+y\omega^2+z\omega} + \frac{\omega^2}{x+y\omega+z\omega^2} \right)$$

$$= -\frac{1}{(x+y+z)^2} - \frac{1}{\left(x+y\omega^2+z\omega\right)^2} - \frac{1}{\left(x+y\omega+z\omega^2\right)^2}$$

$$- \frac{1}{(x+y+z)^2} - \frac{\omega^4}{\left(x+y\omega^2+z\omega\right)^2} - \frac{\omega^2}{\left(x+y\omega+z\omega^2\right)^2}$$

$$- \frac{1}{(x+y+z)^2} - \frac{\omega^2}{\left(x+y\omega^2+z\omega\right)^2} - \frac{\omega^4}{\left(x+y\omega+z\omega^2\right)^2}$$

$$= -\frac{3}{(x+y+z)^2} \quad \left(\because \omega^4 = \omega, \ 1+\omega+\omega^2 = 0 \right).$$

So option (c) is true.

67. (b)

$V = \log(x^2 + y^2 + z^2)$ (given)

$$\therefore \frac{\partial V}{\partial x} = \frac{2x}{x^2+y^2+z^2},$$

$$\frac{\partial V}{\partial y} = \frac{2y}{x^2+y^2+z^2},$$

$$\frac{\partial V}{\partial z} = \frac{2z}{x^2+y^2+z^2}.$$

Now $\dfrac{\partial^2 V}{\partial x^2} = \dfrac{\partial}{\partial x}\left(\dfrac{\partial V}{\partial x} \right) = \dfrac{\partial}{\partial x}\left(\dfrac{2x}{x^2+y^2+z^2} \right),$

$$= \frac{\left(x^2+y^2+z^2\right)\times 2 - 2x\times 2x}{\left(x^2+y^2+z^2\right)^2}$$

$$= \frac{2y^2+2z^2-2x^2}{\left(x^2+y^2+z^2\right)^2}.$$

Similarly we can get,

$$\frac{\partial^2 V}{\partial y^2} = \frac{2x^2+2z^2-2y^2}{\left(x^2+y^2+z^2\right)^2}, \quad \frac{\partial^2 V}{\partial z^2} = \frac{2x^2+2y^2-2z^2}{x^2+y^2+z^2}.$$

Hence, $\dfrac{\partial^2 V}{\partial x^2} + \dfrac{\partial^2 V}{\partial y^2} + \dfrac{\partial^2 V}{\partial z^2}$

$$=\frac{4\left(x^2+y^2+z^2\right)-2\left(x^2+y^2+z^2\right)}{\left(x^2+y^2+z^2\right)^2}$$

$$=\frac{2}{\left(x^2+y^2+z^2\right)}.$$

68. (c) Given $x = r\cos\theta$, $y = r\sin\theta$, $x^2 + y^2 = r^2$.

$$\therefore 2x=2r\frac{\partial r}{\partial x},\ \ 2y=2r\frac{\partial r}{\partial y}\ i.e;\ \frac{x}{r}=\frac{\partial r}{\partial x},\ \frac{y}{r}=\frac{\partial r}{\partial y}.$$

So $\dfrac{\partial^2 r}{\partial x^2}=\dfrac{r\times 1-x\times\dfrac{\partial r}{\partial x}}{r^2}=\dfrac{r-\dfrac{x^2}{r}}{r^2}\left(\because\ \dfrac{\partial r}{\partial x}=\dfrac{x}{r}\right),$

So $\dfrac{\partial^2 r}{\partial x^2}=\dfrac{r\times 1-x\times\dfrac{\partial r}{\partial x}}{r^2}\dfrac{r-x^2}{r^2}\left(\because\ \dfrac{\partial r}{\partial x}=\dfrac{x}{r}\right),$

$\dfrac{\partial^2 r}{\partial y^2}=\dfrac{r\times 1-y\times\dfrac{\partial r}{\partial y}}{r^2}\dfrac{r-\dfrac{y^2}{r}}{r^2}\left(\because\ \dfrac{\partial r}{\partial y}=\dfrac{y}{r}\right).$

Now $u = f(r)$

$$\Rightarrow\frac{\partial u}{\partial x}=\frac{\partial}{\partial x}f(r)=\frac{\partial}{\partial r}\{f(r)\}\times\frac{\partial r}{\partial x}=f'(r)\times\frac{\partial r}{\partial x}.$$

Similarly, it can be shown that $\dfrac{\partial u}{\partial y}=f'(r)\times\dfrac{\partial r}{\partial y}$.

Then,

$$\frac{\partial^2 u}{\partial x^2}=\frac{\partial}{\partial x}\left(\frac{\partial u}{\partial x}\right)=\frac{\partial}{\partial x}\left\{f'(r)\frac{\partial r}{\partial x}\right\}$$

$$=f'(r)\times\frac{\partial^2 r}{\partial x^2}+\frac{\partial r}{\partial x}\times\frac{\partial}{\partial r}\{f'(r)\}\times\frac{\partial r}{\partial x}$$

$$=f'(r)\times\frac{\partial^2 r}{\partial x^2}+f''(r)\times\left(\frac{\partial r}{\partial x}\right)^2.$$

Similarly,

$$\frac{\partial^2 u}{\partial y^2}+f'(r)\times\frac{\partial^2 r}{\partial y^2}f''(r)\times\left(\frac{\partial r}{\partial y}\right)^2.$$

So, $\dfrac{\partial^2 u}{\partial x^2}+\dfrac{\partial^2 u}{\partial y^2}$

$$=f'(r)\times\left\{\frac{\partial^2 r}{\partial x^2}+\frac{\partial^2 r}{\partial y^2}\right\}+f''(r)\times\left\{\left(\frac{\partial r}{\partial x}\right)^2+\left(\frac{\partial r}{\partial y}\right)^2\right\}$$

$$f'(r)\times\frac{2r-\dfrac{1}{r}(x^2+y^2)}{r^2}+f''(r)\times\left\{\frac{x^2}{r^2}+\frac{y^2}{r^2}\right\}$$

$$=\frac{1}{r}f'(r)+f''(r)\left(\because\ x^2+y^2=r^2\right)$$

69. (a)

$$u=\sin^{-1}\left(\frac{x^2+y^2}{x+y}\right)\ \Rightarrow\sin u=\frac{x^2+y^2}{x+y}$$

$$=\frac{x^2\left\{1+\left(\dfrac{y}{x}\right)^2\right\}}{x\left\{1+\dfrac{y}{x}\right\}}=x\left\{\frac{1+\left(\dfrac{y}{x}\right)^2}{1+\dfrac{y}{x}}\right\}$$

$$=x^1 f\left(\frac{y}{x}\right)\ \text{(say)}$$

$\therefore$ $\sin u$ is a homogeneous function of degree "1." So by Euler's theorem,

$$x\frac{\partial}{\partial x}(\sin u)+y\frac{\partial}{\partial y}(\sin u)=1\times\sin u$$

or, $x\left[\dfrac{\partial}{\partial u}(\sin u)\right]\dfrac{\partial u}{\partial x}+y\left[\dfrac{\partial}{\partial u}(\sin u)\right]\dfrac{\partial u}{\partial y}=\sin u$

or, $x\cos u\dfrac{\partial u}{\partial x}+y\cos u\dfrac{\partial u}{\partial y}=\sin u$

or, $x\dfrac{\partial u}{\partial x}+y\dfrac{\partial u}{\partial y}=\tan u.$

70. (b)

$$x = e^{\varphi} + e^{-\psi} \Rightarrow \frac{\partial x}{\partial\varphi}=e^{\varphi},\ \frac{\partial x}{\partial\psi}=-e^{-\psi},$$

$$y = e^{-\varphi} - e^{\psi} \Rightarrow \frac{\partial y}{\partial\varphi}=-e^{-\varphi},\ \frac{\partial y}{\partial\psi}=-e^{\psi}.$$

Then $\dfrac{\partial u}{\partial\varphi}=\dfrac{\partial u}{\partial x}\times\dfrac{\partial x}{\partial\varphi}+\dfrac{\partial u}{\partial y}\times\dfrac{\partial y}{\partial\varphi}$

$$=\frac{\partial u}{\partial x}\times e^{\varphi}+\frac{\partial u}{\partial y}\times\left(-e^{-\varphi}\right)\ldots\ldots(i)$$

$$\frac{\partial u}{\partial\psi}=\frac{\partial u}{\partial x}\times\frac{\partial x}{\partial\psi}+\frac{\partial u}{\partial y}\times\frac{\partial y}{\partial\psi}$$

$$=\frac{\partial u}{\partial x}\times\left(-e^{-\psi}\right)+\frac{\partial u}{\partial y}\times\left(-e^{\psi}\right)\ldots\ldots(ii)$$

$\therefore$ $(i)-(ii)\Rightarrow$

$$\frac{\partial u}{\partial\varphi}-\frac{\partial u}{\partial\psi}=\frac{\partial u}{\partial x}\left(e^{\varphi}+e^{-\psi}\right)+\frac{\partial u}{\partial y}\left(e^{\psi}-e^{-\varphi}\right)$$

$$=x\frac{\partial u}{\partial x}-y\frac{\partial u}{\partial y}$$

$$\left[\because x = e^{\varphi} + e^{-\psi},\ y = e^{-\varphi} - e^{\psi}\right]$$

71. (c)

$u = f(2x - 3y, 3y - 4z, 4z - 2x)$ (given)

Let, $p = 2x - 3y$, $q = 3y - 4z$, $r = 4z - 2x$

Then $\dfrac{\partial p}{\partial x} = 2$, $\dfrac{\partial p}{\partial y} = -3$, $\dfrac{\partial p}{\partial z} = 0$;

$\dfrac{\partial q}{\partial x} = 0$, $\dfrac{\partial q}{\partial y} = 3$, $\dfrac{\partial q}{\partial z} = -4$;

$\dfrac{\partial r}{\partial x} = -2$, $\dfrac{\partial r}{\partial y} = 0$, $\dfrac{\partial r}{\partial z} = 4$.

Again, $u = f(p, q, r)$ and so

$\dfrac{\partial u}{\partial x} = \dfrac{\partial u}{\partial p} \times \dfrac{\partial p}{\partial x} + \dfrac{\partial u}{\partial q} \times \dfrac{\partial q}{\partial x} + \dfrac{\partial u}{\partial r} \times \dfrac{\partial r}{\partial x}$

$= \dfrac{\partial u}{\partial p} \times 2 + \dfrac{\partial u}{\partial q} \times 0 + \dfrac{\partial u}{\partial r} \times (-2)$

$= 2\left(\dfrac{\partial u}{\partial p} - \dfrac{\partial u}{\partial r} \right)$

$\dfrac{\partial u}{\partial y} = \dfrac{\partial u}{\partial p} \times \dfrac{\partial p}{\partial y} + \dfrac{\partial u}{\partial q} \times \dfrac{\partial q}{\partial y} + \dfrac{\partial u}{\partial r} \times \dfrac{\partial r}{\partial y}$

$= \dfrac{\partial u}{\partial p} \times (-3) + \dfrac{\partial u}{\partial q} \times 3 + \dfrac{\partial u}{\partial r} \times 0$

$= 3\left(\dfrac{\partial u}{\partial q} - \dfrac{\partial u}{\partial p} \right)$

$\dfrac{\partial u}{\partial z} = \dfrac{\partial u}{\partial p} \times \dfrac{\partial p}{\partial z} + \dfrac{\partial u}{\partial q} \times \dfrac{\partial q}{\partial z} + \dfrac{\partial u}{\partial r} \times \dfrac{\partial r}{\partial z}$

$= \dfrac{\partial u}{\partial p} \times 0 + \dfrac{\partial u}{\partial q} \times (-4) + \dfrac{\partial u}{\partial r} \times 4$

$= 4\left(\dfrac{\partial u}{\partial r} - \dfrac{\partial u}{\partial q} \right)$

$\therefore \dfrac{1}{2}u_x + \dfrac{1}{3}u_y + \dfrac{1}{4}u_z$

$= \left(\dfrac{\partial u}{\partial p} - \dfrac{\partial u}{\partial r} \right) + \left(\dfrac{\partial u}{\partial q} - \dfrac{\partial u}{\partial p} \right) + \left(\dfrac{\partial u}{\partial r} - \dfrac{\partial u}{\partial q} \right) = 0$.

72. (d)

$\dfrac{\partial(u,v)}{\partial(x,y)} = \begin{vmatrix} \dfrac{\partial u}{\partial x} & \dfrac{\partial u}{\partial y} \\ \dfrac{\partial v}{\partial x} & \dfrac{\partial v}{\partial y} \end{vmatrix}$

$= \begin{vmatrix} \dfrac{(1-xy)\times 1 - (x+y)\times(-y)}{(1-xy)^2} & \dfrac{(1-xy)\times 1 - (x+y)\times(-x)}{(1-xy)^2} \\ \dfrac{1}{1+x^2} & \dfrac{1}{1+y^2} \end{vmatrix}$

$= \begin{vmatrix} \dfrac{1+y^2}{(1-xy)^2} & \dfrac{1+x^2}{(1-xy)^2} \\ \dfrac{1}{1+x^2} & \dfrac{1}{1+y^2} \end{vmatrix}$

$= \dfrac{(1+y^2)}{(1-xy)^2} \times \dfrac{1}{(1+y^2)} - \dfrac{(1+x^2)}{(1-xy)^2} \times \dfrac{1}{(1+x^2)} = 0$.

$\therefore$ u and v are functionally related.

Now, $v = \tan^{-1} x + \tan^{-1} y = \tan^{-1}\left(\dfrac{x+y}{1-xy} \right) = \tan^{-1} u$

$\therefore u = \tan v$.

73. (a)

Let $v = xf\left(\dfrac{y}{x} \right)$ and $w = g\left(\dfrac{y}{x} \right)$.

Then clearly v is a homogeneous function of degree "1" and w is a homogeneous function of degree "0." So by Euler's theorem, we have

$x\dfrac{\partial v}{\partial x} + y\dfrac{\partial v}{\partial y} = 1 \times v$ ....(i)

and $x\dfrac{\partial w}{\partial x} + y\dfrac{\partial w}{\partial y} = 0 \times w = 0$ ....(ii)

Now (i) + (ii) $\Rightarrow$

$x\left(\dfrac{\partial v}{\partial x} + \dfrac{\partial w}{\partial x} \right) + y\left(\dfrac{\partial v}{\partial y} + \dfrac{\partial w}{\partial y} \right) = v + 0$

or, $x\dfrac{\partial u}{\partial x} + y\dfrac{\partial u}{\partial y} = xf\left(\dfrac{y}{x} \right)$

$\because$ $u = v + w \Rightarrow$

$\left[\because u = v + w \Rightarrow \dfrac{\partial u}{\partial x} = \dfrac{\partial v}{\partial x} + \dfrac{\partial w}{\partial x} \text{ and } \dfrac{\partial u}{\partial y} = \dfrac{\partial v}{\partial y} + \dfrac{\partial w}{\partial y} \right]$

74. (b)

Let $f(x,y) = x^3 + y^3 + 3xy - 1$.

Then $\dfrac{\partial f}{\partial x} = 3x^2 + 3y$ and $\dfrac{\partial f}{\partial y} = 3y^2 + 3x$.

$\therefore \dfrac{dy}{dx} = -\dfrac{\partial f/\partial x}{\partial f/\partial y} = \dfrac{-(3x^2 + 3y)}{3y^2 + 3x} = \dfrac{-(x^2 + y)}{(x + y^2)}$

75. (d)

Let $f(x,y) = x^2 + y^2 + 2xy - 1$.

Then $\dfrac{\partial f}{\partial x} = 2x + 2y$ and $\dfrac{\partial f}{\partial y} = 2y + 2x$.

$\therefore \dfrac{dy}{dx} = -\dfrac{\partial f/\partial x}{\partial f/\partial y} = \dfrac{-(2x + 2y)}{2y + 2x} = -1$.

Hence, $\dfrac{du}{dx}=\dfrac{\partial u}{\partial x}+\dfrac{\partial u}{\partial y}\times\dfrac{dy}{dx}$

$=\left\{\log(xy)+x\times\dfrac{1}{xy}\times y\right\}+x\times\dfrac{1}{xy}\times x\times(-1)$

$=\log(xy)+1-\dfrac{x}{y}.$

76. (d)

Here, $\dfrac{\partial f}{\partial x}=y(1-x-y)+xy\times(-1)$

$\qquad\qquad =y-2xy-y^2,$

$\qquad \dfrac{\partial f}{\partial y}=x(1-x-y)+xy\times(-1)$

$\qquad\qquad =x-2xy-x^2,$

$\dfrac{\partial^2 f}{\partial x^2}=-2y,\ \dfrac{\partial^2 f}{\partial x\,\partial y}=1-2y-2x$

and $\dfrac{\partial^2 f}{\partial y^2}=-2x.$

$\therefore \dfrac{\partial f}{\partial x}=0,\ \dfrac{\partial f}{\partial y}=0$

$\Rightarrow y-2xy-y^2=0.............(i)$

$\qquad x-2xy-x^2=0...............(ii)$

Subtracting (i) from (ii) we get,

$x-y-(x^2-y^2)=0$

or, $(x-y)(1-x-y)=0$

or, either $x=y$ or $x+y=1$

Case-I: $x=y$

Equation (i)$\Rightarrow y-2y^2-y^2=0$

or, $3y^2=y$

or, $y(3y-1)=0$

or, $y=0,\dfrac{1}{3}$

$y=0\Rightarrow x=0\ \&\ y=\dfrac{1}{3}\Rightarrow x=\dfrac{1}{3}\ (\because x=y)$

Therefore, in this case the points are (0,0) and $\left(\dfrac{1}{3},\dfrac{1}{3}\right)$

At (0,0):

$A=\dfrac{\partial^2 f}{\partial x^2}\bigg|_{(0,0)}=0,\ B=\dfrac{\partial^2 f}{\partial x\,\partial y}\bigg|_{(0,0)}=1$

and $C=\dfrac{\partial^2 f}{\partial y^2}\bigg|_{(0,0)}=0.$

$\therefore AC-B^2=-1<0.$

Therefore, $f(x,y)$ has neither a maximum nor a minimum at (0, 0).

At $\left(\dfrac{1}{3},\dfrac{1}{3}\right)$:

$A=\dfrac{\partial^2 f}{\partial x^2}\bigg|_{\left(\frac{1}{3},\frac{1}{3}\right)}=-\dfrac{2}{3},\ B=\dfrac{\partial^2 f}{\partial x\,\partial y}\bigg|_{\left(\frac{1}{3},\frac{1}{3}\right)}=-\dfrac{1}{3}$

and $C=\dfrac{\partial^2 f}{\partial y^2}\bigg|_{\left(\frac{1}{3},\frac{1}{3}\right)}=-\dfrac{2}{3}.$

$\therefore AC-B^2=\dfrac{4}{9}-\dfrac{1}{9}>0$ and $A<0$

Therefore, $f(x,y)$ has a maximum value at $\left(\dfrac{1}{3},\dfrac{1}{3}\right).$

Case-II: $x+y=1$

Equation (i) $\Rightarrow y-2y(1-y)-y^2=0$

or, $y^2-y=0$

or, $y(y-1)=0$

or, $y=0,1$

$y=0\Rightarrow x=1\ \&\ y=1\Rightarrow x=0\ (\because x+y=1)$

Therefore, in this case the points are (1, 0) and (0,1)

At (1,0):

$A=\dfrac{\partial^2 f}{\partial x^2}\bigg|_{(1,0)}=0,\ B=\dfrac{\partial^2 f}{\partial x\,\partial y}\bigg|_{(1,0)}=-1$

and $C=\dfrac{\partial^2 f}{\partial y^2}\bigg|_{(1,0)}=-2.$

$\therefore AC-B^2=-1<0.$

Therefore, $f(x,y)$ has neither a maximum nor a minimum at (1, 0).

At $(0, 1)$:

$$A = \frac{\partial^2 f}{\partial x^2}\bigg|_{(0,1)} = -2, \ B = \frac{\partial^2 f}{\partial x\,\partial y}\bigg|_{(0,1)} = -1$$

and $C = \dfrac{\partial^2 f}{\partial y^2}\bigg|_{(0,1)} = 0$.

$\therefore AC - B^2 = -1 < 0$.

Therefore, $f(x, y)$ has neither a maximum nor a minimum at $(0, 1)$.

Hence, $f(x, y)$ has a maximum at $\left(\dfrac{1}{3}, \dfrac{1}{3}\right)$ only but it does not have any minimum value.

77. (a)

Here $f(x, y, z) = x^2 + y^2 + z^2$,

$\phi(x, y, z) = x + 2y + 3z - 4$.

$\therefore F(x, y, z)$

$= f(x, y, z) + \lambda\phi(x, y, z)$

$= x^2 + y^2 + z^2 + \lambda(x + 2y + 3z - 4)$

Then, $\dfrac{\partial F}{\partial x} = 0, \ \dfrac{\partial F}{\partial y} = 0$ and $\dfrac{\partial F}{\partial z} = 0$

$\Rightarrow 2x + \lambda = 0, 2y + 2\lambda = 0, 2z + 3\lambda = 0$

$\Rightarrow -\lambda = \dfrac{2x}{1} = \dfrac{y}{1} = \dfrac{2z}{3}$

$\Rightarrow \dfrac{2x}{1} = \dfrac{y}{1} = \dfrac{2z}{3} \Rightarrow \dfrac{2x}{1} = \dfrac{4y}{4} = \dfrac{6z}{9}$

$\Rightarrow \dfrac{2x}{1} = \dfrac{4y}{4} = \dfrac{6z}{9} = \dfrac{2x + 4y + 6z}{1 + 4 + 9}$

$\qquad = \dfrac{2(x + 2y + 3z)}{14}$

$\qquad = \dfrac{8}{14} = \dfrac{4}{7}$

$\qquad [\because x + 2y + 3z = 4]$

$\Rightarrow x = \dfrac{2}{7}, y = \dfrac{4}{7}$ and $z = \dfrac{6}{7}$.

Therefore, minimum value of $x^2 + y^2 + z^2$ is $\left(\dfrac{2}{7}\right)^2 + \left(\dfrac{4}{7}\right)^2 + \left(\dfrac{6}{7}\right)^2$ i.e; $\dfrac{8}{7}$.

78. (a) Let $P(x, y, z)$ be any point on the given sphere and $A(1, 2, 3)$ be the given point. Then,

$\overline{AP} = \sqrt{(x-1)^2 + (y-2)^2 + (z-3)^2}$.

$\therefore \overline{AP}^2 = (x-1)^2 + (y-2)^2 + (z-3)^2$

$\qquad = f(x, y, z) \ (say)$

Let $\phi(x, y, z) = x^2 + y^2 + z^2 - 1$.

$\therefore F(x, y, z)$

$= f(x, y, z) + \lambda\phi(x, y, z)$

$= (x-1)^2 + (y-2)^2 + (z-3)^2$

$\quad + \lambda\left(x^2 + y^2 + z^2 - 1\right)$

Then, $\dfrac{\partial F}{\partial x} = 0, \ \dfrac{\partial F}{\partial y} = 0$ and $\dfrac{\partial F}{\partial z} = 0$

$\Rightarrow 2(x-1) + 2\lambda x = 0, 2(y-2) + 2\lambda y = 0,$

$\quad 2(z-3) + 2\lambda z = 0$

$\Rightarrow -\lambda = \dfrac{x-1}{x} = \dfrac{y-2}{y} = \dfrac{z-3}{z}$

$\qquad = \dfrac{\sqrt{(x-1)^2 + (y-2)^2 + (z-3)^2}}{\sqrt{x^2 + y^2 + z^2}}$

$\qquad = \dfrac{\sqrt{f(x, y, z)}}{1} = \sqrt{f(x, y, z)}$

$\Rightarrow x = \dfrac{1}{1 + \sqrt{f(x, y, z)}}, y = \dfrac{2}{1 + \sqrt{f(x, y, z)}},$

$\quad z = \dfrac{3}{1 + \sqrt{f(x, y, z)}}$.

Then, $x^2 + y^2 + z^2 = 1$

$\Rightarrow \dfrac{14}{\left(1 + \sqrt{f(x, y, z)}\right)^2} = 1$

$\Rightarrow 1 + \sqrt{f(x, y, z)} = \pm\sqrt{14}$

Then, $x^2 + y^2 + z^2 = 1$

$\Rightarrow \dfrac{14}{\left(1 + \sqrt{f(x, y, z)}\right)^2} = 1$

$\Rightarrow 1 + \sqrt{f(x, y, z)} = \pm\sqrt{14}$

$$\Rightarrow \sqrt{f(x,y,z)} = \pm\sqrt{14} - 1$$

$$\Rightarrow \left|\sqrt{f(x,y,z)}\right| = \sqrt{14} - 1, \sqrt{14} + 1$$

Hence, the minimum and maximum values of $\overline{AP}$ are, respectively, $\sqrt{14} - 1$ and $\sqrt{14} + 1$.

79. (b)

$$\iiint\limits_{V} \left(x^2 + y^2 + z^2\right) dx\,dy\,dz$$

$$= \int\limits_{x=0}^{2} \left[\int\limits_{y=0}^{2} \left\{ \int\limits_{z=0}^{2} \left(x^2 + y^2 + z^2\right) dz \right\} dy \right] dx$$

$$= \int\limits_{x=0}^{2} \left(\int\limits_{y=0}^{2} \left\{ \left[x^2 z + y^2 z + \frac{z^3}{3} \right]_{z=0}^{2} \right\} dy \right) dx$$

$$= \int\limits_{x=0}^{2} \left\{ \int\limits_{y=0}^{2} \left(2x^2 + 2y^2 + \frac{8}{3} \right) dy \right\} dx$$

$$= \int\limits_{x=0}^{2} \left[2x^2 y + \frac{2y^3}{3} + \frac{8}{3} y \right]_{y=0}^{2} dx$$

$$= \int\limits_{0}^{2} \left(4x^2 + \frac{16}{3} + \frac{16}{3} \right) dx$$

$$= \left[\frac{4x^3}{3} + \frac{16}{3} x + \frac{16x}{3} \right]_{0}^{2}$$

$$= \frac{4}{3} \times 8 + \frac{2 \times 16}{3} \times 2 = 32.$$

80. (c)

$$\int\limits_{0}^{1} \int\limits_{0}^{x} \int\limits_{0}^{y} x^3 y^2 z \, dz\,dy\,dx$$

$$= \int\limits_{x=0}^{1} \left\{ \int\limits_{y=0}^{x} \left(\int\limits_{z=0}^{y} x^3 y^2 z \, dz \right) dy \right\} dx$$

$$= \int\limits_{x=0}^{1} \left\{ \int\limits_{y=0}^{x} \left[x^3 y^2 \frac{z^2}{2} \right]_{z=0}^{y} dy \right\} dx$$

$$= \int\limits_{x=0}^{1} \left\{ \int\limits_{y=0}^{x} x^3 y^2 \times \frac{y^2}{2} dy \right\} dx$$

$$= \int\limits_{x=0}^{1} \frac{x^3}{2} \left[\frac{y^5}{5} \right]_{0}^{x} dx = \int\limits_{0}^{1} \frac{x^3}{10} \times x^5 dx$$

$$= \frac{1}{10} \left[\frac{x^9}{9} \right]_{0}^{1} = \frac{1}{90}.$$

81. (a)

Here, V = $\{(x,y,z) : x = 0, y = 0, z = 0,$

$x + y + z = 1\}$.

Let us put, $x + y + z = u$, $x + y = uv$, $y = uvw$.

Then, $y = uvw$, $x = uv - y = uv - uvw = uv(1 - w)$.

$z = u - x - y = u(1 - v)$.

$$\therefore \text{ Jacobian, } J = \begin{vmatrix} \dfrac{\partial x}{\partial u} & \dfrac{\partial x}{\partial v} & \dfrac{\partial u}{\partial w} \\[2mm] \dfrac{\partial y}{\partial u} & \dfrac{\partial y}{\partial v} & \dfrac{\partial y}{\partial w} \\[2mm] \dfrac{\partial z}{\partial u} & \dfrac{\partial z}{\partial v} & \dfrac{\partial z}{\partial w} \end{vmatrix}$$

$$= \begin{vmatrix} v(1-w) & u(1-w) & -uv \\ vw & uw & uv \\ 1-v & -u & 0 \end{vmatrix}$$

$$= v(1-w)\{0 + u^2 v\} - u(1-w)\{0 - uv(1-v)\}$$
$$\qquad - uv\{-uvw - uw(1-v)\}$$

$$= u^2 v^2 (1-w) + u^2 v (1-w) - u^2 v^2 (1-w)$$
$$\qquad + u^2 v^2 w + u^2 vw(1-v)$$

$$= u^2 v - u^2 vw + u^2 v^2 w + u^2 vw - u^2 v^2 w$$

$$= u^2 v$$

Now, $x + y + z = 1 \Rightarrow u = 1$,

$x = 0 \Rightarrow uv(1 - w) = 0$
$\qquad\qquad \Rightarrow u = 0, v = 0, w = 1$

$y = 0 \Rightarrow uvw = 0$
$\qquad\qquad \Rightarrow u = 0, v = 0, w = 0$

$z = 0 \Rightarrow u(1 - v) = 0 \Rightarrow u = 0, v = 1$

Thus, the new region of integration is given by $R' = \{(u,v,w) : 0 \leqslant u \leqslant 1, 0 \leqslant v \leqslant 1, 0 \leqslant w \leqslant 1\}$.

Hence, the given integral

$$= \int\limits_{u=0}^{1} \int\limits_{v=0}^{1} \int\limits_{w=0}^{1} u \times uv(1-w) \times uvw \times u(1-v) \times u^2 v \, du\,dv\,dw$$

$$(\because (\because dx\,dy\,dz = J\,du\,dv\,dw = u^2 v \, du\,dv\,dw)$$

$$= \int\limits_{u=0}^{1} u^6 du \times \int\limits_{v=0}^{1} v^3 (1-v) dv \times \int\limits_{w=0}^{1} w(1-w) dw$$

$$=\left[\frac{u^7}{7}\right]_0^1\times\left[\frac{v^4}{4}-\frac{v^5}{5}\right]_0^1\times\left[\frac{w^2}{2}-\frac{w^3}{3}\right]_0^1$$

$$=\frac{1}{7}\times\left(\frac{1}{4}-\frac{1}{5}\right)\times\left(\frac{1}{2}-\frac{1}{3}\right)=\frac{1}{7}\times\frac{1}{20}\times\frac{1}{6}=\frac{1}{840}.$$

82. (d) We know that

$$\iiint_V x^{m-1}y^{n-1}z^{l-1}\,dxdydz=\frac{\Gamma(m)\Gamma(n)\Gamma(l)}{\Gamma(m+n+l)},$$

where $V=\{(x,y,z):x\geqslant0,\ y\geqslant0,\ z\geqslant0,\ x+y+z\leqslant1\}$

∴ Given integral

$$=\iiint_V x^{4-1}y^{5-1}z^{3-1}\,dxdydz$$

$$=\frac{\Gamma(4)\Gamma(5)\Gamma(3)}{\Gamma(4+5+3)}\quad(\text{here, }m=4,n=5,l=3)$$

$$=\frac{3!\times4!\times2!}{11!}\quad\left(\because\overline{|n}=(n-1)!\text{ for }n\in Z^+\right)$$

$$=\frac{2!\times3!\times4!}{(11)!}.$$

83. (d) $2x+y+z=4\Rightarrow z=4-2x-y$

Again, $2x+y+z=4\Rightarrow y=4-2x$ (for $z=0$)

$2x+y+z=4\Rightarrow 2x=4$ (for $y=0,z=0$)

$\Rightarrow x=2$

$$\therefore\int_0^2\int_0^{\lambda(x)}\int_0^{\mu(x,y)}dzdydx=\int_{x=0}^2\int_{y=0}^{4-2x}\int_{z=0}^{4-2x-y}dzdydx$$

Hence, $\lambda(x)=4-2x,\ \mu(x,y)=4-2x-y$.

So, $\lambda(x)-\mu(x,y)=y$.

84. (b)

Given integral $=\int_{x=0}^1\left\{\int_{y=0}^{x^2}\left[yz+z^2\right]_{z=0}^y dy\right\}dx$

$$=\int_{x=0}^1\left\{\int_{y=0}^{x^2}(y^2+y^2)dy\right\}dx$$

$$=\int_{x=0}^1\left[\frac{2y^3}{3}\right]_{y=0}^{x^2}dx=\int_0^1\frac{2}{3}x^6dx=\frac{2}{3}\left[\frac{x^7}{7}\right]_0^1$$

$$=\frac{2}{3}\times\frac{1}{7}=\frac{2}{21}.$$

85. (b) Required volume of the solid of revolution

$$=\pi\int_{-\infty}^\infty y^2dx=\pi\int_{-\infty}^\infty\frac{64}{(x^2+4)^2}dx$$

$$=2\pi\int_0^\infty\frac{64}{(x^2+4)^2}dx$$

[∵ the integrand is an even function]

$$=2\pi\int_0^{\frac{\pi}{2}}\frac{64}{(4\tan^2\theta+4)^2}\times2\sec^2\theta\,d\theta$$

[putting $x=2\tan\theta$ so that $dx=2\sec^2\theta d\theta$;
$x=0\Rightarrow\theta=0; x=\infty\Rightarrow\theta=\pi/2$]

$$=256\pi\int_0^{\frac{\pi}{2}}\frac{\sec^2\theta d\theta}{16\sec^4\theta}=16\pi\int_0^{\frac{\pi}{2}}\cos^2\theta d\theta$$

$$=8\pi\int_0^{\frac{\pi}{2}}2\cos^2\theta d\theta=8\pi\int_0^{\frac{\pi}{2}}(1+\cos2\theta)\,d\theta$$

$$=8\pi\left[\theta+\frac{\sin2\theta}{2}\right]_0^{\frac{\pi}{2}}$$

$$=8\pi\left[\frac{\pi}{2}+\sin\pi-0-0\right]=4\pi^2\quad(\because\sin\pi=0)$$

86. (b) Clearly x varies from 0 to 2.

Hence, required volume of the solid of revolution

$$=\pi\int_0^2 y^2dx=\pi\int_0^2\frac{x^2(2-x)}{2+x}dx$$

$$=\pi\int_0^2\frac{x^2\{(2+x)-2x\}}{2+x}dx$$

$$=\pi\int_0^2\frac{x^2(2+x)}{(2+x)}dx-2\pi\int_0^2\frac{x^3}{x+2}dx$$

$$=\pi\left\{\int_0^2 x^2dx-2\int_0^2\frac{(x^3+2^3)-8}{(x+2)}dx\right\}$$

$$=\pi\left\{\left[\frac{x^3}{3}\right]_0^2-2\int_0^2\frac{(x+2)(x^2-2x+4)}{(x+2)}dx+16\int_0^2\frac{dx}{x+2}\right\}$$

$$= \frac{8\pi}{3} - 2\pi \int_0^2 \left(x^2 - 2x + 4\right) dx + 16\pi \left[\log(x+2)\right]_0^2$$

$$= \frac{8\pi}{3} - 2\pi \left[\frac{x^3}{3} - x^2 + 4x\right]_0^2 + 16\pi\left(\log 4 - \log 2\right)$$

$$= \frac{8\pi}{3} - 2\pi\left[\frac{8}{3} - 4 + 8\right] + \left(32\log 2 - 16\log 2\right)\pi$$

$$= -\frac{32\pi}{3} + 16\pi \log 2$$

$$= 16\pi\left(\log 2 - \frac{2}{3}\right)$$

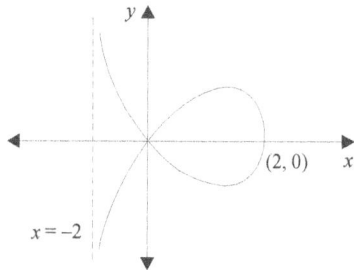

87. (a) Required volume of solid of revolution

$$= \frac{2\pi}{3} \int_{\theta=0}^{\pi} r^3 \sin\theta\, d\theta$$

$$= \frac{2\pi}{3} \int_{\theta=0}^{\pi} \left\{a(1+\cos\theta)\right\}^3 \sin\theta\, d\theta = -\frac{2\pi a^2}{3}\int_2^0 z^3\, dz$$

(putting $1 + \cos\theta = z$, so that $-\sin\theta d\theta = dz$ and $\theta = 0 \Rightarrow z = 1 + 1 = 2$, $\theta = \pi \Rightarrow z = 1 - 1 = 0$)

$$= -\frac{2\pi a^2}{3}\left[\frac{z^4}{4}\right]_2^0$$

$$= -\frac{2\pi a^2}{3} \times \left(0 - \frac{16}{4}\right) = \frac{8\pi a^2}{3}.$$

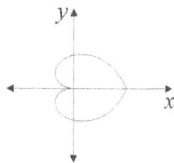

88. (c)

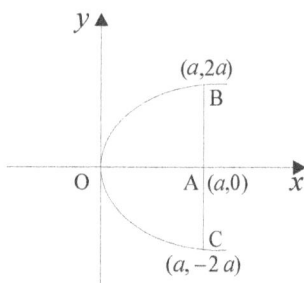

Here the end points of the latus rectum arc $(a, 2a)$ and $(a, -2a)$. So x varies from "0" to "a."

Hence, the required surface area of the solid of revolution

$$= 2\pi \int_0^a y \sqrt{1 + \left(\frac{dy}{dx}\right)^2}\, dx$$

$$= 2\pi \int_0^a y \sqrt{1 + \frac{a}{x}}\, dx$$

$$\left(\because y^2 = 4ax \Rightarrow y = 2\sqrt{a}\sqrt{x} \Rightarrow \frac{dy}{dx} = \frac{\sqrt{a}}{\sqrt{x}}\right)$$

$$= 2\pi \int_0^a 2\sqrt{a}\sqrt{x}\frac{\sqrt{a+x}}{\sqrt{x}}dx$$

$$= 4\pi\sqrt{a}\int_0^a (x+a)^{\frac{1}{2}}\, dx = 4\pi\sqrt{a}\left[\frac{(x+a)^{\frac{3}{2}}}{\frac{3}{2}}\right]_0^a$$

$$= 4\pi\sqrt{a} \times \frac{2}{3}\left\{(2a)^{\frac{3}{2}} - a^{\frac{3}{2}}\right\}$$

$$= \frac{8\pi\sqrt{a}}{3}\left(2\sqrt{2}\,a\sqrt{a} - a\sqrt{a}\right) = \frac{8\pi a^2}{3}\left(2\sqrt{2} - 1\right).$$

89. (a)

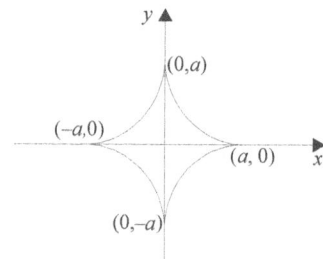

Clearly x varies from 0 to a in first quadrant. Then the surface area of the solid of revolution

$$= 2 \times 2\pi \int_0^a y \sqrt{1 + \left(\frac{dy}{dx}\right)^2}\, dx$$

$$= 4\pi \int_0^a \left(a^{\frac{2}{3}} - x^{\frac{2}{3}}\right)^{\frac{3}{2}} \sqrt{1 + \frac{y^{\frac{2}{3}}}{x^{\frac{2}{3}}}}\, dx$$

$$\left(\because x^{\frac{2}{3}} + y^{\frac{2}{3}} = a^{\frac{2}{3}} \Rightarrow \frac{2}{3}x^{-\frac{1}{3}} + \frac{2}{3}y^{-\frac{1}{3}}\frac{dy}{dx} = 0\right.$$

$$\Rightarrow \frac{dy}{dx} = -\frac{y^{1/3}}{x^{1/3}} \Big)$$

$$= 4\pi \int_0^a \left(a^{\frac{2}{3}} - x^{\frac{2}{3}}\right)^{\frac{3}{2}} \frac{\sqrt{x^{\frac{2}{3}} + y^{\frac{2}{3}}}}{x^{\frac{1}{3}}} dx$$

$$= 4\pi \int_0^a \left(a^{\frac{2}{3}} - x^{\frac{2}{3}}\right)^{\frac{3}{2}} \frac{a^{\frac{1}{3}}}{x^{\frac{1}{3}}} dx \quad \left(\because x^{\frac{2}{3}} + y^{\frac{2}{3}} = a^{\frac{2}{3}}\right)$$

$$= 4\pi \int_0^{\frac{\pi}{2}} \left(a^{\frac{2}{3}} - a^{\frac{2}{3}} \sin^2 \theta\right)^{\frac{3}{2}} \frac{a^{\frac{1}{3}}}{a^{\frac{1}{3}} \sin\theta} \times 3a\left(\sin^2\theta\right)\cos\theta d\theta$$

$$\begin{bmatrix} \text{putting } x^{\frac{2}{3}} = a^{\frac{2}{3}} \sin^2\theta \text{ so that } x = a\left(\sin^3\theta\right); \\ dx = 3a\left(\sin^2\theta\right)\cos\theta d\theta, \ x = 0 \Rightarrow \theta = 0; \\ x = a \Rightarrow \theta = \frac{\pi}{2} \end{bmatrix}$$

$$= 12\pi a \int_0^{\frac{\pi}{2}} \left\{a^{\frac{2}{3}}\left(1 - \sin^2\theta\right)\right\}^{\frac{3}{2}} \sin\theta\cos\theta d\theta$$

$$= 12\pi a^2 \int_0^{\frac{\pi}{2}} \cos^3\theta \times \sin\theta\cos\theta d\theta$$

$$= 12\pi a^2 \int_1^0 t^4 (-dt)$$

(putting $\cos\theta = t$ so that $-\sin\theta d\theta = dt$ and
$\theta = 0 \Rightarrow t = 1, \theta = \frac{\pi}{2} \Rightarrow t = 0$)

$$= -12\pi a^2 \left[\frac{t^5}{5}\right]_1^0 = \frac{12\pi a^2}{5}.$$

90. (d)

Here, x-axis is the initial line and θ varies from 0 to π.

Also, $r = a\,(1 + \cos\theta) \Rightarrow \frac{dr}{d\theta} = -a\sin\theta \cdot$

∴ Required surface area of the solid of revolution

$$= 2\pi \int_{\theta=0}^{\pi} r\sin\theta \sqrt{r^2 + \left(\frac{dr}{d\theta}\right)^2} \, d\theta$$

$$= 2\pi \int_0^{\pi} a(1+\cos\theta)\sin\theta\sqrt{a^2(1+\cos\theta)^2 + a^2\sin^2\theta}\,d\theta$$

$$= 2\pi a^2 \int_0^{\pi} (1+\cos\theta)\sin\theta\sqrt{1+1+2\cos\theta}\,d\theta$$

$$= 2\pi a^2 \int_0^{\pi} \sqrt{2}(1+\cos\theta)^{\frac{3}{2}} \sin\theta d\theta$$

$$= 2\pi a^2 \int_2^0 \sqrt{2}\,t^{\frac{3}{2}}(-dt)$$

(putting $1+\cos\theta = t$ so that $-\sin\theta d\theta = dt$ and
$\theta = 0 \Rightarrow t = 2, \theta = \pi \Rightarrow t = 0$)

$$= -2\sqrt{2}\pi a^2 \left[\frac{t^{\frac{5}{2}}}{\frac{5}{2}}\right]_2^0 = 2\sqrt{2}\pi a^2 \times \frac{2}{5} \times 2^{\frac{5}{2}}$$

$$= \frac{32\pi a^2}{5}.$$

Previous Years Solved Papers (2000-2018)

1. Limit of the function

$f(x) = \dfrac{1-a^4}{x^4}$ as $x \to \infty$ is given by

(a) 1   (b) e^{-a^4}

(c) ∞ (d) 0

[GATE 2000]

2. Limit of the following series as x approaches $\dfrac{\pi}{2}$ is

$$f(x) = x - \frac{x^3}{3!} + \frac{x^5}{5!} - \frac{x^7}{7!} + \ldots\ldots\infty$$

(a) $\dfrac{2\pi}{3}$ (b) $\dfrac{\pi}{2}$ (c) $\dfrac{\pi}{3}$ (d) 1

[CE GATE 2001]

3. $\displaystyle \lim_{x \to \frac{\pi}{4}} \left[\frac{\sin 2\left(x - \dfrac{\pi}{4}\right)}{x - \dfrac{\pi}{4}} \right] = ?$

(a) 0   (b) $\dfrac{1}{2}$ (c) 1   (d) 2

[IN GATE 2001]

4. $\lim\limits_{x\to 0} \dfrac{\sin^2 x}{x} = ?$

    (a) 0   (b) ∞ (c) 1   (d) -1

                        [GATE 2003]

5. The value of the function,

$$f(x) = \lim\limits_{x\to 0} \dfrac{x^3 + x}{2x^3 - 7x^2} \text{ is}$$

    (a) 0   (b) $\dfrac{-1}{7}$ (c) $\dfrac{1}{7}$ (d) ∞

                        [GATE 2004]

6. $\lim\limits_{\theta\to 0} \dfrac{\sin\left(\dfrac{\theta}{2}\right)}{\theta}$ is

    (a) 0.5   (b) 1   (c) 2

    (d) not defined

                  [EC GATE 2007]

7. $\lim\limits_{x\to 0} \dfrac{e^x - \left(1 + x + \dfrac{x^2}{2}\right)}{x^3} = ?$

    (a) 0   (b) $\dfrac{1}{6}$ (c) $\dfrac{1}{3}$ (d) 1

                  [ME GATE 2007]

8. If $y = x + \sqrt{x + \sqrt{x + \sqrt{x + \ldots\infty}}}$ then $y(2) = ?$

    (a) 4 or 1   (b) 4 only

    (c) 1 only   (d) Undefined

                  [ME GATE 2007]

9. What is the value of $\lim\limits_{x\to\frac{\pi}{4}} \dfrac{\cos x - \sin x}{x - \dfrac{\pi}{4}}$?

    (a) $\sqrt{2}$ (b) 0

    (c) $-\sqrt{2}$ (d) limit doesn't exist

                  [PI GATE 2007]

10. $\lim\limits_{x\to\infty} \dfrac{x - \sin x}{x + \cos x} = ?$

    (a) 1   (b) -1   (c) ∞ (d) $-\infty$

                  [CS GATE 2008]

11. $\lim\limits_{x\to 0} \dfrac{\sin x}{x}$ is

    (a) indeterminate   (b) 0

    (c) 1   (d) ∞

                  [IN GATE 2008]

12. The expression $e^{-\ln x}$ for $x > 0$ is equal to

    (a) $-x$ (b) x (c) x^{-1} (d) $-x^{-1}$

                  [IN GATE 2008]

13. The value of $\lim\limits_{x\to 8} \dfrac{x^{\frac{1}{3}} - 2}{x - 8}$ is

    (a) $\dfrac{1}{16}$ (b) $\dfrac{1}{12}$ (c) $\dfrac{1}{8}$ (d) $\dfrac{1}{4}$

                  [ME GATE 2008]

14. The $\lim\limits_{x\to 0} \dfrac{\sin\left(\dfrac{2}{3}x\right)}{x}$ is

    (a) $\dfrac{2}{3}$ (b) 1   (c) $\dfrac{3}{2}$ (d) ∞

                  [CE GATE 2010]

15. What is the value of $\lim\limits_{n\to\infty} \left(1 - \dfrac{1}{n}\right)^{2n}$?

    (a) 0   (b) e^{-2} (c) $e^{-\frac{1}{2}}$ (d) 1

                  [CS GATE 2010]

16. What is $\lim\limits_{\theta\to 0} \dfrac{\sin\theta}{\theta}$ equal to?

    (a) θ (b) $\sin\theta$ (c) 0   (d) 1

                  [ME GATE 2011]

17. The value of $\lim\limits_{x\to\infty} \left(x + \dfrac{1}{x}\right)^x$ is

    (a) ln2   (b) 1.0   (c) e   (d) ∞

                  [EC GATE 2014]

18. $\lim\limits_{x\to\infty} \left(\dfrac{x + \sin x}{x}\right)$ equals to

    (a) $-\infty$ (b) 0   (c) 1   (d) ∞

                  [CE GATE 2014]

19. Given

    $x(t) = 3\sin(1000\pi t)$ and

    $y(t) = 5\cos\left(1000\pi t + \dfrac{\pi}{4}\right)$

    The x-y plot will be

    (a) a circle

    (b) a multiloop closed curve

    (c) a hyperbola   (d) an ellipse

                  [IN GATE 2014]

20. $\underset{x\to\infty}{\mathrm{Lt}}\left(1+\dfrac{1}{x}\right)^{2x}$ is equal to

(a) e^{-2} (b) e (c) 1   (d) e^2

[CE GATE 2015]

21. A function $f(x)$ is linear and has a value of 29 at $x=-2$ and 39 at $x=3$. Find its value $x=5$.

(a) 59   (b) 45

(c) 43   (d) 35

[CE GATE 2015]

22. Choose the most appropriate equation for the function drawn as a thick line in the plot below:

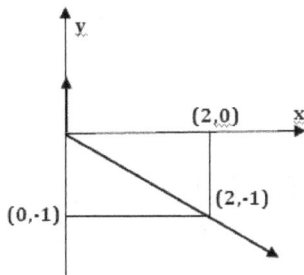

(a) $x=y-|y|$ (b) $x=-(y-|y|)$

(c) $x=y+|y|$ (d) $x=-(y+|y|)$

[CS GATE 2015]

23. $\underset{n\to\infty}{\mathrm{Lt}}\left(\sqrt{n^2+n}-\sqrt{n^2+1}\right)$ is ____?

[IN GATE 2016]

24. $\underset{x\to\infty}{\mathrm{Lt}}\left(\sqrt{x^2+x-1}-x\right)$ is

(a) 0   (b) ∞ (c) $\dfrac{1}{2}$ (d) $-\infty$

[ME GATE 2016]

25. $\underset{x\to0}{\mathrm{Lt}}\left(\dfrac{e^{5x}-1}{x}\right)^2$ is equal to ____?

[PI GATE 2016]

26. $\underset{x\to4}{\mathrm{Lt}}\dfrac{\sin(x-4)}{x-4}=$?

[CS GATE 2016]

27. How many distinct values of x satisfy the equation $\sin x=\dfrac{x}{2}$, where x is in radians?

(a) 1   (b) 2

(c) 3   (d) 4 or more

[EC GATE 2016]

28. $\underset{x\to0}{\mathrm{Lt}}\dfrac{\log_e(1+4x)}{e^{3x}-1}$ is equal to?

(a) 0   (b) $\dfrac{1}{12}$ (c) $\dfrac{4}{3}$ (d) 1

[ME GATE 2016]

29. The value of $\underset{x\to0}{\lim}\left(\dfrac{x^3-\sin x}{x}\right)$ is

(a) 0   (b) 3   (c) 1   (d) -1

[EE GATE 2017]

30. $\underset{x\to0}{\lim}\left(\dfrac{\tan x}{x^2-x}\right)$ is equal to ____?

[CE GATE 2017]

31. Which of the following function(s) is an accurate description of the graph for the range(s) indicated?

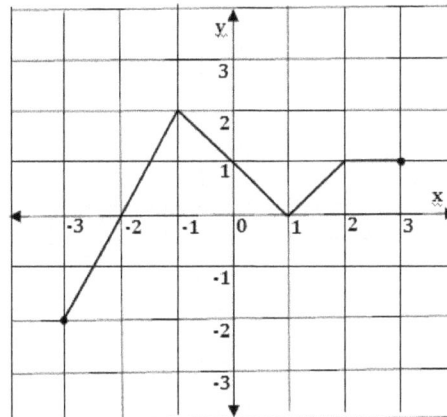

(i) $y=2x+4$ for $-3\le x\le-1$

(ii) $y=|x-1|$ for $-1\le x\le2$

(iii) $y=\big||x|-1\big|$ for $-1\le x\le2$

(iv) $y=1$ for $2\le x\le3$

(a) (i), (ii) and (iii) only

(b) (i), (ii) and (iv) only

(c) (i) and (iv) only

(d) (ii) and (iv) only

[CE GATE 2018]

32. Consider two functions $f(x)=(x-2)^2$ and $g(x)=2x-1$, where x is real. The smaller value of x for which $f(x)=g(x)$ is ____?

[IN GATE 2018]

33. Which of the following functions describe the graph shown in the below figure?

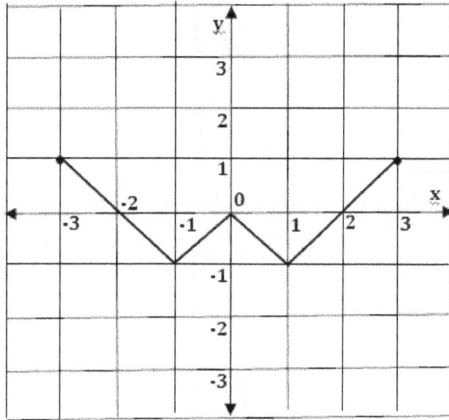

(a) $y = \left\| x \right| + 1 \right| - 2$

(b) $y = \left\| x \right| - 1 \right| - 1$

(c) $y = \left\| x \right| + 1 \right| - 1$

(d) $y = |x - 1| - 1$

[PI GATE 2018]

34. Which of the following functions is not differentiable in the domain $[-1,1]$?

(a) $f(x) = x^2$

(b) $f(x) = x - 1$

(c) $f(x) = 2$

(d) $f(x) = \text{maximum } (x, -x)$

[GATE 2002]

35. If $x = a(\theta + \sin\theta)$ and $y = a(1 - \cos\theta)$, then $\dfrac{dy}{dx} = ?$

(a) $\sin\dfrac{\theta}{2}$

(b) $\cos\dfrac{\theta}{2}$

(c) $\tan\dfrac{\theta}{2}$

(d) $\cot\dfrac{\theta}{2}$

[GATE 2004]

36. Consider the function $f(x) = |x|^3$, where x is real. Then the function $f(x)$ at $x = 0$ is

(a) continuous but not differentiable

(b) once differentiable but not twice

(c) twice differentiable but not thrice

(d) thrice differentiable

[IN GATE 2007]

37. Given $y = x^2 + 2x + 10$ the value of $\left. \dfrac{dy}{dx} \right|_{x=1}$ is

(a) 0

(b) 4

(c) 12

(d) 13

[IN GATE 2008]

38. If $f(x) = \sin|x|$, then the value of $\dfrac{df}{dx}$ at $x = \dfrac{-\pi}{4}$ is

(a) 0

(b) $\dfrac{1}{\sqrt{2}}$

(c) $-\dfrac{1}{\sqrt{2}}$

(d) 1

[PI GATE 2010]

39. The function $y = |2 - 3x|$

(a) is continuous $\forall x \in R$ and differentiable $\forall x \in R$

(b) is continuous $\forall x \in R$ and differentiable $\forall x \in R$ except at $x = \dfrac{3}{2}$

(c) is continuous $\forall x \in R$ and differentiable $\forall x \in R$ except at $x = \dfrac{2}{3}$

(d) is continuous $\forall x \in R$ and except at $x = 3$ and differentiable $\forall x \in R$

[ME GATE 2010]

40. At $t = 0$, the function $f(t) = \dfrac{\sin t}{t}$ has

(a) a minimum

(b) a discontinuity

(c) a point of inflection

(d) a maximum

[EE GATE 2010]

41. What should be the value of λ such that the function defined below is continuous at $x\dfrac{\pi}{2}$? $x = \dfrac{\pi}{2}$?

$$f(x) = \begin{cases} \dfrac{\lambda \cos x}{\dfrac{\pi}{2} - x}, & \text{if } x \neq \dfrac{\pi}{2} \\ 1, & \text{if } x = \dfrac{\pi}{2} \end{cases}$$

(a) 0

(b) 2π

(c) 1

(d) $\dfrac{\pi}{2}$

[CE GATE 2011]

42. Consider the function $f(x) = |x|$ in the interval $-1 \leq x \leq 1$. At the point $x = 0$, $f(x)$ is

(a) continuous and differentiable

(b) non-continuous and differentiable

(c) continuous and non-differentiable

(d) neither continuous nor differentiable

[ME, PI GATE 2012]

43. A function $y = 5x^2 + 10x$ is defined over an open interval $x = (1, 2)$. At least at one point in this interval, $\frac{dy}{dx}$ exactly

(a) 20 (b) 25 (c) 30 (d) 35

[EE GATE 2013]

44. Which of the following functions is continuous at $x = 3$?

(a) $f(x) = \begin{cases} 2 & \text{if } x = 3 \\ x-1 & \text{if } x > 3 \\ \dfrac{x+3}{3} & \text{if } x < 3 \end{cases}$

(b) $f(x) = \begin{cases} 4 & \text{if } x = 3 \\ 8-x & \text{if } x \neq 3 \end{cases}$

(c) $f(x) = \begin{cases} x+3 & \text{if } x \leq 3 \\ x-4 & \text{if } x > 3 \end{cases}$

(d) $f(x) = \dfrac{1}{x^3 - 27}$ if $x \neq 3$

[CS GATE 2013]

45. If a function is continuous at a point,

(a) the limit of the function may not exist at the point

(b) the function must be derivable at the point

(c) the limit of the function at the point tends to infinity

(d) the limit must exist the point and the value of limit should be same as the value of function at the point

[ME GATE 2014]

46. The function $f(x) = x \sin x$ satisfies the following equation:

$f''(x) + f(x) + t \cos x = 0$. The value of t is ___?

[CS GATE 2014]

47. The values of for which the function

$f(x) = \dfrac{x^2 - 3x - 4}{x^2 + 3x - 4}$ is NOT continuous are

(a) 4 and −1 (b) 4 and 1

(c) −4 and 1 (d) −1 and −4

[ME GATE 2016]

48. Given the following statements about a function $f : \mathbb{R} \to \mathbb{R}$, select the right option:

P: If $f(x)$ is continuous at $x = x_0$, then it is also differentiable at $x = x_0$

Q: If $f(x)$ is continuous at $x = x_0$, then it may not be differentiable at $x = x_0$

R: If $f(x)$ is differentiable at $x = x_0$, then it is also continuous at $x = x_0$

(a) P is true, Q is false, R is false

(b) P is false, Q is true, R is true

(c) P is false, Q is true, R is false

(d) P is true, Q is false, R is true

[EC GATE 2016]

49. At $x = 0$, the function is

$f(x) = \left| \sin \dfrac{2\pi x}{L} \right|$ $(-\infty < x < \infty, L > 0)$

(a) continuous and differentiable

(b) not continuous and not differentiable

(c) not continuous but differentiable

(d) continuous but not differentiable

[PI GATE 2016]

50. A function $f(x)$ is defined as

$f(x) = \begin{cases} e^x, & x < 1 \\ \ln x + ax^2 + bx, & x \geq 1 \end{cases}$

Which one of the following statements is TRUE?

(a) $f(x)$ is not differentiable at $x = 1$ for any values of a and b.

(b) $f(x)$ is differentiable at $x = 1$ for the unique values of a and b.

(c) $f(x)$ is differentiable at $x = 1$ for all values of a and b such that $a + b = e$.

(d) $f(x)$ is differentiable at $x = 1$ for all values of a and b.

[EE GATE 2017]

51. Let $g(x) = \begin{cases} -x, & x \leq 1 \\ x+1, & x \geq 1 \end{cases}$ and

$f(x) = \begin{cases} 1-x, & x \leq 0 \\ x^2, & x > 0 \end{cases}$

Consider the composition of f and g, i.e., $(f \circ g)(x) = f(g(x))$. Then the number of discontinuities in $(f \circ g)(x)$ present in the interval $(-\infty, 0)$ is

(a) 0   (b) 1   (c) 2   (d) 4

[EE GATE 2017]

52. The tangent to the curve represented by $y = x \ln x$ is required to have inclination $45°$ with the x-axis. The coordinates of the tangent point would be

(a) $(1,0)$ (b) $(0,1)$
(c) $(1, 1)$ (d) $\left(\sqrt{2}, \sqrt{2}\right)$

[CE GATE 2017]

53. A real-valued function y of real variable x is such that $y = 5|x|$. At $x = 0$, the function is

(a) discontinuous but differentiable
(b) both continuous and differentiable
(c) discontinuous and not differentiable
(d) discontinuous but not differentiable

[PI GATE 2018]

54. Let f be a real valued function of a real variable defined as $f(x) = x^2$ for $x \geq 0$, and $f(x) = -x^2$ for $x < 0$. Which one of the following statements is true?

(a) $f(x)$ is discontinuous at $x = 0$

(b) $f(x)$ is continuous but not differentiable at $x = 0$

(c) $f(x)$ is differentiable but its first derivative is not continuous at $x = 0$

(d) $f(x)$ is differentiable but its first derivative is not differentiable at $x = 0$

[EE GATE 2018]

55. Limit of the following sequence as $n \to \infty$ is ___?

$$x_n = n^{\frac{1}{n}}$$

(a) 0   (b) 1   (c) ∞ (d) $-\infty$

[CE GATE 2002]

56. What is the value of $\underset{x \to \frac{\pi}{4}}{Lim} \dfrac{\cos x - \sin x}{x - \dfrac{\pi}{4}}$

(a) $\sqrt{2}$ (b) 0
(c) $-\sqrt{2}$ (d) Limit does not exist

[PI GATE 2007]

57. The value of the expression $\underset{x \to 0}{Lt} \dfrac{\sin x}{x e^x}$ is

(a) 0   (b) $\dfrac{1}{2}$ (c) 1   (d) $\dfrac{1}{1+e}$

[PI GATE 2008]

58. $\underset{x \to 0}{Lt} \left(\dfrac{1 - \cos x}{x^2} \right)$ is

(a) $\dfrac{1}{4}$ (b) $\dfrac{1}{2}$ (c) 1   (d) 2

[ME, PI GATE 2012]

59. $\underset{x \to 0}{Lt} \dfrac{x - \sin x}{1 - \cos x}$ is

(a) 0   (b) 1
(c) 3   (d) not defined

[ME GATE 2014]

60. $\underset{x \to 0}{Lt} \left(\dfrac{e^{2x} - 1}{\sin 4x} \right)$ is

(a) 0   (b) 0.5   (c) 1   (d) 2

[ME GATE 2014]

61. The expression $\underset{a \to 0}{Lt} \dfrac{x^a - 1}{a}$ is equal to

(a) $\log x$ (b) 0

(c) $x \log x$ (d) ∞

[CE GATE 2014]

62. The value of

$$\underset{x \to 0}{Lt} \left(\dfrac{-\sin x}{2 \sin x + x \cos x} \right) \text{ is } \_\_\_\_?$$

[ME GATE 2015]

63. The value of $\underset{x \to 0}{Lt} \dfrac{1 - \cos x^2}{2x^4}$ is ___?

(a) 0   (b) $\dfrac{1}{2}$

(c) $\dfrac{1}{4}$ (d) undefined

[ME GATE 2015]

64. $\underset{x \to 0}{Lt} \dfrac{\log(1 + 4x)}{e^{3x} - 1}$ is equal to

(a) 0   (b) $\dfrac{1}{12}$ (c) $\dfrac{4}{3}$ (d) 1

[ME GATE 2016]

65. The value of $\underset{x \to 1}{\lim} \dfrac{x^7 - 2x^5 + 1}{x^3 - 3x^2 + 2}$ is

(a) 0   (b) -1
(c) 1   (d) does not exist

[CS/IT GATE 2017]

66. Let the function

$$f(\theta) = \begin{vmatrix} \sin\theta & \cos\theta & \tan\theta \\ \sin\left(\dfrac{\pi}{6}\right) & \cos\left(\dfrac{\pi}{6}\right) & \tan\left(\dfrac{\pi}{6}\right) \\ \sin\left(\dfrac{\pi}{3}\right) & \cos\left(\dfrac{\pi}{3}\right) & \tan\left(\dfrac{\pi}{3}\right) \end{vmatrix}$$

where $\theta \in \left[\dfrac{\pi}{6}, \dfrac{\pi}{3}\right]$ and $f'(\theta)$ denotes the derivative of f with respect to θ. Which of the following statement/statements is/are true?

(I) There exists $\theta \in \left(\dfrac{\pi}{6}, \dfrac{\pi}{3}\right)$ such that

$f'(\theta) = 0$

(II) There exists $\theta \in \left(\dfrac{\pi}{6}, \dfrac{\pi}{3}\right)$ such that

$f'(\theta) \neq 0$

(a) I only   (b) II only
(c) Both I and II (d) neither I nor II
[CS GATE 2014]

67. A function $f(x)$ is continuous in the interval $[0, 2]$. It is given that $f(0)=-1=f(2)$ and $f(1)=1$. Which one of the following statements must be true?

(a) there exist a "y" in the interval $(0,1)$ such that $f(y)=f(y+1)$

(b) for every "y" in the interval $(0, 1)$, $f(y)=f(2-y)$

(c) the maximum value of the function in the interval $(0, 2)$ is 1

(d) there exists a "y" in the interval $(0, 1)$ such that $f(y)=-f(2+y)$

[CS GATE 2014]

68. A function $f(x) = 1 - x^2 + x^3$ is defined in the closed interval $[-1,1]$. The value of x, in the open interval $(-1,1)$. for which the mean value theorem is satisfied, is

(a) $\dfrac{-1}{2}$ (b) $\dfrac{-1}{3}$ (c) $\dfrac{1}{3}$ (d) $\dfrac{1}{2}$

[CE GATE 2015]

69. According to the Mean Value Theorem, for a continuous function $f(x)$ in the interval $[a,b]$, there exists a value ξ in this interval such that $\displaystyle\int_a^b f(x)dx =$

(a) $f(\xi)(b-a)$ (b) $f(b)(\xi-a)$

(c) $f(a)(b-\xi)$ (d) 0
[ME GATE 2018]

70. As x varies from −1 to 3, which of the following describes the behavior of the function $f(x) = x^3 - 3x^2 + 1$?

(a) $f(x)$ Increases monotonically

(b) $f(x)$ Increases, then decreases and increases again

(c) $f(x)$ decreases, then increases and decreases again

(d) $f(x)$ Increases and then decreases
[EC GATE 2016]

71. The following function has local minima at which value of x,

$$f(x) = x\sqrt{5-x^2}$$

(a) $-\dfrac{\sqrt{5}}{2}$ (b) $\sqrt{5}$ (c) $\sqrt{\dfrac{5}{2}}$ (d) $-\sqrt{\dfrac{5}{2}}$

[CE GATE 2002]

72. The function $f(x) = 2x^3 - 3x^2 - 36x + 2$ has its maxima at

(a) $x = -2$ only (b) $x = 0$ only

(c) $x = 3$ only

(d) both $x = -2$ and $x = 3$
[GATE 2004]

73. For the function $f(x) = x^2 e^{-x}$, the maximum occurs when x is equal to

(a) 2 (b) 1 (c) 0 (d) −1
[EE GATE 2005]

74. For real x the maximum value of $\dfrac{e^{\sin x}}{e^{\cos x}}$ is

(a) 1 (b) e (c) $e^{\sqrt{2}}$ (d) ∞
[IN GATE 2007]

75. The minimum value of function $y = x^2$ in the interval $[1, 5]$ is

(a) 0   (b) 1

(c) 25   (d) undefined

[ME GATE 2007]

76. Consider the function $f(x) = x^2 - x - 2$. The maximum value of $f(x)$ in the closed interval $[-4, 4]$ is

(a) 18   (b) 10

(c) −2.25   (d) indeterminate

[EC GATE 2007]

77. Consider the function $f(x) = (x^2 - 4)^2$ where x is a real number. Then the function has

(a) Only one minimum

(b) Only two minima

(c) Three minima   (d) Three maxima

[EE GATE 2007]

78. Consider the function $f(x) = x^2 - x - 2$. Then the maximum value of $f(x)$ in the closed interval $[-4, 4]$ is

(a) 18   (b) 10

(c) -2.25   (d) indeterminate

[EC GATE 2007]

79. A point on the curve is said to be an extremum if it is a local minimum (or) a local maximum. The number of distinct extrema for the curve $3x^4 - 16x^3 + 24x^2 + 37$ is

(a) 0   (b) 1   (c) 2   (d) 3

[CS GATE 2008]

80. Consider the function $y = x^2 - 6x + 9$. The maximum value of y obtained when x varies over the interval 2 to 5 is

(a) 1   (b) 3   (c) 4   (d) 9

[IN GATE 2008]

81. For the real value x, the minimum value of the function $f(x) = e^x + e^{-x}$ is

(a) 2   (b) 1   (c) 0.5   (d) 0

[EC GATE 2008]

82. If $e^y = x^{\frac{1}{x}}$, then y has a

(a) maximum at $x = e$

(b) minimum at $x = e$

(c) maximum at $x = e^{-1}$

(d) minimum at $x = e^{-1}$

[EC GATE 2010]

83. The function $f(x) = 2x - x^2 + 3$ has

(a) a maxima at $x = 1$ and minima at $x = 5$

(b) a maxima at $x = 1$ and minima at $x = -5$

(c) only a maxima at $x = 1$

(d) only a minima at $x = 5$

[EE GATE 2011]

84. The maximum value of $f(x) = x^3 - 9x^2 + 24x + 5$ in the interval $[1, 6]$ is

(a) 21   (b) 25   (c) 41   (d) 46

[EE, EC, IN GATE 2012]

85. At $x = 0$, the function $f(x) = x^3 + 1$ has

(a) a maximum value   (b) a minimum value

(c) a singularity

(d) a point of inflection

[ME, PI GATE 2012]

86. A political party orders an arch for the entrance to the ground in which the annual convention is being held. The profile of the arch follows the equation $y = 2x - 0.1x^2$ where y is the height of the arch in meters. The maximum possible height of the arch is

(a) 8 meters   (b) 10 meters

(c) 12 meters   (d) 14 meters

[PI, ME GATE 2012]

87. For $0 \le t < \infty$, the maximum value of the function $f(t) = e^{-t} - 2e^{-2t}$ occurs at

(a) $t = \log_e 4$ (b) $t = \log_e 2$

(c) $t = 0$ (d) $t = \log_e 8$

[EC GATE 2014]

88. The maximum value of the function $f(x) = \ln(1 + x) - x$ (where $x > -1$) occurs at $x = ?$

[EC GATE 2014]

89. The maximum value of $f(x) = 2x^3 - 9x^2 + 12x - 3$ in the interval $0 \le x \le 3$ is ____?

[EC GATE2014]

90. For a right-angled triangle, if the sum of the lengths of the hypotenuse and a side is kept constant, in order to have maximum area of the triangle, the angle between the hypotenuse and the side is

(a) $12°$ (b) $36°$ (c) $60°$ (d) $45°$

[EC GATE 2014]

91. Let $f(x) = xe^{-x}$. The maximum value of the function in the interval $(0, \infty)$ is

(a) e^{-1} (b) e
(c) $1 - e^{-1}$ (d) $1 + e^{-1}$

[EE GATE 2014]

92. Minimum of the real valued function $f(x) = (x-1)^{\frac{2}{3}}$ occurs at x equal to

(a) $-\infty$ (b) 0 (c) 1 (d) ∞

[EE GATE 2014]

93. The minimum value of the function $f(x) = x^3 - 3x^2 - 24x + 100$ in the interval $[-3, 3]$ is

(a) 20 (b) 28 (c) 16 (d) 32

[EE GATE 2014]

94. The maximum area (in square units) of a rectangle whose vertices lie on the ellipse $x^2 + 4y^2 = 1$ is____?

[EC GATE 2015]

95. Which one of the following graphs describes the function?
$f(x) = e^{-x}(x^2 + x + 1)$?

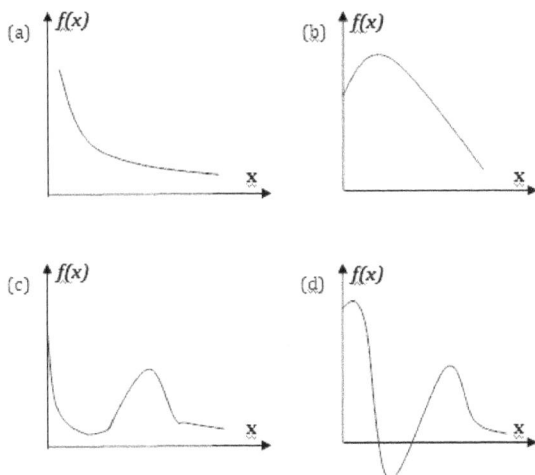

[EC GATE 2015]

96. The maximum value attained by the function $f(x) = x(x-1)(x-2)$ in the interval $[1, 2]$ is____?

[EE GATE 2016]

97. Let $f : [-1, 1] \to R$, where $f(x) = 2x^3 - x^4 - 10$. The minimum value of $f(x)$ is_____?

[IN GATE 2016]

98. Consider the function $f(x) = 2x^3 - 3x^2$ in the domain $[-1, 2]$. The global minimum of $f(x)$ is_____?

[ME GATE 2016]

99. The range of values of k for which the function $f(x) = (k^2 - 4)x^2 + 6x^3 + 8x^4$ has local maxima at point $x = 0$ is

(a) $k < -2$ or $k > 2$ (b) $k \le -2$ or $k \ge 2$
(c) $-2 < k < 2$ (d) $-2 \le k \le 2$

[PI GATE 2016]

100. At the point $x = 0$, the function $f(x) = x^3$ has
(a) local maximum
(b) local minimum
(c) both local maximum and minimum
(d) neither local maximum local minimum

[CE GATE 2018]

101. Let $f(x) = 3x^3 - 7x^2 + 5x + 6$. The maximum value of $f(x)$ over the interval $[0, 2]$ is____ ? (up to 1 decimal places).

[EE GATE 2018]

102. The Taylor series expansion of $\sin x$ about $x = \dfrac{\pi}{6}$ is given by

(a)
$$\frac{1}{2} + \frac{\sqrt{3}}{2}\left(x - \frac{\pi}{6}\right) - \frac{1}{4}\left(x - \frac{\pi}{6}\right)^2 - \frac{\sqrt{3}}{12}\left(x - \frac{\pi}{6}\right)^3 + ...\infty$$

(b) $x - \dfrac{x^3}{3!} + \dfrac{x^5}{5!} - \dfrac{x^7}{7!} + ...$

(c) $\dfrac{\left(x - \frac{\pi}{6}\right)}{1!} - \dfrac{\left(x - \frac{\pi}{6}\right)^3}{3!} + \dfrac{\left(x - \frac{\pi}{6}\right)^5}{5!} - \dfrac{\left(x - \frac{\pi}{6}\right)^7}{7!} + ...\infty$

(d) $\dfrac{1}{2}$

[CE GATE 2000]

103. For the function e^{-x}, the linear approximation around $x = 2$ is

(a) $(3-x)e^{-2}$ (b) $1-x$

(c) $\left[3 + 2\sqrt{2} - (1+\sqrt{2}\,x)\right]e^{-2}$

(d) e^{-2}

103. (a) Let $f(x) = e^{-x}$.

Then $f'(x) = -e^{-x}, f''(x) = e^{-x}, \ldots\ldots$

So by the Taylor series expansion around $x = 2$, we have,

$$f(x) = f(2) + (x-2)f'(2) + \frac{(x-2)^2}{2!}f''(2) + \ldots\infty$$

Therefore, around $x = 2$, the linear approximation of e^{-x}

$$= f(2) + (x-2)f'(2)$$
$$= e^{-2} + (x-2)(-e^{-2}) = (3-x)e^{-2}.$$

104. For $|x| \ll 1, \coth(x)$ can be approximated as

(a) x (b) x^2 (c) $\dfrac{1}{x}$ (d) $\dfrac{1}{x^2}$

105. Which of the following function would have only the odd powers of x in its Taylor series expansion about the point $x = 0$?

(a) $\sin(x^3)$ (b) $\sin(x^2)$

(c) $\cos(x^3)$ (d) $\cos(x^2)$

106. In the Taylor series expansion of $e^x + \sin x$ about the point $x = \pi$, the coefficient of $(x-\pi)^2$ is

(a) e^π (b) $0.5\,e^\pi$

(c) $e^\pi + 1$ (d) $e^\pi - 1$

107. In the Taylor series expansion of e^x about $x = 2$, the coefficient of $(x-2)^4$ is

(a) $\dfrac{1}{4!}$ (b) $\dfrac{2^4}{4!}$ (c) $\dfrac{e^2}{4!}$ (d) $\dfrac{e^4}{4!}$

108. The Taylor series expansion of $\dfrac{\sin x}{x - \pi}$ at $x = \pi$ is given by

(a) $1 + \dfrac{(x-\pi)^2}{3!} + \ldots\ldots\infty$

(b) $-1 - \dfrac{(x-\pi)^2}{3!} + \ldots\ldots\infty$

(c) $1 - \dfrac{(x-\pi)^2}{3!} + \ldots\ldots\infty$

(d) $-1 + \dfrac{(x-\pi)^2}{3!} + \ldots\ldots\infty$

109. The infinite series

$$f(x) = x - \frac{x^3}{3!} + \frac{x^5}{5!} - \frac{x^7}{7!} + \ldots\ldots\infty \text{ converges to}$$

(a) $\cos x$ (b) $\sin x$ (c) $\sinh x$ (d) e^x

110. A series expansion for the function $\sin\theta$ is

(a) $1 - \dfrac{\theta^2}{2!} + \dfrac{\theta^4}{4!} - \ldots\ldots$

(b) $\theta - \dfrac{\theta^3}{3!} + \dfrac{\theta^5}{5!} - \ldots\ldots$

(c) $1 + \theta + \dfrac{\theta^2}{2!} + \dfrac{\theta^3}{3!} + \ldots\ldots$

(d) $\theta + \dfrac{\theta^3}{3!} + \dfrac{\theta^5}{5!} + \ldots\ldots$

111. The series $\displaystyle\sum_{m=0}^{\infty}\frac{1}{4^m}(x-1)^{2m}$ converges for

(a) $-2 < x < 2$ (b) $-1 < x < 3$

(c) $-3 < x < 1$ (d) $x < 3$

112. The infinite series

$$1 + x + \frac{x^2}{2!} + \frac{x^3}{3!} + \frac{x^4}{4!} + \ldots\ldots \text{ corresponds to}$$

(a) $\sec x$ (b) e^x

(c) $\cos x$ (d) $1 + \sin^2 x$

113. The Taylor series expansion of $3\sin x + 2\cos x$ is

(a) $2 + 3x - x^2 - \dfrac{x^3}{2} + \ldots\ldots\infty$

(b) $2 - 3x + x^2 - \dfrac{x^3}{2} + \ldots\ldots\infty$

(c) $2 + 3x + x^2 + \dfrac{x^3}{2} + \ldots\ldots\ldots\infty$

(d) $2 - 3x - x^2 + \dfrac{x^3}{2} + \ldots\ldots\ldots\infty$

[EC GATE 2014]

114. The series $\displaystyle\sum_{n=0}^{\infty} \dfrac{1}{n!}$ converges to

(a) $2\ln 2$ (b) $\sqrt{2}$ (c) 2 (d) e

[EC GATE 2014]

115. The value of $\displaystyle\sum_{n=0}^{\infty} n\left(\dfrac{1}{2}\right)^n$ is _____?

[EC GATE 2015]

116. Let $S = \displaystyle\sum_{n=0}^{\infty} na^n$ where $|a| < 1$. The value of a in the range $0 < a < 1$, such that $S = 2a$ is _____?

[EE GATE 2016]

117. The quadratic approximation of $f(x) = x^3 - 3x^2 - 5$ at the point $x = 0$ is

(a) $3x^2 - 6x - 5$ (b) $-3x^2 - 5$

(c) $-3x^2 + 6x - 5$ (d) $3x^2 - 5$

[CE GATE 2016]

118. Let $f(x) = e^{x+x^2}$ for real x. From among the following chose the Taylor series approximation of $f(x)$ around $x = 0$, which includes all powers of x less than or equal to 3.

(a) $1 + x + x^2 + x^3$ (b) $1 + x + \dfrac{3}{2}x^2 + x^3$

(c) $1 + x + \dfrac{3}{2}x^2 + \dfrac{7}{6}x^3$ (d) $1 + x + 3x^2 + 7x^3$

[EC GATE 2017]

119. The Taylor series expansion of $f(x) = \displaystyle\int_0^x e^{-\left(\frac{t^2}{2}\right)}dt$ around $x = 0$ has the form $f(x) = a_0 + a_1 x + a_2 x^2 + \ldots$ the coefficient a_2 (correct to two decimal places) is equal to _____?

[EC GATE 2018]

120. The value of the integral is $I = \displaystyle\int_0^{\pi/4} \cos^2 x\, dx$

(a) $\dfrac{\pi}{8} + \dfrac{1}{4}$ (b) $\dfrac{\pi}{8} - \dfrac{1}{4}$

(c) $-\dfrac{\pi}{8} - \dfrac{1}{4}$ (d) $-\dfrac{\pi}{8} + \dfrac{1}{4}$

[GATE 2001]

121. The value of the following definite integral in $\displaystyle\int_{-\frac{\pi}{2}}^{\frac{\pi}{2}} \dfrac{\sin 2x}{1 + \cos x} dx = ?$

(a) $-2\log 2$ (b) 2

(c) 0 (d) none

[GATE 2002]

122. The area enclosed between the parabola $y = x^2$ and the straight line $y = x$ is

(a) $\dfrac{1}{8}$ (b) $\dfrac{1}{6}$ (c) $\dfrac{1}{3}$ (d) $\dfrac{1}{2}$

[GATE 2004]

123. $\displaystyle\int_{-a}^{a}\left[\sin^6 x + \sin^7 x\right]dx$ is equal to

(a) $2\displaystyle\int_0^a \sin^6 x\, dx$ (b) $2\displaystyle\int_0^a \sin^7 x\, dx$

(c) $2\displaystyle\int_0^a\left(\sin^6 x + \sin^7 x\right)dx$

(d) zero

[ME GATE 2005]

124. The expression $V = \displaystyle\int_0^H \pi R^2\left(1 - \dfrac{h}{H}\right)^2 dh$ for the volume of a cone is equal to

(a) $\displaystyle\int_0^R \pi R^2\left(1 - \dfrac{h}{H}\right)^2 dr$

(b) $\displaystyle\int_0^R \pi R^2\left(1 - \dfrac{h}{H}\right)^2 dh$

(c) $\displaystyle\int_0^H 2\pi rH\left(1 - \dfrac{r}{R}\right)dh$

(d) $\displaystyle\int_0^R 2\pi rH\left(1 - \dfrac{r}{R}\right)dr$

[EE GATE 2006]

125. The following plot shows a function y which varies linearly with x. The value of the integral $I = \int_1^2 y\, dx$

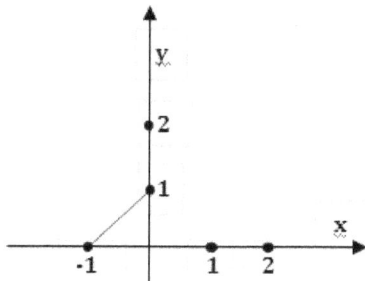

(a) 1   (b) 2.5   (c) 4   (d) 5

[EC GATE 2007]

126. The integral $\dfrac{1}{2\pi}\int_0^{2\pi} \sin(t-\tau)\cos\tau\, d\tau$ equals

(a) $\sin t \cos t$ (b) 0

(c) $\dfrac{1}{2}\cos t$ (d) $\dfrac{1}{2}\sin t$

[EE GATE 2007]

127. The value of the integral $\int_{-\frac{\pi}{2}}^{\frac{\pi}{2}} x\cos x\, dx$ is

(a) 0   (b) $\pi - 2$ (c) π (d) $\pi + 2$

[PI GATE 2008]

128. The area enclosed between the curves $y^2 = 4x$ and $x^2 = 4y$ is

(a) $\dfrac{16}{3}$ (b) 8   (c) $\dfrac{32}{3}$ (d) 16

[ME GATE 2009]

129. The value of the quantity, where

$P = \int_0^1 x e^x\, dx$ is

(a) 0   (b) 1   (c) e (d) $\dfrac{1}{e}$

[EE GATE 2010]

130. What is the value of the definite integral $\int_0^a \dfrac{\sqrt{x}}{\sqrt{x}+\sqrt{a-x}}\, dx$?

(a) 0   (b) $\dfrac{a}{2}$ (b) a (d) $2a$

[CE GATE 2011]

131. If $f(x)$ is even function and a is a positive real number, then $\int_{-a}^{a} f(x)\, dx$ is

(a) 0   (b) a

(c) $2a$ (d) $2\int_0^a f(x)\, dx$

[ME GATE 2011]

132. Given $i = \sqrt{-1}$, what will be the evalution of the definite integral $\int_0^{\frac{\pi}{2}} \dfrac{\cos x + i\sin x}{\cos x - i\sin x}\, dx$?

(a) 0   (b) 2   (c) $-i$ (d) i

[CS GATE 2011]

133. The area enclosed between the straight line $y = x$ and the parabola $y = x^2$ in the xy plane

(a) $\dfrac{1}{6}$ (b) $\dfrac{1}{4}$ (c) $\dfrac{1}{3}$ (d) $\dfrac{1}{2}$

[ME, PI GATE 2012]

134. The value of the definite integral $\int_1^e \sqrt{x}\ln(x)\, dx$ is

(a) $\dfrac{4}{9}\sqrt{e^3} + \dfrac{2}{9}$ (b) $\dfrac{2}{9}\sqrt{e^3} - \dfrac{4}{9}$

(c) $\dfrac{2}{9}\sqrt{e^3} + \dfrac{4}{9}$ (d) $\dfrac{4}{9}\sqrt{e^3} - \dfrac{2}{9}$

[ME GATE 2013]

135. A particle, starting from origin at $t = 0$ sec, is traveling along $x-$ axis with velocity $v = \dfrac{\pi}{2}\cos\left(\dfrac{\pi}{2}t\right) m/s$. At $t = 3$ sec, the difference between the distance covered by the particle and the magnitude of displacement from the origin is _____?

[EE GATE 2014]

136. The value of the integral $\int_0^2 \dfrac{(x-1)^2 \sin(x-1)}{(x-1)^2 + \cos(x-1)}\, dx$ is

(a) 3   (b) 0

(c) -1 (d) -2

[ME GATE 2014]

137. If $\int\limits_{0}^{2\pi} |x\sin x|\,dx = k\pi$, then the value of k is

equal to _____ ?

[CS GATE 2014]

138. The value of the integral given below is

$\int\limits_{0}^{\pi} x^2 \cos x\,dx$

(a) -2π (b) π (c) $-\pi$ (d) 2π

[CS GATE 2014]

139. Given $i = \sqrt{-1}$, the value of the definite integral $I = \int\limits_{0}^{\frac{\pi}{2}} \dfrac{\cos x + i\sin x}{\cos x - i\sin x}\,dx$ is : is:

(a) 1 (b) −1 (c) i (d) $-i$

[CE GATE 2015]

140. If for nonzero x, $af(x) + bf\left(\dfrac{1}{x}\right) = \dfrac{1}{x} - 25$

where $a \neq b$, then $\int\limits_{1}^{2} f(x)\,dx$ is

(a) $\dfrac{1}{a^2 - b^2}\left[a(\ln 2 - 25) + \dfrac{47}{2}b \right]$

(b) $\dfrac{1}{a^2 - b^2}\left[a(2\ln 2 - 25) - \dfrac{47}{2}b \right]$

(c) $\dfrac{1}{a^2 - b^2}\left[a(2\ln 2 - 25) + \dfrac{47}{2}b \right]$

(d) $\dfrac{1}{a^2 - b^2}\left[a(\ln 2 - 25) - \dfrac{47}{2}b \right]$

141. The integral $\int\limits_{0}^{1} \dfrac{dx}{\sqrt{1-x}}$ is equal to _____ ?

[EC GATE 2016]

142. Consider the following definite integral

$I = \int\limits_{0}^{1} \dfrac{\left(\sin^{-1} x\right)^2}{\sqrt{1-x^2}}\,dx$

The value of the integral is

(a) $\dfrac{\pi^3}{24}$ (b) $\dfrac{\pi^3}{12}$ (c) $\dfrac{\pi^3}{48}$ (d) $\dfrac{\pi^3}{64}$

[CE GATE 2017]

143. Let X be a continuous variable defined over the interval $(-\infty, \infty)$ and $f(x) = e^{-x - e^{-x}}$.

The integral $g(x) = \int f(x)\,dx$ is equal to

(a) $e^{e^{-x}}$ (b) $e^{-e^{-x}}$ (c) $e^{-e^{x}}$ (d) e^{-x}

[CE GATE 2017]

144. If $f(x) = R\sin\left(\dfrac{\pi x}{2}\right) + S$, $f'\left(\dfrac{1}{2}\right) = \sqrt{2}$

and $\int\limits_{0}^{1} f(x)\,dx = \dfrac{2R}{\pi}$, then the constant R and S, respectively are

(a) $\dfrac{2}{\pi}$ and $\dfrac{16}{\pi}$ (b) $\dfrac{2}{\pi}$ and 0

(c) $\dfrac{4}{\pi}$ and 0 (d) $\dfrac{4}{\pi}$ and $\dfrac{16}{\pi}$

[CS/IT GATE 2017]

145. The value of the integral $\int\limits_{0}^{\pi} x\cos^2 x\,dx$ is

(a) $\dfrac{\pi^2}{8}$ (b) $\dfrac{\pi^2}{4}$ (c) $\dfrac{\pi^2}{2}$ (d) π^2

[CE GATE 2018]

146. The value of $\int\limits_{0}^{\frac{\pi}{4}} x\cos(x^2)\,dx$ correct to three-decimal places (assuming that $\pi = 3.14$) is

[CS/IT GATE 2018]

147. Let f be a real valued function of a real variable defined as $f(x) = x - [x]$, where $[x]$ denotes the largest integer less than or equal to x. The value of $\int\limits_{0.25}^{1.25} f(x)\,dx$ is _____?

(up to 2 decimal places)

[EE GATE 2018]

148. Consider the following integral

$\underset{a\to\infty}{Lim} \int\limits_{1}^{a} x^{-4}\,dx$. It

(a) diverges (b) converges to $\dfrac{1}{3}$

(c) converges to $\dfrac{-1}{a^3}$ (d) converges to 0

[GATE 2000]

149. The value of the following improper integral is

$$\int_0^1 x \log x \, dx$$

(a) $\dfrac{1}{4}$ (b) 0   (c) $-\dfrac{1}{4}$ (d) 1

[GATE 2002]

150. The value of the integral

$$I = \dfrac{1}{\sqrt{2\pi}} \int_0^\infty e^{-\frac{x^2}{8}} \, dx \text{ is}$$

(a) 1   (b) π (c) 2   (d) 2π

[EC GATE 2005]

151. If $S = \int_1^\infty x^{-3} \, dx$ then S has the value

(a) $\dfrac{-1}{3}$ (b) $\dfrac{1}{4}$ (c) $\dfrac{1}{2}$ (d) 1

[EE GATE 2005]

152. Which of the following integrals is unbounded?

(a) $\displaystyle\int_0^{\frac{\pi}{4}} \tan x \, dx$ (b) $\displaystyle\int_0^\infty \dfrac{1}{1+x^2} \, dx$

(c) $\displaystyle\int_0^\infty x e^{-x} \, dx$ (d) $\displaystyle\int_0^1 \dfrac{1}{1-x} \, dx$

[ME GATE 2008]

153. The integral $\dfrac{1}{\sqrt{2\pi}} \displaystyle\int_{-\infty}^\infty e^{\frac{-x^2}{2}} \, dx$ is equal to

(a) $\dfrac{1}{2}$ (b) $\dfrac{1}{\sqrt{2}}$

(c) 1   (d) ∞

[PI GATE 2010]

154. The value of the integral $\displaystyle\int_{-\infty}^\infty \dfrac{dx}{1+x^2}$

(a) $-\pi$ (b) $\dfrac{-\pi}{2}$ (c) $\dfrac{\pi}{2}$ (d) π

[ME GATE 2010]

155. The solution for $\displaystyle\int_0^{\frac{\pi}{6}} \cos^4 3\theta \, \sin^3 6\theta \, d\theta$ is:

(a) 0   (b) $\dfrac{1}{15}$ (c) 1   (d) $\dfrac{8}{3}$

[CE GATE 2013]

156. The value of the integral

$$\int_{-\infty}^\infty 12 \cos(2\pi t) \dfrac{\sin(4\pi t)}{4\pi t} \, dt \text{ is } \rule{1.5cm}{0.4pt} ?$$

[EC GATE 2015]

157. The value of the integral

$$2\int_{-\infty}^\infty \left(\dfrac{\sin 2\pi t}{\pi t} \right) dt \text{ is equal to}$$

(a) 0   (b) 0.5   (c) 1   (d) 2

[EE GATE 2016]

158. If $f(x, y, z) = (x^2 + y^2 + z^2)^{-\frac{1}{2}}$ then

$\dfrac{\partial^2 f}{\partial x^2} + \dfrac{\partial^2 f}{\partial y^2} + \dfrac{\partial^2 f}{\partial z^2}$ is equal to

(a) 0   (b) 1

(c) 2   (d) $-3(x^2 + y^2 + z^2)^{\frac{-5}{2}}$

[GATE 2000]

159. I f $f = a_0 x^n + a_1 x^{n-1} y + \ldots\ldots + a_{n-1} x y^{n-1} + a_n y^n$

where $a_i \, (i = 0, 1, 2, \ldots, n)$ are constants, then

$x \dfrac{\partial f}{\partial x} + y \dfrac{\partial f}{\partial y}$ is

(a) $\dfrac{f}{n}$ (b) $\dfrac{n}{f}$

(c) nf (d) $n\sqrt{f}$

[IN GATE 2005]

160. The value of the integral of the function $g(x, y) = 4x^3 + 10y^4$ along the straight line segment from the point $(0,0)$ to the $(1, 2)$ in the xy-plane is

(a) 33   (b) 35   (c) 40   (d) 56

[EC GATE 2008]

161. The total derivative of the function xy is

(a) $x \, dy + y \, dx$ (b) $x \, dx + y \, dy$

(c) $dx + dy$ (d) $dx \, dy$

[PI GATE 2009]

162. If $z = xy \ln(xy)$, then

(a) $x \dfrac{\partial z}{\partial x} + y \dfrac{\partial z}{\partial y} = 0$ (b) $y \dfrac{\partial z}{\partial x} = x \dfrac{\partial z}{\partial y}$

(c) $x \dfrac{\partial z}{\partial x} = y \dfrac{\partial z}{\partial y}$ (d) $y \dfrac{\partial z}{\partial x} + x \dfrac{\partial z}{\partial y} = 0$

[EC GATE 2014]

163. The contour on the $x-y$ plane, where the partial derivative of x^2+y^2 with respect to y is equal to the partial derivative of $6x+4y$ with respect to x is

(a) $y=2$ (b) $x=2$

(c) $x+y=4$ (d) $x-y=0$

[EC GATE 2015]

164. For the two functions

$f(x,y)=x^3-3xy^2$ and $g(x,y)=3x^2y-y^3$

which one of the following options is correct?

(a) $\dfrac{\partial f}{\partial x}=\dfrac{\partial g}{\partial x}$ (b) $\dfrac{\partial f}{\partial x}=-\dfrac{\partial g}{\partial y}$

(c) $\dfrac{\partial f}{\partial y}=-\dfrac{\partial g}{\partial x}$ (d) $\dfrac{\partial f}{\partial y}=\dfrac{\partial g}{\partial x}$

[PI GATE 2016]

165. Consider a function $f(x,y,z)$ given by

$f(x,y,z) = (x^2 + y^2 -2z^2)(y^2 + z^2)$. The partial derivative of this function with respect to x at the point x = 2, y = 1 and z = 3 is _____?

[EE GATE 2017]

166. Let $W=f(x,y)$, where x and y are functions of t. Then according to the chain rule, $\dfrac{dw}{dt}$ is equal to

(a) $\dfrac{dw}{dx}\dfrac{dx}{dt}+\dfrac{dw}{dy}\dfrac{dy}{dt}$ (b) $\dfrac{\partial w}{\partial x}\dfrac{\partial x}{\partial t}+\dfrac{\partial w}{\partial y}\dfrac{\partial y}{\partial t}$

(c) $\dfrac{\partial w}{\partial x}\dfrac{dx}{dt}+\dfrac{\partial w}{\partial y}\dfrac{dy}{dt}$ (b) $\dfrac{dw}{dx}\dfrac{\partial x}{\partial t}+\dfrac{dw}{dy}\dfrac{\partial y}{\partial t}$

[CE GATE 2017]

167. Let $f(x,y)=\dfrac{ax^2+by^2}{xy}$, where a and b are constants. If $\dfrac{\partial f}{\partial x}=\dfrac{\partial f}{\partial y}$ at $x=1$ and $y=2$, then the relation between a and b is

(a) $a=\dfrac{b}{4}$ (b) $a=\dfrac{b}{2}$

(c) $a=2b$ (d) $a=4b$

[EC GATE 2018]

168. Consider the following equations:

$\dfrac{\partial V(x,y)}{\partial x}=px^2+y^2+2xy$ and

$\dfrac{\partial V(x,y)}{\partial y}=x^2+qy^2+2xy$

where p and q are constants, $V(x,y)$ that satisfies the above equation is

(a) $p\dfrac{x^3}{3}+q\dfrac{y^3}{3}+2xy+6$

(b) $p\dfrac{x^3}{3}+q\dfrac{y^3}{3}+5$

(c) $p\dfrac{x^3}{3}+q\dfrac{y^3}{3}+x^2y+xy^2+xy$

(d) $p\dfrac{x^3}{3}+q\dfrac{y^3}{3}+x^2y+xy^2$

[IN GATE 2018]

169. Let $r=x^2+y-z$ and $z^3-xy+yz+y^3=1$. Assume that x and y are independent variables. At $(x,y,z)=(2,-1,1)$, the value (correct to two decimal places) of $\dfrac{\partial r}{\partial x}$ is _____ ?

[EC GATE 2018]

170. The function $f(x,y)=2x^2+2xy-y^3$ has

(a) Only one stationary point at $(0,0)$

(b) Two stationary points at $(0,0)$ and $\left(\dfrac{1}{6},-\dfrac{1}{3}\right)$

(c) Two stationary points at $(0,0)$ and $(1,-1)$

(d) no stationary point.

[GATE 2002]

171. For the function $f(x,y)=x^2-y^2$ defined on R^2, the point $(0,0)$ is

(a) a local minimum

(b) neither a local minimum nor a local maximum.

(c) a local maximum

(d) both a local minimum and a local maximum

[PI GATE 2007]

172. Let $f=y^x$. What is $\dfrac{\partial^2 f}{\partial x\partial y}$ at $x=2$, $y=1$?

(a) 0   (b) ln2

(c) 1   (d) $\dfrac{1}{\ln 2}$

[ME GATE 2008]

173. The distance between the origin and the point nearest to it on the surface $z^2 = 1 + xy$ is

(a) 1   (b) $\dfrac{\sqrt{3}}{2}$ (c) $\sqrt{3}$ (d) 2

[ME GATE 2009]

174. Given a function
$f(x, y) = 4x^2 + 6y^2 - 8x - 4y + 8,$

The optimal values of $f(x, y)$ is

(a) a minimum equal to $\dfrac{10}{3}$

(b) a maximum equal to $\dfrac{10}{3}$

(c) a minimum equal to $\dfrac{8}{3}$

(d) a maximum equal to $\dfrac{8}{3}$

[CE GATE 2010]

175. Changing the order of integration in the double integral $I = \int_0^8 \int_{\frac{x}{4}}^2 f(x, y)\,dy\,dx$ leads to

$I = \int_r^s \int_p^q f(x, y)\,dy\,dx$. What is q?

(a) $4y$ (b) $16y^2$ (c) x (d) 8

[GATE 2005]

176. The double integral
$\int_0^a \int_0^y f(x, y)\,dx\,dy$ is equivalent to

(a) $\int_0^x \int_0^y f(x, y)\,dx\,dy$ (b) $\int_0^a \int_x^y f(x, y)\,dx\,dy$

(c) $\int_0^a \int_x^a f(x, y)\,dy\,dx$ (d) $\int_0^a \int_0^a f(x, y)\,dx\,dy$

[IN GATE 2015]

177. $\int_0^{\frac{\pi}{2}} \int_0^{\frac{\pi}{2}} \sin(x + y)\,dx\,dy$

(a) 0   (b) π

(c) $\pi/2$ (d) 2

[GATE 2000]

178. The volume of an object expressed in spherical co-ordinates is given by

$V = \int_0^{2\pi} \int_0^{\frac{\pi}{3}} \int_0^1 r^2 \sin\phi \, dr\, d\phi\, d\theta$. The value of the integral

(a) $\dfrac{\pi}{3}$ (b) $\dfrac{\pi}{6}$ (c) $\dfrac{2\pi}{3}$ (d) $\dfrac{\pi}{4}$

[GATE 2004]

179. By a change of variables

$x(u,v) = uv$, $y(u,v) = \dfrac{u}{v}$ in a double integral, the integral $f(x, y)$ changes to $f\left(uv, \dfrac{v}{u}\right)$. Then, $\phi(u,v)$ is _____?

(a) $\dfrac{2v}{u}$ (b) $2uv$ (c) v^2 (d) 1

[GATE 2005]

180. The value of $\int_0^\infty \int_0^\infty e^{-x^2} e^{-y^2}\,dx\,dy$ is

(a) $\dfrac{\sqrt{\pi}}{2}$ (b) $\sqrt{\pi}$ (c) π (d) $\dfrac{\pi}{4}$

[IN GATE 2007]

181. The value of $\int_0^3 \int_0^x (6 - x - y)\,dx\,dy$ is

(a) 13.5   (b) 27.0

(c) 40.5   (d) 54.0

[CS GATE 2008]

182. Consider the shaded triangular region P shown in the figure below. What is $\iint_P xy\,dx\,dy$?

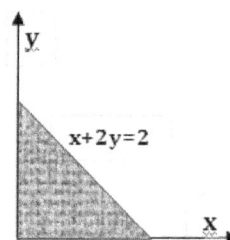

(a) $\dfrac{1}{6}$ (b) $\dfrac{2}{9}$ (c) $\dfrac{7}{16}$ (d) 1

[ME GATE 2008]

183. If $f(x,y)$ is continuous function defined over $(x,y)\in[0,1]\times[0,1]$. Given two constraints, $x>y^2$ and $y>x^2$, the volume under $f(x,y)$ is

(a) $\displaystyle\int_{y=0}^{y=1}\int_{x=y^2}^{x=\sqrt{y}} f(x,y)\,dx\,dy$

(b) $\displaystyle\int_{y=x^2}^{y=1}\int_{x=y^2}^{x=1} f(x,y)\,dx\,dy$

(c) $\displaystyle\int_{y=x^2}^{y=1}\int_{x=0}^{x=1} f(x,y)\,dx\,dy$

(d) $\displaystyle\int_{y=0}^{y=\sqrt{x}}\int_{x=0}^{x=\sqrt{y}} f(x,y)\,dx\,dy$

[EE GATE 2009]

184. The volume under the surface $Z(x,y)=x+y$ and above the triangle in the xy plane defined by
$\{0\le y\le x$ and $0\le x\le 12\}$ is _____?

[EC GATE 2014]

185. To evaluate the double integral $\displaystyle\int_0^8\left(\int_{\frac{y}{2}}^{\frac{y}{2}+1}\left(\frac{2x-y}{2}\right)dx\right)dy$ we make the substitution $u=\left(\frac{2x-y}{2}\right)$ and $v=\frac{y}{2}$. The integral will reduce to

(a) $\displaystyle\int_0^4\left(\int_0^2 2u\,du\right)dv$ (b) $\displaystyle\int_0^4\left(\int_0^1 2u\,du\right)dv$

(c) $\displaystyle\int_0^4\left(\int_0^1 u\,du\right)dv$ (b) $\displaystyle\int_0^4\left(\int_0^{21} 2u\,du\right)dv$

[EE GATE 2014]

186. The value of the integral $\displaystyle\int_0^2\int_0^x e^{x+y}\,dy\,dx$ is

(a) $\frac{1}{2}(e-1)$ (b) $\frac{1}{2}(e^2-1)^2$

(c) $\frac{1}{2}(e^2-e)$ (d) $\frac{1}{2}\left(e-\frac{1}{e}\right)^2$

[ME GATE 2014]

187. The volume enclosed by the surface $f(x,y)=e^x$ over the triangle bounded by the line $x=y,x=0,y=1$ in the xy plane is ___?

[EE GATE 2015]

188. The integral $\frac{1}{2\pi}\iint_D(x+y+10)\,dx\,dy$ where D denotes the disc: $x^2+y^2\le 4$, evaluates to _____?

[EC GATE 2016]

189. The region specified by
$$\left\{(\rho,\varphi,z):3\le\rho\le5,\frac{\pi}{8}\le\varphi\le\frac{\pi}{4},3\le z\le4.5\right\}$$
in cylindrical coordinates has volume of _____?

[EC GATE 2016]

190. A triangle in the xy-plane is bounded by the straight lines $2x=3y$, $y=0$, and $x=3$. The volume above the triangle and the under the plane $x+y+z=6$ is _____?

[EC GATE 2016]

191. The values of the integrals
$$\int_0^1\left(\int_0^1\frac{x-y}{(x+y)^3}\,dy\right)dx \text{ and}$$
$$\int_0^1\left(\int_0^1\frac{x-y}{(x+y)^3}\,dx\right)dy \text{ are}$$

(a) and equal to 0.5

(b) same and equal to -0.5

(c) 0.5 and -0.5, respectively

(d) -0.5 and -0.5, respectively

[EC GATE 2017]

192. Let $I=c\iint_R xy^2\,dx\,dy$, where R is the region shown in the figure and $c=6\times10^{-4}$ The value of I equals _____?

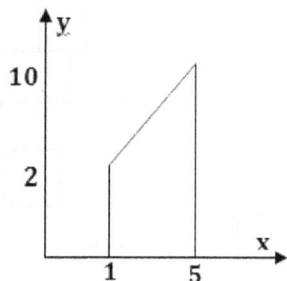

[EE GATE 2017]

193. The length of the curve $y = \dfrac{2}{3} x^{\frac{3}{2}}$ between $x = 0 \,\&\, x = 1$ is

(a) 0.27   (b) 0.67

(c) 1   (d) 1.22

[MEGATE 2008]

194. A parabolic cable is held between two supports at the same level. The horizontal span between the supports is L. The sag at the mid–span is h. The equation of the parabola is $y = 4h\dfrac{x^2}{L^2}$, where x is the horizontal coordinate and y is the vertical coordinate with the origin at the center of the cable. The expression for the total length of the cable is

(a) $\displaystyle\int_0^L \sqrt{1 + 64\frac{h^2 x^2}{L^4}}\,dx$ (b) $2\displaystyle\int_0^{\frac{L}{2}} \sqrt{1 + 64\frac{h^3 x^2}{L^4}}\,dx$

(c) $\displaystyle\int_0^{\frac{L}{2}} \sqrt{1 + 64\frac{h^2 x^2}{L^4}}\,dx$ (d) $2\displaystyle\int_0^{\frac{L}{2}} \sqrt{1 + 64\frac{h^2 x^2}{L^4}}\,dx$

[CE GATE 2010]

195. Consider a spatial curve in three-dimensional space given in parametric form by $x(t) = \cos t, y(t) = \sin t,\ \ z(t) = \dfrac{2}{\pi} t,\ \ 0 \le t \le \dfrac{\pi}{2}.$ The length of the curve is _____?

[ME GATE 2015]

196. The parabolic arc $y = \sqrt{x}$ is revolved around the x-axis. The volume of the solid of revolution is

(a) $\dfrac{\pi}{4}$ (b) $\dfrac{\pi}{2}$ (c) $\dfrac{3\pi}{4}$ (d) $\dfrac{3\pi}{2}$

[ME GATE 2010]

197. A three-dimensional region R of finite volume is described by $x^2 + y^2 \le z^3, 0 \le z \le 1$, where x, y, z are real. The volume of R correct to two decimal places is _____?

[EC GATE 2017]

198. A parametric curve defined by $x = \cos\left(\dfrac{\pi u}{2}\right),\ y = \sin\left(\dfrac{\pi u}{2}\right)$ in the range $0 \le u \le 1$ is rotated about x-axis by 360 degrees. Area of the surface generated is

(a) $\dfrac{\pi}{2}$ (b) π (c) 2π (d) 4π

[EE GATE 2017]

Answer key				
1. (d)	2. (d)	3. (d)	4. (a)	5. (b)
6. (a)	7. (b)	8. (b)	9. (c)	10. (a)
11. (c)	12. (c)	13. (b)	14. (a)	15. (b)
16. (d)	17. (c)	18. (c)	19. (d)	20. (d)
21. (c)	22. (b)	23. $\frac{1}{2}$.	24. (c)	25. 25
26. 1.	27. (c)	28. (c)	29. (d)	30. −1
31. (b)	32. 1.	33. (b)	34. (d)	35. (c)
36. (c)	37. (b)	38. (c)	39. (c)	40. (b)
41. (c)	42. (c)	43. (a)	44. (a)	45. (d)
46. −2	47. (c)	48. (b)	49. (d)	50. (b).
51. (a)	52. (a)	53. (d)	54. (d)	55. (b)
56. (c)	57. (c)	58. (b)	59. (a)	60. (b)
61. (a)	62. −0.333		63. (c)	64. (c)
65. (c)	66. (c)	67. (a)	68. (b)	69. (a)
70. (b)	70. (b)	71. (d)	72. (a)	73. (a)
74. (c)	75. (b)	76. (a)	77. (b)	78. (a)
79. (b)	80. (c)	81. (a)	82. (a)	83. (c)
84. (c)	85. (d)	86. (b)	87. (a)	88. 0
89. 6	90. (c)	91. (a)	92. (c)	93. (b)
94. 1.	95. (b)	96. 0	97. −13	98. −5
99. (c)	100. (d)	101. 12	102. (a)	103. (a)
104. (c)	105. (a)	106. (b)	107. (c)	108. (d)
109. (b)	110. (b)	111. (b)	112. (b)	113. (a)
114. (d)	115. 2	116. 0.293	117. (b)	118. (c)
119. 0	120. (a)	121. (c)	122. (b)	123. (a)
124. (d)	125. (b)	126. (d)	127. (a)	128. (a)
129. (b)	130. (b)	131. (d)	132. (d)	133. (a)

134. (c)	**135.** 1.	**136.** (b)	**137.** 4.	**138.** (a)
139. (c)	**140.** (a)	**141.** 2	**142.** (a)	**143.** (b)
144. (c)	**145.** (b)	**146.** 0.005.	**147.** 0.5	**148.** (b)
149. (c)	**150.** (a)	**151.** (c)	**152.** (d)	**153.** (c)
154. (d)	**155.** (b)	**156.** 3	**157.** (d)	**158.** (a)
159. (c)	**160.** (a)	**161.** (a)	**162.** (c)	**163.** (a)
164. (c)	**165.** 40.	**166.** (c)	**167.** (d)	**168.** (d)
169. 4.5	**170.** (b)	**171.** (b)	**172.** (c)	**173.** (a)
174. (a)	**175.** (a)	**176.** (c)	**177.** (d)	**178.** (a)
179. (a)	**180.** (d)	**181.** (a)	**182.** (a)	**183.** (a)
184. 864	**185.** (a)	**186.** (b)	**187.** e-2.	**188.** 20
189. 4.71	**190.** 10	**191.** (c)	**192.** 1666.13	
193. (d)	**194.** (d)	**195.** 1.86	**196.** (d)	**197.** $\pi/4$
198. (c)				

Explanation

1. (d)

$$\lim_{x \to \infty} \frac{1-a^4}{x^4} = \left(1-a^4\right) \lim_{x \to \infty} \frac{1}{x^4} = 0$$

$$\left(\because \lim_{x \to \infty} \frac{1}{x} = 0 \right)$$

2. (d)

$$f(x) = x - \frac{x^3}{3!} + \frac{x^5}{5!} - \frac{x^7}{7!} + \ldots \ldots \infty$$

$$\Rightarrow f(x) = \sin x$$

$$\therefore \lim_{x \to \frac{\pi}{2}} f(x) = \lim_{x \to \frac{\pi}{2}} \sin x = \sin \frac{\pi}{2} = 1.$$

3. (d)

$$\lim_{x \to \frac{\pi}{4}} \frac{\sin 2\left(x - \frac{\pi}{4}\right)}{x - \frac{\pi}{4}}$$

$$= \lim_{x - \frac{\pi}{4} \to 0} \frac{\sin 2\left(x - \frac{\pi}{4}\right)}{x - \frac{\pi}{4}} = \lim_{t \to 0} \frac{\sin 2t}{t}$$

$$\left(\text{putting } x - \frac{\pi}{4} = t \text{ so that } t \to 0 \text{ when } x \to \frac{\pi}{4} \right)$$

$$= \lim_{t \to 0} \left[\frac{\sin 2t}{2t} \times 2 \right] = 2 \lim_{2t \to 0} \frac{\sin 2t}{2t} = 2 \times 1 = 2.$$

4. (a)

$$\lim_{x \to 0} \frac{\sin^2 x}{x}$$

$$= \lim_{x \to 0} \left(\frac{\sin x}{x} \times \sin x \right)$$

$$= \lim_{x \to 0} \frac{\sin x}{x} \times \lim_{x \to 0} \sin x = 1 \times \sin 0 = 0.$$

5. (b)

$$f(x)$$

$$= \lim_{x \to 0} \frac{x^3 + x}{2x^3 - 7x^2} = \lim_{x \to 0} \frac{x^2(x+1)}{x^2(2x-7)}$$

$$= \lim_{x \to 0} \frac{x+1}{2x-7} = \frac{0+1}{2 \times 0 - 7} = -\frac{1}{7}.$$

6. (a)

$$\underset{\theta \to 0}{\text{Lt}} \frac{\sin\left(\frac{\theta}{2}\right)}{\theta}$$

$$= \left[\underset{\frac{\theta}{2} \to 0}{\text{Lt}} \frac{\sin\left(\frac{\theta}{2}\right)}{\frac{\theta}{2}} \right] \times \frac{1}{2} = 1 \times \frac{1}{2} = \frac{1}{2} \left(\because \lim_{x \to 0} \frac{\sin x}{x} = 1 \right)$$

7. (b)

$$\lim_{x \to 0} \frac{e^x \left(1 + x + \frac{x^2}{2}\right)}{x^3}$$

$$= \lim_{x \to 0} \frac{\left(1 + x + \frac{x^2}{2!} + \frac{x^3}{3!} + \frac{x^4}{4!} + \ldots \infty\right) - \left(1 + x + \frac{x^2}{2}\right)}{x^3}$$

$$\lim_{x \to 0} \frac{e^x - \left(1 + x + \frac{x^2}{2}\right)}{x^3}$$

$$= \lim_{x \to 0} \frac{\left(1 + x + \frac{x^2}{2!} + \frac{x^3}{3!} + \frac{x^4}{4!} + \ldots \infty\right) - \left(1 + x + \frac{x^2}{2}\right)}{x^3}$$

$$= \lim_{x \to 0} \frac{\frac{x^3}{3!} + \frac{x^4}{4!} + \ldots \infty}{x^3} = \lim_{x \to 0} \frac{\frac{1}{3!} + \frac{x}{4!} + \ldots \infty}{1}$$

$$= \frac{1}{3!} + \frac{0}{4!} + \ldots \ldots \infty = \frac{1}{6}.$$

8. (b)

$$y = x + \sqrt{x + \sqrt{x + \sqrt{x + ... \infty}}}$$

$$\Rightarrow y = x + \sqrt{y}$$

$$\Rightarrow (y - x)^2 = y$$

$$\Rightarrow y^2 - 2xy + x^2 = y$$

At $x = 2$,

$$y^2 - 4y + 4 = y$$

or, $y^2 - 5y + 4 = 0$

or, $y = 1, 4$

But at $x = 2$,

$$y = 2 + \sqrt{2 + \sqrt{2 + ... \infty}} > 1$$

$$\therefore y = 4 \text{ only.}$$

9. (c)

$$\underset{x \to \frac{\pi}{4}}{\text{Lt}} \frac{\cos x - \sin x}{x - \frac{\pi}{4}}$$

$$= \sqrt{2} \underset{x \to \frac{\pi}{4}}{\text{Lt}} \left[\frac{\frac{1}{\sqrt{2}}(\cos x - \sin x)}{x - \frac{\pi}{4}} \right]$$

$$= \sqrt{2} \underset{x \to \frac{\pi}{4}}{\text{Lt}} \left[\frac{\frac{1}{\sqrt{2}}\cos x - \frac{1}{\sqrt{2}}\sin x}{x - \frac{\pi}{4}} \right]$$

$$= \sqrt{2} \underset{x \to \frac{\pi}{4}}{\text{Lt}} \left[\frac{\sin\left(\frac{\pi}{4} - x\right)}{x - \frac{\pi}{4}} \right] = -\sqrt{2} \underset{x - \frac{\pi}{4} \to 0}{\text{Lt}} \left[\frac{\sin\left(x - \frac{\pi}{4}\right)}{x - \frac{\pi}{4}} \right]$$

$$= -\sqrt{2} \times 1 = -\sqrt{2} \quad \left(\because \lim_{x \to 0} \frac{\sin x}{x} = 1 \right).$$

10. (a)

$-1 \; \sin x \le 1$ and $-1 \le \cos x \le 1$

$\Rightarrow 1 \ge -\sin x \ge -1$ and $-1 \le \cos x \le 1$

$\Rightarrow x - 1 \le x - \sin x \le x + 1$ and

$x - 1 \le x + \cos x \le x + 1$

$\Rightarrow x - 1 \le x - \sin x \le x + 1$ and

$-1 \le \sin x \le 1$ and $-1 \le \cos x \le 1$

$\Rightarrow 1 \ge -\sin x \ge -1$ and $-1 \le \cos x \le 1$

$\Rightarrow x - 1 \le x - \sin x \le x + 1$ and

$x - 1 \le x + \cos x \le x + 1$

$\Rightarrow x - 1 \le x - \sin x \le x + 1$ and

$$\frac{1}{x+1} \le \frac{1}{x + \cos x} \le \frac{1}{x-1}$$

$$\Rightarrow \frac{x-1}{x+1} \le \frac{x - \sin x}{x + \cos x} \le \frac{x+1}{x-1}$$

Now $\underset{x \to \infty}{\text{Lt}} \dfrac{x-1}{x+1} = \underset{x \to \infty}{\text{Lt}} \dfrac{1 - \frac{1}{x}}{1 + \frac{1}{x}} = \dfrac{1-0}{1+0} = 1$

and $\underset{x \to \infty}{\text{Lt}} \dfrac{x+1}{x-1} = \underset{x \to \infty}{\text{Lt}} \dfrac{1 + \frac{1}{x}}{1 - \frac{1}{x}} = \dfrac{1+0}{1-0} = 1.$

Therefore, by the Sandwich theorem, we get

$$\underset{x \to \infty}{\text{Lt}} \frac{x - \sin x}{x + \cos x} = 1.$$

11. (b) Result follows from the fundamental formulas.

12. (c) $e^{-\ln x} = e^{-\log_e x} = e^{\log_e x^{-1}} = x^{-1} = \dfrac{1}{x}.$

13. (b)

$$\underset{x \to 8}{\text{Lt}} \frac{x^{\frac{1}{3}} - 2}{x - 8}$$

$$= \underset{x \to 8}{\text{Lt}} \frac{x^{\frac{1}{3}} - 8^{\frac{1}{3}}}{x - 8} = \frac{1}{3}(8)^{\frac{1}{3} - 1}$$

$$\left(\text{using } \underset{x \to a}{\text{Lt}} \frac{x^n - a^n}{x - a} = n a^{n-1} \right)$$

$$= \frac{1}{3}(8)^{\frac{-2}{3}} = \frac{1}{12}.$$

14. (a)

$$\underset{x \to 0}{\text{Lt}} \frac{\sin\left(\frac{2}{3} x\right)}{x}$$

$$= \underset{\frac{2}{3}x \to 0}{\text{Lt}} \left[\frac{\sin\left(\frac{2}{3} x\right)}{\frac{2}{3} x} \right] \times \frac{2}{3} = 1 \times \frac{2}{3} = \frac{2}{3} \left(\because \underset{x \to 0}{\text{Lt}} \frac{\sin x}{x} = 1 \right)$$

15. $\left(\underset{n\to\infty}{Lt}\left(1-\dfrac{1}{n}\right)^{2n} = \left[\underset{n\to\infty}{Lt}\left(1-\dfrac{1}{n}\right)^{n}\right]^{2} = \left(e^{-1}\right)^{2} = e^{-2}. \right.$

$\left[\because \underset{n\to\infty}{Lt}\left(1-\dfrac{1}{n}\right)^{n} = e^{-1}\right]$

16. (d) Follows from fundamental formulas.

17. (c)

$\underset{x\to\infty}{Lt}\left(1+\dfrac{1}{x}\right)^{x} = \underset{x\to 0}{Lt}\left(1+x\right)^{\frac{1}{x}} = e.$

18. (c)

$\underset{x\to\infty}{Lt}\left(\dfrac{x+\sin x}{x}\right)$

$= \underset{x\to\infty}{Lt}\left(1+\dfrac{\sin x}{x}\right) = 1 + \underset{x\to\infty}{Lt}\dfrac{\sin x}{x} = 1 + 0 = 1.$

$\left[\because \sin x \text{ is bounded and } \underset{x\to\infty}{Lt}\dfrac{1}{x} = 0\right]$

Remember:

If $f(x)$ is bounded and $\underset{x\to\infty}{Lt}\, g(x) = 0$, then $\underset{x\to\infty}{Lt}\left[f(x)g(x)\right] = 0.$

19. (d) Let $\theta = 1000\pi t$. Then

$x = 3\sin\theta \text{ i.e; } \dfrac{x}{3} = \sin\theta \dots\dots\dots(1)$

$y = 5\cos\left[\theta + \dfrac{\pi}{4}\right] \text{ i.e;}$

$\dfrac{y}{5} = \cos\left[\theta + \dfrac{\pi}{4}\right]$

$= \cos\theta\cos\dfrac{\pi}{4} - \sin\theta\sin\dfrac{\pi}{4}$

$= \dfrac{1}{\sqrt{2}}\left(\cos\theta - \sin\theta\right)\dots\dots\dots(2)$

Using (1) in (2), we get

$\dfrac{y}{5} = \dfrac{1}{\sqrt{2}}\left(\cos\theta - \dfrac{x}{3}\right)$

or, $\left(\dfrac{y\sqrt{2}}{5} + \dfrac{x}{3}\right) = \cos\theta \dots\dots\dots(3)$

Squaring and adding equations (1) and (3), we get,

$\dfrac{x^2}{9} + \left(\dfrac{2}{25}y^2 + \dfrac{2\sqrt{2}}{15}xy + \dfrac{x^2}{9}\right)$

$= \sin^2\theta + \cos^2\theta = 1$

i.e., $\dfrac{2x^2}{9} + \dfrac{2\sqrt{2}}{15}xy + \dfrac{2y^2}{25} - 1 = 0\dots\dots\dots(4)$

Now comparing equation (4) with

$ax^2 + 2hxy + by^2 + 2gx + 2fy + c = 0$, we get

$a = \dfrac{2}{9}, 2h = \dfrac{2\sqrt{2}}{15} \text{ i.e; } h = \dfrac{\sqrt{2}}{15}, b = \dfrac{2}{25}, g = 0,$

$f = 0$ and $c = -1$.

Therefore, $ab - h^2 = \left(\dfrac{4}{225} - \dfrac{2}{225}\right) = \dfrac{2}{225} > 0,$

$\Delta = \begin{vmatrix} \dfrac{2}{9} & \dfrac{\sqrt{2}}{15} & 0 \\ \dfrac{\sqrt{2}}{15} & \dfrac{2}{25} & 0 \\ 0 & 0 & -1 \end{vmatrix} = -\dfrac{2}{225} \neq 0$

Hence, equation (4) represents an ellipse.

Remember:

$ax^2 + 2hxy + by^2 + 2gx + 2fy + c = 0$ represents

an ellipse if $ab - h^2 > 0$ and $\Delta = \begin{vmatrix} a & h & g \\ h & b & f \\ g & f & c \end{vmatrix} \neq 0.$

20. (d) $\underset{x\to\infty}{Lt}\left(1+\dfrac{1}{x}\right)^{2x} = \left[\underset{x\to\infty}{Lt}\left(1+\dfrac{1}{x}\right)^{x}\right]^{2} = e^{2}.$

21. (c) Let $f(x) = ax + b\dots\dots\dots(i)$
Then,

$f(-2) = 29 \Rightarrow -2a + b = 29\dots\dots(i)$
and $f(3) = 39 \Rightarrow 3a + b = 39\dots\dots\dots(ii)$

Solving equations (i) and (ii), we get,
a=2 and b=33.

Hence, $f(x) = 2x + 33$ and so $f(5) = 2 \times 5 + 33 = 43.$

22. (b) The point (2, -1) lies on the curve (b) only.

23. $\dfrac{1}{2}$.

$$\underset{n\to\infty}{\text{Lt}}\left(\sqrt{n^2+n}-\sqrt{n^2+1}\right)$$

$$=\underset{n\to\infty}{\text{Lt}}\frac{\left(\sqrt{n^2+n}-\sqrt{n^2+1}\right)\left(\sqrt{n^2+n}+\sqrt{n^2+1}\right)}{\left(\sqrt{n^2+n}+\sqrt{n^2+1}\right)}$$

$$=\underset{n\to\infty}{\text{Lt}}\frac{(n^2+n)-(n^2+1)}{\left(\sqrt{n^2+n}+\sqrt{n^2+1}\right)}$$

$$=\underset{x\to\infty}{\text{Lt}}\frac{(n-1)}{\left(\sqrt{1+\dfrac{1}{n}}+\sqrt{1+\dfrac{1}{n^2}}\right)}$$

$$=\underset{x\to\infty}{\text{Lt}}\frac{\left(1-\dfrac{1}{n}\right)}{\left(\sqrt{1+\dfrac{1}{n}}+\sqrt{1+\dfrac{1}{n^2}}\right)}=\frac{1-0}{\sqrt{1+0}+\sqrt{1+0}}=\frac{1}{2}.$$

24. (c)

$$\underset{x\to\infty}{\text{Lt}}\left(\sqrt{x^2+x-1}-x\right)$$

$$=\underset{x\to\infty}{\text{Lt}}\frac{\left(\sqrt{x^2+x-1}-x\right)\left(\sqrt{x^2+x-1}+x\right)}{\left(\sqrt{x^2+x-1}+x\right)}$$

$$=\underset{x\to\infty}{\text{Lt}}\frac{(x^2+x-1-x^2)}{\sqrt{x^2+x-1}+x}=\underset{x\to\infty}{\text{Lt}}\frac{x\left(1-\dfrac{1}{x}\right)}{x\left(\sqrt{1+\dfrac{1}{x}-\dfrac{1}{x^2}}+1\right)}$$

$$=\underset{x\to\infty}{\text{Lt}}\frac{\left(1-\dfrac{1}{x}\right)}{\left(\sqrt{1+\dfrac{1}{x}-\dfrac{1}{x^2}}+1\right)}=\frac{1-0}{\sqrt{1+0-0}+1}=\frac{1}{2}.$$

25. 25

$$\underset{x\to0}{\text{Lt}}\left(\frac{e^{5x}-1}{x}\right)^2$$

$$=\underset{x\to0}{\text{Lt}}\left\{\left(\frac{e^{5x}-1}{5x}\right)\times5\right\}^2=25\times\left\{\underset{5x\to0}{\text{Lt}}\left(\frac{e^{5x}-1}{5x}\right)\right\}^2$$

$$=25\times1^2=25.\quad\left(\because\underset{x\to0}{\text{Lt}}\left(\frac{e^x-1}{x}\right)=1\right)$$

26. 1.

$$\underset{x\to4}{\text{Lt}}\frac{\sin(x-4)}{x-4}$$

$$=\underset{x-4\to0}{\text{Lt}}\frac{\sin(x-4)}{x-4}=\underset{y\to0}{\text{Lt}}\frac{\sin y}{y}=1$$

(taking $x-4=y$).

27. (c) Let $f(x)=\sin x$ & $g(x)=\dfrac{x}{2}$.

We know that $-1\le\sin x\le1$.

$\therefore-1\le\dfrac{x}{2}\le1$ *i.e*; $-2\le x\le2$

Let us draw both curves for $x\in[-2,2]$.

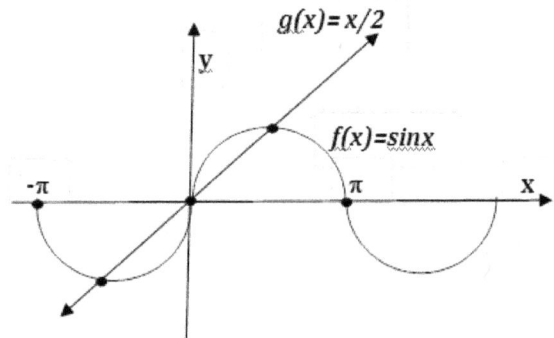

From the figure above it is clear that $f(x)=\sin x$ and $g(x)=\dfrac{x}{2}$ intersects at 3 points. Hence, three solutions exist.

28. (c)

$$\underset{x\to0}{\text{Lt}}\frac{\log_e(1+4x)}{e^{3x}-1}$$

$$=\underset{x\to0}{\text{Lt}}\left\{\frac{\dfrac{\log_e(1+4x)}{x}}{\dfrac{e^{3x}-1}{x}}\right\}=\frac{\underset{x\to0}{\text{Lt}}\dfrac{\log_e(1+4x)}{x}}{\underset{x\to0}{\text{Lt}}\dfrac{e^{3x}-1}{x}}$$

$$=\frac{\left\{\underset{4x\to0}{\text{Lt}}\dfrac{\log_e(1+4x)}{4x}\right\}\times4}{\left\{\underset{3x\to0}{\text{Lt}}\dfrac{e^{3x}-1}{3x}\right\}\times3}=\frac{1\times4}{1\times3}=\frac{4}{3}.$$

29. (d)

$$\lim_{x \to 0}\left(\frac{x^3 - \sin x}{x}\right)$$

$$= \lim_{x \to 0}\left(x^2 - \frac{\sin x}{x}\right) = \lim_{x \to 0} x^2 - \lim_{x \to 0}\frac{\sin x}{x}$$

$$= 0 - 1 = -1.$$

30. –1.

$$\lim_{x \to 0}\left(\frac{\tan x}{x^2 - x}\right)$$

$$= \lim_{x \to 0}\left(\frac{\dfrac{\tan x}{x}}{\dfrac{x^2 - x}{x}}\right) = \frac{\lim\limits_{x \to 0}\dfrac{\tan x}{x}}{\lim\limits_{x \to 0}(x-1)} = \frac{1}{0-1} = -1.$$

31. (b) The equation of the line joining the points $(-3,-2)$ and $(-1,2)$ is given by:

$$\frac{y+2}{x-(-3)} = \frac{2-(-2)}{-1-(-3)}$$

or, $y + 2 = 2(x+3) = 2x + 6$

or, $y = 2x + 4$ for $-3 \le x \le -1$

∴ (i) is correct.

The equation of the line joining the points $(-1,2)$ and $(1,0)$ is given by:

$$\frac{y-2}{x+1} = \frac{0-2}{1-(-1)}$$

or, $y - 2 = -(x+1)$

or, $y = 1 - x = -(x-1)$ for $-1 \le x \le 1$..........(1)

The equation of the line joining the points $(1,0)$ and $(2,1)$ is given by:

$$\frac{y-0}{x-1} = \frac{1-0}{2-1}$$

or, $y = x - 1$ for $1 \le x \le 2$..........(2)

Combining (1) and (2), we get,

$y = |x-1|$ for $-1 \le x \le 2$

∴ (ii) is correct.

Again from the graph, we have, $y = 1$ for $2 \le x \le 3$.

So (iv) is also correct.

Consequently, option (b) is correct.

Remember:

The equation of a line passing through the points (x_1, y_1) and (x_2, y_2) is given by $\dfrac{y - y_1}{x - x_1} = \dfrac{y_2 - y_1}{x_2 - x_1}$.

32. 1.

$f(x) = g(x)$

$\Rightarrow x^2 - 4x + 4 = 2x - 1$

$\Rightarrow x^2 - 6x + 5 = 0$

$\Rightarrow (x-1)(x-5) = 0 \Rightarrow x = 1, 5$.

Hence, the smaller value of x for which $f(x) = g(x)$ is 1.

33. (b) From the given graph we have the following points:

$(0,0), (\pm 1, -1), (\pm 2, 0)$.

It is easy to verify that all these points lie on the curve (b) only.

34. (d) Since all polynomial functions and all constant functions are differentiable, so the functions $f(x) = x^2, f(x) = x - 1$ and $f(x) = 2$ are all differentiable in $[-1,1]$.

$$f(x) = \max(x,-x) = \begin{cases} x, & x > 0 \\ -x, & x < 0 \\ 0, & x = 0 \end{cases} = |x|$$

which is not differentiable at $x = 0 \in [-1, 1]$.

35. (c) Given that

$$x = a(\theta + \sin\theta) \Rightarrow \frac{dx}{d\theta} = a(1 + \cos\theta),$$

$$y = a(1 - \cos\theta) \Rightarrow \frac{dy}{d\theta} = a\sin\theta.$$

$$\therefore \frac{dy}{dx} = \frac{\dfrac{dy}{d\theta}}{\dfrac{dx}{d\theta}} = \frac{a\sin\theta}{a(1+\cos\theta)}$$

$$= \frac{2\sin\dfrac{\theta}{2}\cos\dfrac{\theta}{2}}{2\cos^2\dfrac{\theta}{2}} = \tan\frac{\theta}{2}$$

36. (c) $f(x) = |x|^3 = \begin{cases} x^3, & x > 0 \\ -x^3, & x < 0 \\ 0, & x = 0 \end{cases}$

R.H.L at $x = 0$

$= \lim_{x \to 0+} f(x) = \lim_{x \to 0+} x^3 = 0^3 = 0.$

L.H.L at $x = 0$

$= \lim_{x \to 0-} f(x) = \lim_{x \to 0-} \left(-x^3\right) = -0^3 = 0$

Thus $\lim_{x \to 0+} f(x) = \lim_{x \to 0-} f(x) = 0 = f(0).$

$\therefore f(x)$ is continuous at $x = 0.$

$Rf'(0)$

$= \lim_{x \to 0+} \dfrac{f(x) - f(0)}{x}$

$= \lim_{x \to 0+} \dfrac{x^3 - 0}{x} = \lim_{x \to 0+} x^2 = 0.$

$Lf'(0)$

$= \lim_{x \to 0-} \dfrac{f(x) - f(0)}{x}$

$= \lim_{x \to 0-} \dfrac{-x^3 - 0}{x} = \lim_{x \to 0-} \left(-x^2\right) = 0.$

Hence, $Rf'(0) = Lf'(0)$ and so $f(x)$ is differentiable at $x = 0.$

Thus $f'(x) = \begin{cases} 3x^2, & x > 0 \\ -3x^2, & x < 0 \\ 0, & x = 0 \end{cases}$

Similarly, $f''(x) = \begin{cases} 6x, & x > 0 \\ -6x, & x < 0 \\ 0, & x = 0 \end{cases}$

and $f'''(x) = \begin{cases} 6, & x > 0 \\ -6, & x < 0 \end{cases}$

But $f'''(x)$ does not exist if $x = 0$.

Hence $f(x)$ is twice differentiable only.

37. (b) $y = x^2 + 2x + 10 \Rightarrow \dfrac{dy}{dx} = 2x + 2$

$\therefore \left.\dfrac{dy}{dx}\right|_{x=1} = 2 \times 1 + 2 = 4.$

38. (c)

$f(x) = \sin|x| \Rightarrow \dfrac{df}{dx} = \dfrac{|x|}{x} \cos|x|$

$\therefore \left.\dfrac{df}{dx}\right|_{x=-\frac{\pi}{4}} = \dfrac{\left|-\dfrac{\pi}{4}\right|}{-\dfrac{\pi}{4}} \cos\left|-\dfrac{\pi}{4}\right| = -\cos\dfrac{\pi}{4} = -\dfrac{1}{\sqrt{2}}.$

39. (c) We know that if $y = |ax + b|$, then y is continuous for all "x" and $\dfrac{dy}{dx} = a\dfrac{|ax+b|}{(ax+b)}$, which exist for $x \neq \dfrac{-b}{a}$.

$\therefore y = |2 - 3x| \Rightarrow \dfrac{dy}{dx} = -3\dfrac{|2-3x|}{(2-3x)}.$

Thus, $y = |2 - 3x|$ is continuous $\forall x \in R$ and differentiable $\forall x \in R$ except at $x = \dfrac{2}{3}$.

40. (b) Given that $f(t) = \dfrac{\sin t}{t}$

$\therefore R.H.L = \underset{t \to 0+}{\text{Lt}}\, f(t) = \underset{t \to 0+}{\text{Lt}}\, \dfrac{\sin t}{t} = 1,$

$L.H.L = \underset{t \to 0-}{\text{Lt}}\, f(t) = \underset{t \to 0-}{\text{Lt}}\, \dfrac{\sin t}{t} = 1.$

Thus $R.H.L = L.H.L$, but $f(0)$ is not given.

Hence, the function $f(t)$ has a discontinuity at $t = 0$

41. (c) $f(x)$ is continuous at $x = \dfrac{\pi}{2}$

$\Rightarrow \underset{x \to \frac{\pi}{2}}{\text{Lt}}\, f(x) = f\left(\dfrac{\pi}{2}\right) \Rightarrow \underset{x \to \frac{\pi}{2}}{\text{Lt}}\left(\dfrac{\lambda \cos x}{\dfrac{\pi}{2} - x}\right) = 1$

$\Rightarrow \underset{x - \frac{\pi}{2} \to 0}{\text{Lt}}\left(\dfrac{-\lambda \sin\left(x - \dfrac{\pi}{2}\right)}{-\left(x - \dfrac{\pi}{2}\right)}\right) = 1$

$\Rightarrow \lambda \underset{x - \frac{\pi}{2} \to 0}{\text{Lt}}\left(\dfrac{\sin\left(x - \dfrac{\pi}{2}\right)}{\left(x - \dfrac{\pi}{2}\right)}\right) = 1 \Rightarrow \lambda \times 1 = 1 \Rightarrow \lambda = 1.$

42. (c) We know that if $f(x) = |ax + b|$, then $f(x)$ is continuous for all "x" and $f'(x) = a\dfrac{|ax+b|}{(ax+b)}$, which exist for $x \neq \dfrac{-b}{a}$.

Here a = 1 and b = 0. So $f(x) = |x|$ is continuous and non-differentiable at x = 0.

43. (a)

$$y = 5x^2 + 10x \Rightarrow \frac{dy}{dx} = 10x + 10.$$

$$\therefore \frac{dy}{dx}\Big|_{x=1} = 10 \times 1 + 10 = 20.$$

44. (a) Consider the function f(x) defined by:

$$f(x) = \begin{cases} 2 & \text{if } x = 3 \\ x - 1 & \text{if } x > 3 \\ \dfrac{x+3}{3} & \text{if } x < 3 \end{cases}$$

Then,

$$L.H.L = \underset{x \to 3-}{\text{Lt}}\, f(x) = \underset{x \to 3-}{\text{Lt}}\left(\frac{x+3}{3}\right) = \frac{3+3}{3} = 2,$$

$$R.H.L = \underset{x \to 3+}{\text{Lt}}\, f(x) = \underset{x \to 3+}{\text{Lt}}\,(x-1) = 3 - 1 = 2$$

and $f(3) = 2$.

$$\therefore L.H.L = R.H.L = f(3).$$

Hence, $f(x)$ is continuous at x=3.

45. (d) If a function $f(x)$ is continuous at $x = a$

$\Rightarrow \underset{x \to a}{\text{Lt}}\, f(x) = f(a)$ i.e; limit exists and is equal to the value of the function at that point.

46. −2

Given $f(x) = x \sin x$.

$\therefore f'(x) = x \cos x + \sin x,$

$f''(x) = -x \sin x + 2 \cos x.$

Then, $f''(x) + f(x) + t \cos x = 0$

$\Rightarrow -x \sin x + 2 \cos x + x \sin x + t \cos x = 0$

$\Rightarrow (2+t) \cos x = 0 \Rightarrow t + 2 = 0$

$\therefore t = -2.$

47. (c) The function $f(x) = \dfrac{x^2 - 3x - 4}{x^2 + 3x - 4}$ is not defined at $x = 1$ and $x = -4$.

$\therefore$ The function $f(x)$ is not continuous at $x = -4, 1$.

48. (b) A continuous function may not be differentiable where as a differentiable function is always continuous.

49. (d) Since any modulus function is continuous,

so $f(x) = \left|\sin \dfrac{2\pi x}{L}\right|$ is continuous for all values of x.

Now, $Rf'(0)$

$$= \underset{x \to 0+}{\text{Lt}}\, \frac{f(x) - f(0)}{x}$$

$$= \underset{x \to 0+}{\text{Lt}}\, \frac{\left|\sin\left(\dfrac{2\pi x}{L}\right)\right|}{x} = \underset{x \to 0+}{\text{Lt}}\, \frac{\sin\left(\dfrac{2\pi x}{L}\right)}{x}$$

$$= \left\{\underset{\frac{2\pi x}{L} \to 0+}{\text{Lt}}\, \frac{\sin\left(\dfrac{2\pi x}{L}\right)}{\dfrac{2\pi x}{L}}\right\} \times \frac{2\pi}{L} = 1 \times \frac{2\pi}{L} = \frac{2\pi}{L}.$$

$Lf'(0)$

$$= \underset{x \to 0-}{\text{Lt}}\, \frac{f(x) - f(0)}{x}$$

$$= \underset{x \to 0-}{\text{Lt}}\, \frac{\left|\sin\left(\dfrac{2\pi x}{L}\right)\right|}{x} = \underset{x \to 0-}{\text{Lt}}\, \frac{-\sin\left(\dfrac{2\pi x}{L}\right)}{x}$$

$$= -\left\{\underset{\frac{2\pi x}{L} \to 0-}{\text{Lt}}\, \frac{\sin\left(\dfrac{2\pi x}{L}\right)}{\dfrac{2\pi x}{L}}\right\} \times \frac{2\pi}{L} = -1 \times \frac{2\pi}{L} = -\frac{2\pi}{L}.$$

Since $Rf'(0) \neq Lf'(0)$, so $f(x)$ is not differentiable at $x = 0$.

50. (b).

$$f(x) = \begin{cases} e^x, & x < 1 \\ \ln x + ax^2 + bx, & x \geq 1 \end{cases}$$

$$\Rightarrow f'(x) = \begin{cases} e^x, & x < 1 \\ \dfrac{1}{x} + 2ax + b, & x \geq 1 \end{cases}$$

Now, if $f(x)$ is differentiable at x=1, then

$$\underset{x \to 1+}{\text{Lt}}\, f'(x) = \underset{x \to 1-}{\text{Lt}}\, f'(x)$$

or, $\dfrac{1}{1} + 2a + b = e$

or, $2a + b = e - 1$ ....(1)

Again, if $f(x)$ is differentiable at x=1, then $f(x)$ is continuous at x=1. This implies,

$$\underset{x \to 1+}{Lt} f(x) = \underset{x \to 1-}{Lt} f(x)$$

or, $\ln 1 + a + b = e$

or, $a + b = e \ldots \ldots \ldots (2)$

Solving (1) and (2) we get, $a = -1, b = e + 1$.

Thus, we can conclude that $f(x)$ is differentiable at $x = 1$ for the unique values of a and b.

51. (a) In $(-\infty, 0)$, $g(x) = -x$.

Then, $(f \circ g)(x)$

$= f(g(x)) = f(-x) = (-x)^2 = x^2$

$\left(x \in (-\infty, 0) \Rightarrow -x \in (0, \infty) \Rightarrow f(-x) = (-x)^2 \right)$

Thus, $(f \circ g)(x)$ is a polynomial function and so it has no points of discontinuity in $(-\infty, 0)$.

52. (a)

$y = x \ln x$

$\Rightarrow \dfrac{dy}{dx} = x \times \dfrac{1}{x} + \ln x = 1 + \ln x$

$\Rightarrow \tan 45° = 1 + \ln x \left(\because \text{slope} = \dfrac{dy}{dx} = \tan 45° \right)$

$\Rightarrow \ln x = 0 = \ln 1 \Rightarrow x = 1$.

At $x = 1, y = 1 \times \ln 1 = 0$.

Hence, the required point is $(1, 0)$.

53. (d) Since any modulus function is continuous, so $y = 5|x|$. is continuous at $x = 0$.

$Lf'(0) = \underset{x \to 0-}{Lt} \dfrac{f(x) - f(0)}{x} = \underset{x \to 0-}{Lt} \dfrac{-5x - 0}{x} = -5$

$\left(\because |x| = -x \text{ for } x < 0 \right)$,

$Rf'(0) = \underset{x \to 0+}{Lt} \dfrac{f(x) - f(0)}{x} = \underset{x \to 0+}{Lt} \dfrac{5x - 0}{x} = 5$

$\left(\because |x| = x \text{ for } x > 0 \right)$

$\therefore Lf'(0) \neq Rf'(0)$.

Hence, $f(x)$ is not differentiable at $x = 0$.

54. (d)

Given that $f(x) = \begin{cases} x^2, & x \geq 0 \\ -x^2, & x < 0 \end{cases}$

$Lf'(0) = \underset{x \to 0-}{Lt} \dfrac{f(x) - f(0)}{x}$

$= \underset{x \to 0-}{Lt} \dfrac{-x^2 - 0}{x - 0} = \underset{x \to 0}{Lt}(-x) = 0,$

$Rf'(0) = \underset{x \to 0+}{Lt} \dfrac{f(x) - f(0)}{x}$

$= \underset{x \to 0+}{Lt} \dfrac{x^2 - 0}{x} = \underset{x \to 0}{Lt}(x) = 0.$

$\because Lf'(0) = Rf'(0), \quad f(x)$ is differentiable at $x = 0$.

Then, $f'(x) = \begin{cases} 2x, & x \geq 0 \\ -2x, & x < 0 \end{cases} = g(x) (\text{say})$

Now

$Lg'(0) = \underset{x \to 0-}{Lt} \dfrac{g(x) - g(0)}{x} = \underset{x \to 0-}{Lt} \dfrac{-2x - 0}{x} = -2,$

$Rg'(0) = \underset{x \to 0+}{Lt} \dfrac{g(x) - g(0)}{x} = \underset{x \to 0+}{Lt} \dfrac{2x - 0}{x} = 2.$

Thus, $Lg'(0) \neq Rg'(0)$.

Hence, $g(x)$ i.e; $f'(x)$ is not differentiable at $x = 0$.

55. (b) Let $y = \underset{n \to \infty}{Lim} n^{\frac{1}{n}}$.

Taking logarithm on both sides we get,

$\log y = \underset{n \to \infty}{Lim} \log n^{\frac{1}{n}}$

$\Rightarrow \log y = \underset{n \to \infty}{Lim} \dfrac{1}{n} \log n \quad \left(\dfrac{\infty}{\infty} \text{ form} \right)$

Using the L'Hospital rule, we get,

$\log y = \underset{n \to \infty}{Lim} \dfrac{\dfrac{d}{dn} \log n}{\dfrac{d}{dn} n} = \underset{n \to \infty}{Lim} \dfrac{\dfrac{1}{n}}{1} = 0 = \log 1$

$\therefore y = 1 \ i.e; \underset{n \to \infty}{Lim} n^{\frac{1}{n}} = 1.$

56. (c)

$\underset{x \to \frac{\pi}{4}}{Lim} \dfrac{\cos x - \sin x}{x - \dfrac{\pi}{4}} \left(\dfrac{0}{0} \text{form} \right)$

$= \underset{x \to \frac{\pi}{4}}{Lim} \dfrac{\dfrac{d}{dx}(\cos x - \sin x)}{\dfrac{d}{dx}\left(x - \dfrac{\pi}{4} \right)}$ (by the L'Hospital Rule)

$= \underset{x \to \frac{\pi}{4}}{Lim} \dfrac{-\sin x - \cos x}{1}$

$= -\sin \dfrac{\pi}{4} - \cos \dfrac{\pi}{4} = -\dfrac{1}{\sqrt{2}} - \dfrac{1}{\sqrt{2}} = -\sqrt{2}.$

57. (c)

$$\underset{x \to 0}{Lt} \frac{\sin x}{e^x x} \quad \left(\text{form } \frac{0}{0}\right)$$

$$= \underset{x \to 0}{Lt} \frac{\dfrac{d}{dx}(\sin x)}{\dfrac{d}{dx}(e^x x)} \quad \text{(by the L'Hospital rule)}$$

$$= \underset{x \to 0}{Lt} \frac{\cos x}{xe^x + e^x} = \frac{\cos 0}{0 + e^0} = 1.$$

58. (b)

$$\underset{x \to 0}{Lt} \left(\frac{1 - \cos x}{x^2}\right) \quad \left(\frac{0}{0} \text{ form}\right)$$

$$= \underset{x \to 0}{Lt} \left(\frac{\dfrac{d}{dx}\{1 - \cos x\}}{\dfrac{d}{dx}\{x^2\}}\right) \quad \text{(by the L'Hospital rule)}$$

$$= \underset{x \to 0}{Lt} \frac{\sin x}{2x} = \frac{1}{2} \times \underset{x \to 0}{Lt} \frac{\sin x}{x} = \frac{1}{2} \left[\because \underset{x \to 0}{Lt} \frac{\sin x}{x} = 1\right]$$

59. (a)

$$\underset{x \to 0}{Lt} \frac{x - \sin x}{1 - \cos x} \quad \left(\frac{0}{0} \text{ form}\right)$$

$$= \underset{x \to 0}{Lt} \frac{\dfrac{d}{dx}(x - \sin x)}{\dfrac{d}{dx}(1 - \cos x)} \quad \text{(by the L'Hospital rule)}$$

$$= \underset{x \to 0}{Lt} \frac{1 - \cos x}{\sin x} \quad \left(\frac{0}{0} \text{ form}\right)$$

$$= \underset{x \to 0}{Lt} \frac{\dfrac{d}{dx}(1 - \cos x)}{\dfrac{d}{dx}(\sin x)} \quad \text{(by the L'Hospital rule)}$$

$$= \underset{x \to 0}{Lt} \frac{\sin x}{\cos x} = \frac{\sin 0}{\cos 0} = 0 \ (\sin 0 = 0).$$

60. (b)

$$\underset{x \to 0}{Lt} \frac{e^{2x} - 1}{\sin 4x} \quad \left(\frac{0}{0} \text{ form}\right)$$

$$= \underset{x \to 0}{Lt} \frac{\dfrac{d}{dx}(e^{2x} - 1)}{\dfrac{d}{dx}(\sin 4x)} \quad \text{(by the L'Hospital rule)}$$

$$= \underset{x \to 0}{Lt} \frac{2e^{2x}}{4 \cos 4x} = \frac{2e^0}{4 \cos 0} = \frac{2}{4} = 0.5.$$

61. (a)

$$\underset{a \to 0}{Lt} \frac{x^a - 1}{a} \quad \left(\frac{0}{0} \text{ form}\right)$$

$$= \underset{a \to 0}{Lt} \frac{\dfrac{d}{da}(x^a - 1)}{\dfrac{d}{da}(a)}$$

$$= \underset{a \to 0}{Lt} \frac{x^a \log x}{1} \quad \text{[using the L'Hospital rule]}$$

$$= \log x \times \underset{a \to 0}{Lt} x^a = \log x \times 1 = \log x.$$

62. -0.333

$$\underset{x \to 0}{Lt} \left(\frac{-\sin x}{2 \sin x + x \cos x}\right) \quad \left(\text{form } \frac{0}{0}\right)$$

$$= \underset{x \to 0}{Lt} \frac{\dfrac{d}{dx}(-\sin x)}{\dfrac{d}{dx}(2 \sin x + x \cos x)} \quad \text{(by the L'Hospital rule)}$$

$$= \underset{x \to 0}{Lt} \frac{-\cos x}{2 \cos x + \cos x - x \sin x}$$

$$= \underset{x \to 0}{Lt} \frac{-\cos x}{3 \cos x - x \sin x} = \frac{-\cos 0}{3 \cos 0 - 0} = \frac{-1}{3} = -0.333.$$

63. (c) $\underset{x \to 0}{Lt} \dfrac{1 - \cos(x^2)}{2x^4} \quad \left(\text{form } \dfrac{0}{0}\right)$

$$= \underset{x \to 0}{Lt} \frac{\dfrac{d}{dx}(1 - \cos x^2)}{\dfrac{d}{dx}(2x^4)} \quad \text{(by the L'Hospital rule)}$$

$$= \underset{x \to 0}{Lt} \frac{2x \sin x^2}{8x^3} = \frac{1}{4} \times \underset{x^2 \to 0}{Lt} \frac{\sin(x^2)}{x^2} = \frac{1}{4} \times 1 = \frac{1}{4}.$$

64. (c) $\underset{x \to 0}{Lt} \dfrac{\log(1 + 4x)}{e^{3x} - 1} \quad \left(\dfrac{0}{0} \text{ form}\right)$

$$= \underset{x \to 0}{Lt} \left\{\frac{\dfrac{d}{dx}(\log(1 + 4x))}{\dfrac{d}{dx}(e^{3x} - 1)}\right\} \quad \text{(by the L'Hospital rule)}$$

$$= \underset{x \to 0}{Lt} \left\{\frac{\dfrac{1}{1 + 4x} \times 4}{3e^{3x}}\right\} = \frac{\dfrac{4}{1 + 0}}{3e^0} = \frac{4}{3}.$$

65. :

$$\lim_{x\to 1}\frac{x^7-2x^5+1}{x^3-3x^2+2}\quad\left(\text{form }\frac{0}{0}\right)$$

$$=\lim_{x\to 1}\frac{\dfrac{d}{dx}\left(x^7-2x^5+1\right)}{\dfrac{d}{dx}\left(x^3-3x^2+2\right)}\quad\text{(by the L'Hospital rule)}$$

$$=\lim_{x\to 1}\frac{7x^6-10x^4}{3x^2-6x}=\lim_{x\to 1}\frac{7x^5-10x^3}{3x-6}$$

$$=\frac{7-10}{3-6}=1.$$

66. (c) It is easy to verify that $f(\theta)$ is continuous

on $\left[\dfrac{\pi}{6},\dfrac{\pi}{3}\right]$ and differentiable on $\left(\dfrac{\pi}{6},\dfrac{\pi}{3}\right)$.

Also $f\left(\dfrac{\pi}{6}\right)=0=f\left(\dfrac{\pi}{3}\right)$.

∴ By Rolle's theorem, there exists at least one

value of $\theta\in\left(\dfrac{\pi}{6},\dfrac{\pi}{3}\right)$ such that $f'(\theta)=0$

Thus, (I) is true.

Again $f'(\theta)$ is always zero if $f(\theta)$ is a constant function. Since $f(\theta)$ is not a constant function, so (II) is also true.

67. () Let $g(y)=f(y)-f(y+1)$ for $y\in[0, 1]$. Then we have,

$g(0)=f(0)-f(0+1)=-1-1=-2<0$ and

$g(1)=f(1)-f(1+1)=1+1=2>0$.

Thus, $g(0)\times g(1)<0$.

Hence, $g(y)=0$ for some $y\in(0, 1)$. This gives,

$f(y)-f(y+1)=0$ i.e; $f(y)=f(y+1)$ for some $y\in(0, 1)$.

Remember:

If $f:[a,b]\to R$ be a function such that f is continuous and $f(a)f(b)<0$, then $f(c)=0$ for some $c\in(a,b)$

68. (b) $f(x)=1-x^2+x^3\Rightarrow f'(x)=-2x+3x^2$.

Clearly, f is continuous on [-1, 1] and differentiable on (-1, 1). Hence, by Lagrange's Mean Value Theorem, we have,

$$f'(c)=\frac{f(1)-f(-1)}{1-(-1)}$$

or, $-2c+3c^2=\dfrac{(1-1+1)-(1-1-1)}{2}=1$

or, $3c^2-2c-1=0$

or, $c=1,-\dfrac{1}{3}$

$\therefore c=-\dfrac{1}{3}\in(-1,1)\ \left[\because c=1\notin(-1,1)\right]$

69. (a) According to the Mean Value Theorem of Integral Calculus, we have,

If a function $f(x)$ is continuous on $[a,b]$, then ∃ a point $\xi\in(a,b)$ such that

$$\int_a^b f(x)\,dx=f(\xi)(b-a).$$

70. (b)

$f(x)=x^3-3x^2+1$

$\Rightarrow f'(x)=3x^2-6x$ and $f''(x)=6x-6$.

Now, $f'(x)=0$

$\Rightarrow 3x^2-6x=0\Rightarrow x=0,2$.

At $x=0$, $f''(x)=0-6<0$.

Therefore, $f(x)$ has a maximum at x=0.

At $x=2$, $f''(x)=12-6>0$.

Therefore, $f(x)$ has a minimum at x=0.

∴ The function $f(x)$ increases from -1 to 0 and decreases from 0 to 2 and then increases from 2 to 3.

71. (d)

$f(x)=x\sqrt{5-x^2}$ gives,

$$f'(x)=\frac{-x^2}{\sqrt{5-x^2}}+\sqrt{5-x^2}=\frac{5-2x^2}{\sqrt{5-x^2}},$$

$$f''(x)=\frac{(5x-2x^3)}{(5-x^2)\sqrt{5-x^2}}-\frac{4x}{\sqrt{5-x^2}}.$$

Now, $f'(x)=0\Rightarrow 5-2x^2=0\Rightarrow x^2=\dfrac{5}{2}$

$$\Rightarrow x=\sqrt{\frac{5}{2}},-\sqrt{\frac{5}{2}}$$

$$f''\left(\sqrt{\frac{5}{2}}\right)$$

$$= \frac{5\sqrt{\frac{5}{2}} - 2\left(\sqrt{\frac{5}{2}}\right)^3}{\left[5 - \left(\sqrt{\frac{5}{2}}\right)^2\right]\sqrt{5 - \left(\sqrt{\frac{5}{2}}\right)^2}} - \frac{4\sqrt{\frac{5}{2}}}{\sqrt{5 - \left(\sqrt{\frac{5}{2}}\right)^2}}$$

$$= \frac{\sqrt{\frac{5}{2}}\left\{5 - 2\left(\sqrt{\frac{5}{2}}\right)^2\right\}}{\left[5 - \frac{5}{2}\right]\sqrt{5 - \frac{5}{2}}} - \frac{4\sqrt{\frac{5}{2}}}{\sqrt{5 - \frac{5}{2}}} = 0 - \frac{4\sqrt{\frac{5}{2}}}{\sqrt{5 - \frac{5}{2}}} < 0.$$

$\therefore f(x)$ has a local maxima at $x = \sqrt{\frac{5}{2}}$.

Again $f''\left(-\sqrt{\frac{5}{2}}\right)$

$$= \frac{-5\sqrt{\frac{5}{2}} - 2\left(-\sqrt{\frac{5}{2}}\right)^3}{\left[5 - \left(-\sqrt{\frac{5}{2}}\right)^2\right]\sqrt{5 - \left(-\sqrt{\frac{5}{2}}\right)^2}} - \frac{\left(-4\sqrt{\frac{5}{2}}\right)}{\sqrt{5 - \left(-\sqrt{\frac{5}{2}}\right)^2}}$$

$$= 0 + \frac{4\sqrt{\frac{5}{2}}}{\sqrt{5 - \frac{5}{2}}} > 0.$$

$\therefore f(x)$ has local minimum at $x = -\sqrt{\frac{5}{2}}$

72. (a)

$$f(x) = 2x^3 - 3x^2 - 36x + 2$$

$$\Rightarrow f'(x) = 6x^2 - 6x - 36, \; f''(x) = 12x - 6.$$

Then, $f'(x) = 0$

$$\Rightarrow 6x^2 - 6x - 36 = 0$$

$$\Rightarrow x^2 - x - 6 = 0$$

$$\Rightarrow (x - 3)(x + 2) = 0$$

$$\Rightarrow x = -2, 3$$

$\therefore x = 3, -2$ are stationary points.

Now, $f''(3) = 12 \times 3 - 6 = 6 > 0$.

Thus, $f(x)$ has a minimum value at $x = 3$.

Also, $f''(-2) = 12 \times (-2) - 6 < 0$.

So, $f(x)$ has a maximum value at $x = -2$.

73. (a)

$$f(x) = x^2 e^{-x}$$

$$\Rightarrow f'(x) = 2xe^{-x} - x^2 e^{-x} = (2x - x^2)e^{-x},$$

$$f''(x) = (x^2 - 4x + 2)e^{-x}.$$

Now, $f'(x) = 0 \Rightarrow (2x - x^2)e^{-x} = 0$

$$\Rightarrow x(2 - x)e^{-x} = 0$$

$x = 0, 2$ are the stationary points.

Now, $f''(0) = (0^2 - 4 \times 0 + 2)e^{-0} = 2 > 0.$

$\therefore f(x)$ has a minimum value at $x = 0$.

$f''(2) = (2^2 - 4 \times 2 + 2)e^{-2} = -2e^{-2} < 0.$

So, $f(x)$ has a maximum value at $x = 2$.

74. (c)

Given $f(x) = \dfrac{e^{\sin x}}{e^{\cos x}} = e^{\sin x - \cos x}$.

So $f'(x) = e^{\sin x - \cos x}(\sin x + \cos x).$

$\therefore f'(x) = 0$

$$\Rightarrow e^{\sin x - \cos x}(\sin x + \cos x) = 0$$

$$\Rightarrow \sin x + \cos x = 0$$

$$\Rightarrow \sin x = -\cos x$$

$$\Rightarrow \tan x = -1 = -\tan\frac{\pi}{4} = \begin{cases} \tan\left(-\dfrac{\pi}{4}\right) \\ \tan\left(\pi - \dfrac{\pi}{4}\right) \end{cases}$$

$$\Rightarrow x = -\frac{\pi}{4}, \frac{3\pi}{4}$$

So, the stationary points are $x = -\dfrac{\pi}{4}, \dfrac{3\pi}{4}$.

Now, we have

$$f''(x)$$

$$= e^{\sin x - \cos x}(\cos x - \sin x)$$

$$\quad + (\sin x + \cos x)e^{\sin x - \cos x}(\cos x + \sin x)$$

$$= e^{\sin x - \cos x}\left\{(\sin x + \cos x)^2 + (\cos x - \sin x)\right\}$$

$$\therefore f''\left(-\frac{\pi}{4}\right)$$

$$= e^{-\frac{1}{\sqrt{2}} - \frac{1}{\sqrt{2}}}\left\{\left(-\frac{1}{\sqrt{2}} + \frac{1}{\sqrt{2}}\right)^2 + \left(\frac{1}{\sqrt{2}} + \frac{1}{\sqrt{2}}\right)\right\} > 0$$

$$\left(\because \sin\left(-\frac{\pi}{4}\right) = -\frac{1}{\sqrt{2}}, \; \cos\left(-\frac{\pi}{4}\right) = \frac{1}{\sqrt{2}}\right)$$

So $f(x)$ has a minimum value at $-\dfrac{\pi}{4}$.

Since, $\sin\left(\dfrac{3\pi}{4}\right)=\sin\left(\pi-\dfrac{\pi}{4}\right)=\sin\dfrac{\pi}{4}=\dfrac{1}{\sqrt{2}}$

and $\cos\left(\dfrac{3\pi}{4}\right)=\cos\left(\pi-\dfrac{\pi}{4}\right)=-\cos\dfrac{\pi}{4}=-\dfrac{1}{\sqrt{2}}$,

we have,

$$f''\left(\dfrac{3\pi}{4}\right)$$
$$=e^{\frac{1}{\sqrt{2}}+\frac{1}{\sqrt{2}}}\left\{\left(\dfrac{1}{\sqrt{2}}-\dfrac{1}{\sqrt{2}}\right)^2+\left(-\dfrac{1}{\sqrt{2}}-\dfrac{1}{\sqrt{2}}\right)\right\}<0$$

So, $f(x)$ has a maximum at $x=\dfrac{3\pi}{4}$.

Hence, maximum value of $f(x)$ is

$f\left(\dfrac{3\pi}{4}\right)$ i.e; $e^{\sin\frac{3\pi}{4}-\cos\frac{3\pi}{4}}$.

$\because e^{\sin\frac{3\pi}{4}-\cos\frac{3\pi}{4}}=e^{\frac{1}{\sqrt{2}}+\frac{1}{\sqrt{2}}}=e^{\frac{2}{\sqrt{2}}}=e^{\sqrt{2}}$,

so the maximum value of $f(x)$ is $e^{\sqrt{2}}$.

75. (b)

Given $y=x^2$ in $[1,5]$. So $\dfrac{dy}{dx}=2x>0$ in $[1,5]$.

This shows that y is an increasing function in $[1,5]$. Hence, y is minimum when x is minimum in $[1,5]$. Consequently, minimum value of y is $(1)^2$, i.e, 1.

76. (a)

$f(x)=x^2-x-2$

$\Rightarrow f'(x)=2x-1<0$ for $x\in[-4,4]$

$\therefore f(x)$ is a decreasing function on $[-4,4]$.

Hence, the maximum value of $f(x)$ occurs at $x=-4$.

$\therefore$ the maximum value of $f(x)$ is $f(-4)$ i.e; 18.

77. (b)

$f(x)=(x^2-4)^2$

$\Rightarrow f'(x)=4x(x^2-4)$,

$f''(x)=4(x^2-4)+4x\times2x=4(3x^2-4)$.

$\therefore f'(x)=0$

$\Rightarrow 4x(x^2-4)=0\Rightarrow x=0,2,-2$.

Now $f''(0)=-16<0, f''(2)=4(12-4)>0$

and $f''(-2)=4(12-4)>0$.

$\therefore f(x)$ has a maximum value at $x=0$ and minimum values at $x=2$ and $x=-2$.

78. (a)

$f(x)=x^2-x-2$

$\Rightarrow f'(x)=2x-1, f''(x)=2$.

$\therefore f'(x)=0\Rightarrow x=\dfrac{1}{2}$.

At $x=\dfrac{1}{2}, f''(x)=2>0$.

$\therefore x=\dfrac{1}{2}$ is a point of local minimum.

As $f(4)=4^2-4-2=10, f(-4)=4^2+4-2=18$, so the maximum value of $f(x)$ in the closed interval $[-4,4]$ is max$\{10,18\}$ i.e; 18.

79. (b)

Let $f(x)=3x^4-16x^3+24x^2+37$.

$\therefore f'(x)=12x^3-48x^2+48x$,

$f''(x)=36x^2-96x+48$.

Now $f'(x)=0$

$\Rightarrow 12x^3-48x^2+48x=0$

$\Rightarrow x(x^2-4x+4)=0$

$\Rightarrow x(x-2)^2=0\Rightarrow x=0,2$.

$\therefore x=0,2$ are the stationary points.

Now $f''(0)=0-0+48>0$ and so $f(x)$ has a minimum at $x=0$.

$f''(2)=36\times4-96\times2+48=0$ and so $f(x)$ has no extremum at $x=2$.

80. (c)

$y=x^2-6x+9\Rightarrow\dfrac{dy}{dx}=2x-6, \dfrac{d^2y}{dx^2}=2$.

$\therefore\dfrac{dy}{dx}=0\Rightarrow 2x-6=0\Rightarrow x=3$.

So, $x=3$ is the only stationary point.

$\because\left.\dfrac{d^2y}{dx^2}\right|_{x=3}=2>0, f(x)$ has a minimum at $x=3$.

Since $y(2)=2^2-6\times2+9=1, y(5)=5^2-6\times5+9=4$,

$\therefore$ the maximum value of y is max$\{f(2),f(5)\}$ i.e; 4.

81. (a) $f(x) = e^x + e^{-x}$

$\Rightarrow f'(x) = e^x - e^{-x}$ and $f''(x) = e^x + e^{-x}$.

$f'(x) = 0 \Rightarrow e^x - e^{-x} = 0 \Rightarrow e^{2x} = 1$

$\Rightarrow e^{2x} = e^0 \Rightarrow 2x = 0$

$\Rightarrow x = 0$

$\therefore x = 0$ is a stationary point.

Then, $f''(0) = e^0 + e^0 = 1 + 1 = 2 > 0$.

$\therefore f(x)$ has a minimum value at $x = 0$.

Minimum value $= f(0) = e^0 + e^{-0} = 2$.

82. (a)

$e^y = x^{\frac{1}{x}} \Rightarrow \log e^y = \log x^{\frac{1}{x}} \Rightarrow y = \frac{1}{x} \log x$.

$\therefore \frac{dy}{dx} = \frac{x \times \frac{1}{x} - \log x}{x^2} = \frac{1 - \log x}{x^2}$ and

$\frac{d^2y}{dx^2} = \frac{x^2 \frac{d}{dx}(1 - \log x) - (1 - \log x)\frac{d}{dx}(x^2)}{\{x^2\}^2}$

$= \frac{x^2\left(0 - \frac{1}{x}\right) - (1 - \log x)2x}{x^4}$

$= \frac{-1 - 2(1 - \log x)}{x^3} = \frac{-3 + 2\log x}{x^3}$

Now, $\frac{dy}{dx} = 0$

$\Rightarrow \frac{1 - \log x}{x^2} = 0 \Rightarrow 1 - \log x = 0$

$\Rightarrow \log x = 1 = \log e \Rightarrow x = e$.

$\therefore x = e$ is the stationary point

At $x = e$, $\frac{d^2y}{dx^2} = \frac{-3 + 2\log e}{e^3} = \frac{-3 + 2}{e^3} < 0$.

$\therefore y$ has a maximum value at $x = e$.

83. (c)

$f(x) = 2x - x^2 + 3$

$\Rightarrow f'(x) = 2 - 2x$ and $f''(x) = -2$.

$f'(x) = 0$

$\Rightarrow 2 - 2x = 0$

$\Rightarrow x = 1$ is a stationary point.

Now $f''(1) = -2 < 0$

$\therefore f(x)$ has a (local) maximum value at $x = 1$.

84. (c) $f(x) = x^3 - 9x^2 + 24x + 5$

$\Rightarrow f'(x) = 3x^2 - 18x + 24$, $f''(x) = 6x - 18$.

Now, $f'(x) = 0$

$\Rightarrow 3x^2 - 18x + 24 = 0 \Rightarrow x^2 - 6x + 8 = 0$

$\Rightarrow x = 2, 4$ are the stationary points.

$f''(2) = 12 - 18 < 0$, $f''(4) = 24 - 18 > 0$.

$\therefore x = 2$ is a local maxima and $x = 4$ is a local minima.

Again, $f(1) = 1 - 9 + 24 + 5 = 21$,

$f(6) = 6^3 - 9 \times 6^2 + 24 \times 6 + 5 = 41$,

$f(2) = 2^3 - 9 \times 2^2 + 24 \times 2 + 5 = 25$,

and $f(4) = 4^3 - 9 \times 4^2 + 24 \times 4 + 5 = 21$.

$\therefore$ Maximum value of $f(x)$ in $[1, 6]$ is

$\max\{f(2), f(1), f(6)\}$ i.e; $\max\{25, 21, 41\}$ i.e; 41.

85. (d) Given $f(x) = x^3 + 1$.

$\therefore f'(x) = 3x^2$, $f''(x) = 6x$, $f'''(x) = 6$.

$f'(x) = 0$

$\Rightarrow 3x^2 = 0$

$\Rightarrow x = 0$ is the stationary point.

Now, $f''(0) = 0$ & $f'''(0) \neq 0$

$\therefore f(x)$ has point of inflection at $x = 0$.

86. (b) Given $y = 2x - 0.1x^2$.

$\therefore \frac{dy}{dx} = 2 - 2(0.1)x$, $\frac{d^2y}{dx^2} = -2 \times 0.1 = -0.2$.

$\frac{dy}{dx} = 0 \Rightarrow 2 - 2(0.1)x = 0 \Rightarrow x = 10$

$\therefore x = 10$ is the stationary point.

Since, $\left.\frac{d^2y}{dx^2}\right|_{x=10} = -0.2 < 0$,

so y is maximum at $x = 10$.

$\therefore$ Max. height $= y(10) = 2 \times 10 - (0.1) \times 10^2 = 10$ m.

87. (a)

$f(t) = e^{-t} - 2e^{-2t} \Rightarrow f'(t) = -e^{-t} + 4e^{-2t}.$

$f'(t) = 0$

$\Rightarrow e^{-t}(-1 + 4e^{-t}) = 0 \Rightarrow e^{-t} = \dfrac{1}{4}$

$\Rightarrow -t = \log\left(\dfrac{1}{4}\right) = -\log 4 \Rightarrow t = \log 4.$

Now, $f''(t) = e^{-t} - 8e^{-2t}.$

At $t = \log 4,$

$f''(t) = e^{-\log 4} - 8e^{-2\log 4} = e^{\log\frac{1}{4}} - 8e^{\log\frac{1}{16}}$

$= \dfrac{1}{4} - 8 \times \dfrac{1}{16} = -\dfrac{1}{4} < 0.$

$\therefore f(t)$ has a maximum value at $t = \log 4.$

88. 0

$f(x) = \ln(1+x) - x \Rightarrow f'(x) = \dfrac{1}{1+x} - 1.$

So $f'(x) = 0 \Rightarrow \dfrac{1}{1+x} = 1 \Rightarrow 1+x = 1 \Rightarrow x = 0.$

Thus, $x = 0$ is a stationary point.

Now, $f''(x) = \dfrac{-1}{(1+x)^2} = -1 < 0$ at $x = 0$

$\therefore f(x)$ is maximum at $x = 0$ and maximum value of the function $= \ln(1+0) - 0 = 0.$

89. 6

$f(x) = 2x^3 - 9x^2 + 12x - 3$

$\Rightarrow f'(x) = 6x^2 - 18x + 12,\ f''(x) = 12x - 18.$

$f'(x) = 0$

$\Rightarrow 6x^2 - 18x + 12 = 0 \Rightarrow x = 1, 2 \in [0, 3].$

At $x = 1$ $f''(x) = 12 \times 1 - 18 = -6 < 0.$

$\therefore x = 1$ is a point of local maximum.

At $x = 2$ $f''(x) = 12 \times 2 - 18 = 6 > 0$

$\therefore x = 2$ is a point of local minimum.

Here, $f(0) = 0 - 0 + 0 - 3 = -3,$

$f(1) = 2 - 9 + 12 - 3 = 2$ and

and $f(3) = 2 \times 27 - 9 \times 9 + 36 - 3 = 6.$

Hence, maximum of $f(x)$ in $[0, 3]$

$= \max\{f(0), f(1), f(3)\}$

$= \max\{-3, 2, 6\} = 6.$

90. (c) Let base=x and height=y. Then, hypotenuse $= \sqrt{x^2 + y^2}$.

Given that sum of the lengths of the hypotenuse and a side is constant

so $x + \sqrt{x^2 + y^2} = c,$ where c is constant

or, $\sqrt{x^2 + y^2} = c - x$

or, $x^2 + y^2 = (c-x)^2$

or, $y^2 = c^2 - 2cx.........(1)$

Let the area of the triangle be

A. Then, $A = \dfrac{1}{2}xy.$

$\therefore A^2 = \dfrac{x^2 y^2}{4} = \dfrac{x^2}{4}(c^2 - 2cx)$

$= \dfrac{1}{4}(c^2 x^2 - 2cx^3) = f(x)$ (say).

$\therefore f'(x) = \dfrac{1}{4}(2c^2 x - 6cx^2),$

$f''(x) = \dfrac{1}{4}(2c^2 - 12cx).$

$f'(x) = 0$

$\Rightarrow 2c^2 x - 6cx^2 = 0$

$\Rightarrow 2cx(c - 3x) = 0 \Rightarrow x = 0, \dfrac{c}{3}.$

$\therefore x = 0, \dfrac{c}{3}$ are stationary points.

$f''(x) = \dfrac{1}{4}(2c^2 - 12cx).$

$f'(x) = 0$

$\Rightarrow 2c^2 x - 6cx^2 = 0$

$\Rightarrow 2cx(c - 3x) = 0 \Rightarrow x = 0, \dfrac{c}{3}.$

$\therefore x = 0, \dfrac{c}{3}$ are stationary points.

Clearly, $f''(0) = \dfrac{1}{4}(2c^2 - 0) = \dfrac{c^2}{2} > 0.$

$\therefore f(x)$ has a minimum value at $x = 0.$

Again, $f''\left(\dfrac{c}{3}\right) = \dfrac{1}{4}\left\{2c^2 - 12c \times \dfrac{c}{3}\right\} = -\dfrac{c^2}{2} < 0.$

$\therefore f(x)$ has a maximum value at $x = \dfrac{c}{3}.$

Now $y^2 = c^2 - 2cx = c^2 - 2c \times \dfrac{c}{3} = \dfrac{c^2}{3}$.

$\therefore y = \dfrac{c}{\sqrt{3}}$.

If θ be the angle between the hypotenuse and the side, then we have

$\cos\theta = \dfrac{x}{\sqrt{x^2+y^2}} = \dfrac{\dfrac{c}{3}}{\sqrt{\dfrac{c^2}{9}+\dfrac{c^2}{3}}} = \dfrac{1}{2} = \cos 60°$

$\therefore \theta = 60°$.

91. (a) Given $f(x) = xe^{-x}$.

$\Rightarrow f'(x) = -xe^{-x} + e^{-x} = (1-x)e^{-x}$,

$f''(x) = -(1-x)e^{-x} - e^{-x} = (x-2)e^{-x}$.

Then, $f'(x) = 0 \Rightarrow (1-x)e^{-x} = 0 \Rightarrow x = 1$.

$\therefore x = 1$ is the only stationary point.

Now, $f''(1) = (1-2)e^{-1} = -e^{-1} < 0$

$\therefore f(x)$ has a maximum value at $x = 1$.

Hence, the maximum value of $f(x) = f(1) = e^{-1}$.

92. (c)

$f(x) = (x-1)^{\frac{2}{3}} \Rightarrow f'(x) = \dfrac{2}{3}(x-1)^{-\frac{1}{3}}$.

$\therefore f'(x) = 0$

$\Rightarrow \dfrac{2}{3}(x-1)^{-\frac{1}{3}} = 0 \Rightarrow \dfrac{1}{(x-1)^{\frac{1}{3}}} = 0$,

which is never possible for any value of x.

Now $f(x) = (x-1)^{\frac{2}{3}} = \left\{(x-1)^{\frac{1}{3}}\right\}^2 \geq 0$.

Thus, the minimum value of $f(x)$ is "0" which occurs at $x = 1$.

93. (b) $f(x) = x^3 - 3x^2 - 24x + 100$

$\Rightarrow f'(x) = 3x^2 - 6x - 24, f''(x) = 6x - 6$.

$f'(x) = 0$

$\Rightarrow 3x^2 - 6x - 24 = 0 \Rightarrow x = -2, 4$.

$\therefore x = -2$ and $x = 4$ are stationary points.

$f''(-2) = 6 \times (-2) - 6 = -18 < 0$

$\therefore f(x)$ has a maximum value at $x = 2$. -2.

$f''(4) = 6 \times 4 - 6 = 18 > 0$.

$\therefore f(x)$ has a minimum value at $x = 4$.

Now, $f(4) = 4^3 - 3 \times 4^2 - 24 \times 4 + 100 = 20$,

$f(-3) = -3^3 - 3 \times 3^2 + 24 \times 3 + 100 = 118$,

$f(3) = 3^3 - 3 \times 3^2 - 24 \times 3 + 100 = 28$.

Since $4 \in [-3, 3]$. Therefore, required minimum value of f(x) = min{f(-3), f(3)}=min{118, 28} = 28.

94. 1.

Let $2x$ and $2y$ be the length and breadth of the rectangle.

If A denotes the area of the rectangle, then

$A = 2x \times 2y = 4xy$.

Then, $A^2 = 16x^2y^2 = 4x^2(1-x^2) = 4x^2 - 4x^4$

$(\because x^2 + 4y^2 = 1)$

Let $f(x) = 4x^2 - 4x^4$.

Then, $f'(x) = 8x - 16x^3$ and $f''(x) = 8 - 48x^2$.

Now $f'(x) = 0$

$\Rightarrow 8x(1 - 2x^2) = 0$

$\Rightarrow x = 0, \dfrac{1}{\sqrt{2}}, -\dfrac{1}{\sqrt{2}}$.

Here $f''\left(\dfrac{1}{\sqrt{2}}\right) = 8 - 48 \times \dfrac{1}{2} < 0$.

Thus, $f(x)$ is maximum at $x = \dfrac{1}{\sqrt{2}}$ and so area is maximum at $x = \dfrac{1}{\sqrt{2}}$.

$\therefore$ Maximum Area

$= 4xy = 4x \times \dfrac{\sqrt{1-x^2}}{2} = 2x\sqrt{1-x^2}$

$= 2 \times \dfrac{1}{\sqrt{2}} \times \sqrt{1 - \dfrac{1}{2}} = 1$.

95. (b) Given $f(x) = e^{-x}(x^2 + x + 1)$

$\Rightarrow f'(x) = (2x+1)e^{-x} - e^{-x}(x^2+x+1)$

$\qquad = (-x^2 + x)e^{-x}.$

Then, $f'(x) = 0$

$\Rightarrow (-x^2 + x)e^{-x} = 0 \Rightarrow x(1-x) = 0$

$\therefore x = 0, 1$ are stationary points.

Now, $f''(x)$

$= (-2x+1)e^{-x} - e^{-x}(-x^2+x) = e^{-x}(x^2 - 3x + 1).$

$\therefore f''(0) = 1 > 0$ and $f''(1) = -e^{-1} - 0 < 0.$

Hence, $f(x)$ has minimum at $x = 0$ and maximum at $x = 1$ which is true for the graph (b) only.

96. 0 $f(x) = x(x-1)(x-2)$ in $[1, 2]$

$\Rightarrow f'(x) = 3x^2 - 6x + 2.$

Then, $f'(x) = 0$

$\Rightarrow x = \dfrac{6 \pm \sqrt{36 - 4 \times 3 \times 2}}{2 \times 3} = \dfrac{3 \pm \sqrt{3}}{3}.$

These two stationary points lie outside the given interval $[1, 2]$.

Therefore, the maximum value of $f(x)$ exists at one of the end points of the interval.

We have $f(1) = 1(1-1)(1-2) = 0$ and

$f(2) = 2(2-1)(2-2) = 0.$

$\therefore$ maximum value of $f(x) = \max\{f(1), f(2)\} = 0.$

97. −13

$f(x) = 2x^3 - x^4 - 10$

$\Rightarrow f'(x) = 6x^2 - 4x^3, f''(x) = 12x - 12x^2.$

$\therefore f'(x) = 0$

$\Rightarrow 6x^2 - 4x^3 = 0 \Rightarrow 2x^2(3 - 2x)$

$\Rightarrow x = 0, \dfrac{3}{2}$ are stationary points.

But $x = \dfrac{3}{2}$ lies outside the interval $[-1,1]$.

Now $f(-1) = -2 - 1 - 10 = -13,$

$f(1) = 2 - 1 - 10 = -9, f(0) = 0 - 0 - 10 = -10.$

$\therefore$ Minimum value of $f(x)$ in $[-1,1]$

$= \min\{f(-1), f(1), f(0)\} = -13.$

98. −5

$f(x) = 2x^3 - 3x^2$

$\Rightarrow f'(x) = 6x^2 - 6x, f''(x) = 12x - 6.$

Then, $f'(x) = 0$

$\Rightarrow 6x^2 - 6x = 0 \Rightarrow 6x(x-1) = 0$

$\Rightarrow x = 0, 1$ are stationary points.

Now, $f''(1) = 12 - 6 > 0.$

$\therefore f(x)$ has a minimum at $x = 1$ and

$f(1) = 2 - 3 = -1.$

Again, $f(-1) = -2 - 3 = -5, f(2) = 16 - 12 = 4.$

$\therefore$ Global minimum

$= \min\{f(1), f(-1), f(2)\} = \min\{-1, -5, 4\} = -5.$

99. (c)

$f(x) = (k^2 - 4)x^2 + 6x^3 + 8x^4$

$\Rightarrow f'(x) = 32x^3 + 18x^2 + 2x(k^2 - 4)$

$\Rightarrow f''(x) = 96x^2 + 36x + 2(k^2 - 4).$

Now, $f(x)$ has a local maxima at $x = 0$

$\Rightarrow f''(0) < 0$

$\Rightarrow 2(k^2 - 4) < 0 \Rightarrow k^2 < 4$

$\Rightarrow |k| < 2 \Rightarrow -2 < k < 2.$

100. (d)

$f(x) = x^3 \Rightarrow f'(x) = 3x^2 \quad f''(x) = 6x, f'''(x) = 6$

Then, $f'(x) = 0 \Rightarrow 3x^2 = 0 \Rightarrow x = 0.$

$\therefore x = 0$ is a stationary point.

Now, $f''(0) = 0$ and $f'''(0) = 6 \neq 0.$

$\therefore x = 0$ is a point of inflection (i.e. neither a maximum nor a minimum).

101. 12

$f(x) = 3x^3 - 7x^2 + 5x + 6$

$\Rightarrow f'(x) = 9x^2 - 14x + 5, f''(x) = 18x - 14.$

Then, $f'(x) = 0$

$\Rightarrow 9x^2 - 14x + 5 = 0$

$\Rightarrow (x-1)(9x - 5) = 0$

$\Rightarrow x = 1, \dfrac{5}{9}$ are stationary points.

Now, $f''(1) = 18 - 14 > 0$ and

$$f''\left(\frac{5}{9}\right) = 18\left(\frac{5}{9}\right) - 14 < 0.$$

Thus, $f(x)$ has a local minimum at $x = 1$

and local maximum $x = \frac{5}{9}$.

Now,

$f(0) = 6$, $f(2) = 3 \times 8 - 7 \times 4 + 5 \times 2 + 6 = 12$ and

$$f\left(\frac{5}{9}\right) = 3 \times \left(\frac{5}{9}\right)^3 - 7 \times \left(\frac{5}{9}\right)^2 + 5 \times \frac{5}{9} + 6 = 7.13.$$

$\therefore$ The maximum value of $f(x)$ in $[0, 2]$

$$= \max\left\{f(0), f(2), f\left(\frac{5}{9}\right)\right\}$$

$$= \max\{6, 12, 7.13\} = 12.$$

102. (a)

Let $f(x) = \sin x$. Then $f'(x) = \cos x$

$f''(x) = -\sin x$, $f'''(x) = -\cos x$. Therefore,

$$f\left(\frac{\pi}{6}\right) = \frac{1}{2}, f'\left(\frac{\pi}{6}\right) = \frac{\sqrt{3}}{2}, f''\left(\frac{\pi}{6}\right) = -\frac{1}{2},$$

$$f'''\left(\frac{\pi}{6}\right) = -\frac{\sqrt{3}}{2}.$$

By the Taylor series expansion of $f(x)$ about $x = \frac{\pi}{6}$, we have,

$$f(x) = f\left(\frac{\pi}{6}\right) + \left(x - \frac{\pi}{6}\right) f'\left(\frac{\pi}{6}\right) + \frac{\left(x - \frac{\pi}{6}\right)^2}{2!} f''\left(\frac{\pi}{6}\right)$$

$$+ \frac{\left(x - \frac{\pi}{6}\right)^3}{3!} f'''\left(\frac{\pi}{6}\right) + \dots \infty$$

$$= \frac{1}{2} + \frac{\sqrt{3}}{2}\left(x - \frac{\pi}{6}\right) - \frac{1}{4}\left(x - \frac{\pi}{6}\right)^2 - \frac{\sqrt{3}}{12}\left(x - \frac{\pi}{6}\right)^3 + \dots \infty$$

103. (a)

Let $f(x) = e^{-x}$.

Then, $f'(x) = -e^{-x}, f''(x) = e^{-x}, \dots\dots$

So by the Taylor series expansion around $x = 2$, we have

$$f(x) = f(2) + (x - 2) f'(2) + \frac{(x-2)^2}{2!} f''(2) + \dots \infty$$

Therefore, around $x = 2$ the linear approximation of e^{-x}

$$= f(2) + (x - 2) f'(2)$$

$$= e^{-2} + (x - 2)(-e^{-2}) = (3 - x)e^{-2}.$$

104. (c)

$$\coth x = \frac{\cosh x}{\sinh x} = \frac{\dfrac{e^x + e^{-x}}{2}}{\dfrac{e^x - e^{-x}}{2}} = \frac{e^x + e^{-x}}{e^x - e^{-x}}$$

$$= \frac{2\left[1 + \dfrac{x^2}{2!} + \dfrac{x^4}{4!} + \dots\dots \infty\right]}{2\left[x + \dfrac{x^3}{3!} + \dfrac{x^5}{5!} + \dots\dots \infty\right]}$$

$$= \frac{\left[1 + \dfrac{x^2}{2!} + \dfrac{x^4}{4!} + \dots\dots \infty\right]}{x\left[1 + \dfrac{x^2}{3!} + \dfrac{x^4}{5!} + \dots\dots \infty\right]}$$

$$= \frac{1}{x}$$

[by neglecting x^2 and higher power of x as $|x| \ll 1$]

105. (a)

$$\sin\left(x^3\right) = x^3 - \frac{\left(x^3\right)^3}{3!} + \frac{\left(x^3\right)^5}{5!} - \dots\dots \infty$$

$$= x^3 - \frac{x^9}{3!} + \frac{x^{15}}{5!} - \dots\dots \infty$$

106. (b)

$f(x) = e^x + \sin x$

$\Rightarrow f'(x) = e^x + \cos x$, $f''(x) = e^x - \sin x$.

$\therefore f(\pi) = e^\pi + \sin \pi = e^\pi$,

$f'(\pi) = e^\pi + \cos \pi = e^\pi - 1$,

$f''(\pi) = e^\pi - \sin \pi = e^\pi$.

Then using the Taylor series expansion of $f(x)$ about $x = \pi$, we can write,

$$f(x) = f(\pi) + (x - \pi) f'(\pi) + \frac{(x - \pi)^2}{2!} f''(\pi) + \dots\dots \infty$$

Therefore, the coefficient of $(x - \pi)^2$ is

$\frac{f''(\pi)}{2}$ i.e, $\frac{e^\pi}{2}$.

107. (c)

$$f(x) = e^x$$

$$\Rightarrow f'(x) = e^x, f''(x) = e^x, f'''(x) = e^x,$$

$$f^{iv}(x) = e^x, \ldots\ldots\ldots$$

$$\therefore f'(2) = e^2, f''(2) = e^2, f'''(2) = e^2,$$

$$f^{iv}(2) = e^2, \ldots\ldots\ldots$$

Then using the Taylor series expansion of $f(x)$ about $x = 2$, we can write

$$f(x) = f(2) + (x-2)f'(2) + \frac{(x-2)^2}{2!}f''(2)$$

$$+ \frac{(x-2)^3}{3!}f'''(2) + \frac{(x-2)^4}{4!}f^{iv}(2) + \ldots\ldots\infty$$

Therefore, the coefficient of $(x-2)^4$ is $\frac{f^{iv}(2)}{4!}$ i.e; $\frac{e^2}{4!}$.

108. (d)

$$\frac{\sin x}{x - \pi}$$

$$= \frac{\sin\{\pi + (x-\pi)\}}{x - \pi} = \frac{-\sin(x-\pi)}{x - \pi}$$

$$(\because \sin(\pi + \theta) = -\sin\theta)$$

$$= \frac{-\left\{(x-\pi) - \frac{(x-\pi)^3}{3!} + \frac{(x-\pi)^5}{5!} - \ldots\ldots\ldots\infty\right\}}{x - \pi}$$

$$= -1 + \frac{(x-\pi)^2}{3!} - \frac{(x-\pi)^4}{5!} - \ldots\ldots\ldots\infty$$

109. (b) We know that expansion of $\sin x$ is

$$x - \frac{x^3}{3!} + \frac{x^5}{5!} - \frac{x^7}{7!} + \ldots\ldots\infty$$

110. (b) We know that the Taylor series expansion of $\sin\theta$ about $\theta = 0$ is given by

$$\sin\theta = \theta - \frac{\theta^3}{3!} + \frac{\theta^5}{5!} - \ldots\ldots\ldots\infty.$$

111. (b) Let $\sum_{m=1}^{\infty} u_m$ be the given series. Then,

$$u_m = \frac{1}{4^m}(x-1)^{2m}.$$

$$\therefore u_{m+1} = \frac{1}{4^{m+1}}(x-1)^{2m+2}.$$

Hence, $\underset{m\to\infty}{\text{Lt}} \frac{u_{m+1}}{u_m} = \underset{m\to\infty}{\text{Lt}} \frac{(x-1)^2}{4} = \frac{(x-1)^2}{4}.$

By the D'Alembert's ratio test, the series $\sum_{m=1}^{\infty} u_m$ converges if $\frac{(x-1)^2}{4} < 1$ i.e; if $|x-1| < 2$

i.e; if $-1 < x < 3$.

$$\left[\because |x-1| < 2 \Rightarrow -2 < x-1 < 2 \Rightarrow -1 < x < 3\right]$$

Alternative method:

$$\sum_{m=0}^{\infty} \frac{1}{4^m}(x-1)^{2m}$$

$$= \sum_{m=0}^{\infty} \left(\frac{x-1}{2}\right)^{2m} = 1 + \left(\frac{x-1}{2}\right)^2 + \left(\frac{x-1}{2}\right)^4 + \ldots\ldots\infty$$

$$= \left[1 - \left(\frac{x-1}{2}\right)^2\right]^{-1}$$

which converges for $\frac{(x-1)^2}{4} < 1$ i.e; for $|x-1| < 2$

i.e; for $-1 < x < 3$.

$$\left[\begin{array}{l}\therefore (1-x)^{-1} = 1 + x + x^2 + \ldots\ldots\infty \text{ converges} \\ \text{for } |x| < 1\end{array}\right]$$

112. (b)

We know that

$$e^x = 1 + x + \frac{x^2}{2!} + \frac{x^3}{3!} + \frac{x^4}{4!} + \ldots\ldots\infty$$

113. (a)

$$3\sin x + 2\cos x$$

$$= 3\left(x - \frac{x^3}{3!} + \frac{x^5}{5!} - \ldots\ldots\right) + 2\left(1 - \frac{x^2}{2!} + \frac{x^4}{4!} - \ldots\ldots\right)$$

$$= 2 + 3x - x^2 - \frac{x^3}{2} + \ldots\ldots\ldots\infty$$

114. (d) We know that $e^x = 1 + x + \frac{x^2}{2!} + \frac{x^3}{3!} + \ldots\ldots$

$$= \sum_{n=0}^{\infty} \frac{x^n}{n!}$$

For $x = 1$, we get, $e^1 = e = \sum_{n=0}^{\infty} \frac{1}{n!}$.

115. 2

$$\sum_{n=0}^{\infty} n\left(\frac{1}{2}\right)^n = \frac{1}{2} + 2\left(\frac{1}{2}\right)^2 + 3\left(\frac{1}{2}\right)^3 + 4\left(\frac{1}{2}\right)^4 + \ldots\ldots$$

$$= \frac{1}{2}\left[1 + 2\frac{1}{2} + 3\left(\frac{1}{2}\right)^2 + 4\left(\frac{1}{2}\right)^3 + \ldots\ldots\right]$$

$$= \frac{1}{2}\left[1 - \frac{1}{2}\right]^{-2} = \frac{1}{2} \times 2^2 = 2$$

$$\left[\because (1-x)^{-2} = 1 + 2x + 3x^2 + 4x^3 + \ldots\ldots\right]$$

116. 0.293

$$S = \sum_{n=0}^{\infty} n a^n$$

$$\Rightarrow S = 0 + 1a + 2a^2 + 3a^3 + \ldots\ldots\ldots\infty$$

$$\Rightarrow S = a + 2a^2 + 3a^3 + \ldots\ldots\ldots\infty$$

$$\Rightarrow S = a(1 + 2a + 3a^2 + \ldots\ldots\ldots\infty)$$

$$\Rightarrow S = a(1-a)^{-2}$$

$$\Rightarrow S = \frac{a}{(1-a)^2} \quad [\because 0 < a < 1]$$

But given that $S = 2a$.

$$\therefore \frac{a}{(1-a)^2} = 2a$$

or, $\frac{1}{2} = (1-a)^2$

or, $(1-a) = \pm\frac{1}{\sqrt{2}}$

or, $a = 1 - \frac{1}{\sqrt{2}}, 1 + \frac{1}{\sqrt{2}}$

or, $a = 1 - 0.707, 1 + 0.707$

$$\Rightarrow a = 0.293 \quad (0 < a < 1)$$

117. (b)

$$f(x) = x^3 - 3x^2 - 5$$

$$\Rightarrow f'(x) = 3x^2 - 6x, \ f''(x) = 6x - 6.$$

Then using the Taylor series expansion of $f(x)$ about $x = 0$, we can write

$$f(x) = f(0) + (x-0)f'(0) + \frac{(x-0)^2}{2!}f''(0)$$

So the quadratic approximation of $f(x)$ about $x = 0$

$$= f(0) + (x-0)f'(0) + \frac{(x-0)^2}{2!}f''(0)$$

$$= -5 + x \times 0 + \frac{x^2}{2} \times (-6) = -3x^2 - 5.$$

118. (c)

$$f(x) = e^{x+x^2}$$

$$= 1 + (x+x^2) + \frac{(x+x^2)^2}{2!} + \frac{(x+x^2)^3}{3!} + \ldots\ldots$$

$$= 1 + x + x^2 + \frac{x^2 + 2x^3 + x^4}{2} + \frac{x^3 + 3x^4 + 3x^5 + x^6}{6} + \ldots\ldots$$

$$= 1 + x + \frac{3}{2}x^2 + \frac{7}{6}x^3 + \ldots\ldots\ldots$$

Therefore, the third order approximation of $f(x)$ around $x = 0$

$$= 1 + x + \frac{3}{2}x^2 + \frac{7}{6}x^3.$$

119. 0 The Taylor series expansion of $f(x)$ around $x = a$ is given by

$$f(x) = f(a) + (x-a)f'(a) + \frac{(x-a)^2}{2!}f''(a) + \ldots..$$

i.e; $f(x) = a_0 + a_1(x-a) + a_2(x-a)^2 + \ldots….$

where $a_n = \dfrac{f^{(n)}(a)}{n!}$.

Given $f(x) = \int_0^x e^{-\left(\frac{t^2}{2}\right)}dt$ and $a = 0$.

$$\therefore f'(x) = e^{-\frac{x^2}{2}} \text{ and } f''(x) = -xe^{-\frac{x^2}{2}}.$$

So $a_2 = \dfrac{1}{2}f''(0) = \dfrac{1}{2} \times 0 = 0$.

120. (a) Given,

$$\int_0^{\frac{\pi}{4}} \cos^2 x \, dx$$

$$= \frac{1}{2}\int_0^{\frac{\pi}{4}} 2\cos^2 x \, dx = \frac{1}{2}\int_0^{\frac{\pi}{4}} (1 + \cos 2x) \, dx$$

$$= \frac{1}{2}\left[x + \frac{\sin 2x}{2}\right]_0^{\frac{\pi}{4}}$$

$$= \frac{1}{2}\left[\frac{\pi}{4} + \frac{\sin\frac{\pi}{2}}{2} - 0 - 0\right] = \frac{\pi}{8} + \frac{1}{4}.$$

$$= \frac{1}{2}\left[\frac{\pi}{4} + \frac{\sin\frac{\pi}{2}}{2} - 0 - 0\right] = \frac{\pi}{8} + \frac{1}{4}.$$

121. (c)

Let $f(x) = \dfrac{\sin 2x}{1 + \cos 2x}$. Then,

$$f(-x) = \frac{\sin(-2x)}{1 + \cos(-2x)} = -\frac{\sin 2x}{1 + \cos 2x} = -f(x)$$

$$(\because \sin(-x) = -\sin x, \cos(-x) = \cos x)$$

$\therefore f(x)$ is an odd function.

So $\displaystyle\int_{-\frac{\pi}{2}}^{\frac{\pi}{2}} \frac{\sin 2x}{1 + \cos x}\, dx = 0.$

122. (b) The given curves are $y = x^2$ and $y = x$.

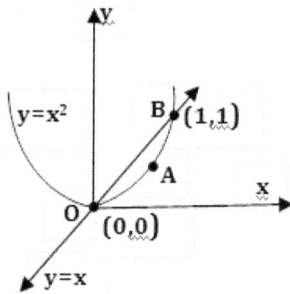

Therefore, required area

= area of the region OABO

$$= \int_0^1 (x - x^2)\,dx = \left[\frac{x^2}{2} - \frac{x^3}{3}\right]_0^1 = \frac{1}{2} - \frac{1}{3} = \frac{1}{6}.$$

123. (a) $\displaystyle\int_{-a}^{a} \left(\sin^6 x + \sin^7 x\right) dx$

$$= \int_{-a}^{a} \sin^6 x\, dx + \int_{-a}^{a} \sin^7 x\, dx$$

$$= 2\int_0^a \sin^6 x\, dx + 0 = 2\int_0^a \sin^6 x\, dx$$

(Since $\sin^6 x$ is even function and $\sin^7 x$ is odd

function, so $\displaystyle\int_{-a}^{a} \sin^6 x\, dx = 2\int_0^a \sin^6 x\, dx$ and

$\displaystyle\int_{-a}^{a} \sin^7 x\, dx = 0) \cdot$

124. (d)

$$\int_o^R 2\pi\, rH\left(1 - \frac{r}{R}\right) dr$$

$$= 2\pi\, H\int_o^R \left(r - \frac{r^2}{R}\right) dr$$

$$= 2\pi\, H\left[\frac{r^2}{2} - \frac{r^3}{3R}\right]_0^R = 2\pi\, H\left[\frac{R^2}{2} - \frac{R^3}{3R}\right]$$

$$= 2\pi\, H \times \frac{R^2}{6} = \frac{1}{3}\pi R^2\, H$$

= volume of the cone

125. (b) From the given diagram, it is clear that
the equation of the line is $\dfrac{x}{-1} + \dfrac{y}{1} = 1.$

[since the line cuts x-axis at (-1, 0) and y-axis at (0, 1)]

This gives $y = 1 + x.$

$$\therefore I = \int_1^2 (1 + x)\,dx$$

$$= \left[x + \frac{x^2}{2}\right]_1^2 = (2 + 2) - \left(1 + \frac{1}{2}\right) = \frac{5}{2} = 2.5.$$

126. (d)

$$\frac{1}{2\pi}\int_0^{2\pi} \sin(t - \tau)\cos\tau\, d\tau$$

$$= \frac{1}{4\pi}\int_0^{2\pi} 2\sin(t - \tau)\cos\tau\, d\tau$$

$$= \frac{1}{4\pi}\int_0^{2\pi} \left[\sin t + \sin(t - 2\tau)\right] d\tau$$

$$(\text{using } \sin(x + y) + \sin(x - y) = 2\sin x \cos y)$$

$$= \frac{1}{4\pi}\left\{[\tau]_0^{2\pi} \sin t + \left[\frac{-\cos(t - 2\tau)}{-2}\right]_0^{2\pi}\right\}$$

$$= \frac{1}{4\pi}\left\{2\pi \sin t + \left[\frac{\cos(t - 4\pi)}{2} - \frac{\cos t}{2}\right]\right\}$$

$$= \frac{1}{4\pi}\left\{2\pi \sin t + \left[\frac{\cos t}{2} - \frac{\cos t}{2}\right]\right\}$$

$$(\because \cos(t - 4\pi) = \cos(4\pi - t) = \cos t)$$

$$= \frac{2\pi \sin t}{4\pi} = \frac{1}{2}\sin t.$$

127. (a)

We know that $\int_{-a}^{a} f(x)\,dx = 0$ if $f(x)$ is an odd function.

Here, $a = \dfrac{\pi}{2}$, $f(x) = x\cos x$.

So $f(-x) = -x\cos(-x) = -x\cos x = -f(x)$.

Hence, $f(x) = x\cos x$ is an odd function.

$\therefore \displaystyle\int_{\frac{-\pi}{2}}^{\frac{\pi}{2}} x\cos x\,dx = 0.$

128. (a)

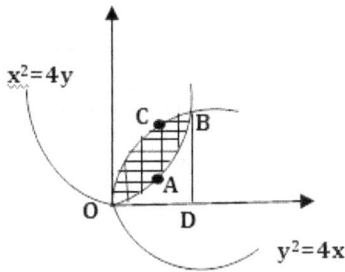

Required Area

$= area\,OCBD - area\,OABD$

$= \displaystyle\int_{0}^{4} 2\sqrt{x}\,dx - \int_{0}^{4} \frac{x^2}{4}\,dx$

$= \left[\dfrac{4}{3} x^{\frac{3}{2}} - \dfrac{x^3}{12} \right]_{0}^{4}$

$\begin{bmatrix} \because x^2 = 4y, y^2 = 4x \Rightarrow x^4 = 16 \times 4x \\ \Rightarrow x(x^3 - 64) = 0 \Rightarrow x = 0, 4 \end{bmatrix}$

$= \left[\dfrac{4}{3} \times 4^{\frac{3}{2}} - \dfrac{4^3}{12} \right] = \dfrac{16}{3}.$

129. (b)

$P = \displaystyle\int_{0}^{1} x e^x\,dx = \left[x e^x \right]_{0}^{1} - \int_{0}^{1} 1 \times e^x\,dx$

$= (e - 0) - \left[e^x \right]_{0}^{1} = e - (e - 1) = 1.$

130. (b)

$I = \displaystyle\int_{0}^{a} \frac{\sqrt{x}}{\sqrt{x} + \sqrt{a - x}}\,dx\,............(1)$

Again, $I = \displaystyle\int_{0}^{a} \frac{\sqrt{a - x}}{\sqrt{a - x} + \sqrt{a - (a - x)}}\,dx............(2)$

$\left[\because \displaystyle\int_{a}^{b} f(x)\,dx = \int_{a}^{b} f(a + b - x)\,dx \right]$

Adding (1) & (2) we get,

$2I = \displaystyle\int_{0}^{a} \frac{\sqrt{x} + \sqrt{a - x}}{\sqrt{a - x} + \sqrt{x}}\,dx = \int_{0}^{a} dx = \left[x \right]_{0}^{a} = a$

$\therefore \quad I = \dfrac{a}{2}.$

131. (d) By properties of definite integral,

$\displaystyle\int_{-a}^{a} f(x)\,dx = \begin{cases} 2\displaystyle\int_{0}^{a} f(x)\,dx, & \text{if } f(x) \text{ is even} \\ 0, & \text{if } f(x) \text{ is odd} \end{cases}$

132. (d)

$\displaystyle\int_{0}^{\frac{\pi}{2}} \frac{\cos x + i\sin x}{\cos x - i\sin x}\,dx$

$= \displaystyle\int_{0}^{\frac{\pi}{2}} \frac{e^{ix}}{e^{-ix}}\,dx = \int_{0}^{\frac{\pi}{2}} e^{2ix}\,dx$

$\left[\because \cos x + i\sin x = e^{ix}, \cos x - i\sin x = e^{-ix} \right]$

$= \left[\dfrac{e^{2ix}}{2i} \right]_{0}^{\frac{\pi}{2}} = \dfrac{1}{2i} \left[e^{i\pi} - e^0 \right]$

$= \dfrac{1}{2i} \left[\cos\pi + i\sin\pi - 1 \right] = \dfrac{-1}{i} = \dfrac{-i}{i^2} = i.$

133. (a)

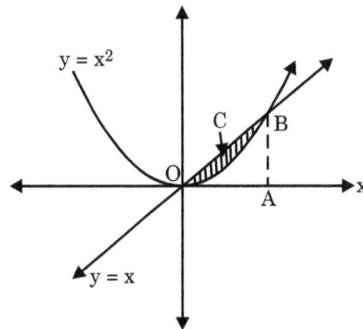

The area bounded by the parabola $y = x^2$ and line $y = x$ is given by the area OBCO, where

area OBCO

= area OAB − area OABCO

area OBCO

$= area\,OABCO - area\,OAB$

$$= \int_0^1 x^2\,dx - \int_0^1 x\,dx$$

$$\left[\because y = x^2, y = x \Rightarrow x^2 = x \Rightarrow x = 0,1\right]$$

$$= \int_0^1 (x^2 - x)\,dx = \left[\frac{x^3}{3} - \frac{x^2}{2}\right]_0^1 = \frac{1}{3} - \frac{1}{2} = -\frac{1}{6}.$$

$$= \int (x - x^2)\,dx = \left[\frac{x^2}{2} - \frac{x^3}{3}\right]_0^1 = \frac{1}{2} - \frac{1}{3} = \frac{1}{6}.$$

Therefore, magnitude of the area $= \frac{1}{6}$.

134. (c)

$$I = \int_1^e \sqrt{x}\ln(x)\,dx$$

$$= \left[\ln x \int \sqrt{x}\,dx\right]_1^e - \int_1^e\left[\frac{1}{x}\int\sqrt{x}\,dx\right]dx$$

$$= \left[\frac{2}{3}x^{\frac{3}{2}}\ln x\right]_1^e - \int_1^e\left[\frac{2}{3}\left(\frac{1}{x}\right)x^{\frac{3}{2}}\right]dx$$

$$= \left[\frac{2}{3}e^{\frac{3}{2}}\ln e - \frac{2}{3}\ln 1\right] - \frac{2}{3}\int_1^e \sqrt{x}\,dx$$

$$= \frac{2}{3}e^{\frac{3}{2}} - 0 - \frac{2}{3}\left[\frac{2}{3}x^{\frac{3}{2}}\right]_1^e = \frac{2}{3}e^{\frac{3}{2}} - \frac{4}{9}\left[e^{\frac{3}{2}} - 1\right]$$

$$= \frac{2}{9}\sqrt{e^3} + \frac{4}{9}$$

135. 1.

Displacement "x" is given by

$$x = \int v\,dt = \int \frac{\pi}{2}\cos\left(\frac{\pi t}{2}\right)dt = \sin\left(\frac{\pi t}{2}\right) + C \quad \dots(1)$$

At $t = 0, x = 0$

$\therefore$ (1) gives $C = 0$

So $x = \sin\left(\frac{\pi t}{2}\right)$.

When $t = 3\sec$,

$$x = \sin\left(\frac{3\pi}{2}\right) = \sin\left(\pi + \frac{\pi}{2}\right) = -\sin\frac{\pi}{2} = -1.$$

$\therefore$ The required distance $= |-1| - 0 = 1$ unit.

136. (b) Let us put $(x-1) = t$.

The $x = t + 1$ and $dx = dt$.

Also, $x \to 0 \Rightarrow t \to -1$ and $x \to 2 \Rightarrow t \to 1$.

$$\therefore \int_0^2 \frac{(x-1)^2\sin(x-1)}{(x-1)^2 + \cos(x-1)}\,dx$$

Then, $x = t + 1$ and $dx = dt$.

Also, $x \to 0 \Rightarrow t \to -1$ and $x \to 2 \Rightarrow t \to 1$.

$$\therefore \int_0^2 \frac{(x-1)^2\sin(x-1)}{(x-1)^2 + \cos(x-1)}\,dx$$

$$= \int_{-1}^1\left(\frac{t^2\sin t}{t^2 + \cos t}\right)dt = 0$$

$$\left[\begin{array}{l}\because \int_{-a}^a f(t)\,dt = 0 \text{ if } f(-t) = -f(t);\\[2mm] \text{Here } f(t) = \frac{t^2\sin t}{t^2 + \cos t} \text{ and so}\\[2mm] f(-t) = \frac{(-t)^2\sin(-t)}{(-t)^2 + \cos(-t)} = \frac{-t^2\sin t}{t^2 + \cos t} = -f(t)\end{array}\right]$$

137. 4.

$$\int_0^{2\pi} |x\sin x|\,dx = k\pi$$

$$\Rightarrow \int_0^{2\pi} x|\sin x|\,dx = k\pi$$

$$\Rightarrow \pi\int_0^{2\pi} |\sin x|\,dx = k\pi$$

$$\left(\because \int_0^a xf(x)\,dx = \frac{a}{2}\int_0^a f(x)\,dx \text{ if } f(a-x) = f(x)\right)$$

$$\Rightarrow 2\pi\int_0^{\pi} |\sin x|\,dx = k\pi$$

$$\left[\because \int_0^{2a} f(x)\,dx = 2\int_0^a f(x)\,dx \text{ if } f(2a-x) = f(x)\right]$$

$$\Rightarrow 2\pi\int_0^{\pi}\sin x\,dx = k\pi \Rightarrow 2\pi\left[-\cos x\right]_0^{\pi} = k\pi$$

$$\Rightarrow 2\pi\left[-\cos\pi + \cos 0\right] = k\pi \Rightarrow 4\pi = k\pi \Rightarrow k = 4.$$

138. (a)

$$\int_0^{\pi} x^2 \cos x \, dx$$

$$= \left[x^2 \int \cos x \, dx \right]_0^{\pi} - \int_0^{\pi} \left\{ \frac{d}{dx}(x^2) \int \cos x \, dx \right\} dx$$

$$= \left[x^2 \sin x \right]_0^{\pi} - \int_0^{\pi} 2x \sin x \, dx$$

$$= \pi^2 \sin \pi - 0 - 2 \left\{ \left[x(-\cos x) \right]_0^{\pi} - \int_0^{\pi} (-\cos x) \, dx \right\}$$

$$= 2\pi \cos \pi - 0 - 2 \left[\sin x \right]_0^{\pi} = -2\pi.$$

139. (d)

$$\int_0^{\frac{\pi}{2}} \frac{\cos x + i \sin x}{\cos x - i \sin x} \, dx$$

$$= \int_0^{\frac{\pi}{2}} \frac{e^{ix}}{e^{-ix}} \, dx = \int_0^{\frac{\pi}{2}} e^{2ix} \, dx$$

$$= \left[\frac{e^{2ix}}{2i} \right]_0^{\frac{\pi}{2}} = \frac{1}{2i} \left[e^{\pi i} - e^0 \right]$$

$$= \frac{1}{2i} \left[\cos \pi + i \sin \pi - 1 \right] = \frac{-1}{i} = \frac{-i}{i^2} = i.$$

140. (a)

Given $af(x) + bf\left(\dfrac{1}{x}\right) = \dfrac{1}{x} - 25$ ...(1)

Replacing x by $\dfrac{1}{x}$ in (1) we get,

$af\left(\dfrac{1}{x}\right) + bf(x) = x - 25$ ...(2)

Solving (1) and (2) we get,

$$f(x) = \frac{1}{a^2 - b^2} \left\{ a\left(\frac{1}{x} - 25\right) - b(x - 25) \right\}.$$

$$\therefore \int_1^2 f(x) \, dx$$

$$= \int_1^2 \frac{1}{a^2 - b^2} \left\{ a\left(\frac{1}{x} - 25\right) - b(x - 25) \right\} dx$$

$$= \frac{1}{a^2 - b^2} \left[a(\ln x - 25x) - b\left(\frac{x^2}{2} - 25x\right) \right]_1^2$$

$$= \frac{1}{a^2 - b^2} \left[\begin{array}{l} a(\ln 2 - 50) - b(2 - 50) \\ \quad - a(\ln 1 - 25) + b\left(\frac{1}{2} - 25\right) \end{array} \right]$$

$$= \frac{1}{a^2 - b^2} \left[a(\ln 2 - 25) + \frac{47}{2} b \right].$$

141. 2

$$\int_0^1 \frac{dx}{\sqrt{1-x}} = \left\{ -2\sqrt{1-x} \right\}_0^1 = -2[0-1] = 2.$$

142. ()

$$I = \int_0^1 \frac{\left(\sin^{-1} x\right)^2}{\sqrt{1-x^2}} \, dx$$

$$= \int_0^{\frac{\pi}{2}} y^2 \, dy$$

$$\left[\begin{array}{l} \text{putting } \sin^{-1} x = y \text{ so that } \dfrac{dx}{\sqrt{1-x^2}} = dy; \\[2mm] \text{also } x = 0 \Rightarrow y = 0 \ \& \ x = 1 \Rightarrow y = \dfrac{\pi}{2} \end{array} \right]$$

$$= \left[\frac{y^3}{3} \right]_0^{\frac{\pi}{2}} = \frac{\left(\frac{\pi}{2}\right)^3}{3} = \frac{\pi^3}{24}.$$

143. (b)

$g(x)$

$$= \int f(x) \, dx = \int e^{-x - e^{-x}} \, dx = \int e^{-x} \, e^{-e^{-x}} \, dx$$

$$= -\int e^{-t} \, dt \ \left[\text{putting } t = e^{-x} \text{ so that } dt = -e^{-x} \right]$$

$$= e^{-t} = e^{-e^{-x}}.$$

144. (c)

$$f(x) = R \sin\left(\frac{\pi x}{2}\right) + S \Rightarrow f'(x) = \frac{\pi R}{2} \cos\left(\frac{\pi x}{2}\right).$$

$$\therefore f'\left(\frac{1}{2}\right) = \sqrt{2} \text{ gives}$$

$$\frac{\pi R}{2} \cos\left(\frac{\pi}{4}\right) = \sqrt{2}$$

or, $\dfrac{\pi R}{2} \times \dfrac{1}{\sqrt{2}} = \sqrt{2}$ i.e; $R = \dfrac{4}{\pi}$.

Then, $\displaystyle \int_0^1 f(x) \, dx = \frac{2R}{\pi}$

$$\Rightarrow \int_0^1 \left\{ \frac{4}{\pi} \sin\left(\frac{\pi x}{2}\right) + S \right\} dx = \frac{2R}{\pi}$$

$$\Rightarrow \left[-\frac{4}{\pi} \times \frac{2}{\pi} \cos\left(\frac{\pi x}{2}\right) + Sx \right]_0^1 = \frac{2R}{\pi}$$

$$\Rightarrow S + \frac{8}{\pi^2} = \frac{2}{\pi} \times \frac{4}{\pi} \Rightarrow S = 0.$$

145. (b)

$$\int_0^\pi x\cos^2 x\,dx$$

$$= \int_0^\pi x\frac{(1+\cos 2x)}{2}\,dx$$

$$= \frac{1}{2}\int_0^\pi x\,dx + \frac{1}{2}\int_0^\pi x\cos 2x\,dx$$

$$= \frac{1}{2}\left[\frac{x^2}{2}\right]_0^\pi + \frac{1}{2}\left[x\frac{\sin 2x}{2}+\frac{\cos 2x}{4}\right]_0^\pi$$

$$= \frac{1}{4}\left[\pi^2-0\right]+\frac{1}{2}\left(0+\frac{\cos 2\pi}{4}-0-\frac{\cos 0}{4}\right)$$

$$= \frac{\pi^2}{4}.$$

146. 0.005.

Let us put $x^2 = t$.

Then, $2x\,dx = dt$ i.e; $x\,dx = \dfrac{dt}{2}$.

Also $x=0 \Rightarrow t=0$ and $x=\dfrac{\pi}{4} \Rightarrow t=\dfrac{\pi^2}{16}$.

$$\therefore \int_0^{\frac{\pi}{4}} x\cos(x^2)\,dx$$

$$= \frac{1}{2}\int_0^{\frac{\pi^2}{16}}\cos t\,dt = \frac{1}{2}\left[\sin t\right]_0^{\frac{\pi^2}{16}}$$

$$= \frac{1}{2}\left[\sin\left(\frac{\pi^2}{16}\right)-\sin 0\right]=\frac{1}{2}\times\sin(0.6162)$$

$$= \frac{1}{2}\times 0.01075 = 0.005.$$

147. 0.5

$$\int_{0.25}^{1.25}\left(x-[x]\right)dx$$

$$= \int_{0.25}^{1.25} x\,dx - \left(\int_{0.25}^{1.25}[x]\,dx\right)$$

$$= \left[\frac{x^2}{2}\right]_{0.25}^{1.25} - \left(\int_{0.25}^{1}[x]\,dx+\int_{1}^{1.25}[x]\,dx\right)$$

$$= \left[\frac{(1.25)^2}{2}-\frac{(0.25)^2}{2}\right]-\left(\int_{0.25}^{1}0\,dx+\int_{1}^{1.25}1\,dx\right)$$

$$= 0.75-\left[0+[x]_1^{1.25}\right]=0.75-0.25=0.5.$$

$$= 0.75 - [0+[x]_1^{1.25}] = 0.75-0.25 = 0.5.$$

148. (b)

$$\underset{a\to\infty}{Lim}\int_1^a x^{-4}\,dx$$

$$= \underset{a\to\infty}{Lim}\left[\frac{x^{-3}}{-3}\right]_1^a = \frac{-1}{3}\underset{a\to\infty}{Lim}\left[\frac{1}{a^3}-1\right]=\frac{-1}{3}(0-1)=\frac{1}{3}.$$

So the given improper integral converges to $\dfrac{1}{3}$.

149. (c)

$$\int_0^1 x\log x\,dx$$

$$= \left[\log x\times\frac{x^2}{2}\right]_0^1 - \int_0^1\frac{x^2}{2}\times\frac{1}{x}\,dx$$

$$= \log 1\times\frac{1}{2}-\underset{\varepsilon\to 0}{lim}\left(\log\varepsilon\times\frac{\varepsilon^2}{2}\right)-\left[\frac{x^2}{4}\right]_0^1$$

$$= 0-\frac{1}{2}\underset{\varepsilon\to 0}{lim}\left(\frac{\log\varepsilon}{\frac{1}{\varepsilon^2}}\right)-\frac{1}{4}\ldots\ldots\ldots(1)$$

Now $\underset{\varepsilon\to 0}{lim}\left(\dfrac{\log\varepsilon}{\frac{1}{\varepsilon^2}}\right)\left(\text{form }\dfrac{\infty}{\infty}\right)$

$$= \underset{\varepsilon\to 0}{lim}\frac{\frac{1}{\varepsilon}}{-\frac{2}{\varepsilon^3}}\quad(\text{by L' Hospital rule})$$

$$= -\underset{\varepsilon\to 0}{lim}\left(-\frac{1}{2}\varepsilon^2\right)=0.$$

$\therefore$ (1) gives,

$$\int_0^1 x\log x\,dx = -\frac{1}{2}\times 0-\frac{1}{4}=-\frac{1}{4}.$$

150. (a)

Let us put $x^2 = 8t$. Then $x=\sqrt{8t}$.

$$\therefore dx = \frac{\sqrt 8}{2\sqrt t}\,dt = \frac{2\sqrt 2}{2\sqrt t}\,dt = \frac{\sqrt 2}{\sqrt t}\,dt.$$

Then $I = \dfrac{1}{\sqrt{2\pi}}\displaystyle\int_0^\infty e^{-\frac{x^2}{8}}\,dx$

$$\frac{1}{\sqrt{2\pi}}\int_0^\infty e^{-t}\frac{\sqrt 2}{\sqrt t}\,dt$$

$(\because\quad x=0\ t=0\text{ and } x=\infty \Rightarrow t=\infty)$

$$\frac{1}{\sqrt{2\pi}} \int_0^\infty e^{-t} \frac{\sqrt{2}}{\sqrt{t}}\, dt$$

$$(\because x = 0 \Rightarrow t = 0 \text{ and } x = \infty \Rightarrow t = \infty)$$

$$= \frac{1}{\sqrt{\pi}} \int_0^\infty e^{-t}\, t^{\frac{1}{2}-1}\, dt$$

$$= \frac{1}{\sqrt{\pi}} \Gamma\left(\frac{1}{2}\right) \left[\because \Gamma(n) = \int_0^\infty e^{-x}\, x^{n-1}\, dx\right]$$

$$= \frac{1}{\sqrt{\pi}} \times \sqrt{\pi} = 1 \quad \left[\because \Gamma\left(\frac{1}{2}\right) = \sqrt{\pi}\right]$$

151. (c) $\displaystyle\int_1^\infty x^{-3}\, dx$

$$= \lim_{X \to \infty} \int_1^X x^{-3}\, dx = \lim_{X \to \infty} \left[\frac{1}{-2x^2}\right]_1^X$$

$$= \frac{-1}{2}\left[\lim_{X \to \infty} \frac{1}{x^2} - 1\right] = -\frac{1}{2}(0-1) = \frac{1}{2}.$$

152. (d) $\displaystyle\int_0^{\frac{\pi}{4}} \tan x\, dx$

$$= \left[\log(\sec x)\right]_0^{\frac{\pi}{4}} = \log\left(\sec \frac{\pi}{4}\right) - \log(\sec 0)$$

$$= \log\sqrt{2} - \log 1 = \log\sqrt{2}.$$

$$\int_0^\infty \frac{1}{1+x^2}\, dx = \left[\tan^{-1}(x)\right]_0^\infty$$

$$= \tan^{-1}(\infty) - \tan^{-1}(0)$$

$$= \frac{\pi}{2} - 0 = \frac{\pi}{2}.$$

$$\int_0^\infty x\, e^{-x}\, dx$$

$$= \left[x(-e^{-x})\right]_0^\infty - \int_0^\infty 1\times(-e^{-x})\, dx$$

$$= -\lim_{x \to \infty}(xe^{-x}) + 0 + \left[\frac{e^{-x}}{(-1)}\right]_0^\infty = -0 + \left[\frac{e^{-\infty}}{(-1)} - \frac{e^{-0}}{(-1)}\right] = 1$$

$$\left(\begin{array}{l} \because \lim_{x \to \infty}(xe^{-x}) \\[2mm] = \lim_{x \to \infty} \frac{x}{e^x}\ \left[\text{form } \frac{\infty}{\infty}\right] \\[2mm] = \lim_{x \to \infty} \frac{1}{e^x}\ \text{(by the L'Hospital Rule)} \\[2mm] = 0 \end{array}\right)$$

$$\int_0^1 \frac{1}{1-x}\, dx = \left[\frac{\log(1-x)}{-1}\right]_0^1, \text{which is unbounded}$$

$$(\text{since } \log(1-1) = \log 0 \text{ is not defined})$$

153. (c)

$$f(x) = \frac{1}{\sqrt{2\pi}} \int_{-\infty}^\infty e^{\frac{-x^2}{2}}\, dx$$

$$= \frac{1}{\sqrt{2\pi}} \times 2\int_0^\infty e^{\frac{-x^2}{2}}\, dx \dots\dots\dots\dots(1)$$

$$(\because e^{\frac{-x^2}{2}} \text{ is an even function of } 'x')$$

Let $\dfrac{x^2}{2} = t.$

Then, $\dfrac{x^2}{2} = t$

$$\Rightarrow x\, dx = dt \Rightarrow dx = \frac{1}{x}\, dt \Rightarrow dx = \frac{1}{\sqrt{2}} t^{\frac{-1}{2}}\, dt.$$

Also, $x = \infty \Rightarrow t = \infty$ and $x = 0 \Rightarrow t = 0.$

Therefore from (1), given integral

$$= \frac{1}{\sqrt{2\pi}} \times 2\int_0^\infty e^{-t} \frac{1}{\sqrt{2}} t^{\frac{-1}{2}}\, dt$$

$$= \frac{1}{\sqrt{\pi}} \int_0^\infty e^{-t}\, t^{\frac{1}{2}-1}\, dt$$

$$= \frac{1}{\sqrt{\pi}} \Gamma\left(\frac{1}{2}\right) = \frac{\sqrt{\pi}}{\sqrt{\pi}} = 1 \left(\because \Gamma\left(\frac{1}{2}\right) = \sqrt{\pi}\right)$$

154. (d)

$$\int_{-\infty}^\infty \frac{dx}{1+x^2} = 2\int_0^\infty \frac{dx}{1+x^2}$$

$$\left(\because \frac{1}{1+x^2} \text{ is an even function of "x"}\right)$$

$$= \left[2\tan^{-1}x\right]_0^\infty = 2\left[\frac{\pi}{2} - 0\right] = \pi.$$

155. (b)

Let $I = \displaystyle\int_0^{\frac{\pi}{6}} \cos^4 3\theta\ \sin^3 6\theta\ d\theta \qquad \dots(1)$

Let us put $3\theta = t$. Then, $d\theta = \dfrac{dt}{3}$.

Also $\theta = 0 \Rightarrow t = 0$ and $\theta = \dfrac{\pi}{6} \Rightarrow t = \dfrac{\pi}{2}$.

$\therefore$ (1) gives,

$$I = \int_0^{\frac{\pi}{2}} \cos^4 t \, \sin^3 2t \, \frac{dt}{3} = \frac{1}{3} \int_0^{\frac{\pi}{2}} \cos^4 t \, (2\sin t \cos t)^3 \, dt$$

$$= \frac{8}{3} \int_0^{\frac{\pi}{2}} \cos^7 t \sin^3 t \, dt = \frac{8}{3} \times \frac{\Gamma\left(\dfrac{7+1}{2}\right) \times \Gamma\left(\dfrac{3+1}{2}\right)}{2\Gamma\left(\dfrac{7+3+2}{2}\right)}$$

$$\left(\text{using} \int_0^{\frac{\pi}{2}} \cos^m x \sin^n x \, dx = \frac{\Gamma\left(\dfrac{m+1}{2}\right) \times \Gamma\left(\dfrac{n+1}{2}\right)}{2\Gamma\left(\dfrac{m+n+2}{2}\right)} \right)$$

$$= \frac{8}{3} \times \frac{\Gamma(4) \times \Gamma(2)}{2\Gamma(6)} = \frac{8}{3} \times \frac{3! \times 1!}{2 \times 5!}$$

$$= \frac{8}{3} \times \frac{3! \times 1!}{2 \times 5 \times 4 \times 3!} = \frac{1}{15}.$$

$$(\because \Gamma(n) = (n-1)! \text{ when n = positive integer})$$

156. 3

$$\int_{-\infty}^{\infty} 12\cos(2\pi t)\frac{\sin(4\pi t)}{4\pi t} dt$$

$$= 12 \times 2\int_0^{\infty} \frac{\cos(2\pi t)\,\sin(4\pi t)}{4\pi t} dt$$

$$\left[\because \frac{\cos(2\pi t)\,\sin(4\pi t)}{4\pi t} \text{ is an even function} \right]$$

$$= \frac{6}{\pi} \int_0^{\infty} \frac{1}{2}\frac{[\sin(6\pi t) + \sin(2\pi t)]}{t} dt$$

$$= \frac{3}{\pi} \left[\int_0^{\infty} \frac{\sin(6\pi t)}{t} dt + \int_0^{\infty} \frac{\sin(2\pi t)}{t} dt \right]$$

$$= \frac{3}{\pi} \left[\frac{\pi}{2} + \frac{\pi}{2} \right] = \frac{3}{\pi} \times \pi = 3 \left[\because \int_0^{\infty} \frac{\sin(at)}{t} dt = \frac{\pi}{2} \right]$$

157. (d)

$$2\int_{-\infty}^{\infty} \left(\frac{\sin 2\pi t}{\pi t} \right) dt$$

$$= \frac{2}{\delta} \times 2 \int_0^{\infty} \frac{\sin 2\pi t}{t} dt$$

$$2\int_{-\infty}^{\infty} \left(\frac{\sin 2\pi t}{\pi t} \right) dt$$

$$= \frac{2}{\pi} \times 2 \int_0^{\infty} \frac{\sin 2\pi t}{t} dt$$

$$\left(\because \frac{\sin 2\pi t}{t} \text{ is an even function} \right)$$

$$= \frac{4}{\pi} \times \frac{\pi}{2} = 2 \left[\because \int_0^{\infty} \left(\frac{\sin a t}{t} \right) = \frac{\pi}{2} \right]$$

158. (a)

Given that $f(x, y, z) = (x^2 + y^2 + z^2)^{-\frac{1}{2}}$.

$$\therefore \frac{\partial f}{\partial x} = -x(x^2 + y^2 + z^2)^{\frac{-3}{2}},$$

$$\frac{\partial f}{\partial y} = -y(x^2 + y^2 + z^2)^{\frac{-3}{2}},$$

$$\frac{\partial f}{\partial z} = -z(x^2 + y^2 + z^2)^{\frac{-3}{2}},$$

$$\frac{\partial^2 f}{\partial x^2} = 3x^2(x^2 + y^2 + z^2)^{\frac{-5}{2}} - (x^2 + y^2 + z^2)^{\frac{-3}{2}},$$

$$\frac{\partial^2 f}{\partial y^2} = 3y^2(x^2 + y^2 + z^2)^{\frac{-5}{2}} - (x^2 + y^2 + z^2)^{\frac{-3}{2}},$$

$$\frac{\partial^2 f}{\partial z^2} = 3z^2(x^2 + y^2 + z^2)^{\frac{-5}{2}} - (x^2 + y^2 + z^2)^{\frac{-3}{2}}.$$

Hence,

$$\frac{\partial^2 f}{\partial x^2} + \frac{\partial^2 f}{\partial y^2} + \frac{\partial^2 f}{\partial z^2}$$

$$= \left(3x^2 + 3y^2 + 3z^2 \right)(x^2 + y^2 + z^2)^{\frac{-5}{2}}$$

$$\quad - 3(x^2 + y^2 + z^2)^{\frac{-3}{2}}$$

$$= 3(x^2 + y^2 + z^2)^{\frac{-3}{2}} - 3(x^2 + y^2 + z^2)^{\frac{-3}{2}} = 0.$$

159. (c) $f = a_0 x^n + a_1 x^{n-1} y + \ldots\ldots + a_{n-1} x y^{n-1} + a_n y^n$

$\Rightarrow f$ is a homogeneous polynomial in x and y of degree n

So by Euler's theorem (for homogeneous function), we have

$$x\frac{\partial f}{\partial x} + y\frac{\partial f}{\partial y} = nf.$$

160. (a) Given $g(x,y) = 4x^3 + 10y^4$.

The equation of the straight line joining $(0,0)$ and $(1,2)$ is $y = 2x$.

The line integral of the function $g(x,y)$ along the line segment from $(0,0)$ to $(1,2)$ is

$\int_C g(x,y)\,dx$

$= \int_0^1 (4x^3 + 10y^4)$

$= \int_0^1 \left[4x^3 + 10(2x)^4\right]dx = \int_0^1 \left[4x^3 + 160x^4\right]dx$

$= \left[4 \times \dfrac{x^4}{4} + 160 \times \dfrac{x^5}{5}\right]_0^1 = 1 + \dfrac{160}{5} = 33.$

161. (a) Let $f(x,y) = xy.$ Then, $\dfrac{\partial f}{\partial x} = y$ and $\dfrac{\partial f}{\partial x} = x.$

$\therefore df = \dfrac{\partial f}{\partial x}dx + \dfrac{\partial f}{\partial y}dy = y\,dx + x\,dy.$

162. (c)

$z = xy\ln(xy)$

$\Rightarrow \dfrac{\partial z}{\partial x} = y\left[\dfrac{x}{xy}y + \ln(xy) \times 1\right] = y(1 + \ln xy),$

$\dfrac{\partial z}{\partial y} = x\left[y\dfrac{1}{xy}x + \ln(xy).1\right] = x(1 + \ln xy).$

$\therefore x\dfrac{\partial z}{\partial x} = y\dfrac{\partial z}{\partial y}.$

163. (a) Given $\dfrac{\partial}{\partial y}(x^2 + y^2) = \dfrac{\partial}{\partial x}(6y + 4x).$ This

gives $2y = 4$ i.e; $y = 2.$

164. (c) The given functions are

$f(x,y) = x^3 - 3xy^2$ & $g(x,y) = 3x^2y - y^3$.

Then, $\dfrac{\partial f}{\partial y} = -6xy,\ \dfrac{\partial f}{\partial x} = 3x^2 - 3y^2,$

$\dfrac{\partial g}{\partial x} = 6xy,\ \dfrac{\partial g}{\partial y} = 3x^2 - 3y^2.$

$\therefore \dfrac{\partial f}{\partial y} = -\dfrac{\partial g}{\partial x}.$

165. 40.

$f(x,y,z) = (x^2 + y^2 - 2z^2)(y^2 + z^2)$

$\Rightarrow \dfrac{\partial f}{\partial x} = 2x(y^2 + z^2)$

$\therefore$ At $x = 2, y = 1, z = 3; \dfrac{\partial f}{\partial x} = 2 \times 2(1^2 + 3^2) = 40.$

166. (c) By chain rule, $\dfrac{\partial w}{\partial x}\dfrac{dx}{dt} + \dfrac{\partial w}{\partial y}\dfrac{dy}{dt} = \dfrac{dw}{dt}.$

167. (d)

$f(x,y) = \dfrac{ax^2 + by^2}{xy} = \dfrac{ax}{y} + \dfrac{by}{x}$

$\Rightarrow \dfrac{\partial f}{\partial x} = \dfrac{a}{y} - \dfrac{by}{x^2}$ and $\dfrac{\partial f}{\partial y} = \dfrac{-ax}{y^2} + \dfrac{b}{x}$

$\Rightarrow \left(\dfrac{\partial f}{\partial x}\right)_{(1,2)} = \dfrac{a}{2} - 2b$ and $\left(\dfrac{\partial f}{\partial y}\right)_{(1,2)} = \dfrac{-a}{4} + b.$

Now, $\left(\dfrac{\partial f}{\partial x}\right)_{(1,2)} = \left(\dfrac{\partial f}{\partial y}\right)_{(1,2)}$ (given)

$\Rightarrow \dfrac{a}{2} - 2b = -\dfrac{a}{4} + b \Rightarrow \dfrac{3a}{4} = 3b \Rightarrow a = 4b.$

168. (d) For $V(x,y) = \dfrac{px^3}{3} + \dfrac{qy^3}{3} + x^2y + xy^2,$

$\dfrac{\partial V}{\partial x} = px^2 + y^2 + 2xy$ and

$\dfrac{\partial V}{\partial y} = x^2 + qy^2 + 2xy.$

169. 4.5

Given $r = x^2 + y - z$............(i)

and $z^3 - xy + yz + y^3 = 1$..........(ii)

Differentiating (i) partially w.r.t x, we get

$\dfrac{\partial r}{\partial x} = 2x - \dfrac{\partial z}{\partial x}$..........(iii)

Differentiating (ii) partially w.r.t x, we get

$3z^2\dfrac{\partial z}{\partial x} - y + y\dfrac{\partial z}{\partial x} = 0$

or, $(3z^2 + y)\dfrac{\partial z}{\partial x} = y$

or, $\dfrac{\partial z}{\partial x} = \dfrac{y}{3z^2 + y}$.............(iv)

Using (iv), (iii) becomes

$$\frac{\partial r}{\partial x} = 2x - \frac{y}{(3z^2 + y)}.$$

$$\therefore \left(\frac{\partial r}{\partial x}\right)_{(2,-1,1)} = 4 - \frac{(-1)}{(3-1)} = 4.5$$

170. (b)

$$f(x, y) = 2x^2 + 2xy - y^3$$

$$\Rightarrow f_x = 4x + 2y, \ f_y = 2x - 3y^2.$$

Then, $f_x = 0 \Rightarrow 4x + 2y = 0 \Rightarrow y = -2x,$

$f_y = 0 \Rightarrow 2x - 3y^2 = 0 \Rightarrow 2x - 3(-2x)^2 = 0$

$$\Rightarrow 2x(1 - 6x) = 0 \Rightarrow x = 0, \frac{1}{6}.$$

Now for $x = 0, \ y = -2 \times 0 = 0$ and

for $x = \frac{1}{6}, \ y = -2 \times \frac{1}{6} = -\frac{1}{3}.$

Thus, the stationary points are $(0,0)$ and $\left(\frac{1}{6}, -\frac{1}{3}\right).$

171. (b)

$$f(x, y) = x^2 - y^2$$

$$\Rightarrow f_x = 2x, f_y = -2y, f_{xx} = 2, f_{xy} = 0, f_{yy} = -2.$$

Then, $f_x = 0, f_y = 0$

$$\Rightarrow 2x = 0, -2y = 0 \Rightarrow x = 0, y = 0.$$

$\therefore (0,0)$ is the only stationary point.

At $(0,0), \ f_{xx}f_{yy} - \left(f_{xy}\right)^2 = -4 < 0.$

$\therefore f(x, y)$ has neither maxima nor minima at $(0,0).$

172. (c)

$$f = y^x$$

$$\Rightarrow \frac{\partial f}{\partial y} = xy^{x-1}$$

$$\Rightarrow \frac{\partial}{\partial x}\left(\frac{\partial f}{\partial y}\right) = \frac{\partial^2 f}{\partial x \partial y} = y^{x-1} + xy^{x-1}\log y$$

$$\therefore \frac{\partial^2 f}{\partial x \partial y}\bigg|_{x=2, y=1} = 1^{2-1} + 2 \times 1^{2-1}\log 1 = 1.$$

173. (a) Let $P(x, y, z)$ be point lying on $z^2 = 1 + xy.$ Then,

$$\overline{OP} = \sqrt{x^2 + y^2 + z^2} = \sqrt{x^2 + y^2 + 1 + xy}$$

Let $f(x, y) = x^2 + y^2 + 1 + xy.$ Then,

$$\frac{\partial f}{\partial x} = 2x + y \text{ and } \frac{\partial f}{\partial y} = 2y + x.$$

$$\therefore \frac{\partial f}{\partial x} = 0 \text{ and } \frac{\partial f}{\partial y} = 0$$

$$\Rightarrow 2x + y = 0 \text{ and } 2y + x = 0$$

$$\Rightarrow x = 0, y = 0$$

Thus, the stationary point is $(0, 0).$

Now at $(0,0), \dfrac{\partial^2 f}{\partial x^2}\dfrac{\partial^2 f}{\partial y^2} - \left(\dfrac{\partial^2 f}{\partial x \partial y}\right)^2 = 2 \times 2 - 1 > 0.$

Also $\dfrac{\partial^2 f}{\partial x^2} = 2 > 0.$

Thus, $f(x, y)$ has a minimum at $(0,0).$

$\therefore$ Minimum distance $= \sqrt{0 + 0 + 1 + 0} = 1$ (since

for x = 0, y=0, $z = \sqrt{1 + xy} = \sqrt{1 + 0} = 1.$

174. (a)

$$f(x, y) = 4x^2 + 6y^2 - 8x - 4y + 8$$
$$\Rightarrow f_x = 8x - 8, f_y = 12y - 4, f_{xx} = 8,$$
$$f_{xy} = 0, f_{yy} = 12.$$

$$\therefore f_x = 0 \Rightarrow 8x - 8 = 0 \Rightarrow x = 1,$$

$$f_y = 0 \Rightarrow 12y - 4 = 0 \Rightarrow y = \frac{1}{3}.$$

Thus, $\left(1, \dfrac{1}{3}\right)$ is the only stationary point.

At $\left(1, \dfrac{1}{3}\right), f_{xx}f_{yy} - f_{xy}^2 = 8 \times 12 - 0 = 96 > 0$

and $f_{xx} = 8 > 0.$

$\therefore f(x, y)$ has a minimum at $\left(1, \dfrac{1}{3}\right).$

$\therefore$ Minimum value

$$= f\left(1, \frac{1}{3}\right) = 4 + \frac{2}{3} - 8 - \frac{4}{3} + 8 = \frac{10}{3}.$$

175. (a) Here the limits of integration are:

$$x = 0 \text{ to } x = 8, \ y = \frac{x}{4} \text{ to } y = 2.$$

Now changing the order of integration, the new limits will be as follows:

$$x = 0 \text{ to } x = 4y, \ y = 0 \text{ to } y = 2.$$

$\therefore I = \int\limits_{0}^{2}\int\limits_{0}^{4y} f(x,y)\,dx\,dy = \int\limits_{r}^{s}\int\limits_{p}^{q} f(x,y)\,dx\,dy$

Comparing we get, $q = 4y$.

176. (c) The limits of the double integral are:
$y = 0$ to $y = a$ and $x = 0$ to $x = y$.

After changing the order of integration, the new limits of integration will be:

$x = 0$ to $x = a$ and $y = x$ to $y = a$.

$\therefore \int\limits_{0}^{a}\int\limits_{0}^{y} f(x,y)\,dx\,dy = \int\limits_{0}^{a}\int\limits_{x}^{a} f(x,y)\,dy\,dx$.

177. (d)

$\int\limits_{0}^{\frac{\pi}{2}}\int\limits_{0}^{\frac{\pi}{2}} \sin(x+y)\,dx\,dy$

$= \int\limits_{0}^{\frac{\pi}{2}} [-\cos(x+y)]_{0}^{\frac{\pi}{2}}\,dy$

$= -\int\limits_{0}^{\frac{\pi}{2}} [\cos\left(\frac{\pi}{2}+y\right) - \cos y]\,dy$

$= -\int\limits_{0}^{\frac{\pi}{2}} [-\sin y - \cos y]\,dy = [-\cos y + \sin y]_{0}^{\pi/2} = 2.$

178. (a)

$V = \int\limits_{0}^{2\pi}\int\limits_{0}^{\frac{\pi}{3}}\int\limits_{0}^{1} r^2 \sin\phi\, dr\, d\phi\, d\theta$

$= \int\limits_{0}^{2\pi}\int\limits_{0}^{\frac{\pi}{3}} \left[\frac{r^3}{3}\right]_{0}^{1} \sin\varphi\, d\phi\, d\theta = \frac{1}{3}\int\limits_{0}^{2\pi}\int\limits_{0}^{\frac{\pi}{3}} \sin\varphi\, d\phi\, d\theta$

$= \frac{1}{3}\int\limits_{0}^{2\pi} [-\cos\phi]_{0}^{\frac{\pi}{3}}\, d\theta = \frac{1}{3}\int\limits_{0}^{2\pi} \left[-\cos\frac{\pi}{3} + \cos 0\right] d\theta$

$= \frac{1}{3}\left(-\frac{1}{2}+1\right)\int\limits_{0}^{2\pi} d\theta = \frac{1}{6}[\theta]_{0}^{2\pi} = \frac{1}{6}\times 2\pi = \frac{\pi}{3}.$

179. (a)

$\iint f(x,y)\,dx\,dy = \iint f\left(uv,\frac{v}{u}\right)\phi(u,v)\,du\,dv$

where $\phi(u,v) = J\left(\dfrac{x,y}{u,v}\right) = \begin{vmatrix} \dfrac{\partial x}{\partial u} & \dfrac{\partial x}{\partial v} \\ \dfrac{\partial y}{\partial u} & \dfrac{\partial y}{\partial v} \end{vmatrix}$ ...(i)

Now $x(u,v) = uv \Rightarrow \dfrac{\partial x}{\partial u} = v, \dfrac{\partial x}{\partial v} = u;$

$y(u,v) = \dfrac{v}{u} \Rightarrow \dfrac{\partial y}{\partial u} = -\dfrac{v}{u^2}, \dfrac{\partial y}{\partial v} = \dfrac{1}{u}.$

Then (i) gives, $\phi(u,v) = \begin{vmatrix} v & u \\ -\dfrac{v}{u^2} & \dfrac{1}{u} \end{vmatrix} = \dfrac{v}{u} + \dfrac{v}{u} = \dfrac{2v}{u}.$

180. (d)

Let $x = \sqrt{t}$. Then, $dx = \dfrac{1}{2\sqrt{t}}\,dt.$

Then, $x = 0 \Rightarrow t = 0$ and $x = \infty \Rightarrow t = \infty$

$\therefore \int\limits_{x=0}^{\infty} e^{-x^2}\,dx$

$= \int\limits_{0}^{\infty} e^{-t}\dfrac{1}{2\sqrt{t}}\,dt = \dfrac{1}{2}\int\limits_{0}^{\infty} e^{-t}\, t^{-\frac{1}{2}}\,dt$

$= \dfrac{1}{2}\int\limits_{0}^{\infty} e^{-t}\, t^{\frac{1}{2}-1}\,dt = \dfrac{1}{2}\Gamma\left(\dfrac{1}{2}\right) = \dfrac{\sqrt{\pi}}{2}.$

In a similar manner, it can be shown that

$\int\limits_{0}^{\infty} e^{-y^2}\,dy = \dfrac{1}{2}\Gamma\left(\dfrac{1}{2}\right) = \dfrac{\sqrt{\pi}}{2}.$

Then given double integral

$= \int\limits_{0}^{\infty}\int\limits_{0}^{\infty} e^{-x^2} e^{-y^2}\,dx\,dy$

$= \left(\int\limits_{0}^{\infty} e^{-y^2}\,dy\right)\left(\int\limits_{0}^{\infty} e^{-x^2}\,dx\right) = \dfrac{\sqrt{\pi}}{2}\times\dfrac{\sqrt{\pi}}{2} = \dfrac{\pi}{4}$.

181. (a)

$\int\limits_{x=0}^{3}\int\limits_{y=0}^{x} (6-x-y)\,dx\,dy$

$= \int\limits_{x=0}^{3}\left\{\int\limits_{y=0}^{x} (6-x-y)\,dy\right\}dx$

$= \int\limits_{x=0}^{3}\left[6y - xy - \dfrac{y^2}{2}\right]_{y=0}^{x} dx$

$= \int\limits_{0}^{3}\left[6x - x^2 - \dfrac{x^2}{2}\right]dx$

$= \int\limits_{0}^{3}\left[6x - \dfrac{3x^2}{2}\right]dx = \left[6\dfrac{x^2}{2} - \dfrac{3x^3}{2\times 3}\right]_{0}^{3}$

$= \left[3x^2 - \dfrac{x^3}{2}\right]_{0}^{3} = 27 - \dfrac{27}{2} = 13.5.$

182. (a) $x + 2y = 2 \Rightarrow \dfrac{x}{2} + \dfrac{y}{1} = 1.$

The above line cuts x-axis at $(2, 0)$ and y-axis at $(0,1)$.

So it is clear that x varies from 0 to 2 and y varies from 0 to $\dfrac{2-x}{2}$.

$\therefore \iint\limits_{P} xy\,dx\,dy$

$= \int\limits_{x=0}^{2} \int\limits_{0}^{\frac{2-x}{2}} xy\,dx\,dy = \int\limits_{0}^{2} x\left[\dfrac{y^2}{2}\right]_{0}^{\frac{2-x}{2}} dx$

$= \dfrac{1}{2}\int\limits_{0}^{2} x\left(\dfrac{2-x}{2}\right)^2 dx = \dfrac{1}{8}\int\limits_{0}^{2} x\left(4-4x+x^2\right)dx$

$= \dfrac{1}{8}\int\limits_{0}^{2}\left[4x - 4x^2 + x^3\right]dx$

$= \dfrac{1}{8}\left[4\dfrac{x^2}{2} - \dfrac{4x^3}{3} + \dfrac{x^4}{4}\right]_{0}^{2}$

$= \dfrac{1}{8}\left[4\times\dfrac{4}{2} - \dfrac{4\times 8}{3} + \dfrac{16}{4}\right] = \dfrac{1}{6}.$

183. (a)

$x = y^2$ and $y = x^2$

$\Rightarrow y^4 = x^2 = y \Rightarrow y^4 - y = 0$

$\Rightarrow y(y^3 - 1) = 0 \Rightarrow y = 0, 1$

Thus, y varies from 0 to 1.

Also, x varies from y^2 to $\sqrt{y}$.

Hence, the required volume $= \int\limits_{y=0}^{1} \int\limits_{x=y^2}^{\sqrt{y}} f(x, y)\,dx\,dy.$

184. 864

Required volume

$= \int\limits_{0}^{12}\int\limits_{0}^{x} z\,dy\,dx = \int\limits_{0}^{12}\int\limits_{0}^{x}(x+y)\,dy\,dx$

$= \int\limits_{0}^{12}\left[xy + \dfrac{y^2}{2}\right]_{0}^{x} dx = \int\limits_{0}^{12}\left(x^2 + \dfrac{x^2}{2}\right)dx$

$= \dfrac{3}{2}\left[\dfrac{x^3}{3}\right]_{0}^{12} = \dfrac{3}{2}\times\dfrac{12^3}{3} = 864.$

185. (a)

$u = \dfrac{2x - y}{2}$ and $v = \dfrac{y}{2}$

$\Rightarrow du = dx$ and $dv = \dfrac{dy}{2}.$

$\therefore dy = 2dv.$

$x = \dfrac{y}{2} \Rightarrow u = 0; \; x = \dfrac{y}{2} + 1 \Rightarrow u = 1.$

$y = 0 \Rightarrow v = 0; \; y = 8 \Rightarrow v = 4.$

$\therefore \int\limits_{0}^{8}\left(\int\limits_{\frac{y}{2}}^{\left(\frac{y}{2}+1\right)}\left(\dfrac{2x-y}{2}\right)dx\right)dy = \int\limits_{v=0}^{4}\int\limits_{u=0}^{1} 2u\,du\,dv.$

186. (b)

$\int\limits_{0}^{2}\int\limits_{0}^{x} e^{x+y}\,dy\,dx$

$= \int\limits_{0}^{2} e^x \left[e^y\right]_{0}^{x} dx = \int\limits_{0}^{2} e^x\left(e^x - 1\right)dx$

$= \int\limits_{0}^{2}\left(e^{2x} - e^x\right)dx = \left[\dfrac{e^{2x}}{2} - e^x\right]_{0}^{2}$

$= \left[\left(\dfrac{e^4}{2} - e^2\right) - \left(\dfrac{1}{2} - 1\right)\right] = \dfrac{e^4}{2} - e^2 + \dfrac{1}{2} = \dfrac{1}{2}(e^2 - 1)^2.$

187. e-2.

Let R be the region bounded by the lines $x = y, x = 0, y = 1$ in the xy plane.

Then the required volume

$= \iint\limits_{R} f(x, y)\,dx\,dy$

$= \int\limits_{x=0}^{1}\int\limits_{y=x}^{1} e^x\,dy\,dx = \int\limits_{0}^{1}\left[y\right]_{x}^{1} e^x\,dx$

$= \int\limits_{0}^{1}(1-x)e^x\,dx = \int\limits_{0}^{1} e^x\,dx - \int\limits_{0}^{1} x e^x\,dx$

$= \left[e^x\right]_{0}^{1} - \left[xe^x - e^x\right]_{0}^{1}$

$= (e-1) - \left[(e-e) - (0-1)\right] = e - 2.$

188. 20

Let $x = r\cos\theta, y = r\sin\theta$. Then Jacobian $= r$.

Also $x^2 + y^2 \le 4 \Rightarrow r^2 \le 4 \Rightarrow 0 \le r \le 2.$

$$\therefore \frac{1}{2\pi} \iint_D (x+y+10)\,dx\,dy$$

$$= \frac{1}{2\pi} \int_{r=0}^{2} \int_{\theta=0}^{2\pi} (r\cos\theta + r\sin\theta +10)\, r\,dr\,d\theta$$

$$= \frac{1}{2\pi} \int_{r=0}^{2} \int_{\theta=0}^{2\pi} (r^2\cos\theta + r^2\sin\theta +10r)\, dr\,d\theta$$

$$= \frac{1}{2\pi} \int_{\theta=0}^{2\pi} \left[\frac{r^3}{3}\cos\theta + \frac{r^3}{3}\sin\theta + 5r^2 \right]_0^2 d\theta$$

$$= \frac{1}{2\pi} \int_{\theta=0}^{2\pi} \left\{ \frac{8}{3}\cos\theta + \frac{8}{3}\sin\theta + 20 \right\} d\theta$$

$$= \frac{1}{2\pi} \left[\frac{8}{3}\sin\theta - \frac{8}{3}\cos\theta + 20\theta \right]_0^{2\pi}$$

$$= \frac{1}{2\pi} \left[\frac{8}{3}\sin 2\pi - \frac{8}{3}\cos 2\pi + 20\times 2\pi - 0 + \frac{8}{3} - 0 \right]$$

$$= \frac{1}{2\pi} \left\{ -\frac{8}{3} + 40\pi + \frac{8}{3} \right\} = 20.$$

189. 4.71

Required volume

$$= \int_{r=3}^{5} \int_{\theta=\frac{\pi}{8}}^{\frac{\pi}{4}} \int_{z=3}^{4.5} r\,dr\,d\theta\,dz$$

$$= \left[\frac{r^2}{2} \right]_3^5 [\theta]_{\frac{\pi}{8}}^{\frac{\pi}{4}} [z]_3^{4.5}$$

$$= \left(\frac{25}{2} - \frac{9}{2} \right)\left(\frac{\pi}{4} - \frac{\pi}{8} \right)(4.5-3) = 8\times\frac{\pi}{8}\times 1.5 = 4.71$$

190. 10

Required volume

$$= \int_{x=0}^{3} \int_{y=0}^{\frac{2}{3}x} (6-x-y)\,dx\,dy$$

$$= \int_0^3 \left[6y - xy - \frac{y^2}{2} \right]_0^{\frac{2x}{3}} dx$$

$$= \int_0^3 \left[6\times\frac{2x}{3} - \frac{2x}{3}\times x - \frac{1}{2}\times\left(\frac{2x}{3} \right)^2 - 0 \right] dx$$

$$= \int_0^3 \left[4x - \frac{2x^2}{3} - \frac{2x^2}{9} \right] dx$$

$$= \left[2x^2 - \frac{2x^3}{9} - \frac{2x^3}{27} \right]_0^3 = [18 - 6 - 2] = 10.$$

191. (c)

$$\int_0^1 \left(\int_0^1 \frac{x-y}{(x+y)^3} dy \right) dx$$

$$= \int_0^1 \left(\int_0^1 \frac{2x-(x+y)}{(x+y)^3} dy \right) dx$$

$$\int_0^1 \left(\int_0^1 \frac{x-y}{(x+y)^3} dy \right) dx$$

$$= \int_0^1 \left(\int_0^1 \frac{2x-(x+y)}{(x+y)^3} dy \right) dx$$

$$= \int_0^1 \left(\int_0^1 \left\{ \frac{2x}{(x+y)^3} - \frac{1}{(x+y)^2} \right\} dy \right) dx$$

$$= \int_0^1 \left(\left[\frac{2x}{-2(x+y)^2} - \frac{1}{(-1)(x+y)} \right]_0^1 \right) dx$$

$$= \int_0^1 \left(\left[-\frac{x}{(x+y)^2} + \frac{1}{(x+y)} \right]_0^1 \right) dx$$

$$= \int_0^1 \left[-\frac{x}{(x+1)^2} + \frac{1}{(x+1)} \right] dx$$

$$= \int_0^1 \left[-\frac{(x+1)-1}{(x+1)^2} + \frac{1}{(x+1)} \right] dx$$

$$= \int_0^1 \left[-\frac{1}{(x+1)} + \frac{1}{(x+1)^2} + \frac{1}{(x+1)} \right] dx$$

$$= \left[-\frac{1}{(x+1)} \right]_0^1 = -\frac{1}{2} + 1 = 0.5$$

$$\int_0^1 \left(\int_0^1 \frac{x-y}{(x+y)^3} dx \right) dx$$

$$= \int_0^1 \left(\int_0^1 \left\{ \frac{(x+y)-2y}{(x+y)^3} \right\} dx \right) dy$$

$$= \int_0^1 \left(\int_0^1 \left\{ \frac{1}{(x+y)^2} - \frac{2y}{(x+y)^3} \right\} dx \right) dy$$

$$= \int_0^1 \left(\left[\frac{1}{-(x+y)} - \frac{2y}{(-2)(x+y)^2} \right]_0^1 \right) dy$$

$$= \int_0^1 \left(\left[-\frac{1}{(x+y)} + \frac{y}{(x+y)^2} \right]_0^1 \right) dy$$

$$= \int_0^1 \left[-\frac{1}{(1+y)} + \frac{y}{(1+y)^2} \right] dy$$

$$= \int_0^1 \left[-\frac{1}{(1+y)} + \frac{(1+y)-1}{(1+y)^2} \right] dy$$

$$= \int_0^1 \left[-\frac{1}{(1+y)} + \frac{1}{(1+y)} - \frac{1}{(1+y)^2} \right] dy$$

$$= -\int_0^1 \frac{1}{(1+y)^2} dy = \left[\frac{1}{1+y} \right]_0^1 = \frac{1}{2} - 1 = -0.5$$

192. 1666.13

Here the region of integration R is divided into two sub regions as shown below:

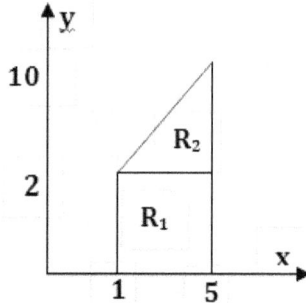

$$\iint_R xy^2 \, dxdy$$

$$= \iint_{R_1} xy^2 \, dxdy + \iint_{R_2} xy^2 \, dxdy$$

$$= \int_{y=0}^2 \int_{x=1}^5 xy^2 \, dx \, dy + \int_{y=2}^{10} \int_{y/2}^5 xy^2 \, dx \, dy$$

$$= \int_{y=0}^2 y^2 \left[\frac{x^2}{2} \right]_1^5 dy + \int_{y=2}^{10} y^2 \left[\frac{x^2}{2} \right]_{y/2}^5 dy$$

$$= \int_{y=0}^2 y^2 \left[\frac{25}{2} - \frac{1}{2} \right] dy + \int_{y=2}^{10} y^2 \left[\frac{25}{2} - \frac{y^2}{8} \right] dy$$

$$= 12 \int_{y=0}^2 y^2 \, dy + \frac{25}{2} \int_{y=2}^{10} y^2 \, dy - \frac{1}{8} \int_{y=2}^{10} y^4 \, dy$$

$$= 12 \left[\frac{y^3}{3} \right]_0^2 + \frac{25}{2} \left[\frac{y^3}{3} \right]_2^{10} - \frac{1}{8} \left[\frac{y^5}{5} \right]_2^{10}$$

$$= 12 \left[\frac{8}{3} - 0 \right] + \frac{25}{2} \left[\frac{1000}{3} - \frac{8}{3} \right] - \frac{1}{8} \left[\frac{10^5}{5} - \frac{2^5}{5} \right]$$

$$= 32 + \frac{25000}{6} - \frac{100}{3} - 2500 + \frac{4}{5} = 1666.13$$

193. (d)

$$y = \frac{2}{3} x^{\frac{3}{2}} \Rightarrow \frac{dy}{dx} = x^{\frac{1}{2}} \Rightarrow \left(\frac{dy}{dx} \right)^2 = x$$

So length of the curve

$$= \int_0^1 \sqrt{1 + \left(\frac{dy}{dx} \right)^2}$$

$$= \int_0^1 \sqrt{1+x} \, dx = \left[\frac{2}{3} (1+x)^{\frac{3}{2}} \right]_0^1 = \frac{2}{3} \left[2^{\frac{2}{3}} - 1 \right] = 1.22$$

194. (d) The equation of the parabola is $y = 4h \dfrac{x^2}{L^2}$.

$$\therefore \frac{dy}{dx} = \frac{8hx}{L^2}$$

So total length

$$= 2 \int_0^{\frac{L}{2}} \sqrt{1 + \left(\frac{dy}{dx} \right)^2} \, dx = 2 \int_0^{\frac{L}{2}} \sqrt{1 + 64 \frac{h^2 x^2}{L^4}} \, dx$$

195. 1.86

$$x(t) = \cos t, \, y(t) = \sin t, \, z(t) = \frac{2}{\pi} t,$$

$$\Rightarrow \frac{dx}{dt} = -\sin t, \frac{dy}{dt} = \cos t \text{ and } \frac{dz}{dt} = \frac{2}{\pi}.$$

Then, the length of the curve

$$= \int_\alpha^\beta \sqrt{\left(\frac{dx}{dt} \right)^2 + \left(\frac{dy}{dt} \right)^2 + \left(\frac{dz}{dt} \right)^2} \, dt$$

(where α, β are limits of t)

$$= \int_0^{\frac{\pi}{2}} \sqrt{(-\sin t)^2 + (\cos t)^2 + \left(\frac{2}{\pi} \right)^2} \, dt$$

$$= \int_0^{\frac{\pi}{2}} \sqrt{\frac{\pi^2 + 4}{\pi^2}} \, dt = \frac{1}{\pi} \int_0^{\frac{\pi}{2}} \sqrt{\pi^2 + 4} \, dt$$

$$= \frac{\sqrt{\pi^2 + 4}}{\pi} [t]_0^{\frac{\pi}{2}} = \frac{\sqrt{\pi^2 + 4}}{\pi} \times \frac{\pi}{2} = 1.86.$$

196. (d) Volume of revolution

$$= \int_{x_1}^{x_2} \pi y^2 \, dx$$

$$= \int_1^2 \pi (\sqrt{x})^2 \, dx = \pi \int_1^2 x \, dx = \pi \left[\frac{x^2}{2} \right]_1^2 = \frac{3\pi}{2}.$$

197. $\pi/4$

In polar co-ordinates,

$x = r\cos\theta, y = r\sin\theta, z = z.$

So $x^2 + y^2 = r^2\cos^2\theta + r^2\sin^2\theta = r^2.$

Hence, the required volume is

$$= \int_0^1 \pi r^2 dz = \pi\int_0^1 z^3 dz = \pi\left[\frac{z^4}{4}\right]_0^1 = \frac{\pi}{4}.$$

198. (c)

$$x = \cos\left(\frac{\pi u}{2}\right), y = \sin\left(\frac{\pi u}{2}\right)$$

$$\Rightarrow \frac{dx}{du} = -\frac{\pi}{2}\sin\left(\frac{\pi u}{2}\right), \frac{dy}{du} = \frac{\pi}{2}\cos\left(\frac{\pi u}{2}\right).$$

Therefore, the required surface area

$$= \int_a^b 2\pi y\left(\frac{ds}{du}\right)du$$

$$= \int_0^1 2\pi y\sqrt{\left(\frac{dx}{du}\right)^2 + \left(\frac{dx}{du}\right)^2}\, du \; (\because a = 0, b = 1)$$

$$= \int_0^1 2\pi y\sqrt{\left\{-\frac{\pi}{2}\sin\left(\frac{\pi u}{2}\right)\right\}^2 + \left\{\frac{\pi}{2}\cos\left(\frac{\pi u}{2}\right)\right\}^2}\, du$$

$$= 2\pi\int_0^1 \sin\left(\frac{\pi u}{2}\right)\times\frac{\pi}{2}du$$

$$= \pi^2\left[-\frac{2}{\pi}\cos\left(\frac{\pi u}{2}\right)\right]_0^1 = -2\pi\left(\cos\frac{\pi}{2} - \cos 0\right) = 2\pi.$$

Questions for Practice

1. If $f : R \to R$ satisfies $f(x+y) = f(x) + f(y)$

$\forall x, y \in R$ and $f(1) = 7$, then $\sum_{r=1}^{n} f(r)$ is equal to

(a) $\dfrac{7n}{2}$ (b) $\dfrac{7(n+1)}{2}$

(c) $7n(n+1)$ (d) $\dfrac{7n(n+1)}{2}$

2. A real valued function $f(x)$ satisfies the functional equation

$f(x-y) = f(x)f(y) - f(a-x)\times f(a+y),$

where "a" is a constant and $f(0) = 1$. Then, $f(2a-x) = ?$

(a) $-f(x)$ (b) $f(x)$

(c) $f(a) + f(a-x)$ (d) $f(-x)$

3. Let $f(x)$ be defined $\forall x > 0$ and be continuous. If $f(x)$ satisfies

$f\left(\dfrac{x}{y}\right) = f(x) - f(y)\,\forall x, y$ and $f(e) = 1,$

then which of the following is true?

(a) $f(x)$ is bounded

(b) $f(x) = \log_e x$

(c) $f\left(\dfrac{1}{x}\right) \to 0$ as $x \to 0$

(d) none of these

4. If $f(x) = |x-1|$, then

(a) $f(x+y) = f(x) + f(y)$

(b) $f(x^2) = [f(x)]^2$

(c) $f(|x|) = |f(x)|$

(d) none of these

5. If $f(x) = \cos(\log_e x)$, then the simplest value

of $f(x)f(y) - \dfrac{1}{2}\left[f\left(\dfrac{x}{y}\right) + f(xy)\right]$ is

(a) 0   (b) −1   (c) 1   (d) $\dfrac{1}{2}$

6. $\lim\limits_{n\to\infty}\left[\dfrac{1}{1-n^2} + \dfrac{2}{1-n^2} + ... + \dfrac{n}{1-n^2}\right] = ?$

(a) −1   (b) $-\dfrac{1}{2}$ (c) 1   (d) 0

7. $\lim\limits_{x\to 1}(1-x)\tan\dfrac{\pi x}{2} = ?$

(a) $\dfrac{\pi}{2}$ (b) $\dfrac{2}{\pi}$ (c) $\dfrac{1}{\pi}$ (d) π

8. $\lim\limits_{x\to\infty}\sqrt{\dfrac{x-\sin x}{x+\cos^2 x}} = ?$

(a) ∞ (b) 0

(c) 1   (d) −1

9. $\lim\limits_{x \to 1} \dfrac{\sqrt{1 - \cos 2(x-1)}}{x-1}$

(a) exists and equals to $\sqrt{2}$
(b) exists and equals to $-\sqrt{2}$
(c) exists and equals to 1
(d) does not exist

10. $\lim\limits_{x \to 0} \dfrac{\sqrt{1 - \cos 2x}}{\sqrt{2}\,x} = ?$

(a) 1　　　　　　　　(b) 0
(c) -1　　　　　　　(d) does not exist

11. If $f(x) = \begin{cases} xe^{-\left(\frac{1}{|x|} + \frac{1}{x}\right)}, & x \neq 0 \\ 0, & x = 0 \end{cases}$; then $f(x)$ is

(a) continuous as well as differentiable for all x
(b) continuous for all x but not differentiable at $x = 0$
(c) neither continuous nor differentiable at $x = 0$
(d) discontinuous everywhere

12. Let $f(x) = \begin{cases} (x-1)\sin\left(\dfrac{1}{x-1}\right), & x \neq 1 \\ 0, & x = 1 \end{cases}$. Then which of the following is true?

(a) f is differentiable at $x = 0$ and $x = 1$

(b) f is differentiable at $x = 0$ but not at $x = 1$

(c) f is differentiable at $x = 1$ but not at $x = 0$

(d) f is differentiable neither at $x = 0$ nor at $x = 1$

13. Suppose f is differentiable at $x = 1$ and $\lim\limits_{h \to 0} \dfrac{1}{h} f(1 + h) = 5$, then $f'(1)$ is equal to

(a) -5　　　　　　　(b) 5
(c) 4　　　　　　　　(d) -4

14. The set of points where $f(x) = \dfrac{x}{1 + |x|}$ is differentiable is

(a) $(-\infty, 0) \cup (0, \infty)$　　(b) $(-\infty, -1) \cup (-1, \infty)$

(c) $(-\infty, \infty)$　　　　　　(d) $(0, \infty)$

15. If $\lim\limits_{x \to 1} \dfrac{\sqrt{f(x)} - 1}{\sqrt{x} - 1} = ?$ $f(1) = 1$ and $f'(1) = 2$, then

(a) 2　　　(b) 1　　　(c) 0　　　(d) -1

16. If $f(x) = x - [x]$, then $f'\left(\dfrac{1}{2}\right)$ is equal to

(a) $\dfrac{1}{2}$　　(b) 1　　(c) 0　　(d) $-\dfrac{1}{2}$

17. If $f(x) = 1 + x + \dfrac{x^2}{2} + \dots + \dfrac{x^{1000}}{1000}$, then $f'(1) = ?$

(a) 1000　　　　　　(b) 5000

(c) $\dfrac{101}{2}$　　　　　　(d) 1

18. Let $f(x) = \begin{cases} \dfrac{\sqrt{1 + kx} - \sqrt{1 - kx}}{x}, & -1 \leq x < 0 \\ \dfrac{2x + 1}{x - 2}, & 0 \leq x \leq 1 \end{cases}$

If $f(x)$ is continuous $\forall x$, then $k = ?$

(a) $\dfrac{1}{2}$　　　　　　(b) $-\dfrac{1}{2}$

(c) 1　　　　　　　　(d) 0

19. If $f(x) = \begin{cases} 5, & \text{if} & x \leq 2 \\ ax + b, & \text{if} & 2 < x < 10 \\ 21, & \text{if} & x \geq 10 \end{cases}$

and $f(x)$ is continuous $\forall x \in R$, then the values of a and b are, respectively:

(a) 2, 1　　(b) 1, 2　　(c) $-1, 2$　　(d) $2, -1$

20. Let $f(x) = \begin{cases} \lambda x^2 + \mu, & |x| < 1 \\ \dfrac{1}{|x|}, & |x| \geq 1 \end{cases}$

If f is differentiable $\forall x \in R$, then which of the following is true?

(a) $\lambda = \dfrac{1}{2}, \mu = \dfrac{3}{2}$　　(b) $\lambda = -\dfrac{1}{2}, \mu = \dfrac{3}{2}$

(c) $\lambda = -\dfrac{1}{2}, \mu = -\dfrac{3}{2}$　　(d) $\lambda = \dfrac{1}{2}, \mu = -\dfrac{3}{2}$

21. Which of the following is true?

(a) Lagrange's Mean Value Theorem is applicable to $f(x) = |x|$ in $[-1, 1]$

(b) Lagrange's Mean Value Theorem is applicable to $f(x) = \dfrac{1}{x}$ in $[-1, 3]$

(c) Rolle's theorem is applicable to $f(x) = |x|$ in $[-1, 1]$

(d) Rolle's theorem is not applicable to $f(x) = \dfrac{1}{x}$ in $[-1, 3]$

22. A stone is dropped into a quiet lake and waves moves in circles at a speed of 4 cm/sec. At the instant when the radius of the circular wave is 10 cm, how fast is the enclosed area increasing?

(a) $40\pi \ cm^2 / \sec$ (b) $80\pi \ cm^2 / \sec$

(c) $100\pi \ cm^2 / \sec$ (d) $120\pi \ cm^2 / \sec$

23. The surface area of a spherical bubble is increasing at the rate of $2 cm^2 / \sec$. When the radius of the bubble is 6 cm, at what rate is the volume of the bubble increasing?

(a) $6 cm^3 / \sec$ (b) $8 cm^3 / \sec$

(c) $10 cm^3 / \sec$ (d) $12 cm^3 / \sec$

24. The function $f(x) = 2x^3 - 9x^2 + 12x + 15$ is increasing on

(a) $(-\infty, 1) \cup (2, \infty)$ (b) $(1, \infty)$

(c) $(-\infty, 2)$ (d) $(1, 2) \cup (2, 4)$

25. The function $f(x) = \log(1+x) - \dfrac{x}{1+x}$ is increasing on

(a) $(1, \infty)$ (b) $(-\infty, \infty)$

(c) $(0, \infty)$ (d) $(-\infty, 0)$

26. The minimum value of the function $f(x) = 2x^3 - 15x^2 + 36x + 1$ in $[1, 5]$ is

(a) 24   (b) 28
(c) 40   (d) 56

27. The height of a a cylinder of maximum volume that can be inscribed in a sphere of radius R is

(a) $\dfrac{R}{\sqrt{3}}$ (b) $\dfrac{R}{\sqrt{2}}$

(c) $\dfrac{2R}{\sqrt{3}}$ (d) $\dfrac{2R}{3}$

28. A tank with rectangular base and rectangular sides, open at the top is to be constructed so that its depth is 2 m and volume is $8 m^3$. If building of tank costs Rs. 70 per square meter for the base and Rs. 45 per square meter for sides, what will be the cost of the least expensive tank?

(a) Rs. 500   (b) Rs. 800
(c) Rs. 1000   (d) Rs. 1200

29. The point on the curve $y^2 = 2x$ which is at a minimum distance from the point (1, 4) is

(a) (2, 2)   (b) (2, 4)
(c) (4, 2)   (d) 2, 6)

30. The value of $\displaystyle\int_0^{\frac{\pi}{2}} \dfrac{\sin 2x}{\sin^4 x + \cos^4 x} dx$ is equal to

(a) π (b) $\dfrac{\pi}{2}$ (c) 2π (d) $\dfrac{\pi}{3}$

31. The value of $\displaystyle\int_1^2 [2x] dx$ is

(a) 5   (b) $\dfrac{3}{2}$ (c) $\dfrac{5}{2}$ (d) 3

32. $\displaystyle\int_{-\frac{\pi}{2}}^{\frac{\pi}{2}} \left(\sin|x| + \cos|x|\right) dx = ?$

(a) 1   (b) 2   (c) 3   (d) 4

33. The value of $\displaystyle\int_0^{\pi} \log(1 + \cos x) dx = ?$

(a) $-\pi \log_e 2$ (b) $\log_e 2$

(c) $\pi \log_e 2$ (d) 2π

34. Which of the following is true?

(a) $\displaystyle\int_0^{\frac{\pi}{2}} \log(\sin x) dx \neq \int_0^{\frac{\pi}{2}} \log(\cos x) dx$

(b) $\displaystyle\int_0^{\frac{\pi}{2}} \log(\sin x) dx = \dfrac{\pi}{2} \log 2,$

$$\int_0^{\frac{\pi}{2}} \log(\cos x)\,dx = -\frac{\pi}{2}\log 2$$

(c) $\int_0^{\frac{\pi}{2}} \log(\cos x)\,dx = \frac{\pi}{2}\log 2,$

$$\int_0^{\frac{\pi}{2}} \log(\sin x)\,dx = -\frac{\pi}{2}\log 2$$

(d) $\int_0^{\frac{\pi}{2}} \log(\sin x)\,dx = \int_0^{\frac{\pi}{2}} \log(\cos x)\,dx = -\frac{\pi}{2}\log 2$

35. If A_n be the area bounded by the curve $y = \tan^n x$ and the lines $y=0, x=0$ and $x = \frac{\pi}{4}$; then which of the following is correct?

(a) $A_n + A_{n-2} = \frac{1}{n-1}$ for $n > 2$

(b) $A_n + A_{n-1} = \frac{1}{n}$ for $n > 1$

(c) $A_{n+1} + A_{n-1} = \frac{1}{n}$ for $n > 1$

(d) both of (a) and (c

36. The area bounded by the parabolas $x^2 = 8y$ and $y^2 = 8x$ is

(a) $\frac{10}{3}$ sq.unit   (b) $\frac{14}{3}$ sq.unit

(c) $\frac{16}{3}$ sq.unit   (d) $\frac{17}{3}$ sq.unit

37. The value of $\lim_{x \to 0}\left\{ \dfrac{\int_0^x \sin(t^2)\,dt}{x} \right\}$ is

(a) -1 (b) 0 (c) 1 (d) 2

38. If $\int_0^{f(x)} t\,dt = 2\sin\pi x,$ then $f'\left(\dfrac{1}{2}\right) = ?$

(a) $\dfrac{2}{3}$ (b) $-\dfrac{1}{2}$ (c) $\dfrac{1}{2}$ (d) 0

39. The function $f(x,y) = x^2y - 3xy + 2y + x$ has
(a) No local maximum
(b) One local maximum but no local minimum
(c) One local minimum but no local maximum
(d) One local minimum and one local maximum

40. The volume generated by revolving the area bounded by parabola $y^2 = 8x$ and the line $x = 2$ about $y-axis$ is

(a) $\dfrac{128\pi}{5}$ (b) $\dfrac{5}{128\pi}$

(c) $\dfrac{127}{5\pi}$ (d) None of these

41. $\lim_{x \to 0} \dfrac{x(e^x - 1) + 2(\cos x - 1)}{x(1 - \cos x)} = ?$

(a) 0 (b) 1 (c) $\dfrac{1}{2}$ (d) $\dfrac{3}{2}$

42. $\lim_{x \to \infty} \dfrac{x^3 - \cos x}{x^2 + \sin^2 x} = ?$

(a) 0 (b) 2 (c) ∞ (d) -1

43. The value of $\displaystyle\int_0^1 \int_x^{\frac{1}{x}} \dfrac{x}{1+y^2}\,dx\,dy = ?$

(a) 1 (b) $1 + \dfrac{\pi}{4}$

(c) $1 - \dfrac{\pi}{4}$ (d) $\dfrac{\pi}{4}$

44. The area bounded by the parabola $x^2 = 2y$ and the line $x = y - 4$ is
(a) 18   (b) 14   (c) 12   (d) 10

45. The function $y = x^2 + \dfrac{250}{x}$ at $x = 5$ attians
(a) Maximum   (b) Minimum
(c) Neither   (d) 1

46. The maximum value of the function $f(x) = 2x^2 - 2x + 6$ in $[0,2]$ is
(a) 6   (b) 10
(c) 12   (d) 14

47. The function $f(x) = x^3 - 6x^2 + 9x + 25$ has

(a) maximun at $x = 1$ and a minimum at $x = 3$

(b) a maximum at $x = 3$ and a minimum at $x = 1$

(c) no maximum but a minimum at $x = 3$

(d) a maximum at $x = 1$, but no minimum

48. The third term in the Taylor series expression of e^x about "a" would be

(a) $e^a (x-a)$

(b) $\dfrac{e^a}{2}(x-a)^2$

(c) $\dfrac{e^a}{2}$

(d) $\dfrac{e^a}{6}(x-a)^3$

49. The value of "c" in the mean value theorem of $f(b) - f(a) = (b-a) f'(c)$ for the function $f(x) = Ax^2 + Bx + C$ in (a, b) is

(a) $a + b$

(b) $b - a$

(c) $\dfrac{b+a}{2}$

(d) $\dfrac{b-a}{2}$

50. By reversing the order of integration $\int_0^2 \int_{x^2}^{2x} f(x, y)\, dy\, dx$ may be represented as

(a) $\int_0^2 \int_{x^2}^{2x} f(x, y)\, dy\, dx$

(b) $\int_0^2 \int_y^{\sqrt{y}} f(x, y)\, dx\, dy$

(c) $\int_0^4 \int_{\frac{y}{2}}^{\sqrt{y}} f(x, y)\, dx\, dy$

(d) $\int_{x^2}^{2x} \int_0^2 f(x, y)\, dy\, dx$

51. $\lim\limits_{x \to \infty} \dfrac{n}{\sqrt{n^2 + n}} = ?$

(a) $\dfrac{1}{2}$

(b) 0

(c) ∞

(d) 1

52. $\lim\limits_{x \to \infty} (x-a)^{x-a} = ?$

(a) 0

(b) 1

(c) a

(d) ∞

53. The Taylor series expansion of $\sin x$ is

(a) $1 - \dfrac{x^2}{2!} + \dfrac{x^4}{4!} - \ldots\ldots\infty$

(b) $1 - \dfrac{x^2}{4!} + \dfrac{x^4}{6!} + \ldots\ldots\infty$

(c) $x - \dfrac{x^3}{3!} + \dfrac{x^5}{5!} - \ldots\ldots\infty$

(d) $x + \dfrac{x^3}{3!} + \dfrac{x^5}{5!} + \ldots\ldots\infty$

54. If $y = \int_0^{x^2} \sqrt{t}\, dt$, then $\dfrac{dy}{dx} = ?$

(a) $2x^2$

(b) $\sqrt{x}$

(c) 0

(d) 1

55. The curve $x^2 + y^2 = 3axy$ is

(a) Symmetrical about x-axis

(b) Symmetrical about y-axis

(c) Symmetrical about the line $y = x$

(d) Tangential to $x = y = \dfrac{a}{3}$

56. Area bounded by the curve $y = x^2$ and the lines $x = 4$ and $y = 0$ is

(a) 64

(b) $\dfrac{64}{3}$

(c) $\dfrac{128}{3}$

(d) $\dfrac{128}{4}$

57. By changing the order of integration, $\int_0^a \int_y^a \dfrac{x\, dx\, dy}{x^2 + y^2}$ becomes

(a) $\int_0^a \int_0^x \dfrac{x\, dy\, dx}{x^2 + y^2}$

(b) $\int_{-a}^a \int_0^x \dfrac{dy\, dx}{x^2 + y^2}$

(c) $\int_0^a \int_0^a \dfrac{x\, dy\, dx}{x^2 + y^2}$

(d) None of these

58. The linear approximation of the function $f(x) = e^{-x}$ around $x = 2$ is given by

(a) $(3-x)e^{-2}$

(b) $1 - x$

(c) $1 - x + x^2$

(d) $(2-x)e^2$

59. The value of $\int_{x=1}^2 \int_{y=3}^4 (xy + e^y)\, dy\, dx$ is

(a) $\dfrac{21}{4} + e^4 - e^3$

(b) $\dfrac{9}{4} - e^3 + e^4$

(c) e^3

(d) $\dfrac{1}{2}(e^3 - e^4)$

60. By changing the order of integration $\int\limits_{-a}^{a}\int\limits_{0}^{\sqrt{a^2-y^2}} g(x,y)\,dxdy$ becomes

(a) $\int\limits_{-a}^{a}\int\limits_{-\sqrt{a^2-x^2}}^{\sqrt{a^2-x^2}} g(x,y)\,dydx$

(b) $\int\limits_{0}^{a}\int\limits_{-\sqrt{a^2-x^2}}^{\sqrt{a^2-x^2}} g(x,y)\,dydx$

(c) $\int\limits_{0}^{a}\int\limits_{0}^{\sqrt{a^2-x^2}} g(x,y)\,dydx$

(d) None of these

61. The value of triple integral $\int\limits_{0}^{1}\int\limits_{0}^{x}\int\limits_{0}^{x+y} (x+y+z)\,dxdydz$

(a) $\dfrac{2}{7}$ (b) $\dfrac{3}{8}$

(c) $\dfrac{7}{8}$ (d) $\dfrac{5}{7}$

Answer key				
1. (d)	**2.** (a)	**3.** (b)	**4.** (d)	**5.** (a)
6. (b)	**7.** (b)	**8.** (c)	**9.** (d)	**10.** (d)
11. (b)	**12.** (d)	**13.** (b)	**14.** (c)	**15.** (a)
16. (b)	**17.** (a)	**18.** (b)	**19.** (a)	**20.** (b)
21. (d)	**22.** (b)	**23.** (a)	**24.** (a)	**25.** (c)
26. (a)	**27.** (c)	**28.** (c)	**29.** (a)	**30.** (b)
31. (c)	**32.** (d)	**33.** (a)	**34.** (d)	**35.** (d)
36. (c)	**37.** (b)	**38.** (d)	**39.** (a)	**40.** (a)
41. (b)	**42.** (c)	**43.** (c)	**44.** (a)	**45.** (b)
46. (b)	**47.** (a)	**48.** (b)	**49.** (c)	**50.** (c)
51. (d)	**52.** (b)	**53.** (c)	**54.** (a)	**55.** (c)
56. (b)	**57.** (a)	**58.** (a)	**59.** (a)	**60.** (b)
61. (c)				

Explanation

1. (d) Take $f(x)=Kx$. Then $f(1)=7\Rightarrow K=7$]

2. (a)

$f(0-0)=f(0)=f^2(0)-f^2(a)\Rightarrow f(a)=0$

$(\because f(0)=1)$

Now take $x=a$ and $y=x-a$ and then proceed]

3. (b)

$f(x)=\log_e x$ satisfies all the given conditions]

4. (d)

5. (a)

6. (b)

$\dfrac{1}{1-n^2}+\dfrac{2}{1-n^2}+\dots+\dfrac{n}{1-n^2}$

$=\dfrac{1+2+\dots+n}{1-n^2}=\dfrac{n(n+1)}{2(1-n^2)}=\dfrac{n}{2(1-n)}$

7. (b)

Put $x=1+t$ so that $x\to1\Rightarrow t\to0$]

8. (c)

9. (d)

10. (d)

$\lim\limits_{x\to0}\dfrac{\sqrt{1-\cos 2x}}{\sqrt{2}x}=\lim\limits_{x\to0}\dfrac{\sqrt{2\sin^2 x}}{\sqrt{2}x}=\lim\limits_{x\to0}\dfrac{|\sin x|}{x}$

Now show that $L.H.L\neq R.H.L$]

11. (b)

12. (d)

13. (b)

14. (c)

15. (a)

16. (b)

$f(x)=x-[x]=x-0$ for $0\le x<1$]

17. (a)

18. (b)

19. (a)

20. (b)

21. (d)

22. (b) Let r be the radius and A be the area of the circular region.

Then $A=\pi r^2\Rightarrow\dfrac{dA}{dt}=2\pi r\dfrac{dr}{dt}$.

Given that $\dfrac{dr}{dt} = 4\,cm/\sec$

Now calculate $\dfrac{dA}{dt}$ for $r = 10$]

23. (a)

24. (a)

25. (c)

26. (a)

27. (c)

28. (c) Let x, y be the length and breadth of the tank, respectively. Given height of the tank = 2 m. So, volume $(V) = 2xy = 8$ (given)

$$\therefore y = \dfrac{4}{x}$$

If C be the total cost, then

$$C = 70xy + 45 \times (2 \times 2x + 2 \times 2y)$$

$$= 70x \times \dfrac{4}{x} + 45 \left(4x + 4 \times \dfrac{4}{x} \right)]$$

30. (b)

$$\int_0^{\frac{\pi}{2}} \dfrac{\sin 2x}{\sin^4 x + \cos^4 x}\,dx = \int_0^{\frac{\pi}{2}} \dfrac{2\sin x \cos x\,dx}{\left(\sin^2 x\right)^2 + \left(1 - \sin^2 x\right)^2}$$

Now put $\sin^2 x = t$]

31. (c) $1 \le x \le 2 \Rightarrow 2 \le 2x \le 4$

$$\therefore [2x] = \begin{cases} 2, & 2 \le 2x < 3 \\ 3, & 3 \le 2x < 4 \end{cases}$$

$$= \begin{cases} 2, & 1 \le x < \dfrac{3}{2} \\ 3, & \dfrac{3}{2} \le x < 2 \end{cases}$$

$$\therefore \int_1^2 [2x]\,dx = \int_1^{\frac{3}{2}} 2\,dx + \int_{\frac{3}{2}}^2 3\,dx]$$

32. (d)

$$\int_{-\frac{\pi}{2}}^{\frac{\pi}{2}} \left(\sin|x| + \cos|x| \right)dx$$

$$= \int_{-\frac{\pi}{2}}^{0} \left(-\sin x + \cos x \right)dx + \int_0^{\frac{\pi}{2}} \left(\sin x + \cos x \right)dx]$$

33. (a)

34. (d)

35. (d)

$$A_n = \int_0^{\frac{\pi}{4}} \tan^n x\,dx = \int_0^{\frac{\pi}{4}} \tan^{n-2} x \times \tan^2 x\,dx$$

$$= \int_0^{\frac{\pi}{4}} \tan^{n-2} x \left(\sec^2 x - 1 \right)dx$$

$$= \int_0^{\frac{\pi}{4}} \tan^{n-2} x \sec^2 x\,dx - \int_0^{\frac{\pi}{4}} \tan^{n-2} x\,dx$$

$$= \int_0^1 t^{n-2}\,dt - A_{n-2}$$

(Putting $\tan x = t$ in the first integral so that $\sec^2 x\,dx = dt$ and $x = 0 \Rightarrow t = 0; x = \dfrac{\pi}{4} \Rightarrow t = 1$)

$$= \dfrac{1}{n-1} - A_{n-2}$$

$$\therefore A_n + A_{n-2} = \dfrac{1}{n-1} \quad (\text{for } n > 2)$$

Now replacing n by $n+1$, we get

$$A_{n+1} + A_{n-1} = \dfrac{1}{n}]$$

36. (c)

37. (b)

38. (d)

39. (a)

40. (a)

41. (b)

42. (c)

43. (c)

44. (a)

45. (b)

46. (b)

47. (a)

48. (b)

49. (c)

50. (c)

51. (d)

52. (b)

53. (c)

54. (a)

55. (c) The equation of the curve remains unaltered if we interchange x and y]

56. (b)

57. (a)

58. (a)

59. (a)

60. (b)

61. (c)

CHAPTER 3

VECTORS

3.1 BASIC CONCEPTS

3.1.1 Scalars and Vectors

A quantity which is characterized solely by its magnitude is called a scalar.

Example: length, height, temperature, etc.

A vector is a physical quantity which has a magnitude as well as definite direction.

Example: Speed, velocity, acceleration, etc.

If A and B be two points, then the directed line segment from A to B, will represent a vector and it will be denoted by $\overline{AB}$. In this case, the length of $\overline{AB}$ will be called the magnitude of the vector $\overline{AB}$ and is denoted by $\left|\overline{AB}\right|$.

Generally, a vector is represented by the symbols $\vec{\alpha}, \vec{\beta}, \vec{\gamma}$, etc.

3.1.2 Position Vector

If O be the origin, then $\overline{OA}, \overline{OB}$ are called the position vectors of the points A and B, respectively. Then $\overline{AB} = \overline{OB} - \overline{OA}$.

3.1.3 Equal Vectors

If two vectors $\vec{\alpha}$ and $\vec{\beta}$ have equal magnitude and same direction, then they are said to be equal.

3.1.4 Negative of a Vector

A vector whose magnitude is equal to that $\vec{\alpha}$ and whose direction is opposite to that of $\vec{\alpha}$, is called the negative of $\vec{\alpha}$ and is denoted by $-\vec{\alpha}$. Thus, $\overline{AB} = -\overline{BA}$.

3.1.5 Unit Vectors

Any vector with magnitude 1 is called an unit vector. It is denoted by the symbols $\hat{\alpha}, \hat{\beta}, \hat{\gamma}$, etc. $\hat{i}, \hat{j}, \hat{k}$ are unit vectors along the three mutually perpendicular axis, namely x-axis, y-axis, z-axis, respectively. Let $P(x, y, z)$ be any point in the space. The we use the symbol $\vec{r}$ to denote $\overline{OP}$; where $\overline{OP} = x\hat{i} + y\hat{j} + z\hat{k}$ Here $\left|\vec{r}\right| = \left|\overline{OP}\right|$ = distance between the points O and $P = \sqrt{x^2 + y^2 + z^2}$. We use the symbol r to denote $\left|\vec{r}\right|$.

Remember: An unit vector in the direction of $\vec{\alpha}$ is given by $\hat{\alpha} = \dfrac{\vec{\alpha}}{\left|\vec{\alpha}\right|}$.

Thus, if $\vec{\alpha} = a\hat{i} + b\hat{j} + c\hat{k}$, then $\left|\vec{\alpha}\right| = \sqrt{a^2 + b^2 + c^2}$ and so $\hat{\alpha} = \dfrac{a\hat{i} + b\hat{j} + c\hat{k}}{\sqrt{a^2 + b^2 + c^2}}$.

Remark: A vector whose magnitude is zero is called a null vector or a zero vector. It is denoted by $\vec{0}$. For a null vector its initial and terminal point coincide. Thus, a zero vector has no definite direction.

3.1.6 Sum and Difference of Two Vectors

If $\vec{\alpha} = a_1\hat{i} + b_1\hat{j} + c_1\hat{k}$, $\vec{\beta} = a_2\hat{i} + b_2\hat{j} + c_2\hat{k}$ then

(i) the sum of $\vec{\alpha}$ and $\vec{\beta}$ is denoted by $\vec{\alpha} + \vec{\beta}$ and is defined by $\vec{\alpha} + \vec{\beta} = (a_1 + a_2)\hat{i} + (b_1 + b_2)\hat{j} + (c_1 + c_2)\hat{k}$.

(ii) the difference of $\vec{\alpha}$ and $\vec{\beta}$ is denoted by $\vec{\alpha} - \vec{\beta}$ and is defined by $\vec{\alpha} - \vec{\beta} = (a_1 - a_2)\hat{i} + (b_1 - b_2)\hat{j} + (c_1 - c_2)\hat{k}$.

3.1.7 Triangle Law of Addition

If A, B, C are the vertices of a triangle, then $\overrightarrow{AB} + \overrightarrow{BC} + \overrightarrow{CA} = \vec{0}$, i.e; $\overrightarrow{AB} + \overrightarrow{BC} = \overrightarrow{AC}$.

The single vector $\overrightarrow{AC}$ whose effect is equal to the combined effect of two vectors $\overrightarrow{AB}$ and $\overrightarrow{BC}$, is called the resultant of the vectors $\overrightarrow{AB}$ and $\overrightarrow{BC}$.

3.1.8 Product of a Vector with a Scalar

The product of a vector $\vec{\alpha}$ and a real number "p" is denoted by $p\vec{\alpha}$ and is defined as the vector whose modulus is "λ" times that of $\vec{\alpha}$.

Thus, if $\vec{\alpha} = a_1\hat{i} + b_1\hat{j} + c_1\hat{k}$ and p be any scalar, then $p\vec{\alpha} = pa_1\hat{i} + pb_1\hat{j} + pc_1\hat{k}$.

3.1.9 Collinear Vectors

Two vectors $\vec{\alpha}$ and $\vec{\beta}$ are said to be parallel or collinear if $\vec{\alpha} = \lambda\vec{\beta}$ for some scalar λ.

Remark:

The necessary and sufficient condition for three vectors $\vec{\alpha}, \vec{\beta}, \vec{\gamma}$ to be collinear is that there exists three scalars x, y, z not all zero such that $x\vec{\alpha} + y\vec{\beta} + z\vec{\gamma} = \vec{0}$ where $x + y + z = 0$.

3.1.10 Coplanar Vectors

Three vectors $\vec{\alpha}, \vec{\beta}, \vec{\gamma}$ are said to be coplanar if they are parallel to some plane.

Remark:

(i) Suppose $\vec{\alpha}$ and $\vec{\beta}$ be two non-collinear vectors. Then every vector $\vec{\gamma}$ coplanar with $\vec{\alpha}$ and $\vec{\beta}$ can be expressed uniquely as $\vec{\gamma} = x\vec{\alpha} + y\vec{\beta}$ for some scalars x and y.

(ii) The necessary and sufficient condition for four vectors $\vec{\alpha}, \vec{\beta}, \vec{\gamma}, \vec{\delta}$ to be coplanar is that there exists four scalars x, y, z, t not all zero such that $x\vec{\alpha} + y\vec{\beta} + z\vec{\gamma} + t\vec{\delta} = \vec{0}$ where $x + y + z + t = 0$.

3.1.11 Section Formula

Let the position vectors of the points A and B be $\vec{\alpha}$ and $\vec{\beta}$, respectively, i.e; $\overrightarrow{OA} = \vec{\alpha}$ and $\overrightarrow{OB} = \vec{\beta}$. Also let $\vec{\gamma}$ be the position vector of the point C which divides the line segment joining A to B in the ratio $m: n$. Then $\vec{\gamma} = \dfrac{m\vec{\beta} + n\vec{\alpha}}{m + n}$.

3.2 PRODUCT OF VECTORS

3.2.1 Scalar Product (Dot Product)

The scalar product (dot product) of two vectors $\vec{a}$ and $\vec{b}$ is denoted by $\vec{a}.\vec{b}$ and is defined by $\vec{a}.\vec{b} = |\vec{a}||\vec{b}|\cos\theta$, where θ is the angle between $\vec{a}$ and $\vec{b}$.

If θ is acute then $\vec{a}.\vec{b} > 0$ and if θ is obtuse, then $\vec{a}.\vec{b} < 0$.

Remember:

(i) $\vec{a}.\vec{b} = \vec{b}.\vec{a}$ i.e; dot product is commutative.

(ii) $\theta = \cos^{-1}\dfrac{\vec{a}.\vec{b}}{|\vec{a}||\vec{b}|}$.

(iii) $\vec{a}.\vec{a} = |\vec{a}|^2$.

(iv) $(\vec{a}+\vec{b})^2 = (\vec{a}+\vec{b}).(\vec{a}+\vec{b}) = |\vec{a}|^2 + 2\vec{a}.\vec{b} + |\vec{b}|^2$

(v) $(\vec{a}-\vec{b})^2 = (\vec{a}-\vec{b}).(\vec{a}-\vec{b}) = |\vec{a}|^2 - 2\vec{a}.\vec{b} + |\vec{b}|^2$.

(vi) $(\vec{a}+\vec{b}).(\vec{a}-\vec{b}) = |\vec{a}|^2 - |\vec{b}|^2$.

(vii) $\vec{a}.(\vec{b}+\vec{c}) = (\vec{b}+\vec{c}).\vec{a} = \vec{a}.\vec{b} + \vec{a}.\vec{c}$.

(viii) If $\vec{a} = a_1\hat{i} + a_2\hat{j} + a_3\hat{k}$ and $\vec{b} = b_1\hat{i} + b_2\hat{j} + b_3\hat{k}$, then $\vec{a}.\vec{b} = a_1b_1 + a_2b_2 + a_3b_3$ and
$$\cos\theta = \dfrac{a_1b_1 + a_2b_2 + a_3b_3}{\sqrt{a_1^2 + a_2^2 + a_3^2}\sqrt{b_1^2 + b_2^2 + b_3^2}}.$$
Further we have, $\vec{a}$ is perpendicular to $\vec{b}$
$\Leftrightarrow \vec{a}.\vec{b} = 0$
$\Leftrightarrow a_1b_1 + a_2b_2 + a_3b_3 = 0$.

(ix) $\hat{i}.\hat{i} = |\hat{i}|^2 = 1, \hat{j}.\hat{j} = |\hat{j}|^2 = 1, \hat{k}.\hat{k} = |\hat{k}|^2 = 1,$
$\hat{i}.\hat{j} = \hat{j}.\hat{i} = 0, \hat{j}.\hat{k} = \hat{k}.\hat{j} = 0, \hat{k}.\hat{i} = \hat{i}.\hat{k} = 0$.

(x) If a constant force $\vec{F}$ acts on a particle and displaces it from point A to point B, then work done by the force $= \vec{F}.\overrightarrow{AB}$.

3.2.2 Vector Product (Cross Product)

The vector product (cross product) of two vectors $\vec{a}$ and $\vec{b}$ is denoted by $\vec{a}\times\vec{b}$ and is defined by $\vec{a}\times\vec{b} = |\vec{a}||\vec{b}|\sin\theta\hat{n}$, where θ is the angle between

$\vec{a}$ and $\vec{b}$ and $\hat{n}$ is a unit vector which is perpendicular to the plane containing the vectors $\vec{a}$ and $\vec{b}$.

Remember:

(i) $\vec{a} \times \vec{b} \neq \vec{b} \times \vec{a}$ (in general).

(ii) $\vec{a}$ and $\vec{b}$ are collinear (or parallel) $\Leftrightarrow$ $\vec{a} \times \vec{b} = \vec{0}$.

(iii) $\left(\vec{a} \times \vec{b}\right)^2 + \left(\vec{a}.\vec{b}\right)^2 = \left|\vec{a}\right|^2 \left|\vec{b}\right|^2$.

(iv) A vector of magnitude "d" unit which is perpendicular to both $\vec{a}$ and $\vec{b}$ is given by
$$d\frac{\left(\vec{a} \times \vec{b}\right)}{\left|\vec{a} \times \vec{b}\right|}.$$

(v) $\hat{i} \times \hat{i} = \hat{j} \times \hat{j} = \hat{k} \times \hat{k} = \vec{0}, \hat{i} \times \hat{j} = \hat{k}, \hat{j} \times \hat{i} = -\hat{k},$

$\hat{j} \times \hat{k} = \hat{i}, \hat{k} \times \hat{j} = -\hat{i}, \hat{k} \times \hat{i} = \hat{j}, \hat{i} \times \hat{k} = -\hat{j}$.

(vii) The area of the parallelogram whose two adjacent sides are $\vec{a}$ and $\vec{b}$ is $\left|\vec{a} \times \vec{b}\right|$.

(viii) The area of a triangle ABC is given by
$ar(\Delta ABC) = \frac{1}{2}\left|\overrightarrow{AB} \times \overrightarrow{AC}\right|$. and the vector area is given by $\frac{1}{2}\left(\overrightarrow{AB} \times \overrightarrow{AC}\right)$.

(ix) The moment of the force $\vec{F}$ acting at the point P about the point A is given by $\overrightarrow{AP} \times \vec{F}$

3.2.3 Scalar Triple Product

The scalar quantity $\vec{a}.\left(\vec{b} \times \vec{c}\right)$ is called the scalar triple product of three vectors $\vec{a}, \vec{b}, \vec{c}$ and is denoted by $\left[\vec{a}, \vec{b}, \vec{c}\right]$.

Remember:

(i) The scalar triple product of three vectors $\vec{a}, \vec{b}, \vec{c}$ is equal in magnitude to the volume of the parallelepiped whose three co-terminus edges are $\vec{a}, \vec{b}, \vec{c}$.

(ii) $\left[\vec{a}, \vec{b}, \vec{c}\right] = \left[\vec{b}, \vec{c}, \vec{a}\right] = \left[\vec{c}, \vec{a}, \vec{b}\right]$

(iii) If $\vec{a} = a_1\hat{i} + a_2\hat{j} + a_3\hat{k}$, $\vec{b} = b_1\hat{i} + b_2\hat{j} + b_3\hat{k}$ and $\vec{c} = c_1\hat{i} + c_2\hat{j} + c_3\hat{k}$, then
$$\left[\vec{a}, \vec{b}, \vec{c}\right] = \begin{vmatrix} a_1 & a_2 & a_3 \\ b_1 & b_2 & b_3 \\ c_1 & c_2 & c_3 \end{vmatrix}.$$

(iv) If three vectors $\vec{a}, \vec{b}, \vec{c}$ are coplanar, then $\left[\vec{a}, \vec{b}, \vec{c}\right] = 0$.

3.2.4 Vector Triple Product

The vector quantity $\left(\vec{a}.\vec{c}\right)\vec{b} - \left(\vec{a}.\vec{b}\right)\vec{c}$ is called the vector triple product of three vectors $\vec{a}, \vec{b}, \vec{c}$ and is denoted by $\vec{a} \times \left(\vec{b} \times \vec{c}\right)$.

3.3 VECTOR DIFFERENTIATION AND INTEGRATION

3.3.1 Derivative of a Vector Function

Let $\vec{f}(t)$ be a vector function of the scalar variable t, if $\lim_{\delta t \to 0} \frac{\vec{f}(t + \delta t) - \vec{f}(t)}{\delta t}$ exist, then this limit is called the derivative of $\vec{f}(t)$ w.r.t t and is denoted by $\vec{f}(t)$ or by $\frac{d\vec{f}(t)}{dt}$.

3.3.2 General Rules for Vector Differentiation

1. If the derivative of $\vec{f}(t) = f_1(t)\hat{i} + f_2(t)\hat{j} + f_3(t)\hat{k}$ exists, where t is a scalar variable, then
$$\frac{d\vec{f}}{dt} = \frac{df_1}{dt}\hat{i} + \frac{df_2}{dt}\hat{j} + \frac{df_3}{dt}\hat{k}.$$

2. If $\vec{u}(t)$ and $\vec{v}(t)$ be two differentiable vector functions of the scalar variable t, then
$$\frac{d}{dt}(\vec{u}(t) \pm \vec{v}(t)) = \frac{d\vec{u}}{dt} \pm \frac{d\vec{v}}{dt}.$$

3. If $\vec{u}(t)$ be a differentiable vector function of the scalar variable t and $\phi(t)$ be a differentiable scalar function of the scalar variable t, then
$$\frac{d(\phi\vec{u})}{dt} = \phi\frac{d\vec{u}}{dt} + \frac{d\phi}{dt}\vec{u}.$$

4. If $\vec{u}(t)$ and $\vec{v}(t)$ be two differentiable vector functions of the scalar variable t, then
$$\frac{d(\vec{u}.\vec{v})}{dt} = \vec{u}.\frac{d\vec{v}}{dt} + \vec{v}.\frac{d\vec{u}}{dt}.$$

Remark:

The necessary and sufficient condition for the vector $\vec{u}(t)$ to have a constant magnitude is
$$\vec{u}.\frac{d\vec{u}}{dt} = 0.$$

252 • ENGINEERING MATHEMATICS EXAM PREP

5. If $\vec{u}(t)$ and $\vec{v}(t)$ be two differentiable vector functions of the scalar variable t, then $\frac{d(\vec{u}\times\vec{v})}{dt}=\vec{u}\times\frac{d\vec{v}}{dt}+\frac{d\vec{u}}{dt}\times\vec{v}$.

Remark:

The necessary and sufficient condition for the vector $\vec{u}(t)$ to have a constant direction is $\vec{u}\times\frac{d\vec{u}}{dt}=\vec{0}$.

6. If $\vec{u}(t)$ is a differentiable vector function of the scalar variable t, then $\frac{d}{dt}(\vec{u}\times\frac{d\vec{u}}{dt})=\vec{u}\times\frac{d^2\vec{u}}{dt^2}$.

7. If $\vec{u},\vec{v},\vec{w}$ be three differentiable vector functions of the scalar variable t, then $\frac{d}{dt}[\vec{u},\vec{v},\vec{w}]=\left[\frac{d\vec{u}}{dt},\vec{v},\vec{w}\right]+\left[\vec{u},\frac{d\vec{v}}{dt},\vec{w}\right]+\left[\vec{u},\vec{v},\frac{d\vec{w}}{dt}\right]$.

3.3.3 Velocity and Acceleration

Let $\vec{r}$ be the position vector of a particle at $P(x, y, z)$ at time t. If $\vec{v}$ and $\vec{f}$ be the velocity and acceleration, respectively at the particle at time t, then $\vec{v}=\frac{d\vec{r}}{dt}$ and $\vec{f}=\frac{d\vec{v}}{dt}=\frac{d^2\vec{r}}{dt^2}$.

3.3.4 Vector Integration

Let $\vec{f}(t)$ be a vector function of a scalar variable t. If there exist a vector function $\vec{F}(t)$ and a constant vector function $\vec{C}$ such that $\frac{d}{dt}(\vec{F}(t)+\vec{C})=\vec{f}(t)$, then we write $\int\vec{f}(t)dt=\vec{F}(t)+\vec{C}$.

Remark:

If $\vec{u}(t)$ and $\vec{v}(t)$ be differentiable vector functions of the scalar variable t and $\phi(t)$ be a differentiable scalar function of the scalar variable t, then

(i) $\int(\vec{u}.\frac{d\vec{v}}{dt}+\frac{d\vec{u}}{dt}.\vec{v})dt=\vec{u}.\vec{v}+\vec{C}$

(ii) $\int(\vec{u}.\frac{d\vec{v}}{dt}+\frac{d\vec{u}}{dt}.\vec{v})dt=\vec{u}.\vec{v}+\vec{C}$

(iii) $\int(2\vec{v}.\frac{d\vec{v}}{dt})dt=\vec{v}^2+\vec{C}$

(iv) $\int(\vec{u}\times\frac{d\vec{v}}{dt}+\frac{d\vec{u}}{dt}\times\vec{v})dt=\vec{u}\times\vec{v}+\vec{C}$

(v) $\int(\vec{u}\times\frac{d^2\vec{u}}{dt^2})dt=\vec{u}\times\frac{d\vec{u}}{dt}+\vec{C}$

(vi) If $\vec{f}(t)=f_1(t)\hat{i}+f_2(t)\hat{j}+f_3(t)\hat{k}$,

then $\int\vec{f}(t)dt=\hat{i}\int f_1(t)dt+\hat{j}\int f_2(t)dt+\hat{k}\int f_3(t)dt$

(vii) $\int_{t_1}^{t_2}\vec{f}(t)\,dt=\vec{F}(t_2)-\vec{F}(t_1)$ if $\int\vec{f}(t)dt=\vec{F}(t)+\vec{C}$.

3.4 GRADIENT, DIVERGENCE AND CURL

3.4.1 Del Operator

The vector operator $\vec{\nabla}$ (read as del or nabla) defined by

$$\vec{\nabla}=\frac{\partial}{\partial x}\hat{i}+\frac{\partial}{\partial y}\hat{j}+\frac{\partial}{\partial z}\hat{k}.$$

3.4.2 Gradient of a Scalar Point Function

Let $\phi(x,y,z)$ be a continuously differentiable scalar point function. Then the gradient of ϕ is denoted by grad ϕ or $\vec{\nabla}\phi$ and is defined as

$grad\phi=\vec{\nabla}\phi=\frac{\partial\phi}{\partial x}\hat{i}+\frac{\partial\phi}{\partial y}\hat{j}+\frac{\partial\phi}{\partial z}\hat{k}.$

Remark:

1. The necessary and sufficient condition for a scalar point function $\phi(x,y,z)$ to be constant is that $\vec{\nabla}\phi=\vec{0}$.

2. $\vec{\nabla}(c\phi)=c\vec{\nabla}\phi$, for any constant "$c$" and for any scalar point function ϕ.

3.4.3 Divergence of a Vector Point Function

If $\vec{f}(x,y,z)=f_1(x,y,z)\hat{i}+f_2(x,y,z)\hat{j}+f_3(x,y,z)\hat{k}$ be a continuously differentiable vector point function, then the divergence of $\vec{f}$ is denoted by $div\vec{f}$ or $\vec{\nabla}.\vec{f}$ and is defined by

$$\vec{\nabla}.\vec{f}=div\vec{f}=\frac{\partial f_1}{\partial x}+\frac{\partial f_2}{\partial y}+\frac{\partial f_3}{\partial z}.$$

Remark:

1. If $\vec{f}$ is a constant vector then $div\vec{f}=0$.

2. $\vec{f}(x,y,z)$ is said to be *solenoidal* if $div\vec{f}=0$.

3.4.4 Curl of a Vector Point Function

If $\vec{f}(x,y,z)=f_1(x,y,z)\hat{i}+f_2(x,y,z)\hat{j}+f_3(x,y,z)\hat{k}$ be a continuously differentiable vector point function, then the curl of $\vec{f}$ is denoted by $curl\vec{f}$ or $\vec{\nabla}\times\vec{f}$ and is defined by

$$curl\vec{f}=\vec{\nabla}\times\vec{f}=\begin{vmatrix}\hat{i}&\hat{j}&\hat{k}\\\frac{\partial}{\partial x}&\frac{\partial}{\partial y}&\frac{\partial}{\partial z}\\f_1&f_2&f_3\end{vmatrix}$$

Remark:

1. If $\vec{f}$ is a constant vector then $curl\ \vec{f} = \vec{0}$

2. $\vec{f}(x, y, z)$ is said to be *irrotational* if $curl\ \vec{f} = \vec{0}$.

3.4.5 Vector Identities

(i) $grad\ (\phi + \psi) = grad\ \phi + grad\ \psi$ *i.e;*

$$\vec{\nabla}\ (\phi + \psi) = \vec{\nabla}\ \phi + \vec{\nabla}\ \psi.$$

(ii) $div\ (\vec{f} + \vec{g}) = div\ \vec{f} + div\ \vec{g}$ *i.e;*

$$\vec{\nabla}.(\vec{f} + \vec{g}) = \vec{\nabla}.\vec{f} + \vec{\nabla}.\vec{g}.$$

(iii) $curl\ (\vec{f} + \vec{g}) = curl\ \vec{f} + curl\ \vec{g}$ *i.e;*

$$\vec{\nabla}\times(\vec{f} + \vec{g}) = \vec{\nabla}\times\vec{f} + \vec{\nabla}\times\vec{g}.$$

(iv) $grad\ (\phi\psi) = \psi(grad\ \phi) + \phi(grad\ \psi)$

i.e; $\vec{\nabla}\ (\phi\psi) = \psi\vec{\nabla}\phi + \phi\ \vec{\nabla}\psi.$

(v) $div\ (\phi\vec{f}) = \phi div\vec{f} + \vec{f}.\ grad\phi$ *i.e;*

$$\vec{\nabla}.\ (\phi\vec{f}) = \phi\ (\vec{\nabla}.\vec{f}) + \vec{f}.\ (\vec{\nabla}\phi).$$

(vi) $curl\ (\phi\vec{f}) = \phi\ curl\ \vec{f} + grad\ \phi\times\vec{f}$ *i.e;*

$$\vec{\nabla}\times(\phi\vec{f}) = \phi\ (\vec{\nabla}\times\vec{f}) + (\vec{\nabla}\phi)\times\vec{f}$$

(vii) $div\ (\vec{f}\times\vec{g}) = \vec{g}.(curl\vec{f}) - \vec{f}.(curl\vec{g})$

i.e; $\vec{\nabla}.(\vec{f}\times\vec{g}) = \vec{g}.(\vec{\nabla}\times\vec{f}) - \vec{f}.(\vec{\nabla}\times\vec{g}).$

(viii) $curl(\vec{f}\times\vec{g}) = \vec{f}\ div\vec{g} - \vec{g}\ div\vec{f} - (\vec{g}.\vec{\nabla})\vec{f}$

$-(\vec{f}.\vec{\nabla})\vec{g}$

i.e; $\vec{\nabla}\times(\vec{f}\times\vec{g}) = \vec{f}\ (\vec{\nabla}.\vec{g}) - \vec{g}\ (\vec{\nabla}.\vec{f})$

$$- (\vec{g}.\vec{\nabla})\vec{f} - (\vec{f}.\vec{\nabla})\vec{g}$$

(ix) $grad(\vec{f}.\vec{g}) = \vec{f}\times curl\ \vec{g} - \vec{g}\times curl\ \vec{f} + (\vec{f}.\vec{\nabla})\vec{g}$

$+(\vec{g}.\vec{\nabla})\vec{f}$ *i.e;*

$\vec{\nabla}(\vec{f}.\vec{g}) = \vec{f}\times(\vec{\nabla}\times\vec{g}) + \vec{g}\times(\vec{\nabla}\times\vec{f}) + (\vec{f}.\vec{\nabla})\vec{g}$

$$- (\vec{g}.\vec{\nabla})\vec{f}$$

(x) $curl\ grad\ \phi = \vec{0}$ *i.e;* $\vec{\nabla}\times(\vec{\nabla}\phi) = \vec{0}$

(xi) $div\ curl\ \vec{f} = \vec{0}$ *i.e;* $\vec{\nabla}.(\vec{\nabla}\times\vec{f}) = \vec{0}$

(xii) $curl\ curl\ \vec{f} = grad\ div\ \vec{f} - \vec{\nabla}^2 f$

i.e; $\vec{\nabla}\times(\vec{\nabla}\times\vec{f}) = \vec{\nabla}(\vec{\nabla}.\vec{f}) - \vec{\nabla}^2\vec{f}$

(xiii) $grad\ div\ \vec{f} = curl\ curl\ \vec{f} + \vec{\nabla}^2\vec{f}$

i.e; $\vec{\nabla}(\vec{\nabla}.\vec{f}) = \vec{\nabla}\times(\vec{\nabla}\times\vec{f}) + \vec{\nabla}^2\vec{f}$

3.4.6 Directional Derivative

The directional derivative of a scalar function $\phi(x, y, z)$ at a point $P(x, y, z)$ in the direction of a unit vector $\hat{a}$ is given by $\vec{\nabla}\phi.\hat{a}$.

Remark:

(i) The directional derivative of a scalar point function $\phi(x, y, z)$ in the direction of x, y, z axes are $\dfrac{\partial\phi}{\partial x}$, $\dfrac{\partial\phi}{\partial y}$, $\dfrac{\partial\phi}{\partial z}$, respectively.

(ii) The maximum value of the directional derivative of ϕ is $\left|\vec{\nabla}\phi\right|$ and it is in the direction of $\vec{\nabla}\phi$.

3.5 LINE, SURFACE, AND VOLUME INTEGRALS

3.5.1 Line Integral

Let $\vec{F}(x, y, z)$ a vector function defined on a closed curve C. Then $\int_C \vec{F}.d\vec{r}$ is called the line integral of $\vec{F}$ along the curve C

Remember:

(i) Sometimes we use the symbol $\oint_C \vec{F}.d\vec{r}$ to denote the line integral of $\vec{F}$.

(ii) If the closed curve C is composed of a finite number of curves $C_1, C_2, C_3, ..., C_n$ joined end to end, then

$$\int_C \vec{F}.d\vec{r} = \int_{C_1} \vec{F}.d\vec{r} + \int_{C_2} \vec{F}.d\vec{r} + ... + \int_{C_n} \vec{F}.d\vec{r}$$

(iii) If $\vec{F} = f_1\ \hat{i} + f_2\ \hat{j} + f_3\ \hat{K}$, then

$$\int_C \vec{F}.d\vec{r} = \int_C \left(f_1 dx + f_2 dy + f_3 dz\right)$$

(iv) If the curve C is given by $x = x(t)$, $y = y(t)$, $z = z(t)$;

then $\int_C \vec{F}.d\vec{r} = \int_{t_1}^{t_2}\left[f_1(t)dx + f_2(t)dy + f_3(t)dz\right]$

$$= \int_{t_1}^{t_2}\left[f_1(t)\frac{dx}{dt} + f_2(t)\frac{dy}{dt} + f_3(t)\frac{dz}{dt}\right]dt$$

where $\vec{F} = \vec{F}(t) = f_1(t)\hat{i} + f_2(t)\hat{j} + f_3(t)\hat{k}$ and t_1, t_2

are the values of the parameter "t" at the two points of the curve.

Example:

(i) Let $\vec{F} = \vec{F}(t) = (x^2 + y^2)\hat{i} - 2xy\hat{j}$ and C be the rectangle in xy-Plane bounded by $y = 0$, $x = 1$, $y = 2$, $x = 0$. Suppose we need to find $\int_C \vec{F}.d\vec{r}$.

Here $\vec{F}.d\vec{r} = (x^2 + y^2)dx - 2xydy$.

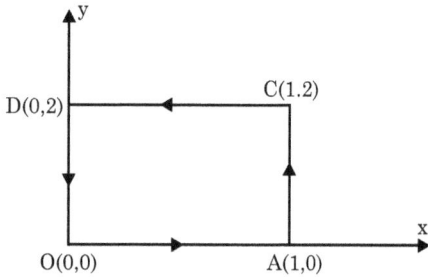

Along OA: $y = 0$ and $dy = 0$. x varies from 0 to 1.

$$\therefore \int_{OA} \vec{F}.d\vec{r} = \int_0^1 \left[(x^2 + 0)dx - 0\right] = \left[\frac{x^3}{3}\right]_0^1 = \frac{1}{3}.$$

Along AC: $x = 1$ and $dx = 0$. y varies from 0 to 2.

$$\therefore \int_{AC} \vec{F}.d\vec{r} = \int_0^2 \left[0 - 2ydy\right] = -2\left[\frac{y^2}{2}\right]_0^2 = -4.$$

Along CD: $y = 2$ and $dy = 0$. x varies from 1 to 0.

$$\therefore \int_{CD} \vec{F}.d\vec{r} = \int_1^0 \left[(x^2 + 4)dx - 0\right]$$

$$= \left[\frac{x^3}{3} + 4x\right]_1^0 = -\frac{1}{3} - 4 = -\frac{13}{3}.$$

Along DO: $x = 0$ and $dx = 0$. y varies from 2 to 0.

$$\therefore \int_{DO} \vec{F}.d\vec{r} = \int_2^0 \left[0 - 0dy\right] = 0.$$

Hence, $\int_C \vec{F}.d\vec{r}$

$$= \int_{OA} \vec{F}.d\vec{r} + \int_{AC} \vec{F}.d\vec{r} + \int_{CD} \vec{F}.d\vec{r} + \int_{DO} \vec{F}.d\vec{r}$$

$$= \frac{1}{3} - 4 - \frac{13}{3} + 0 = \frac{-24}{3} = -8.$$

(ii) Let $\vec{F} = y\hat{i} - x\hat{j}$ and C be the curve given by $x = t, y = t^2$. Let us find $\int_C \vec{F}.d\vec{r}$ from $(0,0)$ to $(1, 1)$.

Here $\vec{F}.d\vec{r}$

$$= ydx - xdy = t^2dt - t(2tdt) = -t^2dt.$$

$$[\because x = t \Rightarrow dx = dt \quad \text{and}$$

$$y = t^2 \Rightarrow dy = 2tdt]$$

As x varies from 0 to 1

So, $x = 0 \Rightarrow t = 0 \quad$ and $\quad x = 1 \Rightarrow t = 1.$

$$\therefore \int_C \vec{F}.d\vec{r} = \int_0^1 (-t^2)dt = -\left[\frac{t^3}{3}\right]_0^1 = -\frac{1}{3}.$$

3.5.2 Surface Integral

Let $\vec{F} = \vec{F}(x,y,z)$ be a vector function defined over a closed surface S. Then the surface integral of $\vec{F}$ over S is denoted by $\iint_S \vec{F}.\hat{n}\,ds$.

Case-I: Let R be the orthogonal projection of S on xy-plane. Then

$$\iint_S \vec{F}.\hat{n}\,ds = \iint_R \vec{F}.\hat{n}\,\frac{dxdy}{|\hat{n}.\hat{k}|}$$

Case-II: Let R be the orthogonal projection of S on yz-plane. Then

$$\iint_S \vec{F}.\hat{n}\,ds = \iint_R \vec{F}.\hat{n}\,\frac{dydz}{|\hat{n}.\hat{i}|}$$

Case-III: Let R be the orthogonal projection of S on xz-plane. Then

$$\iint_S \vec{F}.\hat{n}\,ds = \iint_R \vec{F}.\hat{n}\,\frac{dx\,dz}{|\hat{n}.\hat{j}|}$$

Remark:

Other forms of surface integrals are

$$\iint_S (\vec{\nabla} \times \vec{F}).\hat{n}\,ds, \int_S \vec{F} \times \hat{n}\,ds, \text{ etc.}$$

Example:

Let us find $\iint_S (\vec{\nabla} \times \vec{F}).\hat{n}\,ds$ where

$\vec{F} = \vec{F}(x,y,z) = yz\hat{i} + xz\hat{j} + xy\hat{k}$ and S is the surface of the sphere $x^2 + y^2 + z^2 = 1$ which lies in first octant.

Let $\phi(x,y,z) = x^2 + y^2 + z^2 - 1.$

Then $\vec{\nabla}\phi$

$$= \hat{i}\frac{\partial}{\partial x}\left(x^2 + y^2 + z^2 - 1\right) + \hat{j}\frac{\partial}{\partial y}\left(x^2 + y^2 + z^2 - 1\right)$$

$$+ \hat{k}\frac{\partial}{\partial z}\left(x^2 + y^2 + z^2 - 1\right)$$

$$= 2x\hat{i} + 2y\hat{j} + 2z\hat{k}.$$

$\therefore \hat{n}$

= Unit normal vector to the surface S

$$= \frac{\vec{\nabla}\phi}{\left|\vec{\nabla}\phi\right|} = \frac{2x\hat{i} + 2y\hat{j} + 2z\hat{k}}{\sqrt{(2x)^2 + (2y)^2 + (2z)^2}}$$

$$= \frac{x\hat{i} + y\hat{j} + z\hat{k}}{\sqrt{x^2 + y^2 + z^2}} = x\hat{i} + y\hat{j} + z\hat{k}.$$

So $\hat{F}.\hat{n} = \left(yz\hat{i} + xz\hat{j} + xy\hat{k}\right).\left(x\hat{i} + y\hat{j} + z\hat{k}\right)$

$$= xyz + xyz + xyz = 3xyz.$$

Also $\hat{n}.\hat{k} = \left(x\hat{i} + y\hat{j} + z\hat{k}\right).\hat{k} = z.$

Let us project the surface S on the xy-plane.

The projection of S on the xy-plane is a circle R whose equation will be $x^2 + y^2 = 1$.

Then the required surface integral

$$= \iint_R \vec{F}.\hat{n}\,\frac{dxdy}{\left|\hat{n}.\hat{k}\right|}$$

$$= \iint_R 3xyz\,\frac{dxdy}{\left|\hat{n}.\hat{k}\right|} = 3\iint_R xy\,dxdy$$

$$= 3\int_{x=0}^{1}\int_{y=0}^{\sqrt{1-x^2}} xy\,dy\,dx$$

$$= 3\int_{x=0}^{1} x\left[\frac{y^2}{2}\right]_0^{y=\sqrt{1-x^2}} dx = \frac{3}{2}\int_{x=0}^{1} x\left[(1-x^2) - 0\right]dx$$

$$= \frac{3}{2}\left[\frac{x^2}{2} - \frac{x^4}{4}\right]_0^1 = \frac{3}{2}\left[\frac{1}{2} - \frac{1}{4}\right] = \frac{3}{8}.$$

3.5.3 Volume Integral

Let $\vec{F}(\vec{r})$ be a continuous vector point function and let a volume V is enclosed by a smooth surface S. Then the volume integral of $\vec{F}(\vec{r})$ over V and is denoted by $\iiint_V \vec{F}\,dV$.

Remember:

The other forms of volume integral are $\iiint_V \vec{\nabla}\times\vec{F}\,dV$ and $\iiint_V \vec{\nabla}.\vec{F}\,dV$.

Example:

Let us find $\iiint_V \vec{\nabla}.\vec{F}\,dV$, taken over the rectangular parallelopiped $0 \le x \le 1, 0 \le y \le 2, \quad 0 \le z \le 3$ where $\vec{F} = (x^2 - yz)\hat{i} + (y^2 - xz)\hat{j} + (z^2 - xy)\hat{k}$.

Here, $\vec{\nabla}.\vec{F}$

$$= \frac{\partial}{\partial x}(x^2 - yz) + \frac{\partial}{\partial y}(y^2 - xz) + \frac{\partial}{\partial z}(z^2 - xy)$$

$$= 2x + 2y + 2z = 2(x + y + z)$$

$$\therefore \iiint_V \vec{\nabla}.\vec{F}\,dV$$

$$= \iiint_V 2(x + y + z)\,dV$$

$$= 2\int_{z=0}^{3}\int_{y=0}^{2}\int_{x=0}^{1}(x + y + z)\,dx\,dy\,dz$$

$$= 2\int_{z=0}^{3}\int_{y=0}^{2}\left[\frac{x^2}{2} + xy + xz\right]_{x=0}^{1} dy\,dz$$

$$= 2\int_{z=0}^{3}\int_{y=0}^{2}\left[\frac{1}{2} + y + z\right]dy\,dz$$

$$= 2\int_{z=0}^{3}\left[\frac{1}{2}y + \frac{y^2}{2} + zy\right]_{y=0}^{2} dz$$

$$= 2\int_{z=0}^{3}\left[\frac{1}{2}\times 2 + \frac{2^2}{2} + 2z\right]dz$$

$$= 2\int_{z=0}^{3}\left[1 + 2 + 2z\right]dz = 2\left[3z + z^2\right]_{z=0}^{3}$$

$$= 2\left[3\times 3 + 3^2\right] = 36.$$

3.6 GREEN'S, STOKES', AND GAUSS DIVERGENCE THEOREM

3.6.1 Greens Theorem (in a Plane)

If $F_1(x, y)$ and $F_2(x, y)$ be two continuous functions of x and y having continuous partial derivatives $\frac{\partial F_1}{\partial y}$ and $\frac{\partial F_2}{\partial x}$ in a region R on xy plane bounded by a closed curve C, then

$$\oint_C (F_1\,dx + F_2\,dy) = \iint_R \left(\frac{\partial F_2}{\partial x} - \frac{\partial F_1}{\partial y}\right)dx\,dy, \quad \text{where}$$

the closed curve is traversed in a counterclockwise direction.

Remember:

By Green's theorem, we can transform the line integral into the double integral and vice versa.

Application:

Let us find the value of the integral

$\oint_C\left[\left(2xy^3+y\right)dx+\left(3x^2y^2+2x\right)dy\right]$ where C is the circle $x^2+y^2=1$ taken in the counterclockwise direction.

$\oint_C\left[\left(2xy^3+y\right)dx+\left(3x^2y^2+2x\right)dy\right]$

$=\iint_R\left[\frac{\partial}{\partial x}\left(3x^2y^2+2x\right)-\frac{\partial}{\partial y}\left(2xy^3+y\right)\right]dx\,dy$

(by Green's theorem)

$=\iint_R\left[\left(6xy^2+2\right)-\left(6xy^2+1\right)\right]dx\,dy$

$=\iint_R dx\,dy$

$=$ area of the circle $x^2+y^2=1$

$=\pi\times1^2=\pi.$

3.6.2 Stokes' Theorem

Let S be a surface having the closed curve C as its boundary and $\overline{F}$ be any vector point function having continuous first order partial derivatives. Then $\oint_C\overline{F}.d\vec{r}=\iint_S curl\overline{F}.\hat{n}\,dS$, where $\hat{n}$ is a unit outward normal drawn at any point of S.

Remember:

(i) By Stokes' theorem, we can transform the line integral into the surface integral and vice versa.

(ii) Green's theorem in a plane is a special case of Stokes' theorem.

Application:

Let us find the value of $\oint_C\overline{F}.d\vec{r}$ where $\overline{F}=y^2\hat{i}+x^2\hat{j}-2(x+z)\hat{k}$ and C is the boundary of the triangle with vertices at $(0,0,0),(1,0,0)$ and $(1,1,0)$.

Then clearly the projection of the triangle formed by the points $(0,0,0),(1,0,0)$ and $(1,1,0)$ in xy plane is a triangle formed by the points $(0,0)$, $(1,0)$ and $(1,1)$.

Here $curl\overline{F}$

$=\begin{vmatrix}\hat{i}&\hat{j}&\hat{k}\\\frac{\partial}{\partial x}&\frac{\partial}{\partial y}&\frac{\partial}{\partial z}\\y^2&x^2&-2(x+z)\end{vmatrix}$

$=\hat{i}\left\{\frac{\partial}{\partial y}(-2(x+z))-\frac{\partial}{\partial z}(x^2)\right\}$

$-\hat{j}\left\{\frac{\partial}{\partial x}(-2(x+z))-\frac{\partial}{\partial z}(y^2)\right\}$

$+\hat{k}\left\{\frac{\partial}{\partial x}(x^2)-\frac{\partial}{\partial y}(y^2)\right\}$

$=2\hat{j}+2(x-y)\hat{k}$

$\therefore curl\overline{F}.\hat{n}$

$=\left\{2\hat{j}+2(x-y)\hat{k}\right\}.\hat{k}$

$\left(\because\hat{n}=\text{unit vector perpendicular to }xy\text{ plane}\atop=\hat{k}\right)$

$=2(x-y)\ (\because\hat{j}.\hat{k}=0,\hat{k}.\hat{k}=1)$

Hence, by Stokes' theorem,

$\oint_C\overline{F}.d\vec{r}$

$=\iint_S curl\overline{F}.\hat{n}\,dS$

$=\int_{x=0}^1\int_{y=0}^x 2(x-y)dy\,dx=\int_{x=0}^1\left[2\left(xy-\frac{y^2}{2}\right)\right]_{y=0}^x dx$

$=\int_{x=0}^1\left[2\left(x^2-\frac{x^2}{2}\right)-0\right]dx=\left[\frac{x^3}{3}\right]_0^1=\frac{1}{3}.$

3.6.3 Gauss Divergence Theorem

Let $\overline{F}=\overline{F}(x,y,z)$ be a vector point function having continuous first-order partial derivatives in the volume V bounded by a closed surface S. Then $\iiint_V\vec{\nabla}.\overline{F}\,dV=\iint_S\overline{F}.\hat{n}\,dS$, where $\hat{n}$ is a unit outward normal drawn at any point of S.

Remember:

(i) By the Gauss Divergence theorem, we can transform the volume integral into surface integral and vice versa.

(ii) If $\overline{F}=F_1\hat{i}+F_2\hat{j}+F_3\hat{k}$, then according to the Gauss Divergence theorem we have,

$$\iiint_V \left(\frac{\partial F_1}{\partial x} + \frac{\partial F_2}{\partial y} + \frac{\partial F_3}{\partial z} \right) dx\, dy\, dz$$

$$= \iint_S \left(F_1\, dy\, dz + F_2\, dx\, dz + F_3\, dx\, dy \right).$$

Application:

If $\vec{F} = x\hat{i} + 2y\hat{j} + 3z\hat{k}$ and S be the surface of the sphere $x^2 + y^2 + z^2 = 1$, then

$$\iint_S \vec{F}.\hat{n}\, dS$$

$$= \iiint_V \vec{\nabla}.\vec{F}\, dV$$

$$= \iiint_V (1+2+3)\, dV$$

$$\left(\begin{array}{l} \because \vec{\nabla}.\vec{F} \\ = \left(\hat{i}\dfrac{\partial}{\partial x} + \hat{j}\dfrac{\partial}{\partial y} + \hat{k}\dfrac{\partial}{\partial z} \right).\left(x\hat{i} + 2y\hat{j} + 3z\hat{k} \right) \\ = \dfrac{\partial}{\partial x}(x) + \dfrac{\partial}{\partial y}(2y) + \dfrac{\partial}{\partial z}(3z) = 1+2+3 \end{array} \right)$$

$$= 6\iiint_V dV$$

$$= 6 \times \text{volume of the sphere } x^2 + y^2 + z^2 = 1$$

$$= 6 \times \frac{4}{3}\pi \times 1^2 = 8\pi.$$

Fully Solved MCQs

1. If $\vec{\alpha}, \vec{\beta}, \vec{\gamma}$ are three vectors such that $\vec{\alpha} + \vec{\beta} + \vec{\gamma} = \vec{0}$, $\left|\vec{\alpha}\right| = 1, \left|\vec{\beta}\right| = 2, \left|\vec{\gamma}\right| = 3$ then $\vec{\alpha}.\vec{\beta} + \vec{\beta}.\vec{\gamma} + \vec{\gamma}.\vec{\alpha} = ?$

(a) –14   (b) –7   (c) 6   (d) –6

2. A unit vector which is perpendicular to both $\vec{\alpha} = \hat{i} + \hat{j} + \hat{k}$ and $\vec{\beta} = 2\hat{i} - \hat{j} + \hat{k}$ is

(a) $\dfrac{1}{\sqrt{6}}\left(\hat{i} - \hat{j} + 2\hat{k} \right)$ (b) $\dfrac{1}{\sqrt{14}}\left(2\hat{i} + \hat{j} - 3\hat{k} \right)$

(c) $\dfrac{1}{\sqrt{14}}\left(\hat{i} + \hat{j} - 2\hat{k} \right)$ (d) None of these

3. If $\left|\vec{\alpha}\right| = 5, \left|\vec{\beta}\right| = 1, \vec{\alpha}.\vec{\beta} = 3$, then $\left|\vec{\alpha} \times \vec{\beta}\right| = ?$

(a) 4   (b) 3   (c) 6   (d) 7

4. A vector of magnitude 5 unit which is perpendicular to both $\vec{\alpha} = 2\hat{i} + \hat{j} - 3\hat{k}$ and $\vec{\beta} = \hat{i} - 2\hat{j} + \hat{k}$ is

(a) $-\dfrac{5}{\sqrt{3}}\left(\hat{i} + \hat{j} + \hat{k} \right)$ (b) $\dfrac{5}{3}\left(\hat{i} + \hat{j} + \hat{k} \right)$

(c) $-\dfrac{5}{3}\left(\hat{i} + \hat{j} + \hat{k} \right)$ (d) None of these

5. The angle between the vectors $\hat{i} - 2\hat{j} - 2\hat{k}$ and $2\hat{i} + \hat{j} - 2\hat{k}$ is

(a) $\cos^{-1}\dfrac{2}{9}$ (b) $\cos^{-1}\dfrac{4}{9}$

(c) $\cos^{-1}\dfrac{4}{3}$ (d) $\cos^{-1}\dfrac{9}{2}$

6. The vectors $\vec{\alpha} = 2\hat{i} + \hat{j} + a\hat{k}$ and $\vec{\beta} = 3\hat{i} + b\hat{j} - \hat{k}$ are mutually perpendicular. If $\left|\vec{\alpha}\right| = \left|\vec{\beta}\right|$, then The value of a and b are:

(a) $\dfrac{41}{12}, -\dfrac{31}{12}$ (b) $\dfrac{31}{12}, -\dfrac{41}{12}$

(c) $-\dfrac{31}{12}, \dfrac{41}{12}$ (d) $-\dfrac{13}{12}, \dfrac{41}{12}$

7. The value of λ for which the vectors $\vec{\alpha} = \lambda\hat{i} - 4\hat{j} + 3\hat{k}$ and $\vec{\beta} = 3\hat{i} + \lambda\hat{j} - 2\hat{k}$ are perpendicular, is

(a) 8   (b) –6   (c) 7   (d) 4

8. The area of ΔABC, the position vectors of whose vectors are $\hat{i} + \hat{j} - \hat{k}$, $\hat{i} + \hat{j} + 2\hat{k}$ and $3\hat{i} - \hat{j} - \hat{k}$ is

(a) $\dfrac{1}{2}\sqrt{21}$ (b) $3\sqrt{2}$ (c) $5\sqrt{3}$ (d) 5

9. If $\vec{\alpha}, \vec{\beta}, \vec{\gamma}$ are three vectors such that $\vec{\alpha} + \vec{\beta} + \vec{\gamma} = \vec{0}$ and $\left|\vec{\alpha}\right| = 3, \left|\vec{\beta}\right| = 5$ and $\left|\vec{\gamma}\right| = 7$;then the angle between $\vec{\alpha}$ and $\vec{\beta}$ is

(a) 30°   (b) 60°   (c) 45°   (d) 90°

10. The value of $\left(\vec{r}.\hat{i}\right)\hat{i} + \left(\vec{r}.\hat{j}\right)\hat{j} + \left(\vec{r}.\hat{k}\right)\hat{k}$ is

(a) $\vec{r}$ (b) $\dfrac{\vec{r}}{r}$ (c) $r\vec{r}$ (d) $\dfrac{1}{r^2}\vec{r}$

11. The value of $\hat{i}.\left(\hat{j} \times \hat{k}\right) + \hat{j}.\left(\hat{k} \times \hat{i}\right) + \hat{k}.\left(\hat{i} \times \hat{j}\right)$ is

(a) 1   (b) 2   (c) 3   (d) 0

12. $\left[\vec{a}-\vec{b},\vec{b}-\vec{c},\vec{c}-\vec{a}\right]=$?

(a) $\left[\vec{a},\vec{b},\vec{c}\right]$ (b) 1   (c) 0   (d) 2

13. $\dfrac{d}{dt}\left\{\vec{r}.\left(\dfrac{d\vec{r}}{dt}\times\dfrac{d^2\vec{r}}{dt^2}\right)\right\}=$?

(a) $\dfrac{d\vec{r}}{dt}$ (b) $\dfrac{d^2\vec{r}}{dt^2}$ (c) $\left[\vec{r},\dfrac{d\vec{r}}{dt},\dfrac{d^3\vec{r}}{dt^3}\right]$

(d) None of these

14. If, $\vec{r}=\sin t\hat{i}+\cos t\hat{j}+t\hat{k}$, then the value of $\left|\dfrac{d^2\vec{r}}{dt^2}\right|$ is

(a) 0   (b) 3   (c) 2   (d) 1

15. If $\vec{\alpha}=t^2\hat{i}-t\hat{j}+(2t+1)\hat{k}$ and $\vec{\beta}=(2t-3)\hat{i}+\hat{j}-t\hat{k}$, then at $t=2$, $\dfrac{d}{dt}\left(\vec{\alpha}\times\dfrac{d\vec{\beta}}{dt}\right)$ is

(a) $\hat{i}+\hat{j}+\hat{k}$ (b) $\hat{i}+4\hat{j}+2\hat{k}$

(c) $\hat{i}+8\hat{j}+2\hat{k}$ (d) $4\hat{i}+2\hat{j}+\hat{k}$

16. A particle moves along the curve $x=t^3+1$, $y=t^2$, $z=2t+5$. Then the velocity and acceleration at $t=1$ are, respectively

(a) $3\hat{i}+2\hat{j}+2\hat{k}, 6\hat{i}+2\hat{j}$

(b) $\hat{i}+\hat{j}+2\hat{k}, 4\hat{i}+2\hat{j}$

(c) $3\hat{i}+\hat{j}+2\hat{k}, \hat{i}+2\hat{j}$

(d) None of these

17. If $\vec{\alpha}=2t\hat{i}-t^2\hat{j}+t^3\hat{k}$ and $\vec{\beta}=-t\hat{i}+t^2\hat{k}$ and $\vec{\gamma}=t^3\hat{j}-2t\hat{k}$, then $\dfrac{d}{dt}\left[\vec{\alpha}\,\vec{\beta}\,\vec{\gamma}\right]$ at $t=1$ is

(a) 11   (b) –11

(c) 3   (d) –3

18. If $\vec{f}(t)=2\cos t\hat{i}+2\sin t\hat{j}+2t\tan\theta\hat{k}$, then the value of $\left|\dfrac{d\vec{f}}{dt}\times\dfrac{d^2\vec{f}}{dt^2}\right|$ is

(a) $4\sec\theta$ (b) $4\tan\theta$

(c) $2\sec\theta$ (d) $2\tan\theta$

19. If $\vec{f}(t)=2t^2\hat{i}+t\hat{j}+3t^3\hat{k}$, then $\displaystyle\int_1^2\vec{f}\times\dfrac{d^2\vec{f}}{dt^2}dt=$?

(a) $-42\hat{i}+90\hat{j}-6\hat{k}$ (b) $40\hat{i}+85\hat{j}-5\hat{k}$

(c) $42\hat{i}-80\hat{j}+9\hat{k}$ (d) None of these

20. The maximum value of the directional derivative of $\phi(x,y,z)=x^2-y^2+2z^2$ at $(1,2,3)$ is

(a) $\sqrt{41}$ (b) $2\sqrt{41}$ (c) $\sqrt{33}$ (d) $\sqrt{46}$

21. If $\vec{r}=x\hat{i}+y\hat{j}+z\hat{k}$ and $r=|\vec{r}|$; then

(i) $\vec{\nabla}\left(\dfrac{1}{r}\right)=$?

(a) $\dfrac{\vec{r}}{r^2}$ (b) $-\dfrac{\vec{r}}{r^2}$ (c) $-\dfrac{\vec{r}}{r^3}$ (d) $\dfrac{\vec{r}}{r^3}$

(ii) $\vec{\nabla}\left(r^n\right)=$?

(a) $nr^{n-2}\vec{r}$ (b) $nr^{n-1}\vec{r}$

(c) $n(n+1)\vec{r}$ (d) $n(n-1)\vec{r}$

22. If $\vec{\nabla}\phi=(y^2-2xyz^3)\hat{i}+(3+2xy-x^2z^3)\hat{j}+(6z^3-3x^2yz^2)\hat{k}$, then the value of $\phi=$?

(a) $3y+xy^2-x^2yz^3+\dfrac{3z^4}{2}$

(b) $3y-2x^2y-x^2yz^2+\dfrac{3z^4}{2}$

(c) $3y+x^2y+x^2yz^2+\dfrac{3z^4}{2}$

(d) None of these

23. If $\vec{f}(x,y,z)=x^2y\hat{i}-2xz\hat{j}+2yz\hat{k}$, then $curl(curl\vec{f})$ is

(a) $x\hat{i}+y\hat{j}$ (b) $2(x+1)\hat{j}$

(c) $z\hat{k}$ (d) None of these

24. The value of "λ" so that the vector function $\vec{f}(x,y,z)=(x+3y)\hat{i}+(y-2z)\hat{j}+(x+\lambda z)\hat{k}$ is solenoidal

(a) 2   (b) –2   (c) 1   (d) –1

25. If $\vec{f}=grad(x^3+y^3+z^3-3xyz)$, then $\vec{f}$ is

(a) Solenoidal   (b) Irrotational

(c) Both irrotational and solenoidal

(d) None of these

26. If $\phi(x,y,z)=x^2y+2xy+z^2$, then $curl\ grad\ \phi$ is

(a) $x^2\hat{i}+y^2\hat{j}+z^2\hat{k}$ (b) $x\hat{i}+2\hat{j}+z\hat{k}$

(c) $\vec{0}$ (d) $2xy\hat{i}+2z\hat{k}$

27. If $\vec{r}=x\hat{i}+y\hat{j}+z\hat{k}$, then $\vec{\nabla}.\vec{r}=$?

(a) 3   (b) 2   (c) 1   (d) 0

28. The unit normal to the surface $x^2y + 2xz = 4$ at the point $(2, -2, 3)$ is

(a) $\hat{i} + 2\hat{j} + 2\hat{k}$
(b) $-\dfrac{1}{3}\left(\hat{i} - 2\hat{j} - 2\hat{k}\right)$

(c) $\dfrac{1}{3}\left(\hat{i} + 2\hat{j} + 2\hat{k}\right)$
(d) $\hat{i} - 2\hat{j} - 2\hat{k}$

29. The directional derivative of $\phi\,(x, y, z) = xy + yz + zx$ in the direction $\vec{\alpha} = \hat{i} + 2\hat{j} + 2\hat{k}$ at the point $(1, 2, 0)$ is

(a) $\dfrac{12}{\sqrt{14}}$
(b) $\dfrac{1}{3}$
(c) $\dfrac{9}{\sqrt{14}}$
(d) $\dfrac{10}{3}$

30. If $\vec{f}(x, y, z) = (mxy - z^3)\hat{i} + (m-2)x^2\hat{j} + (1-m)xz^2\hat{k}$ is irrotational, then the value of m

(a) 4
(b) -4
(c) 0
(d) 2

31. If $\vec{r} = x\hat{i} + y\hat{j} + z\hat{k}$ and $r = |\vec{r}| = \sqrt{x^2 + y^2 + z^2}$; then $\vec{\nabla}e^{r^2} = ?$

(a) $e^{r^2}\vec{r}$
(b) $2e^{r^2}\vec{r}$

(c) $\vec{r}$
(d) $e^{r^2}\left(\vec{r} \times \vec{r}\right)$

32. If $\vec{u} = e^{xyz}\left(\hat{i} + \hat{j} + \hat{k}\right)$, then $curl\,\vec{u}$ at $(1, 1, 1)$ is

(a) $(\hat{i} + \hat{j} + \hat{k})\text{e}$
(b) $\hat{i} + \hat{j} + \hat{k}$

(c) $\vec{0}$
(d) None of these

33. If $\vec{f} = (x + 2y + az)\hat{i} + (bx - 3y - z)\hat{j} + (4x + cy + 2z)\hat{k}$ is irrotational, then

(a) $a = 4, b = 2, c = -1$
(b) $a = 2, b = 3, c = -1$
(c) $a = 3, b = 2, c = 1$
(d) $a = 3, b = 2, c = -1$

34. If $\vec{f} = \left(xyz\right)^b\left(x^a\hat{i} + y^a\hat{j} + z^a\hat{k}\right)$ is irrotational, then

(a) $a = -1$ or $b = 0$
(b) $a = 1$ or $b = 0$
(c) $a = 1$ or $b = 0$
(d) $a = -1$ or $b = 2$

35. If $\vec{F} = (3 + 2xy)\hat{i} + (x^2 - 3y^2)\hat{j}$ and let C be the curve $\vec{r}(t) = e^t \sin t\,\hat{i} + e^t \cos t\,\hat{j}$, where $0 \le t \le \pi$, then $\int_C \vec{F}.d\vec{r} = ?$

(a) $e^{3\pi} + 1$
(b) $e^{2\pi} - 1$

(c) $-e^{3\pi} + 1$
(d) 0

36. Consider the vector field $\vec{F} = r^\beta(y\hat{i} - x\hat{j})$, where β is a real number, $\vec{r} = x\hat{i} + y\hat{j}$ and $r = |\vec{r}|$.

If the absolute value of the line integral $\int_C \vec{F}.d\vec{r}$ along the closed curve $C : x^2 + y^2 = a^2$ (oriented counterclockwise) is 2π, then β is

(a) 3
(b) 1
(c) 6
(d) -2

37. The value of $\iint_S \left(\vec{\nabla} \times \vec{F}\right).\hat{n}\,dS$ where $\vec{F} = \left(x^2 + y^2\right)\hat{i} - 2xy\,\hat{j}$ and S represents the surface bounded by the lines $x = \pm 1, y = 0$ and $y = 3$, is

(a) 20
(b) -36
(c) 32
(d) 24

38. The value of the integral $\oint_C\left[\left(x^2 - \cosh y\right)dx + (y + \sin x)dy\right]$, where C is the rectangle with vertices $(0,0), (\pi, 0), (\pi, 1)$ and $(0, 1)$ is

(a) $\pi(\cos h\,1 - 1)$
(b) $\pi(\sin 1 - 1)$
(c) $\pi \cosh 1$
(d) $\pi \sin h\,1$

39. The value of the integral $\oint_C\left[(x + y)dx + x^2dy\right]$, where C is the triangle with vertices $(0,0), (2,0)$ and $(2, 4)$ is

(a) $\dfrac{1}{4}$
(b) $\dfrac{3}{10}$
(c) $\dfrac{20}{3}$
(d) $\dfrac{5}{21}$

40. Let C be the boundary of the region $R = \left\{(x, y) : -1 \le y \le 1, 0 \le x \le 1 - y^2\right\}$ oriented in the counterclockwise direction. Then the value of the integral $\oint_C\left[ydx + 2xdy\right]$ is

(a) $\dfrac{3}{4}$
(b) $\dfrac{4}{3}$
(c) $\dfrac{1}{3}$
(d) $\dfrac{3}{2}$

41. The value of the integral $\oint_C\left[(\cos x \sin y - xy)dx + \sin x \cos y\,dy\right]$, where C is the circle $x^2 + y^2 = 1$, is

(a) 2
(b) 0
(c) 1
(d) 3

42. The value of $\oint_C \vec{F}.d\vec{r}$ where

$\vec{F} = (2x - y)\hat{i} - yz^2\,\hat{j} - y^2z\,\hat{k}$ and S is the upper half surface of the sphere $x^2 + y^2 + z^2 = 1$ with C as its boundary, is

(a) π (b) 2π (c) 3π (d) $\dfrac{\pi}{2}$

43. Let $\vec{F}$ be a vector field given by $\vec{F}(x,y,z) = -y\hat{i} + 2xy\,\hat{j} + z^3\,\hat{k}$. If C is the curve of intersection of the surfaces $x^2 + y^2 = 1$ and $y + z = 2$, then which of the following is equal to $\left|\oint_C \vec{F}.d\vec{r}\right|$?

(a) $\displaystyle\int_0^{2\pi}\left(\dfrac{1}{2} + \dfrac{1}{2}\sin\theta\right)d\theta$ (b) $\displaystyle\int_0^{2\pi}(1 + 2\sin\theta)d\theta$

(c) $\displaystyle\int_0^{2\pi}\left(\dfrac{1}{2} + \dfrac{2}{3}\sin\theta\right)d\theta$ (d) $\displaystyle\int_0^{2\pi}\left(\dfrac{1}{2} + \dfrac{1}{3}\sin\theta\right)d\theta$

44. The value of $\displaystyle\iint_S \vec{F}.\hat{n}\,dS$ where $\vec{F} = 4xz\hat{i} - y^2\hat{j} + yz\,\hat{k}$ and S represents the surface of the cube bounded by $x = 0, x = 1, y = 0, y = 1, z = 0$ and $z = 1$, is

(a) $\dfrac{3}{2}$ (b) $\dfrac{1}{2}$ (c) 1 (d) 0

Answer Key				
1. (b)	**2.** (b)	**3.** (a)	**4.** (a)	**5.** (b)
6. (a)	**7.** (b)	**8.** (b)	**9.** (b)	**10.** (a)
11. (c)	**12.** (c)	**13.** (c)	**14.** (d)	**15.** (c)
16. (a)	**17.** (b)	**18.** (a)	**19.** (a)	**20.** (b)
21. (i)-(c); (ii)-(a).	**22.** (a)	**23.** (b)	**24.** (b)
25. (b)	**26.** (c)	**27.** (a)	**28.** (b)	**29.** (d)
30. (a)	**31.** (b)	**32.** (c)	**33.** (a)	**34.** (a)
35. (a)	**36.** (d)	**37.** (b)	**38.** (a)	**39.** (c)
40. (b)	**41.** (b)	**42.** (a)	**43.** (c)	**44.** (a)

Explanation

1. (b)

$\vec{\alpha} + \vec{\beta} + \vec{\gamma} = \vec{0} \Rightarrow (\vec{\alpha} + \vec{\beta} + \vec{\gamma})^2 = \vec{0}^2$

$\Rightarrow |\vec{\alpha}|^2 + |\vec{\beta}|^2 + |\vec{\gamma}|^2 + 2(\vec{\alpha}.\vec{\beta} + \vec{\beta}.\vec{\gamma} + \vec{\gamma}.\vec{\alpha}) = 0$

$\Rightarrow 1^2 + 2^2 + 3^2 + 2(\vec{\alpha}.\vec{\beta} + \vec{\beta}.\vec{\gamma} + \vec{\gamma}.\vec{\alpha}) = 0$

$\Rightarrow \vec{\alpha}.\vec{\beta} + \vec{\beta}.\vec{\gamma} + \vec{\gamma}.\vec{\alpha} = -7.$

2. (b)

$\vec{\alpha} \times \vec{\beta} = \begin{vmatrix} \hat{i} & \hat{j} & \hat{k} \\ 1 & 1 & 1 \\ 2 & -1 & 1 \end{vmatrix}$

$= (1+1)\hat{i} - (1-2)\hat{j} + (-1-2)\hat{k} = 2\hat{i} + \hat{j} - 3\hat{k}$

Then, $|\vec{\alpha} \times \vec{\beta}| = \sqrt{2^2 + 1^2 + (-3)^2} = \sqrt{14}$.

$\therefore$ required unit vector

$= \dfrac{\vec{\alpha} \times \vec{\beta}}{|\vec{\alpha} \times \vec{\beta}|} = \dfrac{1}{\sqrt{14}}(2\hat{i} + \hat{j} - 3\hat{k})$.

3. (a)

$|\vec{\alpha} \times \vec{\beta}|^2 + (\vec{\alpha}.\vec{\beta})^2 = |\vec{\alpha}|^2 |\vec{\beta}|^2$

$\Rightarrow |\vec{\alpha} \times \vec{\beta}|^2 + 9 = 25 \times 1$

$\Rightarrow |\vec{\alpha} \times \vec{\beta}|^2 = 25 - 9 = 16 \Rightarrow |\vec{\alpha} \times \vec{\beta}| = 4.$

4. (a)

$\vec{\alpha} \times \vec{\beta} = \begin{vmatrix} \hat{i} & \hat{j} & \hat{k} \\ 2 & 1 & -3 \\ 1 & -2 & 1 \end{vmatrix}$

$= (1-6)\hat{i} - (2+3)\hat{j} + (-4-1)\hat{k} = -5\hat{i} - 5\hat{j} - 5\hat{k}.$

Then, $|\vec{\alpha} \times \vec{\beta}| = \sqrt{(-5)^2 + (-5)^2 + (-5)^2} = 5\sqrt{3}$.

$\therefore$ Required vector

$= 5 \dfrac{(\vec{\alpha} \times \vec{\beta})}{|\vec{\alpha} \times \vec{\beta}|}$

$= \dfrac{5}{5\sqrt{3}}(-5\hat{i} - 5\hat{j} - 5\hat{k}) = -\dfrac{5}{\sqrt{3}}(\hat{i} + \hat{j} + \hat{k}).$

5. (b)

Let $\vec{\alpha} = \hat{i} - 2\hat{j} - 2\hat{k}$ and $\vec{\beta} = 2\hat{i} + \hat{j} - 2\hat{k}$.

Let θ be the angle between them.

Then, $\vec{\alpha}.\vec{\beta} = |\vec{\alpha}||\vec{\beta}|\cos\theta$

$\Rightarrow 1 \times 2 + (-2) \times 1 + (-2) \times (-2)$

$= \sqrt{1^2 + (-2)^2 + (-2)^2}\sqrt{2^2 + 1^2 + (-2)^2}\cos\theta$

$\Rightarrow 4 = 9\cos\theta \Rightarrow \cos\theta = \dfrac{4}{9} \Rightarrow \theta = \cos^{-1}\dfrac{4}{9}.$

6. (a)

$\vec{\alpha}$ & $\vec{\beta}$ are perpendicular

$\Rightarrow \vec{\alpha}.\vec{\beta} = 0$

$\Rightarrow \left(2\hat{i} + \hat{j} + a\hat{k}\right).\left(3\hat{i} + b\hat{j} - \hat{k}\right) = 0$

$\Rightarrow 2 \times 3 + 1 \times b + a \times (-1) = 0$

$\Rightarrow a = b + 6$

$|\vec{\alpha}| = |\vec{\beta}|$

$\Rightarrow \sqrt{2^2 + 1^2 + a^2} = \sqrt{3^2 + b^2 + (-1)^2}$

$\Rightarrow a^2 + 5 = b^2 + 10$

$\Rightarrow (b+6)^2 + 5 = b^2 + 10$

$\Rightarrow b^2 + 12b + 36 + 5 = b^2 + 10$

$\Rightarrow 12b = -31 \Rightarrow b = -\dfrac{31}{12}$

$\therefore a = -\dfrac{31}{12} + 6 = \dfrac{41}{12}.$

7. (b)

$\vec{\alpha}$ & $\vec{\beta}$ are perpendicular

$\Rightarrow \vec{\alpha}.\vec{\beta} = 0$

$\Rightarrow \left(\lambda\hat{i} - 4\hat{j} + 3\hat{k}\right).\left(3\hat{i} + \lambda\hat{j} - 2\hat{k}\right) = 0$

$\Rightarrow 3 \times \lambda + (-4) \times \lambda + 3 \times (-2) = 0$

$\Rightarrow \lambda = -6.$

8. (b)

Here, $\overrightarrow{OA} = \hat{i} + \hat{j} - \hat{k}$, $\overrightarrow{OB} = \hat{i} + \hat{j} + 2\hat{k}$ and

$\overrightarrow{OC} = 3\hat{i} - \hat{j} - \hat{k}.$

$\overrightarrow{AB} = \overrightarrow{OB} - \overrightarrow{OA} = 3\hat{k},$

$\overrightarrow{AC} = \overrightarrow{OC} - \overrightarrow{OA} = 2\hat{i} - 2\hat{j}.$

Then, $\overrightarrow{AB} \times \overrightarrow{AC} = \begin{vmatrix} \hat{i} & \hat{j} & \hat{k} \\ 0 & 0 & 3 \\ 2 & -2 & 0 \end{vmatrix}$

$= (0+6)\hat{i} - (0-6)\hat{j} + (0-0)\hat{k}$

$= 6\hat{i} + 6\hat{j}$

Hence, area (ΔABC) is

$= \dfrac{1}{2}\left|\overrightarrow{AB} \times \overrightarrow{AC}\right|$

$= \dfrac{1}{2}\sqrt{6^2 + 6^2} = 3\sqrt{2} \text{ unit}^2$

9. (b)

$\vec{\alpha} + \vec{\beta} + \vec{\gamma} = \vec{0}$

$\Rightarrow \vec{\alpha} + \vec{\beta} = -\vec{\gamma} \Rightarrow \left(\vec{\alpha} + \vec{\beta}\right)^2 = \left(-\vec{\gamma}\right)^2$

$\Rightarrow |\vec{\alpha}|^2 + 2\vec{\alpha}.\vec{\beta} + |\vec{\beta}|^2 = |\vec{\gamma}|^2$

$\Rightarrow 9 + 2|\vec{\alpha}||\vec{\beta}|\cos\theta + 25 = 49$

$\Rightarrow 2 \times 3 \times 5 \cos\theta = 15$

$\Rightarrow \cos\theta = \dfrac{15}{30} = \dfrac{1}{2} = \cos 60^o \Rightarrow \theta = 60^o.$

10. (a)

Here, $\vec{r} = x\hat{i} + y\hat{j} + z\hat{k}.$

Then, $\left(\vec{r}.\hat{i}\right)\hat{i} + \left(\vec{r}.\hat{j}\right)\hat{j} + \left(\vec{r}.\hat{k}\right)\hat{k} = x\hat{i} + y\hat{j} + z\hat{k} = \vec{r}.$

11. (c)

$\hat{i}.\left(\hat{j} \times \hat{k}\right) + \hat{j}.\left(\hat{k} \times \hat{i}\right) + \hat{k}.\left(\hat{i} \times \hat{j}\right)$

$= \hat{i}.\hat{i} + \hat{j}.\hat{j} + \hat{k}.\hat{k} = 1 + 1 + 1 = 3.$

12. (c)

$\left[\vec{a} - \vec{b}, \vec{b} - \vec{c}, \vec{c} - \vec{a}\right]$

$= \left(\vec{a} - \vec{b}\right).\left\{\left(\vec{b} - \vec{c}\right) \times \left(\vec{c} - \vec{a}\right)\right\}$

$= \left(\vec{a} - \vec{b}\right).\left(\vec{b} \times \vec{c} - \vec{c} \times \vec{c} - \vec{b} \times \vec{a} + \vec{c} \times \vec{a}\right)$

$= \left(\vec{a} - \vec{b}\right).\left(\vec{b} \times \vec{c} + \vec{a} \times \vec{b} + \vec{c} \times \vec{a}\right) \quad \left(\because \vec{c} \times \vec{c} = \vec{0}\right)$

$= \vec{a}.\left(\vec{b} \times \vec{c}\right) + \vec{a}.\left(\vec{a} \times \vec{b}\right) + \vec{a}.\left(\vec{c} \times \vec{a}\right) - \vec{b}.\left(\vec{b} \times \vec{c}\right)$

$\quad - \vec{b}.\left(\vec{a} \times \vec{b}\right) - \vec{b}.\left(\vec{c} \times \vec{a}\right)$

$= \left[\vec{a}, \vec{b}, \vec{c}\right] + \left[\vec{a}, \vec{a}, \vec{b}\right] + \left[\vec{a}, \vec{c}, \vec{a}\right] - \left[\vec{b}, \vec{b}, \vec{c}\right]$

$\quad - \left[\vec{b}, \vec{a}, \vec{b}\right] - \left[\vec{b}, \vec{c}, \vec{a}\right]$

$= \left[\vec{a}, \vec{b}, \vec{c}\right] + 0 + 0 - 0 - 0 - \left[\vec{a}, \vec{b}, \vec{c}\right] = 0.$

13. (c)

$\dfrac{d}{dt}\left\{\vec{r}.\left(\dfrac{d\vec{r}}{dt} \times \dfrac{d^2\vec{r}}{dt^2}\right)\right\}$

$= \dfrac{d\vec{r}}{dt}.\left(\dfrac{d\vec{r}}{dt} \times \dfrac{d^2\vec{r}}{dt^2}\right) + \vec{r}.\dfrac{d}{dt}\left(\dfrac{d\vec{r}}{dt} \times \dfrac{d^2\vec{r}}{dt^2}\right)$

$= \left[\dfrac{d\vec{r}}{dt}, \dfrac{d\vec{r}}{dt}, \dfrac{d^2\vec{r}}{dt^2}\right] + \vec{r}.\left(\dfrac{d^2\vec{r}}{dt^2} \times \dfrac{d^2\vec{r}}{dt^2} + \dfrac{d\vec{r}}{dt} \times \dfrac{d^3\vec{r}}{dt^3}\right)$

$$= 0 + \vec{r}.\left(\frac{d^2\vec{r}}{dt^2} \times \frac{d^2\vec{r}}{dt^2}\right) + \vec{r}.\left(\frac{d\vec{r}}{dt} \times \frac{d^3\vec{r}}{dt^3}\right)$$

$$\left[\because \vec{a}.(\vec{b} \times \vec{c}) = [\vec{a}\ \vec{b}\ \vec{c}] \text{ and } [\vec{a}\ \vec{a}\ \vec{b}] = 0\right]$$

$$= \left[\vec{r}, \frac{d^2\vec{r}}{dt^2}, \frac{d^2\vec{r}}{dt^2}\right] + \left[\vec{r}, \frac{d\vec{r}}{dt}, \frac{d^3\vec{r}}{dt^3}\right]$$

$$= \left[\vec{r}, \frac{d\vec{r}}{dt}, \frac{d^3\vec{r}}{dt^3}\right] \quad \left(\text{using } [\vec{a},\vec{b},\vec{b}] = 0\right)$$

14. (d)

$$\frac{d\vec{r}}{dt} = \cos t\hat{i} - \sin t\hat{j} + \hat{k},$$

$$\frac{d^2\vec{r}}{dt^2} = -\sin t\hat{i} - \cos t\hat{j}.$$

$$\therefore \left|\frac{d^2\vec{r}}{dt^2}\right| = \sqrt{(-\sin t)^2 + (-\cos t)^2} = 1.$$

15. (c)

$$\frac{d\vec{\beta}}{dt} = 2\hat{i} - \hat{k}.$$

$$\therefore \vec{\alpha} \times \frac{d\vec{\beta}}{dt} = \begin{vmatrix} \hat{i} & \hat{j} & \hat{k} \\ t^2 & -t & 2t+1 \\ 2 & 0 & -1 \end{vmatrix}$$

$$= (t-0)\hat{i} - (-t^2 - 4t - 2)\hat{j} + (0+2t)\hat{k}$$

$$= t\hat{i} + (t^2 + 4t + 2)\hat{j} + 2t\hat{k}$$

Hence, $\dfrac{d}{dt}\left(\vec{\alpha} \times \dfrac{d\vec{\beta}}{dt}\right)$

$$= \frac{d}{dt}(t)\hat{i} + \frac{d}{dt}(t^2 + 4t + 2)\hat{j} + \frac{d}{dt}(2t)\hat{k}$$

$$= \hat{i} + (2t+4)\hat{j} + 2\hat{k}$$

$$\therefore \left[\frac{d}{dt}\left(\vec{\alpha} \times \frac{d\vec{\beta}}{dt}\right)\right]_{t=2} = \hat{i} + 8\hat{j} + 2\hat{k}.$$

16. (a)

$$\vec{r} = x\hat{i} + y\hat{j} + z\hat{k} = (t^3 + 1)\hat{i} + t^2\hat{j} + (2t+5)\hat{k}$$

$$\therefore \frac{d\vec{r}}{dt} = 3t^2\hat{i} + 2t\hat{j} + 2\hat{k},$$

$$\frac{d^2\vec{r}}{dt^2} = 6t\hat{i} + 2\hat{j}.$$

So $\left[\dfrac{d\vec{r}}{dt}\right]_{t=1} = 3\hat{i} + 2\hat{j} + 2\hat{k}$ and

$$\left[\frac{d^2\vec{r}}{dt^2}\right]_{t=1} = 6\hat{i} + 2\hat{j}.$$

17. (b)

$$[\vec{\alpha}\ \vec{\beta}\ \vec{\gamma}] = \begin{vmatrix} 2t & -t^2 & t^3 \\ -t & 0 & t^2 \\ 0 & t^3 & -2t \end{vmatrix}$$

$$= 2t(0 - t^5) + t^2(2t^2 - 0) + t^3(-t^4 - 0)$$

$$= -2t^6 + 2t^4 - t^7$$

$$\therefore \frac{d}{dt}[\vec{\alpha}\ \vec{\beta}\ \vec{\gamma}] = -12t^5 + 8t^3 - 7t^6.$$

Hence, $\left[\dfrac{d}{dt}[\vec{\alpha}\ \vec{\beta}\ \vec{\gamma}]\right]_{t=1} = -12 + 8 - 7 = -11.$

18. (a)

$$\frac{d\vec{f}}{dt}$$

$$= \hat{i}\frac{d}{dt}(2\cos t) + \hat{j}\frac{d}{dt}(2\sin t) + \hat{k}\frac{d}{dt}(2t\tan\theta)$$

$$= -2\sin t\hat{i} + 2\cos t\hat{j} + 2\tan\theta\hat{k},$$

$$\frac{d^2\vec{f}}{dt^2}$$

$$= \hat{i}\frac{d}{dt}(-2\sin t) + \hat{j}\frac{d}{dt}(2\cos t) + \hat{k}\frac{d}{dt}(2\tan\theta)$$

$$= -2\cos t\hat{i} - 2\sin t\hat{j}.$$

$$\therefore \frac{d\vec{f}}{dt} \times \frac{d^2\vec{f}}{dt^2} = \begin{vmatrix} \hat{i} & \hat{j} & \hat{k} \\ -2\sin t & 2\cos t & 2\tan\theta \\ -2\cos t & -2\sin t & 0 \end{vmatrix}$$

$$= (0 + 2\tan\theta \times 2\sin t)\hat{i} - (0 + 2\tan\theta \times 2\cos t)\hat{j}$$
$$\quad + (2\sin t \times 2\sin t + 2\cos t \times 2\cos t)\hat{k}$$

$$= 4\tan\theta\sin t\hat{i} - 4\tan\theta\cos t\hat{j} + 4\hat{k}$$

Hence, $\left|\dfrac{d\vec{f}}{dt} \times \dfrac{d^2\vec{f}}{dt^2}\right|$

$$= \sqrt{(4\tan\theta\sin t)^2 + (-4\tan\theta\cos t)^2 + 4^2}$$

$$= \sqrt{16\tan^2\theta(\sin^2 t + \cos^2 t) + 16}$$

$$= \sqrt{16\tan^2\theta + 16} = \sqrt{16(\tan^2\theta + 1)} = 4\sec\theta.$$

19. (a)

$$\frac{d\vec{f}}{dt} = \hat{i}\frac{d}{dt}(2t^2) + \hat{j}\frac{d}{dt}(t) + \hat{k}\frac{d}{dt}(3t^3)$$

$$= 4t\hat{i} + \hat{j} + 9t^2\hat{k}$$

$$\therefore \frac{d^2\vec{f}}{dt^2} = 4\hat{i} + 18t\hat{k}.$$

Then, $\vec{f} \times \dfrac{d^2 \vec{f}}{dt^2} = \begin{vmatrix} \hat{i} & \hat{j} & \hat{k} \\ 2t^2 & t & 3t^3 \\ 4 & 0 & 18t \end{vmatrix}$

$= (-18t \times t - 0)\hat{i} - (18t \times 2t^2 - 4 \times 3t^3)\hat{j}$

$\quad + (0 - 4t)\hat{k}$

$= -18t^2 \hat{i} + 24t^3 \hat{j} - 4t\hat{k}$

$\therefore \displaystyle\int_1^2 \vec{f} \times \dfrac{d^2\vec{f}}{dt^2} dt$

$= \displaystyle\int_1^2 \left(-18t^2\hat{i} + 24t^3\hat{j} - 4t\hat{k}\right) dt$

$= -18\hat{i}\displaystyle\int_1^2 t^2 dt + 24\hat{j}\int_1^2 t^3 dt - 4\hat{k}\int_1^2 t\, dt$

$= -18\hat{i}\left[\dfrac{t^3}{3}\right]_1^2 + 24\hat{j}\left[\dfrac{t^4}{4}\right]_1^2 - 4\hat{k}\left[\dfrac{t^2}{2}\right]_1^2$

$= -18\hat{i}\left[\dfrac{8}{3} - \dfrac{1}{3}\right] + 24\hat{j}\left[\dfrac{16}{4} - \dfrac{1}{4}\right] - 4\hat{k}\left[\dfrac{4}{2} - \dfrac{1}{2}\right]$

$= -42\hat{i} + 90\hat{j} - 6\hat{k}.$

20. (b)

$\vec{\nabla}\phi = \hat{i}\dfrac{\partial\phi}{\partial x} + \hat{j}\dfrac{\partial\phi}{\partial y} + \hat{k}\dfrac{\partial\phi}{\partial z}$

$= \hat{i}\dfrac{\partial}{\partial x}\left(x^2 - y^2 + 2z^2\right) + \hat{j}\dfrac{\partial}{\partial y}\left(x^2 - y^2 + 2z^2\right)$

$\quad + \hat{k}\dfrac{\partial}{\partial z}\left(x^2 - y^2 + 2z^2\right)$

$= 2x\hat{i} - 2y\hat{j} + 4z\hat{k}$

$\therefore At\ (1,2,3), \vec{\nabla}\phi = 2\hat{i} - 4\hat{j} + 12\hat{k}.$

The maximum value of the directional derivative, $\left|\vec{\nabla}\phi\right| = \sqrt{2^2 + (-4)^2 + (-12)^2} = \sqrt{164} = 2\sqrt{41}.$

21. (i)-(c); (ii)-(a).

$r = |\vec{r}| = \sqrt{x^2 + y^2 + z^2}$

$\Rightarrow r^2 = x^2 + y^2 + z^2$

$\therefore 2r\dfrac{\partial r}{\partial x} = 2x, 2r\dfrac{\partial r}{\partial y} = 2y, 2r\dfrac{\partial r}{\partial z} = 2z$

i.e; $\dfrac{\partial r}{\partial x} = \dfrac{x}{r}, \dfrac{\partial r}{\partial y} = \dfrac{y}{r}, \dfrac{\partial r}{\partial z} = \dfrac{z}{r}.$

(I) $\vec{\nabla}\left(\dfrac{1}{r}\right)$

$= \hat{i}\dfrac{\partial}{\partial x}\left(\dfrac{1}{r}\right) + \hat{j}\dfrac{\partial}{\partial y}\left(\dfrac{1}{r}\right) + \hat{k}\dfrac{\partial}{\partial z}\left(\dfrac{1}{r}\right)$

$= \hat{i}\dfrac{\partial}{\partial r}\left(\dfrac{1}{r}\right)\dfrac{\partial r}{\partial x} + \hat{j}\dfrac{\partial}{\partial r}\left(\dfrac{1}{r}\right)\dfrac{\partial r}{\partial y} + \hat{k}\dfrac{\partial}{\partial r}\left(\dfrac{1}{r}\right)\dfrac{\partial r}{\partial z}$

$= \hat{i}\left(-\dfrac{1}{r^2}\right)\dfrac{x}{r} + \hat{j}\left(-\dfrac{1}{r^2}\right)\dfrac{y}{r} + \hat{k}\left(-\dfrac{1}{r^2}\right)\dfrac{z}{r}$

$= -\dfrac{1}{r^3}\left(x\hat{i} + y\hat{j} + z\hat{k}\right) = -\dfrac{\vec{r}}{r^3}.$

(II) $\vec{\nabla}\left(r^n\right)$

$= \hat{i}\dfrac{\partial}{\partial x}\left(r^n\right) + \hat{j}\dfrac{\partial}{\partial y}\left(r^n\right) + \hat{k}\dfrac{\partial}{\partial z}\left(r^n\right)$

$= \hat{i}\dfrac{\partial}{\partial r}\left(r^n\right)\dfrac{\partial r}{\partial x} + \hat{j}\dfrac{\partial}{\partial r}\left(r^n\right)\dfrac{\partial r}{\partial y} + \hat{k}\dfrac{\partial}{\partial r}\left(r^n\right)\dfrac{\partial r}{\partial z}$

$= \hat{i}\left(nr^{n-1}\right)\dfrac{x}{r} + \hat{j}\left(nr^{n-1}\right)\dfrac{y}{r} + \hat{k}\left(nr^{n-1}\right)\dfrac{z}{r}$

$= nr^{n-2}\left(x\hat{i} + y\hat{j} + z\hat{k}\right) = nr^{n-2}\vec{r}.$

22. (a)

$\vec{\nabla}\phi = \hat{i}\dfrac{\partial\phi}{\partial x} + \hat{j}\dfrac{\partial\phi}{\partial y} + \hat{k}\dfrac{\partial\phi}{\partial z}$

$\quad = (y^2 - 2xyz^3)\hat{i} + (3 + 2xy - x^2z^3)\hat{j}$

$\qquad + (6z^3 - 3x^2yz^2)\hat{k}$

$\therefore \dfrac{\partial\phi}{\partial x} = y^2 - 2xyz^3 \ldots\ldots\ldots\ldots(1)$

$\quad \dfrac{\partial\phi}{\partial y} = 3 + 2xy - x^2z^3 \ldots\ldots\ldots(2)$

$\quad \dfrac{\partial\phi}{\partial z} = 6z^3 - 3x^2yz^2 \ldots\ldots\ldots\ldots(3)$

Integrating both sides of (1) w.r.t. x, we get

$\phi = \displaystyle\int (y^2 - 2xyz^3)dx = y^2x - x^2yz^3$

Integrating both sides of (2) w.r.t. y, we get

$\phi = \displaystyle\int (3 + 2xy - x^2z^3)dy = 3y + xy^2 - x^2yz^3.$

Integrating both sides of (3) w.r.t. z, we get

$\phi = \displaystyle\int (6z^3 - 3x^2yz^2)dz = \dfrac{3z^4}{2} - x^2yz^3$

Hence, $\phi = \phi(x,y,z) = 3y + xy^2 - x^2yz^3 + \dfrac{3z^4}{2}$

$$\left(\begin{array}{l} \because d\phi = \dfrac{\partial\phi}{\partial x}dx + \dfrac{\partial\phi}{\partial y}dy + \dfrac{\partial\phi}{\partial z}dz, \text{ so} \\[2mm] \phi = \int\dfrac{\partial\phi}{\partial x}dx + \int\dfrac{\partial\phi}{\partial y}dy + \int\dfrac{\partial\phi}{\partial z}dz \end{array} \right)$$

23. (b)

$$curl\vec{f} = \begin{vmatrix} \hat{i} & \hat{j} & \hat{k} \\ \dfrac{\partial}{\partial x} & \dfrac{\partial}{\partial y} & \dfrac{\partial}{\partial z} \\ x^2y & -2xz & 2yz \end{vmatrix}$$

$$= \hat{i}\left\{\dfrac{\partial}{\partial y}(2yz) - \dfrac{\partial}{\partial z}(-2xz)\right\}$$

$$\quad - \hat{j}\left\{\dfrac{\partial}{\partial x}(2yz) - \dfrac{\partial}{\partial z}(x^2y)\right\}$$

$$\quad + \hat{k}\left\{\dfrac{\partial}{\partial x}(-2xz) - \dfrac{\partial}{\partial y}(x^2y)\right\}$$

$$= \hat{i}(2z + 2x) - (x^2 + 2z)\hat{k}$$

$\therefore curl(curl\vec{f})$

$$= \begin{vmatrix} \hat{i} & \hat{j} & \hat{k} \\ \dfrac{\partial}{\partial x} & \dfrac{\partial}{\partial y} & \dfrac{\partial}{\partial z} \\ 2(z+x) & 0 & -(x^2+2z) \end{vmatrix}$$

$$= \hat{i}\left\{\dfrac{\partial}{\partial y}(-x^2 - 2z) - 0\right\}$$

$$\quad - \hat{j}\left\{\dfrac{\partial}{\partial x}(-x^2 - 2z) - \dfrac{\partial}{\partial z}(2x + 2z)\right\}$$

$$\quad + \hat{k}\left\{0 - \dfrac{\partial}{\partial y}(2x + 2z)\right\}$$

$$= \hat{i}(0-0) - \hat{j}(-2x - 2) - \hat{k}(0-0) = 2(x+1)\hat{j}.$$

24. (b)

$\vec{f}$ is solenoidal

$\Rightarrow div\vec{f} = 0$

$\Rightarrow \dfrac{\partial}{\partial x}(x + 3y) + \dfrac{\partial}{\partial y}(y - 2z) + \dfrac{\partial}{\partial z}(x + \lambda z) = 0$

$\Rightarrow 1 + 1 + \lambda = 0 \Rightarrow \lambda = -2.$

25. (b)

$\vec{f} = grad(x^3 + y^3 + z^3 - 3xyz)$

$$= \hat{i}\dfrac{\partial}{\partial x}(x^3 + y^3 + z^3 - 3xyz)$$

$$\quad + \hat{j}\dfrac{\partial}{\partial y}(x^3 + y^3 + z^3 - 3xyz)$$

$$\quad + \hat{k}\dfrac{\partial}{\partial z}(x^3 + y^3 + z^3 - 3xyz)$$

$$= (3x^2 - 3yz)\hat{i} + (3y^2 - 3xz)\hat{j} + (3z^2 - 3xy)\hat{k}$$

$\therefore curl\vec{f}$

$$= \begin{vmatrix} \hat{i} & \hat{j} & \hat{k} \\ \dfrac{\partial}{\partial x} & \dfrac{\partial}{\partial y} & \dfrac{\partial}{\partial z} \\ 3x^2 - 3yz & 3y^2 - 3xz & 3z^2 - 3xy \end{vmatrix}$$

$$= \hat{i}\left\{\dfrac{\partial}{\partial y}(3z^2 - 3xy) - \dfrac{\partial}{\partial z}(3y^2 - 3xz)\right\}$$

$$\quad - \hat{j}\left\{\dfrac{\partial}{\partial x}(3z^2 - 3xy) - \dfrac{\partial}{\partial z}(3x^2 - 3yz)\right\}$$

$$\quad + \hat{k}\left\{\dfrac{\partial}{\partial x}(3y^2 - 3xz) - \dfrac{\partial}{\partial y}(3x^2 - 3yz)\right\}$$

$$= (-3x + 3x)\hat{i} - (-3y + 3y)\hat{j} + (-3z + 3z)\hat{k}$$

$$= \vec{0}.$$

Hence, $\vec{f}$ is Solenoidal.

26. (c)

$grad\ \phi$

$$= \hat{i}\dfrac{\partial\phi}{\partial x} + \hat{j}\dfrac{\partial\phi}{\partial y} + \hat{k}\dfrac{\partial\phi}{\partial z}$$

$$= \hat{i}\dfrac{\partial}{\partial x}(x^2y + 2xy + z^2) + \hat{j}\dfrac{\partial}{\partial y}(x^2y + 2xy + z^2)$$

$$\quad + \hat{k}\dfrac{\partial}{\partial z}(x^2y + 2xy + z^2)$$

$$= (2xy + 2y)\hat{i} + (x^2 + 2x)\hat{j} + (2z)\hat{k}$$

$\therefore curl\ grad\ \phi$

$$= \begin{vmatrix} \hat{i} & \hat{j} & \hat{k} \\ \dfrac{\partial}{\partial x} & \dfrac{\partial}{\partial y} & \dfrac{\partial}{\partial z} \\ 2xy + 2y & x^2 + 2x & 2z \end{vmatrix}$$

$$= \hat{i}\left\{\dfrac{\partial}{\partial y}(2z) - \dfrac{\partial}{\partial z}(x^2 + 2x)\right\}$$

$$\quad - \hat{j}\left\{\dfrac{\partial}{\partial x}(2z) - \dfrac{\partial}{\partial z}(2xy + 2y)\right\}$$

$$+ \hat{k} \left\{ \frac{\partial}{\partial x}(x^2 + 2x) - \frac{\partial}{\partial y}(2xy + 2y) \right\}$$

$$= (0 - 0)\hat{i} - (0 - 0)\hat{j} + (2x + 2 - 2x - 2)\hat{k}$$

$$= \vec{0}.$$

27. (a)

$$\vec{\nabla}.\vec{r} = \frac{\partial}{\partial x}(x) + \frac{\partial}{\partial y}(y) + \frac{\partial}{\partial z}(z) = 1 + 1 + 1 = 3.$$

28. (b) We know that the unit normal to the surface $\phi(x, y, z) = c$ is $\dfrac{\vec{\nabla}\phi}{|\vec{\nabla}\phi|}$.

Hence, $\vec{\nabla}\phi$

$$= \hat{i} \frac{\partial \phi}{\partial x} + \hat{j} \frac{\partial \phi}{\partial y} + \hat{k} \frac{\partial \phi}{\partial z}$$

$$= \hat{i} \frac{\partial}{\partial x}(x^2 y + 2xz) + \hat{j} \frac{\partial}{\partial y}(x^2 y + 2xz)$$

$$+ \hat{k} \frac{\partial}{\partial z}(x^2 y + 2xz)$$

$$= (2xy + 2z)\hat{i} + (x^2 + 0)\hat{j} + (0 + 2x)\hat{k}$$

$$\therefore |\vec{\nabla}\phi| = \sqrt{(2xy + 2z)^2 + (x^2)^2 + (2x)^2}.$$

Hence, the required unit normal at the point $(2, -2, 3)$ is

$$\left(\frac{\vec{\nabla}\phi}{|\vec{\nabla}\phi|} \right)_{(2,-2,3)} = \frac{-2\hat{i} + 4\hat{j} + 4\hat{k}}{\sqrt{4 + 16 + 16}} = -\frac{1}{3}(\hat{i} - 2\hat{j} - 2\hat{k}).$$

29. (d)

$$\vec{\nabla}\phi$$

$$= \hat{i} \frac{\partial \phi}{\partial x} + \hat{j} \frac{\partial \phi}{\partial y} + \hat{k} \frac{\partial \phi}{\partial z}$$

$$= \hat{i} \frac{\partial}{\partial x}(xy + yz + zx) + \hat{j} \frac{\partial}{\partial y}(xy + yz + zx)$$

$$+ \hat{k} \frac{\partial}{\partial z}(xy + yz + zx)$$

$$= (y + z)\hat{i} + (x + z)\hat{j} + (y + x)\hat{k}.$$

$$\therefore \text{At } (1, 2, 0), \vec{\nabla}\phi = 2\hat{i} + \hat{j} + 3\hat{k}.$$

The required directional derivative

$$= \frac{\vec{\alpha}}{|\vec{\alpha}|}.\vec{\nabla}\phi$$

$$= \frac{1}{\sqrt{9}}(\hat{i} + 2\hat{j} + 2\hat{k}).(2\hat{i} + \hat{j} + 3\hat{k})$$

$$= \frac{1}{3}(2 \times 1 + 2 \times 1 + 2 \times 3) = \frac{10}{3}.$$

30. (a)

$\vec{f}$ is irrotational $\Rightarrow \operatorname{curl} \vec{f} = \vec{0}$

$$\Rightarrow \begin{vmatrix} \hat{i} & \hat{j} & \hat{k} \\ \dfrac{\partial}{\partial x} & \dfrac{\partial}{\partial y} & \dfrac{\partial}{\partial z} \\ mxy - z^3 & (m-2)x^2 & (1-m)xz^2 \end{vmatrix} = \vec{0}$$

$$\Rightarrow \hat{i} \left\{ \frac{\partial}{\partial y} \left[(1-m)xz^2 \right] - \frac{\partial}{\partial z} \left[(m-2)x^2 \right] \right\}$$

$$- \hat{j} \left\{ \frac{\partial}{\partial x} \left[(1-m)xz^2 \right] - \frac{\partial}{\partial z}(mxy - z^3) \right\}$$

$$+ \hat{k} \left\{ \frac{\partial}{\partial x} \left[(m-2)x^2 \right] - \frac{\partial}{\partial y}(mxy - z^3) \right\} = \vec{0}$$

$$\Rightarrow (0 - 0)\hat{i} - \left\{ (1-m)z^2 + 3z^2 \right\}\hat{j}$$

$$+ \left\{ 2x(m-2) - mx \right\}\hat{k} = 0\hat{i} + 0\hat{j} + 0\hat{k}$$

$$\Rightarrow (1-m)z^2 + 3z^2 = 0 \Rightarrow 1 - m + 3 = 0 \Rightarrow m = 4.$$

31. (b)

$$r = |\vec{r}| = \sqrt{x^2 + y^2 + z^2}$$

$$\Rightarrow r^2 = x^2 + y^2 + z^2$$

$$\therefore 2r \frac{\partial r}{\partial x} = 2x, 2r \frac{\partial r}{\partial y} = 2y, 2r \frac{\partial r}{\partial z} = 2z$$

$$i.e; \frac{\partial r}{\partial x} = \frac{x}{r}, \frac{\partial r}{\partial y} = \frac{y}{r}, \frac{\partial r}{\partial z} = \frac{z}{r}.$$

$$\therefore \vec{\nabla} e^{r^2}$$

$$= \hat{i} \frac{\partial}{\partial x}\left(e^{r^2} \right) + \hat{j} \frac{\partial}{\partial y}\left(e^{r^2} \right) + \hat{k} \frac{\partial}{\partial z}\left(e^{r^2} \right)$$

$$= \hat{i} \frac{\partial}{\partial r}\left(e^{r^2} \right)\frac{\partial r}{\partial x} + \hat{j} \frac{\partial}{\partial r}\left(e^{r^2} \right)\frac{\partial r}{\partial y} + \hat{k} \frac{\partial}{\partial r}\left(e^{r^2} \right)\frac{\partial r}{\partial z}$$

$$= \hat{i}\left(2re^{r^2} \right)\frac{x}{r} + \hat{j}\left(2re^{r^2} \right)\frac{y}{r} + \hat{k}\left(2re^{r^2} \right)\frac{z}{r}$$

$$= 2e^{r^2}\left(x\hat{i} + y\hat{j} + z\hat{k} \right) = 2e^{r^2}\vec{r}.$$

32. (c)

$curl\vec{f}$

$$= \begin{vmatrix} \hat{i} & \hat{j} & \hat{k} \\ \dfrac{\partial}{\partial x} & \dfrac{\partial}{\partial y} & \dfrac{\partial}{\partial z} \\ e^{xyz} & e^{xyz} & e^{xyz} \end{vmatrix}$$

$$= \hat{i}\left\{ \frac{\partial}{\partial y}(e^{xyz}) - \frac{\partial}{\partial z}(e^{xyz}) \right\}$$

$$\quad - \hat{j}\left\{ \frac{\partial}{\partial x}(e^{xyz}) - \frac{\partial}{\partial z}(e^{xyz}) \right\}$$

$$\quad + \hat{k}\left\{ \frac{\partial}{\partial x}(e^{xyz}) - \frac{\partial}{\partial y}(e^{xyz}) \right\}$$

$$= \hat{i}\left(xze^{xyz} - xye^{xyz}\right) - \hat{j}\left(yze^{xyz} - xye^{xyz}\right)$$

$$\quad + \hat{k}\left(yze^{xyz} - xze^{xyz}\right)$$

Hence, at (1, 1, 1), curl $\vec{u}$

$$= \hat{i}(e-e) - \hat{j}(e-e) + \hat{k}(e-e) = \vec{0}.$$

33. (a)

$\vec{f}$ is irrotational

$\Rightarrow curl\ \vec{f} = \vec{0}$

$$\Rightarrow \begin{vmatrix} \hat{i} & \hat{j} & \hat{k} \\ \dfrac{\partial}{\partial x} & \dfrac{\partial}{\partial y} & \dfrac{\partial}{\partial z} \\ x+2y+az & bx-3y-z & 4x+cy+2z \end{vmatrix} = \vec{0}$$

$$\Rightarrow \hat{i}\left\{ \frac{\partial}{\partial y}(4x+cy+2z) - \frac{\partial}{\partial z}(bx-3y-z) \right\}$$

$$\quad - \hat{j}\left\{ \frac{\partial}{\partial x}(4x+cy+2z) - \frac{\partial}{\partial z}(x+2y+az) \right\}$$

$$\quad + \hat{k}\left\{ \frac{\partial}{\partial x}(bx-3y-z) - \frac{\partial}{\partial y}(x+2y+az) \right\} = \vec{0}$$

$$\Rightarrow (c+1)\hat{i} - (4-a)\hat{j} + (b-2)\hat{k} = 0\hat{i} + 0\hat{j} + 0\hat{k}$$

$$\Rightarrow c+1 = 0, -(4-a) = 0, b-2 = 0$$

$$\Rightarrow c = -1, a = 4, b = 2.$$

34. (a)

Here, $\vec{f} = x^{a+b}y^b z^b\hat{i} + x^b y^{a+b}z^b\hat{j} + x^b y^b z^{a+b}\hat{k}$.

$\vec{f}$ is irrotational $\Rightarrow curl\ \vec{f} = \vec{0}$

$$\Rightarrow \begin{vmatrix} \hat{i} & \hat{j} & \hat{k} \\ \dfrac{\partial}{\partial x} & \dfrac{\partial}{\partial y} & \dfrac{\partial}{\partial z} \\ x^{a+b}y^b z^b & x^b y^{a+b}z^b & x^b y^b z^{a+b} \end{vmatrix} = \vec{0}$$

$$\Rightarrow \hat{i}\left\{ \frac{\partial}{\partial y}\left(x^b y^b z^{a+b}\right) - \frac{\partial}{\partial z}\left(x^b y^{a+b}z^b\right) \right\}$$

$$\quad - \hat{j}\left\{ \frac{\partial}{\partial x}\left(x^b y^b z^{a+b}\right) - \frac{\partial}{\partial z}\left(x^{a+b}y^b z^b\right) \right\}$$

$$\quad + \hat{k}\left\{ \frac{\partial}{\partial x}\left(x^b y^{a+b}z^b\right) - \frac{\partial}{\partial y}\left(x^{a+b}y^b z^b\right) \right\} = \vec{0}$$

$$\Rightarrow \hat{i}\left(bx^b y^{b-1}z^{a+b} - bx^b y^{a+b}z^{b-1}\right)$$

$$\quad - \hat{j}\left(bx^{b-1}y^b z^{a+b} - bx^{a+b}y^b z^{b-1}\right)$$

$$\quad + \hat{k}\left(bx^{b-1}y^{a+b}z^b - bx^{a+b}y^{b-1}z^b\right) = 0\hat{i} + 0\hat{j} + 0\hat{k}$$

$$\Rightarrow \hat{i}\left(y^{-1}z^a - y^a z^{-1}\right)b - \hat{j}\left(x^{-1}z^a - x^a z^{-1}\right)b$$

$$\quad + \hat{k}\left(x^{-1}y^a - x^a y^{-1}\right)b = 0\hat{i} + 0\hat{j} + 0\hat{k},$$

[on dividing both sides by $x^b y^b z^b$]

which is true if $a = -1$ or $b = 0$.

35. (a) Here, $x = e^t \sin t$ and $y = e^t \cos t$.

$\therefore t = 0 \Rightarrow x = 0$ and $t = \pi \Rightarrow x = 0$;

$t = 0 \Rightarrow y = 1$ and $t = \pi \Rightarrow y = -e^\pi$.

Then, $\displaystyle\int_C \vec{F}.d\vec{r}$

$$= \int_C \left[(3+2xy)\,dx + (x^2-3y^2)\,dy\right]$$

$$= \int_C d\left(3x + x^2 y - y^3\right)$$

$$= \int_{(0,1)}^{(0,-e^\pi)} d\left(3x + x^2 y - y^3\right)$$

$$= \left[3x + x^2 y - y^3\right]_{(0,1)}^{(0,-e^\pi)}$$

$$= e^{3\pi} - (-1) = 1 + e^{3\pi}.$$

36. (d)

$\vec{r} = x\hat{i} + y\hat{j}$ and $r = |\vec{r}| \Rightarrow r = \sqrt{x^2+y^2}$.

So, $\vec{F} = r^\beta(y\hat{i} - x\hat{j}) = \left(\sqrt{x^2+y^2}\right)^\beta (y\hat{i} - x\hat{j})$

$$= a^\beta(y\hat{i} - x\hat{j}) \quad (\because x^2 + y^2 = a^2)$$

$$= ya^\beta\hat{i} - xa^\beta\hat{j}$$

Then, $\displaystyle\int_C \vec{F}.d\vec{r}$

$$= \int_C \left[ya^\beta dx - xa^\beta dy\right] = a^\beta\int_C \left[y\,dx - x\,dy\right] \dots\dots(i)$$

Let $x = a\cos\theta$ and $y = a\sin\theta$.

Then $dx = -a\sin\theta\,d\theta$ and $dy = a\cos\theta\,d\theta$.

Then (i) $\Rightarrow$

$$\int_C \vec{F}.d\vec{r}$$

$$= a^\beta \int_{\theta=0}^{2\pi} \left[(a\sin\theta)(-a\sin\theta\,d\theta) - (a\cos\theta)(a\cos\theta\,d\theta)\right]$$

$$= a^{\beta+2} \int_{\theta=0}^{2\pi} \left[-\sin^2\theta - \cos^2\theta\right] d\theta = -a^{\beta+2} \int_{\theta=0}^{2\pi} d\theta$$

$$= -a^{\beta+2}\left[\theta\right]_0^{2\pi} = -2\pi a^{\beta+2}.$$

$$\therefore \left|\int_C \vec{F}.d\vec{r}\right| = 2\pi$$

$$\Rightarrow 2\pi a^{\beta+2} = 2\pi \Rightarrow a^{\beta+2} = 1 = a^0$$

$$\Rightarrow \beta + 2 = 0 \Rightarrow \beta = -2.$$

37. (b)

Here, $\vec{\nabla} \times \vec{F} = \operatorname{curl}\vec{F}$

$$= \begin{vmatrix} \hat{i} & \hat{j} & \hat{k} \\ \dfrac{\partial}{\partial x} & \dfrac{\partial}{\partial y} & \dfrac{\partial}{\partial z} \\ x^2+y^2 & -2xy & 0 \end{vmatrix}$$

$$= \hat{i}\left\{\frac{\partial}{\partial y}(0) - \frac{\partial}{\partial z}(-2xy)\right\}$$

$$- \hat{j}\left\{\frac{\partial}{\partial x}(0) - \frac{\partial}{\partial z}(x^2+y^2)\right\}$$

$$+ \hat{k}\left\{\frac{\partial}{\partial x}(-2xy) - \frac{\partial}{\partial y}(x^2+y^2)\right\}$$

$$= -4y\hat{k}$$

$$\therefore \operatorname{curl}\vec{F}.\hat{n} = \left\{-4y\hat{k}\right\}.\hat{k} = -4y.$$

So, $\iint_S \operatorname{curl}\vec{F}.\hat{n}\,dS$

$$= \int_{y=0}^{3} \int_{x=-1}^{1} (-4y)\,dx\,dy = -4\int_{y=0}^{3} y[\mathrm{x}]_{x=-1}^{x=1}\,dy$$

$$= -4\int_{y=0}^{3} 2y\,dy = -8\left[\frac{y^2}{2}\right]_0^3 = -36.$$

38. (a)

$$\oint_C \left[\left(x^2 - \cosh y\right)dx + \left(y + \sin x\right)dy\right],$$

$$= \iint_R \left[\frac{\partial}{\partial x}(y + \sin x) - \frac{\partial}{\partial y}\left(x^2 - \cosh y\right)\right]dx\,dy$$

(by Green's theorem)

$$= \int_{y=0}^{1} \int_{x=0}^{\pi} \left[\cos x + \sinh y\right]dx\,dy$$

$$= \int_{y=0}^{1} \left[\sin x + x\sinh y\right]_{x=0}^{\pi}\,dy$$

$$= \int_{y=0}^{1} \left[\sin\pi + \pi\sinh y - 0\right]dy$$

$$= \pi \int_{y=0}^{1} \sinh y\,dy = \pi\left[\cosh y\right]_{y=0}^{1} = \pi\left[\cosh 1 - 1\right].$$

39. (c)

$$\oint_C \left[(x+y)dx + x^2 dy\right]$$

$$= \iint_R \left[\frac{\partial}{\partial x}(x^2) - \frac{\partial}{\partial y}(x+y)\right]dx\,dy$$

(by Green's theorem)

$$= \int_{x=0}^{2} \int_{y=0}^{2x} \left[2x-1\right]dy\,dx = \int_{x=0}^{2} \left[2xy - y\right]_{y=0}^{2x}\,dx$$

$$= \int_{x=0}^{2} \left[(2x\times 2x - 2x) - 0\right]dx = \int_{x=0}^{2} \left[4x^2 - 2x\right]dx$$

$$= \left[4\frac{x^3}{3} - 2\frac{x^2}{2}\right]_0^2 = 4\times\frac{2^3}{3} - 2\times\frac{2^2}{2} = \frac{20}{3}.$$

40. (b)

$$\oint_C \left[y\,dx + 2x\,dy\right]$$

$$= \iint_R \left[\frac{\partial}{\partial x}(2x) - \frac{\partial}{\partial y}(y)\right]dx\,dy$$

(by Green's theorem)

$$= \int_{y=-1}^{1} \int_{x=0}^{1-y^2} dx\,dy = \int_{y=-1}^{1} \left[x\right]_{x=0}^{1-y^2}\,dy$$

$$= \int_{y=-1}^{1} \left[1 - y^2\right] dy = 2 \int_{y=0}^{1} \left[1 - y^2\right] dy$$

$$\left(\because f(y) = 1 - y^2 \text{ is an even function}\right)$$

$$= 2\left[y - \frac{y^3}{3}\right]_{y=0}^{1} = 2\left(1 - \frac{1}{3}\right) = \frac{4}{3}.$$

41. (b)

$$\oint_C \left[(\cos x \sin y - xy)dx + \sin x \cos y \, dy\right]$$

$$= \iint_R \left[\frac{\partial}{\partial x}(\sin x \cos y) - \frac{\partial}{\partial y}(\cos x \sin y - xy)\right] dx \, dy$$

(by Green's theorem)

$$= \iint_R \left[\cos x \cos y - \cos x \cos y + x\right] dx \, dy$$

$$= \iint_R x \, dx \, dy = \int_{\theta=0}^{2\pi} \int_{r=0}^{1} r\cos\theta \times r \, dr \, d\theta$$

$$\begin{bmatrix} \text{putting } x = r\cos\theta, y = r\sin\theta \text{ so that} \\ \text{Jacobian} = r; \therefore dx\,dy = r\,dr\,d\theta. \\ \text{Also } 0 \le \theta \le 2\pi \text{ and } 0 \le r \le 1 \text{ in the} \\ \text{circle } x^2 + y^2 = 1 \end{bmatrix}$$

$$= \int_{\theta=0}^{2\pi} \cos\theta \, d\theta \times \int_{r=0}^{1} r^2 \, dr$$

$$= \left[\sin\theta\right]_0^{2\pi} \times \left[\frac{r^3}{3}\right]_0^{1} = 0 \, (\because \sin 0 = 0 = \sin 2\pi)$$

42. (a)

Here, $curl\,\vec{F}$

$$= \begin{vmatrix} \hat{i} & \hat{j} & \hat{k} \\ \dfrac{\partial}{\partial x} & \dfrac{\partial}{\partial y} & \dfrac{\partial}{\partial z} \\ 2x-y & -yz^2 & -y^2 z \end{vmatrix}$$

$$= \hat{i}\left\{\frac{\partial}{\partial y}(-y^2 z) - \frac{\partial}{\partial z}(-yz^2)\right\}$$

$$- \hat{j}\left\{\frac{\partial}{\partial x}(-y^2 z) - \frac{\partial}{\partial z}(2x - y)\right\}$$

$$+ \hat{k}\left\{\frac{\partial}{\partial x}(-yz^2) - \frac{\partial}{\partial y}(2x - y)\right\}$$

$$= \hat{i}\{-2yz + 2yz\} - \hat{j}\{0 - 0\} + \hat{k}\{0 - (-1)\}$$

$$= \hat{k}$$

$$\therefore curl\,\vec{F}.\hat{n} = \hat{k}.\hat{k} = 1.$$

So by Stokes' theorem,

$$\oint_C \vec{F}.d\vec{r}$$

$$= \iint_S curl\,\vec{F}.\hat{n}\,dS = \iint_{S_1} 1\,dx\,dy$$

$$\begin{bmatrix} \text{where } S_1 \text{ is the plane region bounded by} \\ \text{the circle } x^2 + y^2 = 1 \end{bmatrix}$$

$$= \text{area of the circle } x^2 + y^2 = 1 = \pi \times 1^2 = \pi.$$

43. (c)

$$\oint_C \vec{F}.d\vec{r}$$

$$= \oint_C \left(-y\hat{i} + 2xy\hat{j} + z^3\hat{k}\right).\left(dx\hat{i} + dy\hat{j} + dz\hat{k}\right)$$

$$= \oint_C \left(-y\,dx + 2xy\,dy + z^3\,dz\right)$$

$$= \oint_C \left(-y\,dx + 2xy\,dy + (2-y)^3\,dy\right)$$

$$(\because y + z = 2)$$

$$= \oint_C \left[-y\,dx + \left(2xy + (2-y)^3\right)dy\right]$$

$$= \iint_R \left[\frac{\partial}{\partial x}\left(2xy - (2-y)^3\right) - \frac{\partial}{\partial y}(-y)\right] dx\,dy$$

(by Green's theorem)

$$= \iint_R \left[2y + 1\right] dx\,dy$$

$$= \int_{\theta=0}^{2\pi} \int_{r=0}^{1} (1 + 2r\sin\theta)\,r\,dr\,d\theta$$

$$\begin{pmatrix} \text{putting } x = r\cos\theta, y = r\sin\theta \text{ so that} \\ dx\,dy = r\,dr\,d\theta \\ \text{Also } 0 \le \theta \le 2\pi \text{ and } 0 \le r \le 1 \text{ in the} \\ \text{circle } x^2 + y^2 = 1 \end{pmatrix}$$

$$= \int_{\theta=0}^{2\pi} \int_{r=0}^{1} (r + 2r^2\sin\theta)\,dr\,d\theta$$

$$= \int_{\theta=0}^{2\pi} \left[\frac{r^2}{2} + \frac{2r^3}{3}\sin\theta\right]_{r=0}^{1} d\theta = \int_{\theta=0}^{2\pi} \left[\frac{1}{2} + \frac{2}{3}\sin\theta\right]d\theta.$$

44. (a) $\vec{\nabla}.\vec{F}$

$$= \frac{\partial}{\partial x}(4xz) + \frac{\partial}{\partial y}(-y^2) + \frac{\partial}{\partial z}(yz)$$

$$= 4z - 2y + y = 4z - y$$

$$\therefore \iint_S \vec{F}.\hat{n}\,dS$$

$$= \iiint_V \vec{\nabla}.\vec{F}\,dV \text{ (by the Gauss Divegrence theorem)}$$

$$= \iiint_V (4z - y)\,dV$$

$$= \int_{z=0}^{1} \int_{y=0}^{1} \int_{x=0}^{1} (4z - y)\,dx\,dy\,dz$$

$$= \int_{z=0}^{1} \int_{y=0}^{1} \left[4zx - xy\right]_{x=0}^{1} dy\,dz$$

$$= \int_{z=0}^{1} \int_{y=0}^{1} \left[4z - y\right]dy\,dz$$

$$= \int_{z=0}^{1} \left[4zy - \frac{y^2}{2}\right]_{y=0}^{1} dz = \int_{z=0}^{1} \left[4z - \frac{1}{2}\right]dz$$

$$= \left[4\frac{z^2}{2} - \frac{1}{2}z\right]_{z=0}^{1} = 2 - \frac{1}{2} = \frac{3}{2}.$$

Previous Years Solved Papers (2000-2018)

1. If a vector $\overline{R}(t)$ has a constant magnitude, then

(a) $\overline{R}.\dfrac{d\overline{R}}{dt} = 0$ (b) $\overline{R} \times \dfrac{d\overline{R}}{dt} = 0$

(c) $\overline{R}.\overline{R} = \dfrac{d\overline{R}}{dt}$ (d) $\overline{R} \times \overline{R} = \dfrac{d\overline{R}}{dt}$

[GATE 2002]

2. The directional derivative of the following function at $(1, 2)$ in the direction of $4\hat{i} + 3\hat{j}$ is:
$$f(x, y) = x^2 + y^2$$
(a) 4/5 (b) 4 (c) 2/5 (d) 1

[GATE 2002]

3. The vector field $\vec{F} = x\vec{i} - y\vec{j}$ (where $\vec{i}$ and $\vec{j}$ are unit vectors) is
(a) divergence free but not irrotational
(b) irrotational but not divergence free
(c) divergence free and irrotational
(d) neither divergence free nor irrotational

[GATE 2003]

4. The angle between two unit-magnitude co-planar vectors $P(0.866, 0.500, 0)$ and $Q(0.259, 0.966, 0)$ will be
(a) 0° (b) 30° (c) 45° (d) 60°

[ME GATE-2004]

5. For the scalar field $u = \dfrac{x}{2} + \dfrac{y^2}{3}$, magnitude of the gradient at the point $(1, 3)$ is

(a) $\sqrt{\dfrac{13}{9}}$ (b) $\sqrt{\dfrac{9}{2}}$

(c) $\sqrt{5}$ (d) $\dfrac{9}{2}$

[EE GATE 2005]

6. A scalar field is given by $f = x^{2/3} + y^{2/3}$, where x and y are the Cartesian co-ordinates. The derivative of "f" along the line $y = x$ directed away from the origin at the point $(8, 8)$ is

(a) $\dfrac{\sqrt{2}}{3}$ (b) $\dfrac{\sqrt{3}}{2}$

(c) $\dfrac{2}{\sqrt{3}}$ (d) $\dfrac{3}{\sqrt{2}}$

[IN GATE 2005]

7. Value of the integral $\oint_C \left(xy\,dy - y^2\,dx\right)$, where C is the square cut from the first quadrant by the lines $x = 1$ and $y = 1$ will be (Use Green's theorem to change the line integral into double integral)

(a) $\dfrac{1}{2}$ (b) 1 (c) $\dfrac{3}{2}$ (d) $\dfrac{5}{3}$

[CE GATE 2005]

8. Stokes' theorem connects
(a) a line integral and a surface integral
(b) a surface integral and a volume integral
(c) a line integral and a volume integral
(d) gradient of a function and its surface integral

[ME GATE 2005]

9. Which one of the following is not associated with vector calculus?
(a) Strokes' theorem
(b) Gauss Divergence theorem
(c) Green's theorem
(d) Kennedy's theorem

[PI GATE 2005]

10. The line integral $\int \vec{V}.\vec{dr}$ of the vector $\vec{V}(\vec{r}) = 2xyz\hat{i} + x^2z\hat{j} + x^2y\hat{k}$ from the origin to the point $P(1, 1, 1)$

(a) is 1

(b) is zero

(c) is –1

(d) cannot be determined without specifying path

[ME GATE 2005]

11. $\iint (\nabla \times P).ds$ where P is a vector, is equal to

(a) $\oint P.dl$

(b) $\oint \nabla \times \nabla \times P.dl$

(c) $\oint \nabla \times P.dl$

(d) $\iiint \nabla \times P.dv$

[EC GATE 2006]

12. The directional derivative of $f(x, y, z) = 2x^2 + 3y^2 + z^2$ at the point $P(2, 1, 3)$ in the direction of the vector $\vec{a} = \hat{i} - 2\hat{k}$ is

(a) –2.785

(b) –2.145

(c) –1.789

(d) 1

[CE GATE 2006]

13. $\vec{\nabla} \times \vec{\nabla} \times \vec{P}$ where $\vec{P}$ is a vector is equal to

(a) $\vec{P} \times \vec{\nabla} \times \vec{P} - \vec{\nabla}^2 \vec{P}$

(b) $\vec{\nabla}^2 \vec{P} + \vec{\nabla}\left(\vec{\nabla} \times \vec{P}\right)$

(c) $\vec{\nabla}^2 \vec{P} + \vec{\nabla} \times \vec{P}$

(d) $\vec{\nabla}\left(\vec{\nabla}.\vec{P}\right) - \vec{\nabla}^2 \vec{P}$

[EC GATE 2006]

14. Divergence of the vector field

$\vec{V}(x, y, z) = -(x\cos xy + y)\vec{i} + (y\cos xy)\vec{j}$

$+ \left(\sin z^2 + x^2 + y^2\right)\vec{k}$ is

(a) $2z \cos z^2$

(b) $\sin xy + 2z\cos z^2$

(c) $x\sin xy - \cos z$

(d) none of these

[EE GATE 2007]

15. A velocity vector is given as $\vec{V} = 5xy\vec{i} + 2y^2\vec{j} + 3yz^2\vec{k}$. The divergence of this velocity vector at $(1, 1, 1)$ is

(a) 9

(b) 10

(c) 14

(d) 15

[CE GATE 2007]

16. Let x and y be two vectors in a 3 dimensional space and $<x, y>$ denote their dot product. Then the determinant $\det \begin{bmatrix} <x,x> & <x,y> \\ <y,x> & <y,y> \end{bmatrix}$

(a) is zero when x and y are linearly independent

(b) is positive when x and y are linearly independent

(c) is nonzero for all nonzero x and y

(d) is zero only when either x or y is zero

[EE GATE 2007]

17. The area of a triangle formed by the tips of vectors $\bar{a}, \bar{b}$ and $\bar{c}$ is

(a) $\frac{1}{2}(\bar{a} - \bar{b}).(\bar{a} - \bar{c})$

(b) $\frac{1}{2}|(\bar{a} - \bar{b}).(\bar{a} - \bar{c})|$

(c) $\frac{1}{2}|\bar{a} \times \bar{b} \times \bar{c}|$

(d) $\frac{1}{2}(\bar{a} \times \bar{b}).\bar{c}$

[ME GATE 2007]

18. The angle (in degrees) between two planar vector $\vec{a} = \frac{\sqrt{3}}{2}\hat{i} + \frac{1}{2}\hat{j}$ and $\vec{b} = -\frac{\sqrt{3}}{2}\hat{i} + \frac{1}{2}\hat{j}$ is

(a) 30°

(b) 60°

(c) 90°

(d) 120°

[PI GATE 2007]

19. Consider the points P and Q in the x-y plane, with $P = (1,0)$ and $Q = (0,1)$.

The line integral $2\int_P^Q (xdx + ydy)$ along the semicircle with the line segment PQ as its diameter

(a) is –1

(b) is 0

(c) is 1

(d) depends on the direction (clockwise or counterclockwise) of the semicircle

[EC GATE 2008]

20. If $\vec{r}$ is the position vector of any point on a closed surface S that encloses the volume V, then $\iint_S \vec{r}.\vec{ds}$ is equal to

(a) $\frac{1}{2}V$ 　(b) V 　(c) $2V$ 　(d) $3V$

[PI GATE 2008]

21. The inner (dot) product of two non zero vectors $\vec{P}$ and $\vec{Q}$ is zero. The angle (degrees) between the two vectors is

(a) 0   (b) 30   (c) 90   (d) 120

[CE GATE 2008]

22. The divergence of the vector field
$(x-y)\hat{i}+(y-x)\hat{j}+(x+y+z)\hat{k}$ is

(a) 0   (b) 1   (c) 2   (d) 3

[ME GATE 2008]

23. The directional derivative of the scalar function $f(x, y, z) = x^2 + 2y^2 + z$ at the point $P = (1, 1, 2)$ in the direction of the vector $\vec{a} = 3\hat{i}-4\hat{j}$ is

(a) –4   (b) –2   (c) –1   (d) 1

[ME GATE 2008]

24. A sphere of unit radius is centered at the origin. The unit normal at a point (x, y, z) on the surface of the sphere is the vector

(a) (x, y, z) (b) $\left(1/\sqrt{3}, 1/\sqrt{3}, 1/\sqrt{3}\right)$

(c) $\left(x/\sqrt{3}, y/\sqrt{3}, z/\sqrt{3}\right)$

(d) $\left(x/\sqrt{2}, y/\sqrt{2}, z/\sqrt{2}\right)$

[IN GATE 2009]

25. If a vector field $\vec{V}$ is related to another field $\vec{A}$ through $\vec{V} = \vec{\nabla} \times \vec{A}$, which of the following is true?

Note *C and SC refer to any closed contour and any surface whose boundary is C*

(a) $\oint_C \vec{V}.\overrightarrow{dl} = \iint_{S_c} \vec{A}.\overrightarrow{ds}$ (b) $\oint_C \vec{A}.\overrightarrow{dl} = \iint_{S_c} \vec{V}.\overrightarrow{ds}$

(c) $\oint_C \vec{\nabla} \times \vec{V}.\overrightarrow{dl} = \iint_{S_c} \vec{\nabla} \times \vec{A}.\overrightarrow{ds}$

(d) $\oint_C \vec{\nabla} \times \vec{A}.\overrightarrow{dl} = \iint_{S_c} \vec{V}.\overrightarrow{ds}$

[EC GATE 2009]

26. $F(x,y) = \left(x^2 + xy\right)\hat{a}_x + \left(y^2 + xy\right)\hat{a}_y$.

Its line integral over the straight line from $(x,y) = (0,2)$ to $(x,y) = (2,0)$ evaluates to

(a) –8   (b) 4   (c) 8   (d) 0

[EE GATE 2009]

27. For a scalar function $f(x, y, z) = x^2 + 3y^2 + 2z^2$, the gradient at the point $P(1, 2, -1)$ is

(a) $2\vec{i} + 6\vec{j} + 4\vec{k}$ (b) $2\vec{i} + 12\vec{j} - 4\vec{k}$

(c) $2\vec{i} + 12\vec{j} - 4\vec{k}$ (d) $\sqrt{56}$

[CE GATE 2009]

28. For a scalar function $f(x, y, z) = x^2 + 3y^2 + 2z^2$, the directional derivative at the point $P(1, 2, -1)$ in the direction of a vector $\vec{i} - \vec{j} + 2\vec{k}$ is

(a) –18   (b) $-3\sqrt{6}$ (c) $3\sqrt{6}$ (d) 18

[CE GATE 2009]

29. The divergence of the vector field $3xz\hat{i} + 2xy\hat{j} - yz^2\hat{k}$ at a point $(1,1,1)$ is equal to

(a) 7   (b) 4   (c) 3   (d) 0

[ME GATE 2009]

30. If $\vec{A} = xy\hat{a}_x + x^2\hat{a}_y$, then $\oint_C \vec{A}.\overrightarrow{dl}$ over the path shown in the figure below is

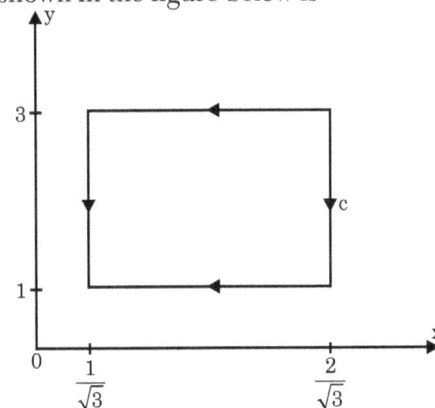

(a) 0   (b) $\dfrac{2}{\sqrt{3}}$ (c) 1   (d) $2\sqrt{3}$

[EC GATE 2010]

31. Velocity vector of a flow field is given as $\vec{V} = 2xy\hat{i} - x^2z\hat{j}$. The vorticity vector at $(1, 1, 1)$ is

(a) $4\hat{i} - \hat{j}$ (b) $4\hat{i} - \hat{k}$

(c) $\hat{i} - 4\hat{j}$ (d) $\hat{i} - 4\hat{k}$

[ME GATE 2010]

32. Divergence of the three-dimensional radial vector field is

(a) 3   (b) 1/r

(c) $\hat{i} + \hat{j} + \hat{k}$ (d) $3\left(\hat{i} + \hat{j} + \hat{k}\right)$

[EE GATE 2010]

33. If $T(x,y,z) = x^2 + y^2 + 2z^2$ defines the temperature at any location (x,y,z), then the magnitude of the temperature gradient at the point $P(1, 1, 1)$ is _____.

[PI GATE 2011]

(a) $2\sqrt{6}$ (b) 4   (c) 24   (d) $\sqrt{6}$

34. Consider a closed surface S surrounding a volume V. If $\vec{r}$ is the position vector of a point inside S with $\hat{n}$ (the unit normal on S), then the value of the integral $\oiint 5\vec{r}.\hat{n}ds$ is

(a) 3V   (b) 5V   (c) 10V   (d) 15V

[EC GATE 2011]

35. If $\vec{a}$ and $\vec{b}$ are two arbitrary vectors with magnitudes a and b, respectively, $\left|\vec{a}\times\vec{b}\right|^2$ will be equal to

(a) $a^2b^2 - (\vec{a}.\vec{b})^2$ (b) $ab - \vec{a}.\vec{b}$

(c) $a^2b^2 + (\vec{a}.\vec{b})^2$ (d) $ab + \vec{a}.\vec{b}$

[CE GATE 2011]

36. The two vectors $[1, 1, 1]$ and $[1, a, a^2]$, where $a = -\dfrac{1}{2} + \dfrac{\sqrt{3}}{2}j$, are

(a) orthonormal   (b) orthogonal
(c) parallel   (d) collinear

[EE GATE 2011]

37. The line integral $\int\limits_{P_1}^{P_2}(ydx + xdy)$ from $P_1(x_1, y_1)$ to $P_2(x_3\ y_2)$ along the semicircle P_1P_2 shown in the figure is?

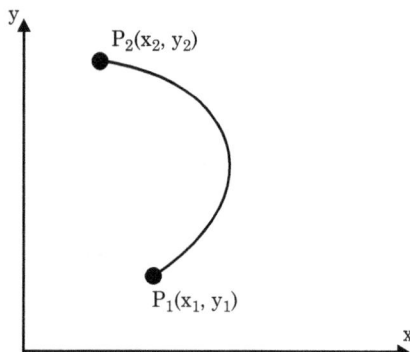

(a) $x_2y_2 - x_1y_1$

(b) $\left(y_2^2 - y_1^2\right) + \left(x_2^2 - x_1^2\right)$

(c) $(x_2 - x_1)(y_2 - y_1)$

(d) $\left(y_2 - y_1\right)^2 + \left(x_2 - x_1\right)^2$

[PI GATE 2011]

38. For the parallelogram OPQR shown in the sketch $\overline{OP} = a\hat{i} + b\hat{j}$ and $\overline{OR} = c\hat{i} + d\hat{j}$.

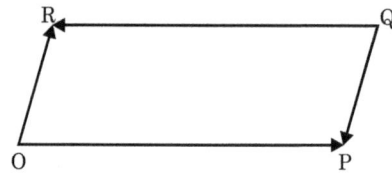

The area of the parallelogram is

(a) $ad - bc$ (b) $ac + bd$
(c) $ad + bc$ (d) $ad - cd$

[CE GATE 2012]

39. For the spherical surface $x^2 + y^2 + z^2 = 1$, the unit outward normal vector at the point $\left(\dfrac{1}{\sqrt{2}}, \dfrac{1}{\sqrt{2}}, 0\right)$ is given by

(a) $\dfrac{1}{\sqrt{2}}\hat{i} + \dfrac{1}{\sqrt{2}}\hat{j}$ (b) $\dfrac{1}{\sqrt{2}}\hat{i} - \dfrac{1}{\sqrt{2}}\hat{j}$

(c) $\hat{k}$ (d) $\dfrac{1}{\sqrt{3}}\hat{i} + \dfrac{1}{\sqrt{2}}\hat{j} + \dfrac{1}{\sqrt{3}}\hat{k}$

[ME GATE 2012]

40. The direction of vector $\vec{A}$ is radially outward from the origin, with $\left|\vec{A}\right| = kr^n$ where $r^2 = x^2 + y^2 + z^2$ and k is a constant. The value of n for which $\vec{\nabla}.\vec{A} = 0$ is

(a) –2   (b) 2   (c) 1   (d) 0

[IN GATE 2012]

41. The following surface integral is to be evaluated over a sphere for the given steady velocity vector field $\vec{F} = x\hat{i} + y\hat{j} + z\hat{k}$ defined with respect to a Cartesian coordinate system having $\hat{i}, \hat{j}, \hat{k}$ as unit base vectors.

$$\iint\limits_S \frac{1}{4}\left(\vec{F}.\hat{n}\right)dA\ ,$$

where S is the sphere, $x^2 + y^2 + z^2 = 1$ and $\hat{n}$ is the outward unit normal vector to the sphere. The value of the surface integral is

(a) π (b) 2π (c) $3\pi/4$ (d) 4π

[ME GATE 2013]

42. Consider a vector field $\vec{A}(\vec{r})$. The closed loop line integral $\oint \vec{A}.\vec{dl}$ can be expressed as

   (a) $\oiint (\nabla \times \vec{A}).\vec{dl}$ over the closed surface bounded by the loop

   (b) $\oiiint (\nabla \times \vec{A})dv$ over the closed volume bounded by the top

   (c) $\iiint (\nabla.\vec{A})dv$ over the open volume bounded by the loop

   (d) $\iint (\nabla \times \vec{A}).\vec{ds}$ over the open surface bounded by the loop

   [EC GATE 2013]

43. Given a vector field $\vec{F} = y^2 x \hat{a}_x - yz \hat{a}_y - x^2 \hat{a}_z$,

   the line integral $\int \vec{F}.\vec{dl}$ evaluated along a segment on the x-axis from $x = 1$ to $x = 2$ is

   (a) –2.33   (b) 0   (c) 2.33   (d) 7

   [EE GATE 2013]

44. The curl of the gradient of the scalar field defined by $V = 2x^2 y + 3y^2 z + 4z^2 x$ is

   (a) $4xyax + 6yzay + 8zxaz$
   (b) $4ax + 6ay + 8az$
   (c) $(4xy + 4z^2)ax + (2x^2 + 6yz)ay + (3y^2 + 8zx)az$
   (d) 0

   [EE GATE 2013]

45. The divergence of the vector field $\vec{A} = x \hat{a}_x + y \hat{a}_y + z \hat{a}_z$ is

   (a) 0   (b) 1/3   (c) 1   (d) 3

   [EC GATE 2013]

46. For a vector E, which one of the following statements is NOT TRUE?

   (a) If $\nabla.E = 0$, E is called solenoidal
   (b) If $\nabla \times E = 0$, E is called conservative
   (c) If $\nabla \times E = 0$, E is called irrotational
   (d) If $\nabla.E = 0$, E is called irrotational

   [IN GATE 2013]

47. A vector is defined as $\vec{f} = y\hat{i} + x\hat{j} + z\hat{k}$, where $\hat{i}, \hat{j}, \hat{k}$ are unit vectors in the Cartesian (x, y, z) co-ordinate system. The surface integral $\oiint \vec{f}.\vec{ds}$ over the closed surface S of a cube with vertices having the following co-ordinates:

$(0,0,0), (1,0,0), (0,1,0), (0,0,1), (1,0,1), (1,1,1),$

$(0,1,1), (1,1,0)$ is _____

   [IN GATE 2014]

48. The integral $\oint_C (ydx - xdy)$ is evaluated along the circle $x^2 + y^2 = \frac{1}{4}$ traversed in counterclockwise direction. The integral is equal to

   (a) 0   (b) $-\frac{\pi}{4}$ (c) $-\frac{\pi}{2}$ (d) $\frac{\pi}{4}$

   [ME GATE 2014]

49. The line integral of the function $\vec{F} = yz\hat{i}$ in the counterclockwise direction along the circle $x^2 + y^2 = 1$ at $z = 1$ is

   (a) -2π (b) $-\pi$ (c) π (d) 2π

   [EE GATE 2014]

50. Which one of the following describes the relationship among the three vectors,

   $\hat{i} + \hat{j} + \hat{k}, 2\hat{i} + 3\hat{j} + \hat{k}$ and $5\hat{i} + 6\hat{j} + 4\hat{k}$?

   (a) The vectors are mutually perpendicular
   (b) The vectors are linearly dependent
   (c) The vectors are linearly independent
   (d) The vectors are unit vectors

   [ME GATE 2014]

51. A particle moves along a curve whose parametric equations are: $x = t^3 + 2t, y = -3e^{-2t}$ and $z = 2\sin 5t$, where x, y, z show variations of the distance covered by the particle (in cm) with time t (in sec). The magnitude of the acceleration of the particle (in cm/sec^2) at $t = 0$ is _____?

   [CE GATE 2014]

52. If $\vec{r} = x\hat{a}_x + y\hat{a}_y + z\hat{a}_z$ and $|\vec{r}| = r$, then

   $div\left(r^2 \vec{\nabla}(\ln r)\right) = $ ____?

   [EC GATE 2014]

53. The directional derivative of $f(x,y) = \dfrac{xy}{\sqrt{2}}(x+y)$ at $(1, 1)$ in the direction of the unit vector at an angle of $\dfrac{\pi}{4}$ with y-axis, is given by _____.

[EC GATE 2014]

54. The curl of vector $\vec{F} = x^2 z^2 \hat{i} - 2xy^2 z \hat{j} + 2y^2 z^3 \hat{k}$ is

(a) $\left(4yz^3 + 2xy^2\right)\hat{i} + 2x^2 z\hat{j} - 2y^2 z\hat{k}$

(b) $\left(4yz^3 + 2xy^2\right)\vec{i} + 2x^2 z\hat{j} - 2y^2 z\hat{k}$

(c) $2xz^2 \hat{i} - 4xyz \hat{j} - 6y^2 z^2 \hat{k}$

(d) $2xz^2 \hat{i} - 4xyz \hat{j} - 6y^2 z^2 \hat{k}$

[ME GATE 2014]

55. The divergence of the vector field $x^2 z\hat{i} + xy\hat{j} - yz^2 \hat{k}$ at $(1, -1, 1)$ is

(a) 0   (b) 3   (c) 5   (d) 6

[ME GATE 2014]

56. Let $\vec{\nabla}\cdot\left(f\vec{V}\right) = x^2 y + y^2 z + z^2 x$, where f and $\vec{V}$ are scalar and vector fields, respectively. If $\vec{V} = y\hat{i} + z\hat{j} + x\hat{k}$, then $\vec{V}\cdot\vec{\nabla}f$ is

(a) $x^2 y + y2z + z2x$ (b) $2xy + 2yz + 2zx$

(c) $x + y + z$ (d) 0

[EE GATE 2014]

57. The directional derivative of $\phi = 2xz - y^2$ at the point $(1, 3, 2)$ becomes maximum in the direction of

(a) $4\hat{i} + 2\hat{j} - 3\hat{k}$ (b) $4\hat{i} - 6\hat{j} + 2\hat{k}$

(c) $2\hat{i} - 6\hat{j} + 2\hat{k}$ (d) $4\hat{i} - 6\hat{j} - 2\hat{k}$

[PI GATE 2014]

58. If $\phi = 2x^3 y^2 z^4$, then $\vec{\nabla}^2 \phi$ is

(a) $12xy^2 z^4 + 4x^2 z^4 + 20x^3 y^2 z^3$

(b) $2x^2 y^2 z + 4x^3 z^4 + 24x^3 y^2 z^2$

(c) $12xy^2 z^4 + 4x^3 z^4 + 24x^3 y^2 z^2$

(d) $4xy^2 z + 4x^2 z^4 + 24x^3 y^2 z^2$

[PI GATE 2014]

59. The value of $\displaystyle\int_C\left[\left(3x - 8y^2\right)dx + \left(4y - 6xy\right)dy\right]$, (where C is the boundary of the region boundary by $x = 0$, $y = 0$ and $x + y = 1$) is _____.

[ME GATE 2015]

60. The surface integral $\displaystyle\iint_S \frac{1}{\pi}\left(9x\hat{i} - 3y\hat{j}\right)\cdot\hat{n}dS$ over the sphere given by $x^2 + y^2 + z^2 = 9$ is _____.

[ME GATE 2015]

61. A vector $\vec{P}$ is given by $\vec{P} = x^3 y\vec{a}_x - x^2 y^2 \vec{a}_y - x^2 yz\vec{a}_z$. Which one of the following statements is TRUE?

(a) $\vec{P}$ is solenoidal, but not irrotational

(b) $\vec{P}$ is irrotational but not solenoidal

(c) $\vec{P}$ is neither solenoidal nor irrotational

(d) $\vec{P}$ is both solenoidal and irrotational

[EC GATE 2015]

62. The directional derivative of the field $u(x, y, z) = x^2 - 3yz$ in the direction of the vector $\left(\vec{i} + \vec{j} - 2\vec{k}\right)$ at point $(2, -1, 4)$ is _____.

[CE GATE 2015]

63. The curl of vector $\vec{V}(x, y, z) = 2x^2 i + 3z^2 j + y^3 k$ at $x = y = z = 1$ is

(a) $-3i$ (b) $3i$

(c) $3i - 4j$ (d) $3i - 6k$

[ME GATE 2015]

64. Let ϕ be an arbitrary smooth real valued scalar function and V be an arbitrary smooth vector valued function in a three-dimensional space. Which one of the following is an identity?

(a) $\text{Curl}\left(\phi\vec{V}\right) = \nabla\left(\phi Div\vec{V}\right)$

(b) $\text{Div } \vec{V} = 0$

(c) $\text{Div Curl } \vec{V} = 0$

(d) $\text{Div}\left(\phi\vec{V}\right) = \phi Div\vec{V}$

[ME GATE 2015]

65. The magnitude of the directional derivative of the function $f(x, y) = x^2 + 3y^2$ in a direction normal to the circle $x^2 + y^2 = 2$, at the point $(1, 1)$, is

(a) $4\sqrt{2}$ (b) $5\sqrt{2}$ (c) $7\sqrt{2}$ (d) $9\sqrt{2}$

[IN GATE 2015]

66. The value of the line integral $\oint_C \bar{F}.\vec{r}'\,ds$, where C is a circle of radius $\dfrac{4}{\sqrt{\pi}}$ units is _____

Here, $\bar{F}(x,y) = y\hat{i} + 2x\,\hat{j}$ and $\vec{r}'$ is the UNIT tangent vector on the curve C at an arc length "s" from a reference point on the curve. $\hat{i}$ and $\hat{j}$ are the basis vectors in the x-y Cartesian reference. In evaluating the line integral, the curve has to be traversed in the counterclockwise direction.

[ME GATE 2016]

67. Suppose C is the closed curve defined as the circle $x^2 + y^2 = 1$ with C oriented counterclockwise. The value of $\oint (xy^2 dx + x^2 y dy)$ over the curve C equals _____.

[EC GATE 2016]

68. A scalar φ potential has the following gradient: $\vec{\nabla}\phi = yz\hat{i} + xz\hat{j} + xy\hat{k}.$ Consider the integral $\int_C \vec{\nabla}\phi.d\vec{r}$ on the curve $\vec{r} = x\hat{i} + y\hat{j} + z\hat{k}$. The curve C is parameterized as follows: $x = t, y = t^2, z = 3t^2$ and $1 \le t \le 3$. The value of the integral is _____.

[ME GATE 2016]

69. The value of the line integral $\int_C \left(2xy^2 dx + 2x^2 y dy + dz\right)$ along a path joining the origin $(0, 0, 0)$ and the point $(1, 1, 1)$ is

(a) 0 (b) 2 (c) 4 (d) 6

[EE GATE 2016]

70. The line integral of the vector field $\vec{F} = 5xz\hat{i} + \left(3x^2 + 2y\right)\hat{j} + x^2 z\hat{k}$ along a path from $(0, 0, 0)$ to $(1, 1, 1)$ parameterized by (t, t^2, t) is _____.

[EE, GATE-2016]

71. Consider the time-varying vector $I = \hat{x}15\cos(\omega t) + \hat{y}5\sin(\omega t)$ in Cartesian coordinates, where $\omega > 0$ is a constant. When the vector magnitude $|I|$ is at its minimum value, the angle θ that I makes with the x axis (in degrees, such that $0 \le \theta \le 180$ is _____.

[EC GATE 2016]

72. The vector that is NOT perpendicular to the vectors $(i + j + k)$ and $(i + 2j + 3k)$ Is _____.

(a) $(i - 2j + k)$ (b) $(-i + 2j - k)$
(c) $(0i + 0j + 0k)$ (d) $(4i + 3j + 5k)$

[IN GATE 2016]

73. The surface integral $\iint_S \vec{F}.\hat{n}ds$ over the surface S of the sphere $x^2 + y^2 + z^2 = 9$, where $\vec{F} = (x + y)\hat{i} + (x + z)\hat{j} + (y + z)\hat{k}$ and $\hat{n}$ is the unit outward surface normal, yields _____.

[ME GATE 2017]

74. Let $I = \int_C (2zdx + 2ydy + 2xdz)$ where x, y, z are *real*, and let C be the straight line segment from point A: $(0, 2, 1)$ to point B: $(4,1,-1)$. The value of I is _____.

[EC GATE 2017]

75. The divergence of the vector field $\vec{V} = x^2\hat{i} + 2y^3\hat{j} + z^4\hat{k}$ at $x = 1$, $y = 2$, $z = 3$ is _____.

[CE GATE 2017]

76. The divergence of the vector $-yi + xj$ is _____.

[ME GATE 2017]

77. For the vector $\vec{V} = 2yz\hat{i} + 3xz\hat{j} + 4xy\hat{k}$, the value of $\vec{\nabla}.\left(\vec{\nabla}\times\vec{V}\right)$ is

[ME GATE 2017]

78. If the vector function $\vec{F} = \hat{a}_x\left(3y - k_1 z\right) + \hat{a}_y\left(k_2 x - 2z\right) - \hat{a}_z\left(k_3 y + z\right)$ is irrotational, then the values of the constants k_1, k_2 and k_3, respectively, are

(a) $0.3, -2.5, 0.5$ (b) $0.0,3.0,2.0$
(c) $0.3,0.33,0.5$ (d) $4.0,3.0,2.0$

[EC GATE 2017]

79. The divergence of the vector field $\vec{u} = e^x(\cos y\hat{i} + \sin y\hat{j})$ is

[ME GATE 2018]

(a) 0 (b) $e^x \cos y + e^x \sin y$
(c) $2e^x \cos y$ (d) $2e^x \sin y$

80. The value (up to two decimal places) of a line integral $\int_C \vec{F}(\vec{r}).d\vec{r}$, for $\vec{F}(\vec{r}) = x^2\hat{i} + y^2\hat{j}$ along C which is a straight line joining $(0, 0)$ to $(1, 1)$ is _____ **[CE GATE 2018]**

81. As shown in the figure, C is the arc from the point $(3, 0)$ to the point $(0, 3)$ on the circle $x^2 + y^2 = 9$. The value of the integral $\int_C \{(y^2 + 2xy)dx + (2xy + x^2)dy\}$ is _____ (up to 2 decimal places).

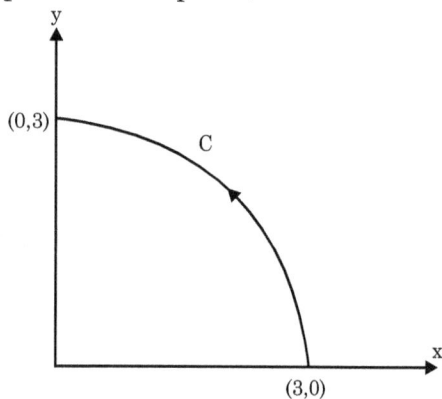

[EE GATE 2018]

82. Given $\vec{F} = (x^2 - 2y)\hat{i} - 4yz\hat{j} + 4xz^2\hat{k}$, the value of the line integral $\int_C \vec{F}.d\vec{l}$ along the straight line C from $(0, 0, 0)$ to $(1, 1, 1)$ is **[IN GATE 2018]**
(a) 3/16 (b) 0 (c) – 5/12 (d) –1

83. The value of the integral $\oiint_S \vec{r}.\hat{n}\,dS$ over the closed surface S bounding a volume V, where $\vec{r} = x\hat{i} + y\hat{j} + z\hat{k}$ is the position **[ME GATE 2018]**
(a) V (b) 2V (c) 3V (d) 4V

84. The value of the surface integral $\iint_S (9x\hat{i} - 2y\hat{j} - z\hat{k}).\hat{n}\,ds$ over the surface S of the sphere $x^2 + y^2 + z^2 = 9$, where $\hat{n}$ is the unit outward normal to the surface element dS, is _____ **[PI GATE 2018]**

85. Vector triple product $\vec{a} \times (\vec{b} \times \vec{c})$ of three vectors $\vec{a}, \vec{b}$ and $\vec{c}$ is given by **[PI GATE 2018]**
(a) $(\vec{a} \bullet \vec{c})\vec{b} - (\vec{a} \bullet \vec{b})\vec{c}$
(b) $(\vec{b} \bullet \vec{c})\vec{a} - (\vec{a} \bullet \vec{c})\vec{b}$
(c) $(\vec{a} \bullet \vec{b})\vec{c} - (\vec{a} \bullet \vec{c})\vec{b}$
(d) $(\vec{b} \bullet \vec{c})\vec{a} - (\vec{a} \bullet \vec{b})\vec{c}$

86. The value of the directional derivative of the function $\phi(x,y,z) = xy^2 + yz^2 + zx^2$ at the point $(2,-1, 1)$ in the direction of the vector $\vec{P} = \hat{i} + 2\hat{j} + 2\hat{k}$ is **[CE GATE 2018]**
(a) 1 (b) 0.95 (c) 0.93 (d) 0.9

87. For a position vector $\vec{r} = x\hat{i} + y\hat{j} + z\hat{k}$ the norm of the vector can be defined as $|\vec{r}| = \sqrt{x^2 + y^2 + z^2}$. Given a function $\phi = \ln|\vec{r}|$, its gradient $\vec{\nabla}\phi$ is **[ME GATE 2018]**
(a) $\vec{r}$ (b) $\dfrac{\vec{r}}{|\vec{r}|}$ (c) $\dfrac{\vec{r}}{\vec{r}.\vec{r}}$ (d) $\dfrac{\vec{r}}{|\vec{r}|^3}$

Answer Key				
1. (a)	2. (b)	3. (c)	4. (c)	5. (c)
6. (a)	7. (c)	8. (a)	9. (d)	10. (a)
11. (a)	12. (c)	13. (d)	14. (a)	15. (d)
16. (b)	17. (b)	18. (d)	19. (b)	20. (d)
21. (c)	22. (d)	23. (b)	24. (a)	25. (b)
26. (d)	27. (b)	28. (b)	29. (c)	30. (c)
31. (d)	32. (a)	33. (a)	34. (d)	35. (a)
36. (b)	37. (a)	38. (a)	39. (a)	40. (a)
41. (a)	42. (d)	43. (b)	44. (d)	45. (d)
46. (d)	47. 1	48. (c)	49. (b)	50. (b)
51. 12	52. 3.	53. 3.	54. (a)	55. (c)
56. (a)	57. (b)	58. (c)	59. 5/3.	60. 216.
61. (a)	62. $7\sqrt{6}$		63. (a)	64. (c)
65. (a)	66. 16.	67. 0	68. 726	69. (b)
70. 53/12		71. 90°	72. (d)	73. 72p
74. –11.	75. 134.	76. 0	77. 0.	78. (b)
79. (c)	80. 0.67	81. 0	82. (d)	83. (c)
84. 678.58		85. (a)	86. (a)	87. (c)

Explanation

1. (a)

$\left|\overline{R}(t)\right| = R\,(say) = $ constant (given)

Now, $\overline{R}.\overline{R} = \left|\overline{R}\right|^2 = R^2$

$\Rightarrow \dfrac{d}{dt}\left(\overline{R}.\overline{R}\right) = 0 \Rightarrow \overline{R}.\dfrac{d\overline{R}}{dt} + \dfrac{d\overline{R}}{dt}.\overline{R} = 0$

$\Rightarrow 2\overline{R}.\dfrac{d\overline{R}}{dt} = 0 \Rightarrow \overline{R}.\dfrac{d\overline{R}}{dt} = 0$

2. (b)

$f(x,y) = x^2 + y^2$

$\Rightarrow \vec{\nabla}f = \hat{i}\dfrac{\partial f}{\partial x} + \hat{j}\dfrac{\partial f}{\partial y} = 2x\hat{i} + 2y\hat{j}.$

$\therefore$ At $(1,2), \vec{\nabla}f = 2\hat{i} + 4\hat{j}.$

Let $\vec{\alpha} = 4\hat{i} + 3\hat{j}.$ Then $\left|\vec{\alpha}\right| = \sqrt{4^2 + 3^2} = 5.$

Hence, the required directional derivative

$= \vec{\nabla}f.\dfrac{\vec{\alpha}}{\left|\vec{\alpha}\right|} = \left(2\hat{i} + 4\hat{j}\right).\dfrac{\left(4\hat{i} + 3\hat{j}\right)}{5} = \dfrac{4\times 2 + 3\times 4}{5} = 4.$

3. (c)

$div\,\vec{F} = \dfrac{\partial}{\partial x}(x) + \dfrac{\partial}{\partial y}(-y) = 1 + (-1) = 0$

$\therefore \vec{F}$ is divergence free, *i.e.*, solenoidal.

Now curl $\vec{F} = \begin{vmatrix} \vec{i} & \vec{j} & \vec{k} \\ \dfrac{\partial}{\partial x} & \dfrac{\partial}{\partial y} & \dfrac{\partial}{\partial z} \\ x & -y & 0 \end{vmatrix}$

$= \vec{i}\left\{\dfrac{\partial}{\partial y}(0) - \dfrac{\partial}{\partial z}(-y)\right\} - \vec{j}\left\{\dfrac{\partial}{\partial x}(0) - \dfrac{\partial}{\partial z}(x)\right\}$

$+ \vec{k}\left\{\dfrac{\partial}{\partial x}(-y) - \dfrac{\partial}{\partial y}(x)\right\} = 0\vec{i} - 0\vec{j} + 0\vec{k} = \vec{0}.$

$\therefore \vec{F}$ is irrational.

4. (c) Here,

$\overrightarrow{OP} = 0.866\hat{i} + 0.500\hat{j} + 0\hat{k} = 0.866\hat{i} + 0.500\hat{j},$

$\overrightarrow{OQ} = 0.259\hat{i} + 0.966\hat{j} + 0\hat{k} = 0.259\hat{i} + 0.966\hat{j}.$

Let θ be the angle between $\overrightarrow{OP}$ and $\overrightarrow{OQ}$.

Then, $\overrightarrow{OP}.\overrightarrow{OQ} = \left|\overrightarrow{OP}\right|\left|\overrightarrow{OQ}\right|\cos\theta$

$\Rightarrow \left(0.866\hat{i} + 0.5\hat{j}\right).\left(0.259\hat{i} + 0.966\hat{j}\right)$

$= \sqrt{(0.866)^2 + (0.5)^2}\sqrt{(0.259)^2 + (0.966)^2}\cos\theta$

$\Rightarrow \cos\theta = \dfrac{(0.866\times 0.259) + (0.5\times 0.966)}{\sqrt{(0.866)^2 + (0.5)^2}\sqrt{(0.259)^2 + (0.966)^2}}$

$\simeq 0.7 = \cos 45°$

$\therefore \theta = 45°$

5. (c)

$u = \dfrac{x^2}{2} + \dfrac{y^2}{3}$

$\Rightarrow grad\,u = \hat{i}\dfrac{\partial u}{\partial u} + \hat{j}\dfrac{\partial u}{\partial y} = \hat{i}x + \hat{j}\dfrac{2}{3}y.$

At $(1,3,), grad\,u = \hat{i} + \hat{j}\left(\dfrac{2}{3}\times 3\right) = \hat{i} + 2\hat{j}.$

So $\left|grad\,u\right| = \sqrt{1^2 + 2^2} = \sqrt{5}.$

6. (a)

Let $\hat{a} = $ unit vector along the line $y = x$

Then, $\hat{a} = x\hat{i} + x\hat{j} = x\left(\hat{i} + \hat{j}\right).$

$\hat{a}$ is a unit vector

$\Rightarrow \left|\hat{a}\right| = 1 \Rightarrow \sqrt{x^2 + x^2} = 1 \Rightarrow x = \dfrac{1}{\sqrt{2}}.$

$\therefore \hat{a} = \dfrac{1}{\sqrt{2}}\left(\hat{i} + \hat{j}\right).$

Now, $f = x^{2/3} + y^{2/3}$

$\Rightarrow \vec{\nabla}f = \hat{i}\dfrac{\partial f}{\partial x} + \hat{j}\dfrac{\partial f}{\partial y} = \hat{i}\left(\dfrac{2}{3}x^{-1/3}\right) + \hat{j}\left(\dfrac{2}{3}y^{-1/3}\right)$

At $(8, 8),$

$\vec{\nabla}f = \left(\dfrac{2}{3}\times 8^{-1/3}\right)\hat{i} + \left(\dfrac{2}{3}\times 8^{-1/3}\right)\hat{j} = \dfrac{1}{3}\hat{i} + \dfrac{1}{3}\hat{j}.$

$\therefore$ Required directional derivative

$= \vec{\nabla}f.\dfrac{\hat{a}}{\left|\hat{a}\right|}$

$= \left(\dfrac{1}{3}\hat{i} + \dfrac{1}{3}\hat{j}\right).\dfrac{1}{\sqrt{2}}\left(\hat{i} + \hat{j}\right) = \dfrac{1}{3\sqrt{2}} + \dfrac{1}{3\sqrt{2}} = \dfrac{\sqrt{2}}{3}.$

7. (c) Comparing $\oint_C \left(xy\,dy - y^2\,dx \right)$ with

$\oint_C \left[\phi(x,y)\,dx + \psi(x,y)\,dy \right]$, we get $\phi(x,y) = -y^2$

and $\psi(x,y) = xy$.

$\therefore \dfrac{\partial \psi}{\partial x} = y$ and $\dfrac{\partial \phi}{\partial y} = -2y$.

Clearly in region R bounded by the closed curve C, "x" varies from 0 to 1 and "y" varies from 0 to 1.

Now by Green's theorem, we have

$$\oint_C \left[\phi(x,y)\,dx + \psi(x,y)\,dy \right] = \iint_R \left(\frac{\partial \psi}{\partial x} - \frac{\partial \phi}{\partial y} \right) dx\,dy$$

or, $\oint_C \left(-y^2\,dx + xy\,dy \right) = \displaystyle\int_{y=0}^{1} \int_{x=0}^{1} (y + 2y)\,dx\,dy$

$= \displaystyle\int_{y=0}^{1} 3y\,dy \times \int_{x=0}^{1} dx$

$= \left[3\dfrac{y^2}{2} \right]_0^1 \times [x]_0^1 = \dfrac{3}{2} \times 1 = \dfrac{3}{2}.$

8. (a)

9. (d)

10. (a)

$\vec{V}.\vec{dr} = \left(2xyz\hat{i} + x^2 z\hat{j} + x^2 y\hat{k} \right).\left(dx\hat{i} + dy\hat{j} + dz\hat{k} \right)$

$= 2xyz\,dx + x^2 z\,dy + x^2 y\,dz$

$= d\left(x^2 yz \right)$

$\therefore$ Required line integral

$= \displaystyle\int_{(0,0,0)}^{(1,1,1)} \vec{V}.\vec{dr} = \int_{(0,0,0)}^{(1,1,1)} d\left(x^2 yz \right)$

$= \left[x^2 yz \right]_{(0,0,0)}^{(1,1,1)} = 1 \times 1 \times 1 - 0 = 1$

11. (a)

12. (c)

$f(x,y,z) = 2x^2 + 3y^2 + z^2$

$\Rightarrow \vec{\nabla} f = \hat{i}\dfrac{\partial f}{\partial x} + \hat{j}\dfrac{\partial f}{\partial y} + \hat{k}\dfrac{\partial f}{\partial z}$

$= 4x\hat{i} + 6y\hat{j} + 2z\hat{k}$

At $P(2, 1, 3)$,

$\vec{\nabla} f = (4 \times 2)\hat{i} + (6 \times 1)\hat{j} + (2 \times 3)\hat{k} = 8\hat{i} + 6\hat{j} + 6\hat{k}.$

Given $\vec{a} = \hat{i} - 2\hat{k}$. Then $|\vec{a}| = \sqrt{1^2 + (-2)^2} = \sqrt{5}.$

Hence, direction derivative of $f(x, y, z)$ at the point $P(2, 1, 3)$ in the direction of $\vec{a}$.

$= \dfrac{\vec{a}}{|\vec{a}|}.\vec{\nabla} f = \dfrac{\left(\hat{i} - 2\hat{k} \right)}{\sqrt{5}}.\left(8\hat{i} + 6\hat{j} + 6\hat{k} \right)$

$= \dfrac{1}{\sqrt{5}} \left\{ 1 \times 8 + 0 \times 6 + (-2) \times 6 \right\} = -\dfrac{4}{\sqrt{5}} = -1.789.$

13. (d)

We know that $\vec{\alpha} \times (\vec{\beta} \times \vec{\gamma}) = (\vec{\alpha}.\vec{\gamma})\vec{\beta} - (\vec{\alpha}.\vec{\beta})\vec{\gamma}.$

Now replacing $\vec{\alpha}$ by $\vec{\nabla}, \vec{\beta}$ by $\vec{\nabla}$ and $\vec{\gamma}$ by $\vec{P}$, we get,

$\vec{\nabla} \times (\vec{\nabla} \times \vec{P}) = (\vec{\nabla}.\vec{P})\vec{\nabla} - (\vec{\nabla}.\vec{\nabla})\vec{P}$

$= (\vec{\nabla}.\vec{P})\vec{\nabla} - \vec{\nabla}^2 \vec{P}.$

14. (a)

$div(\vec{V}) = \dfrac{\partial}{\partial x} \left\{ -(x \cos xy + y) \right\} + \dfrac{\partial}{\partial y} (y \cos xy)$

$+ \dfrac{\partial}{\partial z} \left(\sin z^2 + x^2 + y^2 \right)$

$= -(\cos xy - xy \sin xy) + \left[\cos xy - xy \sin xy \right]$

$+ 2z \cos z^2$

$= 2z \cos z^2.$

15. (d)

$\vec{V} = 5xy\hat{i} + 2y^2\hat{j} + 3yz^2\hat{k}$

$\Rightarrow div\,\vec{V} = \dfrac{\partial}{\partial x}(5xy) + \dfrac{\partial}{\partial y}(2y^2) + \dfrac{\partial}{\partial z}(3yz^2)$

$= 5y + 4y + 6yz.$

$\therefore div\,\vec{V}\Big|_{(1,1,1)} = (5 \times 1) + (4 \times 1) + (6 \times 1 \times 1) = 15.$

16. (b)

Let $x = x_1\hat{i} + x_2\hat{j} + x_3\hat{k}$ and $y = y_1\hat{i} + y_2\hat{j} + y_3\hat{k}.$

Then $x.x = |x|^2 = x_1^2 + x_2^2 + x_3^2,$

$y.y = \left|y\right|^2 = y_1^2 + y_2^2 + y_3^2$ and

$x.y = y.x = x_1y_1 + x_2y_2 + x_3y_3.$

Then, $\det \begin{bmatrix} \langle x,x \rangle & \langle x,y \rangle \\ \langle y,x \rangle & \langle y,y \rangle \end{bmatrix}$

$= \begin{bmatrix} x.x & x.y \\ y.x & y.y \end{bmatrix}$

$= \begin{vmatrix} x_1^2 + x_2^2 + x_3^2 & x_1y_1 + x_2y_2 + x_3y_3 \\ x_1y_1 + x_2y_2 + x_3y_3 & y_1^2 + y_2^2 + y_3^2 \end{vmatrix}$

$= \left(x_1^2 + x_2^2 + x_3^2\right)\left(y_1^2 + y_2^2 + y_3^2\right)$

$- \left(x_1y_1 + x_2y_2 + x_3y_3\right)^2$

$= x_1^2y_1^2 + x_1^2y_2^2 + x_1^2y_3^2 + x_2^2y_1^2 + x_2^2y_2^2 + x_2^2y_3^2$

$+x_3^2y_1^2 + x_3^2y_2^2 + x_3^2y_3^2 - \left(x_1^2y_1^2 + x_2^2y_2^2 + x_3^2y_3^2\right.$

$\left.+2x_1x_2y_1y_2 + 2x_2x_3y_2y_3 + 2x_1x_3y_1y_3\right)$

$= \left(x_1^2y_2^2 - 2x_1x_2y_1y_2 + x_2^2y_1^2\right)$

$+ \left(x_1^2y_3^2 - 2x_1x_3y_1y_3 + x_3^2y_1^2\right)$

$+ \left(x_2^2y_3^2 - 2x_2x_3y_2y_3 + x_3^2y_2^2\right)$

$= \left(x_1y_2 - x_2y_1\right)^2 + \left(x_1y_3 - x_3y_1\right)^2$

$+ \left(x_2y_3 - x_3y_2\right)^2 \geq 0$

If x and y are linearly dependent, then

$\dfrac{x_1}{y_1} = \dfrac{x_2}{y_2} = \dfrac{x_3}{y_3}$ and so $x_1y_2 = y_1x_2, x_2y_3 = x_3y_2$ and

$x_1y_3 = x_3y_1.$

Then determinant $= 0^2 + 0^2 + 0^2 = 0.$

Hence, x and y are linearly independent $\Rightarrow$ determinant is positive.

17. (b) Let ΔABC be the triangle and position vectors of the vertices, A, B, C be, respectively, $\vec{a}, \vec{b}, \vec{c}$. Then, $\overrightarrow{OA} = \vec{a}, \overrightarrow{OB} = \vec{b}$ and $\overrightarrow{OC} = \vec{c}.$

Then area (ΔABC)

$= \dfrac{1}{2}\left|\overrightarrow{AB} \times \overrightarrow{AC}\right| = \dfrac{1}{2}\left(\overrightarrow{OB} - \overrightarrow{OA}\right) \times \left(\overrightarrow{OC} - \overrightarrow{OA}\right)$

$= \dfrac{1}{2}\left|\left(\vec{b} - \vec{a}\right) \times \left(\vec{c} - \vec{a}\right)\right| = \dfrac{1}{2}\left|\left(\vec{a} - \vec{b}\right) \times \left(\vec{a} - \vec{c}\right)\right|.$

18. (d)

$\vec{a} = \dfrac{\sqrt{3}}{2}\hat{i} + \dfrac{1}{2}\hat{j} \Rightarrow \left|\vec{a}\right| = \sqrt{\left(\dfrac{\sqrt{3}}{2}\right)^2\left(\dfrac{1}{2}\right)^2} = 1$ and

$\vec{b} = \sqrt{\left(-\dfrac{\sqrt{3}}{2}\right)^2\left(\dfrac{1}{2}\right)^2} = 1.$

Now, $\cos\theta = \dfrac{\vec{a}.\vec{b}}{\left|\vec{a}\right|\left|\vec{b}\right|} = \dfrac{\dfrac{\sqrt{3}}{2} \times \left(-\dfrac{\sqrt{3}}{2}\right) + \dfrac{1}{2} \times \dfrac{1}{2}}{1 \times 1}$

$= -\dfrac{1}{2} = \cos 120°$

$\therefore \theta = 120°.$

19. (b)

$2\displaystyle\int_P^Q (xdx + ydy)$

$= 2\displaystyle\iint_R \left[\dfrac{\partial}{\partial x}(y) - \dfrac{\partial}{\partial y}(x)\right]dx\,dy$ (by Green's theorem)

$= 2\displaystyle\iint_R [0 - 0]dx\,dy = 0.$

20. (d)

$\displaystyle\iint_S \vec{r}.d\vec{s}$

$= \displaystyle\iiint_V (\vec{\nabla}.\vec{r})dV$ (by the Gauss Divergence theorem)

$= \displaystyle\iiint_V 3\,dV$

$\left(\because \vec{\nabla}.\vec{r} = \left(\hat{i}\dfrac{\partial}{\partial x} + \hat{j}\dfrac{\partial}{\partial y} + \hat{k}\dfrac{\partial}{\partial z}\right).\left(x\hat{i} + y\hat{j} + z\hat{k}\right)\right.$

$= \dfrac{\partial}{\partial x}(x) + \dfrac{\partial}{\partial y}(y) + \dfrac{\partial}{\partial z}(z)$

$\left. = 1 + 1 + 1 = 3\right)$

$= 3\displaystyle\iiint_V dV$

$= 3V \left(\because \displaystyle\iiint_V dV \text{ represents the volume } V\right)$

21. (c)

$$\vec{P}.\vec{Q} = |\vec{P}||\vec{Q}|\cos\theta = 0$$

$$\Rightarrow \cos\theta = 0$$

$(\because \vec{P}\,\&\,\vec{Q}$ are nonzero vectors, and so on $|\vec{P}| \neq 0; |\vec{Q}| \neq 0)$

$$\Rightarrow \theta = 90°.$$

22. (d)

Let $\vec{\alpha} = (x-y)\hat{i} + (y-x)\hat{j} + (x+y+z)\hat{k}.$
Then,

$$div\vec{\alpha} = \frac{\partial}{\partial x}(x-y) + \frac{\partial}{\partial y}(y-x) + \frac{\partial}{\partial z}(x+y+z)$$

$$= 1+1+1 = 3.$$

23. (b)

$$f(x,y,z) = x^2 + 2y^2 + z$$

$$\Rightarrow \vec{\nabla}f = \hat{i}\frac{\partial f}{\partial y} + \hat{j}\frac{\partial f}{\partial y} + \hat{k}\frac{\partial f}{\partial z} = 2x\hat{i} + 4y\hat{j} + \hat{k}.$$

$\therefore$ At $P(1,1,2), \vec{\nabla}f = 2\hat{i} + 4\hat{j} + \hat{k}.$

Again, $\vec{\alpha} = 3\hat{i} - 4\hat{j} \Rightarrow |\vec{\alpha}| = \sqrt{3^2 + (-4)^2} = 5.$

$\therefore$ Required directional derivative at $P(1, 1, 2)$

$$= \frac{\vec{\alpha}}{|\vec{\alpha}|}.\vec{\nabla}f = \frac{1}{5}(3\hat{i} - 4\hat{j}).(2\hat{i} - 4\hat{j} + \hat{k})$$

$$= \frac{1}{5}\{3\times2 + (-4)\times4 + 0\times1\} = -2.$$

24. (a) Equation of the sphere of unit radius is
$$x^2 + y^2 + z^2 = 1$$

Let, $\phi(x,y,z) = x^2 + y^2 + z^2 = 1.$

Now normal at (x, y, z) on the surface of the sphere

$$= \vec{\nabla}\phi = \hat{i}\frac{\partial\phi}{\partial x} + \hat{j}\frac{\partial\phi}{\partial y} + \hat{k}\frac{\partial\phi}{\partial z} = 2x\hat{i} + 2y\hat{j} + 2z\hat{k}$$

$\therefore$ unit normal at (x, y, z) on the surface of the sphere

$$= \frac{\vec{\nabla}\phi}{|\vec{\nabla}\phi|} = \frac{2x\hat{i} + 2y\hat{j} + 2z\hat{k}}{\sqrt{(2x)^2 + (2y)^2 + (2z)^2}} = \frac{2x\hat{i} + 2y\hat{j} + 2z\hat{k}}{\sqrt{4(x^2 + y^2 + z^2)}}$$

$$= x\hat{i} + y\hat{j} + z\hat{k} \,(\because x^2 + y^2 + z^2 = 1), \text{ which rep-}$$
resents the position vector of the point $(x, y, z).$

25. (b) By Stokes' theorem,

$$\oint_C \vec{A}.\vec{dl} = \iint_{S_c}(\vec{\nabla}\times\vec{A}).\vec{ds} = \iint_{S_c}\vec{V}.\vec{ds}$$

$$(\because \vec{V} = \vec{\nabla}\times\vec{A})$$

26. (d) The equation of the line passing through the points $(0, 2)$ and $(2, 0)$ is
$$\frac{x-0}{2-0} = \frac{y-2}{0-2}, \text{ i.e., } y = 2-x \text{ and so } dy = -dx.$$
Clearly, x varies from 0 to 2.

$\therefore$ Required line integral $= \int \vec{F}.\vec{dr}$

$$= \int\{(x^2 + xy)\hat{a}_x + (y^2 + xy)\hat{a}_y\}.(dx\hat{a}_x + dy\hat{a}_y)$$

$$= \int[(x^2 + xy)dx + (y^2 + xy)dy]$$

$$= \int_0^2[\{x^2 + x(2-x)\}dx + \{(2-x)^2 + x(2-x)\}(-dx)]$$

$$= \int_0^2[x^2 + x - x^2 - (4 - 4x + x^2 + 2x - x^2)]dx$$

$$= \int_0^2(4x - 4)dx = [2x^2 - 4x]_0^2 = 8 - 8 - 0 = 0.$$

27. (b)

$$f(x,y,z) = x^2 + 3y^2 + 2z^2$$

$$\Rightarrow grad\,f = \hat{i}\frac{\partial f}{\partial x} + \hat{j}\frac{\partial f}{\partial y} + \hat{k}\frac{\partial f}{\partial z} = 2x\hat{i} + 6y\hat{j} + 4z\hat{k}$$

$\therefore$ At $P(1,2,-1),$

$$grad\,f = (2\times1)\hat{i} + (6\times2)\hat{j} + 4\times(-1)\hat{k}$$

$$= 2\hat{i} + 12\hat{j} - 4\hat{k}.$$

28. (b)

Given, $f(x,y,z) = x^2 + 3y^2 + 2z^2.$
Then, $\vec{\nabla}f = \hat{i}\frac{\partial f}{\partial x} + \hat{j}\frac{\partial f}{\partial y} + \hat{k}\frac{\partial f}{\partial z} = 2x\hat{i} + 6y\hat{j} + 4z\hat{k}.$

$$\therefore \vec{\nabla}f\Big|_{(1,2,-1)} = (2\times1)\hat{i} + (6\times2)\hat{j} + 4\times(-1)\hat{k}$$

$$= 2\hat{i} + 12\hat{j} - 4\hat{k}.$$

Let $\vec{\alpha} = \hat{i} - \hat{j} + 2\hat{k}.$ Then,
$$|\vec{\alpha}| = \sqrt{1^2 + (-1)^2 + 2^2} = \sqrt{6}.$$

∴ Required directional derivative

$$= \frac{\vec{\alpha}}{|\vec{\alpha}|} \cdot \vec{\nabla} f \Big|_{(1,2,-1)}$$

$$= \frac{1}{\sqrt{6}} \left(\hat{i} - \hat{j} + 2\hat{k} \right) \cdot \left(2\hat{i} + 12\hat{j} - 4\hat{k} \right)$$

$$= \frac{1}{\sqrt{6}} \left\{ 1 \times 2 + (-1) \times 12 + 2 \times (-4) \right\}$$

$$= \frac{1}{\sqrt{6}} \times (-18) = -\frac{18\sqrt{6}}{\sqrt{6} \times \sqrt{6}} = \frac{-18\sqrt{6}}{6} = -3\sqrt{6}.$$

29. (c) Let $\vec{f}(x,y,z) = 3xz\hat{i} + 2xy\hat{j} - yz^2\hat{k}$

Then, $div\vec{f} = \dfrac{\partial}{\partial x}(3xz) + \dfrac{\partial}{\partial y}(2xy) + \dfrac{\partial}{\partial z}(-yz^2)$

$$= 3z + 2x - 2yz.$$

∴ At $(1,1,1), div\vec{f} = 3 + 2 - 2 = 3.$

30. (c) Here, $\vec{A} = xy\hat{a}_x + x^2\hat{a}_y$ and $\vec{l} = x\hat{a}_x + y\hat{a}_y.$

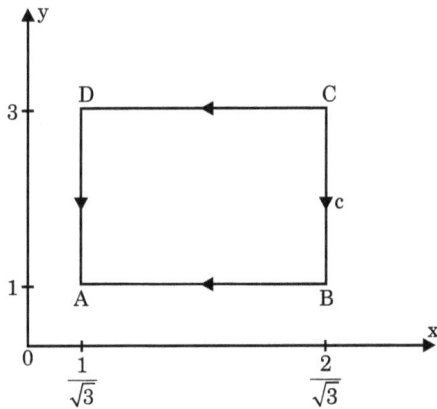

Then, $\displaystyle\oint_C \vec{A} \cdot d\vec{l}$

$$= \oint_C \left(xy\hat{a}_x + x^2\hat{a}_y \right) \cdot \left(dx\hat{a}_x + dy\hat{a}_y \right)$$

$$= \oint_C \left(xy\,dx + x^2\,dy \right)$$

$$= \oint_{AB} \left(xy\,dx + x^2\,dy \right) + \oint_{BC} \left(xy\,dx + x^2\,dy \right)$$

$$+ \oint_{CD} \left(xy\,dx + x^2\,dy \right) + \oint_{DA} \left(xy\,dx + x^2\,dy \right) \quad \dots (i)$$

Along AB:

x varies from $1/\sqrt{3}$ to $2/\sqrt{3}, y = 1$ and so $dy = 0.$

Hence, $\displaystyle\oint_{AB} \left(xy\,dx + x^2\,dy \right)$

$$= \int_{1/\sqrt{3}}^{2/\sqrt{3}} x\,dx = \left[\frac{x^2}{2} \right]_{1/\sqrt{3}}^{2/\sqrt{3}} = \frac{1}{2} \left(\frac{4}{3} - \frac{1}{3} \right) = \frac{1}{2}.$$

Along BC:

y varies from 1 to 3, $x = \dfrac{2}{\sqrt{3}}$ and so $dx = 0.$

Hence, $\displaystyle\oint_{BC} \left(xy\,dx + x^2\,dy \right)$

$$= \int_1^3 \frac{4}{3}\,dy = \frac{4}{3}[y]_1^3 = \frac{4}{3}(3-1) = \frac{8}{3}.$$

Along CD:

x varies from $2/\sqrt{3}$ to $1/\sqrt{3}, y = 3$ and so $dy = 0.$

$$\therefore \oint_{CD} \left(xy\,dx + x^2\,dy \right)$$

$$= \int_{2/\sqrt{3}}^{1/\sqrt{3}} 3x\,dx = 3\left[\frac{x^2}{2} \right]_{2/\sqrt{3}}^{1/\sqrt{3}} = \frac{3}{2} \left(\frac{1}{3} - \frac{4}{3} \right) = -\frac{3}{2}.$$

Along DA:

y varies from 3 to 1, $x = \dfrac{1}{\sqrt{3}}$ and so $dx = 0.$

Hence, $\displaystyle\oint_{DA} \left(xy\,dx + x^2\,dy \right)$

$$= \int_3^1 \frac{1}{3}\,dy = \frac{1}{3}[y]_3^1 = \frac{1}{3}(1-3) = -\frac{2}{3}.$$

Hence, (i) gives,

$$\oint_C \vec{A} \cdot d\vec{l} = \frac{1}{2} + \frac{8}{3} + \left(-\frac{3}{2} \right) + \left(-\frac{2}{3} \right) = 1.$$

31. (d) Given, velocity vector, $\vec{V} = 3xy\hat{i} - x^2z\hat{j}.$

Then vorticity vector, curl $\vec{V}$

$$= \begin{vmatrix} \hat{i} & \hat{j} & \hat{k} \\ \dfrac{\partial}{\partial x} & \dfrac{\partial}{\partial y} & \dfrac{\partial}{\partial z} \\ 2xy & -x^2z & 0 \end{vmatrix}$$

$$= \hat{i}\left\{ 0 + \frac{\partial}{\partial z}(x^2z) \right\} - \hat{j}\left\{ 0 - \frac{\partial}{\partial z}(2xy) \right\}$$

$$+ \hat{k}\left\{ \frac{\partial}{\partial x}(-x^2z) - \frac{\partial}{\partial y}(2xy) \right\}$$

$$= x^2\hat{i} + \vec{0} - (2xz + 2x)\hat{k}$$

$$\therefore \quad \text{At } (1,1,1), curl\,\vec{V} = 1\hat{i} - (2+2)\hat{k} = \hat{i} - 4\hat{k}.$$

32. (a)

We know that $\vec{r} = x\hat{i} + y\hat{j} + z\hat{k}$

$$\therefore div\left(\vec{r}\right) = \frac{\partial}{\partial x}(x) + \frac{\partial}{\partial y}(y) + \frac{\partial}{\partial z}(z) = 1+1+1 = 3.$$

33. (a) Given that temperature,

$T(x,y,z) = x^2 + y^2 + 2z^2$. Then temperature

gradient $= \vec{\nabla}T = \hat{i}\dfrac{\partial T}{\partial x} + \hat{j}\dfrac{\partial T}{\partial y} + \hat{k}\dfrac{\partial T}{\partial z}$

$$= 2x\hat{i} + 2y\hat{j} + 4z\hat{k}.$$

$\therefore$ At $P(1,1,1)$, temperature gradient $= 2\hat{i} + 2\hat{j} + 4\hat{k}$ and so its magnitude

$$= \sqrt{2^2 + 2^2 + 4^2} = 2\sqrt{6}.$$

34. (d)

$$\oiint 5\vec{r}.\hat{n}ds$$

$$= 5\oiint \vec{r}.\hat{n}ds = 5\iiint_V \left(\vec{\nabla}.\vec{r}\right)dV$$

(by the Gauss Divergence theorem)

$$= 5\iiint_V 3dV = 15\iiint_V dV = 15V.$$

$$\left[\because \vec{\nabla}.\vec{r} = \left(\hat{i}\frac{\partial}{\partial x} + \hat{j}\frac{\partial}{\partial y} + \hat{k}\frac{\partial}{\partial z} \right).\left(x\hat{i} + y\hat{j} + z\hat{k} \right) \right.$$

$$\left. = \frac{\partial}{\partial x}(x) + \frac{\partial}{\partial y}(y) + \frac{\partial}{\partial z}(z) = 1+1+1 = 3 \right]$$

35. (a) If θ be the angle between $\vec{a}$ and $\vec{b}$, then

$$\vec{a}.\vec{b} = |\vec{a}||\vec{b}|\cos\theta \text{ and } \vec{a}\times\vec{b} = |\vec{a}||\vec{b}|\sin\theta\,\hat{n}.$$

$$\therefore \left(\vec{a}.\vec{b}\right)^2 = |\vec{a}|^2|\vec{b}|^2\cos^2\theta \dots\dots(1)$$

$$\left|\vec{a}\times\vec{b}\right|^2 = |\vec{a}|^2|\vec{b}|^2\sin^2\theta \dots\dots\dots(2)$$

$$\left(\because (\hat{n})^2 = |\hat{n}|^2 = 1 \right)$$

$$(1) + (2) \Rightarrow \left(\vec{a}.\vec{b}\right)^2 + \left|\vec{a}\times\vec{b}\right|^2$$

$$= |\vec{a}|^2|\vec{b}|^2\left(\cos^2\theta + \sin^2\theta\right) = |\vec{a}|^2|\vec{b}|^2 = a^2b^2$$

$$\Rightarrow \left|\vec{a}\times\vec{b}\right|^2 = a^2b^2 - \left(\vec{a}.\vec{b}\right)^2.$$

36. (b)

$$a = -\frac{1}{2} + \frac{\sqrt{3}}{2}j \Rightarrow a^2 = -\frac{1}{2} - \frac{\sqrt{3}}{2}j.$$

$$\therefore [1,1,1].[1,a,a^2]$$

$$= \left(\hat{i} + \hat{j} + \hat{k}\right).\left(\hat{i} + a\hat{j} + a^2\hat{k}\right)$$

$$= 1 + a + a^2$$

$$= 1 + \left(-\frac{1}{2} + j\frac{\sqrt{3}}{2}\right) + \left(-\frac{1}{2} - j\frac{\sqrt{3}}{2}\right) = 0.$$

Hence, the given vectors are orthogonal (perpendicular).

37. (a)

$$\int_{P_1}^{P_2} (ydx + xdy)$$

$$= \int_{P_1}^{P_2} d(xy) = \left[xy\right]_{P_1}^{P_2} = \left[xy\right]_{(x_1,y_1)}^{(x_2,y_2)} = x_2y_2 - x_1y_1.$$

38. (a) The area of the parallelogram $OPQR$

$$= \left|\overrightarrow{OP}\times\overrightarrow{OQ}\right|$$

$$= ad - bc$$

$$\left[\because \overrightarrow{OP}\times\overrightarrow{OQ} = \begin{vmatrix} \hat{i} & \hat{j} & \hat{k} \\ a & b & 0 \\ c & d & 0 \end{vmatrix} = \hat{k}(ad - bc) \right.$$

$$\left. \text{so } \left|\overrightarrow{OP}\times\overrightarrow{OQ}\right| = \left|\hat{k}\right|(ad - bc) = ad - bc\left(\because \left|\hat{k}\right| = 1 \right) \right]$$

39. (a)

Let $f(x,y,z) = x^2 + y^2 + z^2 - 1$.

Then,

$$grad\,f = \hat{i}\frac{\partial f}{\partial x} + \hat{j}\frac{\partial f}{\partial y} + \hat{k}\frac{\partial f}{\partial z} = 2x\hat{i} + 2y\hat{j} + 2z\hat{k}.$$

At $\left(\dfrac{1}{\sqrt{2}}, \dfrac{1}{\sqrt{2}}, 0\right)$,

$$grad\,f = \frac{2}{\sqrt{2}}\hat{i} + \frac{2}{\sqrt{2}}\hat{j} + 0\hat{k} = \sqrt{2}\hat{i} + \sqrt{2}\hat{j}.$$

so $|grad\,f| = \sqrt{\left(\sqrt{2}\right)^2 + \left(\sqrt{2}\right)^2} = \sqrt{4} = 2.$

Hence, the required outward normal vector is

$$= \frac{grad\, f}{|grad\, f|} = \frac{\sqrt{2}\hat{i} + \sqrt{2}\hat{j}}{2} = \frac{1}{\sqrt{2}}\hat{i} + \frac{1}{\sqrt{2}}\hat{j}.$$

40. (a)

$$\left|\vec{A}\right| = kr^n \Rightarrow \vec{A} = kr^n \frac{\vec{r}}{|\vec{r}|} = \frac{kr^n \vec{r}}{r} = kr^{n-1}\vec{r}$$

$$k\left(x^2 + y^2 + z^2\right)^{\frac{n-1}{2}}\left(x\hat{i} + y\hat{j} + z\hat{k}\right)$$

$$= k\left(x^2 + y^2 + z^2\right)^{\frac{n-1}{2}} x\hat{i} + k\left(x^2 + y^2 + z^2\right)^{\frac{n-1}{2}} y\hat{j}$$

$$+ k\left(x^2 + y^2 + z^2\right)^{\frac{n-1}{2}} z\hat{k}$$

Then, $\vec{\nabla}.\vec{A} = 0 \Rightarrow div\vec{A} = 0$

$$\Rightarrow \frac{\partial}{\partial x}\left\{kx\left(x^2 + y^2 + z^2\right)^{\frac{n-1}{2}}\right\}$$

$$+ \frac{\partial}{\partial y}\left\{ky\left(x^2 + y^2 + z^2\right)^{\frac{n-1}{2}}\right\}$$

$$+ \frac{\partial}{\partial z}\left\{kz\left(x^2 + y^2 + z^2\right)^{\frac{n-1}{2}}\right\} = 0$$

$$\Rightarrow k\left(x^2 + y^2 + z^2\right)^{\frac{n-1}{2}}$$

$$+ kx \times \left(\frac{n-1}{2}\right) \times 2x\left(x^2 + y^2 + z^2\right)^{\frac{n-3}{2}}$$

$$+ k\left(x^2 + y^2 + z^2\right)^{\frac{n-1}{2}}$$

$$+ ky \times \left(\frac{n-1}{2}\right) \times 2y\left(x^2 + y^2 + z^2\right)^{\frac{n-3}{2}}$$

$$+ k\left(x^2 + y^2 + z^2\right)^{\frac{n-1}{2}}$$

$$+ kz \times \left(\frac{n-1}{2}\right) \times 2z\left(x^2 + y^2 + z^2\right)^{\frac{n-3}{2}} = 0$$

$$\Rightarrow 3k\left(x^2 + y^2 + z^2\right)^{\frac{n-1}{2}}$$

$$+ (n-1)k\{\ x^2\left(x^2 + y^2 + z^2\right)^{\frac{n-3}{2}}$$

$$+ y^2\left(x^2 + y^2 + z^2\right)^{\frac{n-3}{2}} + z^2\left(x^2 + y^2 + z^2\right)^{\frac{n-3}{2}}\right\} = 0$$

$$\Rightarrow 3\left(x^2 + y^2 + z^2\right)^{\frac{n-1}{2}} + (n-1)\times$$

$$\left\{\left(x^2 + y^2 + z^2\right)^{\frac{n-3}{2}}\left(x^2 + y^2 + z^2\right)\right\}$$

$$\Rightarrow 3\left(x^2 + y^2 + z^2\right)^{\frac{n-1}{2}}$$

$$+ (n-1)\left(x^2 + y^2 + z^2\right)^{\frac{n-1}{2}} = 0$$

$$\Rightarrow 3 + n - 1 = 0 \Rightarrow n = -2.$$

41. (a)

$$\iint_S \frac{1}{4}\left(\vec{F}.\vec{n}\right)dA$$

$$= \frac{1}{4}\iiint_V \left(\vec{\nabla}.\vec{F}\right)dV \text{ (by the Gauss Divergence theorem)}$$

$$= \frac{1}{4}\iiint_V 3\, dV$$

$$\left[\because \vec{\nabla}.\vec{F} = \frac{\partial}{\partial x}(x) + \frac{\partial}{\partial y}(y) + \frac{\partial}{\partial x}(z) = 1 + 1 + 1 = 3\right]$$

$$= \frac{3}{4}\iiint_V dV$$

$$= \frac{3}{4}\times \text{ volume of the sphere } x^2 + y^2 + z^2 = 1$$

$$= \frac{3}{4}\times\frac{4}{3}\pi\times 1^3 = \pi.$$

42. (d) By Stokes' theorem,

$$\oint \vec{A}.d\vec{l} = \iint_S (\vec{\nabla}\times\vec{A}).d\vec{s}.$$

43. (b)

$$\vec{l} = x\hat{a}_x + y\hat{a}_y + z\hat{a}_z \text{ and so}$$

$$d\vec{l} = \hat{a}_x dx + \hat{a}_y dy + \hat{a}_z dz.$$

$$\therefore \vec{F}.d\vec{l}$$

$$= \left(y^2 x\hat{a}_x - yz\hat{a}_y - x^2\hat{a}_z dz\right).\left(\hat{a}_x dx + \hat{a}_y dy + \hat{a}_z dz\right)$$

$$= y^2 x dx - yz dy - x^2 dz.$$

Then the required integral

$$= \int_{x=1}^{2} \left(y^2 x\, dx - yz\, dy - x^2 dz \right) = \int_{1}^{2} 0\, dx = 0.$$

($\therefore$ along $x-axis, y = 0, z = 0$ and so $dy = 0, dz = 0$)

44. (d) We know that for a function $\phi(x, y, z)$,
curl $(grad\phi) = 0.$

45. (d)

$$\vec{A} = x\hat{a}_x + y\hat{a}_y + z\hat{a}_z$$

$$\Rightarrow div\vec{A} = \left(\hat{a}_x \frac{\partial}{\partial x} + \hat{a}_y \frac{\partial}{\partial y} + \hat{a}_z \frac{\partial}{\partial z} \right).\left(x\hat{a}_x + y\hat{a}_y + z\hat{a}_z \right)$$

$$\Rightarrow div\vec{A} = \frac{\partial}{\partial x}(x) + \frac{\partial}{\partial y}(y) + \frac{\partial}{\partial z}(z) = 1 + 1 + 1 = 3.$$

46. (d) We know that a vector $\vec{\alpha}$ is called

(i) solenoidal if $div\vec{\alpha} = 0$, i.e., if $\vec{\nabla}.\vec{\alpha} = 0$

(ii) irrotational if curl $\vec{\alpha} = \vec{0}$, i.e., if $\vec{\nabla} \times \vec{\alpha} = \vec{0}$

(iii) conservative if curl $\vec{\alpha} = \vec{0}$, i.e., if $\vec{\nabla} \times \vec{\alpha} = \vec{0}.$

47. 1

$$\oiint \vec{f}.d\vec{s}$$

$$= \iiint_V (\vec{\nabla}.\vec{f})dV$$

(by the Gauss Divergence theorem)

$$= \int_{x=0}^{1} \int_{y=0}^{1} \int_{z=0}^{1} \left[\frac{\partial}{\partial x}(y) + \frac{\partial}{\partial y}(x) + \frac{\partial}{\partial z}(z) \right] dz\, dy\, dx$$

$$= \int_{x=0}^{1} \int_{y=0}^{1} \int_{z=0}^{1} (0 + 0 + 1)dz\, dy\, dx$$

$$= [x]_0^1 \times [y]_0^1 \times [z]_0^1 = 1 \times 1 \times 1 = 1.$$

48. (c)

$$\oint_C (y\, dx - x\, dy)$$

$$= \iint_R \left[\frac{\partial}{\partial x}(-x) - \frac{\partial}{\partial y}(y) \right] dx\, dy$$

(by Green's theorem)

$$= \iint_R (-1-1)dx\, dy = -2\iint_R dx\, dy$$

$$= (-2) \times \text{ area of the circle } x^2 + y^2 = \left(\frac{1}{2}\right)^2$$

$$= (-2) \times \pi \times \left(\frac{1}{2}\right)^2 = -\frac{\pi}{2}.$$

49. (b)

$$\vec{F} = yz\hat{i} \Rightarrow curl\vec{F} = \begin{vmatrix} \hat{i} & \hat{j} & \hat{k} \\ \frac{\partial}{\partial z} & \frac{\partial}{\partial y} & \frac{\partial}{\partial z} \\ yz & 0 & 0 \end{vmatrix}$$

$$= -\hat{j}\left[0 - \frac{\partial}{\partial z}(yz) \right] + \hat{k}\left[0 - \frac{\partial}{\partial y}(yz) \right]$$

$$= y\hat{j} - z\hat{k} = y\hat{j} - \hat{k} \qquad (\because z = 1)$$

By Stokes' theorem,

$$\int_C \vec{F}.d\vec{r}$$

$$= \int_S curl\vec{F}.\hat{n}\, ds = \iint_S (y\hat{j} - \hat{k}).\hat{k}\, ds$$

$$= \iint_s (-1)ds = -\iint_s ds$$

$$= - \text{ surface area of } x^2 + y^2 = 1 = -\pi \times 1^2 = -\pi.$$

50. (b) $5\hat{i} + 6\hat{j} + 4\hat{k} = 3(\hat{i} + \hat{j} + \hat{k}) + 1(2\hat{i} + 3\hat{j} + \hat{k})$

Thus, a vector is expressed as a liner combination of the rest of the vectors. So the given vectors are linearly dependent.

51. 12

Here,

$$\vec{r} = x\hat{i} + y\hat{j} + z\hat{k} = (t^3 + 2t)\hat{i} - 3e^{-2t}\hat{j} + 2\sin 5t\hat{k}$$

$\therefore$ velocity

$$= \frac{d\vec{r}}{dt} = (3t^2 + 2)\hat{i} + 6e^{-2t}\hat{j} + 10\cos 5t\hat{k}.$$

So acceleration $= \dfrac{d^2\vec{r}}{dt^2} = 6t\hat{i} - 12e^{-2t}\hat{j} - 50\sin 5t\hat{k}$

At $t = 0, \dfrac{d^2\vec{r}}{dt^2} = \vec{0} - 12\hat{j} - \vec{0} = -12\hat{j}$

Hence, at $t = 0, \left|\dfrac{d^2\vec{r}}{dt^2}\right| = \sqrt{(-12)^2} = 12.$

52. 3.

We know that $\vec{\nabla}(f(r)) = \dfrac{f'(r)}{r}\vec{r}$ ...(i)

Now take $f(r) = \ln r.$. Then, $f'(r) = \dfrac{1}{r}$.

Then (i) gives, $\vec{\nabla}(\ln r) = \dfrac{1}{r^2}\vec{r}$.

$\therefore div\left(r^2\vec{\nabla}(\ln r)\right)$

$= div\left(\vec{r}\right) = \vec{\nabla}.\vec{r}$

$= \left(\hat{i}\dfrac{\partial}{\partial x} + \hat{j}\dfrac{\partial}{\partial y} + \hat{k}\dfrac{\partial}{\partial z}\right).\left(x\hat{i} + y\hat{j} + z\hat{k}\right)$

$= \dfrac{\partial}{\partial x}(x) + \dfrac{\partial}{\partial y}(y) + \dfrac{\partial}{\partial z}(z)$

$= 1 + 1 + 1 = 3.$

53. 3.

$f(x, y) = \dfrac{x^2 y}{\sqrt{2}} + \dfrac{xy^2}{\sqrt{2}} \Rightarrow \vec{\nabla}f = \hat{i}\dfrac{\partial f}{\partial x} + j\dfrac{\partial f}{\partial y}$

$= \left(\dfrac{2xy}{\sqrt{2}} + \dfrac{y^2}{\sqrt{2}}\right)\hat{i} + \left(\dfrac{x^2}{\sqrt{2}} + \dfrac{2xy}{\sqrt{2}}\right)\hat{j}$

$\therefore$ At $(1,1), \vec{\nabla}f = \left(\dfrac{2}{\sqrt{2}} + \dfrac{1}{\sqrt{2}}\right)\hat{i} + \left(\dfrac{1}{\sqrt{2}} + \dfrac{2}{\sqrt{2}}\right)\hat{j}$

$= \dfrac{3}{\sqrt{2}}\hat{i} + \dfrac{3}{\sqrt{2}}\hat{j}.$

We know that $\hat{n} = \cos\theta\hat{i} + \sin\theta\hat{j}$ is a unit vector that makes an angle θ with x-axis.

Here, $\theta = \dfrac{\pi}{2} - \dfrac{\pi}{4} = \dfrac{\pi}{4}$ and so

$\hat{n} = \cos\dfrac{\pi}{4}\hat{i} + \sin\dfrac{\pi}{4}\hat{j} = \dfrac{1}{\sqrt{2}}\hat{i} + \dfrac{1}{\sqrt{2}}\hat{j}$

Hence, the required direction derivative

$= \vec{\nabla}f.\dfrac{\hat{n}}{|\hat{n}|} = \left(\dfrac{3}{\sqrt{2}}\hat{i} + \dfrac{3}{\sqrt{2}}\hat{j}\right)\left(\dfrac{1}{\sqrt{2}}\hat{i} + \dfrac{1}{\sqrt{2}}\hat{j}\right)$

$\left(\because |\hat{n}| = 1\right)$

$= \dfrac{3}{\sqrt{2}} \times \dfrac{1}{\sqrt{2}} + \dfrac{3}{\sqrt{2}} \times \dfrac{1}{\sqrt{2}} = \dfrac{3+3}{2} = 3.$

54. (a)

$curl\vec{F} = \begin{vmatrix} \hat{i} & \hat{j} & \hat{k} \\ \dfrac{\partial}{\partial x} & \dfrac{\partial}{\partial y} & \dfrac{\partial}{\partial z} \\ x^2 z^2 & -2xy^2 z & 2y^2 z^3 \end{vmatrix}$

$= \hat{i}\left\{\dfrac{\partial}{\partial y}\left(2y^2 z^3\right) - \dfrac{\partial}{\partial z}\left(-2xy^2 z\right)\right\}$

$\quad - \hat{j}\left\{\dfrac{\partial}{\partial x}\left(2y^2 z^3\right) - \dfrac{\partial}{\partial z}\left(x^2 z^2\right)\right\}$

$\quad + \hat{k}\left\{\dfrac{\partial}{\partial x}\left(-2xy^2 z\right) - \dfrac{\partial}{\partial y}\left(x^2 z^2\right)\right\}$

$= \left(4yz^3 + 2xy^2\right)\hat{i} + 2x^2 z\hat{j} - 2y^2 z\hat{k}.$

55. (c) Let $\vec{f}(x, y, z) = x^2 z\hat{i} + xy\hat{j} - yz^2\hat{k}$.

Then, $div\left(\vec{f}\right) = \dfrac{\partial}{\partial x}\left(x^2 z\right) + \dfrac{\partial}{\partial y}(xy) + \dfrac{\partial}{\partial z}\left(-yz^2\right)$

$= 2xz + x - 2yz.$

$\therefore$ At $(1, -1, 1)$,

$div\left(\vec{f}\right) = 2 \times 1 \times 1 + 1 - 2(-1) \times 1 = 5.$

56. (a) We know that

$\vec{\nabla}.\left(f\vec{V}\right) = \left(\vec{\nabla}f\right).\vec{V} + f\left(\vec{\nabla}.\vec{V}\right)$

$\therefore \vec{V}.\left(\vec{\nabla}f\right)$

$= \vec{\nabla}.\left(f\vec{V}\right) - f\left(\vec{\nabla}.\vec{V}\right)$

$= \left(x^2 y + y^2 z + z^2 x\right) - \left[\dfrac{\partial}{\partial x}(y) + \dfrac{\partial}{\partial y}(z) + \dfrac{\partial}{\partial z}(x)\right]f$

$= x^2 y + y^2 z + z^2 x - (0 + 0 + 0)f$

$= x^2 y + y^2 z + z^2 x.$

57. (b)

$\phi = 2xz - y^2$

$\Rightarrow \vec{\nabla}\phi = \hat{i}\dfrac{\partial\phi}{\partial x} + \hat{j}\dfrac{\partial\phi}{\partial y} + \hat{k}\dfrac{\partial\phi}{\partial z} = 2z\hat{i} - 2y\hat{j} + 2x\hat{k}.$

$\therefore$ Maximum directional derivative at $(1, 3, 2)$

$= \vec{\nabla}\phi\Big|_{(1,3,2)} = (2 \times 2)\hat{i} - (2 \times 3)\hat{j} + (2 \times 1)\hat{k}$

$= 4\hat{i} - 6\hat{j} + 2\hat{k}.$

58. (c)

$\vec{\nabla}^2\phi = \dfrac{\partial^2\phi}{\partial x^2} + \dfrac{\partial^2\phi}{\partial y^2} + \dfrac{\partial^2\phi}{\partial z^2}$

$= \dfrac{\partial}{\partial x}\left(\dfrac{\partial\phi}{\partial x}\right) + \dfrac{\partial}{\partial y}\left(\dfrac{\partial\phi}{\partial y}\right) + \dfrac{\partial}{\partial z}\left(\dfrac{\partial\phi}{\partial z}\right)$

$$= \frac{\partial}{\partial x}\left(6x^2y^2z^4\right) + \frac{\partial}{\partial y}\left(4x^3yz^4\right) + \frac{\partial}{\partial z}\left(8x^3y^2z^3\right)$$

$$= 12xy^2z^4 + 4x^3z^4 + 24x^3y^2z^2.$$

59. 5/3.

By Green's theorem,

$$\int_C \left(\varphi dx + \psi dy\right) = \iint_R \left(\frac{\partial \psi}{\partial x} - \frac{\partial \varphi}{\partial y}\right) dxdy \qquad \ldots(i)$$

Here $\varphi = 3x - 8y^2, \psi = 4y - 6xy.$

$\therefore$ (i) gives, $\displaystyle\int_C \left[\left(3x - 8y^2\right)dx + \left(4y - 6xy\right)dy\right]$

$$= \iint_R \left[\frac{\partial}{\partial x}\left(4y - 6xy\right) - \frac{\partial}{\partial y}\left(3x - 8y^2\right)\right]dxdy$$

$$= \iint_R \left(0 - 6y - 0 + 16y\right)dxdy$$

$$= \int_{x=0}^{1}\int_{y=0}^{1-x} 10y\,dy\,dx$$

($\therefore$ the region R is bounded by
$x = 0$, $y = 0$, and $x + y = 1$)

$$= 10\int_{x=0}^{1}\left[\frac{y^2}{2}\right]_{y=0}^{1-x} dx$$

$$= 10\int_{x=0}^{1}\frac{(1-x)^2}{2}dx = -\frac{5}{3}\left[(1-x)\right]_{x=0}^{1}$$

$$= -\frac{5}{3}(0-1) = \frac{5}{3}.$$

60. 216.

$$\iint_S \frac{1}{\pi}\left(9x\hat{i} - 3y\hat{j}\right).\hat{n}ds$$

$$= \frac{1}{\pi}\iiint_V div\left(9x\hat{i} - 3y\hat{j}\right)dV$$

(by the Gauss Divergence theorem)

$$= \frac{1}{\pi}\iiint_V \left[\frac{\partial}{\partial x}(9x) + \frac{\partial}{\partial y}(-3y)\right]dV$$

$$= \frac{1}{\pi}\iiint_V (9-3)dV = \frac{6}{\pi}\iiint_V dV$$

$$= \frac{6}{\pi}\times \text{volume of the sphere } x^2 + y^2 + z^2 = 3^2$$

$$= \frac{6}{\pi}\times\frac{4}{3}\pi(3)^3 = 216.$$

61. (a)
$$\vec{P} = x^3y\vec{a}_x - x^2y^2\vec{a}_y - x^2yz\vec{a}_z$$

$$\Rightarrow div\left(\vec{P}\right) = \frac{\partial}{\partial x}\left(x^3y\right) + \frac{\partial}{\partial y}\left(-x^2y^2\right) + \frac{\partial}{\partial z}\left(-x^2yz\right)$$

$$= 3x^2y - 2x^2y - x^2y = 0.$$

$\therefore \vec{P}$ is solenoidal.

Now curl $\vec{P} = \begin{vmatrix} \vec{a}_x & \vec{a}_y & \vec{a}_z \\ \dfrac{\partial}{\partial x} & \dfrac{\partial}{\partial y} & \dfrac{\partial}{\partial z} \\ x^3y & -x^2y^2 & -x^2yz \end{vmatrix}$

$$= \vec{a}_x\left\{\frac{\partial}{\partial y}\left(-x^2yz\right) - \frac{\partial}{\partial z}\left(-x^2y^2\right)\right\}$$

$$-\vec{a}_y\left\{\frac{\partial}{\partial x}\left(-x^2yz\right) - \frac{\partial}{\partial z}\left(x^3y\right)\right\}$$

$$+\vec{a}_z\left\{\frac{\partial}{\partial x}\left(-x^2y^2\right) - \frac{\partial}{\partial y}\left(x^3y\right)\right\}$$

$$= \left(-x^2z + 0\right)\vec{a}_x + \left(2xyz + 0\right)\vec{a}_y$$

$$-\left(2xy^2 + x^3\right)\vec{a}_z \neq \vec{0}$$

$\therefore \vec{P}$ is not irrotational.

62. $-\dfrac{7\sqrt{6}}{3}.$

$$u(x,y,z) = x^2 - 3yz$$

$$\Rightarrow \vec{\nabla}u = \hat{i}\frac{\partial u}{\partial x} + \hat{j}\frac{\partial u}{\partial y} + \hat{k}\frac{\partial u}{\partial z} = 2x\hat{i} - 3z\hat{j} - 3y\hat{k}.$$

$\therefore$ At $(2,-1,4)$,

$$\vec{\nabla}u = (2\times 2)\hat{i} - (3\times 4)\hat{j} - 3\times(-1)\hat{k}$$

$$= 4\hat{i} - 12\hat{j} + 3\hat{k}.$$

Let, $\vec{\alpha} = \vec{i} + \vec{j} - 2\vec{k}.$

Then, $|\vec{\alpha}| = \sqrt{1^2 + 1^2 + (-2)^2} = \sqrt{6}.$

Hence, the required directional derivative is

$= \vec{\nabla} u. \dfrac{\vec{\alpha}}{\left|\vec{\alpha}\right|}$

$= \left(4\hat{i} - 12\hat{j} + 3\hat{k}\right). \dfrac{1}{\sqrt{6}}\left(\hat{i} + \hat{j} - 2\hat{k}\right)$

$= \dfrac{1}{\sqrt{6}}\left\{4 \times 1 - 12 \times 1 + 3 \times (-2)\right\} = -\dfrac{14}{\sqrt{6}} = \dfrac{-14\sqrt{6}}{\sqrt{6} \times \sqrt{6}}$

$= \dfrac{-14\sqrt{6}}{6} = \dfrac{-7\sqrt{6}}{3}.$

63. (a)

$curl\,\vec{V} = \begin{vmatrix} \hat{i} & \hat{j} & \hat{k} \\ \dfrac{\partial}{\partial x} & \dfrac{\partial}{\partial y} & \dfrac{\partial}{\partial z} \\ 2x^2 & 3z^2 & y^3 \end{vmatrix}$

$= \hat{i}\left\{\dfrac{\partial}{\partial y}\left(y^3\right) - \dfrac{\partial}{\partial z}\left(3z^2\right)\right\} - \hat{j}\left\{\dfrac{\partial}{\partial x}\left(y^3\right) - \dfrac{\partial}{\partial z}\left(2x^2\right)\right\}$

$+ \hat{k}\left\{\dfrac{\partial}{\partial x}\left(3z^2\right) - \dfrac{\partial}{\partial y}\left(2x^2\right)\right\}$

$= \left(3y^2 - 6z\right)\hat{i} - (0 - 0)\hat{j} + (0 - 0)\hat{k}$

$= \left(3y^2 - 6z\right)\hat{i}.$

$\therefore$ At $x = y = z = 1$, $curl\,\vec{V} = (3 - 6)\hat{i} = -3\hat{i}.$

64. (c) We know that for any vector $\vec{V}$, $div\,curl\,\vec{V} = 0.$

65. (a) Let $\varphi(x, y) = x^2 + y^2 - 2$. Then, $\vec{\nabla}\varphi$ is a normal to the circle $x^2 + y^2 = 2$.

Now, $\vec{\nabla}\varphi = \hat{i}\dfrac{\partial\varphi}{\partial x} + \hat{j}\dfrac{\partial\varphi}{\partial y} = 2x\hat{i} + 2y\hat{j}.$

$\therefore \left.\vec{\nabla}\varphi\right|_{(1,1)} = 2\hat{i} + 2\hat{j}$ & $\left|\vec{\nabla}\varphi\right| = \sqrt{2^2 + 2^2} = 2\sqrt{2}.$

Also, $f(x, y) = x^2 + 3y^2.$

So, $\vec{\nabla}f = \hat{i}\dfrac{\partial f}{\partial x} + \hat{j}\dfrac{\partial f}{\partial y} = 2x\hat{i} + 6y\hat{j}.$

$\therefore \left.\vec{\nabla}f\right|_{(1,1)} = 2\hat{i} + 6\hat{j}.$

Hence, the required directional derivative is

$= \vec{\nabla}f. \dfrac{\vec{\nabla}\varphi}{\left|\vec{\nabla}\varphi\right|} = \left(2\hat{i} + 6\hat{j}\right). \dfrac{1}{2\sqrt{2}}\left(2\hat{i} + 2\hat{j}\right)$

$= \dfrac{1}{2\sqrt{2}}\left(2 \times 2 + 6 \times 2\right) = \dfrac{16}{2\sqrt{2}} = 4\sqrt{2}.$

66. 16.

$\overline{F}.\overline{dr} = \left(y\hat{i} + 2x\hat{j}\right).\left(dx\hat{i} + dy\hat{j}\right) = ydx + 2xdy$

$\therefore \oint_C \overline{F}.\overline{dr} = \oint_C (ydx + 2xdy)$

$= \iint_R \left[\dfrac{\partial}{\partial x}(2x) - \dfrac{\partial}{\partial y}(y)\right]dxdy$ (by Green's theorem)

$= \iint_R (2 - 1)dxdy$

$= \iint_R dxdy =$ area of the circle of radius $\dfrac{4}{\sqrt{\pi}}$ unit

$= \pi \times \left(\dfrac{4}{\sqrt{\pi}}\right)^2 = 16$

67. 0

$\oint_C \left(xy^2dx + x^2ydy\right)$

$= \iint_R \left[\dfrac{\partial}{\partial x}\left(x^2y\right) - \dfrac{\partial}{\partial y}\left(xy^2\right)\right]dxdy$

(by Green's theorem)

$= \iint_R (2xy - 2xy)\,dxdy = 0.$

68. 726

$\int_C \vec{\nabla}\phi.\overline{dr}$

$= \int_C \left(yz\hat{i} + xz\hat{j} + xy\hat{k}\right).\left(dx\hat{i} + dy\hat{j} + dz\hat{k}\right)$

$= \int_C (yzdx + xzdy + xydz)$

$$= \int_1^3 \left[t^2 \times 3t^2 dt + t \times 3t^2 \times 2t dt + t \times t^2 \times 6t dt \right]$$

$$\left[\because x = t, y = t^2, z = 3t^2 \text{ and so} \right.$$

$$\left. dx = dt, dy = 2t\, dt, dz = 6t dt \right]$$

$$= \int_1^3 15t^4 dt = \frac{15}{5} \left[t^5 \right]_1^3 = 3\left(3^5 - 1 \right) = 726.$$

69. (b)

$$\int_C \left(2xy^2 dx + 2x^2 y\, dy + dz \right)$$

$$= \int_C \left\{ d\left(x^2 y^2 \right) + dz \right\}$$

$$= \int_C d\left(x^2 y^2 \right) + \int_C dz = \int_{(0,0)}^{(1,1)} d\left(x^2 y^2 \right) + \int_0^1 dz$$

$$= \left[x^2 y^2 \right]_{(0,0)}^{(1,1)} + \left[z \right]_0^1 = 1 + 1 = 2.$$

70. 53/12

Here, $x = t, y = t^2$ & $z = t$.

$\therefore dx = dt, dy = 2t dt, dz = dt.$

clearly "t" varies from 0 to 1.

$$\therefore \int_C \vec{F}.d\vec{r}$$

$$= \int_C \left[5xz dx + \left(3x^2 + 2y \right) dy + x^2 z dz \right]$$

$$= \int_0^1 \left[\left(5 \times t \times t \right) dt + \left(3t^2 + 2t^2 \right) 2t dt + \left(t^2 \times t \right) dt \right]$$

$$\int_0^1 \left(5t^2 + 11t^3 \right) dt = \frac{5}{3} \left[t^3 \right]_0^1 + \frac{11}{4} \left[t^4 \right]_0^1$$

$$= \frac{5}{3} + \frac{11}{4} = \frac{53}{12}.$$

71. 90°

$$I = 15 \cos wt\, \hat{x} + 5 \sin wt\, \hat{y}$$

$$\Rightarrow |I| = \sqrt{\left(15 \cos wt \right)^2 + \left(5 \sin wt \right)^2}$$

$$= \sqrt{225 \cos^2 wt + 25 \sin^2 wt}$$

$$= \sqrt{25 \left(1 - \cos^2 wt \right) + 225 \cos^2 wt}$$

$$= \sqrt{25 + 200 \cos^2 wt}$$

So $|I|$ is minimum when $25 + 200 \cos^2 wt$ is minimum, *i.e.*, when $\cos wt = 0$, *i.e.*, when $wt = 90° = \theta$.

72. (d)

$$\left(\hat{i} + \hat{j} + \hat{k} \right).\left(4\hat{i} + 3\hat{j} + 5\hat{k} \right)$$

$$= 4 \times 1 + 3 \times 1 + 5 \times 1 \neq 0$$

$\therefore 4\hat{i} + 3\hat{j} + 5\hat{k}$ is not perpendicular to $\hat{i} + \hat{j} + \hat{k}$.

Again, $\left(\hat{i} + 2\hat{j} + 3\hat{k} \right).\left(4\hat{i} + 3\hat{j} + 5\hat{k} \right)$

$$= 4 \times 1 + 3 \times 2 + 5 \times 3 \neq 0$$

$\therefore 4\hat{i} + 3\hat{j} + 5\hat{k}$ is not perpendicular to $\hat{i} + 2\hat{j} + 3\hat{k}$.

73. 72π

$$\vec{\nabla}.\vec{F}$$

$$= \frac{\partial}{\partial x} (x + y) + \frac{\partial}{\partial y} (x + z) + \frac{\partial}{\partial z} (y + z)$$

$$= 1 + 0 + 1 = 2.$$

By the Gauss Divergence theorem,

$$\iint_S \vec{F}.\hat{n} ds = \iiint_V \left(\vec{\nabla}.\vec{F} \right) dV = \iiint_V 2 dV$$

$$= 2 \iiint_V dV$$

$= 2 \times$ volume of the sphere $x^2 + y^2 + z^2 = 9$

$$= 2 \times \frac{4}{3} \pi (3)^3 = 72\pi.$$

74. −11.

$$I = \int_C \left(2z dx + 2y dy + 2x dz \right)$$

$$= \int_C \left[2d(xz) + 2y dy \right] = 2 \int_{(0,2,1)}^{(4,1,-1)} \left[d(xz) + y dy \right]$$

$$= 2\left\{ \left[xz\right]_{(0,2,1)}^{(4,1,-1)} + \left[\frac{y^2}{2}\right]_{(0,2,1)}^{(4,1,-1)} \right\}$$

$$= 2\left\{ \left[-4-0\right] + \left[\frac{1}{2}-\frac{4}{2}\right] \right\} = -11.$$

75. 134.

$$\vec{V} = x^2\hat{i} + 2y^3\hat{j} + z^4\hat{k}$$

$$\Rightarrow div\vec{V} = \frac{\partial}{\partial x}\left(x^2\right) + \frac{\partial}{\partial y}\left(2y^3\right) + \frac{\partial}{\partial z}\left(z^4\right)$$

$$= 2x + 6y^2 + 4z^3.$$

$$\therefore \quad \text{At } x=1, \, y=2, \, z=3;$$

$$div\vec{V} = 2\times 1 + 6\times 2^2 + 4\times 3^3 = 134.$$

76. 0

$$div\left(-y\hat{i} + x\hat{j}\right)$$

$$= \frac{\partial}{\partial x}\left(-y\right) + \frac{\partial}{\partial y}\left(x\right) + \frac{\partial}{\partial z}\left(0\right) = 0 + 0 + 0 = 0.$$

77. 0.

For any vector $\vec{V}$, we know that div curl $\vec{V} = 0$,

i.e., $\vec{\nabla}.\left(\vec{\nabla}\times\vec{V}\right) = 0.$

78. (b)

$$curl\,\vec{F} = \begin{vmatrix} \hat{a}_x & \hat{a}_y & \hat{a}_z \\ \dfrac{\partial}{\partial x} & \dfrac{\partial}{\partial y} & \dfrac{\partial}{\partial z} \\ 3y-k_1z & k_2x-2z & -\left(k_3y+z\right) \end{vmatrix}$$

$$= \hat{a}_x\left\{ \frac{\partial}{\partial y}\left(-k_3y-z\right) - \frac{\partial}{\partial z}\left(k_2x-2z\right) \right\}$$

$$-\hat{a}_y\left\{ \frac{\partial}{\partial x}\left(-k_3y-z\right) - \frac{\partial}{\partial z}\left(3y-k_1z\right) \right\}$$

$$+\hat{a}_z\left\{ \frac{\partial}{\partial x}\left(k_2x-2z\right) - \frac{\partial}{\partial y}\left(3y-k_1z\right) \right\}$$

$$= \left(-k_3+2\right)\hat{a}_x - k_1\hat{a}_y + \left(k_2-3\right)\hat{a}_z$$

Now, $\vec{F}$ is irrotational

$$\Rightarrow curl\,\vec{F} = \vec{0}$$

$$\Rightarrow \left(-k_3+2\right)\hat{a}_x + \left(-k_1\right)\hat{a}_y + \left(k_2-3\right)\hat{a}_z$$

$$= 0\hat{a}_x + 0\hat{a}_y + 0\hat{a}_z$$

$$\Rightarrow -k_3+2=0, -k_1=0, k_2-3=0$$

$$\Rightarrow k_3=2, k_2=3, k_1=0.$$

79. (c)

$$\vec{u} = e^x\cos y\hat{i} + e^x\sin y\hat{j} + 0\hat{k}$$

$$\Rightarrow div\,\vec{u} = \frac{\partial}{\partial x}(e^x\cos y) + \frac{\partial}{\partial y}(e^x\sin y) + \frac{\partial}{\partial z}(0)$$

$$= e^x\cos y + e^x\cos y + 0.$$

$$= 2e^x\cos y.$$

80. 0.67

$$\vec{F}(\vec{r}) = x^2\hat{i} + y^2\hat{j}$$

$$\Rightarrow \vec{F}(\vec{r}).d\vec{r} = x^2dx + y^2dy$$

$$\therefore \int_C \vec{F}(\vec{r}).d\vec{r}$$

$$= \int_{(0,0)}^{(1,1)}\left[x^2dx + y^2dy\right] = \left(\frac{x^3}{3} + \frac{y^3}{3}\right)\Bigg|_{(0,0)}^{(1,1)}$$

$$= \left(\frac{1}{3} + \frac{1}{3}\right) - (0+0) = \frac{2}{3} = 0.67.$$

81. 0

$$\int_C\left\{(y^2 + 2xy)dx + (2xy + x^2)dy\right\}$$

$$= \int_{(3,0)}^{(0,3)}\left\{(y^2 + 2xy)dx + (2xy + x^2)dy\right\}$$

$$= \int_{(3,0)}^{(0,3)} d(xy^2 + x^2y) = \left[xy^2 + x^2y\right]_{(3,0)}^{(0,3)} = 0.$$

82. (d) $\vec{F}.d\vec{l} = (x^2 - 2y)dx - 4yzdy + 4xz^2dz.$

The equation of straight line passing through $(0, 0, 0)$ to $(1, 1, 1)$ is

$$\Rightarrow \frac{x}{1} = \frac{y}{1} = \frac{z}{1}$$

$$\Rightarrow x = y = z = t \text{ (say)}$$

$$\therefore dx = dt, dy = dt, dz = dt.$$

Now, $\int_C \vec{F}.d\vec{l}$

$= \int_C \left\{ (x^2 - 2y)dx - 4yz dy + 4xz^2 dz \right\}$

$= \int_{t=0}^{1} \left(t^2 - 2t - 4t^2 + 4t^3 \right) dt$

$= \int_{t=0}^{1} \left(4t^3 - 3t^2 - 2t \right) dt = \left[t^4 - t^3 - t^2 \right]_0^1 = -1.$

83. (c) By the Gauss Divergence theorem, we have,

$$\oiint_S \vec{r}.\vec{n}\, dS = \iiint_V div\ \vec{r}\, dV$$

$$\Rightarrow \oiint_S \vec{r}.\vec{n}\, dS = \iiint_V \left(\frac{\partial}{\partial x}(x) + \frac{\partial}{\partial y}(y) + \frac{\partial}{\partial z}(z) \right) dV$$

$$(\because \vec{r} = x\hat{i} + y\hat{j} + z\hat{k})$$

$$= \iiint_V (1+1+1)\, dx\, dy\, dz$$

$$= 3\iiint_V dx\, dy\, dz = 3V.$$

84. 678.58

By the Gauss Divergence theorem, we have

$$\oiint_S \vec{f}.\vec{n}\, dS = \iiint_V div\ \vec{f}\, dV$$

$$\therefore \iint_S \left(9x\hat{i} - 2y\hat{j} - z\hat{k} \right).\hat{n}\, ds$$

$$= \iiint_V div \left(9x\hat{i} - 2y\hat{j} - z\hat{k} \right).\hat{n}\, dV$$

$$= \iiint_V \left[\frac{\partial}{\partial x}(9x) + \frac{\partial}{\partial y}(-2y) + \frac{\partial}{\partial z}(-z) \right] dx\, dy\, dz$$

$$= \iiint_V (9-2-1)\, dx\, dy\, dz$$

$$= 6\iiint_V dx\, dy\, dz$$

$= 6 \times$ volume of the sphere $x^2 + y^2 + z^2 = 9$

$$= 6 \times \frac{4\pi}{3} \times 3^3 = 216\pi = 678.58.$$

85. (a) It follows from the definition of vector triple product.

86. (a)

$\phi(x,y,z) = xy^2 + yz^2 + zx^2$

$\Rightarrow \vec{\nabla}\phi = \frac{\partial\phi}{\partial x}\hat{i} + \frac{\partial\phi}{\partial y}\hat{j} + \frac{\partial\phi}{\partial z}\hat{k}$

$= \left(y^2 + 2xz \right)\hat{i} + \left(z^2 + 2xy \right)\hat{j} + \left(x^2 + 2yz \right)\hat{k}$

$\Rightarrow \vec{\nabla}\phi\Big|_{(2,-1,1)} = 5\hat{i} - 3\hat{j} + 2\hat{k}$

Again, $\vec{P} = \hat{i} + 2\hat{j} + 2\hat{k} \Rightarrow |\vec{P}| = \sqrt{1^2 + 2^2 + 2^2} = 3.$

Hence, the directional derivative is

$= \vec{\nabla}\phi.\frac{\vec{P}}{|\vec{P}|} = \left(5\hat{i} - 3\hat{j} + 2\hat{k} \right).\left(\frac{\hat{i} + 2\hat{j} + 2\hat{k}}{3} \right)$

$= \frac{5-6+4}{3} = 1.$

87. (c)

$|\vec{r}| = \sqrt{x^2 + y^2 + z^2}$

$\Rightarrow \frac{\partial}{\partial x}\ln|\vec{r}| = \frac{\partial}{\partial x}\ln\left(\sqrt{x^2 + y^2 + z^2} \right)$

$= \frac{1}{\sqrt{x^2 + y^2 + z^2}} \times \frac{2x}{2\sqrt{x^2 + y^2 + z^2}}$

$= \frac{x}{x^2 + y^2 + z^2}$

Similarly, $\frac{\partial}{\partial y}\ln|\vec{r}| = \frac{y}{x^2 + y^2 + z^2}$ and

$\frac{\partial}{\partial z}\ln|\vec{r}| = \frac{z}{x^2 + y^2 + z^2}.$

$\therefore \vec{\nabla}\ln|\vec{r}|$

$= \frac{\partial}{\partial x}\ln|\vec{r}|\hat{i} + \frac{\partial}{\partial y}\ln|\vec{r}|\hat{j} + \frac{\partial}{\partial z}\ln|\vec{r}|\hat{k}$

$= \frac{x}{x^2 + y^2 + z^2}\hat{i} + \frac{y}{x^2 + y^2 + z^2}\hat{j} + \frac{z}{x^2 + y^2 + z^2}\hat{k}$

$= \frac{1}{x^2 + y^2 + z^2}\left(x\hat{i} + y\hat{j} + z\hat{k} \right) = \frac{\vec{r}}{|\vec{r}|^2} = \frac{\vec{r}}{\vec{r}.\vec{r}}.$

Questions for Practice

1. If $\vec{\alpha}$ and $\vec{\beta}$ be two non zero non-collinear vectors and x, y be two scalars such that $x\vec{\alpha} + y\vec{\beta} = \vec{0}$, then which of the following is true?

(a) $x = 0 = y$ (b) $x \neq y$

(c) $x, y \neq 0$ (d) $y \neq 0$

2. The vectors $3\vec{a}-7\vec{b}-\vec{c}, 3\vec{a}-2\vec{b}+\vec{c}, \vec{a}+\vec{b}+2\vec{c}$ (given that $\vec{a},\vec{b},\vec{c}$, are any three non-coplanar vectors) are

(a) coplanar
(b) non-coplanar
(c) coplanar if $\vec{a}=\vec{0}$
(d) coplanar if and only if $\vec{a}=\vec{b}=\vec{c}=\vec{0}$

3. The vectors $2\hat{i}-\hat{j}+\hat{k}, \hat{i}-3\hat{j}-5\hat{k}$ and

$3\hat{i}-4\hat{j}-4\hat{k}$ form the side of a

(a) equilateral triangle
(b) right angled triangle
(c) scalene triangle
(d) right angled scalene triangle

4. The values of λ and μ for which the vectors $-3\hat{i}+4\hat{j}+\lambda\hat{k}$ and $\mu\hat{i}+8\hat{j}+6\hat{k}$ are collinear, are

(a) $\lambda=3, \mu=-6$ (b) $\lambda=6, \mu=-3$
(c) $\lambda=\mu=5$ (d) none of these

5. If $\vec{a},\vec{b},\vec{c}$ be unit vectors satisfying $\vec{a}+\vec{b}+\vec{c}=\vec{0}$, then the value of $\vec{a}.\vec{b}+\vec{b}.\vec{c}+\vec{c}.\vec{a}$ is

(a) $\dfrac{3}{2}$ (b) $-\dfrac{3}{2}$ (c) $-\dfrac{1}{2}$ (d) $\dfrac{1}{2}$

6. If $\vec{\alpha}$ and $\vec{\beta}$ are vectors with magnitude 3 units and 4 units, respectively, and if the vectors $\vec{\alpha}+\lambda\vec{\beta}$ and $\vec{\alpha}-\lambda\vec{\beta}$ are perpendicular to each other, then l = ?

(a) $\dfrac{1}{2}$ (b) $-\dfrac{1}{2}$ (c) $\pm\dfrac{3}{4}$ (d) $\pm\dfrac{1}{2}$

7. The vector $\vec{\delta}$ which is perpendicular to both $\vec{\alpha}=4\hat{i}+5\hat{j}-\hat{k}$ and $\vec{\beta}=\hat{i}-4\hat{j}+5\hat{k}$ and which satisfies $(3\hat{i}+\hat{j}-\hat{k}).\vec{\delta}=21$ is given by

(a) $7\hat{i}+7\hat{j}-7\hat{k}$ (b) $7\hat{i}-7\hat{j}+7\hat{k}$

(c) $7\hat{i}-7\hat{j}-7\hat{k}$ (d) none of these

8. Which of the following is true?

(a) $\left[\vec{a},\vec{a},\vec{c}\right]=0$

(b) $\left[\vec{a}\times\vec{b},\vec{b}\times\vec{c},\vec{c}\times\vec{a}\right]=\left[\vec{a},\vec{b},\vec{c}\right]^2$

(c) $\left[\vec{a}+\vec{b},\vec{b}+\vec{c},\vec{c}+\vec{a}\right]=2\left[\vec{a},\vec{b},\vec{c}\right]$

(d) all of the above

9. A force $\vec{F}=3\hat{i}+2\hat{j}-4\hat{k}$ is applied at the point $(1, -1, 2)$. Then the moment of the force about the point $(2, -1, 3)$ is

(a) $2\hat{i}-7\hat{j}-2\hat{k}$ (b) $4\hat{i}-7\hat{j}+5\hat{k}$

(c) $7\hat{i}-2\hat{j}+8\hat{k}$ (d) $\hat{i}-\hat{j}-2\hat{k}$

10. A particle is acted on by constant forces $\vec{F}_1=3\hat{i}+2\hat{j}-4\hat{k}$ and $\vec{F}_2=\hat{i}-\hat{j}+3\hat{k}$ and is displaced from the point $(1, 1, 2)$ to the point $(2, 4, 5)$. Then the work done by the forces is

(a) 5 unit   (b) 10 unit   (c) 4 unit   (d) 8 unit

11. If $\vec{r}=(\sinh t)\hat{i}+(\cosh t)\hat{j}$, then $\dfrac{d^2 r}{dt^2}=$?

(a) $4\vec{r}$ (b) $3\vec{r}$ (c) $\vec{r}$ (d) $2\vec{r}$

12. If $\vec{r}(t)=5t^2\hat{i}+t\hat{j}-t^3\hat{k}$, then $\displaystyle\int_1^2 \vec{r}\times\dfrac{d^2\vec{r}}{dt^2}dt=$?

(a) $-14\hat{i}+75\hat{j}-15\hat{k}$ (b) $20\hat{i}+15\hat{j}-5\hat{k}$

(c) $16\hat{i}-17\hat{j}+9\hat{k}$ (d) None of these

13. If $\vec{F}=x^2 y\hat{i}-2xz\hat{j}+2yz\hat{k}$, then $div\vec{F}$

(a) $2xy$ (b) $2x(y+1)$
(c) $2y(x+1)$ (d) 0

14. Consider $\vec{F}=(2x+3y)\hat{i}+(4y-3z)\hat{j}-6z\hat{k}$. Then,

(a) $\vec{F}$ is solenoidal   (b) $\vec{F}$ irrotational

(c) $\vec{\nabla}.\vec{F}\neq 0$ (d) $\vec{\nabla}\times\vec{F}\neq\vec{0}$

15. If the vectors $\vec{f}$ and $\vec{g}$ are irrotational, then which of the followings is true?

(a) $\vec{f}\times\vec{g}$ is irrotational

(b) $\vec{f}\times\vec{g}$ is solenoidal

(c) $\vec{f}$ is solenoidal

(d) $\vec{g}$ is solenoidal

16. The unit normal vector to the surface $x^2+y-z=1$ at the point $(1, 0, 0)$ is

(a) $\dfrac{1}{\sqrt{6}}\left(2\hat{i}+\hat{j}-\hat{k}\right)$ (b) $\dfrac{1}{\sqrt{6}}\left(2\hat{i}-\hat{j}-\hat{k}\right)$

(c) $\dfrac{1}{\sqrt{6}}\left(2\hat{i}-\hat{j}+\hat{k}\right)$ (d) $\dfrac{1}{\sqrt{6}}\left(2\hat{i}+\hat{j}+\hat{k}\right)$

17. The directional derivative of $\phi(x,y,z) = x^2yz + 4xz^2$ at the point $(1, 2, -1)$ in the direction of the vector $2\hat{i} - \hat{j} - 2\hat{k}$ is

(a) $\dfrac{17}{3}$ (b) $\dfrac{13}{3}$ (c) $\dfrac{7}{3}$ (d) $\dfrac{19}{3}$

18. If $\vec{f}(x,y,z) = x^2y\hat{i} + y^2z\hat{j} + z^2\hat{k}$, then $curl(curl\vec{f})$ at $(1, -1, 0)$ is

(a) $2(\hat{i} + \hat{j})$ (b) $2(\hat{j} - \hat{k})$

(c) $2(\hat{i} + \hat{j} - \hat{k})$ (d) $2(\hat{i} - \hat{j} + \hat{k})$

19. The value of $\int_C \vec{F}.d\vec{r}$ where $\vec{F} = yz\hat{i} + zx\hat{j} + xy\hat{k}$ and C is the portion of the curve $\vec{r} = 4\cos t\,\hat{i} + 2\sin t\,\hat{j} + 2t\,\hat{k}$ from $t = 0$ to $t = \dfrac{\pi}{2}$, is

(a) 4   (b) 2   (c) 0   (d) 1

20. Evaluate $\int_C \vec{F} \times d\vec{r}$ where $\vec{F} = xy\hat{i} - z\hat{j} + x^2\hat{k}$ and C is the curve $x = t^2, y = 2t, z = t^3$ from $t = 0$ to $t = 1$, is

(a) $-\dfrac{9}{10}\hat{i} - \dfrac{2}{3}\hat{j} + \dfrac{7}{5}\hat{k}$ (b) $-\dfrac{3}{10}\hat{i} + \dfrac{1}{3}\hat{j} + \dfrac{1}{5}\hat{k}$

(c) $\dfrac{1}{10}\hat{i} + \dfrac{2}{3}\hat{j} - \dfrac{3}{5}\hat{k}$ (d) $-\dfrac{1}{10}\hat{i} - \dfrac{7}{3}\hat{j} + \dfrac{8}{5}\hat{k}$

21. Using Stokes' theorem, the value of $\iint_S curl\vec{F}.\hat{n}\,dS$ where $\vec{F} = (2x - y)\hat{i} - yz^2\hat{j} - y^2z\hat{k}$

and S is the upper half surface of the sphere $x^2 + y^2 + z^2 = 1$ from $t = 0$ to $t = 1$ and C its boundary, is

(a) π (b) $\dfrac{3\pi}{4}$

(c) 3π (d) 0

22. Using the Gauss Divergence theorem, the value of $\iint_S \vec{F}.\hat{n}\,dS$ where $\vec{F} = 4xz\hat{i} + xyz^2\hat{j} + 3z\hat{k}$ above the xy plane bounded by the cone $z^2 = x^2 + y^2$ and the plane $z = 4$, is

(a) 345π (b) 319π (c) 334π (d) 320π

23. Using Green's theorem, the value of the integral $\int_C \left[(xy + y^2)dx + x^2dy\right]$ where C is the closed curve of the region bounded by $y = x^2$ and $y = x$, is

(a) $\dfrac{1}{5}$ (b) $-\dfrac{1}{5}$ (c) $-\dfrac{1}{20}$ (d) $\dfrac{1}{20}$

24. The value of $\iint_S (xdy\,dz + ydz\,dx + zdx\,dy)$ where S is the surface of the cube $\{(x,y,z) \in \mathbb{R}^3 : 0 \le x \le 1, 0 \le y \le 1, 0 \le z \le 1\}$ is

(a) 2   (b) 1   (c) 3   (d) 0

Answer Key				
1. (a)	2. (a)	3. (d)	4. (a)	5. (b)
6. (c)	7. (c)	8. (d)	9. (a)	10. (c)
11. (c)	12. (a)	13. (c)	14. (a)	15. (b)
16. (a)	17. (b)	18. (b)	19. (c)	20. (a)
21. (a)	22. (d)	23. (c)	24. (c)	

CHAPTER 4

ORDINARY DIFFERENTIAL EQUATIONS

4.1 BASIC CONCEPTS

4.1.1 Definition of a Differential Equation

A differential equation is an equation involving at least one differential or differential coefficient with or without variables.

Examples:

(i) $dy = (2x - e^y)dx$ In this equation the differentials are dx and dy; dependent variable is "y" and independent variable is "x."

(ii) $\dfrac{d^3y}{dx^3} + 4\left(\dfrac{d^2y}{dx^2}\right)^3 = e^x$

In this equation, the differential coefficients are $\dfrac{d^3y}{dx^3}$ and $\dfrac{d^2y}{dx^2}$; dependent variables is "y" and independent variable is "x."

(iii) $\dfrac{d^2y}{dx^2} = 0$

In this equation, the only differential coefficient is $\dfrac{d^2y}{dx^2}$; dependent variables is "y" and independent variable is "x."

(iv) $x\dfrac{\partial u}{\partial x} + e^y\dfrac{\partial^2 u}{\partial y^2} = 0$

In this above equation, differential coefficients are $\dfrac{\partial u}{\partial x}$ and $\dfrac{\partial^2 u}{\partial y^2}$; "$u$" is the dependent variable, where as "x" and "y" are independent variables.

(v) $2\dfrac{\partial^2 u}{\partial x^2} + 3\dfrac{\partial^2 u}{\partial y^2} + 4\dfrac{\partial^2 u}{\partial z^2} = 0$

In this above equation, differential coefficients are $\dfrac{\partial^2 u}{\partial x^2}, \dfrac{\partial^2 u}{\partial y^2}$ and $\dfrac{\partial^2 u}{\partial z^2}$; dependent variable is "u" and independent variables are "x," "y," and "z."

4.1.2 Classification of Differential Equations

(a) **Ordinary differential equation:**
An ordinary differential equation is an equation involving derivatives or differentials with respect to a single independent variable.

Examples:

(i) $dy = (2x - e^y)dx$

(ii) $\dfrac{d^2y}{dx^2} + 3x\dfrac{dy}{dx} = x^2$

(b) **Partial differential equation:**
A partial differential equation is an equation involving partial derivatives or differentials w.r.t at least two independent variables.

Examples:

(i) $x\dfrac{\partial u}{\partial x} + y\dfrac{\partial u}{\partial y} = 0$

(ii) $xy\dfrac{\partial^2 u}{\partial x^2} + z\dfrac{\partial^2 u}{\partial y^2} + y^2\dfrac{\partial^2 u}{\partial z^2} = 0$

4.1.3 Order of a Differential Equation

The order of a differential equation is the order of the highest derivative or highest differential occurring in the equation.

Examples:

(i) The order of the differential equation $\dfrac{d^2y}{dx^2} + x^2\dfrac{dy}{dx} = e^x$ is "2" (since the highest order derivative occurring in the equation is $\dfrac{d^2y}{dx^2}$)

(ii) The order of the differential equation $9\dfrac{d^3y}{dx^3} + 5\dfrac{d^2y}{dx^2} + 2\dfrac{dy}{dx} = \log(x+1)$ is 3 (since the highest order derivative occurring in the equation is $\dfrac{d^3y}{dx^3}$)

(*iii*) The order of the differential equation $x^2 dy = (\sin x + y)dx$ is "1" (since the equation can be written as $\dfrac{dy}{dx} = \dfrac{\sin x + y}{x^2}$ and so highest order derivative occurring in the equation is $\dfrac{dy}{dx}$)

4.1.4 Degree of a Differential Equation

The degree of a differential equation is the power of the derivative (or differential) of the highest order in the equation, after the equation is freed from radicals and fractions.

Examples:

(*i*) The equation $\left(\dfrac{d^2y}{dx^2}\right)^2 + 5x\dfrac{dy}{dx} = x^2$ has degree "2"

[Since power of highest order derivative, *i.e*, the power of $\dfrac{d^2y}{dx^2}$ is "2"]

(*ii*) The equation $y\dfrac{d^3y}{dx^3} + 6\left(\dfrac{d^2y}{dx^2}\right)^4 = \sin x$ has degree "3." [Since power of highest order derivative, *i.e*, the power of $\left(\dfrac{d^3y}{dx^3}\right)$ is "1"]

(*iii*) Consider the equation $2\dfrac{d^2y}{dx^2} = \sqrt{5 + \left(\dfrac{dy}{dx}\right)^3}$.

This equation can be re-written as $5 + \left(\dfrac{dy}{dx}\right)^3 = 4\left(\dfrac{d^2y}{dx^2}\right)^2$ Hence, degree of this equation = power of highest order derivative *i.e*; power of $\left(\dfrac{d^2y}{dx^2}\right) = 1$

4.1.5 Formation of a Differential Equation

Consider a relation
$f(x,y,c_1,c_2,....c_n) = 0$
where c_1, c_2, ...c_n are n-number of arbitrary constants.

Then differentiating the relation n-times, we will get "n" number of equation. Now between these "n" equations and the given equation, *i.e.* (n + 1) equations in total, if we eliminate the n-arbitrary constants c_1, c_2,...c_n; then finally we will get the differential equation of nth order.

Examples:

(*a*) Let $y = mx + 3$ ...(i)

Here, "m" is the only arbitrary constant. Differentiating w.r.t "x," we get $\dfrac{dy}{dx} = m$ ...(ii)

Now eliminating "m" between (i) and (ii) we get
$$y = x\dfrac{dy}{dx} + 3$$
which is the required differential equation.

(*b*) Let $x = A\cos t + B\sin t$ ...(i)
Here the arbitrary constants are "A" and "B"
Differentiating, w.r.t x we get
$$\dfrac{dx}{dt} = -A\sin t + B\cos t \quad ...(ii)$$
Now differentiating (*ii*) w.r.t "x" we get,
$$\dfrac{d^2x}{dt^2} = -A\cos t - B\sin t = -x$$
i.e; $\dfrac{d^2x}{dt^2} + x = 0$,

which is the required differential equation.

4.1.6 Solution of a Differential Equation

Any relation between the dependent and independent variables from which the differential equation can be derived, is called the solution of the differential equation.

Example:

Consider the differential equation
$$\dfrac{d^2x}{dt^2} + x = 0 \quad ...(i)$$
Now consider a relation $x = A\cos t + B\sin t$ ...(ii)

Then from the Subsection 4.1.5 (Example-(*b*)), it is clear that from the relation (*ii*), the differential equation (*i*) can be derived.

Hence, (*ii*) is the solution of the differential equation (*i*).

Remember

(I) A general solution of a differential equation is a solution in which number of arbitrary constants = order of the differential equation.

Example:

Let $y = ae^{-x} + be^x$ is a solution of the differential equation $\dfrac{d^2y}{dx^2} - y = 0$;

Here the no. of arbitrary constants = 2 = order of the differential equation.

So $y = ae^{-x} + be^x$ is a general solution of the equation $\dfrac{d^2 y}{dx^2} - y = 0$

(II) A particular solution of a differential equation is a solution which is obtained by putting particular values of the arbitrary constants.

Example:

$y = ae^{-x} + be^x$ is a solution of the differential equation $\dfrac{d^2 y}{dx^2} - y = 0$;

For $a = 2, b = 3$, we get one particular solution, namely $y = 2e^{-x} + 3e^x$.

4.2 LINEARLY DEPENDENT AND LINEARLY INDEPENDENT SOLUTIONS

4.2.1 Wronskian

1. If $y_1(x)$ and $y_2(x)$ are two solutions of second-order ordinary differential equation $a_0 \dfrac{d^2 y}{dx^2} + a_1 \dfrac{dy}{dx} + a_2 y = 0$, then their Wronskian is denoted by W or $W(x)$ or $W(y_1, y_2)$ and is defined by

$$W(x) = W(y_1, y_2) = \begin{vmatrix} y_1(x) & y_2(x) \\ y_1'(x) & y_2'(x) \end{vmatrix}$$

2. If $y_1(x)$, $y_2(x)$, and $y_3(x)$ are three solutions of second-order ordinary differential equation $a_0 \dfrac{d^3 y}{dx^3} + a_1 \dfrac{d^2 y}{dx^2} + a_2 \dfrac{dy}{dx} + a_3 y = 0$, then their Wronskian is denoted by W or $W(x)$ or $W(y_1, y_2, y_3)$ and is defined by

$$W(x) = W(y_1, y_2, y_3) = \begin{vmatrix} y_1(x) & y_2(x) & y_3(x) \\ y_1'(x) & y_2'(x) & y_3'(x) \\ y_1''(x) & y_2''(x) & y_3''(x) \end{vmatrix}$$

Example:

Let $y_1(x) = 1, y_2(x) = x, y_3(x) = x^2$.
Then $y'_1(x) = 0, y''_1(x) = 0, y'_2(x) = 1, y''_2(x) = 0$
$y'_3(x) = 2x, y''_3(x) = 2$.

$$\therefore \quad W(y_1, y_2, y_3) = \begin{vmatrix} 1 & x & x^2 \\ 0 & 1 & 2x \\ 0 & 0 & 2 \end{vmatrix} = 2.$$

4.2.2 Linearly Dependent Solutions

1. Let $y_1(x)$ and $y_2(x)$ be two solutions of a second order ordinary differential equation

$$a_0 \dfrac{d^2 y}{dx^2} + a_1 \dfrac{dy}{dx} + a_2 y = 0$$

Then $y_1(x)$ and $y_2(x)$ are said to be linearly dependent if there exist constants c_1 and c_2, not all equal to zero such that $c_1 y_1(x) + c_2 y_2(x) = 0$.

2. Let $y_1(x)$, $y_2(x)$, and $y_3(x)$ be three solutions of a 3rd order differential equation

$$a_0 \dfrac{d^3 y}{dx^3} + a_1 \dfrac{d^2 y}{dx^2} + a_2 \dfrac{dy}{dx} + a_3 y = 0.$$

Then $y_1(x)$, $y_2(x)$, and $y_3(x)$ are said to be linearly dependent if there exist constants c_1, c_2 and c_3 not all equal to zero such that $c_1 y_1(x) + c_2 y_2(x) + c_3 y_3(x) = 0$.

Remember

1. If $W(y_1, y_2) = 0$, then the solutions $y_1(x)$ and $y_2(x)$ are linearly dependent.
2. If $W(y_1, y_2, y_3) = 0$, then the solutions $y_1(x)$, $y_2(x)$ and $y_3(x)$ are linearly dependent.

Example:

Let $y_1(x) = 1 - x$, $y_2(x) = 2x - 2$.

Then $W(y_1, y_2) = \begin{vmatrix} y_1(x) & y_2(x) \\ y_1'(x) & y_2'(x) \end{vmatrix} = \begin{vmatrix} 1 - x & 2x - 2 \\ -1 & 2 \end{vmatrix}$

$= 2(1 - x) - (-1)(2x - 2) = 0$

$\therefore$ The solutions $y_1(x)$ and $y_2(x)$ are linearly dependent.

4.2.3 Linearly Independent Solutions

1. Let $y_1(x)$ and $y_2(x)$ be two solutions of a second order ordinary differential equation

$$a_0 \dfrac{d^2 y}{dx^2} + a_1 \dfrac{dy}{dx} + a_2 y = 0$$

Then $y_1(x)$ and $y_2(x)$ are said to be linearly independent if there exist constants c_1 and c_2, such that $c_1 y_1(x) + c_2 y_2(x) = 0$ is satisfied only for $c_1 = c_2 = 0$.

2. Let $y_1(x)$, $y_2(x)$ and $y_3(x)$ be three solutions of a 3rd order differential equation

$$a_0 \dfrac{d^3 y}{dx^3} + a_1 \dfrac{d^2 y}{dx^2} + a_2 \dfrac{dy}{dx} + a_3 y = 0.$$

Then $y_1(x)$, $y_2(x)$ and $y_3(x)$ are said to be linearly independent if there exist constants c_1, c_2, and c_3 such that $c_1 y_1(x) + c_2 y_2(x) + c_3 y_3(x) = 0$ is satisfied only for $c_1 = c_2 = c_3 = 0$.

Remember

1. If $W(y_1, y_2) \neq 0 \; \forall \; x$, then $y_1(x)$ and $y_2(x)$ are linearly independent.
2. If $W(y_1, y_2, y_3) \neq 0 \; \forall \; x$, then $y_1(x)$, $y_2(x)$ and y_3 are linearly independent.

Example:

Let $y_1(x) = x^2$ and $y_2(x) = x^2 \log x$.

Then $y_1'(x) = 2x$ and $y_2'(x) = 2x \log x + x^2 \times \dfrac{1}{x}$

$= 2x \log x + x$.

$\therefore W(y_1, y_2) = \begin{vmatrix} x^2 & x^2 \log x \\ 2x & 2x \log x + x \end{vmatrix}$

$= x^2 (2x \log x + x) - 2x \times x^2 \log x$

$= x^3 \neq 0 \ \forall \ x$

Thus, $W(y_1, y_2)$ does not vanish identically. So, $y_1(x)$ and $y_2(x)$ are linearly independent.

4.3 DIFFERENTIAL EQUATIONS OF 1ST ORDER AND 1ST DEGREE

4.3.1 General Form

An ordinary differential equation of first order and first degree can be put in one of the following forms:

(i) $\dfrac{dy}{dx} = f(x, y)$ (ii) $\dfrac{dx}{dy} = f(x, y)$

(iii) $M dx + N dy = 0$, where M and N are functions of x and y or constants.

Examples:

(i) $x^2 dx + e^y dy = 0$ (ii) $\dfrac{dy}{dx} = \dfrac{x\sqrt{y+1}}{y\sqrt{x+1}}$ etc

4.3.2 Solution by Separation of Variables

Put the equation $M dx + N dy = 0$ in the form

$f_1(x) dx + f_2(y) dy = 0$ (if possible)

Then the equation can be solved by integrating each term separately. Hence, the solution of the equation is $\int f_1(x) dx + \int f_2(y) dy = C$, where C is the constant of integration.

Example:

Let us find the solution of $x\dfrac{dy}{dx} - y = 1$.

$x\dfrac{dy}{dx} - y = 1$

$\Rightarrow x\dfrac{dy}{dx} = 1 + y$

$\Rightarrow \dfrac{dy}{1+y} = \dfrac{dx}{x}$

$\Rightarrow \int \dfrac{dy}{1+y} = \int \dfrac{dx}{x} + C$

$\Rightarrow \log(1 + y) = \log x + C \ldots\ldots\ldots\ldots(i)$

which is the required solution.

Again (i)

$\Rightarrow \log(1 + y) - \log x = \log C_1$ (where $C = \log C_1$)

$\Rightarrow \log\left(\dfrac{1+y}{x}\right) = \log C_1$

$\Rightarrow \dfrac{1+y}{x} = C_1 \qquad \Rightarrow 1 + y = C_1 x$

4.3.3 Homogeneous Differential Equation

If a differential equation of first order and first degree can be put in the form $\dfrac{dy}{dx} = f\left(\dfrac{y}{x}\right)$ or in the form $\dfrac{dx}{dy} = f\left(\dfrac{x}{y}\right)$, then the equation is said to be homogeneous.

Remember:

A. Consider the equation $\dfrac{dy}{dx} = f\left(\dfrac{y}{x}\right)$. This type of homogeneous equation can be easily solved by putting $y = vx$.

B. Consider the equation $\dfrac{dx}{dy} = f\left(\dfrac{x}{y}\right)$, This type of homogeneous equation can be easily solved by putting $x = vy$.

Example:

Consider $(x^2 + y^2) dx - 2xy dy = 0$.

Then, $\dfrac{dy}{dx} = \dfrac{x^2 + y^2}{2xy} = \dfrac{1 + \left(\dfrac{y}{x}\right)^2}{2\dfrac{y}{x}} = f\left(\dfrac{y}{x}\right) \qquad \ldots(i)$

So the given differential equation is homogeneous.

Let us put $y = vx$. Then $\dfrac{dy}{dx} = v + x\dfrac{dv}{dx}$ and $v = \dfrac{y}{x}$

$\therefore (i) \Rightarrow v + x\dfrac{dv}{dx} = \dfrac{1+v^2}{2v}$

$or, \ x\dfrac{dv}{dx} = \dfrac{1+v^2}{2v} - v = \dfrac{1-v^2}{2v}$

$or, \ \dfrac{2v dv}{v^2 - 1} = -\dfrac{dx}{x}$

$or, \ \int \dfrac{2v}{v^2 - 1} dv = -\int \dfrac{dx}{x} + \log C$

(taking $\log C$ as the constant of integration)

$or, \ \log(v^2 - 1) = -\log x + \log C$

$or, \ \log(v^2 - 1) + \log x = \log C$

$or, \ \log\{x(v^2 - 1)\} = \log C$

$or, \ x\left\{\left(\dfrac{y}{x}\right)^2 - 1\right\} = C$

$or, \ y^2 - x^2 = Cx$, which is the required solution.

C. Let us consider a nonhomogeneous equation of the form:

$$\frac{dy}{dx} = \frac{a_1 x + b_1 y + c_1}{a_2 x + b_2 y + c_2} \qquad \left(\frac{a_1}{a_2} \neq \frac{b_1}{b_2}\right) \qquad \ldots(i)$$

To solve (i), put $x = x' + h$ and $y = y' + k$.

Here, the constants h and k are so chosen that

$a_1 h + b_1 k + c_1 = 0 \quad$ and $\quad a_2 h + b_2 k + c_2 = 0$

Then (i) becomes,

$$\frac{dy'}{dx'} = \frac{a_1 x' + b_1 y'}{a_2 x' + b_2 y'} = \frac{a_1 + b_1\left(\dfrac{y'}{x'}\right)}{a_2 + b_2\left(\dfrac{y'}{x'}\right)} = f\left(\frac{y'}{x'}\right)$$

which is a homogeneous equation.

So put $y' = v\,x'$ and proceed.

If $\dfrac{a_1}{a_2} = \dfrac{b_1}{b_2} = k$ (say), then

$$\frac{dy}{dx} = \frac{a_2 kx + b_2 ky + c_1}{a_2 x + b_2 y + c_2} = \frac{k(a_2 x + b_2 y) + c_1}{(a_2 x + b_2 y)},$$

which can be solved by putting $a_2 x + b_2 y = u$

so that $a_2 + b_2 \dfrac{dy}{dx} = \dfrac{du}{dx}$.

Example:

Let us solve the equation $\dfrac{dy}{dx} = \dfrac{x + y + 1}{2x + 2y + 1}$.

Here, $\dfrac{a_1}{a_2} = \dfrac{1}{2} = \dfrac{b_1}{b_2}$.

So let us put $x + y = u$. So $1 + \dfrac{dy}{dx} = \dfrac{du}{dx}$.

Then, $\dfrac{dy}{dx} = \dfrac{x + y + 1}{2(x + y) + 1}$

$\Rightarrow \dfrac{du}{dx} - 1 = \dfrac{u + 1}{2u + 1}$

$\Rightarrow \dfrac{du}{dx} = \dfrac{u + 1}{2u + 1} + 1 = \dfrac{3u + 2}{2u + 1}$

$\Rightarrow \left(\dfrac{2u + 1}{3u + 2}\right) du = dx$

$\Rightarrow \left\{\dfrac{\dfrac{2}{3}(3u + 2) - \dfrac{1}{3}}{(3u + 2)}\right\} du = dx$

$\Rightarrow \left(\dfrac{2}{3} - \dfrac{1}{3} \times \dfrac{1}{3u + 2}\right) du = dx$

$\Rightarrow \int\left[\dfrac{2}{3} - \dfrac{1}{3(3u + 2)}\right] du = \int dx + C$

$\Rightarrow \dfrac{2}{3}u - \dfrac{1}{9}\log(3u + 2) = x + C$

$\Rightarrow \dfrac{2}{3}(x + y) - \dfrac{1}{9}\log\{3(x + y) + 2\} = x + C$

$\Rightarrow \dfrac{2}{3}y - \dfrac{1}{3}x - \dfrac{1}{9}\log(3x + 3y + 2) = C$

4.3.4 Exact Differential Equations

If the differential equation $Mdx + Ndy = 0$ can be expressed in the form $du = 0$, where $u = u(x, y)$, then the equation is called exact and its solution is given by $u(x, y) = C$, where C is an arbitrary constant. In other words, a differential equation $Mdx + Ndy = 0$ is said to be exact if $\dfrac{\partial M}{\partial y} = \dfrac{\partial N}{\partial x}$.

Working procedure to solve an exact differential equation:

Step-1: Compare the given differential equation with $Mdx + Ndy = 0$ and find out M and N.

Step-2: Show that $\dfrac{\partial M}{\partial y} = \dfrac{\partial N}{\partial x}$.

Step-3: Find $\int M\,dx$ considering "y" as a constant.

Step-4: Find $\int N\,dy$ considering "x" as a constant.

Step-5: The required solution is given by the sum $\int M\,dx + \int N\,dy = C$ (the common terms in $\int M\,dx$ and $\int N\,dy$ come once in the sum).

Example:

(i) Consider the equation $ydx + xdy = 0$

Then, $ydx + xdy = 0$

$\Rightarrow d(xy) = 0$,

which shows that the equation is exact.

Integrating we get $xy = C$, which is the required solution.

(ii) Consider $(x^3 + 4y)dx + (4x + y + 2)dy = 0$.

Here, $M = x^3 + 4y$ and $N = 4x + y + 2$. So we have,

$\dfrac{\partial M}{\partial y} = 4 = \dfrac{\partial N}{\partial x}$.

Hence, the given differential equation is exact.

Now $\int M\,dx = \int (x^3 + 4y)dx = \dfrac{x^4}{4} + 4xy$,

$\int N\,dy = \int (4x + y + 2)dy = 4xy + \dfrac{y^2}{2} + 2y$.

Therefore, the solution is given by $\int M\,dx + \int N\,dy = C$

$i.e;\ \dfrac{x^4}{4} + 4xy + \dfrac{y^2}{2} + 2y = C.$

Note: If the given equation is not exact, then it can be made exact by multiplying some function of x or some function of y or some function of x and y.

This multiplier is called an ***Integrating factor (I.F.)*** of the given differential equation. The number of integrating factors (if exist) is infinite.

Remember:

(i) $ydx + xdy = d(xy)$

(ii) $xdx + ydy = \dfrac{1}{2}d\left(x^2 + y^2\right)$

(iii) $\dfrac{-xdy + ydx}{y^2} = d\left(\dfrac{x}{y}\right)$

(iv) $\dfrac{xdy - ydx}{x^2} = d\left(\dfrac{y}{x}\right)$

(v) $\dfrac{xdy - ydx}{x^2 + y^2} = d\left(\tan^{-1}\dfrac{y}{x}\right)$

(vi) $\dfrac{ydx - xdy}{x^2 + y^2} = d\left(\tan^{-1}\dfrac{x}{y}\right)$

4.3.5 Linear Differential Equations

(i) An equation of the form $\dfrac{dy}{dx} + Py = Q$, where P and Q are functions of x alone or constants

is called a linear equation of first order and first degree in y.

The solution of this equation is given by

$$y\left(e^{\int Pdx}\right) = \int\left[Q\times\left(e^{\int Pdx}\right)\right]dx + C$$

where C is a constant of integration.
Here integrating factor (I.F.) $= e^{\int Pdx}$.

Example:

Consider the equation $\dfrac{dy}{dx} + y\tan x = \sec x$.

Then, $P = \tan x$ and $Q = \sec x$

$\therefore\ e^{\int Pdx} = e^{\int \tan x dx} = e^{\log(\sec x)} = \sec x$.

So required solution is given by

$$y\times\sec x = \int(\sec x\times\sec x)dx + C$$

or, $y\sec x = \int\sec^2 x dx + C$

or, $y\sec x = \tan x + C$

(ii) An equation of the form $\dfrac{dx}{dy} + Px = Q$, where P and Q are functions of y alone or constants

is called a linear equation of first order and first degree in x.

The solution of this equation is given by

$$x\left(e^{\int Pdy}\right) = \int\left[Q\times\left(e^{\int Pdy}\right)\right]dy + C,$$

where C is a constant of integration.
Here integrating factor (I.F) $= e^{\int Pdy}$.

Example:

Consider the equation $(y - x - 1)dy = dx$

Then $(y - x - 1)\,dy = dx$

$\Rightarrow \dfrac{dx}{dy} = y - x - 1$

$\Rightarrow \dfrac{dx}{dy} + 1\times x = y - 1$, Which is of the form $\dfrac{dx}{dy} + Px = Q$

$\therefore\ e^{\int Pdy} = e^{\int 1 dy} = e^{y}$.

So the required solution is given by

$$xe^{y} = \int\left[(y - 1)\times e^{y}\right]dy + C$$

or, $xe^{y} = (y - 1)e^{y} - \int 1\times e^{y}dy + C$

or, $xe^{y} = (y - 1)e^{y} - e^{y} + C$

or, $xe^{y} = (y - 2)e^{y} + C$

Remember:

The equation $\dfrac{dy}{dx} + Py = Qy^{n}$ (Known as Bernoulli's equation; where P, Q are functions of x alone or constants) can be reduced to a linear equation by substituting $y^{1-n} = v$.

Example:

Consider $x\dfrac{dy}{dx} + y = xy^{2}$

Then, $x\dfrac{dy}{dx} + y = xy^{2}$

$\Rightarrow \dfrac{1}{y^2}\dfrac{dy}{dx} + \dfrac{1}{xy} = 1$ ...(i)

Let us put $y^{1-2} = v$ i.e; $\dfrac{1}{y} = v$.

Then, $-\dfrac{1}{y^2}\dfrac{dy}{dx} = \dfrac{dv}{dx}$

$\therefore$ (i) gives,

$-\dfrac{dv}{dx} + \dfrac{1}{x}v = 1$

or, $\dfrac{dv}{dx} + \left(-\dfrac{1}{x}\right)v = -1$ ...(ii)

Here,

$e^{\int Pdx} = e^{\int\left(-\frac{1}{x}\right)dx} = e^{-\int\frac{dx}{x}} = e^{-\log x} = e^{\log\left(\frac{1}{x}\right)} = \dfrac{1}{x}$.

So the required solution of (ii) is given by

$$v\times\dfrac{1}{x} = \int(-1)\times\dfrac{1}{x}dx + C = -\log x + C$$

or, $\dfrac{1}{xy} = -\log x + C$

4.4 LINEAR DIFFERENTIAL EQUATIONS OF 2ND ORDER

4.4.1 General Form

An ordinary differential equation of nth order has the following form:

$$a_0 \frac{d^n y}{dx^n} + a_1 \frac{d^{n-1} y}{dx^{n-1}} + a_2 \frac{d^{n-2} y}{dx^{n-2}} + \ldots\ldots + a_{n-1} \frac{dy}{dx} + a_n y = X$$

where X and $a_0, a_1, a_2, \ldots\ldots, a_{n-1}, a_n$ are constants or functions of x only.

When $n = 2$, we have

$$a_0 \frac{d^2 y}{dx^2} + a_1 \frac{dy}{dx} + a_2 y = X \qquad \ldots(i)$$

(i) is called the general form of a linear differential equation of order "2."

4.4.2 Complementary Function (C.F)

Let us consider

$$a_0 \frac{d^2 y}{dx^2} + a_1 \frac{dy}{dx} + a_2 y = X \qquad \ldots(i)$$

where a_0, a_1, a_2 are constants.

Let $y = e^{mx}$ be a trial solution of the equation

$$a_0 \frac{d^2 y}{dx^2} + a_1 \frac{dy}{dx} + a_2 y = 0 \qquad \ldots(ii)$$

Then $\dfrac{dy}{dx} = m e^{mx}$ and $\dfrac{d^2 y}{dx^2} = m^2 e^{mx}$.

∴ (ii) gives,

$$a_0 m^2 e^{mx} + a_1 m e^{mx} + a_2 e^{mx} = 0$$

or, $a_0 m^2 + a_1 m + a_2 = 0 \qquad \left(\because e^{mx} \neq 0 \right)\ldots\ldots\ldots(iii)$

Equation (iii) is called the auxiliary equation of (ii). Let m_1 and m_2 be two roots of the equation (iii). The solution of (ii) is called complementary function (C.F).

CASE-I: Let m_1 and m_2 are distinct real numbers.

In this case, the complementary function (C.F.) is given by

$$y_{CF} = c_1 e^{m_1 x} + c_2 e^{m_2 x},$$

where c_1 and c_2 are arbitrary constants.

CASE-II: Let m_1 and m_2 be equal real numbers.

In this case, the complementary function (C.F.) is given by $y_{CF} = (c_1 + c_2 x) e^{m_1 x}$

CASE-III: Let m_1 and m_2 be complex conjugates and $m_1 = \alpha + i\beta$, $m_2 = \alpha - i\beta$.

In this case, the complementary function (C.F.) is given by $y_{CF} = e^{\alpha x}(c_1 \cos \beta x + c_2 \sin \beta x)$

where c_1, c_2 are arbitrary constants.

Examples:

(i) Let us consider the equation

$$\frac{d^2 y}{dx^2} + 3\frac{dy}{dx} + 2y = x \qquad \ldots(i)$$

Let $y = e^{mx}$ be a trial solution of

$$\frac{d^2 y}{dx^2} + 3\frac{dy}{dx} + 2y = 0 \qquad \ldots(ii)$$

Then (ii) gives $m^2 e^{mx} + 3m e^{mx} + 2e^{mx} = 0$

or, $m^2 - 3m + 2 = 0 \quad (\because e^{mx} \neq 0)$

or, $m = -2, -1$

Thus, values of m are real and distinct.

So, $y_{CF} = c_1 e^{-2x} + c_2 e^{-x}$

(ii) Consider the equation

$$\frac{d^2 y}{dx^2} - 2\frac{dy}{dx} + y = e^x \qquad \ldots(i)$$

Let $y = e^{mx}$ be a trial solution of

$$\frac{d^2 y}{dx^2} - 2\frac{dy}{dx} + y = 0 \qquad \ldots(ii)$$

Then (ii) gives,

$$m^2 e^{mx} - 2m e^{mx} + e^{mx} = 0$$

or, $m^2 - 2m + 1 = 0 \quad (\because e^{mx} \neq 0)$

or, $(m-1)^2 = 0 \qquad$ or, $m = 1, 1$

Thus, the values of m are real and equal.

So, $y_{CF} = (c_1 + c_2 x)e^{1x} = (c_1 + c_2 x)e^x$

(iii) Consider the equation

$$\frac{d^2 y}{dx^2} + \frac{dy}{dx} + y = 2x \qquad \ldots(i)$$

Let $y = e^{mx}$ be a trial solution of

$$\frac{d^2 y}{dx^2} + \frac{dy}{dx} + y = 0 \qquad \ldots(ii)$$

Then (ii) gives,

$$m^2 e^{mx} + m e^{mx} + e^{mx} = 0$$

or $m^2 + m + 1 = 0 \quad (\because e^{mx} \neq 0)$

or, $m = \dfrac{-1 \pm \sqrt{1^2 - 4 \times 1 \times 1}}{2 \times 1}$

$= \dfrac{-1 \pm \sqrt{3}i}{2}$, where $i = \sqrt{-1}$

$$\therefore \quad m = -\frac{1}{2} + \frac{\sqrt{3}}{2}i, \quad -\frac{1}{2} - \frac{\sqrt{3}}{2}i$$

Thus, the values of m are complex conjugates.

So, $y_{CF} = e^{-\frac{1}{2}x} \left\{ c_1 \cos\left(\frac{\sqrt{3}}{2}x\right) + c_2 \sin\left(\frac{\sqrt{3}}{2}x\right) \right\}$

Remember:

We use the notation D for $\frac{d}{dx}$ and D^2 for $\frac{d^2}{dx^2}$.

Thus, $\frac{dy}{dx} = Dy$ and $\frac{d^2y}{dx^2} = D^2y$

Using these notations, the equation

$\frac{d^2y}{dx^2} + 3\frac{dy}{dx} + 4y = x^2$ can be re-written as

$D^2y + 3Dy + 4y = x^2$ or $(D^2 + 3D + 4)y = x^2$

4.4.3 Particular Integral (P.I)

Let us consider the differential equation

$$a_0 \frac{d^2y}{dx^2} + a_1 \frac{dy}{dx} + a_2 y = X \qquad \ldots(i)$$

where a_0, a_1, a_2, and X are constants or functions of x only

The equation (i) can be re-written as

$a_0D^2y + a_1Dy + a_2y = X$

or $(a_0D^2 + a_1D + a_2)y = X$

or $f(D)y = X$

where $f(D) = a_0D^2 + a_1D + a_2$

Then we define the particular integral (P.I) of the equation (i) by

$$y_{PI} = \frac{1}{f(D)}X$$

Methods for finding P.I

Case-I: Let $X = x^n$

In order to get P.I. in this case, expand $\frac{1}{f(D)}$
i.e; $\{f(D)\}^{-1}$ in ascending powers of D and operate on x^n with the result. Then the result will follows. Use the following formula to get the result quickly:

$$D^m\left(x^n\right) = \begin{cases} n! & , if \ n = m \\ 0 & , if \ m > n \\ \dfrac{n!}{(n-m)!}x^{n-m} & , if \ n > m \end{cases}$$

Thus, $D^2(x^2) = 2! = 2$, $D^3(x^2) = 0$,

$D^2\left(x^5\right) = \frac{5!}{(5-2)!}x^{5-2} = 20x^3$ etc.

Example:

Let us find the particular integral of
$(D^2 - 2D + 2)y = x$

Here, $y_{PI} = \dfrac{1}{D^2 - 2D + 2}(x) = \dfrac{1}{2\left\{1 + \left(\dfrac{D^2 - 2D}{2}\right)\right\}}(x)$

$= \dfrac{1}{2}\left\{1 + \left(\dfrac{D^2 - 2D}{2}\right)\right\}^{-1}(x)$

$= \dfrac{1}{2}\left\{1 - \left(\dfrac{D^2 - 2D}{2}\right) + \left(\dfrac{D^2 - 2D}{2}\right)^2 \ldots\ldots\infty\right\}(x)$

$= \dfrac{1}{2}\left\{1 - \dfrac{1}{2}D^2 + D\right\}(x)$

$= \dfrac{1}{2}\left\{x - \dfrac{1}{2}D^2(x) + D(x)\right\}$

$= \dfrac{1}{2}\left(x - \dfrac{1}{2} \times 0 + 1!\right) = \dfrac{1}{2}(x + 1).$

Case-II: Let $X = e^{ax}$

In this case, $y_{PI} = \dfrac{1}{f(D)}e^{ax}$

$$= \begin{cases} \dfrac{1}{f(a)}e^{ax} & provided \ f(a) \neq 0 \\ \dfrac{xe^{ax}}{f'(a)} & if \ f(a) = 0 \end{cases}$$

Example:

Let us find the P.I of $y'' - 2y' + 3y = e^{4x}$

The given equation can be re-written as:
$(D^2 - 2D + 3)y = e^{4x}$

Then $y_{PI} = \dfrac{1}{D^2 - 2D + 3}e^{4x}$

$= \dfrac{e^{4x}}{4^2 - 2 \times 4 + 3} \quad \left[\because f(D) = D^2 - 2D + 3 \right.$

$\left. \text{and } f(4) = 4^2 - 2 \times 4 + 3 \neq 0\right]$

$= \dfrac{e^{4x}}{11}$

Case-III: Let $X = \sin(ax + b)$ or $\cos(ax + b)$.

Assume that $f(D)$ contains only even powers of D.

Then $f(D) = \phi(D^2)$

In this case,

$$y_{PI} = \frac{1}{\phi(-a^2)}\sin(ax + b) \quad or \quad \frac{1}{\phi(-a^2)}\cos(ax + b)$$

[provided $f(-a^2) \neq 0$]

Example:

(*i*) Let us find the P.I. of $y'' + y = \sin^2 x$.

Here, the given differential equation can be re-written as: $(D^2 + 1)y = \sin 2x$

$$\therefore y_{PI} = \frac{1}{D^2 + 1}\sin 2x$$

$$= \frac{1}{-2^2 + 1}\sin 2x = -\frac{1}{3}\sin 2x$$

(*ii*) Let us find the P.I of $(D^2 + D + 1)y = \cos x$

Here $y_{PI} = \dfrac{1}{D^2 + D + 1}\cos x$

$$= \frac{1}{-1^2 + D + 1}\cos x$$

$$= \frac{1}{D}\cos x = \int \cos x\,dx = \sin x$$

$$\left(\because \frac{1}{D} \text{ stands for integration}\right)$$

Remember:

(*i*) If $f(-a^2) = 0$, then

$$y_{PI} = \frac{1}{f(D)}\sin ax = \frac{x\sin ax}{f'(-a^2)}$$

$$[\text{provided } f'(-a^2) \neq 0]$$

(*ii*) If $f(-a^2) = 0$, then

$$y_{PI} = \frac{1}{f(D)}\cos ax = \frac{x\cos ax}{f'(-a^2)}$$

$$[\text{provided } f'(-a^2) \neq 0]$$

Case-IV: Let $X = e^{ax}V$, where V is a function of x.

Here, $y_{PI} = \dfrac{1}{f(D)}\left(e^{ax}V\right) = e^{ax} \times \dfrac{1}{f(D+a)}(V)$

Example:

Let us consider, $\dfrac{d^2y}{dx^2} - 2\dfrac{dy}{dx} + y = xe^x$

This equation can be written as

$(D^2 - 2D + 1)y = xe^x$

$$\therefore y_{PI} = \frac{1}{D^2 - 2D + 1}\left(xe^x\right) = \frac{1}{(D-1)^2}xe^x$$

$$= e^x \frac{1}{\left[\{(D+1)-1\}\right]^2}(x) = e^x \frac{1}{D^2}(x) = e^x \frac{1}{D}\int x\,dx$$

$$= e^x \frac{1}{D}\left(\frac{x^2}{2}\right) = e^x \int \frac{x^2}{2}\,dx = e^x \times \frac{x^3}{2\times 3} = \frac{e^x x^3}{6}.$$

Remember:

If $X = xV$, where V is any function of x, then

$$y_{PI} = \frac{1}{f(D)}(xV) = x\frac{1}{f(D)}V + \left[\frac{d}{dD}\left\{\frac{1}{f(D)}\right\}\right](V)$$

$$= x\frac{1}{f(D)}V - \frac{f'(D)}{\left[f(D)\right]^2}(V)$$

4.4.4 Complete (General) Solution

Let us consider the second order linear differential equation

$$a_0 \frac{d^2y}{dx^2} + a_1 \frac{dy}{dx} + a_2 y = X \qquad \ldots(i)$$

where a_0, a_1, a_2, and X are functions of x or constants.

For the equation (*i*), the complete (general) solution is given by

$y = y_{CF} + y_{PI}$, which contains two arbitrary constants in y_{CF} part.

Example:

Let us solve $\dfrac{d^2y}{dx^2} + 4\dfrac{dy}{dx} + 4 = e^x \qquad \ldots(i)$

Consider $\dfrac{d^2y}{dx^2} + 4\dfrac{dy}{dx} + 4 = 0 \qquad \ldots(ii)$

Let $y = e^{mx}$ be a trial solution of (*ii*)

Then (*ii*) gives

$m^2 e^{mx} - 4me^{mx} + 4e^{mx} = 0$

$or,\ m^2 - 4m + 4 = 0 \quad \left[\because e^{mn} \neq 0\right]$

$or,\ (m-2)^2 = 0 \quad or,\ m = 2,\ 2$

Here, the values of m are real and equal.

So, $y_{CF} = (c_1 + c_2 x)e^{2x}$

Now, equation (i) can be re-written as

$(D^2 - 4D + 4)y = e^x$

$$\therefore y_{PI} = \frac{1}{\left(D^2 - 4D + 4\right)}e^x = \frac{1}{(D-2)^2}e^x$$

$$= \frac{e^x}{(1-2)^2} = e^x$$

Hence, the complete solution is given by

$y = y_{CF} + y_{PI} = (c_1 + c_2 x)e^{2x} + e^x$.

4.4.5 Homogeneous Linear Differential Equations of Order Two

An equation of the form

$x^2 \dfrac{d^2y}{dx^2} + a_1 x\dfrac{dy}{dx} + a_2 y = X$, where a_1, a_2 are constants and X is a function of x, is called a linear homogeneous differential equation of order two.

To solve this type of equation, put $\log x = z$.

Then $\dfrac{dy}{dz} = x\dfrac{dy}{dx}$ and $x^2 \dfrac{d^2y}{dx^2} = \dfrac{d^2y}{dz^2} - \dfrac{dy}{dz}$

Using these results the given equation reduces to a linear differential equations with constant co-efficients which can be solved by finding out its CF and PI.

Example:

Let us solve $x^2 \dfrac{d^2y}{dx^2} - x\dfrac{dy}{dx} + y = 2\log x$ ...(i)

The given differential equation is a linear homogeneous equation of order 2.

So put $z = \log x$

Then $\dfrac{dy}{dz} = x\dfrac{dy}{dx}$ and $\dfrac{d^2y}{dz^2} - \dfrac{dy}{dz} = x^2\dfrac{d^2y}{dx^2}$

∴   (i) gives

$\dfrac{d^2y}{dz^2} - \dfrac{dy}{dz} - \dfrac{dy}{dz} + y = 2z$

or $\dfrac{d^2y}{dz^2} - 2\dfrac{dy}{dz} + y = 2z$ ...(ii)

Consider $\dfrac{d^2y}{dz^2} - 2\dfrac{dy}{dz} + y = 0$ ...(iii)

Let $y = e^{mz}$ be a trial solution of (iii)

Then, $(iii) \Rightarrow m^2 e^{mz} - 2me^{mz} + e^{mz} = 0$

$\Rightarrow m^2 - 2m + 1 = 0 \ (\because e^{mz} \neq 0)$

$\Rightarrow m = 1, 1$

∴ $y_{CF} = (c_1 + c_2 z)e^{1z} = (c_1 + c_2 \log x)x \ (\because z = \log x)$

Now equation (ii) can be written as:

$(D^2 - 2D + 1)y = 2z$, where $D \equiv \dfrac{d}{dz}$

∴ $y_{PI} = \dfrac{1}{D^2 - 2D + 1}(2z) = \left[1 + \left(D^2 - 2D\right)\right]^{-1}(2z)$

$= \left[1 - \left(D^2 - 2D\right) + \left(D^2 - 2D\right)^2 \dots\dots\right](2z)$

$\left(\text{using } (1+x)^{-1} = 1 - x + x^2 - x^3 + \dots\dots\dots\infty\right)$

$= \left[1 - \left(D^2 - 2D\right)\right](2z)$

$= 2z - D^2(2z) + 2D(2z)$

$= 2z - 0 + 4 = 2(\log x + 2)$

Hence, the complete solution of (i) is given by

$y = y_{CF} + y_{PI} = x(c_1 + c_2 \log x) + 2(\log x + 2)$

Remember:

An equation of the form

$(ax + b)^2 \dfrac{d^2y}{dx^2} + (ax + b)P_1\dfrac{dy}{dx} + P_2 y = X,$

where P_1 and P_2 are constants and X is a function of x alone, can be reduced to a linear homogeneous differential equation of order 2 by substituting $ax + b = u$.

1. The differential equation of all straight lines passing through the origin is given by
   (a) $\dfrac{dy}{dx} = \dfrac{y}{x}$ (b) $\dfrac{dy}{dx} = \dfrac{x}{y}$
   (c) $xdy + ydx = 0$ (d) none of these

2. The order and degree of the following differential equation are, respectively:
   $$\left[1 + \left(\dfrac{dy}{dx}\right)^2\right]^{\frac{4}{5}} = \dfrac{d^2y}{dx^2}$$
   (a) 2,1   (b) 2,2   (c) 2,5   (d) 5,2

3. The general solution of a differential equation is given as $y = a \cos(bx + c)$, here "b" is the only parameter. Then the differential equation is giver by:
   (a) $y_2 = b^2 y$ (b) $y_2 + b^2 y = 0$
   (c) $y_2 + y = 0$ (d) $y_2 - y = 0$

4. $y = a + \dfrac{b}{x}$ is a general solution of a differential equation given by
   (a) $x\dfrac{d^2y}{dx^2} + 2\dfrac{dy}{dx} = 0$ (b) $x\dfrac{d^2y}{dx^2} - 2\dfrac{dy}{dx} = 0$
   (c) $x\dfrac{d^2y}{dx^2} = 2\dfrac{dy}{dx}$ (d) none of these

5. $xy = ae^x + be^{-x}$ is a general solution of a differential equation given by
   (a) $x^2 y_2 + 2xy_1 = y$ (b) $xy_2 + 2y_1 = xy$
   (c) $y_2 + y_1 = y$ (d) none of these

6. $xy = Ae^x + Be^{-x} + x^2$ is a general solution of a differential equation given by
   (a) $x^2 y_2 + xy_1 - xy + y^2 - 2 = 0$
   (b) $x^2 y_2 + xy_1 + y^2 - 2 = 0$
   (c) $xy_2 + 2y_1 - xy + x^2 - 2 = 0$
   (d) None of these

7. The solution $(1 + x^2)\tan^{-1} xdy + y^2 dx = 0$ is
   (a) $\log\left(\tan^{-1} x\right) - \dfrac{1}{y} = C$

(b) $\log(\tan^{-1}x) + y = C$

(c) $\log\left(\cot^{-1}x\right) - \dfrac{1}{y^2} = C$

(d) none of these

8. $(xy^2 + x)dx + (yx^2 + y)dy = 0$ has the solution

(a) $\log\left(\dfrac{x^2+1}{y^2+1}\right) = C$

(b) $\log(x^2 + 1) = C \log(y^2 + 1)$

(c) $\dfrac{1}{2}\log\left\{\left(x^2+1\right)\left(y^2+1\right)\right\} = C$

(d) none the these

9. The solution of $\log\left(\dfrac{dy}{dx}\right) = 2x + 3y$ is given by

(a) $2e^{-3y} + 3e^{2x} = C$ (b) $3e^{-3y} + 2e^{2x} = C$
(c) $2 + e^{2x+3y} = Cxy$ (d) none of these

10. Solve: $x(ydx + xdy) = y\tan\dfrac{y}{x}(xdy - ydx)$

(a) $\sec\dfrac{y}{x} = cxy$ (d) $\cos\dfrac{y}{x} = \dfrac{cx}{y}$

(c) $\sec\dfrac{y}{x} = \dfrac{cy}{x}$ (d) none of these

11. Solve: $xydx = (x^2 - y^2)dy$

(a) $xe^{\frac{x^2}{2y^2}} = C$ (b) $ye^{\frac{x^2}{2y^2}} = C$

(c) $xe^{\frac{x^2}{y}} = C$ (d) $ye^{\frac{y^2}{x}} = C$

12. The solution of $x\dfrac{dy}{dx} + y = y^2\log x$ is given by

(a) $y(1 + \log x) = 1 + Cxy$
(b) $x(1 + \log x) = 1 + Cxy$
(c) $y\log x = Cx$ (d) $x\log x = Cy$

13. Solve: $\left(x^2 + 1\right)\dfrac{dy}{dx} + 2xy = x^2$

(a) $y\left(x^2+1\right) = \dfrac{x^3}{3} + c$ (b) $y\left(x^2+1\right) = \dfrac{x^2}{2} + c$

(c) $y(x^2 - 1) = xy + c$ (d) none of these

14. The solution of $(x + y + 1)dy - dx = 0$ is given by

(a) $x - y = ce^{-y}$ (b) $x + y + 2 = ce^y$
(c) $2x + y = ce^y$ (d) $2x - y = ce^{-y}$

15. The solution of $x\dfrac{dy}{dx} + \sin 2y = x^2\cos^2 y$ is

(a) $x^2\tan y = \dfrac{x^4}{4} + c$ (b) $x^2\tan y = \dfrac{x^2}{2} + c$

(c) $x^2\cot y = \dfrac{x^4}{4} + c$ (d) $x^2\sec y = -\dfrac{x^4}{4} + c$

16. If $y(x)$ satisfies the initial value problem $(x^2 + y)dx = xdy$ with $y(1) = 2$, then $y(2) = ?$
(a) 4   (b) 5   (c) 6   (d) 8

17. The homogeneous equation
$(ax + by)\,dx + (cx + fy)\,dy$ is exact if and only if
(a) $a = c$ (b) $b = f$ (c) $b = c$ (d) $a = f$

18. The differential equation
$\left|\dfrac{dy}{dx}\right| + |y| = 0,\ y(0) = 2$ has

(a) a unique solution
(b) no solution
(c) finite number of solutions
(d) infinite number of solutions

19. Solve: $\dfrac{d^2y}{dx^2} + \dfrac{dy}{dx} - 2y = 0$, when $x = 0, y = 2$ and

$\dfrac{dy}{dx} = 0,\ y = \dfrac{4}{3}e^x + \dfrac{2}{3}e^{-2x}$

(a) $y = \dfrac{1}{3}e^x + \dfrac{2}{3}e^{-2x}$ (b) $y = \dfrac{1}{3}e^x + \dfrac{1}{3}e^{-2x}$

(c) $y = \dfrac{4}{3}e^{-x} + \dfrac{2}{3}e^{2x}$ (d) $y = \dfrac{4}{3}e^x + \dfrac{2}{3}e^{-2x}$.

20. Solve: $\dfrac{d^2y}{dx^2} = \sin 3x$

(a) $y = -\dfrac{1}{9}\sin 3x + c_1x + c_2$

(b) $y = \dfrac{1}{4}\sin 3x + c_1x + c_2$

(c) $y = -\dfrac{1}{4}\sin 3x + c_1x + c_2$

(d) none of these

21. Solve: $\dfrac{d^2y}{dx^2} + 2\dfrac{dy}{dx} + y = e^x - e^{-2x}$

(a) $(c_1 + c_2x)e^x + e^x + e^{-2x}$

(b) $(c_1 + c_2x)e^{-x} + \dfrac{1}{4}e^x - e^{-2x}$

(c) $(c_1 + c_2x)e^{-x} + e^x - \dfrac{1}{2}e^{-2x}$

(d) none of these

22. Solve: $\left(D^2 + 1\right)y = \sin 2x;\ y = 0\ \&\ \dfrac{dy}{dx} = 0$ at $x = 0$

(a) $y = \dfrac{2}{3}\sin x - \dfrac{1}{3}\sin 2x$

(b) $y = \dfrac{4}{3}\sin x - \dfrac{1}{3}\sin 2x$

(c) $y = \dfrac{1}{3}\sin x + \dfrac{2}{3}\sin 2x$

(d) $y = \dfrac{1}{3}\sin x - \dfrac{4}{3}\sin 2x$

23. If $y = 3e^{2x} + e^{-2x} - \alpha x$ is the solution of the initial value problem

$$\frac{d^2y}{dx^2} + \beta y = 4a\,x,\ y(0) = 4 \text{ and } \left.\frac{dy}{dx}\right|_{x=0} = 1, \text{ then}$$

(a) $\alpha = 3, \beta = -4$ (b) $\alpha = 3, \beta = 4$
(c) $\alpha = -3, \beta = -4$ (d) $\alpha = -3, \beta = 4$

24. Solve: $y'' + 4y = x^2$

(a) $c_1 \cos 2x + c_2 \sin 2x + \frac{1}{4}\left(x^2 - \frac{1}{2}\right)$

(b) $c_1 \cos 2x + c_2 \sin 2x + \frac{1}{2}\left(x^2 + \frac{1}{2}\right)$

(c) $c_1 \cos 2x + c_2 \sin 2x - \frac{1}{2}\left(x^2 - \frac{1}{2}\right)$

(d) none of these

25. Solve: $y'' - 4y = 2\sin 3x + e^x$

(a) $c_1 e^{2x} + c_2 e^{-2x} + \frac{1}{13}\sin 3x - \frac{1}{3}e^x$

(b) $c_1 e^{2x} + c_2 e^{-2x} - \frac{2}{13}\sin 3x - \frac{1}{3}e^x$

(c) $(c_1 + c_2 x)e^{2x} + \frac{1}{13}\cos 3x + \frac{1}{2}e^x$

(d) $(c_1 + c_2 x)e^{-2x} - \frac{2}{13}\cos 3x - \frac{1}{2}e^x$

26. The Wronskian of e^{2x}, e^{-4x} and e^{4x} is
(a) $48e^{-2x}$ (b) $12e^{-2x}$ (c) $24e^{2x}$ (d) $-96e^{2x}$

27. For three solutions $x^2 - 1$, $x^2 + x + 1$, $2x^2 + x$; which of the following is true?
(a) $W(x) \neq 0$
(b) The solutions are linearly dependent
(c) The solutions are linearly independent
(d) $W(x) = 2x^2 - x$

28. For the solutions 1, $\sin^2 x$, $\cos^2 x$ which of the following is true?
(a) $W(x) \neq 0$
(b) The solutions are linearly dependent
(c) The solutions are linearly independent
(d) None of these.

29. Let $y_1(x)$ and $y_2(x)$ be two linearly independent solutions of

$y'' + y \sin x = 0, 0 \leq x \leq 1$.

Let $g(x) = W(y_1, y_2)$. Then which of the following is true?
(a) $g'(x) > 0$ on $[0, 1]$ (b) $g'(x) < 0$ on $[0, 1]$
(c) $g'(x)$ vanishes at only one point of $[0,1]$
(d) $g'(x)$ vanishes at only all points of $[0,1]$

30. Let $W(y_1, y_2)$ be Wronskian of two linearly independent solutions y_1 and y_2 of the equation $y'' + P(x)y' + Q(x)y = 0$. Then the product $W(y_1, y_2) \times P(x)$ is equal to_____?
(a) $y_1'' y_2 - y_1 y_2''$ (b) $y_1 y_2' - y_2 y_1'$
(c) $y_1' y_2'' - y_1'' y_2'$ (d) $y_2' y_1'' - y_1' y_2''$

Answer key				
1. (a)	**2.** (c)	**3.** (b)	**4.** (a)	**5.** (b)
6. (c)	**7.** (a)	**8.** (c)	**9.** (a)	**10.** (a)
11. (b)	**12.** (a)	**13.** (a)	**14.** (b)	**15.** (a)
16. (c)	**17.** (c)	**18.** (b)	**19.** (d)	**20.** (a)
21. (b)	**22.** (a)	**23.** (a)	**24.** (a)	**25.** (b)
26. (d)	**27.** (b)	**28.** (b)	**29.** (d)	**30.** (a)

Explanations

1. (*a*) Let, $y = mx$ be the straight line passing through the origin.

Then, $\frac{dy}{dx} = m$ and so $y = \frac{dy}{dx} \times x$

i.e; $\frac{dy}{dx} = \frac{y}{x}$

2. (*c*) $\left[1 + \left(\frac{dy}{dx}\right)^2\right]^{\frac{4}{5}} = \frac{d^2y}{dx^2}$

$\Rightarrow \left(\frac{d^2y}{dx^2}\right)^5 = \left[1 + \left(\frac{dy}{dx}\right)^2\right]^4$

Here highest order derivative occurring in the equation is $\frac{d^2y}{dx^2}$. So order = 2.

Again, power of highest order derivative *i.e;* power of $\frac{d^2y}{dx^2}$ is 5. Hence, degree = 5.

3. (*b*) $y = a \cos (bx + c) \Rightarrow y_1 = -ab \sin (b + c)$

$\Rightarrow y_2 = -ab^2 \cos (bx + c)$

$\Rightarrow y_2 = -b^2 y$

$\Rightarrow y_2 + b^2 y = 0$.

4. (*a*) $y = a + \frac{b}{x} \Rightarrow \frac{dy}{dx} = \frac{d}{dx}\left(a + \frac{b}{x}\right)$

$\Rightarrow \frac{dy}{dx} = 0 - \frac{b}{x^2}$

$\Rightarrow x^2 \dfrac{dy}{dx} = -b$

$\Rightarrow \dfrac{d}{dx}\left(x^2 \dfrac{dy}{dx}\right) = \dfrac{d}{dx}(-b)$

$\Rightarrow 2x\dfrac{dy}{dx} + x^2 \dfrac{d^2 y}{dx^2} = 0$

$\Rightarrow x\dfrac{d^2 y}{dx^2} + 2\dfrac{dy}{dx} = 0.$

5. (b) $xy = ae^x + be^{-x}$

$\Rightarrow \dfrac{d}{dx}(xy) = \dfrac{d}{dx}\left(ae^x + be^{-x}\right)$

$\Rightarrow x\dfrac{dy}{dx} + y = ae^x - be^{-x}$

$\Rightarrow \dfrac{d}{dx}\left(x\dfrac{dy}{dx} + y\right) = \dfrac{d}{dx}(ae^x - be^{-x})$

$\Rightarrow xy_2 + 2y_1 = ae^x + be^{-x} = xy.$

6. (c) $xy = Ae^x + Be^{-x} + x^2$

$\Rightarrow \dfrac{d}{dx}(xy) = \dfrac{d}{dx}\left(Ae^x + Be^{-x} + x^2\right)$

$\Rightarrow xy_1 + y = Ae^x - Be^{-x} + 2x \qquad \ldots(i)$

Again, differentiating both sides of (i) w.r.t "x" we get,

$xy_2 + y_1 + y_1 = Ae^x + Be^{-x} + 2$

or, $xy_2 + 2y_1 = (Ae^x + Be^{-x} + x^2) + (2 - x^2)$

or, $xy_2 + 2y_1 = xy + 2 - x^2$

$(\because \quad xy = Ae^x + Be^{-x} + x^2)$

or, $xy_2 + 2y_1 - xy + x^2 - 2 = 0$

7. (a) $(1 + x^2)\tan^{-1} x\, dy + y^2 dx = 0$

$\Rightarrow \dfrac{dx}{\left(1+x^2\right)\tan^{-1} x} + \dfrac{dy}{y^2} = 0$

$\Rightarrow \int \dfrac{dx}{\left(1+x^2\right)\tan^{-1} x} + \int \dfrac{dy}{y^2} = C$

$\Rightarrow \int \dfrac{dt}{t} - \dfrac{1}{y} = C$

$\left[\text{putting } \tan^{-1} x = t \text{ so that } \dfrac{1}{1+x^2}dx = dt\right]$

$\Rightarrow \log t - \dfrac{1}{y} = C \quad \Rightarrow \log\left(\tan^{-1} x\right) - \dfrac{1}{y} = C.$

8. (c) $(xy^2 + x)dx + (yx^2 + y)dy = 0$

$\Rightarrow x\left(y^2 + 1\right)dx + y\left(x^2 + 1\right)dy = 0$

$\Rightarrow \dfrac{xdx}{x^2 + 1} + \dfrac{ydy}{y^2 + 1} = 0$

$\Rightarrow \int \dfrac{xdx}{x^2 + 1} + \int \dfrac{ydy}{y^2 + 1} = C$

$\Rightarrow \dfrac{1}{2}\int \dfrac{2xdx}{x^2 + 1} + \dfrac{1}{2}\int \dfrac{2ydy}{y^2 + 1} = C$

$\Rightarrow \dfrac{1}{2}\log\left(x^2 + 1\right) + \dfrac{1}{2}\log\left(y^2 + 1\right) = C$

$\Rightarrow \dfrac{1}{2}\left[\log\left(x^2 + 1\right)\left(y^2 + 1\right)\right] = C.$

9. (a) $\log\left(\dfrac{dy}{dx}\right) = 2x + 3y$

$\Rightarrow \dfrac{dy}{dx} = e^{2x+3y} = e^{2x} \times e^{3y}$

$\Rightarrow \dfrac{dy}{e^{3y}} = e^{2x}dx \Rightarrow e^{-3y}dy = e^{2x}dx$

Integrating we get,

$\int e^{-3y}dy = \int e^{2x}dx - \dfrac{C}{6}$

(where $-\dfrac{C}{6}$ is the constant of integration)

or, $\dfrac{e^{-3y}}{-3} = \dfrac{e^{2x}}{2} - \dfrac{C}{6}$

or, $\dfrac{e^{2x}}{2} + \dfrac{e^{-3y}}{3} = \dfrac{C}{6}$

or, $3e^{2x} + 2e^{-3y} = C.$

10. (a)

$x(ydx + xdy) = y\tan\dfrac{y}{x}(xdy - ydx)$

$\Rightarrow \dfrac{x}{x^2}d(xy) = y\tan\dfrac{y}{x}\left(\dfrac{xdy - ydx}{x^2}\right)$

$\Rightarrow \dfrac{1}{xy}d(xy) = \tan\dfrac{y}{x}d\left(\dfrac{y}{x}\right)$

Integrating we get,

$\int \dfrac{d(xy)}{xy} = \int \tan\dfrac{y}{x}d\left(\dfrac{y}{x}\right) - \log C$

(where $-\log C$ is the constant of integration)

or, $\log(xy) = \log\left(\sec\dfrac{y}{x}\right) - \log C$

or, $\log\left(\sec\dfrac{y}{x}\right) = \log xy + \log C = \log(Cxy)$

or, $\sec\dfrac{y}{x} = Cxy$

11. (b) $xy\,dx = (x^2 - y^2)\,dy$

$$\Rightarrow \frac{dy}{dx} = \frac{xy}{x^2 - y^2} = \frac{\dfrac{y}{x}}{1 - \left(\dfrac{y}{x}\right)^2} = f\left(\frac{y}{x}\right) \ (say) \qquad \ldots(i)$$

$\therefore$ the given differential equation is homogeneous. So let us put $y = vx$.

Then, $\dfrac{dy}{dx} = v + x\dfrac{dv}{dx}$.

$\therefore$ (i) gives $v + x\dfrac{dv}{dx} = \dfrac{v}{1 - v^2}$

$or,\ x\dfrac{dv}{dx} = \dfrac{v}{1-v^2} - v = \dfrac{v - v + v^3}{1 - v^2}$

$or,\ \left(\dfrac{1-v^2}{v^3}\right)dv = \dfrac{dx}{x}$

$or,\ \left(\dfrac{1}{v^3} - \dfrac{1}{v}\right)dv = \dfrac{dx}{x}$

Integrating we get,

$$-\frac{1}{2v^2} - \log v = \log x - \log C$$

(where $-\log C$ is the constant of integration)

$or,\ \log C - \log v - \log x = \dfrac{1}{2v^2}$

$or,\ \log C - \log xv = \dfrac{1}{2v^2}$

$or,\ \log\left(\dfrac{C}{xv}\right) = \dfrac{1}{2v^2}$

$or,\ \dfrac{C}{xv} = e^{\frac{1}{2v^2}}$

$or,\ \dfrac{C}{y} = e^{\frac{x^2}{2y^2}} \qquad \left(\because v = \dfrac{y}{x}\right)$

$or,\ ye^{\frac{x^2}{2y^2}} = C$

12. (a)

$$x\frac{dy}{dx} + y = y^2 \log x$$

$$\Rightarrow xdy + ydx = y^2 \log x\,dx$$

$$\Rightarrow d(xy) = y^2 \log x\,dx$$

$$\Rightarrow \frac{d(xy)}{x^2 y^2} = \frac{\log x}{x^2}\,dx$$

Integrating we get,

$$\int \frac{d(xy)}{x^2 y^2} = \int \frac{\log x}{x^2} + C$$

$or,\ -\dfrac{1}{xy} = \log x \times \left(-\dfrac{1}{x}\right) - \int\dfrac{1}{x}\times\left(-\dfrac{1}{x}\right)dx + C$

$or,\ -\dfrac{1}{xy} = -\dfrac{\log x}{x} - \dfrac{1}{x} + C$

$or,\ \dfrac{1}{xy} = \dfrac{\log x}{x} + \dfrac{1}{x} - C$

$or,\ 1 = y(1 + \log x) - Cxy$

$or,\ y(1 + \log x) = 1 + Cxy$

Alternative Method:

$$x\frac{dy}{dx} + y = y^2 \log x$$

$$\Rightarrow \frac{1}{y^2}\frac{dy}{dx} + \frac{y}{xy^2} = \frac{\log x}{x}$$

[on dividing both sides by xy^2]

$$\Rightarrow \frac{1}{y^2}\frac{dy}{dx} + \frac{1}{xy} = \frac{\log x}{x} \qquad \ldots(i)$$

Let $\dfrac{1}{y} = z$, then $-\dfrac{1}{y^2}\dfrac{dy}{dx} = \dfrac{dz}{dx}$

Then (i) becomes,

$$-\frac{dz}{dx} + \frac{z}{x} = \frac{\log x}{x}$$

$or,\ \dfrac{dz}{dx} + \left(-\dfrac{1}{x}\right)z = -\dfrac{\log x}{x} \qquad \ldots(ii)$

which is a linear equation in z.
So the solution of (ii) is given by

$$z \times \left\{ e^{\int\left(-\frac{1}{x}\right)dx} \right\} = \int\left[\left(\frac{-\log x}{x}\right) \times e^{\int\left(-\frac{1}{x}\right)dx}\right]dx - C$$

$or,\ ze^{-\log x} = \int\left(-\dfrac{\log x}{x}\right)\times e^{-\log x}dx - C$

$or,\ z\times\dfrac{1}{x} = -\int\dfrac{\log x}{x}\times\dfrac{1}{x}dx - C$

$\left(\because e^{-\log x} = e^{\log x^{-1}} = x^{-1} = \dfrac{1}{x}\right)$

$or,\ \dfrac{z}{x} = -\int\dfrac{\log x}{x^2}dx - C$

$or,\ \dfrac{1}{xy} = -\left[-\dfrac{1}{x}\log x - \int\left(-\dfrac{1}{x}\right)\times\dfrac{1}{x}dx\right] - C$

$or,\ \dfrac{1}{xy} = \dfrac{1}{x}\log x + \dfrac{1}{x} - C$

$or,\ y(1 + \log x) = 1 + Cxy.$

13. (a)

$$\left(x^2 + 1\right)\frac{dy}{dx} + 2xy = x^2$$

$or,\ \dfrac{dy}{dx} + \left(\dfrac{2x}{x^2 + 1}\right)y = \dfrac{x^2}{x^2 + 1}$

which is linear in y.

$\therefore e^{\int Pdx} = e^{\int \frac{2x}{x^2+1}dx} = e^{\log(x^2+1)} = x^2 + 1.$

Then, the solution is given by

$$y \times (x^2 + 1) = \int \left(\frac{x^2}{x^2+1}\right) \times (x^2+1)dx + c$$

$or,\ y(x^2+1) = \int x^2 dx + c = \frac{x^3}{3} + c$

14. (b) $(x + y + 1)dy - dx = 0$

$\Rightarrow \dfrac{dx}{dy} = x + y + 1$

$\Rightarrow \dfrac{dx}{dy} + (-1)x = y + 1,$ which is linear in x

$\therefore e^{\int Pdy} = e^{\int (-1)dy} = e^{-y}$

So, the solution is given by

$x \times e^{-y} = \int (y+1) \times e^{-y} dy + c$

$\qquad = -e^{-y}(y+1) + \int (e^{-y} \times 1)dy + c$

$\qquad = -e^{-y}(y+1) - e^{-y} + c$

$\qquad = -(y+2)e^{-y} + c$

$or,\ x = -(y+2) + ce^{y}$

$or,\ x + y + 2 = ce^{y}.$

15. (a)

$$x\frac{dy}{dx} + \sin 2y = x^2 \cos^2 y$$

$\Rightarrow \sec^2 y \dfrac{dy}{dx} + \dfrac{2\sin y \cos y}{x\cos^2 y} = x$

$\Rightarrow \sec^2 y \dfrac{dy}{dx} + \dfrac{2}{x}\tan y = x$

Let $\tan y = z.$ So $\sec^2 y \dfrac{dy}{dx} = \dfrac{dz}{dx}.$
Then (i) gives

$\dfrac{dz}{dx} + \dfrac{2z}{x} = x,$ which is linear in $z.$

Then $e^{\int Pdx} = e^{\int \frac{2}{x}dx} = e^{2\log x} = e^{\log x^2} = x^2$

$\therefore$ the required solution is,

$z \times x^2 = \int x \times x^2 dx + c$

$or,\ x^2 \tan y = \dfrac{x^4}{4} + c.$

16. (c) $(x^2 + y)dx = xdy$

$\Rightarrow \dfrac{dy}{dx} = \dfrac{x^2 + y}{x} = \dfrac{y}{x} + x$

$\Rightarrow \dfrac{dy}{dx} + \left(-\dfrac{1}{x}\right)y = x,$

Which is lirear in $y.$

Here, I.F $= e^{\int \left(-\frac{1}{x}\right)dx} = e^{-\log x} = e^{\log x^{-1}} = x^{-1} = \dfrac{1}{x}.$

$\therefore$ the solution is given by

$y \times (I.F) = \int (x \times I.F)dx + c$

$or,\ y \times \dfrac{1}{x} = \int x \times \dfrac{1}{x}dx + c$

$or,\ \dfrac{y}{x} = x + c$

$or,\ y = x^2 + cx$

Now, $y(1) = 2 \Rightarrow 2 = 1 + c \Rightarrow c = 1$

$\therefore\ y(x) = x^2 + x$
Hence, $y(2) = 2^2 + 2 = 6.$

17. (c) The equation $Mdx + Ndy = 0$ is exact if
$\dfrac{\partial M}{\partial y} = \dfrac{\partial N}{\partial x}$

Here, $M = ax + by$ and $N = cx + fy.$

So, $\dfrac{\partial M}{\partial y} = b, \dfrac{\partial N}{\partial x} = c.$

$\therefore$ the given differential equation is exact if
$\dfrac{\partial M}{\partial y} = \dfrac{\partial N}{\partial x}$ i.e; if $b = c.$

18. (b) $\left|\dfrac{dy}{dx}\right| + |y| = 0 \Rightarrow \dfrac{dy}{dx} = 0$ and $y = 0$

$\Rightarrow y = c$ and $y = 0$ (c is a constant)
But this contradicts the fact that $y(0) = 2.$
Hence, the given differential equation has no solution.

19. (d) Since the R.H.S. of the given equation is
"0," so let us assume that $y = e^{mx}$ be a solution of the equation.

Then $\dfrac{d^2y}{dx^2} + \dfrac{dy}{dx} - 2y = 0$

$\Rightarrow m^2 e^{mx} + me^{mx} - 2e^{mx} = 0$

$\Rightarrow m^2 + m - 2 = (\because e^{mx} \neq 0)$

$\Rightarrow m^2 + 2m - m - 2 = 0$

$\Rightarrow m(m + 2) - 1(m + 2) = 0$

$\Rightarrow (m + 2)(m - 1) = 0 \Rightarrow m = 1, -2$

$\therefore$ the required solution is given by

$y = c_1 e^{1x} + c_2 e^{-2x} = c_1 e^x + c_2 e^{-2x}$ ...(i)

So, $\dfrac{dy}{dx} = c_1 e^x - 2c_2 e^{-2x}$ ...(ii)

When $x = 0, y = 2;$
So (i) $\Rightarrow c_1 + c_2 = 2$ ...(iii)

When $x = 0, \dfrac{dy}{dx} = 0;$

So, $(iii) \Rightarrow c_1 - 2c_2 = 0$ ...(iv)

Solving (iii) and (iv) we get, $c_1 = \dfrac{4}{3}$ and $c_2 = \dfrac{2}{3}$

∴   The required solution is given by

$$y = \frac{4}{3}e^x + \frac{2}{3}e^{-2x}.$$

20. (a) Given that $\dfrac{d^2y}{dx^2} = \sin 3x$

Integrating both sides w.r.t "x" we get,

$$\int \frac{d^2y}{dx^2}dx = \int \sin 3x\, dx + c_1$$

$$or, \int \frac{d}{dx}\left(\frac{dy}{dx}\right)dx = -\frac{1}{3}\cos 3x + c_1$$

$$or, \int d\left(\frac{dy}{dx}\right) = -\frac{1}{3}\cos 3x + c_1$$

$$or, \frac{dy}{dx} = -\frac{1}{3}\cos 3x + c_1$$

Again, integrating both sides w.r.t "x" we get

$$\int \frac{dy}{dx}dx = -\frac{1}{3}\int \cos 3x\, dx + \int c_1 dx + c_2$$

$$or, \int \frac{d}{dx}(y)dx = -\frac{1}{9}\sin 3x + c_1 x + c_2$$

$$or, \int dy = -\frac{1}{9}\sin 3x + c_1 x + c_2$$

$$or, y = -\frac{1}{9}\sin 3x + c_1 x + c_2.$$

21. (b) Given equation is

$$\frac{d^2y}{dx^2} + 2\frac{dy}{dx} + y = e^x - e^{-2x} \qquad ...(i)$$

Consider $\dfrac{d^2y}{dx^2} + 2\dfrac{dy}{dx} + y = 0$ ...(ii)

Let $y = e^{mx}$ be trial solution of (ii)

Then, $(ii) \Rightarrow m^2 e^{mx} + 2m e^{mx} + e^{mx} = 0$

$\Rightarrow m^2 + 2m + 1 = 0 \ (\because e^{mx} \neq 0)$

$\Rightarrow m\ -1, -1$

∴ $y_{CF} = (c_1 + c_2 x)e^{-x}$.

Now (i) can be expressed as:

$(D^2 + 2D + 1)y = e^x - e^{-2x}$

$$\therefore y_{PI} = \frac{1}{(D^2 + 2D + 1)}\left(e^x - e^{-2x}\right)$$

$$= \frac{1}{D^2 + 2D + 1}(e^x) - \frac{1}{D^2 + 2D + 1}(e^{-2x})$$

$$= \frac{e^x}{1 + 2\times 1 + 1} - \frac{e^{-2x}}{(-2)^2 + 2\times(-2) + 1}$$

$$= \frac{1}{4}e^x - e^{-2x}.$$

Hence, the complete solution is given by

$$y = y_{CF} + y_{PI} = (c_1 + c_2 x)e^{-x} + \frac{1}{4}e^x - e^{-2x}.$$

22. (a) $\left(D^2 + 1\right)y = \operatorname{Sin} 2x$

$$\Rightarrow D^2 y + y = \operatorname{Sin} 2x \Rightarrow \frac{d^2y}{dx^2} + y = \sin 2x$$

Consider $\dfrac{d^2y}{dx^2} + y = 0$..........$(1)$

Let $y = e^{mx}$ be a trial solution of (1)

Then(1)

$\Rightarrow m^2 e^{mx} + e^{mx} = o \Rightarrow m^2 + 1 = 0 \ \left(\because e^{mx} \neq 0\right)$

$\Rightarrow m^2 = -1 = i^2 \Rightarrow m = \pm i = 0 + i, 0 - i$

∴ $y_{CF} = e^{0x}(c_1 \cos x + c_2 \sin x) = c_1 \cos x + c_2$ $\sin x$.

$$y_{PI} = \frac{1}{D^2 + 1}\sin 2x = \frac{\sin 2x}{-2^2 + 1} = -\frac{1}{3}\sin 2x$$

So, the complete solution is given by

$$y = y_{CF} + y_{PI} = c_1 \cos x + c_2 \sin x - \frac{1}{3}\sin 2x \qquad ...(ii)$$

$$\therefore \frac{dy}{dx} = -c_1 \sin x + c_2 \cos x - \frac{2}{3}\cos 2x \qquad ...(iii)$$

When $x = 0$, $y = 0$;

So, (ii) gives $0 = c_1 + c_2 \times 0 - \dfrac{1}{3}\times 0$ or $c_1 = 0$.

Again for $x = 0$, $\dfrac{dy}{dx} = 0$; so (iii) gives

$$0 = 0 + c_2 - \frac{2}{3} \ i.e; \ c_2 = \frac{2}{3}.$$

Hence, the required solution is given by

$$y = 0 \times \cos x + \frac{2}{3}\sin x - \frac{1}{3}\sin 2x$$

$$= \frac{2}{3}\sin x - \frac{1}{3}\sin 2x.$$

23. (a) $y = 3e^{2x} + e^{-2x} - \alpha x$

$$\Rightarrow \frac{dy}{dx} = 6e^{2x} - 2e^{-2x} - a \ ....(i)$$

$$\Rightarrow \frac{d^2y}{dx^2} = 12e^{2x} + 4e^{-2x}$$

$$= 4\left(3e^{2x} + e^{-2x} - a\,x\right) + 4a\,x$$

$$= 4y + 4a\,x$$

$$\Rightarrow \frac{d^2y}{dx^2} - 4y = 4a\,x ....(ii)$$

Comparing (ii) with $\dfrac{d^2y}{dx^2} + \beta\, y = 4a\,x$,

we get $\beta = -4$.

Now $\dfrac{dy}{dx}\Big|_{x=0} = 1 \qquad \Rightarrow 1 = 6-2-a \qquad \big[\text{using (i)}\big]$

$$\Rightarrow a = 3.$$

24. (a) The given equation is $y'' + 4y = x^2 \qquad \ldots(i)$

Consider $y'' + 4y = 0 \qquad\qquad \ldots(ii)$

Let $y = e^{mx}$ be a trial solution of (ii)

Then (ii) gives $m^2 e^{mx} + 4e^{mx} = 0$

or, $m^2 + 4 = 0$

or, $m^2 = -4 = i^2 \times 2^2$

or, $m = 2i = 0 + 2i,\ 0 - 2i$

$\therefore \quad y_{CF} = c_1 \cos 2x + c_2 \sin 2x.$

Now (i) can be written as: $(D^2 + 4)y = x^2$

$\therefore y_{PI} = \dfrac{1}{D^2 + 4}(x^2) = \dfrac{1}{4\left(1 + \dfrac{D^2}{4}\right)}(x^2)$

$= \dfrac{1}{4}\left(1 + \dfrac{D^2}{4}\right)^{-1}(x^2)$

$= \dfrac{1}{4}\left[1 - \dfrac{D^2}{4} + \left(\dfrac{D^2}{4}\right)^2 - \ldots\ldots\infty\right](x^2)$

$= \dfrac{1}{4}\left[x^2 - \dfrac{D^2}{4}(x^2) + 0 - 0 + \ldots\ldots\right]$

$= \dfrac{1}{4}\left[x^2 - \dfrac{1}{4}\times 2!\right] = \dfrac{1}{4}\left(x^2 - \dfrac{1}{2}\right)$

Hence, the complete solution is given by

$y = c_1 \cos 2x + c_2 \sin 2x + \dfrac{1}{4}\left(x^2 - \dfrac{1}{2}\right).$

25. (b) Given $y'' - 4y = 2\sin 3x + e^x$

Consider $y'' - 4y = 0 \qquad\qquad \ldots(i)$

Let $y = e^{mx}$ be a trial solution of (i)

Then (i) gives, $m^2 e^{mx} - 4e^{mx} = 0$

or, $m^2 = 4$ or, $m = \pm 2$

$\therefore \quad y_{CF} = c_1 e^{2x} + c_2 e^{-2x}$

Now, the given differential equation can be re-written as: $(D^2 - 4)y = 2\sin 3x + e^x$

$\therefore y_{PI} = \dfrac{1}{D^2 - 4}\left(2\sin 3x + e^x\right)$

$= \dfrac{1}{D^2 - 4}(2\sin 3x) + \dfrac{1}{D^2 - 4}(e^x)$

$= \dfrac{1}{(-3^2 - 4)}2\sin 3x + \dfrac{1}{(1^2 - 4)}e^x$

$= -\dfrac{2}{13}\sin 3x - \dfrac{1}{3}e^x$

Hence, the complete solution is given by

$y = y_{CF} + y_{PI}$

$= c_1 e^{2x} + c_2 e^{-2x} - \dfrac{2}{13}\sin 3x - \dfrac{1}{3}e^x.$

26. (d)

Let $y_1(x) = e^{2x},\ y_2(x) = e^{-4x},\ y_3(x) = e^{4x}$

Then, $y'_1(x) = 2e^{2x},\ y'_2(x) = -4e^{-4x},\ y'_3(x) = 4e^{4x}$,

$y''_1(x) = 4e^{2x},\ y''_2(x) = 16e^{-4x}$ and $y''_3(x) = 16e^{4x}$

$\therefore W(y_1, y_2, y_3) = \begin{vmatrix} e^{2x} & e^{-4x} & e^{4x} \\ 2e^{2x} & -4e^{-4x} & 4e^{4x} \\ 4e^{2x} & 16e^{-4x} & 16e^{4x} \end{vmatrix}$

$= e^{2x} \times e^{-4x} \times e^{4x} \begin{vmatrix} 1 & 1 & 1 \\ 2 & -4 & 4 \\ 4 & 16 & 16 \end{vmatrix} = -96e^{2x}.$

27. (b)

Let $y_1(x) = x^2 - 1,\ y_2(x) = x^2 + x + 1$

and $y_3(x) = 2x^2 + x.$

Then $1 \times y_1(x) + 1 \times y_2(x) + (-1) \times y_3(x) = 0.$

So clearly the solutions are linearly dependent.

Here, $y'_1(x) = 2x,\ y''_1(x) = 2,\ y'_2(x) = 2x + 1,$

$y''_2(x) = 2,\ y'_3(x) = 4x + 1,\ y_3''(x) = 4.$

$\therefore W(x) = W(y_1, y_2, y_3) = \begin{vmatrix} x^2 - 1 & x^2 + x + 1 & 2x^2 + x \\ 2x & 2x + 1 & 4x + 1 \\ 2 & 2 & 4 \end{vmatrix}$

$= (x^2 - 1)\{4(2x + 1) - 2(4x + 1)\}$

$\quad - (x^2 + x + 1)\{4 \times 2x - 2(4x + 1)\}$

$\quad + (2x^2 + x)\{2 \times 2x - 2 \times (2x + 1)\}$

$= 2(x^2 - 1) + 2(x^2 + x + 1) - 2(2x^2 + x) = 0.$

28. (b)

Let $f(x) = 1,\ g(x) = \sin^2 x,\ h(x) = \cos^2 x$

Then, $W(x) = \begin{vmatrix} f(x) & g(x) & h(x) \\ f'(x) & g'(x) & h'(x) \\ f''(x) & g''(x) & h''(x) \end{vmatrix}$

$= \begin{vmatrix} 1 & \sin^2 x & \cos^2 x \\ 0 & 2\sin x\cos x & -2\cos x\sin x \\ 0 & 2\cos 2x & -2\cos 2x \end{vmatrix}$

$= \sin 2x \times (-2\cos 2x) - (-\sin 2x) \times 2\cos 2x = 0$

$\therefore$ the functions $f(x), g(x)$ and $h(x)$ are linearly dependent.

29. (d)

∵ $y_1(x)$ and $y_2(x)$ are solutions of $y'' + y \sin x = 0$

So, $y_1'' + y_1 \sin x = 0$ and $y_2'' + y_2 \sin x = 0$

i.e; $y_1'' = -y_1 \sin x$ and $y_2'' = -y_2 \sin x$

Then, $g(x) = W(y_1, y_2) = \begin{vmatrix} y_1 & y_2 \\ y_1' & y_2' \end{vmatrix} = y_1 y_2' - y_1' y_2.$

$\therefore g'(x) = y_1 y_2'' + y_1' y_2' - y_1'' y_2 - y_1' y_2'$

$\qquad = y_1 y_2'' - y_1'' y_2$

$\qquad = y_1(-y_2 \sin x) - (-y_1 \sin x) y_2$

$\qquad = 0 \ \forall \ x \in [0,1]$

30. (a)

$W(y_1, y_2) = \begin{vmatrix} y_1 & y_2 \\ y_1' & y_2' \end{vmatrix} = y_1 y_2' - y_1' y_2.$

∵ y_1 and y_2 are solutions of

$y'' + P(x)y' + Q(x)y = 0.$ So

$y_1'' + P(x)y_1' + Q(x)y_1 = 0 \qquad \ldots(i)$

$y_2'' + P(x)y_2' + Q(x)y_2 = 0 \qquad \ldots(ii)$

$(i) \times y_2 - (ii) \times y_1 \Rightarrow$

$\quad y_1'' y_2 + P(x)y_1' y_2 - y_2'' y_1 - P(x)y_2' y_1 = 0$

or, $(y_1' y_2 - y_2' y_1)P(x) = y_2'' y_1 - y_1'' y_2$

or, $-W(y_1, y_2) \times P(x) = y_1 y_2'' - y_2 y_1''$

or, $W(y_1, y_2) \times P(x) = y_1'' y_2 - y_1 y_2''.$

Fully Solved MCQs

1. The differential equation corresponding to the solution $y = e^{-2x}(A \cos x + B \sin x)$ is given by
(a) $y'' + 4y' + 3y = 0$ (b) $y'' + 4y' + 5y = 0$
(c) $y'' - 4y' + 3y = 0$ (d) $y'' - 4y' + 5y = 0$

2. The differential equation of the system of circles having a constant radius "a" and having their centers on the x-axis is given by
(a) $(1 + y_1^2)y^2 = a^2$ (b) $yy_1^2 = a^2$
(c) $1 - y_1^2 = a^2 y^2$ (d) none of these

3. The differential equation of all circles touching y-axis at origin is given by
(a) $(x^2 + y^2)dy - 2xydx = 0$
(b) $(y^2 - x^2)dx + 2xydy = 0$
(c) $(x^2 + y^2)dy + 2xydx = 0$
(d) $(y^2 - x^2)dx - 2xydy = 0$

4. The differential equation of all circles touching both the axes is given by
(a) $y^2 - 2xy - 2xy\dfrac{dy}{dx} + \left(x^2 - 2xy\right)\left(\dfrac{dy}{dx}\right)^2 = 0$

(b) $x^2 + y^2 - 2xy\dfrac{dy}{dx} = 0$

(c) $x^2 + y^2 - 2xy\dfrac{dy}{dx} + \left(\dfrac{dy}{dx}\right)^2 = 0$

(d) none of these

5. Solve: $y - x\dfrac{dy}{dx} = 1 + x^2\dfrac{dy}{dx}$; given that $y = 2$ when $x = 1$
(a) $xy + x + y + 1 = 0$ (b) $xy - y + 2x = 1$
(c) $xy + y - 3x - 1 = 0$ (d) none of these

6. The solution of $(3x + 2y - 6)dx + (2x + 3y - 6)dy = 0$ is
(a) $x^2 + y^2 + xy - 6x - 6y = c$
(b) $x^2 + y^2 + 6xy - 3x - 3y = c$

(c) $\dfrac{1}{2}\left(x^2 + y^2\right) + 3xy - 2x - 3y = c$

(d) $\dfrac{3}{2}\left(x^2 + y^2\right) + 2xy - 6x - 6y = c$

7. Consider the differential equation

$p\dfrac{dy}{dx} + qy = re^{-kx}$

where p, q, r are positive constants and k is a non-negative constant. Then every solution of the given differential equation approaches to $\dfrac{r}{q}$ as $x \to \infty$ when

(a) $k = 0$ (b) $k > 0$ (c) $k = \dfrac{p}{q}$ (d) $k = \dfrac{q}{r}$

8. The solution of the differential equation $yy' + y^2 - x = 0$ is given by

(a) $y^2 = x + ce^{-2x}$ (b) $y^2 = x + ce^{-2x} - \dfrac{1}{2}$

(c) $y^2 = x + ce^{2x} + \dfrac{1}{2}$ (d) $y^2 = 2x + ce^{2x} + 1$

9. The general solution of $yy'' - (y')^2 = 0$ is given by

(a) $y = c_2 e^{c_1 x^2}$ (b) $y = (c_2 - x)e^{c_1 x^2}$

(c) $y = (c_2 + x)e^{c_1 x^2}$ (d) $y = c_2 e^{c_1 x}$

10. Consider the differential equation

$\dfrac{dy}{dx} - y = -y^2$. Then $\lim\limits_{y \to \infty} y(x) = ?$

(a) 0   (b) 2

(c) 1   (d) –1

11. If $g(x, y)dx + (x + y)dy = 0$ is an exact differential equation and if $g(x, 0) = x^2$, then the general solution of the differential equation is given by

(a) $x^3 + 2xy + y^3 = c$ (b) $2x^3 + 6xy + 3y^3 = c$

(c) $2x^3 + 4xy + y^3 = c$ (d) $x^2 + 6xy + 3y^3 = c$

12. If λ be a constant such that $xy + k = e^{\frac{(x-1)^2}{2}}$ satisfies the differential equation

$$x\frac{dy}{dx} = \left(x^2 - x - 1\right)y + x - 1, \text{ then "}k\text{" is equal to}$$

(a) 0   (b) –1

(c) 1   (d) 2

13. If y^k is an integrating factor of the differential equation $2xy\,dx - (3x^2 - y^2)dy = 0$, then k = ?

(a) –4   (b) 4

(c) –1   (d) 1

14. Consider the differential equation

$$\frac{dy}{dx} = ay - by^2, \text{ where } a, b > 0 \text{ and } y(0) = y_0.$$

When $x \to \infty$, $y(x)$ tends to–

(a) 1   (b) 0

(c) $\dfrac{a}{b}$ (d) $\dfrac{b}{a}$

15. Let α and β be two real numbers such that every solution of $\dfrac{d^2y}{dx^2} + 2a\dfrac{dy}{dx} + \beta y = 0$

satisfies $\lim\limits_{x\to\infty} y(x) = 0$. Then,

(a) $3\alpha^2 + \beta < 0$ and $\alpha > 0$

(b) $\alpha^2 + \beta > 0$ and $\alpha < 0$

(c) $\alpha^2 - \beta > 0$ and $\alpha > 0$

(d) $\alpha^2 - \beta > 0$, $\alpha > 0$ and $\beta > 0$

16. Let $y(x)$ be the solution of the differential equation $\dfrac{d}{dx}\left(x\dfrac{dy}{dx}\right) = x$; $y(1) = 0$ and $\dfrac{dy}{dx}\Big|_{x=1} = 0$

Then, $y(2) = ?$

(a) $\dfrac{3}{4} - \dfrac{1}{2}\log_e 2$ (b) $\dfrac{3}{4} + \dfrac{1}{2}\log_e 2$

(c) $\dfrac{1}{2} + \log_e 2$ (d) none of these

17. Let $y_1(x)$ and $y_2(x)$ be two linearly independent solution of the differential equation $x^2y'' - 2xy' - 4y = 0$ for $x \in [1, 9]$. If $W(x) = y_1(x)$

$y'_2(x) - y_2(x)y'(x)$ and $W(1) = 5$ Then what will be the value of $W(2) = $?

(a) 16   (b) $\dfrac{1}{16}$

(c) $\dfrac{5}{16}$ (d) 0

18. If $\dfrac{1}{x}(A + B\log_e x)$ is the general solution of the differential equation

$$x^2\frac{d^2y}{dx^2} + \lambda x\frac{dy}{dx} + y = 0, \ x > 0; \text{ then } \lambda = ?$$

(a) 3   (b) 3

(c) 2   (d) 1

19. Consider the initial value problem

$x^2y'' - 6y = 0$, $y(1) = k$ and $y'(1) = 6$

If $y(x) \to 0$ as $x \to 0 +$, then "k" is equal to–

(a) 1   (b) 2   (c) 4   (d) 0

20. Solve: $\cos^2 y\dfrac{d^2y}{dx^2} = \tan y$; given that $\dfrac{dy}{dx} = 0$ when $y = 0$

(a) $(\sin y + ce^x)(\sin y + ce^{-x}) = 0$

(b) $(\sin y + c)(\cos y - e^x) = 0$

(c) $(\cos x + ce^y)(\sin y + ce^{-x}) = 0$

(d) none of these

21. Solve: $(x^2D^2 + xD + 1)y = \sin(2\log x)$

(a) $y = c_1\cos(\log x) + c_2\sin(\log x) - \dfrac{1}{2}\sin(\log x^2)$

(b) $y = c_1\cos(\log x) + c_2\sin(\log x) + \dfrac{2}{3}\sin(\log x)$

(c) $y = c_1\cos(\log 2x) + c_2\sin(\log 2x) - \dfrac{1}{3}\sin(\log x)$

(d) $y = c_1\cos(\log x) + c_2\sin(\log x) - \dfrac{1}{3}\sin(2\log x)$

Answer key				
1. (b)	**2.** (a)	**3.** (d)	**4.** (a)	**5.** (c)
6. (d)	**7.** (a)	**8.** (b)	**9.** (d)	**10.** (c)
11. (b)	**12.** (c)	**13.** (a)	**14.** (c)	**15.** (d)
16. (a)	**17.** (c)	**18.** (a)	**19.** (b)	**20.** (a)
21. (d)				

Explanations				

1. (b) $y = e^{-2x}(A\cos x + B\sin x)$

$ye^{2x} = A\cos x + B\sin x$ ...(i)

Differentiating both sides of (i) w.r.t "x," we get

$$e^{2x}\frac{dy}{dx} + 2ye^{2x} = -A\sin x + B\cos x \quad ...(ii)$$

Now differentiating both sides of (*ii*) w.r.t "*x*," we get

$$2e^{2x}\frac{dy}{dx} + e^{2x}\frac{d^2y}{dx^2} + 2\left[e^{2x}\frac{dy}{dx} + 2ye^x\right]$$

$$= -A\cos x - B\sin x$$

or $e^{2x}\frac{d^2y}{dx^2} + 4e^{2x}\frac{dy}{dx} + 4ye^{2x} = -ye^{2x}$ [using (i)]

or $\frac{d^2y}{dx^2} + 4\frac{dy}{dx} + 4y = -y$

or $y'' + 4y' + 5y = 0$

2. (*a*) Let the equation of the circle be

$(x-a)^2 + y^2 = \alpha^2$ ...(i)

Here, radius of the circle = α and center of the circle = $(a, 0)$

In equation (*i*), "*a*" is an arbitrary constant and α is a parameter.

Differentiating (*i*) w.r.t "*x*" we get,

$2(x-a) + 2y\frac{dy}{dx} = 0$ *i.e*; $x - a = -y\frac{dy}{dx}$

Then equation (*i*) becomes,

$\left(-y\frac{dy}{dx}\right)^2 + y^2 = \alpha^2$

or $y^2\left(\frac{dy}{dx}\right)^2 + y^2 = \alpha^2$

or $y^2\left(1 + y_1^2\right) = \alpha^2$

3. (*d*) Let the equation of the circle be

$(x-a)^2 + y^2 = a^2$ *i.e.*; $x^2 - 2ax + y^2 = 0$ ...(i)

Now differentiating (*i*) w.r.t "*x*" we get,

$$2x - 2a + 2y\frac{dy}{dx} = 0$$

or, $a = x + y\frac{dy}{dx}$

∴   (*i*) gives

$$x^2 - 2x\left(x + y\frac{dy}{dx}\right) + y^2 = 0$$

or, $y^2 - x^2 - 2xy\frac{dy}{dx} = 0$

or, $\left(y^2 - x^2\right)dx - 2xydy = 0$

4. (*a*) Let the equation of the circle be

$(x-a)^2 + (y-a)^2 = a^2$

i.e.; $x^2 + y^2 - 2ax - 2ay + a^2 = 0$ ...(i)

Differentiating (*i*) w.r.t "*x*" we get,

$$2x + 2y\frac{dy}{dx} - 2a - 2a\frac{dy}{dx} + 0 = 0$$

or, $a = \dfrac{x + y\dfrac{dy}{dx}}{1 + \dfrac{dy}{dx}}$

Then, (*i*) gives

$$x^2 + y^2 - 2x\left(\frac{x + y\frac{dy}{dx}}{1 + \frac{dy}{dx}}\right) - 2y\left(\frac{x + y\frac{dy}{dx}}{1 + \frac{dy}{dx}}\right)$$

$$+ \left(\frac{x + y\frac{dy}{dx}}{1 + \frac{dy}{dx}}\right)^2 = 0$$

or, $\left(x^2 + y^2\right)\left(1 + \frac{dy}{dx}\right)^2 - 2(x+y)\left(x + y\frac{dy}{dx}\right)\left(1 + \frac{dy}{dx}\right)$

$+\left(x + y\frac{dy}{dx}\right)^2 = 0$

or, $\left(x^2 + y^2\right)\left[1 + 2\frac{dy}{dx} + \left(\frac{dy}{dx}\right)^2\right]$

$-2(x+y)\left[x + (x+y)\frac{dy}{dx} + y\left(\frac{dy}{dx}\right)^2\right]$

$+\left[x^2 + 2xy\frac{dy}{dx} + y^2\left(\frac{dy}{dx}\right)^2\right] = 0$

or, $x^2 + y^2 + 2\left(x^2 + y^2\right)\frac{dy}{dx} + \left(x^2 + y^2\right)\left(\frac{dy}{dx}\right)^2$

$+\left(x^2 + y^2\right)\left(\frac{dy}{dx}\right)^2 - 2x^2 - 2xy$

$-2\left(x^2 + 2xy + y^2\right)\frac{dy}{dx} - \left(2xy + 2y^2\right)\left(\frac{dy}{dx}\right)^2$

$+x^2 + 2xy\frac{dy}{dx} + y^2\left(\frac{dy}{dx}\right)^2 = 0$

or, $y^2 - 2xy - 2xy\frac{dy}{dx} + \left(x^2 - 2xy\right)\left(\frac{dy}{dx}\right)^2 = 0$.

5. (*c*) $y - x\frac{dy}{dx} = 1 + x^2\frac{dy}{dx}$

$\Rightarrow y - 1 = x(1+x)\frac{dy}{dx}$

$\Rightarrow \dfrac{dx}{x(1+x)} = \dfrac{dy}{y-1}$

Integrating we get,

$$\int \frac{dx}{x(1+x)} = \int \frac{dy}{y-1} + C$$

$$\Rightarrow \int \frac{(x+1)-x}{x(1+x)} dx = \log(y-1) + C$$

$$\Rightarrow \int \frac{dx}{x} - \int \frac{dx}{x+1} = \log(y-1) + C$$

$$\Rightarrow \log x - \log(x+1) = \log(y-1) + C \dots\dots\dots(i)$$

Putting $y = 2$ and $x = 1$, we get from (i),

$\log 1 - \log(1+1) = \log(2-1) + C$

or, $C = \log 1 - \log 2 - \log 1 = -\log 2$

∴ (i) gives

$\log x - \log(x+1) = \log(y-1) - \log 2$

or, $\log \dfrac{x}{x+1} = \log \dfrac{y-1}{2}$

or, $\dfrac{x}{x+1} = \dfrac{y-1}{2}$

or, $2x = (y-1)(x+1)$

or, $xy + y - 3x - 1 = 0$

6. (d) $(3x + 2y - 6)dx + (2x + 3y - 6)dy = 0$

$\Rightarrow \ 3x\,dx + 3y\,dy + 2y\,dx + 2x\,dy - 6dx - 6dy = 0$

$\Rightarrow \ 3(x\,dx + y\,dy) + 2(y\,dx + x\,dy) - 6(dx + dy) = 0$

Integrating we get,

$3\int(x\,dx + y\,dy) + 2\int(y\,dx + x\,dy) - 6\int d(x+y) = c$

or, $3\int d\left\{\dfrac{1}{2}\left(x^2 + y^2\right)\right\} + 2\int d(xy) - 6(x + y) = c$

or, $\dfrac{3}{2}\left(x^2 + y^2\right) + 2xy - 6x - 6y = c$

Alternative Method:

The given equation can be re-written as

$$\frac{dy}{dx} = \frac{-(3x + 2y - 6)}{(2x + 3y - 6)} \qquad \dots(i)$$

Let us put $x = x' + h$ and $y = y' + k$

Then $dx = dx'$ and $dy = dy'$

∴ (i) gives,

$$\frac{dy'}{dx'} = \frac{-\{3(x'+h) + 2(y'+k) - 6\}}{\{2(x'+h) + 3(y'+k) - 6\}}$$

$$= \frac{-\{3x' + 2y' + (3h + 2k - 6)\}}{\{2x' + 3y' + (2h + 3k - 6)\}} \dots\dots(ii)$$

Let us consider

$3h + 2k - 6 = 0 \qquad \dots(iii)$ and

$2h + 3k - 6 = 0 \qquad\qquad (iv)$

Solving (iii) and (iv) we get, $h = \dfrac{6}{5}$ & $k = \dfrac{6}{5}$.

Now from (ii) we have

$$\frac{dy'}{dx'} = \frac{-(3x' + 2y')}{2x' + 3y'} = \frac{-\left(3 + 2\dfrac{y'}{x'}\right)}{2 + 3\dfrac{y'}{x'}} \qquad \dots(v)$$

which is a homogeneous equation.

Let us put $y' = vx'$.

So $\dfrac{dy'}{dx'} = v + x'\dfrac{dv}{dx'}$.

Then $(v) \Rightarrow v + x'\dfrac{dv}{dx'} = \dfrac{-(3 + 2v)}{2 + 3v}$

or, $x'\dfrac{dv}{dx'} = \dfrac{-3 - 2v}{2 + 3v} - v$

or, $x'\dfrac{dv}{dx'} = \dfrac{-3 - 4v - 3v^2}{2 + 3v}$

or, $\dfrac{(2 + 3v)dv}{3v^2 + 4v + 3} = \dfrac{dx'}{x'}$

Integrating we, get,

$\int \dfrac{(2 + 3v)dv}{3v^2 + 4v + 3} = -\int \dfrac{dx'}{x'} + \log c'$

or, $\dfrac{1}{2}\int \dfrac{(6v + 4)dv}{3v^2 + 4v + 3} = -\log x' + \log c'$

or, $\dfrac{1}{2}\log(3v^2 + 4v + 3) = -\log x' + \log c'$

or, $\log(3v^2 + 4v + 3) = -2\log x' + 2\log c'$

or, $\log(3v^2 + 4v + 3) = \log\left\{(x')^{-2}\right\} + \log\left\{(c')^2\right\}$

or, $\log(3v^2 + 4v + 3) - \log\left\{\dfrac{1}{(x')^2}\right\} = \log c_1$

$$\left[\text{where } (c')^2 = c_1\right]$$

or, $\log\left\{\dfrac{3v^2 + 4v + 3}{\dfrac{1}{(x')^2}}\right\} = \log c_1$

or, $(x')^2(3v^2 + 4v + 3) = c_1$

or, $(x')^2\left\{\dfrac{3(y')^2}{(x')^2} + \dfrac{4y'}{x'} + 3\right\} = c_1$

or, $3(y')^2 + 4x'y' + 3(x')^2 = c_1$

or, $3(y-k)^2 + 4(x-h)(y-k) + 3(x-h)^2 = c_1$

or, $3\left(y - \dfrac{6}{5}\right)^2 + 4\left(x - \dfrac{6}{5}\right)\left(y - \dfrac{6}{5}\right) + 3\left(x - \dfrac{6}{5}\right)^2 = c_1$

or, $3y^2 - \dfrac{36y}{5} + \dfrac{108}{25} + 4xy - \dfrac{24x}{5} - \dfrac{24y}{5}$

$+ \dfrac{144}{25} + 3x^2 - \dfrac{36x}{5} + \dfrac{108}{25} = c_1$

$or,\ 3(x^2 + y^2) + 4xy - 12x - 12y = c_1 - \dfrac{216}{25}$

$or,\ \dfrac{3}{2}(x^2 + y^2) + 2xy - 6x - 6y = c,$

where $c = \dfrac{1}{2}\left(c_1 - \dfrac{216}{25}\right).$

7. (a) $p\dfrac{dy}{dx} + qy = re^{-kx}$ $\Rightarrow \dfrac{dy}{dx} + \left(\dfrac{q}{p}\right)y = \dfrac{r}{p}e^{-kx}$

which is linear in y.

Here, $e^{\int Pdx} = e^{\int \frac{q}{p}dx} = e^{\frac{qx}{p}}.$

$\therefore$ the solution is given by,

$$y \times e^{\frac{qx}{p}} = \int e^{\frac{qx}{p}} \times \dfrac{r}{p}e^{-kx}dx + C$$

$or,\ ye^{\frac{qx}{p}} = \dfrac{r}{p}\int e^{\left(\frac{q}{p} - k\right)x}dx + C$

$or,\ ye^{\frac{qx}{p}} = \dfrac{r}{p} \times \dfrac{e^{\left(\frac{q}{p} - k\right)x}}{\left(\dfrac{q}{p} - k\right)} + C$

$or,\ y = \left(\dfrac{r}{q - pk}\right)e^{-kx} + Ce^{-\frac{qx}{p}}$ ...(i)

When $k = 0$, (i) gives,

$y = \dfrac{r}{q}e^{-0x} + Ce^{-\frac{qx}{p}} = \dfrac{r}{q} + Ce^{-\frac{qx}{p}}$, which approaches

to $\dfrac{r}{q}$ when $x \to \infty$ $\left(\because e^{-\infty} = \dfrac{1}{e^{\infty}} = 0\right).$

8. (b) Let us put $y^2 = z$.

Then, $2y\dfrac{dy}{dx} = \dfrac{dz}{dx}$ $i.e;\ yy' = \dfrac{1}{2}\dfrac{dz}{dx}$

Then, $yy' + y^2 - x = 0$

$\Rightarrow \dfrac{1}{2}\dfrac{dz}{dx} + z - x = 0$

$\Rightarrow \dfrac{dz}{dx} + 2z = 2x$, which is linear in z.

$\therefore$ $e^{\int Pdx} = e^{\int 2dx} = e^{2x}.$

Hence, the solution is given by

$$z \times e^{2x} = \int e^{2x} \times 2xdx + c$$

$\Rightarrow y^2e^{2x} = 2x \times \dfrac{e^{2x}}{2} - \int 2 \times \dfrac{e^{2x}}{2}dx + c$

$\Rightarrow y^2e^{2x} = xe^{2x} - \dfrac{e^{2x}}{2} + c$

$\Rightarrow y^2 = x - \dfrac{1}{2} + ce^{-2x}.$

9. (d) $yy'' - (y')^2 = 0$ $\Rightarrow \dfrac{y''}{y'} = \dfrac{y'}{y}$

Integrating we get,

$$\int \dfrac{y''}{y'} = \int \dfrac{y'}{y} + \log c_1$$

$or,\ \log(y') = \log(y) + \log c_1$

$\left[\text{using } \int \dfrac{f'(x)}{f(x)}dx = \log(f(x)) + c\right]$

$or,\ \log y' - \log y = \log c_1$

$or,\ \log\left(\dfrac{y'}{y}\right) = \log c_1$ $or,\ \dfrac{y'}{y} = c_1$

$or,\ \dfrac{dy}{y} = c_1 dx$

Integrating we get, $\int \dfrac{dy}{y} = \int c_1 dx + \log c_2$

$\log y = c_1 x + \log c_2$

$or,\ \log\left(\dfrac{y}{c_2}\right) = c_1 x$

$or,\ \dfrac{y}{c_2} = e^{c_1 x}$

$or,\ y = c_2 e^{c_1 x}$

10. (c) $\dfrac{dy}{dx} - y = -y^2$ $\Rightarrow \dfrac{dy}{dx} = y(1 - y)$

$\Rightarrow \dfrac{dy}{y(1 - y)} = dx$ $\Rightarrow \dfrac{\{(1 - y) + y\}dy}{y(1 - y)} = dx$

$\Rightarrow \left(\dfrac{1}{y} + \dfrac{1}{1 - y}\right)dy = dx$

Integrating we get, $\int\left(\dfrac{1}{y} + \dfrac{1}{1 - y}\right)dy = \int dx + \log c_1$

$or,\ \log y - \log(1 - y) = x + \log c_1$

$or,\ \log\left(\dfrac{y}{1 - y}\right) - \log c_1 = x$

$or,\ \log\left\{\dfrac{y}{c_1(1 - y)}\right\} = x$

$or,\ \dfrac{y}{c_1(1 - y)} = e^x$

$or,\ \dfrac{y}{1 - y} = c_1 e^x$

$or,\ \dfrac{y}{1 - y} + 1 = c_1 e^x + 1$

$or,\ \dfrac{1}{1 - y} = 1 + c_1 e^x$

$or,\ 1 - y = \dfrac{1}{1 + c_1 e^x}$

$or,\ y = 1 - \dfrac{1}{1 + c_1 e^x} = \dfrac{c_1 e^x}{1 + c_1 e^x} = \dfrac{c_1}{e^{-x} + c_1}$

Now $x \to \infty$, $y(x) = \dfrac{c_1}{e^{-\infty} + c_1} = \dfrac{c_1}{0 + c_1} = 1.$

$\therefore \quad \lim_{y \to \infty} y(x) = 1.$

11. (b) $g(x, y)dx + (x + y)dy = 0$ is exact

$\Rightarrow \dfrac{\partial}{\partial x}(x + y) = \dfrac{\partial g}{\partial y} \quad \Rightarrow \dfrac{\partial g}{\partial y} = 1$

Given that $g(x, 0) = x^2.$

So we can take $g(x, y) = x^2 + y,$

which satisfies $g(x, 0) = x^2$ and $\dfrac{\partial g}{\partial y} = 1$

Now, $\int M dx = \int g(x, y)dx = \int (x^2 + y)dx$

(taking y as constant)

$= \dfrac{x^3}{3} + xy$

$\int N dy = \int (x + y)dy = xy + \dfrac{y^2}{2}$

(taking x as constant)

Hence, the general solution is given by

$\dfrac{x^3}{3} + xy + \dfrac{y^2}{2} = c_1$

or, $2x^3 + 6xy + 3y^2 = 6c_1 = c$ (taking $c = 6c_1$).

12. (c) $x\dfrac{dy}{dx} = (x^2 - x - 1)y + x - 1$

$\Rightarrow \dfrac{dy}{dx} = \left(x - 1 - \dfrac{1}{x}\right)y + \left(1 - \dfrac{1}{x}\right)$

$\Rightarrow \dfrac{dy}{dx} + \left(1 + \dfrac{1}{x} - x\right)y = 1 - \dfrac{1}{x},$ which is linear in y.

$\therefore$ I.F $= e^{\int P dx} = e^{\int \left(1 + \frac{1}{x} - x\right)dx} = e^{\int \left[\frac{1}{x} - (x-1)\right]dx}$

$= e^{\log_e x - \frac{(x-1)^2}{2}} = e^{\log_e x} \, e^{-\frac{(x-1)^2}{2}} = xe^{-\frac{(x-1)^2}{2}}$

$\therefore$ The solution is given by

$y \times \left\{xe^{-\frac{(x-1)^2}{2}}\right\} = \int \left[\left(1 - \dfrac{1}{x}\right) \times xe^{-\frac{(x-1)^2}{2}}\right]dx + c$

or, $xye^{-\frac{(x-1)^2}{2}} = \int (x-1)e^{-\frac{(x-1)^2}{2}}dx + c$

$= \int (e^{-z})dz + c$

$\left(\text{by putting } \dfrac{(x-1)^2}{2} = z \text{ so that } (x-1)dx = dz\right)$

$= -e^{-z} + c = -e^{-\frac{(x-1)^2}{2}} + c$

or, $(xy+1)e^{-\frac{(x-1)^2}{2}} = c$

or, $xy + 1 = ce^{\frac{(x-1)^2}{2}}$ ....(i)

Comparing (i) with $xy + k = e^{\frac{(x-1)^2}{2}}$, we get,

$k = 1$ and $c = 1.$

13. (a) $2xydx - (3x^2 - y^2)dy = 0$

$\Rightarrow 2xy \times y^k dx - (3x^2 - y^2) \times y^k dy = 0 \times y^k$

$\Rightarrow 2xy^{k+1} dx - (3x^2 y^k - y^{k+2})dy = 0,$ which is exact.

$\therefore \dfrac{\partial}{\partial x}\{-(3x^2 y^k - y^{k+2})\} = \dfrac{\partial}{\partial y}(2xy^{k+1})$

or, $-6xy^k + 0 = 2x(k+1)y^k$

or, $-3 = k+1$

or, $k = -4.$

14. (c)

$\dfrac{dy}{dx} = ay - by^2$

$\Rightarrow \dfrac{1}{y^2}\dfrac{dy}{dx} - \dfrac{a}{y} = -b$ ....(i)

Put, $\dfrac{1}{y} = z$, so that $-\dfrac{1}{y^2}\dfrac{dy}{dx} = \dfrac{dz}{dx}.$

Then (i) gives,

$-\dfrac{dz}{dx} - az = -b$

or, $\dfrac{dz}{dx} + az = b,$ which is linear in z.

$\therefore$ I.F $= e^{\int P dx} = e^{\int a dx} = e^{ax}$

So, the solution is given by

$z \times e^{ax} = \int e^{ax} \times b dx + c$

or, $\dfrac{1}{y}e^{ax} = \dfrac{b}{a}e^{ax} + c$

or, $\dfrac{1}{y} = \dfrac{b}{a} + ce^{-ax}$ ....(ii)

Now, $y(0) = y_0 \Rightarrow \dfrac{1}{y_0} = \dfrac{b}{a} + c \Rightarrow c = \dfrac{1}{y_0} - \dfrac{b}{a}$

Then from (ii) we get,

$\dfrac{1}{y} = \dfrac{b}{a} + \left(\dfrac{1}{y_0} - \dfrac{b}{a}\right) \times e^{-ax}$

when $x \to \infty$, $e^{-ax} \to e^{-\infty} = 0$

$\therefore \dfrac{1}{y} = \dfrac{b}{a} + \left(\dfrac{1}{y_0} - \dfrac{b}{a}\right) \times 0 \quad i.e; \quad y = \dfrac{a}{b}.$

15. (*d*) Let $y = e^{mx}$ be a solution of

$$\frac{d^2y}{dx^2} + 2a\frac{dy}{dx} + \beta y = 0 \qquad ...(i)$$

Then $(i) \Rightarrow m^2 e^{mx} + 2am e^{mx} + \beta e^{mx} = 0$

$\Rightarrow m^2 + 2am + \beta = 0 \quad (\because e^{mx} \neq 0)$

$$\Rightarrow m = \frac{-2a \pm \sqrt{(2a)^2 - 4 \times 1 \times \beta}}{2 \times 1}$$

$$= -a \pm \sqrt{a^2 - \beta}$$

Let $\alpha^2 - \beta > 0$

In this case, values of m are distinct real numbers.

So, the solution is given by

$$y(x) = c_1 e^{\left(-\alpha + \sqrt{\alpha^2 - \beta}\right)x} + c_2 e^{\left(-\alpha - \sqrt{\alpha^2 - \beta}\right)x}$$

Now, $\lim\limits_{x \to \infty} y(x) = 0 \quad (given)$,

which is possible

if $e^{\left(-\alpha + \sqrt{\alpha^2 - \beta}\right)x} \to 0$ and $e^{\left(-\alpha - \sqrt{\alpha^2 - \beta}\right)x} \to 0$ as $x \to \infty$

i.e; if $-\alpha + \sqrt{\alpha^2 - \beta} < 0$ and $-\alpha - \sqrt{\alpha^2 - \beta} < 0$

i.e; if $\alpha > 0$ and $\beta > 0$

$\therefore$ option (*d*) is correct.

16. (*a*) $\dfrac{d}{dx}\left(x\dfrac{dy}{dx}\right) = x$

$\Rightarrow \displaystyle\int \frac{d}{dx}\left(x\frac{dy}{dx}\right)dx = \int x\,dx + c_1$

$\Rightarrow \displaystyle\int d\left(x\frac{dy}{dx}\right) = \frac{x^2}{2} + c_1$

$\Rightarrow x\dfrac{dy}{dx} = \dfrac{x^2}{2} + c_1$

$\Rightarrow dy = \left(\dfrac{x}{2} + \dfrac{c_1}{x}\right)dx$

$\Rightarrow \displaystyle\int dy = \int\left(\frac{x}{2} + \frac{c_1}{x}\right)dx + c_2$

$\Rightarrow y = \dfrac{x^2}{4} + c_1 \log_e x + c_2 \,.................(i)$

$\therefore \dfrac{dy}{dx} = \dfrac{x}{2} + \dfrac{c_1}{x}\,..................(ii)$

When, $x = 1$, $y = 0$;

So, $(i) \Rightarrow 0 = \dfrac{1}{4} + c_1 \log_e 1 + c_2$

$\Rightarrow c_2 = -\dfrac{1}{4} \quad (\because \log_e 1 = 0)$

When, $x = 1$, $\dfrac{dy}{dx} = 0$;

So, $(ii) \Rightarrow 0 = \dfrac{1}{2} + c_1 \Rightarrow c_1 = -\dfrac{1}{2}$

Thus, we have, $y = \dfrac{x^2}{4} - \dfrac{1}{2}\log_e x - \dfrac{1}{4}$

$\therefore y(2) = \dfrac{2^2}{4} - \dfrac{1}{2}\log_e 2 - \dfrac{1}{4} = \dfrac{3}{4} - \dfrac{1}{2}\log_e 2.$

17. (*c*) The given equation is $x^2 y'' - 2xy' - 4y = 0$
.......(i) which is a linear homogeneous differential equation of order "2."

Let us put $\log x = z$.

Then, $\dfrac{dy}{dz} = x\dfrac{dy}{dx} = xy'$ and

$\dfrac{d^2y}{dz^2} - \dfrac{dy}{dz} = x^2\dfrac{d^2y}{dx^2} = x^2 y''$

$\therefore (i) \Rightarrow \dfrac{d^2y}{dz^2} - \dfrac{dy}{dz} - 2\dfrac{dy}{dz} - 4y = 0$

or, $\dfrac{d^2y}{dz^2} - 3\dfrac{dy}{dz} - 4y = 0 ....(ii)$

Let $y = e^{mz}$ be a trial solution of (ii)

Then, $(ii) \Rightarrow m^2 e^{mz} - 3m e^{mz} - 4e^{mz} = 0$

or, $m^2 - 3m - 4 = 0 \quad (\because e^{mz} \neq 0)$

or, $m = -4, 1$

$\therefore$ complete solution is given by

$y(x) = c_1 e^{-4z} + c_2 e^z = \dfrac{c_1}{x^4} + c_2 x$

$\left(\because e^z = x \text{ and } e^{-z} = \dfrac{1}{x}\right)$

Let, $y_1(x) = \dfrac{1}{x^4}$ and $y_2(x) = x$. Then clearly $y_1(x)$ and $y_2(x)$ are an independent solution of (i)

Now, $W(x)$

$= \begin{vmatrix} y_1(x) & y_2(x) \\ y_1'(x) & y_2''(x) \end{vmatrix} = \begin{vmatrix} \dfrac{1}{x^4} & x \\ \dfrac{-4}{x^5} & 1 \end{vmatrix}$

$= \dfrac{1}{x^4} \times 1 - x \times \left(\dfrac{-4}{x^5}\right) = \dfrac{5}{x^4}$, which satisfies $W(1) = 5.$

$\therefore W(2) = \dfrac{5}{2^4} = \dfrac{5}{16}$

18. (a) The equation $x^2 \dfrac{d^2y}{dx^2} + \lambda x \dfrac{dy}{dx} + y = 0$ is a linear homogeneous differential equation of order "2."

So, let us put $z = \log_e x$

Then, $\dfrac{dy}{dz} = x\dfrac{dy}{dx}, \dfrac{d^2y}{dz^2} - \dfrac{dy}{dz} = x^2\dfrac{d^2y}{dx^2}$

∴ the given differential equation becomes

$\dfrac{d^2y}{dz^2} - \dfrac{dy}{dz} + \lambda\dfrac{dy}{dz} + y = 0$

or, $\dfrac{d^2y}{dz^2} + (\lambda - 1)\dfrac{dy}{dz} + y = 0$ ...(i)

Now, $\dfrac{1}{x}(A + B\log_e x)$ is the general solution of (i)

$\Rightarrow \dfrac{1}{e^z}(A + Bz)$ is the general solution of (i)

$\Rightarrow (A + Bz)e^{-z}$ is the general solution of (i)

$\Rightarrow m = -1, -1$ are the roots of the auxiliary equation.

$\Rightarrow y = e^{-z}$ is a trial solution of (i)

Then, (i) gives $e^{-z} - (\lambda - 1) \times (e^{-z}) + e^{-z} = 0$

or, $1 + (1 - \lambda) + 1 = 0$ ∴ $\lambda = 3$

19. (b) The given differential equation is a linear homogeneous differential equation of order 2.

So, let us put $\log x = z$

Then, $\dfrac{dy}{dz} = x\dfrac{dy}{dx}, \quad \dfrac{d^2y}{dz^2} - \dfrac{dy}{dz} = x^2\dfrac{d^2y}{dx^2}$

∴ the given differential equation becomes

$\dfrac{d^2y}{dz^2} - \dfrac{dy}{dz} - 6y = 0$ ...(i)

Let $y = e^{mz}$ be a trial solution of (i)

Then (i) becomes,

$m^2e^{mz} - me^{mz} - 6e^{mz} = 0$

or $m^2 - m - 6 = 0$

or $m = -3, 2$

Hence, the solution of (i) is given by

$y = c_1e^{-3z} + c_2e^{2z} = \dfrac{c_1}{x^3} + c_2x^2$ ....(ii)

$\left(\because \log x = z \Rightarrow x = e^z, \dfrac{1}{x} = e^{-z}\right)$

Then, $y(1) = k \Rightarrow k = c_1 + c_2$ [from (ii)] ...(iii)

Also (ii) gives, $\dfrac{dy}{dx} = -\dfrac{3c_1}{x^4} + 2c_2x$

∴ $y'(1) = 6 \Rightarrow 6 = -3c_1 + 2c_2$ ...(iv)

Solving (iii) and (iv), we get,

$c_1 = 6 - 2k, c_2 = 3k - 6$

Then (ii) becomes, $y(x) = \dfrac{6 - 2k}{x^3} + (3k - 6)x^2$

Now, $y(x) \to 0$ as $x \to 0 +$

$\Rightarrow \lim_{x \to 0+} y(x) = 0$

$\Rightarrow \lim_{x \to 0+}\left[\dfrac{6 - 2k}{x^3} + (3k - 6)x^2\right] = 0$

$\Rightarrow \lim_{x \to 0+}\dfrac{6 - 2k}{x^3} + (3k - 6) \times 0 = 0$

$\Rightarrow \lim_{x \to 0+}\left(\dfrac{6 - 2k}{x^3}\right) = 0$

$\Rightarrow 6 - 2k = 0 \qquad \Rightarrow k = 3.$

20. (a)

$\cos^2 y\dfrac{d^2y}{dx^2} = \tan y$

$\Rightarrow \dfrac{d^2y}{dx^2} = \sec^2 y \times \tan y$

$\Rightarrow 2\dfrac{dy}{dx}\dfrac{d^2y}{dx^2} = 2\sec^2 y \times \tan y\dfrac{dy}{dx}$

$\Rightarrow \dfrac{d}{dx}\left(\dfrac{dy}{dx}\right)^2 = 2\tan y\sec^2 y\dfrac{dy}{dx}$

Now integrating both sides w.r.t "x" we get

$\left(\dfrac{dy}{dx}\right)^2 = 2\int \tan y \times \sec^2 y\, dy + c_1$

$= 2 \times \dfrac{(\tan y)^2}{2} + c_1 = (\tan y)^2 + c_1$

$\left(\text{using } \int f(y) \times f'(y)dy = \dfrac{[f(y)]^2}{2}\right)$

or, $\left(\dfrac{dy}{dx}\right)^2 = \tan^2 y + c_1$ ....(i)

Now when $y = 0, \dfrac{dy}{dx} = 0$

∴ (i) gives, $0 = \tan^2 0 + c_1$

or, $c_1 = 0$

Thus, we have $\left(\dfrac{dy}{dx}\right)^2 = \tan^2 y$

or, $\dfrac{dy}{dx} = \pm\tan y$

or, $\cot y\, dy = \pm dx$

Integrating we get,

$$\int \cot y \, dy = \pm \int dx + \log c'$$

or, $\log(\sin y) = \pm x + \log c'$

or, $\log(\sin y) - \log c' = \pm x$

or, $\log\left(\dfrac{\sin y}{c'}\right) = \pm x$

or, $\dfrac{\sin y}{c'} = e^{\pm x}$

or, $\sin y = c'e^{x}$ and $\sin y = c'e^{-x}$

or, $\sin y - c'e^{x} = 0$ and $\sin y - c'e^{-x} = 0$

or, $\sin y + ce^{x} = 0$ and $\sin y + ce^{-x} = 0$

(taking $c = -c'$)

Hence, the required solution is

$(\sin y + ce^{x})(\sin y + ce^{-x}) = 0$

21. (d) $(x^2 D^2 + xD + 1)y = \sin(2 \log x)$

$\Rightarrow x^2 D^2 y + xDy + y = \sin(2 \log x)$

$\Rightarrow x^2 \dfrac{d^2 y}{dx^2} + x\dfrac{dy}{dx} + y = \sin(2\log x)$ ...(i)

Equation (i) is a linear homogeneous differential equation of order "2"

So let us put $\log x = z$

Then, $\dfrac{dy}{dz} = x\dfrac{dy}{dx}$ and $\dfrac{d^2 y}{dz^2} - \dfrac{dy}{dz} = x^2\dfrac{d^2 y}{dx^2}$

$\therefore$ (i) gives,

$\dfrac{d^2 y}{dz^2} - \dfrac{dy}{dz} + \dfrac{dy}{dz} + y = \sin 2z$

or, $\dfrac{d^2 y}{dz^2} + y = \sin 2z$ ....(ii)

Consider $\dfrac{d^2 y}{dz^2} + y = 0$ ...(iii)

Let $y = e^{mz}$ be a trial solution of (iii)

Then (iii) gives

$m^2 e^{mz} + e^{mz} = 0$

or, $m^2 + 1 = 0$ or, $m^2 = -1 = i^2$

or, $m = \pm i = 0 + i,\ 0 - i$

$\therefore$ $y_{CF} = e^{0z}(c_1 \cos z + c_2 \sin z) = c_1 \cos z + c_2 \sin z$

$= c_1 \cos(\log x) + c_2 \sin(\log z)$

Now equation (ii) can be re-written as:

$\left(D^2 + 1\right)y = \sin 2z$ where $D \equiv \dfrac{d}{dz}$

$\therefore y_{PI} = \dfrac{1}{D^2 + 1}(\sin 2z) = \dfrac{1}{-2^2 + 1}\sin 2z = -\dfrac{1}{3}\sin 2z$

$= -\dfrac{1}{3}\sin(2\log x)$

Hence, the complete solution is given by

$y = y_{CF} + y_{PI}$

$= c_1 \cos(\log x) + c_2 \sin(\log x) - \dfrac{1}{3}\sin(2\log x).$

PREVIOUS YEARS QUESTIONS (2000-18)

1. The solution of the differential equation $\dfrac{d^2 y}{dx^2} = 3x - 2$ with boundary conditions $y(0) = 2$ and $y'(1) = -3$ is

(a) $y = \dfrac{x^3}{3} - \dfrac{x^2}{2} - 3x - 2$

(b) $y = 3x^3 - \dfrac{x^2}{2} - 5x + 2$

(c) $y = \dfrac{x^3}{2} - x^2 - \dfrac{5x}{2} + 2$

(d) $y = x^3 - \dfrac{x^2}{2} + 5x + \dfrac{3}{2}$ **[CE GATE 2001]**

2. The differential equation
$$\left\{1 + \left(\dfrac{dy}{dt}\right)^2\right\}^3 = C^2\left(\dfrac{d^2 y}{dt^2}\right)^2 \text{ is of}$$

(a) Second order and third degree
(b) Third order and Second degree
(c) Second order and Second degree
(d) Third order and Third degree
[PI GATE-2005]

3. The following differential equation has
$$3\left(\dfrac{d^2 y}{dt^2}\right) + 4\left(\dfrac{dy}{dt}\right)^3 + y^2 + 2 = t$$

(a) degree = 2, order = 1
(b) degree = 1, order = 2
(c) degree = 4, order = 3
(d) degree = 2, order = 3 **[EC GATE 2005]**

4. The degree of the differential equation $\dfrac{d^2 x}{dt^2} + 2x^3 = 0$ is

(a) 0   (b) 1   (c) 2   (d) 3
[CE GATE 2007]

5. The order of the differential equation
$$\dfrac{d^2 y}{dt^2} + \left(\dfrac{dy}{dt}\right)^3 + y^4 = e^{-t}$$

(a) 1   (b) 2   (c) 3   (d) 4
[EC GATE-2009]

6. The order and degree of the differential equation $\dfrac{d^3y}{dx^3}+4\sqrt{\left(\dfrac{dy}{dx}\right)^3+y^2}=0$ are, respectively

(a) 3 and 2   (b) 2 and 3
(c) 3 and 3   (d) 3 and 1

[CE GATE 2010]

7. Which of the following differential equations has a solution given by the function?

$$y = 5\sin\left(3x+\dfrac{\pi}{3}\right)$$

(a) $\dfrac{dy}{dx}-\dfrac{5}{3}\cos 3x = 0$ (b) $\dfrac{dy}{dx}-\dfrac{5}{3}\cos 3x = 0$

(c) $\dfrac{d^2y}{dx^2}+9y = 0$ (d) $\dfrac{d^2y}{dx^2}-9y = 0$

[PI GATE 2010]

8. The Blasius equation, $\dfrac{d^3f}{dn^3}+\dfrac{f}{2}\dfrac{d^2f}{dn^2}=0$, is a

(a) second-order nonlinear ordinary differential equation
(b) third-order nonlinear ordinary differential equation
(c) third-order linear ordinary differential equation
(d) mixed order nonlinear ordinary differential equation

[ME GATE-2010]

9. The solution of the differential equation $\dfrac{dy}{dx}+y^2 = 0$ is

(a) $y = \dfrac{1}{x+c}$ (b) $y = \dfrac{-x^3}{3}+c$ (c) ce^x

(d) unsolvable as equation is non-liner

[ME GATE 2003]

10. Biotransformation of an organic compound having concentration "x" can be modeled using an ordinary differential equation $\dfrac{dx}{dt}+kx^2 = 0$, where k is the reaction rate constant. If $x = a$ at $t = 0$, the solution of the equation is

(a) $x = ae^{-kt}$ (b) $\dfrac{1}{x}=\dfrac{1}{a}+kt$
(c) $x = a(1-e^{-kt})$ (d) $x = a + kt$

[CE GATE 2004]

11. Transformation to linear form by substituting $v = y^{1-n}$ of the equation

$$\dfrac{dy}{dt}+p(t)y = q(t)y^n; n > 0 \text{ will be}$$

(a) $\dfrac{dv}{dt}+(1-n)pv = (1-n)q$

(b) $\dfrac{dv}{dt}+(1+n)pv = (1+n)q$

(c) $\dfrac{dv}{dt}+(1+n)pv = (1-n)q$

(d) $\dfrac{dv}{dt}+(1-n)pv = (1+n)q$

[CE GATE 2005]

12. The solution of the first-order differential equation $x'(t) = -3x(t)$, $x(0) = x_0$ is
(a) $x(t) = x_0e^{-3t}$ (b) $x(t) = x_0e^{-3}$
(c) $x(t) = x_0e^{-1/3}$ (d) $x(t) = x_0e^{-1}$

[EE GATE 2005]

13. A spherical naphthalene ball exposed to the atmosphere loses volume at a rate proportional to its instantaneous surface area due to evaporation. If the initial diameter of the ball is 2 cm and the diameter reduces to 1 cm after 3 months, the ball completely evaporates in
(a) 6 months   (b) 9 months
(c) 12 months   (d) infinite time

[CE GATE2006]

14. The solution of the differential equation $\dfrac{dy}{dx}+2xy = e^{-x^2}$ with $y(0) = 1$ is

(a) $(1+x)e^{+x^2}$ (b) $(1+x)e^{-x^2}$

(c) $(1-x)e^{+x^2}$ (d) $(1-x)e^{-x^2}$

[ME GATE 2006]

15. The solution of $dy/dx = y^2$ with initial value $y(0) = 1$ is bounded in the interval
(a) $-\infty \le x \le \infty$ (b) $-\infty \le x \le 1$
(c) $x < 1, x > 1$ (d) $-2 \le x \le 2$

[ME GATE 2007]

16. The solution for the differential equation $\dfrac{dy}{dx}=x^2y$ with the condition that $y = 1$ at $x = 0$ is

(a) $y = e^{\frac{1}{2x}}$ (b) $In(y) = \dfrac{x^3}{3}+4$

(c) $In(y) = \dfrac{x^3}{2}$ (d) $y = e^{\frac{x^3}{3}}$

[CE GATE 2007]

17. A body originally at 60°C cools down to 40°C in 15 minutes when kept in air at a temperature of 25°C. What will be the temperature of the body at the end of 30 minutes?
    (a) 35.2°C   (b) 31.5°C
    (c) 28.7°C   (d) 15°C

    [CE GATE 2007]

18. Solution of $\frac{dy}{dx} = -\frac{x}{y}$ at $x = 1$ and $y = \sqrt{3}$ is
    (a) $x - y^2 = 2$ (b) $x + y^2 = 4$
    (c) $x^2 - y^2 = -2$ (d) $x^2 + y^2 = 4$

    [CE GATE 2008]

19. Consider the differential equations $\frac{dy}{dx} = 1 + y^2$ which one of the following can be particular solution of this differentials equation?
    (a) $y = \tan(x + 3)$ (b) $y = \tan x + 3$
    (c) $y = \tan(y + 3)$ (d) $x = \tan y + 3$

    [IN GATE 2008]

20. Which of the following is a solution to the differential equation $\frac{dx(t)}{dt} + 3x(t) = 0$?
    (a) $x(t) = 3e^{-t}$ (b) $x(t) = 2e^{-3t}$
    (c) $x(t) = \frac{3}{2}t^2$ (d) $x(t) = 3t^2$

    [EC GATE 2008]

21. Match List-I with List-II and select the correct answer using the codes given below:

List-I	List-II
A. $\frac{dy}{dx} = \frac{y}{x}$	1. Circles
B. $\frac{dy}{dx} = -\frac{y}{x}$	2. Straight lines
C. $\frac{dy}{dx} = \frac{x}{y}$	3. Hyperbolas
D. $\frac{dy}{dx} = -\frac{x}{y}$	

    Codes:

	A	B	C	D
(a)	2	3	3	1
(b)	1	3	2	1
(c)	2	1	3	3
(d)	3	2	1	2

    [EC GATE 2009]

22. Solution of the differential equation
    $3y\frac{dy}{dx} + 2x = 0$ represents a family of

    (a) ellipses   (b) circles
    (c) parabolas   (d) hyperbolas

    [CE GATE 2009]

23. The solution of $x\frac{dy}{dx} + y = x^4$ with the condition $y(1) = \frac{6}{5}$ is
    (a) $y = \frac{x^4}{5} + \frac{1}{x}$ (b) $y = \frac{4x^4}{5} + \frac{4}{5x}$
    (c) $y = \frac{x^4}{5} + 1$ (d) $y = \frac{x^5}{5} + 1$

    [ME GATE 2009]

24. Consider the differential equation
    $\frac{dy}{dx} + y = e^x$ with $y(0) = 1$. Then the value of $y(1)$ is
    (a) $e + e^{-1}$ (b) $\frac{1}{2}[e - e^{-1}]$
    (c) $\frac{1}{2}[e + e^{-1}]$ (d) $2[e - e^{-1}]$

    [IN GATE 2010]

25. The solution of the differential equation $\frac{dy}{dx} - y^2 = 1$ satisfying the conditions $y(0) = 1$ is
    (a) $y = e^{x^2}$ (b) $y = \sqrt{x}$
    (c) $y = \cot\left(x + \frac{\pi}{4}\right)$ (d) $y = \tan\left(x + \frac{\pi}{4}\right)$

    [PI GATE 2010]

26. Consider the differential equation $\frac{dy}{dx} = (1 + y^2)x$. The general solution with constant c is
    (a) $y = \tan\frac{x^2}{2} + \tan c$ (b) $y = \tan^2\left(\frac{x}{2} + c\right)$
    (c) $y = \tan^2\left(\frac{x}{2}\right) + c$ (d) $y = \tan\left(\frac{x^2}{2} + c\right)$

    [ME GATE 2011]

27. The solution of the differential equation $\frac{dy}{dx} = ky, y(0) = c$ is
    (a) $x = ce^{-ky}$ (b) $x = ke^{cy}$
    (c) $y = ce^{kx}$ (d) $y = ce^{-kx}$

    [EC GATE 2011]

28. The solution of the differential equation $\frac{dy}{dx} + \frac{y}{x} = x$, with the condition that $y = 1$ at $x = 1$, is

(a) $y = \dfrac{2}{3x^2} + \dfrac{x}{3}$ (b) $y = \dfrac{x}{2} + \dfrac{1}{2x}$

(c) $y = \dfrac{2}{3} + \dfrac{x}{3}$ (d) $y = \dfrac{2}{3x} + \dfrac{x^2}{3}$

[CE GATE 2011]

29. With K as a constant, the possible solution for the first-order differential equation $\dfrac{dy}{dx} = e^{-3x}$ is

(a) $-\dfrac{1}{3}e^{-3x} + K$ (b) $-\dfrac{1}{3}e^{3x} + K$

(c) $-3e^{-3x} + K$ (d) $-3e^{-x} + K$

[EE GATE 2011]

30. With initial condition $x(1) = 0.5$, the solution the differential equation, $t\dfrac{dx}{dt} + x = t$ is

(a) $x = t - \dfrac{1}{2}$ (b) $x = t^2 - \dfrac{1}{2}$

(c) $x = \dfrac{t^2}{2}$ (d) $x = \dfrac{t}{2}$

[EC, EE, IN GATE2012]

31. The solution of the ordinary differential equation $\dfrac{dy}{dx} + 2y = 0$ for the boundary condition. $y = 5$ at $x = 1$ is

(a) $y = e^{-2x}$ (b) $y = 2e^{-2x}$
(c) $y = 10.95e^{-2x}$ (d) $y = 36.95e^{-2x}$

[CE GATE 2012]

32. The matrix form of the linear system $\dfrac{dx}{dt} = 3x - 5y$ and $\dfrac{dy}{dt} = 4x + 8y$ is

(a) $\dfrac{d}{dt}\begin{Bmatrix}x\\y\end{Bmatrix} = \begin{bmatrix}3 & -5\\4 & 8\end{bmatrix}\begin{Bmatrix}x\\y\end{Bmatrix}$

(b) $\dfrac{d}{dt}\begin{Bmatrix}x\\y\end{Bmatrix} = \begin{bmatrix}3 & 8\\4 & -5\end{bmatrix}\begin{Bmatrix}x\\y\end{Bmatrix}$

(c) $\dfrac{d}{dt}\begin{Bmatrix}x\\y\end{Bmatrix} = \begin{bmatrix}4 & -5\\3 & 8\end{bmatrix}\begin{Bmatrix}x\\y\end{Bmatrix}$

(d) $\dfrac{d}{dt}\begin{Bmatrix}x\\y\end{Bmatrix} = \begin{bmatrix}4 & 8\\3 & -5\end{bmatrix}\begin{Bmatrix}x\\y\end{Bmatrix}$

[ME GATE 2014]

33. The general solution of the differential equation $\dfrac{dy}{dx} = \cos(x+y)$, with "$c$" as a constant, is

(a) $y + \sin(x + y) = x + c$

(b) $\tan\left(\dfrac{x+y}{2}\right) = y + c$

(c) $\cos\left(\dfrac{x+y}{2}\right) = x + c$

(d) $\tan\left(\dfrac{x+y}{2}\right) = y + c$

[ME GATE 2014]

34. Which ONE of the following is a linear non-homogeneous differential equation, where x and y are the independent and dependent variables, respectively?

(a) $\dfrac{dy}{dx} + y = e^{-x}$ (b) $\dfrac{dy}{dx} + xy = 0$

(c) $\dfrac{dy}{dx} + xy = e^{-y}$ (d) $\dfrac{dy}{dx} + e^{-y} = 0$

[EC GATE 2014]

35. The solution of the initial value problem $\dfrac{dy}{dx} = -2xy; y(0) = 2$ is

(a) $1 + e^{-x^2}$ (b) $2e^{-x^2}$

(c) $1 + e^{x^2}$ (d) $2e^{x^2}$

[ME GATE 2014]

36. Consider the following difference equation $x(y\,dx + x\,dy)\cos\dfrac{y}{x} = y(x\,dy - y\,dx)\sin\dfrac{y}{x}$ Which of the following is the solution of the above equation (C is an arbitrary constant)?

(a) $\dfrac{x}{y}\cos\dfrac{y}{x} = C$ (b) $\dfrac{x}{y}\sin\dfrac{y}{x} = C$

(c) $xy\cos\dfrac{y}{x} = C$ (d) $xy\sin\dfrac{y}{x} = C$

[CE GATE 2015]

37. The solution to $6yy' - 25x = 0$ represents a
(a) family of circles (b) family of ellipses
(c) family of parabolas
(d) family of hyperbolas

[PI GATE 2015]

38. Consider the following differential equation: $\dfrac{dy}{dt} = -5y$; initial condition: $y = 2$ at $t = 0$ The value of y at $t = 3$ is

(a) $-5e^{-10}$ (b) $2e^{-10}2e^{-10}$
(c) $2e^{-15}$ (d) $-15e^2$

[ME GATE 2015]

39. The general solution of the differential equation $\dfrac{dy}{dx} = \dfrac{1+\cos 2y}{1-\cos 2x}$ is
 (a) $\tan y - \cot x = c$ (c is a constant)
 (b) $\tan x - \cot y = c$ (c is a constant)
 (c) $\tan y + \cot x = c$ (c is a constant)
 (d) $\tan x + \cot y = c$ (c is a constant)

[EC GATE 2015]

40. Consider the differential equation
 $\dfrac{dx}{dt} = 10 - 0.2x$ with initial condition
 $x(0) = 1$. The response $x(t)$ for $t > 0$ is
 (a) $2 - e^{-0.2t}$ (b) $2 - e^{0.2t}$
 (c) $50 - 49e^{-0.2t}$ (d) $50 - 49e^{0.2t}$

[EC GATE 2015]

41. A differential equation $\dfrac{di}{dt} - 0.2i = 0$ is applicable over $-10 < t < 10$. If $i(4) = 10$, then $i(-5)$ is _____.

[EE GATE 2015]

42. Which one of the following is the general solution of the first-order differential equation $\dfrac{dy}{dx} = (x+y-1)^2$ where x, y are real?
 (a) $y = 1 + x + \tan^{-1}(x + c)$, where c is a constant.
 (b) $y = 1 + x + \tan(x + c)$, where c is a constant.
 (c) $y = 1 - x + \tan^{-1}(x + c)$, where c is a constant.
 (d) $y = 1 - x + \tan(x + c)$, where c is a constant.

[EC GATE 2017]

43. Consider the differential equation
 $(t^2 - 81)\dfrac{dy}{dt} + 5ty = \sin(t)$ with $y(1) = 2\pi$. There exists a unique solution for this differential equation when t belongs to the interval
 (a) $(-2, 2)$ (b) $(-10, 10)$
 (c) $(-10, 2)$ (d) $(0, 10)$

[EE GATE 2017]

44. A particle of mass 2 kg is traveling at a velocity of 1.5 m/s. A force $f(t) = 3t^2$ (in N) is applied to it in the direction of motion for a duration of 2 seconds, where t denotes time in seconds. The velocity (in m/s, up to one decimal place) of the particle immediately after the removal of the force is _____.

[CE GATE 2017]

45. The solution of the equation $\dfrac{dQ}{dt} + Q = 1$ with $Q = 0$ at $t = 0$ is
 (a) $Q(t) = e^{-t} - 1$ (b) $Q(t) = 1 + e^{-t}$
 (c) $Q(t) = 1 - e^{t}$ (d) $Q(t) = 1 - e^{-t}$

[CE GATE 2017]

46. The solution of the equation $x\dfrac{dy}{dx} + y = 0$ passing through the point $(1, 1)$ is
 (a) x (b) x^2 (c) x^{-1} (d) x^{-2}

[CE GATE 2018]

47. A curve passes through the point $(1, 0)$ and satisfies the differential equation $\dfrac{dy}{dx} = \dfrac{x^2 + y^2}{2y} + \dfrac{y}{x}$. The equation that describes the curve is

 (a) $\ln\left(1 + \dfrac{y^2}{x^2}\right) = x - 1$ (b) $\dfrac{1}{2}\ln\left(1 + \dfrac{y^2}{x^2}\right) = x - 1$

 (c) $\ln\left(1 + \dfrac{y}{x}\right) = x - 1$ (d) $\dfrac{1}{2}\ln\left(1 + \dfrac{y}{x}\right) = x - 1$

[EC GATE 2018]

48. If y is a solution of the differential equation $y^3\dfrac{dy}{dx} + x^3 = 0$, $y(0) = 1$, the value of $y(-1)$ is _____?
 (a) -2 (b) -1
 (c) 0 (d) 1

[ME GATE 2018]

49. The solution for the following differential equation with boundary conditions $y(0) = 2$ and $y'(1) = -3$ is
 $\dfrac{d^2y}{dx^2} = 3x - 2$

 (a) $y = \dfrac{x^3}{3} - \dfrac{x^2}{2} + 3x - 6$

 (b) $y = 3x^3 - \dfrac{x^2}{2} - 5x + 2$

 (c) $y = \dfrac{x^3}{2} - x^2 - \dfrac{5x}{2} + 2$

 (d) $y = x^3 - \dfrac{x^2}{2} - 5x + \dfrac{3}{2}$

[GATE 2001]

50. Solve the differential equation,

$$\frac{d^2 y}{dx^2} + y = x$$

With the following conditions:

(i) at $x = 0$, $y = 1$

(ii) at $x = \frac{\pi}{2}$, $y = \frac{\pi}{2}$

[GATE 2001]

51. The solution of $\frac{d^2 y}{dx^2} + 2\frac{dy}{dx} + 17y = 0$;

$y(0) = 1, \frac{dy}{dx}\left(\frac{\pi}{4}\right) = 0$ in the range $0 < x < \frac{\pi}{4}$ is

given by

(a) $e^{-x}\left(\cos 4x + \frac{1}{4}\sin 4x\right)$

(b) $e^{x}\left(\cos 4x - \frac{1}{4}\sin 4x\right)$

(c) $e^{-4x}\left(\cos x - \frac{1}{4}\sin x\right)$

(d) $e^{-4x}\left(\cos 4x - \frac{1}{4}\sin 4x\right)$

[CE GATE 2005]

52. The complete solution of the ordinary differential equation

$$\frac{d^2 y}{dx^2} + p\frac{dy}{dx} + qy = 0 \text{ is } y = c_1 e^{-x} + c_2 e^{-3x}.$$

Then, p and q are

(a) $p = 3, q = 3$ (b) $p = 3, q = 4$

(c) $p = 4, q = 3$ (d) $p = 4, q = 4$

[ME GATE 2005]

53. The general solution of the differential equation $(D^2 - 4D + 4)y = 0$ is of the form

$\left(\text{given } D = \frac{d}{dx} \text{ and } C_1, C_2 \text{ are constants}\right)$

(a) $C_1 e^{2x}$

(b) $C1e2x + C_2 e^{-2x}$

(c) $C_1 e^{2x} + C_2 e^{-2x}$

(d) $C_1 e^{2x} + C_2 x e^{2x}$

[IN GATE 2005]

54. For the equation,

$x''(t) + 3x'(t) + 2x(t) = 5$, the solution $x(t)$ approaches to the following value as $t \to \infty$

(a) 0

(b) $\frac{5}{2}$

(c) 5

(d) 10

[EE GATE 2005]

55. Which of the following is a solution of the differential equation $\frac{d^2 y}{dx^2} + p\frac{dy}{dx} + (q+1)y = 0$?

(a) e^{-3x}

(b) xe^{-x}

(c) xe^{-2x}

(d) $x^2 e^{-2x}$

[ME GATE 2005]

56. A solution of the following differential equation is given by $\frac{d^2 y}{dx^2} - 5\frac{dy}{dx} + 6y = 0$

(a) $y = e^{2x} + e^{-3x}$

(b) $y = e^{2x} + e^{3x}$

(c) $y = e^{-2x} + e^{3x}$

(d) $y = e^{-2x} + e^{-3x}$

[EC GATE 2005]

57. For $\frac{d^2 y}{dx^2} + 4\frac{dy}{dx} + 3y = 3e^{2x}$, the particular integral is

(a) $\frac{1}{15}e^{2x}$

(b) $\frac{1}{5}e^{2x}$

(c) $3e^{2x}$

(d) $C_1 e^{-x} + C_2 e^{-3x}$

[ME GATE 2006]

58. For the initial value problem

$y'' + 2y' + 101y = 10.4e^x$, $y(0) = 1.1$ and

$y'(0) = -0.9$;

various solutions are written in the following groups. Match the type of solution with the correct expression.

Group -1

P. General solution of homogeneous equations

Q. Particular integral

R. Total solution satisfying boundary conditions

Group-II

(1) $0.1e^x$

(2) $e^{-x}[A \cos 10x + B \sin 10x]$

(3) $e^{-x} \cos 10x + 0.1e^x$

(a) $P - 2, Q - 1, R - 3$

(b) $P - 1, Q - 3, R - 2$

(c) $P - 1, Q - 2, R - 3$

(d) $P - 3, Q - 2, R - 1$

[IN GATE 2006]

59. The general solution $\frac{d^2 y}{dx^2} + y = 0$ is

(a) $y = P \cos x + Q \sin x$

(b) $y = P \cos x$

(c) $y = P \sin x$

(d) $y = P \sin^2 x$

[EC GATE 2007]

60. Given that $\ddot{x} + 3x = 0$, and $x(0) = 1$, $\dot{x}(0) = 0$, what is $x(1)$?

   (a) –0.99   (b) –0.16

   (c) 0.16   (d) 0.99

                          [ME GATE 2008]

61. It is given that $y'' + 2y' + y = 0$, $y(0) = 0$, $y(1) = 0$ what is y?

   (a) 0   (b) 0.37

   (c) 0.62   (d) 1.13

                          [ME GATE 2008]

62. The solutions of the differential equations

$$\frac{d^2y}{dx^2} + 2\frac{dy}{dx} + 2y = 0 \text{ are}$$

   (a) $e^{-(1+i)x}, e^{-(1-i)x}$ (b) $e^{(1+i)x}, e^{(1-i)x}$

   (c) $e^{-(1+i)x}, e^{(1-i)x}$ (d) $e^{(1+i)x}, e^{-(1-i)x}$

                          [PI GATE 2008]

63. The homogeneous part of the differential equation

$$\frac{d^2y}{dx^2} + P\frac{dy}{dx} + qy = 0 \text{ (p and q, r are contants) has}$$

   real distinct roots if

   (a) $P^2 - 4q > 0$ (b) $P^2 - 4q < 0$

   (c) $P^2 - 4q = 0$ (d) $P^2 - 4q = r$

                          [PI GATE 2009]

64. The solution to the ordinary differential equation $\dfrac{d^2y}{dx^2} + \dfrac{dy}{dx} - 6y = 0$

   (a) $y = c_1e^{3x} + c_2e^{-2x}$ (b) $y = c_1e^{3x} + c^2e^{2x}$

   (c) $y = c_1e^{-3x} + c_2e^{2x}$ (d) $y = c_1e^{-3x} + c_2e^{-2x}$

                          [EE GATE 2010]

65. The function $n(x)$ satisfies the differential equation $\dfrac{d^2n(x)}{dx^2} - \dfrac{n(x)}{L^2} = 0$ where L is a constant. The boundary conditions are: $n(0) = K$ and $n(\infty) = 0$. The solution to this equation is

   (a) $n(x) = K \exp(x/L)$

   (b) $n(x) = K \exp(-x/\sqrt{L})$

   (c) $n(x) = K^2 \exp(-x/L)$

   (d) $n(x) = K \exp(-x/L)$

                          [EC GATE 2010]

66. For the differential equation $\dfrac{d^2x}{dt^2} + 6\dfrac{dx}{dt} + 8x = 0$ with initial conditions $x(0)$

   $= 1$ and $\dfrac{dx}{dt}\bigg|_{t=0} = 0$, the solution is

   (a) $x(t) = 2e^{-6t} - e^{-2t}$

   (b) $x(t) = 2e^{-2t} - e^{-4t}$

   (c) $x(t) = -e^{-6t} + 2e^{-4t}$

   (d) $x(t) = e^{-2t} + 2e^{-4t}$

                          [EE GATE 2010]

67. Consider the differential equation

$$\frac{d^2y}{dx^2} + 2\frac{dy}{dx} + y = 0 \text{ with boundary conditions}$$

   $y(0)=1$ & $y(1)=0$. The value of $y(2)$ is

   (a) –1   (b) $-e^{-1}$

   (c) $-e^{-2}$ (d) e^2

                          [IN GATE 2011]

68. The solution of the differential equation

$$\frac{d^2y}{dx^2} + 6\frac{dy}{dx} + 9y = 9x + 6 \text{ with } C_1 \text{ and } C_2 \text{ as}$$

   constant is

   (a) $y = (C_1x + C_2)e^{-3x}$

   (b) $y = C_1e^{3x} + C_2e^{-3x}$

   (c) $y = (C_1x + C_2)e^{-3x} + x$

   (d) $y = (C_1x + C_2)e^{3x} + x$

                          [PI GATE 2011]

69. Consider the differential equation

$$x^2\frac{d^2y}{dx^2} + x\frac{dy}{dx} - 4y = 0$$

   With the boundary condition of $y(0) = 0$ and $y(1) = 1$. The complete solution of the differential equation is

   (a) x^2 (b) $\sin\left(\dfrac{\pi x}{2}\right)$

   (c) $e^x \sin\left(\dfrac{\pi x}{2}\right)$ (d) $e^{-x} \sin\left(\dfrac{\pi x}{2}\right)$

                   [ME, PI GATE 2012]

70. The solution to the differential equation $\dfrac{d^2u}{dx^2} - k\dfrac{du}{dx} = 0$ where k is constant, subjected to the boundary conditions $u(0) = 0$ and $u(L) = U$ is

   (a) $u = U\dfrac{x}{L}$ (b) $u = U\left(\dfrac{1-e^{kx}}{1-e^{kL}}\right)$

   (c) $u = U\left(\dfrac{1-e^{-kx}}{1-e^{-kL}}\right)$ (d) $u = U\left(\dfrac{1+e^{kx}}{1+e^{kL}}\right)$

                         [ME GATE 2013]

71. The maximum value of the solution $y(t)$ of the differential equation $y(t) + \ddot{y}(t) = 0$ with initial conditions $\dot{y}(0) = 1$ and $y(0) = 1$, for $t \geq 0$ is

(a) 1 (b) 2

(c) π (d) $\sqrt{2}$

[IN GATE 2013]

72. The solution for the differential equation

$\dfrac{d^2x}{dt^2}+9x=0$ with initial conditions $x(0)=1$

and $\dfrac{dx}{dt}\Big|_{t=0}=1$, is

(a) t^2+t+1 (b) $\sin 3t+\dfrac{1}{3}\cos 3t+\dfrac{2}{3}$

(c) $\dfrac{1}{3}\sin 3t+\cos 3t$ (d) $\cos 3t+t$

[EE GATE 2014]

73. Consider two solutions $x(t)=x_1(t)$ and $x(t)=x_2(t)$ of the differential equation $\dfrac{d^2x(t)}{dt^2}+x(t)=0, t>0$, such that at $t=0,\ x_2=0,\dfrac{dx_1(t)}{dt}=0$. The Wronskian

$W(t)=\begin{vmatrix} x_1(t) & x_2(t) \\ \dfrac{dx_1(t)}{dt} & \dfrac{dx_2(t)}{dt} \end{vmatrix}$ at $t=\pi/2$ is

(a) 1 (b) –1

(c) 0 (d) $\pi/2$

[ME GATE 2014]

74. If $y=f(x)$ is the solution of $\dfrac{d^2y}{dx^2}=0$

with the boundary conditions $y=5$ at $x=0$ and $\dfrac{dy}{dx}=2$ at $x=10$, $f(15)=$?

[ME GATE 2014]

75. If a and b are constants, the most general solution of the differential equation $\dfrac{d^2x}{dt^2}+2\dfrac{dx}{dt}+x=0$ is

(a) ae^{-t} (b) $ae^{-t}+bte^{-t}$

(c) ae^t+bte^{-t} (d) ae^{-2t}

[EC GATE 2014]

76. Consider the differential equation

$x^2\dfrac{d^2y}{dx^2}+x\dfrac{dy}{dx}-y=0$. Which of the following is a solution to this differential equation for $x=0$?

(a) e^x (b) x^2

(c) $1/x$ (d) $\ln x$

[EE GATE 2014]

77. If the characteristic equation of the differential equation $\dfrac{d^2y}{dx^2}+2a\dfrac{dy}{dx}+y=0$ has two equal roots, then the values of α are

(a) ±1 (b) 0, 0

(c) $\pm j$ (d) $\pm1/2$

[EC GATE 2014]

78. Consider the following second-order linear differential equation $\dfrac{d^2y}{dx^2}=-12x^2+24x-20$. The boundary conditions are at $x=0$, $y=5$ and $x=2$, $y=21$. The value of y at $x=1$ is _____.

[CE GATE 2015]

79. A solution of the ordinary differential equation $\dfrac{d^2y}{dt^2}+5\dfrac{dy}{dt}+6y=0$ is such that $y(0)=2$ and $y(1)=-\dfrac{1-3e}{e^3}$. The value of $\dfrac{dy}{dt}(0)$ is _____.

[EE GATE 2015]

80. The solution of the differential equation $\dfrac{d^2y}{dt^2}+2\dfrac{dy}{dt}+y=0$ with $y(0)=y'(0)=1$

(a) $(2-t)e^t$ (b) $(1+2t)e^{-t}$

(c) $(2+t)e^{-t}$ (d) $(1-2t)e^t$

[EC GATE 2015]

81. The solution to $x^2y''+xy'-y=0$ is

(a) $y=C_1x^2+C_2x^{-3}$ (b) $y=C_1+C_2x^{-2}$

(c) $y=C_1x+\dfrac{C_2}{x}$ (d) $y=C_1x+C_2x^4$

[PI GATE 2015]

82. Find the solution of $\dfrac{d^2y}{dx^2}=y$ which passes through the origin and the point $\left(\ln 2,\dfrac{3}{4}\right)$

(a) $y=\dfrac{1}{2}e^x-e^{-x}$ (b) $y=\dfrac{1}{2}\left(e^x+e^{-x}\right)$

(c) $y=\dfrac{1}{2}\left(e^x-e^{-x}\right)$ (d) $y=\dfrac{1}{2}e^x+e^{-x}$

[ME GATE 2015]

83. Consider the differential equation

$\frac{d^2x(t)}{dt^2} + 3\frac{dx(t)}{dt} + 2x(t) = 0.$ Given $x(0) = 20$ and $x(1) = 10/e$, where $e = 2.718$, then the value of $x(2)$ is _____?

[EC GATE 2015]

84. The particular solution of the initial value problem given below is

$\frac{d^2y}{dx^2} + 12\frac{dy}{dx} + 36y = 0$

with $y(0) = 3$ and $\frac{dy}{dx}\big|_{x=0} = -36$

(a) $(3 - 18x)e^{-6x}$ (b) $(3 + 25x)e^{-6x}$
(c) $(3 + 20x)e^{-6x}$ (d) $(3 - 12x)e^{-6x}$

[EC GATE 2016]

85. If $y = f(x)$ satisfies the boundary value problem $y'' + 9y = 0, y(0) = 0, y\left(\frac{\pi}{2}\right) = \sqrt{2}$, then $y\left(\frac{\pi}{4}\right)$ is _____.

[ME GATE 2016]

86. Let $y(x)$ be the solution of the differential equation $\frac{d^2y}{dx^2} - 4\frac{dy}{dx} + 4y = 0$ with initial conditions $y(0) = 0$ and $\frac{dy}{dx}\big|_{x=0} = 1$. Then the value of $y(1)$ is _____.

[EE GATE 2016]

87. The respective expressions for complimentary function and particular integral part of the solution of the differential equation $\frac{d^4y}{dx^4} + 3\frac{d^2y}{dx^2} = 108x^2$ are

(a) $\left[c_1 + c_2 x + c_3 \sin(\sqrt{3}x) + c_4 \cos(\sqrt{3}x)\right]$ and

$\left[3x^4 - 12x^2 + c\right]$

(b) $\left[c_2 + c_3 \sin(\sqrt{3}x) + c_4 \cos(\sqrt{3}x)\right]$

and $\left[5x^4 - 12x^2 + c\right]$

(c) $\left[c_1 + c_3 \sin(\sqrt{3}x) + c_4 \cos(\sqrt{3}x)\right]$

and $\left[3x^4 - 12x^2 + c\right]$

(d) $\left[c_1 + c_2 x + c_3 \sin(\sqrt{3}x) + c_4 \cos(\sqrt{3}x)\right]$

and $\left[5x^4 - 12x^2 + c\right]$

[CE GATE 2016]

88. A function $y(t)$, such that $y(0) = 1$ and $y(1) = 3e^{-1}$, is a solution of the differential equation $\frac{d^2y}{dt^2} + 2\frac{dy}{dt} + y = 0$. Then $y(2)$ is

(a) $5e^{-1}$ (b) $5e^{-2}$
(c) $7e^{-1}$ (d) $7e^{-2}$

[EE GATE 2016]

89. The differential equation $\frac{d^2y}{dx^2} + 16y = 0$ for $y(x)$ with the two boundary conditions $\frac{dy}{dx}\big|_{x=0} = 1$ and $\frac{dy}{dx}\big|_{x=\frac{\pi}{2}} = -1$ has

(a) no solution   (b) exactly two solution
(c) exactly one solution
(d) infinitely many solutions

[ME GATE 2017]

90. Consider the following second-order differential equation:

$y'' - 4y' + 3y = 2t - 3t^2.$ The particular solution of the differential equation is

(a) $-2 - 2t - t^2$ (b) $-2t - t^2$
(c) $2t - t^2$ (d) $-2 - 2t - 3t^2$

[CE GATE 2017]

91. The general solution of the differential equation $\frac{d^2y}{dx^2} + 2\frac{dy}{dx} - 5y = 0$ in terms of arbitrary constants K_1 and K_2 is

(a) $K_1 e^{(-1+\sqrt{6})x} + K_2 e^{(-1-\sqrt{6})x}$

(b) $K_1 e^{(-1+\sqrt{8})x} + K_2 e^{(-1-\sqrt{8})x}$

(c) $K_1 e^{(-2+\sqrt{6})x} + K_2 e^{(-2-\sqrt{6})x}$

(d) $K_1 e^{(-2+\sqrt{8})x} + K_2 e^{(-2-\sqrt{8})x}$

[EC GATE-2017]

92. Consider the differential equation $3y''(x) + 27y(x) = 0$ with initial conditions $y(0) = 0$ and $y'(0) = 2000$. The value of y at $x = 1$ is _____.

[ME GATE 2017]

93. The solution (up to three decimal places) at $x = 1$ of the differential equation $\frac{d^2y}{dx^2} + 2\frac{dy}{dx} + y = 0$ subject to boundary conditions $y(0) = 0$ and $\frac{dy}{dx}\big|_{x=0} = -1$ is _____.

[CE GATE 2018]

94. The position of a particle $y(t)$ is described by the differential equation $\dfrac{d^2y}{dt^2} = -\dfrac{dy}{dt} + \dfrac{5y}{4}$.

The initial conditions are $y(0) = 1$ and $\dfrac{dy}{dt}\Big|_{t=0} = 0$.

The position (accurate to two decimal places) of the particle at $t = \pi$ is _____.

[EC GATE 2018]

95. Consider the differential equation $2\dfrac{d^2y}{dt^2} + 8y = 0$ with initial conditions $y(0) = 0$ and $\dfrac{dy}{dt}\Big|_{t=0} = 10$. Then the value of y (correct upto two decimal places) at $t = 1$ is _____.

[PI GATE 2018]

Answer Key

1. (c)	2. (c)	3. (b)	4. (b)	5. (b)
6. (a)	7. (c)	8. (b)	9. (a)	10. (b)
11. (a)	12. (a)	13. (a)	14. (b)	15. (c)
16. (d)	17. (b)	18. (d)	19. (a)	20. (b)
21. (a)	22. (a)	23. (a)	24. (c)	25. (d)
26. (d)	27. (c)	28. (d)	29. (a)	30. (d)
31. (d)	32. (a)	33. (d)	34. (a)	35. (b)
36. (c)	37. (d)	38. (c)	39. (c)	40. (c)
41. 1.652		42. (d)	43. (a)	44. 5.5
45. (d)	46. (c)	47. (a)	48. (c)	49. (c)
50. $\cos x + x$	51. (a)	52. (c)	53. (d)	54. (b)
55. (c)	56. (b)	57. (b)	58. (a)	59. (a)
60. (d)	61. (a)	62. (a)	63. (a)	64. (c)
65. (d)	66. (b)	67. (c)	68. (c)	69. (a)
70. (b)	71. (d)	72. (c)	73. (a)	74. 35
75. (b)	76. (c)	77. (a)	78. 18	79. –3
80. (b)	81. (c)	82. (c)	83. (0.8566)	
84. (a)	85. (–1)	86. (7.38)	87. (a)	88. (b)
89. (a)	90. (a)	91. (a)	92. (94.08)	
93. (–0.37)		94. (–0.21)		95. 4.54

Explanations

1. (c)

$$\frac{d^2y}{dx^2} = 3x - 2$$

$$\Rightarrow \int \frac{d^2y}{dx^2}dx = \int (3x-2)\,dx + C$$

$$\Rightarrow \int \frac{d}{dx}\left(\frac{dy}{dx}\right)dx = 3\frac{x^2}{2} - 2x + C$$

$$\Rightarrow \int d\left(\frac{dy}{dx}\right) = 3\frac{x^2}{2} - 2x + C$$

$$\Rightarrow \frac{dy}{dx} = \frac{3x^2}{2} - 2x + C \quad\text{............(i)}$$

Now $y'(1) = -3 \Rightarrow -3 = \dfrac{3}{2} - 2 + C$ (from (i))

$$\Rightarrow C = -\frac{5}{2}$$

$\therefore$ (i) gives, $\dfrac{dy}{dx} = \dfrac{3x^2}{2} - 2x - \dfrac{5}{2}$

Integrating both sides with respect to "x," we get,

$$\int \frac{dy}{dx}dx = \int \left\{\frac{3x^2}{2} - 2x - \frac{5}{2}\right\}dx + K$$

or, $\int dy = \dfrac{x^3}{2} - x^2 - \dfrac{5x}{2} + K$

or, $y = \dfrac{x^3}{2} - x^2 - \dfrac{5x}{2} + K \quad\text{............(ii)}$

Now $y(0) = 2$ gives,

$2 = 0 - 0 - 0 + K$, i.e, $k = 2$.

So from (ii) we have, $y = \dfrac{x^3}{2} - x^2 - \dfrac{5x}{2} + 2$.

2. (c) The highest order derivative occurring in the equation is $\dfrac{d^2y}{dy^2}$. So order of the equation = 2. Also the power of the highest order derivative, i.e, the power of $\dfrac{d^2y}{dy^2}$ is "2." Hence, degree of the equation = 2.

3. (b) The highest order derivative occurring in the equation is $\dfrac{d^2y}{dy^2}$. So order of the equation = 2. Also the power of the highest order derivative, i.e, the power of $\dfrac{d^2y}{dy^2}$ is "1." Hence, degree of the equation = 1.

4. (b) The highest order derivative occurring in the equation is $\dfrac{d^2x}{dt^2}$. So degree of the equation = power of the highest order derivative = power of $\dfrac{d^2x}{dt^2}$ = 1.

5. (b) The highest order derivative occurring in the equation is $\dfrac{d^2y}{dt^2}$. So order of the equation = 2.

6. (*a*)

$$\frac{d^3y}{dx^3} + 4\sqrt{\left(\frac{dy}{dx}\right)^3 + y^2} = 0$$

$$\Rightarrow \frac{d^3y}{dx^3} + 16\left\{\left(\frac{dy}{dx}\right)^3 + y^2\right\}, \text{ which is radical free.}$$

∴ order of the equation = 3 (since highest order derivative occurring in the equation is $\frac{d^3y}{dx^3}$).

The degree of the equation is "2" (since power of highest order derivative is "2").

7. (*c*)

$$y = 5\sin\left(3x + \frac{\pi}{3}\right)$$

$$\Rightarrow \frac{dy}{dx} = 15\cos\left(3x + \frac{\pi}{3}\right)$$

$$\Rightarrow \frac{d^2y}{dx^2} = -45\sin\left(3x + \frac{\pi}{3}\right)$$

$$\Rightarrow \frac{d^2y}{dx^2} = -9y \Rightarrow \frac{d^2y}{dx^2} + 9y = 0$$

8. (*b*)

$$\frac{d^3f}{dn^3} + \frac{f}{2}\frac{d^2f}{dn^2} = 0, \text{ is third-order ordinary differ-}$$

ential equation (since highest order derivative occurring in the equation is $\frac{d^3f}{dn^3}$) and it is non-linear (since the product $f \times \frac{d^2f}{dn^2}$ is not allowed in case of linear differential equation).

9. (*a*)

$$\frac{dy}{dx} + y^2 = 0 \Rightarrow \frac{dy}{y^2} = -dx$$

(by separation of variables)

Integrating, we get

$$\int \frac{dy}{y^2} = -\int dx + K$$

or, $-\dfrac{1}{y} = -x + K$

or, $\dfrac{1}{y} = x - K = x + C$ (where $C = -K$)

or, $y = \dfrac{1}{x + C}$.

10. (*b*) $\dfrac{dx}{dt} + kx^2 = 0 \Rightarrow \dfrac{dx}{x^2} = -k\,dt$

(by separation of variables)

Integrating, we get

$$\int \frac{dx}{x^2} = -k\int dt + C$$

or, $-\dfrac{1}{x} = -kt + C$ ....(i)

At $x = a, t = 0$. So (*i*) gives,

$$-\frac{1}{a} = -k \times 0 + C$$

or, $C = -\dfrac{1}{a}$

Thus, we get,

$$-\frac{1}{x} = -kt - \frac{1}{a} \quad (from\ (i))$$

or, $\dfrac{1}{x} = kt + \dfrac{1}{a}$

11. (*a*)

$$\frac{dy}{dt} + p(t)y = q(t)y^n$$

$$\Rightarrow \frac{1}{y^n}\frac{dy}{dt} + \frac{1}{y^{n-1}}p(t) = q(t)$$

$$\Rightarrow y^{-n}\frac{dy}{dt} + y^{1-n}p(t) = q(t) \text{........(i)}$$

Let us put $v = y^{1-n}$. Then we have

$$\frac{dv}{dt} = (1-n)y^{-n}\frac{dy}{dt}$$

or, $y^{-n}\dfrac{dy}{dt} = \dfrac{1}{(1-n)}\dfrac{dv}{dt}$

Substituting this in (*i*), we get,

$$\frac{1}{(1-n)}\frac{dv}{dt} + vp(t) = q(t)$$

or, $\dfrac{dv}{dt} + (1-n)vp(t) = (1-n)q(t)$

12. (*a*)

$$x'(t) = -3x(t)$$

$$\Rightarrow \frac{dx}{dt} = -3x \Rightarrow \frac{dx}{x} = -3dt$$

Integrating, we get,

$$\int \frac{dx}{x} = -3\int dt + \ln C$$

(where $\ln C$ is the constant of integration)

or, $\ln x = -3t + \ln C$

or, $\ln x - \ln C = -3t$

or, $\ln\dfrac{x}{C} = -3t$ *i.e;* $x = Ce^{-3t}$ ....(i)

Given that $x(0) = x_0$. so (i) gives,

$$x_0 = Ce^0 \text{ } i.e; \text{ } C = x_0$$

Hence, $x = x_0e^{-3t}$ ..

13. (*a*) If V be the volume and A be the surface area of the ball, then ATQ,

$$\frac{dV}{dt} = -kA \qquad \qquad ...(i)$$

Here $V = \frac{4}{3}\pi r^3$, $A = 4\pi r^2$.

$$\therefore \frac{dV}{dt} = 4\pi r^2 \frac{dr}{dt}$$

Substituting this in (i) we get,

$$4\pi r^2 \frac{dr}{dt} = -k \times 4\pi r^2$$

$$or, \frac{dr}{dt} = -k$$

$$or, \ dr = -kdt$$

Integrating, we get,

$$r = -kt + C \qquad \qquad ...(ii)$$

At $t = 0$, $r = 1$. So (ii) gives,

$$1 = -k \times 0 + C \quad i.e., C = 1.$$

So $r = -kt + 1 \qquad \qquad ...(iii)$

Now at $t = 3$ months, $r = 0.5$ cm; we get from (iii),

$$0.5 = -k \times 3 + 1 \quad i.e; \ k = \frac{1}{6}.$$

Now substituting this value of r in equation (iii) we get,

$$r = -\frac{1}{6}t + 1 \qquad \qquad ...(iv)$$

When the ball completely evaporates, we have $r = 0$. In this case, we get from (iv),

$$0 = -\frac{1}{6}t + 1, \ i.e, t = 6 \ months$$

14. (*b*) The given equation is

$$\frac{dy}{dx} + 2xy = e^{-x^2}, \text{ which is linear in } y.$$

Integrating factor, I.F $= e^{\int Pdx} = e^{\int 2xdx} = e^{x^2}$.

Hence, the solution is given by

$$y \times e^{x^2} = \int e^{x^2} \times e^{-x^2} dx + C$$

$$or, ye^{x^2} = x + C..........(i)$$

Given that at $x = 0$, $y = 1$. So (i) gives,

$$1e^0 = 0 + C \ i.e., C = 1$$

Hence, the required particlular solution is

$$ye^{x^2} = x + 1 \ i.e; \ y = e^{-x^2}(x+1).$$

15. (*c*)

$$\frac{dy}{dx} = y^2$$

$$\Rightarrow \int \frac{dy}{y^2} = \int dx + C$$

$$\Rightarrow -\frac{1}{y} = x + C$$

$$\therefore y = -\frac{1}{x + C} \qquad \qquad ...(i)$$

When $x = 0$, $y = 1$; So (i) gives

$$1 = -\frac{1}{0 + C} \ i.e; C = -1$$

Hence, $y = -\frac{1}{x - 1}$

Now y is bounded when $x - 1 \neq 0$, $i.e$, when $x \neq 1$, $i.e$, when $x < 1$ or $x > 1$.

16. (*d*)

$$\frac{dy}{dx} = x^2 y$$

$$\Rightarrow \frac{dy}{y} = x^2 dx$$

$$\Rightarrow \int \frac{dy}{y} = \int x^2 dx + C_1$$

$$\Rightarrow \log_e y = \frac{x^3}{3} + C_1$$

$$\Rightarrow y = e^{\frac{x^3}{3} + C_1} = e^{C_1} \times e^{\frac{x^3}{3}}$$

$$\Rightarrow y = C \times e^{\frac{x^3}{3}} \ (\text{where } C = e^{C_1}) \qquad ...(i)$$

Given $x = 0$, $y = 1$; so (i) gives

$$1 = C \times e^{\frac{0}{3}}$$

$$i.e; C = 1$$

$$\therefore \quad y = e^{\frac{x^3}{3}} \text{ is the required solution.}$$

17. (*b*) By Newton's law of cooling, we have

$$\frac{d\theta}{dt} = -k(\theta - \theta_0)$$

$$or, \frac{d\theta}{\theta - \theta_0} = -kdt$$

$$or, \int \frac{d\theta}{\theta - \theta_0} = \int -kdt + C_1$$

$$or, ln \ (\theta - \theta_0) = -kt + C_1$$

$$or, \theta - \theta_0 = Ce^{-kt} \ (\text{where } C = e^{C_1})$$

$$or, \theta = \theta_0 + Ce^{-kt}$$

Given that, $\theta_0 = 25°C$

Now at $t = 0$, $\theta = 60°$; so from (i) we get

$$60 = 25 + C.e^0$$

$$\Rightarrow C = 35$$

$$\therefore \quad \theta = 25 + 35e^{-kt} \qquad \qquad ...(ii)$$

Again at $t = 15$ minutes, $\theta = 40°C$. So from (ii) we have

$$\therefore \quad 40 = 25 + 35e^{(-k \times 15)}$$

$$\Rightarrow e^{-15k} = \frac{3}{7}$$

Now at $t = 30$ minutes, we get

$$\theta = 25 + 35e^{-30k} = 25 + 35(e^{-15k})^2$$

$$= 25 + 35 \times \left(\frac{3}{7}\right)^2 = 31.4°C \approx 31.5°C$$

18. (d)

$$\frac{dy}{dx} = \frac{-x}{y}$$

$$\Rightarrow y\,dy = -x\,dx$$

$$\Rightarrow \int y\,dy = -\int x\,dx + C$$

$$\Rightarrow \frac{y^2}{2} = -\frac{x^2}{2} + C \qquad \qquad \dots(i)$$

Given $x = 1$, $y = \sqrt{3}$; so (i) given

$$\frac{\left(\sqrt{3}\right)^2}{2} = \frac{-1^2}{2} + C \ i.e; \ C = 2$$

The required solution is $\dfrac{y^2}{2} = \dfrac{-x^2}{2} + 2$,

i.e., $x^2 + y^2 = 4$.

19. (a)

$$\frac{dy}{dx} = 1 + y^2$$

$$\Rightarrow \frac{dy}{1 + y^2} = dx$$

Integrating, we get,

$\tan^{-1}(y) = x + c$

or, $y = \tan(x + c)$,

where c can take any arbitrary value.

20. (b)

$$\frac{dx}{dt} = -3x$$

$$\Rightarrow \frac{dx}{x} = -3dt$$

$$\Rightarrow \int \frac{dx}{x} = \int -3dt + C$$

$$\Rightarrow x = t + c$$

$$\Rightarrow x = e^{-3t+c}$$

$$\Rightarrow x = e^c.e^{-3t} = c_1e^{-3t}(c_1 = e^c)$$

For $C_1 = 2$, $x = 2e^{-3t}$.

21. (a)

A. $\dfrac{dy}{dx} = \dfrac{y}{x}$

$$\Rightarrow \frac{dy}{y} = \frac{dx}{x}$$

$$\Rightarrow \int \frac{dy}{y} = \int \frac{dx}{x} + \log c$$

$$\Rightarrow \log y = \log x + \log c = \log cx$$

$\Rightarrow y = cx$, which represents a family of straight-lines.

B. $\dfrac{dy}{dx} = \dfrac{-y}{x}$

$$\Rightarrow \frac{dy}{y} = \frac{-dx}{x} \Rightarrow \int \frac{dy}{y} = -\int \frac{dx}{x} + \log c$$

$$\Rightarrow \log y = -\log x + \log c$$

$$\Rightarrow \log y + \log x = \log c$$

$$\Rightarrow \log yx = \log c$$

$$\Rightarrow yx = c$$

$\Rightarrow y = c/x$, which represents a family of hyperbolas.

C. $\dfrac{dy}{dx} = \dfrac{x}{y}$

$$\Rightarrow ydy = xdx$$

$$\Rightarrow \int ydy = \int xdx + \frac{C^2}{2}$$

$$\Rightarrow \frac{y^2}{2} = \frac{x^2}{2} + \frac{C^2}{2}$$

$$\Rightarrow y^2 - x^2 = e^2$$

$$\Rightarrow \frac{y^2}{c^2} - \frac{x^2}{c^2} = 1 \text{ which represents a family of hyperbolas.}$$

D. $\dfrac{dy}{dx} = \dfrac{-x}{y}$

$$\Rightarrow \int ydy = -\int xdx + \frac{c^2}{2}$$

$$\Rightarrow \frac{y^2}{2} + \frac{x^2}{2} = \frac{c^2}{2}$$

$\Rightarrow x^2 + y^2 = c^2$ which represents a family of a circle.

22. (a)

$$3y\frac{dy}{dx} + 2x = 0$$

$$\Rightarrow 3\,ydy = -2x\,dy$$

$$\Rightarrow \int 3y\,dy = -\int 2x\,dx + \frac{C}{2}$$

$$\Rightarrow \frac{3}{2}y^2 = -2\frac{x^2}{2} + \frac{C}{2}$$

$$\Rightarrow 3\,y^2 + 2x^2 = C$$

$$\Rightarrow \frac{x^2}{\left(\sqrt{\frac{C}{2}}\right)^2} + \frac{y^2}{\left(\sqrt{\frac{C}{3}}\right)^2} = 1, \text{ which represents a}$$

family of ellipse.

23. (a) $x\dfrac{dy}{dx}+y=x^4$

$\Rightarrow \dfrac{dy}{dx}+\left(\dfrac{y}{x}\right)=x^3$ ...(i)

which is a liner equation of the form

$\dfrac{dy}{dx}+Py=Q$

Here, $P=\dfrac{1}{x}$ and $Q=x^3$

∴ integrating factor (I.F.) $=e^{\int Pdx}$

$=e^{\int\frac{1}{x}dx}=e^{Inx}=x$

So the solution is given

$y\times I.F.=\int Q\times(I.F.)dx+C$

or, $y\times x=\int x^3\times x\ dx+C$

or, $yx=\dfrac{x^5}{5}+c$ ...(ii)

Given that $y(1)=\dfrac{6}{5}$. So we get from (ii)

$\dfrac{6}{5}\times 1=\dfrac{1}{5}+C$

$\Rightarrow C=\dfrac{6}{5}-\dfrac{1}{5}=1$

Hence, we have $yx=\dfrac{x^5}{5}+1$ i.e; $y=\dfrac{x^4}{5}+\dfrac{1}{x}$.

24. (c)

Given equation is $\dfrac{dy}{dx}+y=e^x$,

which is linear in y.

∴ I. F $=e^{\int dx}=e^x$

Hence, the general solution of the equation is

$y\times I.F=\int(I.F)e^x dx+C$

or, $ye^x=\int e^x e^x dx+C=\int e^{2x}dx+C$

or, $ye^x=\dfrac{e^{2x}}{2}+C$ ...(i)

Now using $y(0)=1$, we get from (i) $C=\dfrac{1}{2}$.

Thus, the solution is given by

$ye^x=\dfrac{e^{2x}}{2}+\dfrac{1}{2}$.

∴ $y(1)=\dfrac{e}{2}+\dfrac{e^{-1}}{2}=\dfrac{e+e^{-1}}{2}$.

25. (d) $\dfrac{dy}{dx}-y^2=1$

$\Rightarrow \dfrac{dy}{dx}=1+y^2$

$\Rightarrow \dfrac{dy}{1+y^2}=dx$

$\Rightarrow \int\dfrac{dy}{1+y^2}=\int dx+c$

$\Rightarrow \tan^{-1}(y)=x+c$ ...(i)

Given $y(0)=1$, so (i) gives, $c=\dfrac{\pi}{4}$.

∴ required solution is $y=\tan\left(x+\dfrac{\pi}{4}\right)$

26. (d)

$\dfrac{dy}{dx}=(1+y^2)x$

$\Rightarrow \int\dfrac{dy}{1+y^2}=\int xdx+C$

$\Rightarrow \tan^{-1}y=\dfrac{x^2}{2}+C$

$\Rightarrow y=\tan\left(\dfrac{x^2}{2}+C\right)$.

27. (c)

$\dfrac{dy}{dx}=ky$

$\Rightarrow \dfrac{dy}{y}=kdx$

$\Rightarrow \int\dfrac{dy}{y}=k\int dx+\log C$

$\Rightarrow \log y=kx+\log C$ ...(i)

Which satisfies $y(0)=C$

Now, (i) $\Rightarrow$

$\log y-\log C=kx$

or, $\log\dfrac{y}{C}=kx$

or, $\dfrac{y}{C}=e^{kx}$

or, $y=Ce^{kx}$

28. (d)

The equation $\dfrac{dy}{dx}+\dfrac{y}{x}=x$ is a liner differential

equation of the form $\dfrac{dy}{dx}+Py=Q$ with $P=\dfrac{1}{x}$ and $Q=x..$

Here, integrating factor (I. F)

$=e^{\int Pdx}=e^{\int\frac{1}{x}dx}=e^{lnx}=x$

Therefore, the solution is given by

$$y \times (I.F.) = \int Q \times (I.F.) dx + C$$

$$\Rightarrow y.x = \int (x.x) dx + c \quad C$$

$$\Rightarrow yx = \int x^2 dx + C$$

$$\Rightarrow yx = \frac{x^2}{3} + C$$

$$\Rightarrow y = \frac{x^3}{3} + \frac{C}{x} \qquad \ldots(i)$$

For $y(1) = 1$, we get from (i),

$$\frac{1^2}{3} + \frac{C}{1} = 1$$

$i.e; \ C = 2/3$

So, the required is solution is $y = \frac{x^2}{3} + \frac{2}{3x}$.

29. (a)

$$\frac{dy}{dx} = e^{-3x}$$

$$\Rightarrow \int dy = \int e^{-3x} dx + k \quad K$$

$$\Rightarrow y = \frac{e^{-3x}}{-3} + K = -\frac{1}{3} e^{-3x} + K$$

30. (d)

$$t\frac{dx}{dt} + x = t$$

$$\Rightarrow \frac{dx}{dt} + \frac{x}{t} = 1$$

which is a liner differential equation of the form $\frac{dx}{dt} + Px = Q$ where $P = \frac{1}{t}$ and $Q = 1$.

∴ Integrating factor (I. F.)

$$= e^{\int Pdt = e^{\int \frac{1}{t} dt}} = e^{\log_e t} = t.$$

Hence, the solution is given by

$$x \times (I.F.) = \int Q \times (I.F.) dt + C$$

or, $x.t = \int 1.t.dt + C$

or, $xt = \frac{t^2}{2} + C$

or, $x = \frac{t}{2} + \frac{C}{t} \qquad \ldots(i)$

Given $x(1) = \frac{1}{2}$; so (i) gives

$$\frac{1}{2} + \frac{C}{1} = \frac{1}{2}$$

$$\Rightarrow C = 0$$

Hence, $x = \frac{t}{2}$ is the required solution.

31. (d)

$$\frac{dy}{dx} + 2y = 0$$

$$\Rightarrow \frac{dy}{dx} = -2y$$

$$\Rightarrow \int \frac{dy}{y} = \int -2 dx + \log C$$

$$\Rightarrow \log y = -2x + \log C$$

$$\Rightarrow \log y - \log C = -2x$$

$$\Rightarrow \log \frac{y}{C} = -2x$$

$$\Rightarrow \frac{y}{C} = e^{-2x}$$

$$\Rightarrow y = Ce^{-2x} \qquad \ldots(i)$$

∴ $y(1) = 5 \Rightarrow 5 = Ce^{-2}$

$i.e., \ C = 5e^2 = 36.95$

Hence, (i) gives $y = 36.95e^{-2x}$.

32. (a)

$$\frac{dx}{dt} = 3x - 5y \text{ and } \frac{dx}{dt} = 4x + 8y$$

So $\dfrac{d}{dt}\begin{Bmatrix} x \\ y \end{Bmatrix} = \begin{bmatrix} 3 & -5 \\ 4 & 8 \end{bmatrix}\begin{Bmatrix} x \\ y \end{Bmatrix}$.

33. (d)

Let us put $z = x + y$

Then, $\dfrac{dz}{dx} = 1 + \dfrac{dy}{dx}$.

∴ $\dfrac{dy}{dx} = \cos(x + y)$

$$\Rightarrow \frac{dz}{dx} - 1 = \cos z$$

$$\Rightarrow \int \frac{dz}{1 + \cos z} = \int dx + C$$

$$\Rightarrow \frac{1}{2} \int \sec^2\left(\frac{z}{2}\right) dz = x + c$$

$$\Rightarrow \tan\left(\frac{z}{2}\right) = x + c$$

$$\Rightarrow \tan\left(\frac{x + y}{2}\right) = x + c.$$

34. (a) We know that the general form of linear differential equation in "y" is $\dfrac{dy}{dx} + Py = Q$, where P and Q are functions of x or constants. Here only option (a) is in the above form.

35. (b)

$$\frac{dy}{dx} = -2xy \Rightarrow \frac{dy}{dx} + 2xy = 0.$$

I.F. $= e^{\int 2x\,dx} = e^{x^2}$

So, the solution is given by

$y \times I.F = \int I.F \times 0\,dx + C$

or, $ye^{x^2} = 0 + C = C$

Now $y(0) = 2 \Rightarrow 2 \times e^0 = C \Rightarrow C = 2$

Hence, $ye^{x^2} = 2$ *i.e*; $y = 2e^{-x^2}$.

36. (c) $x(y\,dx + x\,dy)\cos\dfrac{y}{x} = y(x\,dy - y\,dx)\sin\dfrac{y}{x}$

$\Rightarrow \dfrac{y\,dx + x\,dy}{x\,dy - y\,dx} = \dfrac{y}{x}\tan\dfrac{y}{x}$ ...(i)

Let $y = v\,.\,x$ Then $dy = v\,dx + x\,dv$. So (i) gives,

$\dfrac{v x\,dx + v x\,dx + x^2\,dv}{v x\,dx + x^2\,dv - v x\,dx} = v\tan v$

or, $\dfrac{x\,dv + 2v\,dx}{x\,dv} = v\tan v$

or, $1 + \dfrac{2v}{x}\dfrac{dx}{dv} = v\tan v$

or, $\dfrac{2v}{x}\dfrac{dx}{dv} = v\tan v - 1$

or, $2\dfrac{dx}{x} = \left(\tan v - \dfrac{1}{v}\right)dv$

or, $2\int\dfrac{dx}{x} = \int\left(\tan v - \dfrac{1}{v}\right)dv + \log c$

or, $2\log x = \log|\sec v| - \log v + \log c$

or, $\log x^2 = \log\left(\dfrac{c\sec v}{v}\right)$

$\Rightarrow x^2 = \dfrac{c\sec v}{v}$

$\Rightarrow x^2 = \dfrac{cx}{y} = \sec\left(\dfrac{y}{x}\right)$ (since $v = y/x$)

$\Rightarrow xy\cos\dfrac{y}{x} = c$

37. (d) $6yy' - 25x = 0$

$\Rightarrow 6y\dfrac{dy}{dx} = 25x$

$\Rightarrow y\,dy = \dfrac{25}{6}x\,dx \Rightarrow \int y\,dy = \dfrac{25}{6}\int x\,dx - \dfrac{C^2}{12}$

$\Rightarrow \dfrac{y^2}{2} = \dfrac{25}{6}.\dfrac{x^2}{2} - \dfrac{C^2}{12}$

$\Rightarrow 6y^2 = 25x^2 - C^2$

$\Rightarrow 25x^2 - 6y^2 = C^2$

$\Rightarrow \dfrac{x^2}{\left(\dfrac{C}{5}\right)^2} - \dfrac{y^2}{\left(\dfrac{C}{\sqrt{6}}\right)^2} = 1,$

which represents the family of hyperbolas.

38. (c)

$\dfrac{dy}{dt} = -5y \Rightarrow \int\dfrac{dy}{y} = -\int 5\,dt \Rightarrow \ln y = -5t + C$ ...(i)

Given that at $t = 0, y = 2$; so (i) gives, $\ln 2 = C$.

Thus, we get,

$\ln y = -5t + \ln 2$

or, $\ln\dfrac{y}{2} = -5t$ or, $\dfrac{y}{2} = e^{-5t}$ or, $y = 2e^{-5t}$.

Hence, at $t = 3$, $y = 2e^{-15}$.

39. (c)

$\dfrac{dy}{dx} = \dfrac{1 + \cos 2y}{1 - \cos 2x}$

$\Rightarrow \dfrac{dy}{1 + \cos 2y} = \dfrac{dx}{1 - \cos 2x}$

$\Rightarrow \dfrac{dy}{2\cos^2 y} = \dfrac{dx}{2\sin^2 x}$

$\Rightarrow \sec^2 y\,dy = \cos ec^2 x\,dx$

Integrating, we get

$\tan y = -\cot x + c$

or, $\tan y + \cot x = c$.

40. (c)

$\dfrac{dx}{dt} = 10 - 0.2x$

$\Rightarrow \dfrac{dx}{10 - 0.2x} = dt$

$\Rightarrow \int\dfrac{dx}{10 - 0.2x} = \int dt + c$

$\Rightarrow \dfrac{\log(10 - 0.2x)}{(-0.2)} = t + c$

$\therefore x(0) = 1 \Rightarrow c = \dfrac{\log 9.8}{(-0.2)}$

Hence, $\dfrac{\log(10 - 0.2x)}{(-0.2)} - \dfrac{\log 9.8}{(-0.2)} = t$

$\Rightarrow \log\left(\dfrac{10 - 0.2x}{9.8}\right) = -0.2t$

$\Rightarrow \dfrac{10 - 0.2x}{9.8} = e^{-0.2t}$

$\Rightarrow \dfrac{5 - 0.1x}{4.9} = e^{-0.2t} \Rightarrow \dfrac{50 - x}{49} = e^{-0.2t}$

$\Rightarrow x = x(t) = 50 - 49e^{-0.2t}$.

41. 1.652

$\dfrac{di}{dt} - 0.2i = 0$

$\Rightarrow \dfrac{di}{i} = 0.2\,dt$

$$\Rightarrow \int \frac{di}{i} = \int 0.2dt + \log C$$

$$\Rightarrow \log i = 0.2t + \log C$$

$$\Rightarrow \log i - \log C = 0.2t$$

$$\Rightarrow \log\left(\frac{i}{C}\right) = 0.2t$$

$$\Rightarrow \frac{i}{C} = e^{0.2t} \Rightarrow i = Ce^{0.2t} \qquad \ldots(i)$$

Given $i(4) = 10$; so (i) gives,

$$10 = Ce^{(0.2)4}$$

or, $10 = C(2.225)$

$\therefore C = 4.493$

Hence, $i = (4.493)e^{0.2t}$ and so

at $t = -5$, $i = 4.493e^{-10} = 1.652$.

42. (d)

$$\frac{dy}{dx} = (x + y - 1)^2 \qquad \ldots(i)$$

Let $x + y - 1 = t$ $\ldots(ii)$

Then, $1 + \dfrac{dy}{dx} = \dfrac{dt}{dx}$ and so

$$\frac{dy}{dx} = \frac{dt}{dx} - 1 \qquad \ldots(iii)$$

Now using (ii) and (iii) , we get from (i),

$$\frac{dt}{dx} - 1 = t^2$$

or, $\dfrac{dt}{t^2 + 1} = dx$

Integrating, we get,

$$\int \frac{1}{t^2 + 1} dt = \int dx + C$$

or, $\tan^{-1} t = x + c$

or, $\tan^{-1}(x + y - 1) = x + c$

or, $x + y - 1 = \tan(x + c)$

or, $y = 1 - x + \tan(x + c)$

43. (a)

$$\left(t^2 - 81\right)\frac{dy}{dt} + 5ty = \sin t$$

$$\Rightarrow \frac{dy}{dt} + \frac{5t}{t^2 - 81} y = \frac{\sin t}{t^2 - 81} \qquad \ldots(i)$$

which is of the form $\dfrac{dy}{dt} + Py = Q$.

Here, $P = \dfrac{5t}{t^2 - 81}$ and $Q = \dfrac{\sin t}{t^2 - 81}$.

So I.F

$$= e^{\int Pdt} = e^{\int \frac{5t}{t^2 - 81} dt} = e^{\frac{5}{2}\int \frac{2t\,dt}{t^2 - 81}} = e^{\frac{5}{2}\log(t^2 - 81)}$$

$$= e^{In(t^2 - 81)^{5/2}} = (t^2 - 81)^{\frac{5}{2}}$$

Hence, the solution is given by

$$y \times I.F = \int I.F \times Q\,dt + C$$

$$y(t^2 - 81)^{\frac{5}{2}} = \int \frac{\sin t}{t^2 - 81} \cdot (t^2 - 81)^{\frac{5}{2}} dt + C$$

or, $y = \dfrac{\int (\sin t)(t^2 - 81)^{3/2} dt}{(t^2 - 81)^{5/2}} + \dfrac{C}{(t^2 - 81)^{5/2}}$,

which exists for $t \neq -9, 9$.

Hence, option (a) is correct.

44. 5.5

Here $m = 2$ kg, $V_0 = 1.5$ m/sec. Then,

$$\int_0^t f(t)dt = m(V - V_0)$$

$$\Rightarrow \int_0^2 3t^2 dt = 2(V - 1.5)$$

$$\Rightarrow \left[t^3\right]_0^2 = 2(V - 1.5)$$

$$\Rightarrow (8 - 0) = 2(V - 1.5)$$

$$\Rightarrow V = 5.5 \text{ m/s}$$

45. (d)

The given equation is $\dfrac{dQ}{dt} + Q = 1$

$\therefore$ I.F. $= e^{\int 1dt} = e^t$

So the solution is given by

$$Q.e^t = \int 1.e^t dt + C = e^t + C$$

or, $Q = 1 + Ce^{-t}$ $\ldots(i)$

When $t = 0, Q = 0$; so equation $(i) \Rightarrow 0 = 1 + C$

$\Rightarrow C = -1$

Hence, $Q(t) = 1 - e^{-t}$.

46. (c)

$$x\frac{dy}{dx} + y = 0$$

$$\Rightarrow xdy + ydx = 0 \Rightarrow d(xy) = 0.$$

Integrating both sides, we get, $xy = C$ $\ldots(i)$

Given that the solution passes through the point $(1, 1)$. So (i) gives $C = 1$.

Hence, the required solution is $xy = 1$ *i.e*;

$y = \dfrac{1}{x} = x^{-1}$.

47. (a)

The given differential equation is

$$\frac{dy}{dx} = \frac{x^2 + y^2}{2y} + \frac{y}{x}$$

or, $2y\dfrac{dy}{dx} - \left(1 + \dfrac{2}{x}\right)y^2 = x^2 \ldots\ldots\ldots(i)$

Let us put $y^2 = z$. So $2y\dfrac{dy}{dx} = \dfrac{dz}{dx}$. Then (i) becomes,

$$\frac{dz}{dx} - \left(1 + \frac{2}{x}\right)z = x^2 \qquad \ldots(ii)$$

Clearly, equation (ii) is linear and it is integrating factor is given by

$$\text{I. F} = e^{-\int\left(1+\frac{2}{x}\right)dx} = e^{-x-2\log x} = e^{-x} \times e^{-2\log x}$$

$$= e^{-x} \times e^{\log x^{-2}} = \frac{e^{-x}}{x^2}.$$

Hence, the general solution of the equation (ii) is

$$z \times \text{I.F} = \int (\text{I.F})x^2 dx + C$$

or, $z\dfrac{e^{-x}}{x^2} = \int \dfrac{e^{-x}}{x^2} \times x^2 dx + C = \int e^{-x}dx + C$

or, $y^2\dfrac{e^{-x}}{x^2} = -e^{-x} + C$

or, $\left(\dfrac{y^2}{x^2} + 1\right)e^{-x} = C$ ....(iii)

Now since the curve passes through $(1, 0)$, so $y(1) = 0$ and hence from (iii) we get,

$$\left(\frac{0}{1^2} + 1\right)e^{-1} = C \text{ i.e; } C = e^{-1}.$$

Consequently, the required solution $i.e$; the curve is

$$\left(\frac{y^2}{x^2} + 1\right)e^{-x} = e^{-1} \text{ i.e; } \frac{y^2}{x^2} + 1 = e^{x-1}$$

$i.e$; $\ln\left(\dfrac{y^2}{x^2} + 1\right) = x - 1.$

48. (c)

$$y^3\frac{dy}{dx} + x^3 = O$$

$$\Rightarrow y^3 dy = -x^3 dx \Rightarrow \int y^3 dy = -\int x^3 dx + \frac{C}{4}$$

$$\Rightarrow \frac{y^4}{4} = -\frac{x^4}{4} + \frac{C}{4} \Rightarrow y^4 = -x^4 + C........(i)$$

Now given that $y(0) = 1$. So (i) gives,

$1^4 = -0^4 + C \text{ i.e., } C = 1.$

Hence, (i) becomes, $y^4 = -x^4 + 1 \text{ i.e; } y = \left(1 - x^4\right)^{\frac{1}{4}}.$

$\therefore y(-1) = \left(1 - 1\right)^{\frac{1}{4}} = 0.$

49. (c) Given $\dfrac{d^2 y}{dx^2} = 3x - 2$

Integrating both sides w.r.t x, we get,

$$\frac{dy}{dx} = \frac{3}{2}x^2 - 2x + A........(i)$$

At $x = 1, \dfrac{dy}{dx} = -3$;

so $(i) \Rightarrow -3 = \dfrac{3}{2} - 2 + A \Rightarrow A = -\dfrac{5}{2}$

$\therefore \dfrac{dy}{dx} = \dfrac{3}{2}x^2 - 2x - \dfrac{5}{2}$

Again, integrating both sides w.r.t x we get,

$$y = \frac{x^3}{2} - x^2 - \frac{5}{2}x + B........(ii)$$

At $x = 0, y = 2$;

So $(ii) \Rightarrow = 2 = 0 - 0 - 0 + B \text{ i.e., } B = 2$

Thus, $y = \dfrac{x^3}{2} - x^2 - \dfrac{5x}{2} + 2.$

50. $y = \cos x + x$

The given equation is $\dfrac{d^2 y}{dx^2} + y = x$ $\qquad \ldots(i)$

Consider, $\dfrac{d^2 y}{dx^2} + y = 0$ $\qquad \ldots(ii)$

Let $y = e^{mx}$ be a trial solution of (ii). Then (ii) gives,

$$m^2 e^{mx} + e^{mx} = 0$$

$$\Rightarrow m^2 + 1 = 0$$

$$\Rightarrow m = 0 \pm i$$

Then, $y_{C.F} = c_1 \cos x + c_2 \sin x.$

$$y_{P.I.} = \frac{1}{(D^2 + 1)}x = (1 + D^2)^{-1}x$$

$$= (1 - D^2 + D^4 - .....)x$$

$$= x - D^2 x + D^4 x - .....$$

$$= x - 0 + 0 - ....$$

$$= x$$

Hence, the complete solution is

$y = c_1 \cos x + c_2 \sin x + x$

Now, $y(0) = 1 \Rightarrow 1 = c_1 + 0 + 0 \Rightarrow c_1 = 1$

Again, $y\left(\dfrac{\pi}{2}\right) = \dfrac{\pi}{2} \Rightarrow 0 + c_2 + \dfrac{\pi}{2} = \dfrac{\pi}{2} \Rightarrow c_2 = 0$

$\therefore \quad y = \cos x + x.$

51. (a)

$$\frac{d^2 y}{dx^2} + 2\frac{dy}{dx} + 17y = 0$$

$$\Rightarrow m^2 e^{mx} + 2me^{mx} + 17e^{mx} = 0$$

(taking $y = e^{mx}$ as a trial solution)

$$\Rightarrow m^2 + 2m + 17 = 0$$

$$\Rightarrow m = \frac{-2 \pm \sqrt{4 - 68}}{2} = -1 \pm 4i$$

Hence, the solution is given by

$$y = e^{-x}(c_1 \cos 4x + c_2 \sin 4x) \qquad \ldots(i)$$

$$\therefore \frac{dy}{dx} = -e^{-x}(c_1 \cos 4x + c_2 \sin 4x)$$

$$+ e^{-x}(-4c_1 \sin 4x + 4c_2 \cos 4x)\ldots\ldots\ldots(ii)$$

Since, $y(0) = 1$, so (i) gives,

$1 = e^0(c_1 \cos 0 + c_2 \sin 0) = c_1$ i.e., $c_1 = 1$.

Since, $\dfrac{dy}{dx}\left(\dfrac{\pi}{4}\right) = 0$, so (ii) gives,

$$0 = -e^{-\frac{\pi}{4}}(c_1 \cos \pi + c_2 \sin \pi)$$

$$+ e^{-\frac{\pi}{4}}(-4c_1 \sin \pi + 4c_2 \cos \pi)$$

or, $-c_1 \cos \pi + 4c_2 \cos \pi = 0$

or, $c_2 = \dfrac{c_1}{4} = \dfrac{1}{4}$.

Hence, the required solution is,

$$y = e^{-x}\left(\cos 4x + \frac{1}{4}\sin 4x\right)$$

52. (c)

$y = c_1 e^{-x} + c_2 e^{-3x}$ is the solution

$\Rightarrow (m+1)(m+3)=0$ is the auxiliary equation

$\Rightarrow m^2 + 4m + 3 = 0$ is the auxiliary equation

$\Rightarrow m^2 e^{mx} + 4m e^{mx} + 3e^{mx} = 0$

$\Rightarrow \dfrac{d^2 y}{dx^2} + 4\dfrac{dy}{dx} + 3y = 0$ $(taking\ y = e^{mx})$

Now comparing $\dfrac{d^2 y}{dx^2} + 4\dfrac{dy}{dx} + 3y = 0$ with

$\dfrac{d^2 y}{dx^2} + p\dfrac{dy}{dx} + qy = 0$, we get $p = 4$ and $q = 3$.

53. (d)

Consider $(D^2 - 4D + 4)y = 0$ $\qquad \ldots(i)$

Let $y = e^{mx}$ be a trial solution of (i). Then (i) gives,

$$m^2 e^{mx} - 4m e^{mx} + 4e^{mx} = 0$$

$$\Rightarrow m^2 - 4m + 4 = 0$$

$$\Rightarrow m = 2, 2.$$

Then, $y = y(x) = (c_1 + c_2 x)e^{2x}$

54. (b)

Given $x''(t) + 3x'(t) + 2x(t) = 5$ $\qquad \ldots(i)$

Consider $x''(t) + 3x'(t) + 2x(t) = 0$ $\qquad \ldots(ii)$

Let $x = e^{mt}$ be a trial solution of (ii). Then (ii) gives,

$$m^2 e^{mt} + 3m e^{mt} + 2e^{mt} = 0$$

$$\Rightarrow m^2 + 3m + 2 = 0$$

$$\Rightarrow m = -1, -2.$$

Then, $x = x(t) = c_1 e^{-t} + c_2 e^{-2t}$

$$x_{P.I.} = \frac{1}{(D^2 + 3D + 2)}5$$

$$= \frac{5}{2}\frac{1}{\left(1 + \dfrac{D^2 + 3D}{2}\right)} \quad (1)$$

$$= \frac{5}{2}\left(1 + \frac{D^2 + 3D}{2}\right)^{-1} \quad (1)$$

$$= \frac{5}{2}\left(1 - \frac{D^2 + 3D}{2}\right)(1) = \frac{5}{2}\left(1 - \frac{0+0}{2}\right) = \frac{5}{2}$$

$$\therefore x(t) = x_{C.F} + x_{P.I.} = c_1 e^{-t} + c_2 e^{-2t} + \frac{5}{2}$$

$$\Rightarrow x(t)\big|_{t \to \infty} = c_1 \times 0 + c_2 \times 0 + \frac{5}{2} = \frac{5}{2}.$$

55. (c)

The given equation is

$$\frac{d^2 y}{dx^2} + p\frac{dy}{dx} + (q+1) = 0 \qquad \ldots(i)$$

Let us take $p = 4$ and $q = 3$. Then (i) becomes,

$$\frac{d^2 y}{dx^2} + 4\frac{dy}{dx} + 4y = 0 \qquad \ldots(ii)$$

Let $y = e^{mx}$ be a trial solution of (ii). Then (ii) gives,

$$m^2 e^{mx} + 4m e^{mx} + 4e^{mx} = 0$$

$$\Rightarrow m^2 + 4m + 4 = 0$$

$$\Rightarrow m = -2, -2.$$

Then the solution is given by,

$y = (c_1 + c_2 x)e^{-2x}$

Therefore, e^{-2x} and xe^{-2x} are two solutions of (ii).

The solution $y = xe^{-2x}$ satisfies option (c).

56. (b)

The given equation is

$$\frac{d^2 y}{dx^2} - 5\frac{dy}{dx} + 6y = 0 \qquad \ldots(i)$$

Let $y = e^{mx}$ be a trial solution of (i). Then (i) gives,

$$m^2 e^{mx} - 5m e^{mx} + 6e^{mx} = 0$$

$$\Rightarrow m^2 - 5m + 6 = 0$$

$$\Rightarrow m = 2, 3.$$

Then the solution is given by,

$y = c_1 e^{2x} + c_2 e^{3x} = e^{2x} + e^{3x}$ for $C_1 = C_2 = 1$

57. (b)

$$\frac{d^2 y}{dx^2} + 4\frac{dy}{dx} + 3y = 3e^{2x} \quad (given)$$

$(D^2 + 4D + 3)y = 3e^{2x}$

Hence,

$$P.I = \frac{1}{(D^2 + 4D + 3)} 3e^{2x} = 3\frac{1}{(D^2 + 4D + 3)} e^{2x}$$

$$= 3\frac{1}{(2^2 + 4 \times 2 + 3)} e^{2x} = \frac{e^{2x}}{5}.$$

$\left[\text{Using } \frac{1}{f(D)} e^{ax} = \frac{1}{f(a)} e^{ax}\right]$

58. (a) The given equation is $y'' + 2y' + 101y = 10.4e^x$ ...(i)

Consider $y'' + 2y' + 101 y = 0$ ...(ii)

Let $y = e^{mx}$ be a trial solution of (ii). Then (ii) gives,

$m^2 e^{mx} + 2me^{mx} + 101e^{mx} = 0$

$\Rightarrow m^2 + 2m + 101 = 0$

$\Rightarrow m = \frac{-2 \pm \sqrt{4 - 404}}{2} = \frac{-2 \pm 20i}{2} = -1 \pm 10i$

Then $y_{C.F} = e^{-x}(c_1 \cos 10x + c_2 \sin 10x)$.

$y_{P.I.} = \frac{1}{(D^2 + 2D + 101)}(10.4e^x)$

$= \frac{10.4}{(1^2 + 2 + 101)} e^x = 0.1e^x$

So, the complete solution is

$y = y_{C.F} + y_{P.I}$

$= e^{-x}[c_1 \cos 10x + c_2 \sin 10x] + 0.1e^x$ ...(iii)

$\therefore \frac{dy}{dx} = -e^{-x}[c_1 \cos 10x + c_2 \sin 10x] + 0.1e^x$

$+ e^{-x}[-10c_1 \sin 10x + 10c_2 \cos 10x]$ ...(iv)

Now using $y(0) = 1.1$ and $y'(0) = -0.9$ we get from (iii) and (iv), $c_1 = 1$ and $c_2 = 0$

Hence, $y = e^{-x} \cos 10x + 0.1e^x$

So $P - 2, Q - 1, R - 3$ is the correct option.

59. (a) The given equation is $\frac{d^2y}{dx^2} + y = 0$ ...(i)

Let $y = e^{mx}$ be a trial solution of (i).

Then, (i)

$\Rightarrow m^2 e^{mx} + e^{mx} = 0 \Rightarrow m^2 + 1 = 0 \Rightarrow m = 0 \pm i$

Hence, the solution is given by

$y = e^0(P \cos x + Q \sin x) = P \cos x + Q \sin x$, where P and Q are constants.

60. (d) Given $\ddot{x} + 3x = 0$ i.e; $\frac{d^2x}{dt^2} + 3x = 0$ ...(i)

Then taking $x = e^{mt}$, we get from (i),

$\Rightarrow m^2 e^{mt} + 3e^{mt} = 0 \Rightarrow m^2 + 3 = 0 \Rightarrow m = 0 \pm i\sqrt{3}$

Hence, the solution is given by

$x = e^0(A \cos \sqrt{3}t + B \sin \sqrt{3}t)$

$= A \cos(\sqrt{3}t) + B \sin(\sqrt{3}t)$ ....(ii)

$\therefore \frac{dx}{dt} = \dot{x} = -A\sqrt{3} \sin(\sqrt{3}t) + B\sqrt{3} \cos(\sqrt{3}t)$ ....(iii)

At $t = 0, x = 1$; so from (ii) we get, $A = 1$.

At $t = 0, \dot{x} = 0$; so from (ii) we get, $B = 0$.

Thus, from (ii), we have $x = \cos \sqrt{3}t$.

Therefore, $x(1) = \cos \sqrt{3} = 0.99$.

61. (a)

Given $y'' + 2y' + y = 0$ ...(i)

Let $y = e^{mx}$ be a trial solution of (i). Then (i) gives,

$m^2 e^{mx} + 2me^{mx} + e^{mx} = 0$

$\Rightarrow m^2 + 2m + 1 = 0$

$\Rightarrow m = -1, -1.$

Then, the solution is given by

$y = (C_1 + C_2x)e^{-x}.$

Then,

$y(0) = 0 \Rightarrow 0 = (C_1 + C_2(0))e^{-0} \Rightarrow C_1 = 0$ and

$y(1) = 0 \Rightarrow 0 = (C_1 + C_2)e^{-1} \Rightarrow C_1 + C_2 = 0$

$\Rightarrow C_2 = 0$

$\therefore y = (0 + 0x)e^{-x} = 0$ is the solution. Consequently $y(0.5) = 0$.

62. (a)

Given equation is $\frac{d^2y}{dx^2} + 2\frac{dy}{dx} + 2y = 0$ ...(i)

Let $y = e^{mx}$ be a trial solution of (i).

Then, (i)

$\Rightarrow m^2 e^{mx} + 2me^{mx} + 2e^{mx} = 0$

$\Rightarrow m^2 + 2m + 2 = 0$

$\Rightarrow m = \frac{-2 \pm \sqrt{4 - 8}}{2} = -1 \pm i$

Hence, the general solution is given by

$y = C_1 e^{(-1+i)x} + C_2 e^{(-1-i)x}.$

So two solutions are $e^{(-1+i)x}$ and $e^{(-1-i)x}$.

63. (a)

The given equation is $\frac{d^2y}{dx^2} + P\frac{dy}{dx} + qy = 0$ ...(i)

Let $y = e^{mx}$ be a trial solution of (i).
Then, (i)
$$\Rightarrow m^2 e^{mx} + Pm e^{mx} + q e^{mx} = 0$$
$$\Rightarrow m^2 + Pm + q = 0$$
$$\Rightarrow m = \frac{-P \pm \sqrt{P^2 - 4q}}{2}$$

Thus if $P^2 - 4q > 0$, then we have real and distinct roots.

64. (c)

Given $\dfrac{d^2 y}{dx^2} + \dfrac{dy}{dx} - 6y = 0$ ...(i)

Let $y = e^{mx}$ be a trial solution of (i). Then (i) gives,

$$m^2 e^{mx} + m e^{mx} - 6 e^{mx} = 0$$
$$\Rightarrow m^2 + m - 6 = 0$$
$$\Rightarrow m = -3, 2.$$

$\therefore$ the required solution is $y = c_1 e^{-3x} + c_2 e^{2x}$.

65. (d)

Given $\dfrac{d^2 n(x)}{dx^2} - \dfrac{n(x)}{L^2} = 0$ ...(i)

Let $n = e^{mx}$ be a trial solution of (i). Then (i) gives,

$$m^2 e^{mx} - \frac{e^{mx}}{L^2} = 0$$
$$\Rightarrow m^2 - \frac{1}{L^2} = 0$$
$$\Rightarrow m = \pm \frac{1}{L}.$$

$\therefore$ The solution is given by $n(x) = C_1 e^{-\frac{1}{L}x} + C_2 e^{\frac{1}{L}x}$.
Then, $n(0) = K \Rightarrow C_1 + C_2 = K$ ...(ii)
Also, $n(\infty) = 0 \Rightarrow C_2 e^{\infty} = 0 \Rightarrow C_2 = 0$
$\therefore \quad C_1 = K$ (using (ii))
Hence, the solution is $n(x) = K e^{-\frac{1}{L}x}$.

66. (b)

Given, $\dfrac{d^2 x}{dt^2} + 6 \dfrac{dx}{dt} + 8x = 0$ ...(i)

Let $x = e^{mt}$ be a trial solution of (i). Then (i) gives,

$$m^2 e^{mt} + 6m e^{mt} + 8 e^{mt} = 0$$
$$\Rightarrow m^2 + 6m + 8 = 0$$
$$\Rightarrow m = -4, -2.$$

So the solution is $x = A e^{-2t} + B e^{-4t}$
Therefore, $\dfrac{dx}{dt} = -2A e^{-2t} - 4B e^{-4t}$

Now $x(0) = 1 \Rightarrow A + B = 1$ ...(ii)
Also $\left.\dfrac{dx}{dt}\right|_{t=0} = 0, \Rightarrow -2A - 4B = 0$ ...(iii)
Solving (ii) and (iii) we get, $A = 2$ and $B = -1$.
Hence, the solution is $x(t) = 2e^{-2t} - e^{-4t}$.

67. (c)

The given equation is $\dfrac{d^2 y}{dx^2} + 2\dfrac{dy}{dx} + y = 0$ ...(i)
Let $y = e^{mx}$ be a trial solution of (i).
Then, (i)
$$\Rightarrow m^2 e^{mx} + 2m e^{mx} + e^{mx} = 0$$
$$\Rightarrow m^2 + 2m + 1 = 0$$
$$\Rightarrow m = -1, -1$$
Hence the general solution is given by
$y = (C_1 + C_2 x)e^{-x}$.
Then, $y(0) = 1 \Rightarrow C_1 = 1$.
Again, $y(1) = 0 \Rightarrow C_2 = -1$.
Hence, $y = y(x) = (1 - x)e^{-x}$ and so $y(2) = -e^{-2}$.

68. (c)

The given equation is $\dfrac{d^2 y}{dx^2} + 6\dfrac{dy}{dx} + 9y = 9x + 6$ ...(i)

Consider $\dfrac{d^2 y}{dx^2} + 6\dfrac{dy}{dx} + 9y = 0$ ...(ii)
Let $y = e^{mx}$ be a trial solution of (ii).
Then, (ii)
$$\Rightarrow m^2 e^{mx} + 6m e^{mx} + 9 e^{mx} = 0$$
$$\Rightarrow m^2 + 6m + 9 = 0 \Rightarrow m = -3, -3$$
Hence, $y_{C.F} = (C_1 x + C_2)e^{-3x}$

$$y_{P.I} = \frac{1}{D^2 + 6D + 9}(9x + 6)$$
$$= \frac{1}{9\left(1 + \dfrac{D^2 + 6D}{9}\right)}(9x + 6)$$
$$= \frac{1}{9}\left(1 + \frac{D^2 + 6D}{9}\right)^{-1}(9x + 6)$$
$$= \frac{1}{9}\left(1 - \frac{D^2 + 6D}{9}\right)(9x + 6)$$
$$= \frac{1}{9}\left(9x + 6 - \frac{D^2(9x+6) + 6D(9x+6)}{9}\right)$$
$$= \frac{1}{9}\left(9x + 6 - \frac{0 + 54 + 0}{9}\right) = x$$

Hence, the complete solution is
$y = y_{C.F} + y_{P.I} = (C_1 x + C_2)e^{-3x} + x$.

69. (a)

Given equation is $x^2\dfrac{d^2y}{dx^2}+x\dfrac{dy}{dx}-4y=0$ ...(i)

which is a linear homogeneous differential equation of order "2."

Let us put $logx = z$.

Then $\dfrac{dy}{dz}=x\dfrac{dy}{dx}$ and $\dfrac{d^2y}{dz^2}-\dfrac{dy}{dz}=x^2\dfrac{d^2y}{dx^2}$

$\therefore$ (i) $\Rightarrow \dfrac{d^2y}{dz^2}-\dfrac{dy}{dz}+\dfrac{dy}{dz}-4y=0$

or, $\dfrac{d^2y}{dz^2}-4y=0$............(ii)

Let $u = e^{mz}$ be a trial solution of (ii)

Then, (ii) $\Rightarrow m^2e^{mz}-4e^{mz}=0$

or $m^2-4=0 (\because e^{mz}\Rightarrow 0)$

or $m=-2,2$

$\therefore$ the complete solution is given by

$y(x)=c_1e^{-2z}+c_2e^{2z}=\dfrac{c_1}{x^2}+c_2x^2$

$\left(\because e^z=x \text{ and } e^{-z}=\dfrac{1}{x}\right)$

Now, $y(0)=0\Rightarrow c_1=0.$

Again, $y(1)=1\Rightarrow 0+c_2=1\Rightarrow c_2=1$

Thus, $y=y(x)=x^2.$

70. (b)

The given differential equation is

$\dfrac{d^2u}{dx^2}-k\dfrac{du}{dx}=0$ ...(i)

Let $u=e^{mx}$ be a trial solution of (i). Then (i) gives,

$m^2e^{mx}-kme^{mx}=0$

$\Rightarrow m^2-km=0$

$\Rightarrow m=0,k.$

So the solution is given by $u=u(x)=Ae^{ox}+Be^{kx}$
$=A+Be^{kx}$

Then, $u(0)=0\Rightarrow A+B=0$ ...(ii) and

$u(L)=U$

$\Rightarrow A+Be^{kx}=U\Rightarrow A-Ae^{kL}=U\Rightarrow A=\dfrac{U}{1-e^{kL}}$

So (ii) gives $B=-\dfrac{U}{1-e^{kL}}.$

$u=\dfrac{U}{1-e^{kL}}-\dfrac{U}{1-e^{kL}}e^{kx}$ i.e; $u=U\left[\dfrac{1-e^{kx}}{1-e^{kL}}\right].$

71. (d)

Given, $y(t)+y(t)=0$ ...(i)

Let $y=e^{mt}$ be a trial solution of (i). Then (i) gives,

$m^2e^{mt}+e^{mt}=0$

$\Rightarrow m^2+1=0$

$\Rightarrow m=0\pm i$

So the solution is $y=A\cos t+B\sin t$ and
$\dot y=-A\sin t+B\cos t.$

Since $y(0)=1,$ $\therefore$ $1=A\times 1+B\times 0$ i.e;
$A=1.$

Since $y(0)=1,\therefore$ $1=-A\times 0+B\times 1$ i.e;
$B=1.$

Hence, $y=\cos t+\sin t$ and $\dot y=-\sin t+\cos t.$

For y to be maximum, we must have

$\dot y=0$ or $-\sin t+\cos t=0$ or $\tan t=1$ or

$t=\dfrac{\pi}{4}=45°$

Since $\ddot y=-\cos x-\sin x$, so at $t=45°, \ddot y=\dfrac{-2}{\sqrt2}$
$<0.$

Hence, y is maximum at $t=45°.$

So the maximum value of y is $\cos 45°+\sin 45°;$
$\sqrt2$.

72. (c)

Given, $\dfrac{d^2x}{dt^2}+9x=0$ ...(i)

Let $x=e^{mt}$ be a trial solution of (i). Then (i) gives,

$m^2e^{mt}+9e^{mt}=0$

$\Rightarrow m^2+9=0$

$\Rightarrow m=\pm 3i$

$\therefore$ $x=C_1\cos 3t+C_2\sin 3t$ and
$\dfrac{dx}{dt}=-3C_1\sin 3t+3C_2\cos 3t.$

Now, $x(0)=1\Rightarrow C_1=1.$

Also, $\dfrac{dx}{dt}\bigg|_{t=0}=1\Rightarrow 3C_2=1\Rightarrow C_2=\dfrac{1}{3}.$

Hence, $x=\cos 3t+\dfrac{1}{3}\sin 3t$

73. (a)

The given differential equation is

$\dfrac{d^2x(t)}{dt^2}+x(t)=0$ ...(i)

Let $x=e^{mt}$ be a trial solution of (i). Then (i)

gives,

$$m^2 e^{mt} + e^{mt} = 0$$

$$\Rightarrow m^2 + 1 = 0$$

$$\Rightarrow m = 0 \pm i$$

So $x = x(t) = C_1 \cos t + C_2 \sin t$ and $\frac{dx}{dt} = -C_1 \sin t + C_2 \cos t$.

Let $x_1(t) = \cos t$ and $x_2(t) = \sin t$. Then the conditions $x_2(0) = 0, \dfrac{dx_1(t)}{dt}\bigg|_{t=0} = 0$ are satisfied.

Then, $W(t) = \begin{vmatrix} x_1(t) & x_2(t) \\ \dfrac{dx_1(t)}{dt} & \dfrac{dx_2(t)}{dt} \end{vmatrix} = \begin{vmatrix} \cos t & \sin t \\ -\sin t & \cos t \end{vmatrix}$

$= \cos^2 t + \sin^2 t = 1$ for all "t."

74. 35

Given equation is $\dfrac{d^2 y}{dx^2} = 0$

Integrating both sides we get, $\dfrac{dy}{dx} = A$

Again, integrating both sides we get,

$y = y(x) = Ax + B$

$\therefore y(0) = 5 \Rightarrow B = 5$.

Again $\dfrac{dy}{dx}\bigg|_{x=10} = 2 \Rightarrow A = 2$

$\therefore y(x) = 2x + 5$ and so $y(15) = 30 + 5 = 35$.

75. (b)

Given $\dfrac{d^2 x}{dt^2} + 2\dfrac{dx}{dt} + x = 0$ ...(i)

Let $x = e^{mt}$ be a trial solution of (i). Then (i) gives,

$$m^2 e^{mt} + 2m e^{mt} + e^{mt} = 0$$

$$\Rightarrow m^2 + 2m + 1 = 0$$

$$\Rightarrow m = -1, -1$$

$\therefore x = x(t) = (a + bt)e^{-t} = ae^{-t} + bte^{-t}$.

76. (c)

Given $x^2 \dfrac{d^2 y}{dx^2} + x\dfrac{dy}{dx} - y = 0$ ...(i)

Let us put $z = \log x$

Then, $\dfrac{dy}{dz} = x\dfrac{dy}{dx}$ and $\dfrac{d^2 y}{dz^2} - \dfrac{dy}{dz} = x^2 \dfrac{d^2 y}{dx^2}$

$\therefore$ (i) gives

$$\dfrac{d^2 y}{dz^2} - \dfrac{dy}{dz} + \dfrac{dy}{dz} - y = 0$$

or, $\dfrac{d^2 y}{dz^2} - y = 0$ ......(ii)

Let $y = e^{mz}$ be a trial solution of (ii)

Then, (ii) $\Rightarrow m^2 e^{mz} - e^{mz} = 0$

$\Rightarrow m^2 - 1 = 0 \ (\because \ e^{mz} \neq 0)$

$\Rightarrow m = -1, 1$

$\therefore y = c_1 e^z + c_2 e^z = c_1 x + \dfrac{c_2}{x}$

Hence, $\dfrac{1}{x}$ and "x" are two solutions.

77. (a)

Given $\dfrac{d^2 y}{dx^2} + 2\alpha \dfrac{dy}{dx} + y = 0$ ...(i)

Let $y = e^{mx}$ be a trial solution of (i).

Then the auxiliary equation is given by

$(m^2 + 2\alpha m + 1) = 0$

$\Rightarrow m = \dfrac{-2\alpha \pm \sqrt{4\alpha^2 - 4}}{2}$

Now both roots *i.e*; values of "m" are equal

$\Rightarrow 4\alpha^2 - 4 = 0$

$\Rightarrow \alpha = \pm 1$

78. 18

Given $\dfrac{d^2 y}{dx^2} = -12x^2 + 24x - 20$.

Integrating both sides w.r.t. x, we get

$\dfrac{dy}{dx} = -4x^3 + 12x^2 - 20x + c_1$

Again, integrating both sides w.r.t. x, we get

$y = -x^4 + 4x^3 - 10x^2 + c_1 x + c_2$ ...(i)

At $x = 0$, $y = 5$; so (i) gives, $c_2 = 5$.

Again at $x = 2$, $y = 21$;

so (i) $\Rightarrow 2c_1 + c_2 = 45$ ...(ii)

Solving (i) and (ii) we get, $c_2 = 5$ and $c_1 = 20$.

So, (i) $\Rightarrow y = -x^4 + 4x^3 - 10x^2 + 20x + 5$

So for $x = 1$, (i) $\Rightarrow y = -1 + 4 - 10 + 20 + 5$

$= 18$.

79. –3

Given $\dfrac{d^2 y}{dt^2} + 5\dfrac{dy}{dt} + 6y = 0$ ...(i)

Let $y = e^{mt}$ be a trial solution of (i). Then (i) gives,

$$m^2 e^{mt} + 5m e^{mt} + 6 e^{mt} = 0$$

$$\Rightarrow m^2 + 5m + 6 = 0$$

$$\Rightarrow m = -2, -3$$

$\therefore y = y(t) = c_1 e^{-2t} + c_2 e^{-3t}$

Then, $y(0) = 2 \Rightarrow c_1 + c_2 = 2$ ...(ii)

Again, $y(1) = -\left(\dfrac{1-3e}{e^3}\right) \Rightarrow \dfrac{c_1}{e^2} + \dfrac{c_2}{e^3} = -\left(\dfrac{1-3e}{e^3}\right)$ ...(iii)

Now solving equation (ii) and (iii), we get

$c_1 = 3$ and $c_2 = -1$.

So $y(t) = 3e^{-2t} - e^{-3t}$.

Now, $\dfrac{dy}{dt} = -6e^{-2t} + 3e^{-3t}$

Therefore, $\left(\dfrac{dy}{dt}\right)_{t=0} = -6 + 3 = 3\ -3.$

80. (b)

Given $\dfrac{d^2y}{dt^2} + 2\dfrac{dy}{dt} + y = 0$ ...(i)

Let $y = e^{mt}$ be a trial solution of (i). Then (i) gives,

$m^2 e^{mt} + 2m e^{mt} + e^{mt} = 0$

$\Rightarrow m^2 + 2m + 1 = 0$

$\Rightarrow m = -1, -1$

$\therefore y = y(t) = (c_1 + c_2 t)e^{-t}$

and $y' = y'(t) = -(c_1 + c_2 t)e^{-t} + c_2 e^{-t}$

Then, $y(0) = 1 \Rightarrow c_1 = 1$.

Again, $y'(0) = 1 \Rightarrow -c_1 + c_2 = 1 \Rightarrow c_2 = 2$

Hence, $y(t) = (1 + 2t)e^{-t}$.

81. (c) The given equation is $x^2 y'' + xy' - y = 0$... (i) which is a linear homogeneous differential equation of order "2."

Let us put $\log x = z$.

Then, $\dfrac{dy}{dz} = x\dfrac{dy}{dx}$ and $\dfrac{d^2y}{dz^2} - \dfrac{dy}{dz} = x^2\dfrac{d^2y}{dx^2}$

$\therefore$ (i) $\Rightarrow \dfrac{d^2y}{dz^2} - \dfrac{dy}{dz} + \dfrac{dy}{dz} - y = 0$

or, $\dfrac{d^2y}{dz^2} - y = 0$............(ii)

Let $y = e^{mz}$ be a trial solution of (ii)

Then, (ii) $\Rightarrow m^2 e^{mz} - e^{mz} = 0$

or, $m^2 - 1 = 0$ $(\because e^{mz} \neq 0)$

or, $m = -1, 1$

$\therefore$ the complete solution is given by

$y(x) = c_2 e^{-z} + c_1 e^z = \dfrac{c_2}{x} + c_1 x. \left(\because e^z = x \text{ and } e^{-z} = \dfrac{1}{x}\right)$

82. (c)

Given $\dfrac{d^2y}{dx^2} = y$ ...(i)

Let $y = e^{mx}$ be a trial solution of (i). Then (i) gives,

$m^2 e^{mx} - e^{mx} = 0$

$\Rightarrow m^2 - 1 = 0$

$\Rightarrow m = -1, 1$

So $y = C_1 e^x + C_2 e^{-x}$ ...(ii)

Since (ii) passes through origin, so $C_1 + C_2 = 0$ ...(iii)

Again, (ii) passes through (ln 2, 3/4); so

$C_1 e^{\ln 2} + C_2 e^{-\ln 2} = \dfrac{3}{4}$

$\Rightarrow \dfrac{3}{4} 2C_1 + \dfrac{C_2}{2} = \dfrac{3}{4}$

$\Rightarrow C_2 + 4C_1 = 1.5$ ...(iv)

Solving (iii) and (iv) we get, $C_1 = 0.5$, $C_2 = -0.5$.

Hence, from (ii), we get, $y = \dfrac{e^x - e^{-x}}{2}$.

83. 0.8566

Given $\dfrac{d^2x(t)}{dt^2} + 3\dfrac{dx(t)}{dt} + 2x(t) = 0.$ ...(i)

Let $x = e^{mt}$ be a trial solution of (i). Then (i) gives,

$m^2 e^{mt} + 3m e^{mt} + 2e^{mt} = 0$

$\Rightarrow m^2 + 3m + 2 = 0$

$\Rightarrow m = -1, -2$

$\therefore x = x(t) = C_1 e^{-t} + C_2 e^{-2t}$

Then, $x(0) = 20 \Rightarrow C_1 + C_2 = 20$ ...(ii)

$x(1) = 10/e \Rightarrow C_1 + C_2 e^{-1} = 10$ ...(iii)

Solving (ii) and (iii), we get

$C_1 = \dfrac{10e - 20}{e-1}; C_2 = \left(\dfrac{10e}{e-1}\right).$

So $x = x(t) = \left\{\dfrac{10e-20}{e-1}\right\}e^{-t} + \left\{\dfrac{10e}{e-1}\right\}e^{-2t}.$

Hence, $x(2) = \left(\dfrac{10e-20}{e-1}\right)e^{-2} + \left(\dfrac{10e}{e-1}\right)e^{-4}$

$= 0.8566$

84. (a)

Given $\dfrac{d^2y}{dx^2} + 12\dfrac{dy}{dx} + 36y = 0$ ...(i)

Let $y = e^{mx}$ be a trial solution of (i). Then (i) gives,

$m^2 e^{mx} + 12m e^{mx} + 36 e^{mx} = 0$

$\Rightarrow m^2 + 12m + 36 = 0$

$\Rightarrow m = -6, -6$

So $y = (C_1 + C_2 x)e^{-6x}$. Then $y(0) = 3 \Rightarrow C_1 = 3$.

$y' = -6C_1e^{-6x} + C_2e^{-6x} - 6C_2xe^{-6x}$

So $y'(0) = -36 \Rightarrow -36 = -6C_1 + C_2$

$\Rightarrow -36 = -18 + C_2$

$\Rightarrow C_2 = -18$

$\therefore \quad y = 3e^{-6x} - 18xe^{-6x}$ *i.e.*, $y = (3 - 18x)e^{-6x}$

85. -1.

Given $y'' + 9y = 0$ ...(i)

Let $y = e^{mx}$ be a trial solution of (i). Then (i) gives,

$m^2e^{mx} + 9e^{mx} = 0$

$\Rightarrow m^2 + 9 = 0$

$\Rightarrow m = \pm 3i$

So $y = C_1 \cos 3x + C_2 \sin 3x$ ...(i)

Then, $y(0) = 0 \Rightarrow C_1 = 0$

Again $y\left(\dfrac{\pi}{2}\right) = \sqrt{2} \Rightarrow C_1 \cos \dfrac{3\pi}{2} + C_2 \sin \dfrac{3\pi}{2} = \sqrt{2}$

$\Rightarrow C_2 = -\sqrt{2}$

Therefore, $y = -\sqrt{2} \sin 3x$ and so

$y\left(\dfrac{\pi}{4}\right) = -\sqrt{2} \sin \dfrac{3\pi}{4} = -\sqrt{2} \times \dfrac{1}{\sqrt{2}} = -1$

$\left(\because \sin \dfrac{3\pi}{4} = \sin\left(\pi - \dfrac{\pi}{4}\right) = \sin \dfrac{\pi}{4} = \dfrac{1}{\sqrt{2}} \right)$

86. 7.38

Given $\dfrac{d^2y}{dx^2} - 4\dfrac{dy}{dx} + 4y = 0$ ...(i)

Let $y = e^{mx}$ be a trial solution of (i). Then (i) gives,

$m^2e^{mx} - 4me^{mx} + 4e^{mx} = 0$

$\Rightarrow m^2 - 4m + 4 = 0$

$\Rightarrow m = 2, 2$

$\therefore \quad y = (C_1 + C_2x)e^{2x}$

Now, $y(0) = 0 \Rightarrow C_1 = 0$

So, $y = C_2xe^{2x}$.

Now, $\dfrac{dy}{dx} = C_2e^{2x} + 2C_2 xe^{2x}$.

Then, $\dfrac{dy}{dx}\Big|_{x=0} = 1 \Rightarrow C_2 = 1$

Thus, $y = xe^{2x}$ and so $y(1) = e^2 = 7.38$.

87. (a)

Given $\dfrac{d^4y}{dx^4} + 3\dfrac{d^2y}{dx^2} = 108x^2$ ...(i)

Consider $\dfrac{d^4y}{dx^4} + 3\dfrac{d^2y}{dx^2} = 0$ ...(ii)

Let $y = e^{mx}$ be a trial solution of (ii). Then (ii) gives,

$m^4e^{mx} + 3m^2e^{mx} = 0$

$\Rightarrow m^4 + 3m^2 = 0$

$\Rightarrow m^2(m^2 + 3) = 0 \Rightarrow m = 0, 0, 0 \pm \sqrt{3}i$

$\therefore \quad CF = (C_1 + C_2x) + C_3 \sin(\sqrt{3}x) + C_4 \cos(\sqrt{3}x)$

Now, P.I $= \dfrac{1}{D^4 + 3D^2}(108x^2)$

$= 108 \dfrac{1}{3D^2\left(1 + \dfrac{D^2}{3}\right)}x^2$

$= 36\dfrac{1}{D^2}\left(1 + \dfrac{D^2}{3}\right)^{-1}x^2$

$= 36\dfrac{1}{D^2}\left(1 - \dfrac{D^2}{3} + \dfrac{D^4}{9} - \cdots\cdots\right)x^2$

$= 36\dfrac{1}{D^2}\left(x^2 - \dfrac{D^2x^2}{3} + \dfrac{D^4x^2}{9} - \cdots\cdots\right)$

$= 36\dfrac{1}{D^2}\left(x^2 - \dfrac{2}{3}\right) = 36\dfrac{1}{D}\int\left(x^2 - \dfrac{2}{3}\right)dx$

$= 36\dfrac{1}{D}\left(\dfrac{x^3}{3} - \dfrac{2}{3}x\right) = 36\int\left(\dfrac{x^3}{3} - \dfrac{2}{3}x\right)dx$

$= 36\left(\dfrac{x^4}{12} - \dfrac{1}{3}x^2\right) = 3x^4 - 12x^2$

88. (b)

Given $\dfrac{d^2y}{dt^2} + 2\dfrac{dy}{dt} + y = 0$ ...(i)

Let $y = e^{mt}$ be a trial solution of (i). Then (i) gives,

$m^2e^{mt} + 2me^{mt} + e^{mt} = 0$

$\Rightarrow m^2 + 2m + 1 = 0$

$\Rightarrow m = -1, -1$

$\therefore \quad y(c_1 + c_2t)e^{-t}$

Now, $y(0) = 1 \Rightarrow c_1 = 1$

So, $y = (1 + c_2t)e^{-t}$

Again, $y(1) = 3e^{-1} \Rightarrow (1 + C_2)e^{-1} = 3e^{-1} \Rightarrow c_2 = 2$

Hence, $y = (1 + 2t)e^{-t}$ and so $y(2) = 5e^{-2}$.

89. (*a*)

Given $\dfrac{d^2y}{dx^2}+16y=0$ ...(*i*)

Let $y=e^{mx}$ be a trial solution of (*i*)

So the auxiliary equation is $m^2+16=0$ which gives $m=\pm 4i$

Therefore, $y=c_1\cos 4x+c_2\sin 4x$ and

$y'=-4c_1\sin 4x+4c_2\cos 4x$.

Now, $y'(0)=1\Rightarrow 1=4c_2\Rightarrow c_2=1/4$

Again, $y'(\pi/2)=-1$

$\Rightarrow -1=-4c_1\sin 2\pi+4c_2\cos 2\pi$

$\Rightarrow -1=04c_2$

$\Rightarrow c_2=-1/4$, which contradicts the fact that $c_2=1/4$.

Hence, the given differential equation has no solution.

90. (*a*)

The equation $y''-4y'+3y=2t-3t^2$ can be re-written as: $(D^2-4D+3)y=2t-3t^2$

So P.I.

$=\dfrac{1}{D^2-4D+3}(2t-3t^2)$

$=\dfrac{1}{(1-D)(3-D)}(2t-3t^2)$

$=\dfrac{(3-D)-(1-D)}{2(3-D)(1-D)}(2t-3t^2)$

$=\dfrac{1}{2(1-D)}(2t-3t^2)-\dfrac{1}{2(3-D)}(2t-3t^2)$

$=\dfrac{1}{2(1-D)}(2t-3t^2)-\dfrac{1}{6\left(1-\dfrac{D}{3}\right)}(2t-3t^2)$

$=\dfrac{1}{2}(1-D)^{-1}(2t-3t^2)-\dfrac{1}{6}\left(1-\dfrac{D}{3}\right)^{-1}(2t-3t^2)$

$=\dfrac{1}{2}(1+D+D^2)(2t-3t^2)-\dfrac{1}{6}\left(1+\dfrac{D}{3}+\dfrac{D^2}{9}\right)(2t-3t^2)$

$=\dfrac{1}{2}\left\{(2t-3t^2)+D(2t-3t^2)+D^2(2t-3t^2)\right\}$

$\qquad -\dfrac{1}{6}\left\{(2t-3t^2)+\dfrac{D}{3}(2t-3t^2)+\dfrac{D^2}{9}(2t-3t^2)\right\}$

$=\dfrac{1}{2}\left[2t-3t^2+2-6t-6\right]-\dfrac{1}{6}\left[2t-3t^2+\dfrac{2-6t}{3}+\dfrac{-6}{9}\right]$

$=\dfrac{1}{2}(-4-4t-3t^2)-\dfrac{1}{6}(-3t^2)$

$=-2-2t-t^2$

91. (*a*)

Given $\dfrac{d^2y}{dx^2}+2\dfrac{dy}{dx}-5y=0$ ...(*i*)

Let $y=e^{mx}$ be a trial solution of (*i*)

So the auxiliary equation is $m^2+2m-5=0$ which gives,

$m=\dfrac{-2\pm\sqrt{4+20}}{2}=\dfrac{-2\pm 2\sqrt{6}}{2}=-1\pm\sqrt{6}.$

Hence, $y=K_1e^{(-1+\sqrt{6})x}+K_2e^{(-1-\sqrt{6})x}$.

92. 94.08

Given $3y''(x)+27y(x)=0$ ...(*i*)

Let $y=e^{mx}$ be a trial solution of (*i*)

Then the auxiliary equation is

$3m^2+27=0$

or, $m^2+9=0$

or, $m=\pm 3i$

So the solution is $y(x)=y=c_1\cos 3x+c_2\sin 3x$ ...(*ii*)

Given $y(0)=0$. Therefore (*ii*) gives, $C_1=0$. So, $y=C_2\sin 3x$

Now $y'=3c_2\cos 3x$.

So $y'(0)=2000\Rightarrow 2000=0+3c_2\Rightarrow c_2=\dfrac{2000}{3}$

$\therefore y=\dfrac{2000}{3}\sin 3x$.

Hence, $y(1)=\dfrac{2000}{3}\sin 3=94.08$.

93. −0.37.

The given differential equation is:

$\dfrac{d^2y}{dx^2}+2\dfrac{dy}{dx}+y=0$ ...(*i*)

Let $y=e^{mx}$ be a trial solution of (*i*)

So the auxiliary equation is $m^2+2m+1=0$ which gives, $(m+1)^2=1$ *i.e*; $m=-1,-1$.

So the general solution of (*i*) is given by

$y=(A+Bx)e^{-x}$ ...(*ii*)

$\therefore y(0)=0\Rightarrow 0=(A+0)e^{-0}\Rightarrow A=0$.

Again, $\dfrac{dy}{dx}\bigg|_{x=0}=-1\Rightarrow\left[-(A+Bx)e^{-x}+Be^{-x}\right]_{x=0}=-1$

$\Rightarrow -(0+0)e^{-0}+Be^0=-1\Rightarrow B=-1$.

Now putting the values of A and B we get from (*ii*), $y=-xe^{-x}$.

$\therefore y(1)=-e^{-1}=-0.37$.

94. –0.21

The given equation is

$$\frac{d^2y}{dt^2} = -\frac{dy}{dt} + \frac{5y}{4} \qquad \ldots(i)$$

Let $y = e^{mt}$ be a trial solution of (i)

So the auxiliary equation is

$$m^2 = -m - \frac{5}{4} \ i.e; 4m^2 + 4m + 5 = 0 \text{ which gives,}$$

$$m = \frac{-4 \pm \sqrt{16 - 80}}{2 \times 4} = \frac{-4 \pm 8i}{8} = -\frac{1}{2} \pm i.$$

So the general solution of (i) is given by

$$y = (A\cos t + B\sin t)e^{-\frac{t}{2}} \qquad \ldots(ii)$$

$$\therefore \quad y(0) = 1 \Rightarrow (A\cos 0 + B\sin 0)e^{-0} = 1 \Rightarrow A = 1.$$

Also

$$\frac{dy}{dt} = (-A\sin t + B\cos t)e^{-\frac{t}{2}} - \frac{1}{2}(A\cos t + B\sin t)e^{-\frac{t}{2}}$$

So $\left.\dfrac{dy}{dx}\right|_{t=0} = 0$

$$\Rightarrow (-0 + B)e^0 - \frac{1}{2}(1 + B \times 0)e^0 = 0$$

$$\Rightarrow B = \frac{1}{2}.$$

Now putting the values of A and B we get from (ii),

$$y = e^{-\frac{t}{2}}\left(\cos t + \frac{1}{2}\sin t\right).$$

$$\therefore y(\pi) = e^{-\frac{\pi}{2}}\left(\cos \pi + \frac{1}{2}\sin \pi\right) = -e^{-\frac{\pi}{2}} = -0.21.$$

95. 4.54.

The given equation is

$$2\frac{d^2y}{dt^2} + 8y = 0 \qquad \ldots(i)$$

Let $y = e^{mt}$ be a trial solution of (i)

So the auxiliary equation is $2m^2 + 8 = 0$ which gives, $m = 0 \pm 2i$.

So the general solution of (i) is given by

$$y = (A\cos 2t + B\sin 2t)e^0$$
$$= A\cos 2t + B\sin 2t \qquad \ldots(ii)$$

$$\therefore \quad y(0) = 0 \Rightarrow (A\cos 0 + B\sin 0)e^{-0} = 0 \Rightarrow A = 0.$$

Also, $\dfrac{dy}{dt} = -2A\sin 2t + 2B\cos 2t = 2B\cos 2t.$

So $\left.\dfrac{dy}{dt}\right|_{t=0} = 10 \Rightarrow 2B\cos 0 = 10 \Rightarrow B = 5.$

Now putting the values of A and B we get from (ii), $y = 10\sin 2t.$

$$\therefore \quad y(1) = 10\sin 2 = 10 \times 0.9092 = 4.54.$$

1. Let $y(x)$ be the solution of the differential equation

$$(xy + y + e^{-x})dx + (x + e^{-x})dy = 0 \text{ satisfying}$$

$y(0) = 1$. Then $y(-1) = $?

(a) $\dfrac{e}{e-1}$ (b) $\dfrac{2e}{e-1}$

(c) $\dfrac{e}{1-e}$ (d) 0

2. Let $y(x)$ be the solution of the differential equation $\dfrac{dy}{dx} = (y-1)(y-3)$ satisfying $y(0) = 2$. Then which of the following is not true?

(a) $\lim\limits_{x\to\infty} y(x) = 1$ (b) $\lim\limits_{x\to-\infty} y(x) = 3$

(c) $y(x)$ is bounded

(d) $y(x)$ is not bounded above

3. The nonzero value of n for which the differential equation

$$(3xy^2 + n^2x^2y)dx + (nx^3 + 3x^2y)dy = 0 \text{ becomes}$$

exact is

(a) –3 (b) –2 (c) 2 (d) 3

4. An integrating factor of the differential equation $\dfrac{dy}{dx} = \dfrac{2xy^2 + y}{x - 2y^3}$ is

(a) $\dfrac{1}{y}$ (b) $\dfrac{1}{y^2}$ (c) y (d) y^2

5. If $y(x)$ is a solution of the differential equation $y'' + 4y = 2e^x$, then $\lim\limits_{x\to\infty} e^{-x}y(x) = $?

(a) $\dfrac{1}{5}$ (b) $\dfrac{2}{5}$ (c) $\dfrac{1}{7}$ (d) $\dfrac{2}{7}$

6. The differential equation

$$(1 + x^2y^3 + \lambda x^2y^2)dx + (5 + x^3y^2 + x^3y)dy = 0 \text{ is}$$

exact if $\lambda = $?

(a) $\dfrac{3}{2}$ (b) $\dfrac{1}{2}$ (c) $\dfrac{5}{3}$ (d) $\dfrac{2}{5}$

7. If $y(x)$ is a solution of $\dfrac{d^2y}{dx^2}+4\dfrac{dy}{dx}+4y=0$ sat-

isfying $y(0) = 4$ and $\left.\dfrac{dy}{dx}\right|_{x=0}=8$; then $y(x)$ is given by

(a) $4e^{2x}$ (b) $(16x + 4)e^{-2x}$
(c) $4e^{-2x} + 16x$ (d) $4e^{-2x} + 16e^{2x}$

8. Consider the following differential equation
$(x + y +1)dx + (2x +2y + 1)dy = 0.$

Then which of the following statements is true?
(a) The differential equation is homogeneous
(b) The differential equation is exact
(c) e^{x+y} is an integrating factor
(d) A suitable substitution transforms the differential equation to the variable seperable form.

9. Consider the differential equation

$2\cos(y^2)dx - xy\sin(y^2)\,dy = 0$. Then which of the following is true?
(a) e^x is an integrating factor.
(b) e^{-x} is an integrating factor.
(c) x^2 is an integrating factor.
(d) x^3 is an integrating factor.

10. Let $y_1(x)$ and $y_2(x)$ be two linearly independent solutions of $y'' + P(x)y' + Q(x)y = 0$.

Then, $y_3(x) = ay_1(x) + by_2(x)$
and $y_4(x) = cy_1(x) + dy_2(x)$
are linearly independent if
(a) $ad = bc$ (b) $ad \ne bc$
(c) $ac = bd$ (d) $ab \ne cd$

11. An integrating factor of

$x\dfrac{dy}{dx}+(3x+1)y=e^{-2x}$ is

(a) xe^{3x} (b) $3xe^x$
(c) xe^x (d) x^3e^x

[**Hint:** Express the equation in the form
$\dfrac{dy}{dx}+Py=Q$, Then I.F $=e^{\int Pdx}$]

12. The general solution of $x^2y' - 5xy' + 9y = 0$ is given by
(a) $(c_1 + c_2x)e^{3x}$ (b) $(c_1 + c_2 \ln x)x^3$
(c) $(c_1 + c_2x)x^3$ (d) $(c_1 + c_2 \ln x)x^2$

13. If x^3y^2 is an integrating factor of
$(6y^2 + \alpha xy)dx + (6xy + \beta x^2)dy,$

(a) $3\alpha - 5\beta = 0$ (b) $2\alpha - \beta = 0$
(c) $3\alpha + 5\beta = 0$ (d) $2\alpha = \beta + 1$

14. Let $y(x)$ be the solution of the initial value problem $x^2y'' + xy' + y = x$, $y(1) = y'(1) = 1$.

Then the value of $y\left(e^{\frac{\pi}{2}}\right)$ is

(a) $\dfrac{1}{2}\left(1-e^{\frac{\pi}{2}}\right)$ (b) $\dfrac{1}{2}\left(1+e^{\frac{\pi}{2}}\right)$

(c) $\dfrac{1}{2}+\dfrac{\pi}{4}$ (d) $\dfrac{1}{2}-\dfrac{\pi}{4}$

15. If the integrating factor of
$(x^7y^2 + 3y)dx + (3x^8y - x)dy = 0$ is x^my^n, then
(a) $m = -7, n = 1$ (b) $m = 1, n = -7$
(c) $m = n = 0$ (d) $m = n = 1$

16. The particular solution of the equation $y\sin x = y$ $\ln y$ satisfying the initial condition $y\left(\dfrac{\pi}{2}\right) = e$ is

(a) $e^{\tan\left(\frac{x}{2}\right)}$ (b) $e^{\cot\left(\frac{x}{2}\right)}$

(c) $\ln\left\{\tan\left(\dfrac{x}{2}\right)\right\}$ (d) $\ln\left\{\cot\left(\dfrac{x}{2}\right)\right\}$

17. The differential equation

$\dfrac{dy}{dx} = k(a - y)(b - y)$ when solved with the condition $y(0) = 0$, yields the result

(a) $\dfrac{b(a - y)}{a(b - y)} = e^{(a-b)kx}$ (b) $\dfrac{b(a - x)}{a(b - x)} = e^{(b-a)ky}$

(c) $\dfrac{a(b - y)}{b(a - y)} = e^{(a-b)kx}$ (d) $xy = ke$

18. Solve: $y'' - 4y' + 4y = x^2$; when $x = 0$, $y = \dfrac{3}{8}$ and $\dfrac{dy}{dx} = 1$

(a) $xe^{2x} + x + \dfrac{x^2}{2} + \dfrac{1}{8}$ (b) $\dfrac{xe^{2x}}{2} + \dfrac{x^2}{2} + x + \dfrac{1}{4}$

(c) $y = \dfrac{xe^{2x}}{2} + \dfrac{x^2}{4} + \dfrac{x}{2} + \dfrac{3}{8}$

(d) none of these

19. Solve: $(D^2 - 1)y = 2$; given that $\dfrac{dy}{dx} = 3$ when y $= 1$; and $x = 2$ when $y = -1$
(a) $y + 4 = e^{2x}$ (b) $y + 2 = e^{x-2}$
(c) $y = e^{2x} + 2$ (d) $y - 2 = e^{x+2}$

20. The equation of a curve passing through the point $(-2, 3)$ given that the slope of the tangent to the curve at any point (x, y) is $\dfrac{2x}{y^2}$, is

(a) $y = \left(3x^2 + 15\right)^{\frac{1}{3}}$

(b) $y^2 = 2x^2 + 15$

(c) $y = \left(x^2 + 5\right)^{\frac{3}{2}}$

(d) $y^2 = \left(3x^2 - 4\right)^{\frac{1}{3}}$

21. Solve: $\dfrac{dy}{dx} + y\cot x = 4x\,\mathrm{cosec}\,x;\ y\left(\dfrac{\pi}{2}\right) = 0$

(a) $y\cos x = x^2 + \dfrac{\pi^2}{2}$

(b) $y\cos x = x^3 - \dfrac{\pi^2}{6}$

(c) $y\sin x = 2x^2 - \dfrac{\pi^2}{2}$

(d) $y\sin x = x^3 - \dfrac{2\pi^2}{3}$

22. Solve: $\dfrac{dy}{dx} + 2y\tan x = \sin x;\ y = 0$ when $x = \dfrac{\pi}{3}$

(a) $y = \cos x - 2\cos^2 x$ (b) $y = \cos x + \sin^2 x$
(c) $y = \sin x - 2\sin^2 x$ (d) $y = \sin x + \cos^2 x$

23. Solve: $\dfrac{dy}{dx} + x\cot y = 2y + y^2\cot y$; given that $x = 0$, when $y = \dfrac{\pi}{2}$

(a) $x\cos y = y^2\sin y + \dfrac{\pi^2}{4}$

(b) $x\sin y = y^2\sin y - \dfrac{\pi^2}{4}$

(c) (c) $x(\sin y + \cos y) = \dfrac{\pi^2}{4}$

(d) none of these

24. What will be the solution of $\dfrac{dy}{dx} - \dfrac{y}{x} = 2x^2$

(a) $y = x^3 + cx$ (b) $y = x^2 + 2cx$

(c) $y = \dfrac{x^3}{3} + cx^2$ (d) $y = \dfrac{x^2}{4} + cx$

25. Solve: $(x + y + 1)^2 dy = dx;\ y(-1) = 0$
(a) $x + y = \tan y + 2$ (b) $(x + y)^2 = \cot y + 1$
(c) $x + y + 1 = \tan y$ (d) $x + y - 1 = \tan y$

26. A particular solution of the differential equation $\log\left(\dfrac{dy}{dx}\right) = 3x + 4y$ given that $y = 0$ when $x = 0$, is given by
(a) $4e^{-3x} + 3e^{3y} = 5$ (b) $4e^{-3x} + 3e^{4y} = 7$
(c) $3e^{3x} + 4e^{4y} = 5$ (d) $4e^{3x} + 3e^{-4y} = 7$

27. Solve: $(x+1)\dfrac{dy}{dx} = 2e^{-y} - 1;\ y(0) = 0$

(a) $y = \log\left(2 - \dfrac{1}{x+1}\right)$ (b) $y = \log\left(1 + \dfrac{1}{x}\right)$

(c) $\dfrac{1}{x} = c + e^y$ (d) none of these

28. Solve: $\dfrac{dy}{dx} = y\sin 2x;\ y(0) = 1$

(a) $y = e^{\sin^2 x}$ (b) $y = e^{\cos^2 x}$
(c) $y = e^{\sin x + \cos x}$ (b) $y = e^{\cos x - \sin x}$

29. Solve: $(1 + e^{2x})dy + (1 + y^2)e^x dx = 0;\ y = 1$ when $x = 0$
(a) $y = e^x$ (b) $y = e^{-x}$
(c) $xy = e^x$ (d) $xy = e^{-x}$

30. Solve: $(1 + y^2)(1 + \log x)dx + xdy = 0;\ y = 1$ when $x = 1$

(a) $y = \tan\left\{\dfrac{\pi}{2} - \dfrac{1}{2} + \dfrac{1}{2}(1 + \log x)^2\right\}$

(b) $y = \cot\left\{\dfrac{\pi}{4} + \dfrac{1}{2} + \dfrac{1}{2}(1 + \log x)^2\right\}$

(c) $y = \tan\left\{\dfrac{\pi}{2} + \dfrac{1}{2} - \dfrac{1}{2}(1 + \log x)^2\right\}$

(d) none of these

31. The solution of the differential equation $K^2\dfrac{d^2y}{dx^2} = y - y_2$ under the boundary conditions
(i) $y = y_1$ and $x = x_0$ and
(ii) $y = y_2$ and $x = \infty$, where K, y_1 and y_2 are constant, is

(a) $y = (y_1 - y_2)\exp\left(-\dfrac{x}{k^2}\right) + y_2$

(b) $y = (y_2 - y_1)\exp\left(-\dfrac{x}{k}\right) + y_1$

(c) $y = (y_1 - y_2)\sinh\left(\dfrac{x}{k}\right) + y_1$

(d) $y = (y_1 - y_2)\exp\left(-\dfrac{x}{k}\right) + y_2$

Answer key				
1. (b)	2. (d)	3. (d)	4. (b)	5. (b)
6. (a)	7. (b)	8. (d)	9. (d)	10. (b)
11. (a)	12. (b)	13. (a)	14. (b)	15. (a)
16. (a)	17. (a)	18. (c)	19. (b)	20. (a)
21. (c)	22. (a)	23. (b)	24. (a)	25. (c)
26. (d)	27. (a)	28. (a)	29. (b)	30. (c)
31. (d)				

Hints	

1. (b) Express the differential equation in the form $\dfrac{dy}{dx} + Py = Q$. Here $I.F. = xe^x + 1$]

3. (d) The given equation is exact

$$\Rightarrow \frac{\partial}{\partial x}\left(nx^3 + 3x^2 y\right) = \frac{\partial}{\partial y}\left(3xy^2 + n^2 x^2 y\right)$$

4. (b) Multiply numerator and denominator by $\dfrac{1}{y^2}$ and then express the equation in the form $Mdx + Ndy = 0$. Then show that $\dfrac{\partial M}{\partial y} = \dfrac{\partial N}{\partial x}$

6. (a) The given differential equation is exact

$$\Rightarrow \frac{\partial}{\partial x}\left(5 + x^3 y^2 + x^3 y\right) = \frac{\partial}{\partial y}\left(1 + x^2 y^3 + \lambda x^2 y^2\right)$$

8. (d) Use the transformation $x + y = u$ so that $1 + \dfrac{dy}{dx} = \dfrac{du}{dx}$ and then proceed.

10. (b) y_1 and y_2 are linearly independent solutions

$$\Rightarrow W(y_1, y_2) \neq 0$$

$$\Rightarrow \begin{vmatrix} y_1 & y_2 \\ y_1' & y_2' \end{vmatrix} \neq 0 \quad \Rightarrow y_1 y_2' - y_1' y_2 \neq 0 \dots\dots(i)$$

Now, $W(y_3, y_4) = \begin{vmatrix} y_3 & y_4 \\ y_3' & y_4' \end{vmatrix}$

$$= \begin{vmatrix} ay_1 + by_2 & cy_1 + dy_2 \\ ay_1' + by_2' & cy_1' + dy_2' \end{vmatrix}$$

$$= (ay_1 + by_2)(cy_1' + dy_2') - (ay_1' + by_2')(cy_1 + dy_2)$$

$$= acy_1 y_1' + bcy_1' y_2 + ady_1 y_2' + bdy_2 y_2'$$

$$\quad - acy_1 y_1' - ady_1' y_2 - bcy_2' y_1 - bdy_2 y_2'$$

$$= (bc - ad)(y_1' y_2 - y_2' y_1)$$

But from (i) we get $y'_1 y_2 - y'_2 y_1 \neq 0$

$\therefore \quad W(y_3, y_4) \neq 0$ if $bc - ad \neq 0$ i.e; if $ad \neq bc$. Thus, y_3 and y_4 are L.I. solutions if $ad \neq bc$

CHAPTER 5

PARTIAL DIFFERENTIAL EQUATIONS

5.1 BASIC CONCEPTS

5.1.1 Introduction

Partial differential equations (PDE for short) have applications in many practical situations such as Brownian motion, population growth, traffic flow along a highway, stochastic process, etc.

A partial differential equation is an equation involving partial differential coefficients of a function of two or more variables.

Let us, consider that x and y are independent variables and $z = f(x, y)$ [or $u = f(x, y)$] be a function of two variables x and y. We shall use the following notations throughout this chapter:

$$p = \frac{\partial z}{\partial x} \left(or \ p = \frac{\partial u}{\partial x} \right), \quad q = \frac{\partial z}{\partial y} \left(or \ q = \frac{\partial u}{\partial y} \right),$$

$$r = \frac{\partial^2 z}{\partial x^2} \left(or \ r = \frac{\partial^2 u}{\partial x^2} \right), \quad s = \frac{\partial^2 z}{\partial x \partial y} \left(or \ s = \frac{\partial^2 u}{\partial x \partial y} \right)$$

and $t = \frac{\partial^2 z}{\partial y^2} \left(or \ t = \frac{\partial^2 u}{\partial y^2} \right)$.

A few examples of partial differential equations are

(i) $x^2 \frac{\partial^2 u}{\partial x^2} + xy \frac{\partial^2 u}{\partial x \partial y} + y^2 \frac{\partial^2 u}{\partial y^2} - (2x+1) \frac{\partial u}{\partial x} = 3y$

(ii) $x \frac{\partial z}{\partial x} + y \frac{\partial z}{\partial y} = 3z$

5.1.2 Order and Degree

The highest order derivative occurring in the partial differential equation is called the order of the equation.

The degree of a partial differential equation is the greatest exponent of the highest order when the equation is made free from all radicals.

Examples:

(i) The order and degree of the PDE

$2\frac{\partial z}{\partial x} + 5x\frac{\partial z}{\partial y} = xy^2$ are respectively 1 and 1.

(ii) The order and degree of the PDE

$\frac{\partial^2 u}{\partial x^2} + \frac{\partial^2 u}{\partial y^2} + x\frac{\partial u}{\partial x} + y\frac{\partial u}{\partial y} = 3u$ are respectively 2 and 1.

5.1.3 Linear and No-Linear Partial Differential Equations

A partial differential equation is called linear if it is of first degree in the dependent variable and its partial derivatives, *i.e.*, if the powers or the products of the dependent variable and its partial derivatives remain absent in the equation.

A PDE which is not linear is called a nonlinear partial differential equation.

Examples:

(i) $x\frac{\partial u}{\partial x} + y\frac{\partial u}{\partial z} = 2u$ is a linear PDE.

(ii) $\left(\frac{\partial z}{\partial x}\right)^2 + \left(\frac{\partial z}{\partial y}\right)^2 = 1$ is a non-linear PDE.

(iii) $x^2 \frac{\partial^2 z}{\partial x^2} + 2\frac{\partial^2 z}{\partial x \partial y} + \frac{\partial z}{\partial x} - e^y \frac{\partial z}{\partial y} = x + y$ is a non-linear PDE.

5.1.4 Formation of Partial Differential Equations

Let us consider an equation

$$F(x,y,z,a,b) = 0 \qquad \ldots(i)$$

where a & b are arbitrary constants and z is a function of two independent variables x and y.

Now differentiating (i) partially w.r.t x and y, respectively, we get,

$$\frac{\partial F}{\partial x} + p\frac{\partial F}{\partial z} = 0 \qquad \ldots(ii)$$

and $\dfrac{\partial F}{\partial y} + q\dfrac{\partial F}{\partial z} = 0 \qquad \ldots(iii)$

Then eliminating the arbitrary constants a and b from equation (i), (ii), and (iii), we get a partial differential equation of the form $f(x, y, z, p, q) = 0$.

5.2 CLASSIFICATION OF 2ND ORDER PARTIAL DIFFERENTIAL EQUATION

Consider the second-order partial differential equation:

$$A\frac{\partial^2 u}{\partial x^2} + B\frac{\partial^2 u}{\partial x\partial y} + C\frac{\partial^2 u}{\partial y^2} + D\frac{\partial u}{\partial x} + E\frac{\partial u}{\partial y} + Fu = G,$$

where A, B, C, D, E, F, G are functions of x and y or constants.

Then the equation is called

(i) elliptic if $B^2 - 4AC < 0$

(ii) parabolic if $B^2 - 4AC = 0$

(iii) hyperbolic if $B^2 - 4AC > 0$

5.3 HEAT, WAVE, AND LAPLACE EQUATIONS

5.3.1 Solution by Separation of Variables

In this method, we assume that the dependent variable is the product of two functions each of which involves only one of the independent variables.

Example:

Let us solve the following:

$\dfrac{\partial u}{\partial x} = 4\dfrac{\partial u}{\partial y}$, given that $u(0, y) = 8e^{-3y}$

Here, $u = u(x,y) = X(x)Y(y) = XY$ (say)

Then, $\dfrac{\partial u}{\partial x} = X'Y$ and $\dfrac{\partial u}{\partial y} = XY'$

$$\left(\text{Here } X' = \frac{dX}{dx}, Y' = \frac{dY}{dy}\right)$$

$$\therefore \frac{\partial u}{\partial x} = 4\frac{\partial u}{\partial y} \Rightarrow X'Y = 4XY'$$

$$\Rightarrow \frac{X'}{X} = \frac{4Y'}{Y} = k\text{(say)}$$

Then, $\dfrac{X'}{X} = k \Rightarrow \dfrac{1}{X}\dfrac{dX}{dx} = k \Rightarrow \dfrac{dX}{X} = k\,dx$

Integrating we get,

$\log X = kx + \log c_1$

$or, \log X - \log c_1 = kx$

$or, \log \dfrac{X}{c_1} = kx$

$or, X = c_1 e^{kx}$

Again, $\dfrac{4Y'}{Y} = k \Rightarrow \dfrac{1}{Y}\dfrac{dY}{dy} = \dfrac{k}{4} \Rightarrow \dfrac{dY}{Y} = \dfrac{k}{4}dy$

Integrating we get,

$\log Y = \dfrac{ky}{4} + \log c_2$

$or, \log Y - \log c_2 = \dfrac{ky}{4}$

$or, \log \dfrac{Y}{c_2} = \dfrac{ky}{4}$

$or, Y = c_2 e^{\frac{ky}{4}}$

$\therefore u(x,y) = XY = c_1 c_2 e^{kx}e^{\frac{ky}{4}} = c_1 c_2 e^{k\left(x+\frac{y}{4}\right)}$

Then, $u(0,y) = 8e^{-3y}$

$\Rightarrow c_1 c_2 e^{k\left(x+\frac{y}{4}\right)} = 8e^{-3y}$

$\therefore c_1 c_2 = 8, \dfrac{k}{4} = -3$ i.e.,$c_1 c_2 = 8, k = -12$

Hence, $u(x,y) = 8e^{-12\left(x+\frac{y}{4}\right)} = 8e^{-12x-3y}$, which is the required solution.

5.3.2 One-Dimensional Heat (Diffusion) Equation and Its Solution

The one dimensional heat equation is given by

$$\frac{\partial u}{\partial t} = c^2\frac{\partial^2 u}{\partial x^2}, \quad 0 < x < l$$

with boundary conditions $u(0, t) = 0$, $u(l, t) = 0$ and initial condition $u(x,0) = f(x)$.

The general solution is given by

$$u(x,t) = \sum_{n=1}^{\infty} b_n e^{\frac{-c^2 n^2 \pi^2 t}{l^2}} \sin\left(\frac{n\pi x}{l}\right)$$

where $b_n = \dfrac{2}{l}\displaystyle\int_0^l f(x)\sin\left(\dfrac{n\pi x}{l}\right)dx.$

5.3.3 One-Dimensional Wave Equation and Its Solution

The one-dimensional wave equation is given by

$$\frac{\partial^2 u}{\partial t^2} = c^2\frac{\partial^2 u}{\partial x^2},\ \ 0 < x < l$$

with boundary conditions $u(0, t) = 0$, $u(l, t) = 0$ and initial condition $u(x,0) = f(x)$, $\dfrac{\partial u}{\partial t}\Big|_{t=0} = g(x)$.

The general solution is given by

$$u(x,t) = \sum_{n=1}^{\infty}\left[a_n\cos\left(\frac{n\pi ct}{l}\right)+b_n\sin\left(\frac{n\pi ct}{l}\right)\right]\sin\left(\frac{n\pi x}{l}\right)$$

where $a_n = \dfrac{2}{l}\displaystyle\int_0^l f(x)\sin\dfrac{n\pi x}{l}dx$ and

$$b_n = \frac{2}{n\pi c}\int_0^l g(x)\sin\frac{n\pi x}{l}dx.$$

Remember:

1. The D'Alembert's solution of the following problem:

$$\frac{\partial^2 u}{\partial t^2} = c^2\frac{\partial^2 u}{\partial x^2},\ \ -\infty < x < \infty,\ t > 0$$

with initial condition $u(x,0) = f(x)$, and

$\dfrac{\partial u}{\partial t}\Big|_{t=0} = g(x)$ is given by

$$u(x,t) = \frac{1}{2}\left[f(x+ct)+f(x-ct)\right]+\frac{1}{2c}\int_{x-ct}^{x+ct} g(\xi)d\xi.$$

2. The D'Alembert's solution of the following problem:

$$\frac{\partial^2 u}{\partial t^2} = c^2\frac{\partial^2 u}{\partial x^2},\ \ -\infty < x < \infty,\ t > 0$$

with initial condition $u(x,0) = f(x)$, and

$\dfrac{\partial u}{\partial t}\Big|_{t=0} = 0$ is given by

$$u(x,t) = \psi(x+ct) + \psi(x-ct).$$

5.3.4 The Laplace Equation and its Solution

The Laplace equation is given by

$$\frac{\partial^2 u}{\partial x^2}+\frac{\partial^2 u}{\partial y^2} = 0\ \text{with}$$

u(0,y) = 0, u(a,y) = 0, u(x,b) = 0 and
u(x,0) = f(x).

The general solution is given by

$$u(x,y) = \sum_{n=1}^{\infty}a_n\sin\left(\frac{n\pi x}{a}\right)\sinh\left\{\frac{n\pi(b-y)}{a}\right\},$$

where, $a_n = \dfrac{2}{a\sinh\left(\dfrac{n\pi b}{a}\right)}\displaystyle\int_0^a f(x)\sin\left(\dfrac{n\pi x}{a}\right)dx.$

Fully Solved MCQs

1. The partial differential equation that can be formed from $u = ax + qy + a^2 + b^2$ has the form
   (a) $z = px + qy + p^2 + q^2$
   (b) $z = px + qy$
   (c) $z = 2px + 3qy + p^2 - q^2$
   (d) none of these

2. The partial differential equation that can be formed from $z = (x + a)(y + b)$ has the form
   (a) $z = \dfrac{\partial z}{\partial x}\times\dfrac{\partial z}{\partial y}$ (b) $z = \dfrac{\partial z}{\partial x}+\dfrac{\partial z}{\partial y}$
   (c) $z = \dfrac{\partial z}{\partial x}-\dfrac{\partial z}{\partial y}$ (d) none of these

3. The partial differential equation that can be formed from $u = (x^2 + a)(y^2 + b)$ has the form
   (a) $\dfrac{\partial u}{\partial x}+\dfrac{\partial u}{\partial y} = 4\dfrac{xy}{u}$ (b) $2\dfrac{\partial u}{\partial x}-\dfrac{\partial u}{\partial y} = 4xy+u$
   (c) $\dfrac{\partial u}{\partial x}\dfrac{\partial u}{\partial y} = 4xyu$ (d) none of these

4. The PDE corresponding to $y = f(x - at) + g(x + at)$ is
   (a) $a^2\dfrac{\partial^2 y}{\partial t^2} = \dfrac{\partial^2 y}{\partial x^2}$ (b) $\dfrac{\partial^2 y}{\partial t^2} = xy\dfrac{\partial^2 y}{\partial x^2}$
   (c) $\dfrac{\partial^2 y}{\partial t^2} = a^2\dfrac{\partial^2 y}{\partial x^2}$ (d) none of these

5. Classify the following PDE:
   $$\frac{\partial^2 u}{\partial x^2}-2\sin x\frac{\partial^2 u}{\partial x\partial y}-\cos^2 x\frac{\partial^2 u}{\partial y^2}-4\sin x = 0$$
   (a) parabolic   (b) hyperbolic
   (c) elliptic   (d) none of these

6. $yu_{xx} + (x - y)u_{xy} - xu_{yy} = 0$ is
   (a) parabolic   (b) hyperbolic
   (c) elliptic   (d) none of these

7. $y^2u_{xx} - 2xyu_{xy} + x^2u_{yy} = \dfrac{2y^2}{x}u_x + \dfrac{x^2}{2y}u_y$
   (a) parabolic   (b) hyperbolic
   (c) elliptic   (d) none of these

8. $\sin^2 x \dfrac{\partial^2 u}{\partial x^2} - \sin 2x \dfrac{\partial^2 u}{\partial x \partial y} + \cos^2 x \dfrac{\partial^2 u}{\partial y^2} = 3u$ is

   (a) parabolic   (b) hyperbolic
   (c) elliptic   (d) none of these

9. Classify the following PDE:

   $(1+x^2)\dfrac{\partial^2 u}{\partial x^2} + (1+y^2)\dfrac{\partial^2 u}{\partial y^2} + y\dfrac{\partial u}{\partial x} + x\dfrac{\partial u}{\partial y} = 0$

   (a) parabolic   (b) hyperbolic
   (c) elliptic   (d) none of these

10. The solution of the PDE $u_{tt} = 4u_{xx}$, $t > 0$, $-\infty < x < \infty$ satisfying the conditions $u(x, 0) = x$, $u_t(x, 0) = 0$ is

    (a) x (b) $\dfrac{1}{2}$
    (c) $2t$ (d) $2x$

11. Let $u(x, t)$ be the solution of $\dfrac{\partial^2 u}{\partial t^2} = \dfrac{\partial^2 u}{\partial x^2}$, $-\infty < x < \infty$, $t > 0$ with $u(x, 0) = x - x^2$ and $u_t(x, 0) = 0$. Then $u(1,1) = ?$

    (a) 0   (b) –1
    (c) 1   (d) 2

12. Consider the initial value problem:

    $\dfrac{\partial^2 u}{\partial t^2} - \dfrac{\partial^2 u}{\partial x^2} = 0$, $u(x,0) = \sin x$, $\dfrac{\partial u}{\partial t}(x,0) = 1$

    Then $u\left(\pi, \frac{\pi}{2}\right) = ?$

    (a) $\dfrac{\pi}{2}$ (b) $1 - \dfrac{\pi}{2}$
    (c) 1   (d) $1 + \pi$

13. Let $u = \Psi(x, t)$ be the solution to the initial value problem $u_{tt} = u_{xx}$ for $-\infty < x < \infty$, $t > 0$ with $u(x, 0) = \sin x$, $u_t(x, 0) = \cos x$

    Then the value of $\psi\left(\dfrac{\pi}{2}, \dfrac{\pi}{6}\right)$ is

    (a) $\dfrac{\sqrt{3}}{2}$ (b) $\dfrac{1}{2}$
    (c) $\dfrac{1}{\sqrt{2}}$ (d) 1

Answer key				
1. (a)	2. (a)	3. (c)	4. (c)	5. (b)
6. (b)	7. (a)	8. (a)	9. (c)	10. (a)
11. (b)	12. (a)	13. (a)		

Explanation

1. (a) $u = ax + by + a^2 + b^2$ ...(i)
   Then differentiating (i) partially w.r.t x and y, we get

$\dfrac{\partial u}{\partial x} = a$ and $\dfrac{\partial u}{\partial y} = b$

Substituting these values of a and b in (i), we

get $u = x\dfrac{\partial u}{\partial x} + y\dfrac{\partial u}{\partial y} + \left(\dfrac{\partial u}{\partial x}\right)^2 + \left(\dfrac{\partial x}{\partial y}\right)^2$

i.e. $u = px + qy + p^2 + q^2$.

2. (a) $z = (x + a)(y + b)$ ...(i)

   $\Rightarrow \dfrac{\partial z}{\partial x} = y + b, \dfrac{\partial z}{\partial y} = x + a$

   Substituting these values of $(x + a)$ and $(y + b)$ in (i), we get $z = \dfrac{\partial z}{\partial x} \times \dfrac{\partial z}{\partial y}$.

3. (c) $u = (x^2 + a)(y^2 + b)$ ...(i)
   Then differentiating (i) partially w.r.t x and y, we get,

   $\dfrac{\partial u}{\partial x} = 2x(y^2 + b)$ i.e. $y^2 + b = \dfrac{1}{2x}\dfrac{\partial u}{\partial x}$ ...(ii)

   and $\dfrac{\partial u}{\partial y} = 2y(x^2 + a)$ i.e. $x^2 + a = \dfrac{1}{2y}\dfrac{\partial u}{\partial y}$ ...(iii)

   Substituting these values of $x^2 + a$ and $y^2 + b$ in

   (i), we get $u = \dfrac{1}{2y}\dfrac{\partial u}{\partial y} \times \dfrac{1}{2x}\dfrac{\partial u}{\partial x}$ i.e; $\dfrac{\partial u}{\partial x}\dfrac{\partial u}{\partial y} = 4xyu$.

4. (c) $y = f(x - at) + g(x + at)$

   $\Rightarrow \dfrac{\partial y}{\partial t} = (-a)f'(x - at) + ag'(x + at)$

   and $\dfrac{\partial^2 y}{\partial t} = af''(x - at) + a^2 g''(x + at)$

   Also $y = f(x - at) + g(x + at)$

   $\Rightarrow \dfrac{\partial y}{\partial x} = f'(x - at) + g'(x + at)$

   and $\dfrac{\partial^2 y}{\partial x^2} = f''(x - at) + g''(x + at)$

   $\therefore \dfrac{\partial^2 y}{\partial t^2} = a^2 \dfrac{\partial^2 y}{\partial x^2}$.

5. (b) Comparing the given equation with

   $A\dfrac{\partial^2 u}{\partial x^2} + B\dfrac{\partial^2 u}{\partial x \partial y} + C\dfrac{\partial^2 u}{\partial y^2} + D\dfrac{\partial u}{\partial x} + E\dfrac{\partial u}{\partial y} + Fu = G$,

   we get $A = 1$, $B = -2\sin x$, $C = -\cos^2 x$
   $\therefore$ $B^2 - 4AC = 4\sin^2 x - 4 \times 1 \times (-\cos^2 x)$
   $= 4(\sin^2 x + \cos^2 x) = 4 > 0$
   Hence, the given equation is hyperbolic.

6. (b) Comparing the given equation with

   $A\dfrac{\partial^2 u}{\partial x^2} + B\dfrac{\partial^2 u}{\partial x \partial y} + C\dfrac{\partial^2 u}{\partial y^2} + D\dfrac{\partial u}{\partial x} + E\dfrac{\partial u}{\partial y} + Fu = G$,

   we get $A = y$, $B = x - y$, $C = -x$.
   So $B^2 - 4AC = (x - y)^2 - 4y(-x)$

$= (x - y)^2 + 4xy = (x + y)^2 > 0$

Hence, the given equation is hyperbolic.

7. (*a*) Comparing the given equation with

$$A\frac{\partial^2 u}{\partial x^2} + B\frac{\partial^2 u}{\partial x \partial y} + C\frac{\partial^2 u}{\partial y^2} + D\frac{\partial u}{\partial x} + E\frac{\partial u}{\partial y} + Fu = G,$$

we get $A = y^2$, $B = -2xy$, $C = x^2$.

$B^2 - 4AC = (-2xy)^2 - 4y^2x^2$

$= 4x^2y^2 - 4x^2y^2 = 0$

Hence, the given equation is parabolic.

8. (*a*) Comparing the given equation with

$$A\frac{\partial^2 u}{\partial x^2} + B\frac{\partial^2 u}{\partial x \partial y} + C\frac{\partial^2 u}{\partial y^2} + D\frac{\partial u}{\partial x} + E\frac{\partial u}{\partial y} + Fu = G,$$

we get $A = \sin^2 x$, $B = -\sin 2x$, $C = \cos^2 x$.

∴ $B^2 - 4AC = (-\sin 2x)^2 - 4\sin^2 x \cos^2 x = 0$.

Hence, the given equation is parabolic.

9. (*c*) Comparing the given equation with

$$A\frac{\partial^2 u}{\partial x^2} + B\frac{\partial^2 u}{\partial x \partial y} + C\frac{\partial^2 u}{\partial y^2} + D\frac{\partial u}{\partial x} + E\frac{\partial u}{\partial y} + Fu = G,$$

we get $A = 1 + x^2$, $B = 1 + y^2$, $C = 0$.

∴ $B^2 - 4AC = 0 - (1 + x^2)(1 + y^2) < 0$.

Hence, the given equation is elliptic.

10. (*a*)

Here, $c = 2$, $f(x) = x$, $g(x) = 0$.

Hence, $u(x,t) = \dfrac{1}{2}\left[f(x + ct) + f(x - ct)\right] + \dfrac{1}{2}\displaystyle\int_{x-ct}^{x+ct} 0\, d\xi$

$= \dfrac{1}{2}\left[(x + t) + (x - t)\right] = x$.

11. (*b*)

Here, $c = 1$, $f(x) = x - x^2$, $g(x) = 0$. Hence,

$u(x,t) = \dfrac{1}{2}\left[f(x + t) + f(x - t)\right] + \dfrac{1}{2}\displaystyle\int_{x-t}^{x+t} 0\, d\xi$

$= \dfrac{1}{2}\left[\left\{(x + t) - (x + t)^2\right\} + \left\{(x - t) - (x - t)^2\right\}\right]$

$= \dfrac{1}{2}\left[2x - 2\left(x^2 + t^2\right)\right] = x - \left(x^2 + t^2\right)$

∴ $u(1,1) = 1 - (1 + 1) = -1$.

12. (*a*) Here, $c = 1$, $f(x) = \sin x$, $g(x) = 1$. Hence,

$u(x,t) = \dfrac{1}{2}\left[f(x - t) + f(x + t)\right] + \dfrac{1}{2}\displaystyle\int_{x-t}^{x+t} g(\xi)\, d\xi$

$= \dfrac{1}{2}\left[\sin(x + t) + \sin(x - t)\right] + \dfrac{1}{2}\displaystyle\int_{x-t}^{x+t} 1\, d\xi$

$= \dfrac{1}{2} \times (2\sin x \cos t) + \dfrac{1}{2}\left\{(x + t) - (x - t)\right\}$

$= \sin x \cos t + t$

∴ $u\left(\pi, \dfrac{\pi}{2}\right) = \sin \pi \cos \dfrac{\pi}{2} + \dfrac{\pi}{2} = \dfrac{\pi}{2}$.

13. (*a*) Here, $c = 1$, $f(x) = \sin x$, $g(x) = \cos x$. Hence,

$\psi(x,t) = \dfrac{1}{2}\left[\sin(x - t) + \sin(x + t)\right] + \dfrac{1}{2}\displaystyle\int_{x-t}^{x+t} \cos \xi\, d\xi$

$= \dfrac{1}{2}\left[\sin(x - t) + \sin(x + t)\right] + \dfrac{1}{2}\left[\sin \xi\right]_{x-t}^{x+t}$

$= \dfrac{1}{2}\left[\sin(x - t) + \sin(x + t)\right] + \dfrac{1}{2}\left[\sin(x + t) - \sin(x - t)\right]$

$= \sin(x + t)$

∴ $\psi\left(\dfrac{\pi}{2}, \dfrac{\pi}{6}\right) = \sin\left(\dfrac{\pi}{2} + \dfrac{\pi}{6}\right) = \sin \dfrac{2\pi}{3}$

$= \sin\left(\pi - \dfrac{\pi}{3}\right) = \cos \dfrac{\pi}{3} = \dfrac{\sqrt{3}}{2}$.

Fully Solved MCQs

1. The PDE corresponding to

$ax + by + cz = f(x^2 + y^2 + z^2)$ is

(*a*) $(cy - bz)\dfrac{\partial z}{\partial x} + (az - cx)\dfrac{\partial z}{\partial y} = bx - ay$

(*b*) $(cy + bz)\dfrac{\partial z}{\partial x} + (az + cx)\dfrac{\partial z}{\partial y} = bx + ay$

(*c*) $(by - cz)\dfrac{\partial z}{\partial x} + (cz - ax)\dfrac{\partial z}{\partial y} = ax - by$

(*d*) none of these

2. The PDE corresponding to

$u = xy + g(x^2 + y^2)$ is

(*a*) $px - qy = x^2 - y^2$ (*b*) $py + qx = 2y^2 - 2x^2$

(*c*) $py - qx = x^2 - y^2$ (*d*) $py + qx = y^2 + x^2$

3. The PDE corresponding to

$u = F(xy) + G(x/y)$ is

(*a*) $y^2\dfrac{\partial^2 u}{\partial x^2} + x^2\dfrac{\partial^2 u}{\partial y^2} + y\dfrac{\partial u}{\partial x} - x\dfrac{\partial u}{\partial y} = 0$

(*b*) $x^2\dfrac{\partial^2 u}{\partial x^2} - y^2\dfrac{\partial^2 u}{\partial y^2} + x\dfrac{\partial u}{\partial x} - y\dfrac{\partial u}{\partial y} = 0$

(*c*) $y^2\dfrac{\partial^2 u}{\partial x^2} + x^2\dfrac{\partial^2 u}{\partial y^2} + x\dfrac{\partial u}{\partial x} - y\dfrac{\partial u}{\partial y} = 0$

(*d*) none of these

4. The PDE corresponding to

$u = y\varphi(x) + x\Psi(y)$ is

(*a*) $xy\dfrac{\partial^2 u}{\partial x \partial y} = x\dfrac{\partial u}{\partial x} + y\dfrac{\partial u}{\partial y} - u$

(*b*) $\dfrac{\partial^2 u}{\partial x \partial y} = y\dfrac{\partial u}{\partial x} + x\dfrac{\partial u}{\partial y} - u$

(c) $(x+y)\dfrac{\partial^2 u}{\partial x \partial y} = x\dfrac{\partial u}{\partial x} - y\dfrac{\partial u}{\partial y} - u$

(d) $xy\dfrac{\partial^2 u}{\partial x \partial y} = \dfrac{\partial u}{\partial x} - xy\dfrac{\partial u}{\partial y} + 2u$

5. The PDE $x^2\dfrac{\partial^2 u}{\partial x^2} - (y^2-1)x\dfrac{\partial^2 u}{\partial x \partial y}$

$+y(y-1)^2\dfrac{\partial^2 u}{\partial y^2} + x\dfrac{\partial u}{\partial x} + y\dfrac{\partial u}{\partial y} = 0$

is hyperbolic in a region in the XY-plane if
(a) $x \neq 0$ and $y = 1$ (b) $x = 0$ and $y \neq 1$
(b) $x \neq 0$ and $y \neq 1$ (d) $x = 0$ and $y = 1$

6. In the region $x > 0$, $y > 0$, the PDE
$(x^2 - y^2)z_{xx} + 2(x^2 + y^2)z_{xy} + (x^2 - y^2)z_{yy} = 0$
(a) is elliptic   (b) is parabolic
(c) is hyperbolic   (d) changes its type

7. The PDE

$x\dfrac{\partial^2 u}{\partial x^2} + 2xy\dfrac{\partial^2 u}{\partial x \partial y} + y\dfrac{\partial^2 u}{\partial y^2} + \dfrac{\partial u}{\partial y} + \dfrac{\partial u}{\partial x} = 0$ is

(a) elliptic in the region $x < 0$, $y < 0$, $xy < 1$
(b) elliptic in the region $x > 0$, $y > 0$, $xy > 1$
(c) parabolic in the region $x < 0$, $y < 0$, $xy < 1$
(d) hyperbolic in the region $x < 0$, $y < 0$, $xy < 1$

8. If $u(x, y)$ be a solution of the initial value problem $\dfrac{\partial^2 u}{\partial t^2} = \dfrac{\partial^2 u}{\partial x^2}$, with $u(x,0) = x^2$, $\dfrac{\partial u}{\partial t}\Big|_{t=0} = 0$, then $u(1,0) = ?$
(a) 0   (b) 5
(c) 4   (d) 1

9. Consider the diffusion problem:
$u_{xx} = u_t$, $0 < x < p$, $t > 0$ with
$u(0, t) = 0 = u(\pi, t)$ and $u(x, 0) = 3\sin 2x$.
Then its solution is given by
(a) $3e^{-t}\sin 2x$ (b) $3e^{-4t}\sin 2x$
(c) $3e^{-9t}\sin 2x$ (d) $3e^{-2t}\sin 2x$

10. Consider the one dimensional wave equation:
$u_{tt} = 4u_{xx}$, $-\infty < x < \infty, t > 0$

with $u(x,0) = \begin{cases} 16 - x^2, & \text{if } |x| \leq 4 \\ 0, & \text{otherwise} \end{cases}$

and $u_t(x,0) = \begin{cases} 1, & \text{if } |x| \leqslant 2 \\ 0, & \text{otherwise} \end{cases}$

Then for $1 < t < 3$, $u(2, t)$ is equal to

(a) $\dfrac{1}{2}\left[16 - (2-2t)^2\right] + \dfrac{1}{2}\left[1 - \min\{1, \, t-1\}\right]$

(b) $\dfrac{1}{2}\left[32 - (2-2t)^2 - (2+2t)^2\right] + t$

(c) $\dfrac{1}{2}\left[32 - (2-2t)^2 - (2+2t)^2\right] + 1$

(d) $\dfrac{1}{2}\left[16 - (2-2t)^2\right] + \dfrac{1}{2}\left[1 - \max\{1-t, -1\}\right]$

11. A function $u(x, t)$ satisfies the

wave equation $\dfrac{\partial^2 u}{\partial t^2} = \dfrac{\partial^2 u}{\partial x^2}$, $0 < x < 1$, $t > 0$

If $u\left(\dfrac{1}{2}, 0\right) = \dfrac{1}{4}$, $u\left(1, \dfrac{1}{2}\right) = 1$ and $u\left(0, \dfrac{1}{2}\right) = \dfrac{1}{2}$

Then, $u\left(\dfrac{1}{2}, 1\right) = ?$

(a) $\dfrac{7}{4}$ (b) $\dfrac{5}{4}$

(c) $\dfrac{4}{5}$ (d) $\dfrac{4}{7}$

Answer key				
1. (a)	**2.** (c)	**3.** (b)	**4.** (a)	**5.** (c)
6. (c)	**7.** (a)	**8.** (d)	**9.** (b)	**10.** (b)
11. (b)				

Explanation

1. (a) Given that
$ax + by + cz = f(x^2 + y^2 + z^2)$ ...(i)
Differentiating (i) partially w.r.t x and y, we get

$a + c\dfrac{\partial z}{\partial x} = f'(x^2 + y^2 + z^2) \times \left[2x + 2z\dfrac{\partial z}{\partial x}\right]$ ...(ii)

And

$b + c\dfrac{\partial z}{\partial y} = f'(x^2 + y^2 + z^2) \times \left[2y + 2z\dfrac{\partial z}{\partial y}\right]$ ...(iii)

Dividing (ii) by (iii) we get,

$\dfrac{a + c\dfrac{\partial z}{\partial x}}{b + c\dfrac{\partial z}{\partial y}} = \dfrac{x + z\dfrac{\partial z}{\partial x}}{y + z\dfrac{\partial z}{\partial y}}$

or, $\left(a + c\dfrac{\partial z}{\partial x}\right)\left(y + z\dfrac{\partial z}{\partial y}\right) = \left(b + c\dfrac{\partial z}{\partial y}\right)\left(x + z\dfrac{\partial z}{\partial x}\right)$

or, $(cy - bz)\dfrac{\partial z}{\partial x} + (az - cx)\dfrac{\partial z}{\partial y} = bx - ay$.

2. (c) $u = xy + g(x^2 + y^2)$ ...(i)
Differentiating (i) partially w.r.t x and y, we get

$\dfrac{\partial u}{\partial x} = y + 2xg'(x^2 + y^2)$ ...(ii) and $\frac{\partial u}{\partial y} = x + 2yg'(x^2 + y^2)$ ...(iii)

From (ii) and (iii) we get,

$$\frac{\frac{\partial u}{\partial x} - y}{2x} = \frac{\frac{\partial u}{\partial y} - x}{2y}$$

or, $y\frac{\partial u}{\partial x} - y^2 = x\frac{\partial u}{\partial y} - x^2$

or, $x\frac{\partial u}{\partial y} - y\frac{\partial u}{\partial x} = y^2 - x^2$

or, $py - qx = x^2 - y^2$.

3. (b) $u = F(xy) + G(x/y)$ ...(i)

So $\frac{\partial u}{\partial x} = yF'(xy) + \frac{1}{y}G'(x/y)$ ...(ii)

$\frac{\partial u}{\partial y} = xF'(xy) - \frac{x}{y^2}G'(x/y)$ ...(iii)

Differentiating (ii) partially w.r.t x partially we get,

$$\frac{\partial^2 u}{\partial x^2} = y^2 F''(xy) + \frac{1}{y^2}G''(x/y)$$

Again, Differentiating (iii) partially w.r.t y we get,

$$\frac{\partial^2 u}{\partial y^2} = x^2 F''(xy) + \left[\frac{x^2}{y^4}G''(x/y) + \frac{2x}{y^3}G'(x/y)\right]$$

$$\therefore x^2\frac{\partial^2 u}{\partial x^2} - y^2\frac{\partial^2 u}{\partial y^2}$$

$$= x^2 y^2 F''(xy) + \frac{x^2}{y^2}G''(x/y)$$

$$-x^2 y^2 F''(xy) - \frac{x^2}{y^2}G''(x/y) - \frac{2x}{y}G'(x/y)$$

$$= -\frac{2x}{y}G'(x/y)$$ ...(iv)

Also, $x\frac{\partial u}{\partial x} - y\frac{\partial u}{\partial y}$

$$= xyF'(xy) + \frac{x}{y}G'(x/y) - xyF'(xy) + \frac{x}{y}G'(x/y)$$

$$= \frac{2x}{y}G'(x/y)$$ ...(v)

Using (v), we get from (iv),

$$x^2\frac{\partial^2 u}{\partial x^2} - y^2\frac{\partial^2 u}{\partial y^2} = -\left[x\frac{\partial u}{\partial x} - y\frac{\partial u}{\partial y}\right]$$

or, $x^2\frac{\partial^2 u}{\partial x^2} - y^2\frac{\partial^2 u}{\partial y^2} + x\frac{\partial u}{\partial x} - y\frac{\partial u}{\partial y} = 0$.

4. (a) Given that $u = y\varphi(x) + x\Psi(y)$ ...(i)

$\therefore \frac{\partial u}{\partial x} = y\varphi(x) + x\psi(y)$ ...(ii) and

$\frac{\partial u}{\partial y} = \varphi(x) + x\psi'(y)$ ...(iii)

Differentiating (iii) w.r.t. x, we have

$\frac{\partial^2 u}{\partial x\partial y} = \varphi'(x) + \psi'(y)$ ...(iv)

From (ii) and (iii) we get,

$$\frac{\partial^2 u}{\partial x\partial y} = \frac{1}{y}\left[\frac{\partial u}{\partial x} - \psi(x)\right] + \frac{1}{x}\left[\frac{\partial u}{\partial y} - \varphi(x)\right]$$

or, $xy\frac{\partial^2 u}{\partial x\partial y} = x\frac{\partial u}{\partial x} - x\psi(y) + y\frac{\partial u}{\partial y} - y\varphi(x)$

or $xy\frac{\partial^2 u}{\partial x\partial y} = x\frac{\partial u}{\partial x} + y\frac{\partial u}{\partial y} - u$ $[\because \ u = y\varphi(x) + x\Psi(y)]$

5. (c) Comparing the given equation with

$$A\frac{\partial^2 u}{\partial x^2} + B\frac{\partial^2 u}{\partial x\partial y} + C\frac{\partial^2 u}{\partial y^2} + D\frac{\partial u}{\partial x} + E\frac{\partial u}{\partial y} + Fu = G,$$

we get $A = x^2$, $B = -(y^2 - 1)x$, $C = y(y - 1)^2$.

$\therefore \ B^2 - 4AC = \{-x(y^2 - 1)\} - 4x^2 y(y - 1)^2$

$= (y - 1)^2 x^2\{(y + 1)^2 - 4y\}$

$= (y - 1)^2 x^2 (y - 1)^2$

$= x^2(y - 1)^4 > 0$ (if $\ x \neq 0$ and $y \neq 1$)

6. (c) Comparing the given equation with

$$A\frac{\partial^2 u}{\partial x^2} + B\frac{\partial^2 u}{\partial x\partial y} + C\frac{\partial^2 u}{\partial y^2} + D\frac{\partial u}{\partial x} + E\frac{\partial u}{\partial y} + Fu = G,$$

we get $A = x^2 - y^2$, $B = 2(x^2 + y^2)$, $C = x^2 - y^2$.

$\therefore \ B^2 - 4AC$

$= \{2(x^2 + y^2)\}^2 - 4(x^2 - y^2) \times (x^2 - y^2)$

$= 4(x^2 + y^2)^2 - 4(x^2 - y^2)^2$

$= 4\left\{\left(x^2 + y^2\right)^2 - \left(x^2 - y^2\right)^2\right\}$

$= 4 \times 2x^2 y^2 = 8x^2 y^2 > 0$ (since $x > 0, y > 0$)

Hence, the given PDE is hyperbolic in the given region $x > 0, y > 0$.

7. (a) Comparing the given equation with

$$A\frac{\partial^2 u}{\partial x^2} + B\frac{\partial^2 u}{\partial x\partial y} + C\frac{\partial^2 u}{\partial y^2} + D\frac{\partial u}{\partial x} + E\frac{\partial u}{\partial y} + Fu = G,$$

we get $A = x$, $B = 2xy$, $C = y$.

$\therefore \ B^2 - 4AC = (2xy)^2 - 4xy$

$= 4xy\,(xy - 1)$ ...(i)

Now if $x < 0, y < 0$ and $xy < 1$, then
$xy > 0$ and $xy - 1 < 0$ i.e; $xy(xy - 1) < 0$.
So from (i), $B^2 - 4AC = 4xy\,(xy - 1) < 0$.
Hence, the given PDE is elliptic.

8. (d) Here, $u(x,t) = \varphi_1(x+ct) + \varphi_2(x-ct)$

$\qquad = \varphi_1(x+t) + \varphi_2(x-t)$

Then, $u(x, 0) = x^2$

$\Rightarrow \varphi_1(x) + \varphi_2(x) = x^2$ ...(i)

Here, $\dfrac{\partial u}{\partial t} = \varphi_1(x+t) - \varphi_2(x-t)$

$\therefore \dfrac{\partial u}{\partial t}\Big|_{t=0} = 0 \Rightarrow \varphi_1(x) - \varphi_2(x) = 0$ ...(ii)

$(i) + (ii) \Rightarrow 2\varphi_1(x) = x^2 \Rightarrow \varphi_1(x) = \dfrac{x^2}{2}$

$(i) - (ii) \Rightarrow 2\varphi_2(x) = x^2 \Rightarrow \varphi_2(x) = \dfrac{x^2}{2}$

Hence, $u(x,t) = \dfrac{(x+t)^2}{2} + \dfrac{(x-t)^2}{2} = x^2 + t^2$.

$\therefore u(1,0) = 1^2 + 0^2 = 1$.

9. (b) The solution is given by

$u(x,t) = \sum_{n=1}^{\infty} b_n e^{\frac{-c^2 n^2 \pi^2 t}{l^2}} \sin\left(\dfrac{n\pi x}{l}\right)$

$\qquad = \sum_{n=1}^{\infty} b_n e^{\frac{-n^2 \pi^2 t}{\pi^2}} \sin\left(\dfrac{n\pi x}{\pi}\right)$

$\qquad\qquad [\because \text{ here } c=1, \ l=\pi]$

$\qquad = \sum_{n=1}^{\infty} b_n e^{-n^2 t} \sin nx$ ....(i)

$b_n = \dfrac{2}{l} \int_0^l f(x) \sin\left(\dfrac{n\pi x}{l}\right) dx$

$\quad = \dfrac{2}{\pi} \int_0^\pi 3 \sin 2x \times \sin\left(\dfrac{n\pi x}{\pi}\right) dx$

$\quad = \dfrac{3}{\pi} \int_0^\pi 2\sin 2x \sin nx\, dx$

$\quad = \dfrac{3}{\pi} \int_0^\pi \left[\cos(nx-2x) - \cos(nx+2x)\right] dx$

$\quad = \dfrac{3}{\pi} \int_0^\pi \left[\cos\{(n-2)x\} - \cos\{(n+2)x\}\right] dx$

$\quad = \dfrac{3}{\pi}\left[\dfrac{\sin\{(n-2)x\}}{n-2} - \dfrac{\sin\{(n+2)x\}}{n+2}\right]_0^\pi$

$\quad = \dfrac{3}{\pi}\left[\dfrac{\sin\{(n-2)\pi\}}{n-2} - \dfrac{\sin\{(n+2)\pi\}}{n+2}\right]$

$\quad = \dfrac{3}{\pi} \times (0-0) = 0 \quad [for\ n \neq 2]$

$(i) \Rightarrow u(x,0) = \sum_{n=1}^{\infty} b_n \sin nx$

$\Rightarrow 3\sin 2x = b_1 \sin x + b_2 \sin 2x + b_3 \sin 3x + \dots\dots\infty$

$\Rightarrow 3\sin 2x = 0 + b_2 \sin 2x + 0 + 0 + \dots\dots\infty$

$\qquad\qquad (\because b_n = 0 \ \forall\ n(\neq 2) \in N)$

$\Rightarrow b_2 = 3$

Hence, (i) gives

$u(x,t) = b_2 e^{-4t} \sin 2x \ (\because b_1 = b_3 = b_4 = b_5 = \dots = 0)$

$\qquad = 3e^{-4t} \sin 2x$

10. (b) Comparing $u_{tt} = 4u_{xx}$ with $u_{tt} = c^2 u_{xx}$, we get

$c = 2$, Here $u(x, 0) = f(x)$.

$\therefore u(x,t) = \dfrac{1}{2}\left[f(x-2t) + f(x+2t)\right] + \dfrac{1}{2\times 2}\int_{x-2t}^{x+2t} g(\xi)d\xi$

So, $u(2,t) = \dfrac{1}{2}\left[f(2-2t) + f(2+2t)\right] + \dfrac{1}{4}\int_{2-2t}^{2+2t} g(\xi)d\xi$

$= \dfrac{1}{2}\left[\{16-(2-2t)^2\} + \{16-(2+2t)^2\}\right] + \dfrac{1}{4}\int_{2-2t}^{2+2t} 1 d\xi$

$= \dfrac{1}{2}\left[32 - (2-2t)^2 - (2+2t)^2\right]$

$+ \dfrac{1}{4}\left[(2+2t) - (2-2t)\right]$

$= \dfrac{1}{2}\left[32 - (2-2t)^2 - (2+2t)^2\right] + t$.

11. (b)

$u(x,t) = \varphi(x+t) + \psi(x-t)$

$\therefore u\left(\dfrac{1}{2},0\right) = \dfrac{1}{4} \Rightarrow \varphi\left(\dfrac{1}{2}\right) + \psi\left(\dfrac{1}{2}\right) = \dfrac{1}{4}$ ...(i)

$u\left(1,\dfrac{1}{2}\right) = 1 \Rightarrow \varphi\left(1+\dfrac{1}{2}\right) + \psi\left(1-\dfrac{1}{2}\right) = 1$

$\Rightarrow \varphi\left(\dfrac{3}{2}\right) + \psi\left(\dfrac{1}{2}\right) = 1$ ...(ii)

$u\left(0,\dfrac{1}{2}\right) = \dfrac{1}{2} \Rightarrow \varphi\left(\dfrac{1}{2}\right) + \psi\left(-\dfrac{1}{2}\right) = \dfrac{1}{2}$ ...(iii)

Eqn (ii) + Eqn (iii) − Eqn (i) ⇒

$\varphi\left(\dfrac{3}{2}\right) + \cancel{\psi\left(\dfrac{1}{2}\right)} + \cancel{\varphi\left(\dfrac{1}{2}\right)} + \psi\left(-\dfrac{1}{2}\right) - \cancel{\varphi\left(\dfrac{1}{2}\right)} - \cancel{\psi\left(\dfrac{1}{2}\right)}$

$= 1 + \dfrac{1}{2} - \dfrac{1}{4}$

$\Rightarrow \varphi\left(\dfrac{3}{2}\right) + \psi\left(-\dfrac{1}{2}\right) = \dfrac{5}{4}$

$\Rightarrow \varphi\left(\dfrac{1}{2}+1\right) + \psi\left(\dfrac{1}{2}-1\right) = \dfrac{5}{4} \Rightarrow u\left(\dfrac{1}{2},1\right) = \dfrac{5}{4}$.

1. The number of boundary conditions required to solve the differential equation $\dfrac{\partial^2 \varphi}{\partial x^2} + \dfrac{\partial^2 \varphi}{\partial y^2} = 0$ is

(a) 2   (b) 0
(c) 4   (d) 1

[CE GATE 2001]

2. The partial differential equation

$\dfrac{\partial^2 \varphi}{\partial x^2} + \dfrac{\partial^2 \varphi}{\partial y^2} + \dfrac{\partial \varphi}{\partial x} + \dfrac{\partial \varphi}{\partial y} = 0$ has

(a) degree 1 and order 2
(b) degree 1 and order 1
(c) degree 2 and order 1
(d) degree 2 and order 2

[ME GATE 2007]

3. The partial differential equation that can be formed from $z = ax + by + ab$ has the form $\left(\text{with } p = \dfrac{\partial z}{\partial x}, q = \dfrac{\partial z}{\partial y} \right)$ is

(a) $z = px + qy$ (b) $z = px + pq$
(c) $z = px + qy + pq$ (d) $z = qy + py$

[CE GATE 2010]

4. The partial differential equation

$\dfrac{\partial u}{\partial t} + u\dfrac{\partial u}{\partial x} = \dfrac{\partial^2 u}{\partial x^2}$ is a

(a) linear equation of order 2
(b) non-linear equation of order 1
(c) linear equation of order 1
(d) non-linear equation of order 2

[ME GATE 2013]

5. The type of partial differential equation

$\dfrac{\partial f}{\partial t} = \dfrac{\partial^2 f}{\partial x^2}$ is

(a) parabolic   (b) elliptic
(c) hyperbolic   (d) non-linear

[IN GATE 2013]

6. The type of the partial differential equation

$\dfrac{\partial^2 P}{\partial x^2} + \dfrac{\partial^2 P}{\partial y^2} + 3\dfrac{\partial^2 P}{\partial x \partial y} + 2\dfrac{\partial P}{\partial x} - \dfrac{\partial P}{\partial y} = 0$ is

(a) elliptic   (b) parabolic
(c) hyperbolic   (d) none of these

[CE GATE 2016]

7. Which one of the followings a property of the solutions to the Laplace equation $\nabla^2 f = 0$?

(a) The solutions have neither maxima nor minima anywhere except at the boundaries
(b) The solutions are not separable in the co-ordinates
(c) The solutions are not continuous
(d) The solutions are not dependent on boundary conditions

[ME GATE 2016]

8. The solution of the partial differential equation $\dfrac{\partial u}{\partial t} = \alpha \dfrac{\partial^2 u}{\partial x^2}$ is of the form

(a) $C \cos kt \left[C_1 e^{\left(\sqrt{k/\alpha}\right)x} + C_2 e^{-\left(\sqrt{k/\alpha}\right)x} \right]$

(b) $C e^{kt} \left[C_1 e^{\left(\sqrt{k/\alpha}\right)x} + C_2 e^{-\left(\sqrt{k/\alpha}\right)x} \right]$

(c) $C e^{kt} \left[C_1 \cos\left\{\left(\sqrt{k/\alpha}\right)x\right\} + C_2 \sin\left\{-\left(\sqrt{k/\alpha}\right)x\right\} \right]$

(d) $C \sin kt \left[C_1 \cos\left\{\left(\sqrt{k/\alpha}\right)x\right\} + C_2 \sin\left\{-\left(\sqrt{k/\alpha}\right)x\right\} \right]$

[CE GATE 2016]

9. Consider the following partial differential equation $\dfrac{\partial u}{\partial y} + c\dfrac{\partial u}{\partial x} = 0.$

Solution of this equation is
(a) $u(x, y) = f(x + cy)$ (b) $u(x, y) = f(x - cy)$
(c) $u(x, y) = f(cx + y)$ (d) $u(x, y) = f(cx - y)$

[ME GATE 2017]

10. Consider the following partial differential equation $3\dfrac{\partial^2 \varphi}{\partial x^2} + B\dfrac{\partial^2 \varphi}{\partial x \partial y} + 3\dfrac{\partial^2 \varphi}{\partial y^2} + 4\varphi = 0.$

For the equation to be classified as parabolic, the value of B^2 must be_____?

[CE GATE 2017]

11. Consider a function u which depends on position "x" and time "t." The partial differential equation $\dfrac{\partial u}{\partial t} = \dfrac{\partial^2 u}{\partial x^2}$ is known as the

[ME GATE 2018]

(a) Wave equation
(b) Heat equation
(c) Laplace's equation
(d) Elasticity equation

12. The solution at $x = 1$, $t = 1$ of the partial differential equation $\dfrac{\partial^2 u}{\partial x^2} = 25\dfrac{\partial^2 u}{\partial t^2}$ subject to initial conditions of $u(0) = 3x$ and $\dfrac{\partial u}{\partial t}\Big|_{t=0} = 3$ is _____

[CE GATE 2018]

(a) 1   (b) 2
(c) 4   (d) 6

Answer key				
1. (c)	**2.** (a)	**3.** (c)	**4.** (d)	**5.** (a)
6. (c)	**7.** (a)	**8.** (b)	**9.** (b)	**10.** 36
11. (b)	**12.** (d)			

Explanation

1. (c) We know that the general solution of the Laplace equation $\dfrac{\partial^2 \varphi}{\partial x^2} + \dfrac{\partial^2 \varphi}{\partial y^2} = 0$ contains four arbitrary constants. Hence, "4" boundary conditions are needed in this case.

2. (a) The order of the equation = order of highest order derivative in the equation = 2;
The degree of the equation = exponent of the highest order derivative in the equation = 1.

3. (c) The given equation is $z = ax + by + ab$ ...(i)
Differentiating (i) partially w.r.t x and y, we get

$$p = \frac{\partial z}{\partial x} = a \text{ and } q = \frac{\partial z}{\partial y} = b.$$

Substituting the values of "a" and "b" we get from (i),
$z = px + qy + pq$.

4. (d) Order of the equation = order of highest order derivative in the equation = 2.

Again, the product term $u\dfrac{\partial u}{\partial x}$ is present in the equation.
Hence, the given equation is a non-linear equation of order 2.

5. (a) Comparing the given equation with

$$A\frac{\partial^2 u}{\partial x^2} + B\frac{\partial^2 u}{\partial x\partial y} + C\frac{\partial^2 u}{\partial y^2} + D\frac{\partial u}{\partial x} + E\frac{\partial u}{\partial y} + Fu = G,$$

we get $A = 1$, $B = 0$, $C = 0$.
$\therefore \quad B^2 - 4AC = 0 - 0 = 0$
Hence, the given *PDE* is parabolic.

6. (c) Comparing the given equation with

$$A\frac{\partial^2 P}{\partial x^2} + B\frac{\partial^2 P}{\partial x\partial y} + C\frac{\partial^2 P}{\partial y^2} + D\frac{\partial P}{\partial x} + E\frac{\partial P}{\partial y} + FP = G,$$

we get $A = 1$, $B = 3$, $C = 1$.
$\therefore \quad B^2 - 4AC = 9 - 4 > 0$.
Hence, the given PDE is hyperbolic.

7. (a) For a Laplace equation, solutions are always continuous and are separable in any coordinate system. The arbitrary constants occurring in the solution can be determined using the boundary conditions.

8. (b)
Here, $u = u(x,t) = X(x)T(t) = XT$ (say)

Then, $\dfrac{\partial u}{\partial x} = X'T, \dfrac{\partial^2 u}{\partial x^2} = X''T$ and $\dfrac{\partial u}{\partial t} = XT'$

$$\left(\text{Here } X' = \frac{dX}{dx}, \ T' = \frac{dY}{dt}\right)$$

$$\therefore \frac{\partial u}{\partial t} = \alpha\frac{\partial^2 u}{\partial x^2} \Rightarrow XT' = \alpha X''T$$

$$\Rightarrow \frac{\alpha X''}{X} = \frac{T'}{T} = k(\text{say})$$

Then, $\dfrac{\alpha X''}{X} = k \Rightarrow X'' - \dfrac{k}{\alpha}X = 0$ ....(i)

Auxiliary equation of (i) is given by:

$$m^2 - \frac{k}{\alpha} = 0 \ \ i.e; \ \ m = \pm\sqrt{\frac{k}{\alpha}}.$$

Hence, the solution of (i) is given by

$$X = X(x) = C_1 e^{\left(\sqrt{\frac{k}{\alpha}}\right)x} + C_2 e^{-\left(\sqrt{\frac{k}{\alpha}}\right)x}.$$

$$\frac{T'}{T} = k \Rightarrow \frac{1}{T}\frac{dT}{dt} = k \Rightarrow \frac{dT}{T} = k\,dt$$

Integrating we get,
$\log T = kt + \log C$
$or, \log T - \log C = kt$
$or, \log\dfrac{T}{C} = kt$
$or, T = Ce^{kt}$

$$\therefore u(x,t) = XT = \left\{C_1 e^{\left(\sqrt{\frac{k}{\alpha}}\right)x} + C_2 e^{-\left(\sqrt{\frac{k}{\alpha}}\right)x}\right\}Ce^{kt}.$$

9. (b) Let $u(x, y) = f(x - cy)$.

$$\therefore \quad \frac{\partial u}{\partial x} = f'(x - cy) \text{ and } \frac{\partial u}{\partial y} = -cf'(x - cy).$$

Then, $\dfrac{\partial u}{\partial y} + c\dfrac{\partial u}{\partial x} = -cf'(x - cy) + cf'(x - cy) = 0.$

10. 36

Comparing the given equation with

$$A\frac{\partial^2 u}{\partial x^2} + B\frac{\partial^2 u}{\partial x\partial y} + C\frac{\partial^2 u}{\partial y^2} + D\frac{\partial u}{\partial x} + E\frac{\partial u}{\partial y} + Fu = G,$$

we get $A = 3$, $B = B$, $C = 3$.
Then, equation is parabolic

$$\Rightarrow B^2 - 4AC = 0$$
$$\Rightarrow B^2 - 4\times3\times3 = 0$$
$$\Rightarrow B^2 = 36.$$

11. (b) We know that the general form of one dimensional heat flow equation is $\frac{\partial u}{\partial t} = c^2\frac{\partial^2 u}{\partial x^2}$.

For $c = 1$, the equation becomes $\frac{\partial u}{\partial t} = \frac{\partial^2 u}{\partial x^2}$.

Hence, the given partial differential equation is a heat equation.

12. (d) We know that the D'Alembert's solution of the following problem:

$$\frac{\partial^2 u}{\partial t^2} = c^2\frac{\partial^2 u}{\partial x^2}, \quad -\infty < x < \infty, \ t > 0$$

with initial condition $u(x,0) = f(x)$, and

$$\frac{\partial u}{\partial t}\Big|_{t=0} = g(x) \text{ is given by}$$

$$u(x,t) = \frac{1}{2}\left[f(x+ct) + f(x-ct)\right] + \frac{1}{2c}\int_{x-ct}^{x+ct} g(\xi)d\xi.$$

Here, $c^2 = \frac{1}{25}$ i.e; $c = \frac{1}{5}$, $f(x) = 3x$ and $g(x) = 3$.

$$\therefore u(1,1) = \frac{1}{2}\left[f(1+c) + f(1-c)\right] + \frac{1}{2c}\int_{1-c}^{1+c} g(\xi)d\xi$$

$$= \frac{1}{2}\left[f\left(1+\frac{1}{5}\right) + f\left(1-\frac{1}{5}\right)\right] + \frac{5}{2}\int_{1-\frac{1}{5}}^{1+\frac{1}{5}} g(\xi)d\xi$$

$$= \frac{1}{2}\left[3\times\left(1+\frac{1}{5}\right) + 3\times\left(1-\frac{1}{5}\right)\right] + \frac{5}{2}\int_{1-\frac{1}{5}}^{1+\frac{1}{5}} 3\,d\xi$$

$$= \frac{1}{2}\left[\frac{18}{5} + \frac{12}{5}\right] + \frac{15}{2}\times[\xi]_{1-\frac{1}{5}}^{1+\frac{1}{5}}$$

$$= 3 + \frac{15}{2}\times\left[\left(1+\frac{1}{5}\right) - \left(1-\frac{1}{5}\right)\right] = 6.$$

1. Solve by the method of separation of variables,

$$\frac{\partial u}{\partial x} = 2\frac{\partial u}{\partial t} + u, \text{ where } u(x, 0) = 6e^{-3x}$$

(a) $u(x, 0) = 6e^{-3x}$
(b) $u = 6e^{3x}-4t$
(c) $u = 6e^{3x+4t}$
(d) $u = 8e^{-(2x+3t)}$

2. Solve: $3\frac{\partial u}{\partial x} + 2\frac{\partial u}{\partial y} = 0; u(x,o) = 4e^{-x}$

(a) $u = 6e^{x-2y}$
(b) $u = 4e^{-x+\frac{3y}{2}}$
(c) $u = 4e^{-x+2y}$
(d) none of these.

3. The equation $t\frac{\partial^2 u}{\partial t^2} + 3\frac{\partial^2 u}{\partial x\partial t} + x\frac{\partial^2 u}{\partial x^2} + \frac{\partial u}{\partial y} = 0$ is

(a) Hyperbolic if $4xt > 9$
(b) Parabolic if $4xt \le 9$
(c) Elliptic if $4xt > 9$
(d) Parabolic if $2xt = 3$

4. The equation $\frac{\partial^2 u}{\partial x^2} = \frac{\partial u}{\partial y}$ is

(a) Parabolic
(b) Hyperbolic
(c) Elliptic
(d) None of these

5. Classify the following equation in second quadrant:

$$\sqrt{x^2+y^2}\frac{\partial^2 u}{\partial x^2} + 2(x-y)\frac{\partial^2 u}{\partial x\partial y} + \sqrt{y^2+x^2}\frac{\partial^2 u}{\partial y^2}$$

$$+x\frac{\partial u}{\partial y} + y\frac{\partial u}{\partial y} = 0$$

(a) Elliptic for $x > 0$
(b) Parabolic for $y < 0$
(c) Hyperbolic
(d) Parabolic for $x = 0$

6. The partial differential equation that can be formed from $z = ax + by + ab$ is given by

(a) $z = xz_x + yz_y + z_xz_y$
(b) $z = yz_x + xz_y + z_xz_y$
(c) $z = z_x + z_y + xyz_xz_y$
(d) None of these.

7. The PDE corresponding to $2z = (ax + y)^2 + b$ is given by

(a) $x\frac{\partial z}{\partial y} + y\frac{\partial z}{\partial y} = 0$

(b) $x\frac{\partial z}{\partial x} + y\frac{\partial z}{\partial y} = \left(\frac{\partial z}{\partial y}\right)^2$

(c) $y\frac{\partial z}{\partial x} + x\frac{\partial z}{\partial y} = \frac{\partial z}{\partial x}\frac{\partial z}{\partial y}$

(d) None of these

8. Let $u(x, t)$ be the D'Alembert's solution of the initial value problem for the wave equation

$u_{tt} - c^2 u_{xx} = 0; u(x,0) = f(x), u_t(x,0) = g(x);$ where c is a positive real number and $f(x)$, $g(x)$ are add functions. Then, $u(0,1)$ is equal to

(a) –1 (b) 1
(c) 0 (d) 2

9. Let $u(x, t)$ be the solution of the initial value problem:

$u_{tt} = u_{xx}, u(x,0) = x, u_t(x,0) = 1,$

Then, $u(2, 2)$.is equal to

(a) 4 (b) 6
(c) –2 (d) 0

10. Let $u(x, y) = 2\cos(x - 2y) \times f(y)$ be a solution of the initial value problem

$2u_x + u_y = u; \; u(x,0) = \cos x$

Then, $f(1) =$

(a) $\dfrac{1}{2}$ (b) e

(c) $\dfrac{e}{2}$ (d) 1

Answer key				
1. (a)	**2.** (b)	**3.** (c)	**4.** (a)	**5.** (c)
6. (a)	**7.** (b)	**8.** (c)	**9.** (a)	**10.** (c)

CHAPTER 6

LAPLACE TRANSFORMS

6.1 BASICS OF LAPLACE TRANSFORMS

6.1.1 Definition of the Laplace Transform

Let $f(t)$ be defined for all real $t \geq 0$. Then the Laplace transform of $f(t)$, denoted by $L[f(t)]$ or $F(s)$ or $\bar{f}(s)$, is defined by $L[f(t)] = F(s) = \int_0^\infty e^{-st} f(t) dt$,

provided the integral on the right-hand side exists (here s is *a* real or complex variable).

6.1.2 Linear Property of the Laplace Transform

Let $f(t)$ and $g(t)$ be two functions of t and a, b be two constants. Then,

$$L\{af(t) + bg(t)\} = a\,L\{f(t)\} + bL\{g(t)\}$$

Example:

$$L(2e^t - 3\sin t) = 2L(e^t) - 3L(\sin t)$$

6.1.3 Fundamental Formulas of the Laplace Transform

(i) $L(1) = \dfrac{1}{s}$ (provided $s > 0$)

(ii) $L(t^n) = \begin{cases} \dfrac{\Gamma(n+1)}{s^{n+1}} & \text{if } s > 0 \quad \text{(in general)} \\[2ex] \dfrac{n!}{s^{n+1}} & \text{if } s > 0 \text{ and } n \in Z^+ \end{cases}$

Example:

$$L(t^{\frac{1}{2}}) = \frac{\Gamma\left(\dfrac{1}{2}+1\right)}{s^{\frac{1}{2}+1}} = \frac{\Gamma\left(\dfrac{3}{2}\right)}{s^{\frac{3}{2}}}$$

and $L(t^2) = \dfrac{2!}{s^{2+1}} = \dfrac{2}{s^3}$.

(iii) $L(e^{at}) = \dfrac{1}{s-a}$ provided $s > a$

Example:

$$L(e^{5t}) = \frac{1}{s-5}, \ L(e^{-2t}) = \frac{1}{s+2}.$$

(iv) $L(\sin at) = \dfrac{a}{s^2 + a^2}$ provided $s > 0$

Example:

$$L(\sin 3t) = \frac{3}{s^2 + 3^2} = \frac{3}{s^2 + 9}$$

(v) $L(\cos at) = \dfrac{s}{s^2 + a^2}$ provided $s > 0$

Example:

$$L(\cos 2t) = \frac{s}{s^2 + 2^2} = \frac{s}{s^2 + 4}$$

(vi) $L(\sinh at) = \dfrac{a}{s^2 - a^2}$ provided $|s| > a$

$$\left[\underline{\text{Remember}} : \sinh x = \frac{e^x - e^{-x}}{2} \right]$$

Example:

$$L(\sinh t) = \frac{1}{s^2 - 1}$$

(vii) $L(\cosh at) = \frac{s}{s^2 - a^2}$ provided $|s| > a$

$$\left[\underline{\text{Remember}} : \cosh x = \frac{e^x + e^{-x}}{2} \right]$$

Example:

$$L(\cosh 3t) = \frac{s}{s^2 - 9}$$

6.1.4 First Shifting Theorem

If $L\{f(t)\} = F(s)$, then $L\{e^{at} f(t)\} = F(s-a)$, "$a$" being a real or complex constant.

Application:

$$L\{e^t \sin t\} = L\{e^{at} f(t)\}$$

Here, $a = 1$ and $f(t) = \sin t$.

$$\therefore F(s) = L\{f(t)\} = L\{\sin t\} = \frac{1}{s^2 + 1}.$$

Hence, by 1st shifting theorem,

$$L\{e^t \sin t\} = L\{e^{at} f(t)\} = F(s-a)$$

$$= F(s-1) = \frac{1}{(s-1)^2 + 1}.$$

6.1.5 Some Advanced Formulas of the Laplace Transform

(i) $L(e^{at} \cos bt) = \frac{s-a}{(s-a)^2 + b^2}$

Example:

$$L(e^t \cos 2t) = \frac{s-1}{(s-1)^2 + 4}$$

(ii) $L(e^{at} \sin bt) = \frac{b}{(s-a)^2 + b^2}$

Example:

$$L(e^{-t} \sin 3t) = \frac{3}{(s+1)^2 + 9}$$

(iii) $L(e^{at} \cosh bt) = \frac{s-a}{(s-a)^2 - b^2}$

Example:

$$L(e^{-2t} \cosh t) = \frac{s+2}{(s+2)^2 - 1}$$

(iv) $L(e^{at} \sinh bt) = \frac{b}{(s-a)^2 - b^2}$

Example:

$$L(e^{3t} \sinh 2t) = \frac{2}{(s-3)^2 - 4}$$

(v) $L(e^{at} t^n) = \frac{\lfloor n}{(s-a)^{n+1}}, \quad n \in Z^+$

Example:

$$L(e^t t^3) = \frac{\lfloor 3}{(s-1)^4}$$

6.1.6 Change of Scale Property

If $L\{f(t)\} = F(s)$, then $L\{f(at)\} = \frac{1}{a} F(s/a)$, "$a$" being a real or complex constant.

Application:

We know that $L(e^{2t}) = \frac{1}{s-2}$.

Take $f(t) = e^{2t}$. Then $F(s) = \frac{1}{s-2}$.

$$\therefore L\{f(3t)\} = \frac{1}{3} F(s/3) = \frac{1}{3} \times \frac{1}{\left(\frac{s}{3} - 2\right)}$$

$$= \frac{1}{s-6}$$

6.2 LAPLACE TRANSFORM ON DERIVATIVES

Let $L\{f(t)\} = F(s)$. Also let $f(t), f'(t), f''(t), ..., f^{(n-1)}(t)$ are all continuous for $t \geq 0$. Then the Laplace transform of nth order derivative of $f(t)$ is denoted by $L\{f^{(n)}(t)\}$ and is defined by

$$L\{f^{(n)}(t)\} = s^n F(s) - s^{n-1} f(0) - s^{n-2} f'(0)$$
$$- .... - f^{(n-1)}(0)$$

In particular,

$$L\{f'(t)\} = s F(s) - f(0),$$
$$L\{f''(t)\} = s^2 F(s) - sf(0) - f'(0),$$
$$L\{f'''(t)\} = s^3 F(s) - s^2 f(0) - sf'(0) - f''(0).$$

Let $f(t) = \sin t$. Then $L\{f(t)\} = F(s) = \frac{1}{s^2 + 1}$.

Here $f(0) = \sin 0 = 0$. $f'(t) = \cos t$ and so $f'(0) = \cos 0 = 1$. Therefore,

$$L\{f''(t)\} = s^2 F(s) - sf(0) - f'(0)$$

$$= s^2 \times \frac{1}{s^2 + 1} - (s \times 0) - 1$$

$$= \frac{s^2}{s^2 + 1} - 1 = \frac{-1}{s^2 + 1}.$$

Some important results:

If $L\{f(t)\} = F(s)$, then

(i) $L\{t\,f(t)\} = -\dfrac{d}{ds}F(s)$

(ii) $L\{t^n f(t)\} = (-1)^n \dfrac{d^n}{ds^n}F(s)$

(for $n = 0,1,2,3......$)

6.3 LAPLACE TRANSFORM ON INTEGRALS

If $L\{f(t)\} = F(s)$, then $L\left[\int_0^t f(t)dt\right] = \dfrac{F(s)}{s}$

Remember:

If $L\{f(t)\} = F(s)$, then

$L\left[\dfrac{f(t)}{t}\right] = \int_s^\infty F(s)ds$, provided the integral is exists.

6.4 LAPLACE TRANSFORM ON PERIODIC FUNCTIONS

If $L\{f(t)\} = F(s)$ and $f(t)$ be a periodic function with period "T," then

$L\{f(t)\} = F(s) = \dfrac{\int_0^T e^{-sT}f(t)dt}{1-e^{-sT}}.$

Remember:

(i) $f(t)$ is periodic with period $T \Leftrightarrow f(t + T) = f(t)$.

(ii) Let $f(t)$ be defined as follows:

$f(t) = \begin{cases} f_1(t), & 0 < t < k_1 \\ f_2(t), & k_1 < t < k_2 \end{cases}$

OR

$f(t) = \begin{cases} f_1(t), & 0 \le t < k_1 \\ f_2(t), & k_1 < t \le k_2 \end{cases}$

OR

$f(t) = \begin{cases} f_1(t), & 0 < t \le k_1 \\ f_2(t), & k_1 < t < k_2 \end{cases}$

OR

$f(t) = \begin{cases} f_1(t), & 0 \le t \le k_1 \\ f_2(t), & k_1 \le t \le k_2 \end{cases}$

Then, the period of $f(t)$ is $k2$.

6.5 EVALUATION OF INTEGRALS USING LAPLACE TRANSFORMS

If the range of integration are "0" and "∞," then the definition of Laplace transform can be used to find the value of the integral. This technique actually reduces the number of steps for obtaining the value of integration.

Example:

$\int_0^\infty e^{-3t}\sin t\, dt$

$= \int_0^\infty e^{-st}f(t)dt$ [where $s = 3, f(t) = \sin t$]

$= F(s) = L(f(t)) = L(\sin t)$

$= \dfrac{1}{s^2+1} = \dfrac{1}{3^2+1} = \dfrac{1}{10}$ [∵ $s = 3$]

6.6 INITIAL AND FINAL VALUE THEOREMS

6.6.1 Initial Value Theorem

If $L(f(t)) = F(s)$, then $\lim\limits_{t\to 0} f(t) = \lim\limits_{s\to\infty} sF(s)$

(provided the limit exists)

6.6.2 Final Value Theorem

If $L(f(t)) = F(s)$, then $\lim\limits_{t\to\infty} f(t) = \lim\limits_{s\to 0} sF(s)$

(provided the limit exists)

6.7 FUNDAMENTALS OF INVERSE LAPLACE TRANSFORM

6.7.1 Definition of Inverse Laplace Transform

If $L[f(t)] = F(s)$, then $f(t)$ is called inverse Laplace transform of $F(s)$, and is written as $f(t) = L^{-1}[F(s)]$.

$L(e^{2t}) = \dfrac{1}{s-2} \Rightarrow L^{-1}\left(\dfrac{1}{s-2}\right) = e^{2t}$

Remember:

(Linearity property)

If $L[f_1(t)] = F_1(s), L[f_2(t)] = F_2(s)$ and c_1, c_2 be two constants, then

$L^{-1}[c_1f_1(s) + c_2f_2(s)] = c_1L^{-1}[F_1(s)] + c_2L^{-1}[F_2(s)]$
$= c_1f_1(t) + c_2f_2(t).$

6.7.2 Useful formulas on Inverse Laplace Transforms

(i) $L^{-1}\left(\dfrac{1}{s}\right) = 1, \quad s > 0$

(ii) $L^{-1}\left(\dfrac{1}{s^n}\right) = \begin{cases} \dfrac{t^{n-1}}{\Gamma(n)}, & \text{provided } s > 0, n > 0 \\ \dfrac{t^{n-1}}{(n-1)!}, & \text{if } n \in Z^+ \end{cases}$

Example:

$$L^{-1}\left(\frac{1}{s^3}\right) = \frac{t^{3-1}}{\underline{|3-1}} = \frac{t^2}{\underline{|2}} = \frac{t^2}{2},$$

$$L^{-1}\left(\frac{1}{s^{3/2}}\right) = \frac{t^{3/2-1}}{\underline{\left|\frac{3}{2}\right.}} = \frac{t^{1/2}}{\frac{1}{2}\underline{\left|\frac{1}{2}\right.}} \quad \left(\because \underline{|n+1} = n\underline{|n} \text{ and } \underline{\left|\frac{1}{2}\right.} = \sqrt{\pi}\right)$$

$$= \frac{2t^{1/2}}{\underline{\left|\frac{1}{2}\right.}} = \frac{2t^{1/2}}{\sqrt{\pi}}$$

(iii) $L^{-1}\left(\frac{1}{s-a}\right) = e^{at}$, provided $s > a$

Example:

$$L^{-1}\left(\frac{1}{s-5}\right) = e^{5t}$$

(iv) $L^{-1}\left(\frac{1}{s^2+a^2}\right) = \frac{1}{a}\sin at$, provided $s > 0$

Example:

$$L^{-1}\left(\frac{1}{s^2+4}\right) = L^{-1}\left(\frac{1}{s^2+2^2}\right) = \frac{1}{2}\sin 2t$$

(v) $L^{-1}\left(\frac{s}{s^2+a^2}\right) = \cos at$, provided $s > 0$

Example:

$$L^{-1}\left(\frac{s}{s^2+9}\right) = L^{-1}\left(\frac{s}{s^2+3^2}\right) = \cos 3t$$

(vi) $L^{-1}\left(\frac{1}{s^2-a^2}\right) = \frac{1}{a}\sinh at; \ L^{-1}\left(\frac{s}{s^2-a^2}\right) = \cosh at$

Example:

$$L^{-1}\left(\frac{1}{s^2-25}\right) = L^{-1}\left(\frac{1}{s^2-5^2}\right) = \frac{1}{5}\sinh 5t$$

(vii) $L^{-1}\left[\frac{1}{(s-a)^n}\right] = e^{at}\frac{t^{n-1}}{\underline{|n-1}}, \quad$ if $n \in Z^+$

Example:

$$L^{-1}\left[\frac{1}{(s-2)^3}\right] = e^{2t}\frac{t^{3-1}}{\underline{|3-1}} = e^{2t}\frac{t^2}{\underline{|2}} = e^{2t}\frac{t^2}{2}$$

(viii) $L^{-1}\left[\frac{1}{(s-a)^2+b^2}\right] = \frac{1}{b}e^{at}\sin bt$

Example:

$$L^{-1}\left[\frac{1}{(s-1)^2+7^2}\right] = \frac{1}{7}e^t\sin 7t$$

(ix) $L^{-1}\left[\frac{s-a}{(s-a)^2+b^2}\right] = e^{at}\cos bt$

Example:

$$L^{-1}\left[\frac{s-2}{(s-2)^2+1^2}\right] = e^{2t}\cos t$$

(x) $L^{-1}\left[\frac{s}{(s^2+a^2)^2}\right] = \frac{1}{2a}t\sin at$

Example:

$$L^{-1}\left[\frac{s}{(s^2+4)^2}\right] = L^{-1}\left[\frac{s}{(s^2+2^2)^2}\right] = \frac{1}{4}t\sin 2t$$

(xi) $L^{-1}\left[\frac{1}{(s^2+a^2)^2}\right] = \frac{1}{2a^3}(\sin at - at\cos at)$

Example:

$$L^{-1}\left[\frac{1}{(s^2+4)^2}\right] = L^{-1}\left[\frac{1}{(s^2+2^2)^2}\right]$$

$$= \frac{1}{2\times 2^3}(\sin 2t - 2t\cos 2t)$$

$$= \frac{1}{16}(\sin 2t - 2t\cos 2t)$$

6.8 IMPORTANT THEOREMS ON INVERSE LAPLACE TRANSFORMS

(i) If $L^{-1}[F(s)] = f(t)$, then,

$$tf(t) = -L^{-1}\left\{\frac{d}{ds}F(s)\right\}$$

(ii) (Convolution theorem) if $L^{-1}[F(s)] = f(t)$ and $L^{-1}[G(s)] = g(t)$, then

$$f * g = L^{-1}[F(s)G(s)] = \int_0^t f(u)g(t-u)du$$

6.9 UNIT STEP FUNCTION AND UNIT IMPULSE FUNCTION

6.9.1 Unit Step Function

The unit step function (Heaviside's unit step function) $u(t-a)$ is defined by

$$u(t-a) = \begin{cases} 0, & \text{for } t < a \\ 1, & \text{for } t \geq a \end{cases}$$

[a being a positive real number]

Remember:

(i) $L\{u(t-a)\} = \frac{1}{s}e^{-as}$

(ii) If $f(t) = \begin{cases} f_1(t), & a < t < c \\ f_2(t), & c < t < b \end{cases}$, then

$$f(t) = f_1(t)\{u(t-a) - u(t-c)\} + f_2(t)\{u(t-c) - u(t-b)\}$$

(iii) If $f(t) = \begin{cases} f_1(t), \ a < t < b \\ f_2(t), \ t > b \end{cases}$, then

$$f(t) = f_1(t)\{u(t-a) - u(t-b)\} + f_2(t) \ u(t-b)\}$$

(iv) If $f(t) = \begin{cases} f_1(t), \ a < t < c \\ f_2(t), \ c < t < b \\ f_3(t), \ t > b \end{cases}$, then

$$f(t) = f_1(t)\{u(t-a) - u(t-c)\} + f_2(t)\{u(t-c) - u(t-b)\}$$

$$+ f_3(t)u(t-b)$$

Example:

If $f(t) = \begin{cases} 0, \ 0 < t < 1 \\ 1, \ 1 < t < 2 \\ 2, \ \ \ \ t > 2 \end{cases}$, then

$$f(t) = 0 \times \{u(t-0) - u(t-1)\}$$

$$+ 1 \times \{u(t-1) - u(t-2)\} + 2 \times u(t-2)$$

6.9.2 Second Shifting Theorem

If $L^{-1}[F(s)] = f(t)$, then $L[f(t-a) \times u(t-a)] = e^{-as} F(s) = e^{-as} L(f(t))$

Example:

$$L\{(t-1)u(t-1)\} = L\{f(t-1)u(t-1)\}$$

$$(\text{here } f(t-1) = t-1; \text{ so} f(t) = t)$$

$$= e^{-1s} L(f(t)) = e^{-s} L(t) = e^{-s} \frac{1}{s^2}$$

6.9.3 Unit Impulse Function

The unit impulse function (Dirac Delta function) $\delta(t - a)$ is defined by

$$\delta(t-a) = \begin{cases} 0, \text{ for } t \neq a \\ \infty, \text{ for } t = a \end{cases}$$

$$\left(\text{provided } \int_{-\infty}^{\infty} \delta(t-a)dt = 1 \right)$$

Remember:

(i) $L\{\delta(t-a)\} = e^{-as}$

(ii) $\int_0^{\infty} f(t)\delta(t-a)dt = f(a)$

6.10 SOLVING ORDINARY DIFFERENTIAL EQUATIONS

The Laplace transform method can be used to solve an ordinary differential equation without finding the general solution.

Suppose a given ordinary differential equation contains the variable $y = y(t)$ and its derivatives. Assume that $L[y(t)] = Y(s)$. Then to solve the equation (*i.e.*, to get $y(t)$) the following steps can be followed:

Step-1: Take the Laplace transform on both sides of the given differential equation and apply the given conditions.

Step-2: Express $Y(s)$ as a function of "s" and then resolve the R.H.S. part as a sum of partial fractions.

Step-3: Take inverse Laplace transform on both sides and put $L^{-1}(Y(s)) = y(t)$.

Fully Solved MCQs (Level-I)

1. The Laplace transform of $f(t) = \dfrac{e^{at}-1}{a}$ is

(a) $\dfrac{1}{s-a}$

(b) $\dfrac{1}{s(s+a)}$

(c) $\dfrac{1}{s(s-a)}$

(d) $\dfrac{1}{s+a}$

2. The Laplace transform of $(\sin t + \cos t)^2$ is

(a) $\dfrac{1}{s} + \dfrac{2}{s^2+4}$

(b) $\dfrac{1}{s} - \dfrac{1}{s^2+4}$

(c) $\dfrac{1}{2s} + \dfrac{1}{s^2+4}$

(d) $\dfrac{1}{s^2} + \dfrac{2}{s^2+4}$

3. $L(\sin 3t \cos t) = ?$

(a) $\dfrac{1}{s^2+16} + \dfrac{2}{s^2+4}$

(b) $\dfrac{1}{s^2+16} + \dfrac{1}{s^2+4}$

(c) $\dfrac{2}{s^2+16} + \dfrac{1}{s^2+4}$

(d) None of these

4. $L(\sin 3t \sin 2t) = ?$

(a) $\dfrac{12s}{(s^2+1)(s^2+25)}$

(b) $\dfrac{2s^2+6}{(s^2+1)(s^2+5)}$

(c) $\dfrac{1}{(s^2+1)(s^2+5)}$

(d) None of these

5. $L(e^{-2t} - e^{-3t}) = ?$

(a) $\dfrac{5}{s^2+5s+6}$

(b) $\dfrac{s}{s^2+5s+6}$

(c) $\dfrac{1}{s^2+5s+6}$

(d) None of these

6. $L(\sin^2 2t) = ?$

(a) $\dfrac{1}{2}\left(\dfrac{1}{s} - \dfrac{s}{s^2+16} \right)$

(b) $\dfrac{1}{2}\left(\dfrac{1}{s} + \dfrac{s}{s^2+16} \right)$

(c) $\dfrac{1}{2}\left(\dfrac{1}{s} - \dfrac{4}{s^2+16} \right)$

(d) $\dfrac{1}{2}\left(\dfrac{1}{s} + \dfrac{4}{s^2+16} \right)$

7. $L(\sin^3 t) = ?$

(a) $\dfrac{4s}{(s^2+1)(s^2+9)}$ (b) $\dfrac{6}{(s^2+1)(s^2+9)}$

(c) $\dfrac{6s}{(s^2+1)(s^2+9)}$ (d) None of these

8. $L\left(1/\sqrt{\pi t}\right) = ?$

(a) $1/\sqrt{\pi s}$ (b) $1/\sqrt{s}$

(c) $\sqrt{\pi}/\sqrt{s}$ (d) $\dfrac{1}{2\sqrt{s}}$

9. $L(t^2 e^{-2t}) = ?$

(a) $\dfrac{1}{(s+2)^3}$ (b) $\dfrac{s}{(s+2)^3}$

(c) $\dfrac{2}{(s+2)^3}$ (d) $\dfrac{3}{(s+2)^3}$

10. $L(e^{-at} \sin h\, bt) = ?$

(a) $\dfrac{b}{(s+a)^2 - b^2}$ (b) $\dfrac{b}{(s+a)^2 + b^2}$

(c) $\dfrac{s}{(s+a)^2 - b^2}$ (d) $\dfrac{s}{(s+a)^2 + b^2}$

11. Find $L(f(t))$ if

$$f(t) = \begin{cases} e^t, & 0 < t < 1 \\ 0, & t > 1 \end{cases}$$

(a) $\dfrac{1 - e^{1-s}}{1-s}$ (b) $\dfrac{e^{1-s} - 1}{1-s}$

(c) $\dfrac{e^{1-s} - 1}{s}$ (d) None of this

12. The Laplace transform of $\sin(2t + 5)$ is

(a) $\dfrac{1}{(s^2-4)}(2\cos 5 - s\sin 5)$

(b) $\dfrac{1}{(s^2+4)}(2\cos 5 + s\sin 5)$

(c) $\dfrac{1}{(s^2-4)}(s\cos 5 - 5\sin 5)$

(d) $\dfrac{1}{(s^2+4)}(s\cos 5 + 5\sin 5)$

13. If $L\{f(t)\} = \dfrac{1}{s+1}$, then $L\{f(4t)\} = ?$

(a) $\dfrac{1}{s-1}$ (b) $\dfrac{1}{s+2}$

(c) $\dfrac{1}{s-3}$ (d) $\dfrac{1}{s+4}$

14. $L(t \sin 2t) = ?$

(a) $\dfrac{4}{(s^2+4)^2}$ (b) $\dfrac{4s}{(s^2+4)^2}$

(c) $\dfrac{2}{s^2+4}$ (d) $\dfrac{4}{s^2+4}$

15. $L(te^{2t}) = ?$

(a) $\dfrac{1}{(s-2)^2}$ (b) $\dfrac{1}{(s+2)^2}$

(c) $\dfrac{-1}{(s-2)^2}$ (d) $\dfrac{-1}{(s+2)^2}$

16. $L\left(\dfrac{1-e^t}{t}\right) = ?$

(a) $\log\left(\dfrac{s-1}{s}\right)$ (b) $\log\left(\dfrac{s}{s-1}\right)$

(c) $\log\left(\dfrac{s}{s+1}\right)$ (d) $\log\left(\dfrac{s+1}{s}\right)$

17. $L\left(\dfrac{e^{-at} - e^{-bt}}{t}\right) = ?$

(a) $\log\left(\dfrac{s+b}{s+a}\right)$ (b) $-\log\left(\dfrac{s+b}{s+a}\right)$

(c) $2\log\left(\dfrac{s+a}{s+b}\right)$ (d) None of these

18. The Laplace transform of $\dfrac{\sin t}{t}$ is

(a) $\cot^{-1} s$ (b) $\tan^{-1} s$

(c) $\cot^{-1} \dfrac{1}{s^2}$ (d) $\tan^{-1} \dfrac{1}{s^2}$

19. The Laplace transform of $\dfrac{\sinh t}{t}$ is

(a) $\log\sqrt{\dfrac{s+1}{s-1}}$ (b) $\log\sqrt{\dfrac{s-1}{s+1}}$

(c) $\log\left(\dfrac{s+1}{s-1}\right)$ (d) $\log\left(\dfrac{s-1}{s+1}\right)$

20. Find the Laplace transform of the periodic function $f(t)$ which is defined as follows:

$$f(t) = \frac{ct}{T} \text{ for } 0 < t < T \text{ and } f(t+T) = f(t).$$

(a) $\dfrac{1}{sT} - \dfrac{ce^{-sT}}{s(1-e^{-sT})}$ (b) $\dfrac{c}{sT} - \dfrac{e^{-sT}}{s(1+e^{-sT})}$

(c) $\dfrac{c}{s^2T} - \dfrac{ce^{-sT}}{s(1-e^{-sT})}$ (d) 0

21. Find the Laplace transform of the periodic function $f(t)$ where:

$$f(t) = e^t \text{ for } 0 < t < 2\pi$$

(a) $\dfrac{e^{2(1-s)\pi}-1}{(1-s)(1-e^{-2\pi s})}$ (b) $\dfrac{e^{2s\pi}-1}{(1-s)(1-e^{-2\pi s})}$

(c) $\dfrac{e^{2(1-s)\pi}-1}{(1-e^{-2\pi s})}$ (d) 1/s

22. $\displaystyle\int_0^\infty te^{-t}\, dt = ?$

(a) 1   (b) $\dfrac{1}{2}$ (c) $\dfrac{1}{3}$ (d) 0

23. $\displaystyle\int_0^\infty e^{-3t}\sin t\cos t\, dt = ?$

(a) $\dfrac{1}{13}$ (b) $\dfrac{3}{13}$ (c) $\dfrac{4}{13}$ (d) $\dfrac{5}{13}$

24. $\displaystyle\int_0^\infty e^{-t}\dfrac{\sin t}{t}\, dt = ?$

(a) $\dfrac{\pi}{2}$ (b) $\dfrac{\pi}{3}$ (c) $\dfrac{\pi}{4}$ (d) $-\dfrac{\pi}{2}$

25. $\displaystyle\int_0^\infty \dfrac{\sin 4t}{t}\, dt = ?$

(a) $\dfrac{\pi}{2}$ (b) $-\dfrac{\pi}{4}$ (c) $\dfrac{\pi}{4}$ (d) $-\dfrac{\pi}{2}$

26. $L^{-1}\left\{\dfrac{3}{(s+2)(s+5)}\right\} = ?$

(a) $e^{2t}-e^{5t}$ (b) $e^{-2t}-e^{-5t}$

(c) $e^{-2t}+e^{-5t}$ (d) None of these

27. $L^{-1}\left\{\dfrac{1}{(s^2+1)(s^2+4)}\right\} = ?$

(a) $\dfrac{1}{3}\left(\sin t - \dfrac{1}{2}\sin 2t\right)$

(b) $\dfrac{1}{3}\left(\dfrac{1}{2}\sin t - \sin 2t\right)$

(c) $\sin t + \sin 2t$ (d) None of these

28. $L^{-1}\left\{\dfrac{1}{(s^2+9)^2}\right\} = ?$

(a) $\dfrac{1}{54}(\sin 3t - 3t\cos 3t)$

(b) $\dfrac{1}{27}(\sin 3t - t\sin 3t)$

(c) $\dfrac{1}{27}(\sin 3t - 3\cos 3t)$

(d) None of these

29. $L^{-1}\left\{\dfrac{s+2}{(s^2+1)^2}\right\} = ?$

(a) $\left(1-\dfrac{t}{2}\right)\sin t + t\cos t$

(b) $\dfrac{t}{2}\sin t + t\cos t$

(c) $\left(1+\dfrac{t}{2}\right)\sin t - t\cos t$

(d) None of these

30. $L^{-1}\left\{\dfrac{1}{(s+2)^3}\right\} = ?$

(a) $\dfrac{1}{2}e^{-2t}t^2$ (b) $\dfrac{1}{2}e^{2t}t^2$

(c) $e^{-2t}t^3$ (d) $e^{-2t}t^2$

31. $L^{-1}\left\{\dfrac{s+4}{(s+2)^3}\right\} = ?$

(a) $te^{2t}(1+t)$ (b) $te^{-2t}(1+t)$

(c) $e^{-2t}(2+t)$ (d) None of these

32. $L^{-1}\left\{\log\left(\dfrac{s+1}{s-1}\right)\right\} = ?$

(a) $\dfrac{2}{t}\sinh t$ (b) $\dfrac{2}{t}\cosh t$

(c) $\dfrac{t}{2}\sin t$ (d) $\dfrac{t}{2}\cos t$

33. $L^{-1}\left[\dfrac{1}{s(s+1)}\right] = ?$

(a) $1 - e^{-t}$

(b) $1 + e^{-t}$

(c) $1 + e^{t}$

(d) $1 - e^{t}$

34. $\displaystyle\int_0^{\infty} e^{2t}\,\delta(t-2)\,dt = ?$

(a) e^{-3}

(b) e^4

(c) e^{-4}

(d) None of this

35. $L^{-1}\left[\dfrac{e^{-2s}}{s^2+1}\right] = ?$

(a) $u(t-2) \times \sin t$

(b) $e^{-2t} \times \sin t$

(c) $u(t-2) \times \sin(t-2)$

(d) $e^{2t} \times \sin(t-2)$

36. $L\{e^{t-2}\,u(t-2)\} = ?$

(a) $\dfrac{1}{(s-1)}e^{-2s}$

(b) $\dfrac{1}{(s+1)}e^{-2s}$

(c) $\dfrac{1}{(s-1)}e^{2s}$

(d) $\dfrac{1}{(s+1)}e^{2s}$

37. Find $L(f(t))$ if $f(t)=\begin{cases} e^{t-2}, & t>2 \\ 0, & t<2 \end{cases}$ is

(a) $\dfrac{e^{-s}}{s-1}$

(b) $\dfrac{e^{-2s}}{s-1}$

(c) $\dfrac{e^{-3s}}{s-1}$

(d) $\dfrac{e^{-s}}{s+1}$

38. Find $L(f(t))$ if $f(t)=\begin{cases} \sin\left(t-\dfrac{\pi}{3}\right), & t>\dfrac{\pi}{3} \\ 0, & t<\dfrac{\pi}{3} \end{cases}$

(a) $\dfrac{e^{-\pi s}}{s^2-1}$

(b) $\dfrac{e^{-\frac{\pi s}{3}}}{s^2-1}$

(c) $\dfrac{e^{-\frac{\pi s}{3}}}{s^2+1}$

(d) $\dfrac{e^{-\pi s}}{s^2+1}$

39. Solve $y' = 1$; given $y(0) = 0$

(a) t

(b) $t+1$

(c) $2t$

(d) t^2

40. Solve $y' + y = 1$; given $y(0) = 0$

(a) e^t

(b) e^{-t}

(c) $1 + e^{-t}$

(d) $1 - e^{-t}$

Answer key				
1. (c)	**2.** (a)	**3.** (c)	**4.** (a)	**5.** (c)
6. (a)	**7.** (b)	**8.** (b)	**9.** (c)	**10.** (a)
11. (b)	**12.** (b)	**13.** (d)	**14.** (b)	**15.** (a)
16. (a)	**17.** (a)	**18.** (a)	**19.** (a)	**20.** (c)
21. (a)	**22.** (a)	**23.** (a)	**24.** (c)	**25.** (a)
26. (b)	**27.** (a)	**28.** (a)	**29.** (c)	**30.** (a)
31. (b)	**32.** (a)	**33.** (a)	**34.** (b)	**35.** (c)
36. (a)	**37.** (b)	**38.** (c)	**39.** (a)	**40.** (d)

Explanation

1. (c)

$$L\{f(t)\} = L\left(\frac{e^{at}-1}{a}\right)$$
$$= \frac{1}{a}L(e^{at}-1)$$
$$= \frac{1}{a}\{L(e^{at}) - L(1)\}$$
$$= \frac{1}{a}\left(\frac{1}{s-a} - \frac{1}{s}\right) = \frac{1}{s(s-a)}$$

2. (a)

$$L\{f(t)\}$$
$$= L\{(\sin t + \cos t)^2\}$$
$$= L(1 + \sin 2t) = L(1) + L(\sin 2t)$$
$$= \frac{1}{s} + \frac{2}{s^2+4}$$

3. (c)

$$\sin 3t \cos t$$
$$= \frac{1}{2}(2\sin 3t \cos t)$$
$$= \frac{1}{2}\{\sin(3t+t) + \sin(3t-t)\}$$
$$= \frac{1}{2}(\sin 4t + \sin 2t)$$
$$\therefore L(\sin 3t \cos t)$$
$$= L\left\{\frac{1}{2}(\sin 4t + \sin 2t)\right\}$$
$$= \frac{1}{2}\{L(\sin 4t) + L(\sin 2t)\}$$
$$= \frac{1}{2}\left(\frac{4}{s^2+16} + \frac{2}{s^2+4}\right) = \frac{2}{s^2+16} + \frac{1}{s^2+4}$$

4. (*a*)

$$\sin 3t \sin 2t$$

$$= \frac{1}{2}(2\sin 3t \sin 2t)$$

$$= \frac{1}{2}\{\cos(3t - 2t) - \cos(3t + 2t)\}$$

$$= \frac{1}{2}(\cos t - \cos 5t)$$

$$\therefore L(\sin 3t \cos t)$$

$$= L\left\{\frac{1}{2}(\cos t - \cos 5t)\right\}$$

$$= \frac{1}{2}\{L(\cos t) - L(\cos 5t)\}$$

$$= \frac{1}{2}\left(\frac{s}{s^2 + 1} - \frac{s}{s^2 + 5^2}\right) = \frac{12s}{(s^2 + 1)(s^2 + 25)}$$

5. (*c*)

$$L(e^{-2t} - e^{-3t}) = L(e^{-2t}) - L(e^{-3t})$$

$$= \frac{1}{s+2} - \frac{1}{s+3}$$

$$= \frac{1}{(s+2)(s+3)} = \frac{1}{s^2 + 5s + 6}$$

6. (*a*)

$$L(\sin^2 2t) = \frac{1}{2}L(2\sin^2 2t) = \frac{1}{2}L(1 - \cos 4t)$$

$$= \frac{1}{2}\{L(1) - L(\cos 4t)\}$$

$$= \frac{1}{2}\left(\frac{1}{s} - \frac{s}{s^2 + 16}\right)$$

7. (*b*)

$$\sin^3 t = \frac{1}{4}(4\sin^3 t) = \frac{1}{4}\{3\sin t - \sin 3t\}$$

$$[\text{Using } \sin 3x = 3\sin x - 4\sin^3 x]$$

$$= \frac{3}{4}\sin t - \frac{1}{4}\sin 3t$$

$$\therefore L(\sin^3 t)$$

$$= L\left(\frac{3}{4}\sin t - \frac{1}{4}\sin 3t\right)$$

$$= \frac{3}{4}L(\sin t) - \frac{1}{4}L(\sin 3t)$$

$$= \frac{3}{4}\times\frac{1}{s^2 + 1} - \frac{1}{4}\times\frac{3}{s^2 + 9}$$

$$= \frac{3}{4}\left(\frac{1}{s^2 + 1} - \frac{1}{s^2 + 9}\right)$$

$$= \frac{6}{(s^2 + 1)(s^2 + 9)}$$

8. (*b*)

$$L(1/\sqrt{\pi t}) = \frac{1}{\sqrt{\pi}}L(1/\sqrt{t}) = \frac{1}{\sqrt{\pi}}L(t^{-1/2})$$

$$= \frac{1}{\sqrt{\pi}}\frac{\Gamma\left(-\frac{1}{2}+1\right)}{s^{(-\frac{1}{2}+1)}} = \frac{1}{\sqrt{\pi}}\frac{\Gamma\left(\frac{1}{2}\right)}{s^{\frac{1}{2}}}$$

$$= \frac{1}{\sqrt{\pi}}\frac{\sqrt{\pi}}{\sqrt{s}} = \frac{1}{\sqrt{s}}\quad\left[\text{since } \Gamma\left(\frac{1}{2}\right) = \sqrt{\pi}\right]$$

9. (*c*)

$$L(t^2) = \frac{\lfloor 2}{s^3} = \frac{2}{s^3}$$

$$\therefore L(t^2 e^{-2t}) = \frac{2}{\{s - (-2)\}^3}$$

$$[\text{by first shifting theorem}]$$

$$= \frac{2}{(s+2)^3}$$

10. (*a*)

$$e^{-at}\sinh bt = e^{-at}\left(\frac{e^{bt} - e^{-bt}}{2}\right)$$

$$= \frac{1}{2}\{e^{(b-a)t} - e^{-(b+a)t}\}$$

$$L(e^{-at}\sinh bt)$$

$$= L\left[\frac{1}{2}\{e^{(b-a)t} - e^{-(b+a)t}\}\right]$$

$$= \frac{1}{2}\{L(e^{(b-a)t}) - L(e^{-(b+a)t})\}$$

$$= \frac{1}{2}\left(\frac{1}{s - (b-a)} - \frac{1}{s + (b+a)}\right)$$

$$= \frac{1}{2}\left(\frac{s + a + b - s - a + b}{(s + a - b)(s + a + b)}\right)$$

$$= \frac{b}{(s+a)^2 - b^2}$$

Alternative Method

$$L(\sinh bt) = \frac{b}{s^2 - b^2}$$

By 1st shifting theorem,

$$L(e^{-at}\sin bt) = \frac{b}{(s+a)^2 - b^2}$$

11. (*b*)

$$L\{f(t)\}$$

$$= \int_0^\infty e^{-st}f(t)dt$$

$$= \int_0^1 e^{-st}e^t dt + \int_1^\infty e^{-st}0\,dt$$

$$= \int_0^1 e^{(1-s)t} dt = \left[\frac{e^{(1-s)t}}{1-s} \right]_0^1$$

$$= \frac{e^{1-s}}{1-s} - \frac{e^0}{1-s} = \frac{e^{1-s}-1}{1-s}$$

12. (b)

$$L\{\sin(2t+5)\}$$

$$= L(\cos 5 \times \sin 2t + \sin 5 \times \cos 2t)$$

$$= \cos 5 \times L(\sin 2t) + \sin 5 \times L(\cos 2t)$$

$$= \cos 5 \times \frac{2}{(s^2+4)} + \sin 5 \times \frac{s}{(s^2+4)}$$

$$= \frac{1}{(s^2+4)}(2\cos 5 + s\sin 5)$$

13. (d)

We know that if $L\{f(t)\} = F(s)$, then,

$$L\{f(at)\} = \frac{1}{a} F\left(\frac{s}{a}\right).$$

$$\therefore L\{f(4t)\} = \frac{1}{4} F\left(\frac{s}{4}\right) = \frac{1}{4} \times \frac{1}{1+\frac{s}{4}} = \frac{1}{s+4}$$

$$\left(\because F(s) = \frac{1}{s+1} \text{ (given)} \right)$$

14. (b)

$$L(t\sin 2t) = L(t^n f(t)) = (-1)^n \frac{d^n}{ds^n} F(s)$$

$$[\text{where } n = 1, f(t) = \sin 2t]$$

$$= (-1)^1 \frac{d}{ds} L\{f(t)\}$$

$$= -\frac{d}{ds} L(\sin 2t)$$

$$= -\frac{d}{ds}\left(\frac{2}{s^2+4}\right)$$

$$= (-2) \times \frac{(-1) \times 2s}{(s^2+4)^2} = \frac{4s}{(s^2+4)^2}$$

15. (a)

$$L(te^{2t}) = L(t^n f(t)) = (-1)^n \frac{d^n}{ds^n} F(s)$$

$$[\text{where } n = 1, f(t) = e^{2t}]$$

$$= (-1)^1 \frac{d}{ds} L\{f(t)\}$$

$$= -\frac{d}{ds} L(e^{2t})$$

$$= -\frac{d}{ds}\left(\frac{1}{s-2}\right)$$

$$= (-1) \times \frac{(-1)}{(s-2)^2} = \frac{1}{(s-2)^2}$$

16. (a)

$$L\left(\frac{1-e^t}{t}\right)$$

$$= L\left[\frac{f(t)}{t}\right], \quad \left[\text{where } f(t) = 1-e^t\right]$$

$$= \int_s^\infty F(s)ds = \int_s^\infty L[f(t)]ds = \int_s^\infty L(1-e^t)ds$$

$$= \int_s^\infty \{L(1) - L(e^t)\}ds \ [\text{Using linear property}]$$

$$= \int_s^\infty \left(\frac{1}{s} - \frac{1}{s-1}\right)ds = \left[\log s - \log(s-1)\right]_s^\infty$$

$$= \left[\log\left(\frac{s}{s-1}\right)\right]_s^\infty = \left[\log\left(\frac{1}{1-(1/s)}\right)\right]_s^\infty$$

$$= \log\left(\frac{1}{1-0}\right) - \log\left(\frac{1}{1-(1/s)}\right)$$

$$\left[\because s \to \infty \Rightarrow \frac{1}{s} \to 0\right]$$

$$= -\log\left(\frac{s}{s-1}\right) = \log\left(\frac{s-1}{s}\right) \ (\because \log 1 = 0)$$

17. (a)

$$L(e^{-at} - e^{-bt}) = L(e^{-at}) - L(e^{-bt})$$

$$= \frac{1}{s+a} - \frac{1}{s+b} = F(s)$$

$$\therefore L\left(\frac{e^{-at} - e^{-bt}}{t}\right)$$

$$= L\left[\frac{f(t)}{t}\right], \quad \left[\text{where } f(t) = e^{-at} - e^{-bt}\right]$$

$$= \int_s^\infty F(s)ds = \int_s^\infty \left(\frac{1}{s+a} - \frac{1}{s+b}\right)ds$$

$$= \left[\log(s+a) - \log(s+b)\right]_s^\infty$$

$$= \left[\log\left(\frac{s+a}{s+b}\right)\right]_s^\infty$$

$$= \left[\log\left(\frac{1+(a/s)}{1+(b/s)}\right)\right]_s^\infty$$

$$= \log\left(\frac{1+0}{1+0}\right) - \log\left(\frac{1+(a/s)}{1+(b/s)}\right)$$

$$\left[\because s \to \infty \Rightarrow \frac{1}{s} \to 0\right]$$

$$= \log 1 - \log\left(\frac{s+a}{s+b}\right)$$

$$= \log\left(\frac{s+b}{s+a}\right)$$

18. (a)

$$L\left(\frac{\sin t}{t}\right) = \int_s^\infty L(\sin t)\,ds = \int_s^\infty \frac{1}{s^2+1}\,ds$$

$$= \left[\tan^{-1} s\right]_s^\infty = \tan^{-1}\infty - \tan^{-1} s$$

$$= \frac{\pi}{2} - \tan^{-1} s = \cot^{-1} s$$

19. (a)

$$L\left(\frac{\sinh t}{t}\right)$$

$$= \int_s^\infty L(\sinh t)\,ds$$

$$= \int_s^\infty \frac{1}{s^2-1}\,ds$$

$$= \frac{1}{2}\left[\log\left(\frac{s-1}{s+1}\right)\right]_s^\infty = \frac{1}{2}\left[\log\left(\frac{1-\frac{1}{s}}{1+\frac{1}{s}}\right)\right]_s^\infty$$

$$= \frac{1}{2}\left[\log\left(\frac{1-0}{1+0}\right) - \log\left(\frac{1-\frac{1}{s}}{1+\frac{1}{s}}\right)\right]$$

$$\left(\because s\to\infty \Rightarrow \frac{1}{s}\to 0\right)$$

$$= \frac{1}{2}\left[\log 1 - \log\left(\frac{1-\frac{1}{s}}{1+\frac{1}{s}}\right)\right]$$

$$= \frac{1}{2}\left[0 - \log\left(\frac{1-\frac{1}{s}}{1+\frac{1}{s}}\right)\right]$$

$$= -\frac{1}{2}\log\left(\frac{s-1}{s+1}\right)$$

$$= \frac{1}{2}\log\left(\frac{s+1}{s-1}\right) = \log\sqrt{\frac{s+1}{s-1}}.$$

20. (c)

$$L\{f(t)\}$$

$$= \frac{\int_0^T e^{-st} f(t)\,dt}{1-e^{-sT}} = \frac{\int_0^T e^{-st} \times \frac{ct}{T}\,dt}{1-e^{-sT}}$$

$$= \frac{c}{T\left(1-e^{-sT}\right)} \int_0^T e^{-st} t\,dt$$

$$= \frac{c}{T\left(1-e^{-sT}\right)}\left[t\left(\frac{e^{-st}}{-s}\right) - \int 1\times\left(\frac{e^{-st}}{-s}\right)dt\right]_0^T$$

(integrating by parts)

$$= \frac{c}{T\left(1-e^{-sT}\right)}\left[t\left(\frac{e^{-st}}{-s}\right) - \frac{e^{-st}}{s^2}\right]_0^T$$

$$= \frac{c}{T\left(1-e^{-sT}\right)}\left[\left\{T\left(\frac{e^{-sT}}{-s}\right) - \frac{e^{-sT}}{s^2}\right\} - \left\{0 - \frac{e^{-0}}{s^2}\right\}\right]$$

$$= \frac{c}{T\left(1-e^{-sT}\right)}\left[\frac{-Te^{-sT}}{s} - \frac{e^{-sT}}{s^2} + \frac{1}{s^2}\right]$$

$$= \frac{c}{T\left(1-e^{-sT}\right)}\left[\frac{-Te^{-sT}}{s} + \frac{1}{s^2}\left(1-e^{-sT}\right)\right]$$

$$= -\frac{ce^{-sT}}{s\left(1-e^{-sT}\right)} + \frac{c}{s^2 T}$$

21. (a) Here, $f(t)$ is a periodic function with period 2π.

$$L\{f(t)\}$$

$$= \frac{\int_0^T e^{-st} f(t)\,dt}{1-e^{-sT}} = \frac{\int_0^{2\pi} e^{-st}\times e^t\,dt}{1-e^{-2\pi s}}$$

$$= \frac{1}{\left(1-e^{-2\pi s}\right)} \int_0^{2\pi} e^{(1-s)t}\,dt = \frac{1}{\left(1-e^{-2\pi s}\right)}\left[\frac{e^{(1-s)t}}{(1-s)}\right]_0^{2\pi}$$

$$= \frac{1}{\left(1-e^{-2\pi s}\right)}\left[\frac{e^{(1-s)2\pi}}{(1-s)} - \frac{e^0}{(1-s)}\right]$$

$$= \frac{1}{\left(1-e^{-2\pi s}\right)}\left[\frac{e^{(1-s)2\pi}}{(1-s)} - \frac{1}{(1-s)}\right]$$

$$= \frac{e^{2(1-s)\pi} - 1}{(1-s)\left(1-e^{-2\pi s}\right)}$$

22. (a)

$$\int_0^\infty te^{-t}\,dt$$

$$= \int_0^\infty e^{-st} f(t)\,dt \quad [\text{where } s=1,\ f(t)=t]$$

$$= F(s) = L(f(t)) = L(t) = \frac{1}{s^2} = 1 \quad [\because s=1]$$

23. (a) $\int\limits_0^\infty e^{-3t} \sin t \cos t \, dt$

$= \int\limits_0^\infty e^{-st} f(t) \, dt$

$\left(\text{here } s = 3 \text{ and } f(t) = \sin t \cos t \right)$

$= F(s) \text{ (by definition of Laplace transform)}$

$= L[f(t)]$

$= L[\sin t \cos t] = \dfrac{1}{2}L[2\sin t \cos t] = \dfrac{1}{2}L[\sin 2t]$

$= \dfrac{1}{2} \times \dfrac{2}{s^2+2^2} = \dfrac{1}{s^2+4} = \dfrac{1}{13} \quad (\because s = 3)$

24. (c)

$\int\limits_0^\infty e^{-t} \dfrac{\sin t}{t} \, dt$

$= \int\limits_0^\infty e^{-st} f(t) \, dt$

$\left(\text{here } s = 1 \text{ and } f(t) = \dfrac{\sin t}{t}\right)$

$= F(s) = L[f(t)] = L\left[\dfrac{\sin t}{t}\right]$

$= \int\limits_s^\infty L(\sin t) \, ds = \int\limits_s^\infty \dfrac{1}{s^2+1} \, ds$

$= \left[\tan^{-1} s\right]_s^\infty = \tan^{-1}\infty - \tan^{-1} s$

$= \dfrac{\pi}{2} - \tan^{-1} s = \cot^{-1} s = \cot^{-1} 1 = \dfrac{\pi}{4}$

$\quad (\because s = 1)$

25. (a) $\int\limits_0^\infty \dfrac{\sin 4t}{t} \, dt$

$= \int\limits_0^\infty e^{-0t} \dfrac{\sin 4t}{t} \, dt$

$= \int\limits_0^\infty e^{-st} f(t) \, dt$

$\left(\text{here } s = 0 \text{ and } f(t) = \dfrac{\sin 4t}{t}\right)$

$= F(s) = L[f(t)] = L\left[\dfrac{\sin 4t}{t}\right]$

$= \int\limits_s^\infty L(\sin 4t) \, ds = \int\limits_s^\infty \dfrac{4}{s^2+4^2} \, ds$

$= \dfrac{4}{4}\left[\tan^{-1}\dfrac{s}{4}\right]_s^\infty = \tan^{-1}\infty - \tan^{-1}\dfrac{s}{4}$

$= \dfrac{\pi}{2} - \tan^{-1}\dfrac{s}{4} = \dfrac{\pi}{2} - \tan^{-1}\dfrac{0}{4} = \dfrac{\pi}{2}$

$\quad (\because s = 0)$

26. (b)

$L^{-1}\left\{\dfrac{3}{(s+2)(s+5)}\right\}$

$= L^{-1}\left\{\dfrac{(s+5)-(s+2)}{(s+2)(s+5)}\right\}$

$= L^{-1}\left(\dfrac{1}{s+2} - \dfrac{1}{s+5}\right)$

$= L^{-1}\left(\dfrac{1}{s+2}\right) - L^{-1}\left(\dfrac{1}{s+5}\right)$

$= e^{-2t} - e^{-5t}$

27. (a)

$L^{-1}\left\{\dfrac{1}{(s^2+1)(s^2+4)}\right\}$

$= \dfrac{1}{3}L^{-1}\left\{\dfrac{3}{(s^2+1)(s^2+4)}\right\}$

$= \dfrac{1}{3}L^{-1}\left(\dfrac{(s^2+4)-(s^2+1)}{(s^2+1)(s^2+4)}\right)$

$= \dfrac{1}{3}\left[L^{-1}\left(\dfrac{1}{s^2+1}\right) - L^{-1}\left(\dfrac{1}{s^2+4}\right)\right]$

$= \dfrac{1}{3}\left(\sin t - \dfrac{1}{2}\sin 2t\right)$

28. (a) $L^{-1}\left\{\dfrac{1}{(s^2+9)^2}\right\} = L^{-1}\left\{\dfrac{1}{(s^2+3^2)^2}\right\}$

$= \dfrac{1}{2\times 3^3}(\sin 3t - 3t\cos 3t)$

$= \dfrac{1}{54}(\sin 3t - 3t\cos 3t)$

29. (c)

$L^{-1}\left\{\dfrac{s+2}{(s^2+1)^2}\right\}$

$= L^{-1}\left\{\dfrac{s}{(s^2+1^2)^2} + 2\dfrac{1}{(s^2+1^2)^2}\right\}$

$= L^{-1}\left\{\dfrac{s}{(s^2+1^2)^2}\right\} + 2L^{-1}\left\{\dfrac{1}{(s^2+1^2)^2}\right\}$

$= \dfrac{t}{2}\sin t + \dfrac{2}{2\times 1^3}(\sin t - t\cos t)$

$= \left(1 + \dfrac{t}{2}\right)\sin t - t\cos t$

30. (a) $L^{-1}\left\{\dfrac{1}{(s+2)^3}\right\} = L^{-1}\left[\dfrac{1}{\{s-(-2)\}^3}\right]$

$$= \frac{e^{-2t}t^{3-1}}{\underline{|3-1}} = \frac{1}{2}e^{-2t}t^2$$

31. (b)

$L^{-1}\left\{\dfrac{s+4}{(s+2)^3}\right\} = L^{-1}\left\{\dfrac{(s+2)+2}{(s+2)^3}\right\}$

$$= L^{-1}\left\{\frac{1}{(s+2)^2} + 2\times\frac{1}{(s+2)^3}\right\}$$

$$= L^{-1}\left\{\frac{1}{(s+2)^2}\right\} + 2L^{-1}\left\{\frac{1}{(s+2)^3}\right\}$$

$$= \frac{e^{-2t}t^{2-1}}{(2-1)!} + 2\frac{e^{-2t}t^{3-1}}{(3-1)!}$$

$$= e^{-2t}(t+t^2)$$

$$= te^{-2t}(1+t)$$

32. (a)

Let $F(s) = \log\left(\dfrac{s+1}{s-1}\right)$.

Then $f(t) = L^{-1}(F(s)) = L^{-1}\left(\dfrac{s+1}{s-1}\right)$

$tf(t)$

$$= -L^{-1}\frac{d}{ds}\{F(s)\}$$

$$= -L^{-1}\frac{d}{ds}\left\{\log\left(\frac{s+1}{s-1}\right)\right\}$$

$$= -L^{-1}\left[\frac{d}{ds}\{\log(s+1)-\log(s-1)\}\right]$$

$$= -L^{-1}\left(\frac{1}{s+1} - \frac{1}{s-1}\right)$$

$$= -L^{-1}\left(\frac{-2}{s^2-1}\right) = 2L^{-1}\left(\frac{1}{s^2-1}\right) = 2\sinh t$$

$$\Rightarrow f(t) = \frac{2}{t}\sinh t$$

$$\Rightarrow L^{-1}\left\{\log\left(\frac{s+1}{s-1}\right)\right\} = \frac{2}{t}\sinh t$$

33. (a) $L^{-1}\left[\dfrac{1}{s(s+1)}\right]$

$$= L^{-1}\left[\frac{1}{(s+1)}\times\frac{1}{s}\right]$$

$$= L^{-1}[F(s)\times G(s)] = \int_0^t f(u)\times g(t-u)\,du$$

$$\ldots\ldots\ldots\ldots(1) \quad \text{(by the Convolution theorem)}$$

$$\left[\text{where, } F(s) = \frac{1}{s}, G(s) = \frac{1}{s+1}\right]$$

Here, $f(t) = L^{-1}[F(s)] = L^{-1}\left[\dfrac{1}{s}\right] = 1$,

$g(t) = L^{-1}[G(s)] = L^{-1}\left[\dfrac{1}{s+1}\right] = e^{-t}$.

$\therefore f(u) = 1, \ g(t-u) = e^{-(t-u)} = e^{u-t}$

So (1) gives

$$L^{-1}\left[\frac{1}{s(s+1)}\right] = \int_0^t 1\times e^{u-t}\,du = e^{-t}\int_0^t e^u\,du$$

$$= e^{-t}\left[e^u\right]_0^t = e^{-t}\left[e^t - e^0\right]$$

$$= e^{-t}\left[e^t - 1\right] = 1 - e^{-t}$$

34. (b)

Here, $f(t) = e^{2t}$ and $a = 2$.

$$\therefore \int_0^\infty e^{2t}\,\delta(t-2)\,dt = f(2) = e^{2\times 2} = e^4.$$

35. (c)

$L[u(t-a)f(t-a)] = e^{-as}F(s)$

$\Rightarrow u(t-a)f(t-a) = L^{-1}[e^{-as}F(s)]$

(by second shifting theorem)

Now comparing $e^{-as}F(s)$ with $\dfrac{e^{-2s}}{s^2+1}$, we get

$$a = 2, \quad F(s) = \frac{1}{s^2+1}$$

$\therefore f(t) = L^{-1}[F(s)] = L^{-1}\left[\dfrac{1}{s^2+1}\right] = \sin t$

Hence, $L^{-1}\left[\dfrac{e^{-2s}}{s^2+1}\right] = L^{-1}[e^{-as}F(s)]$

$$= u(t-a)f(t-a)$$

$$= u(t-2)f(t-2)$$

$$= u(t-2)\times \sin(t-2)$$

36. (a)

$$L\{e^{t-2}u(t-2)\} = e^{-2s}L(e^t) = \frac{1}{(s-1)}e^{-2s}.$$

(by the second shifting theorem)

37. (b) We know that if $f(t) = \begin{cases} g(t-a), t > a \\ 0, t < a \end{cases}$, then,

$$L[f(t)] = e^{-as}\times L[g(t)].$$

Here, $a = 2$ and $g(t-2) = e^{t-2}$. So $g(t) = e^t$.

$\therefore L[f(t)] = e^{-2s} \times L[g(t)] = e^{-2s} \times L[e^t]$

$= e^{-2s} \times \dfrac{1}{(s-1)}$.

38. (c)

We know that if $f(t) = \begin{cases} g(t-a), t > a \\ 0, t < a \end{cases}$, then,

$L[f(t)] = e^{-as} \times L[g(t)]$.

Here, $a = \dfrac{\pi}{3}$ and $g(t - \dfrac{\pi}{3}) = \sin\left(t - \dfrac{\pi}{3}\right)$.

So $g(t) = \sin t$.

$\therefore L[f(t)] = e^{-\frac{\pi s}{3}} \times L[g(t)] = e^{-\frac{\pi s}{3}} \times L[\sin t]$

$= e^{-\frac{\pi s}{3}} \times \dfrac{1}{(s^2+1)}$.

39. (a)

$y' = 1$

$\Rightarrow y'(t) = 1 \Rightarrow L[y'(t)] = L(1)$

$\Rightarrow sY(s) - y(0) = \dfrac{1}{s}$

$\Rightarrow sY(s) = \dfrac{1}{s} \quad (\because y(0) = 0)$

$\Rightarrow Y(s) = \dfrac{1}{s^2}$

$\Rightarrow L^{-1}[Y(s)] = L^{-1}\left(\dfrac{1}{s^2}\right) = \dfrac{t^{2-1}}{\lfloor 2-1} = t$

$\Rightarrow y(t) = t$

40. (d)

$y' + y = 1 \Rightarrow y'(t) + y(t) = 1$

$\Rightarrow L[y'(t)] + L[y(t)] = L(1)$

$\Rightarrow sY(s) - y(0) + Y(s) = \dfrac{1}{s}$

$\Rightarrow (s+1)Y(s) = \dfrac{1}{s} \quad (\because y(0) = 0)$

$\Rightarrow Y(s) = \dfrac{1}{s(s+1)} = \dfrac{(s+1)-s}{s(s+1)} = \dfrac{1}{s} - \dfrac{1}{s+1}$

$\Rightarrow L^{-1}[Y(s)] = L^{-1}\left(\dfrac{1}{s} - \dfrac{1}{s+1}\right)$

$= L^{-1}\left(\dfrac{1}{s}\right) - L^{-1}\left(\dfrac{1}{s+1}\right)$

$\Rightarrow y(t) = 1 - e^{-t}$

Fully Solved MCQs (Level-II)

1. Find $L(f(t))$

if $f(t) = \begin{cases} \sin t, & 0 < t < \pi \\ 0, & t > \pi \end{cases}$

(a) $\dfrac{1 - e^{-s\pi}}{s^2 + 1}$ (b) $\dfrac{1}{s^2 + 1}$

(c) $\dfrac{1 + e^{-s\pi}}{s^2 + 1}$ (d) $\dfrac{1}{s}$

2. Find $L(f(t))$

if $f(t) = \begin{cases} 0, & 0 < t < 1 \\ (t-1)^2, & t > 1 \end{cases}$

(a) $\dfrac{2e^{-s}}{s^3}$ (b) $\dfrac{2e^s}{s^3}$

(c) $\dfrac{2e^{-s}}{s^2}$ (d) $\dfrac{2e^s}{s^2}$

3. $L(e^{-t}\sin^2 t) = ?$

(a) $\dfrac{1}{(s+1)(s^2+2s+5)}$ (b) $\dfrac{s+1}{s^2+2s+5}$

(c) $\dfrac{2}{(s+1)(s^2+2s+5)}$ (d) None of these

4. $L\{(1 + te^{-t})^3\} = ?$

(a) $\dfrac{1}{s} + \dfrac{3}{(s+1)^2} + \dfrac{6}{(s+2)^3} + \dfrac{6}{(s+3)^4}$

(b) $\dfrac{1}{s} + \dfrac{2}{(s+1)^2} + \dfrac{3}{(s+2)^3} + \dfrac{4}{(s+3)^4}$

(c) $-\dfrac{1}{s} + \dfrac{2}{(s+1)^2} + \dfrac{6}{(s+2)^3} + \dfrac{6}{(s+3)^4}$

(d) None of this

5. The Laplace transform of $e^{-t}(4\sin 2t - 5\cosh 2t)$ is

(a) $\dfrac{8}{s^2+2s+5} - \dfrac{5(s+1)}{s^2+2s-3}$

(b) $\dfrac{8}{s^2-2s+5} - \dfrac{5(s+1)}{s^2+2s-3}$

(c) $\dfrac{4}{s^2-2s+5} - \dfrac{10(s+1)}{s^2+2s-3}$

(d) none of these

6. The Laplace transform of $\cosh^2 2t$ is

(a) $\dfrac{1}{2s} + \dfrac{s}{2(s^2-16)}$ (b) $\dfrac{1}{2s} - \dfrac{s}{2(s^2+16)}$

(c) $\dfrac{1}{s} + \dfrac{s}{2(s^2 - 4)}$ (d) $\dfrac{1}{s} - \dfrac{s}{(s^2 - 4)}$

7. The Laplace transform of $e^{-t}(t + 2)^2$ is

(a) $\dfrac{3}{(s-1)^3} + \dfrac{4}{(s-1)^2} + \dfrac{2}{(s-1)}$

(b) $\dfrac{2}{(s-1)^3} + \dfrac{4}{(s-1)^2} + \dfrac{3}{(s-1)}$

(c) $\dfrac{1}{(s+1)^3} + \dfrac{3}{(s+1)^2} + \dfrac{2}{(s+1)}$

(d) $\dfrac{2}{(s+1)^3} + \dfrac{4}{(s+1)^2} + \dfrac{4}{(s+1)}$

8. If $L\{f(t)\} = \dfrac{e^{-s}}{s}$, $L\{e^t f(3t)\} = ?$

(a) $\dfrac{1}{(s-1)} e^{\left(\frac{s+1}{3}\right)}$ (b) $\dfrac{1}{(s+1)} e^{\left(\frac{s-1}{3}\right)}$

(c) $\dfrac{1}{(s-1)} e^{-\left(\frac{s-1}{3}\right)}$ (d) $\dfrac{1}{(s+1)} e^{-\left(\frac{s+1}{3}\right)}$

9. $L(te^{-t} \cos 2t) = ?$

(a) $\dfrac{s^2 - 3 + 2s}{(s^2 + 2s + 5)^2}$ (b) $\dfrac{1}{(s^2 + 2s + 5)^2}$

(c) $\dfrac{5 - s^2}{(s^2 + 2s + 5)^2}$ (d) None of this

10. If $L\left(2\sqrt{\dfrac{t}{\pi}}\right) = \dfrac{1}{s^{\frac{3}{2}}}$, then $L\left(\dfrac{1}{\sqrt{\pi t}}\right) = ?$

(a) $\dfrac{1}{s}$ (b) $\dfrac{1}{\sqrt{s}}$

(c) $\dfrac{-1}{\sqrt{s}}$ (d) None of this

11. If $L(f(t)) = \tan^{-1}\left(\dfrac{1}{s}\right)$, then $L(tf(t)) = ?$

(a) $\tan^{-1}\left(\dfrac{1}{s^2}\right)$ (b) $\dfrac{1}{s^2 + 1}$

(c) $\dfrac{s}{s^2 + 1}$ (d) None of this

12. $L\left(\dfrac{e^{at} - \cos bt}{t}\right) = ?$

(a) $\log\left(\dfrac{s - a}{\sqrt{s^2 + b^2}}\right)$ (b) $-\log\left(\dfrac{s - a}{\sqrt{s^2 + b^2}}\right)$

(c) $\log\left(\dfrac{s - a}{s^2 + b^2}\right)$ (d) None of these

13. $L\left(e^{-t}\displaystyle\int_0^t \dfrac{\sin t}{t} dt\right) = ?$

(a) $\dfrac{1}{s+1} \tan^{-1}(s+1)$ (b) $\dfrac{1}{s+1} \cot^{-1}(s+1)$

(c) $\dfrac{1}{s} \tan^{-1} s$ (d) $\dfrac{1}{s} \cot^{-1} s$

14. $L\left(\displaystyle\int_0^t e^{-t} \cosh t \, dt\right) = ?$

(a) $\dfrac{(s+1)}{s^2(s+2)}$ (b) $\dfrac{(s-1)}{s^2(s-2)}$

(c) $\dfrac{(s-1)}{s(s+2)}$ (d) $\dfrac{(s+1)}{s(s-2)}$

15. Find the Laplace transform of the periodic function $f(t)$ where

$$f(t) = \begin{cases} \sin kt, & 0 < t < \dfrac{\pi}{k} \\ 0, & \dfrac{\pi}{k} < t < \dfrac{2\pi}{k} \end{cases}$$

(a) $\dfrac{1}{\left(1 + e^{-\frac{\pi}{k}s}\right)\left(s^2 + k^2\right)}$

(b) $\dfrac{1}{\left(1 + e^{-\frac{2\pi}{k}s}\right)\left(s^2 + k^2\right)}$

(c) $\dfrac{k}{\left(1 - e^{-\frac{2\pi}{k}s}\right)\left(s^2 + k^2\right)}$

(d) $\dfrac{k}{\left(1 - e^{-\frac{\pi}{k}s}\right)\left(s^2 + k^2\right)}$

16. If $f(t)$ is a periodic function with period 2π and $f(t)$ is given by

$$f(t) = \begin{cases} 1, & 0 < t < \pi \\ -1, & \pi < t < 2\pi \end{cases}$$

then $L[f(t)] = ?$

(a) $\dfrac{1}{s(e^{-2\pi s} - 1)}$ (b) $\dfrac{e^{-\pi s}}{s(e^{-2\pi s} - 1)}$

(c) $\dfrac{1}{s(e^{-2\pi s} - 1)}\{2e^{-\pi s} - e^{-2\pi s} - 1\}$

(d) $\dfrac{e^{-\pi s}}{s}$

17. $\displaystyle\int_0^\infty \frac{e^{-at} - e^{-bt}}{t}\,dt = ?$

    (a) $\log\left(\dfrac{b}{a}\right)$ (b) $\log\left(\dfrac{a}{b}\right)$
    (c) $\log(a + b)$ (d) $\log(ab)$

18. $\displaystyle\int_0^\infty e^{-2t}\frac{\sinh t}{t}\,dt = ?$

    (a) $\dfrac{1}{2}\log 2$ (b) $\dfrac{1}{2}\log 3$
    (c) 1 (d) 0

19. If $L\big(f(t)\big) = \dfrac{4}{s(s+4)}$, then $\lim\limits_{t\to\infty} f(t) = ?$

    (a) 0 (b) 1
    (c) -1 (d) 4

20. If $L\big(f(t)\big) = \dfrac{4s^2 - 3s + 8}{s(s^2 + 4)}$, then $\lim\limits_{t\to\infty} f(t) = ?$

    (a) 0 (b) 1
    (c) -1 (d) 2

21. $L^{-1}\left(\dfrac{s+2}{s^2 - 4s + 13}\right) = ?$

    (a) $e^{2t}\left(\cos 3t + \dfrac{4}{3}\sin 3t\right)$

    (b) $e^{-2t}\left(\cos 3t + \dfrac{4}{3}\sin 3t\right)$

    (c) $e^{t}\left(\cos 2t - \dfrac{4}{3}\sin 3t\right)$

    (d) None of these

22. $L^{-1}\left(\dfrac{3s+4}{s^2 + 2s + 5}\right) = ?$

    (a) $e^{-t}\cos 2t + \dfrac{1}{3}e^{-t}\sin 2t$

    (b) $3e^{-t}\cos 2t + \dfrac{1}{2}e^{-t}\sin 2t$

    (c) $e^{t}\cos 2t + \dfrac{1}{2}e^{t}\sin 2t$

    (d) None of these

23. Find $L^{-1}\left\{\dfrac{1}{s^2(s^2 + 2)}\right\} = ?$

    (a) $\dfrac{1}{2}\left[t - \dfrac{1}{\sqrt{2}}\sin(\sqrt{2}t)\right]$

    (b) $\dfrac{1}{2}\left[t + \dfrac{1}{2}\sin(2t)\right]$

    (c) $\dfrac{1}{2}\left[1 - \dfrac{1}{2}\sin(2t)\right]$

    (d) $\dfrac{1}{2}\left[t + \dfrac{1}{\sqrt{2}}\sin(\sqrt{2}t)\right]$

24. $L^{-1}\left\{\dfrac{1}{s^2(s+1)^2}\right\} = ?$

    (a) $t - 2 + (t + 2)e^{-t}$ (b) $t + 2 + (t - 2e^{-t})$
    (c) $t + 2 + (t - 2)e^{t}$ (d) $t - 2 + (t + 2)e^{-2t}$

25. $L^{-1}\left[\dfrac{s}{(s+1)(s^2 + 1)}\right] = ?$

    (a) $\dfrac{1}{2}\left\{-\cos t + \sin t + e^{t}\right\}$

    (b) $\dfrac{1}{2}\left\{\cos t - \sin t - e^{-t}\right\}$

    (c) $\dfrac{1}{2}\left\{\cos t - \sin t + e^{-t}\right\}$

    (d) $\dfrac{1}{2}\left\{\cos t + \sin t - e^{-t}\right\}$

26. $L^{-1}\left\{\dfrac{s+1}{(s^2 + 2s + 5)^2}\right\} = ?$

    (a) $\dfrac{1}{2}te^{t}\sin 2t$ (b) $\dfrac{1}{4}te^{-t}\sin 2t$
    (c) $\dfrac{1}{4}te^{t}\sin t$ (d) $\dfrac{1}{2}te^{-t}\sin 2t$

27. $L^{-1}\left[\log\left\{\dfrac{s^2 + 1}{s(s+1)}\right\}\right] = ?$

    (a) $\dfrac{1}{t}\left(1 + e^{t} - 2\cos t\right)$ (b) $\dfrac{1}{t}\left(1 - e^{-t} + 2\cos t\right)$
    (c) $\dfrac{1}{t}\left(1 + e^{-t} - 2\cos t\right)$ (d) None of these

28. The Laplace transform of $f(t)$, where z
    $$f(t) = \begin{cases} 0, & 0 < t < 1 \\ 1, & 1 < t < 2 \\ 2, & t > 2 \end{cases} \text{ is}$$

    (a) $\dfrac{1}{s}\left(e^{-s} - e^{-2s}\right)$ (b) $\dfrac{1}{s}\left(e^{-s} + e^{-2s}\right)$
    (c) $\dfrac{1}{s}\left(e^{-2s} - e^{-s}\right)$ (d) $\dfrac{1}{s}\left(e^{-3s} + e^{-s}\right)$

29. The Laplace transform of $f(t)$, where

$$f(t) = \begin{cases} t, & 0 < t < \dfrac{1}{2} \\ t-1, & \dfrac{1}{2} < t < 1 \\ 0, & t > 1 \end{cases} \text{ is}$$

(a) $\dfrac{1}{s}\left(e^{-\frac{s}{2}} - \dfrac{1}{s}e^{-s} + \dfrac{1}{s}\right)$

(b) $-\dfrac{1}{s}\left(e^{-\frac{s}{2}} + \dfrac{1}{s}e^{-s} + \dfrac{1}{s}\right)$

(c) $-\dfrac{1}{s}\left(e^{-\frac{s}{2}} + \dfrac{1}{s}e^{-s} - \dfrac{1}{s}\right)$

(d) None of these

30. The Laplace transform of $f(t)$, where

$$f(t) = \begin{cases} \sin 2t, & 0 < t < \pi \\ 0, & t > \pi \end{cases} \text{ is}$$

(a) $\dfrac{2\left(1 - e^{-\pi s}\right)}{s^2 + 4}$

(b) $\dfrac{2\left(1 + e^{-\pi s}\right)}{s^2}$

(c) $\dfrac{1 - e^{-\pi s}}{s^2 + 4}$

(d) $\dfrac{1 + e^{-\pi s}}{s^2 + 4}$

31. The Laplace transform of $f(t)$, where

$$f(t) = \begin{cases} 0, & 0 < t < 1 \\ (t-1)^3, & t > 1 \end{cases} \text{ is}$$

(a) $\dfrac{6e^s}{s^4}$

(b) $\dfrac{-6e^s}{s^3}$

(c) $\dfrac{6e^{-s}}{s^4}$

(d) $\dfrac{2e^s}{s^3}$

32. The Laplace transform of $f(t)$, where

$$f(t) = \begin{cases} 1, & 0 < t < 3 \\ t, & t > 3 \end{cases} \text{ is}$$

(a) $\dfrac{1}{s} + \dfrac{e^{-3s}}{s^2} + \dfrac{2}{3}e^{-3s}$

(b) $\dfrac{1}{s} + \dfrac{e^{-3s}}{s^2} + \dfrac{2}{s}e^{-3s}$

(c) $\dfrac{1}{s} - \dfrac{2e^{-3s}}{s^3} + e^s$

(d) $\dfrac{2e^{-3s}}{s^3}$

33. Find $L(f(t))$ if $f(t) = \begin{cases} 2t, & 0 \le t \le 1 \\ t, & t > 1 \end{cases}$ is

(a) $\dfrac{2}{s^2} - \left(\dfrac{1}{s^2} + \dfrac{1}{s}\right)e^{-s}$

(b) $\dfrac{2}{s^2} + \left(\dfrac{1}{s^2} - \dfrac{1}{s}\right)e^{-s}$

(c) $\dfrac{2}{s^2} + \left(\dfrac{1}{s^2} + \dfrac{1}{s}\right)e^{-s}$

(d) $\dfrac{-2}{s^2} - \left(\dfrac{1}{s^2} + \dfrac{1}{s}\right)e^{-s}$

34. $L\{t^2 u(t-1)\} = ?$

(a) $\left(\dfrac{1}{s^3} - \dfrac{2}{s^2} + \dfrac{1}{s}\right)e^{-s}$

(b) $\left(\dfrac{1}{s^3} - \dfrac{1}{s^2} + \dfrac{1}{s}\right)e^{-s}$

(c) $\left(\dfrac{2}{s^3} + \dfrac{2}{s^2} + \dfrac{1}{s}\right)e^{-s}$

(d) $\left(\dfrac{2}{s^3} - \dfrac{2}{s^2} - \dfrac{1}{s}\right)e^{-s}$

35. Solve: $y'' + 2y' + y = 1$; given $y(0) = 0 = y'(0)$

(a) $1 - e^{-t}(1 + t)$

(b) $1 + e^t(t - 1)$

(c) $1 - te^{-t}$

(d) te^t

36. Solve: $y'' - 4y' = e^t$; given $y(0) = 0 = y'(0)$

(a) $\dfrac{1}{4}e^{-2t} - \dfrac{1}{3}e^t + \dfrac{1}{12}e^{2t}$

(b) $\dfrac{1}{4}e^{2t} - \dfrac{1}{3}e^t + \dfrac{1}{12}e^{-2t}$

(c) $\dfrac{1}{4}e^{-t} - \dfrac{1}{3}e^t + \dfrac{1}{12}e^{-2t}$

(d) $\dfrac{1}{12}e^{2t} - \dfrac{1}{3}e^t + \dfrac{1}{12}e^{-2t}$

Answers key				
1. (c)	**2.** (a)	**3.** (c)	**4.** (a)	**5.** (a)
6. (a)	**7.** (d)	**8.** (d)	**9.** (a)	**10.** (b)
11. (b)	**12.** (b)	**13.** (b)	**14.** (a)	**15.** (d)
16. (c)	**17.** (a)	**18.** (b)	**19.** (b)	**20.** (d)
21. (a)	**22.** (b)	**23.** (a)	**24.** (a)	**25.** (d)
26. (b)	**27.** (c)	**28.** (b)	**29.** (c)	**30.** (a)
31. (c)	**32.** (b)	**33.** (a)	**34.** (c)	**35.** (a)
36. (b)				

Explanation

1. (c)

$L\{f(t)\}$

$$= \int_0^\infty e^{-st} f(t)\, dt$$

$$= \int_0^\pi e^{-st} f(t)\, dt + \int_\pi^\infty e^{-st} f(t)\, dt$$

$$= \int_0^\pi e^{-st} \sin t\, dt + \int_\pi^\infty e^{-st} \times 0\, dt$$

$$= \int_0^\pi e^{-st} \sin t\, dt$$

$$= \left[\frac{e^{-st}}{s^2+1}(-s\sin t - \cos t)\right]_0^\pi$$

$$\left[\text{Using}\int e^{ax}\sin bx\,dx\right.$$

$$\left.= \frac{e^{ax}}{a^2+b^2}(a\sin bx - b\cos bx)\right]$$

$$= \frac{e^{-s\pi}}{s^2+1}(-s\sin\pi - \cos\pi) - \frac{e^0}{s^2+1}(-s\sin 0 - \cos 0)$$

$$= \frac{e^{-s\pi}}{s^2+1} + \frac{1}{s^2+1} = \frac{e^{-s\pi}+1}{s^2+1}$$

$$(\because \sin\pi = 0, \cos 0 = 1, \cos\pi = -1)$$

2. (a)

$$L\{f(t)\}$$

$$= \int_0^\infty e^{-st}f(t)dt$$

$$= \int_0^1 e^{-st}f(t)dt + \int_1^\infty e^{-st}f(t)dt$$

$$= \int_1^\infty e^{-st}(t-1)^2\,dt$$

$$\left(\begin{array}{l}\text{Putting}\quad t-1=q\text{ so that }dt=dq\\\text{and }t=1\Rightarrow q=0; t\to\infty\Rightarrow q\to\infty\end{array}\right)$$

$$= \int_0^\infty e^{-s(1+q)}q^2\,dq = e^{-s}\int_0^\infty e^{-sq}q^{3-1}\,dq$$

$$= e^{-s}\frac{\Gamma(3)}{s^3}\quad\left[\because \int_0^\infty e^{-aq}q^{n-1}\,dq = \frac{\Gamma(n)}{a^n}\right]$$

$$= e^{-s}\frac{\underline{|2}}{s^3}\quad\left[\because \Gamma(n) = \underline{|n-1}\text{ if }n\in Z^+\right]$$

$$= e^{-s}\frac{2}{s^3}$$

3. (c)

$$L\left(e^{-t}\sin^2 t\right)$$

$$= \frac{1}{2}L\left\{e^{-t}(2\sin^2 t)\right\}$$

$$= \frac{1}{2}L\left\{e^{-t}(1-\cos 2t)\right\}$$

$$= \frac{1}{2}L(e^{-t}) - \frac{1}{2}L(e^{-t}\cos 2t)$$

$$= \frac{1}{2}\left(\frac{1}{s+1} - \frac{s+1}{(s+1)^2+4}\right)$$

$$= \frac{1}{2}\left(\frac{s^2+2s+5-(s^2+2s+1)}{(s+1)(s^2+2s+5)}\right)$$

$$= \frac{2}{(s+1)(s^2+2s+5)}$$

4. (a)

$$\left(1+te^{-t}\right)^3 = 1 + 3te^{-t} + 3\left(te^{-t}\right)^2 + \left(te^{-t}\right)^3$$

$$= 1 + 3te^{-t} + 3t^2e^{-2t} + t^3e^{-3t}$$

$$L\left[\left(1+te^{-t}\right)^3\right]$$

$$= L\left(1 + 3te^{-t} + 3t^2e^{-2t} + t^3e^{-3t}\right)$$

$$= L(1) + 3L\left(te^{-t}\right) + 3L\left(t^2e^{-2t}\right) + L\left(t^3e^{-3t}\right)$$

$$\ldots\ldots\ldots\ldots\ldots(1)$$

Now, $L(t) = \dfrac{\Gamma(2)}{s^2} = \dfrac{\underline{|1}}{s^2} = \dfrac{1}{s^2}$,

$$L\left(t^2\right) = \frac{\Gamma(3)}{s^3} = \frac{\underline{|2}}{s^3} = \frac{2}{s^3},$$

$$L\left(t^3\right) = \frac{\Gamma(4)}{s^4} = \frac{\underline{|3}}{s^4} = \frac{6}{s^4}$$

So by the first shifting theorem,

$$L\left(te^{-t}\right) = \frac{1}{\{s-(-1)\}^2} = \frac{1}{(s+1)^2},$$

$$L\left(t^2e^{-2t}\right) = \frac{2}{\{s-(-2)\}^3} = \frac{2}{(s+2)^3},$$

$$L\left(t^3e^{-3t}\right) = \frac{6}{\{s-(-3)\}^4} = \frac{6}{(s+3)^4}$$

Hence, from (1) we get,

$$L\left[\left(1+te^{-t}\right)^3\right] = \frac{1}{s} + \frac{3}{(s+1)^2} + \frac{6}{(s+2)^3} + \frac{6}{(s+3)^4}$$

$$\left(\text{since } L(1) = 1/s\right)$$

5. (a)

$$L\left\{e^{-t}\left(4\sin 2t - 5\cosh 2t\right)\right\}$$

$$= 4L\left(e^{-t}\sin 2t\right) - 5L\left(e^{-t}\cosh 2t\right)$$

$$= 4\times\frac{2}{(s+1)^2+2^2} - 5\times\frac{(s+1)}{(s+1)^2-2^2}$$

$$= \frac{8}{s^2+2s+5} - \frac{5(s+1)}{s^2+2s-3}$$

6. (a)

$$\cosh^2 2t = (\cosh 2t)^2 = \left(\frac{e^{2t}+e^{-2t}}{2}\right)^2$$

$$\left(\because \cosh x = \frac{e^x+e^{-x}}{2}\right)$$

$$= \frac{e^{4t}+2+e^{-4t}}{4} = \frac{e^{4t}}{4} + \frac{1}{2} + \frac{e^{-4t}}{4}$$

$\therefore L\left(\cosh^2 2t\right)$

$= L\left(\dfrac{e^{4t}}{4}\right) + L\left(\dfrac{1}{2}\right) + L\left(\dfrac{e^{-4t}}{4}\right)$

$= \dfrac{1}{4}L\left(e^{4t}\right) + \dfrac{1}{2}L(1) + \dfrac{1}{4}L\left(e^{-4t}\right)$

$= \dfrac{1}{4}\times\dfrac{1}{(s-4)} + \dfrac{1}{2}\times\dfrac{1}{s} + \dfrac{1}{4}\times\dfrac{1}{(s+4)}$

$= \dfrac{1}{2s} + \dfrac{1}{4}\times\dfrac{2s}{\left(s^2-16\right)} = \dfrac{1}{2s} + \dfrac{s}{2\left(s^2-16\right)}$

7. (d)

$L\left(e^{-t}(t+2)^2\right)$

$= L\left[e^{-t}\left(t^2+4t+4\right)\right]$

$= L\left(e^{-t}t^2\right) + 4L\left(e^{-t}t\right) + 4L\left(e^{-t}\right)$

$= \dfrac{2!}{(s+1)^3} + 4\times\dfrac{1!}{(s+1)^2} + 4\times\dfrac{1}{(s+1)}$

(by the first shifting theorem)

$= \dfrac{2}{(s+1)^3} + \dfrac{4}{(s+1)^2} + \dfrac{4}{(s+1)}$

8. (d)

We know that

$L\{f(at)\} = \dfrac{1}{a}F\left(\dfrac{s}{a}\right)$ if $L\{f(t)\} = F(s)$.

Therefore, $L\{f(3t)\} = \dfrac{1}{3}F\left(\dfrac{s}{3}\right)$ and so

$L\{e^{-t}f(3t)\} = \dfrac{1}{3}F\left(\dfrac{s+1}{3}\right)$ (by the first shifting theorem)

Given $L\{f(t)\} = F(s) = \dfrac{e^{-s}}{s}$.

$\therefore L\{e^{-t}f(3t)\} = \dfrac{1}{3}F\left(\dfrac{s+1}{3}\right)$

$= \dfrac{1}{3}\dfrac{e^{-\left(\frac{s+1}{3}\right)}}{\left(\dfrac{s+1}{3}\right)} = \dfrac{1}{(s+1)}e^{-\left(\frac{s+1}{3}\right)}$

9. (a) $L(\cos 2t) = \dfrac{s}{s^2+4}$

$\therefore$ By first shifting theorem,

$L\left(e^{-t}\cos 2t\right) = \dfrac{s+1}{(s+1)^2+4} = \dfrac{s+1}{s^2+2s+5}$

$L\left(te^{-t}\cos 2t\right)$

$= L\{t(e^{-t}\cos 2t)\} = (-1)^1\dfrac{d}{ds}L\left(e^{-t}\cos 2t\right)$

$= -\dfrac{d}{ds}\left(\dfrac{s+1}{s^2+2s+5}\right)$

$= -\dfrac{\{(s^2+2s+5)\times 1 - (s+1)(2s+2)\}}{(s^2+2s+5)^2}$

$= \dfrac{s^2+2s-3}{(s^2+2s+5)^2}$

10. (b)

Let $f(t) = 2\sqrt{\dfrac{t}{\pi}}$. Then,

$F(s) = L(f(t)) = L\left(2\sqrt{\dfrac{t}{\pi}}\right) = \dfrac{1}{s^{\frac{3}{2}}}$ (given)

Also, $f'(t) = \dfrac{2}{\sqrt{\pi}}\times\dfrac{1}{2\sqrt{t}} = \dfrac{1}{\sqrt{\pi t}}$ and $f(0) = 0$.

Hence, $L(f'(t)) = sF(s) - f(0)$

$\Rightarrow L\left(\dfrac{1}{\sqrt{\pi t}}\right) = s\times\dfrac{1}{s^{\frac{3}{2}}} - 0 = \dfrac{1}{s^{\frac{1}{2}}}$.

11. (b) $L(f(t)) = \tan^{-1}\left(\dfrac{1}{s}\right)$

$\Rightarrow L(tf(t)) = -\dfrac{d}{ds}L(f(t))$

$= -\dfrac{d}{ds}\tan^{-1}\left(\dfrac{1}{s}\right)$

$= -\dfrac{1}{\left\{1+\left(\dfrac{1}{s}\right)^2\right\}}\times\left(\dfrac{-1}{s^2}\right)$

$= \dfrac{1}{s^2+1}$

12. (b) $L\left(e^{at} - \cos bt\right) = L\left(e^{at}\right) - L(\cos bt)$

$= \dfrac{1}{s-a} - \dfrac{s}{s^2+b^2} = F(s)$

$\therefore L\left(\dfrac{e^{at}-\cos bt}{t}\right)$

$= L\left[\dfrac{f(t)}{t}\right], \quad$ where $f(t) = e^{at} - \cos bt$

$$= \int_s^\infty F(s)ds = \int_s^\infty \left(\frac{1}{s-a} - \frac{s}{s^2+b^2} \right) ds$$

$$= \left[\log(s-a) - \frac{1}{2}\log(s^2+b^2) \right]_s^\infty$$

$$= \left[\log\left(\frac{s-a}{\sqrt{s^2+b^2}} \right) \right]_s^\infty = \left[\log\left(\frac{1-(a/s)}{\sqrt{1+(b/s)^2}} \right) \right]_s^\infty$$

$$= \log\left(\frac{1-0}{\sqrt{1+0}} \right) - \log\left(\frac{1-(a/s)}{\sqrt{1+(b/s)^2}} \right)$$

$$\left[\because s \to \infty \Rightarrow \frac{1}{s} \to 0 \right]$$

$$= \log 1 - \log\left(\frac{s-a}{\sqrt{s^2+b^2}} \right) = -\log\left(\frac{s-a}{\sqrt{s^2+b^2}} \right)$$

13. (b)

$$L\left(\int_0^t \frac{\sin t}{t} dt \right)$$

$$= L\left(\int_0^t f(t)dt \right) \quad \left[\text{where } f(t) = \frac{\sin t}{t} \right]$$

$$= \frac{1}{s} F(s) = \frac{1}{s} L\{f(t)\} = \frac{1}{s} L\left[\frac{\sin t}{t} \right]$$

$$= \frac{1}{s} \int_s^\infty L(\sin t) ds = \frac{1}{s} \int_s^\infty \frac{1}{s^2+1} ds$$

$$= \frac{1}{s} \left[\tan^{-1} s \right]_s^\infty = \frac{1}{s} \left[\tan^{-1} \infty - \tan^{-1} s \right]$$

$$= \frac{1}{s} \left[\frac{\pi}{2} - \tan^{-1} s \right] = \frac{1}{s} \cot^{-1} s$$

By first shifting theorem,

$$L\left(e^{-t} \int_0^t \frac{\sin t}{t} dt \right) = \frac{1}{s-(-1)} \cot^{-1}\{s-(-1)\}$$

$$= \frac{\cot^{-1}(s+1)}{s+1}$$

14. (a)

Here, $f(t) = e^{-t} \cosh t$ and

$$F(s) = L(f(t)) = L(e^{-t} \cosh t)$$

$$= \frac{s-(-1)}{\{s-(-1)\}^2 - 1}$$

$$= \frac{s+1}{s^2+2s} = \frac{s+1}{s(s+2)}$$

$$\therefore L\left(\int_0^t e^{-t} \cosh t \, dt \right) = L\left(\int_0^t f(t)dt \right)$$

$$= \frac{F(s)}{s} = \frac{s+1}{s^2(s+2)}.$$

15. (d)

Here, $f(t)$ is a periodic function with period $2\pi/k$.

$$L[f(t)] = \frac{\int_0^T e^{-st} f(t) dt}{1-e^{-sT}}$$

$$= \frac{1}{1-e^{-\frac{2\pi}{k}s}} \int_0^{\frac{2\pi}{k}} e^{-st} f(t)dt$$

$$= \frac{1}{\left(1-e^{-\frac{2\pi}{k}s}\right)} \left[\int_0^{\frac{\pi}{k}} e^{-st} \times \sin kt \, dt + \int_{\frac{\pi}{k}}^{\frac{2\pi}{k}} e^{-st} \times 0 \, dt \right]$$

$$= \frac{1}{\left(1-e^{-\frac{2\pi}{k}s}\right)} \int_0^{\frac{\pi}{k}} e^{-st} \times \sin kt \, dt$$

$$= \frac{1}{\left(1-e^{-\frac{2\pi}{k}s}\right)} \left[\frac{e^{-st}(-s\sin kt - k\cos kt)}{s^2+k^2} \right]_0^{\frac{\pi}{k}}$$

$$= \frac{1}{\left(1-e^{-\frac{2\pi}{k}s}\right)} \left[\frac{e^{-\frac{\pi}{k}s}(-s\sin \pi - k\cos \pi) - (-k)}{s^2+k^2} \right]$$

$$= \frac{1}{\left(1-e^{-\frac{2\pi}{k}s}\right)} \left[\frac{ke^{-\frac{\pi}{k}s} + k}{s^2+k^2} \right] \quad (\because \sin \pi = 0, \cos \pi = -1)$$

$$= \frac{k}{\left(1-e^{-\frac{\pi}{k}s}\right)\left(1+e^{-\frac{\pi}{k}s}\right)} \left[\frac{e^{-\frac{\pi}{k}s} + 1}{s^2+k^2} \right]$$

$$= \frac{k}{\left(1-e^{-\frac{\pi}{k}s}\right)(s^2+k^2)}$$

16. (c) We know that if $f(t)$ is a periodic function with period T, then

$$L[f(t)] = \frac{1}{\left(1-e^{-sT}\right)}\int_0^T e^{-st}f(t)dt$$

$$= \frac{1}{\left(1-e^{-2\pi s}\right)}\int_0^{2\pi} e^{-st}f(t)dt \quad \left(\text{here } T=2\pi\right)$$

$$= \frac{1}{\left(1-e^{-2\pi s}\right)}[\int_0^\pi e^{-st}\times 1dt + \int_\pi^{2\pi} e^{-st}(-1)dt]$$

$$= \frac{1}{(1-e^{-2\pi s})}\left\{\left[\frac{e^{-st}}{-s}\right]_0^\pi - \left[\frac{e^{-st}}{-s}\right]_\pi^{2\pi}\right\}$$

$$= \frac{1}{(1-e^{-2\pi s})}\left\{-\frac{1}{s}\left[e^{-\pi s}-e^0\right]+\frac{1}{s}\left[e^{-2\pi s}-e^{-\pi s}\right]\right\}$$

$$= \frac{1}{s(1-e^{-2\pi s})}\{e^{-2\pi s}-2e^{-\pi s}+1\}$$

$$= \frac{1}{s(e^{-2\pi s}-1)}\{2e^{-\pi s}-e^{-2\pi s}-1\}$$

17. (a)

$$L\left(e^{-at}-e^{-bt}\right)=L\left(e^{-at}\right)-L\left(e^{-bt}\right)$$

$$=\frac{1}{s+a}-\frac{1}{s+b}=F(s)$$

$$\therefore L\left(\frac{e^{-at}-e^{-bt}}{t}\right)$$

$$=L\left[\frac{f(t)}{t}\right], \quad \text{where } f(t)=e^{-at}-e^{-bt}$$

$$=\int_s^\infty F(s)ds = \int_s^\infty\left(\frac{1}{s+a}-\frac{1}{s+b}\right)ds$$

$$=\left[\log(s+a)-\log(s+b)\right]_s^\infty$$

$$=\left[\log\left(\frac{s+a}{s+b}\right)\right]_s^\infty = \left[\log\left(\frac{1+(a/s)}{1+(b/s)}\right)\right]_s^\infty$$

$$=\log\left(\frac{1+0}{1+0}\right)-\log\left(\frac{1+(a/s)}{1+(b/s)}\right)$$

$$\left[\because s\to\infty\Rightarrow\frac{1}{s}\to 0\right]$$

$$=\log 1-\log\left(\frac{s+a}{s+b}\right)=\log\left(\frac{s+b}{s+a}\right)$$

Let $g(t)=\dfrac{e^{-at}-e^{-bt}}{t}$. Then

$$L\{g(t)\}=L\left(\frac{e^{-at}-e^{-bt}}{t}\right)=\log\left(\frac{s+b}{s+a}\right)$$

$$\Rightarrow G(s)=\log\left(\frac{s+b}{s+a}\right) \quad (\text{if } L\{g(t)\}=G(s))$$

Now comparing $\int_0^\infty\dfrac{e^{-at}-e^{-bt}}{t}dt$ with

$$\int_0^\infty e^{-st}g(t)dt \text{ we get, } s=0.$$

$$\therefore\int_0^\infty\frac{e^{-at}-e^{-bt}}{t}dt = \int_0^\infty e^{-st}g(t)dt = G(s)=G(0)$$

$$=\log\left(\frac{0+b}{0+a}\right)=\log\left(\frac{b}{a}\right)$$

18. (b)

$$\int_0^\infty e^{-2t}\frac{\sinh t}{t}dt = \int_0^\infty e^{-st}f(t)dt$$

$$\left(\text{here } s=2 \text{ and } f(t)=\frac{\sinh t}{t}\right)$$

$$=F(s)=L[f(t)]=L\left[\frac{\sinh t}{t}\right]$$

$$=\int_s^\infty L(\sinh t)ds = \int_s^\infty\frac{1}{s^2-1}ds$$

$$=\frac{1}{2}\left[\log\left(\frac{s-1}{s+1}\right)\right]_s^\infty = \frac{1}{2}\left[\log\left(\frac{1-\frac{1}{s}}{1+\frac{1}{s}}\right)\right]_s^\infty$$

$$=\frac{1}{2}\left[\log\left(\frac{1-0}{1+0}\right)-\log\left(\frac{1-\frac{1}{s}}{1+\frac{1}{s}}\right)\right]$$

$$=\frac{1}{2}\left[\log 1-\log\left(\frac{s-1}{s+1}\right)\right]=\frac{1}{2}\left[0-\log\left(\frac{s-1}{s+1}\right)\right]$$

$$=-\frac{1}{2}\log\left(\frac{2-1}{2+1}\right)=-\frac{1}{2}\log\left(\frac{1}{3}\right)=\frac{1}{2}\log 3 \;(\because s=2)$$

19. (b)

$$\lim_{t\to\infty}f(t)=\lim_{s\to 0}sF(s)=\lim_{s\to 0}sL(f(t))$$

$$=\lim_{s\to 0}\left\{s\times\frac{4}{s(s+4)}\right\}$$

$$=\lim_{s\to 0}\left\{\frac{4}{(s+4)}\right\}=\frac{4}{(0+4)}=1.$$

20. (d)

$$\lim_{t\to\infty} f(t) = \lim_{s\to 0} sF(s) = \lim_{s\to 0} sL\left(f(t)\right)$$

$$= \lim_{s\to 0}\left\{s\times\frac{\left(4s^2 - 3s + 8\right)}{s\left(s^2 + 4\right)}\right\}$$

$$= \lim_{s\to 0}\left\{\frac{4s^2 - 3s + 8}{\left(s^2 + 4\right)}\right\} = \frac{0 - 0 + 8}{(0 + 4)} = 2.$$

21. (a)

$$L^{-1}\left(\frac{s+2}{s^2 - 4s + 13}\right)$$

$$= L^{-1}\left\{\frac{(s-2)+4}{(s-2)^2 + 3^2}\right\}$$

$$= L^{-1}\left\{\frac{s-2}{(s-2)^2 + 3^2} + 4\frac{1}{(s-2)^2 + 3^2}\right\}$$

$$= L^{-1}\left\{\frac{s-2}{(s-2)^2 + 3^2}\right\} + 4L^{-1}\left\{\frac{1}{(s-2)^2 + 3^2}\right\}$$

$$= e^{2t}\cos 3t + e^{2t}\frac{4}{3}\sin 3t$$

$$= e^{2t}(\cos 3t + \frac{4}{3}\sin 3t)$$

22. (b)

$$L^{-1}\left(\frac{3s+4}{s^2 + 2s + 5}\right)$$

$$= L^{-1}\left\{\frac{3(s+1)+1}{(s+1)^2 + 2^2}\right\}$$

$$= L^{-1}\left\{\frac{3(s+1)}{(s+1)^2 + 2^2} + \frac{1}{(s+1)^2 + 2^2}\right\}$$

$$= 3L^{-1}\left\{\frac{(s+1)}{(s+1)^2 + 2^2}\right\} + L^{-1}\left\{\frac{1}{(s+1)^2 + 2^2}\right\}$$

$$= 3e^{-t}\cos 2t + \frac{1}{2}e^{-t}\sin 2t$$

23. (a)

$$L^{-1}\left\{\frac{1}{s^2(s^2 + 2)}\right\}$$

$$= L^{-1}\left[\frac{1}{2}\times\frac{2}{s^2(s^2 + 2)}\right]$$

$$= \frac{1}{2}L^{-1}\left[\frac{(s^2 + 2) - s^2}{s^2(s^2 + 2)}\right]$$

$$= \frac{1}{2}L^{-1}\left[\frac{1}{s^2} - \frac{1}{(s^2 + 2)}\right]$$

$$= \frac{1}{2}\left[L^{-1}\left(\frac{1}{s^2}\right) - L^{-1}\left(\frac{1}{s^2 + 2}\right)\right]$$

$$= \frac{1}{2}\left[L^{-1}\left(\frac{1}{s^2}\right) - L^{-1}\left(\frac{1}{s^2 + \left(\sqrt{2}\right)^2}\right)\right]$$

$$= \frac{1}{2}\left[t - \frac{1}{\sqrt{2}}\sin(\sqrt{2}t)\right]$$

24. (a)

$$L^{-1}\left\{\frac{1}{s^2(s+1)^2}\right\}$$

$$= L^{-1}\left\{\frac{(s+1)-s}{s^2(s+1)^2}\right\}$$

$$= L^{-1}\left\{\frac{(s+1)}{s^2(s+1)^2} - \frac{s}{s^2(s+1)^2}\right\}$$

$$= L^{-1}\left\{\frac{1}{s^2(s+1)} - \frac{1}{s(s+1)^2}\right\}$$

$$= L^{-1}\left\{\frac{(s+1)-s}{s^2(s+1)} - \frac{(s+1)-s}{s(s+1)^2}\right\}$$

$$= L^{-1}\left\{\frac{(s+1)}{s^2(s+1)} - \frac{s}{s^2(s+1)} - \frac{(s+1)}{s(s+1)^2} + \frac{s}{s(s+1)^2}\right\}$$

$$= L^{-1}\left\{\frac{1}{s^2} - \frac{1}{s(s+1)} - \frac{1}{s(s+1)} + \frac{1}{(s+1)^2}\right\}$$

$$= L^{-1}\left\{\frac{1}{s^2} - \frac{(s+1)-s}{s(s+1)} - \frac{(s+1)-s}{s(s+1)} + \frac{1}{(s+1)^2}\right\}$$

$$= L^{-1}\left\{\frac{1}{s^2} - \frac{1}{s} + \frac{1}{s+1} - \frac{1}{s} + \frac{1}{s+1} + \frac{1}{(s+1)^2}\right\}$$

$$= L^{-1}\left\{\frac{1}{s^2} - \frac{2}{s} + \frac{2}{s+1} + \frac{1}{(s+1)^2}\right\}$$

$$= L^{-1}\left\{\frac{1}{s^2}\right\} - 2L^{-1}\left\{\frac{1}{s}\right\} + 2L^{-1}\left\{\frac{1}{s+1}\right\}$$

$$\quad + L^{-1}\left\{\frac{1}{(s+1)^2}\right\}$$

$$= t - 2 + 2e^{-t} + te^{-t}$$

$$= t - 2 + (t+2)e^{-t}$$

25. (d)

$$L^{-1}\left[\frac{s}{(s+1)(s^2+1)}\right]$$

$$= L^{-1}\left[\frac{1}{2}\times\frac{2s}{(s+1)(s^2+1)}\right]$$

$$= \frac{1}{2}L^{-1}\left[\frac{2s}{(s+1)(s^2+1)}\right]$$

$$= \frac{1}{2}L^{-1}\left[\frac{(s+1)^2-(s^2+1)}{(s+1)(s^2+1)}\right]$$

$$= \frac{1}{2}L^{-1}\left[\frac{(s+1)^2}{(s+1)(s^2+1)}-\frac{(s^2+1)}{(s+1)(s^2+1)}\right]$$

$$= \frac{1}{2}L^{-1}\left[\frac{s+1}{s^2+1}-\frac{1}{s+1}\right]$$

$$= \frac{1}{2}\left\{L^{-1}\left[\frac{s}{s^2+1}\right]+L^{-1}\left[\frac{1}{s^2+1}\right]-L^{-1}\left[\frac{1}{s+1}\right]\right\}$$

$$= \frac{1}{2}\left\{\cos t+\sin t-e^{-t}\right\}$$

26. (b)

$$L^{-1}\left(\frac{1}{s^2+2s+5}\right)=L^{-1}\left\{\frac{1}{(s+1)^2+2^2}\right\}$$

$$\Rightarrow f(t)=L^{-1}\{F(s)\}=\frac{1}{2}e^{-t}\sin 2t$$

$$\left(\text{where } F(s)=\frac{1}{s^2+2s+5}\right)$$

Then, $tf(t)=-L^{-1}\left\{\frac{d}{ds}\left(\frac{1}{s^2+2s+5}\right)\right\}$

$$= -L^{-1}\left\{\frac{(-1)(2s+2)}{(s^2+2s+5)^2}\right\}$$

$$= 2L^{-1}\left\{\frac{s+1}{(s^2+2s+5)^2}\right\}$$

$$\Rightarrow L^{-1}\left\{\frac{s+1}{(s^2+2s+5)^2}\right\}=\frac{1}{2}tf(t)$$

$$= \frac{1}{2}t\times\frac{1}{2}e^{-t}\sin 2t$$

$$= \frac{1}{4}te^{-t}\sin 2t$$

27. (c) Let $F(s)=\log\left\{\frac{s^2+1}{s(s+1)}\right\}$

Then, $f(t)=L^{-1}\{F(s)\}=L^{-1}\left[\log\left\{\frac{s^2+1}{s(s+1)}\right\}\right]$

$\therefore tf(t)$

$$= -L^{-1}\frac{d}{ds}\{F(s)\}$$

$$= -L^{-1}\frac{d}{ds}\left[\log\left\{\frac{s^2+1}{s(s+1)}\right\}\right]$$

$$= -L^{-1}\left[\frac{d}{ds}\{\log\left(s^2+1\right)-\log s-\log\left(s+1\right)\}\right]$$

$$= -L^{-1}\left(\frac{2s}{s^2+1}-\frac{1}{s}-\frac{1}{s+1}\right)$$

$$= -2L^{-1}\left(\frac{s}{s^2+1}\right)+L^{-1}\left(\frac{1}{s}\right)+L^{-1}\left(\frac{1}{s+1}\right)$$

$$= -2\cos t+1+e^{-t}$$

$$\Rightarrow f(t)=\frac{1}{t}\left(-2\cos t+1+e^{-t}\right)$$

$$\Rightarrow L^{-1}\{F(s)\}=\frac{1}{t}\left(-2\cos t+1+e^{-t}\right)$$

28. (b) Using the concept of the unit step function, we can write:

$$f(t)=0\times\{u(t-0)-u(t-1)\}$$
$$+1\times\{u(t-1)-u(t-2)\}+2\times u(t-2)$$

$\therefore L\{f(t)\}$

$$= L\left[0\times\{u(t-0)-u(t-1)\}\right]$$
$$+L\left[1\times\{u(t-1)-u(t-2)\}\right]+L\left[2\times u(t-2)\right]$$

$$= L(0)+L\{u(t-1)-u(t-2)\}+2L\{u(t-2)\}$$

$$= 0+L\{u(t-1)\}-L\{u(t-2)\}+2L\{u(t-2)\}$$

$$= L\{u(t-1)\}+L\{u(t-2)\}=\frac{1}{s}e^{-s}+\frac{1}{s}e^{-2s}$$

$$\left(\text{using } L\{u(t-a)\}=\frac{1}{s}e^{-as}\right)$$

29. (c) Using the concept of the unit step function, we can write:

$$f(t)=t\times\left\{u(t-0)-u(t-\frac{1}{2})\right\}$$
$$+(t-1)\times\left\{u(t-\frac{1}{2})-u(t-1)\right\}+0\times u(t-1)$$

$\therefore L\{f(t)\}$

$$= L\left[t\times\left\{u(t-0)-u(t-\frac{1}{2})\right\}\right]$$
$$+L\left[(t-1)\times\left\{u(t-\frac{1}{2})-u(t-1)\right\}\right]+L\left[0\times u(t-1)\right]$$

$$= L\{tu(t-0)\}-L\left\{tu\left(t-\frac{1}{2}\right)\right\}+L\left\{(t-1)u\left(t-\frac{1}{2}\right)\right\}$$

$$-L\{(t-1)u(t-1)\}+L(0)$$

$$= L\{(t-0)u(t-0)\} - L\left\{u\left(t-\frac{1}{2}\right)\right\} - L\{(t-1)u(t-1)\}$$

$$= e^{-0s}L(t) - \frac{1}{s}e^{-\frac{s}{2}} - e^{-1s}L(t)$$

$$\left(\text{using the second shifting theorem \& } L\{u(t-a)\} = \frac{1}{s}e^{-as}\right)$$

$$= \frac{1}{s^2} - \frac{1}{s}e^{-\frac{s}{2}} - \frac{1}{s^2}e^{-s} = -\frac{1}{s}\left(e^{-\frac{s}{2}} + \frac{1}{s}e^{-s} - \frac{1}{s}\right).$$

30. (a) Using the concept of the unit step function, we can write:

$$f(t) = \sin 2t \times \{u(t-0) - u(t-\pi)\} + 0 \times u(t-\pi)$$
$$= \sin 2t \times u(t-0) - \sin 2t \times u(t-\pi) + 0$$

$$\therefore L\{f(t)\}$$
$$= L[\sin 2t \times u(t-0)] - L[\sin 2t \times u(t-\pi)]$$

$$= L[2\sin t \cos t \times u(t-0)] - L[2\sin t \cos t \times u(t-\pi)]$$

$$= L[2\sin(t-0)\cos(t-0) \times u(t-0)]$$
$$\quad - L[2\sin(t-\pi)\cos(t-\pi) \times u(t-\pi)]$$

$$\left(\begin{array}{l} \because \sin(t-\pi) = -\sin(\pi-t) = -\sin t, \\ \cos(t-\pi) = \cos(\pi-t) = -\cos t \end{array}\right)$$

$$= e^{-0s}L(2\sin t \cos t) - e^{-\pi s}L(2\sin t \cos t)$$

(by the second shifting theorem)

$$= L(\sin 2t) - e^{-\pi s}L(\sin 2t)$$

$$= \frac{2}{s^2 + 2^2} - e^{-\pi s} \times \frac{2}{s^2 + 2^2} = \frac{2(1 - e^{-\pi s})}{s^2 + 4}$$

31. (c) Using the concept of the unit step function, we can write:

$$f(t) = 0 \times \{u(t-0) - u(t-1)\} + (t-1)^3 \times u(t-1)$$

$$= (t-1)^3 \times u(t-1)$$

$$\therefore L\{f(t)\} = L[(t-1)^3 \times u(t-1)] = e^{-1s} \times L(t^3)$$

(by the second shifting theorem of Laplace transform)

$$= e^{-s} \times \frac{3!}{s^4} = \frac{6e^{-s}}{s^4}$$

32. (b) Using the concept of the unit step function, we can write:

$$f(t) = 1 \times \{u(t-0) - u(t-3)\} + t \times u(t-3)$$

$$= u(t-0) + (t-1) \times u(t-3)$$

$$= u(t-0) + (t-3) \times u(t-3) + 2u(t-3)$$

$$\therefore \quad L\{f(t)\}$$

$$= L[u(t-0)] + L[(t-3) \times u(t-3)] + 2L[u(t-3)]$$

$$f(t) = 1 \times \{u(t-0) - u(t-3)\} + t \times u(t-3)$$
$$= u(t-0) + (t-1) \times u(t-3)$$
$$= u(t-0) + (t-3) \times u(t-3) + 2u(t-3)$$

$$\therefore L\{f(t)\}$$
$$= L[u(t-0)] + L[(t-3) \times u(t-3)] + 2L[u(t-3)]$$

$$= \frac{1}{s}e^{-0s} + e^{-3s} \times L(t) + 2 \times \frac{1}{s}e^{-3s}$$

$$\left(\begin{array}{l} \text{by second shifting theorem and using} \\ L(u(t-a)) = \frac{1}{s}e^{-as} \end{array}\right)$$

$$= \frac{1}{s} + \frac{e^{-3s}}{s^2} + \frac{2}{s}e^{-3s}$$

33. (a) $f(t)$
$$= 2t \times \{u(t-0) - u(t-1)\} + t \times u(t-1)$$
$$= 2t\, u(t) - t \times u(t-1)$$

$$= 2\{(t-0) \times u(t-0)\} - (t-1) \times u(t-1) - u(t-1)$$

$$\therefore L[f(t)] = 2L\{(t-0) \times u(t-0)\} - L[(t-1) \times u(t-1)]$$
$$\quad - L[u(t-1)]$$

$$= 2 \times e^{-0s}L(t) - e^{-1s}L(t) - \frac{e^{-1s}}{s}$$

$$= 2 \times \frac{1}{s^2} - e^{-s} \times \frac{1}{s^2} - \frac{e^{-1s}}{s}$$

$$= \frac{2}{s^2} - \left(\frac{1}{s^2} + \frac{1}{s}\right)e^{-s}$$

34. (c)

$$L\{t^2 u(t-1)\}$$

$$= L\left[\{(t-1)+1\}^2 u(t-1)\right]$$

$$= L\left[\{(t-1)^2 + 2(t-1) + 1\}^2 u(t-1)\right]$$

$$= L[(t-1)^2 u(t-1)] + 2L[(t-1)u(t-1)] + L[u(t-1)]$$

$$= e^{-1s}L(t^2) + 2e^{-1s}L(t) + \frac{1}{s}e^{-s}$$

$$= e^{-s}\frac{2!}{s^3} + 2e^{-s}\frac{1}{s^2} + \frac{1}{s}e^{-s}$$

$$= \left(\frac{2}{s^3} + \frac{2}{s^2} + \frac{1}{s}\right)e^{-s}$$

35. (a)

$y'' + 2y' + y = 1$

$\Rightarrow y''(t) + 2y'(t) + y(t) = 1$

$\Rightarrow L[y''(t)] + 2L[y'(t)] + L[y(t)] = L(1)$

$\Rightarrow s^2 Y(s) - sy(0) - y'(0) + 2\{sY(s) - y(0)\}$

$\quad + Y(s) = \dfrac{1}{s}$

$\Rightarrow s^2 Y(s) - 0 - 0 + 2sY(s) + Y(s) = \dfrac{1}{s}$

$\Rightarrow (s^2 + 2s + 1)Y(s) = \dfrac{1}{s}$

$\Rightarrow Y(s) = \dfrac{1}{s(s^2 + 2s + 1)} = \dfrac{1}{s(s+1)^2}$

$\quad = \dfrac{(s+1) - s}{s(s+1)^2} = \dfrac{1}{s(s+1)} - \dfrac{1}{(s+1)^2}$

$\quad = \dfrac{1}{s} - \dfrac{1}{s+1} - \dfrac{1}{(s+1)^2}$

$\Rightarrow L^{-1}[Y(s)] = L^{-1}\left(\dfrac{1}{s} - \dfrac{1}{s+1} - \dfrac{1}{(s+1)^2}\right)$

$\quad = L^{-1}\left(\dfrac{1}{s}\right) - L^{-1}\left(\dfrac{1}{s+1}\right) - L^{-1}\left\{\dfrac{1}{(s+1)^2}\right\}$

$\Rightarrow y(t) = 1 - e^{-t} - \dfrac{e^{-t} \times t^{2-1}}{\underline{|2-1}} = 1 - e^{-t} - te^{-t}$

$\quad = 1 - e^{-t}(1+t)$

36. (b)

$y'' - 4y' = e^t$

$\Rightarrow y''(t) - 4y'(t) = e^t$

$\Rightarrow L[y''(t)] - 4L[y'(t)] = L[e^t]$

$\Rightarrow [s^2 Y(s) - sy(0) - y'(0)] - 4[sY(s) - y(0)]$

$\quad = \dfrac{1}{s-1}$

$\Rightarrow s^2 Y(s) - 4Y(s) = \dfrac{1}{s-1} \qquad [\because y(0) = 0 = y'(0)]$

$\Rightarrow (s^2 - 4)Y(s) = \dfrac{1}{s-1}$

$\Rightarrow Y(s) = \dfrac{1}{(s-1)(s^2-4)} = \dfrac{1}{(s-1)(s-2)(s+2)}$

$\quad = \dfrac{(s+2) - (s-2)}{4(s-1)(s-2)(s+2)}$

$\quad = \dfrac{1}{4(s-1)(s-2)} - \dfrac{1}{4(s-1)(s+2)}$

$\quad = \dfrac{(s-1)-(s-2)}{4(s-1)(s-2)} - \dfrac{[(s+2)-(s-1)]}{12(s-1)(s+2)}$

$\quad = \dfrac{1}{4(s-2)} - \dfrac{1}{4(s-1)} - \dfrac{1}{12(s-1)} + \dfrac{1}{12(s+2)}$

$\Rightarrow L^{-1}[Y(s)] = \dfrac{1}{4}L^{-1}\left[\dfrac{1}{(s-2)}\right] - \dfrac{1}{4}L^{-1}\left[\dfrac{1}{(s-1)}\right]$

$\quad - \dfrac{1}{12}L^{-1}\left[\dfrac{1}{(s-1)}\right] + \dfrac{1}{12}L^{-1}\left[\dfrac{1}{(s+2)}\right]$

$\Rightarrow y(t) = \dfrac{1}{4}e^{2t} - \dfrac{1}{4}e^t - \dfrac{1}{12}e^t + \dfrac{1}{12}e^{-2t}$

$\quad = \dfrac{1}{4}e^{2t} - \dfrac{1}{3}e^t + \dfrac{1}{12}e^{-2t}$

Previous Years Questions (2000-2018)

1. If $L\{f(t)\} = \dfrac{s+2}{s^2+1}$, $L\{g(t)\} = \dfrac{s^2+1}{(s+3)(s+2)}$

and $h(t) = \int_0^t f(T)g(t-T)\,dT$, then $L\{h(t)\}$ is

(a) $\dfrac{s^2+1}{s+3}$ $\qquad$ (b) $\dfrac{1}{s+3}$

(c) $\dfrac{s^2+1}{(s+3)(s+2)} + \dfrac{(s+2)}{s^2+1}$

(d) None of these

$\hfill$ **(EC GATE 2000)**

2. Let $F(s) = L(f(t))$ denotes the Laplace transform of the function $f(t)$. Which of the following statements is true?

(a) $L\left(\dfrac{df}{dt}\right) = \dfrac{1}{s}F(s); \ L\left(\int_0^t f(t)dt\right) = sF(s) - f(0)$

(b) $L\left(\dfrac{df}{dt}\right) = sF(s) - F(0); \ L\left(\int_0^t f(t)dt\right) = -\dfrac{dF}{ds}$

(c) $L\left(\dfrac{df}{dt}\right) = sF(s) - F(0); \ L\left(\int_0^t f(t)dt\right) = F(s-a)$

(d) $L\left(\dfrac{df}{dt}\right) = sF(s) - f(0); \ L\left(\int_0^t f(t)dt\right) = \dfrac{1}{s}F(s)$

$\hfill$ **(GATE 2000)**

3. The Laplace transform of the function $L(\sin^2 2t)$ is

(a) $\dfrac{1}{2s} - \dfrac{s}{2(s^2+16)}$

(b) $\dfrac{s}{s^2+16}$

(c) $\dfrac{1}{s} - \dfrac{s}{s^2+4}$

(d) $\dfrac{s}{s^2+4}$

(GATE 2002)

4. The Laplace transform of the following function is

$$f(t) = \begin{cases} \sin t, & 0 \le t \le \pi \\ 0, & t > \pi \end{cases}$$

(a) $\dfrac{1}{(s^2+1)}$ for $s > 0$

(b) $\dfrac{1}{(s^2+1)}$ for $s < \pi$

(c) $\dfrac{1+e^{-\pi s}}{(s^2+1)}$ for $s > 0$

(d) $\dfrac{e^{-\pi s}}{(s^2+1)}$ for $s > 0$

(GATE 2002)

5. If L defines the Laplace transform of a function, then $L(\sin at)$ will be equal to

(a) $\dfrac{1}{s^2-a^2}$

(b) $\dfrac{a}{s^2+a^2}$

(c) $\dfrac{s}{s^2+a^2}$

(d) $\dfrac{s}{s^2-a^2}$

(CE GATE 2003)

6. The Laplace transform of a function $\sin \omega t$ is

(a) $\dfrac{s}{s^2+\omega^2}$

(b) $\dfrac{\omega}{s^2+\omega^2}$

(c) $\dfrac{s}{s^2-\omega^2}$

(d) $\dfrac{\omega}{s^2-\omega^2}$

(ME GATE 2003)

7. A delayed unit step function is defined as
$$u(t-a) = \begin{cases} 0, & \text{for } t < a \\ 1, & \text{for } t \ge a \end{cases},$$

Its Laplace transform is

(a) ae^{-as}

(b) $\dfrac{1}{s}e^{-as}$

(c) $\dfrac{1}{s}e^{as}$

(d) $\dfrac{1}{a}e^{as}$

(ME GATE 2004)

8. The Dirac delta $\delta(t)$ is defined as

(a) $\delta(t) = \begin{cases} 1, t = 0 \\ 0, \text{otherwise} \end{cases}$

(b) $\delta(t) = \begin{cases} \infty, t = 0 \\ 0, \text{otherwise} \end{cases}$

(c) $\delta(t) = \begin{cases} 1, t = 0 \\ 0, \text{otherwise} \end{cases}$ and $\displaystyle\int_{-\infty}^{\infty} \delta(t)\, dt = 1$

(d) $\delta(t) = \begin{cases} \infty, t = 0 \\ 0, \text{otherwise} \end{cases}$ and $\displaystyle\int_{-\infty}^{\infty} \delta(t)\, dt = 1$

(EC GATE 2005)

9. A solution for the differential equation $x'(t) + 2x(t) = \delta(t)$ with initial condition $x(0) = 0$ is

(a) $e^{-2t}\, u(t)$

(b) $e^{2t}\, u(t)$

(c) $e^{-t}\, u(t)$

(d) $e^{t}\, u(t)$

($u(t)$ denotes the unit step function)

(GATE 2006)

10. Evaluate $\displaystyle\int_{0}^{\infty} \dfrac{\sin t}{t}\, dt$

(a) π

(b) $\dfrac{\pi}{2}$

(c) $\dfrac{\pi}{4}$

(d) $\dfrac{\pi}{8}$

(GATE 2007)

11. The Laplace transform of the function $8t^3$ is

(a) $\dfrac{8}{s^4}$

(b) $\dfrac{16}{s^4}$

(c) $\dfrac{24}{s^4}$

(d) $\dfrac{48}{s^4}$

(PI GATE 2008)

12. The Laplace transform for the function $f(x) = \cos h\,(ax)$ is

(a) $\dfrac{a}{s^2-a^2}$

(b) $\dfrac{s}{s^2-a^2}$

(c) $\dfrac{a}{s^2+a^2}$

(d) $\dfrac{s}{s^2+a^2}$

(CE GATE 2009)

13. If $F(s)$ is the Laplace transform of the $f(t)$, then the Laplace transform of $\displaystyle\int_{0}^{t} f(\tau)\, d\tau$ is

(a) $\dfrac{1}{s}F(s)$

(b) $\dfrac{1}{s}F(s) - f(0)$

(c) $sF(s) - f(0)$

(d) $\displaystyle\int F(s)\, ds$

(ME GATE 2007, EC GATE 2009)

14. The inverse Laplace transform of $\dfrac{1}{s^2+s}$ is

(a) $1 + e^t$ (b) $1 - e^t$

(c) $1 - e^{-t}$ (d) $1 + e^{-t}$

(ME GATE 2009)

15. The Laplace transform of a function $f(t)$ is $\dfrac{1}{s^2(s+1)}$. Then $f(t)$ is

(a) $t - 1 + e^{-t}$ (b) $t + 1 + e^{-t}$

(c) $-1 + e^{-t}$ (d) $2t + e^t$

(ME GATE 2010)

16. Given $L^{-1}\left[\dfrac{3s+1}{s^3+4s^2+(k-3)s}\right] = f(\mathrm{t}).$

If $\lim\limits_{t\to\infty} f(t) = 1,$ then the value of "k" is

(a) 1   (b) 2   (c) 3   (d) 4

(EE GATE 2010)

17. If $u(t)$ represents the unit step function, then the Laplace transform of $u(t - \tau)$ is

(a) $\dfrac{1}{s\tau}$ (b) $\dfrac{1}{s-\tau}$ (c) $\dfrac{e^{-s\tau}}{s}$ (d) $e^{-s}\tau$

(IN GATE 2010)

18. The integral $\displaystyle\int_{-\infty}^{\infty} \delta\left(t - \dfrac{\pi}{6}\right) \times 6\sin t\, dt$ evalutes to

(a) 6   (b) 3   (c) 1.5   (d) 0

(IN GATE 2010)

19. Given $f(t)$ and $g(t)$ shown below:

$g(t)$ can be expressed as

(a) $g(t) = f(2t - 3)$ (b) $g(t) = f\left(\dfrac{t}{2} - 3\right)$

(c) $g(t) = f\left(2t - \dfrac{3}{2}\right)$ (d) $g(t) = f\left(\dfrac{t}{2} - \dfrac{3}{2}\right)$

(EE GATE 2011)

20. If $F(s) = L\{f(\mathrm{t})\} = \dfrac{2(s+1)}{s^2+4s+7}$ then the initial and final values of $f(t)$ are, respectively

(a) 0, 2   (b) 2, 0

(c) $0, \dfrac{2}{7}$ (d) $\dfrac{2}{7}, 0$

(GATE 2011)

21. Given two continuous time signal $x(t) = e^{-t}$ and $y(t) = e^{-2t}$ which exist for $t > 0$ then the convolution $z(t) = x(t) \circ y(t)$ is

(a) $e^{-t} - e^{-2t}$ (b) e^{-2t}

(c) e^{-t} (d) $e^{-t} + e^{-3t}$

(GATE 2011)

22. The unilateral Laplace transform of $y(t)$ is $\dfrac{1}{s^2+s+1}$. The unilateral Laplace transform of $tf(t)$ is

(a) $-\dfrac{s}{\left(s^2+s+1\right)^2}$ (b) $-\dfrac{(2s+1)}{\left(s^2+s+1\right)^2}$

(c) $\dfrac{s}{\left(s^2+s+1\right)^2}$ (d) $\dfrac{(2s+1)}{\left(s^2+s+1\right)^2}$

(EE, EC GATE 2012)

23. The inverse Laplace transform of the function $F(s) = \dfrac{1}{s(s+1)}$ is given by

(a) $f(t) = \sin t$ (b) $f(t) = e^{-t} \sin t$

(c) $f(t) = e^{-t}$ (d) $f(t) = 1 - e^{-t}$

(ME GATE 2012)

24. Consider the differential equation

$$\dfrac{d^2 y(t)}{dt^2} + 2\dfrac{dy(t)}{dt} + y(t) = \delta(t) \text{ with } y(t)\,|_{t=0} = -2 \text{ and}$$

$\dfrac{dy}{dt}\,|_{t=0} = 0.$ Then the numerical value of $\dfrac{dy}{dt}\,|_{t=0}$ is

(a) -2 (b) -1

(c) 0   (d) 1

(EE, IN GATE 2012)

25. The function $f(t)$ satisfies the differential equation $\dfrac{d^2 f}{dt^2} + f = 0$ and the auxiliary conditions are $f(0) = 0, \dfrac{df}{dt}(0) = 4$. The Laplace transform of $f(t)$ is given by

(a) $\dfrac{2}{s+1}$ (b) $\dfrac{4}{s+1}$

(c) $\dfrac{4}{s^2+1}$ (d) $\dfrac{2}{s^4+1}$

(ME GATE 2013)

26. Let $\dfrac{3s-15}{s^2+10s+21}$ be the Laplace transform of a signal $x(t)$. Then $x(0)$ is

(a) 0   (b) 3
(c) 5   (d) 21

(EE GATE 2014)

27. The Laplace transform of $\cos \omega t$ is $\dfrac{s}{s^2+\omega^2}$.

The Laplace transform of $e^{-2t}\cos 4t$ is

(a) $\dfrac{s-2}{(s+2)^2+16}$ (b) $\dfrac{s+2}{(s-2)^2+16}$

(c) $\dfrac{s-2}{(s+2)^2+16}$ (d) $\dfrac{s+2}{(s+2)^2+16}$

(ME GATE 2014)

28. The Laplace transform of e^{i5t}, where $i = \sqrt{-1}$ is

(a) $\dfrac{s-5i}{s^2-25}$ (b) $\dfrac{s+5i}{s^2+25}$

(c) $\dfrac{s+5i}{s^2-25}$ (d) $\dfrac{s-5i}{s^2+25}$

(ME GATE 2015)

29. The bilateral Laplace transform of a function $f(t)$, where

$$f(t) = \begin{cases} 1, & \text{if } a \le t \le b \\ 0, & \text{otherwise} \end{cases}$$

(a) $\dfrac{a-b}{s}$ (b) $\dfrac{e^s(a-b)}{s}$

(c) $\dfrac{e^{-as}-e^{-bs}}{s}$ (d) $\dfrac{e^{s(a-b)}}{s}$

(EC GATE 2015)

30. The Laplace transform of the function $f(t)$ is given by

$$F(s) = L\{f(t)\} = \int_0^\infty f(t)e^{-st}dt .$$

The Laplace transform of the function shown below is given by

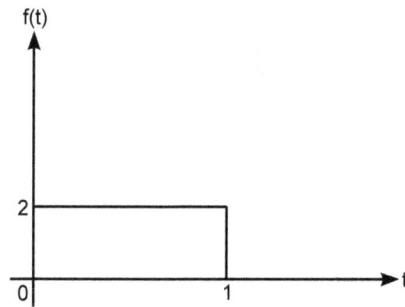

(a) $\dfrac{1-e^{-2s}}{s}$ (b) $\dfrac{1-e^{-s}}{2s}$

(c) $\dfrac{2-2e^{-s}}{s}$ (d) $\dfrac{1-2e^{-s}}{s}$

(GATE 2015)

31. The value of $\displaystyle\int_0^\infty \dfrac{1}{(1+x^2)}dx + \int_0^\infty \dfrac{\sin x}{x}dx$ is

(a) $\dfrac{\pi}{2}$ (b) π

(c) $\dfrac{3\pi}{2}$ (d) 1

(CE GATE 2016)

32. The value of the integral $2\displaystyle\int_{-\infty}^\infty \dfrac{\sin 2\pi t}{\pi t}dt$ is

(a) 0   (b) 0.5
(c) 1   (d) 2

(EE GATE 2016)

33. The Laplace transform of the causal periodic square wave of period T shown in the figure below is

(a) $F(s) = \dfrac{1}{1+e^{-\frac{sT}{2}}}$ (b) $F(s) = \dfrac{1}{s(1-e^{-sT})}$

(c) $F(s) = \dfrac{1}{s(1+e^{-\frac{sT}{2}})}$ (d) $F(s) = \dfrac{1}{1-e^{sT}}$

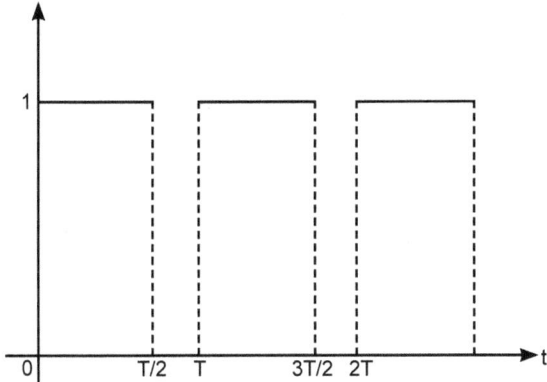

(GATE 2016)

34. Let $y(x)$ be the solution of the differential equation $\dfrac{d^2y}{dx^2} - 4\dfrac{dy}{dx} + 4y = 0$ with initial conditions $y(0) = 0$ and $\dfrac{dy}{dx}\Big|_{x=0} = 1$. Then the value of $y(1)$ is _____.

(GATE 2016)

35. The solution of the differential equation, for $t > 0$, $y''(t) + 2y'(t) + y(t) = 0$ with initial conditions $y(0) = 0$ and $y'(0) = 1$, is ($u(t)$ denotes the unit step function).
(a) $te^{-t} u(t)$ (b) $(e^{-t} - te^{-t})u(t)$
(c) $(-e^{-t} + te^{-t})u(t)$ (d) $e^{-t}u(t)$

(GATE 2016)

36. Let a causal *LTI* system be characterized by the following differential equation, with initial rest condition

$$\frac{d^2y(t)}{dt^2} + 7\frac{dy(t)}{dt} + 10y(t) = 4x(t) + 5\frac{dx(t)}{dt}$$

Where $x(t)$ and $y(t)$ are the input an output, respectively. The impulse response of the system is ($u(t)$ is the unit step function)
(a) $2e^{-2t}u(t) - 7e^{-5t}u(t)$ (b) $-2e^{-2t}u(t) + 7e^{-5t}u(t)$
(c) $7e^{-2t}u(t) - 2e^{-5t}u(t)$ (d) $-7e^{-2t}u(t) + 2e^{-5t}u(t)$

(EE GATE 2017)

37. Consider the state space realization
$$\begin{bmatrix} \dot{x}_1(t) \\ \dot{x}_2(t) \end{bmatrix} = \begin{bmatrix} 0 & 0 \\ 0 & -9 \end{bmatrix}\begin{bmatrix} x_1(t) \\ x_2(t) \end{bmatrix} + \begin{bmatrix} 0 \\ 45 \end{bmatrix}u(t)$$

With the initial condition $\begin{bmatrix} x_1(t) \\ x_2(t) \end{bmatrix} = \begin{bmatrix} 0 \\ 0 \end{bmatrix}$

Where $u(t)$ denotes the unit step function.
Then the value of $\lim\limits_{t\to\infty} \left|\sqrt{x_1^2(t) + x_2^2(t)}\right|$

is _____

(EC GATE 2017)

38. If the Laplace transform of a function is $\dfrac{s+1}{s(s+2)}$, then the initial and final values of $f(t)$ are, respectively
(a) 0, 1 (b) 1, 1/2
(c) 1/2, 1 (d) $\dfrac{1}{2}$, 0

(GATE 2017)

39. The Laplace transform of a casual signal $y(t)$ is $Y(s) = \dfrac{s+2}{s+6}$. Then the value of the signal $y(t)$ at $t = 0.1$ is _____ unit.
(a) ∞ (b) –2.19
(c) 1 (d) non-existant

(IN GATE 2017)

40. The Laplace transform $F(s)$ of the exponential function $f(t) = e^{at}$ when $t \geq 0$ where "a" is a constant and $s - a > 0$ is
(a) $\dfrac{1}{s+a}$ (b) $\dfrac{1}{s-a}$
(c) $\dfrac{1}{a-s}$ (d) ∞

(CE GATE 2018)

41. If $F(s)$ is the Laplace transform of the function $f(t) = 2t^2e^{-t}$, then the value of $F(1)$ is _____?

(ME GATE 2018)

Answers key				
1. (b)	**2.** (d)	**3.** (a)	**4.** (c)	**5.** (b)
6. (b)	**7.** (b)	**8.** (d)	**9.** (a)	**10.** (b)
11. (d)	**12.** (b)	**13.** (a)	**14.** (c)	**15.** (a)
16. (d)	**17.** (c)	**18.** (b)	**19.** (d)	**20.** (b)
21. (a)	**22.** (d)	**23.** (d)	**24.** (d)	**25.** (c)
26. (b)	**27.** (d)	**28.** (b)	**29.** (c)	**30.** (c)
31. (b)	**32.** (d)	**33.** (b)	**34.** 7.389	
35. (a)	**36.** (b)	**37.** 5	**38.** (b)	**39.** (b)
40. (b)	**41.** $\dfrac{1}{2}$.			

Explanations

1. (b)

$$L\{f(t)\} = \frac{s+2}{s^2+1} = F(s)$$

and $L\{g(t)\} = \frac{s^2+1}{(s+3)(s+2)} = G(s)$

By the convolution theorem,

$$L^{-1}[F(s)G(s)] = \int_0^t f(T)\,g(t-T)\,\mathrm{d}T$$

$$\Rightarrow L^{-1}[F(s)G(s)] = h(t)$$
$$\Rightarrow L\{h(t)\} = F(s)\,G(s)$$

$$= \left(\frac{s+2}{s^2+1}\right)\left(\frac{s^2+1}{(s+3)(s+2)}\right)$$

$$= \frac{1}{(s+3)}$$

2. (d) Using the Laplace transform on derivative and integral.

3. (a)

$$L(\sin^2 2t) = \frac{1}{2}L(2\sin^2 2t) = \frac{1}{2}L(1-\cos 4t)$$

$$= \frac{1}{2}\{L(1) - L(\cos 4t)\}$$

$$= \frac{1}{2}\left(\frac{1}{s} - \frac{s}{s^2+16}\right)$$

4. (c)

$$L(f(t))$$

$$= \int_0^\infty e^{-st} f(t)\,dt$$

$$= \int_0^\pi e^{-st} \times \sin t\,dt + \int_\pi^\infty e^{-st} \times 0\,dt$$

$$= \int_0^\pi e^{-st} \times \sin t\,dt$$

$$= \left[\frac{e^{-st}(-s\sin t - \cos t)}{s^2+1}\right]_0^\pi$$

$$= \frac{e^{-s\pi}(-s\sin\pi - \cos\pi)}{s^2+1} - \frac{e^{-0}(-s\sin 0 - \cos 0)}{s^2+1}$$

$$= \frac{1}{s^2+1}\left[e^{-s\pi}\{0-(-1)\} + 0 + 1\right]$$

$$= \frac{e^{-s\pi}+1}{s^2+1}$$

5. (b) By fundamental formulas of Laplace transform.

6. (b) Use the fundamental formula.

7. (b) $L[u(t-a)] = \frac{1}{s}e^{-as}$

8. (d) Use the definition of the Dirac Delta function.

9. (a) $x'(t) + 2x(t) = \delta(t)$

$$\Rightarrow L[x'(t)] + 2L[x(t)] = L[\delta(t)]$$
$$\Rightarrow sX(s) - x(0) + 2X(s) = 1$$
$$(\text{assuming } L(x(t) = X(s))$$
$$\Rightarrow X(s)(s+2) = 1$$
$$\Rightarrow X(s) = \frac{1}{s+2}$$
$$\Rightarrow L^{-1}[X(s)] = L^{-1}\left[\frac{1}{s+2}\right]$$
$$\Rightarrow x(t) = e^{-2t}\,u(t)$$

10. (b)

$$\int_s^\infty \frac{\sin t}{t}\,dt = \int_s^\infty e^{-st} f(t)\,dt \left(\text{here } s = 0, f(t) = \frac{\sin t}{t}\right)$$

$$= L(f(t)) = L\left(\frac{\sin t}{t}\right)$$

$$= \int_S^\infty L(\sin t)\,ds = \int_S^\infty \frac{1}{s^2+1}\,ds$$

$$= \left[\tan^{-1} s\right]_s^\infty = \tan^{-1}\infty - \tan^{-1} s$$

$$= \frac{\pi}{2} - \tan^{-1} s = \frac{\pi}{2} \ (\because s = 0)$$

11. (d) $L(8t^3) = 8L(t^3) = 8 \times \frac{3!}{s^{3+1}} = \frac{48}{s^4}$.

12. (b),

13. (a)

14. (c)

$$L^{-1}\left(\frac{1}{s^2+s}\right) = L^{-1}\left\{\frac{1}{s(s+1)}\right\}$$

$$= L^{-1}\left\{\frac{(s+1)-s}{s(s+1)}\right\}$$

$$= L^{-1}\left(\frac{1}{s} - \frac{1}{s+1}\right)$$

$$= L^{-1}\left(\frac{1}{s}\right) - L^{-1}\left(\frac{1}{s+1}\right)$$

$$= 1 - e^{-t}$$

15. (a)

$f(t)$

$= L^{-1}\left[\dfrac{1}{s^2(s+1)}\right]$

$= L^{-1}\left[\dfrac{1}{s^2(s+1)}\right]$

$= L^{-1}\left[\dfrac{(s+1)-s}{s^2(s+1)}\right]$

$= L^{-1}\left[\dfrac{1}{s^2}-\dfrac{1}{s(s+1)}\right]$

$= L^{-1}\left[\dfrac{1}{s^2}-\dfrac{(s+1)-s}{s(s+1)}\right]$

$= L^{-1}\left(\dfrac{1}{s^2}-\dfrac{1}{s}+\dfrac{1}{s+1}\right)$

$= L^{-1}\left(\dfrac{1}{s^2}\right)-L^{-1}\left(\dfrac{1}{s}\right)+L^{-1}\left(\dfrac{1}{s+1}\right)$

$= t-1+e^{-t}$

16. (d) $\lim\limits_{t\to\infty} f(t)=\lim\limits_{s\to 0} sF(s)$

$\qquad$ (using final value theorem)

$\Rightarrow 1=\lim\limits_{s\to 0} s\left[\dfrac{3s+1}{s^3+4s^2+(k-3)s}\right]$

$\Rightarrow \lim\limits_{s\to 0}\left[\dfrac{3s+1}{s^2+4s+(k-3)}\right]=1$

$\Rightarrow \dfrac{0+1}{0+0+(k-3)}=1$

$\Rightarrow k=4$

17. (c) Use $L\{u(t-a)\}=\dfrac{e^{-as}}{s}$

18. (b) $\displaystyle\int_{-\infty}^{\infty} f(t)\,\delta(t-a)\,dt=f(a)$

$\Rightarrow \displaystyle\int_{-\infty}^{\infty}\delta\left(t-\dfrac{\pi}{6}\right)\times 6\sin t\,dt=6\sin\dfrac{\pi}{6}=3$

19. (d)

Here, $f(t)=\begin{cases} 0, & t<0 \\ 1, & 0<t<1 \text{ and} \\ 0, & t>0 \end{cases}$

$g(t)=\begin{cases} 0, & 0<t<3 \\ 1, & 3<t<5 \\ 0, & t>5 \end{cases}$

$\therefore f\left(\dfrac{t}{2}-\dfrac{3}{2}\right)=f\left(\dfrac{t-3}{2}\right)=\begin{cases} 0, & \dfrac{t-3}{2}<0 \\ 1, & 0<\dfrac{t-3}{2}<1 \\ 0, & \dfrac{t-3}{2}>0 \end{cases}$

$=\begin{cases} 0, & 0<t<3 \\ 1, & 3<t<5 \\ 0, & t>5 \end{cases}$

$=g(t)$

20. (b)

Initial value of $f(t)=\lim\limits_{t\to 0} f(t)$

$=\lim\limits_{s\to\infty} sF(s)$

$=\lim\limits_{s\to\infty} s\left[\dfrac{2(s+1)}{s^2+4s+7}\right]$

$=\lim\limits_{s\to\infty}\dfrac{2\left(1+\dfrac{1}{s}\right)}{1+\dfrac{4}{s}+\dfrac{7}{s^2}}$

$=\dfrac{2(1+0)}{1+0+0}=2$

Final value of $f(t)=\lim\limits_{t\to\infty} f(t)$

$=\lim\limits_{s\to 0} sF(s)$

$=\lim\limits_{s\to 0} s\left[\dfrac{2(s+1)}{s^2+4s+7}\right]$

$=\lim\limits_{s\to 0}\left[\dfrac{2(s^2+s)}{s^2+4s+7}\right]$

$=\dfrac{2(0+0)}{0+0+7}=0.$

21. (a)

Here, $x(t)=e^{-t}$ and $y(t)=e^{-2t}$

$\therefore x*y=\displaystyle\int_0^t x(u)\times y(t-u)\,du$

$\Rightarrow z(t)=\displaystyle\int_0^t e^{-u}\,e^{-2(t-u)}\,du=e^{-2t}\int_0^t e^u\,du$

$=e^{-2t}[e^t-1]=e^{-t}-e^{-2t}$

22. (d)

Let, $F(s)=\dfrac{1}{s^2+s+1}$

Here, $L^{-1}\{F(s)\}=f(t)$

Then, $tf(t)$

$$= -L^{-1}\frac{d}{ds}\{F(s)\} = -L^{-1}\frac{d}{ds}\left(\frac{1}{s^2+s+1}\right)$$

$$= -L^{-1}\left[\frac{(-1)(2s+1)}{\left(s^2+s+1\right)^2}\right]$$

$$= L^{-1}\left[\frac{(2s+1)}{\left(s^2+s+1\right)^2}\right]$$

$$\therefore L[tf(t)] = \frac{2s+1}{\left(s^2+s+1\right)^2}$$

23. (d)

$$L^{-1}\{F(s)\} = L^{-1}\left\{\frac{1}{s(s+1)}\right\}$$

$$= L^{-1}\left\{\frac{(s+1)-s}{s(s+1)}\right\}$$

$$= L^{-1}\left\{\frac{1}{s}-\frac{1}{s+1}\right\}$$

$$= L^{-1}\left\{\frac{1}{s}\right\} - L^{-1}\left\{\frac{1}{s+1}\right\}$$

$$\Rightarrow f(t) = 1 - e^{-t}$$

24. (d)

$$\frac{d^2y(t)}{dt^2} + 2\frac{dy(t)}{dt} + y(t) = \delta(t)$$

$$\Rightarrow L\left[\frac{d^2y(t)}{dt^2}\right] + 2L\left[\frac{dy(t)}{dt}\right] + L[y(t)] = 1$$

$$\Rightarrow \left[s^2Y(s) - sy(0) - y'(0)\right] + 2[sY(s) - y(0)] + Y(s) = 1$$

$$\Rightarrow \left[s^2Y(s) + 2s - 0\right] + 2[sY(s) + 2] + Y(s) = 1$$

$$\Rightarrow (s^2 + 2s + 1)Y(s) = -3 - 2s$$

$$\Rightarrow Y(s) = -\frac{(2s+3)}{(s^2+2s+1)} = -\frac{\{2(s+1)+1\}}{(s+1)^2}$$

$$= -\frac{2}{s+1} - \frac{1}{(s+1)^2}$$

$$\Rightarrow L^{-1}[Y(s)] = -2L^{-1}\left[\frac{1}{s+1}\right] - L^{-1}\left[\frac{1}{(s+1)^2}\right]$$

$$\Rightarrow y(t) = -2e^{-t} - te^{-t}$$

$$\Rightarrow \frac{dy}{dt} = 2e^{-t} + te^{-t} - e^{-t}$$

$$\Rightarrow \frac{dy}{dt}\bigg|_{t=0} = 2 + 0 - 1 = 1$$

25. (c)

$$\frac{d^2f}{dt^2} + f = 0$$

$$\Rightarrow f''(t) + f(t) = 0$$

$$\Rightarrow L[f''(t)] + L[f(t)] = 0$$

$$\Rightarrow \left[s^2 F(s) - sf(0) - f'(0)\right] + F(s) = 0$$

$$\Rightarrow s^2 F(s) - 0 - 4 + F(s) = 0$$

$$\Rightarrow (s^2 + 1)F(s) = 4$$

$$\Rightarrow F(s) = \frac{4}{s^2+1}$$

26. (b)

$$L(x(t)) = \frac{3s-15}{s^2+10s+21}$$

$$\Rightarrow x(t) = L^{-1}\left(\frac{3s-15}{s^2+10s+21}\right)$$

$$= L^{-1}\left[\frac{3(s-5)}{s^2+10s+25-4}\right]$$

$$= L^{-1}\left[\frac{3(s+5)-20}{(s+5)^2-2^2}\right]$$

$$= 3L^{-1}\left[\frac{(s+5)}{(s+5)^2-2^2}\right] - 20L^{-1}\left[\frac{1}{(s+5)^2-2^2}\right]$$

$$= 3e^{-5t}\cosh 2t - \frac{20}{2}e^{-5t}\sinh 2t$$

$$\therefore x(0) = 3e^0\cosh 0 - 10e^0\sinh 0 = 3$$

$$(\because \cosh 0 = 1; \sinh 0 = 0)$$

27. (d)

$$L(\cos\omega t) = \frac{s}{s^2+\omega^2}$$

$$\Rightarrow L(\cos 4t) = \frac{s}{s^2+16}$$

$$\Rightarrow L\left(e^{-2t}\cos 4t\right) = \frac{s-(-2)}{[s-(-2)]^2+16}$$

[by first shifting property]

$$= \frac{s+2}{(s+2)^2+16}$$

28. (b)

$$L\left(e^{i5t}\right) = \frac{1}{s-5i} = \frac{s+5i}{(s-5i)(s+5i)}$$

$$= \frac{s+5i}{s^2-(5i)^2} = \frac{s+5i}{s^2+25}.$$

29. (c)

$\therefore L[f(t)]$

$= \int\limits_0^\infty e^{-st} f(t)\, dt$

$= \int\limits_0^a e^{-st} \times 0\, dt + \int\limits_a^b e^{-st} \times 1\, dt + \int\limits_b^\infty e^{-st} \times 0\, dt$

$= \int\limits_a^b e^{-st}\, dt = \left[\dfrac{e^{-st}}{-s}\right]_a^b = \dfrac{e^{-as} - e^{-bs}}{s}$

30. (c)

Here, $f(t) = \begin{cases} 0, & t < 0 \\ 2, & 0 < t < 1 \\ 0, & t > 1 \end{cases}$

$\therefore L(f(t)) = \int\limits_0^\infty e^{-st} f(t)\, dt$

$= \int\limits_0^1 e^{-st} \times 2\, dt + \int\limits_1^\infty e^{-st} \times 0\, dt$

$= 2\int\limits_0^1 e^{-st}\, dt = 2\left[\dfrac{e^{-st}}{-s}\right]_0^1$

$= \dfrac{-2\left(e^{-s} - 1\right)}{s} = \dfrac{2 - 2e^{-s}}{s}$

31. (b) $\int\limits_0^\infty \dfrac{1}{\left(1 + x^2\right)}\, dx = \left[\tan^{-1} x\right]_0^\infty$

$= \tan^{-1}\infty - \tan^{-1} 0$

$= \dfrac{\pi}{2} - 0 = \dfrac{\pi}{2}$

$\int\limits_0^\infty \dfrac{\sin x}{x}\, dx = \int\limits_0^\infty \dfrac{\sin t}{t}\, dt = \dfrac{\pi}{2}$

$\therefore \int\limits_0^\infty \dfrac{1}{\left(1 + x^2\right)}\, dx + \int\limits_0^\infty \dfrac{\sin x}{x}\, dx = \dfrac{\pi}{2} + \dfrac{\pi}{2} = \pi.$

32. (d)

$2\int\limits_{-\infty}^\infty \dfrac{\sin 2\pi t}{\pi t}\, dt$

$= 4\int\limits_{-\infty}^\infty \dfrac{\sin 2\pi t}{2\pi t}\, dt$

$= 4 \times \dfrac{1}{2\pi}\int\limits_{-\infty}^\infty \dfrac{\sin x}{x}\, dx \quad (\text{putting } x = 2\pi t)$

$= \dfrac{2}{\pi} \times 2\int\limits_0^\infty \dfrac{\sin x}{x}\, dx$

$\begin{pmatrix} \because f(x) = \dfrac{\sin x}{x} \\ \Rightarrow f(-x) = \dfrac{\sin(-x)}{(-x)} = \dfrac{-\sin x}{-x} = f(x) \\ \text{and so } f(x) \text{ is even} \end{pmatrix}$

$= \dfrac{4}{\pi} \times \dfrac{\pi}{2} = 2 \quad \left(\because \int\limits_0^\infty \dfrac{\sin x}{x}\, dx = \dfrac{\pi}{2}\right)$

33. (b)

Here, $f(t) = \begin{cases} 1 & 0 < 1 \le T/2 \\ 0 & T/2 < t \le T \end{cases}$

$F(s)$

$= \dfrac{1}{1 - e^{-sT}}\int\limits_0^T f(t) e^{-st}\, dt$

$= \dfrac{1}{1 - e^{sT}}\int\limits_0^{\frac{T}{2}} 1.e^{-st}\, dt + \dfrac{1}{1 - e^{sT}}\int\limits_{\frac{T}{2}}^T 0.e^{-st}\, dt$

$= \dfrac{1}{1 - e^{-sT}}\left[\dfrac{e^{-st}}{-s}\right]_0^{T/2}$

$= \dfrac{-1}{\left(1 - e^{-sT}\right)} \cdot \dfrac{1}{s} \cdot (e^{-sT/2} - 1)$

$= \dfrac{\left(1 - e^{-sT}/2\right)}{s(1 - e^{-\frac{sT}{2}})}$

$= \dfrac{(1 - e^{-\frac{sT}{2}})}{s(1 - e^{-\frac{sT}{2}})(1 + e^{-sT/2})}$

$= \dfrac{1}{s\left(1 + e^{-\frac{sT}{2}}\right)}$

34. 7.389

$\dfrac{d^2 y}{dx^2} - 4\dfrac{dy}{dx} + 4y = 0$

$\Rightarrow L\left[\dfrac{d^2 y}{dx^2}\right] - 4L\left[\dfrac{dy}{dx}\right] + 4L[y] = 0$

$\Rightarrow s^2 Y(s) - sy(0) - y'(o) - 4(sY(s) - y(0)) + 4Y(s) = 0$

$(\text{taking } L(y(x)) = Y(s))$

$$\Rightarrow \quad s^2 Y(s) - 0 - 1 - 4sY(s) + 4Y(s) = 0$$

$$\Rightarrow \quad (s^2 - 4s + 4)Y(s) = 1$$

$$\Rightarrow \quad Y(s) = \frac{1}{s^2 - 4s + 4} = \frac{1}{(s-2)^2}$$

$$\Rightarrow \quad L^{-1}[Y(s)] = L^{-1}\left[\frac{1}{(s-2)^2}\right]$$

$$\Rightarrow \quad y(x) = xe^{2x}$$

So $y(1) = e^2 = 7.389$

35. (a) $L[y''] + 2L[y'] + L[y] = 0$

$$\Rightarrow \quad s^2Y(s) - sy(0) - y'(0) + 2(sY(s) - y(0))$$

$$+ \; Y(s) = 0 \quad \text{(Where } Y(s) = L[y(x)])$$

$$\Rightarrow \quad s^2Y(s) - 0 - 1 + 2sY(s) - 0 + Y(s) = 0$$

$$\Rightarrow \quad Y(s) = \frac{1}{(s^2 + 2s + 1)} = \frac{1}{(s+1)^2}$$

$$\Rightarrow \quad L^{-1}[Y(s)] = L^{-1}\left[\frac{1}{(s+1)^2}\right]$$

$$\Rightarrow \quad y(t) = te^{-t} = te^{-t}u(t)$$

36. (b) Let $L(x(t)) = X(s)$ and $L(y(t)) = Y(s)$. Then the initial rest conditions are

$x(0) = 0 = y(0); x'(0) = 0 = y'(0)$. then

$$\frac{d^2 y(t)}{dt^2} + 7\frac{dy(t)}{dt} + 10y(t) = 4x(t) + \frac{dx(t)}{dt}$$

$$\Rightarrow L\left[\frac{d^2 y(t)}{dt^2}\right] + 7L\left[\frac{dy(t)}{dt}\right] + 10L[y(t)]$$

$$= 4L[x(t)] + 5L\left[\frac{dx(t)}{dt}\right]$$

$$\Rightarrow \quad [s^2Y(s) - sy(0) - y'(0)] + 7[sY(s) - y(0)]$$

$$+ \; 10Y(s) = 4X(s) + 5[sX(s) - x(0)]$$

$$\Rightarrow \quad s^2Y(s) + 7sY(s) + 10Y(s) = 4X(s) + 5sX(s)$$

$$\Rightarrow \frac{Y(s)}{X(s)} = \frac{4+5s}{s^2 + 7s + 10} = \frac{4+5s}{(s+2)(s+5)}$$

Then, the impulse response

$$= L^{-1}\left[\frac{Y(s)}{X(s)}\right]$$

$$= L^{-1}\left[\frac{4+5s}{(s+2)(s+5)}\right]$$

$$= L^{-1}\left[\frac{5(s+2) - 6}{(s+2)(s+5)}\right]$$

$$= 5L^{-1}\left[\frac{1}{(s+5)}\right] - 6L^{-1}\left[\frac{1}{(s+2)(s+5)}\right]$$

$$= 5L^{-1}\left[\frac{1}{(s+5)}\right] - 2L^{-1}\left[\frac{(s+5) - (s+2)}{(s+2)(s+5)}\right]$$

$$= 5L^{-1}\left[\frac{1}{(s+5)}\right] - 2L^{-1}\left[\frac{1}{(s+2)}\right] + 2L^{-1}\left[\frac{1}{(s+5)}\right]$$

$$= 7L^{-1}\left[\frac{1}{(s+5)}\right] - 2L^{-1}\left[\frac{1}{(s+2)}\right]$$

$$= 7e^{-5t}u(t) - 2e^{-2t}u(t)$$

37. 5 The initial conditions are $x_1(0) = 0$, $x_2(0) = 0$. Now

$$\begin{bmatrix} \dot{x}_1(t) \\ \dot{x}_2(t) \end{bmatrix} = \begin{bmatrix} 0 & 0 \\ 0 & -9 \end{bmatrix}\begin{bmatrix} x_1(t) \\ x_2(t) \end{bmatrix} + \begin{bmatrix} 0 \\ 45 \end{bmatrix}u(t)$$

$$\Rightarrow \begin{bmatrix} \dot{x}_1(t) \\ \dot{x}_2(t) \end{bmatrix} = \begin{bmatrix} 0 \\ -9x_2(t) + 45u(t) \end{bmatrix}$$

$$\Rightarrow \dot{x}_1(t) = 0, \dot{x}_2(t) = -9x_2(t) + 45u(t)$$

$$\Rightarrow L[\dot{x}_1(t)] = L[0] = 0,$$

$$L[\dot{x}_2(t)] = -9L[x_2(t)] + 45L[u(t)]$$

$$\Rightarrow sX_1(s) - x_1(0) = 0,$$

$$sX_2(s) - x_2(0) = -9X_2(s) + \frac{45}{s}$$

$$\Rightarrow X_1(s) = 0, (s+9)X_2(s) = \frac{45}{s}$$

$$\Rightarrow L^{-1}[X_1(s)] = L^{-1}[0] = 0,$$

$$L^{-1}[X_2(s)] = L^{-1}\left[\frac{45}{s(s+9)}\right]$$

$$= 5L^{-1}\left[\frac{(s+9) - s}{s(s+9)}\right]$$

$$= 5L^{-1}\left[\frac{1}{s}\right] - 5L^{-1}\left[\frac{1}{(s+9)}\right]$$

$$= 5 - 5e^{-9t}$$

$$\Rightarrow x_1(t) = 0, x_2(t) = 5 - 5e^{-9t}$$

$$\therefore \lim_{t \to \infty}\left|\sqrt{x_1^2(t) + x_2^2(t)}\right|$$

$$= \left|\sqrt{\lim_{t \to \infty}x_1^2(t) + \lim_{t \to \infty}x_2^2(t)}\right|$$

$$= \left|\sqrt{0 + (5-0)^2}\right| = 5$$

38. (b)

Initial value of $f(t) = \lim\limits_{t \to 0} f(t)$

$$= \lim\limits_{s \to \infty} sF(s)$$

$$= \lim\limits_{s \to \infty} s \left[\frac{s+1}{s(s+2)} \right]$$

$$= \lim\limits_{s \to \infty} \left[\frac{1 + \dfrac{1}{s}}{1 + \dfrac{2}{s}} \right]$$

$$= \frac{1+0}{1+0} = 1$$

Final value of $f(t) = \lim\limits_{t \to \infty} f(t)$

$$= \lim\limits_{s \to 0} sF(s)$$

$$= \lim\limits_{s \to 0} s \left[\frac{s+1}{s(s+2)} \right]$$

$$= \lim\limits_{s \to 0} \left[\frac{s+1}{s+2} \right]$$

$$= \frac{0+1}{0+2} = \frac{1}{2}.$$

39. (b)

$y(t)$

$$= L^{-1} \left[\frac{s+2}{s+6} \right]$$

$$= L^{-1} \left[\frac{(s+6) - 4}{s+6} \right]$$

$$= L^{-1} \left[1 - \frac{4}{s+6} \right]$$

$$= L^{-1}[1] - 4L^{-1} \left[\frac{1}{s+6} \right]$$

$$= \delta(t) - 4e^{-6t}$$

$$\therefore y(0.1) = \delta(0.1) - 4e^{-6 \times 0.1}$$

$$= 0 - 4e^{-0.6} = -2.19$$

40. (b)

41. $\dfrac{1}{2}$.

$F(s)$

$$= L(f(t)) = L\left(2t^2 e^{-t} \right) = 2L\{t^2 \, g(t)\} \text{[where } g(t)=e^{-t}]$$

$$= 2(-1)^2 \frac{d^2}{ds^2} G(s) = 2\frac{d^2}{ds^2} L\{g(t)\} = 2\frac{d^2}{ds^2} L\{e^{-t}\}$$

$$= 2\frac{d^2}{ds^2} \left(\frac{1}{s+1} \right) = 4\frac{1}{(s+1)^3}.$$

$$\therefore F(1) = 4\frac{1}{(1+1)^3} = \frac{1}{2}.$$

1. Find the Laplace transform of $(\sin 2t - \cos 2t)^2$

(a) $\dfrac{1}{s} - \dfrac{4}{s^2 + 16}$ (b) $\dfrac{1}{s} + \dfrac{4}{s^2 + 16}$

(c) $\dfrac{1}{s} - \dfrac{2}{s^2 + 4}$ (d) $\dfrac{1}{s} + \dfrac{2}{s^2 + 4}$

2. $L[\cos(2t + 7)] = ?$

(a) $\dfrac{s\cos 7 + \sin 7}{s^2 + 4}$ (b) $\dfrac{s\cos 7 - 2\sin 7}{s^2 + 4}$

(c) $\dfrac{2\cos 7 - s\sin 7}{s^2 + 4}$ (d) $\dfrac{s\cos 7 + \sin 7}{s^2 + 4}$

3. Find the Laplace transform of $f(t)$, where

$$f(t) = \begin{cases} \cos(t - \dfrac{2\pi}{3}), & \text{if } t > \dfrac{2\pi}{3} \\ 0, & \text{if } t < \dfrac{2\pi}{3} \end{cases}$$

(a) $\left(\dfrac{1}{s^2 - 1} \right) e^{\frac{-2\pi s}{3}}$ (b) $\left(\dfrac{s}{s^2 + 1} \right) e^{\frac{\pi s}{3}}$

(c) $\left(\dfrac{s}{s^2 + 1} \right) e^{\frac{-2\pi s}{3}}$ (d) none of these

4. Find $L[e^{-t} f(2t)]$ given that $L[f(t)] = \dfrac{1}{s(s^2 + 1)}$

(a) $\dfrac{4}{(s+1)(s^2 - 2s + 1)}$ (b) $\dfrac{4}{(s+1)(s^2 - 2s + 5)}$

(c) $\dfrac{4}{(s-1)(s^2 - 2s + 5)}$ (d) none of these

5. Find the Laplace transform of $f(t)$, if s

$$f(t) = \begin{cases} 4, \, 0 \le t \le 1 \\ 2, \, t > 1 \end{cases}$$

(a) $\dfrac{2}{s}(2 - e^s)$ (b) $\dfrac{2}{s}(2 + e^s)$

(c) $\dfrac{1}{s}(1 - e^s)$ (d) $\dfrac{1}{s}(1 + e^s)$

6. $L\{e^{-2t}(2\sin 3t - 5\cos 2t)\} = ?$

(a) $\dfrac{5}{s^2 + 4s + 13} + \dfrac{6(s+2)}{s^2 + 4s + 8}$

(b) $\dfrac{2}{s^2 + 4s + 13} - \dfrac{3(s+2)}{s^2 + 4s + 8}$

(c) $\dfrac{5}{s^2 + 4s + 13} - \dfrac{6(s+2)}{s^2 + 4s + 8}$

(d) $\dfrac{6}{s^2 + 4s + 13} - \dfrac{5(s+2)}{s^2 + 4s + 8}$

7. Find the Laplace transform of $e^{-2t} \sin 5t$

(a) $\dfrac{5}{s^2 + 4s + 26}$ (b) $\dfrac{5}{s^2 + 4s + 29}$

(c) $\dfrac{5}{s^2 + 2s + 25}$ (d) none of these

8. $L(\cos^3 t) = ?$

(a) $\dfrac{1}{4}\left[\dfrac{3s}{s^2 + 9} + \dfrac{s}{s^2 + 1}\right]$

(b) $\dfrac{1}{4}\left[\dfrac{s}{s^2 + 9} - \dfrac{2s}{s^2 + 1}\right]$

(c) $\dfrac{1}{4}\left[\dfrac{s}{s^2 + 9} + \dfrac{3s}{s^2 + 1}\right]$

(d) none of these

9. $L(t^3 e^{3t}) =$

(a) $\dfrac{6}{(s-3)^4}$ (b) $\dfrac{5}{(s+3)^4}$

(c) $\dfrac{6}{(s-3)^3}$ (d) none of these

10. Find $L\{f(t)\}$ if $f(t) = \begin{cases} e^{2t}, & 0 < t < 2 \\ 0, & t > 2 \end{cases}$

(a) $\dfrac{e^{4+2s}+1}{2-s}$ (b) $\dfrac{e^{4-2s}-1}{2-s}$

(c) $\dfrac{e^{4-2s}+1}{2+s}$ (d) none of these

11. $L(t \cos 3t) = ?$

(a) $\dfrac{s^2 + 9}{(s^2 - 9)^2}$ (b) $\dfrac{s^2 + 9}{(s^2 + 5)^2}$

(c) $\dfrac{s^2 - 9}{(s^2 + 9)^2}$ (d) $\dfrac{s^2 - 9}{(s^2 + 9)^2}$

12. $\displaystyle\int_0^\infty e^{2t} \cos 3t\, dt = ?$

(a) $\dfrac{-2}{13}$ (b) $\dfrac{2}{13}$

(c) $\dfrac{5}{13}$ (d) $\dfrac{-5}{13}$

13. $\displaystyle\int_0^\infty e^{2t} \cos 3t\, dt = ?$

(a) $\dfrac{-2}{13s}$ (b) $\dfrac{2}{13s}$

(c) $-\dfrac{2}{13}$ (d) $\dfrac{1}{12s}$

14. $L^{-1}\left(\dfrac{s}{s^2 - 2s + 5}\right) = ?$

(a) $e^t \cos 2t - e^t \sin 2t$ (b) $e^t \cos 2t + \dfrac{1}{2}e^t \sin 2t$

(c) $e^t \cos 2t - \dfrac{3}{2}e^t \sin 2t$

(d) $e^t \cos 2t + 2e^t \sin 2t$

15. $L^{-1}\left[\log\left(\dfrac{s+2}{s-2}\right)\right] = ?$

(a) $\dfrac{2}{t}\sinh 2t$ (b) $\dfrac{2}{t}\sinh t.$

(c) $\dfrac{1}{t}\sinh 2t.$ (d) $\dfrac{1}{t}\sinh t$

16. Find the Laplace transform of $f(t)$ where

$$f(t) = \begin{cases} 2t, & 0 < t < \pi \\ 1, & t > \pi \end{cases}$$

(a) $\dfrac{2}{s^2} + \left(\dfrac{1+2\pi}{s} - \dfrac{3}{s^2}\right)e^{-\pi s}$

(b) $\dfrac{2}{s^2} + \left(\dfrac{1}{s} + \dfrac{2}{s^2}\right)e^{-\pi s}$

(c) $\dfrac{2}{s^2} + \left(\dfrac{2\pi}{s} - \dfrac{2}{s^2}\right)e^{-\pi s}$

(d) $\dfrac{2}{s^2} + \left(\dfrac{1-2\pi}{s} - \dfrac{2}{s^2}\right)e^{-\pi s}$

17. If $f(t)$ is defined by

$$f(t) = \begin{cases} t, & 0 < t < \pi \\ \pi - t, & \pi < t < 2\pi, \end{cases}$$

then find the Laplace transform of $f(t)$

(a) $\dfrac{1 + e^{-\pi s}(\pi s - 1)}{(1 + e^{-\pi s})s}$ (b) $\dfrac{e^{-\pi s}(\pi s + 1)}{(1 + e^{-\pi s})s^2}$

(c) $\dfrac{1 - e^{-\pi s}(\pi s + 1)}{(1 + e^{-\pi s})s^2}$ (d) none of these

18. Find $L\{t^2 u(t - 3)\}$

(a) $\dfrac{e^{-3s}}{s^3} + \dfrac{5e^{-3s}}{s^2} + \dfrac{e^{-3s}}{s}$

(b) $\dfrac{e^{-3s}}{s^3}+\dfrac{4e^{-3s}}{s^2}+\dfrac{6e^{-3s}}{s}$

(c) $\dfrac{2e^{-3s}}{s^3}+\dfrac{6e^{-3s}}{s^2}+\dfrac{9e^{-3s}}{s}$

(d) none of these

19. If $L\{g(t)\}=\dfrac{1}{s-2}$, then $L\{g(2t)\}$ = ?

(a) $\dfrac{1}{s-4}$

(b) $\dfrac{1}{s+4}$

(c) $\dfrac{1}{s-3}$

(d) $\dfrac{1}{s+3}$

20. Find the Laplace transform of $f(t)$, if

$f(t)=\begin{cases}0, & t<4\\1, & t\ge 4\end{cases}$

(a) $\dfrac{e^{-4s}}{s}$

(b) $\dfrac{e^{4s}}{s}$

(c) $\dfrac{e^{-2s}}{s}$

(d) $\dfrac{e^{2s}}{s}$

21. $L[\cos(t-2)u(t-2)]$ = ?

(a) $e^{-t}\left(\dfrac{s}{s^2+1}\right)$

(b) $e^{-t}\left(\dfrac{2s}{s^2+1}\right)$

(c) $e^{-2t}\left(\dfrac{s}{s^2+1}\right)$

(d) $e^{2t}\left(\dfrac{s}{s^2+1}\right)$

22. The Laplace transform of is

(a) $\dfrac{5}{s^2-4s+25}$

(b) $\dfrac{5}{s^2-4s+29}$

(c) $\dfrac{5}{s^2-2s+10}$

(d) none of these

23. If $L^{-1}\left(\dfrac{1}{s^3}\right)=x^3$, then x = ?

(a) 4   (b) 3   (c) 2   (d) 1

24. If $L^{-1}\left\{\dfrac{s}{(s^2+4)^2}\right\}=f(t)*g(t)$, then find $f(t)$ and $g(t)$, respectively (by the convolution theorem)

(a) $\cos 2t$ and $\dfrac{\sin 2t}{2}$ (b) $\cos t$ and $\dfrac{\sin 2t}{2}$

(c) $\cos 2t$ and $\dfrac{\sin t}{2}$ (d) none of these

25. $L^{-1}\left\{\dfrac{1}{s(s+4)}\right\}$ = ?

(a) $\dfrac{1}{2}(1-e^{-2t})$

(b) $\dfrac{1}{2}(1+e^{-2t})$

(c) $\dfrac{1}{4}(1-e^{-4t})$

(d) $\dfrac{1}{4}(1+e^{-4t})$

26. Find $L^{-1}\left(\dfrac{s+2}{2s^2+8}\right)$ = ?

(a) $\dfrac{1}{2}(\cos 2t-\sin t)$ (b) $\dfrac{1}{2}(\cos 2t+\sin 2t)$

(c) $\dfrac{1}{2}(\cos t+\sin t)$ (d) none of these

27. If $L(e^{at})=\dfrac{1}{s-a}$, then $L(3e^{5t}\sin h\,5t)$ is equal to _____ ?

(a) $\dfrac{5}{s^2-5s}$

(b) $\dfrac{5}{s^2-2s}$

(c) $\dfrac{15}{s^2-10s}$

(d) $\dfrac{10}{s^2-15s}$

28. If a is a constant, then the value of the integral $a^2\int_0^\infty xe^{-ax}\,dx$ is ?

(a) 4   (b) 3

(c) 2   (d) 1

29. The Laplace transform of $f(t)=\dfrac{1}{\sqrt{t}}$ is _____ ?

(a) $\sqrt{\dfrac{2\pi}{s}}$

(b) $\sqrt{\dfrac{\pi}{2s}}$

(c) $\sqrt{\dfrac{1}{s}}$

(d) $\sqrt{\dfrac{\pi}{s}}$

30. The Laplace transform of the function shown in the figure below is _____ ?

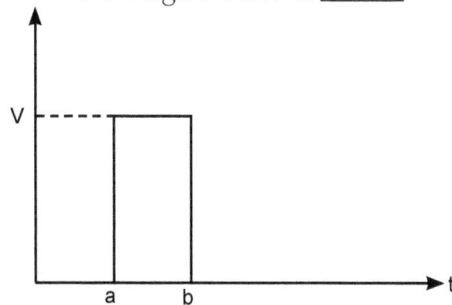

(a) $\dfrac{V}{s}\left(e^{-as}-e^{-bs}\right)$ (b) $\dfrac{V}{s}\left(e^{as}-e^{bs}\right)$

(c) $\dfrac{V}{s}\left(e^{-as}-e^{bs}\right)$ (d) $\dfrac{V}{s}\left(e^{as}-e^{-bs}\right)$

31. For the time-domain function $f(t)=t$, the Laplace transform of $\int_0^t f(t)\,dt$ is _____ ?

(a) $\dfrac{1}{s}$

(b) $\dfrac{1}{s^2}$

(c) $\dfrac{1}{s^3}$

(d) $\dfrac{1}{2s^3}$

32. Given that the Laplace transform of the function below a single period $0 < t < 2$ is $\frac{1}{s^2}\left(1 - e^{-s}\right)^2$. Then the Laplace transform of the periodic function over $0 < t < \infty$ is

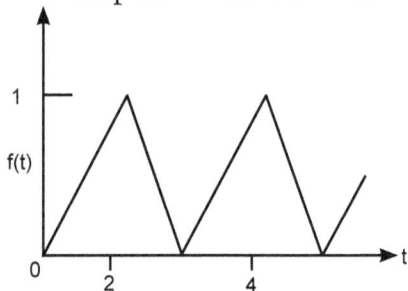

(a) $\dfrac{1}{s^2}\left(\dfrac{1 + e^{-s}}{1 - e^{-s}}\right)$

(b) $\dfrac{1}{s^2}\left(\dfrac{1 - e^{-s}}{1 + e^{-s}}\right)$

(c) $\dfrac{1}{s^2}\left(\dfrac{e^{-s}}{1 + e^{-s}}\right)$

(d) $\dfrac{1}{s^2}\left(\dfrac{1 - e^{-s}}{e^s}\right)$

33. Consider the differential equation $y'' + 2y' + y' = 0$ with boundary condition $y(0) = 1$, $y(1) = 0$. The value of $y(2)$ is

(a) e^2 (b) $-e^2$

(c) e^{-2} (d) $-e^{-2}$

34. The inverse Laplace transform of $\dfrac{1}{2s^2 + 3s + 1}$ is _____ ?

(a) $e^{-\frac{t}{2}} + e^t$

(b) $e^{\frac{t}{2}} - e^{-t}$

(c) $e^{-\frac{t}{2}} - e^{-t}$

(d) $e^{-\frac{t}{2}} + e^{-t}$

35. $L^{-1}\left[\log\left\{\dfrac{s^2 + 1}{(s - 1)^2}\right\}\right] = ?$

(a) $\dfrac{2}{t}\left(e^t - \cos t\right)$

(b) $\dfrac{2}{t}\left(e^t + \cos t\right)$

(c) $\dfrac{2}{t}\left(e^{-t} - \cos t\right)$

(d) $\dfrac{2}{t}\left(e^{-t} + \cos t\right)$

Answers key				
1. (a)	**2.** (b)	**3.** (c)	**4.** (c)	**5.** (a)
6. (d)	**7.** (b)	**8.** (c)	**9.** (a)	**10.** (b)
11. (d)	**12.** (a)	**13.** (c)	**14.** (b)	**15.** (a)
16. (d)	**17.** (c)	**18.** (c)	**19.** (a)	**20.** (a)
21. (c)	**22.** (b)	**23.** (d)	**24.** (a)	**25.** (c)
26. (b)	**27.** (c)	**28.** (d)	**29.** (d)	**30.** (a)
31. (c)	**32.** (b)	**33.** (d)	**34.** (c)	**35.** (a)

1. (a)

$(\sin 2t - \cos 2t)^2$

$= \sin^2 2t - \cos^2 2t - 2\sin 2t \cos 2t$

$= 1 - \sin 4t$

2. (b) $\cos(x + y) = \cos x \cos y - \sin x \sin y$

3. (c), **4.** (c), **5.** (a), **6.** (d), **7.** (b),

8. (c)

$\cos^3 t = \dfrac{1}{4}(4\cos^3 t) = \dfrac{1}{4}(\cos 3t + 3\cos t)$

9. (a), **10.** (b), **11.** (d), **12.** (a), **13.** (c)

14. (b)

$L^{-1}\left(\dfrac{s}{s^2 - 2s + 5}\right)$

$= L^{-1}\left\{\dfrac{(s - 1) + 1}{(s - 1)^2 + 2^2}\right\}$

$= L^{-1}\left\{\dfrac{(s - 1)}{(s - 1)^2 + 2^2}\right\} + L^{-1}\left\{\dfrac{1}{(s - 1)^2 + 2^2}\right\}$

15. (a), **16.** (d) **17.** (c)

18. (c)

$t^2 u(t - 3)$

$= \{(t - 3) + 3\}^2 u(t - 3)$

$= (t - 3)^2 u(t - 3) + 6(t - 3)u(t - 3) + 9u(t - 3)$

Thus,

$L[t^2 u(t - 3)]$

$= L\{(t - 3)^2 u(t - 3)\} + 6L\{(t - 3)u(t - 3)\} + 9L\{u(t - 3)\}$

$= e^{-3s}L(t^2) + 6e^{-3s}L(t) + 9L\{u(t - 3)\}$

19. (a), **20.** (a), **21.** (c), **22.** (b), **23.** (d)

24. (a)

$f(t) * g(t) = L^{-1}\left\{\dfrac{s}{(s^2 + 4)^2}\right\}$

$= L^{-1}\left\{\dfrac{s}{(s^2 + 4)} \times \dfrac{1}{(s^2 + 4)}\right\}$

$= L^{-1}\{F(s) \times G(s)\}$

$\left(\text{where, } F(s) = \dfrac{s}{(s^2 + 4)}, G(s) = \dfrac{1}{(s^2 + 4)}\right)$

$\therefore f(t) = L^{-1}\left(F(s)\right) = L^{-1}\left(\dfrac{s}{(s^2 + 4)}\right) = \cos 2t,$

$g(t) = L^{-1}\left(G(s)\right) = L^{-1}\left(\dfrac{1}{(s^2 + 4)}\right) = \dfrac{1}{2}\sin 2t$

25. (c) **26.** (b) **27.** (c)

28. (d)

$$\int\limits_0^\infty xe^{-ax}\,dx = \int\limits_0^\infty te^{-at}\,dt = \int\limits_0^\infty f(t)e^{-at}\,dt$$

$$[\text{where } s = a,\ f(t) = t]$$

$$= L[f(t)] = L(t) = \frac{1}{s^2} = \frac{1}{a^2}$$

$$\therefore a^2 \int\limits_0^\infty xe^{-ax}\,dx = 1$$

29. (d)

$$L[f(t)] = L\left(\frac{1}{\sqrt{t}}\right) = L\left(t^{-\frac{1}{2}}\right)$$

$$= \frac{\left\lfloor -\dfrac{1}{2}+1 \right.}{s^{-\frac{1}{2}+1}} = \frac{\left\lfloor \dfrac{1}{2} \right.}{s^{\frac{1}{2}}} = \frac{\sqrt{\pi}}{\sqrt{s}}$$

30. (a)

Here, $f(t) = \begin{cases} 0, & t < a \\ V, & a < t < b \\ 0, & t > b \end{cases}$

$$\therefore L[f(t)] = \int\limits_0^\infty e^{-st} f(t)\,dt$$

$$= \int\limits_0^a e^{-st} \times 0\,dt + \int\limits_a^b e^{-st} \times V\,dt + \int\limits_b^\infty e^{-st} \times 0\,dt$$

31. (c)

$$L\left[\int\limits_0^t f(t)\,dt\right] = \frac{1}{s}F(s) = \frac{1}{s}L[f(t)]$$

$$= \frac{1}{s}L(t) = \frac{1}{s} \times \frac{1}{s^2} = \frac{1}{s^3}$$

32. (b) We know that if $f(t)$ be periodic with period T, where $0 < t < \infty$ then,

$$L[f(t)] = \frac{1}{1-e^{-sT}} \int\limits_0^T e^{-st} f(t)\,dt .....(i)$$

$$= \frac{1}{1-e^{-2s}} \int\limits_0^2 e^{-st} f(t)\,dt \ \left(\text{here } T = 2\right)$$

$$= \frac{1}{1-e^{-2s}} \times \frac{1}{s^2}\left(1-e^{-s}\right)^2$$

$$= \frac{1}{1-\left(e^{-s}\right)^2} \times \frac{\left(1-e^{-s}\right)^2}{s^2}$$

$$= \frac{1}{\left(1-e^{-s}\right)\left(1+e^{-s}\right)} \times \frac{\left(1-e^{-s}\right)^2}{s^2}$$

$$= \frac{\left(1-e^{-s}\right)}{s^2\left(1+e^{-s}\right)}$$

33. (d)

$$y'' + 2y' + y = 0$$

$$\Rightarrow y''(t) + 2y'(t) + y(t) = 0$$

$$\Rightarrow L[y''(t)] + 2L[y'(t)] + L[y(t)] = L[0]$$

$$\Rightarrow [s^2Y(s) - sy(0) - y'(0)] + 2[sY(s) - y(0)] + Y(s)$$
$$= 0$$

$$\Rightarrow s^2Y(s) - s - k + 2sY(s) - 2 + Y(s) = O$$
$$[\text{assume } y'(0) = k]$$

$$\Rightarrow (s^2 + 2s + 1)Y(s) = s + 2 + k$$

$$\Rightarrow Y(s) = \frac{s+2+k}{(s+1)^2} = \frac{(s+1)+(k+1)}{(s+1)^2}$$

$$= \frac{1}{s+1} + (k+1) \times \frac{1}{(s+1)^2}$$

$$\Rightarrow L^{-1}[Y(s)] = L^{-1}\left[\frac{1}{s+1}\right] + (k+1)L^{-1}\left[\frac{1}{(s+1)^2}\right]$$

$$\Rightarrow y(t) = e^{-t} + (k+1)e^{-t}t$$

Now, $y(1) = 0 \Rightarrow 0 = e^{-1} + (k+1)e^{-1}$

$$\Rightarrow k + 1 + 1 = 0$$

$$\Rightarrow k = -2$$

$$\therefore y(t) = e^{-t} + (-2+1)e^{-t}t = e^{-t}(1-t)$$

Hence, $y(2) = -e^{-2}$

34. (c)

$$L^{-1}\left(\frac{1}{2s^2 + 3s + 1}\right)$$

$$= L^{-1}\left(\frac{1}{2s^2 + 2s + s + 1}\right)$$

$$= L^{-1}\left\{\frac{1}{2s(s+1) + (s+1)}\right\}$$

$$= L^{-1}\left\{\frac{1}{(s+1)(2s+1)}\right\}$$

$$= 2L^{-1}\left\{\frac{1}{(2s+2)(2s+1)}\right\}$$

$$= 2L^{-1}\left\{\frac{(2s+2)-(2s+1)}{(2s+2)(2s+1)}\right\}$$

$$= 2L^{-1}\left(\frac{1}{2s+1}\right) - 2L^{-1}\left(\frac{1}{2s+2}\right)$$

$$= 2L^{-1}\left\{\frac{1}{2\left(s+\frac{1}{2}\right)}\right\} - 2L^{-1}\left\{\frac{1}{2(s+1)}\right\}$$

$$= L^{-1}\left(\frac{1}{s+\frac{1}{2}}\right) - L^{-1}\left(\frac{1}{s+1}\right)$$

$$= e^{-\frac{t}{2}} - e^{-t}$$

35. (a)

Let, $F(s) = \log\left\{\frac{s^2+1}{(s-1)^2}\right\}$

Then, $L^{-1}\{F(s)\} = f(t)$

$\therefore tf(t)$

$$= -L^{-1}\frac{d}{ds}\{F(s)\} = -L^{-1}\frac{d}{ds}\left[\log\left\{\frac{s^2+1}{(s-1)^2}\right\}\right]$$

$$= -L^{-1}\left[\frac{d}{ds}\left\{\log\left(s^2+1\right) - 2\log\left(s-1\right)\right\}\right]$$

$$= -L^{-1}\left(\frac{2s}{s^2+1} - \frac{2}{s+1}\right)$$

CHAPTER 7

NUMERICAL ANALYSIS

7.1 ERRORS AND APPROXIMATIONS

7.1.1 Rounding off

For computation purpose, we generally cut off some unwanted digits from given numbers. This process of dropping unnecessary digits is called *rounding off*.

The general rules for rounding off a number to n significant figures are as follows:

Discard all digits to the right of the nth digit and if among these discarded digits the digit in the $(n + 1)^{th}$ place is

(*i*) greater than 5 then the digit in the nth place is increased by 1.

(*ii*) less than 5 then the digit in the nth place is left unchanged.

(*iii*) exactly 5 then leave the nth digit unaltered if it is even and increase the nth digit by 1 if it is odd.

For example, the following numbers are rounded off to four significant figures:

7.029886 becomes 7.029

3.5634 becomes 3.563

89.3999 becomes 89.40

8.52854 becomes 8.528

8.42757 becomes 8.428

7.1.2 Errors and Their Computation

Let x_T and x_A denote the true and approximate value, respectively, of a number.

Then the *absolute error*, E_a involved in x_A is defined by $E_a = |x_T - x_A|$.

The relative error, E_r is defined by $E_r = \dfrac{|x_T - x_A|}{x_T}$, provided $x_T \neq 0$.

The percentage error (or relative percentage error), $E_p = E_r \times 100 = \dfrac{|x_T - x_A|}{x_T} \times 100$.

7.2 CALCULUS OF FINITE DIFFERENCES

Let $y = f(x)$ be a real-valued function of x defined in an interval $[a, b]$ and its values are known for $(n + 1)$ equally spaced points $x_i (i = 0, 1, 2, ..., n)$ such that $x_i = x_0 + ih (i = 0, 1, 2, ..., n)$ where $x_0 = a$, $x_n = b$ and h is the spacing (*space length*). Then x_i $(i = 0, 1, 2, ..., n)$ are called nodes and the corresponding values y_i are termed as *entries*.

7.2.1 Forward Difference Operator

The differences $y_1 - y_0, y_2 - y_1, y_3 - y_2, ..., y_n - y_{n-1}$ are denoted by $\Delta y_0, \Delta y_1, \Delta y_2, ..., \Delta y_{n-1}$, respectively.

Thus, we have $\Delta y_i = y_{i+1} - y_i (i = 0, 1, 2, ..., n - 1)$

where Δ is called forward difference operator.

In general, the forward difference operator is defined by $\Delta f(x) = f(x + h) - f(x)$.

Similarly, the higher order forward differences are defined as

$$\Delta^2 y_i = \Delta y_{i+1} - \Delta y_i,$$
$$\Delta^3 y_i = \Delta^2 y_{i+1} - \Delta^2 y_i,$$
$$........ \quad .... \quad ....$$
$$\Delta^r y_i = \Delta^{r-1} y_{i+1} - \Delta^{r-1} y_i.$$

where $i = 0, 1, 2, ..., n - 1$ and $r (1 \leq r \leq n)$ is a positive integer.

Properties of Δ:

If a and b be any two constants, then

(i) $\Delta a = 0$

(ii) $\Delta\{af(x)\} = a\Delta f(x)$

(iii) $\Delta\{af(x) \pm bg(x)\} = a\Delta f(x) \pm b\Delta g(x)$

(iv) $\Delta[f(x)g(x)] = f(x+h)\Delta g(x) + \Delta f(x)g(x+h)$

(v) $\Delta\left\{\dfrac{f(x)}{g(x)}\right\} = \dfrac{g(x)\Delta f(x) - f(x)\Delta g(x)}{g(x)g(x+h)}$

7.2.2 Backward Difference Operator

The differences $y_1 - y_0, y_2 - y_1, y_3 - y_2, ..., y_n - y_{n-1}$

are denoted by $\nabla y_1, \nabla y_2, \nabla y_3, ..., \nabla y_n$, respectively.

Thus, we have $\nabla y_{i+1} = y_{i+1} - y_i (i = 0,1,2,....,n-1)$

where ∇ is called backward difference operator.

In general, the backward difference operator is defined by $\nabla f(x) = f(x) - f(x-h)$.

Similarly, the higher order backward differences are defined as

$\nabla^2 y_i = \nabla y_i - \nabla y_{i-1}$,

$\nabla^3 y_i = \nabla^2 y_i - \nabla^2 y_{i-1}$,

....

$\nabla^r y_i = \nabla^{r-1} y_i - \nabla^{r-1} y_{i-1}$.

where $i = 1, 2, ..., n$ and $r(1 \leq r \leq n)$ is a positive integer.

7.2.3 Shift Operator

The shift operator is denoted by E and is defined as $E\{f(x)\} = f(x+h), h$ being the spacing.

$\therefore E^2 f(x) = Ef(x+h) = f(x+2h)$,

$E^3 f(x) = Ef(x+2h) = f(x+3h)$ and so on.

Thus, in general, $E^n f(x) = f(x+nh)$

The inverse shift operator, E^{-1} is defined by $E^{-1}f(x) = f(x-h)$ and in general, $E^{-n}f(x) = f(x-nh)$.

Remember:

(i) $E \equiv \Delta + 1$ (ii) $E^{-1} \equiv 1 - \nabla$

7.3 INTERPOLATION

Let $f(x)$ be a function of x defined in the interval $[a, b]$ in which it is assumed to be continuous and continuously differentiable.

Suppose the function $y = f(x)$ is not known explicitly, but the values of $f(x)$ are known for $(n + 1)$ distinct values of x, say $x_0, x_1, ...,x_n$ called *arguments* or *nodes*. Let $y_0 = f(x_0), y_1 = f(x_1), ..., y_n = f(x_n)$ be the corresponding entries and there is no other information available about the function.

The problem of interpolation is to compute the value of $f(x)$, at least approximately, for a given argument x, not found in the table. When x lies slightly outside the interval $[x_o, x_n]$, then the process is called *extrapolation*.

Our target is to find a polynomial function $\varphi(x)$ such that $f(x_i) = \varphi(x_i)$ at $i = 0, 1, 2, ..., n$. The polynomial $\varphi(x)$ is called interpolating polynomial.

7.3.1 Newton's Forward Difference Interpolation Formula

Let $y = f(x)$ be a real valued function of x defined in $[a, b]$ and is known for the corresponding $(n + 1)$ equi-spaced arguments $x_i(i = 0, 1, 2, ..., n)$ such that $x_i = x_0 + ih(i = 0, 1, 2, ...,n)$

with $x_0 = a, x_n = b$ and h is the space length.

Let $y_i = f(x_i)$ $(i = 0, 1, 2, ..., n)$

Then,

$y = f(x) \approx \varphi(x)$

$= y_0 + u\Delta y_0 + \dfrac{u(u-1)}{2!}\Delta^2 y_0 + \dfrac{u(u-1)(u-2)}{3!}\Delta^3 y_0$

$+ .... + \dfrac{u(u-1)...(u-n+1)}{n!}\Delta^n y_0$

where $u = \dfrac{x - x_0}{h}$.

7.3.2 Newton's Backward Difference Interpolation Formula

Let $y = f(x)$ be a real valued function of x defined in $[a, b]$ and is known for the corresponding $(n + 1)$ equi-spaced arguments x_i $(i = 0, 1, 2, ..., n)$ such that $x_i = x_0 + ih(i = 0, 1, 2, ..., n)$ with $x_0 = a, x_n = b$ and h is the space length.

Let $y_i = f(x_i)$ $(i = 0, 1, 2, ..., n)$.

Then,

$y = f(x) \approx \varphi(x)$

$= y_n + s\nabla y_n + \dfrac{s(s+1)}{2!}\nabla^2 y_n + \dfrac{s(s+1)(s+2)}{3!}\nabla^3 y_n$

$+ .... + \dfrac{s(s+1)(s+2)...(s+n-1)}{n!}\nabla^n y_n$

where $s = \dfrac{x - x_n}{h}$.

7.3.3 Lagrange's Interpolation Formula

Let $y = f(x)$ be a function of x, continuous and $(n + 1)$ times continuously differentiable in $[a, b]$. Let us divide the interval $[a, b]$ by $(n + 1)$ points $a = x_0, x_1, ...x_n = b$ which are not necessarily equispaced and the corresponding entries are

$y_i = f(x_i)$ $(i = 0, 1, 2, ..., n)$.

Then, $y = f(x) \approx \varphi(x)$

$$= \frac{(x-x_1)(x-x_2)(x-x_3).......(x-x_n)}{(x_0-x_1)(x_0-x_2)(x_0-x_3).......(x_0-x_n)}y_0$$

$$+ \frac{(x-x_0)(x-x_2)(x-x_3).......(x-x_n)}{(x_1-x_0)(x_1-x_2)(x_1-x_3).......(x_1-x_n)}y_1$$

$$+ \frac{(x-x_0)(x-x_1)(x-x_3).......(x-x_n)}{(x_2-x_0)(x_2-x_1)(x_2-x_3).......(x_2-x_n)}y_2$$

$$+ ....$$

$$+ \frac{(x-x_0)(x-x_1)(x-x_2).......(x-x_{n-1})}{(x_n-x_0)(x_n-x_1)(x_n-x_2).......(x_n-x_{n-1})}y_n$$

7.3.4 Error in Interpolation

The error in approximating $f(x)$ by $\varphi(x)$ is given by

$$R_{n+1} = \frac{(x-x_0)(x-x_1)(x-x_2).......(x-x_n)}{(n+1)!}f^{(n+1)}(t)$$

where $a = x_0 < t < x_n = b$

7.4 NUMERICAL DIFFERENTIATION

Let $f(x)$ be a function of x defined in the interval $[a, b]$ in which it is assumed to be continuous and continuously differentiable.

Suppose the function $y = f(x)$ is not known explicitly, but the values of $f(x)$ are known for $(n + 1)$ distinct values of x, say $x_0, x_1, ..., x_n$. Let $y_0 = f(x_0)$, $y_1 = f(x_1)$ ..., $y_n = f(x_n)$ be the corresponding entries.

Our target is to find a polynomial function $\varphi(x)$ such that $f(x_i) = \varphi(x_i)$ at $i = 0, 1, 2, ..., n$ and then differentiate it as many times as we desire.

7.4.1 Differentiation Formula Based on Newton's Forward Difference Formula

(i) $f'(x_0)$

$$\approx \frac{1}{h}\left[\Delta y_0 - \frac{1}{2}\Delta^2 y_0 + \frac{1}{3}\Delta^3 y_0 - \frac{1}{4}\Delta^4 y_0 + ...\right]$$

(ii) $f''(x_0) \approx \frac{1}{h^2}\left[\Delta^2 y_0 - \Delta^3 y_0 + \frac{11}{12}\Delta^4 y_0 - ....\right]$

7.4.2 Differentiation Formula Based on Newton's Backward Difference Formula

(i) $f'(x_n)$

$$\approx \frac{1}{h}\left[\nabla y_n + \frac{1}{2}\nabla^2 y_n + \frac{1}{3}\nabla^3 y_n + \frac{1}{4}\nabla^4 y_n + ...\right]$$

(ii) $f''(x_0) \approx \frac{1}{h^2}\left[\nabla^2 y_n + \nabla^3 y_n + \frac{11}{12}\nabla^4 y_n + ...\right]$

7.5 NUMERICAL INTEGRATION

Consider the definite integral $I = \int_a^b f(x)dx$

and divide the interval $[a, b]$ of integration into n equal subintervals such that $a = x_0 < x_1 < x_2 < ... < x_n = b$ and $x_i = x_0 + ih$ $(i = 0, 1, 2, ..., n)$, h is the space length.

Also, let the function $f(x)$ be known at the nodes x_i, i.e., the values $y_i = f(x_i)$ $(i = 0, 1, 2, ..., n)$ are given.

7.5.1 Trapezoidal Rule

$$\int_{x_0}^{x_n} f(x)dx \approx \frac{h}{2}\left[y_0 + y_n + 2(y_1 + y_2 + ... + y_{n-1})\right]$$

In particular,

$$\int_{x_0}^{x_1} f(x)dx \approx \frac{h}{2}[y_0 + y_1],$$

$$\int_{x_0}^{x_2} f(x)dx \approx \frac{h}{2}[y_0 + y_2 + 2y_1].$$

Remember:

(i) The error in computing the value of the integral using trapezoidal rule is given by

$$E_T = -\frac{nh^3}{12}f''(t) \text{ where } a = x_0 < t < x_n = b$$

(ii) The trapezoidal rule geometrically interprets that the curve $f(x)$ in $[x_0, x_n]$ is replaced by n straight lines joining the points (x_0, y_0), (x_1, y_1), (x_2, y_2),........., (x_n, y_n).

7.5.2 Simpson's 1/3rd Rule

$$\int_{x_0}^{x_n} f(x)dx$$

$$\approx \frac{h}{3}[y_0 + y_n + 4(y_1 + y_3 + ... + y_{n-1})$$

$$+ 2(y_2 + y_4 + ... + y_{n-2})$$

Remember:

(*i*) The error in computing the value of the integral using Simpson's $1/3^{rd}$ rule is given by

$$E_S = -\frac{nh^5}{180} f^{iv}(t) \text{ where } a = x_0 < t < x_n = b$$

(*ii*) Simpson's $1/3^{rd}$ rule gives us exact result for a polynomial of degree less than or equal to three.

(*iii*) Simpson's $1/3^{rd}$ rule geometrically interprets that the curve $f(x)$ in $[x_0, x_2]$ is replaced by a second degree parabola through the points $(x_0, y_0), (x_1, y_1), (x_2, y_2)$.

7.5.3 Weddle's Rule

$$\int_{x_0}^{x_n} f(x)\,dx$$

$$\simeq \frac{3h}{10}\left[y_0 + y_n + (y_2 + y_4 + y_8 + \ldots + y_{n-4})\right.$$

$$+5(y_1 + y_5 + y_7 \ldots + y_{n-1}) + 6(y_3 + y_9 + y_{15} \ldots + y_{n-3})$$

$$\left. + 2(y_6 + y_{12} + y_{18} \ldots + y_{n-6})\right]$$

In particular,

$$\int_{x_0}^{x_6} f(x)\,dx \simeq \frac{3h}{10}\left[y_0 + 5y_1 + y_2 + 6y_3 + y_4 + 5y_5 + y_6\right]$$

7.5.4 Simpson's 3/8th's Rule

$$\int_{x_0}^{x_3} f(x)\,dx \simeq \frac{(b-a)}{8}\left[y_0 + 3y_1 + 3y_2 + y_3\right]$$

7.6 SYSTEM OF LINEAR ALGEBRAIC EQUATIONS

7.6.1 Gauss Elimination Method

In this method, the given system of equations is reduced to an equivalent upper triangular system by a systematic elimination procedure from which the unknowns are found by back substitution.

Let us consider the system of three equations in three unknowns given by

$$\left.\begin{array}{l} a_{11}^{(1)}x_1 + a_{12}^{(1)}x_2 + a_{13}^{(1)}x_3 = b_1^{(1)} \\ a_{21}^{(1)}x_1 + a_{22}^{(1)}x_2 + a_{23}^{(1)}x_3 = b_2^{(1)} \\ a_{31}^{(1)}x_1 + a_{32}^{(1)}x_2 + a_{33}^{(1)}x_3 = b_3^{(1)} \end{array}\right\} \quad \ldots(i)$$

First suppose that $a_{11} \neq 0$, $a_{22} \neq 0$, $a_{33} \neq 0$. Using the Gauss elimination method, the above system (*i*) reduces to:

$$\left.\begin{array}{l} a_{11}^{(1)}x_1 + a_{12}^{(1)}x_2 + a_{13}^{(1)}x_3 = b_1^{(1)} \\ a_{22}^{(2)}x_2 + a_{23}^{(2)}x_3 = b_2^{(2)} \\ a_{33}^{(3)}x_3 = b_3^{(3)} \end{array}\right\} \quad \ldots(ii)$$

where

$$b_2^{(2)} = b_2^{(1)} - \frac{b_1^{(1)}a_{21}^{(1)}}{a_{11}^{(1)}}, \; b_3^{(2)} = b_3^{(1)} - \frac{b_1^{(1)}a_{31}^{(1)}}{a_{11}^{(1)}},$$

$$a_{22}^{(2)} = a_{22}^{(1)} - \frac{a_{21}^{(1)}a_{12}^{(1)}}{a_{11}^{(1)}}, \; a_{23}^{(2)} = a_{23}^{(1)} - \frac{a_{21}^{(1)}a_{13}^{(1)}}{a_{11}^{(1)}},$$

$$a_{32}^{(2)} = a_{32}^{(1)} - \frac{a_{31}^{(1)}a_{12}^{(1)}}{a_{11}^{(1)}}, \; a_{33}^{(2)} = a_{33}^{(1)} - \frac{a_{31}^{(1)}a_{13}^{(1)}}{a_{11}^{(1)}},$$

$$b_3^{(3)} = b_3^{(2)} - \frac{b_2^{(2)}a_{32}^{(2)}}{a_{22}^{(2)}}, \; a_{33}^{(3)} = a_{23}^{(2)} - \frac{a_{32}^{(2)}a_{23}^{(2)}}{a_{22}^{(2)}},$$

The elements a_{11}, a_{22}, a_{33} are called pivots. The solutions of the (*i*) are then obtained from (*ii*) by back substitutions.

7.6.2 LU Decomposition Method

This method is also known as *triangular decomposition* method. This method is based on the fact that every square matrix can be expressed as the product of a lower and an upper triangular matrix.

Let us consider a system of three equations with three unknowns given below:

$$a_{11}x_1 + a_{12}x_2 + a_{13}x_3 = b_1$$
$$a_{21}x_1 + a_{22}x_2 + a_{23}x_3 = b_2$$
$$a_{31}x_1 + a_{32}x_2 + a_{33}x_3 = b_3$$

Here, the coefficient matrix

$$A = \begin{bmatrix} a_{11} & a_{12} & a_{13} \\ a_{21} & a_{22} & a_{23} \\ a_{31} & a_{32} & a_{33} \end{bmatrix} \text{ can be written as } A = LU$$

where $L = \begin{bmatrix} l_{11} & 0 & 0 \\ l_{21} & l_{22} & 0 \\ l_{31} & l_{32} & l_{33} \end{bmatrix}, \; U = \begin{bmatrix} 1 & u_{12} & u_{13} \\ 0 & 1 & u_{23} \\ 0 & 0 & 1 \end{bmatrix}$

$\therefore \quad A = LU$ gives

$$\begin{bmatrix} l_{11} & 0 & 0 \\ l_{21} & l_{22} & 0 \\ l_{31} & l_{32} & l_{33} \end{bmatrix} \begin{bmatrix} 1 & u_{12} & u_{13} \\ 0 & 1 & u_{23} \\ 0 & 0 & 1 \end{bmatrix} = \begin{bmatrix} a_{11} & a_{12} & a_{13} \\ a_{21} & a_{22} & a_{23} \\ a_{31} & a_{32} & a_{33} \end{bmatrix}$$

Then we have,

$l_{11} = a_{11}, l_{21} = a_{21}, l_{31} = a_{31},$

$$l_{11}u_{12} = a_{12} \Rightarrow u_{12} = \frac{a_{12}}{l_{11}}, l_{11}u_{13} = a_{13} \Rightarrow u_{13} = \frac{a_{13}}{l_{11}},$$

$$l_{21}u_{12} + l_{22} = a_{22} \Rightarrow l_{22} = a_{22} - l_{21}u_{12} = a_{22} - \frac{a_{21}a_{12}}{a_{11}},$$

$$l_{31}u_{12} + l_{32} = a_{32} \Rightarrow l_{32} = a_{32} - l_{31}u_{12} = a_{32} - \frac{a_{31}a_{12}}{a_{11}},$$

$$l_{21}u_{13} + l_{22}u_{23} = a_{23} \Rightarrow u_{23} = \frac{a_{23} - l_{21}u_{13}}{l_{22}},$$

and $l_{31}u_{13} + l_{32}u_{23} + l_{33} = a_{33} \Rightarrow$

$l_{33} = a_{33} - l_{31}u_{13} - l_{32}u_{23}$.

Solving all these equations, we obtain the values of l_{11}, l_{21},..... and u_{12}, u_{13}...., and thus, we get the matrices L and U.

7.6.3 Gauss–Seidel Iteration Method

The sufficient condition for the convergence of the Gauss–Seidel iteration method is that the system of equations must be strictly diagonally dominant, *i.e*; for the system of equations $AX = b$ where $A = (a_{ij})_{n \times n}$, we must have $|a_{ii}| > \sum_{j=1, j \neq i}^{n} |a_{ij}|$.

Let us consider the system of three equations in three unknowns given by

$$a_{11}x_1 + a_{12}x_2 + a_{13}x_3 = b_1$$
$$a_{21}x_1 + a_{22}x_2 + a_{23}x_3 = b_2$$
$$a_{31}x_1 + a_{32}x_2 + a_{33}x_3 = b_3$$

Let us consider the initial approximations: $x_1 = x_1^{(0)}$, $x_2 = x_2^{(0)}$, $x_3 = x_3^{(0)}$.

Then the iteration formula of the Gauss–Seidel method are as follows:

$$x_1^{(k+1)} = \frac{1}{a_{11}}\left[b_1 - a_{12}x_2^{(k)} - a_{13}x_3^{(k)}\right],$$

$$x_2^{(k+1)} = \frac{1}{a_{22}}\left[b_2 - a_{21}x_1^{(k)} - a_{23}x_3^{(k)}\right],$$

$$x_3^{(k+1)} = \frac{1}{a_{33}}\left[b_3 - a_{31}x_1^{(k)} - a_{32}x_2^{(k)}\right].$$

7.7 SOLUTION OF ALGEBRAIC AND TRANSCENDAL EQUATIONS

7.7.1 Method of Bisection

This method is based on the well-known theorem which states that if $f(x)$ be a continuous function in $[a, b]$ and $f(a) f(b) < 0$, then there exist at least one real root of the equation $f(x) = 0$ in (a, b).

In this method, we first find out a sufficiently small interval $[a_0, b_0]$ containing the required root α of the equation $f(x) = 0$. Then $f(a_0)f(b_0) < 0$.

To generate the sequence of iterates $\{x_n\}$, each member of which is a successively better approximation of the exact root, say α. We put $x_0 = a_0$ and $x_1 = \frac{1}{2}(a_0 + b_0)$ and find $f(x_1)$. If $f(a_0) f(x_1) < 0$, then set $a_1 = a_0$, $b_1 = x_1$ so that $[a_1, b_1] = [a_1, x_1]$. On the other hand, if $f(x_1)f(b_0) < 0$ a, then put $a_1 = x_1$, $b_1 = b_0$, *i.e*; $[a_1, b_1] = [x_1, b_0]$.

Thus, we see that $[a_1, b_1]$ contains the root α in either case.

Next, set $x_2 = \frac{1}{2}(a_1 + b_1)$ and repeat the above process till we obtain $x_{n+1} = \frac{1}{2}(a_n + b_n)$ with desired accuracy with $x_n \to \alpha$ as $n \to \infty$.

7.7.2 Regula Falsi Method

The Regula Falsi method or method of false position is the oldest method for computing real roots of an equation $f(x) = 0$.

The general iteration formula for the Regula Falsi method is

$$x_{n+1} = x_n - \frac{f(x_n)}{f(x_n) - f(x_{n-1})}(x_n - x_{n-1}),$$

where $n = 0, 1, 2, 3,\ldots\ldots$

7.7.3 Newton–Raphson Method

Let x_0 be an initial approximation of the desired root α of the equation $f(x) = 0$.

The iteration formula for the Newton–Raphson method is

$$x_{n+1} = x_n - \frac{f(x_n)}{f'(x_n)} \text{ (provided } f'(x_n) \neq 0\text{)}$$

where $n = 0, 1, 2, 3.....$

7.8 NUMERICAL SOLUTION OF ORDINARY DIFFERENTIAL EQUATIONS

7.8.1 Euler's Method

It is a simple and single step method for solving an ordinary initial value problem differential equation, where the solution will be obtained as a set of tabulated values of variables x and y.

Let us consider a first-order and first-degree differential equation as

$$\frac{dy}{dx} = f(x, y), \text{ With } y(x_o) = y_o.$$

We divide the range $[x_0, x_n]$ into n-equal subintervals by the points $x_0, x_1, x_2, ..., x_{r-1}, x_r, x_{r+1}, ...,x_n$

where $x_r = x_0 + rh$ $(r = 1, 2, 3,...,n)$ and $h =$ length of each subinterval.

Then, Euler's iteration formula is given by

$$y_{n+1} = y_n + hf(x_n, y_n) = y(x_{n+1}) \text{ for}$$

$n = 0, 1, 2, 3,.....$

7.8.2 Modified Euler's Method

This method is also known as the Euler–Cauchy method. This method gives us a rapid and moderately accurate result up to a desired degree of accuracy.

The modified Euler's iteration formula is

$$y_r^{(n)} = y_{r-1} + \frac{h}{2}\left[f(x_{r-1}, y_{r-1}) + f(x_r, y_r^{(n-1)})\right]$$

where $y_r^{(n)}$ is the nth approximation of y_r and

$$y_r^{(0)} = y_{r-1} + \frac{h}{2}\left[f(x_{r-1}, y_{r-1}) + f(x_r, y_r)\right].$$

7.8.3 Runge–Kutta Method

The Runge–Kutta methods for the numerical solution of an ordinary differential equation give us a greater accuracy.

I. Second-Order Runge–Kutta Method

Consider the differential equation:

$$y' = \frac{dy}{dx} = f(x, y) \text{ with } y(x_0) = y_0.$$

The computational formulae for the Runge–Kutta method of order two is as follows:

$$y_1 = y_0 + k \text{ where}$$

$$k = \frac{1}{2}(k_1 + k_2),$$

$$k_1 = hf(x_0, y_0),$$

$$k_2 = hf(x_0 + h, y_0 + k_1).$$

II. Fourth-Order Runge–Kutta Method

The computational formulae for the Runge–Kutta method of order four is as follows:

$$y_1 = y(x_0 + h) = y_0 + k \text{ where}$$

$$k = \frac{1}{6}(k_1 + 2k_2 + 2k_3 + k_4)$$

and $k_1 = hf(x_0, y_0)$,

$$k_2 = hf\left(x_0 + \frac{h}{2}, y_0 + \frac{k_1}{2}\right),$$

$$k_3 = hf\left(x_0 + \frac{h}{2}, y_0 + \frac{k_1}{2}\right),$$

$$k_4 = hf\left(x_0 + h, y_0 + k_3\right).$$

7.8.4 Predictor-Corrector Method

Let us consider the differential equation

$$\frac{dy}{dx} = f(x, y) \text{ with } y(x_0) = y_0.$$

In order to solve the above differential equation, we first obtain the approximate value of y_{n+1} by predictor formula and then improve this value by means of a corrector formula.

The predictor-corrector formulas are given by

$$y_{n+1}^P = y_n + hf(x_n, y_n),$$

$$y_{n+1}^C = y_n + \frac{h}{2}\left[f(x_n, y_n) + f(x_{n+1}, y_{n+1}^P)\right],$$

where $n = 0, 1, 2, 3\ldots\ldots$

Fully Solved MCQs

1. If $\pi = 3.14$ is used in place of 3.14156, the absolute and relative errors are, respectively
(a) 0.00156, 4.966×10^{-4}
(b) 0.0018, 4.25×10^{-4}
(c) 0.06, 4.2×10^{-4}
(d) 0.05, 4.06×10^{-4}

2. Find the percentage error in approximating $\frac{4}{3}$ to 1.3333 is
(a) 0.45 (b) 0.0036
(c) 0.0025 (d) 0.05

3. If $y = x^2 + 4$ for $y = 1, 3, 5, 7, 9$; find the value of $\Delta^3 f(9)$
(a) 0 (b) 1
(c) 10 (d) 15

3. (a)

x	1	3	5	7	9
y	5	13	29	53	85
$= x^2 + 4$	$(= y_0)$	$(= y_1)$	$(= y_2)$	$(= y_3)$	$(= y_4)$

$$\nabla y_4 = y_4 - y_3 = 85 - 53 = 32,$$
$$\nabla^2 y_4 = \nabla y_4 - \nabla y_3 = \nabla y_4 - (y_3 - y_2)$$
$$= 32 - (53 - 29) = 8,$$
$$\nabla^3 f(9) = \nabla^3 y_4 = \nabla^2 y_4 - \nabla^2 y_3$$
$$= \nabla^2 y_4 - (\nabla y_3 - \nabla y_2)$$
$$= \nabla^2 y_4 - \{(y_3 - y_2) - ((y_2 - y_1))\}$$
$$= 8 - \{(53 - 29) - (29 - 13)\}$$
$$= 0.$$

4. Which of the following relation is true?
(a) $\Delta.\nabla \equiv 2\Delta - \nabla$ (b) $\Delta.\nabla \equiv \Delta - \nabla$
(c) $\Delta.\nabla \equiv \Delta + \nabla$ (d) none of these

5. $\Delta \log f(x) = ?$

(a) $\log\left[\frac{\Delta f(x)}{f(x)}\right]$ (b) $\log\left[1 - \frac{\Delta f(x)}{f(x)}\right]$

(c) $\log\left[1 + \frac{f(x)}{\Delta f(x)}\right]$ (d) $\log\left[1 + \frac{\Delta f(x)}{f(x)}\right]$

6. The simplest value of $\left(\frac{\Delta^2}{E}e^x\right)\frac{Ee^x}{\Delta^2 e^x}$ is

(a) e^x (b) e^{2x}
(c) e^{-x} (d) e^{-2x}

7. Which of the following is true?
(a) $f(4) = f(3) + \Delta f(2) + \Delta^2 f(1) + \Delta^3 f(1)$
(b) $f(4) = f(2) + \Delta f(3) + \Delta^2 f(2) + \Delta^3 f(2)$
(c) $f(4) = f(1) + \Delta f(2) + \Delta^2 f(3) + \Delta^3 f(1)$
(d) none of these

8. The missing terms in the following table are, respectively:

x	0	1	2	3	4	5
y	0	—	8	15	—	35

(a) 1, 8   (b) 2, 12
(c) 3, 24   (d) 5, 25

9. Given the data:

x	1	2	3	4	5
y	−1	2	−3	4	−5

If the derivative of $y(x)$ is approximated as

$$y'(x_k) = \frac{1}{h}\left(\Delta y_k + \frac{1}{2}\Delta^2 y_k - \frac{1}{4}\Delta^3 y_k \right),$$

then the value of $y'(2)$ is
(a) 4   (b) 8
(c) 12   (d) 16

10. The value of $\log 2^{1/3}$ from $\int_0^1 \frac{x^2}{1+x^3}dx$ using

Simpson's 1/3rd rule with $h = 0.25$ is

(a) 0.23108   (b) 0.25251
(c) 0.10234   (d) 0.20501

11. The values of $\int_{0.1}^{0.7}(e^x + 2x)dx$ by the trapezoidal

rule and Simpson's 1/3rd Rule (taking $h = 0.1$, correct to 5-decimal places) are, respectively
(a) 2.32445, 2.38541 (b) 1.25034, 1.00350
(c) 1.38934, 1.38858 (d) none of these

12. If a quadrature formula $\frac{3}{2}f\left(-\frac{1}{3}\right) + Kf\left(\frac{1}{3}\right) +$

$\frac{1}{2}f(1)$ that approximates $\int_{-1}^{1} f(x)dx$ is found

to be exact for quadric polynomials, then the value of K is
(a) 2   (b) 1
(c) 0   (d) − 1

13. Consider the quadrature formula

$$\int_{-1}^{1}|x|f(x)dx = \frac{1}{2}[f(x_0) + f(x_1)]$$

where x_0 and x_1 are quadrature points. Then, the highest degree of the polynomial, for which the above formula of exact, equals
(a) 1   (b) 2
(c) 3   (d) 4

14. The values of the constants α, β for which the quadratic formula

$$\int_0^1 f(x)dx = \alpha f(0) + \beta f(x_1)$$

is exact for polynomials of degree as high as possible are

(a) $\alpha = \frac{2}{3}, \beta = \frac{1}{4}$ (b) $\alpha = \frac{3}{4}, \beta = \frac{1}{4}$

(c) $\alpha = \frac{1}{4}, \beta = \frac{3}{4}$ (d) $\alpha = \frac{2}{3}, \beta = \frac{3}{4}$

15. The root of the equation $xe^x = 1$ between 0 and 1, obtained by using two iterations of bisection method, is
(a) 0.25   (b) 0.50
(c) 0.75   (d) 0.65

16. While solving the equation $x^2 − 3x + 1 = 0$ using the Newton–Raphson method with the initial guess of a root as 1, the value of the root after one iteration is
(a) 1.5   (b) 0.1
(c) 0.5   (d) 0

17. Consider the system of equations

$$\begin{bmatrix} 5 & 2 & 1 \\ -2 & 5 & 2 \\ -1 & 2 & 8 \end{bmatrix}\begin{bmatrix} x_1 \\ x_2 \\ x_3 \end{bmatrix} = \begin{bmatrix} 13 \\ -22 \\ 14 \end{bmatrix}$$

with the initial guess of the solution

$\left[x_1^{(0)}, x_2^{(0)}, x_3^{(0)}\right]^T = [1,1,1]^T$, the approximate

value of the solution $\left[x_1^{(1)}, x_2^{(1)}, x_3^{(1)}\right]^T$ after one

iteration by the Gauss–Seidel method is
(a) $[2, − 4.4, 1.625]^T$ (b) $[2, −4, − 3]^T$
(c) $[2, 4.4, 1.625]^T$ (d) $[2, − 4, 3]^T$

18. The following system of equations by the LU-factorization method has a solution:
$8x_1 − 3x_2 + 2x_3 = 20$
$4x_1 + 11x_2 − 2x_3 = 33$
$6x_1 + 3x_2 + 12x_3 = 36$
(a) 3, 3, 2   (b) 2, 1, 3
(c) 3, 2, 1   (d) 1, 2, 3

19. Using the Euler's method taking step size = 0.1, the approximate value of y obtained corresponding to $x = 0.2$ for the initial value

problem $\frac{dy}{dx} = x^2 + y^2$ and $y(0) = 1$, is

(a) 1.322   (b) 1.122
(c) 1.222   (d) 1.110

20. Given $\frac{dy}{dx} = \frac{y-x}{y+x}$ with initial condition $y = 1$

at $x = 0$, then the value of y for $x = 0.1$ by the Euler's method (correct up to 4 decimal places, taking step length $h = 0.02$) is
(a) 2.0540   (b) 1.0528
(c) 1.0928   (d) 2.0345

21. Given $\dfrac{dy}{dx} + \dfrac{y}{x} = \dfrac{1}{x^2}, y(1) = 1$. Then the value of $y(1.2)$ by the modified Euler's method correct up to 4 decimal places, is

    (a) 0.9858   (b) 1.0528

    (c) 1.8776   (d) 0.0333

22. The value of $y(1.1)$ using the Runge–Kutta method of fourth order, given that $\dfrac{dy}{dx} = y^2 + xy$, $y(1) = 1$ is

    (a) 1.4561   (b) 1.0528

    (c) 1.8776   (d) 1.2415

Answer key				
1. (a)	**2.** (c)	**3.** (a)	**4.** (b)	**5.** (d)
6. (a)	**7.** (a)	**8.** (c)	**9.** (b)	**10.** (a)
11. (c)	**12.** (c)	**13.** (a)	**14.** (c)	**15.** (c)
16. (d)	**17.** (a)	**18.** (c)	**19.** (c)	**20.** (c)
21. (a)	**22.** (d)			

Explanation

1. (a) Here, true value, $x_T = 3.14156$ and approximate value, $x_A = 3.14$.

    $\therefore$ Absolute error, $E_a = |x_T - x_A|$

$$= |3.14156 - 3.14| = 0.00156$$

Relative error, $E_r = \dfrac{|x_T - x_A|}{x_T} = \dfrac{0.00156}{3.14156}$

$$= 4.966 \times 10^{-4}$$

2. (c) True value, $x_T = \dfrac{4}{3}$, approximate value, $x_A = 1.333$.

Therefore, percentage error

$$= \dfrac{|x_T - x_A|}{x_T} \times 100 = \dfrac{\left|\dfrac{4}{3} - 1.3333\right|}{\dfrac{4}{3}} \times 100 = 0.0025.$$

4. (b)

We have $\Delta f(x) = f(x+h) - f(x)$

and $\nabla f(x) = f(x) - f(x-h) \cdot$

$\therefore \Delta . \nabla f(x)$

$$= \Delta\left[f(x) - f(x-h)\right] = \Delta f(x) - \Delta f(x-h)$$

$$= \Delta f(x) - \left[f(x) - (x-h)\right]$$

$$= \Delta f(x) - \nabla f(x) = (\Delta - \nabla) f(x)$$

$\therefore \Delta . \nabla \equiv \Delta - \nabla$

5. (d) We have,

$\Delta \log f(x) = \log f(x+h) - \log f(x)$ (h being the space length)

$$= \log \dfrac{f(x+h)}{f(x)} = \log\left\{\dfrac{\Delta f(x) + f(x)}{f(x)}\right\}$$

$$\left[\because \Delta f(x) = f(x+h) - f(x)\right]$$

$$= \log\left\{1 + \dfrac{\Delta f(x)}{f(x)}\right\}$$

6. (a)

We have,

$$\left(\dfrac{\Delta^2}{E} e^x\right)\dfrac{Ee^x}{\Delta^2 e^x}$$

$$= \left(\Delta^2 E^{-1} e^x\right)\dfrac{e^{x+h}}{\Delta^2 e^x} = \left(\Delta^2 e^{x-h}\right)\dfrac{e^{x+h}}{\Delta^2 e^x} = e^{-h}\Delta^2 e^x \dfrac{e^x . e^h}{\Delta^2 e^x}$$

$$= e^{-h} . e^x . e^h = e^x, h \text{ being the space length.}$$

7. (a)

$\Delta f(3) = f(4) - f(3)$

$\Rightarrow f(4) = f(3) + \Delta f(3)$

$= f(3) + \Delta\{f(2) + \Delta f(2)\}$

$\left[\because \Delta f(2) = f(3) - f(2)\right]$

$= f(3) + \Delta f(2) + \Delta^2 f(2)$

$= f(3) + \Delta f(2) + \Delta^2\{f(1) + \Delta f(1)\}$

$= f(3) + \Delta f(2) + \Delta^2 f(1) + \Delta^3 f(1)$

Thus, $f(4) = f(3) + \Delta f(2) + \Delta^2 f(1) + \Delta^3 f(1)$

8. (c) Since we are given four values, therefore, we take $f(x)$ to be a polynomial of degree 3 in x so that

$\Delta^4 f(x) = 0$ for all values of x.

Now $\Delta^4 f(x) = 0$

$\Rightarrow (E-1)^4 f(x) = 0$

$\Rightarrow E^4 f(x) - 4E^3 f(x) + 6E^2 f(x) - 4Ef(x) + f(x) = 0$

$\Rightarrow f(x+4) - 4f(x+3) + 6f(x+2) - 4f(x+1) + f(x) = 0$ ...(i)

Putting $x = 0$ in (i), we get

$f(4) - 4f(3) + 6f(2) - 4f(1) + f(0) = 0$

or $f(4) - 4 \times 15 + 6 \times 8 - 4f(1) + 0 = 0$

or, $f(4) - 4f(1) = 12$ ...(ii)

Again, putting $x = 1$ in (i), we get

$f(5) - 4f(4) + 6f(3) - 4f(2) + f(1) = 0$

or, $35 - 4f(4) + 6 \times 15 - 4 \times 8 + f(1) = 0$

or, $4f(4) - f(1) = 93$ ...(iii)

Solving (ii) and (iii) we get, $f(1) = 3, f(4) = 24 \cdot$

9. (b)

x	1	2	3	4	5
y	−1 (y_0)	2(y_1)	−3(y_2)	4 (y_3)	−5(y_4)

Here, $\Delta y_1 = y_2 - y_1 = -3 - 2 = -5$,

$$\Delta^2 y_1 = \Delta y_2 - \Delta y_1 = (y_3 - y_2) - (y_2 - y_1)$$
$$= (4 - (-3)) - (-3 - 2) = 12,$$

$$\Delta^3 y_1 = \Delta^2 y_2 - \Delta^2 y_1 = (\Delta y_3 - \Delta y_2) - \Delta^2 y_1$$
$$= (y_4 - y_3) - (y_3 - y_2) - \Delta^2 y_1$$
$$= (-5 - 4) - (4 + 3) - 12 = -28.$$

Now, $x_1 = x_0 + h = 1 + 1 = 2$.

$$\therefore y'(x_1) = \frac{1}{h}\left(\Delta y_1 + \frac{1}{2}\Delta^2 y_1 - \frac{1}{4}\Delta^3 y_1\right)$$

$$\Rightarrow y'(2) = \frac{1}{1}\left[-5 + \frac{1}{2}\times 12 - \frac{1}{4}(-28)\right] = 8.$$

10. (a)

x	0	0.25	0.50	0.75	1
$f(x)$ $= \dfrac{x^2}{1+x^3}$	0 (y_0)	0.06154 (y_1)	0.22222 (y_2)	0.3956 (y_3)	0.5 (y_4)

Here, $h = 0.25$.

∴ By Simpson's 1/3rd rule we get,

$$\int_0^1 \frac{x^2}{1+x^3}dx$$

$$\approx \frac{h}{3}\left[y_0 + y_4 + 4(y_1 + y_3) + 2y_2\right]$$

$$= \frac{0.25}{3}\left[0 + 0.5 + 4(0.06154 + 0.39560) + 2\times 0.22222\right]$$

$$= 0.23108$$

Now, $\int_0^1 \frac{x^2}{1+x^3}dx$

$$= \frac{1}{3}\left[\log(1+x^3)\right]_0^1 = \frac{1}{3}\log 2 = \log 2^{1/3}.$$

Therefore, $\log 2^{1/3} \approx 0.23108$

11. (c)

Here, $f(x) = e^x + 2x, a = 0.1, b = 0.7, h = 0.1$

x	$f(x)$
$x_0 = 0.1$	$y_0 = 1.30517$
$x_1 = 0.2$	$y_1 = 1.62140$
$x_2 = 0.3$	$y_2 = 1.94986$
$x_3 = 0.4$	$y_3 = 2.29182$
$x_4 = 0.5$	$y_4 = 2.64872$
$x_5 = 0.6$	$y_5 = 3.02212$
$x_6 = 0.7$	$y_6 = 3.41375$

Then by the trapezoidal rule, $\int_{0.1}^{0.7}(e^x + 2x)dx$

$$= \frac{h}{2}\left[(y_0 + y_6) + 2(y_1 + y_2 + y_3 + y_4 + y_5)\right]$$

$$= \frac{0.1}{2}\left[(1.30517 + 3.41375) + 2(1.62140 + 1.94986\right.$$
$$\left. + 2.29182 + 2.64872 + 3.02212)\right]$$

$$= 1.389338 \approx 1.38934.$$

Again, by Simpson's 1/3rd rule, $\int_{0.1}^{0.7}(e^x + 2x)dx$

$$= \frac{h}{3}\left[(y_0 + y_6) + 4(y_1 + y_3 + y_5) + 2(y_2 + y_4)\right]$$

$$= \frac{0.1}{3}\left[(1.30517 + 3.41375) + 4(1.62140 + 2.29182\right.$$
$$\left. + 3.02212) + 2(1.94986 + 2.64872)\right]$$

$$= 1.3885813 \approx 1.38858.$$

12. (c) We have,

$$\int_{-1}^1 f(x)dx = \frac{3}{2}f\left(-\frac{1}{3}\right) + Kf\left(\frac{1}{3}\right) + \frac{1}{2}f(1) \qquad ...(i)$$

For $f(x) = 1$, (i) gives,

$$\int_{-1}^1 1dx = \frac{3}{2}\times 1 + K\times 1 + \frac{1}{2}\times 1$$

$$\Rightarrow 2 = \frac{3}{2} + K + \frac{1}{2}$$

$$\Rightarrow k = 0$$

13. (a)

$$\int_{-1}^1 |x| f(x)dx = \frac{1}{2}\left[f(x_0) + f(x_1)\right]$$

$$\Rightarrow -\int_{-1}^0 xf(x)dx + \int_0^1 xf(x)dx = \frac{1}{2}\left[f(x_0) + f(x_1)\right]$$
$$...(i)$$

For $f(x) = 1$, (i) gives,

$$-\int_{-1}^0 xdx + \int_0^1 xdx = \frac{1}{2}[1 + 1]$$

$$\Rightarrow -\left[\frac{x^2}{2}\right]_{-1}^0 + \left[\frac{x^2}{2}\right]_0^1 = 1 \Rightarrow \frac{1}{2} + \frac{1}{2} = 1 \Rightarrow 1 = 1$$

For $f(x) = x$, (i) gives,

$$-\int_{-1}^0 x^2 dx + \int_0^1 x^2 dx = \frac{1}{2}[-1 + 1]$$

$$\Rightarrow -\left[\frac{x^3}{3}\right]_{-1}^0 + \left[\frac{x^3}{3}\right]_0^1 = 0 \Rightarrow -\frac{1}{3} + \frac{1}{3} = 0 \Rightarrow 0 = 0.$$

For $f(x) = x^2$, (i) gives,

$$-\int_{-1}^0 x^3 dx + \int_0^1 x^3 dx = \frac{1}{2}[1 + 1]$$

$$\Rightarrow -\left[\frac{x^4}{4}\right]_{-1}^0 + \left[\frac{x^4}{4}\right]_0^1 = 1 \Rightarrow \frac{1}{4} + \frac{1}{4} = 1 \Rightarrow \frac{1}{2} = 1,$$

a contradiction.

∴ Highest degree of $f(x)$ is 1.

14. (c)

Given that, $\int_0^1 f(x)dx = \alpha f(0) + \beta f(x_1)$ ...(i)

When $f(x) = 1$, we get from (i),

$\int_0^1 1 dx = \alpha \times 1 + \beta \times 1$

$\Rightarrow 1 = \alpha + \beta$ ...(ii)

When $f(x) = x$, we get from (i),

$\int_0^1 x\, dx = \alpha \times 0 + \beta \times x_1$

$\Rightarrow \left[\dfrac{x^2}{2}\right]_0^1 = \beta x_1 \Rightarrow \dfrac{1}{2} = \beta x_1$ ...(iii)

When $f(x) = x^2$, we get from (i),

$\Rightarrow \int_0^1 x^2 dx = \alpha \times 0 + \beta \times x_1^2$

$\Rightarrow \left[\dfrac{x^3}{3}\right]_0^1 = \beta x_1^2 \Rightarrow \dfrac{1}{3} = \beta x_1^2$ ...(iv)

Now, dividing Eq. (iv) by Eq. (iii), we get,

$\dfrac{\beta x_1^2}{\beta x_1} = \dfrac{1/3}{1/2}$ i.e; $x_1 = \dfrac{2}{3}$.

From Eq. (iii), we get, $\dfrac{1}{2} = \beta \times \dfrac{2}{3}$ i.e; $\beta = \dfrac{3}{4}$.

∴ From Eq. (ii), we get,

$\alpha = 1 - \beta = 1 - \dfrac{3}{4} = \dfrac{1}{4}$.

Thus $\alpha = \dfrac{1}{4}$, $\beta = \dfrac{3}{4}$.

15. (c) Let $f(x) = xe^x - 1$.

Now, $f(0) = -1 < 0$ and $f(1) = e - 1 > 0$.

Thus, the exact root lies in (0, 1).

Hence, first approximation of the root,

$x_1 = \dfrac{0+1}{2} = 0.5$.

Then, $f(x_1) = (0.5)e^{0.5} - 1 = -0.175 < 0$.

So second approximation of the root,

$x_2 = \dfrac{0.5+1}{2} = \dfrac{1.5}{2} = 0.75$.

16. (d) We have,

$x_0 = 1, f(x) = x^2 - 3x + 1, f'(x) = 2x - 3$.

∴ $f(1) = 1 - 3 + 1 = -1, f'(1) = 2 - 3 = -1$.

Then by the Newton–Raphson method,

$x_1 = x_0 - \dfrac{f(x_0)}{f'(x_0)} = 1 - \dfrac{f(1)}{f'(1)} = 1 - \dfrac{(-1)}{(-1)} = 0$.

17. (a)

$\begin{bmatrix} 5 & 2 & 1 \\ -2 & 5 & 2 \\ -1 & 2 & 8 \end{bmatrix} \begin{bmatrix} x_1 \\ x_2 \\ x_3 \end{bmatrix} = \begin{bmatrix} 13 \\ -22 \\ 14 \end{bmatrix}$

$\Rightarrow \begin{bmatrix} 5x_1 + 2x_2 + x_3 \\ -2x_1 + 5x_2 + 2x_3 \\ -x_1 + 2x_2 + 8x_3 \end{bmatrix} = \begin{bmatrix} 13 \\ -22 \\ 14 \end{bmatrix}$

$\Rightarrow 5x_1 + 2x_2 + x_3 = 13,$

$-2x_1 + 5x_2 + 2x_3 = -22,$

$-x_1 + 2x_2 + 8x_3 = 14.$

$\Rightarrow x_1 = \dfrac{13}{5} - \dfrac{2}{5}x_2 - \dfrac{1}{5}x_3,$

$x_2 = -\dfrac{22}{5} + \dfrac{2}{5}x_1 - \dfrac{2}{5}x_3,$

$x_3 = \dfrac{14}{8} + \dfrac{1}{8}x_1 - \dfrac{2}{8}x_2.$

Now, the initial guess is $\left[x_1^{(0)}, y_2^{(0)}, z_3^{(0)} \right] = \begin{bmatrix} 1 & 1 & 1 \end{bmatrix}^T$.

∴ $x_1^{(0)} = x_2^{(0)} = x_3^{(0)} = 1$.

∴ $x_1^{(1)} = \dfrac{13}{5} - \dfrac{2}{5} \times 1 - \dfrac{1}{5} \times 1 = 2,$

$x_2^{(1)} = -\dfrac{22}{5} + \dfrac{2}{5} \times 1 - \dfrac{1}{5} \times 1 = 4.4, = -4.4$

$x_3^{(1)} = \dfrac{14}{8} + \dfrac{1}{8} \times 1 - \dfrac{2}{8} \times 1 = 1.625.$

∴ $\begin{bmatrix} x_1^{(1)} & x_2^{(1)} & x_3^{(1)} \end{bmatrix} = \begin{bmatrix} 2.0, -4.4, 1.625 \end{bmatrix}^T$.

18. (c) The given system of equations can be written as

$AX = b$

where $A = \begin{bmatrix} 8 & -3 & 2 \\ 4 & 11 & -1 \\ 6 & 3 & 12 \end{bmatrix}$, $b = \begin{bmatrix} 20 \\ 33 \\ 36 \end{bmatrix}$, $X = \begin{bmatrix} x_1 \\ x_2 \\ x_3 \end{bmatrix}$

Then, $A = LU$

$\Rightarrow \begin{bmatrix} 8 & -3 & 2 \\ 4 & 11 & -1 \\ 6 & 3 & 12 \end{bmatrix} = \begin{bmatrix} l_{11} & 0 & 0 \\ l_{21} & l_{22} & 0 \\ l_{31} & l_{32} & l_{33} \end{bmatrix} \times \begin{bmatrix} 1 & u_{12} & u_{13} \\ 0 & 1 & u_{23} \\ 0 & 0 & 1 \end{bmatrix}$

$\Rightarrow \begin{bmatrix} 8 & -3 & 2 \\ 4 & 11 & -1 \\ 6 & 3 & 12 \end{bmatrix} = \begin{bmatrix} l_{11} & l_{11}u_{12} & l_{11}u_{13} \\ l_{21} & l_{21}u_{12}+l_{22} & l_{21}u_{13}+l_{22}u_{23} \\ l_{31} & l_{31}u_{12}+l_{32} & l_{31}u_{13}+l_{32}u_{23}+l_{33} \end{bmatrix}$

∴ $l_{11} = 8, l_{11}u_{12} = -3 \Rightarrow u_{12} = -\dfrac{3}{8}$,

$l_{11}u_{13} = 2 \Rightarrow u_{13} = \dfrac{2}{8} = \dfrac{1}{4}$, $l_{21} = 4$,

$l_{21}u_{12} + l_{22} = 11$

$\Rightarrow l_{22} = 11 - l_{21}u_{12} \Rightarrow l_{22} = 11 - 4\left(-\dfrac{3}{8}\right) = \dfrac{25}{2}$,

$l_{21}u_{13} + l_{22}u_{23} = -1 \Rightarrow 4 \times \dfrac{1}{4} + \dfrac{25}{2}u_{23} = -1$

$\Rightarrow u_{23} = -\dfrac{4}{25}$,

$l_{31} = 6, l_{31}u_{12} + l_{32} = 3$

$\Rightarrow 6\left(-\dfrac{3}{8}\right) + l_{32} = 3 \Rightarrow l_{32} = \dfrac{21}{4},$

$l_{31}u_{13} + l_{22}u_{23} + l_{33} = 12$

$\Rightarrow 6 \times \dfrac{1}{4} + \dfrac{21}{4} \times \left(\dfrac{-4}{25}\right) + l_{33} = 12 \Rightarrow l_{33} = \dfrac{567}{50}.$

Hence, $L = \begin{bmatrix} 8 & 0 & 0 \\ 4 & \dfrac{25}{2} & 0 \\ 6 & \dfrac{21}{4} & \dfrac{567}{50} \end{bmatrix}$, $U = \begin{bmatrix} 1 & -\dfrac{3}{8} & \dfrac{1}{4} \\ 0 & 1 & \dfrac{-4}{25} \\ 0 & 0 & 1 \end{bmatrix}$

Now, $AX = b$

$\Rightarrow \quad LUX = b \Rightarrow \quad LY = b \qquad \qquad \dots(i)$

where $UX = Y$

$i.e; \begin{bmatrix} 1 & -\dfrac{3}{8} & \dfrac{1}{4} \\ 0 & 1 & \dfrac{-4}{25} \\ 0 & 0 & 1 \end{bmatrix} \begin{bmatrix} x_1 \\ x_2 \\ x_3 \end{bmatrix} = \begin{bmatrix} y_1 \\ y_2 \\ y_3 \end{bmatrix}$

$i.e; \begin{bmatrix} x_1 - \dfrac{3}{8}x_2 + \dfrac{1}{4}x_3 \\ x_2 - \dfrac{4}{25}x_3 \\ x_3 \end{bmatrix} = \begin{bmatrix} y_1 \\ y_2 \\ y_3 \end{bmatrix} \qquad \dots(ii)$

From (i) we get,

$\begin{bmatrix} 8 & 0 & 0 \\ 4 & \dfrac{25}{2} & 0 \\ 6 & \dfrac{21}{4} & \dfrac{567}{50} \end{bmatrix} \begin{bmatrix} y_1 \\ y_2 \\ y_3 \end{bmatrix} = \begin{bmatrix} 20 \\ 33 \\ 36 \end{bmatrix}$

$or, \begin{bmatrix} 8y_1 \\ 4y_1 + \dfrac{25}{2}y_2 \\ 6y_1 + \dfrac{21}{4}y_2 + \dfrac{567}{50}y_3 \end{bmatrix} = \begin{bmatrix} 20 \\ 33 \\ 36 \end{bmatrix}$

$or, 8y_1 = 20, 4y_1 + \dfrac{25}{2}y_2, 6y_1 + \dfrac{21}{4}y_2 + \dfrac{567}{50}y_3 = 36.$

Now, $8y_1 = 20 \Rightarrow y_1 = \dfrac{5}{2},$

$4y_1 + \dfrac{25}{2}y_2 = 33 \Rightarrow y_2 = \left(33 - 4 \times \dfrac{5}{2}\right)\dfrac{2}{25} = \dfrac{46}{25},$

$6y_1 + \dfrac{21}{4}y_2 + \dfrac{567}{50}y_3 = 36$

$\Rightarrow y_3 = \dfrac{50}{567}\left[36 - 15 - \dfrac{21 \times 23}{50}\right] = 1.$

Then from (ii), we have

$\begin{bmatrix} x_1 - \dfrac{3}{8}x_2 + \dfrac{1}{4}x_3 \\ x_2 - \dfrac{4}{25}x_3 \\ x_3 \end{bmatrix} = \begin{bmatrix} \dfrac{5}{2} \\ \dfrac{46}{25} \\ 1 \end{bmatrix}.$

This gives
$x_3 = 1,$

$x_2 - \dfrac{4}{25}x_3 = \dfrac{46}{25} \Rightarrow x_2 = \dfrac{46}{25} + \dfrac{4}{25} = 2,$

$x_1 - \dfrac{3}{8}x_2 + \dfrac{x_3}{4} = \dfrac{5}{2} \Rightarrow x_1 = \dfrac{5}{2} + \dfrac{3}{8} \times 2 - \dfrac{1}{4} = 3.$

19. (c)

We have, $\dfrac{dy}{dx} = x^2 + y^2, x_0 = 0, y_0 = 1, h = 0.1$

$\therefore \quad f(x, y) = x^2 + y^2.$

By the Euler's method,

$y_1 = y(x_1) = y_0 + hf(x_0, y_0)$
$= 1 + 0.1(0^2 + 1^2) = 1.1,$
$y_2 = y(x_2) = y(0.2) = y_1 + h f(x_1, y_1)$
$= 1.1 + 0.1\left[(0.1)^2 + (1.1)^2\right]$
$= 1.1 + 0.1(0.01 + 1.21) = 1.222.$

20. (c)

Here, $f(x, y) = \dfrac{y - x}{y + x}, x_0 = 0, y_0 = 1$ and $h = 0.02.$

$\therefore \quad$ By the Euler's method, we get

$y_1 = y_0 + h f(x_0, y_0) = 1 + 0.02\left(\dfrac{1 - 0}{1 + 0}\right) = 1.02.$

$y(0.04) = y_2 = y_1 + h f(x_1, y_1)$

$= 1.02 + 0.02\left(\dfrac{1.02 - 0.02}{1.02 + 0.02}\right) = 1.039231.$

Similarly,
$y(0.06) = y_3 = y_2 + h f(x_2, y_2)$

$= 1.039231 + 0.02\left(\dfrac{1.039231 - 0.04}{1.039231 + 0.04}\right)$

$= 1.057748$

$y(0.08) = y_4 = y_3 + h f(x_3, y_3)$

$= 1.057748 + 0.02\left(\dfrac{1.057748 - 0.06}{1.057748 + 0.06}\right)$

$= 1.075601$

$$y(0.10) = y_5 = y_4 + h f(x_4, y_4)$$

$$= 1.075601 + 0.02\left(\frac{1.075601 - 0.08}{1.075601 + 0.08}\right)$$

$$= 1.092832 \approx 1.0928$$

21. *(a)*

Here, $f(x,y) = \dfrac{1}{x^2} - \dfrac{y}{x}, x_0 = 1, y_0 = 1.$

Let us take $h = 0.1$ so that $x_1 = 1 + 0.1 = 1.1.$

$\therefore y_1^{(0)} = y_0 + h f(x_0, y_0) = 1 + 0.1 \times (1-1) = 1.$

The modified Euler's iteration formula is

$$y_r^{(n)} = y_{r-1} + \frac{h}{2}\left[f(x_{r-1}, y_{r-1}) + f(x_r, y_r^{(n-1)})\right]$$

$$\therefore y_1^{(1)} = y_0 + \frac{h}{2}\left[f(x_0, y_0) + f\left(x_1, y_1^{(0)}\right)\right]$$

$$= 1 + \frac{0.1}{2}\left[(1-1) + \left\{\frac{1}{(1.1)^2} - \frac{1}{1.1}\right\}\right]$$

$$= 1 + 0.05(-0.08264) = 0.99587,$$

$$y_1^{(2)} = y_0 + \frac{h}{2}\left[f(x_0, y_0) + f\left(x_0, y_1^{(1)}\right)\right]$$

$$= 1 + \frac{0.1}{2}\left[(1-1) + \left(\frac{1}{(1.1)^2} - \frac{0.99587}{1.1}\right)\right]$$

$$= 1 + 0.05\,(-0.078888) = 0.99606,$$

$$y_1^{(3)} = y_0 + \frac{h}{2}\left[f(x_0, y_0) + f\left(x_0, y_2^{(1)}\right)\right]$$

$$= 1 + 0.05\left[(1-1) + \left(\frac{1}{(1.1)^2} - \frac{0.99606}{1.1}\right)\right]$$

$$= 1 + 0.05(-0.079063) = 0.99607.$$

Hence, $y_1 = y(1.1) = 0.9961.$

Thus, $x_1 = 1.1, y_1 = 0.9961.$

$$\therefore f(x_1, y_1) = \frac{1}{(1.1)^2} - \frac{0.9961}{1.1} = -0.079.$$

Now, $y_2^{(0)} = y_1 + h(x_1, y_1)$

$= 0.9961 + 0.1 \times (-0.079) = 0.98819$

Then we have,

$$y_2^{(1)} = y_1 + \frac{h}{2}\left[f(x_1, y_1) + f\left(x_2, y_2^{(0)}\right)\right]$$

$$= 0.9961 + 0.05\left[-0.079 + \frac{1}{(1.2)^2} - \frac{0.98819}{1.2}\right]$$

$$= 0.98569,$$

$$y_2^{(2)} = y_1 + \frac{h}{2}\left[f(x_1, y_1) + f\left(x_2, y_2^{(1)}\right)\right]$$

$$= 0.9961 + 0.05\left[-0.079 + \frac{1}{(1.2)^2} - \frac{0.98569}{1.2}\right]$$

$$= 0.98580,$$

$$y_2^{(3)} = y_1 + \frac{h}{2}\left[f(x_1, y_1) + f\left(x_2, y_2^{(2)}\right)\right]$$

$$= 0.9961 + 0.05\left[-0.079 + \frac{1}{(1.2)^2} - \frac{0.98580}{1.2}\right]$$

$$= 0.985797$$

Thus, $y_2 \approx 0.9858$, correct up to four decimal places.

$\therefore y(1.2) = y_2 \approx 0.9858.$

22. *(d)* Here, $f(x, y) = y^2 + xy, x_0 = 1, y_0 = 1.$

Let us take $h = 0.1$. Then iterative formula of the Runge–Kutta method of order 4 gives:

$$y_1 = y_0 + \frac{1}{6}\left(k_1 + 2k_2 + 2k_3 + k_4\right)$$

where

$$k_1 = hf(x_0, y_0) = 0.1(1^2 + 1 \times 1) = 0.2,$$

$$k_3 = hf\left(x_0 + \frac{1}{2}h, y_0 + \frac{1}{2}k_2\right) = 0.1 \times f(1.05, 1.1)$$

$$= 0.1 \times \{(1.1)^2 + 1.05 \times 1.1\} = 0.2365,$$

$$k_3 = hf\left(x_0 + \frac{1}{2}h, y_0 + \frac{1}{2}k_2\right)$$

$$= 0.1 \times f(1.05, 1.11825)$$

$$= 0.1 \times \{(1.11825)^2 + 1.11825 \times 1.05\} = 0.2425,$$

$$k_4 = hf(x_0 + h, y_0 + k_3) = 0.1 \times f(1.1, 1.2425)$$

$$= 0.1 \times \{(1.2425)^2 + 1.1 \times 1.2425\} = 0.2910556.$$

$\therefore y(1.1) = y_1$

$$= y_0 + \frac{1}{6}\left(k_1 + 2k_2 + 2k_3 + k_4\right)$$

$$= 1 + \frac{1}{6}\left(0.2 + 2 \times 0.2365 + 2 \times 0.2425 + 0.2910556\right)$$

$$\approx 1.2415 \text{ (correct up to four decimal places)}$$

PREVIOUS YEARS SOLVED PAPERS (2000-2018)

1. The trapezoidal rule for integration gives exact result when the integrand is a polynomial of degree

(a) but not 1   *(b)* 1 but not 0

(c) 0 or 1   *(d)* 2

[CS GATE 2002]

2. The accuracy of Simpson's rule quadrature for a step size h is

(a) $O(h^2)$ (b) $O(h^3)$
(c) $O(h^4)$ (d) $O(h^5)$

[ME GATE 2003]

3. Given $a > 0$, we wish to calculate the reciprocal value $\dfrac{1}{a}$ by the Newton–Raphson method for $f(x) = 0$.

(i) The Newton–Raphson algorithm for the function will be

(a) $x_{k+1} = \dfrac{1}{2}\left(x_k + \dfrac{a}{x_k}\right)$

(b) $x_{k+1} = \left(x_k + \dfrac{a}{2}x_k^2\right)$

(c) $x_{k+1} = 2x_k - ax_k^2$

(d) $x_{k+1} = x_k - \dfrac{a}{2}x_k^2$

(ii) For $a = 7$ and starting with $x_0 = 0.2$, the first two iterations will be

(a) 0.11, 0.1299   (b) 0.12, 0.1392
(c) 0.12, 0.1416   (d) 0.13, 0.1428

[CE GATE 2005]

4. Match **List-l** with **List-ll** and select the correct combination:

List-l

A. Newton–Raphson method
B. Runge–Kutta method equations
C. Simpson's 1/3rd rule
D. Gauss elimination

List-ll

1. Solving nonlinear equations
2. Solving linear simultaneous equations
3. Solving ordinary differential equations
4. Numerical integration method
5. Interpolation
6. Calculation of eigenvalues

Codes:

(a) A-6, B-1, C-5, D-3
(b) A-1, B-6, C-4, D-3
(c) A-1, B-3, C-4, D-2
(d) A-5, B-3, C-4, D-1

[EC GATE-2005]

5. Starting from $x_0 = 1$, one step of the Newton–Raphson method in solving the equation $x^3 + 3x - 7 = 0$ gives the next value x_1 as

(a) $x_1 = 0.5$ (b) $x_1 = 1.406$
(c) $x_1 = 1.5$ (d) $x_1 = 2$

[ME GATE 2005]

6. For solving algebraic and transcendental equations which one of the following is used?

(a) Coulomb's theorem
(b) Newton–Raphson method
(c) Euler method
(d) Stokes' theorem

[PI GATE-2005]

7. The real root of the equation $xe^x = 2$ is evaluated using the Newton–Raphson method. If the first approximation of the value of "x" is 0.8679, the second approximation of the value of "x" correct to three decimal places is

(a) 0.865   (b) 0.853
(c) 0.849   (d) 0.838

[PI GATE-2005]

8. A second degree polynomial, $f(x)$ has values of 1, 4, and 15 at $x = 0$, 1, and 2, respectively. The integral $\displaystyle\int_0^2 f(x)dx$ is to be estimated by applying the trapezoidal rule to this data. What is the error in the estimate?

(a) $-\dfrac{4}{3}$ (b) $-\dfrac{2}{3}$

(c) 0   (d) $\dfrac{2}{3}$

[CE GATE 2006]

9. The differential equation $\dfrac{dy}{dx} = 0.25y^2$ is to be solved using the backward (implicit) Euler's method with the boundary condition $y = 1$ at $x = 0$ and with a step size of 1. What would be the value of y at $x = 1$?

(a) 1.33   (b) 1.67
(c) 2.00   (d) 2.33

[CE GATE 2006]

10. The following equation needs to be numerically solved using the Newton–Raphson method:
$x^3 + 4x - 9 = 0$
The iterative equation for this purpose is (k indicates the iteration level)

(a) $x_{k+1} = \dfrac{2x_k^3 + 9}{3x_k^2 + 4}$ (b) $x_{k+1} = \dfrac{3x_k^2 + 4}{3x_k^2 + 9}$

(c) $x_{k+1} = x_k - 3x_k^2 + 4$

(d) $x_{k+1} = \dfrac{4x_k^2 + 3}{9x_k^2 + 2}$

[CE GATE 2007]

11. Given that one root of the equation $x^3 - 10x^2 + 31x - 30 = 0$ is 5, the other two roots are

(a) 2 and 3   (b) 2 and 4
(c) 3 and 4   (d) –2 and –3

[CE GATE 2007]

12. The equation $x^3 - x^2 + 4x - 4 = 0$ is to be solved using the Newton–Raphson method. If $x = 2$ is taken as the initial approximation of the solution, then the next approximation using this method will be

(a) $\frac{2}{3}$ (b) $\frac{4}{3}$

(c) 1 (d) $\frac{3}{2}$

[EC GATE 2007]

13. Consider the series $x_{n+1} = \frac{x_n}{2} + \frac{9}{8x_n}, x_0 = 0.5$ obtained from the Newton–Raphson method. The series converges to

(a) 1.5 (b) $\sqrt{2}$

(c) 1.6 (d) 1.4

[CS GATE 2007]

14. Identify the Newton–Raphson iteration scheme for finding the square root of 2

(a) $x_{n+1} = \frac{1}{2}\left(x_n + \frac{2}{x_n}\right)$ (b) $x_{n+1} = \frac{1}{2}\left(x_n - \frac{2}{x_n}\right)$

(c) $x_{n+1} = \frac{2}{x_n}$ (d) $x_{n+1} = \sqrt{2 + x_n}$

[IN GATE 2007]

15. A calculator has accuracy up to 8 digits after decimal place. The value of $\int_0^{2\pi} \sin x\,dx$ when evaluated using this calculator by trapezoidal method with 8 equal intervals, to 5 significant digits is

(a) 0.00000 (b) 1.0000

(c) 0.00500 (d) 0.00025

[ME GATE 2007]

16. Match **List-l** with **List-ll** and select the correct combination:

List-l

P. Second order differential equations
Q. Non-linear algebraic equations
R. Linear algebraic equations
S. Numerical integration

List-ll

1. Runge–Kutta method
2. Newton–Raphson method
3. Gauss elimination
4. Simpson's rule

Codes:

(a) P-3, Q-2, R-4, S-1
(b) P-2, Q-4, R-3, S-1
(c) P-1, Q-2, R-3, S-4
(d) P-1, Q-3, R-2, S-4

[PI GATE 2007]

17. Equation $e^x - 1 = 0$ is required to be solved using Newton's method with an initial guess $x_0 = -1$. Then, after one step of Newton's method, estimate x_1 of the solution will be given by

(a) 0.71828 (b) 0.36784

(c) 0.20587 (d) 0.00000

[EE GATE 2008]

18. It is known that two roots of the non-linear equation $x^3 - 6x^2 + 11x - 6 = 0$ are 1 and 3. The third root will be

(a) j (b) $-j$

(c) 2 (d) 4

[IN GATE 2008]

19. The Newton–Raphson iteration $x_{n+1} = \frac{1}{2}\left(x_n + \frac{R}{x_n}\right)$ can be used to compute the

(a) square of R (b) reciprocal of R

(c) square root of R (d) logarithm of R

[CS GATE 2008]

20. The recursion relation to solve $x = e^{-x}$ using the Newton–Raphson method is

(a) $x_{n+1} = e^{-x_n}$

(b) $x_{n+1} = x_n - e^{-x_n}$

(c) $x_{n+1} = (1 + x_n)\frac{e^{-x_n}}{1 + e^{-x_n}}$

(d) $x_{n+1} = \frac{x_n^2 - e^{-x_n}(1 + x_n) - 1}{x_n - e^{-x_n}}$

[EC GATE 2008]

21. The minimum number of equal length subintervals needed to approximate $\int_1^2 xe^x dx$ to an accuracy of at least $\frac{1}{3} \times 10^{-6}$ using the trapezoidal rule is

(a) $1000e$ (b) 1000

(c) $100e$ (d) 100

[CS GATE 2008]

22. A differential equation $\frac{dx}{dt} = e^{-2t}u(t)$ has to be solved using trapezoidal rule of integration with a step size $h = 0.01$ sec. Function $u(t)$ indicates a unit step function. If $x(0) = 0$, then the value of "x" at $t = 0.01$ sec will be given by

(a) 0.00099 (b) 0.00495

(c) 0.0099 (d) 0.0198

[EE GATE 2008]

23. Let $x^2 - 117 = 0$. The iterative steps for the solution using the Newton–Raphson method is given by

(a) $x_{k+1} = \dfrac{1}{2}\left(x_k + \dfrac{117}{x_k}\right)$

(b) $x_{k+1} = x_k - \dfrac{117}{x_k}$

(c) $x_{k+1} = x_k - \dfrac{x_k}{117}$

(d) $x_{k+1} = x_k - \dfrac{1}{2}\left(x_k + \dfrac{117}{x_k}\right)$

[EE GATE-2009]

24. During the numerical solution of a first-order differential equation using the Euler (also known as Euler-Cauchy) method with step size h, the local truncation error is of the order of

(a) h^2 (b) h^3 (c) h^4 (d) h^5

[PI GATE 2009]

25. The table below gives values of a function $F(x)$ obtained for values of x at intervals of 0.25:

x	0	0.25	0.50	0.75	1
$F(x)$	1	0.9412	0.8	0.64	0.5

The value of the integral of the function between the times 0 to 1 using Simpson's rule is

(a) 0.7854   (b) 2.3562

(c) 3.1416   (d) 7.5000

[CE GATE 2010]

26. The Newton–Raphson method is used to compute a root of the equation $x^2 - 13 = 0$ with 3.5 as the initial value. The approximation after one iteration is

(a) 3.575   (b) 3.677

(c) 3.667   (d) 3.607

[CS GATE 2010]

27. Consider a differential equation $\dfrac{dy}{dx} - y = x$ with the initial condition $y(0) = 0$. Using Euler's first-order method with a step size of 0.1, the value of $y(0.3)$ is

(a) 0.01   (b) 0.031

(c) 0.0631   (d) 0.1

[EC GATE-2010]

28. Torque exerted on a flywheel over a cycle is listed in the table. Flywheel energy (in J per unit cycle) using Simpson's 1/3rd rule is

Angle (degree)	0	60	120	180	240	300	360
Torque (Nm)	0	1066	–323	0	323	–355	0

(a) 542   (b) 993

(c) 1444   (d) 1986

[ME GATE 2010]

29. The square root of a number N is to be obtained by applying the Newton–Raphson iterations to the equation $x^2 - N = 0$. If i denotes the iteration index, the correct iterative scheme will be

(a) $x_{i+1} = \dfrac{1}{2}\left(x_i + \dfrac{N}{x_i}\right)$ (b) $x_{i+1} = \dfrac{1}{2}\left(x_i^2 + \dfrac{N}{x_i^2}\right)$

(c) $x_{i+1} = \dfrac{1}{2}\left(x_i + \dfrac{N^2}{x_i}\right)$ (d) $x_{i+1} = \dfrac{1}{2}\left(x_i + \dfrac{N}{x_i}\right)$

[CE GATE 2011]

30. Solution of the variables x_1 and x_2 for the following equations is to be obtained by employing the Newton–Raphson iterative method:

Equation (i) $10x_2 \sin x_1 - 0.8 = 0$

Equation (ii) $10x_2^2 - 10x_2 \cos x_1 - 0.6 = 0$

Assuming the initial values $x_1 = 0.0$ and $x_2 = 1.0$, the Jacobian matrix is

(a) $\begin{bmatrix} 10 & -0.8 \\ 0 & -0.6 \end{bmatrix}$ (b) $\begin{bmatrix} 10 & 0 \\ 0 & 10 \end{bmatrix}$

(c) $\begin{bmatrix} 0 & -0.8 \\ 10 & -0.6 \end{bmatrix}$ (d) $\begin{bmatrix} 10 & 0 \\ 10 & -10 \end{bmatrix}$

[EE GATE 2011]

31. A numerical solution of the equation $f(x) = x + \sqrt{x} - 3 = 0$ can be obtained using the Newton–Raphson method. If the starting value is $x = 2$ for the iteration, the value of x that is to be used in the next step is

(a) 0.306   (b) 0.739

(c) 1.694   (d) 2.306

[EC GATE 2011]

32. The matrix $[A] = \begin{bmatrix} 2 & 1 \\ 4 & -1 \end{bmatrix}$ is decomposed into a product of a lower triangular matrix L and an upper triangular matrix U. The properly decomposed L and U matrices, respectively, are

(a) $\begin{bmatrix} 1 & 0 \\ 4 & -1 \end{bmatrix}$ and $\begin{bmatrix} 1 & 1 \\ 0 & -2 \end{bmatrix}$

(b) $\begin{bmatrix} 2 & 0 \\ 4 & -1 \end{bmatrix}$ and $\begin{bmatrix} 1 & 1 \\ 0 & -1 \end{bmatrix}$

(c) $\begin{bmatrix} 1 & 0 \\ 4 & 1 \end{bmatrix}$ and $\begin{bmatrix} 2 & 1 \\ 0 & -1 \end{bmatrix}$

(d) $\begin{bmatrix} 2 & 0 \\ 4 & -3 \end{bmatrix}$ and $\begin{bmatrix} 1 & 0.5 \\ 0 & 1 \end{bmatrix}$

[EE GATE 2011]

33. The integral $\int_1^3 \frac{1}{x} dx$, when evaluated by using Simpson's $1/3^{rd}$ rule on two equal subintervals each of length 1, equals

(a) 1.000 (b) 1.098
(c) 1.111 (d) 1.120

[ME GATE 2011]

34. The estimate of $\int_{0.5}^{1.5} \frac{dx}{x}$ obtained using Simpson's rule with three-point function evaluation exceeds the exact value by

(a) 0.235 (b) 0.068
(c) 0.024 (d) 0.012

[CE GATE 2012]

35. The bisection method is applied to compute a zero of the function $f(x) = x^4 - x^3 - x^2 - 4$ in the interval [1, 9]. The method converges to a solution after _____ iterations.

(a) 1 (b) 3
(c) 5 (d) 7

[CS GATE 2012]

36. Match the correct pairs:

Numerical Integration scheme	Order of Fitting polynomial
P. Simpson's 3/8 Rule	1. First
Q. Trapezoidal Rule	2. Second
R. Simpson's 1/3 Rule	3. Third

(a) P-2, Q-1, R-3 (b) P-3, Q-2, R-1
(c) P-1, Q-2, R-3 (d) P-3, Q-1, R-2

[ME GATE 2013]

37. When the Newton–Raphson method is applied to solve the equation $f(x) = x^3 + 2x - 1 = 0$, the solution at the end of the first iteration with the initial guess value as $x_0 = 1.2$ is

(a) −0.82 (b) 0.49
(c) 0.705 (d) 1.69

[EE GATE 2013]

38. Find the magnitude of the error (correct to two decimal places) in the estimation of the Integral $\int_0^4 (x^4 + 10) dx$ using Simpson $1/3^{rd}$ rule [Take the step length as 1]

[CE GATE 2013]

39. While numerically solving the differential equation $\frac{dy}{dx} + 2xy^2 = 0, y(0) = 1$ using Euler's predictor-corrector (improved Euler-Cauchy) with a step size of 0.2, the value of y after the first step is

(a) 1.00 (b) 1.03
(c) 0.97 (d) 0.96

[IN GATE 2013]

40. Function f is known at the following points:

x	$f(x)$
0	0
0.3	0.09
0.6	0.36
0.9	0.81
1.2	1.44
1.5	2.25
1.8	3.24
2.1	4.41
2.4	5.76
2.7	7.29
3.0	9.00

The value of $\int_0^3 f(x) dx$ computed using the trapezoidal rule is

(a) 8.983 (b) 9.003
(c) 9.017 (d) 9.045

[CS GATE 2013]

41. Match the application to appropriate numerical method.

Application:
P1: Numerical integration
P2: Solution to a transcendental equation
P3: Solution to a system of linear equations
P4: Solution to a differential equation

Numerical method:
M1: Newton–Raphson Method
M2: Runge–Kutta Method
M3: Simpson's 1/3rd rule
M4: Gauss Elimination Method

(a) P1— M3, P2—M2, P3—M4, P4—M1
(b) P1— M3, P2—M1, P3—M4, P4—M2
(c) P1— M4, P2—M1, P3—M3, P4—M2
(d) P1— M2, P2—M1, P3—M3, P4—M4

[EC GATE 2014]

42. The definite integral $\int_1^3 \frac{1}{x} dx$ is evaluated using trapezoidal rule with a step size of 1. The correct answer is _____.

[ME GATE 2014]

43. The function $f(x) = e^x - 1$ is to be solved using the Newton–Raphson method. If the initial value of x_0 is taken as 1.0, then the absolute error observed at second iteration is _____.

[EE GATE 2014]

44. The real root of the equation $5x - 2cosx - 1 = 0$ (up to two decimal accuracy) is _____.

[ME GATE 2014]

45. The value of $\int_{2.5}^{4} In(x)dx$ calculated using the trapezoidal rule with five subintervals is _____.

[ME GATE 2014]

46. In the Newton–Raphson method, an initial guess of $x_0 = 2$ is made and the sequence $x_0, x_1, x_2.....$ is obtained for the function $0.75x^3 - 2x^2 - 2x + 4 = 0$.

Consider the statements

(i) $x_3 = 0$

(ii) The method converges to a solution in a finite number of iterations.

Which of the following is TRUE?

(a) Only (i)   (b) Only (ii)

(c) Both (i) and (ii)   (d) Neither (i) nor (ii)

[CS GATE 2014]

47. Consider an ordinary differential equation $\frac{dx}{dt} = 4t + 4$. If $x = x_0$ at $t = 0$, the increment in x calculated using the Runge–Kutta fourth-order multi-step method with a step size of $\Delta t = 0.2$ is

(a) 0.22   (b) 0.44

(c) 0.66   (d) 0.88

[ME GATE 2014]

48. Using the trapezoidal rule, and dividing the interval of integration into three equal sub-intervals, the definite integral $\int_{-1}^{+1}|x|\,dx$ is _____.

[ME GATE 2014]

49. With respect to the numerical evaluation of the definite integral $K = \int_{a}^{b}x^2dx$, where a and b are given, which of the following statements is/are TRUE?

(i) The value of K obtained using the trapezoidal rule is always greater than or equal to the exact value of the definite integral.

(ii) The value of K obtained using the Simpson's rule is always equal to the exact value of the definite integral

(a) (i) only   (b) (ii) only

(c) Both (i) and (ii)   (d) Neither (i) nor (ii)

[CS GATE 2014]

50. A non-zero polynomial $f(x)$ of degree 3 has roots $x = 1$, $x = 2$, and $x = 3$. Which one of the following must be true?

(a) $f(0)f(4) < 0$ (b) $f(0)\,f(4) > 0$

(c) $f(0) + f(4) > 0$ (d) $f(0) + f(4) < 0$

[CS GATE 2014]

51. If the equation $\sin x = x^2$ is solved by Newton–Raphson's method with the initial guess of $x = 1$, then the value of x after second iteration would be ___?

[PI GATE 2014]

52. The iteration step in order to solve for the cube roots of a given number "N" using the Newton–Raphson method is

(a) $x_{k+1} = x_k + \frac{1}{3}\left(N - x_k^3\right)$

(b) $x_{k+1} = \frac{1}{3}\left(2x_k + \frac{N}{x_k^2}\right)$

(c) $x_{k+1} = x_k - \frac{1}{3}\left(N - x_k^3\right)$

(d) $x_{k+1} = \frac{1}{3}\left(2x_k - \frac{N}{x_k^2}\right)$

[IN GATE 2014]

53. In the LU decomposition of the matrix $\begin{bmatrix} 2 & 2 \\ 4 & 9 \end{bmatrix}$, if the diagonal elements of U are both 1, then the lower diagonal entry l_{22} of L is _____.

[CS GATE 2015]

54. The Newton–Raphson method is used to find the roots of the equation, $x^3 + 2x^2 + 3x - 1 = 0$. If the initial guess is $x_0 = 1$, then the value of x after second iteration is _____.

[ME GATE 2015]

55. The quadratic equation $x^2 - 4x + 4 = 0$ is to be solved numerically, starting with the initial guess $x_0 = 3$. The Newton–Raphson method is applied once to get a new estimate and then the Secant method is applied once using the initial guess and this new estimate. The estimated value of the root after the application of the Secant method is _____.

[CE GATE 2015]

56. The Newton–Raphson method is used to solve the equation $x^3 - 5x^2 + 6x - 8 = 0$. Taking the initial guess as $x = 5$, the solution obtained at the end of the first iteration is_____.

[EC GATE 2015]

57. In the Newton–Raphson iterative method, the initial guess value (x_{ini}) is considered as zero while finding the roots of the equation: $f(x) = -2 + 6x - 4x^2 + 0.5x^3$. The correction, Δx, to be added to x_{ini} in the first iteration is _____.

[CE GATE 2015]

58. The values of function $f(x)$ at 5 discrete points are given below:

x	0	0.1	0.2	0.3	0.4
$y = f(x)$	0	10	40	90	160

Using trapezoidal rule step size of 0.1, the value of $\int_0^{0.4} f(x)dx$ is _____.

[ME GATE 2015]

59. Using a unit step size, the value of integral $\int_1^2 x \ln x \, dx$ by trapezoidal rule is _____.

[ME GATE 2015]

60. Simpson's $1/3^{rd}$ rule is used to integrate the function $f(x) = \frac{3}{5}x^2 + \frac{9}{5}$ between $x = 0$ and $x = 1$ using the least number of equal subintervals. The value of the integral is _____.

[ME GATE 2015]

61. For step-size, $\Delta x = 0.4$, the value of following integral using Simpson's $1/3^{rd}$ rule is _____

$$\int_0^{0.8} \left(0.2 + 25x - 200x^2 + 675x^3 - 900x^4 + 400x^5\right)dx$$

[CE GATE 2015]

62. In numerical integration using Simpson's $1/3^{rd}$ rule, the approximating function in the interval is a

(a) constant (b) straight line
(c) cubic B-spline (d) parabola

[PI GATE 2015]

63. The velocity v in kilometer/minute) of a motorbike which starts from rest is given at fixed intervals of time t (in minutes) as follows:

t	v
2	10
4	18
6	25
6	29
10	32
12	20
14	11
16	5
18	2
20	0

The approximate distance (in kilometers) rounded to two places of decimals covered in 20 minutes using Simpson's $1/3^{rd}$ rule is _____.

[CS GATE 2015]

64. Solve the equation $x = 10 \cos(x)$ using the Newton–Raphson method. The initial guess is $x = \frac{\pi}{4}$. The value of the predicted root after the first iteration, up to second decimal, is _____.

[ME GATE 2016]

65. The ordinary differential equation $\frac{dx}{dt} = -3x + 2$, with $x(0) = 1$ is to be solved using the forward Euler method. The largest time step that can be used to solve the equation without making the numerical solution unstable is _____.

[EC GATE 2016].

66. Numerical integration using trapezoidal rule gives the best result for a single variable function, which is

(a) linear (b) parabolic
(c) logarithmic (d) hyperbolic

[ME GATE 2016]

67. Consider the first-order initial value problem $\frac{dy}{dx} = y + 2x - x^2, y(0) = 1 (0 \le x < \infty)$

with exact solution $y(x) = x^2 + e^x$. For $x = 0.1$, the percentage difference between the exact solution and the solution obtained using a single iteration of the second-order Runge–Kutta method with step-size $h = 0.1$ is _____.

[EC GATE 2016]

68. The root of the function $f(x) = x^3 + x - 1$ obtained after first iteration on application of the Newton–Raphson scheme using an initial guess of $x_0 = 1$ is

(a) 0.682 (b) 0.686
(c) 0.750 (d) 1.000

[ME GATE 2016]

69. The error in numerically computing the integral $\int_0^{\pi} (\sin x + \cos x)dx$ using the trapezoidal rule with three intevals of equal length between 0 and π is _____.

[ME GATE 2016]

70. The Gauss Seidel method is used to solve the following equations (as per the given order):

$x_1 + 2x_2 + 3x_3 = 5$
$2x_1 + 3x_2 + x_3 = 1$
$3x_1 + 2x_2 + x_3 = 3$

Assuming initial guess as $x_1 = x_2 = x_3 = 0$, the value of x_3 after the first iteration is _____ .

[ME GATE 2016]

71. The Newton–Raphson method is to be used to find foot of equation $3x - e^x + \sin x = 0$. If the initial trial value of the roots is taken as 0.333, the next approximation for the root would be _____.

[CE GATE 2016]

72. Only one of the real roots of $f(x) = x^6 - x - 1$ lies in the interval $1 \le x \le 2$ and bisection method is used to find its value. For achieving an accuracy of 0.001, the required minimum number of iterations is _____. (Give the answer up to two decimal places.)

[EE GATE 2017]

73. $P(0, 3)$, $Q(0.5, 4)$, and $R(1, 5)$ are three points on the curve defined by $f(x)$. Numerical integration is carried out using the both trapezoidal rule and Simpson's rule within limits $x = 0$ and $x = 1$ for the curve. The difference between the two results will be

(a) 0   (b) 0.25
(c) 0.5   (d) 1

[ME GATE 2017]

74. Starting with $x = 1$, the solution of the equation $x^3 + x = 1$, after two iterations the Newton–Raphson's method (up to two decimal places) is _____.

[EC GATE 2017]

75. Consider the equation $\frac{du}{dt} = 3t^2 + 1$ with $u = 0$ at $t = 0$. This is numerically solved by using the forward Euler method with a step size $\Delta t = 2$. The absolute error in the solution in the end of the first time step is _____.

[CE GATE 2017]

76. The quadratic equation $2x^2 - 3x + 3 = 0$ is to be solved numerically starting with an initial guess as $x_0 = 2$. The new estimate of x after the first iteration using the Newton–Raphson method is _____.

[CE GATE 2017]

77. In order to evaluate the integral $\int_0^1 e^x dx$ with Simpson's $1/3^{\text{rd}}$ rule, values of the function e^x are used at $x = 0$, 0.5 and 1.0. The absolute value of the error of numerical integration is

(a) 0.000171   (b) 0.000440
(c) 0.000579   (d) 0.002718

[ME GATE 2018]

Answer key				
1. (c)	**2.** (c)	**3.** (i)-(c); (ii)-(b)		**4.** (c)
5. (c)	**6.** (b)	**7.** (b)	**8.** (a)	**9.** (c)
10. (a)	**11.** (a)	**12.** (b)	**13.** (a)	
14. (a)	**15.** (a)	**16.** (c)	**17.** (a)	**18.** (c)
19. (c)	**20.** (c)	**21.** (a)	**22.** (c)	**23.** (a)
24. (a)	**25.** (a)	**26.** (d)	**27.** (b)	
28. (b)	**29.** (a)	**30.** (b)	**31.** (c)	**32.** (d)
33. (c)	**34.** (d)	**35.** (b)	**36.** (d)	**37.** (c)
38. 0.53	**39.** (d)	**40.** (d)	**41.** (b)	**42.** 1.166
43. 0.06	**44.** 0.54	**45.** 1.753.	**46.** (a)	**47.** (d)
48. 1.11	**49.** (c)	**50.** (a)	**51.** 0.88	**52.** (b)
53. 5	**54.** 0.3043	**55.** 2.33	**56.** 4.29	**57.** 0.33
58. 22	**59.** 0.693	**60.** 2	**61.** 1.367	
62. (d)	**63.** **309.33**	**64.** 1.56	**65.** 0.66	**66.** (a)
67. 0.05	**68.** (c)	**69.** 0.186	**70.** –6	
71. 0.3601	**72.** 10	**73.** (a)	**74.** 0.686	
75. 8	**76.** 1	**77.** (c)		

Explanation

1. (c) The Trapezoidal rule gives exact result for a polynomial of degree ≤ 1.

2. (c)

3. (i)-(c); (ii)-(b)

(i) Let $x = \frac{1}{a}$. Then $\frac{1}{x} = a$ $i.e; \frac{1}{x} - a = 0$.

Take $f(x) = \frac{1}{x} - a$. Then $f'(x) = -\frac{1}{x^2}$.

$\therefore f(x_k) = \frac{1}{x_k} - a$ and $f'(x_k) = -\frac{1}{x_k^2}$.

Now by the Newton–Raphson iteration formula,

$x_{k+1} = x_k - \dfrac{f(x_k)}{f'(x_k)}$

or, $x_{k+1} = x_k - \dfrac{(1/x_k - a)}{-\dfrac{1}{x_k^2}}$

or, $x_{k+1} = 2x_k - ax_k^2$

(ii) For $a = 7$ and $x_0 = 0.2$, the iteration formula, becomes $x_{k+1} = 2x_k - 7x_k^2$

$\therefore$ $x_1 = 2x_0 - 7x_0^2 = 2 \times 0.2 - 7(0.2)^2$
$= 0.12$ and $x_2 = 2x_1 - 7x_1^2 = 2 \times 0.12 - 7(0.12)^2$
$= 0.1392$.

4. (c)

5. (c) Given $f(x) = x^3 + 3x - 7$ and $x_0 = 1$.

$\therefore f'(x) = 3x^2 + 3$. Then,
$f(x_0) = f(1) = 1^3 + 3 \times 1 - 7 = -3$ and
$f'(x_0) = f'(1) = 3 \times 1^2 + 3 = 6$.

Now the Newton–Raphson iteration formula gives,

$$x_1 = x_0 - \frac{f(x_0)}{f'(x_0)}$$

or, $x_1 = 1 - \left(\frac{-3}{6}\right) \times 1 = 1.5$.

6. (b)

7. (b) Take $f(x) = xe^x - 2$. Then $f'(x) = xe^x + e^x$.

Given that $x_1 = 0.8679$.

Now the Newton–Raphson iteration formula gives,

$$x_1 = x_0 - \frac{f(x_0)}{f'(x_0)}$$

or, $x_1 = 0.8679 - \left(\dfrac{0.8679 \times e^{0.8679} - 2}{0.8679 \times e^{0.8679} + e^{0.8679}}\right)$

or, $x_1 = 0.8679 - \left(\dfrac{0.8679 \times 2.3819 - 2}{0.8679 \times 2.3819 + e^{0.8679}}\right)$

or $x_1 = 0.853$.

8. (a) Given that $f(0) = 1$, $f(1) = 4$, $f(2) = 15$. Here, $h = 1$.

So by the trapezoidal formula,

$$\int_0^2 f(x)\,dx$$

$$\approx \frac{h}{2}(f(0) + 2f(1) + f(2)) = \frac{1}{2}(1 + 8 + 15) = 12.$$

Since $f(x)$ is second-degree polynomial, let
$f(x) = a_0 + a_1 x + a_2 x^2$.
Then, $f(0) = 1 \Rightarrow a_0 + 0 + 0 = 1 \Rightarrow a_0 = 1$.
$f(1) = 4 \Rightarrow a_0 + a_1 + a_2 = 4$
$\Rightarrow 1 + a_1 + a_2 = 4 \Rightarrow a_1 + a_2 = 3$ ...(i)
$f(2) = 15 \Rightarrow a_0 + a_1 + 4a_2 = 15$
$\Rightarrow 1 + 2a_1 + 4a_2 = 15$
$\Rightarrow a_1 + 2a_2 = 7$ ...(ii)
Solving (i) and (ii), we get, $a_1 = -1$ and $a_2 = 4$.
Thus, $f(x) = 1 - x + 4x^2$.

Now, $\int_0^2 f(x)\,dx$

$$= \int_0^2 (1 - x + 4x^2)\,dx = \left[x - \frac{x^2}{2} + \frac{4x^3}{3}\right]_0^2 = \frac{32}{3}.$$

Hence, error
= Exact – Approximate value $= \dfrac{32}{3} - 12 = -\dfrac{4}{3}$.

9. (c) Given, $\dfrac{dy}{dx} = 0.25y^2$, $y_0 = y(0) = 1$ and $h = 1$.

Iterative equation for the backward (implicit) Euler methods is given by
$y_{k+1} = y_k + hf(x_{k+1}, y_{k+1})$, where $\dfrac{dy}{dx} = f(x, y)$.
Then we have,

$y_{k+1} = y_k + 1 \times 0.25y_{k+1}^2$

or, $0.25y_{k+1}^2 - y_{k+1} + y_k = 0$ ...(i)

Now putting $k = 0$ in (i), we get,

$0.25y_1^2 - y_1 + y_0 = 0$

or, $0.25y_1^2 - y_1 + 1 = 0$

or, $y_1 = \dfrac{1 \pm \sqrt{1 - 4 \times 0.25}}{2 \times 0.25} = 2$.

$\Rightarrow y_1 = 2$

10. (a) Here $f(x) = x^3 + 4x - 9$.

$\therefore f'(x) = 3x^2 + 4$, $f(x_k) = x_k^3 + 4x_k - 9$ and

$f'(x_k) = 3x_k^2 + 4$.

Now iteration formula for the Newton–Raphson method gives,

$$x_{k+1} = x_k - \frac{f(x_k)}{f'(x_k)}$$

or, $x_{k+1} = x_k - \dfrac{x_k^3 + 4x_k - 9}{3x_k^2 + 4}$

$= \dfrac{(3x_k^3 + 4x_k) - (x_k^3 + 4x_k - 9)}{3x_k^2 + 4} = \dfrac{2x_k^3 + 9}{3x_k^2 + 4}$.

11. (a)

Since $x = 5$ is a root, so $x - 5$ is a factor.

$\therefore x^3 - 10x^2 + 31x - 30 = 0$

$\Rightarrow x^2(x - 5) - 5x(x - 5) + 6(x - 5) = 0$

$\Rightarrow (x - 5)(x^2 - 5x + 6) = 0$

$\Rightarrow (x - 5)(x - 2)(x - 3) = 0$

$\Rightarrow x = 5, 2, 3$.

Hence, the other roots of $x^2 - 5x + 6 = 0$ are 2 and 3.

12. (b) Here, $x_0 = 2$, $f(x) = x^3 - x^2 + 4x - 4$.

$\therefore f'(x) = 3x^2 - 2x + 4$,
$f(x_0) = f(2) = 2^3 - 2^2 + 4 \times 2 - 4 = 8$,
$f'(x_0) = f'(2) = 3 \times 2^2 - 2 \times 2 + 4 = 12$.
Now, the Newton–Raphson iteration formula

gives $x_1 = x_0 - \dfrac{f(x_0)}{f'(x_0)}$

or, $x_1 = 2 - \dfrac{8}{12} = \dfrac{4}{3}$.

13. (a)

Given, $x_{n+1} = \dfrac{x_n}{2} + \dfrac{9}{8x_n}, x_0 = 0.5$

When the series converges, we have,
$x_{n+1} = x_n = \alpha =$ root of equation.

Then, $x_{n+1} = \dfrac{x_n}{2} + \dfrac{9}{8x_n}$

$\Rightarrow \alpha = \dfrac{\alpha}{2} + \dfrac{9}{8\alpha} \Rightarrow 8\alpha^2 = 4\alpha^2 + 9$

$\Rightarrow \alpha^2 = \dfrac{9}{4} \Rightarrow \alpha = \dfrac{3}{2} = 1.5$.

14. (a) Let $x = \sqrt{2}$. Then $x^2 - 2 = 0$.

Now let $f(x) = x^2 - 2$. Then $f'(x) = 2x$.
Now iteration formula for the Newton–Raphson method gives,

$x_{n+1} = x_n - \dfrac{f(x_n)}{f'(x_n)}$

or, $x_{n+1} = x_n - \dfrac{x_n^2 - 2}{2x_n} = \dfrac{1}{2}\left\{x_n + \dfrac{2}{x_n}\right\}$.

15. (a) Here, $h = \dfrac{2\pi - 0}{8} = \dfrac{\pi}{4}$ and $y = f(x) = \sin x$.

i	x	$y = \sin x$
0	0	$y_0 = \sin 0 = 0$
1	$\dfrac{\pi}{4}$	$y_1 = \sin\dfrac{\pi}{4} = \dfrac{1}{\sqrt{2}} = 0.7071$
2	$\dfrac{\pi}{2}$	$y_2 = \sin\dfrac{\pi}{2} = 1$
3	$\dfrac{3\pi}{4}$	$y_3 = \sin\dfrac{3\pi}{4} = \sin\left(\pi - \dfrac{\pi}{4}\right)$ $= \sin\dfrac{\pi}{4} = \dfrac{1}{\sqrt{2}} = 0.7071$
4	ϖ	$y_4 = \sin \pi = 0$
5	$\dfrac{5\pi}{4}$	$y_5 = \sin\dfrac{5\pi}{4} = \sin\left(\pi + \dfrac{\pi}{4}\right)$ $= -\sin\dfrac{\pi}{4} = -\dfrac{1}{\sqrt{2}} = -0.7071$
6	$\dfrac{6\pi}{4}$	$y_6 = \sin\dfrac{6\pi}{4} = \sin\dfrac{3\pi}{2}$ $= \sin\left(\pi + \dfrac{\pi}{2}\right) = -\sin\dfrac{\pi}{2} = -1$
7	$\dfrac{7\pi}{4}$	$y_7 = \sin\dfrac{7\pi}{4} = \sin\left(2\pi - \dfrac{\pi}{4}\right)$ $= -\sin\dfrac{\pi}{4} = -\dfrac{1}{\sqrt{2}} = -0.7071$
8	2ϖ	$y_8 = \sin 2\pi = 0$

By the trapezoidal rule, we have,

$\displaystyle\int_{x_0}^{x_8} f(x)dx$

$= \dfrac{h}{2}\left[(y_0 + y_8 + 2(y_1 + y_2 + ... + y_7)\right]$

or, $\displaystyle\int_0^{2\pi} \sin x \, dx$

$= \dfrac{\pi}{8}[2(0.70710 + 1 + 0.70710 - 0.70710 - 1 - 0.70710)]$

$= 0.00000$

16. (c)

17. (a) Here, $f(x) = e^x - 1$. So $f'(x) = e^x$.

∴ $f(x_0) = f(-1) = e^{-1} - 1, f'(x_0) = f'(-1) = e^{-1}$.
Now, the Newton–Raphson iteration formula gives,

$x_1 = x_0 - \dfrac{f(x_0)}{f'(x_0)}$

or, $x_1 = -1 - \dfrac{e^{-1} - 1}{e^{-1}} = \dfrac{-2e^{-1} + 1}{e^{-1}} = -2 + e$

$= -2 + 5.436563 = 0.71828$

18. (c) Here product of the roots
$= -$(constant term)/coefficient of $x^3 = 6/1 = 6$.
If α be the third root, then we have,
$\alpha \times 1 \times 3 = 6 \ i.e; \ \alpha = 2$.

19. (c) Given that the Newton–Raphson iteration formula is

$x_{n+1} = \dfrac{1}{2}\left(x_n + \dfrac{R}{x_n}\right)$ ...(i)

If the above formula converges to the root after n iterations, then $x_{n+1} = x_n = \alpha$(say).
Then (i) gives,

$\alpha = \dfrac{1}{2}\left(\alpha + \dfrac{R}{\alpha}\right)$

or, $2\alpha = \dfrac{\alpha^2 + R}{\alpha}$

or, $2\alpha^2 = \alpha^2 + R$

or, $\alpha^2 = R$, i.e., $\alpha = \sqrt{R}$.

20. (c) Here $f(x) = x - e^{-x}$. So $f'(x) = 1 + e^{-x}$.

Now iteration formula for the Newton–Raphson method gives,

$$x_{n+1} = x_n - \frac{f(x_n)}{f'(x_n)}$$

or, $x_{n+1} = x_n - \dfrac{x_n - e^{-x_n}}{1 + e^{-x_n}} = \dfrac{e^{-x_n}(1 + x_n)}{1 + e^{-x_n}}.$

21. (a)

Here, $f(x) = xe^x$,

$f'(x) = xe^x + e^x = e^x(x + 1)$,

$f''(x) = xe^x + e^x + e^x = e^x(x + 2)$.

Since, both e^x and x are increasing functions of x, maximum value of $f''(t)$ in interval $1 \le t \le 2$, occurs at $t = 2$.

So, $\max |f''(t)| = e^2(2 + 2) = 4e^2$.

Now the truncation error for the trapezoidal rule

$$= \frac{nh^3}{12} \max_{1 \le t \le 2} |f''(t)|$$

(where n is number of subintervals of equal length)

$$= \frac{h^2}{12}(b - a) \max_{1 \le t \le 2} |f''(t)| \quad \left[\because n = \frac{b - a}{h} \right]$$

$$= \frac{h^2}{12}(2 - 1)(4e^2) = \frac{h^2}{3}e^2$$

Now ATQ, $\dfrac{h^2}{3}e^2 = \dfrac{1}{3} \times 10^{-6}$

$$\Rightarrow h^2 = \frac{10^{-6}}{e^2} \Rightarrow h = \frac{10^{-3}}{e}$$

So minimum number of subintervals of equal length

$$= n = \frac{b - a}{h} = \frac{2 - 1}{(10^{-3}/e)} = 1000e.$$

22. (c)

$$\frac{dx}{dt} = e^{-2t}u(t)$$

$$\Rightarrow dx = e^{-2t}u(t)\,dt \Rightarrow x = \int_{t=0}^{0.01} e^{-2t}u(t)\,dt$$

$$\Rightarrow x = \int_{t=0}^{0.01} f(t)\,dt, \text{where } f(t) = e^{-2t}u(t)$$

$$= \frac{h}{2}\big[f(0) + f(0.01)\big] \text{ (by trapezoidal rule)}$$

$$= \frac{0.01}{2}\Big[e^0 u(0) + e^{-2 \times 0.01}u(0)\Big]$$

$$= 0.005 \times \big[1 + e^{-0.02}\big] \quad \left[\because u(t) = \begin{cases} 0, t < 0 \\ 1, t \ge 0 \end{cases} \right]$$

$$= 0.0099$$

23. (a) Here $f(x) = x^2 - 117$. So $f'(x) = 2x$.

Now, the iteration formula for the Newton–Raphson method gives,

$$x_{k+1} = x_k - \frac{f(x_k)}{f'(x_k)}$$

or, $x_{k+1} = x_k - \dfrac{x_k^2 - 117}{2x_k} = \dfrac{x_k^2 + 117}{2x_k} = \dfrac{1}{2}\left(x_k + \dfrac{117}{x_k} \right).$

24. (a)

25. (a) Using Simpson's $1/3^{\text{rd}}$ formula for $n = 4$, we have,

$$\int_0^1 F(x)\,dx$$

$$= \frac{h}{3}\big[(y_0 + y_4) + 4(y_1 + y_3) + 2y_4\big]$$

$$= \frac{0.25}{3}[(1 + 0.5) + 4 \times (0.9412 + 0.64) + 2 \times 0.8]$$

$$= \frac{0.25}{3} \times 9.4248 = 0.7854$$

$$= 0.7854$$

26. (d)

Here, $f(x) = x^2 - 13$ and so $f'(x) = 2x$.

$\therefore f(x_0) = f(3.5) = 3.5^2 - 13 = -0.75,$

$f'(x_0) = f'(3.5) = 2 \times 3.5 = 7.$

The Newton–Raphson iteration formula gives

$$x_1 = x_0 - \frac{f(x_0)}{f'(x_0)}$$

$$or,\ x_1 = 3.5 - \left[\frac{-0.75}{7} \right] = 3.607$$

27. (b) Comparing $\dfrac{dy}{dx} = f(x, y)$ with $\dfrac{dy}{dx} = x + y$,

we get, $f(x, y) = x + y$.

Here, step size, $h = 0.1$, $x_0 = 0$, $y_0 = 0$,

$x_1 = x_0 + h = 0 + 0.1 = 0.1$

$x_2 = x_0 + 2h = 0 + 2 \times 0.1 = 0.2$,

$x_3 = x_0 + 3h = 0 + 3 \times 0.1 = 0.3$.

Euler's first-order formula is given by

$y = y_i + hf(x_i\ y_i)$

$\therefore y_1 = y_0 + hf(x_0, y_0) = 0 + 0.1 \times (0 + 0) = 0,$

$y_2 = y_1 + hf(x_1, y_1) = 0 + 0.1 \times (0.1 + 0) = 0.01,$

$y_3 = y_2 + hf(x_2, y_2) = 0.01 + 0.1 \times (0.2 + 0.01)$
$= 0.031.$

28. (b) If $T(\theta)$ represents the torque exerted, then flywheel energy

$$= \int_0^{2\pi} T(\theta)d\theta$$

$$= \frac{h}{3}\Big[(y_0 + y_6) + 4(y_1 + y_3 + y_5) + 2(y_2 + y_4)\Big]$$

(using Simpson's 1/3rd rule)

$$= \frac{\frac{\pi}{3}}{3}\Big[(0+0) + 4(1066 + 0 - 355) + 2(-323 + 323)\Big]$$

[since $h = 60$ degrees $= \dfrac{\pi}{3}$ radians]

$$= 993.14 = 993 \text{ (approx)}$$

29. (a) Here, $f(x) = x_i^2 - N$ and so $f'(x) = 2x_i$.

The Newton–Raphson iteration formula gives

$$x_{i+1} = x_i - \frac{f(x_i)}{f'(x_i)}$$

or, $x_{i+1} = x_i - \left(\dfrac{x_i^2 - N}{2x_i}\right) = \dfrac{x_i^2 + N}{2x_i} = \dfrac{1}{2}\left[x_i + \dfrac{N}{x_i}\right].$

30. (b) Let $u(x_1, x_2) = 10x_2 \sin x_1 - 0.8$ and

$v(x_1, x_2) = 10x_2^2 - 10x_2 \cos x_1 - 0.6$.

Then the Jacobian matrix is

$$\begin{bmatrix} \dfrac{\partial u}{\partial x_1} & \dfrac{\partial u}{\partial x_2} \\ \dfrac{\partial v}{\partial x_1} & \dfrac{\partial v}{\partial x_2} \end{bmatrix}$$

$$= \begin{bmatrix} 10x_2 \cos x_1 & 10 \sin x_1 \\ 10x_2 \sin x_1 & 20x_2 - 10 \cos x_1 \end{bmatrix}$$

Therefore, at $x_1 = 0.0$ and $x_2 = 1.0$, the Jacobian matrix is

$$= \begin{bmatrix} 10 \cos 0 & 10 \sin 0 \\ 10 \sin 0 & 20 - 10 \cos 0 \end{bmatrix} = \begin{bmatrix} 10 & 0 \\ 0 & 10 \end{bmatrix}.$$

31. (c)

Here, $f(x) = x + \sqrt{x} - 3$ and so $f'(x) = 1 + \dfrac{1}{2\sqrt{x}}.$

$\therefore f(x_0) = f(2) = 2 + \sqrt{2} - 3 = \sqrt{2} - 1,$

$f'(x_0) = f'(2) = 1 + \dfrac{1}{2\sqrt{2}}.$

Now, the Newton–Raphson iteration formula gives

$$x_1 = x_0 - \frac{f(x_0)}{f'(x_0)}$$

or, $x_1 = 2 - \dfrac{\sqrt{2} - 1}{1 + \dfrac{1}{2\sqrt{2}}} = 2 - \dfrac{2\sqrt{2}(\sqrt{2} - 1)}{2\sqrt{2} + 1}$

$$= \frac{6\sqrt{2} - 2}{2\sqrt{2} + 1} = 1.694.$$

32. (d) Let us take $u_{11} = 1$ and $u_{22} = 1$.

Then, $\begin{bmatrix} 2 & 1 \\ 4 & -1 \end{bmatrix} = LU$

$\Rightarrow \begin{bmatrix} 2 & 1 \\ 4 & -1 \end{bmatrix} = \begin{bmatrix} l_{11} & 0 \\ l_{21} & l_{22} \end{bmatrix}\begin{bmatrix} 1 & u_{12} \\ 0 & 1 \end{bmatrix}$

$\Rightarrow \begin{bmatrix} 2 & 1 \\ 4 & -1 \end{bmatrix} = \begin{bmatrix} l_{11} & l_{11}u_{12} \\ l_{21} & l_{21}u_{12} + l_{22} \end{bmatrix}$

$\Rightarrow l_{11} = 2, l_{11}u_{12} = 1, l_{21} = 4, l_{21}u_{12} + l_{22} = -1$

$\Rightarrow l_{11} = 2, u_{12} = 0.5, l_{21} = 4, l_{22} = -3.$

Thus, $\begin{bmatrix} 2 & 1 \\ 4 & -1 \end{bmatrix} = \begin{bmatrix} 2 & 0 \\ 4 & -3 \end{bmatrix}\begin{bmatrix} 1 & 0.5 \\ 0 & 1 \end{bmatrix}$

33. (c)

x	1	2	3
$y = \dfrac{1}{x}$	$1(= y_0)$	$0.5(= y_1)$	$0.3333(= y_2)$

By Simpson's 1/3rd rule, given integral

$$= \frac{h}{3}(y_0 + 4y_1 + y_2)$$

$$= \frac{1}{3}(1 + 4 \times 0.5 + 0.3333) = 1.1111$$

34. (d)

$$\int_{0.5}^{1.5} \frac{dx}{x} = \big[\log x\big]_{0.5}^{1.5} = \log(1.5) - \log(0.5) = 1.0986.$$

Thus, exact value of the integral $= 1.0986$.

x	0.5	1	1.5
$y = \dfrac{1}{x}$	$2(= y_0)$	$1(= y_1)$	$0.666(= y_2)$

Now using Simpson's 1/3rd rule, we have the approximate value of the given integral

$$= \frac{h}{3}(y_0 + 4y_1 + y_2)$$

$$= \frac{0.5}{3}(2 + 4 \times 1 + 0.666) = 1.111.$$

Therefore, approximate value – Exact value
$= 1.1111 - 1.0986$
$= 0.012499 \approx 0.012$

So the estimate exceeds the exact value by 0.012.

35. (b)

n	a_n (+ve)	b_n (−ve)	x_{n+1} $=\dfrac{a_n+b_n}{2}$	$f(x_{n+1})$
0	9	1	$x_1 = 5$	471(+ve)
1	5	1	$x_2 = 3$	41(+ve)
2	3	1	$x_3 = 2$	0

Thus, the method converges exactly to the root in three iterations.

36. (d)

37. (c) Here, $f(x) = x^3 + 2x - 1. \therefore f'(x) = 3x^2 + 2.$

So, $f(x_0) = f(1.2) = (1.2)^3 + 2 \times 1.2 - 1 = 3.128,$

$f'(x_0) = f'(1.2) = 3(1.2)^2 + 2 = 6.32.$

Now the Newton–Raphson iteration formula gives

$$x_1 = x_0 - \frac{f(x_0)}{f'(x_0)}$$

or, $x_1 = 1.2 - \dfrac{3.128}{6.32} = 0.705$

38. 0.53

x	0	1	2	3	4
y $= x^4 + 10$	10 (y_0)	11 (y_1)	26 (y_2)	91 (y_3)	266 (y_4)

Using Simpson's rule, the estimated value of the integral $\int_0^4 (x^4 + 10)\,dx$

$$= \frac{h}{3}\left[(y_0 + y_4) + 4(y_1 + y_3) + 2y_2\right]$$

$$= \frac{1}{3}\left[(10 + 266) + 4 \times (11 + 91) + 2 \times 26\right] = 245.33.$$

The exact value of integral $\int_0^4 (x^4 + 10)\,dx$

$$= \left[\frac{x^5}{5} + 10x\right]_0^4 = \frac{4^5}{5} + 10 \times 4 = 244.8$$

$\therefore$ Magnitude of error
= |exact value − estimated value|
= $|244.8 - 245.33| = 0.53$

39. (d)

$$\frac{dy}{dx} + 2xy^2 = 0 \Rightarrow \frac{dy}{dx} = -2xy^2.$$

$\therefore f(x, y) = -2xy^2.$

$y(0) = 1 \Rightarrow x_0 = 0, y_0 = 1.$

After one iteration,

$y_1^P = y_0 + hf(x_0, y_0)$

or, $y_1^P = 1 + 0.2\left[-2 \times 0 \times 1^2\right] = 1.$

$y_1^C = y_0 + \dfrac{h}{2}\left[f(x_0, y_0) + f(x_0, y_1^P)\right]$

or, $y_1^C = 1 + \dfrac{0.2}{2} \times \left[0 - 2 \times 0.2 \times 1^2\right] = 0.96.$

40. (d) Clearly $h = 0.3$.

Using the trapezoidal rule, the value of the integral $\int_0^3 f(x)\,dx$

$$= \frac{h}{3}\left[(y_0 + y_{10}) + 2(y_1 + y_2 + y_3 + y_4 + y_5 + y_6 + y_7 + y_8 + y_9)\right]$$

$$= \frac{0.3}{3}\Big[(0 + 9) + 2(0.09 + 0.36 + 0.81 + 1.44 + 2.25$$

$$+3.24 + 4.41 + 5.76 + 7.29)\Big]$$

$$= 9.045$$

41. (b)

42. 1.166

x	1	2	3
$y = \dfrac{1}{x}$	$1(= y_0)$	$0.5(= y_1)$	$0.3333(= y_2)$

By the trapezoidal rule,

$$\int_1^3 \frac{1}{x}\,dx$$

$$= \frac{h}{2}\left[(y_0 + y_2) + 2y_1\right]$$

$$= \frac{1}{2}\left[1 + 0.333 + 2 \times 0.5\right] = \frac{2.333}{2} = 1.166$$

43. 0.06

Here, $f(x) = e^x - 1$ and so $f'(x) = e^x$.
Given that $x_0 = 1$.
In the Newton–Raphson method, iteration formula is

$$x_{k+1} = x_k - \frac{f(x_k)}{f'(x_k)}$$

$\therefore x_1 = x_0 - \dfrac{f(x_0)}{f'(x_0)} = 1 - \dfrac{e^1 - 1}{e^1} = e^{-1} = 0.367.$

Now, $x_2 = x_1 - \dfrac{f(x_1)}{f'(x_1)} = 0.367 - \dfrac{e^{0.367} - 1}{e^{0.367}}$

$$= \frac{0.5297 - 1.443 + 1}{1.443} = 0.06$$

Therefore, the absolute error observed at the second iteration = 0.06.

44. 0.54

Here, $f(x) = 5x - 2\cos x - 1,$

$f'(x) = 5 + 2\sin x$.

Let us take $x_0 = 1$ (in radian).

The Newton–Raphson's iteration formula is

$$x_{n+1} = x_n - \frac{f(x_n)}{f'(x_n)}$$

$$\therefore x_1 = x_0 - \frac{f(x_0)}{f'(x_0)} = 1 - \frac{5 \times 1 - 2\cos 1 - 1}{5 + 2\sin 1}$$

$$= 1 - \frac{4 - 2 \times 0.5403}{5 + 2 \times 0.8414} = 0.5632,$$

Again,

$$x_2 = x_1 - \frac{f(x_1)}{f'(x_1)}$$

$$= 0.5632 - \frac{5 \times 0.5632 - 2\cos(0.5632) - 1}{5 + 2\sin(0.5632)}$$

$$= 0.5632 - \frac{2.816 - 2 \times 0.84555 - 1}{5 + 2 \times 0.53389} = 0.5426,$$

$$\therefore x_3 = x_2 - \frac{f(x_2)}{f'(x_2)}$$

$$= 0.5426 - \frac{5 \times 0.5426 - 2\cos(0.5426) - 1}{5 + 2\sin(0.5426)}$$

$$= 0.5426 - \frac{2.713 - 2 \times 0.856369 - 1}{5 + 2 \times 0.53389} = 0.5425.$$

$\therefore$ Real root (up to two decimal accuracy) = 0.54.

45. 1.753.

Here, $h = \dfrac{b-a}{n} = \dfrac{4-2.5}{5} = 0.3$

x	$y = f(x) = ln(x)$
2.5	0.916 (= y_0)
2.8	1.030 (= y_1)
3.1	1.131 (= y_2)
3.4	1.224 (= y_3)
3.7	1.308 (= y_4)
4.0	1.386 (= y_5)

By the trapezoidal rule,

$$I = \int_{2.5}^{4} ln(x)\,dx$$

$$= \frac{h}{2}\left[(y_0 + y_5) + 2(y_1 + y_2 + y_3 + y_4)\right]$$

$$= \frac{0.3}{2}[(0.916 + 1.386) + 2(1.0296 + 1.131$$

$$+ 1.224 + 1.308$$

$$= \frac{0.3}{2} \times 11.688 = 1.753$$

46. (a) Here, $f(x) = 0.75x^3 - 2x^2 - 2x + 4$, $f'(x) = 2.25x^2 - 4x - 2$.

The Newton–Raphson's iteration formula is

$$x_{n+1} = x_n - \frac{f(x_n)}{f'(x_n)}$$

$$\therefore x_1 = x_0 - \frac{f(x_0)}{f'(x_0)} = 2 - \frac{f(2)}{f'(2)}$$

$$= 2 - \frac{0.75 \times 8 - 2 \times 4 - 4 + 4}{2.25 \times 4 - 8 - 2} = 0,$$

$$x_2 = x_1 - \frac{f(x_1)}{f'(x_1)}$$

$$= 0 - \frac{0.75 \times 0 - 2 \times 0 - 0 + 4}{2.25 \times 0 - 0 - 2} = 2,$$

$$x_3 = x_2 - \frac{f(x_2)}{f'(x_2)}$$

$$= 2 - \frac{0.75 \times 8 - 2 \times 4 - 4 + 4}{2.25 \times 4 - 8 - 2} = 0,$$

$$x_4 = x_3 - \frac{f(x_3)}{f'(x_3)}$$

$$= 0 - \frac{0.75 \times 0 - 2 \times 0 - 0 + 4}{2.25 \times 0 - 0 - 2} = 2.$$

Thus, $x_0 = 2$, $x_1 = 0$, $x_2 = 2$, $x_3 = 0$, $x_4 = 2$, ...
So, $x_3 = 0$ is correct but it converges in an infinite number of iterations.

47. (d)

Given, $\dfrac{dx}{dt} = 4t + 4,$ at $t = 0$, $x = x_0$.

So, $f(t, x) = 4t + 4$.

By the Runge–Kutta fourth order multistep method,

$$x_1 = x_0 + k = x_0 + \frac{1}{6}(k_1 + 2k_2 + 2k_3 + k_4), \text{ where,}$$

$$k_1 = hf(t_0, x_0) = 0.2(0 + 4) = 0.8,$$

$$k_2 = hf\left(t_0 + \frac{1}{2}h, x_0 + \frac{k_1}{2}\right) = 0.2f\left(0 + \frac{0.2}{2}, x_0 + \frac{0.8}{2}\right)$$

$$= 0.2f(0.1, x_0 + 0.4) = 0.2 \times (4 \times 0.1 + 4) = 0.88,$$

$$k_3 = hf\left(t_0 + \frac{1}{2}h, x_0 + \frac{k_2}{2}\right)$$

$$= 0.2f\left(0 + \frac{0.2}{2}, x_0 + \frac{0.88}{2}\right) = 0.2f(0.1, x_0 + 0.44)$$

$$= 0.2(4 \times 0.1 + 4) = 0.88,$$

$$k_4 = hf(t_0 + h, x_0 + k_3)$$
$$= 0.2f(0.2, x_0 + 0.88) = 0.2(4 \times 0.2 + 4) = 0.96.$$

$$\therefore k = \frac{1}{6}(k_1 + 2k_2 + 2k_3 + k_4)$$

$= \dfrac{1}{6}(0.8 + 2 \times 0.88 + 2 \times 0.88 + 0.96) = 0.88.$

Hence, the increment in $x = x_1 - x_0 = k = 0.88$

48. 1.11

$h = \dfrac{b-a}{n} = \dfrac{1-(-1)}{3} = \dfrac{2}{3}.$

x	-1	$-\dfrac{1}{3}$	$\dfrac{1}{3}$	1
$y = \|x\|$	$1 \ (= y_0)$	$\dfrac{1}{3} \ (= y_1)$	$\dfrac{1}{3} \ (= y_2)$	$1 \ (= y_3)$

By the trapezoidal rule, given integral

$= \dfrac{h}{2}\left[(y_0 + y_3) + 2(y_1 + y_2)\right]$

$= \dfrac{\frac{2}{3}}{2}\left[(1+1) + 2\left(\dfrac{1}{3} + \dfrac{1}{3}\right)\right] = \dfrac{10}{9} = 1.11$

49. (c) **Case-I:** When we use the trapezoidal rule:

Error $= -\dfrac{nh^3}{12} f''(\xi).$

Here, $f(x) = x^2$, so $f''(x) = 2$ which is always positive. So the sign of the error is always negative. Hence, approximate value always greater than or equal the exact value of the integral.

Case-II: When we use Simpson's 1/3rd rule:

Error $= -\dfrac{nh^5}{90} f^{(iv)}(\xi).$

Here, $f(x) = x^2$, so $f^{(iv)}(x) = 0$. So the error is always "0" which means that the approximate value is always equal to the exact value of the integral.

50. (a) Let $f(x) = k(x-1)(x-2)(x-3), k \neq 0.$
Then $f(0) f(4)$
$= k(0-1)(0-2)(0-3) \times k(4-1)(4-2)$
$(4-3) = -36K^2 < 0$

51. 0.88(approx)

Here, $f(x) = x^2 - \sin x, f'(x) = 2x - \cos x.$
Given $x_0 = 1.$
The Newton Raphson's iteration formula is

$x_{n+1} = x_n - \dfrac{f(x_n)}{f'(x_n)}$

$\therefore x_1 = x_0 - \dfrac{f(x_0)}{f'(x_0)}$

$= 1 - \dfrac{1^2 - \sin 1}{2 \times 1 - \cos 1} = 1 - \dfrac{1 - 0.84147}{2 - 0.54030} = 0.89139,$

$x_2 = x_1 - \dfrac{f(x_1)}{f'(x_1)}$

$= 0.89139 - \dfrac{0.89139^2 - \sin(0.89139)}{2 \times 0.89139 - \cos(0.89139)}$

$= 0.89139 - \dfrac{0.79457 - 0.77794}{1.78278 - 0.62833} = 0.87698.$

52. (b) Here, $f(x) = x^3 - N, f'(x) = 3x^2.$
The Newton Raphson's iteration formula gives:

$x_{k+1} = x_k - \dfrac{f(x_n)}{f'(x_n)}$

or, $x_{k+1} = x_k - \dfrac{x_k^3 - N}{3x_k^2} = \dfrac{1}{3}\left(2x_k + \dfrac{N}{x_k^2}\right).$

53. 5

$\begin{bmatrix} l_{11} & 0 \\ l_{21} & l_{22} \end{bmatrix} \begin{bmatrix} 1 & u_{12} \\ 0 & 1 \end{bmatrix} = \begin{bmatrix} 2 & 2 \\ 4 & 9 \end{bmatrix}$

$\Rightarrow \begin{bmatrix} l_{11} & l_{11}u_{12} \\ l_{21} & l_{21}u_{12} + l_{22} \end{bmatrix} = \begin{bmatrix} 2 & 2 \\ 4 & 9 \end{bmatrix}$

$\Rightarrow l_{11} = 2, l_{11}u_{12} = 2, l_{21} = 4, l_{21}u_{12} + l_{22} = 9$

$\Rightarrow l_{11} = 2, u_{12} = 1, l_{21} = 4, l_{22} = 5.$

54. 0.3043

Here,
$f(x) = x^3 + 2x^2 + 3x - 1,$
$f'(x) = 3x^2 + 4x + 3,$
$x_0 = 1.$
The Newton–Raphson iteration formula gives:

$x_{n+1} = x_n - \dfrac{f(x_n)}{f'(x_n)}$

$\therefore x_1 = x_0 - \dfrac{f(x_0)}{f'(x_0)} = 1 - \dfrac{1+2+3-1}{3+4+3} = 0.5,$

$x_2 = x_1 - \dfrac{f(x_1)}{f'(x_1)}$

$= 0.5 - \dfrac{0.5^3 + 2 \times 0.5^2 + 3 \times 0.5 - 1}{3 \times 0.5^2 + 4 \times 0.5 + 3} = 0.3043$

55. 2.33

Here,
$f(x) = x^2 - 4x + 4, f'(x) = 2x - 4, x_0 = 3.$
$\therefore \ f(3) = 1 - 4 + 4 = 1, f'(3) = 2 \times 3 - 4 = 2.$
By The Newton–Raphson method, we have

$x_1 = x_0 - \dfrac{f(x_0)}{f'(x_0)} = 3 - \dfrac{f(3)}{f'(3)} = 3 - \dfrac{1}{2} = \dfrac{5}{2}.$

Then, $f(x_1) = f\left(\dfrac{5}{2}\right) = \left(\dfrac{5}{2}\right)^2 - 4 \times \dfrac{5}{2} + 4 = \dfrac{1}{4}.$

Now, by the Secant method,

$$x_2 = x_1 - \frac{(x_1 - x_0)f(x_1)}{f(x_1) - f(x_0)}$$

$$= \frac{5}{2} - \frac{\left(\frac{5}{2} - 3\right) \times \frac{1}{4}}{\frac{1}{4} - 1} = \frac{7}{3} = 2.33$$

56. 4.29

Here,

$$f(x) = x^3 - 5x^2 + 6x - 8,$$

$$f'(x) = 3x^2 - 10x + 6,$$

$$x_0 = 5.$$

$$\therefore f(5) = 5^3 - 5 \times 25 + 6 \times 5 - 8 = 22,$$

$$f'(5) = 3 \times 25 - 10 \times 5 + 6 = 31.$$

Now the Newton–Raphson iteration formula gives,

$$x_1 = x_0 - \frac{f(x_0)}{f'(x_0)} = 5 - \frac{f(5)}{f'(5)} = 5 - \frac{22}{31} = 4.29$$

57. 0.33 Here,

$$f(x) = -2 + 6x - 4x^2 + 0.5x^3,$$
$$f'(x) = 6 - 8x + 1.5x^2,$$
$$x_{ini} = 0.$$

By the Newton Raphson method,

$$x_1 = x_{ini} - \frac{f(x_{ini})}{f'(x_{ini})} = 0 - \frac{(-2 + 0 - 0 + 0)}{(6 - 0 - 0)} = 0.33.$$

$$\therefore \quad \Delta x = x_1 - x_{ini} = 0.33 - 0 = 0.33$$

58. 22

$$\int_0^{0.4} f(x)\,dx$$

$$= \frac{h}{2}\left[y_0 + y_4 + 2(y_1 + y_2 + y_3)\right]$$

$$= \frac{0.1}{2}\left[0 + 160 + 2(10 + 40 + 90)\right] = 22.$$

59. 0.693

x	1	2
$f(x) = x \ln x$	0	$2 \ln 2$

By the trapezoidal rule, $\int_1^2 x \ln x\, dx$

$$= \frac{h}{2}\left[y_0 + y_n\right] = \frac{1}{2}\left[0 + 2 \ln 2\right] = \ln 2 = 0.693$$

60. 2

x	1	0.5	1
$f(x) = \frac{3}{5}x^2 + \frac{9}{5}$	$\frac{9}{5}$	$\frac{39}{20}$	$\frac{12}{5}$

By Simpson's 1/3rd rule, $\int_0^1 f(x)$

$$= \frac{h}{3}\left[y_0 + y_2 + 4y_1\right] = \frac{0.5}{3}\left[\frac{9}{5} + \frac{12}{5} + 4 \times \frac{39}{20}\right] = 2$$

61. 1.367

x	0	0.4	0.8
$f(x)$ $= 0.2 + 25x - 200x^2 + 675x^3$ $- 900x^4 + 400x^5$	0.2	2.456	0.232

By Simpson's 1/3rd Rule, given integral

$$= \frac{h}{3}\left[y_0 + 4y_1 + y_2\right]$$

$$= \frac{0.4}{3}(0.2 + 4 \times 2.456 + 0.232) = 1.367$$

62. (*d*)

63. 309.33 km

t	v
0	0 ($= y_0$)
2	10 ($= y_1$)
4	18 ($= y_2$)
6	25 ($= y_3$)
6	29 ($= y_4$)
10	32 ($= y_5$)
12	20 ($= y_6$)
14	11 ($= y_7$)
16	5 ($= y_8$)
18	2 ($= y_9$)
20	0 ($= y_{10}$)

By Simpson's 1/3rd rule, distance (in kilometers)

$$= \int_0^{20} v\,dt$$

$$= \frac{h}{3}\left[(y_0 + y_{10}) + 4(y_1 + y_3 + y_5 + y_7 + y_9)\right.$$
$$\left. + 2(y_2 + y_4 + y_6 + y_8)\right]$$

$$= \frac{2}{3}\left[(0 + 0) + 4(10 + 25 + 32 + 11 + 2)\right.$$
$$\left. + 2(18 + 29 + 20 + 5)\right]$$

$$= 309.33 \; km$$

64. 1.56

Here,

$$f(x) = x - 10\cos x,$$

$$f'(x) = 1 + 10\sin x,$$

$$x_0 = \frac{\pi}{4}.$$

$$\therefore f(x_0) = f\left(\frac{\pi}{4}\right) = \frac{\pi}{4} - 10\cos\frac{\pi}{4} = -6.2856,$$

$$f'(x) = 1 + 10\sin\frac{\pi}{4} = 8.0711.$$

Now, by the Newton–Raphson iteration formula,

$$x_1 = x_0 - \frac{f(x_0)}{f'(x_0)} = \frac{\pi}{4} - \left(\frac{-6.2856}{8.0711}\right)$$

$$= 0.7854 + 0.7788 = 1.5642 \approx 1.56$$

65. 0.66

Given, $\dfrac{dx}{dt} = -3x + 2, \quad y(0) = 1 \cdot$

The solution of differential equations is stable if $|1 - 3h| < 1$

i.e; if $-1 < 1 - 3h < 1$

i.e; if $-2 < -3h < 0$

i.e; if $0 < h < \dfrac{2}{3}.$

Hence, $h_{max} = \dfrac{2}{3} = 0.66$

66. (a)The trapezoidal rule gives the best result when the function is linear, i.e, of degree 1.

67. 0.05

Here, $f(x, y) = y + 2x - x^2, x_0 = 0, y_0 = 1, h = 0.1.$
By the second-order Runge–Kutta method, we have,

$$y_1 = y_0 + \frac{1}{2}(k_1 + k_2) \text{ where}$$

$$k_1 = hf(x_0, y_0) = hf(0,1)$$

$$= 0.1(1 + 2 \times 0 - 0^2) = 0.1,$$

$$k_2 = hf(x_0 + h, y_0 + k_1) = hf(0 + 0.1, 1 + 0.1)$$

$$= 0.1\{(1 + 0.1) + 2 \times 0.1 - (0.1)^2\} = 0.129.$$

$$\therefore y_1 = y_0 + \frac{1}{2}(k_1 + k_2)$$

$$= 1 + \frac{1}{2}(0.1 + 0.129) = 1.1145$$

Again, the exact solution

$$= y(0.1) = (0.1)^2 + e^{0.1} = 1.1151$$

Hence, error

= exact value-approximate value

$$= 1.1151 - 1.1145 = 0.0006.$$

$$\therefore \% \ error = \frac{error}{exact \ value} \times 100\%$$

$$= \frac{0.0006}{1.1151} \times 100\% = 0.05\%$$

68. (c)

Here, $f(x) = x^3 + x - 1, f'(x) = 3x^2 + 1, x_0 = 1.$

$$\therefore f(x_0) = f(1) = 1 + 1 - 1 = 1,$$

$$f'(x_0) = f'(1) = 3 + 1 = 4.$$

Then by the Newton–Raphson iteration formula,

$$x_1 = x_0 - \frac{f(x_0)}{f'(x_0)} = 1 - \frac{1}{4} = 1 - 0.25 = 0.75.$$

69. 0.186

x	0	$\dfrac{\pi}{3}$	$\dfrac{2\pi}{3}$	π
$f(x)$ $= \sin x + \cos x$	$1(= y_0)$	$\dfrac{\sqrt{3}+1}{2}$ $(= y_1)$	$\dfrac{\sqrt{3}-1}{2}$ $(= y_2)$	$-1(= y_3)$

By the trapezoidal rule, $\displaystyle\int_0^\pi (\sin x + \cos x)\,dx$

$$= \frac{h}{2}\{y_0 + y_3 + 2(y_1 + y_2)\}$$

$$= \frac{\pi/3}{2}\left\{1 + (-1) + 2\left(\frac{\sqrt{3}+1}{2} + \frac{\sqrt{3}-1}{2}\right)\right\}$$

$$= 1.813799$$

Again, $\displaystyle\int_0^\pi (\sin x + \cos x)\,dx$

$$= \left[(-\cos x + \sin x)\right]_0^\pi = -\cos\pi + \cos 0 = 2.$$

Then error

= Exact value – approx. value

$$= 2 - 1.813799 = 0.186$$

70. –6

ATQ, $(x_1)_0 = (x_2)_0 = (x_3)_0 = 0.$
According to the Gauss Seidel method, the first iterations are given by

$$(x_1)_1 = 5 - 2(x_2)_0 - 3(x_3)_0 = 5 - 0 - 0 = 5,$$

$$(x_2)_1 = \frac{1}{3}\left[1 - 2(x_1)_1 - (x_3)_0\right]$$

$$= \frac{1}{3}\left[1 - 2 \times 5 - 0\right] = -3,$$

$$(x_3)_1 = 3 - 3(x_1)_1 - 2(x_2)_1 = 3 - 3 \times 5 + 2 \times 3 = -6.$$

71. 0.3601

Here,

$f(x) = 3x - e^x + \sin x$,

$f'(x) = 3 - e^x + \cos x, x_0 = 0.333$.

Now, the Newton–Raphson iteration formula gives:

$$x_1 = x_0 - \frac{f(x_0)}{f'(x_0)}$$

or, $x_1 = x_0 - \dfrac{f(0.333)}{f'(0.333)}$

$$= 0.333 - \frac{3 \times 0.333 - e^{0.333} + \sin(0.333)}{3 - e^{0.333} + \cos(0.333)}$$

$$= 0.333 + \frac{0.0693}{2.5499} = 0.3601$$

72. 10

Here, $f(x) = x^6 - x - 1$, $a = 1, b = 2, \varepsilon = 0.001$
$= 10^{-3}$.

If n be the number of iterations by the Bisection method then,

$$\frac{|b-a|}{2^n} < \varepsilon$$

or, $\dfrac{2-1}{2^n} < 10^{-3}$

or, $2^n > 10^3$

or, $ln\ 2^n > ln\ 10^3$

or, $n > \dfrac{3 ln\ 10}{ln\ 2} = \dfrac{3 \times 2.30258}{0.693147}$

or, $n > 9.96$

Hence, the minimum number of iterations = 10.

73. (a) Here, $h = 0.5$ and $y_0 = 3, y_1 = 4, y_2 = 5$

By the trapezoidal rule, $\displaystyle\int_0^1 f(x)dx$

$$= \frac{h}{2}\left[(y_0 + y_2) + 2y_1\right] = \frac{0.5}{2}[(3+5) + 2 \times 4] = 4.$$

By the Simpson's rule, $\displaystyle\int_0^1 f(x)dx$

$$= \frac{h}{3}\left[(y_0 + y_2) + 4y_1\right] = \frac{0.5}{3}[(3+5) + 4 \times 4] = 4.$$

Thus, the difference between these two results will be zero.

74. 0.686

Here, $f(x) = x^3 + x - 1, f'(x) = 3x^2 + 1, x_0 = 1$.

∴ $f(1) = 1 + 1 - 1 = 1, f'(1) = 3 + 1 = 4$.

By the Newton–Raphson method,

$$x_1 = x_0 - \frac{f(x_0)}{f'(x_0)} = 1 - \frac{1}{4} = 0.75,$$

$$x_2 = x_1 - \frac{f(x_1)}{f'(x_1)} = 0.75 - \frac{f(0.75)}{f'(0.75)}$$

$$= 0.75 - \frac{0.75^3 + 0.75 - 1}{3 \times 0.75^2 + 1}$$

$$= 0.75 - \frac{0.171875}{2.6875} = 0.686$$

75. 8

Given that, $\dfrac{du}{dt} = 3t^2 + 1; u_0 = 0, t_0 = 0; h = \Delta t = 2$.

∴ $f(u, t) = 3t^2 + 1$.

By the Euler's method

$u_1 = u_0 + hf(u_0, t_0)$

$= 0 + 2f(0, 0) = 2 \times (3 \times 0 + 1) = 2$.

Again, $t_1 = t_0 + h = 0 + 2 = 2$.

Thus after the first iteration $u = 2$ when $t = 2$.

Now, $\dfrac{du}{dt} = 3t^2 + 1$

$\Rightarrow du = (3t^2 + 1)dt \Rightarrow \int du = \int (3t^2 + 1)dt$

$\Rightarrow u = 3 \times \dfrac{t^3}{3} + t + C = t^3 + t + C$ ...(i)

Putting $u = 0$ and $t = 0$ in (i), we get, $C = 0$.

So, at $t = 2, u = 8 + 2 = 10$.

Hence, Absolute error

= Exact value – approx. value $= 10 - 2 = 8$.

76. 1

Here, $f(x) = 2x^2 - 3x + 3, f'(x) = 4x - 3, x_0 = 2$.

∴ $f(2) = 8 - 6 + 3 = 5, f'(1) = 8 - 3 = 5$.

By the Newton–Raphson method,

$$x_1 = x_0 - \frac{f(x_0)}{f'(x_0)} = 2 - \frac{5}{5} = 1.$$

77. (c)

x	0	0.5	1
$f(x)$	1	1.64872	2.71828

By the Simpson's $1/3^{rd}$ rule, $\displaystyle\int_0^1 e^x dx$

$$= \frac{h}{2}(y_0 + y_2 + 2y_1)$$

$$= \frac{0.5}{3}\{1 + 2.71828 + 4 \times 1.64872\}$$

$$= 1.71886$$

$$\int_0^1 e^x dx = \left[e^x\right]_0^1 = e - 1 = 1.71828$$

Hence, error
= exact value – approximate value
= 1.71828 – 1.71886 = –0.00058.
So absolute value of error = 0.00058.

Questions for Practice

1. The $(n+1)^{th}$ order forward difference of the nth degree polynomial is
(a) $n!$ (b) n
(c) 0 (d) $(n+1)!$

2. $(\Delta - \nabla)x^2$ is equal to (h being the spacing)
(a) h^2 (b) $-2h^2$
(c) $2h^2$ (d) $-2h$

3. Using the following table, the value of $f(1.1)$ is

x	0	1	2	3
$f(x)$	1	2	11	34

(a) 3.5 (b) 2.4
(c) 5.2 (d) 3.8

4. The Newton–Raphson method is to be used to find the root of the equation $f(x) = 0$ and $f'(x)$ is the derivative of $f(x)$. The method converges
(a) always
(b) only if f is a polynomial
(c) only if $f(x_0) < 0$
(d) none of the above

5. If a continuous function $f(x)$ does not have a root in the interval $[a, b]$, then which one of the following statements is true?
(a) $f(a)f(b) = 0$ (b) $f(a)f(b) < 0$
(c) $f(a)/f(b) \le 0$ (d) $f(a)f(b) > 0$

6. To solve the equation $2 \sin x = x$ by the Newton–Raphson method, the initial guess was chosen to be $x = 2.0$. Consider x in radian only. The value of x (in radian) obtained after one iteration will be closest to
(a) –8.101 (b) 1.901
(c) 2.099 (d) 12.101

7. The integral $\int_{x_1}^{x_2} x^2 \, dx$ with $x_2 > x_1 > 0$ is evaluated analytically as well as numerically using a single application of the trapezoidal rule. If I is the exact value of the integral obtained analytically and J is the approximate value obtained using the trapezoidal rule, which of the following statements is correct about their relationship?
(a) $J > I$ (b) $J < I$
(c) $J = I$ (d) data is insufficient

8. An explicit forward Euler method is used to numerically integrate the differential equation $\frac{dy}{dx} = y$ using a time step of 0.1. With the initial condition $y(0) = 1$, the value of $y(1)$ computed by this method is _____ (correct to two decimal places)
(a) 2.11 (b) 4.34
(c) 2.59 (d) 4.56

9. The Newton–Raphson method is used to find the root of the equation $x^2 - 2 = 0$. If the iterations are started from –1, then the iteration will
(a) converge to –1
(b) converge to $\sqrt{2}$
(c) converge to $-\sqrt{2}$
(d) not converge

10. Simpson's $1/3^{rd}$ rule for integration gives exact result when f(x) is a polynomial function of degree less than or equal to
(a) 1 (b) 2
(c) 3 (d) 4

11. The Newton–Raphson iteration formula for finding $\sqrt[3]{c}$, where $c > 0$ is
(a) $x_{n+1} = \dfrac{2x_n^3 + \sqrt[3]{c}}{3x_n^2}$ (b) $x_{n+1} = \dfrac{2x_n^3 - \sqrt[3]{c}}{3x_n^2}$
(c) $x_{n+1} = \dfrac{2x_n^3 + c}{3x_n^2}$ (d) $x_{n+1} = \dfrac{2x_n^3 - c}{3x_n^2}$

12. Given the differential equation $y' = x - y$ with initial condition $y(0) = 0$. The value of $y(0.1)$ calculated numerically up to the third place of decimal by the second order the Runge–Kutta method with step size $h = 0.1$ is
(a) 1/200 (b) 1/50
(c) 4/35 (d) 1/20

13. Using Simpson's $1/3^{rd}$ rule, the value of the integral $\int_0^6 \dfrac{dx}{(1+x)^2}$ taking six equal subintervals of $[0, 6]$ is

(a) 0.751   (b) 0.849
(c) 0.904   (d) 0.356

14. Using Simpson's $1/3^{rd}$ rule, if we want to find the approximate value of $\int_{12}^{24} f(x)\,dx$, then the error committed is

(a) $-\dfrac{1}{90} f^{iv}(\xi)$ (b) $-\dfrac{1}{15} f^{iv}(\xi)$

(c) $-\dfrac{2}{15} f^{iii}(\xi)$ (d) $-\dfrac{1}{15} f^{iii}(\xi)$

15. A system of equations $AX = b$ where $A = (a_{ij})_{n \times n}$ is said to be diagonally dominant if

(a) $|a_{ii}| > \displaystyle\sum_{j=1, j \neq i}^{n} |a_{ij}|$

(b) $|a_{ii}| < \displaystyle\sum_{j=1, j \neq i}^{n} |a_{ij}|$

(c) $|a_{ii}| > \displaystyle\sum_{j=1}^{n} |a_{ij}|$ (d) $|a_{ii}| < \displaystyle\sum_{j=1}^{n} |a_{ij}|$

16. The condition of convergence of the Newton–Raphson method when applied to an equation $f(x) = 0$ in an interval $[a, b]$ is

(a) $f'(x) \neq 0$

(b) $|f'(x)| < 1$

(c) $\{f'(x)\}^2 > |f''(x) f(x)|$

(d) none of these

17. The Runge–Kutta method is used to solve

(a) a first order ordinary differential equation

(b) a first order partial differential equation

(c) a second order ordinary differential equation

(d) a second order partial differential equation

18. Using the Newton–Raphson method, an approximate real root of $3x - \cos x - 1 = 0$ is (up to three decimal places)

(a) 0.607   (b) 0.809
(c) 0.004   (d) 1.560

19. Using Weddle's rule and taking 12 subintervals, the value of $\int_{0}^{2} \dfrac{1}{1+x^2}\,dx$ correct to four decimal places is

(a) 3.0234   (b) 3.0211
(c) 1.1071   (d) 1.2075

20. Using the Gauss–Seidel iteration method, the solution of the following system is (correct up to four significant figures)

$$3x_1 + 9x_2 - 2x_3 = 11$$
$$4x_1 + 2x_2 + 13x_3 = 24$$
$$4x_1 - 2x_2 + x_3 = -8$$

(a) $x_1 = 1.005, x_2 = 3.879, x_3 = -2.667$

(b) $x_1 = -1.240, x_2 = 2.002, x_3 = 4.123$

(c) $x_1 = 1.205, x_2 = 4.111, x_3 = 3.533$

(d) $x_1 = -1.423, x_2 = 2.131, x_3 = 1.956$

21. If the third order differences of $f(x)$ be constant and $f(-1) = -1, f(0) = 0, f(1) = 1, f(2) = 8$, and $f(3) = 27$, then the value of $f(4)$ is

(a) 15   (b) 64
(c) 45   (d) 45

22. The smallest degree of the polynomial that interpolates the data is

x	−2	−1	0	1	2	3
$f(x)$	−58	−21	−12	−13	−6	27

(a) 3   (b) 4
(c) 5   (d) 6

23. An iterative method of find the n th root $(n \in N)$ of a positive number a is given by $x_{k+1} = \dfrac{1}{2}\left[x_k + \dfrac{a}{x_k^{n-1}}\right]$. A value of n for which this iterative method fails to converge is

(a) 1   (b) 2
(c) 3   (d) 8

24. Solution the equation $x^3 - 9x + 1 = 0$ for the root lying between 2 and 3, correct to 2-significant figures, is

(a) 0.45   (b) 0.94
(c) 0.60   (d) 0.50

25. The following system of linear equations by the Gauss-elimination method has a solution

$$x - 2y + 9z = 8,$$
$$3x + y - z = 3,$$
$$2x - 8y + z = -5.$$

(a) $x = 1, y = 1, z = 1$ (b) $x = 1, y = 2, z = 3$

(c) $x = -1, y = 1, z = 0$ (d) $x = -2, y = -1, z = -3$

26. The following system of equations by LU factorization method has a solution

$2x - 6y + 8z = 24$

$5x + 4y - 8z = 2$

$3x + y + 2z = 16$

(a) $x = 1, y = -1, z = 2$

(b) $x = 3, y = 2, z = 1$

(c) $x = 1, y = 1, z = 2$ (d) $x = 1, y = 3, z = 5$

27. Given $\dfrac{dy}{dx} = x^3 + y, y(0) = 1.$ Then $y(0.02)$, by Euler's method correct up to two decimal places, taking step length $h = 0.01$ is

(a) 2.05   (b) 1.02

(c) 1.25   (d) 2.35

Answer Key				
1. (c)	**2.** (c)	**3.** (b)	**4.** (d)	**5.** (d)
6. (b)	**7.** (a)	**8.** (c)	**9.** (c)	**10.** (c)
11. (c)	**12.** (a)	**13.** (b)	**14.** (b)	**15.** (a)
16. (c)	**17.** (a)	**18.** (a)	**19.** (c)	**20.** (d)
21. (b)	**22.** (a)	**23.** (a)	**24.** (b)	**25.** (a)
26. (d)	**27.** (b)			

CHAPTER 8

COMPLEX ANALYSIS

8.1 BASICS OF COMPLEX ANALYSIS

8.1.1 Complex Number

A number of the form $x + iy$, where $i = \sqrt{-1}$, x and y being real numbers, is called a complex number and is denoted by "z." The real numbers x and y are called the real part of z and imaginary part of z, respectively. Symbolically, we write $z = x + iy$, $Re(z) = x$, $\text{Img}(z) = y$.

Example: $2 + 3i, -\sqrt{2} + \frac{1}{3}i, \sqrt{5}\,i$, etc., all are complex numbers.

Remember:

(*i*) Consider $z = x + iy$.

If $x = 0$, then $z = iy$ and so z is purely imaginary.

If $y = 0$, then $z = x$. In this case, z becomes a real number.

(*ii*) The form $z = x + iy$ is called the Cartesian form of the complex number "z." The complex number z can be regarded as an order pair (x, y).

Thus, every complex number $z = x + iy$ corresponds a point $P(x, y)$ in the xy-plane and vice versa. In the theory of complex numbers, xy-plane is called complex plane or argand plane or Gaussian plane.

Example:

(1) $z = 1 + 2i$ represent a point $(1, 2)$ in the argand plane.

(2) $z = -1 - i$ represents a point $(-1, -1)$ in the argand plane.

8.1.2 Modulus and Amplitude of a Complex Number

Let the complex number $z = x + iy$ corresponds the point $P(x, y)$. Join O(origin) to P. Let the line segment makes an angle θ with the positive direction of x-axis $\overline{OP}$ and $\overline{OP} = \pi$.

Then, $x = r \cos \theta$ and $y = r \sin \theta$ and so

$\tan \theta = y/x$.

Therefore, $z = x + iy = r(\cos \theta + i \sin \theta)$, which is called the polar form or modulus amplitude form of the complex number "z."

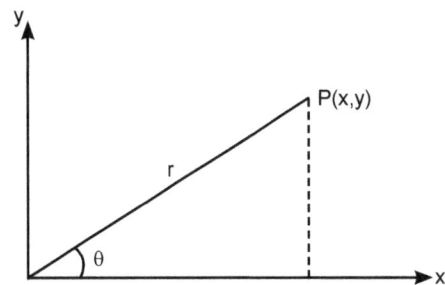

Here $\overline{OP} = r = \sqrt{x^2 + y^2}$ is called the modulus of z and is denoted by $|z|$.

Thus, $|z| = r = \sqrt{x^2 + y^2}$.

θ is called the amplitude (argument) of z and is denoted by amp (z) or arg (z).

Thus, $\text{amp}(z) = \theta = \tan^{-1}(y/x)$.

Remember:

(a) Let $z = r(\cos\theta + i\sin\theta)$........(1) be the polar form of $z = x + iy$

Then amp (z) $(= \theta)$ is not unique because of equation (1) remains unaltered if we replace θ by $\theta + 2\pi$. Thus, θ can take infinite number of values which differ from each other by 2π.

If a value of θ satisfies (i) and lies between $-\pi$ and; π, then that value of θ is called the principle value of the amplitude. In the polar form of the complex number, we need to consider the principle value of the amplitude.

Case-I:

The point (x, y) lies in the first quadrant. In this case, the polar form of z is

$z = r(\cos\theta + i\sin\theta)$, where

$r = \sqrt{x^2 + y^2}$ and $\theta = \tan^{-1}\left|\dfrac{y}{x}\right|$.

Example: Consider $z = 1 + i$ which represents the point $(1, 1)$ that lies in first quadrant.

Comparing $z = 1 + i$ with $z = x + iy$, we get $x =1, y = 1$.

Then, $r = \sqrt{x^2 + y^2} = \sqrt{1^2 + 1^2} = \sqrt{2}$

and $\theta = \tan^{-1}\left|\dfrac{y}{x}\right| = \tan^{-1}\left|\dfrac{1}{1}\right| = \tan^{-1}1 = \dfrac{\pi}{4}$.

Hence, the polar form of z is

$z = \sqrt{2}\left(\cos\dfrac{\pi}{4} + i\sin\dfrac{\pi}{4}\right)$.

Case-II: The point (x, y) lies in the second quadrant. In this case, the polar form of z is

$z = r(\cos\theta + i\sin\theta)$, where

$r = \sqrt{x^2 + y^2}$ and $\theta = \pi - \tan^{-1}\left|\dfrac{y}{x}\right|$.

Example: Consider $z = -1 + i$ which represents the point $(-1, 1)$ that lies in second quadrant.

Comparing $z = -1 + i$ with $z = x + iy$, we get $x = -1, y = 1$.

Then, $r = \sqrt{x^2 + y^2} = \sqrt{(-1)^2 + 1^2} = \sqrt{2}$ and

$\theta = \pi - \tan^{-1}\left|\dfrac{y}{x}\right| = \pi - \tan^{-1}\left|\dfrac{1}{-1}\right| = \pi - \tan^{-1}1$

$= \pi - \dfrac{\pi}{4} = \dfrac{3\pi}{4}$

Hence, the polar form of z is

$z = \sqrt{2}\left(\cos\dfrac{3\pi}{4} + i\sin\dfrac{3\pi}{4}\right)$.

Case-III:

The point (x, y) lies in the third quadrant. In this case, the polar form of z is

$z = r(\cos\theta + i\sin\theta)$, where

$r = \sqrt{x^2 + y^2}$ and $\theta = -\pi + \tan^{-1}\left|\dfrac{y}{x}\right|$.

Example: Consider $z = -1-i$ which represents the point $(-1, -1)$ that lies in third quadrant.

Comparing $z = -1-i$ with $z = x + iy$, we get $x = -1, y = -1$.

Then, $r = \sqrt{x^2 + y^2} = \sqrt{(-1)^2 + (-1)^2} = \sqrt{2}$ and

$\theta = -\pi + \tan^{-1}\left|\dfrac{y}{x}\right| = -\pi + \tan^{-1}\left|\dfrac{-1}{-1}\right|$

$= -\pi + \tan^{-1}1 = -\pi + \dfrac{\pi}{4} = -\dfrac{3\pi}{4}$

Hence, the polar form of z is

$z = \left\{\cos\left(-\dfrac{3\pi}{4}\right) + i\sin\left(-\dfrac{3\pi}{4}\right)\right\}$.

Case-IV: The point (x, y) lies in the fourth quadrant. In this case, the polar form of z is

$z = r(\cos\theta + i\sin\theta)$, where

$r = \sqrt{x^2 + y^2}$ and $\theta = -\tan^{-1}\left|\dfrac{y}{x}\right|$.

Example: Consider $z = 1-i$ which represents the point $(1, -1)$ that lies in fourth quadrant.

Comparing $z = 1 + i$ with $z = x + iy$, we get $x = 1, y = -1$.

Then, $r = \sqrt{x^2 + y^2} = \sqrt{1^2 + (-1)^2} = \sqrt{2}$ and

$\theta = -\tan^{-1}\left|\dfrac{y}{x}\right| = -\tan^{-1}\left|\dfrac{-1}{1}\right| = -\tan^{-1}1 = -\dfrac{\pi}{4}$.

Hence, the polar form of z is

$z = \sqrt{2}\left\{\cos\left(-\dfrac{\pi}{4}\right) + i\sin\left(-\dfrac{\pi}{4}\right)\right\}$.

(b) $z = r(\cos\theta + i\sin\theta)$ can be written as

$z = re^{i\theta}$ $\qquad [\because\ e^{i\theta} = \cos\theta + i\sin\theta]$ and

$z = r(\cos\theta - i\sin\theta)$ can be written as

$z = \pi e^{-i\theta}$ $\qquad [\because\ e^{-i\theta} = \cos\theta - i\sin\theta]$

8.1.3 Conjugate of a Complex Number

The conjugate of the complex number $z = x + iy$ is denoted by $\bar{z}$ and is defined by $\bar{z} = x - iy$.

Example:

The conjugate of $z = 1 - i$ is $\bar{z} = 1 + i$

Remember:

The conjugate of $z = re^{i\theta}$ is $\bar{z} = re^{-i\theta}$.

8.1.4 Properties of Modulus, Argument, and Conjugate

(i) $|z| = 0 \Leftrightarrow z = 0$

(ii) $|z_1 z_2| = |z_1||z_2|$ for any two complex numbers z_1 and z_2

(iii) $\left|\dfrac{z_1}{z_2}\right| = \dfrac{|z_1|}{|z_2|}$

(iv) $|z_1 + z_2| \le |z_1| + |z_2|$

(v) $\big||z_1| - |z_2|\big| \le |z_1 - z_2|$

(vi) $|z_1| \le |z_1 - z_2| + |z_2|$

(vii) $\arg(z_1 z_2) = \arg(z_1) + \arg(z_2)$

(viii) $\arg\left(\dfrac{z_1}{z_2}\right) = \arg(z_1) - \arg(z_2)$

(ix) $\arg(0)$ is not defined.

(x) Arguments of any positive real number is 0.

(xi) Arguments of any negative real number is $\pm \pi$

(xii) $\arg(z) - \arg(-z) = \pm \pi$

(xiii) $z = \bar{z}$ if and only if z is real.

(xiv) $z\bar{z} = (x + iy)(x - iy) = x^2 + y^2 = |z|^2$

(xv) $x = \dfrac{z + \bar{z}}{2}, y = \dfrac{z - \bar{z}}{2}$

(xvi) $\overline{z_1 \pm z_2} = \bar{z}_1 \pm \bar{z}_2$

8.1.5 Sum, Difference, and Product of Two Complex Numbers

If $z_1 = x_1 + iy_1$, $z_2 = x_2 + iy_2$ be two complex numbers, then

(i) The sum of z_1 and z_2 is denoted by $z_1 + z_2$ and is defined by $z_1 + z_2 = (x_1 + x_2) + i(y_1 + y_2)$.

(ii) The difference of z_1 from z_2 is denoted by $z_1 - z_2$ and is defined by $z_1 - z_2 = (x_1 - x_2) + i(y_1 - y_2)$.

(iii) The product of z_1 and z_2 is denoted by $z_1 z_2$ and is defined by $z_1 z_2 = (x_1 x_2 - y_1 y_2) + i(x_1 y_2 + y_1 x_2)$

Remember:

1. $\dfrac{z_1}{z_2} = \dfrac{x_1 + iy_1}{x_2 + iy_2} = \dfrac{(x_1 + iy_1)(x_2 - iy_2)}{(x_2 + iy_2)(x_2 - iy_2)}$

$= \dfrac{(x_1 x_2 + y_1 y_2) - i(x_1 y_2 - y_1 x_2)}{x_2^2 + y_2^2}$.

2. If $z_1 = r(\cos\theta_1 + i\sin\theta_1)$ and $z_2 = r(\cos\theta_2 + i\sin\theta_2)$ be two complex numbers, then

$z_1 z_2 = r[\cos(\theta_1 + \theta_2) + i\sin(\theta_1 + \theta_2)]$

8.1.6 Cube Roots of Unity

The cube roots of unity (*i.e*; 1) are $1, w, w^2$.

where $\omega, \omega^2 = \dfrac{-1 \pm i\sqrt{3}}{2}$.

Remember:

(i) If we take $\omega = \dfrac{-1 + i\sqrt{3}}{2}$, then $\omega^2 = \dfrac{-1 - i\sqrt{3}}{2}$ and

if we take $\omega = \dfrac{-1 - i\sqrt{3}}{2}$, then $\omega^2 = \dfrac{-1 + i\sqrt{3}}{2}$.

(ii) $1 + \omega + \omega^2 = 0$

(iii) $\omega^3 = 1$, $\omega^4 = \omega^3 \times \omega = \omega$, $\omega^5 = \omega^3 \times \omega^2 = \omega^2$, etc.

(iv) $w^{3n} = 1$ where $n \in Z^+$.

8.1.7 De Moivre's Theorem

If $n \in Z$, then $(\cos\theta + i\sin\theta)^n = \cos n\theta + i\sin n\theta$ and if n is a fraction of the form $\dfrac{p}{q}$ (where $p \in Z$, $q(\ne 0) \in Z$), then one of the values of $(\cos\theta + i\sin\theta)^n$ will be $\cos n\theta + i\sin n\theta$.

Application:

Let $x + \dfrac{1}{x} = 2\cos\dfrac{\pi}{5}$.

Then, $x + \dfrac{1}{x} = 2\cos\dfrac{\pi}{5} \Rightarrow x^2 - 2x\cos\dfrac{\pi}{5} + 1 = 0$

$\Rightarrow x = \dfrac{2\cos\dfrac{\pi}{5} \pm \sqrt{4\cos^2\dfrac{\pi}{5} - 4}}{2}$

$= \cos\dfrac{\pi}{5} \pm \sqrt{-\left(1 - \cos^2\dfrac{\pi}{5}\right)} = \cos\dfrac{\pi}{5} \pm \sqrt{-\sin^2\dfrac{\pi}{5}}$

$= \cos\dfrac{\pi}{5} \pm i\sin\dfrac{\pi}{5}$

$\therefore x^5 = \left(\cos\dfrac{\pi}{5} \pm i\sin\dfrac{\pi}{5}\right)^5 = \cos\dfrac{5\pi}{5} \pm i\sin\dfrac{5\pi}{5}$

$= \cos\pi \pm i\sin\pi = -1 \quad (\because \quad \cos\pi = -1, \sin\pi = 0)$

8.1.8 Hyperbolic Functions

Hyperbolic functions are defined as follows

$$\sinh x = \frac{e^x - e^{-x}}{2},$$

$$\cosh x = \frac{e^x + e^{-x}}{2},$$

$$\tanh x = \frac{e^x - e^{-x}}{e^x + e^{-x}} = \frac{\sinh x}{\cosh x},$$

$$\operatorname{sech} x = \frac{2}{e^x + e^{-x}} = \frac{1}{\cosh x},$$

$$\operatorname{cosech} x = \frac{2}{e^x - e^{-x}} = \frac{1}{\sinh x},$$

$$\coth x = \frac{e^x + e^{-x}}{e^x - e^{-x}} = \frac{1}{\tanh x}.$$

Remember:

(i) $\sin(ix) = i \sinh x$

(ii) $\cos(ix) = \cosh x$

(iii) $\tan(ix) = i \tanh x$

(iv) $\sinh x = x + \frac{x^3}{3!} + \frac{x^5}{5!} \ldots \ldots \infty$

(v) $\cosh x = 1 + \frac{x^2}{2!} + \frac{x^4}{4!} \ldots \ldots \infty$

(vi) $\sinh 0 = 0 = \tanh 0; \cosh 0 = 1$

(vii) $\cosh^2 x - \sinh^2 x = 1$

(viii) $\operatorname{sech}^2 x + \tanh^2 x = 1$

(ix) $\coth^2 x - \operatorname{cosech}^2 x = 1$

(x) $\sinh(2x) = 2 \sinh x \cosh x$

8.1.9 Logarithm of a Complex Number

Let us consider the complex number z in polar form given by $z = r(\cos\theta + i\sin\theta) = re^{i\theta}$,

where $r = |z|$ and $\theta = \operatorname{amp}(z)$.

Then we define

$$\operatorname{Log}(z) = \log r + i(2n\pi + \theta), n \in Z$$

$$= \log|z| + i(2n\pi + \operatorname{amp}(z))$$

When $n = 0$, we get the principal value of the logarithm denoted by $\log z$ which is given by

$$\log z = \log|z| + i \operatorname{amp}(z)$$

Remember:

(i) $\operatorname{Log}(z_1 z^2) = \operatorname{Log} z_1 + \operatorname{Log} z_2$, where z_1 and z_2 are non-zero complex numbers

(ii) $\operatorname{Log}\left(\frac{z_1}{z_2}\right) = \operatorname{Log} z_1 - \operatorname{Log} z_2$, where z_1 and z_2 are non-zero complex numbers

(iii) $\operatorname{Log} z_1^{z2} = z_2 \operatorname{Log} z_1 + 2n\pi i$, where $z_1(\neq 0)$ and z_2 are two complex numbers and $n \in Z$.

(iv) $\operatorname{Log}(-x) = \log x + \pi i$, where x is a real number

(v) If $a(\neq 0)$ and z be two complex numbers, then $a^z = e^{\operatorname{Log} a^z} = e^{z\log a} = e^{z(\log a + 2n\pi i)}$

8.2 CALCULUS OF COMPLEX VALUED FUNCTIONS

8.2.1 Function of a Complex Variable

If for each value of a complex variable $z(= x + iy)$, there corresponds one or more than one values of a complex variable w, then we say that w is a function of complex variable z and is denoted by $w = f(z)$.

Example:

$z^3 + 2z + 2$, $\sin z$, $e^z + z$, etc. are functions of complex variable z.

We express $w = f(z)$ by
$$w = f(z) = u(x,y) + iv(x,y) \quad \text{or} \quad w = f(z) = u + iv.$$

8.2.2 Limit of a Complex Valued Function

Let D be the domain of the complex valued function $f(z)$ and $z_0 \in D$. Then we say that l (a complex number) is a limit of $f(z)$ as z tends to z_0 if for a given $\varepsilon > 0$(however small), there exist a number $\delta > 0$ (depends on ε) such that

$$|f(z) - l| < \varepsilon \text{ whenever } |z - z_0| < \delta.$$

Symbolically, we write $\lim_{z \to z_0} f(z) = l$.

Remember:

(i) The limit of a function, if exist, is unique.

(ii) If $\lim_{z \to z_0} f(z) = l_1$ and $\lim_{z \to z_0} g(z) = l_2$ then,

(a) $\lim_{z \to z_0} [f(z) \pm g(z)] = l_1 \pm l_2$

(b) $\lim_{z \to z_0} [f(z)g(z)] = l_1 l_2$

(c) $\lim_{z \to z_0} \left[\frac{f(z)}{g(z)}\right] = \frac{l_1}{l_2}$ provided $l_2 \neq 0$

8.2.3 Continuity of a Complex Valued Function

Let D be a domain of the complex valued function $f(z)$ and $z_0 \in D$. Then $f(z)$ is said to be continuous at $z = z_0$ if for a given $\varepsilon > 0$, there exist a number $\delta > 0$ (depends on ε) such that

$$|f(z) - f(z_0)| < \varepsilon \text{ whenever } |z - z_0| < \delta.$$

Thus, $f(z)$ is continuous at $z = z_0$ if and only if $\lim\limits_{z \to z_0} f(z) = f(z_0)$.

$f(z)$ is said to be continuous on D if $f(z)$ is continuous at every point of D.

Remember:

If $f(z)$ and $g(z)$ are defined on a domain D and both are continuous on D, then

(a) $f(z) \pm g(z)$ is continuous on D.

(b) $f(z)\, g(z)$ is continuous on D.

(c) $\dfrac{f(z)}{g(z)}$ is continuous on D provided $g(z) \neq 0$ for any $z \in D$

8.2.4 Derivative of a Complex Valued Function

Let the complex valued function $f(z)$ is defined on D and $z_0 \in D$. Then the derivative of $w = f(z)$ at $z = z_0$ is denoted by $f'(z)$ or $\dfrac{dw}{dz}$ and is defined by

$$f'(z_0) = \lim_{h \to 0} \frac{f(z_0 + h) - f(z_0)}{h},\ \text{provided the limit}$$

exists and is independent of the path.

Remember:

(i) If $f(z)$ is differentiable at $z = z_0$, then $f(z)$ is continuous at $z = z_0$

(ii) $f'(0) = \lim\limits_{z \to 0} \dfrac{f(z) - f(0)}{z}$

8.2.5 Analytic Function

A complex valued function $f(z)$ defined on a domain D is said to be analytic at $z = z_0$ if there exist a neighborhood $|z - z_0| < \delta$ of z_0 at all points of which the function $f(z)$ is differentiable $i.e;\ f'(z)$ exist.

If $f'(z)$ exist at every point of the domain D, then $f(z)$ is said to be analytic in D.

Example:

The function $f(z) = e^z$ is analytic in every finite region of the complex plane.

Remember:

(i) The term regular and holomorphic are also used as synonyms for analytic.

(ii) If $f(z)$ is analytic at every point of a finite complex plane, then $f(z)$ is called an entire function.

(iii) A point $z = a$ is said to be singular point of $f(z)$ if $f'(a)$ doesn't exist.

(iv) If $f(z) = u + iv$ be analytic and $|z| = $ constant, then $f(z)$ becomes constant.

(v) If $f(z) = u + iv$ be analytic and $u = $ constant, then $f(z)$ becomes constant.

8.2.6 Cauchy Riemann Equations

If $f(z) = u(x, y) + iv(x, y)\,(= u + iv)$ be analytic, then

$$\frac{\partial u}{\partial x} = \frac{\partial v}{\partial y} \text{ and } \frac{\partial v}{\partial x} = -\frac{\partial u}{\partial y}\ i.e;$$

$$u_x = v_y \text{ and } v_x = -u_y$$

The above two equations are called the Cauchy-Riemann equations (CR Equations).

Remember:

(i) If u_x, v_x, u_y, v_y are all continuous and u, v are differentiable for a function $f(z) = u + iv$, and if $u_x = v_y, v_x = -u_y$; then $f(z)$ becomes analytic.

(ii) If $z = r(\cos\theta + i\sin\theta) = re^{i\theta}$, then the polar form of the CR equations are:

$$\frac{\partial u}{\partial r} = \frac{1}{r}\frac{\partial v}{\partial \theta} \text{ and } \frac{\partial v}{\partial r} = -\frac{1}{r}\frac{\partial u}{\partial \theta}.$$

8.2.7 Conjugate Function

If $f(z) = u + iv$ is analytic and if u and v both satisfy the Laplace equation, $i.e.$, $\dfrac{\partial^2 u}{\partial x^2} + \dfrac{\partial^2 u}{\partial y^2} = 0$ and $\dfrac{\partial^2 v}{\partial x^2} + \dfrac{\partial^2 v}{\partial y^2} = 0$.

Then u and v are called conjugate functions (or conjugate to each other).

8.2.8 Harmonic Function

A function $g(x, y)$ is called a harmonic function if $\dfrac{\partial g}{\partial x}, \dfrac{\partial g}{\partial y}, \dfrac{\partial^2 g}{\partial x^2}, \dfrac{\partial^2 g}{\partial y^2}$ are all continuous and

$$\frac{\partial^2 g}{\partial x^2} + \frac{\partial^2 g}{\partial y^2} = 0.$$

Example:

Let $g(x, y) = \dfrac{1}{2}\log(x^2 + y^2)$.

Then, $\dfrac{\partial g}{\partial x} = \dfrac{1}{2} \times \dfrac{2x}{x^2 + y^2} = \dfrac{x}{x^2 + y^2},$

$\dfrac{\partial g}{\partial y} = \dfrac{1}{2} \times \dfrac{2y}{x^2 + y^2} = \dfrac{y}{x^2 + y^2}.$

$\dfrac{\partial^2 g}{\partial x^2} = \dfrac{\left(x^2 + y^2\right) \times 1 - x \times 2x}{\left(x^2 + y^2\right)^2} = \dfrac{y^2 - x^2}{\left(x^2 + y^2\right)^2},$

$\dfrac{\partial^2 g}{\partial y^2} = \dfrac{\left(x^2 + y^2\right) \times 1 - y \times 2y}{\left(x^2 + y^2\right)^2} = \dfrac{x^2 - y^2}{\left(x^2 + y^2\right)^2}.$

Hence, $\dfrac{\partial^2 g}{\partial x^2} + \dfrac{\partial^2 g}{\partial y^2} = 0.$

Clearly $\dfrac{\partial g}{\partial x}, \dfrac{\partial g}{\partial y}, \dfrac{\partial^2 g}{\partial x^2}, \dfrac{\partial^2 g}{\partial y^2}$ are all continuous.

Hence, g is harmonic.

Remember:

If $f(z) = u + iv$ is analytic, then u and v are both harmonic functions.

8.2.9 Construction of an Analytic Function (by Milne Thomson's method)

Case-I: Let $u = u(x, y)$ be given.

Then $f(z) = \int \left\{ \dfrac{\partial u}{\partial x}\bigg|_{(z,0)} - i \dfrac{\partial u}{\partial y}\bigg|_{(z,0)} \right\} + C$

C being the constant of integration.

Case-II:

Let $v = v(x, y)$ be given.

Then $f(z) = \int \left\{ \dfrac{\partial v}{\partial y}\bigg|_{(z,0)} + i \dfrac{\partial v}{\partial x}\bigg|_{(z,0)} \right\} + C$

C being the constant of integration.

8.2.10 Construction of Harmonic Conjugate

Case-I: Suppose $f(z) = u + iv$ is analytic and u is given. We need to find v.

$dv = \dfrac{\partial v}{\partial x}dx + \dfrac{\partial v}{\partial y}dy$

$= \left(-\dfrac{\partial u}{\partial y}\right)dx + \dfrac{\partial u}{\partial x}dy$

$= M_1 dx + N_1 dy$

where $M_1 = -\dfrac{\partial u}{\partial y}, N_1 = \dfrac{\partial u}{\partial x}.$

Then, $v = \int M_1\, dx + \int N_1\, dy + C$

(C being the constant of integration).

Remember:

The common terms in $\int M_1\, dx$ and $\int N_1\, dy$ should appear once while calculating $v = \int M_1\, dx + \int N_1\, dy + C.$

Case-II: Suppose $f(z) = u + iv$ is analytic and v is given. We need to find u.

$du = \dfrac{\partial u}{\partial x}dx + \dfrac{\partial u}{\partial y}dy$

$= \left(\dfrac{\partial v}{\partial y}\right)dx + \left(-\dfrac{\partial v}{\partial x}\right)dy$

$= M_1\, dx + N_1\, dy$

where $M_1 = \dfrac{\partial v}{\partial y}, N_1 = -\dfrac{\partial v}{\partial x}.$

Then $u = \int M_1\, dx + \int N_1\, dy + C$

(C being the constant of integration).

Remember:

The common terms in $\int M_1\, dx$ and $\int N_1\, dy$ should appear once while calculating $u = \int M_1\, dx + \int N_1\, dy + C.$

8.3 COMPLEX INTEGRATION

8.3.1 Curves

Let $z = x + iy$, where $x, y \in R$. The "z" is a complex variable and z represents a point (x, y) which moves on the complex plane and makes a curve, say C. Then we say that the curve C is represented by the complex variable "z."

Further if $x = \varphi(t)$ and $y = \psi(t)$, where "t" is a real variable, then $z = \varphi(t) + i\psi(t)$ represents the curve C.

For example, if $z = x + iy$ and $x = a \cos t$, $y = a \sin t$.

Then z forms a curve $x^2 + y^2 = a^2 \cos^2 t + a^2 \sin^2 t = a^2$, which is a circle. Thus, $z = a \cos t + i(a \sin t)$ represents a circle in the complex plane.

A curve C is called a **simple curve** if does not intersect itself.

Curve – 1 (Fig: 1)   Curve – 2 (Fig: 2)

Here, curve 1 is simple, where as curve 2 is not simple.

A simple curve is called **closed** if the two end points of the curve co-inside. The followings are few examples of closed curves:

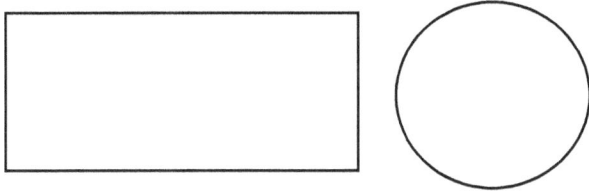

A curve is called **contour** (piecewise smooth) if it is comprised of a finite number of curves each of which has an unique tangent at every point of it. For example, a circle is a contour.

8.3.2 Complex Line Integral

Let $f(z)$ be a complex valued function where z varies over a piecewise smooth and simple curve C. Then $\int_C f(z)dz$ is called the complex line integral along the curve C.

Example:

Let us find the value of the integral $\int_i^{2-i} (3xy + iy^2)dz$ along the straight line joining $z = i$ and $z = 2 - i$.

Here $z = i$ represents the point $A(0, 1)$ and $z = 2 - i$ represents the point $B(2, -1)$.

∴ x varies from 0 to 2 and y varies from 1 to –1.

Now the equation of AB: $\dfrac{y - y_1}{y_2 - y_1} = \dfrac{x - x_1}{x_2 - x_1}$

$\Rightarrow \dfrac{y - 1}{-1 - 1} = \dfrac{x - 0}{2 - 0}$

$\Rightarrow y - 1 = -x \Rightarrow y = 1 - x$

∴ $dy = -dx$ and so

$dz = dx + idy = dx - idx = (1 - i)dx$

Let C represents the straight line AB

Then, $\int_C f(z)dz = \int_{z=i}^{z=2-i} (3xy + iy^2)dz$

$= \int_{x=0}^{2} [3x(1-x) + i(1-x)^2](1-i)dx$

$= (1-i)\left\{ 3\left[\dfrac{x^2}{2} - \dfrac{x^3}{3}\right]_0^2 + i\left[x - x^2 + \dfrac{x^3}{3}\right]_0^2 \right\}$

$= (1-i)\left\{ 3\left[\dfrac{4}{2} - \dfrac{8}{3}\right] + i\left[\left(2^2 - 4 + \dfrac{8}{3}\right) - 0\right]\right\}$

$= (1-i)\left(-2 + \dfrac{8i}{3} - 2i\right)$

$= -2 + 2i + \dfrac{2i}{3} + \dfrac{2}{3} = -\dfrac{4}{3} + \dfrac{8i}{3}$

8.3.3 Cauchy-Goursat Theorem

Statement:

If $f(z)$ be analytic within and on a simple closed curve C, then, $\int_C f(z)dz = 0$.

Example:

Let us find the value of $\int_C \dfrac{5}{z-3}dz$, where $C : |z| = 2$.

Here, $f(z) = \dfrac{5}{z-3}$ and so, $f'(z) = -\dfrac{5}{(z-3)^2}$, which does not exist only for $z = 3$, i.e., at the point $(3, 0)$.

Now, $|z| = 2 \Rightarrow |x + iy| = 2$

$\Rightarrow \sqrt{x^2 + y^2} = 2$

$\Rightarrow (x - 0)^2 + (y - 0)^2 = 2^2$,

which represents a circle with center $(0, 0)$ and radius "2" unit.

Clearly $z = 3$ lies outside the circle C.

Hence, $f(z)$ is analytic within and on C.

∴ $\int_C f(z)dz = \int_C \dfrac{5}{z-3}dz = 0$.

(by the Cauchy-Goursat theorem)

8.3.4 Cauchy's Integral Formula

If $f(z)$ is analytic with and on a simple closed curve C and "$z = a$" be any point lying within C, then $\int_C \dfrac{f(z)dz}{z-a} = 2\pi i \times f(a)$.

Example:

Let us find the value of $\int_C \dfrac{2z+3}{z-1}dz$, where $|z| = 2$.

Comparing the given integral with $\int_C \dfrac{2z+3}{z-1}\,dz$, we get $f(z) = 2z + 3$ and $a = 1$ (which represents the point $(1, 0)$).

$$|z| = 2 \Rightarrow |x + iy| = 2 \Rightarrow \sqrt{x^2 + y^2} = 2$$

$\Rightarrow \quad (x - 0)^2 + (y - 0)^2 = 2^2$, which represents a circle with center $(0, 0)$ and radius "2" unit.

Clearly the point $z = 1$ lies within the circle C.

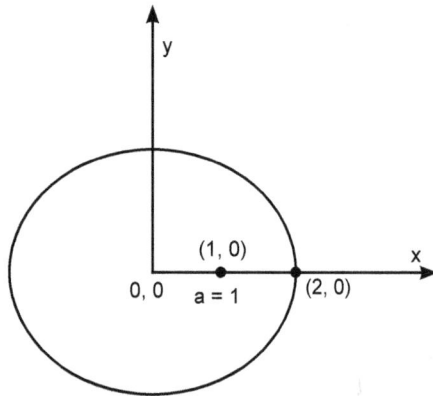

Hence, by Cauchy's integral formula,

$$\int_C \frac{f(z)\,dz}{z-a} = 2\pi i \times f(a)$$

$$\Rightarrow \int_C \frac{(2z+3)}{z-1}\,dz = 2\pi i \times f(1) = 2\pi i \times 5 = 10i$$

$(\because \quad f(1) = 2 \times 1 + 3 = 5).$

8.3.5 Cauchy's Integral Formula on Higher Order Derivatives

Let $f(z)$ be analytic within and on a simple closed curve C and $z = a$ be any point lying within C. If the nth order derivative of $f(z)$ at $z = a$ is denoted by $f^{(n)}(a)$, then

$$\int_C \frac{f(z)\,dz}{(z-a)^{n+1}} = \frac{2\pi i}{n!} \times f^{(n)}(a); \text{ for } n = 0,1,2,....$$

In particular,

$$\int_C \frac{f(z)\,dz}{(z-a)^2} = 2\pi i \times f^{(1)}(a) = 2\pi i \times f'(a) \quad \text{(For } n = 1\text{)}$$

Example:

Let us find the value of $\int_C \dfrac{2z+1}{(z-1)^2}\,dz$, where $C:|z| = 2$.

Comparing the given integral with $\int_C \dfrac{f(z)\,dz}{(z-a)^{n+1}}$ we get $f(z) = 2z + 1$, $a = 1$, $n = 1$. Also in the previ-

ous example, we see that C represents a circle with center $(0, 0)$ & radius "2" unit and $z = 1$ lies within C.

$\therefore$ By Cauchy's integral formula on higher order derivatives, we get,

$$\int_C \frac{2z+1}{(z-1)^2}\,dz = \int_C \frac{f(z)\,dz}{(z-a)^{n+1}} = \frac{2\pi i}{n!} \times f^{(n)}(a)$$

$$= \frac{2\pi i}{1!} \times f^{(1)}(1)$$

$$= 2\pi i \times f'(1) = 2\pi i \times 2 = 4\pi i$$

$$[\because \quad f(z) = 2z + 1 \quad \Rightarrow \quad f'(z) = 2 \Rightarrow f'(1) = 2]$$

8.4 Taylor and Laurent Series

8.4.1 The Taylor Series

Let $f(z)$ be a complex valued function which is analytic within a circle $C : |z - a| = r$. Then at every point "z" lying within C, $f(z) = \sum\limits_{n=0}^{\infty} a_n (z-a)^n$ where $a_n = \dfrac{1}{n!} f^{(n)}(a)$.

8.4.2 The Laurent Series

Let $f(z)$ be analytic in the closed ring shaped region bounded by the concentric circles $C_1 : |z - a| = R_1$ and $C_2 : |z - a| = R_2$ (where $R_1 > R_2$. Then at every point "z" lying in the ring-shaped region,

$$f(z) = \sum_{n=0}^{\infty} a_n (z-a)^n + \sum_{n=1}^{\infty} b_n (z-a)^{-n}$$

where $a_n = \dfrac{1}{2\pi i} \displaystyle\int_{C_1} \dfrac{f(z)\,dz}{(z-a)^{n+1}}$ and $b_n = \dfrac{1}{2\pi i} \displaystyle\int_{C_2} \dfrac{f(z)\,dz}{(z-a)^{-n+1}}$

8.5 Singularities

8.5.1 Singular Point

A point $z = a$ is called a singular point of a complex valued function $f(z)$ if $f(z)$ is not analytic at $z = a$, i.e; if $f'(a)$ does not exist.

Example:

If $f(z) = \dfrac{1}{z-1}$, then $f'(z) = \dfrac{-1}{(z-1)^2}$, which does not exist at $z = 1$. Hence, $z = 1$ is singular point of $f(z) = \dfrac{1}{z-1}$.

Remark: $z = a$ is a singular point of $f(z) \Leftrightarrow f(z)$ has a singularity at $z = a$.

8.5.2 Types of Singularities

8.5.2.1 Isolated singularity

A point $z = a$ is said to be an isolated singularity of $f(z)$ if the following conditions are satisfied:

(a) $f'(a)$ does not exist, $i.e; f(z)$ is not analytic at $z = a$.

(b) $f(z)$ is analytic in the deleted neighborhood of a $i.e; f(z)$ is analytic in $(a - \varepsilon, a + \varepsilon) - \{a\}$, for any pre-assigned $\varepsilon > 0$.

Example:

Consider $f(z) = \dfrac{1}{z-1}$. This function is analytic everywhere except at $z = 1$. Hence, $z = 1$ is an isolated singularity of $f(z)$.

8.5.2.2 Removable singularity

A singularity $z = a$ is said to be a removable singularity of $f(z)$ if $\lim\limits_{z \to a} f(z) = a$ finite quantity.

Example:

Suppose $f(z) = \dfrac{e^z - 1}{z}$.

Then, $f'(z) = \dfrac{ze^z - (e^z - 1) \times 1}{z^2} = \dfrac{e^z(z-1)+1}{z^2}$

$\therefore$ $f'(z)$ does not exist at $z = 0$ and so $f(z)$ is not analytic at $z = 0$. Hence, $z = 0$ is a singularity of $f(z)$.

Again, $\lim\limits_{z \to 0} f(z)$

$= \lim\limits_{z \to 0} \dfrac{e^z - 1}{z} = \lim\limits_{z \to 0} \left\{ \dfrac{\left(1 + z + \dfrac{z^2}{2!} + \dfrac{z^3}{3!} + ...\infty \right) - 1}{z} \right\}$

$= \lim\limits_{z \to 0} \left(1 + \dfrac{z}{2!} + \dfrac{z^2}{3!} + ...\infty \right)$

$= 1 + (0 + 0 + ...\infty) = 1,$ a finite quantity

Hence, $z = 0$ is a removable singularity of $f(z)$

Remember:

If $z = a$ is a removable singularity of $f(z)$, then $f(z)$ can be expressed as:

$$f(z) = \sum_{n=0}^{\infty} a_n (z-a)^n$$

8.5.2.3 Essential singularity:

Consider the Laurent series expansion of $f(z)$ about $z = a$ given by

$$f(z) = \sum_{n=0}^{\infty} a_n (z-a)^n + \sum_{n=1}^{\infty} b_n (z-a)^{-n} \qquad ...(i)$$

If the principal part of (i) $i.e;$ $\sum\limits_{n=1}^{\infty} b_n (z-a)^{-n}$ contains an infinite number of terms, then $z = a$ is called an essential singularity of $f(z)$.

Example:

Consider the function $f(z) = e^{\frac{1}{z^2}}$.

$\because e^{\frac{1}{z^2}} = 1 + \dfrac{1}{z^2} + \left(\dfrac{1}{z^2} \right)^2 \times \dfrac{1}{2!} + \left(\dfrac{1}{z^2} \right)^3 \times \dfrac{1}{3!} + ...\infty,$

so clearly $z = 0$ is a singularity of $f(z)$.

Here, $\sum\limits_{n=1}^{\infty} b_n (z-0)^{-n} = \dfrac{1}{z^2} + \dfrac{1}{2z^4} + \dfrac{1}{6z^6} + ...\infty,$

which contains an infinite number of terms.

Hence, $z = 0$ is an essential singularity of $f(z)$.

8.5.3 Zeros and Poles

(A) Zeros of an analytic function:

$z = a$ is called a zero of an analytic function $f(z)$ if $f(a) = 0$.

An analytic function $f(z)$ is said to have a zero of order "n" if we can express $f(z)$ in the following form:

$f(z) = (z-a)^n \varphi(z),$

where $\varphi(z)$ is analytic and $\varphi(a) \neq 0$

Example:

Consider the function $f(z) = (z-1)^2 \dfrac{1}{z^2 + 4}$.

Then, $f(z) = 0 \Rightarrow (z-1)^2 \times \dfrac{1}{z^2 + 4} = 0$

$\Rightarrow (z-1)^2 = 0 \qquad \Rightarrow z = 1, 1$

$\therefore$ $z = 1$ is a zero of order "2" of $f(z)$.

Remember:

(i) $z = a$ is called a simple zero of $f(z)$ if $z = a$ is a zero of order 1.

(*ii*) If a function $f(z)$ has infinite number of "zeros" and these zeros approaches to a number say "α" (finite or infinite), then "α" is called the limit point of zeros of $f(z)$.

(*iii*) The limit point of zeros is an isolated essential singularity.

Example:

Consider the function $f(z) = \sin\left(\dfrac{1}{z+2}\right)$.

Then zeros of $f(z)$ are given by

$f(z) = 0$ or, $\sin\left(\dfrac{1}{z+2}\right) = 0$

or, $\dfrac{1}{z+2} = n\pi$ $\left(\because \sin x = 0 \Rightarrow x = n\pi,\ n \in Z^+\right)$

or, $z = -2 + \dfrac{1}{n\pi},\ n \in Z^+$

So clearly $z = -2$ is a limit point of zeros, since the zeros $-2+\dfrac{1}{\pi},\ -2+\dfrac{1}{2\pi},\ -2+\dfrac{1}{3\pi}$, etc. approaches to "–2."

Consequently, $z = -2$ is an isolated essential singularity of $f(z)$.

B. Poles

A function $f(z)$ is said to have a pole of order "n" if we can express $f(z)$ in the following way:

$f(z) = \dfrac{\varphi(z)}{(z-a)^n}$, where $\varphi(a) \neq 0$ and $\varphi(z)$ is analytic.

Example:

Consider the function $f(z) = \dfrac{z^2+5}{(z-1)^3}$

Then $z = 1$ is a pole of order "3."

Remember:

(*i*) If $f(z)$ has a pole of order "k" say at $z = a$, then $f(z)$ can be express as

$f(z) = \sum_{n=0}^{\infty} a_n(z-a)^n + \sum_{n=1}^{k} b_n(z-a)^{-n}$

(*ii*) A pole of order "1" is called a simple pole.

(*iii*) If a function $f(z)$ has infinite number of poles and these poles approaches to a number say "β," then β is called the limit point of poles of $f(z)$.

(*iv*) Limit point of poles is a non-isolated essential singularity.

Example:

Consider the function $f(z) = e^z \operatorname{cosec}\dfrac{1}{z}$

Then, $f(z) = \dfrac{e^z}{\sin\dfrac{1}{z}}$.

$\therefore$ The poles of $f(z)$ are given by $\sin\dfrac{1}{z} = 0$

or, $\dfrac{1}{z} = n\pi,\ n \in Z^+$ $\left(\because \sin x = 0 \Rightarrow x = n\pi,\ n \in Z^+\right)$

or, $z = \dfrac{1}{n\pi},\ n \in Z^+$

So, clearly $z = 0$ is a limit point of the poles $\dfrac{1}{\pi}$, $\dfrac{1}{2\pi}, \dfrac{1}{3\pi}, \dfrac{1}{4\pi}$, etc.

Consequently, $z = 0$ is a non-isolated essential singularity of $f(z)$.

(V) $f(z)$ has a pole of order "n" at $z = a$ $\Leftrightarrow$ $\dfrac{1}{f(z)}$ has a zero of order "n" at $z = a$.

8.6 Residues

8.6.1 Residue at a Simple Pole

The residue of $f(z)$ at a simple pole $z = a$ is defined by Res $(z = a) = \lim\limits_{z \to a}(z-a)f(z)$.

Example:

Consider $f(z) = \dfrac{ze^z}{z-1}$. Then clearly $z = 1$ is a simple pole.

$\therefore$ Res $(z = 1) = \lim\limits_{z \to 1}(z-1)f(z)$

$= \lim\limits_{z \to 1}(z-1)\dfrac{ze^z}{z-1}$

$= \lim\limits_{z \to 1} ze^z = 1 \times e^1 = e$.

8.6.2 Residue at a Pole of Order "n"

Let $f(z) = \dfrac{\varphi(z)}{(z-a)^n}$, where $\varphi(a) \neq 0$.

Then $z = a$ is a pole of order "n" and the residue of

$f(z)$ at $z = a$ is defined by

Res $(z = a) = \dfrac{1}{(n-1)!} \times \varphi^{(n-1)}(a)$,

where $\varphi^{(n-1)}(a)$ represents the $(n-1)^{\text{th}}$ derivative of $\varphi(z)$ at $z = a$.

Example:

Consider $f(z) = \dfrac{z^3 + 4}{(z-1)^2}$.

Then clearly $z = 1$ is a pole of order "2." Here, $\varphi(z) = z^3 + 4$. Then $\varphi^{(1)}(z) = \varphi'(z) = 3z^2$ and so $\varphi^{(1)}(1) = 3 \times 1^2 = 3$.

$$\therefore \quad \text{Res } (z = 1) = \frac{1}{(2-1)!}\varphi^{(2-1)}(1) = \varphi^{(1)}(1) = 3.$$

8.6.3 Residue at Infinity

If an analytic function $f(z)$ has a pole at $z = \infty$, then $\text{Res}(z = \infty) = -$ coefficient of $\dfrac{1}{z}$ in the expansion of $f(z)$.

Example:

Consider the function $f(z) = \dfrac{z^2}{(z-2)(z-3)}$

Clearly $z = \infty$ is a pole of $f(z)$.

Now, $f(z) = \dfrac{z^2}{(z-2)(z-3)}$

$$= \frac{z^2}{z^2\left(1 - \dfrac{2}{z}\right)\left(1 - \dfrac{3}{z}\right)} = \left(1 - \frac{2}{z}\right)^{-1}\left(1 - \frac{3}{z}\right)^{-1}$$

$$= \left(1 + \frac{2}{z} + \frac{4}{z^2} + \dots \infty\right) \times \left(1 + \frac{3}{z} + \frac{9}{z^2} + \dots \infty\right)$$

$$= 1 + \left(\frac{2}{z} + \frac{3}{z}\right) + \left(\frac{9}{z^2} + \frac{4}{z^2} + \frac{6}{z^2}\right) + \dots \infty$$

$\therefore$ The coefficient of $\dfrac{1}{z}$ in the expansion of $f(z)$ is 5.

Hence, $\text{Res}(z = \infty) = -(5) = -5$.

Remember:

If $f(z)$ is analytic at $z = \infty$, then

$$\text{Res}(z = \infty) = \lim_{z \to \infty}\{-zf(z)\}.$$

Example:

Consider the function $f(z) = \dfrac{z}{(z-1)(z-2)}$.

Then clearly $f(z)$ in analytic at $z = \infty$.

$\text{Res}(z = \infty)$

$$= \lim_{z \to \infty}\{-z \times f(z)\}$$

$$= -\lim_{z \to \infty}\frac{z^2}{(z-1)(z-2)} = -\lim_{z \to \infty}\frac{1}{\left(1 - \dfrac{1}{z}\right)\left(1 - \dfrac{2}{z}\right)}$$

$$= -\frac{1}{(1-0)(1-0)} = -1.$$

8.6.4 Cauchy's Residue Theorem

Suppose $f(z)$ is analytic within and on a closed curve C except at a finite number of poles, namely, $z_1, z_2, \dots, z_m$ that lies entirely within C. Then

$$\int_C f(z)dz = 2\pi i \times \text{(the sum of residues at the poles)}$$

$$= 2\pi i \times \sum_{i=1}^{m} \text{Res}(z = z_i).$$

Application:

Let us evaluate the integral $\displaystyle\int_C \frac{dz}{(z-1)(z-2)}$, where C is the circle $|z| = 4$.

Here, $f(z) = \dfrac{1}{(z-1)(z-2)}$.

The poles of $f(z)$ are given by

$(z-1)(z-2) = 0$, *i.e;* $z = 1, 2$.

The poles $z = 1$ and $z = 2$ are simple poles and lies within the circle C.

Now $\text{Res } (z = 1) = \lim_{z \to 1}(z-1)f(z)$

$$= \lim_{z \to 1}(z-1) \times \frac{1}{(z-1)(z-2)}$$

$$= \lim_{z \to 1}\frac{1}{(z-2)} = \frac{1}{1-2} = -1$$

$\text{Res } (z = 2) = \lim_{z \to 2}(z-2)f(z)$

$$= \lim_{z \to 2}(z-2) \times \frac{1}{(z-1)(z-2)}$$

$$= \lim_{z \to 2}\frac{1}{(z-1)} = \frac{1}{2-1} = 1.$$

Hence, by Cauchy's residue theorem,

$$\int_C f(z)dz = 2\pi i \times \text{(sum of residues at the poles)}$$

$$= 2\pi i \times (-1 + 1) = 0.$$

Fully Solved MCQ's

1. Let z_1 and z_2 be two complex numbers such that $|z_1| = |z_2|$ and $\arg(z_1) + \arg(z_2) = \pi$. Then, $z_1 = ?$

(a) $\overline{z_2}$ (b) $-\overline{z_2}$ (c) z_2 (d) $-z_2$

2. Let z_1 and z_2 be two complex numbers such that $|z_1| = |z_2| = 1$ and $\arg(z_1) + \arg(z_2) = 0$. Then, $z_1 z_2 = $?
   (a) 0   (b) –1   (c) 1   (d) 2

3. If z be a complex number such that, $|z - 1| = |z - 3| = |z - i|$ then $z = $?
   (a) $2 + 2i$ (b) $-2 + 2i$ (c) $-2 - 2i$ (d) $2 - 2i$

4. If (z_1, z_2) and (z_3, z_4) be the two pair of conjugate complex numbers, then,
   $$\arg\left(\frac{z_1}{z_3}\right) + \arg\left(\frac{z_2}{z_4}\right) = ?$$
   (a) $\dfrac{\pi}{2}$ (b) π (c) 2π (d) 0

5. The amplitude of the complex number
   $$z = \frac{(1 + i\sqrt{3})^2}{4i(1 - i\sqrt{3})} \text{ is}$$
   (a) $\dfrac{\pi}{6}$ (b) $\dfrac{\pi}{4}$ (c) $\dfrac{\pi}{3}$ (d) $\dfrac{\pi}{2}$

6. If z_1 and z_2 be two complex numbers such that $|z_1 + z_2| = |z_1| + |z_2|$, then which of the following is true?
   (a) $\arg(z_1) + \arg(z_2) = 0$
   (b) $\arg(z_1) = \arg(z_2)$
   (c) $\arg(z_1 z_2) = 0$
   (d) $z_1 = -\overline{z_2}$

7. For any complex number z_1 and z_2, given that $|z_1 + z_2|^2 = |z_1|^2 + |z_2|^2$. Then
   (a) $\mathrm{Re}(z_1/z_2) = 0$ (b) $\mathrm{Im}(z_1/z_2) = 0$
   (c) $\mathrm{Re}(z_1 z_2) = 0$ (d) $\mathrm{Im}(z_1 z_2) = 0$

8. The number of solutions of the equation $z^2 = \overline{z}$ is
   (a) 4   (b) 3   (c) 2   (d) 1

9. Let z_1 and z_2 be two complex numbers such that $|z_1| = 12$ and $|z_2 - 3 - 4i| = 5$. Then the minimum value of $|z_1 - z_2|$ is
   (a) 1   (b) 3   (c) 2   (d) 5

10. If $x^2 + y^2 = 1$, then $\dfrac{1 + x + iy}{1 + x - iy} = ?$
    (a) $x + iy$ (b) $x - iy$
    (c) $y + ix$ (d) $y - ix$

11. If $\dfrac{z + 1}{z + i}$ is purely imaginary, then z lies on
    (a) a circle   (b) a parabola
    (c) an ellipse   (d) a straight line

12. The function $f(z) = xy + iy$
    (a) is continuous every where
    (b) is analytic every where
    (c) may be analytic
    (d) satisfies the CR-equations

13. The function $f(z) = x^2 + iy^2$ is
    (a) nowhere analytic.
    (b) everywhere analytic.
    (c) continuous nowhere.
    (d) analytic at finite number of points

14. The function $f(z) = \overline{z}$ is
    (a) nowhere continuous.
    (b) nowhere analytic.
    (c) analytic everywhere.
    (d) none of these.

15. The function $f(z) = \log z$ is
    (a) analytic everywhere except $z = 0$
    (b) nowhere analytic
    (c) analytic everywhere
    (d) none of these

16. The values of the integral $\int_C \overline{z}\, dz$ from $z = 0$ to $z = 4 + 2i$ along the curve C given by $z = t^2 + it$ is
    (a) $10 - \dfrac{8}{3}i$ (b) $-10 + \dfrac{8}{3}i$
    (c) $-10 - \dfrac{8}{3}i$ (d) none of these

17. The value of $\int (x - y + ix^2)\, dz$ along a straight line from $z = 0$ to $z = 1 + i$ is
    (a) $\dfrac{1}{3}(i - 1)$ (b) $\dfrac{1}{3}(i + 1)$
    (c) $-\dfrac{1}{3}(i + 1)$ (d) $\dfrac{i}{3}$

18. The value of $\int_0^{1+i} (x^2 + iy)\, dz$ along a straight line joining $z = 0$ and $z = 1 + i$ is
    (a) $\dfrac{1}{6}(i - 5)$ (b) $\dfrac{1}{6}(1 - 5i)$
    (c) $\dfrac{1}{6}(5i - 1)$ (d) $\dfrac{1}{6}(i - 1)$

19. Let C be any circle enclosing the origin and oriented counterclockwise. Then the value of the integral $\int_C \dfrac{\cos z}{z^2} dz$ is

(a) $2\pi i$ (b) 0

(c) $-2\pi i$ (d) undefined

20. The coefficient of $\dfrac{1}{z}$ in the expansion of $\log\left(\dfrac{z}{z-1}\right)$, valid in $|z| > 1$, is

(a) -1 (b) 1 (c) $-\dfrac{1}{2}$ (d) $\dfrac{1}{2}$

21. Let $f(z) = \dfrac{1}{z^2 - 3z + 2}$

Then, the coefficient of $\dfrac{1}{z^3}$ in the Laurent series expansion of $f(z)$ for $|z| > 2$, is

(a) 0 (b) 1 (c) 3 (d) 5

22. The Laurent series expansion of the function $f(z) = \dfrac{1}{z^2(1-z)}$ about $z = 0$ is given by

(a) $f(z) = -\dfrac{1}{z^2} - \dfrac{2}{z} - 1 + \sum\limits_{n=1}^{\infty} z^n$

(b) $f(z) = \dfrac{1}{z^2} + \dfrac{1}{z} + 1 + \sum\limits_{n=1}^{\infty} z^n$

(c) $f(z) = \dfrac{2}{z^2} - \dfrac{1}{z} + 1 - \sum\limits_{n=1}^{\infty} z^n$

(d) none the these

23. The Taylor series expansion of $f(z) = \dfrac{z}{z^2 + 4}$ about $z = 0$ is given by

(a) $\sum\limits_{n=0}^{\infty} (-1)^n \dfrac{z^{2n+1}}{4^{n+1}}$ (b) $\sum\limits_{n=0}^{\infty} \dfrac{z^{2n-1}}{4^{n+1}}$

(c) $\sum\limits_{n=0}^{\infty} (-1)^n \dfrac{z^{2n+3}}{4^{2n}}$ (d) none of these

24. The Taylor series expansion of $f(z) = \dfrac{1}{z}$ about $z = 1$ is given by

(a) $f(z) = \sum\limits_{n=0}^{\infty} (-1)^n (z-1)^n$

(b) $f(z) = \sum\limits_{n=0}^{\infty} (-1)^n (z-1)^{n+1}$

(c) $f(z) = \dfrac{1}{2} \sum\limits_{n=0}^{\infty} (-1)^n (z-1)^n$

(d) none of these

25. The Taylor series expansion of $f(z) = \dfrac{z+1}{z-1}$ about $z = 0$ is given by

(a) $-1 + 2\sum\limits_{n=0}^{\infty} z^n$ (b) $1 - 2\sum\limits_{n=0}^{\infty} z^n$

(c) $2 - \sum\limits_{n=0}^{\infty} (-1)^n z^n$ (d) none of these

26. The poles of $\dfrac{e^{2z}}{\left(z^2 + 1\right)^2}$ are given as

(a) $-i, i, -i, i$ (b) $-1, 1, i, -i$

(c) $0, 1, -1, i$ (d) none of these

27. Consider $f(z) = \cos z - \sin z$. Then, $z = \infty$ is

(a) a pole of order 1

(b) a zero of order 1

(c) a non-isolated essential singularity

(d) an isolated essential singularity

28. For $f(z) = \dfrac{z}{(z-2)^3 (z+1)^4}$, which of the following is not true?

(a) $f(0) = 0$

(b) $z = 2$ is a pole of, order 3

(c) $z = -1$ is a pole of order 4

(d) $z = 0$ is a zero of order 2

29. The function $f(z) = \begin{cases} \dfrac{\sin z}{z}, & z \neq 0 \\ 0, & z = 0 \end{cases}$ has

(a) a pole of order at $z = 0$

(b) an essential singularity at $z = 0$

(c) a zero of order 2 at $z = 0$

(d) a removable singularity at $z = 0$

30. Consider the function $f(z) = \dfrac{1 - e^{-z}}{z}$. Then which of the following is true?

(a) $z = 0$ is a simple pole

(b) $z = 0$ is an essential singularity

(c) $z = 0$ is a removable singularity

(d) none of these.

31. Consider the function $f(z) = \dfrac{z - \sin z}{z^3}$. Then $z = 0$ is

(a) a pole of order 1   (b) a pole of order 2

(c) a removable singularity

(d) an essential singularity

32. Consider $f(z) = \cos\dfrac{1}{z}$. Then $z = 0$ is

    (a) an isolated singularity

    (b) a non-isolated singularity

    (c) an isolated essential singularity

    (d) a non-isolated essential singularity

33. For the function $f(z) = z^2 e^{\frac{1}{z^2}}$, $z = 0$ is

    (a) a removable singularity

    (b) a pole of order 2

    (c) a simple pole

    (d) an essential singularity

34. The value of $\displaystyle\int_C \dfrac{e^{2z} dz}{z(z-1)}$, where $C: |z| = 2$, is

    (a) $2\pi i e^2$ (b) $2\pi i(e^2 + 1)$

    (c) $-2\pi i e^2$ (d) $2\pi i(e^2 - 1)$

35. The residue of $f(z) = \dfrac{z^3}{1+z^2}$ at $z = \infty$ is given by

    (a) 1   (b) –1   (c) $\dfrac{1}{2}$ (d) $-\dfrac{1}{2}$

36. If $f(z) = \dfrac{\sin 3z}{(z+1)^3}$, then $\mathrm{Res}(z = -1) = ?$

    (a) $\dfrac{1}{4}\sin 3$ (b) $-\dfrac{1}{2}\sin 3$ (c) $\dfrac{9}{2}\sin 3$ (d) 0

37. If $f(z) = \dfrac{e^z}{z\left(z^2 + 4\right)}$, then $\mathrm{Res}(z = -2i) = ?$

    (a) $\dfrac{3}{4}e^{2i}$ (b) $-\dfrac{1}{8}e^{-2i}$

    (c) $\dfrac{1}{2}e^{2i}$ (d) $-\dfrac{1}{4}e^{-2i}$

38. The value of the integral $\displaystyle\int_{|z|=1} \dfrac{e^{-2z}}{z^2} dz$ is

    (a) 0   (b) $-4\pi i$ (c) $4\pi i$ (d) $2\pi i$

39. Let $f : C \to C$ be analytic except for a simple pole at $z = 0$ and let $g : C \to C$ be analytic.

    Then, the value of $\dfrac{\underset{z\to 0}{\mathrm{Res}}\{f(z)g(z)\}}{\underset{z\to 0}{\mathrm{Res}}f(z)}$ is

    (a) $g(0)$ (b) $g'(0)$

    (c) $\lim_{z\to 0} z f(z)$ (d) $\lim_{z\to 0} z f(z)g(z)$

Answer key				
1. (b)	**2.** (c)	**3.** (a)	**4.** (d)	**5.** (d)
6. (b)	**7.** (a)	**8.** (a)	**9.** (c)	**10.** (a)
11. (a)	**12.** (a)	**13.** (a)	**14.** (b)	**15.** (a)
16. (a)	**17.** (a)	**18.** (c)	**19.** (b)	**20.** (a)
21. (c)	**22.** (b)	**23.** (a)	**24.** (a)	**25.** (b)
26. (a)	**27.** (d)	**28.** (d)	**29.** (d)	**30.** (c)
31. (c)	**32.** (c)	**33.** (d)	**34.** (d)	**35.** (a)
36. (c)	**37.** (b)	**38.** (b)	**39.** (a)	

Explanation

1. (b)

Let $z_1 = r_1 e^{i\theta_1}$ and $z_2 = r_2 e^{i\theta_2}$.

Then $|z_1| = |z_2| \Rightarrow r_1 = r_2 = r$ (say)

$\therefore z_1 = re^{i\theta_1}$ and $z_2 = re^{i\theta_2}$.

Now $\arg(z_1) + \arg(z_2) = \pi$

$\Rightarrow \theta_1 + \theta_2 = \pi \Rightarrow \theta_1 = \pi - \theta_2$

$\therefore z_1 = re^{i\theta_1} = re^{i(\pi-\theta_2)} = re^{i\pi}e^{-i\theta_2} = (-1)re^{-i\theta_2} = -\overline{z_2}$

$\left[\because e^{i\pi} = \cos\pi + i\sin\pi = -1 + i\times 0 = -1 \right]$

2. (c)

Let $z_1 = r_1 e^{i\theta_1}$ and $z_2 = r_2 e^{i\theta_2}$.

Then $|z_1| = |z_2| = 1 \Rightarrow r_1 = r_2 = 1$

$\therefore z_1 = e^{i\theta_1}$ and $z_2 = e^{i\theta_2}$

Now $\arg(z_1) + \arg(z_2) = 0$

$\Rightarrow \theta_1 + \theta_2 = 0 \Rightarrow \theta_2 = -\theta_1$.

$\therefore z_1 z_2 = e^{i\theta_1}e^{-i\theta_1} = e^0 = 1$

3. (a)

Let $z = x + iy$

Then $|z - 1| = |z - 3|$

$\Rightarrow |x + iy - 1| = |x + iy - 3|$

$\Rightarrow |(x-1) + iy| = |(x-3) + iy|$

$\Rightarrow \sqrt{(x-1)^2 + y^2} = \sqrt{(x-3)^2 + y^2}$

$\Rightarrow x^2 - 2x + 1 + y^2 = x^2 - 6x + 9 + y^2$

$\Rightarrow 4x = 8 \Rightarrow x = 2$

Also $|z - 1| = |z - i|$

$\Rightarrow |x + iy - 1| = |x + iy - i|$

$\Rightarrow |(x-1) + iy| = |x + i(y-1)|$

$\Rightarrow |1 + iy| = |2 + i(y-1)|$

$\Rightarrow \sqrt{1^2 + y^2} = \sqrt{2^2 + (y-1)^2}$

$\Rightarrow 1 + y^2 = 4 + y^2 - 2y + 1 \Rightarrow 2y = 4 \Rightarrow y = 2$.

$\therefore z = x + iy = 2 + 2i$.

4. (d) Since z_1 and z_2 be two conjugate complex numbers, so let $z_1 = r_1 e^{i\theta_1}$ and $z_2 = r_1^{-i\theta_1}$.

Since z_3 and z_4 be be two conjugate complex numbers, so let $z_3 = r_2 e^{i\theta_2}$ and $z_4 = r_2 e^{-i\theta_2}$.

$$\therefore \frac{z_1}{z_3} = \frac{r_1}{r_2} e^{i(\theta_1 - \theta_2)}, \frac{z_2}{z_4} = \frac{r_1}{r_2} e^{-i(\theta_1 - \theta_2)}$$

Then $\arg\left(\dfrac{z_1}{z_3}\right) - \arg\left(\dfrac{z_2}{z_4}\right)$

$$= (\theta_1 - \theta_2) - \{-(\theta_1 - \theta_2)\} = 0.$$

5. (d)

$$z = \frac{(1 + i\sqrt{3})^2}{4i(1 - i\sqrt{3})} = \frac{1 + 2i\sqrt{3} + (i\sqrt{3})^2}{4i(1 - i\sqrt{3})}$$

$$= \frac{1 + 2i\sqrt{3} - 3}{4i + 4\sqrt{3}} = \frac{-1 + i\sqrt{3}}{2\sqrt{3} + 2i}$$

$$\therefore \arg(z) = \arg\left(\frac{-1 + i\sqrt{3}}{2\sqrt{3} + 2i}\right)$$

$$= \arg\left(-1 + i\sqrt{3}\right) - \arg\left(2\sqrt{3} + 2i\right)$$

$$= \pi - \tan^{-1}\left|\frac{\sqrt{3}}{-1}\right| - \tan^{-1}\left|\frac{2}{2\sqrt{3}}\right|$$

$$\begin{bmatrix} \because \text{ the complex number } -1 + i\sqrt{3} \text{ represents} \\ \text{the point } (-1, \sqrt{3}) \text{ that lies on second quadrant} \\ \text{and the complex number } 2\sqrt{3} + 2i \text{ represents} \\ \text{the point } (2\sqrt{3}, 2) \text{ that lies on first quadrant} \end{bmatrix}$$

$$= \pi - \tan^{-1}\sqrt{3} - \tan^{-1}\frac{1}{\sqrt{3}} = \pi - \frac{\pi}{3} - \frac{\pi}{6} = \frac{\pi}{2}.$$

6. (b)

Let $z_1 = r_1 e^{i\theta_1}$ and $z_2 = r_2 e^{i\theta_2}$

Then $z_1 + z_2 = r_1 e^{i\theta_1} + r_2 e^{i\theta_2}$

$= r_1(\cos\theta_1 + i\sin\theta_1) + r_2(\cos\theta_2 + i\sin\theta_2)$

$= (r_1\cos\theta_1 + r_2\cos\theta_2) + i(r_1\sin\theta_1 + r_2\sin\theta_2)$

$\therefore |z_1 + z_2|$

$$= \sqrt{(r_1\cos\theta_1 + r_2\cos\theta_2)^2 + (r_1\sin\theta_1 + r_2\sin\theta_2)^2}$$

$$= \sqrt{r_1^2 + r_2^2 + 2r_1 r_2(\cos\theta_1\cos\theta_2 + \sin\theta_1\sin\theta_2)}$$

$$= \sqrt{r_1^2 + r_2^2 + 2r_1 r_2\cos(\theta_1 - \theta_2)}$$

$$= \sqrt{r_1^2 + r_2^2 + 2r_1 r_2} \quad \left[\text{if } \cos(\theta_1 - \theta_2) = 1\right]$$

$$= \sqrt{(r_1 + r_2)^2} = r_1 + r_2 = |z_1| + |z_2|,$$

which satisfies the given condition.

Now $\cos(\theta_1 - \theta_2) = 1 = \cos 0^\circ \Rightarrow \theta_1 - \theta_2 = 0$

$\Rightarrow \theta_1 = \theta_2$.

Hence, $\arg(z_1) = \arg(z_2)$.

7. (a) Let $z_1 = r_1 e^{i\theta_1}$ and $z_2 = r_2 e^{i\theta_2}$.

Then $|z_1|^2 = r_1^2$ and $|z_2|^2 = r_2^2$

Then $z_1 + z_2 = r_1 e^{i\theta_1} + r_2 e^{i\theta_2}$

$= r_1(\cos\theta_1 + i\sin\theta_1) + r_2(\cos\theta_2 + i\sin\theta_2)$

$= (r_1\cos\theta_1 + r_2\cos\theta_2) + i(r_1\sin\theta_1 + r_2\sin\theta_2)$

$\therefore |z_1 + z_2|$

$$= \sqrt{(r_1\cos\theta_1 + r_2\cos\theta_2)^2 + (r_1\sin\theta_1 + r_2\sin\theta_2)^2}$$

$$= \sqrt{r_1^2 + r_2^2 + 2r_1 r_2\cos(\theta_1 - \theta_2)}$$

$$\therefore |z_1 + z_2|^2 = |z_1|^2 + |z_2|^2$$

$$\Rightarrow r_1^2 + r_2^2 + 2r_1 r_2\cos(\theta_1 - \theta_2) = r_1^2 + r_2^2$$

$$\Rightarrow \cos(\theta_1 - \theta_2) = 0 = \cos\frac{\pi}{2} \quad [\text{Assuming } r_1, r_2 \neq 0]$$

$$\Rightarrow \theta_1 - \theta_2 = \frac{\pi}{2} \Rightarrow \theta_1 = \frac{\pi}{2} + \theta_2$$

$$\therefore \frac{z_1}{z_2} = \frac{r_1 e^{i\left(\frac{\pi}{2} + \theta_2\right)}}{r_2 e^{i\theta_2}} = \frac{r_1 e^{i\frac{\pi}{2}} e^{i\theta_2}}{r_2 e^{i\theta_2}} = \frac{r_1}{r_2} e^{i\frac{\pi}{2}}$$

$$= \frac{r_1}{r_2}\left(\cos\frac{\pi}{2} + i\sin\frac{\pi}{2}\right) = 0 + i\frac{r_1}{r_2}$$

Hence, $\text{Re}(z_1 / z_2) = 0$.

8. (a) $z^2 = \bar{z} \Rightarrow (x + iy)^2 = x - iy$

$$\Rightarrow x^2 - y^2 + 2ixy = x - iy$$

Comparing the real and imaginary parts from both sides we get,

$x^2 - y^2 = x$ ....(1)

$2xy = -y$ i.e; $y(2x + 1) = 0$...........................(2)

$(2) \Rightarrow 2x = -1$ or $y = 0$

$\Rightarrow x = -1/2$ or $y = 0$.

Case – I : $y = 0$

$\therefore (1) \Rightarrow x^2 - 0^2 = x \Rightarrow x(x - 1) = 0$

$\Rightarrow x = 0, x = 1$

Case – II : $x = -1/2$

$\therefore (1) \Rightarrow y^2 = x^2 - x = (1/4) + (1/2) = 3/4$

$$\Rightarrow y = \pm\frac{\sqrt{3}}{2}.$$

Hence, the solutions are

$x = 0, y = 0; x = 1, y = 0; x = -1/2, y = \sqrt{3}/2;$

$x = -1/2, y = -\sqrt{3}/2.$

Thus, the total number of solutions = 4.

9. (c)

$$|z_1| = |(z_1 - z_2) + (z_2 - 3 - 4i) + (3 + 4i)|$$

$$\Rightarrow |z_1| \le |z_1 - z_2| + |z_2 - 3 - 4i| + |3 + 4i|$$

$$\Rightarrow 12 \le |z_1 - z_2| + 5 + \sqrt{3^2 + 4^2}$$

$$\Rightarrow |z_1 - z_2| \ge 12 - 5 - 5 = 2$$

$$\Rightarrow |z_1 - z_2| \ge 2.$$

Therefore, the minimum value of $|z_1 - z_2|$ is 2.

10. (a)

$$\frac{1 + x + iy}{1 + x - iy}$$

$$= \frac{(1 + x + iy)(1 + x + iy)}{(1 + x - iy)(1 + x + iy)} = \frac{(1 + x + iy)^2}{(1 + x)^2 + y^2}$$

$$= \frac{(1 + x)^2 + 2iy(1 + x) - y^2}{(1 + x)^2 + y^2}$$

$$= \frac{1 + 2x + x^2 + 2iy(1 + x) - y^2}{1 + 2x + x^2 + y^2}$$

$$= \frac{x^2 + y^2 + 2x + x^2 + 2iy(1 + x) - y^2}{1 + 2x + 1}$$

$$[\because x^2 + y^2 = 1]$$

$$= \frac{2x + 2x^2 + 2iy(1 + x)}{2 + 2x}$$

$$= \frac{2x(1 + x) + 2iy(1 + x)}{2(1 + x)} = x + iy.$$

11. (a)

$$\frac{z + 1}{z + i}$$

$$= \frac{x + iy + 1}{x + iy + i} = \frac{(x + 1) + iy}{x + i(y + 1)}$$

$$= \frac{(x + 1 + iy)\{x - i(y + 1)\}}{\{x + i(y + 1)\}\{x - i(y + 1)\}}$$

$$= \frac{x^2 + x + ixy - ix(y + 1) - i(y + 1) + y(y + 1)}{x^2 + (y + 1)^2}$$

$$= \frac{x^2 + x - ix - iy + y^2 + y - i}{x^2 + (y + 1)^2}$$

$$= \frac{x^2 + y^2 + x + y}{x^2 + (y + 1)^2} - i\frac{x + y + 1}{x^2 + (y + 1)^2}$$

Now if $\frac{z + 1}{z + i}$ is purely imaginary, then

$$\operatorname{Re}\left(\frac{z + 1}{z + i}\right) = 0.$$

Now $\operatorname{Re}\left(\dfrac{z + 1}{z + i}\right) = 0$

$$\Rightarrow \frac{x^2 + y^2 + x + y}{x^2 + (y + 1)^2} = 0 \Rightarrow x^2 + y^2 + x + y = 0$$

$$\Rightarrow x^2 + 2 \times \frac{1}{2} \times x + \left(\frac{1}{2}\right)^2 + y^2 + 2 \times \frac{1}{2} \times y + \left(\frac{1}{2}\right)^2$$

$$= \left(\frac{1}{2}\right)^2 + \left(\frac{1}{2}\right)^2 = \frac{1}{2}$$

$$\Rightarrow \left(x + \frac{1}{2}\right)^2 + \left(y + \frac{1}{2}\right)^2 = \left(\frac{1}{\sqrt{2}}\right)^2,$$

which is an equation of a circle.

12. (a) $f(z) = xy + iy = u + iv$.

Therefore, $u = xy$, $v = y$

Since u and v are both polynomials in x and y, so they are continuous, and hence $f(z)$ is continuous everywhere.

Now, $\dfrac{\partial u}{\partial x} = y, \dfrac{\partial u}{\partial y} = x, \dfrac{\partial v}{\partial x} = 0, \dfrac{\partial v}{\partial y} = 1$.

Hence, $\dfrac{\partial u}{\partial x} \ne \dfrac{\partial v}{\partial y}$ & $\dfrac{\partial v}{\partial x} \ne -\dfrac{\partial u}{\partial y}$.

Thus, CR-equations are not satisfied and so the function is not analytic.

13. (a)

$$f(z) = x^2 + iy^2 = u + iv \Rightarrow u = x^2, v = y^2$$

Now $\dfrac{\partial u}{\partial x} = 2x, \dfrac{\partial u}{\partial y} = 0, \dfrac{\partial v}{\partial x} = 0, \dfrac{\partial v}{\partial y} = 2y$.

Hence, $\dfrac{\partial u}{\partial x} \ne \dfrac{\partial v}{\partial y}$.

Thus, CR-equations are not satisfied and so the function is not analytic.

Moreover, $u = x^2$, $v = y^2$ being both polynomials, are continuous and so $f(z)$ is continuous everywhere.

14. (b)

$$f(z) = \bar{z} = \overline{x + iy} = x - iy = u + iv$$

$$\Rightarrow u = x, v = -y.$$

Now, $\dfrac{\partial u}{\partial x} = 1, \dfrac{\partial u}{\partial y} = 0, \dfrac{\partial v}{\partial x} = 0, \dfrac{\partial v}{\partial y} = -1$.

$$\therefore \frac{\partial u}{\partial x} \neq \frac{\partial v}{\partial y}.$$

Thus, *CR*-equations are not satisfied and so the function is nowhere analytic.

15. (*a*)

$f(z) = \log z \Rightarrow f'(z) = \dfrac{1}{z}$ which is not defined at $z = 0$.

So $f'(z)$ exist except at $z = 0$.

Thus, $f(z)$ is analytic everywhere except at $z = 0$.

16. (*a*)

$z = t^2 + it$

$\Rightarrow \begin{cases} t^2 + it = 0 & \text{for} & z = 0 \\ t^2 + it = 4 + 2i & \text{for} & z = 4 + 2i \end{cases}$

$\Rightarrow \begin{cases} t = 0, & \text{for} & z = 0 \\ t = 2, & \text{for} & z = 4 + 2i \end{cases}$

Also $dz = (2t + i)dt$

$\therefore$ the given integral

$= \int_C \bar{z}\, dz = \int_C \left(t^2 - it\right) dz$

$= \int_0^2 \left(t^2 - it\right)(2t + i)\,dt = \int_0^2 \left(2t^3 - 2it^2 + t^2 i + t\right) dt$

$= \left[2\frac{t^4}{4} - 2i \times \frac{t^3}{3} + i\frac{t^3}{3} + \frac{t^2}{2}\right]_0^2$

$= \frac{1}{2} \times 16 - \frac{16i}{3} + \frac{8i}{3} + 2 = 10 - \frac{8i}{3}.$

17. (*a*) $z = 0 = 0 + i \times 0$ represents the point $A(0,0)$ &

$z = 1 + i$ represents the point $B(1, 1)$.

$\therefore$ *x* varies from 0 to 1:

Now, the equation of *AB* is: $\dfrac{y - 0}{1 - 0} = \dfrac{x - 0}{1 - 0}$

i.e., $y = x$ and so $dy = dx$.

$\therefore \quad dz = dx + idy = dx + idx = (1 + i)dx$

Hence, the given integral

$= \int_0^1 \left(x - x + ix^2\right)(1 + i)dx$

$= (1 + i) \int_0^1 x^2 dx = (i - 1)\left[\frac{x^3}{3}\right]_0^1 = \frac{1}{3}(i - 1)$

18. (*c*)

$z = 0 = 0 + i \times 0$ represents the point $A(0,0)$ &

$z = 1 + i$ represents the point $B(1, 1)$.

$\therefore$ *x* varies from 0 to 1:

Now, the equation of *AB* is: $\dfrac{y - 0}{1 - 0} = \dfrac{x - 0}{1 - 0}$

i.e., $y = x$ and so $dy = dx$.

$\therefore \quad dz = dx + idy = dx + idx = (1 + i)dx$

$\therefore$ the given integral

$= \int_0^1 \left(x^2 + ix\right)(1 + i)dx = (1 + i)\left[\frac{x^3}{3} + \frac{ix^2}{2}\right]_0^1$

$= (1 + i)\left(\frac{1}{3} + \frac{i}{2}\right) = \frac{1}{6}(1 + i)(2 + 3i)$

$= \frac{1}{6}\left(2 + 2i + 3i + 3i^2\right) = \frac{1}{6}(5i - 1).$

19. (*b*) Comparing the given integral with

$\int_C \dfrac{f(z)\,dz}{(z - a)^{n+1}}$ we get $f(z) = \cos z$, $a = 0$, $n = 1$.

Clearly, $z = 1$ lies within *C*.

$\therefore$ By Cauchy's integral formula on higher order derivatives, we get,

$\int_C \dfrac{\cos z}{z^2}\,dz = \int_C \dfrac{f(z)\,dz}{(z - a)^{n+1}} = \dfrac{2\pi i}{n!} \times f^{(n)}(a)$

$= \dfrac{2\pi i}{1!} \times f^{(1)}(0)$

$= 2\pi i \times f'(0) = 2\pi i \times 0 = 0.$

$[\because \ f(z) = \cos z \ \Rightarrow \ f'(z) = -\sin z \ \Rightarrow \ f'(0) = 0].$

20. (*a*)

$\log\left(\dfrac{z}{z - 1}\right) = -\log\left(1 - \dfrac{1}{z}\right)$

$= -\left[\dfrac{1}{z} + \dfrac{1}{2z^2} + \dfrac{1}{3z^3} + \cdots\right]$

$\left[\text{using } \log(1 - x) = x + \dfrac{x^2}{2} + \dfrac{x^3}{3} + \cdots\right]$

So, coefficient of $1/z$ in the expansion of

$\log\left(\dfrac{z}{z - 1}\right) = -1$, $|z| > 1$.

21. (c)

$$f(z) = \frac{1}{z^2 - 3z + 2} = \frac{1}{(z-2)(z-1)}$$

$$= \frac{(z-1)-(z-2)}{(z-2)(z-1)} = -\frac{1}{(z-1)} + \frac{1}{(z-2)}.$$

Now, $|z| > 2$

$\Rightarrow |z| > 2$ and $|z| > 1$

$\Rightarrow \dfrac{2}{|z|} < 1$ and $\dfrac{1}{|z|} < 1$

$$f(z) = -\frac{1}{(z-1)} + \frac{1}{(z-2)}$$

$$= -\frac{1}{z(1-1/z)} + \frac{1}{z(1-2/z)}$$

$$= \frac{1}{z}\left[\left(1-\frac{2}{z}\right)^{-1} - \left(1-\frac{1}{z}\right)^{-1}\right]$$

$$= \frac{1}{z}\left[\left(1-\frac{2}{z}+\frac{4}{z^2}+...\right)\right.$$

$$\left. -\left(1+\frac{1}{z}+\frac{1}{z^2}+...\right)\right]$$

$$\left[\text{using } (1-x)^{-1} = 1+x+x^2+x^3+.......\text{ for}|x|<1\right]$$

$$= \frac{1}{z}\left[\frac{1}{z}+\frac{3}{z^2}+...\right] = \frac{1}{z^2}+\frac{3}{z^3}+......$$

So, the coefficient of $1/z^3 = 3$.

22. (b)

$$f(z) = \frac{1}{z^2(1-z)} = \frac{1}{z^2}(1-z)^{-1}$$

$$= \frac{1}{z^2}\left(1+z+z^2+z^3+z^4+.......\infty\right)$$

$$= \frac{1}{z^2}+\frac{1}{z}+1+\left(z+z^2+.......\infty\right)$$

$$= \frac{1}{z^2}+\frac{1}{z}+1+\sum_{n=1}^{\infty}z^n.$$

23. (a)

$$f(z) = \frac{z}{z^2+4} = \frac{z}{4\left(1+\dfrac{z^2}{4}\right)} = \frac{z}{4}\left(1+\frac{z^2}{4}\right)^{-1}$$

$$= \frac{z}{4}\sum_{n=0}^{\infty}(-1)^n\left(\frac{z^2}{4}\right)^n = \sum_{n=0}^{\infty}(-1)^n\frac{z^{2n+1}}{4^{n+1}}$$

$$\left(\text{using } (1+x)^{-1} = \sum_{n=0}^{\infty}(-1)^n x^n\right)$$

24. (a)

$f(z)$

$$= \frac{1}{z} = \frac{1}{1+(z-1)} = \left\{1+(z-1)\right\}^{-1}$$

$$= \sum_{n=0}^{\infty}(-1)^n(z-1)^n \left[\text{using } (1+x)^{-1} = \sum_{n=0}^{\infty}(-1)^n x^n\right]$$

25. (b)

$$f(z) = \frac{z+1}{z-1} = \frac{(z-1)+2}{z-1} = 1+\frac{2}{z-1}$$

$$= 1-\frac{2}{1-z} = 1-2(1-z)^{-1} = 1-2\sum_{n=0}^{\infty}z^n$$

$$\left(\text{using } (1-x)^{-1} = \sum_{n=0}^{\infty}x^n\right)$$

26. (a)

Here, $f(z) = \dfrac{e^{2z}}{\left(z^2+1\right)^2}.$

So the poles of $f(z)$ are given by $(z^2+1)^2 = 0$

or, $\{(z+i)(z-i)\}^2 = 0$

or, $(z+i)^2(z-i)^2 = 0$

or, $z = i, i, -i, -i.$

27. (d) The zeros of $f(z)$ are given by

$f(z) = 0$

or, $\cos z - \sin z = 0$

or, $\tan z = 1 = \tan\dfrac{\pi}{4}$

or, $z = n\pi + \dfrac{\pi}{4},\ n \in Z^+$

$$\left(\because \tan x = \tan\alpha \Rightarrow x = n\pi+\alpha,\ n \in Z^+\right)$$

$\therefore$ The zeros of $f(z)$ are $\pi+\dfrac{\pi}{4},\ 2\pi+\dfrac{\pi}{4},\ 3\pi+\dfrac{\pi}{4},$ and so on.

Clearly $z = \infty$ is a limit point of these zeros. Hence, $z = \infty$ is an isolated essential singularity.

28. (d)

The zeros of $f(z)$ are given by

$f(z) = 0$

or, $\dfrac{z}{(z-2)^3(z+1)^4} = 0$

or, $z = 0$

$\therefore$ $z = 0$ is a zero of order "1."

The poles of $f(z)$ are given by:

$(z-2)^3(z+1)^4 = 0$

or $(z-2)^3 = 0$, $(z+1)^4 = 0$

or $z = 2, 2, 2$ and $z = -1, -1, -1, -1$.

$\therefore$ $z = 2$ is a pole of order "3" and $z = -1$ is a pole of order "4"

29. (d)

$\lim\limits_{z\to 0} f(z) = \lim\limits_{z\to 0} \dfrac{\sin z}{z} = 1$, a finite quantity,

So $z = 0$ is a removable singularity of $f(z)$.

30. (c)

$f(z) = \dfrac{1-e^{-z}}{z} \Rightarrow f'(z) = \dfrac{ze^{-z} - \left(1-e^{-z}\right)\times 1}{z^2}$

$\Rightarrow$ $z = 0$ is a singularity of $f(z)$

$[\because \ f'(0) \text{ does not exist}]$

Now, $\lim\limits_{z\to 0}\dfrac{1-e^{-z}}{z}$ $\left(\text{form } \dfrac{0}{0}\right)$

$= \lim\limits_{z\to 0}\dfrac{0+e^{-z}}{1}$

$= \lim\limits_{z\to 0} e^{-z} = e^{-0} = 1$, a finite quantity

Hence, $z = 0$ is a removable singularity of $f(z)$.

31. (c) Clearly, $z = 0$ is a singularity of $f(z)$.

Now, $\lim\limits_{z\to 0} f(z) = \lim\limits_{z\to 0}\dfrac{z - \sin z}{z^3}$

$= \lim\limits_{z\to 0}\dfrac{z - \left(z - \dfrac{z^3}{3!} + \dfrac{z^5}{5!} - \dfrac{z^7}{7!} + \dots\infty\right)}{z^3}$

$= \lim\limits_{z\to 0}\dfrac{1}{z^3}\left(\dfrac{z^3}{3!} - \dfrac{z^5}{5!} + \dfrac{z^7}{7!} - \dots\infty\right)$

$= \lim\limits_{z\to 0}\left(\dfrac{1}{3!} - \dfrac{z^2}{5!} + \dfrac{z^4}{7!} - \dots\infty\right)$

$= \dfrac{1}{3!} - 0 + 0 - \dots\infty = \dfrac{1}{6}$, a finite quantity

Hence, $z = 0$ is a removable singularity of $f(z)$.

32. (c) The zeros of $f(z)$ are given by

$f(z) = 0$ *or,* $\cos\dfrac{1}{z} = 0$

or, $\dfrac{1}{z} = 2n\pi + \dfrac{\pi}{2}$ $\left(\because \cos x = 0 \Rightarrow x = 2n\pi + \dfrac{\pi}{2}, n \in Z^+\right)$

or, $z = \dfrac{1}{2n\pi + \dfrac{\pi}{2}}$, $n \in Z^+$

Thus, the zeros of $f(z)$ are $\dfrac{1}{2\pi + \dfrac{\pi}{2}}$, $\dfrac{1}{4\pi + \dfrac{\pi}{2}}$, $\dfrac{1}{6\pi + \dfrac{\pi}{2}}$, and so on.

So clearly, $z = 0$ is a limit point of zeros. Hence, $z = 0$ is an isolated essential singularity.

33. (d)

$f(z) = z^2 e^{\frac{1}{z^2}}$

$= z^2\left[1 + \dfrac{1}{z^2} + \left(\dfrac{1}{z^2}\right)^2\times\dfrac{1}{2!} + \left(\dfrac{1}{z^2}\right)^3\times\dfrac{1}{3!} + \dots\infty\right]$

$= z^2 + 1 + \dfrac{1}{2z^2} + \dfrac{1}{6z^4} + \dots\infty$

Since infinite number of items containing the negative power of "$z - 0$" exist, so $z = 0$ is an essential singularity.

34. (d) Here, $f(z) = \displaystyle\int_C \dfrac{e^{2z}dz}{z(z-1)}$. So the poles of $f(z)$

are given by $z(z-1) = 0$ i.e; $z = 0, 1$.

The poles $z = 0$ and $z = 1$ are simple poles and lies within C.

Now, $\text{Res}(z = 0) = \lim\limits_{z\to 0} zf(z) = \lim\limits_{z\to 0}\dfrac{e^{2z}}{z-1} = \dfrac{e^0}{0-1} = -1$.

$\text{Res}(z = 1) = \lim\limits_{z\to 1}(z-1)f(z) = \lim\limits_{z\to 1}\dfrac{e^{2z}}{z} = \dfrac{e^2}{1} = e^2$.

$\therefore$ By the Cauchy's residual theorem,

$\displaystyle\int_C \dfrac{e^{2z}dz}{z(z-1)} = 2\pi i \times\left(-1 + e^2\right) = 2\pi i\left(e^2 - 1\right)$.

35. (a)

$f(z) = \dfrac{z^3}{1+z^2} = \dfrac{z^3}{z^2\left(1+\dfrac{1}{z^2}\right)} = z\left(1+\dfrac{1}{z^2}\right)^{-1}$

$= z\left\{1 - \dfrac{1}{z^2} + \left(\dfrac{1}{z^2}\right)^2 - \dots\infty\right\}$

$= z - \dfrac{1}{z} + \dfrac{1}{z^3} - \dots\infty$

Res$(z = \infty)$

$= -$ coefficient of $\dfrac{1}{z}$ in the expansion of $f(z)$

$= -(-1) = 1.$

36. (c)

$$f(z) = \frac{\sin 3z}{(z+1)^3} = \frac{\varphi(z)}{(z+1)^3} \quad \text{(say)}$$

(where $\varphi(z) = \sin 3z$)

Clearly, $z = -1$ is a pole of order "3."

Here, $\varphi'(z) = 3 \cos 3z$, $\varphi''(z) = -9\sin 3z$

So, $\varphi^{(2)}(-1) = \varphi''(-1) = -9\sin(-3) = 9\sin 3$

$\therefore \quad \text{Res}(z = -1) = \dfrac{1}{(3-1)!} \varphi^{(3-1)}(-1)$

$= \dfrac{1}{2!} \varphi^{(2)}(-1) = \dfrac{9}{2}\sin 3.$

37. (b)

$$f(z) = \frac{e^z}{z(z^2 + 4)} = \frac{e^z}{z(z+2i)(z-2i)}$$

Clearly, $z = -2i$ is a simple pole of $f(z)$.

$\therefore \quad \text{Res}(z = -2i) = \lim\limits_{z \to -2i} \{z - (-2i)\} f(z)$

$= \lim\limits_{z \to -2i} (z + 2i) \times \dfrac{e^z}{z(z+2i)(z-2i)}$

$= \lim\limits_{z \to -2i} \dfrac{e^z}{z(z-2i)} = \dfrac{e^{-2i}}{-2i(-2i-2i)}$

$= \dfrac{e^{-2i}}{8i^2} = -\dfrac{1}{8}e^{-2i}.$

38. (b)

Here, $f(z) = \dfrac{e^{-2z}}{z^2} = \dfrac{e^{-2z}}{(z-0)^2} = \dfrac{\varphi(z)}{(z-0)^2}$ (say)

where $\varphi(z) = e^{-2z}$.

$\therefore \quad z = 0$ is a pole of order 2 and lies within the circle $|z| = 1$.

Now, $\varphi'(z) = -2e^{-2z}$ and so $\varphi'(0) = -2e^0 = -2$.

$\therefore \quad \text{Res}(z = 0) = \dfrac{1}{(2-1)!} \varphi^{(2-1)}(0) = \varphi^{(1)}(0) = -2.$

Hence, by Cauchy's residue theorem,

$$\int\limits_{|z|=1} \frac{e^{-2z}}{z^2} dz = 2\pi i \times \text{Re}\, s(z = 0) = 2\pi i \times (-2) = -4\pi i.$$

39. (a) Let $f: C \to C$ be analytic except for a simple pole at $z = 0$ and $g: C \to C$ be analytic. Then, the value of

$$\frac{\text{Res}\limits_{z \to 0}\{f(z) \times g(z)\}}{\text{Res}\limits_{z \to 0} f(z)}$$

$$= \frac{\lim\limits_{z \to 0}(z - 0) f(z) \times g(z)}{\lim\limits_{z \to 0}(z - 0) f(z)}$$

$$= g(0) \times \frac{\lim\limits_{z \to 0} z f(z)}{\lim\limits_{z \to 0} z f(z)} = g(0).$$

Fully Solved MCQ's (Level-II)

1. The general values of i^i is given by

(a) $e^{-\left(2n+\frac{1}{2}\right)\pi} \quad (n \in Z)$

(b) $e^{-\left(4n+\frac{1}{2}\right)\pi} \quad (n \in Z)$

(c) $e^{-\left(2n-\frac{1}{2}\right)\pi} \quad (n \in Z)$

(d) $e^{-\left(4n-\frac{1}{2}\right)\pi} \quad (n \in Z)$

2. The principal value of $(-1+i)^i$ is given by

(a) $e^{-\frac{\pi}{4}} \left\{ \cos\left(\frac{1}{2}\log 2\right) + i \sin\left(\frac{1}{2}\log 2\right) \right\}$

(b) $e^{-\frac{3\pi}{4}} \left\{ \cos\left(\frac{1}{2}\log 2\right) + i \sin\left(\frac{1}{2}\log 2\right) \right\}$

(c) $e^{\frac{\pi}{4}} \left\{ \cos\left(\frac{1}{2}\log 2\right) + i \sin\left(\frac{1}{2}\log 2\right) \right\}$

(d) $e^{\frac{3\pi}{4}} \left\{ \cos\left(\frac{1}{2}\log 2\right) + i \sin\left(\frac{1}{2}\log 2\right) \right\}$

3. $\sin\left(i \log \dfrac{a - ib}{a + ib}\right) = ?$

(a) $\dfrac{2ab}{a^2 - b^2}$ (b) $\dfrac{ab}{a^2 - b^2}$

(c) $\dfrac{2ab}{a^2 + b^2}$ (d) $\dfrac{ab}{a^2 + b^2}$

4. The equation $\tan\left(i\log\dfrac{x-iy}{x+iy}\right)=2$ represents

    (a) an ellipse   (b) a hyperbola

    (c) a circle

    (d) a rectangular hyperbola

5. If $\sin(a+ib)=x+iy$, then which of the following is true?

    (a) $\dfrac{x^2}{\sin^2 a}+\dfrac{y^2}{\cos^2 a}=1$

    (b) $\dfrac{x^2}{\sin^2 a}-\dfrac{y^2}{\cos^2 a}=1$

    (c) $\dfrac{x^2}{\cosh^2 b}+\dfrac{y^2}{\sinh^2 b}=1$

    (d) $\dfrac{x^2}{\cosh^2 b}-\dfrac{y^2}{\sinh^2 b}=1$

6. If $x+\dfrac{1}{x}=2\cos\dfrac{\pi}{5}$, then $x^5+\dfrac{1}{x^5}=$?

    (a) –2   (b) –1   (c) 0   (d) 1

7. If $z_n=\cos\dfrac{\pi}{2^n}+i\sin\dfrac{\pi}{2^n}$, then

    $z_1 z_2 z_3 \ldots \infty=$?

    (a) 1   (b) –1   (c) 0   (d) 4

8. The polar form of the complex number

    $\sin\dfrac{\pi}{5}+i\left(1-\cos\dfrac{\pi}{5}\right)$ is

    (a) $2\sin\dfrac{\pi}{10}\left(\cos\dfrac{\pi}{10}+i\sin\dfrac{\pi}{10}\right)$

    (b) $2\sin\dfrac{\pi}{5}\left(\cos\dfrac{\pi}{10}+i\sin\dfrac{\pi}{10}\right)$

    (c) $2\cos\dfrac{\pi}{10}\left(\cos\dfrac{\pi}{10}+i\sin\dfrac{\pi}{10}\right)$

    (d) $2\cos\dfrac{\pi}{10}\left(\cos\dfrac{\pi}{10}+i\sin\dfrac{\pi}{10}\right)$

9. Let z be a complex such that $\left|\dfrac{1-iz}{z-i}\right|=1$. Then,

    (a) z is any complex number

    (b) $z=0$

    (c) z is purely real

    (d) $z=\pm i$

10. If α and β be two complex numbers such that $\alpha\ne\beta$ and $|\beta|=1$. Then the value of $\left|\dfrac{\beta-\alpha}{1-\bar{\alpha}\beta}\right|$ is

    (a) 2   (b) 1   (c) 4   (d) $\dfrac{1}{2}$

11. $\sqrt{i}=$?

    (a) $\pm\dfrac{1}{\sqrt{2}}(1+i)$ (b) $\pm\dfrac{1}{\sqrt{2}}(1-i)$

    (c) $\pm\dfrac{1}{\sqrt{2}}(\sqrt{3}+i)$ (d) $\pm i$

12. The product of all the values of $(1+i\sqrt{3})^{\frac{3}{4}}$ is

    (a) 1   (b) 4   (c) 3   (d) 8

13. The function $f(z)=\begin{cases}\dfrac{x^2y^5(x+iy)}{x^4+y^{10}},&z\ne 0\\[2mm]0,&z=0\end{cases}$

    (a) is analytic at $(0,0)$

    (b) not analytic at $(0,0)$

    (c) $f'(0)$ exist   (d) none of these.

14. For the function $f(z)=\begin{cases}\dfrac{xy^2(x+iy)}{x^2+y^4},&z\ne 0\\[2mm]0,&z=0\end{cases}$

    which of the followings is true?

    (a) $f(z)$ is analytic at $z=0$

    (b) $f(z)$ is not analytic at $z=0$

    (c) $f'(0)$ does not exist

    (d) both (b) and (c) are correct

15. Consider $f(z)=\sqrt{|xy|}$. Then,

    (a) CR equations are satisfied at $z=0$

    (b) CR equations are not satisfied at $z=0$

    (c) $f(z)$ is analytic at $z=0$.

    (d) none of these

16. The function $f(z)=|z|$ is

    (a) differentiable for all z

    (b) differentiable except finite number of points

    (c) differentiable except at $z=0$

    (d) not continuous

17. The harmonic conjugate of $u(x,y)=x^3-3xy^2$ is

    (a) $u=y^3+3x^2y$ (b) $u=y^3-3x^2y$

    (c) $u=-y^3+3x^2y$ (d) $u=-y^3-3x^2y$

18. Let $f(z)=u+iu$ be analytic and $u=x^2-y^2$. Then $v=$?

    (a) $2xy$ (b) $-2xy$

    (c) x^2y^2 (d) x^2+2y^2

19. Find an analytic function whose real part is $u(x, y) = e^x(x \cos y - y \sin y)$
(a) $-ze^{-z} + C$ (b) $ze^{-z} + C$
(c) $-ze^z + c$ (d) $ze^z + C$

20. The harmonic conjugate of $v = x^2 - y^2 - 2xy + 2x - 3y$ is ?
(a) $x^2 + y^2 - 2x + 3y$
(b) $x^2 + y^2 - 2xy + 3x + 3y$
(c) $x^2 - 2x + y^2$
(d) $-x^2 + y^2 - 2xy - 3x - 2y$

21. The analytic function $f(z)$ whose imaginary part v is given by $v = e-x(x \cos y + y \sin y)$ is
(a) $ize^{-z} + c$ (b) $ize^z + c$
(c) $ze^{-z} + c$ (d) $ze^z + c$

22. If $f(z) = u + iu$ be analytic and $v - u = e^x(\cos y - \sin y)$, then $f(z) = ?$
(a) $ze^{-z} + c$ (b) $e^{-z} + c$
(c) $e^z + c$ (d) $ze^z + c$

23. Let us consider a function $f(z)$ defined by
$$f(z) = \begin{cases} \dfrac{(\bar{z})^2}{z}, & z \neq 0 \\ 0, & z = 0 \end{cases}$$

Then,
(a) CR equations are not satisfied at $(0, 0)$ and $f'(0)$ exist
(b) CR equations are not satisfied at $(0, 0)$
(c) $f'(0)$ does not exist
(d) $f(z)$ is not analytic at $z = 0$ although CR equations are satisfied at $z = 0$

24. Let $u(x, y) = 2x(1 - y)$ for all real x and y. Then, a function $v(x, y)$, so that $f(z) = u(x, y) + iv(x, y)$ is analytic, is
(a) $x^2 - (y - 1)^2$ (b) $(x - 1)2 - y^2$
(c) $(x - 1)2 + y^2$ (d) $x^2 + (y - 1)^2$

25. The value of $\int_C \dfrac{zdz}{(z^2+9)(z-i)}$, where $c : |z| = 2$ is given by
(a) $\dfrac{\pi}{2}$ (b) $-\dfrac{\pi}{2}$ (c) $\dfrac{\pi}{4}$ (d) $-\dfrac{\pi}{4}$

26. The value of the integral $\int_C \dfrac{dz}{z(z+\pi i)}$, where C is the circle $|z + 3i| = 1$, is
(a) -2 (b) -3 (c) -4 (d) 0

27. The value of $\int_C \dfrac{e^{2z} dz}{z - \pi i}$, where C is the ellipse $|z - 2| + |z + 2| = 6$, is
(a) -2π (b) $2\pi ie^2\pi$
(c) 0 (d) 1

28. The value of integral $\int_C \dfrac{dz}{z^2 + 2z + 2}$ where C is the square having vertices at $(0, 0)$, $(-2, 0)$, $(2, -2)$, and $(0, -2)$, is given by
(a) $\dfrac{\pi}{2}$ (b) π (c) $-\pi$ (d) $\dfrac{2\pi}{3}$

29. The value of $\int_C \dfrac{\sin z dz}{\left(z - \dfrac{\pi}{4}\right)^3}$, where $C : \left| z - \dfrac{\pi}{4} \right| = \dfrac{1}{2}$, is
(a) $-\dfrac{\pi i}{\sqrt{2}}$ (b) $\dfrac{\pi i}{\sqrt{2}}$ (c) $\dfrac{\pi i}{2}$ (d) 0

30. The value of $\int_C \dfrac{e^{-z} dz}{z^2}$, where $c : |z| = 1$, is
(a) πi (b) $-2\pi i$ (c) $2\pi i$ (d) $\dfrac{-\pi i}{2}$

31. The value of $\int_C \dfrac{1}{z} dz$, where C is the circle $z = e^{i\theta}, 0 \leq \theta \leq \pi$, is
(a) $-\dfrac{\pi i}{2}$ (b) $\dfrac{\pi i}{2}$ (c) $-\pi i$ (d) πi

32. The coefficient of $(z - \pi)^2$ in the Taylor series expansion of
$$f(z) = \begin{cases} \dfrac{\sin z}{z - \pi}, & \text{if } z \neq \pi \\ -1, & \text{if } z = \pi \end{cases}$$

around π is
(a) $\dfrac{1}{3}$ (b) $-\dfrac{1}{2}$ (c) $\dfrac{1}{6}$ (d) $-\dfrac{1}{6}$

33. The Taylor series expansion of the function $f(z) = \dfrac{z^2 + 1}{(z+2)(z+3)}$ in the region $|z| < 2$ is given by
(a) $1 + \sum_{n=0}^{\infty} (-1)^n \left\{ \dfrac{5}{2^{n+1}} - \dfrac{10}{3^{n+1}} \right\} z^n$
(b) $1 - \sum_{n=0}^{\infty} (-1)^n \left\{ \dfrac{5}{2^{n+1}} - \dfrac{10}{3^{n+1}} \right\} z^n$

(c) $\dfrac{1}{2}+\sum\limits_{n=0}^{\infty}(-1)^{n}\left\{\dfrac{3}{2^{n}}-\dfrac{2}{3^{n}}\right\}z^{n}$

(d) none of these

34. The Laurent series expansion of $f(z)=\dfrac{1}{z^{2}-5z+6}$ in $2<|z|<3$ is given by

(a) $-\dfrac{1}{3}\sum\limits_{n=0}^{\infty}(-1)^{n}\left(\dfrac{z}{3}\right)^{n}+\dfrac{1}{2}\sum\limits_{n=0}^{\infty}(-1)^{n}\left(\dfrac{z}{2}\right)^{n}$

(b) $\dfrac{1}{3}\sum\limits_{n=0}^{\infty}(-1)^{n}\left(\dfrac{z}{3}\right)^{n}-\dfrac{1}{2}\sum\limits_{n=0}^{\infty}(-1)^{n}\left(\dfrac{z}{2}\right)^{n}$

(c) $-\dfrac{1}{3}\sum\limits_{n=0}^{\infty}\left(\dfrac{z}{3}\right)^{n}-\dfrac{1}{z}\sum\limits_{n=0}^{\infty}\left(\dfrac{2}{z}\right)^{n}$

(d) none of these

35. The Laurent series expansion of $f(z)=\dfrac{z}{(z+1)(z+2)}$ about $z=-1$ is given by

(a) $\dfrac{1}{2}\sum\limits_{z=0}^{\infty}(z+1)^{n}$

(b) $\dfrac{1}{z+1}+\sum\limits_{n=0}^{\infty}(-1)^{n}(z+1)^{n}$

(c) $-\dfrac{1}{z+1}+2\sum\limits_{n=0}^{\infty}(-1)^{n}(z+1)^{n}$

(d) none of these

36. The Taylor series expansion of $f(z)=\cos z$ about $z=\dfrac{\pi}{4}$ is given by

(a) $\sum\limits_{n=0}^{\infty}\cos\left(\dfrac{\pi}{4}+\dfrac{n\pi}{2}\right)\dfrac{\left(z-\dfrac{\pi}{4}\right)^{n}}{n!}$

(b) $\sum\limits_{n=0}^{\infty}\sin\left(\dfrac{\pi}{4}+\dfrac{n\pi}{2}\right)\dfrac{\left(z-\dfrac{\pi}{4}\right)^{n}}{n!}$

(c) $\sum\limits_{n=0}^{\infty}\sin\left(\dfrac{n\pi}{4}-\dfrac{\pi}{4}\right)\left(z-\dfrac{\pi}{4}\right)^{n}$

(d) none of these

37. The Taylor series expansion of $f(z)=\dfrac{1}{(z^{2}+1)(z^{2}+2)}$ is powers of z when $|z|<1$, is given by

(a) $\sum\limits_{n=0}^{\infty}\left(1+\dfrac{1}{2^{n+1}}\right)z^{2n}$ (b) $\sum\limits_{n=0}^{\infty}(-1)^{n}\left(1-\dfrac{1}{2^{n+1}}\right)z^{2n}$

(c) $\sum\limits_{n=0}^{\infty}\left(1-\dfrac{1}{2^{n}}\right)z^{n}$ (d) none of these

38. Find the residue of $f(z)$ at $z=i$ if
$$f(z)=\dfrac{1}{\left(z^{2}+1\right)^{3}}$$

(a) $\dfrac{3}{16i}$ (b) $\dfrac{3i}{16}$ (c) $\dfrac{5i}{16}$ (d) $\dfrac{5}{16i}$

39. The residue of $z^{3}\cos\left(\dfrac{1}{z-2}\right)$ at $z=2$ is given by

(a) $-\dfrac{143}{24}$ (b) $\dfrac{11}{24}$ (c) $-\dfrac{143}{21}$ (d) 0

40. The value of the integral $\displaystyle\int_{C}\dfrac{z\,dz}{(z-1)(z-2)^{2}}$, where C is the circle $|z-2|=\dfrac{1}{2}$, is

(a) 0 (b) πi (c) $2\pi i$ (d) $-2\pi i$

41. Let $f(z)=\sum\limits_{n=0}^{15}z^{n}$ for $z\in C$. If $C:|z-i|=2$, then

$$\oint_{C}\dfrac{f(z)\,dz}{(z-i)^{15}}\text{ is equal to}$$

(a) $2\pi i(1+15i)$ (b) $2\pi i(1-15i)$

(c) $4\pi i(1+15i)$ (d) $2\pi i$

42. Let C be the positively oriented circle given by $|z-3i|=2$. Then, the value of $\displaystyle\int_{C}\dfrac{dz}{z^{2}+4}$ is

(a) $-\dfrac{\pi}{2}$ (b) $\dfrac{\pi}{2}$ (c) $-\dfrac{i\pi}{2}$ (d) $\dfrac{i\pi}{2}$

Answer key				
1. (a)	**2.**(b)	**3.**(c)	**4.**(d)	**5. (c)**
6.(a)	**7.**(b)	**8.**(a)	**9.**(c)	**10. (b)**
11.(a)	**12.**(d)	**13.**(b)	**14.**(d)	**15.** (a)
16.(c)	**17.**(c)	**18.**(a)	**19.**(d)	**20.** (d)
21.(a)	**22.**(c)	**23.**(d)	**24.**(a)	**25.** (d)
26.(a)	**27.**(c)	**28.**(c)	**29.**(a)	**30.** (b)
31.(d)	**32.**(c)	**33.**(a)	**34.**(c)	**35.** (c)
36.(a)	**37.**(b)	**38.**(a)	**39.**(a)	**40.** (d)
41.(a)	**42.** (b)			

Explanation

1. (a)

$$i^i = e^{Log\, i^i} = e^{iLog\, i} = e^{i\{\log i + 2n\pi i\}}$$

$$= e^{i\{\log|i| + iamp(i) + 2n\pi i\}}$$

$$= e^{i\{\log\sqrt{0^2+1^2} + i\tan^{-1}(\infty) + 2n\pi i\}}$$

$$= e^{i\{\log 1 + i\frac{\pi}{2} + 2n\pi i\}}$$

$$= e^{-\frac{\pi}{2} - 2n\pi} \qquad (\because \log 1 = 0)$$

$$= e^{-\left(2n+\frac{1}{2}\right)\pi}, \; n \in Z.$$

2. (b)

Let $z = -1 + i$. Then $r = |z| = \sqrt{(-1)^2 + 1^2} = \sqrt{2}$.

$$amp(z) = \theta = \pi - \tan^{-1}\left|\frac{1}{-1}\right| = \pi - \frac{\pi}{4} = \frac{3\pi}{4}.$$

$\because$ $z = -1 + i$ represents a point $(-1, 1)$ that lies in second quadrant, so

$$amp(z) = \pi - \tan^{-1}\left|\frac{1}{-1}\right| = \pi - \frac{\pi}{4} = \frac{3\pi}{4}.$$

$$\therefore z = r(\cos\theta + i\sin\theta) = \sqrt{2}\left(\cos\frac{3\pi}{4} + i\sin\frac{3\pi}{4}\right).$$

Now, $(-1 + i)^i$

$$= e^{Log(-1+i)^i} = e^{iLog(-1+i)}$$

$$= e^{iLog\,z} = e^{i\{\log z + 2n\pi i\}} \qquad \text{(taking } z = -1 + i\text{)}$$

$$= e^{i\{\log|z| + iamp(z) + 2n\pi i\}}$$

$$= e^{i\{\log\sqrt{2} + i\frac{3\pi}{4} + 2n\pi i\}} = e^{i\log\sqrt{2} - \frac{3\pi}{4} - 2n\pi}, \; n \in z$$

Then the principal value is obtained by

putting $n = 0$ and is given by $e^{i\log\sqrt{2} - \frac{3\pi}{4}}$.

$$i.e., \; e^{-\frac{3\pi}{4}}\left\{\cos\left(\log\sqrt{2}\right) + i\sin\left(\log\sqrt{2}\right)\right\}$$

$$i.e., \; e^{-\frac{3\pi}{4}}\left\{\cos\left(\frac{1}{2}\log 2\right) + i\sin\left(\frac{1}{2}\log 2\right)\right\}.$$

3. (c) Let $a = r\cos\theta$ and $b = r\sin\theta$.

Then, $\dfrac{a - ib}{a + ib} = \dfrac{r(\cos\theta - i\sin\theta)}{r(\cos\theta + i\sin\theta)} = \dfrac{re^{-i\theta}}{re^{i\theta}} = e^{-2i\theta}$

$$\therefore \sin\left(i\log\frac{a-ib}{a+ib}\right) = \sin\left(i\log e^{-2i\theta}\right)$$

$$= \sin\{i \times (-2i\theta)\} = \sin 2\theta$$

$$= \frac{2\tan\theta}{1 + \tan^2\theta} = \frac{2\times\dfrac{b}{a}}{1 + \left(\dfrac{b}{a}\right)^2} = \frac{2ab}{a^2 + b^2}.$$

$$\left(\because \frac{b}{a} = \frac{r\sin\theta}{r\cos\theta} = \tan\theta\right)$$

4. (d)

Let $x = r\cos\theta$, $y = r\sin\theta$. Then, $\tan\theta = \dfrac{y}{x}$.

Then, $\tan\left(i\log\dfrac{x-iy}{x+iy}\right) = 2$

$$\Rightarrow \tan\left[i\log\left\{\frac{r(\cos\theta - i\sin\theta)}{r(\cos\theta + i\sin\theta)}\right\}\right] = 2$$

$$\Rightarrow \tan\left(i\log\frac{e^{-i\theta}}{e^{i\theta}}\right) = 2 \Rightarrow \tan\left(i\log e^{-2i\theta}\right) = 2$$

$$\Rightarrow \tan\{i \times (-2i\theta)\} = 2 \Rightarrow \tan(2\theta) = 2$$

$$\Rightarrow \frac{2\tan\theta}{1 - \tan^2\theta} = 2 \Rightarrow \frac{2\dfrac{y}{x}}{1 - \left(\dfrac{y}{x}\right)^2} = 2$$

$\Rightarrow xy = x^2 - y^2$, which is a rectangular hyperbola.

5. (c)

$\sin(a + ib) = x + iy$

$\Rightarrow \sin a \cos(ib) + \cos a \sin(ib) = x + iy$

$\Rightarrow \sin a \cosh b + i\cos a \sinh b = x + iy$

$\left[\because \cos(ib) = \cosh b \text{ and } \sin(ib) = i\sinh b\right]$

$\Rightarrow x = \sin a \cosh b, \; y = \cos a \sinh b$

$$\therefore \frac{x^2}{\cosh^2 b} + \frac{y^2}{\sinh^2 b}$$

$$= \frac{\sin^2 a \cosh^2 b}{\cosh^2 b} + \frac{\cos^2 a \sinh^2 b}{\sinh^2 b}$$

$$= \sin^2 a + \cos^2 a = 1.$$

6. (a)

$$x + \frac{1}{x} = 2\cos\frac{\pi}{5}$$

$$\Rightarrow x^2 - 2x\cos\frac{\pi}{5} + 1 = 0$$

$$\Rightarrow x = \frac{2\cos\frac{\pi}{5} \pm \sqrt{4\cos^2\frac{\pi}{5} - 4}}{2}$$

$$= \frac{2\cos\frac{\pi}{5} \pm 2\sqrt{-1}\sqrt{1 - \cos^2\frac{\pi}{5}}}{2}$$

$$= \cos\frac{\pi}{5} \pm i\sin\frac{\pi}{5}$$

$$\therefore \quad x^5 = \left(\cos\frac{\pi}{5} \pm i\sin\frac{\pi}{5}\right)^5 = \cos\frac{5\pi}{5} \pm i\sin\frac{5\pi}{5}$$

(by De Moivre's theorem)

$$= \cos\pi \pm i\sin\pi = -1 \pm 0 = -1$$

Hence, $x^5 + \dfrac{1}{x^5} = -1 + \dfrac{1}{(-1)} = -2$.

7. (b)

$$z_1 = \cos\frac{\pi}{2} + i\sin\frac{\pi}{2}, \quad z_2 = \cos\frac{\pi}{2^2} + i\sin\frac{\pi}{2^2},$$

$$z_3 = \cos\frac{\pi}{2^3} + i\sin\frac{\pi}{2^3}, \quad z_4 = \cos\frac{\pi}{2^4} + i\sin\frac{\pi}{2^4},$$

and so on.

$$\therefore \quad z_1 z_2 z_3 z_4 \cdots \text{ up to } \infty$$

$$= \left(\cos\frac{\pi}{2} + i\sin\frac{\pi}{2}\right) \times \left(\cos\frac{\pi}{2^2} + i\sin\frac{\pi}{2^2}\right)$$

$$\times \left(\cos\frac{\pi}{2^3} + i\sin\frac{\pi}{2^3}\right) \times \left(\cos\frac{\pi}{2^4} + i\sin\frac{\pi}{2^4}\right) \times \cdots\infty$$

$$= \cos\left(\frac{\pi}{2} + \frac{\pi}{2^2} + \frac{\pi}{2^3} + \frac{\pi}{2^4} + \cdots\infty\right)$$

$$+ i\sin\left(\frac{\pi}{2} + \frac{\pi}{2^2} + \frac{\pi}{2^3} + \frac{\pi}{2^4} + \cdots\infty\right)$$

[using $(\cos\theta_1 + i\sin\theta_1) \times (\cos\theta_2 + i\sin\theta_2)$

$$\times (\cos\theta_3 + i\sin\theta_3) \times \cdots\infty$$

$$= \cos(\theta_1 + \theta_2 + \cdots\infty) + i\sin(\theta_1 + \theta_2 + \cdots\infty)]$$

$$= \cos\left(\frac{\frac{\pi}{2}}{1 - \frac{1}{2}}\right) + i\sin\left(\frac{\frac{\pi}{2}}{1 - \frac{1}{2}}\right)$$

$$\left[\because a + ar + ar^2 + ar^3 + \cdots\infty = \frac{a}{1-r}\right]$$

$$= \cos\pi + i\sin\pi = -1 + i \times 0 = -1.$$

8. (a)

Let $z = \sin\dfrac{\pi}{5} + i\left(1 - \cos\dfrac{\pi}{5}\right)$

Then, $r = |z| = \sqrt{\sin^2\dfrac{\pi}{5} + \left(1 - \cos\dfrac{\pi}{5}\right)^2}$

$$= \sqrt{\sin^2\frac{\pi}{5} + 1 - 2\cos\frac{\pi}{5} + \cos^2\frac{\pi}{5}}$$

$$= \sqrt{2 - 2\cos\frac{\pi}{5}}$$

$$= \sqrt{2\left(1 - \cos\frac{\pi}{5}\right)} = \sqrt{2 \times 2\sin^2\frac{\pi}{10}}$$

$$= 2\sin\frac{\pi}{10}$$

Let $\text{amp}(z) = \theta$.

Then, $\tan\theta = \dfrac{1 - \cos\dfrac{\pi}{5}}{\sin\dfrac{\pi}{5}} = \dfrac{2\sin^2\dfrac{\pi}{10}}{2\sin\dfrac{\pi}{10}\cos\dfrac{\pi}{10}}$

$$= \tan\frac{\pi}{10}$$

$$\therefore \theta = \frac{\pi}{10}$$

Hence, the required polar form is given by:

$$z = r(\cos\theta + i\sin\theta)$$

$$= 2\sin\frac{\pi}{10}\left(\cos\frac{\pi}{10} + i\sin\frac{\pi}{10}\right).$$

9. (c)

Let $z = x + iy$. Then, $\left|\dfrac{1 - iz}{z - i}\right| = 1$

$$\Rightarrow \left|\frac{1 - i(x + iy)}{x + iy - i}\right| = 1 \Rightarrow \left|\frac{(1+y) - ix}{x + (y-1)i}\right| = 1$$

$$\Rightarrow |(1+y) - ix| = |x + (y-1)i|$$

$$\Rightarrow \sqrt{(y+1)^2 + (-x)^2} = \sqrt{x^2 + (y-1)^2}$$

$$\Rightarrow y^2 + 2y + 1 + x^2 = x^2 + y^2 - 2y + 1$$

or, $4y = 0$ or $y = 0$

$\therefore \quad z = x + iy = x + i \times 0 = x.$

Hence, z is purely real.

10. (b)

$$\left|\frac{\beta-\alpha}{1-\overline{\alpha}\beta}\right| = \left|\frac{\beta-\alpha}{\beta\overline{\beta}-\overline{\alpha}\beta}\right| \quad \left(\because |\beta|=1 \Rightarrow |\beta|^2 = \beta\overline{\beta}=1\right)$$

$$= \left|\frac{\beta-\alpha}{\beta(\overline{\beta}-\overline{\alpha})}\right| = \left|\frac{\beta-\alpha}{\beta(\overline{\beta-\alpha})}\right|$$

$$= \frac{|\beta-\alpha|}{|\beta||\overline{\beta-\alpha}|} = \frac{|\beta-\alpha|}{|\beta||\beta-\alpha|} = \frac{1}{|\beta|} = 1 \quad \left(\because |z| = |\overline{z}| = 1\right)$$

11. (a)

i

$$= \frac{1}{2} \times 2i = \frac{1}{2} \times (1 + 2i - 1)$$

$$= \frac{1}{2}\left\{1^2 + 2i + i^2\right\} = \frac{1}{2}(1+i)^2$$

$$\therefore \sqrt{i} = \pm\frac{1}{\sqrt{2}}(1+i).$$

12. (d) Let $z = 1 + i\sqrt{3}$. Then z represents the point $(1,\sqrt{3})$ that lies in first quadrant.

Now, $r = |z| = \sqrt{1^2 + \left(\sqrt{3}\right)^2} = \sqrt{4} = 2$.

If amp$(z) = \theta$, then $\theta = \tan^{-1}\left|\frac{\sqrt{3}}{1}\right| = \frac{\pi}{3}$.

Hence, $z = r(\cos\theta + i\sin\theta) = 2\left(\cos\frac{\pi}{3} + i\sin\frac{\pi}{3}\right)$

$$\therefore \left(1+i\sqrt{3}\right)^{\frac{3}{4}} = z^{\frac{3}{4}} = \left\{2\left(\cos\frac{\pi}{3} + i\sin\frac{\pi}{3}\right)\right\}^{3/4}$$

$$= 2^{3/4}\left(\cos\frac{3\pi}{3} + i\sin\frac{3\pi}{3}\right)^{1/4}$$

(using De Moivre's theorem)

$$= 2^{3/4}(\cos\pi + i\sin\pi)^{1/4}$$

$$= 2^{3/4}\left\{\cos(2n\pi + \pi) + i\sin(2n\pi + \pi)\right\}^{1/4} \ (n \in Z)$$

$$= 2^{3/4}\left\{\cos\frac{(2n+1)\pi}{4} + i\sin\frac{(2n+1)\pi}{4}\right\},$$

$$n = 0,1,2,3$$

$$= 2^{3/4}e^{\frac{i(2n+1)\pi}{4}}, n = 0,1,2,3$$

So the values of $\left(1+i\sqrt{3}\right)^{\frac{3}{4}}$ are:

$$2^{\frac{3}{4}}e^{\frac{i\pi}{4}}, 2^{\frac{3}{4}}e^{\frac{3i\pi}{4}}, 2^{\frac{3}{4}}e^{\frac{5i\pi}{4}} \text{ and } 2^{\frac{3}{4}}e^{\frac{7i\pi}{4}}.$$

Hence, the product of these values

$$= 2^{\frac{3}{4}\times4} \times e^{\frac{i\pi}{4}(1+3+5+7)} = 2^3 \times e^{4\pi i} = 8$$

$$\left(\because e^{2n\pi i} = 1 \text{ for } n \in N\right)$$

Remember:

$$z^{\frac{1}{k}} = \left\{r(\cos\theta + i\sin\theta)\right\}^{\frac{1}{k}}$$

$$= r^{\frac{1}{k}}(\cos\theta + i\sin\theta)^{\frac{1}{k}}$$

$$= r^{\frac{1}{k}}\left\{\cos(2n\pi + \theta) + i\sin(2n\pi + \theta)\right\}^{\frac{1}{k}}, n \in Z$$

$$= r^{\frac{1}{k}}\left\{\cos\left(\frac{2n\pi+\theta}{k}\right) + i\sin\left(\frac{2n\pi+\theta}{k}\right)\right\},$$

where $n = 0, 1, 2, 3, ..., k-1$.

13. (b)

$f'(0)$

$$= \lim_{z\to0}\frac{f(z) - f(0)}{z}$$

$$= \lim_{z\to0}\frac{\frac{x^2y^5(x+iy)}{x^4+y^{10}} - 0}{x+iy} = \lim_{z\to0}\frac{x^2y^5}{x^4+y^{10}}.........(1)$$

Let us consider a curve $x^2 = my^5$

(m = arbitrary constant).

Then, $z = x + iy = \sqrt{m}y^{\frac{5}{2}} + iy = \left(\sqrt{m}y^{\frac{3}{2}} + i\right)y$.

$\therefore \quad z \to 0 \Rightarrow y \to 0.$

Then, $(1) \Rightarrow f'(0) = \lim_{y\to0}\frac{my^5y^5}{m^2y^{10}+y^{10}}$

$$= \lim_{y\to0}\frac{m}{m^2+1} = \frac{m}{m^2+1}$$

which is not unique as it depends on m.

So $f'(0)$ does not exist, and hence, $f(z)$ is not analytic at $(0, 0)$, *i.e.*, at $z = 0$.

14. (d)

$$f'(0) = \lim_{z \to 0} \frac{f(z) - f(0)}{z} = \lim_{z \to 0} \frac{\frac{xy^2(x+iy)}{x^2+y^4} - 0}{x+iy}$$

$$= \lim_{z \to 0} \frac{xy^2}{x^2+y^4} \quad\ldots\ldots\ldots(1)$$

Let us consider a curve $x = my^2$ (m = arbitrary constant).

So, $z = x + iy = my^2 + iy = (my + i)y.$

$\therefore \quad z \to 0 \Rightarrow y \to 0$

Then, $(1) \Rightarrow f'(0) = \lim_{y \to 0} \frac{my^2 y^2}{m^2 y^4 + y^4}$

$$= \lim_{y \to 0} \frac{m}{m^2+1} = \frac{m}{m^2+1}$$

which is not unique it depends on m.

So $f'(0)$ does not exist and hence $f(z)$ is not analytic at $z = 0$.

15. (a)

$$f(z) = \sqrt{|xy|} = \sqrt{|xy|} + i0 = u(x,y) + iv(x,y).$$

$\therefore u(x,y) = \sqrt{|xy|}, v(x,y) = 0$ and so

$u(x,0) = 0, u(0,y) = 0, v(x,0) = 0, v(0,y) = 0.$

Moreover, $f(0) = 0 \Rightarrow u(0,0) = 0, v(0,0) = 0.$

Then,

$$\frac{\partial u}{\partial x}\bigg|_{(0,0)} = \lim_{x \to 0} \frac{u(x,0) - u(0,0)}{x} = \lim_{x \to 0} \frac{0-0}{x} = 0,$$

$$\frac{\partial u}{\partial y}\bigg|_{(0,0)} = \lim_{y \to 0} \frac{u(0,y) - u(0,0)}{y} = \lim_{y \to 0} \frac{0-0}{y} = 0,$$

$$\frac{\partial v}{\partial x}\bigg|_{(0,0)} = \lim_{x \to 0} \frac{v(x,0) - v(0,0)}{x} = \lim_{x \to 0} \frac{0-0}{x} = 0,$$

$$\frac{\partial v}{\partial y}\bigg|_{(0,0)} = \lim_{y \to 0} \frac{v(0,y) - v(0,0)}{y} = \lim_{y \to 0} \frac{0-0}{y} = 0.$$

Thus, $\frac{\partial u}{\partial x}\bigg|_{(0,0)} = \frac{\partial v}{\partial y}\bigg|_{(0,0)}$ & $\frac{\partial u}{\partial y}\bigg|_{(0,0)} = -\frac{\partial v}{\partial x}\bigg|_{(0,0)}$.

Hence, CR equations are satisfied at $(0, 0)$, i.e., at $z = 0$.

$$f'(0) = \lim_{z \to 0} \frac{f(z) - f(0)}{z} = \lim_{z \to 0} \frac{\sqrt{|xy|} - 0}{x+iy}$$

$$= \lim_{x \to 0} \frac{\sqrt{|x \times mx|}}{x+imx}$$

$\left[\text{Taking } y = mx \text{ so that } z \to 0 \Rightarrow x \to 0\right]$

$$= \lim_{x \to 0} \frac{\sqrt{|m|}}{1+im} = \frac{\sqrt{|m|}}{1+im},$$

which depends on m and so is not unique.

$\therefore \quad f'(z)$ does not exist and so $f(z)$ is not analytic.

16. (c)

$$f'(0) = \lim_{z \to 0} \frac{f(z) - f(0)}{z}$$

$$= \lim_{z \to 0} \frac{|z| - 0}{x+iy} \quad (\because f(0) = |0| = 0)$$

$$= \lim_{z \to 0} \frac{\sqrt{x^2+y^2}}{x+iy}$$

Let us consider a curve $y = mx$ (m = arbitrary constant).

Then, $z = x + iy = x + imx = (1 + im)x.$

$\therefore \quad z \to 0 \Rightarrow x \to 0$

Then, $(1) \Rightarrow f'(0) = \lim_{x \to 0} \frac{\sqrt{x^2 + m^2 x^2}}{x+imx}$

$$= \lim_{x \to 0} \frac{\sqrt{1+m^2}}{1+im} = \frac{\sqrt{1+m^2}}{1+im},$$

which is not unique as it depends on m.

So $f'(0)$ does not exist and hence $f(z)$ is not differentiable at $z = 0$.

17. (c)

Let v be the harmonic conjugate of u.

Then, $dv = \frac{\partial v}{\partial x} dx + \frac{\partial v}{\partial y} dy$

$$= -\left(\frac{\partial u}{\partial y}\right)dx + \left(\frac{\partial u}{\partial x}\right)dy \text{ (by } CR \text{ equations)}$$

$$= -(-6xy)dx + \left(3x^2 - 3y^2\right)dy$$

$$= 6xy\,dx + \left(3x^2 - 3y^2\right)dy$$

$$= M dx + N dy \text{ (say)}$$

Now, $\int M dx = \int 6xy\,dx = 6y\int x\,dx = 6y \times \frac{x^2}{2} = 3x^2 y,$

$\int N dy = \int \left(3x^2 - 3y^2\right)dy = 3x^2 \int dy - 3\int y^2 dy$

$$= 3x^2 y - y^3$$

$\therefore v = \int M dx + \int N dy = 3x^2 y - y^3.$

18. (a)

$$u = x^2 - y^2 \Rightarrow \frac{\partial u}{\partial x} = 2x, \frac{\partial u}{\partial y} = -2y.$$

$\therefore du = \dfrac{\partial u}{\partial x}dx + \dfrac{\partial u}{\partial y}dy$

$= \left(-\dfrac{\partial u}{\partial y}\right)dx + \left(\dfrac{\partial u}{\partial x}\right)dy$ (by the CR equations)

$-(-(-2y))dx + (2x)dy - (-24)$

$= 2ydx + 2xdy$

$= 2(ydx + xdy) = 2d(xy)$

Integrating we get, $v = 2xy$.

19. (d)

$\dfrac{\partial u}{\partial x} = e^x(x\cos y - y\sin y) + e^x(\cos y - 0)$

$\therefore \dfrac{\partial u}{\partial x}\Big|_{(z,0)} = e^z(z\cos 0 - 0) + e^z(\cos 0)$

$\qquad = ze^z + e^z = (z+1)e^z.$

$\dfrac{\partial u}{\partial y} = e^x(-x\sin y - \sin y - y\cos y).$

$\therefore \dfrac{\partial u}{\partial y}\Big|_{(z,0)} = e^z(-z\times 0 - 0 - 0) = 0.$

By the Milne Thomson method,

$f(z) = \int\left\{\dfrac{\partial u}{\partial x}\Big|_{(z,0)} - i\dfrac{\partial u}{\partial y}\Big|_{(z,0)}\right\}dz + C$

$= \int(z+1)e^z dz + C$

$= (z+1)\int e^z dz - \left[\int\dfrac{d}{dz}(z+1)\int e^z dz\right]dz + C$

$= (z+1)e^z - \int e^z dz + C$

$= (zH)e^z - e^z + C = ze^z + C.$

20. (d)

$v = x^2 - y^2 - 2xy + 2x - 3y$

$\Rightarrow \dfrac{\partial v}{\partial x} = 2x - 2y + 2, \dfrac{\partial v}{\partial y} = -2y - 2x - 3.$

$du = \dfrac{\partial u}{\partial x}dx + \dfrac{\partial u}{\partial y}dy$

$= \dfrac{\partial v}{\partial y}dx + \left(-\dfrac{\partial v}{\partial x}\right)dy$

$= (-2y - 2x - 3)dx + \{-(2x - 2y + 2)\}dy$

$-(2x + 2y + 3)dx + (2y - 2x - 2)dy$

$= Mdx + Ndy$ (say)

Then,

$\int Mdx = -\int(2x + 2y + 3)dx$

$\qquad = -2\int xdx - 2y\int dx - 3\int dx$

$= -x^2 - 2xy - 3x,$

$\int Ndx = \int(2y - 2x - 2)$

$\qquad = 2\int ydy - 2x\int dy - 2\int dy$

$= y^2 - 2xy - 2y$

$\therefore u = \int Mdx + Ndy$

$= -x^2 - 2xy - 3x + y^2 - 2y$

21. (a)

Here, $v = e^{-x}(x\cos y + y\sin y)$

$\therefore \dfrac{\partial v}{\partial x} = -e^{-x}(x\cos y + y\sin y) + e^{-x}\cos y$

and so $\dfrac{\partial v}{\partial x}\Big|_{(z,0)} = -e^{-z}(z+0) + e^{-z} = (1-z)e^{-z}.$

Again, $\dfrac{\partial v}{\partial y} = e^{-x}(-x\sin y + y\cos y + \sin y)$

and so $\dfrac{\partial v}{\partial y}\Big|_{(z,0)} = 0.$

Hence, by the Milne Thomson method,

$f(z) = \int\left\{\dfrac{\partial v}{\partial y}\Big|_{(z,0)} + i\dfrac{\partial v}{\partial x}\Big|_{(z,0)}\right\}dz + c$

$= \int\{0 + i(1-z)e^{-z}\}dz + c$

$= i\left[(1-z)\int e^{-z}dz - \int\left\{\dfrac{d}{dz}(1-z)\int e^{-z}dz\right\}dz\right] + c$

$= i\left\{(1-z)(-e^{-z}) - \int(-1)(-1)e^{-z}dz\right\} + c$

$= i\{(z-1)e^{-z} + e^{-z}\} + c$

$= ize^{-z} + c$

22. (c) $f(z) = u + iv \Rightarrow if(z) = iu + i^2v = iu - v$

$\therefore f(z) + if(z) = (u+iv) + (iu-v)$

$\qquad = (u-v) + i(u+v)$

$\Rightarrow \quad (1+i)f(z) = (u-v) + i(u+v)$

$\Rightarrow \quad F(z) = U + iV$ (say), where,

$F(z) = (1 + i)f(z)$, which is analytic, $U = u - v = e^x(\cos y - \sin y)$ (given)

$V = u + v$.

Now, $U = e^x(\cos y - \sin y)$

$\Rightarrow \dfrac{\partial U}{\partial x} = e^x(\cos y - \sin y), \dfrac{\partial U}{\partial y} = e^x(-\sin y - \cos y)$

$\Rightarrow \dfrac{\partial U}{\partial x}\bigg|_{(z,0)} = e^z(1 - 0), \dfrac{\partial U}{\partial y}\bigg|_{(z,0)} = e^z(-0 - 1)$

$\Rightarrow \dfrac{\partial U}{\partial x}\bigg|_{(z,0)} = e^z, \dfrac{\partial U}{\partial y}\bigg|_{(z,0)} = -e^z .$

Then by the Milne Thomson method,

$F(z) = \int \left\{ \dfrac{\partial U}{\partial x}\bigg|_{(z,0)} - i\dfrac{\partial U}{\partial y}\bigg|_{(z,0)} \right\} dz + C'$

$= \int \left\{ e^z - i\left(-e^z \right) \right\} dz + C'$

$= (1 + i)\int e^z dz + (1 + i)C \quad [\text{taking} (1 + i)C = C']$

$\Rightarrow (1 + i)f(z) = (1 + i)e^z + (1 + i)C$

$\Rightarrow f(z) = e^z + C.$

23. (d)

$f'(0) = \lim_{z \to 0} \dfrac{f(z) - f(0)}{z} = \lim_{z \to 0} \dfrac{\dfrac{(\bar{z})^2}{z} - 0}{z}$

$= \lim_{z \to 0} \dfrac{\dfrac{(x - iy)^2}{x + iy}}{(x + iy)} \quad \left(\because z = x + iy, \text{so } \bar{z} = x - iy \right)$

$= \lim_{z \to 0} \dfrac{(x - iy)^2}{(x + iy)^2}$

$= \lim_{z \to 0} \dfrac{(x - imx)^2}{(x + imx)^2} \quad \begin{bmatrix} \text{taking } y = mx \text{ so that } x \to 0 \\ \text{whenever } z \to 0 \end{bmatrix}$

$= \lim_{z \to 0} \dfrac{(1 - im)^2}{(1 + im)^2} = \left(\dfrac{1 - im}{1 + im} \right)^2,$ which depends on "m" and so is not unique.

$\therefore \quad f'(0)$ does not exist and hence, $f(z)$ is not analytic at $z = 0$.

Now, $f(z) = \dfrac{(\bar{z})^2}{z} = \dfrac{(x - iy)^2}{(x + iy)} = u + iv$ (for $z \neq 0$)

$\Rightarrow u + iv = \dfrac{(x - iy)^3}{(x + iy)(x - iy)}$

$= \dfrac{x^3 - 3x^2 iy + 3xi^2 y^2 - i^3 y^3}{x^2 - i^2 y^2}$

$\Rightarrow u + iv = \dfrac{(x^3 - 3xy^2) + i(-3x^2 y + y^3)}{x^2 + y^2}$

$\Rightarrow u = \dfrac{x^3 - 3xy^2}{x^2 + y^2}, v = \dfrac{y^3 - 3x^2 y}{x^2 + y^2}.$

At $z = 0, f(z) = u(x, y) + iv(x, y) = 0.$

$\therefore \quad u(0, 0) + iv(0, 0) = 0 + i0$ at $z = 0.$

So, $u(0, 0) = 0, v(0, 0) = 0.$

Now, $u = u(x, y) = \dfrac{x^3 - 3xy^2}{x^2 + y^2}$

$\Rightarrow u(x, 0) = \dfrac{x^3}{x^2} = x, u(0, y) = \dfrac{0 - 0}{0 + y^2} = 0,$

$v = v(x, y) = \dfrac{y^3 - 3x^2 y}{x^2 + y^2}$

$\Rightarrow v(x, 0) = \dfrac{0 - 0}{0 + x^2} = 0, v(0, y) = \dfrac{y^3 - 0}{y^2 + 0} = y.$

$\therefore \dfrac{\partial u}{\partial x}\bigg|_{(z,0)} = \lim_{x \to 0} \dfrac{u(x,0) - u(0,0)}{x} = \lim_{x \to 0} \left(\dfrac{x - 0}{x} \right) = 1,$

$\dfrac{\partial u}{\partial y}\bigg|_{(z,0)} = \lim_{x \to 0} \dfrac{u(0,y) - u(0,0)}{y} = \lim_{y \to 0} \dfrac{0 - 0}{y} = 0,$

$\dfrac{\partial v}{\partial x}\bigg|_{(z,0)} = \lim_{x \to 0} \dfrac{v(x,0) - v(0,0)}{x} = \lim_{x \to 0} \left(\dfrac{0 - 0}{x} \right) = 0,$

$\dfrac{\partial v}{\partial y}\bigg|_{(z,0)} = \lim_{y \to 0} \dfrac{v(0,y) - v(0,0)}{y} = \lim_{y \to 0} \dfrac{y - 0}{y} = 1.$

Hence, $\dfrac{\partial u}{\partial x}\bigg|_{(z,0)} = \dfrac{\partial v}{\partial y}\bigg|_{(z,0)}$ and $\dfrac{\partial u}{\partial y}\bigg|_{(z,0)} = -\dfrac{\partial v}{\partial x}\bigg|_{(z,0)}.$

So the CR equations are satisfied at $z = 0.$

24. (a) $f(z) = u(x, y) + iv(x, y)$ is analytic

$\Rightarrow \dfrac{\partial u}{\partial x} = \dfrac{\partial v}{\partial y}$ and $\dfrac{\partial u}{\partial y} = -\dfrac{\partial v}{\partial x}.$

Given, $u(x, y) = 2x(1 - y).$

$$\therefore \frac{\partial v}{\partial y} = \frac{\partial u}{\partial x} = 2(1-y)$$

$$\Rightarrow v = \int 2(1-y)\,dy + f(x) = 2\left(y - \frac{y^2}{2}\right) + f(x)$$

$$\Rightarrow \frac{\partial v}{\partial x} = f'(x) = -\frac{\partial u}{\partial y} = 2x$$

$$\Rightarrow f(x) = 2\int x\,dx + C = x^2 + C$$

Hence, $v = 2\left(y - \frac{y^2}{2}\right) + x^2 + C = 2y - y^2 + x^2 + C$

$$= 2y - y^2 + x^2 - 1 \text{ (taking } C = -1)$$
$$= x^2 - (y-1)^2.$$

25. (d) $(z^2 + 9)(z - i) = 0 \Rightarrow (z + 3i)(z - 3i)$
$(z - i) = 0$
$\Rightarrow z = i, -3i, 3i.$

$|z| = 2 \Rightarrow |x + iy| = 2 \Rightarrow (x-0)^2 + (y-0)^2 = 2^2,$

which represents the circle with center $(0, 0)$ and radius "2" unit.

$z = i$ represents the point $(0, 1)$ which lies within C.

$z \pm 3i$ represents the point $(0, 3)$ and $(0, -3)$ which lie outside C.

$$\therefore \int_C \frac{z\,dz}{(z^2+9)(z-i)} = \int_C \frac{\frac{z}{z^2+9}}{z-i}\,dz$$

$$= \int_C \frac{f(z)}{z-a}\,dz \text{ (say)}$$

$$\left(\text{where } f(z) = \frac{z}{z^2+9}, a = i\right)$$

$= 2\pi i \times f(a)$ (by Cauchy's integral formula)

$$= 2\pi i \times f(i) = 2\pi i \times \frac{i}{(i^2+9)} = \frac{-2\pi}{-1+9} = -\frac{\pi}{4}.$$

26. (a) $z(z + \pi i) = 0 \Rightarrow z = 0, z = -\pi i.$
$|z + 3i| = 1 \Rightarrow |x + iy + 3i| = 1$

$$\Rightarrow |x + i(y+3)| = 1$$

$$\Rightarrow \sqrt{x^2 + (y+3)^2} = 1$$

$$\Rightarrow (x-0)^2 + \{y-(-3)\}^2 = 1^2,$$

which represents the circle with the center $(0, -3)$ and radius "1" unit.

$z = 0$ represents the point$(0, 0)$ which lies outside C.

$z = -\pi i$ represents the point $(0, -\pi)$ which lies inside C.

$$\therefore \int_C \frac{dz}{z(z+\pi i)} = \int_C \frac{\frac{1}{z}}{z-(-\pi i)}\,dz$$

$$= \int_C \frac{f(z)\,dz}{z-a} \left(\text{where } f(z) = \frac{1}{z}, a = -\pi i\right)$$

$= 2\pi i \times f(a)$ (by Cauchy's integral formula)

$$= 2\pi i \times f(-\pi i) = 2\pi i \times \frac{1}{(-\pi i)} = -2.$$

27. (c)
$$|z - 2| + |z + 2| = 6$$

$$\Rightarrow |x + iy - 2| + |x + iy + 2| = 6$$

$$\Rightarrow \sqrt{(x-2)^2 + y^2} + \sqrt{(x+2)^2 + y^2} = 6$$

$$\Rightarrow \sqrt{(x-2)^2 + y^2} = 6 - \sqrt{(x+2)^2 + y^2}$$

Squaring both sides we get,

$$(x-2)^2 + y^2 = 36 + (x+2)^2 + y^2 - 12\sqrt{(x+2)^2 + y^2}$$

or, $x^2 + 4 - 4x + y^2 = 36 + x^2 +$

$4 + 4x + y^2 - 12\sqrt{x^2 + 4x + 4 + y^2}$

or, $12\sqrt{x^2 + 4x + 4 + y^2} = 8x + 36$

or, $3\sqrt{x^2 + 4x + 4 + y^2} = 2x + 9$

Again, squaring both sides we get,

$9(x^2 + y^2 + 4x + 4) = 4x^2 + 36x + 81$

or, $5x^2 + 9y^2 = 45$

or, $\dfrac{x^2}{9} + \dfrac{y^2}{5} = 1$ or $\dfrac{x^2}{3^2} + \dfrac{y^2}{(\sqrt{5})^2} = 1$, which repre-

sents the ellipse C.

Now, $z - \pi i = 0 \Rightarrow z = \pi i$ which represents the point $(0, \pi)$ that lies outside C.

So by the Cauchy-Goursat theorem,

$$\int_C \frac{e^{2z}\,dz}{z - \pi i} = 0.$$

28. (c)

$$z^2 + 2z + 2 = 0 \Rightarrow z = \frac{-2 \pm \sqrt{4 - 4 \times 2}}{2} = -1 \pm i$$

$-1 + i$ represents the point $(-1, 1)$ which lies outside the square C.

$-1 - i$ represents the point $(-1, -1)$ which lies inside the square C.

$$\therefore \int_C \frac{dz}{z^2 + 2z + 2} = \int_C \frac{dz}{\{z - (-1+i)\}\{z - (-1-i)\}}$$

$$= \int_C \frac{f(z)dz}{z - a} \text{ (say)}$$

(where $f(z) = \frac{1}{z + 1 - i}, a = -1 - i$)

$= 2\pi i \times f(a)$ (by Cauchy's integral formula)

$= 2\pi i \times f(-1 - i)$

$$= 2\pi i \times \frac{1}{[(-1-i)+1-i]} = -\pi.$$

29. (a)

$$\left| z - \frac{\pi}{4} \right| = \frac{1}{2} \Rightarrow \left| x + iy - \frac{\pi}{4} \right| = \frac{1}{2}$$

$$\Rightarrow \left| \left(x - \frac{\pi}{4} \right) + iy \right| = \frac{1}{2} \Rightarrow \sqrt{\left(x - \frac{\pi}{4} \right)^2 + y^2} = \frac{1}{2}$$

$$\Rightarrow \left(x - \frac{\pi}{4} \right)^2 + (y - 0)^2 = \left(\frac{1}{2} \right)^2,$$

which represents a circle with the center $\left(\frac{\pi}{4}, 0 \right)$ and radius $= \frac{1}{2}$ unit.

$z = \frac{\pi}{4}$ represents the point $A\left(\frac{\pi}{4}, 0 \right)$ which lies inside the circle C.

$$\therefore \int_C \frac{\sin z \, dz}{\left(z - \frac{\pi}{4} \right)^3} = \int_C \frac{\sin z}{\left(z - \frac{\pi}{4} \right)^{2+1}} dz$$

$$= \int_C \frac{f(z)dz}{(z - a)^{n+1}} dz \text{ (say)}$$

[where $f(z) = \sin z, a = \frac{\pi}{4}, n = 2$]

$$= \frac{2\pi i}{n!} \times f^{(n)}(a) \text{ [by Cauchy's integral formula on}$$

higher order derivatives]

$$= \frac{2\pi i}{2!} f^{(2)}\left(\frac{\pi}{4} \right)$$

$$= \pi i \times f''\left(\frac{\pi}{4} \right)$$

$$= \pi i \times \left(-\frac{1}{\sqrt{2}} \right) \left[\because f(z) = \sin z \Rightarrow f''(z) = -\sin z \right]$$

$$= -\frac{\pi i}{\sqrt{2}}.$$

30. (b) $|z| = 1 \Rightarrow |x + iy| = 1 \Rightarrow \sqrt{x^2 + y^2} = 1$

$$\Rightarrow (x - 0)^2 + (y - 0)^2 = 1^2,$$

which represents a circle with the center $(0, 0)$ and radius "1" unit.

$z = 0$ represents the point $O\ (0, 0)$ which lies within the circle C.

$$\therefore \int_C \frac{e^{-z}}{z^2} dz = \int_C \frac{e^{-z} dz}{(z - 0)^{1+1}} = \int_C \frac{f(z) dz}{(z - a)^{n+1}} \text{ (say)}$$

(where $f(z) = e^{-z}, a = 0, n = 1$)

$$= \frac{2\pi i}{n!} \times f^{(n)}(a) \text{ [by Cauchy's integral formula on higher order derivatives]}$$

$$= \frac{2\pi i}{1!} \times f^{(1)}(0) = 2\pi i \times f'(0)$$

$$= 2\pi i \times (-1) = -2\pi i.$$

$$\left[\because f(z) = e^{-z} \Rightarrow f'(z) = -e^{-z} \Rightarrow f'(0) = -e^{-0} = -1 \right]$$

31. (d) $z = e^{i\theta} \Rightarrow dz = ie^{i\theta} d\theta \Rightarrow \frac{dz}{e^{i\theta}} = id\theta$

$$\Rightarrow \frac{dz}{z} = id\theta$$

$$\therefore \text{ given integral } = \int_0^\pi id\theta = i[\theta]_0^\pi = \pi i.$$

32. (c)

$$\frac{\sin z}{z - \pi}$$

$$= \frac{\sin\{\pi + (z - \pi)\}}{z - \pi} = \frac{-\sin(z - \pi)}{z - \pi}$$

$$= -\frac{1}{z - \pi}\left\{(z - \pi) - \frac{(z - \pi)^3}{3!} + \frac{(z - \pi)^5}{5!} + \dots \infty\right\}$$

$$= -1 + \frac{(z - \pi)^2}{6} - \frac{(z - \pi)^4}{120} - \dots \infty$$

Hence, the coefficient of $(z - \pi)^2$ in the Taylor series expansion of $f(z)$ is $\frac{1}{6}$.

33. (a)

$$f(z) = \frac{z^2 + 1}{(z + 2)(z + 3)} = \frac{z^2 - 4 + 5}{(z + 2)(z + 3)}$$

$$= \frac{(z + 2)(z - 2) + 5\{(z + 3) - (z + 2)\}}{(z + 2)(z + 3)}$$

$$= \frac{(z + 2)(z - 2)}{(z + 2)(z + 3)} + \frac{5(z + 3)}{(z + 2)(z + 3)} - \frac{5(z + 2)}{(z + 2)(z + 3)}$$

$$= \frac{z - 2}{z + 3} + \frac{5}{z + 2} - \frac{5}{z + 3}$$

$$= \frac{(z + 3) - 5}{z + 3} + \frac{5}{z + 2} - \frac{5}{z + 3}$$

$$= 1 - \frac{5}{z + 3} + \frac{5}{z + 2} - \frac{5}{z + 3}$$

$$= 1 + \frac{5}{z + 2} - \frac{10}{z + 3} \quad \dots \dots (i)$$

$$|z| < 2 \Rightarrow \frac{|z|}{2} < 1 \text{ and } \frac{|z|}{3} < \frac{|z|}{2} < 1$$

Now, (i) $\Rightarrow f(z) = \frac{5}{2\left(1 + \dfrac{z}{2}\right)} - \frac{10}{3\left(1 + \dfrac{z}{3}\right)} + 1$

$$= 1 + \frac{5}{2}\left(1 + \frac{z}{2}\right)^{-1} - \frac{10}{3}\left(1 + \frac{z}{3}\right)^{-1} \quad \dots(ii)$$

$\because \dfrac{|z|}{2} = \left|\dfrac{z}{2}\right| < 1$ and $\dfrac{|z|}{3} = \left|\dfrac{z}{3}\right| < 1$, so each of the se-

ries $\left(1 + \dfrac{z}{2}\right)^{-1}$ and $\left(1 + \dfrac{z}{3}\right)^{-1}$ can be expanded using

the formula $(1 + x)^{-1} = 1 - x + x^2 - x^3 + \dots \infty$

$$= \sum_{n=0}^{\infty} (-1)^n x^n$$

Then,

$(ii) \Rightarrow f(z) = 1 + \frac{5}{2}\sum_{n=0}^{\infty}(-1)^n\left(\frac{z}{2}\right)^n - \frac{10}{3}\sum_{n=0}^{\infty}(-1)^n\left(\frac{z}{3}\right)^n$

$$= 1 + \sum_{n=0}^{\infty}(-1)^n\left\{\frac{5}{2^{n+1}} - \frac{10}{3^{n+1}}\right\}z^n,$$

which is the required Taylor series expansion

34. (c)

$$f(z) = \frac{1}{z^2 - 5z + 6} = \frac{1}{(z - 2)(z - 3)}$$

$$= \frac{(z - 2) - (z - 3)}{(z - 2)(z - 3)} = \frac{1}{z - 3} - \frac{1}{z - 2}$$

$$= \frac{1}{(-3)\left(1 - \dfrac{z}{3}\right)} - \frac{1}{z\left(1 - \dfrac{2}{z}\right)}$$

$$= -\frac{1}{3}\left(1 - \frac{z}{3}\right)^{-1} - \frac{1}{z}\left(1 - \frac{2}{z}\right)^{-1} \quad \dots \dots \dots (i)$$

$$2 < |z| < 3 \Rightarrow \frac{2}{|z|} < 1, \frac{|z|}{3} < 1$$

$$\Rightarrow \left|\frac{2}{z}\right| < 1, \left|\frac{z}{3}\right| < 1$$

Then, $(i) \Rightarrow$

$$f(z) = -\frac{1}{3}\sum_{n=0}^{\infty}\left(\frac{z}{3}\right)^n - \frac{1}{z}\sum_{n=0}^{\infty}\left(\frac{2}{z}\right)^n$$

$$\left[\text{using } (1 - x)^{-1} = \sum_{n=0}^{\infty} x^n\right]$$

which is the required Laurent series expansion in the annulus $2 < |z| < 3$.

35. (c) $f(z) = \frac{z}{(z + 1)(z + 2)} = \frac{z\{(z + 2) - (z + 1)\}}{(z + 1)(z + 2)}$

$$= \frac{z(z + 2)}{(z + 1)(z + 2)} - \frac{z(z + 1)}{(z + 1)(z + 2)}$$

$$= \frac{z}{z + 1} - \frac{z}{z + 2} = \frac{(z + 1) - 1}{z + 1} - \frac{[(z + 2) - 2]}{z + 2}$$

$$= 1 - \frac{1}{z + 1} - 1 + \frac{2}{z + 2} = -\frac{1}{z + 1} + \frac{2}{z + 2}$$

$$= \frac{2}{\{1 + (z + 1)\}} - \frac{1}{z + 1} = 2\{1 + (z + 1)\}^{-1} - \frac{1}{z + 1}$$

$$= 2\sum_{n=0}^{\infty}(-1)^n(z + 1)^n - \frac{1}{z + 1}$$

$$\left[\text{using } (1 + x)^{-1} = \sum_{n=0}^{\infty}(-1)^n x^n\right]$$

which is the required Taylor series expansion about $z = -1$

36. (a) We know that the Taylor series expansion of $f(z)$ about $z = a$ is given by

$$f(z) = \sum_{n=0}^{\infty} a_n (z-a)^n \text{ where } a_n = \frac{f^n(a)}{n!}.$$

Here, $f(z) = \cos z$ and $a = \frac{\pi}{4}$

$$\therefore f^{(n)}(z) = \cos\left(z + \frac{n\pi}{2}\right)$$

$$\left[u\sin g\ f^{(n)}(x) = \cos\left(\frac{n\pi}{2} + x\right) \text{if } f(x) = \cos x\right]$$

So, $f^{(n)}\left(\frac{\pi}{4}\right) = \cos\left(\frac{\pi}{4} + \frac{n\pi}{2}\right)$

Hence, $f(z) = \sum_{n=0}^{\infty} a_n \left(z - \frac{\pi}{4}\right)^n$

$$= \sum_{n=0}^{\infty} \frac{\left(z - \frac{\pi}{4}\right)^n}{n!} \cos\left(\frac{\pi}{4} + \frac{n\pi}{4}\right).$$

37. (b)

$$f(z) = \frac{1}{(z^2+1)(z^2+2)} = \frac{(z^2+2)-(z^2+1)}{(z^2+1)(z^2+2)}$$

$$= \frac{1}{z^2+1} - \frac{1}{z^2+2} \quad ....(i)$$

$|z| < 1 \Rightarrow |z^2| < 1 \text{ and } \left|\frac{z^2}{2}\right| = \frac{|z^2|}{2} < |z^2| < 1.$

$\therefore$ Each of the series $(1+z^2)^{-1}$ and $\left(1 + \frac{z^2}{2}\right)^{-1}$

can be expanded using the formula

$$(1+x)^{-1} = \sum_{n=0}^{\infty} (-1)^n x^n \text{ for } |x| < 1$$

Then (i) gives,

$$f(z) = \frac{1}{1+z^2} - \frac{1}{2\left(1+\frac{z^2}{2}\right)}$$

$$= \left(1+z^2\right)^{-1} - \frac{1}{2}\left(1+\frac{z^2}{2}\right)^{-1}$$

$$= \sum_{n=0}^{\infty} (-1)^n \left(z^2\right)^n - \frac{1}{2}\sum_{n=0}^{\infty}(-1)^n \left(\frac{z^2}{2}\right)^n$$

$$= \sum_{n=0}^{\infty} (-1)^n z^{2n} - \sum_{n=0}^{\infty} (-1)^n \frac{z^{2n}}{2^{n+1}}$$

$$= \sum_{n=0}^{\infty} (-1)^n \left[1 - \frac{1}{2^{n+1}}\right] z^{2n}.$$

38. (a)

Here, $f(z) = \frac{1}{\left(z^2+1\right)^3} = \frac{1}{\{(z+i)(z-i)\}^3} = \frac{\frac{1}{(z+i)^3}}{(z-i)^3}$

$$= \frac{\varphi(z)}{(z-i)^3} \quad \text{(say) where } \varphi(z) = \frac{1}{(z+i)^3}$$

$$\therefore \varphi'(z) = -\frac{3}{(z+i)^4}, \ \varphi''(z) = \frac{12}{(z+i)^5} = \varphi^{(2)}(z).$$

Then, Res$(z = i)$

$$= \frac{1}{(3-1)!}\varphi^{(3-1)}(i) \ (\because \ z = i) \text{ is a pole of order 3}$$

$$= \frac{1}{2}\varphi^{(2)}(i) = \frac{1}{2}\varphi^{(2)}(z)\big|_{z=i}$$

$$= \frac{1}{2}\times\frac{12}{(i+i)^5} = \frac{6}{(2i)^5} = \frac{6}{32i} = \frac{3}{16i}.$$

39. (a)

$$z^3 \cos\left(\frac{1}{z-2}\right)$$

$$= \{(z-2)+2\}^3 \cos\left(\frac{1}{z-2}\right)$$

$$= \left\{(z-2)^3 + 6(z-2)^2 + 12(z-2) + 8\right\}$$

$$\times\left\{1 - \left(\frac{1}{z-2}\right)^2\times\frac{1}{2!} + 1 + \left(\frac{1}{z-2}\right)^4\times\frac{1}{4!} - ....\infty\right\}$$

$$\left(\text{using } \cos x = 1 - \frac{x^2}{2!} + \frac{x^4}{4!} - ....\infty\right)$$

Here the coefficient of $\frac{1}{z-2}$ in the expansion

of $f(z)$ is $12\times\left(-\frac{1}{2!}\right) + \frac{1}{4!}$ i.e; $-\frac{143}{24}$. So Res$(z = 2) =$

$-\frac{143}{24}.$

Remember:

Res$(z = a)$ = co-efficient of $\frac{1}{z-a}$ in the expansion of $f(z)$.

40. (d)

Here, $f(z) = \dfrac{Z}{(z-1)(z-2)^2}$. Thus, the poles of f (z) are given by:

$(z-1)(z-2)^2 = 0$,   i.e., $z = 1, 2, 2$.

∴ $z = 1$ is a simple pole and $z = 2$ is a pole of order 2.

$|z-2| = \dfrac{1}{2}$ $\Rightarrow |x + iy - 2| = \dfrac{1}{2}$

$\Rightarrow |(x-2) + iy| = \dfrac{1}{2}$

$\Rightarrow \sqrt{(x-2)^2 + y^2} = \dfrac{1}{2}$

$\Rightarrow (x-2)^2 + (y-0)^2 = \left(\dfrac{1}{2}\right)^2$

Hence, C is a circle with the center $(2, 0)$ and radius $\dfrac{1}{2}$ unit.

The pole $z = 1$ represents the point $(1,0)$ which lies outside the circle C.

The point $z = 2$ represents the point $(2,0)$ which lies inside the circle C.

Again, $f(z) = \dfrac{\frac{z}{z-1}}{(z-2)^2} = \dfrac{\varphi(z)}{(z-2)^2} = \dfrac{\varphi(z)}{(z-2)^2}$ (say)

where $\varphi(z) = \dfrac{z}{z-1}$ and so $\varphi'(z) = \dfrac{-1}{(z-1)^2}$.

∴   Res$(z = 2)$

$= \dfrac{1}{(2-1)!}\varphi^{(2-1)}(2) = \varphi^{(1)}(2) = \dfrac{-1}{(2-1)^2} = -1$.

Hence, by Cauchy's residue theorem,

$\displaystyle\int_C f(z)dz = 2\pi i \times \text{Re}\,s(z = 2) = 2\pi i \times(-1) = -2\pi i$.

41. (a) Given, $f(z) = \displaystyle\sum_{n=0}^{15} z^n$ for $z \in C$.

The poles are given by: $(z-i)^{15} = 0$ which means $z = i$ is a pole of order 15.

$|z - i| = 2$

$\Rightarrow |x + iy - i| = 2$

$\Rightarrow |x + i(y-1)| = 2$

$\Rightarrow \sqrt{x^2 + (y-1)^2} = 2$

$\Rightarrow (x-0)^2 + (y-1)^2 = 2^2$

Hence, C is a circle with the center $(0, 1)$ and radius "2" unit.

The pole $z = i$ represents the point $(0,1)$ which lies inside the circle C.

$f(z) = \displaystyle\sum_{n=0}^{15} z^n = 1 + z + z^2 + \ldots\ldots + z^{14} + z^{15}$

$\Rightarrow f^{(14)}(z) = 14! + 15! \times z$

$\Rightarrow f^{(14)}(i) = 14! + 15! \times i$

∴ $\text{Res}(z = i) = \dfrac{f^{(15-1)}(i)}{(15-1)!} = \dfrac{f^{(14)}(i)}{14!} = \dfrac{14! + 15!i}{14!}$

$= 1 + 15i$.

Then, $\displaystyle\oint_C \dfrac{f(z)dz}{(z-i)^{15}} = 2\pi i \,\text{Re}\,s(z = i) = 2\pi i(1 + 15i)$.

42. (b)

Here, $f(z) = \dfrac{1}{z^2 + 4} = \dfrac{1}{(z+2i)(z-2i)}$.

The poles are given by

$z^2 + 4 = 0$ i.e; $z^2 = -4 = 2^2 i^2$ i.e; $z = \pm 2i$.

Again, $|z - 3i| = 2$

$\Rightarrow |x + iy - 3i| = 2 \Rightarrow |x + i(y-3)| = 2$

$\Rightarrow \sqrt{x^2 + (y-3)^2} = 2 \Rightarrow (x-0)^2 + (y-3)^2 = 2^2$

Hence, C is a circle with the center $(0, 3)$ and radius "2" unit.

The pole $z = 2i$ represents the point $(0, 2)$ which lies inside the circle C.

But the pole $z = -2i$ represents the point $(0, -2)$ which lies outside the circle C.

Now, Res$(z = 2i)$

$= \displaystyle\lim_{z \to 2i}(z - 2i)f(z) = \lim_{z \to 2i}\dfrac{(z - 2i)}{(z + 2i)(z - 2i)}$

$= \displaystyle\lim_{z \to 2i}\dfrac{1}{(z + 2i)} = \dfrac{1}{2i + 2i} = \dfrac{1}{4i}$.

∴ $\displaystyle\int_C \dfrac{dz}{z^2 + 4} = 2\pi i \times \text{Re}\,s(z = 2i) = 2\pi i \times \dfrac{1}{4i} = \dfrac{\pi}{2}$.

Previous Years Solved Papers (2000-2018)

1. The bilinear transformation $w = \dfrac{z-1}{z+1}$

(a) Maps the inside of the unit circle in the z plane to the left half of the w-plane

(b) Maps the outside the unit circle in the z-plane to the left half of the w-plane.

(c) Maps the inside of the unit circle in the z-plane to the right half of the w-plane.

(d) Maps the outside the unit circle in the z-plane to the right half of the w-plane.

[IN GATE 2002]

2. Which one of the following is not true for the complex number z_1, z_2?

(a) $\dfrac{z_1}{z_2} = \dfrac{z_1 \bar{z_2}}{|z_2|^2}$

(b) $|z_1 + z_2| \leqslant |z_1| + |z_2|$

(c) $|z_1 - z_2| \leqslant |z_1| - |z_2|$

(d) $|z_1 + z_2|^2 + |z_1 - z_2|^2 = 2|z_1|^2 + 2|z_2|^2$

[CE GATE 2005]

3. The function

$$w = u + iv = \frac{1}{2}\log\left(x^2 + y^2\right) + i\tan^{-1}\left(\frac{y}{x}\right)$$

is not analytic at the point

(a) $(0,0)$ (b) $(0,1)$ (c) $(1,0)$ (d) $(2,\infty)$

[PI GATE 2005]

4. Consider the circle $|z - 5 - 5i| = 2$ in the complex plane (x, y) with $z = x + iy$. Then the minimum distance from the origin to the circle is

(a) $5\sqrt{2} - 2$ (b) $\sqrt{54}$ (c) $\sqrt{34}$ (d) $5\sqrt{2}$

[IN GATE 2005]

5. Consider likely applicability of the Cauchy's integral theorem to evaluate the following integral counterclockwise around the unit circle C:

$I = \oint_C \sec z\, dz, z$ being a complex variable.

Then, the value of I will be

(a) $I = 0$, singularities set $= \phi$

(b) $I = 0$, singularities set

$$= \left\{\pm\left(\frac{2n+1}{2}\right)\pi : n = 0,1,2,\ldots\right\}$$

(c) $I = \dfrac{\pi}{2}$, singularities set $= \{\pm n\pi : n = 0,1,2,\ldots\}$

(d) None of the above

[CE GATE 2005]

6. Let $z^3 = \bar{z}$, where z is a complex number not equal to zero. Then, z is a solution of

(a) $z^2 = 1$ (b) $z^3 = 1$ (c) $z^4 = 1$ (d) $z^9 = 1$

[IN GATE 2005]

7. Using the Cauchy's integral thorem, the value of the integral (integration being taken in counterclockwise direction) $\oint \dfrac{z^3 - 6}{3z - i}dz$ is (within the unit circle)

(a) $\dfrac{2\pi}{81} - 4\pi i$

(b) $\dfrac{\pi}{8} - 6\pi i$

(c) $\dfrac{4\pi}{81} - 6\pi i$

(d) 1

[CE GATE 2006]

8. For the function of a complex variable $w = \ln z$ (where $w = u + jv$ and $z = x + jy$), $u = $ constant lines get mapped in z-plane as

(a) set of radial straight lines
(b) set of concentric circles
(c) set of conformal hyperbolas
(d) set of conformal ellipses

[EC GATE 2006]

9. The value of the contour integral $\oint_{|z-j|=2} \dfrac{1}{z^2 + 4}dz$ in positive sense is

(a) $j\pi/2$ (b) $-\pi/2$ (c) $-j\pi/2$ (d) $\pi/2$

[EC GATE 2006]

10. The value of $\oint_C \dfrac{1}{z^2 + 1}dz$ where C is the contour $\left|z - \dfrac{i}{2}\right| = 1$ is

(a) $2\pi i$

(b) π

(c) $\tan^{-1} z$

(d) $\pi i \tan^{-1} z$

[EC GATE 2007]

11. If the semi-circular contour D of radius "2" is as shown in the following figure, then the value of the integral $\oint_D \dfrac{1}{s^2 - 1}ds$ is

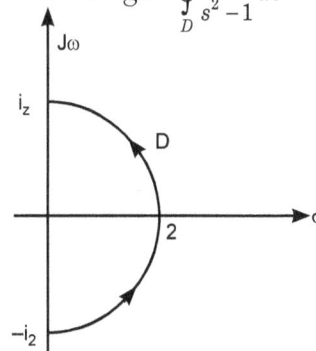

(a) πi (b) $-\pi i$ (c) $-\pi$ (d) π

[EC GATE 2007]

12. Potential function ϕ is given as $\phi = x^2 - y^2$. What will be the stream function ψ with the condition $\psi = 0$ at $x = y = 0$?

(a) $2xy$ (b) $x^2 + y^2$
(c) $x^2 - y^2$ (d) $2x^2y^2$

[CE GATE 2007]

13. If $\phi(x, y)$ and $\psi(x, y)$ are functions with continuous second order derivatives, then $\phi(x, y) + i\psi(x, y)$ can be expressed as an analytic function of $x + iy$ $\left(i = \sqrt{-1}\right)$ when

(a) $\dfrac{\partial \phi}{\partial x} = \dfrac{\partial \psi}{\partial x}, \dfrac{\partial \phi}{\partial y} = \dfrac{\partial \psi}{\partial y}$

(b) $\dfrac{\partial \phi}{\partial y} = -\dfrac{\partial \psi}{\partial x}, \dfrac{\partial \phi}{\partial x} = \dfrac{\partial \psi}{\partial y}$

(c) $\dfrac{\partial^2 \phi}{\partial x^2} + \dfrac{\partial^2 \psi}{\partial y^2} = \dfrac{\partial^2 \psi}{\partial x^2} + \dfrac{\partial^2 \psi}{\partial y^2} = 1$

(d) $\dfrac{\partial \phi}{\partial x} + \dfrac{\partial \phi}{\partial y} = \dfrac{\partial \psi}{\partial x} + \dfrac{\partial \psi}{\partial y} = 0$

[CE GATE 2007]

14. For the function $\dfrac{\sin z}{z^3}$ of a complex variable z, the point $z = 0$ is

(a) a pole of order 3   (b) a pole of order 2
(c) a pole of order 1   (d) not a singularity

[IN GATE 2007]

15. Let $j = \sqrt{-1}$. Then one value of j^j is

(a) $\sqrt{3}$ (b) -1 (c) $\dfrac{1}{\sqrt{2}}$ (d) $e^{-\frac{\pi}{2}}$

[IN GATE 2007]

16. If a complex number $z = \dfrac{\sqrt{3}}{2} + \dfrac{i}{2}$, then z^4 is

(a) $2\sqrt{2} + 2i$ (b) $-\dfrac{1}{2} + \dfrac{i\sqrt{3}}{2}$

(c) $\dfrac{\sqrt{3}}{2} + i\dfrac{1}{2}$ (d) $\dfrac{\sqrt{3}}{8} - \dfrac{i}{8}$

[PI GATE 2007]

17. The residue of the function $f(z) = \dfrac{1}{(z+2)^2 (z-2)^2}$ at $z = 2$ is

(a) $-\dfrac{1}{32}$ (b) $-\dfrac{1}{16}$ (c) $\dfrac{1}{16}$ (d) $\dfrac{1}{32}$

[EC GATE 2008]

18. Given $X(z) = \dfrac{z}{(z-a)^2}$ with $|z| > a$. Then residue of $X(z) \times z^{n-1}$ at $z = a$ for $n \geq 0$ will be

(a) a^{n-1} (b) a^n (c) na^n (d) na^{n-1}

[EE GATE 2008]

19. The value of the expression $\dfrac{-5 + 10i}{3 + 4i}$ is

(a) $1 - 2i$ (b) $1 + 2i$ (c) $2 - i$ (d) $2 + i$

[PI GATE 2008]

20. The integral $\oint f(z)dz$ evaluated around the unit circle on the complex plane for $f(z) = \dfrac{\cos z}{z}$ is?

(a) $2\pi i$ (b) $4\pi i$ (c) $-2\pi i$ (d) 0

[ME GATE 2008]

21. The equation $\sin z = 10$ has

(a) No real (or complex) solution

(b) Exactly two distinct complex solutions

(c) A unique solution

(d) An infinite number of complex solutions

[ME GATE 2008]

22. A complex variable $z = x + (0.1)j$ has its real part x varying in the range $-\infty$ to ∞. Which one of the following is the locus (shown in thick lines) of $\dfrac{1}{z}$ in the complex plane?

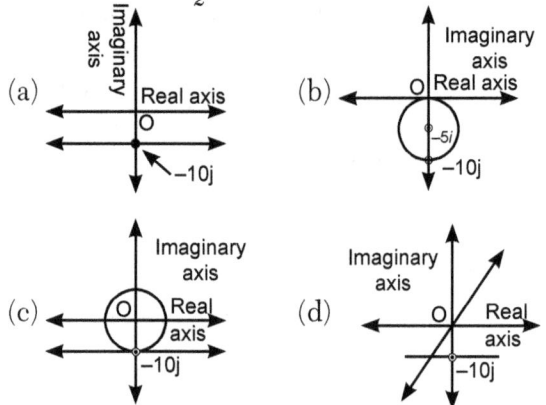

[IN GATE 2008]

23. An analytic function of a complex variable $z = x + iy$ is expressed as $f(z) = u(x, y) + iv(x, y)$, where $i = \sqrt{-1}$. If $u = xy$, the expression for v should be

(a) $\dfrac{(x+y)^2}{2} + k$ (b) $\dfrac{x^2 - y^2}{2} + k$

(c) $\frac{y^2-x^2}{2}+k$ (d) $\frac{(x-y)^2}{2}+k$

[ME GATE 2009]

24. The analytic function $f(z)=\frac{z-1}{z^2+1}$ has singularities at
(a) 1 and –1   (b) 1 and i
(c) 1 and $-i$ (d) i and $-i$

[EC GATE 2009]

25. The value of the integral $\int_c \frac{\cos(\pi z)}{(2z-1)(z-3)}dz$ where c is a closed curve given by $|z| = 1$ is
(a) $-\pi i$ (b) $\frac{\pi i}{5}$ (c) $\frac{2\pi i}{5}$ (d) πi

[CE GATE 2009]

26. If $f(z) = c_0 + \frac{c_1}{z}$, then $\oint_{|z|=1} \frac{1+f(z)}{z}dz$ is
(a) $2\pi c_1$ (b) $2\pi(1+c_0)$
(c) $2\pi j c_1$ (d) $2\pi j(1+c_0)$

[EC GATE 2009]

27. The value of $\oint_C \frac{\sin z}{z}dz$, where the contour of the integration is a simple closed curve around the origin is
(a) 0   (b) $2\pi i$ (c) ∞ (d) $\frac{1}{2\pi i}$

[IN GATE 2009]

28. If $z = x + jy$ where x, y are real numbers, then the value of $|e^{jz}|$ is?
(a) 1   (b) $e^{\sqrt{x^2+y^2}}$ (c) e^y (d) e^{-y}

[IN GATE 2009]

29. One of the roots of the equation $x^3 = j$ (where j is the +ve square root of –1) is–
(a) j (b) $\frac{\sqrt{3}}{2}+\frac{j}{2}$
(c) $\frac{\sqrt{3}}{2}-\frac{j}{2}$ (d) $-\frac{\sqrt{3}}{2}-\frac{j}{2}$

[IN GATE 2009]

30. The product of complex numbers $3 – 2i$ and $3 + 4i$ results in
(a) $1 + 6i$ (b) $9 – 8i$
(c) $9 + 8i$ (d) $17 + 6i$

[PI GATE 2009]

31. If $f(x+iy)=x^3-3xy^2+i\phi(x,y)$, where $i=\sqrt{-1}$ and $f(x+iy)$ is an analytic function, then $\phi(x,y)$ is
(a) y^3-3x^2y (b) $3x^2y-y^3$
(c) x^4-4x^2y (d) $xy-y^2$

[PI GATE 2010]

32. The contour C in the adjoining figure is described by $x^2 + y^2 = 16$. Then the value of $\oint_C \frac{z^2+8}{(0.5)z-(1.5)j}dz$
(a) $-2\pi j$ (b) $2\pi j$ (c) $4\pi j$ (d) $-4\pi j$

[IN GATE 2010]

33. The modulus of the complex number $\frac{3+4i}{1-2i}$ is
(a) 5   (b) $\sqrt{5}$ (c) $\frac{1}{\sqrt{5}}$ (d) $\frac{1}{5}$

[CE GATE 2010]

34. If a complex number ω satisfies the equation $\omega^3 = 1$, then the value of $1+\omega+\frac{1}{\omega}$ is
(a) 0   (b) 1   (c) 2   (d) 4

[PI GATE 2010]

35. For an analytic function $f(x+iy)=u(x,y)+iv(x,y)$, u is given by $u = 3x^2 - 3y^2$. The expression for v, consider k to be a constant is
(a) $3y^2-3x^2+k$ (b) $6x-6y+k$
(c) $6y-6x+k$ (d) $6xy+k$

[ME GATE 2011]

36. The product of two complex numbers $1 + i$ and $2 – 5i$ is
(a) $7 – 3i$ (b) $3 – 4i$
(c) $-3 – 4i$ (d) $7 + 3i$

[ME GATE 2011]

37. A point z has been plotted in the complex plane, as shown in figure below:

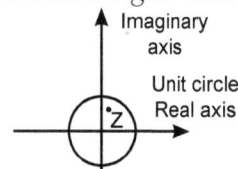

Then, $\frac{1}{z}$ lies in the curve

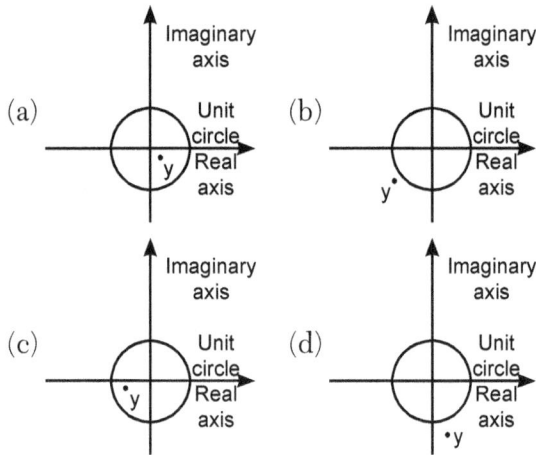

[EE GATE 2011]

38. The contour integral $\int_c e^{\frac{1}{z}}dz$ with C as the counterclockwise unit circle in the z-plane is equal to

(a) 0   (b) 2π (c) $2\pi\sqrt{-1}$ (d) ∞

[IN GATE 2011]

39. The value of the integral $\oint_C \frac{(-3z+4)}{(z^2+4z+5)}dz$, where "$c$" is the circle $|z| = 1$, is given by

(a) 0   (b) $\frac{1}{10}$ (c) $\frac{4}{5}$ (d) 1

[EC GATE 2011]

40. The value of $\oint_C \left(\frac{z^2}{z^4-1}\right)dz$, using the Cauchy's integral around the circle $|z + 1| = 1$ where $z = x + iy$ is

(a) $2\pi i$ (b) $\frac{-\pi i}{2}$ (c) $\frac{-3\pi i}{2}$ (d) πi

[PI GATE 2011]

41. Given $i = \sqrt{-1}$. What will be the value of the definite integral $\int_0^{\pi/2}\left(\frac{\cos x + i\sin x}{\cos x - i\sin x}\right)dx$ = ?

(a) 0   (b) 2   (c) $-i$ (d) i

[CSE GATE 2011]

42. Given $f(z) = \frac{1}{z+1} - \frac{2}{z+3}$. If C is a closed clockwise path in the z-plane such that $|z + 1| = 1$, the value of $\frac{1}{2\pi j}\oint_C f(z)dz$ is

(a) –2   (b) –1   (c) 1   (d) 2

[EE, EC, IN GATE 2012]

43. If $x = \sqrt{-1}$, then the value of x^x is

(a) $e^{-\frac{\pi}{2}}$ (b) $e^{\frac{\pi}{2}}$ (c) x (d) 1

[EE, EC, IN GATE 2012]

44. $\oint_C \left(\frac{z^2-4}{z^2+4}\right)dz$ evaluated   counterclockwise around the circle $C: |z - i| = 2$, where $i = \sqrt{-1}$ is

(a) -4π (b) 0   (c) $2 + \pi$ (d) $2 - \pi$

[EE GATE 2013]

45. Square root of $-i$, where $i = \sqrt{-1}$, are

(a) $i, -i$

(b) $\cos\left(-\frac{\pi}{4}\right) + i\sin\left(-\frac{\pi}{4}\right)$, $\cos\frac{3\pi}{4} + i\sin\frac{3\pi}{4}$

(c) $\cos\frac{\pi}{4} + i\sin\frac{3\pi}{4}$, $\cos\frac{3\pi}{4} + i\sin\frac{\pi}{4}$

(d) $\cos\frac{3\pi}{4} + i\sin\left(-\frac{3\pi}{4}\right)$, $\cos\left(-\frac{3\pi}{4}\right) + i\sin\frac{3\pi}{4}$

[EE GATE 2013]

46. The argument of the complex number $\frac{1+i}{1-i}$ where $i = \sqrt{-1}$, is

(a) $-\pi$ (b) $-\frac{\pi}{2}$ (c) $\frac{\pi}{2}$ (d) π

[ME GATE 2014]

47. An analytic function of a complex variable $z = x + iy$ is expressed as $f(z) = u(x,y) + iv(x,y)$, where $i = \sqrt{-1}$. If $u(x,y) = 2xy$, then $v(x,y)$ must be

(a) $x^2 + y^2$ + constant

(b) $x^2 - y^2$ + constant

(c) $y^2 - x^2$ + constant

(d) $-x^2 - y^2$ + constant

[ME GATE 2014]

48. An analytic function of a complex variable $z = x + iy$ is expressed as $f(z) = u(x,y) + iv(x,y)$, where $i = \sqrt{-1}$. If $u(x,y) = x^2 - y^2$, then expansion for $v(x, y)$ in terms of x and y and a general constant C would be

(a) $xy + C$ (b) $\dfrac{x^2 + y^2}{2} + C$

(c) $2xy + C$ (d) $\dfrac{(x - y)^2}{2} + C$

[ME GATE 2014]

49. The real part of an analytic function $f(z)$, where $z = x + jy$ is given by $e^{-y}\cos x$. The imaginary part of $f(z)$ is

(a) $e^{y}\cos x$.   (b) $e^{-y}\sin x$.

(c) $-e^{y}\sin x$.   (d) $-e^{-y}\sin x$.

[EC GATE 2014]

50. Let S be the set of points in the complex plane corresponding to the unit circle, i.e; $S = \{z : |z| = 1\}$. Consider the function $f(z) = zz^{*}$, where z^{*} denotes the complex conjugate of z. Then $f(z)$ maps S to which one of the followings in the complex plane.

(a) unit circle

(b) horizontal axis line segment from origin to $(1,0)$

(c) the point $(1,0)$

(d) the entire horizontal axis

[EE GATE 2014]

51. C is a closed path in the z-plane given by $|z| = 3$. The value of the integral $\oint_{c} \dfrac{z^2 - z + 4j}{z + 2j}dz$ is

(a) $-4\pi(1 + 2j)$ (b) $4\pi(3 - 2j)$

(c) $-4\pi(3 + 2j)$ (d) $4\pi(1 - 2j)$

[EC GATE 2014]

52. Integration of the complex function $f(z) = \dfrac{z^2}{z^2 - 1}$, in the counterclockwise direction, around $|z - 1| = 1$, is

(a) $-\pi i$ (b) 0 (c) πi (d) $2\pi i$

[EE GATE 2014]

53. If z is a complex variable, then the value of $\int_{5}^{3i} \dfrac{dz}{z}$ is

(a) $-0.511 - 1$.   (b) $-0.511 + 1.57i$

(c) $0.511 - 1.57i$ (d) $0.511 + 1.57i$

[ME GATE 2014]

54. All the values of the multivalued complex function 1^{i}, where $i = \sqrt{-1}$, are

(a) purely imaginary

(b) real and non-negative

(c) on the unit circle

(d) equal in real and imaginary parts

[EE GATE 2014]

55. Given that $f(z) = g(z) + h(z)$, where f, g, h are complex valued functions of a complex variable z. Which one of the following statement is true?

(a) If $f(z)$ is differentiable at z_0, then $g(z)$ and $h(z)$ are also differentiable at z_0

(b) If $g(z)$ and $h(z)$ are differentiable at z_0, then $f(z)$ is also differentiable at z_0.

(c) If $f(z)$ is continuous at z_0, then it is differentiable at z_0

(d) If $f(z)$ is differentiable at z_0, then so are its real and imaginary parts

[EE GATE 2015]

56. If "C" is a circle of radius "r" with center z_0 in the complex "z" plane and if "n" is a non-zero integer, then $\oint_{C} \dfrac{dz}{(z - z_0)^{n+1}}$ equals

(a) $2\pi nj$ (b) 0 (c) $\dfrac{nj}{2\pi}$ (d) $2\pi n$

[EC GATE 2015]

57. The value of $\oint \dfrac{1}{z^2}dz$, where the contour is the unit circle traversed clockwise, is

(a) $-2\pi i$ (b) 0 (c) $2\pi i$ (d) $4\pi i$

[IN GATE 2015]

58. Consider the following complex function $f(z) = \dfrac{9}{(z - 1)(z + 2)^2}$. Which of the following is ONE of the residues of the above function?

(a) -1 (b) $\dfrac{9}{16}$ (c) 2 (d) 9

[CE GATE 2015]

59. Let $z = x + iy$ be a complex variable. Consider that contour integration is performed along the unit circle C in counterclockwise direction. Which one of the following statement is not true?

(a) The residue of $\dfrac{z}{z^2 - 1}$ at $z = 1$ is $\dfrac{1}{2}$

(b) $\oint_{c} z^2 dz = 0$

(c) $\frac{1}{2\pi i}\oint_c \frac{1}{z}dz = 1$

(d) $\bar{z}$ is an analytic function

[EC GATE 2015]

60. If "c" denotes the counterclockwise unit circle. The value of $\frac{1}{2\pi i}\int_c \text{Re}(z)$ is _____.

[EC GATE 2015]

61. Given two complex number $z_1 = 5 + 5\sqrt{3}i$ and $z_2 = \frac{2}{\sqrt{3}} + 2i$, the argument of $\frac{z_1}{z_2}$ in degree is

(a) 0   (b) 30   (c) 60   (d) 90

[ME GATE 2015]

62. Consider the complex valued function $f(z) = 2z^2 + b|z|^3$, where "z" is a complex variable. Then the value of "b" for which the function $f(z)$ is analytic is _____?

[EC GATE 2016]

63. The value of the integral $\int_c \frac{(2z+5)}{\left(z-\frac{1}{2}\right)(z^2-4z+5)}$ over the contour $|z| = 1$, taken in the counterclockwise direction would be

(a) $\frac{24\pi i}{13}$ (b) $\frac{48\pi i}{13}$ (c) $\frac{24}{13}$ (d) $\frac{12}{13}$

[EE GATE 2016]

64. The value of the integral $\frac{1}{2\pi j}\int_c \left(\frac{z^2+1}{z^2-1}\right)dz$ where z is a complex number and "c" is a unit circle with the center at $1 + 0j$ in the complex plane is _____.

[IN GATE 2016]

65. The value of the integral $\frac{1}{2\pi j}\int_c \frac{e^z}{z-1}dz$ along a closed contour C in counterclockwise direction for

(i) the point $z_0 = 2$ inside the contour C and

(ii) the point $z_0 = 2$ outside the contour C, respectively, are

(a) 2.72, 0   (b) 7.39, 0
(c) 0, 2.72   (d) 0, 7.39

[EC GATE 2016]

66. A function f of the complex variable $z = x + iy$, is given as $f(x,y) = u(x,y) + iv(x,y)$, where $u(x,y) = 2kxy$ and $v(x,y) = x^2 - y^2$. The value of k, for which the function is analytic is____?

[ME GATE 2016]

67. For $f(z) = \frac{\sin z}{z^2}$, the residue of the pole at $z = 0$ is _____.

[EC GATE 2016]

68. The value of $\oint_\Gamma \frac{(3z-5)}{(z-1)(z-2)}dz$ along a closed path Γ is equal to $4\pi i$, where $z = x + iy$ and $i = \sqrt{-1}$. The correct path Γ is

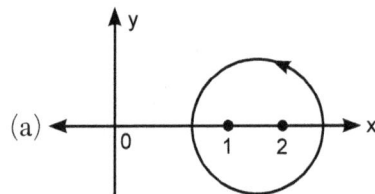

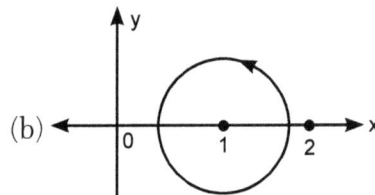

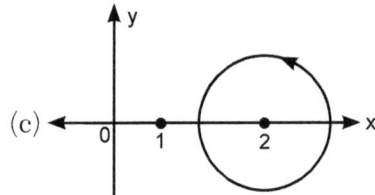

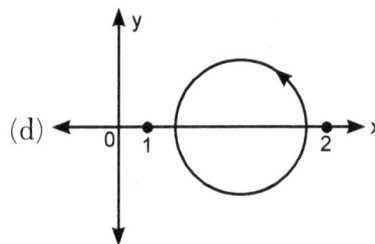

[ME GATE 2016]

69. In the following integral, the contour "c" encloses the points $2\pi j$ and $-2\pi j$. The value of the integral $\frac{1}{\pi}\oint_C \frac{\sin z}{(z-2\pi j)^3}dz$ is _____.

[EC GATE 2016]

70. Consider the function $f(z) = z + z^*$, where z is a complex variable and z^* denotes its complex conjugate. Which one of the following is true?

(a) $f(z)$ is both continuous and analytic

(b) $f(z)$ is continuous but not analytic

(c) $f(z)$ is not continuous but is analytic

(d) $f(z)$ is neither continuous nor analytic

[EE GATE 2016]

71. In the neighborhood of $z = 1$ the function $f(z)$ has a power series expansion of the form $f(z) = 1 + (1-z) + (1-z)^2 + ...\infty$. Then $f(z)$ is

(a) $\dfrac{1}{z}$ (b) $\dfrac{-1}{z-2}$ (c) $\dfrac{z-1}{z+1}$ (d) $\dfrac{1}{2z-1}$

[IN GATE 2016]

72. The residue of a function $f(z) = \dfrac{1}{(z-4)(z+1)^3}$ are

(a) $-\dfrac{1}{27}$ and $-\dfrac{1}{125}$ (b) $\dfrac{1}{125}$ and $-\dfrac{1}{125}$

(c) $-\dfrac{1}{27}$ and $\dfrac{1}{5}$ (d) $\dfrac{1}{125}$ and $-\dfrac{1}{5}$.

[EC GATE 2017]

73. An integral I over a counterclockwise circle c is given by $I = \oint_c \left(\dfrac{z^2-1}{z^2+1}\right) e^z dz$. If c is defined as $|z| = 3$, then the value of I is.

(a) $-\pi i \sin 1$ (b) $-2\pi i \sin 1$

(c) $-3\pi i \sin 1$ (d) $-4\pi i \sin 1$

[EC GATE 2017]

74. The residue of a function $f(z) = \dfrac{1}{(z-4)(z+1)^3}$ are

(a) $-\dfrac{1}{27}$ and $-\dfrac{1}{125}$ (b) $\dfrac{1}{125}$ and $-\dfrac{1}{125}$

(c) $-\dfrac{1}{27}$ and $\dfrac{1}{5}$ (d) $\dfrac{1}{125}$ and $-\dfrac{1}{5}$.

[EC GATE 2017]

75. An integral I over a counterclockwise circle c is given by $I = \oint_c \left(\dfrac{z^2-1}{z^2+1}\right) e^z dz$. If c is defined as $|z| = 3$, then the value of I is.

(a) $-\pi i \sin 1$ (b) $-2\pi i \sin 1$

(c) $-3\pi i \sin 1$ (d) $-4\pi i \sin 1$

[EC GATE 2017]

76. If $f(z) = \left(x^2 + ay^2\right) + ibxy$ is a complex analytic function of $z = x + iy$, where $i = \sqrt{-1}$, then

(a) $a = -1, b = -1$ (b) $a = -1, b = 2$

(c) $a = 1, b = 2$ (d) $a = 2, b = 2$

[ME GATE 2017]

77. Consider the line integral $I = \int_c (x^2 + iy^2)dz$, where $z = x + iy$. The line C is shown in the figure below:

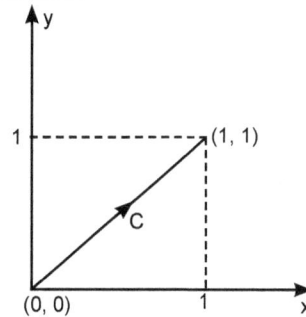

Then the value of I is

(a) $\dfrac{1}{2}i$ (b) $\dfrac{2}{3}i$ (c) $\dfrac{3}{4}i$ (d) $\dfrac{4}{5}i$

[EE GATE 2017]

78. What is the value of "m" for which $2x - x^2 + my^2$ is harmonic?

(a) 1 (b) –1 (c) 2 (d) –2

[EE, ECE GATE 2017]

79. The value of the contour integral in the complex plane $\oint_c \left(\dfrac{z^3 - 2z + 3}{z-2}\right) dz$ along the contour $|z| = 3$, taken counterclockwise is

(a) $-18\pi i$ (b) 0 (c) $14\pi i$ (d) $48\pi i$

[EE GATE 2017]

80. For a complex number z,

$\lim\limits_{z \to i} \dfrac{z^2 + 1}{z^3 + 2z - i\left(z^2 + 2\right)}$ is

(a) $-2i$ (b) $-i$ (c) i (d) $2i$

[EE GATE 2017]

81. Which one of the following options correctly describes the location of the roots of the equation $s^4 + s^2 + 1 = 0$ on the complex plane?
    (a) Four left half plane (LHP) roots
    (b) One right half plane (RHP) root, one LHP root and two roots on the imaginary axis
    (c) Two RHP roots and two LHP roots
    (d) All four roots are on the imaginary axis
    [EC GATE 2017]

82. Let $f_1(z) = z^2$ and $f_2(z) = \bar{z}$ be two complex variable functions. Here $\bar{z}$ is the complex conjugate of z. Choose the correct answer.
    (a) Both $f_1(z)$ and $f_2(z)$ are analytic
    (b) Only $f_1(z)$ is analytic
    (c) Only $f_2(z)$ is analytic
    (d) Both $f_1(z)$ and $f_2(z)$ are not analytic
    [IN GATE 2018]

83. $F(z)$ is a function of the complex variable $z = x + iy$ is given by $F(z) = iz + k \operatorname{Re}(z) + \operatorname{Im}(z)$
    For what value of k will $F(z)$ satisfy the Cauchy-Riemann equations?
    (a) 0   (b) 1   (c) –1   (d) y
    [ME GATE 2018]

84. Consider the analytic function $f(z) = x^2 - y^2 + 2ixy$ of the complex variable $z = x + iy$, where $i = \sqrt{-1}$. The derivative $f'(z)$ is
    (a) $2x + 2iy$ (b) $x^2 + iy^2$
    (c) $x + iy$ (d) $2x - 2iy$
    [PI GATE 2018]

85. The contour C given below is on the complex plane $z = x + jy$, where $j = \sqrt{1}$.

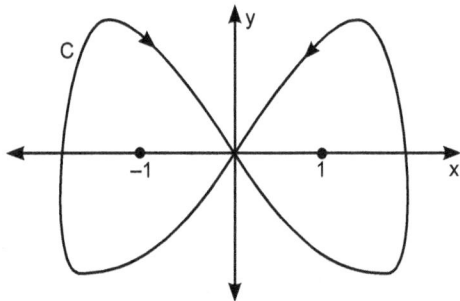

The value of the integral $\dfrac{1}{\pi j}\oint_C \dfrac{dz}{z^2 - 1}$ is
    [EC GATE 2018]

86. The value of the integral $\oint_C \dfrac{z+1}{z^2 - 4}dz$ in the counterclockwise direction around a circle C of radius 1 with center at the point $z = -2$ is
    (a) $\dfrac{\pi i}{2}$ (b) $2\pi i$ (c) $-\dfrac{\pi i}{2}$ (d) $-2\pi i$
    [EE GATE 2018]

87. If C is a circle $|z| = 4$ and $f(z) = \dfrac{z^2}{(z^2 - 3z + 2)^2}$, then $\oint_C f(z)\, dz$ is
    (a) 1   (b) 0   (c) –1   (d) –2
    [EE GATE 2018]

Answer Key				
1. (a)	2. (c)	3. (a)	4. (a)	5. (a)
6. (c)	7. (a)	8. (b)	9. (d)	10. (b)
11. (a)	12. (a)	13. (b)	14. (b)	15. (d)
16. (b)	17. (a)	18. (d)	19. (b)	20. (a)
21. (d)	22. (b)	23. (c)	24. (d)	25. (c)
26. (d)	27. (a)	28. (d)	29. (b)	30. (d)
31. (b)	32. (d)	33. (b)	34. (a)	35. (d)
36. (a)	37. (d)	38. (c)	39. (a)	40. (b)
41. (d)	42. (c)	43. (a)	44. (a)	45. (b)
46. (c)	47. (c)	48. (c)	49. (b)	50. (c)
51. (c)	52. (c)	53. (b)	54. (b)	55. (b)
56. (b)	57. (b)	58. (a)	59. (d)	60. (a)
61. (a)	62. 0.	63. (b)	64. 1.	65. (b)
66. –1	67. 1	68. (b)	69. sin h(2p).	
70. (b)	71. (a)	72. (b)	73. (d)	74. (b)
75. (d)	76. (b)	77. (b)	78. (a)	79. (c)
80. (d)	81. (c)	82. (b)	83. (b)	84. (a)
85. 2	86. (a)	87. (b)		

Explanation

1. (a)
$$w = \frac{z-1}{z+1} \Rightarrow w(z+1) = z - 1$$

$$\Rightarrow\quad wz + w = z - 1 \quad \Rightarrow \quad wz - z = -1 - w$$

$$\Rightarrow z = \frac{1+w}{1-w}$$

In z-plane, the interior of the unit circle is represented by $|z| < 1$

Now, $|z|<1 \Rightarrow \left|\frac{1+w}{1-w}\right|<1 \Rightarrow \left|\frac{1+u+iv}{1-(u+iv)}\right|<1$

[taking $w = u + iv$]

$\Rightarrow \left|\frac{(1+u)+iv}{(1-u)-iv}\right|<1 \Rightarrow \frac{\sqrt{(1+u)^2+v^2}}{\sqrt{(1-u)^2+v^2}}<1$

$\Rightarrow (1+u)^2+v^2 < (1-u)^2+v^2$

$\Rightarrow 1+u^2+2u+v^2 < 1-2u+u^2+v^2$

$\Rightarrow 4u < 0 \Rightarrow u < 0$, which represents the left half of the w-plane.

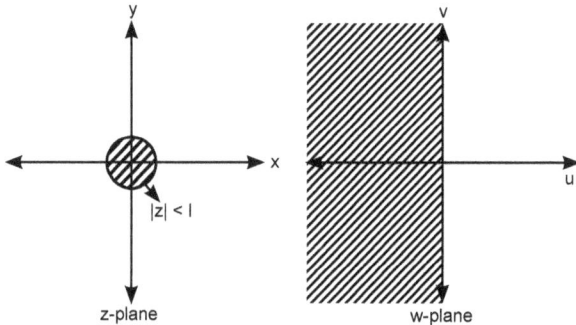

z-plane   w-plane

2. (c) Let, $z_1 = 1+i$ and $z_2 = 1-i$.

Then, $|z_1 - z_2| = |1+i-1+i| = |2i| = 2|i| = 2$

Now, $|z_1| = |1+i| = \sqrt{1^2+1^2} = \sqrt{2}$

and $|z_2| = |1+i| = \sqrt{1^2+(-1)^2} = \sqrt{2}$

So, $|z_1|-|z_2| = \sqrt{2}-\sqrt{2} = 0$

Thus, $|z_1 - z_2| = 2 \nless |z_1|-|z_2|$.

3. (a)

$w = u+iv = \frac{1}{2}\log(x^2+y^2)+i\tan^{-1}\left(\frac{y}{x}\right)$

$= \log\left(\sqrt{x^2+y^2}\right)+i\tan^{-1}\left(\frac{y}{x}\right)$

$= \log|z|+i\,amp(z)$ [where $z = x+iy$]

$= \log z$

$\therefore \frac{dw}{dz} = \frac{1}{z}$, which does not exist at $z = 0$ $i.e;$ at $(0,0)$

Hence, w is not analytic at $(0,0)$.

4. (a)

$|z-5-5i| = |x+iy-5-5i|$

$= |(x-5)+i(y-5)| = \sqrt{(x-5)^2+(y-5)^2}$

$\therefore |z-5-5i| = 2 \Rightarrow \sqrt{(x-5)^2+(y-5)^2} = 2$

$\Rightarrow (x-5)^2+(y-5)^2 = 2^2$ which represents a circle with the center $C(5,5)$ and radius = 2 units.

Let the arbitrary point $z(= x+iy)$ is represented on the circle by the point P. O is the origin.

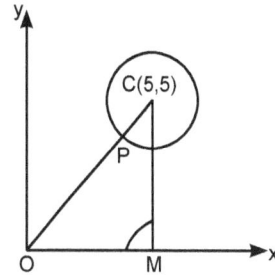

Then, clearly $\overline{PC} = 2$

$\overline{OC}^2 = \overline{OM}^2 + \overline{MC}^2$

$= 5^2+5^2 = 50$

$\therefore \overline{OC} = \sqrt{50} = 5\sqrt{2}$

Hence, $\overline{OP} = \overline{OC} - \overline{PC} = 5\sqrt{2}-2$, which is the required minimum distance from the origin to the circle.

5. (a)

$I = \oint_C \sec z\,dz = \oint_C \frac{dz}{\cos z}$. So poles of $f(z)\left(=\frac{1}{\cos z}\right)$

are given by: $\cos z = 0$ $i.e;$ $\cos z = \cos\frac{\pi}{2}$ and so $z = (2n+1)\frac{\pi}{2}, n = 0,1,2,....$

None of these poles lie within the unit circle $C: |z| = 1$

Hence, by the Cauchy Integral formula $I = 0$ and set all singularities = ϕ.

6. (c) Let $z = re^{i\theta}$ (in polar form).

Then, $\bar{z} = re^{-i\theta}$ and $z^3 = r^3 e^{3i\theta}$.

$\therefore z^3 = \bar{z} \Rightarrow r^3 e^{3i\theta} = re^{-i\theta} \Rightarrow r^2 e^{4i\theta} = 1$

$\Rightarrow r^2 e^{4i\theta} = 1 = 1\times(\cos 2\pi + i\sin 2\pi)$

$\Rightarrow r^2 e^{4i\theta} = 1\times e^{2\pi i}$

Comparing both sides we get $r^2 = 1$ and $4 = 2\pi$ $i.e; r = 1$ and $\theta = \frac{\pi}{2}$.

Hence, $z = 1 \times e^{\frac{i\pi}{2}} = \left(\cos \frac{\pi}{2} + i \sin \frac{\pi}{2} \right) = i$ which satisfies the equation $z^4 = 1$ $(\because i^4 = 1)$.

7. (a)

$$\oint \left(\frac{z^3 - 6}{3z - i} \right) dz$$

$$= \oint \frac{\frac{1}{3}\left(z^3 - 6\right)}{z - (i/3)} dz = \oint \frac{f(z)}{(z-a)} dz$$

$$\left[\text{where } f(z) = \frac{1}{3}\left(z^3 - 6\right), a = \frac{i}{3} \right]$$

$= 2\pi i \times f(a)$ [by the Cauchy's integral formula;

since $a = \frac{i}{3}$ represents the point $\left(0, \frac{1}{3}\right)$ that lies within the unit circle $|z| = 1$]

$$= 2\pi i \times f(i/3) = 2\pi i \times \frac{1}{3}\left\{ (i/3)^3 - 6 \right\}$$

$$= \frac{2\pi i}{3}\left(-\frac{i}{27} - 6 \right) = \frac{2\pi}{81} - 4\pi i.$$

8. (b)

$$w = \ln(z) = \ln(x + jy) = \ln\left(re^{j\theta} \right)$$

$$\left[\because \text{in polar form } z = re^{j\theta} \right]$$

$$= \ln(r) + \ln\left(e^{j\theta} \right) = \ln(r) + j\theta$$

$$= \ln\left(\sqrt{x^2 + y^2} \right) + j \tan^{-1} \frac{y}{x}$$

$$= \frac{1}{2} \ln\left(x^2 + y^2 \right) + j \tan^{-1} \frac{y}{x}$$

$$= u + jv \ (given)$$

$$\therefore u = \frac{1}{2} \ln\left(x^2 + y^2 \right)$$

and so $x^2 + y^2 = e^{2u} = \left(e^u \right)^2$, which is a positive constant since $u = $ constant.

Thus, we get a family of concentric circles

$$x^2 + y^2 = \left(e^u \right)^2.$$

9. (d)

Here, the poles are given by $z^2 + 4 = 0$ $i.e;$

$$(z + 2i)(z - 2i) = 0 \ i.e; \ z = 2i, -2i.$$

$$|z - j| = 2 \Rightarrow |x + jy - j| = 2 \Rightarrow |x + (y - 1)j| = 2$$

$$\Rightarrow \sqrt{x^2 + (y-1)^2} = 2 \Rightarrow (x - 0)^2 + (y - 1)^2 = 2^2,$$

which represents a circle with the center $(0, 1)$ and radius 2 unit.

$z = 2i$ represents the point $(0, 2)$ that lies inside the circle.

$z = -2i$ represents the point $(0, -2)$ that lies outside the circle.

$$\therefore \oint_{|z-j|=2} \frac{1}{z^2 + 4} dz$$

$$= \oint_{|z-j|=2} \frac{1}{(z + 2i)(z - 2i)} dz = \oint_{|z-j|=2} \frac{\left(\frac{1}{z + 2i} \right) dz}{z - 2i}$$

$$= \oint_{|z-j|=2} \frac{f(z)dz}{z - 2i} \left[\text{where } f(z) = \frac{1}{z + 2i} \right]$$

$= 2\pi i \times f(2i)$ [by Cauchy's integral formula]

$$= 2\pi i \times \left(\frac{1}{2i + 2i} \right) = \frac{\pi}{2}.$$

10. (b) $z^2 + 1 = 0 \Rightarrow (z + i)(z - i) = 0 \Rightarrow z = i, -i$

$$\left| z - \frac{i}{2} \right| = 1 \Rightarrow \left| x + iy - \frac{i}{2} \right| = 1 \Rightarrow \left| x + i\left(y - \frac{i}{2} \right) \right| = 1$$

$$\Rightarrow \sqrt{x^2 + \left(y - \frac{1}{2} \right)^2} = 1$$

$$\Rightarrow (x - 0)^2 + \left(y - \frac{1}{2} \right)^2 = 1^2$$

which represents a circle with the center $\left(0, \frac{1}{2} \right)$ and radius = 1 unit.

$z = i = 0 + 1 \times i$ represents a point $(0, 1)$ that lies inside the circle.

$z = -i = 0 + (-1)$ represents a point $(0, -1)$ that lies outside the circle.

$$\therefore \oint_C \frac{1}{z^2 + 1} dz$$

$$= \oint_C \frac{1}{(z + i)(z - i)} dz = \oint_C \frac{\frac{1}{(z + i)}}{z - i} dz$$

$$= \oint_C \frac{f(z)}{z - i} dz \left[\text{where } f(z) = \frac{1}{(z + i)} \right]$$

$= 2\pi i \times f(i)$ [using Cauchy's integral formula]

$= 2\pi i \times \left(\dfrac{1}{i+i}\right) = \pi$

11. $(a)\ s^2 + 1 = 0 \Rightarrow s = -1, 1.$

$s = -1 = -1 + i \times 0$ represents the point $(-1, 0)$ which lies outside the contour D.

$s = 1 = 1 + i \times 0$ represents the point $(1, 0)$ which lies inside the contour D.

Now, $\displaystyle\oint_D \dfrac{1}{s^2-1}\,ds$

$= \displaystyle\oint_D \dfrac{1}{(s-1)(s+1)}\,ds = \oint_D \dfrac{\left(\dfrac{1}{s+1}\right)}{s-1}\,ds$

$= \displaystyle\oint_D \dfrac{f(s)}{s-1}\,ds\ \left[\text{where } f(s) = \dfrac{1}{s+1}\right]$

$= 2\pi i \times f(1)$ [using Cauchy's integral formula]

$= 2\pi i\left(\dfrac{1}{1+1}\right) = \pi i.$

12. $(a)\ \phi(x, y) = x^2 - y^2 \Rightarrow \dfrac{\partial\phi}{\partial x} = 2x$ and $\dfrac{\partial\phi}{\partial y} = -2y.$

Since, $w = \phi(x, y) + i\psi(x, y)$ is analytic, so

$\dfrac{\partial\phi}{\partial x} = \dfrac{\partial\psi}{\partial y}$ and $\dfrac{\partial\phi}{\partial y} = -\dfrac{\partial\psi}{\partial x}$

Then, $\dfrac{\partial\psi}{\partial x} = 2y$ and $\dfrac{\partial\psi}{\partial y} = 2x$

Now, $\dfrac{\partial\psi}{\partial x} = 2y \Rightarrow \psi(x, y) = 2xy + f(y)$ ...(i)

$\therefore \dfrac{\partial\psi}{\partial y} + f'(y) \Rightarrow 2x + f'(y) = 2x$

$\Rightarrow f'(y) = 0 \Rightarrow f(y) = C$ (constant)

Given at $\psi = 0$ at $y = 0$. So (1) gives, $0 = 0 + C$ i.e; $C = 0.$

Thus, we get $\psi(x, y) = 2xy.$

13. (b) Let $w = \phi(x, y) + i\psi(x, y)$. Then w is an analytic function of $z = x + iy$ if $\dfrac{\partial\phi}{\partial x} = \dfrac{\partial\psi}{\partial y}$ and $\dfrac{\partial\phi}{\partial y} = -\dfrac{\partial\psi}{\partial x}.$

14. (b)

$\dfrac{\sin z}{z^3}$

$= \dfrac{1}{z^3}\left(z - \dfrac{z^3}{3!} + \dfrac{z^5}{5!} - \dots \infty\right)$

$= \dfrac{1}{z^2} - \dfrac{1}{3!} + \dfrac{z^2}{5!} - \dfrac{z^4}{7!} + \dots \infty$

$= \dfrac{1}{(z-0)^2} - \dfrac{1}{3!} + \dfrac{(z-0)^2}{5!} - \dfrac{(z-0)^4}{7!} + \dots \infty$

$\therefore\quad z = 0$ is a pole of order 2.

15. (d)

$j = 0 + 1 \times j = \cos\dfrac{\pi}{2} + j\sin\dfrac{\pi}{2} = e^{\frac{j\pi}{2}}$

$\left(\text{using } e^{i\theta} = \cos\theta + i\sin\theta\right)$

$\therefore j^j = \left(e^{\frac{j\pi}{2}}\right)^j = e^{-\frac{\pi}{2}} \left(\because j = \sqrt{-1} \text{ and } j^2 = -1\right)$

16. (b)

$z = \dfrac{\sqrt{3}}{2} + \dfrac{i}{2} = \cos\dfrac{\pi}{6} + i\sin\dfrac{\pi}{6} = e^{\frac{i\pi}{6}}$

$\therefore z^4 = \left(e^{\frac{i\pi}{6}}\right)^4 = e^{\frac{2i\pi}{3}} = \cos\dfrac{2\pi}{3} + i\sin\dfrac{2\pi}{3}$

$= \cos\left(\pi - \dfrac{\pi}{3}\right) + i\sin\left(\pi - \dfrac{\pi}{3}\right) = -\cos\dfrac{\pi}{3} + i\sin\dfrac{\pi}{3}$

$= -\dfrac{1}{2} + \dfrac{i\sqrt{3}}{2}.$

17. (a) Clearly, $z = 2$ is a pole of order "2."

Now $f(z) = \dfrac{1}{(z+2)^2(z-2)^2} = \dfrac{\phi(z)}{(z-2)^2}$ (say)

$\left[\text{where } \phi(z) = \dfrac{1}{(z+2)^2}\right]$

$\therefore \operatorname{Res}(z = 2) = \dfrac{\phi^{(2-1)}(2)}{(2-1)!} = \phi'(z)\big|_{z=2}$

$= -\dfrac{2}{(z+2)^3}\bigg|_{z=2} = -\dfrac{2}{(2+2)^3} = -\dfrac{1}{32}.$

18. (d) Let $g(z) = X(z) \times z^{n-1}.$

Then $g(z) = z^{n-1} \times \dfrac{z}{(z-a)^2} = \dfrac{z^n}{(z-a)^2}$ and so $z = a$ is a pole of order "2" of $g(z).$

Let $\phi(z) = z^n$. Then $g(z) = \dfrac{\phi(z)}{(z-a)^2}$

$\therefore \operatorname{Res}(g(z))\big|_{z=a} = \dfrac{\phi^{(2-1)}(a)}{(2-1)!} = \phi'(z)\big|_{z=a} = na^{n-1}.$

19. (b)

$\dfrac{-5+10i}{3+4i} = \dfrac{(-5+10i)(3-4i)}{(3+4i)(3-4i)}$

$= \dfrac{(-15+40)+i(20+30)}{3^2+4^2}$

[using $(a+ib)(c+id) = (ac-bd)+i(ad+bc)$
and $(a+ib)(a-ib) = a^2+b^2$]

$= \dfrac{25+50i}{25} = 1+2i$

20. (a) Clearly, $z = 0$ is a singular point of the function $f(z) = \dfrac{\cos z}{z}$. Also, $z = 0$ lies inside the unit circle $|z| = 1$.

Hence, by Cauchy's integral formula,

$\oint f(z)dz = \oint_{|z|=1} \dfrac{\cos z\,dz}{(z-0)} = \oint_{|z|=1} \dfrac{f(z)}{z-0}dz$

(where $f(z) = \cos z$)

$= 2\pi i \times f(0) = 2\pi i \times \cos 0 = 2\pi i.$

21. (d)

$\sin z = 10 \Rightarrow \dfrac{e^{iz}-e^{-iz}}{2i} = 10 \Rightarrow e^{iz}-\dfrac{1}{e^{iz}} = 20i$

$\Rightarrow e^{2iz} - 20i\left(e^{iz}\right) - 1 = 0$

$\Rightarrow u^2 - 20iu - 1 = 0 \quad \left(\text{where } u = e^{iz}\right)$

$\Rightarrow u = \dfrac{20i \pm \sqrt{(20i)^2 + 4}}{2} = \dfrac{20i \pm \sqrt{-400+4}}{2}$

$= \dfrac{20i \pm 6i\sqrt{11}}{2} = 10i \pm 3i\sqrt{11}$

$\Rightarrow e^{iz} = 10i \pm 3i\sqrt{11} = ik \ (say)$

$\left(\text{where } k == \dfrac{20 \pm 6\sqrt{11}}{2} = 10 \pm 3\sqrt{11} > 0\right)$

$\Rightarrow \log_e\left(e^{iz}\right) = \log_e(ik) = \log_e i + \log_e k$

$\Rightarrow iz = \log|i| + i\big[2n\pi + amp(i)\big] + \log_e k, \ n \in z^+$

$\Rightarrow iz = \log 1 + i\left(2n\pi + \dfrac{\pi}{2}\right) + \log_e k$

$= i\left(2n\pi + \dfrac{\pi}{2}\right) + \log_e k$

$\Rightarrow z = \left(2n\pi + \dfrac{\pi}{2}\right) + \dfrac{1}{i}\log_e\left(10 \pm 3\sqrt{11}\right)$

$= \left(2n\pi + \dfrac{\pi}{2}\right) - i\log_e\left(10 \pm 3\sqrt{11}\right), \ n \in Z^+ ....(i)$

$\left(\because \dfrac{1}{i} = \dfrac{i}{i^2} = \dfrac{i}{-1} = -i\right)$

Hence, $\sin z = 10$ has an infinite number of complex solutions given by (i).

22. (b)

Let $z = x + iy$ and $w = u + iv = \dfrac{1}{z}$

Then $w = \dfrac{1}{z}$

$\Rightarrow z = \dfrac{1}{w} \Rightarrow x+iy = \dfrac{1}{u+iv} = \dfrac{u-iv}{(u+iv)(u-iv)} = \dfrac{u-iv}{u^2+v^2}$

$\Rightarrow x+iy = \left(\dfrac{u}{u^2+v^2}\right) + i\left(\dfrac{-v}{x^2+y^2}\right)$

$\Rightarrow x = \dfrac{u}{u^2+v^2}, \quad y = \dfrac{-v}{u^2+v^2}$

Then for $y = 0.1 = \dfrac{1}{10}$, we get

$-\dfrac{v}{u^2+v^2} = \dfrac{1}{10}$ or, $u^2 + v^2 + 10v = 0$

or, $u^2 + (v+5)^2 = 5^2$, which represents a circle in u-v plane having the center at $(0,-5)$ and radius $= 5$ unit.

In Figure (b), $-5i$ represents the point $(0,-5)$ and radius of this circle $= |(-5i)-(-10i)| = |5i| = 5$ unit.

23. (c) Since $f(z) = u(x, y) + iv(x, y)$ is analytic, so CR equations are satisfied and hence,

$\dfrac{\partial u}{\partial x} = \dfrac{\partial v}{\partial y}$ and $\dfrac{\partial u}{\partial y} = -\dfrac{\partial v}{\partial x}$

Now, $u = xy \Rightarrow \dfrac{\partial u}{\partial x} = y$ and $\dfrac{\partial u}{\partial y} = x.$

So, $\dfrac{\partial v}{\partial x} = -x$ and $\dfrac{\partial v}{\partial y} = y.$

$$\therefore dv = \frac{\partial v}{\partial x}dx + \frac{\partial v}{\partial y}dy = (-x)dx + ydy$$

Integrating both sides we get,

$$\int dv = \int(-x)dx + ydy + k$$

$$\text{or, } v = \frac{-x^2}{2} + \frac{y^2}{2} + k = \left(\frac{y^2 - x^2}{2}\right) + k.$$

24. (d)

$f(z) = \frac{z-1}{z^2+1} \Rightarrow$ singularities of $f(z)$ are given by $z^2 + 1 = 0$

$\Rightarrow (z+i)(z-i) = 0 \Rightarrow z = i, -i.$

25. (c)

$|z| = 1 \Rightarrow |x + iy| = 1 \Rightarrow \sqrt{x^2 + y^2} = 1$

$\Rightarrow (x-0)^2 + (y-0)^2 = 1^2,$

which represents a circle with the center $(0, 0)$ and radius $=1$ unit.

Now, $(2z-1)(z-3) = 0 \Rightarrow z = \frac{1}{2}, 3.$

$z = \frac{1}{2}$ represents the point $\left(\frac{1}{2}, 0\right)$ which lies inside C.

$z = 3$ represents the point $(3, 0)$ which lies outside C.

$$\therefore \int_C \frac{\cos(2\pi z)}{(2z-1)(z-3)}dz = \int_C \frac{\frac{\cos(2\pi z)}{2(z-3)}}{z-\frac{1}{2}}dz = \int_C \frac{f(z)dz}{z-\frac{1}{2}}$$

$$\left(\text{where } f(z) = \frac{\cos(2\pi z)}{2(z-3)}\right)$$

$= 2\pi i \times f\left(\frac{1}{2}\right)$ (by Cauchy's integral formula)

$$= \frac{2\pi i \times \cos \pi}{-5} = \frac{2\pi i}{5} \qquad (\because \cos \pi = -1).$$

26. (d) Clearly the singularity $z = 0$ lies within $|z| = 1$

$$\therefore \oint_{|z|=1} \frac{1+f(z)}{z}dz$$

$$= \oint_{|z|=1} \frac{\left(1 + c_0 + \frac{c_1}{z}\right)dz}{z}$$

$$= \oint_{|z|=1} \frac{(z + c_0 z + c_1)}{(z-0)^2}dz = \oint_{|z|=1} \frac{g(z)dz}{(z-0)^2}$$

$[\text{where } g(z) = z + c_0 z + c_1]$

$$= \oint_{|z|=1} \frac{g(z)dz}{(z-a)^{n+1}} = \frac{2\pi i \times g'(0)}{1!}$$

$$\left[\text{using} \int_c \frac{g(z)dz}{(z-a)^{n+1}} = \frac{2\pi i}{n!} \times g^{(n)}(a)\right]$$

$= 2\pi i \times g'(z)\big|_{z=0} = 2\pi i \times (1 + C_0)\big|_{z=0} = 2\pi i(1 + C_0)$

$= 2\pi j(1 + C_0)$ [where $j = \sqrt{-1} = i$].

27. (a) Comparing $\oint_C \frac{\sin z}{z}dz$ with $\oint_C \frac{f(z)}{z-a}dz$, we

get $f(z) = \sin z$ and $a = 0$ which lies within C.

$$\therefore \oint_C \frac{\sin z}{z} = \oint_C \frac{f(z)}{z-0} = 2\pi i \times f(0)$$

$$\oint_C \frac{\sin z}{z} = \oint_C \frac{f(z)}{z-0} = 2\pi i \times f(0)$$

(by Cauchy's integral formula)

$= 2\pi i \times \sin 0 = 0.$

28. (d)

$\left|e^{jz}\right| = \left|e^{j(x+jy)}\right| = \left|e^{jx-y}\right| \quad (\because j^2 = -1)$

$= \left|e^{jx}.e^{-y}\right| = \left|e^{jx}\right|\left|e^{-y}\right| = e^{-y}$

$\left(\because \left|e^{jx}\right| = |\cos x + j\sin x| = \sqrt{\cos^2 x + \sin^2 x} = 1\right).$

29. (b) For, $x = \frac{\sqrt{3}}{2} + \frac{j}{2},$

$$x^3 = \left(\frac{\sqrt{3}}{2} + \frac{j}{2}\right)^3 = \left(\cos\frac{\pi}{6} + j\sin\frac{\pi}{6}\right)^3 = \left(e^{j\frac{\pi}{6}}\right)^3$$

$$= e^{j\frac{\pi}{2}} = \cos\frac{\pi}{2} + j\sin\frac{\pi}{2} = 0 + j\times 1 = j.$$

30. (d)

$(3 - 2i)(3 + 4i) = 9 + 12i - 6i - 8i^2 = 17 + 6i.$

31. (b) Given that

$f(x+iy) = f(z) = x^3 - 3xy^2 + i\phi(x, y)$ is an analytic function of $z = x + iy$.

Then by the CR equations, we have

$$\frac{\partial \phi}{\partial y} = \frac{\partial}{\partial x}(x^3 - 3xy^2) = 3x^2 - 3y^2 \text{ and}$$

$$\frac{\partial \phi}{\partial x} = -\frac{\partial}{\partial y}\left(x^3 - 3xy^2\right) = 6xy$$

Now, $d\phi = \dfrac{\partial \phi}{\partial x}dx + \dfrac{\partial \phi}{\partial y}dy$

$$= 6xydx + \left(3x^2 - 3y^2\right)dy = d\left(3x^2y - y^3\right)$$

Integrating we get, $\displaystyle\int d\phi = \int d\left(3x^2y - y^3\right)$

or, $\phi = 3x^2y - y^3$.

32. (d)

$$\oint_c \frac{(z^2 + 8)}{(0.5)z - (1.5)j}dz = \oint_c \frac{(z^2 + 8)}{\left(\dfrac{1}{2}z - \dfrac{3}{2}j\right)}dz$$

$$= 2\oint_c \frac{(z^2 + 8)dz}{(z - 3j)}$$

Clearly $z = 3j$ is a singularly of the integrand $\dfrac{z^2 + 8}{z - 3j}$ and $z = 3j$ represents the point $(0, 3)$ which lies inside the circle $x^2 + y^2 = 16$ (center $= (0, 0)$ & radius $= 4$ unit).

$$\therefore \oint_c \frac{(z^2 + 8)}{(0.5)z - (1.5)j}dz = 2\times\oint_c \frac{f(z)}{z - 3j}dz$$

(where $f(z) = z^2 + 8$)

$= 2 \times 2\pi j \times f(3j)$ (using Cauchy's integral formula)

$= 4j\pi \times (9j^2 + 8) = 4j\pi(-9 + 8) = -4\pi j$ $(j = \sqrt{-1}\,)$.

33. (b)

$$\left|\frac{3 + 4i}{1 - 2i}\right| = \frac{|3 + 4i|}{|1 - 2i|} = \frac{\sqrt{3^2 + 4^2}}{\sqrt{1^2 + (-2)^2}} = \frac{5}{\sqrt{5}} = \sqrt{5}\,.$$

34. (a) $1 + \omega + \dfrac{1}{\omega} = 1 + \omega + \dfrac{\omega^3}{\omega} = 1 + \omega + \omega^2 = 0$

$$[\because \omega^3 = 1 \text{ and } 1 + \omega + \omega^2 = 0]$$

35. (d) Given that the function
$f(z) = f(x + iy) = u(x, y) + iv(x, y)$ is analytic and so the CR equations are satisfied.

$$\therefore \frac{\partial u}{\partial x} = \frac{\partial v}{\partial y} \text{ and } \frac{\partial u}{\partial y} = -\frac{\partial v}{\partial x}.$$

Here, $u = u(x, y) = 3x^2 - 3y^2$ and so $\dfrac{\partial u}{\partial x} = 6x$ and $\dfrac{\partial u}{\partial y} = -6y$.

$$\therefore dv = \frac{\partial v}{\partial x}dx + \frac{\partial v}{\partial y}dy = \left(-\frac{\partial u}{\partial y}\right)dx + \left(\frac{\partial u}{\partial x}\right)dy$$

$$= 6ydx + 6xdy = 6d(xy)$$

Integrating we get,

$$\int dv = \int 6d(xy) + k \quad \text{or, } v = 6xy + k.$$

36. (a) $(1 + i)(2 - 5i) = 2 - 5i + 2i - 5i^2 = 7 - 3i$.

37. (d)

$$z = x + iy \Rightarrow \frac{1}{z} = \frac{1}{x + iy} = \frac{x - iy}{(x + iy)(x - iy)}$$

$$= \frac{x - iy}{x^2 + y^2} = \frac{x}{x^2 + y^2} + i\left(\frac{-y}{x^2 + y^2}\right),$$

which represents the point $\left(\dfrac{x}{x^2 + y^2}, \dfrac{-y}{x^2 + y^2}\right)$.

According to given diagram, $z = x + iy$, i.e; the point (x, y) lies in first quadrant. So $x > 0$ and $y > 0$. Hence, $\dfrac{x}{x^2 + y^2} > 0$ and $\dfrac{-y}{x^2 + y^2} < 0$. So the point

$\left(\dfrac{x}{x^2 + y^2}, \dfrac{-y}{x^2 + y^2}\right)$ i.e; $\dfrac{1}{z}$ lies in fourth quadrant.

38. (c)

$$e^{\frac{1}{z}} = 1 + \frac{1}{z} + \frac{1}{2!}\left(\frac{1}{z}\right)^2 + \frac{1}{3!}\left(\frac{1}{z}\right)^3 + \dots\infty$$

So clearly the integrand $e^{\frac{1}{z}}$ has a singularity at $z = 0$, which lies inside the circle $|z| = 1$.

Here, Res$(z = 0) =$ co-efficient of $\dfrac{1}{z}$ in the expansion of $e^{\frac{1}{z}} = 1$.

Hence, by Cauchy's residue theorem,

$$\int_c e^{\frac{1}{z}}dz = 2\pi i \times (\text{sum of residues}) = 2\pi i = 2\pi\sqrt{-1}.$$

39. (a) The singular points of the integrand are given by the solution of the equation $z^2 + 4z + 5 = 0$.

Now, $z^2 + 4z + 5 = 0$

$$\Rightarrow z = \frac{-4 \pm \sqrt{16 - 4\times 5}}{2} = -2 \pm i$$

$z = -2 + i$ represents the point $(-2, 1)$ which lies outside the circle $|z| = 1$.

$z = -2 - i$ represents the point $(-2, -1)$ which lies outside the circle $|z| = 1$.

Thus, the function $\dfrac{-3z+4}{z^2+4z+5}$ is analytic within and on the circle $|z| = 1$. Hence, by Cauchy's residue theorem,

$$\oint_c \left(\frac{-3z+4}{z^2+4z+5}\right)dz = 0.$$

40. (b) $\displaystyle\oint_c \frac{z^2}{z^4-1}dz = \int_c \frac{z^2}{(z^2+1)(z^2-1)}dz =$

$$= \oint_c \frac{z^2 dz}{(z+i)(z-i)(z+1)(z-1)}.$$

Clearly, $z = \pm i, \pm 1$ are the singularities of the integrand.

Now $|z + 1| = 1$

$\Rightarrow |x+iy+1| = 1 \Rightarrow \sqrt{(x+1)^2 + y^2} = 1$

$\Rightarrow \{x-(-1)\}^2 + (y-0)^2 = 1^2$, which represents a circle with the center $(-1, 0)$ and radius = 1 unit.

$z = i$ represents the point $(0, 1)$ which lies outside the circle.

$z = -i$ represents the point $(0, -1)$ which lies outside the circle.

$z = -1$ represents the point $(-1, 0)$ which lies inside the circle.

$z = 1$ represents the point $(1, 0)$ which lies outside the circle.

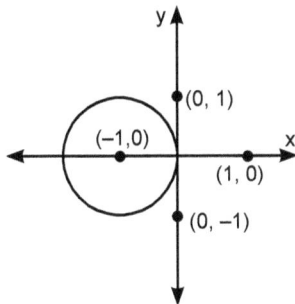

$$\therefore \oint_c \frac{z^2}{z^4-1}dz = \oint_c \frac{\left\{\dfrac{z^2}{(z+i)(z-i)(z-1)}\right\}}{(z+1)}dz$$

$$= \int_c \frac{f(z)}{z-(-1)}dz$$

$\left[\text{where } f(z) = \dfrac{z^2}{(z+i)(z-i)(z-1)} = \dfrac{z^2}{(z^2+1)(z-1)}\right]$

$= 2\pi i \times f(-1)$ (by Cauchy's integral formula)

$$= 2\pi i \times \frac{(-1)^2}{(1+1)(-1-1)} = -\frac{\pi i}{2}.$$

41. (d)

$$\int_0^{\pi/2}\left(\frac{\cos x + i\sin x}{\cos x - i\sin x}\right)dx = \int_0^{\pi/2}\frac{e^{ix}}{e^{-ix}}dx$$

$$= \int_0^{\pi/2} e^{2ix}dx = \frac{1}{2i}\left[e^{2ix}\right]_0^{\pi/2}$$

$$= \frac{1}{2i}\left[e^{i\pi} - e^0\right] = \frac{1}{2i}\{-1-1\} = -\frac{1}{i} = \frac{i^2}{i} = i$$

$\left(\because e^{i\pi} = \cos\pi + i\sin\pi = -1\right)$

42. (c)

$$\frac{1}{2\pi j}\oint_c f(z)dz = \frac{1}{2\pi j}\oint_c \frac{(z+3-2z-2)dz}{(z+1)(z+3)}$$

$$\frac{1}{2\pi j}\oint_c \frac{(1-z)dz}{(z+1)(z+3)}$$

Clearly, $z = -1, -3$ are the singularities of the integrand.

Now, $|z + 1| = 1$

$\Rightarrow |x+jy+1| = 1 \Rightarrow \sqrt{(x+1)^2 + y^2} = 1$

$\Rightarrow \{x-(-1)\}^2 + (y-0)^2 = 1^2$

which is an equation of a circle with the center $(-1, 0)$ and radius = 1 unit.

$z = -1$ represents the point $(-1, 0)$ which lies inside the circle.

$z = -3$ represents the point $(-3, 0)$ which lies outside the circle.

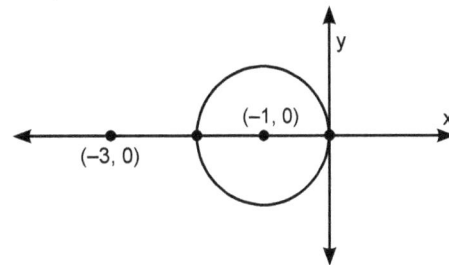

$$\therefore \frac{1}{2\pi j}\oint_c f(z)dz = \frac{1}{2\pi j}\times\left\{\oint_c \frac{\left(\dfrac{1-z}{z+3}\right)dz}{(z-(-1))}\right\}$$

$$= \frac{1}{2\pi j}\oint_c \frac{g(z)}{z-(-1)}\left[\text{where } g(z) = \frac{1-z}{z+3}\right]$$

$$= \frac{1}{2\pi j} \times 2\pi j \times g(-1) = g(-1) = \frac{1-(-1)}{-1+3} = 1.$$

43. (a) Let $z = x^x = \left(\sqrt{-1}\right)^{\sqrt{-1}} = i^i.$

Then, $Log\, z = Log\left(i^i\right) = iLog(i)$

$= i\{\log |i| + [2n\pi + amp(i)]\}$

$= i\left\{\log 1 + i\left(2n\pi + \frac{\pi}{2}\right)\right\}$

$= i\left(0 + \frac{i\pi}{2}\right)$ (for $n = 0$) $= -\frac{\pi}{2}$

$\therefore z = e^{-\frac{\pi}{2}}$ i.e; $x^x = e^{-\frac{\pi}{2}}.$

44. (a) $\oint_c \frac{z^2-4}{z^2+4}dz = \oint_c \frac{z^2-4}{(z+2i)(z-2i)}dz$

Clearly, $z = \pm 2i$ are the singularities of the integrand.

Now, $|z - i| = 2 \Rightarrow |x + iy - i| = 2 \Rightarrow |x + i(y-1)| = 2$

$\Rightarrow \sqrt{x^2(y-1)^2} = 2^2 \Rightarrow (x-0)^2 + (y-1)^2 = 2^2,$

which represents a circle with the center $(0, 1)$ and radius $= 2$ unit.

$z = 2i$ represents the point $(0, 2)$ which lies inside the circle C.

$z = -2i$ represents the point $(0, -2)$ which lies outside the circle C.

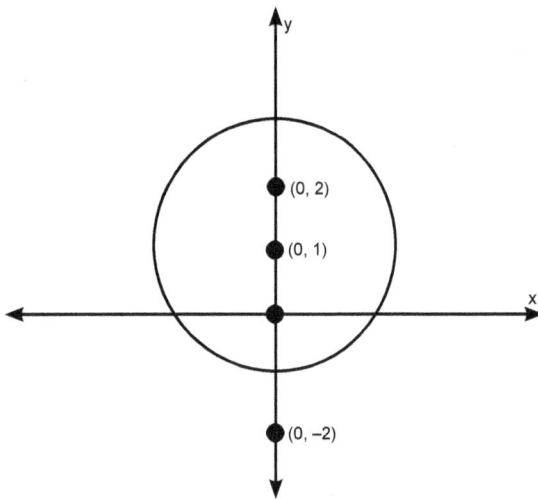

$\therefore \oint_C \left(\frac{z^2-4}{z^2+4}\right)dz = \oint_C \frac{\left(\frac{z^2-4}{z+2i}\right)dz}{z-2i} = \oint_C \frac{f(z)dz}{z-2i}$

(where $f(z) = \frac{z^2-4}{z+2i}$)

$= 2\pi i \times f(2i)$ (by Cauchy's integral formula)

$= 2\pi i \times \left(\frac{4i^2-4}{2i+2i}\right) = \frac{\pi}{2}(-4-4) = -4\pi$

45. (b) Let $z = \sqrt{-i}.$

Then, $z = (-i)^{\frac{1}{2}} \Rightarrow Log\, z = \frac{1}{2}Log(-i)$

$= \frac{1}{2}\{\log |i| + i(2n\pi + amp(-i))\}$ $[n \in Z^+]$

$= \frac{1}{2}\left\{\log 1 + i\left(2n\pi + \left(-\frac{\pi}{2}\right)\right)\right\}$

$= \frac{i}{2}\left(2n\pi - \frac{\pi}{2}\right)$ for $n \in Z^+$

For, $n = 0, 1;$ $\log z = \frac{i}{2}\left(-\frac{\pi}{2}\right), \frac{i}{2}\left(2\pi - \frac{\pi}{2}\right)$

i.e; $\log z = -\frac{i\pi}{4}, \frac{3\pi i}{4}$

So, $z = e^{-\frac{i\pi}{4}}, e^{\frac{3\pi i}{4}}$

$= \cos\left(-\frac{\pi}{4}\right) + i\sin\left(-\frac{\pi}{4}\right), \cos\frac{3\pi}{4} + i\sin\frac{3\pi}{4}.$

46. (c) $\frac{1+i}{1-i} = \frac{(1+i)(1-i)}{(1-i)(1-i)} = \frac{1+2i+i^2}{1^2+1^2} = \frac{2i}{2} = i$

$\therefore \arg\left(\frac{1+i}{1-i}\right) = \arg(i) = \arg(0+1\times i) = \tan^{-1}\left(\frac{1}{0}\right) = \frac{\pi}{2}.$

47. (c) $u(x, y) = 2xy \Rightarrow \frac{\partial u}{\partial x} = 2y$ and $\frac{\partial u}{\partial y} = 2x$

Since $f(z) = u(x,y) + iv(x,y)$ is analytic, so the CR equations are satisfied.

$\therefore \frac{\partial v}{\partial y} = \frac{\partial u}{\partial x} = 2y$ and $\frac{\partial v}{\partial x} = -\frac{\partial u}{\partial y} = -2x$

Then, $dv = \frac{\partial v}{\partial x}dx + \frac{\partial v}{\partial y}dy = -2xdx + 2ydy$

$= d\left(y^2 - x^2\right)$

Integrating we get, $v = y^2 - x^2$ + constant

48. (c)

$u(x, y) = x^2 - y^2 \Rightarrow \frac{\partial u}{\partial x} = 2x$ and $\frac{\partial u}{\partial y} = -2y$

Since $f(z) = u(x, y) + iv(x, y)$ is analytic, so the CR equations are satisfied.

Then, $\dfrac{\partial v}{\partial y} = \dfrac{\partial u}{\partial x} = 2x$ and $\dfrac{\partial v}{\partial x} = -\dfrac{\partial u}{\partial y} = 2y$.

$\therefore dv = \dfrac{\partial v}{\partial x} dx + \dfrac{\partial v}{\partial y} dy = 2y \, dx + 2x \, dy = 2d(xy)$

Integrating we get, $v = 2xy + $ C.

49. (b) Let $f(z) = u + iv$, where $u = e^{-y} \cos x$.

Then, $\dfrac{\partial u}{\partial x} = -e^{-y} \sin x$ and $\dfrac{\partial u}{\partial y} = -e^{-y} \cos x$.

Since $f(z)$ is analytic, so by the CR equations, we have $\dfrac{\partial v}{\partial y} = \dfrac{\partial u}{\partial x} = -e^{-y} \sin x$ and $\dfrac{\partial v}{\partial x} = -\dfrac{\partial u}{\partial y} = e^{-y} \cos x$.

$\therefore dv = \left(\dfrac{\partial v}{\partial x}\right) dx + \left(\dfrac{\partial v}{\partial y}\right) dy$

$= e^{-y} \cos x \, dx + \left(-e^{-y} \sin x\right) dy = d\left(e^{-y} \sin x\right)$

Integrating we get, $v = e^{-y} \sin x$.

50. (c)

$f(z) = zz^* = z\bar{z} = |z|^2 = 1^1 = 1 = 1 + i \times 0$,

which represents the point $(1, 0)$ in the complex plane.

51. (c) Clearly, $z = -2j$ is a singularity of the integrand.

$|z| = 3 \Rightarrow |x + jy| = 3 \Rightarrow \sqrt{x^2 + y^2} = 3$

$\Rightarrow (x - 0)^2 + (y - 0)^2 = 3^2$,

which is a circle with the center $(0, 0)$ and radius = 3 unit.

$z = -2j$ represents a point $(0, -2)$ which lies inside the circle C.

$\therefore \oint_c \dfrac{(z^2 - z + 4j) dz}{z + 2j} = \oint_c \dfrac{f(z) dz}{z - (-2j)}$

[where $f(z) = z^2 - z + 4j$]

$= 2\pi j \times f(-2j)$ (by Cauchy's integral formula)

$= 2\pi j \times \{(-2j)^2 - (-2j) + 4j\}$

$= 2\pi j \, (-4 + 2j + 4j)$

$= -8\pi j - 12\pi \ \left(\because \ j = \sqrt{-1}\right)$

$= -4\pi \, (3 + 2j)$.

52. (c) $z = \pm 1$ are the only singularities of the integrand.

$|z - 1| = 1 \Rightarrow |x + iy - 1| = 1 \Rightarrow |(x - 1) + iy| = 1$

$\Rightarrow \sqrt{(x-1)^2 + y^2} = 1 \Rightarrow (x - 1)^2 + (y - 0)^2 = 1^2$

which is an equation of a circle C with the center $(1, 0)$ and radius = 1 unit.

The singularity $z = 1$ represents the point $(1, 0)$ which lies inside the circle.

The singularity $z = -1$ represents the point $(-1, 0)$ which lies outside the circle.

$\therefore \displaystyle\int_c \dfrac{z^2}{z^2 - 1} dz = \int_c \dfrac{z^2}{(z+1)(z-1)} dz = \int_c \dfrac{\left(\dfrac{z^2}{z+1}\right) dz}{z - 1}$

$= \displaystyle\int_c \dfrac{f(z) dz}{z - 1} \quad \left(\text{where } f(z) = \dfrac{z^2}{z+1}\right)$

$= 2\pi i \times f(1)$ [by Cauchy's integral formula]

$= 2\pi i \times \left(\dfrac{1}{1+1} = \pi i\right)$.

53. (b)

$\displaystyle\int_5^{3i} \dfrac{dz}{z} = \left[\log z\right]_5^{3i} = \log(3i) - \log 5$

$= \log 3 + \log i - \log 5 = \log\left(\dfrac{3}{5}\right) + \log i$

$= -0.511 + \log\left(\cos\dfrac{\pi}{2} + i \sin\dfrac{\pi}{2}\right)$

$= -0.511 + \log\left(e^{\frac{i\pi}{2}}\right)$

$= -0.511 + \dfrac{i\pi}{2} = -0.511 + 1.57i$

54. (b)

$1^i = \left(\cos 0 + i \sin 0\right)^i$

$= \left\{\cos\left(2k\pi + 0\right) + i \sin\left(2k\pi + 0\right)\right\}^i, \ k \in Z$

$= \left(e^{2k\pi i}\right)^i = e^{-2k\pi}, \ k \in Z$

Thus, the values of 1^i are real and non-negative.

55. (b) Sum of two differentiable functions is again a differentiable function.

56. (b) The point $z = z_0$ lies within the circle C: $|z - z_0| = r$.

$$\therefore \oint_c \frac{dz}{(z - z_0)^{n+1}} = \int \frac{f(z)dz}{(z - z_0)^{n+1}} \text{ (where } f(z) = 1\text{)}$$

$$= \frac{2\pi i}{n!} \times f^{(n)}(z_0)$$

(by Cauchy's integral formula for higher order derivaties)

$$= \frac{2\pi i}{n!} \times f^{(n)}(z)\Big|_{z=z_0} = \frac{2\pi i}{n!} \times 0 = 0$$

57. (b) The function $f(z) = \frac{1}{z^2}$ has a pole $z = 0$ of

order 2, which lies within $|z| = 1$.

Now, $f(z) = \frac{1}{z^2} = \frac{\phi(z)}{(z - 0)^2}$, where $\phi(z) = 1$.

$$\therefore \quad \mathrm{Re}\, s(z = 0) = \frac{1}{(2-1)!} \phi^{(2-1)}(0) = \phi'(z)\Big|_{z=0} = 0$$

So by Cauchy's residue theorem,

$$\oint_{|z|=1} \frac{dz}{z^2} = 2\pi i \times \text{ (sum of the residue at the poles)}$$

$$= 2\pi i \times 0 = 0.$$

58. (a)

$$f(z) = \frac{9}{(z - 1)(z + 2)^2} = \frac{\frac{9}{z-1}}{\{z - (-2)\}^2} = \frac{\phi_1(z)}{\{z - (-2)\}^2}$$

where $\phi_1(z) = \frac{9}{z - 1}$.

$\therefore$ $z = -2$ is a pole of order 2 of $f(z)$.

So, $\mathrm{Re}\, s(z = -2) = \frac{1}{(2-1)!} \phi^{(2-1)}(-2) = \phi'(z)\Big|_{z=-2}$

$$= -\frac{9}{(z-1)^2}\Big|_{z=-2} = -\frac{9}{(-2-1)^2} = -1.$$

Again, $f(z) = \frac{9}{(z - 1)(z + 2)^2} = \frac{\frac{9}{(z+2)^2}}{z - 1} = \frac{\phi_2(z)}{(z - 1)^1}$

where $\phi_2(z) = \frac{9}{(z + 2)^2}$.

$\therefore$ $z = 1$ is a pole of order 1 of $f(z)$.

So, $\mathrm{Re}\, s(z = 1) = \lim_{z \to 1}(z - 1)f(z) = \lim_{z \to 1} \phi_2(z)$

$$= \lim_{z \to 1} \frac{9}{(z + 2)^2} = \frac{9}{(1 + 2)^2} = 1.$$

59. (d)

$$f(z) = \frac{z}{z^2 - 1} = \frac{z}{(z + 1)(z - 1)}.$$

So, $z = 1$ is a pole of order 1.

$\therefore$ $\mathrm{Re}\, s(z = 1) = \lim_{z \to 1}(z - 1)f(z)$

$$= \lim_{z \to 1}(z - 1) \times \frac{z}{(z + 1)(z - 1)}$$

$$= \lim_{z \to 1} \frac{z}{z + 1} = \frac{1}{1 + 1} = \frac{1}{2}.$$

$\therefore$ option (a) is a correct statement.

$$\frac{1}{2\pi i}\oint_c \frac{dz}{z} = \frac{1}{2\pi i}\oint_{|z|=1} \frac{g(z)dz}{z - 0} \text{ (where } g(z) = 1 \text{ and}$$

$z = 0$ lies inside C)

$$= \frac{1}{2\pi i} \times (2\pi i \times g(0))$$

(by Cauchy's integral formula)

$$= g(0) = 1.$$

$\therefore$ option (c) is a correct statement.

$$\oint_c z^2 dz = \oint_{|z|=1} z^2 dz = 0 \ (\because \text{ the function } h(z) = z^2 \text{ is analytic within and on } C)$$

$\therefore$ option (b) is a correct statement.

Let $f(z) = \bar{z}$. Then $\phi(z) = x - iy = u + iv$ (say)

$\therefore$ $u = u(x, y) = x$, $v = v(x, y) = -y$.

So $\frac{\partial u}{\partial x} = 1 \neq \frac{\partial v}{\partial y} = -1$. Hence, the CR equations are not satisfied. Consequently, $\phi(z) = \bar{z}$ is not an analytic function of z. So option (d) is an incorrect statement.

60. $\frac{1}{2}$

Here, "c" is the unit circle $|z| = 1$. Let $z = e^{i\theta}$.

Then, $z = \cos\theta + i\sin\theta$ and $dz = (-\sin\theta + i\cos\theta)d\theta$.

$$\therefore \frac{1}{2\pi i}\int \mathrm{Re}(z)dz = \frac{1}{2\pi i}\int_{|z|=1} \mathrm{Re}(z)dz$$

$$= \frac{1}{2\pi i}\int_{\theta=0}^{2\pi} \cos\theta(-\sin\theta + i\cos\theta)d\theta$$

$$= \frac{1}{4\pi i}\int_0^{2\pi} (-2\sin\theta\cos\theta + i \times 2\cos^2\theta)d\theta$$

$$= \frac{1}{4\pi i}\int_0^{2\pi} \left[-\sin 2\theta + i(1 + \cos 2\theta)\right]d\theta$$

$$= \frac{1}{4\pi i}\left[\frac{\cos 2\theta}{2} + i\left(\theta + \frac{\sin 2\theta}{2}\right)\right]_0^{2\pi}$$

$$= \frac{1}{4\pi i}\left[\frac{\cos 4\pi}{2} + i\left(2\pi + \frac{\sin 4\pi}{2}\right) - \frac{\cos o}{2} - i(0 + \sin 0)\right]$$

$$= \frac{1}{4\pi i}\left[\frac{1}{2} + i(2\pi + 0) - \frac{1}{2} - 0\right] = \frac{1}{2}.$$

61. (a)

$$\arg\left(\frac{z_1}{z_2}\right) = \arg(z_1) - \arg(z_2)$$

$$= \arg\left(5 + 5\sqrt{3}i\right) - \arg\left(\frac{2}{\sqrt{3}} + 2i\right)$$

$$= \tan^{-1}\left(\frac{5\sqrt{3}}{5}\right) - \tan^{-1}\left(\frac{2}{\frac{2}{\sqrt{3}}}\right)$$

$$= \tan^{-1}\sqrt{3} - \tan^{-1}\sqrt{3} = 0.$$

62. 0.

Since, $z = x + iy$,

so $f(z) = 2z^2 + b|z|^3 = 2(x + iy)^2 + b\left(x^2 + y^2\right)^{\frac{3}{2}}$

$$= 2\left(x^2 - y^2 + 2ixy\right) + b\left(x^2 + y^2\right)^{\frac{3}{2}}$$

$$= \left\{2x^2 - 2y^2 + b\left(x^2 + y^2\right)^{\frac{3}{2}}\right\} + i(4xy) = u + iv \text{ (say)}$$

Then, $u = 2x^2 - 2y^2 + b\left(x^2 + y^2\right)^{\frac{3}{2}}$ and $v = 4xy$

∵ $f(z)$ is analytic, so by the CR equations,

$$\frac{\partial u}{\partial x} = \frac{\partial v}{\partial y} \text{ or, } 4x + \frac{3b}{2}\left(x^2 + y^2\right)^{\frac{1}{2}} \times 2x = 4x$$

which is true for $b = 0$ only.

63. (b) $|z| = 1$ represents the circle "c" with the center $(0, 0)$ and radius = 1 unit.

The singularities of $f(z) = \dfrac{2z + 5}{\left(z - \frac{1}{2}\right)(z^2 - 4z + 5)}$ are given by.

$$\left(z - \frac{1}{2}\right)(z^2 - 4z + 5) = 0, \text{ or } z = \frac{1}{2}, \frac{4 \pm \sqrt{16 - 4 \times 5}}{2}$$

or $z = \frac{1}{2}, 2 \pm i$.

Now the singular point $z = \frac{1}{2}$ represents the point $\left(\frac{1}{2}, 0\right)$ which lies inside the circle c.

The singular points $z = 2 + i$ and $z = 2 - i$ represent the points $(2, 1)$ and $(2, -1)$, respectively, which lie outside the circle "c"

$$\therefore \oint_c \frac{(2z + 5)\,dz}{\left(z - \frac{1}{2}\right)(z^2 + 4z + 5)}$$

$$= \oint_c \frac{\left(\frac{2z + 5}{z^2 + 4z + 5}\right)dz}{\left(z - \frac{1}{2}\right)}$$

$$= \oint_c \frac{g(z)dz}{z - \frac{1}{2}} \text{ [where } g(z) = \frac{2z + 5}{z^2 + 4z + 5}\text{]}$$

$$= 2\pi i \times g\left(\frac{1}{2}\right) \text{ [by Cauchy's integral formula]}$$

$$= 2\pi i \times \left\{\frac{\left(2 \times \frac{1}{2} + 5\right)}{\left(\frac{1}{2}\right)^2 - \left(4 \times \frac{1}{2}\right) + 5}\right\} = 2\pi i \times \frac{24}{13} = \frac{48\pi i}{13}$$

64. 1.

$$\frac{1}{2\pi j}\int_C\left(\frac{z^2 + 1}{z^2 - 1}\right)dz = \frac{1}{2\pi j}\int_C \frac{(z^2 + 1)}{(z + 1)(z - 1)}\,dz$$

Clearly, $z = \pm 1$ are the singularities of the integrand. The equation of the unit circle with the center at $1 + 0j$ is given by $|z - 1| = 1$.

Now $|z - 1| = 1 \Rightarrow |x + jy - 1| = 1$

$$\Rightarrow \sqrt{(x - 1)^2 + y^2} = 1$$

$$\Rightarrow (x - 1)^2 + (y - 0)^2 = 1^2,$$

which represents the circle "C" with the center $(1, 0)$ and radius = 1 unit.

The singularity $z = 1$ represents the point $(1, 0)$ which lies within "C" and $z = -1$ represents the point $(-1, 0)$ which lies outside "C."

$$\therefore \frac{1}{2\pi j}\int_c\left(\frac{z^2 + 1}{z^2 - 1}\right)dz = \frac{1}{2\pi j}\int_c \frac{\left(\frac{z^2 + 1}{z + 1}\right)dz}{z - 1}$$

$$= \frac{1}{2\pi j}\int_c \frac{f(z)dz}{z-1} \quad [\text{where } f(z) = \frac{z^2+1}{z+1}]$$

$$= \frac{1}{2\pi j}\times\{2\pi j \times f(1)\} \quad \text{(by Cauchy's integral formula)}$$

$$= f(1) = \frac{1^2+1}{1+1} = 1.$$

65. (*b*) We know that if the point $z = z_0$ lies outside the contour C, then $\oint_C \frac{f(z)dz}{z-z_0} = 0.$

So if $z_0 = 2$ lies outside the contour C, then $\oint_C \frac{e^z}{z-2}dz = 0$ and hence, $\frac{1}{2\pi j}\oint_C \frac{e^z}{z-2}dz = 0.$

Now let us assume that $z_0 = 2$ lies inside C.

Then, $\frac{1}{2\pi j}\oint_C \frac{e^z dz}{z-2} = \frac{1}{2\pi j}\int_C \frac{f(z)}{z-2}dz$

(where $f(z) = e^z$)

$$= \frac{1}{2\pi j}\times\{2\pi j\times f(2)\} = f(2) = e^2 = 7.39.$$

(by Cauchy's integral formula)

66. –1.

$f(z)$ is analytic $\Rightarrow$ the *CR* equations are satisfied

$$\Rightarrow \frac{\partial u}{\partial x} = \frac{\partial v}{\partial y} \Rightarrow \frac{\partial}{\partial x}(2kxy) = \frac{\partial}{\partial y}(x^2 - y^2)$$

$$\Rightarrow 2ky = -2y \Rightarrow k = -1.$$

67. 1

$f(z) = \frac{\sin z}{z^2}$ which is of the form $\frac{\phi(z)}{(z-0)^2}$ where $\phi(z) = \sin z.$

So $z = 0$ is a pole of order 2 of $f(z)$.

$\therefore$ Res($z = 0$)

$$= \frac{1}{(2-1)!}\phi^{(2-1)}(0) = \phi'(0) = \phi'(z)\big|_{z=0}$$

$$= \cos z\big|_{z=0} = \cos 0 = 1.$$

68. (*b*) Clearly, $z = 1$ and $z = 2$ are the poles of order 1 of $f(z) = \frac{3z-5}{(z-1)(z-2)}.$

Now assume that the pole $z = 1$ lies within Γ and $z = 2$ lies outside Γ.

$$\text{Res}\, f(z)\big|_{z=1} = \lim_{z\to 1}(z-1)f(z)$$

$$= \lim_{z\to 1}(z-1)\frac{(3z-5)}{(z-1)(z-2)}$$

$$= \lim_{z\to 1}\frac{(3z-5)}{(z-2)} = \frac{3-5}{1-2} = 2.$$

$\therefore$ By Cauchy's residue theorem,

$$\oint_\Gamma \frac{(3z-5)}{(z-1)(z-2)}dz = 2\pi i \times \text{(sum of the residues at the poles which lies inside } \Gamma)$$

$$= 2\pi i \times \text{Res}\, f(z)\big|_{z=1} = 2\pi i \times 2$$

$$= 4\pi i.$$

69. $\sin h(2\pi)$.

Clearly, $z = 2\pi j$ is a pole of order "3" of the integrand.

Now, $f(z) = \frac{\sin z}{(z-2\pi j)^3} = \frac{\phi(z)}{(z-2\pi j)^3},$

where $\phi(z) = \sin z.$

$\therefore \phi'(z) = -\sin z.$

Hence, Res$(z = 2\pi j) = \frac{1}{(3-1)!}\phi''(z) = -\frac{1}{2}\sin(2\pi j).$

$\therefore \oint_C \frac{\sin z}{(z-2\pi j)^3}dz$

$= 2\pi j \times$ sum of the residues at the poles

$$= 2\pi j \times \left\{-\frac{1}{2}\sin(2\pi j)\right\}$$

$$= -\pi j \times j\sinh(2\pi) = \pi\sinh(2\pi).$$

Consequently, $\frac{1}{\pi}\oint \frac{\sin z}{(z-2\pi j)^3}dz = \sinh(2\pi).$

70. (*b*)

$f(z) = z + z^* = (x+iy) + (x-iy) = 2x = u + iv$ (say)

$\therefore$ $u = 2x$ and $v = 0$. Since u and v are both continuous functions of x and y, so $f(z)$ is continuous.

Now, $\frac{\partial u}{\partial x} = 2$ and $\frac{\partial v}{\partial y} = 0$ and so $\frac{\partial u}{\partial x} \neq \frac{\partial v}{\partial y}.$ Thus, *CR* equations are not satisfied. Hence, $f(z)$ is not analytic.

71. (*a*) $f(z) = 1 + (1-z) + (1-z)^2 + \dots\infty$

$$= 1 - (z-1) + (z-1)^2 \dots\infty$$

$$= \{1 + (z-1)\}^{-1} [\text{using } (1+x)^{-1} = 1 - x + x^2 \dots\infty]$$

$$= z^{-1} = \frac{1}{z}.$$

72. (*b*) Clearly, $f(z) = \dfrac{1}{(z-4)(z+1)^3}$ has a pole $z =$ 4 of order 1 and a pole $z = -1$ of order 3.

$$\therefore \ \operatorname{Res} f(z)\big|_{z=4} = \lim_{z \to 4}(z-4)f(z)$$

$$= \lim_{z \to 4}(z-4) \times \frac{1}{(z-4)(z+1)^3}$$

$$= \lim_{z \to 4} \frac{1}{(z+1)^3} = \frac{1}{(4+1)^3} = \frac{1}{125}.$$

Again, $f(z) = \dfrac{1}{(z-4)(z+1)^3} = \dfrac{\phi(z)}{[z-(-1)]^3}$,

where $\phi(z) = \dfrac{1}{z-4}$.

$$\therefore \ \phi'(z) = -\frac{1}{(z-4)^2} \text{ and } \phi''(z) = \frac{2}{(z-4)^3}.$$

So, $\operatorname{Res} f(z)\big|_{z=-1} = \dfrac{1}{(3-1)!}\phi^{(3-1)}(-1) = \dfrac{1}{2}\phi''(z)\big|_{z=-1}$

$$= \frac{1}{2} \times \frac{2}{(z-4)^3}\bigg|_{z=-1} = \frac{1}{(-1-4)^3} = -\frac{1}{125}.$$

73. (*d*)

$$I = \oint_c \left(\frac{z^2-1}{z^2+1}\right)e^z\,dz = \oint_c \frac{(z^2-1)e^z}{(z+i)(z-i)}\,dz \qquad \dots(1)$$

Clearly, each of $z = i$ and $z = -i$ is a pole of order "1" of $f(z) = \dfrac{(z^2-1)e^z}{(z+i)(z-i)}$ and each of these poles lie within the circle $|z| = 3$.

Now, $\operatorname{Res} f(z)\big|_{z=i} = \lim_{z \to i}(z-i)f(z)$

$$= \lim_{z \to i}(z-i) \times \frac{(z^2-1)e^z}{(z+i)(z-i)}$$

$$= \lim_{z \to i}\left(\frac{z^2-1}{z+i}\right)e^z = \left(\frac{i^2-1}{i+i}\right)e^i = -\frac{2e^i}{2i} = ie^i.$$

$\operatorname{Res} f(z)\big|_{z=-i} = \lim_{z \to -i}(z+i)f(z)$

$$= \lim_{z \to -i}(z+i)\frac{(z^2-1)e^z}{(z+i)(z-i)}$$

$$= \lim_{z \to -i}\left(\frac{z^2-1}{z-i}\right)e^z = \left(\frac{i^2-1}{-i-i}\right)e^{-i} = -\frac{2e^{-i}}{-2i} = -ie^i.$$

By Cauchy's residue theorem, we get from (1),

$I = 2\pi i \times$ (sum of the residues at the poles)

$= 2\pi i \times \{ie^i + (-i)e^{-i}\}$

$$= -2\pi(e^i - e^{-i}) = -4i\pi\left(\frac{e^i - e^{-i}}{2i}\right) = -4\pi i \sin 1$$

74. (*b*) Clearly, $f(z) = \dfrac{1}{(z-4)(z+1)^3}$ has a pole $z =$ 4 of order 1 and a pole $z = -1$ of order 3.

$$\therefore \ \operatorname{Res} f(z)\big|_{z=4} = \lim_{z \to 4}(z-4)f(z)$$

$$= \lim_{z \to 4}(z-4) \times \frac{1}{(z-4)(z+1)^3}$$

$$= \lim_{z \to 4} \frac{1}{(z+1)^3} = \frac{1}{(4+1)^3} = \frac{1}{125}.$$

Again, $f(z) = \dfrac{1}{(z-4)(z+1)^3} = \dfrac{\phi(z)}{[z-(-1)]^3}$,

where $\phi(z) = \dfrac{1}{z-4}$.

$$\therefore \ \phi'(z) = -\frac{1}{(z-4)^2} \text{ and } \phi''(z) = \frac{2}{(z-4)^3}.$$

So, $\operatorname{Res} f(z)\big|_{z=-1} = \dfrac{1}{(3-1)!}\phi^{(3-1)}(-1) = \dfrac{1}{2}\phi''(z)\big|_{z=-1}$

$$= \frac{1}{2} \times \frac{2}{(z-4)^3}\bigg|_{z=-1} = \frac{1}{(-1-4)^3} = -\frac{1}{125}.$$

75. (*d*)

$$I = \oint_c \left(\frac{z^2-1}{z^2+1}\right)e^z\,dz = \oint_c \frac{(z^2-1)e^z}{(z+i)(z-i)}\,dz \qquad \dots(1)$$

Clearly, each of $z = i$ and $z = -i$ is a pole of order "1" of $f(z) = \dfrac{(z^2-1)e^z}{(z+i)(z-i)}$ and each of these poles lie within the circle $|z| = 3$.

Now, $\operatorname{Res} f(z)\big|_{z=i} = \lim_{z \to i}(z-i)f(z)$

$$= \lim_{z \to i}(z-i) \times \frac{(z^2-1)e^z}{(z+i)(z-i)}$$

$$= \lim_{z \to i}\left(\frac{z^2-1}{z+i}\right)e^z = \left(\frac{i^2-1}{i+i}\right)e^i = -\frac{2e^i}{2i} = ie^i.$$

$\operatorname{Res} f(z)\big|_{z=-i} = \lim_{z \to -i}(z+i)f(z)$

$$= \lim_{z \to -i}(z+i)\frac{(z^2-1)e^z}{(z+i)(z-i)}$$

$$= \lim_{z \to -i} \left(\frac{z^2 - 1}{z - i} \right) e^z = \left(\frac{i^2 - 1}{-i - i} \right) e^{-i} = -\frac{2e^{-i}}{-2i} = -ie^i.$$

By Cauchy's residue theorem, we set from (1),

$I = 2\pi i \times$ (sum of the residues at the poles)

$= 2\pi i \times \{ie^i + (-i)e^{-i}\}$

$$= -2\pi(e^i - e^{-i}) = -4i\pi \left(\frac{e^i - e^{-i}}{2i} \right) = -4\pi i \sin 1$$

$$\left(\text{using } \sin x = \frac{e^{ix} - e^{-ix}}{2i} \right)$$

76. (b) $f(z) = (x^2 + ay^2) + ibxy$ is analytic

$$\Rightarrow \frac{\partial}{\partial x}(x^2 + ay^2) = \frac{\partial}{\partial y}(bxy) \Rightarrow 2x + 0 = bx \Rightarrow b = 2]$$

Again, $f(z) = (x^2 + ay^2) + ibxy$ is analytic

$$\Rightarrow \frac{\partial}{\partial y}(x^2 + ay^2) = -\frac{\partial}{\partial x}(bxy)$$

$$\Rightarrow 2ay = -by = -2y \Rightarrow a = -1$$

77. (b) Clearly, the equation of the line is $y = x$ and so $dy = dx$. It follows from the given figure that x varies from 0 to 1.

$$\therefore I = \int_c (x^2 + iy^2)dz = \int_c (x^2 + ix^2)(dx + idx)$$

$$= \int_0^1 x^2(1 + i)(dx + idx)$$

$$= (1 + i)^2 \int_0^1 x^2 dx = (1 + 2i + i^2) \times \frac{1}{3} = \frac{2i}{3}.$$

78. (a) Let $u(x, y) = 2x - x^2 + my^2$.

Then, $\frac{\partial u}{\partial x} = 2 - 2x$, $\frac{\partial u}{\partial y} = 2my$

Now, $u(x, y)$ is harmonic $\Rightarrow \frac{\partial^2 u}{\partial x^2} + \frac{\partial^2 u}{\partial x^2} = 0$

$$\Rightarrow \frac{\partial}{\partial x}\left(\frac{\partial u}{\partial x} \right) + \frac{\partial}{\partial y}\left(\frac{\partial u}{\partial y} \right) = 0$$

$$\Rightarrow \frac{\partial}{\partial x}(2 - 2x) + \frac{\partial}{\partial y}(2my) = 0$$

$$\Rightarrow -2 + 2m = 0 \Rightarrow m = 1.$$

79. (c) $z = 2$ is a singularity of the integrand and it lies within the contour $|z| = 3$.

$$\therefore \oint_c \left(\frac{z^3 - 2z + 3}{z - 2} \right) dz = \oint_c \frac{f(z)}{z - 2} dz$$

$= 2\pi i \times f(2)$ [where $f(z) = z^3 - 2z + 3$]

[by Cauchy's integral formula]

$= 2\pi i \times (2^3 - 2 \times 2 + 3) = 14\pi i.$

80. (d)

$$\lim_{z \to i} \frac{z^2 + 1}{z^3 + 2z - i(z^2 + 2)} \quad \left(\text{form } \frac{0}{0} \right)$$

$$= \lim_{z \to i} \frac{\frac{d}{dz}(z^2 + 1)}{\frac{d}{dz}\{z^3 + 2z - i(z^2 + 2)\}} = \lim_{z \to i} \frac{2z}{3z^2 + 2 - 2iz}$$

$$= \frac{2i}{3i^2 + 2 - 2i^2} = \frac{2i}{i^2 + 2} = \frac{2i}{-1 + 2} = 2i.$$

81. (c)

$$s^4 + s^2 + 1 = 0 \Rightarrow t^2 + t + 1 = 0 \ (\text{for } t = s^2)$$

$$\Rightarrow t = \frac{-1 \pm \sqrt{1 - 4}}{2} = \frac{-1 \pm \sqrt{3}i}{2} = \omega, \omega^2$$

$$\Rightarrow s^2 = \omega, \omega^2 \Rightarrow s = \pm\sqrt{\omega}, \pm\omega.$$

Now, $\sqrt{\omega} = \sqrt{\frac{-1 + \sqrt{3}i}{2}} = \sqrt{-\frac{1}{2} + \frac{\sqrt{3}}{2}i}$

$$= \sqrt{\cos\frac{2\pi}{3} + i\sin\frac{2\pi}{3}} = \sqrt{e^{\frac{2\pi i}{3}}} = e^{\frac{\pi i}{3}}$$

$$= \cos\frac{\pi}{3} + i\sin\frac{\pi}{3} = \frac{1}{2} + i\frac{\sqrt{3}}{2}, \text{ which represents the}$$

point $\left(\frac{1}{2}, \frac{\sqrt{3}}{2} \right)$ that lies in first quadrant.

$$\therefore -\sqrt{\omega} = -e^{\frac{\pi i}{3}} = -\left(\cos\frac{\pi}{3} + i\sin\frac{\pi}{3} \right) = -\frac{1}{2} - i\frac{\sqrt{3}}{2},$$

which represents the point $\left(-\frac{1}{2}, -\frac{\sqrt{3}}{2} \right)$ that lies in third quadrant.

$\omega = \frac{-1 + \sqrt{3}i}{2} = -\frac{1}{2} + i\frac{\sqrt{3}}{2}$, which represents the

point $\left(-\frac{1}{2}, \frac{\sqrt{3}}{2} \right)$ that lies in second quadrant.

$\therefore -\omega = \frac{1}{2} - i\frac{\sqrt{3}}{2}$, which represents the point

$\left(\dfrac{1}{2}, -\dfrac{\sqrt{3}}{2}\right)$ that lies on fourth quadrant.

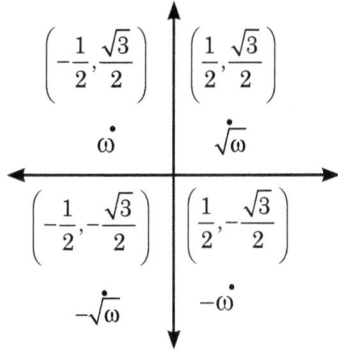

Thus, we can conclude that the given equation has two right-half plane (RHP) roots and two left-half plane (LHP) roots.

82. (*b*) $f_1(z) = z^2 \Rightarrow f_1'(z) = 2z$, which exists finitely $f(z)$ is analytic.

Again, $f_2(z) = \bar{z} = x - iy = u + iv$

$\Rightarrow u = x, v = -y$

$\Rightarrow \dfrac{\partial u}{\partial x} = 1, \dfrac{\partial u}{\partial y} = 0, \dfrac{\partial v}{\partial x} = 0, \dfrac{\partial v}{\partial y} = -1$

Therefore, $\dfrac{\partial u}{\partial x} \neq \dfrac{\partial v}{\partial y}$ and so $f_2(z)$ is not analytic.

83. (*b*) Let $F(z) = u + iv$.
Then, $u + iv = F(z)$
$\Rightarrow u + iv = iz + k\,\mathrm{Re}(z) + i\,\mathrm{Im}(z) = i(x + iy) + kx + iy$
$\Rightarrow u + iv = (kx - y) + i(x + y)$
$\Rightarrow u = kx - y$ and $v = x + y$.

From the *CR* equations, we have,

$\dfrac{\partial u}{\partial x} = \dfrac{\partial v}{\partial y} \Rightarrow k = 1.$

84. (*a*) We have,
$u + iv = f(z) = x^2 - y^2 + i(2xy) = (x^2 - y^2) + i(2xy)$

So, $u = x^2 - y^2$ and $v = 2xy$.

$\therefore \dfrac{\partial u}{\partial x} = 2x, \dfrac{\partial u}{\partial y} = -2y, \dfrac{\partial v}{\partial x} = 2y$ and $\dfrac{\partial v}{\partial y} = 2x.$

Thus, the *CR* equations are satisfied at every point. Also, $u, v, \dfrac{\partial u}{\partial x}, \dfrac{\partial u}{\partial y}, \dfrac{\partial v}{\partial x}, \dfrac{\partial v}{\partial y}$ are continuous at every point.

$\therefore f(z) = x^2 - y^2 + i(2xy)$ is analytic function and

so

$f'(z) = \dfrac{\partial u}{\partial x} + i\dfrac{\partial v}{\partial x} = 2x + 2iy.$

85. 2 Clearly, $f(z) = \dfrac{1}{z^2 - 1}$ has singularities at $z = 1$ and $z = -1$.

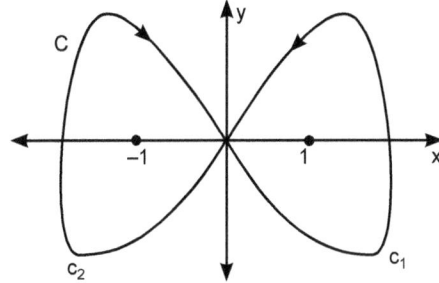

Now, $\displaystyle\oint_C \dfrac{dz}{z^2 - 1}\, dz = \oint_{C_1} \dfrac{dz}{z^2 - 1} + \left(-\oint_{C_2} \dfrac{dz}{z^2 - 1}\right)$

(Since the direction of the curve C_2 is clockwise)

$\Rightarrow \displaystyle\oint_C f(z)\, dz = \oint_{C_1} \dfrac{1}{(z-1)(z+1)}dz - \oint_{C_2} \dfrac{1}{(z-1)(z+1)}dz$

$= \displaystyle\oint_{C_1} \dfrac{\frac{1}{(z+1)}}{[z-1]}dz - \oint_{C_2} \dfrac{\frac{1}{(z-1)}}{[z-(-1)]}dz$

$= \displaystyle\oint_{C_1} \dfrac{g(z)}{[z-1]}dz - \oint_{C_2} \dfrac{h(z)}{[z-(-1)]}dz$

$\left(\text{where } g(z) = \dfrac{1}{(z+1)}, h(z) = \dfrac{1}{(z-1)}\right)$

$= 2\pi j \times g(1) - 2\pi j \times h(-1)$

(by Cauchy's integral formula)

$= 2\pi j\left(\dfrac{1}{2}\right) - 2\pi j\left(\dfrac{1}{-1-1}\right) = 2\pi j.$

$\therefore \dfrac{1}{\pi j}\displaystyle\oint_C \dfrac{1}{(z^2 - 1)}dz = \dfrac{1}{\pi j}(2\pi j) = 2.$

86. (*a*) Clearly, the poles of $f(z) = \dfrac{z+1}{z^2 - 4}$ are $z = 2, -2$.

The pole $z = 2$ lies out-side of C and the pole $z = -2$ lies in-side of C.

Now, $\mathrm{Res}\, f(z)\big|_{z=-2}$

$= \displaystyle\lim_{z\to-2}(z+2)f(z) = \lim_{z\to-2}(z+2)\dfrac{z+1}{(z+2)(z-2)}$

$= \displaystyle\lim_{z\to-2}\dfrac{z+1}{(z-2)} = \dfrac{-2+1}{-2-2} = \dfrac{1}{4}.$

Hence, by Cauchy's residue theorem,

$$\oint_C \frac{z+1}{z^2-4} dz$$

$$= 2\pi i \times \text{sum of residues} = 2\pi i \times \frac{1}{4} = \frac{\pi i}{2}.$$

87. (b) The poles of $f(z) = \dfrac{z^2}{(z^2-3z+2)^2}$ are given by:

$$(z^2-3z+2)^2 = 0 \ i.e; (z-1)^2(z-2)^2 = 0.$$

Thus, $z = 1$ is a pole of order 2 and $z = 2$ is a pole of order 2.

It is easy to verify that both poles of $f(z)$ lie inside of C.

Now, $f(z) = \dfrac{z^2}{(z^2-3z+2)^2}$

$$= \frac{z^2}{(z-2)^2(z-1)^2} = \frac{g(z)}{(z-1)^2},$$

where $g(z) = \dfrac{z^2}{(z-2)^2}$.

$\therefore \text{Res} f(z)\big|_{z=1}$

$$= \frac{g^{(2-1)}(1)}{(2-1)!} = g'(z)\big|_{z=1}$$

$$= \frac{(z-2)^2 \times 2z - z^2 \times 2(z-2)}{(z-2)^4}\bigg|_{z=1} = 4.$$

Again, $f(z) = \dfrac{z^2}{(z^2-3z+2)^2}$

$$= \frac{z^2}{(z-2)^2(z-1)^2} = \frac{h(z)}{(z-2)^2},$$

where $h(z) = \dfrac{z^2}{(z-1)^2}$.

$\therefore \text{Res} f(z)\big|_{z=2}$

$$= \frac{h^{(2-1)}(2)}{(2-1)!} = h'(z)\big|_{z=2}$$

$$= \frac{(z-1)^2 \times 2z - z^2 \times 2(z-1)}{(z-1)^4}\bigg|_{z=2} = -4.$$

Hence, by Cauchy's residue theorem,

$$\oint_C f(z)\, dz$$

$$= 2\pi i \times (\text{sum of residues}) = 2\pi i \times (4-4) = 0.$$

Questions for Practice

1. For any complex number z, $|z-8| + |z+8| = 20$ represents
(a) a circle   (b) a parabola
(c) an ellipse   (d) a hyperbola

2. The square root of $-5 + 12i$ is
(a) $\pm(2+3i)$ (b) $\pm(1-3i)$
(c) $\pm(3+2i)$ (d) $\pm(1-i)$

3. Find the value of $\sqrt{-3+\sqrt{-3+\sqrt{-3+....}}}$

(a) $\dfrac{-1\pm i\sqrt{3}}{2}$ (b) $\dfrac{-2\pm i\sqrt{3}}{2}$

(c) $\dfrac{1\pm i\sqrt{11}}{2}$ (d) $\dfrac{3\pm i\sqrt{11}}{2}$

4. If $2+2i, -2-2i$ and $-2\sqrt{3}+2\sqrt{3}i$ one vertices of a triangle, then area of this triangle is

(a) $4\sqrt{3}$ sq.unit   (b) $8\sqrt{3}$ sq.unit

(c) $10\sqrt{2}$ sq.unit   (d) $\dfrac{13}{3}\sqrt{2}$ sq.unit

5. If ω be an imaginary cube root of unity, then the simplest value of $(1-\omega)(1-\omega^2)(1-\omega^4)(1-\omega^8)$ is
(a) 1   (b) -1 (c) 0   (d) 9

6. If $x + \dfrac{1}{x} = 2\cos\dfrac{\pi}{7}$, then $x^7 + \dfrac{1}{x^7} = ?$
(a) -1 (b) -2 (c) 0   (d) 1

7. $(1+i)^n + (1-i)^n = ?$

(a) $2^{\frac{n}{2}+1} \cos\dfrac{n\pi}{4}$ (b) $2^{\frac{n}{2}+1} \sin\dfrac{n\pi}{4}$

(c) $2^{\frac{n}{2}} \cos\dfrac{n\pi}{4}$ (d) $2^{\frac{n}{2}} \sin\dfrac{n\pi}{4}$

8. If $\alpha_r = \cos\left(\dfrac{\pi}{3^r}\right) + i\sin\left(\dfrac{\pi}{3^r}\right)$, for $r = 1, 2, 3...$ then $\alpha_1\alpha_2\alpha_3...$ upto $\infty = ?$
(a) 0   (b) -1 (c) $-i$ (d) i

9. The values of i^{-i} are given by

(a) $e^{\left(4n+\frac{1}{2}\right)\pi}, n \in Z$ (b) $e^{\left(2n+\frac{1}{2}\right)\pi}, n \in Z$

(c) $-e^{\left(2n+\frac{1}{2}\right)\pi}, n \in Z$ (d) $e^{\left(4n-\frac{1}{2}\right)\pi}, n \in Z$

10. $\cos\left[i\log\left(\dfrac{a-ib}{a+ib}\right)\right] = ?$

(a) $\dfrac{2ab}{a^2+b^2}$ (b) $\dfrac{2ab}{a^2-b^2}$

(c) $\dfrac{a^2-b^2}{a^2+b^2}$ (d) 1

11. If $v(r,\theta) = r^2\cos 2\theta - r\cos\theta + 2$ and $f(z) = u(r,\theta)+iv(r,\theta)$, then $f(z)$ is given by

(a) $i(r^2e^{2i\theta} - re^{i\theta}) + 2i$

(b) $i(re^{2i\theta} + e^{-i\theta}) - 2i$

(c) $i(re^{-2i\theta} + r^2e^{i\theta})$

(d) none of these

12. If $f(z) = \dfrac{1}{2}\log_e(x+y^2)+i\tan^{-1}(kx/y)$ be analytic, then $k = ?$

(a) 2   (b) 1   (c) 0   (d) –1

13. The analytic function whose real part $y + e^x \cos y$, is given by

(a) $e^z + i(c+z)$ (b) $e^{-z} + i(c+z)$

(c) $e^z + i(c-z)$ (d) $e^{-z} - i(c+2z)$

14. The analytic function whose imaginary part is $e^x\sin y$ given by

(a) $f(z) = \sin z$ (b) $f(z) = e^z$

(c) $f(z) = e^{-2z}$ (d) $f(z) = \cos z$

15. The value of the integral $\displaystyle\int_{|z|=\frac{1}{2}}\dfrac{3z^2+7z+1}{z+1}dz$ is

(a) $2\pi i$ (b) $-\pi i$ (c) 0   (d) $-2\pi i$

16. The value of $\displaystyle\int_{|z|=3}\dfrac{e^{-z}}{(z-1)(z-2)^2}dz$ is

(a) 0   (b) πi (c) ie (d) $2\pi ie$

17. The value of $\displaystyle\int_c\dfrac{zdz}{z^2-3z+2}$, where c is the circle $|z-2|=\dfrac{1}{2}$, is

(a) $4\pi i$ (b) $2\pi i$ (c) 0   (d) πi

18. Determine the points where the function $f(z) = z\,\mathrm{Re}\,(z)$ is not differentiable

(a) C (b) $C - \{0\}$

(c) $C - \{0, 1\}$ (d) $C - \{1\}$

19. The singular points of the function $f(z) = \dfrac{z^2+i}{(z-3)(z^2-2z+3)}$ are

(a) $\pm i$, 1   (b) $0, \pm i$

(c) $3, 1\pm\sqrt{2i}$ (d) $\pm i, \pm\sqrt{2i}$

20. The function $f(z) = xy^2 + 1 + iy$ is

(a) analytic at $z = 0$ only

(b) analytic $\forall z$

(c) analytic for $z \in C - \{0\}$

(d) no where analytic

21. If $f(z)$ be an analytic function of z, then $\left(\dfrac{\partial^2}{\partial x^2}+\dfrac{\partial^2}{\partial y^2}\right)|f(z)|^2 = ?$

(a) $4|f'(z)|^2$ (b) $2|f'(z)|^2$

(c) $|f'(z)|^2$ (d) 0

22. The value of the integral $\displaystyle\int_0^{1+2i}\left(3x^2-y+ix^3\right)dz$ along the real axis from $z = 0$ to $z = 1$ is given by

(a) $1-\dfrac{17}{4}i$ (b) $1-\dfrac{3}{4}i$

(c) $1 - 3 + i$ (d) $3 - i$

23. Find the value of $\displaystyle\int_c\bar{z}dz$, where C is the parabola $x^2 = y$ from 0 to $1 + i$

(a) $1+\dfrac{i}{3}$ (b) $-1+\dfrac{i}{3}$ (c) $-3 + i$ (d) $3 - i$

24. If $C = \{z: z = 2e^{-4i\theta}, 0 \leq \theta \leq 4\pi\}$ then the value of the integral $\displaystyle\int_c\dfrac{e^{\pi z}}{z(z+4)}dz$ is

(a) $-2\pi i$ (b) $-4\pi i$ (c) $4\pi i$ (d) $2\pi i$

25. Let $f(z)$ be analytic within and on $C: |z-2| = 3$. Suppose $|f(z)|$ has maximum value "2" on C. Then an upper bound of $|f^{(4)}(2)|$ is

(a) $\dfrac{21}{80}$ (b) $\dfrac{48}{81}$ (c) $\dfrac{4}{81}$ (d) $\dfrac{9}{25}$

26. The value of the integral $\displaystyle\int_{|z|=3}\dfrac{\sin\pi z+\cos\pi z}{(z-1)(z-2)}dz$ is

(a) $4\pi i$ (b) $2\pi i$ (c) 0   (d) $-2\pi i$

27. Let $f(z)$ be analytic in $|z| < 1$ such that $|f(z)| \leq 1$ and $|f(1/3)| = 0$. Then which of the following is true?

(a) $|f(1/7)| \leq \frac{1}{3}$ (b) $|f(z)| \leq \frac{1}{2}$

(c) $|f(z)| \leq \frac{1}{5}$ (d) $|f(1/7)| \leq 1$

28. If $f(z)$ be analytic on $|z| < 1$ and $|f(z)| < 1$, then which one of following is possible?

(a) $f(1/2) = \frac{3}{4}, f'(1/2) = \frac{2}{3}$

(b) $f(1/2) = \frac{3}{4}, f'(1/2) = \frac{5}{3}$

(c) $f(0) = \frac{1}{2}, f'(0) = \frac{2}{3}$

(d) none of these

29. The function $f(z) = \begin{cases} \dfrac{xy(y-ix)}{x^2+y^2}, z \neq 0 \\ 0 \end{cases}$

(a) is analytic at $(0, 0)$
(b) not analytic at $(0, 0)$
(c) $f'(0)$ exist
(d) None of these.

30. The analytic function whose real part $u = e^x \cos y$ is?

(a) $e^{-z} + c$ (b) $-e^z + c$
(c) $e^z + c$ (d) $-e^{-z} + c$

31. The analytic function whose imaginary part is $e^x \sin y - x$, is given by

(a) $f(x) = e^{-z} + iz + c$ (b) $f(z) = e^z - iz + c$
(c) $f(z) = e^z - ize^z$ (d) none of these

32. Consider a function $f(z) = u + iv$ defined on $|z - i| < 1$, where u, v are real-valued functions of x, y. Then, $f(z)$ is analytic for u equals to

(a) $x^2 + y^2$ (b) $\ln(x^2 + y^2)$
(c) e^{xy} (d) x^2y^2

33. Consider the functions $f(z) = x^2 + iy^2$ and $g(z) = z^2 + y^2 + 2ixy$. At $z = 0$,

(a) f is analytic but not g
(b) g is analytic but not f
(c) both f and g are analytic
(d) neither f nor g is analytic

34. The value of the integral $\int_C \frac{dz}{z^2-1}$, $C: |z| = 4$ is equal to

(a) πi (b) 0 (c) $-\pi i$ (d) $2\pi i$

35. Let $I = \int_C \frac{f(z)}{(z-1)(z-2)}$ where

$f(z) = \sin\frac{\pi z}{2} + \cos\frac{\pi z}{2}$ and C is the curve $|z| = 3$ oriented counterclockwise. Then, the value of I is

(a) $4\pi i$ (b) 0 (c) $-2\pi i$ (d) $-4\pi i$

36. e^z is a periodic function with period.

(a) 2π (b) $2\pi i$ (c) π (d) $i\pi$

37. The complex number $z = x + iy$ which satisfies the equation $|z + 1| = 1$ lie on.

(a) a circle with $(1, 0)$ as the center and radius 1
(b) a circle with $(-1, 0)$ as the center and radius 1
(c) y-axis
(d) x-axis

38. $\cos \theta$ can be represented as

(a) $\dfrac{e^{i\theta} - e^{-i\theta}}{2}$ (b) $\dfrac{e^{i\theta} - e^{-i\theta}}{2i}$

(c) $\dfrac{e^{i\theta} + e^{-i\theta}}{i}$ (d) $\dfrac{e^{i\theta} + e^{-i\theta}}{2}$

39. The real part of a complex number $z = x + iy$ is given by

(a) $\mathrm{Re}(z) = z - z^*$ (b) $\mathrm{Re}(z) = \dfrac{z - z^*}{2}$

(c) $\mathrm{Re}(z) = \dfrac{z + z^*}{2}$ (d) $\mathrm{Re}(z) = z + z^*$

(z^* denotes the conjugate of z)

40. If $f(z) = \dfrac{1}{(z^2+1)^3}$ then the residue of $f(z)$ at $z = -i$ is

(a) $\dfrac{-3}{16i}$ (b) $\dfrac{3i}{16}$ (c) $\dfrac{3}{16i}$ (d) $\dfrac{-3i}{16}$

41. Consider the function $f(z) = \begin{cases} \dfrac{xy^2(x+iy)}{2x^2-y^4}, z \neq 0 \\ 0, z = 0 \end{cases}$

Then

(a) $f(z)$ is analytic at $z = 0$
(b) $f(z)$ is not analytic at $z = 0$
(c) $f'(0)$ does not exist
(d) both (b) and (c) are correct

42. Consider $I = \oint_C \left(\dfrac{z^2-4}{z^2+4} \right) e^z dz$. If C is defined as $|z| = 4$ then the value of I is

(a) $-16\pi i \sin 2$ (b) $4\pi i \sin 4$
(c) 0 (d) $8\pi i \sin 8$

43. Evaluate $\oint_C |z| \, dz$, , where C is the circle given by $|z - 1| = 1$ (described in the positive sense)

(a) $\dfrac{2i}{3}$ (b) $\dfrac{4i}{3}$ (c) $\dfrac{8i}{3}$ (d) 0

44. If C is given by the circle $|z - a| = R$, then the value of $\oint_C \dfrac{dz}{z - a}$ is

(a) 0   (b) πi (c) $2\pi i$ (d) $4\pi i$

45. The value of the integral $\oint_C \dfrac{(2z + 3)dz}{z(z^2 + 1)(z + 1)^2}$ where $C: |z| = 3$

(a) πi (b) 0   (c) $2\pi i$ (d) $\dfrac{2i}{3}$

46. Evaluate $\displaystyle\int_c \dfrac{12z - 7}{(z - 1)^2(2z + 3)} dz$, where $C: |z + i| = \sqrt{3}$

(a) 0   (b) πi (c) $2\pi i$ (d) $4\pi i$

47. Suppose a function $f(z)$ has a double pole at $z = 0$ with Res $(f(z))_{z=0} = 2$, a simple pole at $z = 1$ with Res $(f(z))_{z=1} = 2$, and is analytic at all other finite points of the plane and is bounded as $|z| \to \infty$. Also, given that $f(2) = 5$ and $f(-1) = 2$. Then, $f(z)$ is

(a) $\dfrac{1}{z} + \dfrac{2}{z^2} + \dfrac{2}{z - 1} + 1$

(b) $-\dfrac{2}{z} + \dfrac{2}{z^2} + \dfrac{4}{z - 1} + 1$

(c) $\dfrac{2}{z} + \dfrac{4}{z^2} + \dfrac{2}{z - 1} + 1$

(d) none of these

48. Let, $I_r = \displaystyle\int_{C_r} \dfrac{dz}{z(z - 1)(z - 2)}$, where $C_r = \{z \in C: |z| = r, r > 0\}$. Then,

(a) $I_r = 2\pi i$, if $r \in (2, 3)$

(b) $I_r = \dfrac{1}{2}$, if $r \in (0, 1)$

(c) $I_r = -2\pi i$, if $r \in (1, 2)$

(d) $I_r = 0$, if $r > 3$

49. The residue of $f(z) = \dfrac{z^2 - 2z}{(z + 1)^2(z^2 + 4)}$ at $z = -1$ is

(a) $\dfrac{2}{5}$ (b) $\dfrac{-9}{5}$ (c) $\dfrac{-14}{25}$ (d) $\dfrac{5}{9}$

Answer Key-				
1. (c)	**2.** (a)	**3.** (c)	**4.** (b)	**5.** (d)
6. (b)	**7.** (a)	**8.** (d)	**9.** (b)	**10.** (c)
11. (a)	**12.** (d)	**13.** (c)	**14.** (b)	**15.** (c)
16. (d)	**17.** (a)	**18.** (b)	**19.** (c)	**20.** (d)
21. (a)	**22.** (a)	**23.** (a)	**24.** (b)	**25.** (b)
26. (a)	**27.** (c)	**28.** (c)	**29.** (b)	**30.** (c)
31. (b)	**32.** (b)	**33.** (d)	**34.** (b)	**35.** (d)
36. (b)	**37.** (b)	**38.** (d)	**39.** (c)	**40.** (a)
41. (d)	**42.** (a)	**43.** (c)	**44.** (c)	**45.** (b)
46. (d)	**47.** (c)	**48.** (d)	**49.** (c)	

Explanation

1. (c) $|z - 8| + |z + 8| = 20 \Rightarrow |x + iy - 8| + |x + iy + 8| = 20$

$\Rightarrow \sqrt{(x - 8)^2 + y^2} + \sqrt{(x + 8)^2 + y^2} = 20$

$\Rightarrow \sqrt{(x - 8)^2 + y^2} = 20 - \sqrt{(x + 8)^2 + y^2}$

Squaring both sides we get,

$(x - 8)^2 + y^2 = 400 - 40\sqrt{(x + 8)^2} + (x + 8)^2 + y^2$

or, $-16x = 400 - 40\sqrt{(x + 8)^2 + y^2}$

or, $40\sqrt{(x + 8)^2 + y^2} = 400 + 16x$

Again, take the square on both sides

2. (a)

3. (c) Let $x = \sqrt{-3 + \sqrt{-3 + \sqrt{-3 + \ldots}}}$

Then $x = \sqrt{-3 + x}$

or, $x^2 = -3 + x$ or $x^2 - x + 3 = 0$

or, $x = \dfrac{1 \pm \sqrt{1 - 4 \times 3}}{2} = \dfrac{1 \pm i\sqrt{11}}{2}$

4. (b) Let the complex numbers $2 + 2i$, $-2 - 2i$ and $-2\sqrt{3} + 2\sqrt{3}i$ are represented by the points A, B, C whose co-ordinates are, respectively $(2, 2), (-2, -2)$, and $(-2\sqrt{3}, 2\sqrt{3})$

$\therefore \overline{AB} = \sqrt{(2 + 2)^2 + (2 + 2)^2} = 4\sqrt{2}$,

$\therefore \overline{BC} = \sqrt{\left(-2 + 2\sqrt{3}\right)^2 + \left(-2 - 2\sqrt{3}\right)^2} = 4\sqrt{2}$,

$\therefore \overline{CA} = \sqrt{\left(2 + 2\sqrt{3}\right)^2 + \left(-2 - 2\sqrt{3}\right)^2} = 4\sqrt{2}$,

Thus, ΔABC is equilateral with each side $(a) = 4\sqrt{2}$ unit. Hence, area of this triangle $= \dfrac{\sqrt{3}}{4} \times a^2 = \dfrac{\sqrt{3}}{4} \times 32$ sq. unit.

5. (d) **6.** (b) **7.** (a)

$$(1+i)^n = \left[\sqrt{2}\left(\frac{1}{\sqrt{2}} + i \times \frac{1}{\sqrt{2}}\right)\right]^n$$

$$= 2^{\frac{n}{2}}\left(\cos\frac{\pi}{4} + i\sin\frac{\pi}{4}\right)^n$$

$$= 2^{\frac{n}{2}}\left(\cos\frac{n\pi}{4} + i\sin\frac{n\pi}{4}\right) \qquad \ldots(i)$$

(by De Moivre's theorem)

Similarly, $(1-i)^n = 2^{\frac{n}{2}}\left(\cos\dfrac{n\pi}{4} - i\sin\dfrac{n\pi}{4}\right) \qquad \ldots(ii)$

Then add (i) and (ii)

8. (d), **9.** (b) **10.** (c), **11.** (a), **12.** (d), **13.** (c),

14. (b), **15.** (c), **16.** (d), **17.** (a), **18.** (b),

19. (c), **20.** (d)

21. (a)

We know that $\dfrac{\partial^2}{\partial x^2} + \dfrac{\partial^2}{\partial y^2} = 4\dfrac{\partial^2}{\partial z \partial z}$

$$\therefore \left(\frac{\partial^2}{\partial x^2} + \frac{\partial^2}{\partial y^2}\right) |f(z)|^2 = 4\frac{\partial^2}{\partial z \partial z}\left(f(z)\overline{f(z)}\right)$$

$$= 4\frac{\partial}{\partial \overline{z}}\left\{\frac{\partial}{\partial z}\left[f(z)\overline{f}(\overline{z})\right]\right\}$$

$$= 4\frac{\partial}{\partial \overline{z}}\{f'(z)\overline{f}(\overline{z})\} = 4f'(z)\overline{f'}(\overline{z})$$

$$= 4f'(z)\overline{f'(z)} = 4|f'(z)|^2$$

22. (a) **23.** (a) **24.** (b)

25. (b) If $f(z)$ be analytic within and on $C: |z-a| = R$ and $M = \max_{z \in C}|f(z)|$ then an upper bound of $|f^{(n)}(a)|$ is $\dfrac{M \times n!}{R^n}$. Here, $a = 2$, $R = 3$, $M = 2$, $n = 4$

26. (a)

27. (c) We know that if $f(z)$ be analytic in $|z| < 1$ and $|f(z)| \leq 1$ for $|z| < 1$ then $\left|\dfrac{f(z) - f(z_0)}{1 - f(z_0)f(z)}\right| \leq \left|\dfrac{z - z_0}{1 - z_0 z}\right|$.

Here, $z = \dfrac{1}{7}, z_0 = \dfrac{1}{3}, f(z_0) = f\left(\dfrac{1}{3}\right) = 0$.

$$\therefore \left|\frac{f\left(\frac{1}{7}\right) - f\left(\frac{1}{3}\right)}{1 - f\left(\frac{1}{3}\right) \times f\left(\frac{1}{7}\right)}\right| \leq \left|\frac{\frac{1}{7} - \frac{1}{3}}{1 - \frac{1}{3} \times \frac{1}{7}}\right|$$

or, $\left|\dfrac{f(1/7) - 0}{1 - 0 \times f\left(\frac{1}{7}\right)}\right| \leq \dfrac{4}{20}$ or $|f(1/7)| \leq \dfrac{1}{5}$

28. (c) We know that if $f(z)$ be analytic on $|z| < 1$ and $|f(z)| \leq 1$, then $|f'(z)| \leq \dfrac{1 - |f(z)|^2}{1 - |z|^2}$.

Let $f(1/2) = \dfrac{3}{4}$ and $f'(1/2) = \dfrac{2}{3}$.

Then, $|f'(1/2)| \leq \dfrac{1 - |f(1/2)|^2}{1 - |1/2|^2} = \dfrac{1 - \frac{9}{16}}{1 - \frac{1}{4}}$

$\Rightarrow \dfrac{2}{3} \leq \dfrac{7}{12}$, which is not true.

Next, let $f\left(\dfrac{1}{2}\right) = \dfrac{3}{4}$ and $f'\left(\dfrac{1}{2}\right) = \dfrac{5}{3}$

Then, $|f'(1/2)| \leq \dfrac{1 - \left|f\left(\frac{1}{2}\right)\right|}{1 - |1/2|^2} \Rightarrow \dfrac{5}{3} \leq \dfrac{1 - \frac{9}{16}}{1 - \frac{1}{4}}$

$\Rightarrow \dfrac{5}{3} \leq \dfrac{7}{12}$ which is not possible.

Now, let $f(0) = \dfrac{1}{2}$ and $f'(0) = \dfrac{2}{3}$

Then, $|f'(0)| \leq \dfrac{1 - |f(0)|^2}{1 - |0|^2} \Rightarrow \dfrac{2}{3} \leq \dfrac{1 - \frac{1}{4}}{1}$

$\Rightarrow \dfrac{2}{3} \leq \dfrac{3}{4}$, which is true.

29. (b) Consider a curve $y = mx$]

30. (c), **31.** (b), **32.** (b), **33.** (d), **34.** (b),

35. (d), **36.** (b), **37.** (b), **38.** (d), **39.** (c),

40. (a), **41.** (d), **42.** (a), **43.** (c), **44.** (c),

45. (b), **46.** (d)

47. (c) $f(z) = a + \dfrac{2}{(z-0)} + \dfrac{b}{(z-0)^2} + \dfrac{2}{z-1}$

48. (d) **49.** (c)

CHAPTER 9

PROBABILITY AND STATISTICS

9.1 BASICS OF PROBABILITY

The theory of probability is the branch of a Mathematics which studies the influence of "chance." Probability is a concept that numerically measures the degree of certainty or uncertainty of occurrence or non-occurrence.

9.1.1 Experiment

An experiment is an act that gives some definite result.

Example:

Tossing a coin, throwing a die, etc.

9.1.2 Random Experiment

An experiment is called a random experiment if

(*i*) all the possible outcomes are known in advance

(*ii*) which particular outcome will occur, cannot be predicted in advance

(*iii*) the experiment can be repeated under identical conditions.

Example:

Tossing a coin is a random experiment since all the possible outcomes, namely "heads" or "tails" are known in advance but among them which will occur particularly cannot be predicted in advance. Moreover, this experiment can be repeated as many times as we want.

9.1.3 Sample Space (Event Space)

The set of all possible outcomes of a random experiment is called a sample space (or event space) of the random experiment and is generally denoted by the symbol S or Ω.

Example:

(*i*) Consider the random experiment "tossing a coin." Then,

$S = \{H, T\}$ (H stands for heads and T stands for tails).

(*ii*) Consider the random experiment "tossing two coins" or "tossing a coin twice successively." Then,

$S = \{HH, HT, TH, TT\}$.

(*iii*) Consider the random experiment "tossing three coins" or "tossing a coin thrice successively." Then,

$S = \{HHH, HHT, HTH, HTT, THH, THT, TTH, TTT\}$

(*iv*) Consider the random experiment "throwing a die." Then, $S = \{1, 2, 3, 4, 5, 6\}$

(*v*) Consider the random experiment "throwing two dice" or "throwing a die twice." Then,

$S = \{(1,1), (1,2), (1,3), (1,4), (1,5), (1,6), (2,1), (2,2), (2,3), (2,4), (2,5), (2,6), (3,1), \ldots\ldots, (3,6), (4,1), \ldots\ldots, (4,6), (5,1), \ldots, (5,6), (6,1), (6,2), (6,3), (6,4), (6,5), (6,6)\}$.

Remember

We use the symbol $n(S)$ or $|S|$ to denote the total number of outcomes of the random experiment, i.e., total number of elements in the sample space S.

If a simple random experiment (like tossing a coin, throwing a die) has n number of possible outcomes and the experiment is repeated k times, then the total number of outcomes of the compound random experiment will be n^k. Thus if a coin is tossed thrice, the sample space will have 2^3, i.e., 8 outcomes.

9.1.4 Event

Any subset of a sample space is called an event. φ is called impossible event and S is called sure (certain) event.

Example:

(i) Consider the random experiment "tossing two coins." Then,

$S = \{HH, HT, TH, TT\}$

Let $A = \{HH, HT, TH\}$. Then, $A \subset S$ and we say that A is the event of getting at least one heads.

Again, if B is the event of getting no heads, then we write $B = \{TT\}$.

(ii) Consider the random experiment "throwing two dice." Then,

$S = \{(1,1),(1,2),(1,3),(1,4),(1,5),(1,6),(2,1),(2,2),(2,3),$
$(2,4),(2,5),(2,6),(3,1),........,(3,6),(4,1),........,(4,6),$
$(5,1),....,(5,6),(6,1),(6,2),(6,3),(6,4),(6,5),(6,6)\}$.

Here, $n(S) = 6^2 = 36$.

Let E be the event of getting a sum of 10. Then, we can write $E = \{(4,6),(6,4),(5,5)\}$ and $E \subset S$

Again, if F denotes the event of getting a sum more than 10, then we can take $F = \{(5,6),(6,5),(6,6)\}$ and $F \subset S$.

9.1.5 Equally Likely Events

Two events, A & B, associated with the same random experiment are said to be equally likely, if no one of them is expected to occur in preference to other.

Example:

Consider the random experiment "tossing a coin." Then, $S = \{H, T\}$. Let $A = \{H\}$, $B = \{T\}$. Here, A & B are equally likely events.

9.1.6 Mutually Exclusive Events

Two events, $A \& B$, associated with the same random experiment are said to be mutually exclusive if $A \cap B = \varphi$ (null set).

Example:

Consider the random experiment "tossing a coin." Then, $S = \{H, T\}$. Let $A = \{H\}$, $B = \{T\}$. Since $A \cap B = \{\} = \varphi$, so $A \& B$ are mutually exclusive.

9.1.7 Mutually Exhaustive Events

Two events, $A \& B$, associated with the same random experiment are said to be mutually exhaustive if $A \cup B = S$ (sample space).

Example:

Consider the random experiment "tossing a coin." Then, $S = \{H, T\}$. Let $A = \{H\}$, $B = \{T\}$. Here, $A \cap B = \varphi, A \cup B = \{H, T\} = S$. So $A \& B$ are mutually exhaustive.

9.1.8 Classical Definition Of Probability

Let S be the sample space associated with a random experiment, say E. Also let $A \subseteq S$, i.e., A be an event. Then, the probability of the event A is denoted by $P(A)$ and is defined by

$$P(A) = \frac{n(A)}{n(S)},$$

where $n(A)$ = number of element in A,

$n(S)$ = number of element in S.

Remember:

(i) $P(A)$ = probability that the event A occurs

(ii) $P(\varphi) = 0$

(iii) $P(S) = 1$

(iv) $0 \leq P(A) \leq 1$ for any event A.

(v) In the classical definition of probability, it is always assumed that the sample space contain a finite number of outcomes.

(vi) $P(A \cup B)$ = Probability that at least one of the events A and B occur

$= P(A) + P(B) - P(A \cap B)$

(vii) $P(A \cap B)$ = Probability that both the events A and B occur

(viii) $P(A^C)$ = Probability that the event A does not occur

$= 1 - P(A)$

(ix) $P(A^C \cap B^C)$

= Probability that none of the events A and B will occur

= 1 – Probability that at least one of the events will occur

= $1 - P(A \cup B)$

(x) $P(A \cup B \cup C)$

$= P(A) + P(B) + P(C) - P(A \cap B) - P(B \cap C)$

$-(A \cap C) + P(A \cap B \cap C)$

(xi) $P(B) = P(A \cap B) + P(A^C \cap B)$,

$P(A) = P(A \cap B) + P(A \cap B^C)$

(xii) If $E_1, E_2, E_3, \ldots, E_n$ are n mutually exclusive events in a sample space S, then

$P(E_1 \cup E_2 \cup E_3 \cup \ldots \cup E_n)$
$= P(E_1) + P(E_2) + P(E_3) + \ldots + P(E_n)$

$i.e;\ P\left(\bigcup_{i=1}^{n} E_i\right) = \sum_{i=1}^{n} P(E_i)$

(xiii) If $E_1, E_2, E_3, \ldots, E_n$ are n mutually exclusive and exhaustive events, then $P\left(\bigcup_{i=1}^{n} E_i\right) = \sum_{i=1}^{n} P(E_i) = 1.$

Remember:

(ii), (iii), and (xii) are called axioms of probability.

9.1.9 Independent Events

Two events are said to be independent if the occurrence (or non-occurrence) of one event has no influence on the occurrence (or non-occurrence) of the other event.

Example:

(i) Consider the random experiment "tossing a coin." Then, $S = \{H, T\}$. Let $A = \{H\}$, $B = \{T\}$. Then, A & B are both events which denotes event of getting heads and event of getting tails, respectively. Since the occurrence of the outcome heads does not depend on the occurrence of the outcome tails, so the events A & B are independent.

(ii) Suppose a bag contains 10 balls out of which 6 are black balls and 4 are red. One red ball is drawn at random from the bag and is again kept in the bag. Next, a black ball is drawn at random from the bag.

Let A = event of getting a red ball in first withdrawn,

B = event of getting a black ball in second withdrawn.

Then, A and B are independent events.

Remember :

(1) If two events A and B are independent, then

$P(A \cap B) = P(A) P(B)$ and vice versa.

(2) If A and B are independent events, then

(i) B and $\bar{A}$ are independent events.

(ii) A and $\bar{B}$ are independent events.

(iii) $\bar{A}$ and $\bar{B}$ are independent events.

The converse of these three results are also true.

(3) A, B, C are independent events

$\Leftrightarrow P(A \cap B \cap C) = P(A) P(B) P(C)$

9.2 CONDITIONAL PROBABILITY AND BAYES' THEOREM

9.2.1 Conditional Probability

Let A and B be two events associated with the same random experiment. Then, the probability that event A occurs, given that event B is already occurred, is called the conditional probability of A, which is denoted by $P(A/B)$ and is defined by

$$P(A/B) = \frac{P(A \cap B)}{P(B)}, \text{ provided } P(B) \neq 0.$$

Example:

Let us consider a random experiment "throwing two dice." Then, $n(S) = 36$.

Let A = event of getting a sum of 10,

B = event that the first die shows 4.

Then, $B = \{(4,1),(4,2),(4,3),(4,4),(4,5),(4,6)\}$ &

$A = \{(4,6),(6,4,),(5,5)\}$.

$\therefore A \cap B = \{(4,6)\}$.

So prob (getting a sum of 10 when the first die shows the number 4)

$$= P(A/B) = \frac{P(A \cap B)}{P(B)} = \frac{\frac{1}{36}}{\frac{6}{36}} = \frac{1}{6}.$$

Remember:

If A and B are independent events, then $P(A/B) = P(A)$ and $P(B/A) = P(B)$.

9.2.2 Theorem on Total Probability

Let S be the sample space and $A_1, A_2, ..., A_n$ be n events which are pairwise mutually exhaustive, *i.e.*, $A_i \cap A_j = \varphi$ for $i \neq j$ and $S = A_1 \cup A_2 \cup .... \cup A_n$.

Now, if A be an event other than A_i (i = 1, 2,...n) then $P(A) = \sum_{i=1}^{n} P(A_i) P(A/A_i)$

$$= P(A_1) P(A/A_1) + P(A_2) P(A/A_2) + ... + P(A_n) P(A/A_n)$$

9.2.3 Bayes' Theorem

Let A_1, A_2, ... A_n be n mutually exclusive and exhaustive events associated with the sample space S. Let E be another event so that $P(A/A_1)$, $P(A/A_2)$ $P(A/A_n)$ are known. Also, assume that $P(A_1)$, $P(A_2)$,... $P(A_n)$ are known. Then,

$$P(A_i / A)$$

= Prob. of the event A_i if the event A is occurred

$$= \frac{P(A_i) \times P(A/A_i)}{\sum_{i=1}^{n} P(A_i) \times P(A/A_i)}, \quad i = 1,2,3....,n$$

In particular for n = 3,

$$P(A_i / A)$$

$$= \frac{P(A_i) \times P(A/A_i)}{P(A_1) \times P(A/A_1) + P(A_2) \times P(A/A_2) + P(A_3) \times P(A/A_3)},$$

i = 1, 2, 3

9.3 RANDOM VARIABLE AND PROBABILITY DISTRIBUTION

9.3.1 Random Variable

A random variable is a function X given by

$$X : S \text{ (sample space)} \to R \text{ (set of real numbers)}.$$

Remember:

If X_1 and X_2 are random variables and c_1 and c_2 are constants, then $X_1 X_2, c_1 X_1, c_1 X_1 + c_2 X_2$, min (X_1, X_2) max (X_1, X_2), etc., are random variables.

9.3.2 Types of Random Variable

1. **Discrete random variable:**

A random variable is called a discrete random variable if it takes values from a discrete set of values.

Example:

Let us consider the random experiment "tossing two unbiased coins."

Then, S = {HH, HT, TH, TT}.

Let the random variable $X : S \to R$ be defined by

$X(w)$ = number of heads in w for $w \in S$.

Then $X(HH)$ = no. of heads in HH = 2,

$X(HT)$ = no. of heads in HT = 1,

$X(TH)$ = no. of heads in TH = 1,

$X(TT)$ = no. of heads in TT = 0.

Therefore, the random variable X takes values from the set {0, 1, 2}. Hence, X is a discrete random variable.

2. **Continuous random variable:**

A random variable is called a continuous random variable if it takes values from a continuous range of values.

Example:

Let X denotes the weight (in kg) of a group of students. If the minimum and maximum weight of this group are, respectively, 40 and 80, then X assumes every value within the range [40, 80]. Thus, X is a continuous random variable.

9.3.3 Probability Mass Function (P.M.F)

1. Let X be a discrete random variable

which takes the values $x_1, x_2, ..., x_n$ with probabilities $p_1, p_2, ... p_n$, respectively, *i.e.*,

$P(X = x_1) = p_1$, $P(X = x_2) = p_2$, ..., $P(X = x_n) = p_n$.

If we define a function f by $f(x_i) = p_i$ for i = 1, 2, 3,..., n, then f is called the probability mass function (p.m.f) or the probability density function (p.d.f) of the discrete random variable X.

Remember:

(i) $f(x_i) \geq 0 \; \forall x_i$

(ii) $\sum_i f(x_i) = 1, \; i.e., \sum_i p_i = 1$

2. The probability mass function (p.m.f) or the

Probability density function (p.d.f) of the continuous random variable X is a function "f" satisfying the following conditions:

(i) $f(x) \geq 0$ $\forall x$

(ii) $\int_{-\infty}^{\infty} f(x)\,dx = 1$

Remember:

(i) The curve given by $y = f(x)$ is called the probability density curve of the continuous random variable X.

(ii) $P(a < X < b) = \int_a^b f(x)\,dx$

9.3.4 Probability Distribution Function

- Let X be a discrete random variable

which takes the values x_1, x_2, ..., x_n with probabilities P_1, P_2, ...P_n, i.e;

$P(X = x_1) = p_1$, $P(X = x_2) = p_2$,

........,$P(X = x_n) = p_n$ and $x_1 < x_2 < .... < x_n$.

Then, the probability density function is denoted by $F(x)$ and is defined by

$F(x) = P(-\infty < X < x)$. Thus,

$$F(x) = \begin{cases} 0 \text{ if } x < x_1 \\ p_1 \text{ if } x_1 \leq x < x_2 \\ p_1 + p_2 \text{ if } x_2 \leq x < x_3 \\ \text{................................} \\ p_1 + p_2 + ... + p_n = 1 \text{ if } x_n \leq x \end{cases}$$

Example:

Let us consider the random experiment "tossing three unbiased coins."

Then,

$S = \{HHH, HHT, HTH, HTT, TTH, TTT, THH, THT\}$

Let the random variable $X : S \to R$ be defined by

$X(w)$ = number of heads in w for $w \in S$.

Then $X(HHH) = 3$, $X(HHT) = 2$,

$X(HTH) = 2$, $X(HTT) = 1$, $X(TTH) = 1$,

$X(TTT) = 0$, $(X(THH) = 2$, $X(THT) = 1$.

Also, $P(X = 0) = P(\{TTT\}) = \dfrac{1}{8}$,

$P(X = 1) = P(\{HTT, THT, TTH\}) = \dfrac{3}{8}$,

$P(X = 2) = P(\{HHT, THH, HTH\}) = \dfrac{3}{8}$,

$P(X = 3) = P(\{HHH\}) = \dfrac{1}{8}$.

Hence, the probability distribution function is given by

$$F(x) = \begin{cases} 0 \text{ if } x < 0 \\ \dfrac{1}{8} \text{ if } 0 \leq x < 1 \\ \dfrac{1}{8} + \dfrac{3}{8} \text{ if } 1 \leq x < 2 \\ \dfrac{1}{8} + \dfrac{3}{8} + \dfrac{3}{8} \text{ if } 2 \leq x < 3 \\ 1 \text{ if } 3 \leq x \end{cases} = \begin{cases} 0 \text{ if } x < 0 \\ \dfrac{1}{8} \text{ if } 0 \leq x < 1 \\ \dfrac{1}{2} \text{ if } 1 \leq x < 2 \\ \dfrac{7}{8} \text{ if } 2 \leq x < 3 \\ 1 \text{ if } 3 \leq x \end{cases}$$

Properties:

(i) $F(x)$ is monotonic non-decreasing, i.e., if

$x_i \leq x_j$ then $F(x_i) \leq F(x_j)$.

(ii) $F(-\infty) = 0$ and $F(\infty) = 1$

(iii) $0 \leq F(x) \leq 1$

(iv) $P(x_i \leq X \leq x_j) = P(X = x_i) + F(x_j) - F(x_i)$

(v) $f(x_i) = F(x_i) - F(x_{i-1})$

- The probability distribution function of a continuous random variable X is denoted by $F(x)$ and is defined by

$F(x) = P(-\infty < X < x) = \int_{-\infty}^{x} f(t)\,dt.$

Example:

Let $f(x) = \begin{cases} \dfrac{3}{x^4}, & 1 \leq x < \infty \\ 0, & \text{elsewhere} \end{cases}$.

Then, clearly $f(x) \geq 0$ and

$\int_{-\infty}^{\infty} f(x)\,dx = \int_{-\infty}^{1} 0\,dx + \int_{1}^{\infty} \dfrac{3}{x^4}\,dx = \left[\dfrac{-3}{3x^3}\right]_1^{\infty} = 1.$

Hence, $f(x)$ is a p.m.f.

Now for $-\infty < x < 1$,

$$F(x) = \int_{-\infty}^{x} f(t)\,dt = \int_{-\infty}^{x} 0\,dt = 0.$$

Again, for $1 \le x < \infty$

$$F(x) = \int_{-\infty}^{x} f(t)\,dt = \int_{-\infty}^{1} f(t)\,dt + \int_{1}^{x} f(t)\,dt$$

$$= \int_{-\infty}^{1} 0\,dt + \int_{1}^{x} \frac{3}{t^4}\,dt = 0 + \left[-\frac{1}{t^3}\right]_{1}^{x}$$

$$= 1 - \frac{1}{x^3}$$

Thus, $F(x) = \begin{cases} 1 - \dfrac{1}{x^3}, & 1 \le x < \infty \\ 0, & -\infty < x < 1 \end{cases}$.

Remember:

(i) $P(a < X < b) = P(a < X \le b)$

$= P(a \le X < b) = P(a \le X \le b)$

$= \int_{a}^{b} f(x)\,dx$

(ii) $F(-\infty) = 0$ and $F(\infty) = 1$

(iii) $\dfrac{d}{dx} F(x) = f(x)$

(iv) $P(x < X \le x + dx) = F'(x) = f(x)$

9.3.5 Expectation or Mean

• The expectation or mean of a discrete random variable X is denoted by $E(X)$ or by "m" and is defined by $E(X) = \sum_{i=1}^{n} x_i p_i$, if the random variable takes the values $x_1, x_2, ..., x_n$ with probabilities $p_1, p_2, ..., p_n$, respectively.

Example:

Let us consider the random experiment "throwing a die." Then, $S = \{1, 2, 3, 4, 5, 6\}$.

Let X be the random variable denoting the number appearing on the die. Then, the probability distribution of X is given by

X	1	2	3	4	5	6
P(X = xᵢ)	$\frac{1}{6}$	$\frac{1}{6}$	$\frac{1}{6}$	$\frac{1}{6}$	$\frac{1}{6}$	$\frac{1}{6}$

$\therefore E(X)$

$$= 1 \times \frac{1}{6} + 2 \times \frac{1}{6} + 3 \times \frac{1}{6} + 4 \times \frac{1}{6} + 5 \times \frac{1}{6} + 6 \times \frac{1}{6}$$

$$= \frac{7}{2}$$

• The expectation or mean of a continuous random variable X is denoted by $E(X)$ or by "m" and is defined by $E(X) = \int_{-\infty}^{\infty} x f(x)\,dx$.

Example:

Let the p.m.f of a random variable X be as follows:

$$f(x) = \begin{cases} \dfrac{1}{4}, & -2 < x < 2 \\ 0, & \text{otherwise} \end{cases}$$

Then, $E(X) = \int_{-\infty}^{\infty} x f(x)\,dx = \int_{-2}^{2} x \times \frac{1}{4}\,dx$

$$= \int_{-2}^{2} \frac{x}{4}\,dx = \left[\frac{x^2}{8}\right]_{-2}^{2} = 0.$$

Properties of mean:

(i) $E(k) = k$ if "k" is a constant

(ii) $E(kX) = kE(X)$ if "k" is a constant

(iii) $E(X \pm Y) = E(X) \pm E(Y)$

(iv) $E(XY) = E(X)E(Y)$ if X and Y are independent

9.3.6 Variance and Standard Deviation

The variance of a random variable X is denoted by $\text{Var}(X)$ or σ^2 and is defined by $\text{Var}(X) = \sigma^2 = E\{(X - m)^2\} = E(x^2) - m^2$.

The positive root of Var (X) is called the standard deviation (S.D) of X and is denoted by "σ." Thus, $\sigma = +\sqrt{\text{Var}(X)}$.

Example:

Let us consider the random experiment "throwing a die." Then, $S = \{1, 2, 3, 4, 5, 6\}$. Let X be the random variable denoting the number appearing on the die. Then, the probability distribution of X is given by

X	1	2	3	4	5	6
P(X = xᵢ)	$\frac{1}{6}$	$\frac{1}{6}$	$\frac{1}{6}$	$\frac{1}{6}$	$\frac{1}{6}$	$\frac{1}{6}$

$$\therefore m = E(X) = 1 \times \frac{1}{6} + 2 \times \frac{1}{6} + 3 \times \frac{1}{6} + 4 \times \frac{1}{6}$$

$$+ 5 \times \frac{1}{6} + 6 \times \frac{1}{6}$$

$$= \frac{7}{6}$$

$$\text{Also, } E(X^2) = 1^2 \times \frac{1}{6} + 2^2 \times \frac{1}{6} + 3^2 \times \frac{1}{6}$$

$$+ 4^2 \times \frac{1}{6} + 5^2 \times \frac{1}{6} + 6^2 \times \frac{1}{6}$$

$$= \frac{1}{6}(1^2 + 2^2 + 3^2 + 4^2 + 5^2 + 6^2)$$

$$= \frac{91}{6}$$

$$\therefore Var(X) = \sigma^2 = E(x^2) - m^2 = \frac{91}{6} - \left(\frac{7}{6}\right)^2$$

$$= \frac{497}{36}.$$

$$S.D = \sigma^2 = \sqrt{Var(X)} = \sqrt{\frac{497}{36}}.$$

Properties of Variance:

(i) $Var(k) = 0$ if "k" is a constant

(ii) $Var(aX + b) = a^2 Var(X)$ if "a" and "b" are constants

(iii) $Var(aX + bY) = a^2 Var(X) + b^2 Var(Y)$

(if X and Y are independent)

(iv) $Var(aX \pm bY) = a^2 Var(X) + b^2 Var(Y)$

$\pm 2Cov(X, Y)$

9.4 SPECIAL TYPES OF PROBABILITY DISTRIBUTIONS

9.4.1 Binomial Distribution

A discrete random variable X is said to have a Binomial distribution with parameter n (a positive integer) and p ($0 < p < 1$), if its p.m.f. is given by $P(X = r) = {}^nC_r P^r q^{n-r}$, where $r = 0, 1, 2, ..., n$; $p, q > 0$ and $p + q = 1$.

Here constants n and p are called the parameters of the binomial distribution. We generally use the notation $X \sim B(n, p)$ to denote that the random variable X follows binomial distribution with parameters n and p.

Remember:

(i) n = no. of trials

p = prob. of success in each trial,

q = prob. of failure in each trial = $1 - p$,

r = no. of success.

$P(X = r)$ is also read as "prob. of r successes out of n trials."

(ii) Mean of binomial distribution = np and variance = $npq = np(1 - p)$.

(iii) If $X_1 \sim B(n_1, p)$ and $X_2 \sim B(n_2, p)$, then $X_1 + X_2 \sim B(n_1 + n_2, p)$

(iv) Let a random experiment be performed repeatedly; each repetition being called a trial and let the occurrence of an event be called a success and its non-occurence a failure. Let us consider n independent trials in which p = probability of success in any trial and $q = 1 - q$ = probability of failure in any trial. Then, the probability of x successes, i.e., $(n - x)$ failures in n-independent trials is given by

$$P(X = x) = {}^nC_x p^x q^{n-x},$$

where X is the random variable denoting the number of successes.

9.4.2 Poisson Distribution

A discrete random variable is said to follow the Poisson distribution if its p.m.f is given by

$$P(X = x) = f(x) = \frac{e^{-m} m^x}{x!}, \ x = 0, 1, 2, 3 ....\infty$$

Remember:

(i) Mean = variance = m.

(ii) $m = np$, where n is the number of trials and p = probability of success in each trial.

(iii) If X_1 and X_2 are independent Poisson variates with parameters m_1 and m_2, respectively, then $X_1 + X_2$ is also a Poisson variate with the parameter $m_1 + m_2$.

9.4.3 Normal Distribution

A continuous random variable is said to follow the normal distribution with parameters μ and σ^2 if its p.m.f is given by

$$f(x) = \frac{1}{\sigma\sqrt{2\pi}} e^{-\frac{1}{2}\left(\frac{x-\mu}{\sigma}\right)^2}, -\infty < x < \infty$$

Remember:

(i) Mean = μ, variance = σ^2

(*ii*) If a continuous random variable X is normally distributed with mean = μ, variance = σ^2, then we say X is distributed as $N(\mu, \sigma^2)$ and we express this by writing $X \sim N(\mu, \sigma)$.

(*iii*) Suppose a continuous random variable X is normally distributed with mean = μ and variance = σ^2. Let $Z = \dfrac{X - \mu}{\sigma}$. Then, the random variable Z follows normal distribution with mean 0 and variance 1. Z is called a standard normal variate and we write $Z \sim N(0,1)$. The p.m.f of the standard normal variate Z is given by

$$\phi(x) = \frac{1}{\sqrt{2\pi}} e^{-\frac{z^2}{2}}, -\infty < z < \infty$$

(*iv*) Normal distribution is a limiting form of binomial distribution under the following conditions

    (a) $n \to \infty$ (n = number of trials)

    (b) p and q are very small.

9.4.4 Geometric Distribution

A discrete random variable X is said to follow the geometric distribution if its p.m.f is given by

$$f(x) = P(X = x) = q^x p, \, x = 0,1,2,...,\infty; \, p + q = 1$$

where n = no. of failures preceding the first success,

    p = probability of success and

    q = probability of failure.

Remember:

Mean = $\dfrac{q}{p}$ and variance = $\dfrac{q}{p^2}$.

9.4.5 Uniform (Rectangular) Distribution

A continuous random variable X is said to follow a Uniform (rectangular) distribution if its p.m.f is given by

$$f(x) = \begin{cases} \dfrac{1}{b-a}, & a < x < b \\ 0, & \text{otherwise} \end{cases}$$

Remember:

(*i*) The distribution function $F(x)$ of a uniform (rectangular) distribution is given by

$$F(x) = \begin{cases} 0, & -\infty < x < a \\ \dfrac{x-a}{b-a}, & a < x < b \\ 1, & \text{otherwise} \end{cases}$$

(*ii*) mean = $\dfrac{1}{2}(a+b)$; variance = $\dfrac{1}{12}(b-a)^2$

9.4.6 Gamma Distribution

A continuous random variable is said to follow the Gamma distribution with parameter $\lambda(>0)$ if its p.m.f is given by

$$f(x) = \frac{e^{-x} x^{\lambda-1}}{\Gamma(\lambda)} \text{ for } \lambda > 0, \, 0 < x < \infty$$

Remember:

(*i*) Mean = variance = λ

(*ii*) If $X_1, X_2, ..., X_n$ are independent Gamma variates with parameters $\lambda_1, \lambda_2, - \pi_n$, respectively, then $X_1 + X_2 + ... + X_n$ is also a Gamma variate with parameter $\lambda_1 + \lambda_2 + ...+ \lambda_n$.

9.4.7 Exponential Distribution

A continuous random variable is said to follow the exponential distribution with parameter $\lambda(> 0)$ if its p.m.f is given by

$$f(x) = \begin{cases} \lambda e^{-\lambda x}, & \text{for } x > 0 \\ 0, & \text{otherwise} \end{cases}$$

Remember:

(*i*) Mean = $\dfrac{1}{\lambda}$, variance = $\dfrac{1}{\lambda^2}$

(*ii*) If $X_1, X_2, ..., X_n$ are independent exponential variates with parameter $\lambda_1, \lambda_2,..., \lambda_n$, respectively, then the random variable $X = \min \{X_1, X_2, ..., X_n\}$ is also a exponential variate with parameter $\lambda_1 + \lambda_2 + ...+ \lambda_n$.

9.5 INTRODUCTION OF STATISTICS

9.5.1 Statistics

According to Croxton and Cowden, "Statistics may be defined as the science of collection, presentation, analysis, and interpretation of numerical data."

9.5.2 Scopes and limitations of Statistics

Scopes:

Statistical techniques are very useful in economic and business problems such as, wages, credit policy, consumption, and distribution of income Statistical tools like time-series analysis, demand analysis, and forecasting techniques are extensively used for economic development and planning.

Limitation of Statistics:

Some important limitations of statistics are

  (*i*) Statistics does not study individuals. It deals with an aggregate of objects.

  (*ii*) Statistical results are only approximate and not exact.

9.5.3 Frequency Distribution

Let us consider the marks in mathematics of 200 students given in following table:

Marks	No. of students (frequency)
35	5
45	4
60	6
75	3
90	2

The representation of the given data as above is known as "frequency distribution." Here marks are called the variable (denoted by "*x*") and "number of students" against the marks is known as the "frequency" (denoted by "*f*") of the variable.

Remember:

The word "frequency" means – how frequently a variable occurs. In the above example, frequency of 35 is 5, since 5 students got 35 marks in mathematics.

9.5.4 Mean (Arithmetic Mean)

It give us an idea about the concentration of the values in the central part of the distribution.

  (*i*) The arithmetic mean (denoted by $\bar{x}$) of n observation $x_1, x_2, ..., x_n$ is given by

$$\bar{x} = \frac{x_1 + x_2 + ... + x_n}{n}.$$

  (*ii*) If the observations $x_1, x_2, ..., x_n$ occur with the frequencies $f_1, f_2, ..., f_n$, respectively, then the arithmetic mean,

$$\bar{x} = \frac{x_1 f_1 + x_2 f_2 + ... + x_n f_n}{f_1 + f_2 + ... + f_n}.$$

Remember:

If there are n observations $x_1, x_2, ..., x_n$ and $\bar{x}$ be their arithmetic mean, then $\sum\limits_{i=1}^{n} f_i \times (x_i - \bar{x}) = 0.$

Thus, the algebraic sum of the deviations of a set of values from their arithmetic mean is zero.

9.5.5 Median

Median of a distribution is the value of the variable which divides the whole data set into two equal parts. In other words, median is the positional average of the distribution.

Let us arrange n number of observations in ascending or descending order.

Then,

  (*i*) Median (*M*)

$$= \text{value of } \left(\frac{n+1}{2}\right)^{th} \text{ observation if } n = \text{odd.}$$

  (*ii*) Median (*M*)

$$= \frac{1}{2}\left[\text{Sum of values of} \left(\frac{n}{2}\right)^{th} \& \left(\frac{n}{2}+1\right)^{th} \text{observations} \right]$$

if n = even.

Example:

  (*i*) Let us find the median of the value 10, 15, 12, 17, 16.

  Then, arranging them in ascending order we get:

  10, 12, 15, 16, 17.

  Here, $n = 5$ (odd)

  So median = value of $\left(\frac{5+1}{2}\right)^{th}$ observation

  = value of third observation = 15

  (*ii*) Let us find the median of the values 10, 15, 11, 16, 18, and 20.

  Then arranging them in ascending order, we have:

  10, 11, 15, 16, 18, 20.

  Here $n = 6$ (even).

  Here, the value of $\left(\frac{6}{2}\right)^{th}$ observation

  = value of third observation = 15.

  Value of $\left(\frac{6}{2}+1\right)^{th}$ observation

  = value of 4rd observation = 16

  Median

$$= \frac{1}{2}\left[\text{Sum of values of }\left(\frac{6}{2}\right)^{th}\text{ and }\left(\frac{6}{2}+1\right)^{th}\text{ observations}\right]$$

$$= \frac{15+16}{2} = 15.5$$

9.5.6 Mode

Mode is the value that occurs most frequently in a set of observations.

Example:

Consider the following table:

Value	Frequency
10	3
11	1
12	2
15	1

Therefore, 10 is the value that occurs maximum number of times (3).

Therefore, mode = 10.

Remember:

Mean − Mode = 3(Mean − Median).

9.5.7 Standard Deviation (S.D)

Let there be "n" number of observations, say, $x_1, x_2, \ldots x_n$ then the S.D. (standard deviation) is denoted by σ and is defined by

$$\sigma = \sqrt{\frac{1}{N}\sum_{i=1}^{n} f_i(x_i - \bar{x})}, \text{ where } \bar{x} = \text{mean of the}$$

distribution and $N = f_1 + f_2 + \ldots + f_n$. ($f_i$ being the frequency of x_1)

9.5.8 Correlation

If the change in value of one random variable affects the change in the value of the other random variable, then we say that the random variables are correlated.

Example:

Let X and Y be random variables denoting the heights and weights of 10 persons.

Then, X and Y are correlated random variables.

The correlation coefficient between the random variables X and Y, denoted by r or $r(X, Y)$, is a numerical measure of linear relationship between them and is defined as

$$r(X,Y) = \frac{Cov(X,Y)}{\sigma_X \sigma_Y}$$

where $Cov(X,Y) = E[(X-\bar{x})(Y-\bar{y})]$,

σ_X = S.D. of the random variable X,

σ_Y = S.D. of the random variable Y,

$\bar{x}$ = mean of the random variable X,

$\bar{y}$ = mean of the random variable Y.

Remember:

(i) $-1 \le r(X,Y) \le 1$

(ii) $r(aX+b,cY+d) = \dfrac{ac}{|ac|}r(X,Y)$

(iii) The random variables X and Y are uncorrelated $\Leftrightarrow$ cov$(X,Y) = 0 \Leftrightarrow r(X,Y) = 0$.

(iv) Var $(aX + by) = a^2$ var$(X) + b^2$ var$(Y) + 2ab$ cov (X, Y)

(v) If there be n pairs of observations (x_1, y_1), $(x_2, y_2),\ldots, (x_n, y_n)$, then

$$\text{cov}(X,Y) = \frac{1}{n}\sum_{i=1}^{n}(x_i - \bar{x})(y_i - \bar{y}).$$

(v) Covariance is independent of change of origin but not of change of scale.

(vi) Coefficient of correlation depends neither on the change of origin nor on the change of scale.

9.5.9 Regression

The term "regression" means "stepping back toward the average." Basically regression analysis is a mathematical measure of the average relationship between two or more variables in terms of the original units of the data.

Let us consider the probability distribution associated with two random variables X and Y. Then, the equation of line of regression of

(i) Y on X is given by $Y - \bar{y} = r\dfrac{\sigma_Y}{\sigma_X}(X - \bar{x})$

(ii) X on Y is given by $X - \bar{x} = r\dfrac{\sigma_X}{\sigma y}(Y - \bar{y})$

Remember:

(i) $b_{YX} = r\dfrac{\sigma_Y}{\sigma_X}$ and $b_{XY} = r\dfrac{\sigma_X}{\sigma_Y}$ are called regression coefficients.

(ii) If $r = 0$, then the regression lines become perpendicular to each other.

(iii) If $r = \pm 1$, the regression lines coincide or become parallel to each other.

(iv) In order to establish a relation $y = a + bx$ we need to find the constants "a" and "b" be such that

$$E = \sum_{i=1}^{n}(y_i - a - bx_i)^2$$ can be minimized which is possible when $\frac{\partial E}{\partial a} = 0$ and $\frac{\partial E}{\partial b} = 0$.

Fully Solved MCQs

1. A bag contains 16 balls numbered from 1 to 10. If a ball is drawn at random, what is the probability of getting a number which is a multiple of 3 or 4?

(a) 1   (b) $\frac{1}{2}$ (c) $\frac{2}{3}$ (d) $\frac{3}{4}$

2. What is the probability of getting a total of 8 or 12 when a pair of fair dice tossed ?

(a) $\frac{5}{6}$ (b) $\frac{1}{3}$ (c) $\frac{1}{6}$ (d) 1

3. What is the probability that a leap year selected at random will contains 53 Sundays ?

(a) $\frac{1}{7}$ (b) $\frac{2}{7}$ (c) $\frac{3}{7}$ (d) $\frac{4}{7}$

4. When three dice are thrown simultaneously, find the probability that the sum on the three faces is less than 16:

(a) $1 - \frac{10}{6^3}$ (b) $\frac{10}{6^3}$ (c) $1 - \frac{5}{6^3}$ (d) $\frac{5}{6^3}$

5. Given that a box of 10 electric bulbs contain 6 defective bulbs. One bulb is drawn at random from the box. Find the probability that it is nondefective.

(a) 0.40   (b) 0.50   (c) 0.25
(d) none of these

6. In a committee of 15 members, each member is proficient in English or in French or in both. If 10 of them are proficient in English and 8 out of them are proficient in French, then the probability that a person selected at random from the committee is proficient in both of the language is :

(a) $\frac{1}{2}$ (b) $\frac{1}{3}$ (c) $\frac{1}{4}$ (d) $\frac{1}{5}$

7. A bag contains 5 red balls and some black balls. If the probability of drawing a black ball is double that of a red ball, then the number of black balls in the bag is

(a) 12   (b) 10   (c) 9   (d) 6

8. A number x is selected from the numbers 1, 2, 3 and a second number y is randomly selected from the numbers 1,4,9. What is the probability that the product xy will be less than 9?

(a) $\frac{2}{9}$ (b) $\frac{4}{9}$ (c) $\frac{5}{9}$ (d) $\frac{7}{9}$

9. A bag contains 54 marbles each of which is black or green or red. The probability of selecting a black marble at random from the bag is $\frac{1}{3}$ and that of selecting a green marble is $\frac{4}{9}$. Then how many red marbles does the box contain ?

(a) 12   (b) 24   (c) 36   (d) 10

10. Candidates A and B appear for an interview for two posts. The probability of A's selection is $\frac{2}{9}$ and that of B's selection is $\frac{3}{4}$. What is the probability that only one of them will get selected?

(a) $\frac{2}{9}$ (b) $\frac{29}{36}$ (c) $\frac{23}{36}$ (d) $\frac{7}{12}$

11. A university has to select an advisor from a list of 100 persons. Out of them 40 are women, 60 are men, 80 of them knows Spanish, and 75 out of them are college teachers. What is the probability that the university selects a Spanish speaking women college teacher?

(a) $\frac{6}{25}$ (b) $\frac{3}{25}$ (c) $\frac{2}{25}$ (d) $\frac{24}{25}$

12. If A and B be two events such that $P(A) = \frac{1}{4}$, $P(B) = \frac{1}{3}$ and $P(A \cap B) = \frac{1}{5}$, then find $P(\bar{B}/\bar{A})$.

(a) $\frac{43}{45}$ (b) $\frac{7}{45}$ (c) $\frac{3}{45}$ (d) $\frac{37}{45}$

13. In a bolt factory, three machines A, B, C manufactures 25%, 35%, and 40% of the total production, respectively. Of their respective outputs, 5%, 4%, and 2% are defective. A bolt is drawn at random from the total product and is found to be defective. Find the probability that it was manufactured by machine C?

(a) $\frac{16}{69}$ (b) $\frac{13}{25}$ (c) $\frac{2}{25}$ (d) $\frac{23}{69}$

14. Three bags contain 3 white and 2 black balls, 2 white and 3 black balls, 4 white and 1 black ball, respectively. One ball is drawn from a bag chosen at random. What is the probability that a white ball is drawn?

(a) $\frac{1}{5}$ (b) $\frac{3}{5}$ (c) $\frac{2}{5}$ (d) $\frac{2}{25}$

15. Assume that each born child is equally likely to be a boy or a girl. If a family has two children, what is the conditional probability that both are girls given that the youngest is a girl?

(a) $\frac{1}{4}$ (b) $\frac{4}{5}$ (c) $\frac{1}{2}$ (d) $\frac{2}{3}$

16. Let X, Y, Z be events which are mutually independent with probabilities a, b, c, respectively. Let N denotes the number of X, Y, or Z which occur. Then, probability that $N = 2$ is

(a) $ab + bc + ca - abc$ (b) $ab + bc + ca - 3abc$
(c) $2(a + b + c) - abc$ (d) $ab + bc + ca$

17. Let X be a random variable such that $E(X^2) = E(X) = 1$. Then, $E(X^{100})$ is

(a) 0 (b) 1
(c) 2^{100} (d) $2^{100} + 1$

18. If the mean of the random variable X is "4" and $Var(X) = E(X^2) - 2E(X) + 2\lambda$, then $\lambda = ?$

(a) 2 (b) -4 (c) -2 (d) -3

19. The p.d.f of a π.v X is $f(x) = \begin{cases} \frac{1}{20}e^{-\frac{x}{20}}, & x > 0 \\ 0, & x \le 0 \end{cases}$.

Then what will be the value of $P(X \ge 15)$?

(a) $e^{-\frac{3}{4}}$ (b) $e^{\frac{3}{4}}$ (c) $e^{-\frac{5}{4}}$ (d) $e^{\frac{5}{4}}$

20. The p.m.f of a random variable X is given by $f(x) = \begin{cases} e^{-x}, & x > 0 \\ 0, & x \le 0 \end{cases}$. Then, $E[(X-1)^2] = ?$

(a) 0 (b) 1 (c) 2 (d) -2

21. The p.d.f of a random variable X is $f(x) = kx(x-1)$ for $1 \le x \le 2$.
Then what will be the value of "k"?

(a) 2/5   (b) 3/5   (c) 4/5   (d) 6/5

22. If X is uniformly distributed in $[-k, k]$ with $k > 0$, then what will be the value of k so that $P(X > 1) = 1/4$?

(a) 4 (b) 6 (c) 2 (d) 5

23. If X is uniformly distributed in $[-1, 2]$, then find the value of $P\left(|X-1| \ge \frac{1}{2}\right)$.

(a) $\frac{7}{11}$ (b) $\frac{2}{5}$ (c) $\frac{2}{3}$ (d) $\frac{1}{3}$

24. Let X have a binomial distribution with parameters n and p, where n is an integer greater than 1 and $0 < p < 1$. If $P(X = 0) = P(X = 1)$, then the value of p is

(a) $\frac{1}{n-1}$ (b) $\frac{n}{n+1}$ (c) $\frac{1}{n+1}$ (d) $\frac{1}{n}$

25. A die is tossed once. If the random variable X is defined as:

$X = \begin{cases} 1, & \text{if the die results in an even number} \\ 0, & \text{if the die results in an odd number} \end{cases}$

Then, the mean of X is
(a) 3/4   (b) 1/2   (c) 2/5   (d) 3/5

Answers key				
1. (b)	2. (c)	3. (b)	4. (a)	5. (a)
6. (d)	7. (b)	8. (c)	9. (a)	10. (c)
11. (a)	12. (d)	13. (a)	14. (b)	15. (c)
16. (b)	17. (b)	18. (b)	19. (a)	20. (b)
21. (d)	22. (c)	23. (c)	24. (c)	25. (b)

Explanations

1. (b) Let A = event that the ball number is a multiple of 3,

B = event that the ball number is a multiple of 4,

Then, $A = \{3, 6, 9, 12, 15\}, B = \{4, 8, 12, 16\}$.

So $A \cap B = \{12\}$.

Hence, the probability that the ball number is a multiple of 3 or 4

$= P(A \cup B)$
$= P(A) + P(B) - P(A \cap B)$
$= \frac{n(A)}{n(S)} + \frac{n(B)}{n(S)} - \frac{n(A \cap B)}{n(S)}$
$= \frac{5}{16} + \frac{4}{16} - \frac{1}{16} (\because n(S) = 16)$
$= \frac{8}{16} = \frac{1}{2}$.

2. (c) Here $n(S) = 6^2 = 36$.

Let A = event of getting a sum of 8,

B = event of getting a sum of 12.

Then, $A = \{(2,6),(6,2),(3,5),(5,3),(4,4)\}$,

$B = \{(6,6)\}$ and so $A \cap B = \varphi$.

Therefore, $n(A) = 5, n(B) = 1, n(A \cap B) = 0$.

Hence, Prob. (receiving a sum of 8 or 12)

$= P(A \cup B)$

$= P(A) + P(B) - P(A \cap B) = \dfrac{5}{36} + \dfrac{1}{36} - \dfrac{0}{36} = \dfrac{1}{6}$.

3. (*b*) In a leap year, there are 366 days (52 complete weeks and 1 extra day).

The following are the possible combination for these two extra days:

(Sunday, Monday), (Monday, Tuesday), (Tuesday, Wednesday), (Wednesday, Thursday), (Thursday, Friday), (Friday, Saturday), (Saturday, Sunday).

Among these only two cases are favorable, viz (Sunday, Monday) and (Saturday, Sunday).

Hence, the required probability = $\dfrac{2}{7}$.

4. (*a*) Here $n(S) = 6^3$.

Let A = event that the sum of the three faces is greater than 16.

Then,

$A = \{(6,6,6),(6,6,5),(6,5,6),(5,6,6),(6,6,4),$
$(6,4,6),(4,6,6),(5,6,6),(5,6,5),(6,5,5)\}$

Thus, $n(A) = 10$

Hence, Prob.(sum is less than 16)

= 1– Prob. (sum is greater than 16)

$= 1 - P(A) = 1 - \dfrac{10}{6^3}$.

5. (*a*) Let A = event that a bulb drawn is defective.

Then, $P(A) = \dfrac{6}{10} = 0.6$.

Hence, prob. (the drawn bulb is non-defective)

$= P(A^C) = 1 - P(A) = 1 - 0.6 = 0.40$.

6. (*d*) Let A = event that selected person is proficient in English,

B = event that selected person is proficient in French

ATQ,

$P(A) = \dfrac{n(A)}{n(S)} = \dfrac{10}{15}$,

$P(B) = \dfrac{n(B)}{n(S)} = \dfrac{8}{15}$.

Then prob. (the selected person is proficient in both of the language)

$= P(A \cap B)$

$= P(A) + P(B) - P(A \cup B)$

$= \dfrac{10}{15} + \dfrac{8}{15} - 1 \quad (A \cup B = S)$

$= \dfrac{1}{5}$.

7. (*b*) Let there be "*n*" number of black balls in the bag.

Then $n(S)$ = total number of balls in the bag

$= n + 5$.

Let A = event of drawing a black ball,

B = event of drawing a red ball.

Then,

$P(A) = \dfrac{n(A)}{n(S)} = \dfrac{n}{5+n}$, $\quad P(B) = \dfrac{n(B)}{n(S)} = \dfrac{5}{5+n}$

ATQ, $P(A) = 2P(B)$

or, $\dfrac{n}{5+n} = 2 \times \dfrac{5}{5+n}$

or, $n = 10$.

Thus, there are 10 black balls in the bag.

8. (*c*) Two numbers x and y can be selected in $3_{C_1} \times 3_{C_1}$, *i.e.*, 9 ways. Thus, $n(S) = 9$.

Let A = event that the product xy is less than 9. Then,

$A = \{(1,1),(1,4),(2,1),(2,4),(3,1)\}$.

Hence, the required probability = $P(A)$ = $\dfrac{n(A)}{n(S)} = \dfrac{5}{9}$

9. (*a*) Let A = event of selecting a black marble,

B = event of selecting a green marble,

C = event of selecting a red marble.

Given $P(A) = \dfrac{1}{3}$, $P(B) = \dfrac{4}{9}$.

Since the events A, B, C are mutually exclusive and exhaustive, so $S = A \cup B \cup C$. This gives

$P(A \cup B \cup C) = P(S) = 1$

or, $P(A) + P(B) + P(C) = 1$

or, $\dfrac{1}{3} + \dfrac{4}{9} + P(C) = 1$

or, $P(C) = \dfrac{12}{54}$.

Thus, the box contains 12 red marbles.

10. (c) Let E = event that candidate A is selected and

F = event that candidate B is selected.

Then, $P(E) = \dfrac{2}{9}$ & $P(F) = \dfrac{3}{4}$.

$\therefore P(\overline{E}) = 1 - \dfrac{2}{9} = \dfrac{7}{9}$ & $P(\overline{F}) = 1 - \dfrac{3}{4} = \dfrac{1}{4}$.

Therefore, prob. (only one of them will get selected)

= Prob.[(A is selected, B is not selected) or (A is not selected, B is selected)]

$= P\left[(E \cap \overline{F}) \cup (\overline{E} \cap F)\right]$

$= P(E \cap \overline{F}) + P(\overline{E} \cap F)$

[$\because$ the events $E \cap \overline{F}$ & $\overline{E} \cap F$ are mulually exclusive]

$= P(E) \times P(\overline{F}) + P(\overline{E}) \times P(F)$

[$\because$ the events E & $\overline{F}, \overline{E}$ & F are independent]

$= \dfrac{2}{9} \times \dfrac{1}{4} + \dfrac{7}{9} \times \dfrac{3}{4} = \dfrac{23}{36}$.

11. (a) Let A = event that a female advisor is selected,

B = event that a Spanish speaking advisor is selected,

C = event that a college teacher is selected as advisor.

Then, $P(A) = \dfrac{40}{100} = \dfrac{2}{5}$, $P(B) = \dfrac{80}{100} = \dfrac{4}{5}$,

$P(C) = \dfrac{75}{100} = \dfrac{3}{4}$.

Hence, required probability

$= P(A \cap B \cap C) = P(A) \times P(B) \times P(C)$

[$\because A, B, C$ are independent events]

$= \dfrac{2}{5} \times \dfrac{4}{5} \times \dfrac{3}{4} = \dfrac{6}{25}$.

12. (d)

$P(\overline{B}/\overline{A})$

$= \dfrac{P(\overline{B} \cap \overline{A})}{P(\overline{A})} = \dfrac{1 - P(B \cup A)}{1 - P(A)}$

$= \dfrac{1 - \{P(A) + P(B) - P(B \cap A)\}}{1 - P(A)}$

$= \dfrac{1 - \left\{\dfrac{1}{4} + \dfrac{1}{3} - \dfrac{1}{5}\right\}}{1 - \dfrac{1}{4}} = \dfrac{37}{45}$.

13. (a) Let

E = event that a bolt is produced by machine A,

F = event that a bolt is produced by machine B,

G = event that a bolt is produced by machine C.

Then, $P(E) = \dfrac{25}{100}, P(F) = \dfrac{35}{100}$ and $P(G) = \dfrac{40}{100}$.

Let R = event that a defective bolt is drawn.

Then ATQ,

$P(R/E) = \dfrac{5}{100}, P(R/F) = \dfrac{4}{100}$ and $P(R/G) = \dfrac{2}{100}$.

Therefore, the required probability

$= P(G/R)$

$= \dfrac{P(G) \times P(R/G)}{P(E) \times P(R/E) + P(F) \times P(R/F) + P(G) \times P(R/G)}$

(by Bayes' theorem)

$= \dfrac{\dfrac{40}{100} \times \dfrac{2}{100}}{\dfrac{25}{100} \times \dfrac{5}{100} + \dfrac{35}{100} \times \dfrac{4}{100} + \dfrac{40}{100} \times \dfrac{2}{100}} = \dfrac{16}{69}$.

14. (b) Let

A = event that first bag is chosen,

B = event that second bag is chosen and

C = event that third bag is chosen.

Since there is an equal probability of each bag being chosen, so we have, $P(A) = P(B) = P(C) = \dfrac{1}{3}$.

Now, let E = event that a white ball is drawn. Then, ATQ,

$P(E/A) = \dfrac{^3C_1}{^5C_1} = \dfrac{3}{5}, P(E/B) = \dfrac{^2C_1}{^5C_1} = \dfrac{2}{5}$ &

$P(E/C) = \dfrac{^4C_1}{^5C_1} = \dfrac{4}{5}$.

Therefore, the required probability

$= P(E)$

$= P(A)\times P(E/A) + P(B)\times P(E/B) + P(C)\times P(E/B)$

(using the theorem of total probability)

$= \dfrac{1}{3}\times\dfrac{3}{5}+\dfrac{1}{3}\times\dfrac{2}{5}+\dfrac{1}{3}\times\dfrac{4}{5}=\dfrac{3}{5}.$

15. (c) Let "B" and "G" represent the boy and the girl, respectively.

Given that a family has two children. So the sample space, $S = \{(B, B), (G, G), (B, G), (G, B)\}$.

Let A = event that the youngest child is a girl and

B = event that both children are girls.

Then, $A = \{(B, G), (G, G)\}$ and $B = \{(G, G)\}$.

So $A \cap B = \{(G, G)\}$

Then, $P(A)=\dfrac{2}{4}=\dfrac{1}{2}, P(B)=\dfrac{1}{4}, P(A\cap B)=\dfrac{1}{4}.$

Therefore, the conditional probability that both are girls given that the youngest is a girl

$= P(B/A)=\dfrac{P(A\cap B)}{P(A)}=\dfrac{\frac{1}{4}}{\frac{1}{2}}=\dfrac{1}{2}.$

16. (b) Prob. $(N = 2)$

= Prob. (two events occur and one does not occur)

= Prob. [(X and Y occur but Z does not occur) or (Y and Z occur but X does not occur) or (X and Z occur but Y does not occur)]

= Prob. (X and Y occur but Z does not occur) + Prob.

(Y and Z occur but X does not occur) + Prob. (X and Z occur but Y does not occur)

$= ab(1 - c) + (1 - a)bc + a(1 - b)c = ab + bc + ca - 3abc.$

17. (b)

$E(X^2) = E(X) = 1$

$\Rightarrow Var(X) = E(X^2) - E(X) = 1 - 1 = 0$

$\Rightarrow X = c\,(\text{constant})$

Now $E(X)=1 \Rightarrow E(c) = 1 \Rightarrow c=1.$

$\therefore X = 1$ and so $E(X^{100})=E(1)=1.$

18. (b)

$Var(X) = E(X^2) - 2E(X) + 2\lambda$

$\Rightarrow E(X^2) - \{E(X)\}^2 = E(X^2) - 2E(X) + 2\lambda$

$\Rightarrow \{E(X)\}^2 - 2E(X) = -2\lambda$

$\Rightarrow 4^2 - 2\times 4 = -2\lambda$ ($\because$ mean $= E(X) = 4$)

$\Rightarrow \lambda = -4.$

19. (a)

$P(X \geq 15)$

$= 1 - P(X < 15)$

$= 1 - \int_0^{15} f(x)dx = 1 - \int_0^{15}\dfrac{1}{20}e^{-\frac{x}{20}}dx$

$= 1 - \dfrac{1}{20}\left[\dfrac{e^{-\frac{x}{20}}}{-\frac{1}{20}}\right]_0^{15} = 1 + \left[e^{-\frac{x}{20}}\right]_0^{15}$

$= 1 + \left[e^{-\frac{15}{20}} - 1\right] = e^{-\frac{3}{4}}.$

20. (b)

$E\left[(X-1)^2\right]$

$= \int_{-\infty}^{\infty}(x-1)^2 f(x)dx$

$= \int_{-\infty}^{0}(x-1)^2 f(x)dx + \int_0^{\infty}(x-1)^2 f(x)dx$

$= \int_{-\infty}^{0}(x-1)^2\times 0\,dx + \int_0^{\infty}(x-1)^2 e^{-x}dx$

$= 0 + \left[(x-1)^2(-e^{-x})\right]_0^{\infty} - \int_0^{\infty}\left[2(x-1)(-e^{-x})\right]dx$

$= 1 + 2\int_0^{\infty}(x-1)e^{-x}dx$ ($\because e^{-\infty} = 0$)

$= 1 + 2\left\{\left[(x-1)(-e^{-x})\right]_0^{\infty} - \int_0^{\infty}\left[1\times(-e^{-x})\right]dx\right\}$

$= 1 + 2\left\{-0 - 1 + \left[-e^{-x}\right]_0^{\infty}\right\}$ ($\because e^{-\infty} = 0$)

$= 1 + 2\{-1 - [0-1]\} = 1.$

21. (d)

$\int_{-\infty}^{\infty} f(x)dx = 1$

$\Rightarrow k\int_1^2 x(x-1)dx = 1 \Rightarrow k\left[\dfrac{x^3}{3} - \dfrac{x^2}{2}\right]_1^2 = 1$

$\Rightarrow k\left(\dfrac{8}{3} - 2 - \dfrac{1}{3} + \dfrac{1}{2}\right) = 1 \Rightarrow k = \dfrac{6}{5}.$

22. (c)

$$f(x) = \begin{cases} \dfrac{1}{2k}, & -k < x < k \\ 0, & \text{otherwise} \end{cases}$$

Now $P(X > 1) = \dfrac{1}{4}$

$\Rightarrow \displaystyle\int_1^k f(x)\, dx = \dfrac{1}{4} \Rightarrow \int_1^k \dfrac{1}{2k}\, dx = \dfrac{1}{4}$

$\Rightarrow \dfrac{1}{2k}[x]_1^k = \dfrac{1}{4} \Rightarrow k = 2.$

23. (c)

$$f(x) = \begin{cases} \dfrac{1}{3}, & -1 < x < 2 \\ 0, & \text{otherwise} \end{cases}$$

Now $P\left(|X-1| \geq \dfrac{1}{2}\right)$

$= P\left[(X-1) \geq \dfrac{1}{2} \text{ or } -(X-1) \geq \dfrac{1}{2}\right]$

$= P\left[X-1 \geq \dfrac{1}{2}\right] + P\left[X-1 \leq -\dfrac{1}{2}\right]$

$= P\left[X \geq \dfrac{3}{2}\right] + P\left[X \leq \dfrac{1}{2}\right]$

$= \displaystyle\int_{\frac{3}{2}}^{2} f(x)\,dx + \int_{-1}^{\frac{1}{2}} f(x)\,dx = \int_{\frac{3}{2}}^{2}\dfrac{1}{3}dx + \int_{-1}^{\frac{1}{2}}\dfrac{1}{3}dx$

$= \dfrac{1}{3}[x]_{3/2}^{2} + \dfrac{1}{3}[x]_{-1}^{1/2} = \dfrac{2}{3}.$

24. (c)

$P(X = 0) = P(X = 1)$

$\Rightarrow {}^nC_0 p^0 (1-p)^{n-0} = {}^nC_1 p^1 (1-p)^{n-1}$

$\Rightarrow 1 - p = np \Rightarrow p = \dfrac{1}{n+1}.$

25. (b) Here, the sample space = {1, 2, 3, 4, 5, 6}.

$P(X = 1) = $ Prob. (getting an even number) $= \dfrac{3}{6} = \dfrac{1}{2},$

$P(X = 0) = $ Prob. (getting an odd number) $= \dfrac{3}{6} = \dfrac{1}{2}.$

Hence, mean $= E(X) = 1 \times \dfrac{1}{2} + 0 \times \dfrac{1}{2} = \dfrac{1}{2}.$

Fully Solved MCQs

1. A shelf has six mathematics books and four physics books. Find the probability that three particular mathematics books will be together?

(a) $\dfrac{3!5!}{10!}$ (b) $\dfrac{4!6!}{10!}$

(c) $\dfrac{8!3!}{10!}$ (d) none of thes

2. There are three players A, B, C in a game. The probability that of A to win the game is twice that of B and the probability of B to win the game is twice that of C. Find the probability of C to win the game.

(a) $\dfrac{1}{7}$ (b) $\dfrac{6}{7}$ (c) $\dfrac{3}{4}$ (d) $\dfrac{1}{4}$

3. Andrew and Anthony are friends. What is the probability that they will have the same birthday?

(Consider a non-leap year).

(a) $\dfrac{364}{365}$ (b) 1   (c) $\dfrac{1}{365}$ (d) $\dfrac{1}{25}$

4. Two customers are visiting a particular shop during the same week. Each is equally likely to visit the shop on any day. What is the probability that both will visit the shop on different days? (consider six days in a week excluding Sunday)

(a) $\dfrac{1}{6}$ (b) $\dfrac{5}{6}$ (c) $\dfrac{1}{2}$ (d) $\dfrac{2}{3}$

5. Two integers x and y are chosen at random with replacement from the numbers 1,2,3,....,9. Find the probability that $x^2 - y^2$ is divisible by 2.

(a) $\dfrac{1}{81}$ (b) $\dfrac{5}{81}$ (c) $\dfrac{39}{81}$ (d) $\dfrac{41}{81}$

6. Two jars A and B contain, respectively, 2 red, 5 black, 7 green balls and 1 red, 4 black, and 9 green balls. One ball is drawn from each jars. Find the probability that both the balls are of the same color?

(a) $\dfrac{1}{85}$ (b) $\dfrac{139}{140}$ (c) $\dfrac{85}{196}$ (d) $\dfrac{79}{196}$

7. If A and B are two events such that $P(A) = 0.3$, $P(B) = 0.4, P(\bar{A} \cap \bar{B}) = 0.5,$ then the value of $P[B/(A \cup \bar{B})]$ is

(a) $\dfrac{3}{7}$ (b) $\dfrac{1}{4}$ (c) $\dfrac{1}{8}$ (d) $\dfrac{2}{5}$

8. An insurance company insured 2,000 motorcycle drivers, 4,000 car drivers and 6,000 truck drivers. The probability of accidents are 0.01,

0.03, and 0.15, respectively. One of the insured person meets with an accident. What is the probability that he is a motorcycle driver?

(a) $\dfrac{51}{52}$ (b) $\dfrac{1}{52}$ (c) $\dfrac{1}{20}$ (d) $\dfrac{19}{20}$

9. A laboratory blood test is 99% effective in detecting a certain disease, when it is in fact present. However, the test also yields a false positive result for 0.5% of the healthy person tested (that is, if a healthy person is tested, then, with probability 0.005, the test will imply he has the disease). If 0.1% of the population actually has the disease, what is the probability that a person has the disease given that his test result is positive?

(a) 0.165   (b) 0.2   (c) 0.5   (d) 0.874

10. A missile has probability 1/2 of destroying its target and probability 1/2 of missing it. Assume that the firings form independent trials, Determine the least number of missiles that should be fired at a target in order to make the probability of destroying the target at least 0.99.

(a) 6   (b) 7   (c) 8   (d) 9

11. A number is chosen at random from the numbers 1, 2, 3, 4,......,50 and another number is chosen at random from the numbers 1, 2, 3,........, 25. Then the expected value of the product of the two numbers chosen is

(a) 38.5   (b) 35   (c) 45   (d) 48.5

12. A discrete random variable X has the following probability distribution:

X	0	1	2	3	4
P(X = x_i)	0	k	2k	2k	4k

Then,

(i) what is the value of "k"?

(a) $\dfrac{1}{9}$ (b) $\dfrac{1}{11}$ (c) $\dfrac{1}{5}$ (d) 1

(ii) $P(X < 3) = ?$

(a) 0   (b) $\dfrac{1}{2}$ (c) $\dfrac{1}{3}$ (d) 1

(iii) $P(X \geq 2) = ?$

(a) $\dfrac{1}{9}$ (b) $\dfrac{8}{9}$ (c) $\dfrac{1}{4}$ (d) 1

(iv) $P(2 < X \leq 4) = ?$

(a) $\dfrac{2}{3}$ (b) $\dfrac{1}{3}$ (c) $\dfrac{1}{4}$ (d) 1

(v) $P(2 < X | X \leq 4) = ?$

(a) $\dfrac{1}{3}$ (b) $\dfrac{1}{2}$ (c) $\dfrac{2}{3}$ (d) 1

(vi) the minimum value of x so that $P(X \leq x) > \dfrac{1}{2}$ is

(a) 2   (b) 3   (c) 1   (d) 4

13. Let X be a non-negative integer valued random variable with $E(X^2) = 3$ and $E(X) = 1$.

Then, $\sum_{i=1}^{\infty} iP(X \geq i)$ is equal to

(a) 1   (b) 3   (c) 2   (d) 4

14. Let X and Y be two independent random variables. If $E(X) = 3$, $E(Y) = 2$, $\text{Var}(X) = 1$ and $Z = aX - bY$, then the values of "a" and "b" such that

$E(Z) = 0$ and Var $(Z) = 1$ are

(a) $\dfrac{2}{\sqrt{13}}, \dfrac{3}{\sqrt{13}}$ (b) $\dfrac{3}{\sqrt{13}}, \dfrac{2}{\sqrt{13}}$

(c) $\dfrac{1}{\sqrt{13}}, \dfrac{2}{\sqrt{13}}$ (d) $\dfrac{2}{\sqrt{13}}, \dfrac{1}{\sqrt{13}}$

15. In 10 independent throws of a defective die, the probability that an even number will appear 5 times is twice the probability that an even number will appear 4 times. Find the probability that an even number will not appear at all in 10 independent throws of the die.

(a) $\left(\dfrac{3}{8}\right)^{10}$ (b) $\left(\dfrac{5}{8}\right)^{10}$ (c) $\left(\dfrac{1}{8}\right)^{10}$ (d) $\left(\dfrac{7}{8}\right)^{10}$

16. Near a busy street intersection, it is estimated that a jaywalker will be hit by a car with probability 0.01. Assuming that individual trips form independent trials, find the probability of a jaywalker remaining unharmed if he crosses the street twice per day for 50 days.

(a) $(0.9)^{100}$ (b) $(0.05)^{100}$
(c) $(0.99)^{100}$ (d) $(0.01)^{100}$

17. If there is a war every 15 years on the average, then what is the probability that there will be no wars in 25 years?

(a) $e^{-\frac{1}{3}}$ (b) $e^{-\frac{2}{3}}$ (c) $e^{-\frac{4}{3}}$ (d) $e^{-\frac{5}{3}}$

18. A radioactive source emits on average 2.5 particles per second. What is the probability that 3 or more particles will be emitted in an interval of 4 seconds?

(a) $1 - 61e^{-10}$ (b) e^{-9} (c) $1 - e^{-9}$ (d) $61e^{-2}$

19. Some airlines find that each passenger who reserves a seat fails to turn up with probability 0.1 independently of other passengers of these airlines. Airline A always sells 10 tickets for their 9 seat airplane while airline B always sells 20 tickets for their 18 seat airplane. Then, find which one of A and B is more often overbooked.
   (a) all of them   (b) A
   (c) B (d) none of them

20. Assume that 45% of the population favors a certain candidate in an election. If a random sample of size 200 is chosen, then the standard deviation of the number of members of the sample that favors the candidate is
   (a) 6.12   (b) 5.26   (c) 8.18   (d) 7.04

21. A sample of 100 dry battery cells tested to find the length of life produced the following results:
   Mean = 12 hours and standard deviation = 3 hours.

   Assuming the data to be normally distributed, what percentage of battery cells are expected to have life more than 15 hours?
   (a) 15.87%   (b) 18%
   (c) 17.9%   (d) 20.36%

Answers key

1. (c)	2. (a)	3. (c)	4. (b)	5. (d)
6. (c)	7. (b)	8. (b)	9. (a)	10. (b)
11. (a)				
12. (i)-(a), (ii)-(c), (iii)-(b), (iv)-(a), (v)-(c), (vi)-(b).				
13. (c)	14. (a)	15. (a)	16. (c)	17. (d)
18. (a)	19. (c)	20. (d)	21. (a)	

Explanations

1. (c) Here all the books can be arranged themselves in 10! ways. So $n(S) = 10!$.
   Now, considering three mathematics books as one, we have a total of $7 + 1$, i.e., 8 books which can be arranged themselves in 8! ways.

   Again, the three particular mathematics books can be arranged themselves in 3! ways.

   Let A = event that three particular mathematics books will be together.
   Then, $n(A) = 8! \times 3!$.

   Hence, the required probability = $P(A)$ = $\frac{8! \times 3!}{10!}$

2. (a) Let E = event that the player A wins the game,
   F = event that player B wins the game,
   G = event that player C wins the game.
   ATQ, $P(E) = 2P(F)$, $P(F) = 2P(G)$
   Now, $S = E \cup F \cup G$
   (since the events E, F, G are mutually exhaustive)
   $\Rightarrow P(E \cup F \cup G) = P(S) = 1$
   $\Rightarrow P(E) + P(F) + P(G) = 1$
   (since the events E, F, G are mutually exclusive)
   $\Rightarrow 2P(F) + P(F) + P(G) = 1$
   $\Rightarrow 6P(G) + P(G) = 1 \Rightarrow P(G) = \frac{1}{7}$
   Hence, the required prob. = $\frac{1}{7}$.

3. (c) Andrew may have one of the 365 days as his birthday. Similarly, Anthony may have one of the 365 days as his birthday.
   So, total number of ways in which Andrew and Anthony may have their birthday = 365 × 365.
   Clearly, the total number of ways in which Andrew and Anthony will have the same birthday = 365.
   Hence, prob. (Andrew and Anthony will have the same birthday) = $\frac{365}{365 \times 365} = \frac{1}{365}$.

Remember:
Prob. that Andrew and Anthony will have different birthdays
   = 1− Prob.(they will have the same birthday)
   = $1 - \frac{1}{365} = \frac{364}{365}$.

4. (b) Two customers may visit the shop on any of the six days, Monday through Saturdays.
   Thus, total numbers of ways in which both customers will visit the shop = 6 × 6 = 36.
   Again, the total number of ways in which both customers will visit the shop on the same day = 6.
   So the probability (both customers will visit the shop on the same day)
   = $\frac{6}{36} = \frac{1}{6}$.

Hence, the probability (they will visit on different days)

$$= 1 - \frac{1}{6} = \frac{5}{6}.$$

Remember:

The probability that first customers visits the shop $= \frac{6}{36} = \frac{1}{6}.$

The probability that second customers visits the shop on a different day $= 1 - \frac{1}{6} = \frac{5}{6}.$

Hence, the probability that both will visit the shop on consecutive days $= \frac{1}{6} \times \frac{5}{6} = \frac{5}{36}.$

5. (d) Out of 9 integers [from {1, 2,....,9}], an integer can be chosen at random in 9 ways.

So, two integers from the set {1, 2,....,9} can be chosen in 9 × 9, i.e., 81 ways.

Since $x^2 - y^2 = (x + y)(x - y)$, so $x^2 - y^2$ will be divisible by 2 if and only if either both x, y are even or both x, y are odd.

Now there are four even numbers and five odd numbers in the set {1,2,....,9}.

So two even numbers can be chosen in 4 × 4, i.e., 16 ways and two odd numbers can be chosen in 5 × 5, i.e., 25 ways.

Thus, the required probability $= \frac{16 + 25}{81} = \frac{41}{81}.$

Alternative Method:

Here,

$S = \{(1,1),(1,2),(1,3),......,(1,8),(1,9),$
$(2,1),(2,2),(2,3),....,(2,8),(2,9),$
$(3,1),(3,2),(3,3),......,(3,8),(3,9),$
....,
$(9,1),(9,2),(9,3),......,(9,8),(9,9)\}$

Thus, $n(S) = 9 \times 9 = 81$

Let A = event that $x^2 - y^2$ is divisible by 2

Thus,

$A = \{((1,1),(1,3),(1,5),(1,7),(1,9),$
$(2,2),(2,4),(2,6),(2,8),(3,1),$
$(3,3),(3,5),(3,7),(3,9),(4,2),$
$(4,4),(4,6),(4,8),.................,$
$(9,1),(9,3),(9,5),(9,7),(9,9)\}$

So $n(A) = 41$

Hence, $P(A) = \frac{41}{81}.$

6. (c) Let E = event that the drawn balls are red,

F = event that the drawn balls are black,

G = event that the drawn balls are green.

Here each of the jars contain 14 balls.

Case-I: Each of the drawn ball is red.

Thus, P(both of the balls are the color red)

= P(E)

= (prob. of getting a red ball from jar A) × (prob. of getting a red ball from jar B)

$= \frac{2}{14} \times \frac{1}{14} = \frac{2}{196}$

Case-II: Each of the drawn ball is black.

Thus, P(both of the balls are the color black)

$= \frac{5}{14} \times \frac{4}{14} = \frac{20}{196}.$

Case-III: Each of the drawn ball is green.

Thus, P(both of the balls are the color green)

$= \frac{7}{14} \times \frac{9}{14} = \frac{63}{196}.$

Hence, the prob. (both the balls are of same color)

$P(E \cup F \cup G)$
$= P(E) + P(F) + P(G)$
(∵ the events E, F, G are mutually exclusive and exhaustive)
$= \frac{2}{196} + \frac{20}{196} + \frac{63}{196} = \frac{85}{196}.$

7. (b)

$P(\bar{A} \cap \bar{B}) = 0.5$
$\Rightarrow 1 - P(A \cup B) = 0.5$
$\Rightarrow P(A) + P(B) - P(A \cap B) = 0.5$
$\Rightarrow P(A \cap B) = 0.3 + 0.4 - 0.5 = 0.2.$

Now $P[B/(A \cup \bar{B})]$

$= \frac{P[B \cap (A \cup \bar{B})]}{P(A \cup \bar{B})} = \frac{P[(B \cap A) \cup (B \cap \bar{B})]}{P(A) + P(\bar{B}) - P(A \cap \bar{B})}$

$= \frac{P(B \cap A)}{P(A \cap B) + 1 - P(B)} \begin{bmatrix} \because B \cap \bar{B} = \varphi, \\ P(A) = P(A \cap B) + P(A \cap \bar{B}) \end{bmatrix}$

$= \frac{0.2}{0.2 + 1 - 0.4} = \frac{1}{4}.$

8. (*b*) Total number of drivers = 2000 + 4000 + 6000 = 12000.

Let

E = event that the driver is a motorcycle driver,

F = event that the driver is a car driver,

G = event that the driver is a truck driver,

A = event that the person meets with an accident.

Then,

$$P(E) = \frac{2000}{12000} = \frac{1}{6}, P(F) = \frac{4000}{12000} = \frac{1}{3} \text{ and}$$

$$P(G) = \frac{6000}{12000} = \frac{1}{2}.$$

ATQ,

$P(A/E) = 0.01, P(A/F) = 0.03, P(A/G) = 0.15.$

Prob. (the driver is a motorcycle driver given that he is met with an accident)

$= P(E/A)$

$$= \frac{P(E) \times P(A/E)}{P(E) \times P(A/E) + P(F) \times P(A/F) + P(G) \times P(A/G)}$$
(by Bayes' theorem)

$$= \frac{\frac{1}{6} \times 0.01}{\frac{1}{6} \times 0.01 + \frac{1}{3} \times 0.03 + \frac{1}{2} \times 0.15} = \frac{1}{52}.$$

9. (*a*) Let,

E = event that a person has a disease,

F = event that a person has no disease and

A = event that the blood test result is positive.

Then,

$$P(E) = 0.1 \times \frac{1}{100} = 0.001,$$

$$P(F) = 1 - P(E) = 1 - 0.001 = 0.999.$$

ATQ,

$P(A/E) = 99\% = 0.99, P(A/F) = 0.5\% = 0.005.$

Prob. (a person has the disease given that his test result is positive)

$= P(E/A)$

$$= \frac{P(E) \times P(A/E)}{P(E) \times P(A/E) + P(F) \times P(A/F)}$$
(by Bayes' theorem)

$$= \frac{0.001 \times 0.99}{0.001 \times 0.99 + 0.999 \times 0.005} = 0.165.$$

10. (*b*) Let n number of missiles should be fired at a target in order to make the probability of destroying the target at least 0.99.

Let A_r denotes the event "hitting the target at the rth trial and missing in the first $(r-1)$ trials where $r = 1, 2, 3, \dots, n$."

Then, $P(A_r) = \frac{1}{2} \times \frac{1}{2} \times \frac{1}{2} \times \dots \times \frac{1}{2} \times \frac{1}{2} = \frac{1}{2^r}$, for $r = 1, 2, \dots, n$.

Let E = event that the target is destroyed in n trials.

Then, clearly, $P(E)$

$= P(A_1 \cup A_2 \cup \dots \cup A_n)$

$= P(A_1) + P(A_2) + \dots + P(A_n)$

$[\because A_1, A_2, \dots, A_n \text{ are pairwise mutually exclusive}]$

$= \frac{1}{2^1} + \frac{1}{2^2} + \dots + \frac{1}{2^n} = 1 - \frac{1}{2^n}$

(using the sum formula of a finite G.P)

Now ATQ, $P(E) \geq 0.99$

or, $1 - \frac{1}{2^n} \geq 0.99$

or, $2^n \geq \frac{1}{0.01} = 100 = 10^2$

or, $n \log 2 \geq 2 \log 10 = 2$

or, $n \geq \frac{2}{\log 2} \approx 6.4$

Hence, the least numbers of missiles that should be fired = 7.

11. (*a*) Let X be the random variable denoting the number chosen from the numbers 1, 2, 3, 4,……,50. Then, the probability distribution of X is given by

X	1	2	3	....	....	50
P(X = x_i)	$\frac{1}{50}$	$\frac{1}{50}$	$\frac{1}{50}$	....	....	$\frac{1}{50}$

Therefore,

$$E(X) = 1 \times \frac{1}{50} + 2 \times \frac{1}{50} + 3 \times \frac{1}{50} + \dots + 50 \times \frac{1}{50}$$

$$= \frac{1}{50}(1 + 2 + 3 + \dots + 50)$$

$$= \frac{1}{50} \times \frac{50(50+1)}{2} = \frac{51}{2} = 25.5$$

Also let Y be the random variable denoting the number chosen from the numbers 1, 2, 3, 4,…

....,25. Then the probability distribution of Y is given by

Y	1	2	3	....	....	25
$P(Y = y_i)$	$\dfrac{1}{25}$	$\dfrac{1}{25}$	$\dfrac{1}{25}$	....	....	$\dfrac{1}{25}$

$$\therefore E(Y) = 1 \times \frac{1}{25} + 2 \times \frac{1}{25} + 3 \times \frac{1}{25} + ... + 25 \times \frac{1}{25}$$

$$= \frac{1}{25}(1 + 2 + 3 + ... + 25)$$

$$= \frac{1}{25} \times \frac{25(25 + 1)}{2} = 13.$$

Hence, expected value of the product of the two numbers chosen = $E(XY) = E(X) \ E(Y)$ = 25.5 + 13 = 38.5

(Since X and Y are independent, so $E(XY) = E(X) \ E(Y)$).

12. (*i*)-(*a*), (*ii*)-(*c*), (*iii*)-(*b*), (*iv*)-(*a*), (*v*)-(*c*), (*vi*)-(*b*).

(*i*)
$$\sum_i f(x_i) = 1 \Rightarrow k + 2k + 2k + 4k = 1$$

$$\Rightarrow k = \frac{1}{9}.$$

Hence, the probability distribution is given by

X	0	1	2	3	4
$P(X = x_i)$	0	$\dfrac{1}{9}$	$\dfrac{2}{9}$	$\dfrac{2}{9}$	$\dfrac{4}{9}$

Thus, $P(X = 0) = 0, P(X = 1) = \dfrac{1}{9},$

$P(X = 2) = \dfrac{2}{9}, P(X = 3) = \dfrac{2}{9},$

$P(X = 4) = \dfrac{4}{9}.$

(*ii*)
$$P(X < 3)$$
$$= P(X = 0 \text{ or } X = 1 \text{ or } X = 2)$$
$$= P(X = 0) + P(X = 1) + P(X = 2)$$
$$= 0 + \frac{1}{9} + \frac{2}{9} = \frac{1}{3}.$$

(*iii*)
$$P(X \geq 2)$$
$$= 1 - P(X < 2)$$
$$= 1 - P(X = 0 \text{ or } X = 1)$$
$$= 1 - \{P(X = 0) + P(X = 1)\}$$
$$= 1 - \left\{0 + \frac{1}{9}\right\} = \frac{8}{9}.$$

(*iv*)
$$P(2 < X \leq 4)$$
$$= P(X = 3 \text{ or } X = 4)$$
$$= P(X = 3) + P(X = 4)$$
$$= \frac{2}{9} + \frac{4}{9} = \frac{2}{3}.$$

(*v*)
$$P(2 < X | X \leq 4)$$

$$= \frac{P(2 < X \leq 4)}{P(X \leq 4)} = \frac{\dfrac{2}{3}}{1} = \frac{2}{3}$$

$$\left(\begin{array}{l} \because P(X \leq 4) = P(X = 0) + P(X = 1) \\ \quad + P(X = 2) + P(X = 3) \\ \quad + P(X = 4) \\ \quad = \text{sum of probabilities} = 1 \end{array} \right)$$

(*vi*) $P(X \leq 0) = P(X = 0) = 0 < \dfrac{1}{2}$

$$P(X \leq 1) = P(X = 0) + P(X = 1) = 0 + \frac{1}{9} < \frac{1}{2}$$

$$P(X \leq 2) = P(X = 0) + P(X = 1) + P(X = 2)$$
$$= 0 + \frac{1}{9} + \frac{2}{9} = \frac{1}{3} < \frac{1}{2}$$

$$P(X \leq 3) = P(X = 0) + P(X = 1) + P(X = 2)$$
$$+ P(X = 3)$$
$$= 0 + \frac{1}{9} + \frac{2}{9} + \frac{2}{9} = \frac{5}{9} > \frac{1}{2}$$

Thus, the minimum value of x for which $P(X \leq x) > \dfrac{1}{2}$ is "3."

13. (*c*)

$$\sum_{i=1}^{\infty} iP(X \geq i)$$

$$= P(X \geq 1) + 2P(X \geq 2) + 3P(X \geq 3) + ...\infty$$

$$= \{P(X = 1) + P(X = 2) + P(X = 3) + ...\infty\}$$
$$+ 2\{P(X = 2) + P(X = 3) + P(X = 4) + ...\infty\}$$
$$+ 3\{P(X = 3) + P(X = 4) + ...\infty\} + ...\infty$$

$$= 1 \times P(X = 1) + (1 + 2)P(X = 2) + (1 + 2 + 3)P(X = 3)$$
$$+ (1 + 2 + 3 + 4)P(X = 4) + ...\infty$$

$$= \sum_{n=1}^{\infty} (1 + 2 + 3 + ... + n)P(X = n)$$

$$= \frac{1}{2} \sum_{n=1}^{\infty} n(n + 1)P(X = n)$$

$$= \frac{1}{2} \left\{ \sum_{n=1}^{\infty} n^2 \times P(X = n) + \sum_{n=1}^{\infty} n \times P(X = n) \right\}$$

$$= \frac{1}{2} \{E(X^2) + E(X)\} = \frac{1}{2}\{3 + 1\} = 2.$$

14. (*a*)

$E(Z)$

$= E(aX - bY) = aE(X) - bE(Y) = 3a - 2b$

$\therefore E(Z) = 0 \Rightarrow 3a - 2b = 0 \Rightarrow a = \dfrac{2}{3}b.$

$Var(Z)$

$= Var(aX - bY) = a^2 Var(X) + b^2 Var(Y)$

$= a^2 \times 1 + b^2 Var(Y) = \dfrac{4b^2}{9} + b^2 Var(Y)$

Now $Var(Z) = 1$

$\Rightarrow \dfrac{4b^2}{9} + b^2 Var(Y) = 1 \Rightarrow Var(Y) = \dfrac{9 - 4b^2}{9b^2}.$

Again $E(X) - E(Y) = 3 - 2 = 1$

$\Rightarrow E(X - Y) = 1 = E(1) \Rightarrow X - Y = 1$

$\Rightarrow Y = X - 1 \Rightarrow Var(Y) = Var(X) = 1$

$\Rightarrow \dfrac{9 - 4b^2}{9b^2} = 1 \Rightarrow b^2 = \dfrac{9}{13} \Rightarrow b = \dfrac{3}{\sqrt{13}}.$

$\therefore a = \dfrac{2}{3}b = \dfrac{2}{3} \times \dfrac{3}{\sqrt{13}} = \dfrac{2}{\sqrt{13}}.$

15. (*a*) Let X be the random variable denoting the number of times even number occurs.

Then, X follows binomial distribution with $n = 10$ and p = prob. (getting an even number in a single trial).

ATQ, $P(X = 5) = 2 \times P(X = 4)$

or, ${}^{10}C_5 p^5 (1 - p)^{10-5} = 2 \times {}^{10}C_4 p^4 (1 - p)^{10-4}$

or, $\dfrac{10 \times 9 \times 8 \times 7 \times 6}{5!} \times p = 2 \times \dfrac{10 \times 9 \times 8 \times 7}{4!} \times (1 - p)$

or, $\dfrac{3p}{5} = (1 - p)$

or, $p = \dfrac{5}{8}.$

Hence, the probability that an even number will not appear at all in 10 independent throws of the die

$= P(X = 0) = {}^{10}C_0 p^0 (1 - p)^{10-0}$

$= (1 - p)^{10} = \left(1 - \dfrac{5}{8}\right)^{10} = \left(\dfrac{3}{8}\right)^{10}.$

16. (*c*)

Since the jaywalker crosses the street twice per days for 50 days, so he crosses the road 100 times in all.

Let X be the random variable denoting the number of hits by the car.

Then, X follows binomial distribution with $n = 100$ and p = prob. (the jaywalker is hit by the car) = 0.01.

Therefore, the required probability

$= P(X = 0) = {}^{100}C_0 (0.01)^0 (1 - 0.01)^{100-0} = (0.99)^{100}.$

17. (*d*)

Let X be the random variable denoting the number of wars in the interval $(0, 25)$.

Then, clearly X follows the Poisson distribution with the parameter, $m = \dfrac{1}{15} \times 25 = \dfrac{5}{3}.$

Therefore, prob. (there will be no wars in 25 years)

$= P(X = 0) = \dfrac{e^{-m} \times m^0}{0!} = e^{-\frac{5}{3}}.$

18. (*a*)

Let X be the random variable denoting the number of particles emitted in the given interval.

Then, X follows the Poisson distribution with parameter, $m = 2.5 \times 4 = 10$.

Therefore, prob. (3 or more particles will be emitted in an interval of 4 seconds)

$= P(X \geq 3) = 1 - P(X < 3)$

$= 1 - \{P(X = 0) + P(X = 1) + P(X = 2)\}$

$= 1 - \left\{\dfrac{e^{-m} \times m^0}{0!} + \dfrac{e^{-m} \times m^1}{1!} + \dfrac{e^{-m} \times m^2}{2!}\right\}$

$= 1 - e^{-m}\left(1 + m + \dfrac{m^2}{2}\right) = 1 - 61e^{-10}.$

19. (*c*) Let X, Y be the random variables denoting the number of passengers who failed to turn up at the airport A and B, respectively.

Then, X is a Poisson variate with the parameter, $m = 10 \times 0.1 = 1$ and Y is a Poisson variate with the parameter, $m = 20 \times 0.1 = 2$.

So prob. (airline A is overbooked)

$= P(X = 0) = \dfrac{e^{-1} \times 1^0}{0!} = \dfrac{1}{e}.$

Again prob. (airline B is overbooked)

$= P(Y = 0) + P(Y = 1) = \dfrac{e^{-2} \times 2^0}{0!} + \dfrac{e^{-2} \times 2^1}{1!} = \dfrac{3}{e^2}.$

Now, $\dfrac{3}{e^2} - \dfrac{1}{e} = \dfrac{3 - e}{e^2} > 0 (\because 2 < e < 3). \therefore \dfrac{3}{e^2} > \dfrac{1}{e}.$

Hence, airline B is more often overbooked.

20. (d) Let X be the random variable denoting the number of members who favor the candidate.

Then, X follows binomial distribution with $n = 200$ and p = prob. (a member favors the candidate) = 45% = 0.45.

Therefore, mean = $np = 200 \times 0.45 = 90$ and Variance = $np(1-p) = 90 \times (1-0.45) = 49.5$. So standard deviation = $\sqrt{\text{variance}} = \sqrt{49.5} = 7.04$.

21. (a) Given mean = 12 hours and standard deviation = 3 hours.

Let X be the random variable denoting the length of life of dry battery cells.

Then, $Z = \dfrac{X\text{-mean}}{\text{S.D}} = \dfrac{15-12}{3} = 1$.

Therefore, $P(X > 15)$

$= P(Z > 1) = P(0 < Z < \infty) - P(0 < Z < 1)$

$= 0.5 - 0.3413 = 0.1587 = 15.87\%$

Previous Years Solved Papers (2000-2018)

1. E_1 and E_2 are events in a probability space satisfying the following constraints:

$P(E_1) = P(E_2), P(E_1 \cup E_2) = 1$.

If E_1 and E_2 are independent, then $P(E_1)$ =?

(a) 0   (b) $\dfrac{1}{4}$ (c) $\dfrac{1}{2}$ (d) 1

[GATE 2000]

2. In a manufacturing plant, the probability of making a defective bolt is 0.1. The mean and standard deviation of defective bolts in a total of 900 bolts are, respectively

(a) 90 and 9   (b) 9 and 90
(c) 81 and 9   (d) 9 and 81

[GATE 2000]

3. Seven car accidents occurred in a week. What is the probability that they all occurred on the same day?

(a) $\dfrac{1}{7^7}$ (b) $\dfrac{1}{7^6}$ (c) $\dfrac{1}{2^7}$ (d) $\dfrac{7}{2^7}$

[GATE 2001]

4. Four fair coins are tossed simultaneously. The probability that at least one heads and at least one tails turn up is

(a) $\dfrac{1}{16}$ (b) $\dfrac{1}{8}$ (c) $\dfrac{7}{8}$ (d) $\dfrac{15}{16}$

[GATE 2002]

5. A box contains ten screws, three of which are defective. Two screws are drawn at random with the replacement. The probability that none of the two screws is defective will be

(a) 100%   (b) 50%
(c) 49%   (d) None of these

[CE GATE 2003]

6. Let $P(E)$ denote the probability of an event E. Given $P(A) = 1$, $P(B) = 1/2$, the values of $P(A/B)$ and $P(B/A)$, respectively, are

(a) 1/4, 1/2   (b) 1/2, 1/4
(c) 1/2, 1   (d) 1, 1/2

[CS GATE 2003]

7. A program consists of two modules executed sequentially. Let $f_1(t)$ and $f_2(t)$, respectively, denote the probability density functions of time taken to execute the two modules. The probability density function of the overall time taken to execute the program is given by

(a) $f_1(t) + f_2(t)$ (b) $\displaystyle\int_0^t f_1(x) f_2(x)\,dx$

(c) $\displaystyle\int_0^t f_1(x) f_2(t-x)\,dx$ (d) $\max\{f_1(t), f_2(t)\}$

[CS GATE 2003]

8. A box contains five black and five red balls. Two balls are randomly picked one after another from the box without a replacement. The probability for both balls being red is

(a) $\dfrac{1}{90}$ (b) $\dfrac{1}{2}$ (c) $\dfrac{19}{90}$ (d) $\dfrac{2}{9}$

[ME GATE 2003]

9. A hydraulic structure has four gates which operate independently. The probability of failure of each gate is 0.2. Given that gate 1 has failed, the probability that both gates 2 and 3 will fail is

(a) 0.240   (b) 0.200   (c) 0.040   (d) 0.008

[CE GATE 2004]

10. If a fair coin is tossed four times. What is the probability that two heads and two tails will result

(a) 3/8   (b) 1/2   (c) 5/8   (d) 3/4

[CS GATE 2004]

11. An examination paper has 150 multiple-choice questions of one mark each, with each question having four choices. Each incorrect answer fetches – 0.25 marks. Suppose 1000 students choose all their answers randomly with uniform probability. The sum total of the expected marks obtained by all these students is

(a) 0 (b) 2550 (c) 7525 (d) 9375
[CS GATE 2004]

12. Two n bit binary strings, S_1 and S_2 are chosen randomly with uniform probability. The probability that the Hamming distance between these strings (the number of bit positions where the two strings differ) is equal to d is

(a) $\frac{^nC_d}{2^n}$ (b) $\frac{^nC_d}{2^d}$ (c) $\frac{d}{2^n}$ (d) $\frac{1}{2^d}$
[CS GATE 2004]

13. From a pack of regular playing cards, two cards are drawn at random. What is the probability that both cards will be kings, if first card is NOT placed?

(a) $\frac{1}{26}$ (b) $\frac{1}{52}$ (c) $\frac{1}{169}$ (d) $\frac{1}{221}$
[ME GATE 2004]

14. A point is randomly selected with uniform probability in the xy plane within the rectangle with corners at (0, 0), (1, 0), (1, 2), and (0, 2). If p is the length of the position vector of the point, the expected value of p^2 is

(a) 2/3 (b) 1 (c) 4/3 (d) 5/3
[CS GATE 2004]

15. In a population of N families, 50% of the families have three children, 30% of families have two children and the remaining families have one child. What is the probability that a randomly picked child belongs to a family with two children?

(a) 3/23 (b) 6/23 (c) 3/10 (d) 3/5
[IT GATE 2004]

16. The following data about the flow of liquid was observed in a continuous chemical process plant:

Flow rate (liters/sec)	Frequency
7.5-7.7	1
7.7-7.9	5
7.9-8.1	35
8.1-8.3	13
8.3-8.5	12
8.5-8.7	10

Mean flow rate of the liquid is?

(a) 8 liters/sec (b) 8.06 liters/sec
(c) 8.16 liters/sec (d) 8.26 liters/sec
[GATE 2004]

17. If P and Q are two random events, then which of the following is TRUE?

(a) Independence of P and Q implies that probability $(P \cap Q) = 0$
(b) Probability $(P \cup Q) \geq$ Probability (P) + Probability(Q)
(c) If P and Q are mutually exclusive, then they must be independent
(d) Probability $(P \cap Q) \leq$ Probability (P)
[EE GATE 2005]

18. A single die is thrown twice. What is the probability that the sum is neither 8 nor 9?

(a) 1/9 (b) 5/36 (c) 1/4 (d) 3/4
[ME GATE 2005]

19. A fair die is rolled twice. The probability that an odd number will follow an even number is

(a) $\frac{1}{2}$ (b) $\frac{1}{6}$ (c) $\frac{1}{3}$ (d) $\frac{1}{4}$
[EC GATE-2005]

20. A lot has 10% defective items. Ten items are chosen randomly from this lot. The probability that exactly two of the chosen items are defective is

(a) 0.0036 (b) 0.1937
(c) 0.2234 (d) 0.3874
[ME GATE 2005]

21. A bag contains 10 blue marbles, 20 black marbles, and 30 red marbles. A marble is drawn from the bag, its color is recorded and it is then put back in the bag. This process is repeated three times. The probability that no two of the marbles drawn have the same color is?

(a) $\frac{1}{36}$ (b) $\frac{1}{6}$ (c) $\frac{1}{4}$ (d) $\frac{1}{3}$
[IT GATE 2005]

22. A fair coin is tossed three times in succession. If the first toss produces heads, then the probability of getting exactly two heads in three tosses is

(a) 1/8   (b) 1/2   (c) 3/8   (d) 3/4

[EE GATE 2005]

23. The life of a bulb (in hours) is a random variable with an exponential distribution $f(t) = \alpha e^{-\alpha t}, 0 \le t < \infty$. The probability that its value lies between 100 and 200 hours is

(a) $e^{-100\alpha} - e^{-200\alpha}$ (b) $e^{-100} - e^{-200}$

(c) $e^{-100\alpha} + e^{-200\alpha}$ (d) $-e^{-100\alpha} + e^{-200\alpha}$

[PI GATE 2005]

24. Two dice are thrown simultaneously. The probability that the sum of numbers on both exceeds 8 is

(a) $\dfrac{4}{36}$ (b) $\dfrac{7}{36}$ (c) $\dfrac{9}{36}$ (d) $\dfrac{10}{36}$

[PI GATE 2005]

25. Which one of the following statements is NOT true?

(a) The measure of skewness is dependent upon the amount of dispersion

(b) In a symmetric distribution, the values of mean, mode and median are the same

(c) In a positively skewed distribution: mean > median > mode

(d) In a negatively skewed distribution: mode > mean > median

[CE GATE-2005]

26. Let $f(x)$ be the continuous probability density function of a random variable X. The probability that $a < X \le b$ is

(a) $f(b - a)$ (b) $f(b) - (a)$

(c) $\int_{a}^{b} f(x)dx$ (d) $\int_{a}^{b} xf(x)dx$

[CS GATE 2005]

27. Two fair dice are rolled and the sum r of the numbers turned up is considered. Which of the following is correct?

(a) $P(r > 6) = 1/6$

(b) $P(r/3 \text{ is an integer}) = 5/6$

(c) $P(r = 8 \mid r/4 \text{ is an integer}) = 5/9$

(d) $P(r = 6 \mid r/5 \text{ is an integer}) = 1/18$

[EE GATE 2006]

28. There are 25 calculators in a box. Two of them are defective. Suppose five calculators are randomly picked for inspection (*i.e.*, each has the same chance of being selected). What is the probability that only one of the defective calculators will be included in the inspection?

(a) $\dfrac{1}{2}$ (b) $\dfrac{1}{3}$ (c) $\dfrac{1}{4}$ (d) $\dfrac{1}{5}$

[CE GATE 2006]

29. Three companies X, Y, and Z supply computers to a university. The percentage of computers supplied by them and the probability of those being defective are tabulated below:

Company	% of computer	Probability of being supplied defective
X	60	0.01
Y	30	0.02
Z	10	0.03

Given that a computer is defective, the probability that it was supplied by Y is

(a) 0.1   (b) 0.2   (c) 0.3   (d) 0.4

[EC GATE 2006]

30. A box contains 20 defective items and 80 non-defective items. If two items are selected at random without replacement, what will be the probability that both items are defective?

(a) $\dfrac{1}{5}$ (b) $\dfrac{1}{25}$ (c) $\dfrac{20}{99}$ (d) $\dfrac{19}{495}$

[ME GATE 2006]

31. For each element in a set of size $2n$, an unbiased coin is tossed. All the $2n$ coin tossed are independent. An element is chosen if the corresponding coin toss were heads. The probability that exactly n elements are chosen is

(a) $\dfrac{^{2n}C_n}{4^n}$ (b) $\dfrac{^{2n}C_n}{2^n}$ (c) $\dfrac{1}{^{2n}C_n}$ (d) $\dfrac{1}{2}$

[CS GATE 2006]

32. Consider the continuous random variable with probability density function

$f(t) = 1 + t$ for $-1 \le t \le 0$

$= 1 - t$ for $0 \le t \le 1$

The standard deviation of the random variable is

(a) $\dfrac{1}{\sqrt{3}}$ (b) $\dfrac{1}{\sqrt{6}}$ (c) $\dfrac{1}{3}$ (d) $\dfrac{1}{6}$

[ME GATE 2006]

33. A class of first year *B. Tech.* students is composed of four batches *A, B, C,* and *D,* each consisting of 30 students. It is found that the sessional marks of students in Engineering Drawing in batch *C* have a mean of 6.6 and standard deviation of 2.3. The mean and standard deviation of the marks for the entire class are 5.5 and 4.2, respectively. It is decided by the course instructor to normalize the marks of the students of all batches to have the same mean and standard deviation as that of the entire class. Due to this, the marks of a student in batch C are changed from 8.5 to

    (a) 6.0　　(b) 7.0　　(c) 8.0　　(d) 9.0
    [CE GATE 2006]

34. A probability density function is of the form $p(x) = Ke^{-\alpha|x|}$, $x \in (-\infty, \infty)$. The value of K is

    (a) 0.5　　(b) 1　　(c) 0.5α　　(d) α
    [EC GATE 2006]

35. A loaded dice has following probability distribution of occurrences:

Dice value	1	2	3	4	5	6
Probability	1/4	1/8	1/8	1/8	1/8	1/4

    If three identical dice as the above are thrown, the probability of occurrence of values 1, 5, and 6 on the three dice is

    (a) same as that of occurrence of 3, 4, 5
    (b) same as that of occurrence of 1, 2, 5
    (c) 1/128
    (d) 5/8
    [EE GATE 2007]

36. Let *X* and *Y* be two independent random variables. Which one of the relations between Expectation (*E*), Variance (*V*), and Covariance (Cov) given below is false?
    (a) $E(XY) = E(X)E(Y)$
    (b) Cov $(X, Y) = 0$
    (c) $\text{Var}(X + Y) = \text{Var}(X) + \text{Var}(Y)$
    (d) $E(X^2Y^2) = (E(X))^2(E(Y))^2$
    [ME GATE 2007]

37. If the standard deviation of the spot speed of vehicles on a highway is 8.8 kmph and the mean speed of the vehicles is 33 kmph, the coefficient of variation in speed is
    (a) 0.1517　(b) 0.1867
    (c) 0.2666　(d) 0.3646
    [CE GATE 2007]

38. An examination consists of two papers, Paper 1 and Paper 2. The probability of failing in Paper 1 is 0.3 and that in Paper 2 is 0.2. Given that a student has failed in Paper 2, the probability of failing in Paper 1 is 0.6. The probability of a student failing in both the papers is

    (a) 0.5　　(b) 0.18　　(c) 0.12　　(d) 0.06
    [EC GATE 2007]

39. Two cards are drawn at random in succession with replacement from a deck of 52 well shuffled cards. The the probability of getting both "aces" is

    (a) $\dfrac{1}{169}$　(b) $\dfrac{2}{169}$　(c) $\dfrac{1}{13}$　(d) $\dfrac{2}{13}$
    [PI GATE 2007]

40. If *X* is a continuous random variable whose probability density function is given by

    $$f(x) = \begin{cases} k(5x - 2x^2), & 0 \le x \le 2 \\ 0, & \text{otherwise} \end{cases}$$

    Then, $P(X > 1)$ is
    (a) 3/14　(b) 4/5　　(c) 14/17　(d) 17/28
    [PI GATE 2007]

41. The random variable *X* takes on the values 1, 2, and 3 with probabilities $\dfrac{2+5p}{5}, \dfrac{1+3p}{5}$ and $\dfrac{1.5+2p}{5}$, respectively. Then the values of "*p*" and *E(X)* are, respectively
    (a) 0.05, 1.87　　　　(b) 1.90, 5.87
    (c) 0.05, 1.10　　　　(d) 0.25, 1.40
    [PI GATE 2007]

42. A coin is tossed four times. What is the probability of getting heads exactly three times?

    (a) 1/4　　(b) 3/8　　(c) 1/2　　(d) 3/4
    [ME GATE 2008]

43. A random variable is uniformly distributed over the interval (2, 10). Its variance will be

    (a) 16/3　(b) 6　　(c) 256/9　(d) 36
    [IN GATE 2008]

44. *X* is a uniformly distributed random variable that takes values between 0 and 1. The value of $E(X^3)$ will be

    (a) 0　　(b) 1/8　　(c) 1/4　　(d) 1/2
    [EE GATE 2008]

45. If probability density function of a random variable X is given by

$$f(x) = \begin{cases} x^2, & -1 \le x \le 1 \\ 0, & \text{otherwise} \end{cases},$$

then the percentage probability $P\left(-\dfrac{1}{3} \le X \le \dfrac{1}{3}\right)$ is

(a) 0.247　(b) 2.47　(c) 24.7　(d) 247

[CE GATE 2008]

46. Let X be a random variable following normal distribution with mean 1 and variance 4. Let Y be another normal variable with mean -1 and variance unknown. If $P(X \le -1) = P(Y \ge 2)$ the standard deviation of Y is

(a) 3　(b) 2　(c) $\sqrt{2}$　(d) 1

[CS GATE 2008]

47. $P(X) = Me^{-2|x|} + Ne^{-3|x|}$ is the probability density function for the real random variable X, over the entire x-axis where M and N are both positive real numbers. The equation relating M and N is

(a) $M + \dfrac{2}{3}N = 1$　(b) $2M + \dfrac{1}{3}N = 1$

(c) $M + N = 1$　(d) $M + N = 3$

[IN GATE 2008]

48. In a game, two players X and Y toss a coin alternatively. Whoever gets "heads" first, wins the game and the game is terminated. Assuming that player X starts the game, the probability of player X winning the game is

(a) 1/3　(b) 1/2　(c) 2/3　(d) 3/4

[PI GATE 2008]

49. A fair coin is tossed 10 times. What is the probability that ONLY the first two tosses will yield heads?

(a) $\left(\dfrac{1}{2}\right)^2$　(b) $^{10}C_2\left(\dfrac{1}{2}\right)^2$

(c) $\left(\dfrac{1}{2}\right)^{10}$　(d) $^{10}C_2\left(\dfrac{1}{2}\right)^{10}$

[EC GATE 2009]

50. If two fair coins are flipped and at least one of the outcomes is known to be heads, what is the probability that both outcomes are heads?

(a) 1/3　(b) 1/4　(c) 1/2　(d) 2/3

[CS GATE 2009]

51. The standard deviation of a uniformly distributed random variable between 0 and 1 is

(a) $\dfrac{1}{\sqrt{12}}$　(b) $\dfrac{1}{\sqrt{3}}$　(c) $\dfrac{5}{\sqrt{12}}$　(d) $\dfrac{7}{\sqrt{12}}$

[ME GATE 2009]

52. An unbalanced dice (with 6 faces, numbered from 1 to 6) is thrown. The probability that the face value is odd is 90% of the probability that the face value is even. The probability of getting any even numbered face is the same. If the probability that the face is even given that it is greater than 3 is 0.75, which one of the following options is closest to the probability that the face value exceeds 3?

(a) 0.453　(b) 0.468　(c) 0.485　(d) 0.492

[CS GATE 2009]

53. Consider two independent random variables X and Y with identical distributions. The variables X and Y take values 0, 1, and 2 with probability 1/2, 1/4, and 1/4, respectively. What is the conditional probability $P(X + Y = 2 / X - Y = 0)$?

(a) 0　(b) 1/16　(c) 1/6　(d) 1

[EC GATE 2009]

54. If three coins are tossed simultaneously, the probability of getting at least one heads is

(a) 1/8　(b) 3/8　(c) 1/2　(d) 7/8

[ME GATE 2009]

55. A discrete random variable X takes value from 1 to 5 with probabilities as shown in the table below:

k	1	2	3	4	5
$P(X = k)$	0.1	0.2	0.4	0.2	0.1

A student calculates the mean of X as 3.5 and her teacher calculates the variance of X as 1.5. Which of the following statements is true?

(a) Both the student and the teacher are right
(b) Both the student and the teacher are wrong
(c) The student is wrong but the teacher is right
(d) The student is right but the teacher is wrong

[EC GATE 2009]

56. A screening test is carried out to detect a certain disease. It is found that 12% of the positive reports and 15% of the negative reports are incorrect. Assuming that the probability of a person getting positive report is 0.01, find the probability that a person tested gets an incorrect report is
(a) 0.0027   (b) 0.0173
(c) 0.1497   (d) 0.2100

[IN GATE 2009]

57. The standard normal cumulative probability function (probability from $-\infty$ to x_n) can be approximated as

$$F f(x_n) = \frac{1}{1 + \exp\left(-1.7255 x_n \left|x_n\right|^{0.12}\right)}$$

where x_n = standard normal deviate. If mean and standard deviation of annual precipitation are 102 cm and 27 cm, respectively, then the probability that the annual precipitation will be between 90 cm and 102 cm is

(a) 66.7%   (b) 50.0%   (c) 33.3% (d) 16.7%

[CE GATE 2009]

58. Consider a company that assembles computers. The probability of a faulty assembly of any computer is p. The company therefore subjects each computer to a testing process. This testing process gives the correct result for any computer with a probability of q. What is the probability of a computer being declared faulty?
(a) $pq + (1-p)(1-q)$
(b) $(1-q)p$
(c) $(1-p)q$
(d) pq

[CS GATE 2010]

59. A box contains two washers, three nuts, and four bolts. Items are drawn from the box at random one at a time without replacement. The probability of drawing two washers first followed by three nuts and subsequently the four bolts is
(a) 2/315   (b) 1/630
(c) 1/1260   (d) 1/2520

[ME GATE 2010]

60. A fair coin tossed independently four times. The probability of the event "the number of times heads show up is more than the number of times tails show up" is
(a) $\frac{1}{16}$ (b) $\frac{1}{8}$ (c) $\frac{1}{4}$ (d) $\frac{5}{16}$

[EC GATE 2010]

61. A box contains four white balls and three red balls. In succession, two balls are randomly selected and removed from the box. Given that the first removed ball is white, the probability that the second removed ball is red is
(a) 1/3   (b) 3/7   (c) 1/2   (d) 4/7

[EE GATE 2010]

62. If a random variable X satisfies the Poisson's distribution with a mean value of 2, then the probability that $X \geq 2$ is?
(a) $2e^{-2}$ (b) $1 - 2e^{-2}$
(c) $3e^{-2}$ (d) $1 - 3e^{-2}$

[PI GATE 2010]

63. A fair dice is tossed two times. The probability that the second toss results in a value that is higher than the first toss is

(a) 2/36   (b) 2/6   (c) 5/12   (d) 1/2

[EC GATE 2011]

64. There are two containers with one containing four red and three green balls and the other containing three blue and four green balls. One ball is drawn at random from each container. The probability that one of the balls is red and the other is blue will be

(a) 1/7   (b) 9/49   (c) 12/49   (d) 3/7

[CE GATE 2011]

65. If the difference between the expectation of the square of a random variable $[E(X^2)]$ and the square of the expectation of the random variable $(E[x])^2$ is denoted by R, then
(a) $R = 0$ (b) $R < 0$ (c) $R \geq 0$ (d) $R > 0$

[CS GATE 2011]

66. Consider the finite sequence of random values $X = \{x_1, x_2, x_3, \ldots, x_n\}$. Let μ_X be the mean and σ_X be the standard deviation of X. Let another finite sequence Y of equal length be derived from this as, $y_i = ax_i + b$, where a and b are positive constants. Let μ_Y be the mean and σ_Y be the standard deviation of this sequence. Which one of the following statements is INCORRECT?

(a) Index position of mode of X in X is the same as the index position of mode of Y in Y

(b) Index position of median of X in X is the same as the index position of median of X in X

(c) $\mu_y = a\mu_x + b$ \qquad (d) $\sigma_y = a\sigma_x + b$

[CS GATE 2011]

67. An unbiased coin is tossed five times. The outcome of each toss is either heads or tails. The probability of getting at least one heads is

(a) $\dfrac{1}{32}$ \quad (b) $\dfrac{13}{32}$ \quad (c) $\dfrac{16}{32}$ \quad (d) $\dfrac{31}{32}$

[ME GATE 2011]

68. A deck of five cards (each carrying a distinct number from 1 to 5) is shuffled thoroughly. Two cards are then removed one at a time from the deck. What is the probability that the two cards are selected with the number on the first card being one higher than the number on the second card?

(a) 1/5 \quad (b) 4/25 \quad (c) 1.4 \quad (d) 2/5

[CS GATE 2011]

69. It is estimated that the average number of events during a year is three. What is the probability of occurrence of not more than two events over a two year duration? (Assume that the number of events follow a Poisson distribution).

(a) 0.052 \quad (b) 0.062 \quad (c) 0.072 \quad (d) 0.082

[PI GATE 2011]

70. The box 1 contains chips numbered 3, 6, 9, 12, and 15. The box 2 contains chips numbered 6, 11, 16, 21, and 26. Two chips, one from each box are drawn at random. The numbers written on these chips are multiplied. The probability for the product to be an even number is _____?

(a) 6/25 \quad (b) 2/5 \quad (c) 3/5 \quad (d) 19/25

[IN GATE 2011]

71. A fair coin is tossed until heads appears for the first time. The probability that the number of required tosses being odd is

(a) 1/3 \quad (b) 1/2 \quad (c) 2/3 \quad (d) 3/4

[EC, EE, IN GATE 2012]

72. In an experiment, positive and negative values are equally likely to occur. The probability of obtaining at most one negative value in five trials is

(a) $\dfrac{1}{32}$ \quad (b) $\dfrac{2}{32}$ \quad (c) $\dfrac{3}{32}$ \quad (d) $\dfrac{6}{32}$

[CE GATE 2012]

73. Suppose a fair six-sided die is rolled once. If the value on the die is 1, 2, or 3 the die is rolled a second time. What is the probability that the sum of total values that turn up is at least 6?

(a) 10/21 \quad (b) 5/12 \quad (c) 2/3 \quad (d) 1/6

[CS GATE 2012]

74. A box contains four red balls and six black balls. Three balls are selected randomly from the box one after another, without replacement. The probability that the selected set contains one red ball and two black balls is

(a) 1/20 \quad (b) 1/12 \quad (c) 3/10 \quad (d) 1/2

[PI, ME, GATE 2012]

75. Consider a random variable X that takes values 1 and –1 with probability 0.5 each. The values of the cumulative distribution function $F(x)$ at $x = -1$ and 1 are, respectively

(a) 0 and 0.5 \qquad (b) 0 and 1

(c) 0.5 and 1 \qquad (d) 0.25 and 0.75

[CS GATE 2012]

76. An automobile plant contracted to buy shock absorbers from two suppliers X and Y. X supplies 60% and Y supplies 40% of the shock absorbers. All shock absorbers are subjected to a quality test. The one that pass the quality test are considered reliable. Of X's shock absorbers, 96% are reliable; of Y's shock absorbers, 72% are reliable. The probability that a randomly chosen shock absorber, which is found to be reliable, is made by Y is

(a) 0.288 \quad (b) 0.334 \quad (c) 0.667 \quad (d) 0.720

[ME, PI GATE 2012]

77. Find the value of λ such that function $f(x)$ is valid probability density function

$$f(x) = \begin{cases} \lambda(x-1)(2-x) & \text{for } 1 \le x \le 2 \\ 0 & \text{otherwise} \end{cases}$$

[CE GATE 2013]

78. Suppose p is the number of cars per minute passing through a certain road junction between 5 p.m and 6 p.m, and p has a Poisson distribution with mean of 3. What is the probability of observing fewer than 3 cars during any given minute in this interval?

(a) $\frac{8}{2e^3}$ (b) $\frac{9}{2e^3}$ (c) $\frac{17}{2e^3}$ (d) $\frac{26}{2e^3}$

[CS GATE 2013]

79. A continuous random variable X has a probability density function $f(x) = e^{-x}, 0 < x < \infty$. Then, $P\{X > 1\}$ is

(a) 0.368 (b) 0.5 (c) 0.632 (d) 1.0

[IN, EE GATE 2013]

80. Let X be a normal random variable with mean of 1 and variance of 4. The probability $P\{X > 0\}$ is

(a) 0.5
(b) greater than zero and less 0.5
(c) greater than 0.5 and less than 1.0
(d) 1.0

[ME GATE 2013]

81. A nationalized bank has found that the daily balance available in its savings accounts follows a normal distribution with a mean of Rs 500 and a standard deviation of Rs. 50. The percentage of savings account holders, who maintain an average daily balance more than Rs. 500 is _____.

[ME GATE 2014]

82. A box contains 25 parts of which 10 are defective. Two parts are being drawn simultaneously in a random manner from the box. The probability of both the parts being good is

(a) $\frac{7}{20}$ (b) $\frac{45}{125}$ (c) $\frac{25}{29}$ (d) $\frac{5}{9}$

[ME GATE 2014]

83. In a housing society, half of the families have a single child per family, while the remaining half have two children per family. The probability that a child picked at random has a sibling is _____.

[EC GATE 2014]

84. A fair (unbiased) coin was tossed four times in succession and resulted in the following

outcomes: (i) Heads, (ii) Heads, (iii) Heads, (iv) Heads. The probability of obtaining a "Tails," when the coin is tossed again is

(a) 0 (b) $\frac{1}{2}$ (c) $\frac{4}{5}$ (d) $\frac{1}{5}$

[CE GATE 2014]

85. A group consists of equal number of men and women. Of this group 20% of the men and 50% of the women are unemployed. If a person is selected at random from this group, the probability of the selected person being employed is _____.

[ME GATE 2014]

86. An unbiased coin is tossed an infinite number of times. The probability that the fourth heads appears at the tenth toss is

(a) 0.067 (b) 0.073 (c) 0.082 (d) 0.091

[EC GATE 2014]

87. A fair coin is tossed n times. The probability that the difference between the number of heads and tails is $(n-3)$ is

(a) 2^{-2} (b) 0
(c) $^nC_{n-3}2^{-n}$ (d) 2^{-n+3}

[EE GATE 2014]

88. Four fair six-sided dice are rolled. The probability that the sum of the results being 22 is $x/1296$. The value of "x" is _____.

[CS GATE 2014]

89. Consider a die with the property that the probability of a face with n dots showing up is proportional to n. The probability of the face with three dots showing up is _____.

[EE GATE 2014]

90. Let S be a sample space and two mutually exclusive events A and B be such that $A \cup B = S$. If $P(.)$ denotes the probability of the event, the maximum value of $P(A) P(B)$ is _____.

[CS GATE 2014]

91. Parcels from sender S to receiver R pass sequentially through two post-offices. Each post-office has a probability 1/5 of losing an incoming parcel, independently of all other parcels. Given at a parcel is lost, the probability that it was lost by the second post-office is _____.

[EC GATE 2014]

92. A machine produces 0, 1 or 2 defective pieces in a day with associated probability of 1/6, 2/3, and 1/6, respectively. The mean value and the variance of the number of defective pieces produced by the machine in a day, respectively, are

(a) 1 and 1/3   (b) 1/3 and 1
(c) 1 and 4/3   (d) 1/3 and 4/3

[ME GATE 2014]

93. The security system at an IT office is composed of 10 computers of which exactly four are working. To check whether the system is functional, the officials inspect four of the computers picked at random (without replacement). The system is deemed functional if at least three of the four computers inspected are working. Let the probability that the system is deemed functional be denoted by p. Then 100p = _____.

[CS GATE 2014]

94. A traffic office imposes on an average 5 number of penalties daily on traffic violators. Assume that the number of penalties on different days is independent and follows a Poisson distribution. The probability that there will be less than 4 penalties in a day is _____.

[CE GATE 2014]

95. The number of accidents occurring in a plant in a month follows the Poisson distribution with mean as 5.2. The probability of occurrence of less than 2 accidents in the plant during a randomly selected month is

(a) 0.029   (b) 0.034   (c) 0.039   (d) 0.044

[ME GATE 2014]

96. The probability that a given positive integer lying between 1 and 100 (both inclusive) is NOT divisible by 2, 3, or 5 is _____.

[CS GATE 2014]

97. The probability density function of evaporation E on any day during a year in a watershed is given by

$$f(x) = \begin{cases} \dfrac{1}{5}, & 0 \le x \le 5 \\ 0, & \text{otherwise} \end{cases}$$

The probability that E lies in between 2 and 4 mm/day in the watershed is _____?

[CE GATE 2014]

98. Let X be a random variable with probability density function

$$f(x) = \begin{cases} 0.2, & |x| \le 1 \\ 0.1, & 1 < |x| \le 4 \\ 0, & \text{otherwise} \end{cases}$$

The probability $P(0.5 < X < 5)$ is _____.

[EE GATE 2014]

99. If X is a continuous, real valued random variable defined over the interval $(-\infty, +\infty)$ and its occurrence is defined by the density function given as

$$f(x) = \frac{1}{\sqrt{2\pi} * b} e^{-\frac{1}{2}\left(\frac{x-a}{b}\right)^2}$$

where "a" and "b" are the statistical attributes of the random variable X.

The value of the integral $\int_{-\infty}^{a} \frac{1}{\sqrt{2\pi} * b} e^{-\frac{1}{2}\left(\frac{x-a}{b}\right)^2} dx$

(a) 1   (b) 0.5   (c) π (d) $\pi/2$

[CE GATE 2014]

100. In the following table, x is a discrete random variable and $p(x)$ is the probability density.

x	1	2	3
P(x)	0.3	0.6	0.1

The standard deviation of x is

(a) 0.18   (b) 0.36   (c) 0.54   (d) 0.6

[ME GATE 2014]

101. Consider an unbiased cubic dice with opposite faces colored identically and each face colored red, blue, or green such that each color appears only two times on the dice. If the dice is thrown thrice, the probability of obtaining red color on top face of the dice at least twice is _____.

[ME GATE 2014]

102. Each of the nine words in the sentence "The quick brown fox jumps over the lazy dog" is written on a separate piece of paper. These nine pieces of paper are kept in a box. One of the pieces is drawn at random from the box. The expected length of the word drawn is _____?

[CS GATE 2014]

103. Suppose you break a stick of unit length at a point chosen uniformly at random. Then the expected length of the shorter stick is _____?

[CS GATE 2014]

104. A simple random sample of 100 observations was taken from a large population. The sample mean and the standard deviation were determined to be 80 and 12, respectively. The standard error of mean is _____ ?

[PI GATE 2014]

105. In a given day in the rainy season, it may rain 70% of the time. If it rains, chance that a village fair will make a loss on that day is 80%. However, if it does not rain, chance that the fair will make a loss on that day is only 10%. If the fair has not made a loss on a given day in the rainy season, what is the probability that it has not rained on that day?

(a) 3/10   (b) 9/11   (c) 14/17   (d) 27/41

[PI GATE 2014]

106. An observer counts 240 vehicles per hour at a specific highway location. Assume that the vehicle arrived at the location is Poisson distributed, the probability of having one vehicle arriving over a 30 second time interval is _____?

[CE GATE 2014]

107. Marks obtained by 100 students in an examination are given in the following table:

Sl. No.	Marks obtained	No. of students
1	25	20
2	30	20
3	35	40
4	40	20

What would be the mean, median, and mode of the marks obtained by the students?

(a) Mean = 33, Median = 35, Mode = 40
(b) Mean = 35, Median = 32.5, Mode = 40
(c) Mean = 33, Median = 35, Mode = 35

(d) Mean = 35, Median = 32.5, Mode = 35

[PI GATE 2014]

108. Given that X is a random variable in the range $[0, \infty]$ with a probability density function $f(x) = \dfrac{e^{-\frac{x}{2}}}{k}$. Then the value of the constant "k" is _____?

[IN GATE 2014]

109. If calls arrive at a telephone exchange such that the time of arrival of any call is independent of the time of arrival of earlier or future calls, the probability distribution function of the total number of calls in a fixed time interval will be

(a) Poisson   (b) Gaussian
(c) Exponential   (d) Gamma

[EC GATE 2014]

110. Lifetime of a electric bulb is a random variable with probability density function $f(x) = kx^2$, where x is measured in years. If the minimum and maximum lifetimes of bulb are 1 and 2 years, respectively, then the value of "k" is _____?

[EE GATE 2014]

111. Let X_1, X_2, and X_3 be independent and identically distributed random variables with the uniform distribution on $[0, 1]$. Then, Prob. (X_1 is the largest) is _____?

[EC GATE 2014]

112. Let X be a random variable which is uniformly chosen from the set of positive odd numbers less than 100. Then, the expectation, $E(X)$ is _____?

[EC GATE 2014]

113. If X be a zero mean unit variance Gaussian random variable, then $E(|X|)$ is equal to _____?

[EC GATE 2014]

114. Let X be a real valued random variable with $E(X)$ and $E(X^2)$ denoting the mean values of X and X^2, respectively. The relation that always holds true is

(a) $[E(X)]^2 > E(X^2)$ (b) $E(X^2) \geq [E(X)]^2$
(c) $E(X^2) = [E(X)]^2$ (d) $E(X^2) > [E(X)]^2$

[EC GATE 2014]

115. Suppose A and B are two independent events with probabilities $P(A) \neq 0$ and $P(B) \neq 0$. Let $\bar{A}$ and $\bar{B}$ be their complements. Which one of the following statements is FALSE?

(a) $P(A \cap B) = P(A)P(B)$
(b) $P(A/B) = P(A)$
(c) $P(A \cup B) = P(A) + P(B)$
(d) $P(\bar{A} \cup \bar{B}) = P(\bar{A})P(\bar{B})$

[EC GATE 2015]

116. The probability of obtaining at least two "SIX" in throwing a fair dice 4 times is

(a) $\dfrac{425}{432}$ (b) $\dfrac{19}{144}$ (c) $\dfrac{13}{144}$ (d) $\dfrac{125}{432}$

[ME GATE 2015]

117. If $P(X) = \dfrac{1}{4}, P(Y) = \dfrac{1}{3}$ and $P(X \cap Y) = \dfrac{1}{12}$, the value of $P(Y/X)$ is

(a) $\dfrac{1}{4}$ (b) $\dfrac{4}{25}$ (c) $\dfrac{1}{3}$ (d) $\dfrac{29}{50}$

[ME GATE 2015]

118. The probability density function of a random variable, X is

$$f(x) = \begin{cases} \dfrac{x}{4}\left(4 - x^2\right), & 0 \leq x \leq 2 \\ 0, & \text{Otherwise} \end{cases}$$

The mean, μ_x of the random variable is _____.

[CE GATE 2015]

119. Three vendors were asked to supply a very high precision component. The respective probabilities of their meeting the strict design specifications are 0.8, 0.7, and 0.5. Each vendor supplies one component. The probability that out of total three components supplied by the vendors, at least one will meet the design specific is _____.

[ME GATE 2015]

120. Two players, A and B, alternately keep rolling a fair dice. The person to get a six first wins the game. Given that player A starts the game, the probability that A wins the game is

(a) $\dfrac{5}{11}$ (b) $\dfrac{1}{2}$ (c) $\dfrac{7}{13}$ (d) $\dfrac{6}{11}$

[EE GATE 2015]

121. The chance of a student passing an exam is 20%. The chance of a student passing the exam and getting above 90% marks in it is 5%. Given that a student passes the examination, the probability that the student gets above 90% marks is _____.

(a) $\dfrac{1}{18}$ (b) $\dfrac{1}{4}$ (c) $\dfrac{2}{9}$ (d) $\dfrac{5}{18}$

[ME GATE 2015]

122. A random variable X has probability density function $f(x)$ as given below:

$$f(x) = \begin{cases} a + bx, & \text{for } 0 < x < 1 \\ 0, & \text{otherwise} \end{cases}$$

If the expected value $E[X] = 2/3$, then Prob. $[X < 0.5]$ is _____.

[EE GATE 2015]

123. The probability that a thermistor randomly picked up from a production unit is defective is 0.1. The probability that out of ten thermistors randomly picked up, three are defective is

(a) 0.001 (b) 0.057 (c) 0.107 (d) 0.3

[IN GATE 2015]

124. Consider the following probability mass function (p.m.f) of a random variable X:

$$p(x,q) = \begin{cases} q, & \text{if } X = 0 \\ 1-q, & \text{if } X = 1 \\ 0, & \text{otherwise} \end{cases}$$

If $q = 0.4$, the variance of X is _____?

[CE GATE 2015]

125. Let $X \in \{0, 1\}$ and $Y \in \{0, 1\}$ be two independent binary random variables. If $P(X = 0) = p$ and $P(Y = 0) = q$, then $P(X + Y \geq 1)$ is equal to

(a) $pq + (1-p)(1-q)$ (b) pq
(c) $p(1 - q)$ (d) $1-pq$

[EC GATE-2015]

126. The probability density function of a random variable X is given below:

$$P_X(x) = \begin{cases} e^{-x}, & x \geq 0 \\ 0, & \text{otherwise} \end{cases}$$

The expected value of the function $g_X(x) = e^{\frac{3x}{4}}$ is _____ ?

[IN GATE-2015]

127. A fair die with faces {1, 2, 3, 4, 5, 6} is thrown repeatedly until "3" is observed for the first time. Let X denotes the number of times the dice thrown. The expected value of X is _____

[EC GATE 2015]

128. Two coins R and S are tossed. The four joint events $H_R H_S$, $T_R T_S$, $H_R T_S$, $T_R H_S$ have probabilities 0.28, 0.18, 0.30, 0.24, respectively, where H represents heads and T represents tails. Which one of the following is TRUE?
(a) The coin tosses are independent
(b) R is fair, S is not
(c) S is fair, R is not
(d) The coin tosses are dependent

[EE GATE 2015]

129. A product is an assemble of five different components. The product can be sequentially assembled in two possible ways. If the five components are placed in a box and these are drawn at random from the box, then the probability of getting a correct sequence is
(a) $\dfrac{2}{5!}$
(b) $\dfrac{2}{5}$
(c) $\dfrac{2}{(5-2)!}$
(d) $\dfrac{2}{(5-3)!}$

[PI GATE 2015]

130. A coin is tossed thrice. Let X be the event that heads occurs in each of the first tosses. Let Y be the event that tails occurs on the third toss. Let Z be the event that two tails occur in three tosses.

Based on the above information, which of the following statements is true?
(a) X and Y are not independent
(b) Y and Z are dependent
(c) Y and Z are independent
(d) X and Z are independent

[PI,IN, ME GATE 2015]

131. Four cards are randomly selected from a pack of 52 cards. If the first two cards are kings, what is the probability that the third card is a king?
(a) $\dfrac{4}{52}$
(b) $\dfrac{2}{50}$
(c) $\dfrac{1}{52}\times\dfrac{1}{52}$
(d) $\dfrac{1}{52}\times\dfrac{1}{51}\times\dfrac{1}{50}$

[CE GATE 2015]

132. Given set A = {2, 3, 4, 5} and the set B = {11, 12, 13, 14, 15}. Two numbers are randomly selected, one from each set. What is the probability that the sum of these two numbers equals 16?
(a) 0.20 (b) 0.25 (c) 0.30 (d) 0.33

[EE, CS GATE 2015]

133. Robert and Richard appeared in an interview for two vacancies in the same department. The probability of Robert's selection is 1/6 and that of Richard is 1/8. What is the probability that only one of them will be selected?
(a) 47/48 (b) 1/4 (c) 13/48 (d) 35/48

[EC, ME GATE 2015]

134. Suppose X_i for i = 1, 2, 3 are independent and identically distributed random variables whose probability mass functions (p.m.f) are $\Pr(X_i=0)=\Pr(X_i=1)=\dfrac{1}{2}$ for i = 1, 2, 3. Define another random variable $Y = X_1 X_2 \oplus X_3$ where $\oplus$ denotes XOR. Then $\Pr(Y=0/X_3=0)=$ _____?

[CS GATE 2015]

135. Three cards were drawn from a pack of 52 cards. The probability that they are a King, a Queen, and a Jack is
(a) $\dfrac{16}{5525}$
(b) $\dfrac{64}{2197}$
(c) $\dfrac{3}{13}$
(d) $\dfrac{8}{16575}$

[ME GATE 2016]

136. The probability that a screw manufactured by a company is defective is 0.1. The company sells screw in packets containing five screws and gives a guarantee of replacement if one or more screws in the packet are found to be defective. The probability that a packet would have to be replaced is _____.

[ME GATE 2016]

137. X and Y are two random independent events. It is known that $P(X) = 0.40$ and $P(X \cup Y^C) = 0.7$, Which one of the following is the value of $P(X\cup Y)$?
(a) 0.7 (b) 0.5 (c) 0.4 (d) 0.3

[CE GATE 2016]

138. An jar contains 5 red and 7 green balls. A ball is drawn at random and its color is noted. The ball is placed back into the jar along with another ball of the same color. The probability of getting a red ball in the next draw is

(a) $\dfrac{65}{156}$ (b) $\dfrac{67}{156}$ (c) $\dfrac{79}{156}$ (d) $\dfrac{89}{156}$

[IN GATE 2016]

139. The spot speeds (expressed in km/hr) observed at a road section are 66, 62, 45, 79, 32, 51, 56, 60, 53, and 49. The median speed (expressed in km/hr) is _____.

[CE GATE 2016]

140. The probability of getting "heads" in a single toss of a biased coin is 0.3. The coin is tossed repeatedly until "heads" is obtained. If the tosses are independent, then the probability of getting "heads" for the first time in the fifth toss is _____.

[EC GATE 2016]

141. If $f(x)$ and $g(x)$ are two probability density functions,

$$f(x) = \begin{cases} \dfrac{x}{a}+1, & -a \le x < 0 \\ -\dfrac{x}{a}+1, & 0 \le x < a \\ 0, & \text{otherwise} \end{cases}$$

$$g(x) = \begin{cases} -\dfrac{x}{a}, & -a \le x < 0 \\ \dfrac{x}{a}, & 0 \le x < a \\ 0, & \text{otherwise} \end{cases}$$

Which one of the following statements is true?

(a) Mean of $f(x)$ and $g(x)$ are same; variance of $f(x)$ and $g(x)$ are same
(b) Mean of $f(x)$ and $g(x)$ are same; variance of $f(x)$ and $g(x)$ are different
(c) Mean of $f(x)$ and $g(x)$ are different; variance of $f(x)$ and $g(x)$ are same
(d) Mean of $f(x)$ and $g(x)$ are different; variance of $f(x)$ and $g(x)$ are different

[CE GATE 2016]

142. The second moment of a Poisson distributed random variable is 2. The mean of the random variable is _____.

[EC GATE 2016]

143. Consider a Poisson distribution for the tossing of a biased coin. The mean for this is

distribution μ. The standard deviation for this distribution is given by

(a) $\sqrt{\mu}$ (b) μ^2 (c) μ (d) $\dfrac{1}{\mu}$

[ME GATE 2016]

144. Probability density function of a random variable X is given below

$$f(x) = \begin{cases} 0.25, & 1 \le x \le 5 \\ 0, & \text{otherwise} \end{cases}$$

Then, $P(X \le 4) = ?$

(a) $\dfrac{3}{4}$ (b) $\dfrac{1}{2}$ (c) $\dfrac{1}{4}$ (d) $\dfrac{1}{8}$

[CE GATE 2016]

145. A probability density function on the interval [a, 1] is given by $1/x^2$ and outside this interval the value of the function is zero. The value of "a" is _____.

[CS GATE 2016]

146. Let the probability density function of a random variable X, be given as:

$$f_X(x) = \dfrac{3}{2}e^{-3x}u(x) + ae^{4x}u(-x)$$

where u(x) is the unit step function. Then the value of "a" and Prob. {X ≤ 0}, respectively, are

(a) $2, \dfrac{1}{2}$ (b) $4, \dfrac{1}{2}$ (c) $2, \dfrac{1}{4}$ (d) $4, \dfrac{1}{4}$

[EE GATE 2016]

147. Consider the following experiment.

Step 1: Flip a fair coin twice.
Step 2: If the outcomes are (TAILS, HEADS) then output Y and stop.
Step 3: If the outcomes are either (HEADS, HEADS) or (HEADS, TAILS), then output N and stop.
Step 4: If the outcomes are (TAILS, TAILS), then go to Step 1.

The probability that the output of the experiment is Y is _____ (up to two decimal places).

[CS GATE 2016]

148. Suppose that a shop has an equal number of LED bulbs of two different types. The probability of an LED bulb lasting more than 100 hours given that it is of Type 1 is 0.7, and given that it is of Type 2 is 0.4. The probability that an LED bulb chosen uniformly at random lasts more than 100 hours is _____.

[CS GATE 2016]

149. Two random variables X and Y are distributed according to

$$f_{X,Y}(x,y) = \begin{cases} x+y, & 0 \le x \le 1, 0 \le y \le 1 \\ 0, & \text{otherwise} \end{cases}$$

The probability P(X + Y ≤ 1) is _____.

[EC GATE 2016]

150. A fair coin is tossed N times. The probability that heads does not turn up in any of the tosses is

(a) $\left(\frac{1}{2}\right)^{N-1}$ (b) $1-\left(\frac{1}{2}\right)^{N-1}$

(c) $\left(\frac{1}{2}\right)^{N}$ (d) $1-\left(\frac{1}{2}\right)^{N}$

[PI GATE 2016]

151. A normal random variable X has the following probability density function

$$f_X(x) = \frac{1}{\sqrt{8\pi}} e^{-\frac{(x-1)^2}{8}}, \quad -\infty < x < \infty$$

Then, $\int_1^\infty f_X(x) = ?$

(a) 0   (b) $\frac{1}{2}$ (c) $1-\frac{1}{e}$ (d) 1

[PI GATE 2016]

152. A sample of 15 data is as follows:

17, 18, 17, 17.13, 18, 5, 5, 6, 7, 8, 9, 20, 17, 3.

The mode of the data is

(a) 4   (b) 13   (c) 17   (d) 20

[ME GATE 2017]

153. If a random variable X has a Poisson distribution with mean 5, then the expectation $E[(X + 2)^2]$ equals _____.

[CS GATE 2017]

154. For the function f(x) = a + bx, 0 ≤ x ≤ 1, to be a valid probability density function, which one of the following statements is correct?

(a) a = 1, b = 4   (b) a = 0.5, b = 1
(c) a = 0, b = 1   (d) a = 1, b = -1

[CE GATE 2017]

155. Two coins are tossed simultaneously. The probability (up to two decimal points accuracy) of getting at least one heads is _____.

[ME GATE 2017]

156. An jar contains five red balls and five black balls. In the first draw, one ball is picked at random and discarded without noticing its color. The probability to get a red ball in the second draw is

(a) $\frac{1}{2}$ (b) $\frac{4}{9}$ (c) $\frac{5}{9}$ (d) $\frac{6}{9}$

[EE GATE 2017]

157. A six-face fair dice is rolled a large number of times. The mean value of the outcomes is _____.

[ME GATE 2017]

158. The number of parameters in the univariate exponential and Gaussian distributions, respectively, are

(a) 2 and 2   (b) 1 and 2
(c) 2 and 1   (d) 1 and 1

[CE GATE 2017]

159. A two-faced coin has its faces designated as heads (H) and tails (T). This coin is tossed three times in succession to record the following outcomes; H, H, H. If the coin is tossed one more time, the probability (up to one decimal place) of obtaining H again, given the previous realizations of H, H, and H, would be _____.

[CE GATE 2017]

160. Assume that in a traffic junction, the cycle of the traffic signal lights is two minutes of green (vehicle does not stop) and three minutes of red (vehicle stops). Consider that the arrival time of vehicles at the junction is uniformly distributed over five minute cycle. The expected waiting time (in minutes) for the vehicle at the junction is _____.

[EE GATE 2017]

161. Three fair cubical dice are thrown simultaneously. The probability that all three dice have the same number of dots on the faces showing up is (up to third decimal place) _____?

[EC GATE 2017]

162. P and Q are considering to apply for a job. The probability that P applies for the job is $\frac{1}{4}$, the probability that P applies for the job given that Q applies for the job is $\frac{1}{2}$, and the probability that Q applies for the job given that P applies for the job is $\frac{1}{3}$. Then the probability that P does not apply for the job given that Q does not apply for the job is

(a) $\frac{4}{5}$ (b) $\frac{5}{6}$ (c) $\frac{7}{8}$ (d) $\frac{11}{12}$

[CS/IT GATE 2017]

163. Passengers try repeatedly to get a seat reservation in any train running between two stations until they are successful. If there is 40% chance of getting reservation in any attempt by a passenger, then the average number of attempts that passengers need to make to get a seat reserved is ___?

[EC GATE 2017]

164. Probability (up to one decimal place) of consecutive picking three red balls without replacement from a box containing five red balls and one white ball is _____?

[CE GATE 2018]

165. A class of twelve children has two more boys than girls. A group of three children are randomly picked from this class to accompany the teacher on a field trip. What is the probability that the group accompanying the teacher contains more girls than boys?

(a) $\frac{4}{11}$ (b) $\frac{325}{864}$ (c) $\frac{525}{864}$ (d) $\frac{5}{12}$

[EE GATE 2018]

166. Four red balls, four green balls, and four blue balls are put in a box. three balls are pulled out of the box at random one after another without replacement. The probability that all the three balls are red is

(a) $\frac{1}{72}$ (b) $\frac{1}{55}$ (c) $\frac{1}{36}$ (d) $\frac{1}{27}$

[EE GATE 2018]

167. Weights (in kg) of six products are 3, 7, 6, 2, 3, and 4. The median weight (in kg) up to one decimal place is _____.

[PI GATE 2018]

168. A six faced dice is rolled five times. The probability (in %) of obtaining "ONE" at least four times is

(a) 33.3   (b) 3.33
(c) 0.33   (d) 0.0033

[ME GATE 2018]

169. The probabilities of occurrence of events F and G are P(F) = 0.3 and P(G) = 0.4, respectively. The probability that both events occur simultaneously is $P(F \cap G) = 0.2$. The probability of occurrence of at least one event $P(F \cup G)$ is _____.

[PI GATE 2018]

170. Let X_1, X_2, X_3 and X_4, be independent normal random variables with a zero mean and unit variance. The probability that X_4 is the smallest among the four is _____.

[EC GATE 2018]

171. Two bags, A and B, have equal number of balls. Bag A has 20% red balls and 80% green balls. Bag B has 30% red balls, 60% green balls, and 10% yellow balls. Contents of bag A and B are mixed thoroughly and a ball is randomly picked from the mixture. What is the chance that the ball picked is red?

(a) 20%   (b) 25%   (c) 30%   (d) 40%

[IN GATE 2018]

172. Two people, P and Q, decide to independently roll two identical dice, each with six faces, numbered one to six. The person with the lower number wins. In case of a tie, they roll the dice repeatedly until there is no tie. Define a trial as a throw of the dice by P and Q. Assume that all six numbers on each dice are equi-probable and that all trials are independent. The probability (round to three decimal places) that one of them wins on the third trial is _____?

[CS/IT GATE 2018]

173. A cab was involved in a hit and run accident at night. You are given the following data about the cabs in the city and the accident:

(*i*) 85% of cabs in the city are green and the remaining cabs are blue

(*ii*) A witness identified the cab involved in the accident as blue

(*iii*) It is known that a witness can correctly identify the cab color only 80% of the time.

Which of the following options is closest to the probability that the accident was caused by a blue cab?

(a) 12%   (b) 15%   (c) 41%   (d) 80%
[EC GATE 2018]

174. X and Y are two independent random variables with variances 1 and 2, respectively. Let Z = X – Y. The variance of Z is

(a) 0   (b) 1   (c) 2   (d) 3
[IN GATE 2018]

175. An unbiased coin is tossed six times in a row and four different such trials are conducted. One trial implies six tosses of the coin. If H stands for heads and T stands for tails, the following are the observations from the four trials:

(*i*) HTHTHT   (*ii*)   TTHHHT

(*iii*) HTTHHT   (*iv*)   HHHT

Which statement describing the last two coin tosses of the fourth trial has the highest probability of being correct?

(a) Two T will occur
(b) One H and one T will occur
(c) Two H will occur
(d) One H will be followed by one T
[ME GATE 2018]

176. Let X_1 and X_2 be two independent random variables with means μ_1, μ_2 and standard deviations σ_1, σ_2, respectively. Consider Y = $X_1 - X_2$; $\mu_1 = \mu_2 = 1$, $\sigma_1 = 1$, $\sigma_2 = 2$. Then
(a) Y is normally distributed with mean 0 and variance 1
(b) Y is normally distributed with mean 0 and variance 5
(c) Y has mean 0 and and variance 5, but it is not normally distributed

(d) Y has mean 0 and variance 1, but it is not normally distributed
[ME GATE 2018]

177. The probability distribution with right skew is shown in figure below:

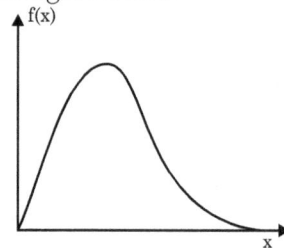

The correct statement for the probability distribution is

(a) Mean is equal to mode
(b) Mean is greater than median but less than mode
(c) Mean is greater than median and mode
(d) Mode is greater than median
[CE GATE 2018]

178. A six-sided unbiased die with four green faces and two red faces is rolled seven times. Which of the following combinations is the most likely outcome of the experiment?

(a) Three green faces and four red faces
(b) Four green faces and three red faces
(c) Five green faces and two red faces
(d) Six green faces and one red face
[CS/IT GATE 2018]

179. The graph of a function f(x) is shown in the figure below:

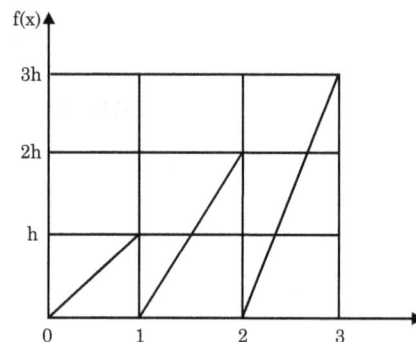

For f(x) to be a valid probability density function, the value of "h" is

(a) 1/3   (b) 2/3   (c) 1   (d) 3
[CE GATE 2018]

Answer key				
1. (d)	**2.** (a)	**3.** (b)	**4.** (c)	**5.** (c)
6. (d)	**7.** (c)	**8.** (d)	**9.** (c)	**10.** (a)
11. (d)	**12.** (a)	**13.** (d)	**14.** (d)	**15.** (b)
16. (c)	**17.** (d)	**18.** (d)	**19.** (d)	**20.** (b)
21. (b)	**22.** (b)	**23.** (a)	**24.** (d)	**25.** (d)
26. (c)	**27.** (c)	**28.** (b)	**29.** (d)	**30.** (d)
31. (a)	**32.** (b)	**33.** (d)	**34.** (c)	**35.** (c)
36. (d)	**37.** (c)	**38.** (c)	**39.** (a)	**40.** (d)
41. (a)	**42.** (a)	**43.** (a)	**44.** (c)	**45.** (b)
46. (a)	**47.** (a)	**48.** (c)	**49.** (c)	**50.** (a)
51. (a)	**52.** (b)	**53.** (c)	**54.** (d)	**55.** (b)
56. (c)	**57.** (d)	**58.** (a)	**59.** (c)	**60.** (d)
61. (c)	**62.** (d)	**63.** (c)	**64.** (c)	**65.** (c)
66. (d)	**67.** (d)	**68.** (a)	**69.** (b)	**70.** (d)
71. (c)	**72.** (d)	**73.** (b)	**74.** (d)	**75.** (c)
76 (b)	**77.** 6.	**78.** (c)	**79.** (a)	**80.** (b)
81. 50%	**82.** (a)	**83.** 0.666	**84.** (b)	**85.** 0.65
86. (c)	**87.** (b)	**88.** 10.	**89.** $\frac{1}{7}$.	**90.** 0.25
91. 0.44	**92.** (a)	**93.** 11.9	**94.** 0.265	**95.** (b)
96. 0.26	**97.** 0.4	**98.** 0.4	**99.** (b)	**100.** (d)
101. 0.259	**102.** 3.88	**103.** 0.25	**104.** 1.2	**105.** (d)
106. 0.27	**107.** (c)	**108.** 2	**109.** (a)	**110.** $\frac{3}{7}$.
111. 0.33	**112.** 50	**113.** $\sqrt{\frac{2}{\pi}}$	**114.** (b)	**115.** (c)
116. (b)	**117.** (c)	**118.** 1.067	**119.** 0.97	**120.** (d)
121. (b)	**122.** 0.25	**123.** (b)	**124.** 0.24	**125.** (d)
126. 4	**127.** 6	**128.** (d)	**129.** (a)	**130.** (b)
131. (b)	**132.** (a)	**133.** (b)	**134.** 0.75	**135.** (a)
136. 0.4095.	**137.** (a)	**138.** (a)	**139.** 54.5	**140.** 0.072
141. (b)	**142.** 1	**143.** (a)	**144.** (a)	**145.** 0.5
146. (a)	**147.** 1/3	**148.** 0.55	**149.** 0.33	**150.** (c)
151. (b)	**152.** (c)	**153.** 54.	**154.** (b)	**155.** $\frac{3}{4}$
156. (a)	**157.** 3.5	**158.** (b)	**159.** 0.5	**160.** 0.9
161. 0.028	**162.** (a)	**163.** 2.5	**164.** $\frac{1}{2}$.	**165.** (a)
166. (b)	**167.** 3.5	**168.** (c)	**169.** 0.5	**170.** 0.25
171. (b)	**172.** 0.0231		**173.** (c)	**174.** (d)
175. (b)	**176.** (b)	**177.** (c)	**178.** (c)	**179.** (a)

Explanation

1. (*d*)

$P(E_1 \cup E_2) = 1$

$\Rightarrow P(E_1) + P(E_2) - P(E_1 \cap E_2) = 1$

$\Rightarrow P(E_1) + P(E_2) - P(E_1) \times P(E_2) = 1$

($\because E_1$ and E_2 are independent events)

$\Rightarrow 2P(E_1) - \{P(E_1)\}^2 = 1 (\because P(E_1) = P(E_2))$

$\Rightarrow \{1 - P(E_1)\}^2 = 0 \Rightarrow P(E_1) = 1.$

2. (*a*) Let X be the random variable denoting the number of defective bolts. Then clearly X follows the binomial distribution.

Here, $n = 900, p = 0.1$.

Therefore, mean $= np = 900 \times 0.1 = 90$,

variance $= np(1-p) = 90 \times 0.9 = 81$.

$\therefore S.D = \sqrt{np(1-p)} = \sqrt{81} = 9$.

3. (*b*) Each of the car accidents can occur on any day of a week. So total number of possible ways $= 7^7$.

Again if all the car accidents occur on the same day, then they all occur on Monday or Tuesday or Wednesday or Thursday or Friday or Saturday or Sunday. Thus, total no. of favorable cases $= 7$.

Hence, prob. (all seven car accidents occur on the same day) $= \dfrac{7}{7^7} = \dfrac{1}{7^6}$

4. (*c*) Here sample space S = {(HHHH), (HHHT), (HHTH),

(HTHH), (THHH), (HHTT), (TTHH), (HTHT), (THTH), (THHT), (HTTH), (HTTT), (TTTH), (THTT), (TTHT), (TTTT)}.

Let A = event that at least one heads and at least one tails turn up.

Then, A = {(HHHT), (HHTH),(HTHH), (THHH), (HHTT), (TTHH), (HTHT), (THTH), (THHT), (HTTH), (HTTT), (TTTH), (THTT), (TTHT)}.

Therefore, Prob. (at least one heads and at least one tails turn up)

$= P(A) = \dfrac{14}{16} = \dfrac{7}{8}$.

5. (*c*)

Let "X" be the random variable denoting the number of defective screws out of two drawn screws.

Clearly X follows binomial distribution with $n = 2$ and p = prob. of getting a defective screw $= \dfrac{3}{10}$.

So prob.(none of the two screws are defective)

$= P(X = 0)$

$= {}^2C_0 \left(\dfrac{3}{10}\right)^0 \left(1 - \dfrac{3}{10}\right)^2 = 0.49 = 49\%$

6. (d) Given, $P(A) = 1$ and $P(B) = 1/2$.

Let us assume that the events A and B are independent.

So, $P(A \cap B) = P(A) \times P(B) = 1 \times \dfrac{1}{2} = \dfrac{1}{2}$.

$\therefore \quad P(A/B) = \dfrac{P(A \cap B)}{P(B)} = \dfrac{1/2}{1/2} = 1$ and

$P(B/A) = \dfrac{P(A \cap B)}{P(A)} = \dfrac{1/2}{1} = 1/2$.

7. (c) Let x and y denote the time taken for first and second modules, respectively.

Also let "t" represents the total time. Then,

$t = x + y$ is a random variable.

So the probability density function of the overall time taken to execute the program

$= g(t) = \int_0^t f(x,y)\,dx = \int_0^t f(x, t-x)\,dx$

$= \int_0^t f_1(x) f_2(t-x)\,dx$.

8. (d) Initially, there are five black and five red balls in the box. Therefore, prob. (a red ball is drawn)

$= \dfrac{5}{5+5} = \dfrac{5}{10} = \dfrac{1}{2}$.

Now if the red ball drawn is not replaced into the box, then number of balls remaining in the box = nine out of which four are red and five are black.

Hence, Prob. (the second ball drawn is red given that first drawn ball is red)

$= \dfrac{4}{4+5} = \dfrac{4}{9}$.

Hence, prob. (both drawn balls are red)

$= \dfrac{1}{2} \times \dfrac{4}{9} = \dfrac{2}{9}$.

9. (c) Let A = event that Gate 1 fails,

B = event that Gate 2 fails,

C = event that Gate 3 fails and

D = event that Gate 4 fails.

Given that $P(A) = P(B) = P(C) = P(D) = 0.2$.

Then, P(gate 2 and gate 3 fail given that gate 1 failed)

$= P((B \cap C)/A) = \dfrac{P(B \cap C \cap A)}{P(A)}$

$= \dfrac{P(B) \times P(C) \times P(A)}{P(A)}$

($\because$ the events A, B, C are independent)

$= P(B) \times P(C) = 0.2 \times 0.2 = 0.04$

10. (a) Here sample space S = {(HHHH), (HHHT), (HHTH),

(HTHH), (THHH), (HHTT), (TTHH), (HTHT), (THTH), (THHT), (HTTH), (HTTT), (TTTH), (THTT), (TTHT), (TTTT)}.

Let A = event that two heads and two tails turn up.

Then, A = {(HHTT), (TTHH), (HTHT), (THTH), (THHT), (HTTH)}.

Therefore, the required probability

$= P(A) = \dfrac{n(A)}{n(S)} = \dfrac{6}{16} = \dfrac{3}{8}$.

11. (d) Let X be the random variable denoting the marks obtained per question.

Then the probability distribution table for X is given by

X	1	–0.25
P(X)	1/4	1– 1/4 (= 3/4)

Expected mark for one question

$= E(X)$

$= 1 \times 1/4 + (-0.25) \times 3/4 = 1/16$.

Total expected marks for 150 questions

$= \dfrac{1}{16} \times 150 = \dfrac{75}{8}$ (for one student).

Total expected marks of 1,000 students

$= \dfrac{75}{8} \times 1000 = 9375$.

12. (a) Clearly we have a binomial distribution with n trials and d successes where probability of success in each trial $= p = 2/4 = 1/2$ (Since out of the four possible outcomes namely (0, 0), (0, 1), (1, 0), (1, 1) ; only two viz (0, 1) and (1, 0) are success.)

Hence, the required probability

$= P(X = d)$

$= {}^nC_d \left(\dfrac{1}{2}\right)^d \left(\dfrac{1}{2}\right)^{n-d} = \dfrac{{}^nC_d}{2^n}$.

13. (d) Prob. (the King card is drawn) = $\frac{^4C_1}{^{52}C_1} = \frac{1}{13}$.

Now if the first King card drawn is not replaced, then no. of cards left = 51 out of which three are Kings.

So, prob. (getting a King in the second drawn) = $\frac{^3C_1}{^{51}C_1} = \frac{1}{17}$.

Hence, prob. (both cards will be Kings, if first card is NOT placed)

= $\frac{1}{13} \times \frac{1}{17} = \frac{1}{221}$.

14. (d) If (x, y) be the coordinates of the point, then ATQ,

$p = \sqrt{x^2 + y^2}$.

$\therefore p^2 = x^2 + y^2$.

Then, $E(p^2) = E(x^2 + y^2) = E(x^2) + E(y^2)$.......(1)

ATQ, x and y are uniformly distributed random variables in $0 \le x \le 1$ and $0 \le y \le 2$, respectively.

The probability density function of $x, f(x) = \frac{1}{1-0} = 1$,

The probability density function of $y, g(y) = \frac{1}{2-0} = \frac{1}{2}$.

Therefore, (1) gives,

$E(p^2)$

$= E(x^2) + E(y^2)$

$= \int_0^1 x^2 f(x) dx + \int_0^2 y^2 g(y) dy$

$= \int_0^1 x^2 \times 1 dx + \int_0^2 y^2 \times \frac{1}{2} dy$

$= \left[\frac{x^3}{3}\right]_0^1 + \left[\frac{y^3}{6}\right]_0^2 = \frac{1}{3} + \frac{4}{3} = \frac{5}{3}$.

15. (b) Here,

The number of children belonging to families with three children = (50% of N) × 3 = $\frac{3N}{2}$,

The number of children belonging to families with two children = (30% of N) × 2 = $\frac{3N}{5}$ and

The number of children belonging to families with one child = (20% of N) × 1 = $\frac{N}{5}$.

So total number of children = $\frac{3N}{2} + \frac{3N}{5} + \frac{N}{5} = \frac{23N}{10}$.

Hence, the required probability

$= \frac{\frac{3N}{5}}{\frac{23N}{10}} = \frac{6}{23}$.

16. (c)

Mid values (x_i)	Frequency (f_i)
7.6	1
7.8	5
8	35
8.2	13
8.4	12
8.6	10

Mean flow rate

$= \dfrac{\sum\limits_{i=1}^{6} f_i x_i}{\sum\limits_{i=1}^{6} f_i}$

$= \dfrac{1 \times 7.6 + 5 \times 7.8 + 35 \times 8 + 13 \times 8.2 + 12 \times 8.4 + 10 \times 8.6}{1 + 5 + 35 + 17 + 12 + 10}$

$= \dfrac{652.8}{80} = 8.16$

17. (d) Option (a) is false since of P and Q are independent implies that

Prob.$(P \cap Q)$ = Prob.$(P) \times$ Prob.(Q)

which may not be zero.

Option (b) is false since

Prob.$(P \cup Q)$ = Prob.(P) + Prob.(Q) − Prob.$(P \cap Q)$

and so Prob.$(P \cup Q) \le$ Prob.(P) + Prob.(Q).

Option (c) is false since independence does not depend on mutually exclusiveness.

Option (d) is true as

$P \cap Q \subseteq P$

$\Rightarrow n(P \cap Q) \le n(P) \Rightarrow$ Prob.$(P \cap Q) \le$ Prob.(P).

18. (*d*) Here, $n(S) = 6^2 = 36$.

Let A = event that sum is either 8 or 9.

Then, $A = \{(2,6),(3,5),(3,6),(4,4),(4,5),(5,3),$ $(5,4),(6,2),(6,3)\}$.

$\therefore \quad P(A) = \dfrac{n(A)}{n(S)} = \dfrac{9}{36} = \dfrac{1}{4}$.

So the prob. that the sum is neither 8 nor 9

$= 1 - P(A) = 1 - \dfrac{1}{4} = \dfrac{3}{4}$.

19. (*d*) Here,

$S = \{(1,1),(1,2),(1,3),(1,4),(1,5),(1,6),(2,1),$ $(2,2),(2,3),(2,4),(2,5),(2,6),(3,1),.......,(3,6),$ $(4,1),.........,(4,6),(5,1),....,(5,6),(6,1),$ $(6,2),(6,3),(6,4),(6,5),(6,6)\}$.

Let A = event that an odd number will follow an even number.

Then,

$A = \{(2,1),(2,3),(2,5),(4,1),(4,3),(4,5),(6,1),$ $(6,3),(6,5)\}$.

Hence, the required probability

$= P(A) = \dfrac{n(A)}{n(S)} = \dfrac{9}{36} = \dfrac{1}{4}$.

20. (*b*) Let X be the random variable denoting the number of defective items.

Clearly, X follows a binomial distribution with $p = 0.1$ and $n = 10$.

Hence, prob. (exactly 2 of the chosen items are defective)

$= {}^{10}C_2\, p^2 (1-p)^{10-2} = 45 \times (0.1)^2 (0.9)^8 = 0.1937$

21. (*b*) Three marbles of different colors can be drawn in 3!, *i.e.*, 6 ways.

Here total number of marbles = 10 + 20 + 30 = 60.

Therefore, the required probability

= 6 × prob. (a blue marble is drawn) × prob. (a black marble is drawn) × prob. (a red marble is drawn)

$= 6 \times \dfrac{10}{60} \times \dfrac{20}{60} \times \dfrac{30}{60} = \dfrac{1}{6}$.

22. (*b*) Here, the sample space, $S = \{$HHH, HTH, HHT, HTT$\}$.

Let A = an event that exactly two heads appear.

Then, $A = \{$HTH, HHT$\}$.

So the required probability $= P(A) = \dfrac{2}{4} = \dfrac{1}{2}$.

23. (*a*) Let X be the random variable denoting the lifetime of bulb in hours.

Then, the required probability

$= P(100 < X < 200)$

$= \displaystyle\int_{100}^{200} f(t)\,dt = \int_{100}^{200} \alpha e^{-\alpha t}\,dt$

$= \alpha \left[-\dfrac{e^{-\alpha t}}{\alpha} \right]_{100}^{200} = e^{-100\alpha} - e^{-200\alpha}$

24. (*d*)

Here,

$S = \{(1,1),(1,2),(1,3),(1,4),(1,5),(1,6),(2,1),$ $(2,2),(2,3),(2,4),(2,5),(2,6),(3,1),.......,(3,6),$ $(4,1),.........,(4,6),(5,1),....,(5,6),(6,1),$ $(6,2),(6,3),(6,4),(6,5),(6,6)\}$.

Let A = an event that the sum of numbers on both exceeds 8.

Then,

$A = \{(3,6),(4,5),(4,6),(5,4),(5,5),(5,6),(6,3),$ $(6,4),(6,5),(6,6)\}$.

Hence, the required probability

$= P(A) = \dfrac{n(A)}{n(S)} = \dfrac{10}{36}$.

25. (*d*) Option (*d*) is not true since for a negatively skewed distribution, we have mode > median > mean.

26. (*c*) We know that for a continuous random variable X with p.m.f $f(x)$,

$P(a < x \le b) = P(a \le x \le b) = \displaystyle\int_a^b f(x)\,dx$.

27. (*c*) Here, $n(S) = 36$.

Let X be the random variable denoting the sum of the numbers on each die.

Then, X takes the values 2, 3, 4, 5,.....,11, 12.

Now $P(X = 2) = P\{(1, 1)\} = 1/36$,

$P(X = 3) = P\{(1, 2), (2, 1)\} = 2/36$,

$P(X = 4) = P\{(1, 3), (2, 2), (3, 1)\} = 3/36$,

$P(X = 5) = P\{(1, 4), (2, 3), (3, 2), (4, 1)\} = 4/36$,

$P(X = 6) = P\{(1, 5), (2, 4), (3, 3), (4, 2), (5, 1)\}$
$= 5/36$,

$P(X = 7) = P\{(1, 6), (2, 5), (3, 4), (4, 3), (5, 2),$
$(6, 1)\} = 6/36$,

$P(X = 8) = P\{(2, 6), (3, 5), (4, 4), (5, 3), (6, 2)\} =$
$5/36$,

$P(X = 9) = P\{(3, 6), (4, 5), (5, 4), (6, 3)\} = 4/36$,

$P(X = 10) = P\{(4, 6), (5, 5), (6, 4)\} = 3/36$,

$P(X = 11) = P\{(5, 6), (6, 5)\} = 2/36$,

$P(X = 12) = P\{(6, 6)\} = 1/36$.

Then, the probability distribution of X is given by

X	2	3	4	5	6	7	8
P(X)	1/36	2/36	3/36	4/36	5/36	6/36	5/36

X	9	10	11	12
P(X)	4/36	3/36	2/36	1/36

Let us denote an arbitrary value of X by "r."

Then, $P(r>6) = P(r \geq 7)$

$= \dfrac{6}{36} + \dfrac{5}{36} + \dfrac{4}{36} + \dfrac{3}{36} + \dfrac{2}{36} + \dfrac{1}{36} = \dfrac{21}{36} = \dfrac{7}{12} \neq \dfrac{1}{6}$.

∴ Option (a) is not correct.

$P(r/3$ is an integer$)$

$= P(r = 3) + P(r = 6) + P(r = 9) + P(r = 12)$

$= \dfrac{2}{36} + \dfrac{5}{36} + \dfrac{4}{36} + \dfrac{1}{36} = \dfrac{12}{36} = \dfrac{1}{3} \neq \dfrac{5}{6}$.

∴ Option (b) is not correct.

$P(r/4$ is an integer$)$

$= P(r = 4) + P(r = 8) + P(r = 12)$

$= \dfrac{3}{36} + \dfrac{5}{36} + \dfrac{1}{36} = \dfrac{9}{36} = \dfrac{1}{4}$.

$P(r = 8$ and $r/4$ is an integer$) = P(r = 8) = \dfrac{5}{36}$.

∴ $P(r = 8 \mid r/4$ is an integer$)$

$= \dfrac{P(r = 8 \text{ and } r/4 \text{ is an integer})}{P(r/4 \text{ is an integer})}$

$= \dfrac{5/36}{1/4} = \dfrac{5}{9}$.

∴ Option (c) is correct.

28. (b) Out of 25 calculators, 2 are defective and 23 are non-defective.

Therefore, the required probability

$= \dfrac{{}^{2}C_1 \times {}^{23}C_4}{{}^{25}C_5}$

$= \dfrac{2 \times 23 \times 22 \times 21 \times 20 \times 5 \times 4 \times 3 \times 2 \times 1}{25 \times 24 \times 23 \times 22 \times 21 \times 4 \times 3 \times 2 \times 1} = \dfrac{1}{3}$.

29. (d) Let A = event that a computer is supplied by Y and

B = event that a computer is defective.

Then, $P(A \cap B) = 0.02 \times 0.03 = 0.006$ and

$P(B) = 60\% \times 0.01 + 30\% \times 0.02 + 10\% \times 0.03$
$= 0.015$.

Hence, prob. (a computer was supplied by Y, given that it was defective)

$= P(A/B)$

$= \dfrac{P(A \cap B)}{P(B)} = \dfrac{0.006}{0.015} = 0.4$.

30. (d) Initially there are 20 + 80, *i.e.*, 100 items in the box out of which 20 are defective.

If one defective item is selected and not re-placed into the box, then after first withdrawn only 99 items will left out inside the box out of which 19 will be defective.

Hence, the required probability $= \dfrac{{}^{20}C_1}{{}^{100}C_1} \times \dfrac{{}^{19}C_1}{{}^{99}C_1} = \dfrac{19}{495}$.

31. (a)

Prob. (exactly n elements are chosen)

= Prob. (getting n heads out of 2n tosses)

$= {}^{2n}C_n \left(\dfrac{1}{2}\right)^n \left(\dfrac{1}{2}\right)^{2n-n}$ (using the binomial distribution)

$= \dfrac{{}^{2n}C_n}{2^{2n}} = \dfrac{{}^{2n}C_n}{4^n}$.

32. (b)

Mean $= E(t) = \int\limits_{-\infty}^{\infty} t f(t) dt$

$$= \int\limits_{-1}^{1} t f(t) dt = \int\limits_{-1}^{0} t(1+t) dt + \int\limits_{0}^{1} t(1-t) dt$$

$$= \left[\frac{t^2}{2} + \frac{t^3}{3}\right]_{-1}^{0} + \left[\frac{t^2}{2} - \frac{t^3}{3}\right]_{0}^{1} = \left(\frac{1}{3} - \frac{1}{2}\right) + \left(\frac{1}{2} - \frac{1}{3}\right) = 0.$$

Now $E(t^2)$

$$= \int\limits_{-\infty}^{\infty} t^2 f(t) dt = \int\limits_{-1}^{0} t^2 (1+t) dt + \int\limits_{0}^{1} t^2 (1-t) dt$$

$$= \int\limits_{-1}^{0} (t^2 + t^3) dt + \int\limits_{0}^{1} (t^2 - t^3) dt$$

$$= \left[\frac{t^3}{3} + \frac{t^4}{4}\right]_{-1}^{0} + \left[\frac{t^3}{3} - \frac{t^4}{4}\right]_{0}^{1}$$

$$= \left(\frac{1}{3} - \frac{1}{4}\right) + \left(\frac{1}{3} - \frac{1}{4}\right) = \frac{1}{6}.$$

So variance $= E(t^2) - \{E(t)\}^2 = \frac{1}{6} - 0^2 = \frac{1}{6}.$

Hence, the standard deviation $= \sqrt{\text{variance}} = \frac{1}{\sqrt{6}}.$

33. (d) Let the mean and standard deviation of the students of batch C be μ_C and σ_C, respectively. Also, let the mean and standard deviation of entire class of first year B.Tech students be μ and σ, respectively.

ATQ, $\mu_C = 6.6$, $\sigma_C = 2.3$, $\mu = 5.5$, and $\sigma = 4.2$.

In order to normalize batch C to entire class, the normalized score (z scores) must be equated.

For the entire class standard normal variate is given by $Z = \frac{x - \mu}{\sigma} = \frac{x - 5.5}{4.2}$, where x denotes the marks of a student.

For batch C, standard normal variate is given by $Z_C = \frac{x_C - \mu_C}{\sigma_C} = \frac{8.5 - 6.6}{2.3}$, where denotes the marks of a student in batch C.

Then, $\frac{8.5 - 6.6}{2.3} = \frac{x - 5.5}{4.2} \Rightarrow x = 8.969 \approx 9.0.$

34. (c) $\int\limits_{-\infty}^{\infty} p(x).dx = 1$

$$\Rightarrow \int\limits_{-\infty}^{\infty} K e^{-\alpha|x|} dx = 1$$

$$\Rightarrow \int\limits_{-\infty}^{0} K e^{-\alpha|x|} dx + \int\limits_{0}^{\infty} K e^{-\alpha|x|} dx = 1$$

$$\Rightarrow \int\limits_{-\infty}^{0} K e^{\alpha x} dx + \int\limits_{0}^{\infty} K e^{-\alpha x} dx = 1$$

$$\Rightarrow \frac{K}{\alpha}\left[e^{\alpha x}\right]_{-\infty}^{0} + \frac{K}{(-\alpha)}\left[e^{-\alpha x}\right]_{0}^{\infty} = 1 \Rightarrow \frac{K}{\alpha} + \frac{K}{\alpha} = 1$$

$$\Rightarrow K = 0.5\alpha.$$

35. (c) The probability of occurrence of values 1, 5, and 6 on the three dice

= Probability of occurrence of value 1 × Probability of occurrence of values 5 × Probability of occurrence of values 6

$$= \frac{1}{4} \times \frac{1}{8} \times \frac{1}{4} = \frac{1}{128}.$$

36. (d) By the properties of expectation, co-variance and variance.

37. (c) Coefficient of variation $= \frac{\sigma}{\mu} = \frac{8.8}{33} = 0.2666.$

38. (c) Let, A and B denote the event of failing in paper 1 and paper 2, respectively.

Given that, $P(A) = 0.3$, $P(B) = 0.2$, and $P(A/B) = 0.6$.

Then, the probability of a student failing in both the papers

$$= P(A \cap B) = P(B) \times P(A/B) = 0.2 \times 0.6 = 0.12.$$

39. (a)

Prob. (drawing an Ace) $= \frac{^4C_1}{^{52}C_1} = \frac{1}{13}.$

Therefore, the probability of getting both "Aces"

$$= \frac{1}{13} \times \frac{1}{13} = \frac{1}{169}.$$

40. (d)

$$\int\limits_{-\infty}^{\infty} f(x) dx = 1$$

$$\Rightarrow \int\limits_{0}^{2} k(5x - 2x^2) dx = 1 \Rightarrow k\left[5\frac{x^2}{2} - 2\frac{x^3}{3}\right]_{0}^{2} = 1$$

$$\Rightarrow k\left(10 - \frac{16}{3}\right) = 1 \Rightarrow k = \frac{3}{14}.$$

$\therefore P(X > 1)$

$$= \int\limits_{1}^{2} f(x) dx = \int\limits_{1}^{2} k(5x - 2x^2) dx = \frac{3}{14}\left[5\frac{x^2}{2} - 2\frac{x^3}{3}\right]_{1}^{2}$$

$$= \frac{3}{14}\left[\frac{5}{2}(4 - 1) - \frac{2}{3}(8 - 1)\right] = \frac{3}{14} \times \frac{17}{6} = \frac{17}{28}.$$

41. (a) Sum of probabilities = 1

$$\Rightarrow \frac{2+5p}{5}+\frac{1+3p}{5}+\frac{1.5+2p}{5}=1 \Rightarrow p=0.05.$$

$E(X)$

$$=1\times\frac{2+5p}{5}+2\times\frac{1+3p}{5}+3\times\frac{1.5+2p}{5}$$

$$=\frac{8.5+17p}{5}=\frac{8.5+17\times0.05}{5}=1.87$$

42. (a) Let X be the random variable denoting the number of heads.

Then clearly X follows the binomial distribution with $n=4$ and p = prob. of getting heads in a single toss = 0.5.

So the probability of getting heads exactly three times

$$=P(X=3)= {}^4C_3(0.5)^3 (1\text{-}0.5)^{4-3}=\frac{1}{4}.$$

43. (a) We know if X follows a uniform distribution in (a,b), then $\text{Var}(X)=\frac{(b-a)^2}{12}$ for $a<x<b$

Here, $a=2, b=10$.

Therefore, $\text{Var}(X)=\frac{(10-2)^2}{12}=\frac{16}{3}.$

44. (c) We know if X follows a uniform distribution in (a,b), then p.d.f, $f(x)=\frac{1}{b-a}$ for $a<x<b$

Here, $a=0$ and $b=1$. So $f(x)=1$.

$\therefore E(X^3)$

$$=\int_0^1 x^3 f(x)dx=\int_0^1 x^3\times1 dx=\left[\frac{x^4}{4}\right]_0^1=\frac{1}{4}.$$

45. (b)

$$P\left(-\frac{1}{3}\le X\le\frac{1}{3}\right)$$

$$=\int_{-\frac13}^{\frac13} f(x)dx=\int_{-\frac13}^{\frac13} x^2 dx=\left[\frac{x^3}{3}\right]_{-\frac13}^{\frac13}=\frac{1}{3}\left(\frac{1}{27}+\frac{1}{27}\right)$$

$$=\frac{2}{81}.$$

Therefore, the probability expressed in percentage

$$p=\frac{2}{81}\times100=2.469\%=2.47\%$$

46. (a)

Given, $\mu_X=1, \sigma_X^2=4, \mu_Y=-1$.

Also given, $P(X\le-1)=P(Y\ge2)$.

Now converting into standard normal variates we get,

$$P\left(Z\le\frac{-1-\mu_X}{\sigma_X}\right)=P\left(Z\ge\frac{2-\mu_Y}{\sigma_Y}\right)$$

or, $P\left(Z\le\frac{-1-1}{2}\right)=P\left(Z\ge\frac{2-(-1)}{\sigma_y}\right)$

or, $P(Z\le-1)=P\left(Z\ge\frac{3}{\sigma_Y}\right)$ ...(i)

We know that in case of standard normal distribution, $P(Z\le-1)=P(Z\ge1)$ ...(ii)

Comparing (i) and (ii) we get,

$\frac{3}{\sigma_Y}=1$, i.e., $\sigma_Y=3$

47. (a)

$$\int_{-\infty}^{\infty} P(X)dx=1$$

$$\Rightarrow \int_{-\infty}^{\infty}\left(Me^{-2|x|}+Ne^{-3|x|}\right)dx=1$$

$$\Rightarrow 2\int_0^{\infty}\left(Me^{-2|x|}+Ne^{-3|x|}\right)dx=1$$

$$\left[\because Me^{-2|x|}+Ne^{-3|x|} \text{ is an even function of } x\right]$$

$$\Rightarrow 2\int_0^{\infty}\left(Me^{-2x}+Ne^{-3x}\right)dx=1$$

$$\Rightarrow 2\left[M\frac{e^{-2x}}{(-2)}+N\frac{e^{-3x}}{(-3)}\right]_0^{\infty}=1$$

$$\Rightarrow 2\left[\left\{M\frac{e^{-\infty}}{(-2)}+N\frac{e^{-\infty}}{(-3)}\right\}-\left\{M\frac{e^0}{(-2)}+N\frac{e^0}{(-3)}\right\}\right]$$

$$\Rightarrow M+\frac{2}{3}N=1\left[\because e^{-\infty}=0, e^0=1\right]$$

48. (c) Let p = probability of getting heads and q = probability of getting tails.

Then, $p=q=1/2$.

Therefore, the probability (player X wins the game if he starts the game)

$$=p+q^2p+q^4p+\ldots\infty$$

$$=p\left(1+q^2+q^4+\ldots\infty\right)$$

$$=p\times\frac{1}{1-q^2}$$

$$\left[\begin{array}{c}\text{using } 1+ar+ar^2+ar^3+\ldots\infty=\frac{a}{1-r};\\ \text{here } a=1, r=q^2\end{array}\right]$$

$$=\frac{1}{2}\times\frac{1}{\left(1-\frac{1}{4}\right)}=\frac{2}{3}.$$

49. (c) p = probability of getting heads = 1/2 and q = probability of getting tails = 1/2.

Since each toss is independent, so the required probability

$= p \times p \times q \times q \times q \times q \times q \times q \times q \times q$

$= \frac{1}{2} \times \frac{1}{2} \times \frac{1}{2} \times \frac{1}{2} \times \frac{1}{2} \times \frac{1}{2} \times \frac{1}{2} \times \frac{1}{2} \times \frac{1}{2} \times \frac{1}{2} = \left(\frac{1}{2}\right)^{10}$.

50. (a) Here, the sample space = {HH, TH, HT}.
Let A = event that both outcomes are heads.

Then, A = {HH}.

Hence, the required probability $= \frac{n(A)}{n(S)} = \frac{1}{3}$.

51. (a) We know if X follows a uniform distribution in (a, b), then its variance $= \frac{(b-a)^2}{12}$.

Here $a = 0$ and $b = 1$. So variance $= \frac{(1-0)^2}{12} = \frac{1}{12}$.

So standard deviation $= \sqrt{\text{variance}} = \frac{1}{\sqrt{12}}$

52. (b) Let A = the event that the face value is odd and

B = the event that the face value is even.

ATQ, $P(A) = 0.9 \times P(B)$.

Now, the sum of the probabilities equals to 1

$\Rightarrow P(A) + P(B) = 1 \Rightarrow 0.9 \times P(B) + P(B) = 1$

$\Rightarrow P(B) = \frac{1}{1.9} = 0.5263$ ...(i)

Given that Prob. (getting any even face) is same i.e; Prob. (getting the face 2) = Prob. (getting the face 4) = Prob. (getting the face 6).

Now Prob. (getting an even face)

= Prob. (getting the face 2 or 4 or 6)

= Prob.(getting the face 2) + Prob.(getting the face 4) +Prob. (getting the face 6)

= 0.5263 (using (i))

∴ Prob. (getting the face 2) = Prob. (getting the face 4) = Prob. (getting the face 6) = $\frac{1}{3} \times 0.5263 = 0.1754$.

Given, Prob. (the face is even given that it is greater than 3) = 0.75.

This implies,

$\frac{\text{P(getting an even face which is greater than 3)}}{\text{P(getting a face which is greater than 3)}} = 0.75$

or, $\frac{\text{P(getting the face 4 or 6)}}{\text{P(getting a face which is greater than 3)}} = 0.75$

or, $\frac{\text{P(getting the face 4)+P(getting the face 6)}}{\text{P(getting a face which is greater than 3)}} = 0.75$

or, P(getting a face which is greater than 3)

$= \frac{0.1754 + 0.1754}{0.75} = 0.4677 \cong 0.468$.

53. (c) ATQ,

$P(X = 0) = P(Y = 0) = 1/2, P(X = 1) = P(Y = 1) = 1/4$,

$P(X = 2) = P(Y = 2) = 1/4$.

∴ P(X+Y=2/X-Y=0)

$= \frac{\text{P(X+Y=2∩X-Y=0)}}{\text{P(X-Y=0)}}$

$= \frac{\text{P(X=1,Y=1)}}{\text{P(X=0,Y=0 or X=1,Y=1 or X=2,Y=2)}}$

$= \frac{\text{P(X=1,Y=1)}}{\text{P(X=0,Y=0) + P(X=1,Y=1)+P(X=2,Y=2)}}$

$= \frac{\frac{1}{4} \times \frac{1}{4}}{\frac{1}{2} \times \frac{1}{2} + \frac{1}{4} \times \frac{1}{4} + \frac{1}{4} \times \frac{1}{4}} = \frac{1}{6}$.

54. (d) Let X be the random variable denoting the number of heads.

Then clearly X follows the binomial distribution with $n = 3$ and $p = \frac{1}{2}$.

So the required probability

$= P(X \geq 1) = 1 - P(X < 1)$

$= 1 - P(X = 0) = 1 - {}^3C_0 \left(\frac{1}{2}\right)^0 \left(1-\frac{1}{2}\right)^3$

$= 1 - \frac{1}{8} = \frac{7}{8}$.

55. (b) Mean = $1 \times 0.1 + 2 \times 0.2 + 3 \times 0.4 + 4 \times 0.2 + 5 \times 0.1 = 3$.

Variance,

$= 1^2 \times 0.1 + 2^2 \times 0.2 + 3^2 \times 0.4 + 4^2 \times 0.2 + 5^2 \times 0.1 - (\text{mean})^2 = 1.2$.

56. (c) Let A = the event that the report is positive, B = the event that the report is negative and E = the event that the report is incorrect.

Then, *ATQ*,

$P(A) = 0.01$, $P(E/A) = 12\% = 0.12$ and $P(E/B) = 15\% = 0.15$.

Therefore, $P(B) = 1 - 0.01 = 0.99$.

Hence, the required probability

$= P(A) \times P(E/A) + P(B) \times P(E/B)$

$= 0.01 \times 0.12 + 0.99 \times 0.15 = 0.1497$.

57. (*d*) Given, $\mu = 102$ cm, $\sigma = 27$ cm.

$\therefore P(90 \le X \le 102)$

$= P\left(\dfrac{90 - \mu}{\sigma} \le \dfrac{X - \mu}{\sigma} \le \dfrac{102 - \mu}{\sigma} \right)$

$= P\left(\dfrac{90 - 102}{27} \le x_n \le \dfrac{102 - 102}{27} \right)$

$= P(-0.44 \le x_n \le 0) = F(0) - F(-0.44)$

$= \dfrac{1}{1 + \exp(0)} - \dfrac{1}{1 + \exp\left(-1.7255 \times (-0.44) \times |0.44|^{0.12} \right)}$

$= 0.5 - 0.3345 = 0.1655 \approx 16.55\%$

Hence, the closest answer is 16.7%.

58. (*a*) Given that, the probability of a faulty assembly of any computer = p and probability that the testing process gives the correct result for any computer = q.

Then the probability of right (correct) assembly of any computer = $1 - p$ and probability that the testing process gives the incorrect result for any computer = $1 - q$.

So the required probability

= Prob. [(faulty assembly of the computer and the testing process gives the correct result) or (correct assembly of the computer and the testing process gives the incorrect result)]

= Prob. (faulty assembly of the computer and the testing process gives the correct result) $+ P$ (correct assembly of the computer and the testing process gives the incorrect result)

$= p \times q + (1 - p) \times (1 - q)$.

59. (*c*) Initially, there are $2 + 3 + 4$, *i.e.*, 9 items in the box.

So the required probability

$= \left(\dfrac{2}{9} \times \dfrac{1}{8} \right) \times \left(\dfrac{3}{7} \times \dfrac{2}{6} \times \dfrac{1}{5} \right) \times \left(\dfrac{4}{4} \times \dfrac{3}{3} \times \dfrac{2}{2} \times \dfrac{1}{1} \right) = \dfrac{1}{1260}$.

60. (*d*) Let X be the random variable denoting the number of heads.

Then X follows the binomial distribution with $n = 4$ and $p = 1/2$, $q = 1/2$.

Hence, Prob. (number of heads > number of tails)

$= P(X = 3) + P(X = 4)$

$= {}^4C_4 \left(\dfrac{1}{2} \right)^4 \left(1 - \dfrac{1}{2} \right)^{4-4} + {}^4C_3 \left(\dfrac{1}{2} \right)^3 \left(1 - \dfrac{1}{2} \right)^{4-3}$

$= 1 \times \dfrac{1}{16} \times 1 + 4 \times \dfrac{1}{8} \times \dfrac{1}{2} = \dfrac{5}{16}$.

61. (*c*) As the first removed ball is of white color, so after first withdrawn only six balls are left in the box out of which three are red.

Hence, Prob. (the second removed ball is red)

$= \dfrac{{}^3C_1}{{}^6C_1} = \dfrac{1}{2}$.

62. (*d*) The p.m.f of a Poisson variate X is given by

$f(x) = P(X = x) = \dfrac{e^{-m} m^x}{x!}, x = 0, 1, 2, 3, \dots\dots, \infty$

where m = mean.

ATQ, $m = 2$.

Therefore, the required probability

$= P(X \ge 2) = 1 - P(X < 2) = 1 - \{P(X = 0) + P(X = 1)\}$

$= 1 - \left(\dfrac{e^{-2} 2^0}{0!} + \dfrac{e^{-2} 2^1}{1!} \right) = 1 - 3e^{-2}$.

63. (*c*) Here,

$S = \{(1,1),(1,2),(1,3),(1,4),(1,5),(1,6),(2,1),$
$(2,2),(2,3),(2,4),(2,5),(2,6),(3,1),\dots\dots,(3,6),$
$(4,1),\dots\dots,(4,6),(5,1),\dots\dots,(5,6),(6,1),(6,2),$
$(6,3),(6,4),(6,5),(6,6)\}$.

Let A = event that second toss results in a value that is higher than the first toss.

Then,

$A = \{(1,2),(1,3),(1,4),(1,5),(1,6),(2,3),(2,4),$
$(2,5),(2,6),(3,4),(3,5),(3,6),(4,5),(4,6),(5,6)\}$.

Hence, the required probability

$= P(A) = \dfrac{n(A)}{n(S)} = \dfrac{15}{36} = \dfrac{5}{12}$.

64. (c) The first container has seven balls out of which four are red and the second container has seven balls out of which three are blue.

Therefore, Prob. (one ball is red and another is blue)

$$= \frac{4}{7} \times \frac{3}{7} = \frac{12}{49}.$$

65. (c)

$$R = Var(X) = E(X^2) - \left[E(X)\right]^2 \geq 0$$

$$(\because Var(X) \geq 0)$$

66. (d) We know that standard deviation is affected by change of scale but not by shifting of origin.

Hence, $y_i = ax_i + b \quad \Rightarrow \sigma_y = a\sigma_x$.

So option (d) is incorrect.

67. (d) Let X be the random variable denoting the number of heads.

Then clearly X follows the binomial distribution with $n = 5$ and $p = q = 1/2$.

Then the probability of getting at least one heads

$$= P(X \geq 1) = 1 - P(x < 1) = 1 - P(X = 0)$$

$$= 1 - {}^5C_0\left(\frac{1}{2}\right)^0\left(1 - \frac{1}{2}\right)^{5-0} = 1 - \frac{1}{32} = \frac{31}{32}.$$

68. (a) Here, the sample space, S

$$= \{(1,2),(1,3),(1,4),(1,5),(2,1),(2,3),(2,4),(2,5),$$
$$(3,1),(3,2),(3,4),(3,5),(4,1),(4,2),(4,3),(4,5),$$
$$(5,1),(5,2),(5,3),(5,4)\}.$$

Let A = the event that the number on the first card is one higher than the number on the second card.

Then, $A = \{(2,1), (3, 2), (4, 3), (5, 4)\}$.

So the required probability = $P(A)$

$$= \frac{n(A)}{n(S)} = \frac{4}{20} = \frac{1}{5}.$$

69. (b) The p.m.f of a Poisson variate X is given by

$$f(x) = P(X = x) = \frac{e^{-m}m^x}{x!}, x = 0,1,2,3,\dots\dots,\infty$$

where m = mean.

Here, $m = 2 \times 3 = 6$ (since mean per year = 3).

Therefore, the required probability

$$= P(X \leq 2) = P(X = 0) + P(X = 1) + P(X = 2)$$

$$= \frac{e^{-6}6^0}{0!} + \frac{e^{-6}6^1}{1!} + \frac{e^{-6}6^2}{2!} = e^{-6} \times 25 = 0.0619 \approx 0.062.$$

70. (d) Here, the sample space, $S = \{(3, 6), (3, 11),$ $(3, 16), (3, 21), (3, 26), (6, 6), (6, 11), (6, 16),$ $(6, 21), (6, 26), (9, 6), (9, 11), (9, 16), (9, 21),$ $(9, 26), (12, 6), (12, 11), (12, 16), (12, 21),$ $(12, 26), (15, 6), (15, 11), (15, 16), (15, 21),$ $(15, 26)\}$.

Let A = the event that the product of the numbers is even.

Then, $A = \{(3,6), (3,16), (3,26), (6,6), (6,11),$ $(6,16), (6,21), (6,26), (9,6), (9,16), (9,26),$ $(12,6), (12,11), (12,16), (12,21), (12,26), (15,6),$ $(15,16), (15,26)\}$.

Then the required probability = $P(A)$

$$= \frac{n(A)}{n(S)} = \frac{19}{25}.$$

71. (c) Prob. (getting heads)

$$= \frac{1}{2} \text{ \& Prob.(getting tails)} = \frac{1}{2}.$$

Then the required probability

= Prob. (getting heads in first toss) + Prob. (getting tails in first two tosses and heads in third toss) + Prob. (getting tails in first four tosses and heads in fifth toss) +........∞

$$= \frac{1}{2} + \left(\frac{1}{2} \times \frac{1}{2} \times \frac{1}{2}\right) + \left(\frac{1}{2} \times \frac{1}{2} \times \frac{1}{2} \times \frac{1}{2} \times \frac{1}{2}\right) + \dots\dots\infty$$

$$= \frac{1}{2}\left[1 + \frac{1}{4} + \frac{1}{4^2} + \dots\right] = \frac{1}{2}\left[\frac{1}{1 - \frac{1}{4}}\right] = \frac{2}{3}.$$

72. (d) Let X be the random variable representing the number of negative values in five trials.

Then X follows the binomial distribution with $n = 5$ and $p = q = 1/2$ (since negative and positive values are equally likely to occur).

Then the probability of obtaining at most one negative value in five trials

$$= P(X \leq 1) = P(X = 0) + P(X = 1)$$

$$= {}^5C_0\left(\frac{1}{2}\right)^0\left(\frac{1}{2}\right)^{5-0} + {}^5C_1\left(\frac{1}{2}\right)^1\left(\frac{1}{2}\right)^{5-1} = \frac{6}{32}$$

73. (b) If first throw shows the number 1, 2, or 3 then the sample space, S

= {(1,1),(1,2),(1,3),(1,4),(1,5),(1,6),(2,1),(2,2),
(2,3),(2,4),(2,5),(2,6),(3,1),(3,2),(3,3),(3,4),
(3,5),(3,6)}.

Out of these, the ordered pairs (1, 5), (1, 6), (2, 4), (2, 5), (2, 6), (3, 3), (3, 4), (3, 5),(3, 6) (in total 9 outcomes) give a sum ≥ 6.

Now if first throw shows the number 4, 5, or 6 then second throw is not made and so the only way to get the sum at least 6, is to consider the outcome "6" only.

Hence, the required probability

$$= \frac{1}{2} \times \frac{9}{18} + \frac{1}{2} \times \frac{1}{3} = \frac{15}{36} = \frac{5}{12}.$$

74. (d) Prob. (the selected set contains one red ball and two black balls)

= Prob. [(first drawn ball is red, second drawn ball is black and third drawn ball is black) or (first drawn ball is black, second drawn ball is black and third drawn ball is red) or (first drawn ball is black, second drawn ball is red, and third drawn ball is black)]

= Prob. (first drawn ball is red, second drawn ball is black and third drawn ball is black) + Prob. (first drawn ball is black, second drawn ball is black and third drawn ball is red) + Prob. (first drawn ball is black, second drawn ball is red, and third drawn ball is black)

= Prob. (first drawn ball is red)×Prob. (second drawn ball is black) × Prob. (third drawn ball is black) + Prob. (first drawn ball is black) × Prob. (second drawn ball is black) × Prob. (third drawn ball is red) + Prob. (first drawn ball is black) × Prob. (second drawn ball is red) × Prob. (third drawn ball is black)

$$= \left(\frac{4}{10} \times \frac{6}{9} \times \frac{5}{8}\right) + \left(\frac{6}{10} \times \frac{5}{9} \times \frac{4}{8}\right) + \left(\frac{6}{10} \times \frac{4}{9} \times \frac{5}{8}\right) = \frac{1}{2}.$$

75. (c) The probability distribution of the random variable X is given by

X	−1	1
P(X = x)	0.5	0.5

Then the cumulative distribution function $F(x)$ is given by

$$F(x) = \begin{cases} 0, \text{ if } x < -1 \\ 0.5, \text{ if } -1 \le x < 1 \\ 0.5 + 0.5 = 1, \text{ if } x \ge 1 \end{cases}$$

So the correct option is (c).

76. (b) Let us consider the following three events:

A = the event that a shock absorber is supplied by X,

B = the event that a shock absorber is supplied by Y and C = the event that a shock absorber is reliable.

So ATQ,

P(A) = 0.60, P(B) = 0.40, P(C/A) = 0.96 & P(C/B) = 0.72.

Then by Bayes' theorem, the required probability

$$= P(B/C) = \frac{P(B) \times P(C/B)}{P(A) \times P(C/A) + P(B) \times P(C/B)}$$

$$= \frac{0.40 \times 0.72}{0.60 \times 0.96 + 0.40 \times 0.72} = 0.334.$$

77. 6.

$$\int_{-\infty}^{\infty} f(x)dx = 1$$

$$\Rightarrow \int_{1}^{2} \lambda(x-1)(2-x)dx = 1 \Rightarrow \int_{1}^{2} \lambda(-x^2 + 3x - 2)dx = 1$$

$$\Rightarrow \lambda \left[-\frac{x^3}{3} + \frac{3x^2}{2} - 2x\right]_{1}^{2} = 1$$

$$\Rightarrow \lambda \left[-\left(\frac{8}{3} - \frac{1}{3}\right) + \frac{3}{2}(4-1) - 2(2-1)\right] = 1$$

$$\Rightarrow \lambda \left(-\frac{7}{3} + \frac{9}{2} - 2\right) = 1 \Rightarrow \lambda = 6.$$

78. (c)

The p.m.f of a Poisson variate X is given by

$$P(X = x) = f(x) = \frac{e^{-\lambda}\lambda^x}{x!}, x = 0,1,2,......\infty$$

where λ = mean of the Poisson distribution = 3 (given).

Hence, the probability of observing fewer than three cars

= P(X < 3) = P(X = 0) + P(X = 1) + P(X = 2)

$$= \frac{e^{-3} \times 3^0}{0!} + \frac{e^{-3} \times 3^1}{1!} + \frac{e^{-3} \times 3^2}{2!} = \frac{17}{2e^3}.$$

79. (a)
P(X > 1)

$$= \int_{1}^{\infty} f(x)dx = \int_{1}^{\infty} e^{-x}dx = \left[-e^{-x}\right]_{1}^{\infty} = 0 - (-e^{-1}) = 0.368.$$

80. (b) Here, $\sigma^2 = 4 \Rightarrow \sigma = 2$. Thus, $\mu = 1$ and $\sigma = 2$.

Now, $P(X > 0)$

$$= P\left(\frac{X-\mu}{\sigma} < \frac{0-\mu}{\sigma}\right) = P\left(Z < \frac{0-1}{2}\right) = P\left(Z < -\frac{1}{2}\right)$$

$$= 1 - P\left(Z < \frac{1}{2}\right)$$

which clearly lies between 0 and 0.5 (which is found from the diagram of the standard normal probability curve).

81. 50%

Let X be the random variable denoting the average daily balance.

Given, $\mu = 500$, $\sigma = 50$.

Therefore, $P(X > 500)$

$$= P\left(\frac{X-\mu}{\sigma} > \frac{500-\mu}{\sigma}\right) = P\left(Z > \frac{500-500}{50}\right)$$

$$= P(Z > 0) = 0.5 = 50\%$$

82. (a) Out of 25 parts 10 are defective and 15 are good.

Hence, the required probability

$$= \frac{^{15}C_2}{^{25}C_2} = \frac{\frac{15 \times 14}{2 \times 1}}{\frac{25 \times 24}{2 \times 1}} = \frac{7}{20}.$$

83. 0.666

Let there be n families in the housing society. Then, $\frac{n}{2}$ families have single child and $\frac{n}{2}$ families have two children.

Hence, the total number of children

$$= \frac{n}{2} \times 1 + \frac{n}{2} \times 2 = \frac{3n}{2}.$$

Now, the total number of cases in which a child picked at random has a sibling $= \frac{n}{2} \times 2 = n$.

Therefore, prob. (a child picked at random has a sibling) $= \frac{n}{\frac{3n}{2}} = \frac{2}{3} = 0.666$

84. (b) Since in any toss, getting tails is independent of the previous tosses, so the required probability

$$= \text{prob. of getting tails} = \frac{1}{2}.$$

85. 0.65

Let M = the event that a man is selected,

W = the event that a woman is selected and

E = the event that an employed person is selected.

Then, ATQ,

$P(M) = 1/2$, $P(W) = 1/2$, $P(E/M) = (100 - 20)\%$ $= 4/5$ and

$P(E/W) = (100 - 50)\% = 1/2$.

Then, the required probability

= Prob. (a man is selected who is employed or a woman is selected who is employed)

= Prob. (a man is selected who is employed)+ Prob. (a woman is selected who is employed)

$= P(E \cap M) + P(E \cap W)$

$= P(M) \times P(E/M) + P(W) \times P(E/W)$

$$= \frac{1}{2} \times \frac{4}{5} + \frac{1}{2} \times \frac{1}{2} = 0.65.$$

86. (c) If the fourth heads appears at the tenth toss, then three heads appear in first nine tosses.

Now, if X be a random variable denoting the number of heads in nine trials, then X must follow binomial distribution with p = prob. of getting heads in a single trial = 1/2.

So Prob. (exactly three heads appear in first nine trials)

$$= P(X=3) = {}^9C_3 \times \left(\frac{1}{2}\right)^3 \left(1-\frac{1}{2}\right)^6 = {}^9C_3 \times \left(\frac{1}{2}\right)^9.$$

In the tenth trial heads must appears and the probability of getting heads in the tenth trial = 1/2.

Hence, the required probability

$$= {}^9C_3 \left(\frac{1}{2}\right)^9 \times \frac{1}{2} = \frac{21}{256} = 0.082.$$

87. (b) Let the number of heads = x. So the number of tails = $n - x$. Then difference between the number of heads and the number of tails $= x - (n - x)$.

Then, $x - (n - x) = n - 3 \Rightarrow x = \frac{2n-3}{2}$, which may not be an integer.

This contradicts the fact that x denotes the number of heads.

∴ required probability = zero.

88. 10.

To get the sum of 22, we have the following two cases:

Case-I: Three 6's and one 4.

No. of ways of getting three 6's and one 4 = $\frac{4!}{3!} = 4$

Case-II: Two 6's and two 5's.

No. of ways of getting two 6's and two 5's = $\frac{4!}{2!2!} = 6$.

Hence, the prob. of getting the sum of 22
$= \frac{6+4}{6^4} = \frac{10}{1296}$
(since the sample space contains 6^4 elements).

ATQ, $\frac{10}{1296} = \frac{x}{1296}$. $\therefore x = 10$.

89. $\frac{1}{7}$.

Let the probability of occurence of one dot be p.

Then, the sum of all probabilities = 1

$\Rightarrow p + 2p + 3p + 4p + 5p + 6p = 1 \Rightarrow p = \frac{1}{21}$.

Hence, the probability of occurence of 3 dots
$= 3 \times p = \frac{3}{21} = \frac{1}{7}$.

90. 0.25

A and B are mutually exclusive and $A \cup B = S$

$\Rightarrow A$ and B are mutually exhaustive events.

Then, $P(A \cup B) = P(S)$

$\Rightarrow P(A) + P(B) = 1 \Rightarrow P(B) = 1 - P(A)$.

Now, we need to maximize $P(A) \times P(B)$.

$P(A)P(B) = P(A)(1 - P(A))$

$= x(1-x) = x - x^2 = y$(say) [taking $P(A) = x$]

Then, $\frac{dy}{dx} = 0 \Rightarrow 1 - 2x = 0 \Rightarrow x = \frac{1}{2}$.

Now $\frac{d^2y}{dx^2}\Big|_{x=\frac{1}{2}} = -2\big|_{x=\frac{1}{2}} = -2 < 0$.

Hence, y, *i.e.*, $P(A)P(B)$ has a maximum value at $x = \frac{1}{2}$.

The maximum value of $P(A)$ $P(B)$
$= \frac{1}{2} - \left(\frac{1}{2}\right)^2 = 0.25$.

91. 0.44

Prob. (a parcel is lost by first post office) $= \frac{1}{5}$.

Prob. (a parcel is lost by second post office)
$= \left(1 - \frac{1}{5}\right) \times \frac{1}{5} = \frac{4}{25}$.

So Prob. (a parcel is lost)

= Prob. (a parcel is lost by first post office or by second post office)

= Prob. (a parcel is lost by first post office) + P (a parcel is lost by second post office)

$= \frac{1}{5} + \frac{4}{25} = \frac{9}{25}$.

Hence, the required probability
$= \frac{4/25}{9/25} = \frac{4}{9} = 0.444$.

92. (*a*) Let X be the random variable denoting the number of defective pieces.

Then, the probability distribution of X is given by

x	0	1	2
$P(X = x)$	$\frac{1}{6}$	$\frac{2}{3}$	$\frac{1}{6}$

So mean = $E(X) = 0 \times \frac{1}{6} + 1 \times \frac{2}{3} + 2 \times \frac{1}{6} = 1$.

Var(X)
$= E(X^2) - \{E(X)\}^2$
$= \left[0^2 \times \frac{1}{6} + 1^2 \times \frac{2}{3} + 2^2 \times \frac{1}{6}\right] - 1^2 = \frac{1}{3}$.

93. 11.9

Prob. (at least three computers are working out of four)

= Prob. (three computers are working and one is non-working) or (all four computers are working)

= Prob. (three computers are working and one is non-working) + P(all four computers are working)

$= \frac{{}^4C_3 \times {}^6C_1}{{}^{10}C_4} + \frac{{}^4C_4}{{}^{10}C_4} = \frac{4}{35} + \frac{1}{210} = \frac{25}{210}$.

$\therefore p = \frac{25}{210}$ and so $100p = 100 \times \frac{25}{210} = 11.9$

94. 0.265

The p.m.f of a Poisson variate X is given by

$$P(X = x) = f(x) = \frac{e^{-\lambda}\lambda^x}{x!}, \; x = 0,1,2,......\infty$$

where λ = mean of the Poisson distribution.

Here the random variable X gives the number of penalties per day and $\lambda = 5$.

Hence, Prob. (there will be less than four penalties in a day)

$= P(X < 4) = P(X = 0) + P(X = 1) + P(X = 2) + P(X = 3)$

$$= \frac{e^{-5}5^0}{0!} + \frac{e^{-5}5^1}{1!} + \frac{e^{-5}5^2}{2!} + \frac{e^{-5}5^3}{3!}$$

$$= e^{-5}\left[1 + 5 + \frac{25}{2} + \frac{125}{6}\right] = e^{-5} \times \frac{118}{3} = 0.265$$

95. (b) The p.m.f of a Poisson variate X is given by

$$P(X = x) = f(x) = \frac{e^{-\lambda}\lambda^x}{x!}, \; x = 0,1,2,......\infty$$

where λ = mean of the Poisson distribution.

Here, the random variable X gives the number of accidents per month and $\lambda = 5.2$.

Hence, the probability of occurrence of less than two accidents in the plant during a randomly selected month

$= P(X < 2) = P(X = 0) + P(X = 1)$

$$= e^{-5.2} \times \frac{5.2^0}{0!} + e^{-5.2} \times \frac{5.2^1}{1!} = e^{-5.2} \times 6.2 = 0.034$$

96. 0.26

Let A = the event that the number is divisible by 2,

B = the event that the number is divisible by 3 and

C = the event that the number is divisible by 5.

Then, $n(A) = \frac{100}{2} = 50, n(B) = \frac{99}{3} = 33, n(C) = \frac{100}{5} = 20.$

Now,

$n(A \cap B)$

= the no. of elements that are divisible by both 2 and 3

= the no. of elements that are divisible by l.c.m (2, 3), i.e., 6

$$= \frac{96}{6} = 16,$$

$n(B \cap C)$

= the no. of elements that are divisible by both 3 and 5

= the no. of elements that are divisible by l.c.m(3,5), i.e.,15

$$= \frac{90}{15} = 6,$$

$n(A \cap C)$

= the no. of elements that are divisible by both 2 and 5

= the no. of elements that are divisible by l.c.m(2, 5), i.e., 10

$$= \frac{100}{10} = 10,$$

$n(A \cap B \cap C)$

= the no. of elements that are divisible by 2, 3, and 5

= the no. of elements that are divisible by l.c.m (2, 3, 5), i.e., 30

$$= \frac{90}{30} = 3.$$

Then, $P(A \cup B \cup C)$

$= P(A) + P(B) + P(C) - P(A \cap B) - P(B \cap C) - P(A \cap C)$

$+ P(A \cap B \cap C)$

$$= \frac{50}{100} + \frac{33}{100} + \frac{20}{100} - \frac{16}{100} - \frac{6}{100} - \frac{10}{100} + \frac{3}{100}$$

$$= \frac{74}{100} = 0.74$$

Therefore, Prob. (the chosen number is not divisible by 2 or 3 or 5)

= 1– Prob. (the chosen number is divisible by at least one of 2, 3, and 5)

$= 1 - P(A \cup B \cup C) = 1 - 0.74 = 0.26$

97. 0.4

$P(2 < E < 4)$

$$= \int_2^4 f(x)\,dx = \int_2^4 \frac{1}{5}\,dx = \frac{1}{5}[x]_2^4 = \frac{2}{5} = 0.4$$

98. 0.4

Probability $(0.5 < X < 5)$

$$= \int_{0.5}^{5} f(x)\,dx = \int_{0.5}^{1} 0.2\,dx + \int_{1}^{4} 0.1\,dx + \int_{4}^{5} 0\,dx$$

$$= 0.2 \times [1 - 0.5] + 0.1 \times [4 - 1] + 0 = 0.4$$

99. (b) We know that in a normal distribution, the area enclosed by the normal curve from $-\infty$ to the mean = 0.5.

Here, "a" is the mean.

Hence, $\int_{-\infty}^{a} \frac{1}{\sqrt{2\pi} * b} e^{-\frac{1}{2}\left(\frac{x-a}{b}\right)^2} dx$

= the area under the normal curve from $-\infty$ to the mean "a" = 0.5

100. (d) Mean

= $E(x)$ = 1 × 0.3 + 2 × 0.6 + 3 × 0.1 = 1.8.

Var(x)

$= E(x^2) - \{E(x)\}^2$

$= (1^2 \times 0.3 + 2^2 \times 0.6 + 3^2 \times 0.1) - (1.8)^2$

$= 3.60 - 3.24 = 0.36$

So, the standard deviation $= \sqrt{Var(x)} = (0.36)^{1/2}$
= 0.6

101. 0.259

Let X be the random variable denoting the number of times red color appeared on the top face of the dice.

Then, X follows the binomial distribution with n = 3 and p = prob. (getting the color red) = $\frac{2}{6} = \frac{1}{3}$.

Hence, Prob. (getting the color red on top face at least twice)

$= P(X \geq 2) = P(X = 2) + P(X = 3)$

$= {}^3C_2 \left(\frac{1}{3}\right)^2 \left(1 - \frac{1}{3}\right)^{3-2} + {}^3C_3 \left(\frac{1}{3}\right)^3 \left(1 - \frac{1}{3}\right)^{3-3}$

$= 3 \times \frac{2}{27} + \frac{1}{27} = \frac{7}{27} = 0.259$

102. 3.88

Let X be the random variable denoting the length of the word drawn.

The X takes the values 3, 4, and 5.

Also $P(X = 3) = P(\{$The, fox, the, dog$\}) = 4/9$,

$P(X = 4) = P(\{$over, lazy$\}) = 2/9$,

$P(X = 5) = P(\{$quick, brown, jumps$\}) = 3/9$.

Then, the probability distribution of X is given by

X	3	4	5
P(x)	4/9	2/9	3/9

Therefore, the expected length

$= E(X) = 3 \times \frac{4}{9} + 4 \times \frac{2}{9} + 5 \times \frac{3}{9} = 3.88.$

103. 0.25

Let X be the random variable denoting the length of the shorter stick. Then, X lies between 0 and $\frac{1}{2}$.

Then, X follows the uniform distribution with p.m.f

$f(x) = \dfrac{1}{\frac{1}{2} - 0} = 2, 0 < x < \frac{1}{2}.$

Therefore, the required expected length

$= \text{mean} = E(X) = \int_{0}^{\frac{1}{2}} x f(x)\,dx = \int_{0}^{\frac{1}{2}} 2x\,dx = 2\left[\frac{x^2}{2}\right]_0^{\frac{1}{2}}$

$= 0.25.$

104. 1.2

Here size of the sample, n = 100 and

Standard deviation, σ = 12.

So the standard error of mean $= \dfrac{\sigma}{\sqrt{n}} = \dfrac{12}{\sqrt{100}} = 1.2.$

105. (d) Let A = event that a chosen day is a rainy day, B = event that a chosen day is a non-rainy day and E = event that the fair will make no loss.

Then, ATQ,

$P(A) = 70\% = \dfrac{7}{10}$, $P(B) = (100 - 70)\% = \dfrac{3}{10}$,

$P(E/A) = (100 - 80)\% = \dfrac{2}{10}$, $P(E/B) = (100 - 10)\% = \dfrac{9}{10}.$

Therefore by Bayes' theorem, the required probability

$= P(B/E) = \dfrac{\frac{3}{10} \times \frac{9}{10}}{\frac{7}{10} \times \frac{2}{10} + \frac{3}{10} \times \frac{9}{10}} = \dfrac{27}{41}.$

106. 0.27

Let X be the random variable denoting the number of vehicles arriving over a 30 second time interval.

Then, X follows the Poisson distribution.

Mean, $\lambda = 240$ vehicles/hour $= \dfrac{240}{60}$ vehicles/min

$= \dfrac{240}{60 \times 2} = 2$ vehicles/30 second

Hence, the required probability

$= P(X = 1) = \dfrac{e^{-2} \times 2^{1}}{1!} = 0.27.$

107. (c)

Mean $= \dfrac{20 \times 25 + 20 \times 30 + 40 \times 35 + 20 \times 40}{20 + 20 + 40 + 20} = 33.$

As there are 100(even no.) observations, so median

$=$ Avg. marks of $\left(\dfrac{100}{2}\right)^{\text{th}}$ and $\left(\dfrac{100}{2} + 1\right)^{\text{th}}$ observations

$=$ Avg. marks of 50th and 51th observations

$= \dfrac{35 + 35}{2} = 35.$

Mode $=$ the mark with the highest frequency $= 35.$

108. 2

$\displaystyle\int_{-\infty}^{\infty} f(x)\,dx = 1$

$\Rightarrow \displaystyle\int_{0}^{\infty} \dfrac{e^{-\frac{x}{2}}}{k}\,dx = 1 \Rightarrow \dfrac{-2}{k}\left[e^{-\frac{x}{2}}\right]_{0}^{\infty} = 1$

$\Rightarrow \dfrac{-2}{k}[0 - 1] = 1 \Rightarrow k = 2 \ (\because e^{-\infty} = 0)$

109. (a) The number of phone calls in a fixed time interval always follow the Poisson's distribution.

110. $\dfrac{3}{7}.$

$\displaystyle\int_{-\infty}^{\infty} f(x)\,dx = 1 \Rightarrow \int_{1}^{2} Kx^{2}\,dx = 1$

$\Rightarrow K\left[\dfrac{x^{3}}{3}\right]_{1}^{2} = 1$

$\Rightarrow K\left(\dfrac{8}{3} - \dfrac{1}{3}\right) = 1$

$\Rightarrow K = \dfrac{3}{7}$

111. 0.33

Since X_1, X_2, and X_3, be independent and identically distributed random variables,

so prob. (X_1 is the largest) $=$ prob. (X_2 is the largest) $=$ prob. (X_3 is the largest) $= 1/3 = 0.33.$

112. 50

Clearly, X takes the values 1, 3, 5, ...,99 and $P(X = 1) = P(X = 3) = ...= P(X = 99) = \dfrac{1}{50}$ (as 50 odd numbers are there in between 1 and 100).

Hence, $E(X)$

$= 1 \times \dfrac{1}{50} + 3 \times \dfrac{1}{50} + \dots\dots\dots + 99 \times \dfrac{1}{50}$

$= \dfrac{1}{50}(1 + 3 + \dots\dots\dots + 99)$

$= \dfrac{1}{50} \times \dfrac{50}{2}\{2 \times 1 + (50 - 1) \times 2\} = 50.$

(Using the sum formula of an arithmetic progression)

113. $\sqrt{\dfrac{2}{\pi}}$

Since X is a zero mean unit variance Gaussian random variable, so X is a standard normal variate and hence its p.m.f is given by

$f(x) = \dfrac{1}{\sqrt{2\pi}} e^{-\frac{x^{2}}{2}}$

Therefore, $E(|X|)$

$= \displaystyle\int_{-\infty}^{\infty} |x| f(x)\,dx = 2\int_{0}^{\infty} xf(x)\,dx$

$\left(\because |x| f(x) = |x| \times \dfrac{1}{\sqrt{2\pi}} e^{-\frac{x^{2}}{2}} \text{ is an even function}\right)$

$= 2\displaystyle\int_{0}^{\infty} x \times \dfrac{1}{\sqrt{2\pi}} e^{-\frac{x^{2}}{2}}\,dx = \sqrt{\dfrac{2}{\pi}}\int_{0}^{\infty} xe^{-\frac{x^{2}}{2}}\,dx = -\sqrt{\dfrac{2}{\pi}}\int_{0}^{\infty} e^{-t}\,dt$

$\left[\begin{array}{l} \text{putting } t = \dfrac{x^{2}}{2} \text{ so that } dt = xdx; \\ \text{Also } x = 0 \Rightarrow t = 0, \ x = \infty \Rightarrow t = \infty \end{array}\right]$

$= \sqrt{\dfrac{2}{\pi}}\left[-e^{-t}\right]_{0}^{\infty} = \sqrt{\dfrac{2}{\pi}}(-0 + 1) = \sqrt{\dfrac{2}{\pi}}.$

114. (b)

$E(X^{2}) - \left[E(X)\right]^{2} = \text{Var}(X) \geq 0$

$\Rightarrow E(X^{2}) \geq \left[E(X)\right]^{2}.$

115. (c) A and B are two independent events

$\Rightarrow \ P(A \cap B) = P(A) \times P(B).$

Therefore, $P(A \cup B)$

$= P(A) + P(B) - P(A \cap B) = P(A) + P(B) - P(A)$
$\times P(B)$

$\neq P(A) + P(B)$.

116. (b) Let X be the random variable denoting the number of times "six" occurs.

Then, X follows the binomial distribution with n = 4 and p = prob. (getting a six in each trial) = $\frac{1}{6}$.

Therefore, the probability of obtaining at least two sixes in throwing a fair dice four times

$= P(X \geq 2) = 1 - P(X < 2)$

$= 1 - \{P(X = 0) + P(X = 1)\}$

$= 1 - \{{}^4C_0 p^0 (1-p)^4 + {}^4C_1 p^1 (1-p)^3\}$

$= 1 - \left\{\left(1 - \frac{1}{6}\right)^4 + 4 \times \frac{1}{6} \times \left(1 - \frac{1}{6}\right)^3\right\} = \frac{19}{144}$.

117. (c)

$P(Y/X) = \dfrac{P(Y \cap X)}{P(X)} = \dfrac{1/12}{1/4} = \dfrac{1}{3}$.

118. 1.067

Mean,

$\mu_X = \displaystyle\int_{-\infty}^{\infty} x f(x) dx = \int_0^2 x \times \frac{x}{4}\left(4 - x^2\right) dx$

$= \dfrac{1}{4} \displaystyle\int_0^2 \left(4x^2 - x^4\right) dx = \dfrac{1}{4}\left[4 \times \dfrac{x^3}{3} - \dfrac{x^5}{5}\right]_0^2$

$= \dfrac{1}{4}\left[4 \times \dfrac{8}{3} - \dfrac{32}{5}\right] = 1.067$

119. 0.97

Let A = the event that first vendor meets the design specifications,

B = the event that second vendor meets the design specifications and

C = the event that third vendor meets the design specifications.

Then, ATQ, $P(A)$ = 0.8, $P(B)$ = 0.7, and $P(C)$ = 0.5.

The probability that at least one meet the specification

= 1–Prob. (none meets the specifications)

$= 1 - P(\bar{A} \cap \bar{B} \cap \bar{C}) = 1 - P(\bar{A}) \times P(\bar{B}) \times P(\bar{C})$

$= 1 - \{(1 - 0.8) \times (1 - 0.7) \times (1 - 0.5)\} = 0.97$

120. (d) Prob. (getting a six) = $\frac{1}{6}$, Prob. (not getting a six) = $\frac{5}{6}$.

Prob. (A wins)

= Prob. (A gets a six in first throw) + Prob. (A does not get a six in first throw, B does not get a six in second throw, A gets a six in third throw) +..............∞

$= \dfrac{1}{6} + \dfrac{5}{6} \times \dfrac{5}{6} \times \dfrac{1}{6} + \ldots\ldots$

$= \dfrac{1}{6}\left\{1 + \left(\dfrac{5}{6}\right)^2 + \left(\dfrac{5}{6}\right)^4 + \ldots\ldots\right\} = \dfrac{1}{6} \times \left\{\dfrac{1}{1 - \left(\dfrac{5}{6}\right)^2}\right\} = \dfrac{6}{11}$

(using the sum formula of an infinite G.P)

121. (b)

Let A = the event that the student pass the exam and B = the event that the student gets above 90% marks.

Then, ATQ, $P(A)$ = 20% = 0.2 and $P(A \cap B)$ = 5% = 0.05.

Therefore, the required probability

$= P(B/A) = \dfrac{P(B \cap A)}{P(A)} = \dfrac{0.05}{0.2} = \dfrac{1}{4}$.

122. 0.25

$E(X) = 2/3$

$\Rightarrow \displaystyle\int_0^1 x f(x) dx = \dfrac{2}{3} \Rightarrow \int_0^1 x(a + bx) dx = \dfrac{2}{3}$

$\Rightarrow a\left[\dfrac{x^2}{2}\right]_0^1 + b\left[\dfrac{x^3}{3}\right]_0^1 = \dfrac{2}{3} \Rightarrow \dfrac{a}{2} + \dfrac{b}{3} = \dfrac{2}{3}$

$\Rightarrow 3a + 2b = 4$ \qquad ...(i)

Now, $\displaystyle\int_{-\infty}^{\infty} f(x) dx = 1$

$\Rightarrow \displaystyle\int_0^1 (a + bx) dx = 1 \Rightarrow \left[ax + \dfrac{bx^2}{2}\right]_0^1 = 1$

$\Rightarrow a + \dfrac{b}{2} = 1 \Rightarrow 2a + b = 2$ \qquad ...(ii)

Solving (i) and (ii), we get,

$a = 0, b = 2$.

So $f(x) = \begin{cases} 2x, & \text{for } 0 < x < 1 \\ 0, & \text{otherwise} \end{cases}$

Now prob. $(x < 0.5)$

$= \int_0^{0.5} f(x)\,dx = \int_0^{1/2} 2x\,dx = \left[x^2\right]_0^{0.5} = 0.25.$

123. (b) Let X be the random variable denoting the number of defective thermistors.

Then, X follows a binomial distribution with $n = 10$ and $p = 0.1$.

So the required probability

$= P(X = 3) = {}^{10}C_3(0.1)^3(1-0.1)^{10-3} = 0.057$

124. 0.24

The probability distribution of X is

X	0	1
P(x = x)	0.4	1−0.4 = 0.6

$E(X) = 0 \times 0.4 + 1 \times 0.6 = 0.6,$

$E(X^2) = 0^2 \times 0.4 + 1^2 \times 0.6 = 0.6.$

$\therefore V(X)$

$= E(X^2) - [E(X)]^2 = 0.6 - (0.6)^2 = 0.24.$

125. (d) $P(X + Y \geq 1)$

$= 1 - P(X + Y < 1) = 1 - P(X = 0, Y = 0) = 1 - P(X) \times P(Y = 0)$

(since X and Y are independent)

$= 1 - pq.$

126. 4

$E[g_X(x)]$

$= E\left[e^{\frac{3x}{4}}\right] = \int_{-\infty}^{\infty} e^{\frac{3x}{4}} \times P_X(x)\,dx$

$= \int_0^{\infty} e^{\frac{3x}{4}} \times e^{-x}\,dx = \int_0^{\infty} e^{-\frac{x}{4}}\,dx = -4\left[e^{-\frac{x}{4}}\right]_0^{\infty}$

$= -4(0-1) = 4.$

127. 6

Prob. (getting the number 3) $= \dfrac{1}{6}$ and Prob.

(not getting the number 3) $= 1 - \dfrac{1}{6} = \dfrac{5}{6}.$

Then, the probability distribution of X is

X	1	2	3	....
P(X)	$\dfrac{1}{6}$	$\dfrac{5}{6} \times \dfrac{1}{6}$	$\dfrac{5}{6} \times \dfrac{5}{6} \times \dfrac{1}{6}$	....

Hence, $E(X)$

$= 1 \times \dfrac{1}{6} + 2 \times \left(\dfrac{5}{6} \times \dfrac{1}{6}\right) + 3 \times \left(\dfrac{5}{6} \times \dfrac{5}{6} \times \dfrac{1}{6}\right) + \ldots\ldots$

$= \dfrac{1}{6}\left[1 + 2 \times \dfrac{5}{6} + 3 \times \left(\dfrac{5}{6}\right)^2 + \ldots\ldots\right]$

$= \dfrac{1}{6}\left[1 - \dfrac{5}{6}\right]^{-2} = \dfrac{1}{6} \times 6^2 = 6.$

128. (d)

If possible, let the tosses are independent.

Then,

$P(H_R H_S) = 0.28 \Rightarrow P(H_R) \times P(H_S) = 0.28 \ldots\ldots(1)$

$P(T_R T_S) = 0.18 \Rightarrow P(T_R) \times P(T_S) = 0.18 \ldots\ldots(2)$

$P(H_R T_S) = 0.30 \Rightarrow P(H_R) \times P(T_S) = 0.30 \ldots\ldots(3)$

$P(T_R H_S) = 0.24 \Rightarrow P(T_R) \times P(H_S) = 0.24 \ldots\ldots(4)$

Now, $(1) \div (3) \Rightarrow \dfrac{P(H_S)}{P(T_S)} = \dfrac{0.28}{0.30} \ldots\ldots(5)$

Again, $(4) \div (2) \Rightarrow \dfrac{P(H_S)}{P(T_S)} = \dfrac{0.24}{0.18} \ldots\ldots(6)$

Thus, the L.H.S of (5) and (6) are equal, but their R.H.S are not equal, a contradiction.

Hence, the coin tosses are not independent and hence they are dependent.

129. (a) Given that, the product can be sequentially assembled in two possible ways.

If the 5 components are placed in a box and these are drawn at random from the box, then the total number of ways of drawing the 5 components sequentially from the box = 5!.

Hence, the required probability $= \dfrac{2}{5!}$.

130. (b) Given that, Y = event that tails occurs on the third toss, i.e., on third attempt and Z = event that two tails occur in three tosses i.e; in three attempts.

Thus, if Y occurs, then only one tails needs to appear in the first two attempts to make Z occurred.

Hence, Y and Z are dependent.

131. (*b*) If two drawn cards are Kings, then only 50 cards are left out of which two are Kings.

Hence, the required probability $= \dfrac{2}{50}$.

132. (*a*) Here $n(S) = n(A) \times n(B) = 4 \times 5 = 20$.

Let E = event that the sum of the numbers is 16.

Then, $E = \{(2,14), (3,13), (4, 12), (5,11)\}$.

So the required probability

$= \dfrac{n(E)}{n(S)} = \dfrac{4}{20} = \dfrac{1}{5} = 0.20.$

133. (*b*) Prob. (one of of them is selected)

= Prob. [(Robert is selected and Richard is not selected) or (Robert is not selected and Richard is selected)] = Prob. (Robert is selected and Richard is not selected) + P(Robert is not selected and Richard is selected) = Prob. (Robert is selected) × Prob. (Richard is not selected) + Prob. (Robert is not selected) × Prob. (Richard is selected)

$= \dfrac{1}{6} \times \left(1 - \dfrac{1}{8}\right) + \left(1 - \dfrac{1}{6}\right) \times \dfrac{1}{8} = \dfrac{12}{48} = \dfrac{1}{4}.$

134. 0.75

$\Pr\left(Y = 0 / X_3 = 0\right)$

$= \dfrac{P\left(X_1 = 1, X_2 = 0, X_3 = 0\right)}{P\left(X_3 = 0\right)}$

$+ \dfrac{P\left(X_1 = 0, X_2 = 1, X_3 = 0\right)}{P\left(X_3 = 0\right)}$

$+ \dfrac{P\left(X_1 = 0, X_2 = 0, X_3 = 0\right)}{P\left(X_3 = 0\right)}$

$= \dfrac{P\left(X_1 = 1\right) \times P\left(X_2 = 0\right) \times P\left(X_3 = 0\right)}{P\left(X_3 = 0\right)}$

$+ \dfrac{P\left(X_1 = 0\right) \times P\left(X_2 = 1\right) \times P\left(X_3 = 0\right)}{P\left(X_3 = 0\right)}$

$+ \dfrac{P\left(X_1 = 0\right) \times P\left(X_2 = 0\right) \times P\left(X_3 = 0\right)}{P\left(X_3 = 0\right)}$

$= \dfrac{\frac{1}{2} \times \frac{1}{2} \times \frac{1}{2}}{\frac{1}{2}} + \dfrac{\frac{1}{2} \times \frac{1}{2} \times \frac{1}{2}}{\frac{1}{2}} + \dfrac{\frac{1}{2} \times \frac{1}{2} \times \frac{1}{2}}{\frac{1}{2}} = 0.75.$

135. (*a*)

Out of 52 cards, there are four Kings, four Queens, and four Jacks.

Therefore, the required probability

$= \dfrac{{}^4C_1 \times {}^4C_1 \times {}^4C_1}{{}^{52}C_3} = \dfrac{4 \times 4 \times 4}{\dfrac{52 \times 51 \times 50}{3 \times 2 \times 1}} = \dfrac{64}{22100} = \dfrac{16}{5525}.$

136. 0.4095.

Let X be the random variable denoting the number of defective screws.

Then, X follows the binomial distribution with $n = 5$,

p = prob. (as the screw is defective) = 0.1.

Therefore, the required probability

$= P(X \geq 1) = 1 - P(X < 1) = 1 - P(X = 0)$

$= 1 - {}^5C_0 (0.1)^0 (1 - 0.1)^{5-0} = 0.4095$

137. (*a*)

$P\left(X \cup Y^c\right) = 0.7$

$\Rightarrow P(X) + P\left(Y^c\right) - P\left(X \cap Y^c\right) = 0.7$

$\Rightarrow P(X) + P\left(Y^c\right) - P(X) P\left(Y^c\right) = 0.7$

($\because$ X and Y are independent events)

$\Rightarrow P(X) + 1 - P(Y) - P(X)\{1 - P(Y)\} = 0.7$

$\Rightarrow P(Y) - P(X \cap Y) = 0.3$

Hence,

$P(X \cup Y) = P(X) + P(Y) - P(X \cap Y)$

$= 0.4 + 0.3 = 0.7.$

138. (*a*)

Case-I: A red ball is drawn and placed back into the jar along with another red ball.

In this case, the total number of balls will be 13 out of which six balls will be red.

Then, the probability of getting a red ball in the next drawn $= \dfrac{5}{12} \times \dfrac{6}{13}.$

Case-II: A green ball is drawn and placed back into the jar along with another green ball.

In this case, the total number of balls will be 13 out of which eight balls will be green.

Then, the probability of getting a red ball in the next drawn $= \dfrac{7}{12} \times \dfrac{5}{13}.$

So the required probability

$= \dfrac{5}{12} \times \dfrac{6}{13} + \dfrac{7}{12} \times \dfrac{5}{13} = \dfrac{65}{156}.$

139. 54.5

The ascending order of spot speed studies are

32, 39, 45, 51, 53, 56, 60, 62, 66, 79.

There are ten (even) observations.

So median speed

= average of $\left(\frac{10}{2}\right)^{\text{th}}$ and $\left(\frac{10}{2}+1\right)^{\text{th}}$ observations

= average of 5th and 6th observations

= $\frac{53+56}{2}$ = 54.5 km/hr

140. 0.072

Here Prob. (getting heads) = 0.3 and Prob. (getting tails) = 1–0.3 = 0.7.

Since all tosses are independent so, the probability of getting heads for the first time in fifth toss

= Prob. (getting tails in first toss) × Prob. (getting tails in second toss) × Prob. (getting tails in third toss) × Prob. (getting tails in fourth toss) × Prob. (getting heads in fifth toss)

= 0.7 × 0.7 × 0.7 × 0.7 × 0.3 = 0.072.

141. (b)

Let $f(x)$ and $g(x)$ represent the p.d.f of the random variables of X and Y, respectively.

Mean of f(x)

$= \int\limits_{-\infty}^{\infty} xf(x)\,dx$

$= \int_{-a}^{0} x\left(\frac{x}{a}+1\right)dx + \int_{0}^{a} x\left(\frac{-x}{a}+1\right)dx$

$= \int_{-a}^{0}\left(\frac{x^2}{a}+x\right)dx + \int_{0}^{a}\left(\frac{-x^2}{a}+x\right)dx$

$= \left[\frac{x^3}{3a}+\frac{x^2}{2}\right]_{-a}^{0} + \left[\frac{-x^3}{3a}+\frac{x^2}{2}\right]_{0}^{a}$

$= \left[\frac{a^3}{3a}-\frac{a^2}{2}\right] + \left[\frac{-a^3}{3a}+\frac{a^2}{2}\right] = 0.$

$E\left(x^2\right)$

$= \int\limits_{-\infty}^{\infty} x^2 f(x)\,dx$

$= \int_{-a}^{0} x^2\left(\frac{x}{a}+1\right)dx + \int_{0}^{a} x^2\left(\frac{-x}{a}+1\right)dx$

$= \int_{-a}^{0}\left(\frac{x^3}{a}+x^2\right)dx + \int_{0}^{a}\left(\frac{-x^3}{a}+x^2\right)dx$

$= \left[\frac{x^4}{4a}+\frac{x^3}{3}\right]_{-a}^{0} + \left[\frac{-x^4}{4a}+\frac{x^3}{3}\right]_{0}^{a}$

$= \left[-\frac{a^4}{4a}+\frac{a^3}{3}\right] + \left[\frac{-a^4}{4a}+\frac{a^3}{3}\right] = \frac{a^3}{6}.$

Therefore, Var(X)

$= E(x)^2 - \left\{E\left(x^2\right)\right\} = \frac{a^3}{6} - 0^2 = \frac{a^3}{6}.$

Mean of g(x)

$= \int\limits_{-\infty}^{\infty} xg(x)\,dx$

$= \int_{-a}^{0} x\left(-\frac{x}{a}\right)dx + \int_{0}^{a} x\left(\frac{x}{a}\right)dx$

$= \int_{-a}^{0}\left(-\frac{x^2}{a}\right)dx + \int_{0}^{a}\left(\frac{x^2}{a}\right)dx$

$= \left[-\frac{x^3}{3a}\right]_{-a}^{0} + \left[\frac{x^3}{3a}\right]_{0}^{a} = -\frac{a^3}{3a}+\frac{a^3}{3a} = 0.$

$E\left(x^2\right)$

$= \int\limits_{-\infty}^{\infty} x^2 g(x)\,dx$

$= \int_{-a}^{0} x^2\left(-\frac{x}{a}\right)dx + \int_{0}^{a} x^2\left(\frac{x}{a}\right)dx$

$= \int_{-a}^{0}\left(-\frac{x^3}{a}\right)dx + \int_{0}^{a}\left(\frac{x^3}{a}\right)dx$

$= \left[-\frac{x^4}{4a}\right]_{-a}^{0} + \left[\frac{x^4}{4a}\right]_{0}^{a}$

$= \frac{a^4}{4a}+\frac{a^4}{4a} = \frac{a^3}{2}.$

Therefore, Var(Y)

$= E\left(x^2\right) - \left\{E(x)^2\right\} = \frac{a^3}{2} - 0^2 = \frac{a^3}{2}.$

∴ Mean of f(x) and g(x) are same but variance of f(x) and g(x) are different.

142. 1

In the Poisson distribution,

Mean = First moment = λ

Second moment = $\lambda^2 + \lambda$

Given, second moment = 2

∴ $\lambda^2 + \lambda = 2$

or, $\lambda^2 + \lambda - 2 = 0$

or, $(\lambda+2)(\lambda-1) = 0$

or, $\lambda = 1$ (since in the Poisson distribution, the mean cannot be negative)

143. (a) We know that in the Poisson distribution, mean = Variance.

Given, Variance = mean = μ.

So, the standard deviation = $\sqrt{\text{Variance}}$ = $\sqrt{\mu}$.

144. (*a*)

$P(X \le 4)$

$= \int_{-\infty}^{4} f(x)dx = \int_{-\infty}^{1} 0\,dx + \int_{1}^{4} 0.25\,dx + \int_{4}^{\infty} 0\,dx$

$= \frac{1}{4}[x]_{1}^{4} = \frac{1}{4}(4-1) = \frac{3}{4}.$

145. 0.5

Given, $f(x) = \begin{cases} \dfrac{1}{x^2}, & a \le x \le 1 \\ 0, & \text{otherwise} \end{cases}$

$\therefore \int_{-\infty}^{\infty} f(x) = 1$

$\Rightarrow \int_{a}^{1} f(x) = 1 \Rightarrow \int_{a}^{1} \frac{1}{x^2} = 1 \Rightarrow \left[\frac{-1}{x}\right]_{a}^{1} = 1$

$\Rightarrow -\left[\frac{1}{1} - \frac{1}{a}\right] = 1 \Rightarrow a = \frac{1}{2} = 0.5.$

146. (*a*) The unit step function u(x) is defined by

$u(x) = \begin{cases} 0, & x < 0 \\ 1, & x \ge 0 \end{cases}$

Then, $u(-x) = \begin{cases} 1, & x < 0 \\ 0, & x \ge 0 \end{cases}$

$\int_{-\infty}^{\infty} f_x(x) = 1$

$\Rightarrow \int_{-\infty}^{\infty} \left(ae^{4x}u(-x) + \frac{3}{2}e^{-3x}u(x) \right)dx = 1$

$\Rightarrow \int_{-\infty}^{0} \left(ae^{4x} \times 1 + \frac{3}{2}e^{-3x} \times 0 \right)dx +$

$+ \int_{0}^{\infty} \left(ae^{4x} \times 0 + \frac{3}{2}e^{-3x} \times 1 \right)dx = 1$

$\Rightarrow \int_{-\infty}^{0} ae^{4x}dx + \frac{3}{2}\int_{0}^{\infty} e^{-3x}dx = 1$

$\Rightarrow \left[\frac{ae^{4x}}{4}\right]_{-\infty}^{0} + \frac{3}{2}\left[\frac{e^{-3x}}{-3}\right]_{0}^{\infty} = 1$

$\Rightarrow \frac{a}{4}(1-0) + \frac{3}{2}\left(0 + \frac{1}{3}\right) = \frac{a}{4} + \frac{1}{2} = 1 \Rightarrow a = 2.$

$\therefore P(X \le 0) = 2\int_{-\infty}^{0} e^{4x}dx = \left[\frac{e^{4x}}{2}\right]_{-\infty}^{0} = \frac{1}{2}.$

147. 1/3

Prob. (getting heads) = $\frac{1}{2}$ = Prob. (getting tails).

Prob. (output is Y)

$= \frac{1}{2} \times \frac{1}{2} + \frac{1}{2} \times \frac{1}{2} \times \frac{1}{2} \times \frac{1}{2} + \dots\dots\dots\infty$

$= \frac{1}{4}\left\{ 1 + \left(\frac{1}{2}\right)^2 + \dots\dots\dots\infty \right\} = \frac{1}{4} \times \frac{1}{\left(1 - \frac{1}{4}\right)} = \frac{1}{3}$

(using the sum formula for an infinite geometric series).

148. 0.55

Let A = the event that an LED bulb chosen uniformly at random lasts more than 100 hours,

E = the event that a LED bulb is of type-I and

F = the event that a LED bulb is of type-II.

Then, P(E) = P(F) = $\frac{1}{2} = 0.5$, P(A/E)=0.7, P(A/F)=0.4.

Therefore,

Prob. (an LED bulb chosen uniformly at random lasts more than 100 hours)

= P(A) = P(E) × P(A/E) + P(F) × P(A/F)

= 0.5×0.7 + 0.5 × 0.4 = 0.55.

149. 0.33

$P(X + Y \le 1)$

$= \int_{x=0}^{1}\int_{y=0}^{1-x} f_{X,Y}(x,y)dx\,dy = \int_{x=0}^{1}\int_{y=0}^{1-x} (x+y)dx\,dy$

$= \int_{x=0}^{1} \left[xy + \frac{y^2}{2} \right]_{0}^{1-x} dx = \int_{x=0}^{1} \left\{ x(1-x) + \frac{(1-x)^2}{2} \right\}dx$

$= \int_{x=0}^{1} \left(\frac{1}{2} - \frac{x^2}{2} \right)dx = \left[\frac{x}{2} - \frac{x^3}{6} \right]_{0}^{1} = \frac{1}{2} - \frac{1}{6} = \frac{1}{3} = 0.33$

150. (*c*) The required probability

= The probability of getting tails in all the N tosses

$= \frac{1}{2} \times \frac{1}{2} \times \dots\dots \times \frac{1}{2} (N \text{ times}) = \left(\frac{1}{2}\right)^N.$

151. (*b*) Clearly X follows normal distribution with mean = 1 and S.D = 2.

$$\therefore \int_1^\infty f_X(x)$$

= area under the normal curve to the right side of the mean

$$= \frac{1}{2}.$$

152. (c) Mode refers the observation which occurs the maximum number of times.

In the given data, 17 occurs four times, i.e., the maximum number of times. So mode is 17.

153. 54. Given, that mean of the Poisson distribution, $\lambda = 5$.

We know that in the Poisson distribution, mean = variance.

So here, $E(X) = V(X) = \lambda = 5$.

Now, $V(X) = E(X^2) - (E(X))^2$

$\Rightarrow 5 = E(X^2) - 5^2 \Rightarrow E(X^2) = 5^2 + 5 = 30.$

$\therefore E[(X+2)^2]$

$= E(X^2 + 4X + 4)$

$= E(X^2) + 4E(X) + 4 = 30 + 4 \times 5 + 4 = 54.$

154. (b)

$$\int_{-\infty}^\infty f(x)dx = 1$$

$$\Rightarrow \int_0^1 (a+bx)dx = 1 \Rightarrow \left[ax + \frac{bx^2}{2}\right]_0^1 = 1 \Rightarrow a + \frac{b}{2} = 1,$$

Which is satisfied for a = 0.5 and b = 1.

155. $\frac{3}{4}$

Here, the sample space, S = {HH, HT, TH, TT}.

Let A = the event that at least one heads occurs.

Then, A = {HH, HT, TH}.

Hence, the required probability = P(A) $= \frac{n(A)}{n(S)} = \frac{3}{4}.$

156. 0.5

Case-I: First drawn ball is red and the second drawn ball is also red.

When the first drawn ball of red color is not replaced, the number of balls left in the jar is nine out of which four are red.

So prob. (first drawn ball is red and the second drawn ball is also red)

$$= \frac{5}{10} \times \frac{4}{9}.$$

Case-II: First drawn ball is black and the second drawn ball is also red.

When the first drawn ball of black color is not replaced, the number of balls left in the jar is nine out of which five are red.

So prob. (first drawn ball is black and the second drawn ball is also red)

$$= \frac{5}{10} \times \frac{5}{9}.$$

Hence, the required probability

$$= \frac{5}{10} \times \frac{4}{9} + \frac{5}{10} \times \frac{5}{9} = \frac{45}{90} = 0.5$$

157. 3.5

Let X be the random variable denoting the outcome of the die.

Then, the probability distribution of X is

X	1	2	3	4	5	6
P(X = x)	$\frac{1}{6}$	$\frac{1}{6}$	$\frac{1}{6}$	$\frac{1}{6}$	$\frac{1}{6}$	$\frac{1}{6}$

So mean

$= E(X)$

$= 1 \times \frac{1}{6} + 2 \times \frac{1}{6} + 3 \times \frac{1}{6} + 4 \times \frac{1}{6} + 5 \times \frac{1}{6} + 6 \times \frac{1}{6}$

$= \frac{1}{6} \times 21 = 3.5$

158. (b) In the exponential distribution, the p.d.f is given by

$f(x) = \lambda e^{-\lambda x}, x > 0; x = 0$

where the only parameter is λ

In the Gaussian distribution, the p.d.f is given by

$$f(x) = \frac{1}{\sigma\sqrt{2\pi}} e^{-\frac{1}{2}\left(\frac{x-\mu}{\sigma}\right)^2}, -\infty < x < \infty$$

where the parameters are μ and σ.

159. 0.5

Since the coin is fair, the outcome of next toss will be independent of the previous toss.

Hence prob. (getting heads in 4th toss)

= Prob. (getting heads) $= \frac{1}{2} = 0.5$

160. 0.9

Let X be the random variable denoting the arrival time of vehicles of the junction which is uniformly distributed in $[0, 5]$.

Then, the p.d.f of X is given by

$$f(x) = \frac{1}{5-0} = \frac{1}{5}, \; 0 \le x \le 5$$

The waiting time at the junction is clearly a function of arrival time "x." Let us denote it by $g(x)$. Then, $g(x)$ is defined by

$$g(x) = \begin{cases} 0, & x < 2 \\ 5-x, & 2 \le x < 5 \end{cases}$$

Then expected waiting time (in minutes) for the vehicle at the junction

$$= E(g(x))$$

$$= \int_{-\infty}^{\infty} f(x) g(x) dx = \int_{0}^{5} \frac{1}{5} \times g(x) dx$$

$$= \frac{1}{5} \int_{0}^{2} 0 \, dx + \frac{1}{5} \int_{2}^{5} (5-x) dx = \frac{1}{5}\left[5x - \frac{x^2}{2} \right]_{2}^{5}$$

$$= \frac{1}{5}\left\{ \left(25 - \frac{25}{2} \right) - \left(10 - \frac{4}{2} \right) \right\}$$

$$= \frac{1}{5}\left(\frac{25}{2} - 8 \right) = 0.9$$

161. 0.028

Here, the number of elements in the sample space, n(S)

= the total number of outcomes when three dice are thrown = 6^3 = 216.

Let A = the event that all three dice have the same number of dots on the faces.

Then, A = {(1,1,1), (2,2,2), (3,3,3), (4,4,4), (5,5,5), (6,6,6)}.

Hence, the required probability

$$= \frac{n(A)}{n(S)} = \frac{6}{216} = 0.028.$$

162. (*a*)

Let A = the event that P applies for the job,

B = the event that Q applies for the job.

Then ATQ,

$P(A) = \frac{1}{4}$, $P(A/B) = \frac{1}{2}$ and $P(B/A) = \frac{1}{3}$.

Now, $P(A/B) = \frac{1}{2} \Rightarrow \dfrac{P(A \cap B)}{P(B)} = \frac{1}{2}$ ....(1)

$P(B/A) = \frac{1}{3}$

$\Rightarrow \dfrac{P(A \cap B)}{P(A)} = \frac{1}{3} \Rightarrow P(A \cap B) = \frac{1}{3} \times \frac{1}{4} = \frac{1}{12}.$

Then, (1) gives, $\dfrac{\frac{1}{12}}{P(B)} = \frac{1}{2}$, *i.e.*, $P(B) = \frac{1}{6}.$

So the required probability

$$= P(\bar{A}/\bar{B}) = \frac{P(\bar{A} \cap \bar{B})}{P(\bar{B})} = \frac{1 - P(A \cup B)}{1 - P(B)}$$

$$= \frac{1 - \{P(A) + P(B) - P(A \cap B)\}}{1 - P(B)}$$

$$= \frac{1 - \left\{ \frac{1}{4} + \frac{1}{6} - \frac{1}{12} \right\}}{1 - \frac{1}{6}} = \frac{4}{5}.$$

163. 2.5

Prob. (a passenger gets a reservation in any attempt)

$$= 40\% = \frac{2}{5},$$

Prob. (a passenger does not get a reservation in any attempt)

$$= (100 - 40)\% = \frac{3}{5}.$$

Let X be the random variable denoting the number of attempts required to get a seat reserved.

Then, X takes the values 1, 2, 3, 4,........

The probability distribution of X is

X	1	2	3	....
P(X = x)	$\dfrac{2}{5}$	$\dfrac{3}{5} \times \dfrac{2}{5}$	$\left(\dfrac{3}{5}\right)^2 \times \dfrac{2}{5}$	....

Then average number of attempts that passengers need to make to get a seat reserved.

= E(X)

$$= \left\{1 \times \frac{2}{5}\right\} + \left\{2 \times \frac{3}{5} \times \frac{2}{5}\right\} + \left\{3 \times \left(\frac{3}{5}\right)^2 \times \frac{2}{5}\right\} + \dots$$

$$= \frac{2}{5}\left\{1 + 2 \times \frac{3}{5} + 3 \times \left(\frac{3}{5}\right)^2 + \dots\right\} = \frac{2}{5}\left(1 - \frac{3}{5}\right)^{-2}$$

$$\left[\text{using } 1 + 2x + 3x^2 + 4x^3 + \dots = (1-x)^{-2}\right]$$

$$= \frac{2}{5} \times \frac{25}{4} = 2.5$$

164. $\frac{1}{2}$.

Initially, there are six balls in the box among which five are red.

After first withdrawn, five balls left out of which four are red.

After second withdrawn only four balls are left out of which three are red.

Hence, the required probability = $\frac{5}{6} \times \frac{4}{5} \times \frac{3}{4} = \frac{1}{2}$.

165. (a)

Given that a class of twelve children has two more boys than girls.

So there are seven boys and five girls in the class.

Then prob. (the group of three children accompanying the teacher contains more girls than boys)

= Prob. (the group contains two girls and one boy or three girls)

= Prob. (the group contains two girls and one boy)+ Prob. (the group contains three girls)

$$= \frac{^5C_2 \times {}^7C_1}{{}^{12}C_3} + \frac{^5C_3}{{}^{12}C_3} = \frac{7}{22} + \frac{1}{22} = \frac{4}{11}.$$

166. (b) The required probability = $\frac{4}{12} \times \frac{3}{11} \times \frac{2}{10} = \frac{1}{55}$.

167. 3.5

The given weights can be arranged in ascending order, as given below:

2, 3, 3, 4, 6, 7.

Here, the total number of observations = 6(even).

Hence, the median weight

= the average of $\left(\frac{6}{2}\right)^{th}$ and $\left(\frac{6}{2}+1\right)^{th}$ observations

= the average of third and fourth observations

$$= \frac{3+4}{2} = 3.5$$

168. (c) Let X be the random variable denoting the number of ones.

Then, X follows the binomial distribution with n = 5 and p = prob. (getting one in a single trial) = $\frac{1}{6}$.

Hence, the required probability

$$= P(X \geq 4) = P(X = 4) + P(X = 5)$$

$$= {}^5C_4 \left(\frac{1}{6}\right)^4 \left(1 - \frac{1}{6}\right)^{5-4} + {}^5C_5 \left(\frac{1}{6}\right)^5 \left(1 - \frac{1}{6}\right)^{5-5}$$

$$= \frac{25}{6^5} + \frac{1}{6^5} = 0.0033 = 0.33\%$$

169. 0.5

$P(F \cup G) = P(F) + P(G) - P(F \cap G) = 0.3 + 0.4 - 0.2 = 0.5$

170. 0.25

Since Let X_1, X_2, X_3, and X_4 are independent random variables, so Prob. (X_4 is the smallest) = Prob. (X_1 is the smallest) = Prob. (X_2 is the smallest) = Prob. (X_3 is the smallest) = 1/4 = 0.25.

171. (b) When the contents of bag A and B are mixed thoroughly, total number of balls.

= 20 + 80 + 30 + 60 + 10 = 200 among with 50 are red balls, 140 are green balls and 10 are yellow balls.

Hence, prob. (the ball picked from the mixture is red)

$\frac{50}{200} = 0.25 = 25\%$.

172. (b) Since two dice are rolled, so there are 36 sample points in total. So n(S) = 36.

A tie happens when both get the same number.

Let A = the event that tie happens.

Then, A = {(1,1), (2,2), (3,3), (4,4), (5,5), (6,6)}.

Therefore, prob. (tie happens)

$= P(A) = \frac{6}{36} = \frac{1}{6}$.

Hence, prob. (one of them wins on the third trial)

= Prob. (tie on the first trial and tie on the second trial and no tie on the third trial)

= Prob. (tie on the first trial)×P(tie on the second trial)×P(no tie on the third trial)

$$=\frac{1}{6}\times\frac{1}{6}\times\left(1-\frac{1}{6}\right)=\frac{5}{216}=0.0231.$$

173. (c) Let A = the event that the color of the cab is green,

B = the event that the color of the cab is blue and

C = the event that a witness identifies the cab color correctly.

Then, ATQ,

$$P(A) = 85\% = \frac{85}{100}, P(B) = (100-85)\% = \frac{15}{100},$$

$$P(C/B) = 80\% = \frac{80}{100}, P(C/A) = (100-80)\% = \frac{20}{100}.$$

Hence, the required probability is

$$=P(B/C)$$

$$=\frac{P(B)\times P(C/B)}{P(A)\times P(C/A)+P(B)\times P(C/B)}$$

$$=\frac{\dfrac{15}{100}\times\dfrac{80}{100}}{\dfrac{85}{100}\times\dfrac{20}{100}+\dfrac{15}{100}\times\dfrac{80}{100}}=\frac{12}{29}=0.41=41\%$$

174. (d)
Var(Z)
= Var(1×X+(-1)×Y)
= 1^2Var(X)+(-1)^{2}Var(Y)=1+2=3.

175. (b)
(a) Prob. (two T will occur)
= Prob. (getting one T)× Prob. (getting one T)
$$=\frac{1}{2}\times\frac{1}{2}=\frac{1}{4}.$$

(b) Prob. (one H and one T will occur)
= Prob. [(getting one H first and then one T)
or (getting one T first and then one H)]
= Prob. (getting one H first and then one T)

+Prob. (getting one T first and then one H)]

= Prob. (getting one H) × Prob. (getting one T)

+ Prob. (getting one H)×Prob. (getting one T)

$$=\frac{1}{2}\times\frac{1}{2}+\frac{1}{2}\times\frac{1}{2}=\frac{1}{2}.$$

(c) Prob. (two H will occur)
= Prob. (getting one H)× Prob. (getting one H)

$$=\frac{1}{2}\times\frac{1}{2}=\frac{1}{4}.$$

(d) Prob. (one H will be followed by one T)
= Prob. (getting one H first and then one T)
= Prob. (getting one H)×Prob. (getting one T)

$$=\frac{1}{2}\times\frac{1}{2}=\frac{1}{4}.$$

Thus in case of option (b), the probability is highest.

176. (b) Since the linear combination of any two independent normal variates, always follow normal distribution, so Y = X_1 – X_2 follows normal distribution.

mean of Y = E(Y) = E(X_1) – E(X_2) = 1 – 1 = 0.

Now Var(Y)

= 1^2Var(X_1) + $(-1)^2$Var(X_2)

(using the properties of properties)

= $\sigma_1^2 + \sigma_2^2 = 1 + 4 = 5$

177. (c) Since the given curve is positively skewed, so mode ≤ median ≤ mean.

178. (c) Prob. (the die shows the green color)
$$=\frac{4}{6}=\frac{2}{3}$$

Prob. (the die shows the red color) $=\frac{2}{6}=\frac{1}{3}$.

Let X be the random variable denoting the number of green faces.

Then, X follows the binomial distribution with n = 7 and p = Prob. (the die shows the green color) $=\frac{2}{3}$

Now,

(a) Prob. (the die shows three green faces and four red faces)

$= P(X = 3) = {}^7C_3\left(\frac{2}{3}\right)^3\left(1-\frac{2}{3}\right)^{7-3} = \frac{280}{3^7}.$

(b) Prob. (four green faces and three red faces)

$= P(X = 4) = {}^7C_4\left(\frac{2}{3}\right)^4\left(1-\frac{2}{3}\right)^{7-4} = \frac{560}{3^7}.$

(c) Prob. (the die shows five green faces and two red faces)

$= P(X = 5) = {}^7C_5\left(\frac{2}{3}\right)^5\left(1-\frac{2}{3}\right)^{7-5} = \frac{672}{3^7}.$

(d) Prob. (the die shows six green faces and one red faces)

$= P(X = 6) = {}^7C_6\left(\frac{2}{3}\right)^6\left(1-\frac{2}{3}\right)^{7-6} = \frac{448}{3^7}.$

Thus in (c), the probability is maximum and so it is the case of most likely outcomes.

179. (a) Let y = f(x).

From the graph, we have the followings:

(i)   a line passes through (0, 0) and (1, h).

(ii) a line passes through (1, 0) and (2, 2h)

(iii) a line passes through (2, 0) and (3, 3h)

Now, the equation of the line passing through (0, 0) and (1, h) is:

$\frac{y-0}{h-0} = \frac{x-0}{1-0}$ i.e; $y = xh$

The equation of the line passing through (1, 0) and (2, 2h) is:

$\frac{y-0}{2h-0} = \frac{x-1}{2-1}$ i.e; $y = 2h(x-1).$

The equation of the line passing through (2, 0) and (3, 3h) is:

$\frac{y-0}{3h-0} = \frac{x-2}{3-2}$ i.e; $y = 3h(x-2).$

Thus, we have,

$y = f(x) = \begin{cases} hx, 0 \le x \le 1 \\ 2h(x-1), 1 \le x \le 2 \\ 3h(x-2), 2 \le x \le 3 \end{cases}$

Then f(x) is a valid probability density function

$\Rightarrow \int_{-\infty}^{\infty} f(x)\,dx = 1$

$\Rightarrow \int_0^1 hx\,dx + \int_1^2 2h(x-1)\,dx + \int_2^3 3h(x-2)\,dx = 1$

$\Rightarrow h\left[\frac{x^2}{2}\right]_0^1 + 2h\left[\frac{x^2}{2}-x\right]_1^2 + 3h\left[\frac{x^2}{2}-2x\right]_2^3 = 1$

$\Rightarrow h\left[\frac{1}{2}-0\right] + 2h\left[\left(\frac{4}{2}-2\right)-\left(\frac{1}{2}-1\right)\right]$

$+ 3h\left[\left(\frac{9}{2}-6\right)-\left(\frac{4}{2}-4\right)\right] = 1$

$\Rightarrow h = \frac{1}{3}.$

Questions For Practice

1. A bag contains five red marbles, eight white marbles, and four green marbles. What is the probability that if one marble is taken out of the bag at random, it will be not green?

(a) $\frac{13}{17}$　(b) $\frac{12}{17}$　(c) $\frac{10}{17}$　(d) $\frac{1}{2}$

2. Two dice are thrown "n" times in succession. What is the probability of obtaining a double six at least once?

(a) $\left(\frac{1}{85}\right)^n$　(b) $1-\left(\frac{35}{36}\right)^n$

(c) $\left(\frac{5}{6}\right)^n$　(d) $1-\left(\frac{5}{6}\right)^n$

3. A bag contains "m" white balls and "n" black balls. If x + y balls are drawn from this bag, what is the probability that among them there will be exactly "x" white and "y" black balls?

(a) $\frac{{}^mC_{x+y} \times {}^nC_{y+x}}{{}^{m+n}C_{x+y}}$　(b) $\frac{{}^mC_{2x} \times {}^nC_{2y}}{{}^{m+n}C_{x+y}}$

(c) $\frac{{}^mC_x \times {}^nC_y}{{}^{m+n}C_{x+y}}$　(d) none of these

4. Katie speaks the truth in 60% of the cases and Rachel in 90% of the cases. In what percentage of cases are they likely to contradict each other in stating the same fact?

(a) 40%　(b) 45%　(c) 42%　(d) 48%

5. A machine operates only when all of its three components function. The probabilities of the failures of the first, second, and third components are 0.14, 0.10, and 0.05, respectively. What is the probability that the machine will fail?

(a) 0.26   (b) 0.29   (c) 0.21   (d) 0.25

6. Let A and B be two switches with the probabilities of working $\frac{2}{3}$ and $\frac{3}{4}$, respectively. Find the probability that the current flows from terminal T_1 to terminal T_2. When the switches A and B are connected parallely as shown below:

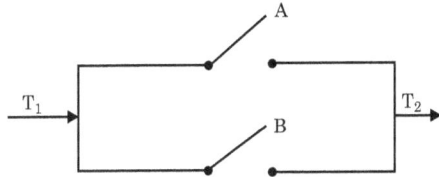

(a) $\frac{11}{12}$ (b) $\frac{2}{17}$ (c) $\frac{5}{12}$ (d) $\frac{1}{2}$

7. A company has two plants to manufacture bicycles. The first plant manufactures 60% of the bicycles and the second plant manufactures 40% of the bicycles. Also 80% of the bicycles are rated of standard quality at the first plant and 90% of standard quality at the second plant. A bicycle is picked up at random and found to be of standard quality. What is the probability that it comes from the second plant?

(a) $\frac{1}{7}$ (b) $\frac{2}{7}$ (c) $\frac{5}{7}$ (d) $\frac{3}{7}$

8. Jar A contains one white, two black, and three red balls; jar B contains two white, one black and one red ball; and jar C contains four white, five black, and three red balls. One jar is chosen at random and two balls are drawn. These happen to be one white and one red. What is the probability that they come from jar A?

(a) $\frac{1}{55}$ (b) $\frac{29}{118}$ (c) $\frac{33}{118}$ (d) $\frac{7}{55}$

9. Assume that each born child is equally likely to be a boy or a girl. If a family has two children, what is the conditional probability that both are girls given that at least one is a girl?

(a) $\frac{1}{3}$ (b) $\frac{4}{5}$ (c) $\frac{1}{2}$ (d) $\frac{2}{3}$

10. A jar contains five red and five black balls. A ball is drawn at random, its color is noted and is returned to the jar. Moreover, two addi-

tional balls of the same color are put in the jar and then a ball drawn at random. What is the probability that the second ball is red?

(a) $\frac{1}{3}$ (b) $\frac{4}{5}$ (c) $\frac{1}{2}$ (d) $\frac{2}{3}$

11. A die is rolled three times. The probability that exactly one odd number turns up among the three outcomes is

(a) 1/6   (b) 3/8   (c) 1/8   (d) 1/2

12. Suppose the expectation of a random variable X is 5. Which of the following statements is true?

(a) There is a sample point at which X has the value 5
(b) There is a sample point at which X has a value > 5
(c) There is a sample point at which X has a value ≥ 5
(d) none of the above

13. Suppose we uniformly and randomly select a permutation from the 20! permutations of 1, 2, 3 ......, 20. What is the probability that 2 appears at an earlier position that any other even number in the selected permutation?

(a) $\frac{1}{2}$ (b) $\frac{1}{10}$

(c) $\frac{9!}{20!}$ (d) None of these

14. What is the probability that a divisor of 10^{99} is a multiple of 10^{96}?

(a) 1/625   (b) 4/625
(c) 12/625   (d) 16/625

15. The probability that it will rain today is 0.5, the probability that it will rain tomorrow is 0.6. The probability that it will rain either today or tomorrow?

(a) 0.3   (b) 0.25   (c) 0.35   (d) 0.40

16. The probabilities that a student passes in Mathematics, Physics, and Chemistry are m, p, c, respectively. Of these subjects, the student has 75% chance of passing in at least one, a 50% chance of passing in at least two and a 40% chance of passing in exactly two. Following relations are drawn in m, p, c:

(*i*)   p + m + c = 27/20

(*ii*)   p + m + c = 13/20

(*iii*) p × m × c = 1/10.

Then

(a) only relation I is true
(b) only relation II is true
(c) relations II and II are true
(d) relations I and III are true

17. The probability that a number selected at random between 100 and 999 (both inclusive) will not contain the digit 7 is

(a) $\frac{16}{25}$ (b) $\left(\frac{9}{10}\right)^3$ (c) $\frac{27}{75}$ (d) $\frac{18}{25}$

18. The probability that a student knows the correct answer to a multiple choice question is $\frac{2}{3}$. If the student does not know the answer, then the student guesses the answer. The probability of the guessed answer being correct is $\frac{1}{4}$. Given that the student has answered the question correctly, the conditional probability that the student knows the correct answer is

(a) $\frac{2}{3}$ (b) $\frac{3}{4}$ (c) $\frac{5}{6}$ (d) $\frac{8}{9}$

19. A person on a trip has a choice between private car and public transport. The probability of using a private car is 0.45. While using the public transport, further choices available are bus and train, out of which the probability of commuting by a bus is 0.55. In such a situation, the probability (rounded up to two decimals) of using a car, bus, and train, respectively, would be

(a) 0.45, 0.30, and 0.25
(b) 0.45, 0.25, and 0.30
(c) 0.45, 0.55, and 0.00
(d) 0.45, 0.35, and 0.20

20. Alexandra studies either computer science or mathematics everyday. If she studies computer science on a day, then the probability that she studies mathematics the next day is 0.6. If she studies mathematics on a day, then the probability that she studies computer science the next day is 0.4. Given that Alexandra studies computer science on Monday, what is the probability that she studies computer science on Wednesday?

(a) 0.24 (b) 0.36 (c) 0.4 (d) 0.6

21. Ryan takes a step forward with probability 0.4 and one step backward with probability 0.6. What is the probability that at the end of eleven steps, he is one step away from the starting point?

(a) $^{11}C_4 \times (0.3)^5$ (b) $^{11}C_5 \times (0.2)^5$
(c) $^{11}C_4 \times (0.25)^5$ (d) $^{11}C_6 \times (0.24)^5$

22. An experiment succeeds twice as often as it fails. Find the probability that in the next six trials there will be at least four successes?

(a) 0.68 (b) 0.45 (c) 0.25 (d) 0.62

23. A and B play a game in which A's chance of winning is 2/3. In a series of eight games, what is the probability that A will win at least six times?

(a) 0.65 (b) 0.45 (c) 0.46 (d) 0.66

24. If there are 480 families each having three children, then how many families can be expected to have at least one girl?

(a) 400 (b) 420 (c) 450 (d) 460

25. Suppose the probability for A to win a game against B is 0.4. If A has an option of playing either a "best of three games" or a "best of five games" match against B, which option should A choose so that the probability of his winning the match is higher?

(a) best of three games
(b) best of five games
(c) none of these games
(d) can't be determined

26. If two coins are tossed five times, what is the probability that an odd number of heads is obtained?

(a) 0 (b) 1/2 (c) 1/3 (d) 1/4

27. Two coins are simultaneously tossed. The probability of two heads simultaneously appearing is

(a) 1/8 (b) 1/6 (c) 1/4 (d) 1/2

28. Using given data points tabulated below, a straight line passing through the origin is fitted using least squares method. The slope of the line is

x	1	2	3
y	1.5	2.2	2.7

(a) 0.9 (b) 1 (c) 1.1 (d) 1.5

29. Three values of x and y are to be fitted in a straight line in the form y = a + bx by the method of least squares. Given $\sum x = 6, \sum y = 21, \sum x^2 = 14,$
$\Sigma xy = 46$. Then, the values of a and b are, respectively
(a) 2, 3   (b) 1, 2   (c) 2, 1   (d) 3, 2

30. Consider a Gaussian distributed random variable with zero mean and standard deviation σ. The value of its cumulative distribution function at the origin will be
(a) 0   (b) 0.5   (c) 1   (d) 10σ

31. For a random variable x (where −∞ < x < ∞) following normal distribution, the mean is μ = 100. If the probability is P = α for x ≥ 110, then the probability of x lying between 90 and 110 is equal to
(a) 1 − 2α   (b) 1 − α
(c) 1 − (α/2)   (d) 2α

32. Two independent random variables X and Y are uniformly distributed in the interval [−1, 1]. The probability that max [X, Y] is less than 1/2 is
(a) 3/4   (b) 9/16   (c) 1/4   (d) 2/3

33. Stephen and Kevin are friends. They decided to meet between 1 p.m and 2 p.m on a given day. There is a condition that whoever arrives first will not wait for the other for more than 15 minutes. The probability that they will meet on that day is
(a) 1/16   (b) 1/4   (c) 7/16   (d) 9/16

34. The annual precipitation data of a city are normally distributed with mean and standard deviation as 1000 mm and 200 mm, respectively. The probability that the annual precipitation will be more than 1200 mm is
(a) < 50%   (b) 50%   (c) 75%   (d) 100%

35. Let U and V be two independent zero mean Gaussian random variables of variances $\frac{1}{4}$ and $\frac{1}{9}$, respectively. Then, the probability P(3V ≥ 2U) is
(a) 4/9   (b) 1/2   (c) 2/3   (d) 5/9

36. A fair coin is tossed repeatedly till both heads and tails appear at least once. The average number of tosses required is
(a) 3   (b) 4   (c) 5   (d) 6

37. Let X_1, X_2, and X_3, be independent and identically distributed random variables with the uniform distribution on [0, 1]. Then prob. ($X_1 + X_2 \leq X_3$) is
(a) 0.45   (b) 0.16   (c) 0.39   (d) 0.56

38. Let the random variable X represents the number of times a fair coin needs to be tossed until two consecutive heads appear for the first time. The expectation of X is
(a) 3   (b) 4   (c) 6   (d) 5

39. Let X_1 and X_2 be two independent exponentially distributed random variables with mean 0.5 and 0.25, respectively. Then $Y = \min(X_1, X_2)$ is
(a) exponentially distributed with mean 1/6
(b) exponentially distributed with mean 2
(c) normally distributed with mean 3/4
(d) normally distributed with mean 1/6

40. Consider two events E_1 and E_2 such that $P(E_1) = 1/2$, $P(E_2) = 1/3$ and $P(E_1 \cap E_2) = 1/5$. Which of the following statements is true?
(a) $P(E_1 \cup E_2) = 2/3$
(b) E_1 and E_2 are independent
(c) E_1 and E_2 are not independent
(d) $P(E_1/E_2) = 4/5$

41. Four arbitrary points (x_1, y_1), (x_2, y_2), (x_3, y_3), (x_4, y_4) are given in the xy-plane using the method of least squares. If regression of y upon x gives the fitted line y = ax + b and regression of x upon y gives the fitted line x = cy + d, then
(a) the two fitted lines must coincide
(b) the two fitted lines need not coincide
(c) it is possible that ac = 0
(d) a = 1/c

42. The probability that of a product by a machine to be defective is 0.01. If 30 products are taken at random, find the probability that exactly two will be defective?
(a) 0.33   (b) 0.5   (c) 0.45   (d) 0.60

43. A jar contains n tickets numbered from 1 to n, from which a ticket is drawn and replaced r times. What is the probability that the greatest number drawn is "i?"
(a) $\dfrac{i^r}{n^r}$ (b) $\dfrac{(i-1)^r}{n^r}$
(c) $\dfrac{(i+1)^r - i^r}{n^r}$ (d) $\dfrac{i^r - (i-1)^r}{n^r}$

44. The probability of a man hitting a target is 1/4. How many times he should fire so that the probability of his hitting the target at least once is greater than 2/3?

(a) 3   (b) 4   (c) 5   (d) 6

45. The distribution function F(x) of a random variable X is given by

$$F(x) = \begin{cases} 1 - \dfrac{1}{2}e^{-x}, & x \geq 0 \\ 0, & \text{elsewhere} \end{cases}$$

Then, find P(X > 1).

(a) e   (b) 1   (c) $\dfrac{1}{2e}$ (d) 2e

46. Suppose that probability of a new born baby a boy is 0.45. In a family of eight children, calculate the probability that there are three or four boys.

(a) 0.519   (b) 0.5   (c) 0.651   (d) 0.245

47. Let F, G and H be pair-wise independent events such that P(F) = P(G) = P(H) = 1/3 and P(F∩G∩H) = 1/4. Then, the probability that at least one event among F, G, H occurs is

(a) 11/12   (b) 7/12   (c) 5/12   (d) 3/4

48. The lifetime of two brands of bulbs X and Y are exponentially distributed with a mean lifetime of 100 hour. Bulb X is switched on 15 hour after bulb Y has been switched on. The probability that the bulb X fails before Y is

(a) 15/100   (b) 1/2   (c) 85/100   (d) 0

49. Let $P(X = n) = \dfrac{\lambda}{n^2(n+1)}$, where λ is an appropriate constant. Then, E(X) is

(a) 2λ + 1   (b) λ   (c) ∞   (d) 2λ

50. Numbers are selected at random, one at a time from the two digit numbers 00, 01, 02, 03,........,99 with replacement. An event E occurs if and only if the product of the two digits of a selected number is 18. If four numbers are selected at random, find the probability that the event E occurs at least three times.

(a) $\dfrac{87}{25^4}$ (b) $\dfrac{3}{25^4}$ (c) $\dfrac{91}{25^4}$ (d) $\dfrac{97}{25^4}$

51. A bag contains a coin of value "M" and a number of other coins whose aggregate value is "m." A person draws one at a time until he draws the coin of value M. Find the value of his expectation.

(a) $M + \dfrac{m}{2}$ (b) $M - \dfrac{m}{2}$

(c) $\dfrac{m}{2}$ (d) 2M + m

52. Two cards are drawn successively with replacement from a well shuffled pack of 52 cards. Then the variance of the number of Kings is

(a) $\dfrac{6}{13}$ (b) $\dfrac{2}{3}$ (c) $\dfrac{29}{169}$ (d) $\dfrac{24}{169}$

53. A pair of dice is thrown four times. If getting a doublet is considered a success, then what will be the mean of the number of successes?

(a) 3/4   (b) 1/2   (c) 2/3   (d) 1/3

54. Students of a particular class appeared for a class test. Their marks were found to be normally distributed with mean 60 and standard deviation 5. What % of students scored more than 60 marks?

(a) 40%   (b) 50%   (c) 60%   (d) 30%

Answer key				
1. (a)	2. (b)	3. (c)	4. (c)	5. (a)
6. (a)	7. (d)	8. (c)	9. (a)	10. (c)
11. (b)	12. (c)	13. (d)	14. (a)	15. (d)
16. (d)	17. (d)	18. (d)	19. (a)	20. (c)
21. (d)	22. (a)	23. (c)	24. (b)	25. (a)
26. (b)	27. (c)	28. (b)	29. (d)	30. (b)
31. (a)	32. (b)	33. (c)	34. (a)	35. (b)
36. (a)	37. (b)	38. (c)	39. (a)	40. (c)
41. (b)	42. (a)	43. (d)	44. (b)	45. (c)
46. (a)	47. (a)	48. (a)	49. (b)	50. (d)
51. (a)	52. (d)	53. (c)	54. (b)	

Explanations

21. (*d*) Here p = prob. that a man takes one step forward = 0.4 and

q = prob. that a man takes one step forward = 0.6.

Next apply the binomial distribution with n = 11.

22. (*a*) Let X be the random variable denoting the number of successes.

Obviously X follows the binomial distribution with n = 6.

ATQ,

p = 2q and so p + q = 1 gives p = 2/3 and q = 1/3.

Then, the required probability

$= P(X \; 4) = P(X = 4) + P(X = 5) + P(X = 6)$

24. (*b*) Here p = probability of having a boy = 1/2 and q = probability of having a girl = 1/2.

Let X be the random variable denoting the number of girl.

Then, number of families having at least one girl
$= 480 \times P(X \geq 1) = 480\{P(X = 1) + P(X = 2) + P(X = 3)\}]$

28. (*b*) Let the line be y = mx.

Then y = mx

$\Rightarrow \sum xy = \sum x \times mx = m \sum x^2 \Rightarrow m = \dfrac{\sum xy}{\sum x^2}$

30. (*b*) [Hint: Normal curve is symmetrical about the mean.]

44. (*b*) Let X be the random variable denoting the number of hitting the target.

Then X follows a binomial distribution with p = prob. (man hits a target) = 1/4.

ATQ, P(X > 1) > 2/3.

Now $P(X > 1)$

$= 1 - P(X = 0) = 1 - {}^nC_0 \left(\dfrac{1}{4}\right)^0 \left(1 - \dfrac{1}{4}\right)^{n-}$

$= 1 - \left(\dfrac{3}{4}\right)^n$.

Then, $P(X > 1) > \dfrac{2}{3}$

if $1 - \left(\dfrac{3}{4}\right)^n > \dfrac{2}{3}$

i.e; if $\left(\dfrac{3}{4}\right)^n < \dfrac{1}{3}$

Now take log on both sides

45. (*c*) $P(X > 1) = 1 - P(X \leq 1) = 1 - F(1).$

53. (*c*)

Prob. (getting a doublet) $= \dfrac{6}{36} = \dfrac{1}{6}$,

Prob. (getting a non-doublet) $= 1 - \dfrac{1}{6} = \dfrac{5}{6}$.

Let X be the random variable denoting the number of doublets when a pair of dice is thrown four times.

Then, X takes the values 0, 1, 2, 3, 4.

Now P(X = 0) $= \dfrac{5}{6} \times \dfrac{5}{6} \times \dfrac{5}{6} \times \dfrac{5}{6} = \dfrac{625}{1296}$,

P(X = 1)

$= \left(\dfrac{1}{6} \times \dfrac{5}{6} \times \dfrac{5}{6} \times \dfrac{5}{6}\right) + \left(\dfrac{5}{6} \times \dfrac{1}{6} \times \dfrac{5}{6} \times \dfrac{5}{6}\right) + \left(\dfrac{5}{6} \times \dfrac{5}{6} \times \dfrac{1}{6} \times \dfrac{5}{6}\right)$
$+ \left(\dfrac{5}{6} \times \dfrac{5}{6} \times \dfrac{5}{6} \times \dfrac{1}{6}\right)$,

P(X = 2)

$= \left(\dfrac{1}{6} \times \dfrac{1}{6} \times \dfrac{5}{6} \times \dfrac{5}{6}\right) + \left(\dfrac{1}{6} \times \dfrac{5}{6} \times \dfrac{1}{6} \times \dfrac{5}{6}\right) + \left(\dfrac{1}{6} \times \dfrac{5}{6} \times \dfrac{5}{6} \times \dfrac{1}{6}\right)$
$+ \left(\dfrac{5}{6} \times \dfrac{1}{6} \times \dfrac{1}{6} \times \dfrac{5}{6}\right) + \left(\dfrac{5}{6} \times \dfrac{1}{6} \times \dfrac{5}{6} \times \dfrac{1}{6}\right) + \left(\dfrac{5}{6} \times \dfrac{5}{6} \times \dfrac{1}{6} \times \dfrac{1}{6}\right)$,

P(X = 3)

$= \left(\dfrac{1}{6} \times \dfrac{1}{6} \times \dfrac{1}{6} \times \dfrac{5}{6}\right) + \left(\dfrac{1}{6} \times \dfrac{1}{6} \times \dfrac{5}{6} \times \dfrac{1}{6}\right) + \left(\dfrac{5}{6} \times \dfrac{1}{6} \times \dfrac{1}{6} \times \dfrac{1}{6}\right)$
$+ \left(\dfrac{1}{6} \times \dfrac{5}{6} \times \dfrac{1}{6} \times \dfrac{1}{6}\right)$,

P(X = 4) $= \dfrac{1}{6} \times \dfrac{1}{6} \times \dfrac{1}{6} \times \dfrac{1}{6}$.

FOURIER SERIES

10.1 BASICS OF THE FOURIER SERIES

10.1.1 Definition of the Fourier Series

The Fourier series for the function $f(x)$ in $(c, c + 2\pi)$ is given by

$$f(x) = \frac{a_0}{2} + \sum_{n=1}^{\infty} (a_n \cos nx + b_n \sin nx)$$

where, $a_0 = \frac{1}{\pi} \int_{c}^{c+2\pi} f(x)dx,$

$$a_n = \frac{1}{\pi} \int_{c}^{c+2\pi} f(x)\cos nx\, dx \quad (n \geq 1),$$

$$b_n = \frac{1}{\pi} \int_{c}^{c+2\pi} f(x)\sin nx\, dx \quad (n \geq 1)$$

a_0, a_n, and b_n are called Fourier coefficients. Here, we have assumed that $f(x)$ is a periodic function with period 2π.

Case – I : When $c = 0$.

In this case, the Fourier series for $f(x)$ in $(0, 2\pi)$ is given by

$$f(x) = \frac{a_0}{2} + \sum_{n=1}^{\infty} (a_n \cos nx + b_n \sin nx)$$

where, $a_0 = \frac{1}{\pi} \int_{0}^{2\pi} f(x)dx,$

$$a_n = \frac{1}{\pi} \int_{0}^{2\pi} f(x)\cos nx\, dx (n \geq 1),$$

$$b_n = \frac{1}{\pi} \int_{0}^{2\pi} f(x)\sin nx\, dx, (n \geq 1)$$

Case – II : when $c = -\pi$

In this case, the Fourier series for $f(x)$ in $(-\pi, \pi)$ is given by

$$f(x) = \frac{a_0}{2} + \sum_{n=1}^{\infty} (a_n \cos nx + b_n \sin nx), \text{ where}$$

$$a_0 = \frac{1}{\pi} \int_{-\pi}^{\pi} f(x)dx,$$

$$a_n = \frac{1}{\pi} \int_{-\pi}^{\pi} f(x)\cos nx\, dx \quad (n \geq 1),$$

$$b_n = \frac{1}{\pi} \int_{-\pi}^{\pi} f(x)\sin nx\, dx \quad (n \geq 1)$$

Remember:

(i) $\int_{-\pi}^{\pi} \cos mx \cos nx\, dx = 0$ for $m \neq n$

(ii) $\int_{-\pi}^{\pi} \sin mx \sin nx\, dx = 0$ for $m \neq n$

(iii) $\int_{-\pi}^{\pi} \cos mx \sin mx\, dx = 0$ for $m \neq 0$

(iv) $\cos n\pi = (-1)^n$, where $n \in Z^+$

(v) $\sin n\pi = 0$, where $n \in Z^+$

(vi) $\int_{-\pi}^{\pi} \sin nx \sin mx\, dx = 0$

(vii) $\int_{-\pi}^{\pi} \sin nx\, dx = 0$ for $n \neq 0$

$(viii)$ $\int_{-\pi}^{\pi} \cos nx\, dx = 0$ for $n \neq 0$

(ix) $\int_{-\pi}^{\pi} \cos^2 nx\, dx = \int_{-\pi}^{\pi} \sin^2 nx = \pi$ for $n \neq 0$

10.1.2 Dirichlet's Condition

Any function $f(x)$ can be expressed as the Fourier series if it satisfies the following conditions:

(a) $f(x)$ is periodic and single valued

(b) $f(x)$ has finite number of discontinuities in any one period

(c) $f(x)$ has finite number of points of local maxima and local minima.

10.2 FOURIER SERIES OF EVEN AND ODD FUNCTIONS

10.2.1 Fourier Series of Even Function

If $f(x)$ be an even periodic function with period 2π, then the Fourier series for $f(x)$ is given by

$$f(x) = \frac{a_0}{2} + \sum_{n=1}^{\infty} a_n \cos nx$$

where, $a_0 = \frac{2}{\pi}\int_0^{\pi} f(x)\,dx$

$$a_n = \frac{2}{\pi}\int_0^{\pi} f(x)\cos nx\,dx\,(n \geq 1)$$

Remember :

A function $f(x)$ is said to be even if $f(-x) = f(x)\ \forall\ x$

10.2.2 Fourier Series for Odd Function

If $f(x)$ be an odd periodic function with period 2π, then the Fourier series for $f(x)$ is given by

$$f(x) = \sum_{n=1}^{\infty} b_n \sin nx \quad \text{where}$$

$$b_n = \frac{2}{\pi}\int_0^{\pi} f(x)\sin nx\,dx$$

Remember :

A function $f(x)$ is said to be odd if $f(-x) = -f(x)\ \forall x$.

10.3 HALF RANGE FOURIER SERIES

10.3.1 Half Range Sine Series

The Fourier sine series for the function $f(x)$ in $(0, c)$ is given by

$$f(x) = \sum_{n=1}^{\infty} b_n \sin\left(\frac{n\pi x}{c}\right)$$

where, $b_n = \frac{2}{c}\int_0^c f(x)\sin\left(\frac{n\pi x}{c}\right)dx\ (n \geq 1)$

10.3.2 Fourier Cosine Series

The Fourier cosine series for the function $f(x)$ in $(0, c)$ is given by

$$f(x) = \frac{a_0}{2} + \sum_{n=1}^{\infty} a_n \cos\left(\frac{n\pi x}{c}\right)$$

where, $a_0 = \frac{2}{c}\int_0^c f(x)\,dx$

$$a_n = \frac{2}{c}\int_0^c f(x)\cos\left(\frac{n\pi x}{c}\right)dx\ (n \geq 1)$$

1. The Fourier series for $f(x) = x$, in $(-\pi, \pi)$ is given by

(a) $\displaystyle\sum_{n=1}^{\infty}\frac{1}{n^2}\cos nx$

(b) $\displaystyle\sum_{n=1}^{\infty}(-1)^n\frac{\cos nx}{n^2}$

(c) $\displaystyle\sum_{n=1}^{\infty}\frac{1}{n}(-1)^n\sin nx$

(d) $\displaystyle\sum_{n=1}^{\infty}\left\{-\frac{2}{n}(-1)^n\right\}\sin nx$

2. Consider the Fourier series for $f(x) = |x|$ in $(-\pi, \pi)$. Then which of the following is true?

(a) $\dfrac{1}{1^2} + \dfrac{1}{3^2} + \dfrac{1}{5^2} + \ldots\infty = \dfrac{\pi^2}{8}$

(b) $\dfrac{1}{1^2} + \dfrac{1}{3^2} + \dfrac{1}{5^2} + \ldots\infty = \dfrac{\pi^2}{12}$

(c) $1 + \dfrac{1}{2} + \dfrac{1}{3} + \ldots\infty = \dfrac{\pi^2}{4}$

(d) none of these

3. For the function $f(x) = x + x^2$, $-\pi < x < \pi$; the value of a_0 in the Fourier series expansion of $f(x)$ is

(a) $\dfrac{2}{3}\pi^2$

(b) $\dfrac{1}{3}\pi^2$

(c) $\dfrac{1}{2}\pi^2$

(d) π

4. Consider the Fourier series of the function $f(x) = x^2 + 4\sin x \cos x$ in $(-\pi, \pi)$. Then find the value of $\left|\displaystyle\sum_{n=0}^{\infty} a_n - \sum_{n=1}^{\infty} b_n\right| = a$?

(a) $\dfrac{\pi}{2}(\pi - 1)$

(b) $\dfrac{\pi^2}{3}(2\pi - 1)$

(c) $\dfrac{\pi^2}{2}(2\pi + 1)$

(d) 0

5. The Fourier series expansion of $f(x)$, where, $f(x) = \begin{cases} 0, & -2 < x < 0 \\ 1, & 0 < x < 2 \end{cases}$ is given by

(a) $\dfrac{1}{2} + \dfrac{2}{\pi}\left\{\sin\dfrac{\pi x}{2} + \dfrac{1}{3}\sin\dfrac{3\pi x}{2} + \dfrac{1}{5}\sin\dfrac{5\pi x}{2} + \ldots\infty\right\}$

(b) $1 + \dfrac{2}{\pi}\displaystyle\sum_{n=1}^{\infty}\sin\dfrac{nx}{2}$

(c) $\dfrac{\pi}{2} - \displaystyle\sum_{n=1}^{\infty}\dfrac{1}{n^2}\sin nx$

(d) none of these

6. Consider the half range cosine series for

$$f(x) = \begin{cases} 2x, & 0 \leq x \leq 2 \\ 2(4-x), & 2 \leq x \leq 4 \end{cases}$$

Then the value of a_0 is _____?

(a) 0   (b) 2

(c) 3   (d) 4

7. Consider the half range sine series of the function $f(x) = 1$ in $0 < x < \pi$. Then which one of the following is true?

(a) $f(x) = \sum\limits_{n=1}^{\infty} \dfrac{2}{n\pi}\left\{(-1)^n - 1\right\}\sin nx$

(b) $f(x) = \sum\limits_{n=1}^{\infty} \dfrac{2}{n\pi}\sin nx$

(c) $f(x) = \sum\limits_{n=1}^{\infty} \dfrac{2}{n\pi}\left\{(-1)^n - 1\right\}\cos nx$

(d) none of these

Answer Key

1. (d)	2. (a)	3. (a)	4. (b)	5. (a)
6. (d)	7. (a)			

Explanation

1. (d)

$$f(x) = x \Rightarrow f(-x) = -x = -f(x)$$

∴ $f(x)$ is an odd function of x.

Hence, the Fourier series for $f(x)$ is given by

$$f(x) = \sum_{n=1}^{\infty} b_n \sin nx$$

where, $b_n = \dfrac{2}{\pi}\int\limits_0^\pi f(x)\sin nx\, dx$

$$= \dfrac{2}{\pi}\int\limits_0^\pi x \sin nx\, dx$$

$$= \dfrac{2}{\pi}\left\{\left[-x \times \dfrac{\cos nx}{n}\right]_0^\pi - \int\limits_0^\pi 1 \times \left(-\dfrac{\cos nx}{n}\right)dx\right\}$$

$$= \dfrac{2}{\pi}\left\{-\dfrac{\pi}{n}\times(-1)^n + \dfrac{1}{n^2}[\sin nx]_0^\pi\right\}\left(\because \cos n\pi = (-1)^n\right)$$

$$= \dfrac{2}{\pi}\left\{-\dfrac{\pi}{n}\times(-1)^n + \dfrac{1}{n^2}(0-0)\right\}$$

$$(\because \sin n\pi = 0 = \sin 0)$$

$$= -\dfrac{2}{n}\times(-1)^n$$

Hence, $f(x) = \sum\limits_{n=1}^{\infty}\left\{-\dfrac{2}{n}(-1)^n\right\}\sin nx.$

2. (a)

$$f(x) = |x| \Rightarrow f(-x) = |x| = |x| = f(x)$$

∴ $f(x)$ is an even function of x.

So the Fourier series for $f(x)$ is given by

$$f(x) = \dfrac{a_0}{2} + \sum_{n=1}^{\infty} a_n \cos nx$$

where, $a_0 = \dfrac{2}{\pi}\int\limits_0^\pi f(x)\,dx = \dfrac{2}{\pi}\int\limits_0^\pi x\,dx$

$$\left[\because\ f(x) = |x| = \begin{cases} x & x \geq 0 \\ -x, & x \leq 0 \end{cases}\right]$$

$$= \dfrac{2}{\pi}\times\left[\dfrac{x^2}{2}\right]_0^\pi = \dfrac{2}{\pi}\times\dfrac{\pi^2}{2} = \pi$$

$$a_n = \dfrac{2}{\pi}\int\limits_0^\pi x\cos nx\,dx$$

$$= \dfrac{2}{\pi}\left\{\left[x\times\dfrac{\sin nx}{n}\right]_0^\pi - \int\limits_0^\pi 1\times\dfrac{\sin nx}{n}dx\right\}$$

$$= \dfrac{2}{\pi}\left\{\pi\times\dfrac{\sin n\pi}{n} - 0\times\dfrac{\sin 0}{n} + \dfrac{1}{n^2}[\cos nx]_0^\pi\right\}$$

$$= \dfrac{2}{\pi n^2}\left\{\cos n\pi - \cos 0\right\}\ (\because \sin n\pi = 0)$$

$$= \dfrac{2}{\pi n^2}\left\{(-1)^n - 1\right\}$$

∴ $f(x) = |x| = \dfrac{\pi}{2} + \sum\limits_{n=1}^{\infty}\dfrac{2}{\pi n^2}\left\{(-1)^n - 1\right\}\cos nx$ ...(i)

Putting $x = 0$ both sides of equation (i) we get,

$$|0| = 0 = \dfrac{\pi}{2} + \sum_{n=1}^{\infty}\dfrac{2}{\pi n^2}\left\{(-1)^n - 1\right\}$$

or, $\dfrac{2}{\pi\times 1^2}\{-1-1\} + \dfrac{2}{\pi\times 2^2}\{1-1\} + \dfrac{2}{\pi\times 3^2}\{-1-1\}$

$$+\dfrac{2}{\pi\times 4^2}\{1-1\} + \dfrac{2}{\pi\times 5^2}\{-1-1\} + ...\infty = -\dfrac{\pi}{2}$$

or, $-\dfrac{4}{\pi\times 1^2} - \dfrac{4}{\pi\times 3^2} - \dfrac{4}{\pi\times 5^2} - ...\infty = -\dfrac{\pi}{2}$

or, $\dfrac{1}{1^2} + \dfrac{1}{3^2} + \dfrac{1}{5^2} + ...\infty = \dfrac{\pi^2}{8}.$

3. (a) The Fourier series for $f(x)$ is given by

$$f(x) = \dfrac{a_0}{2} + \sum_{n=1}^{\infty}\left(a_n\cos nx + b_n\sin nx\right), -\pi < x < \pi$$

Here, $a_0 = \dfrac{1}{\pi}\int\limits_{-\pi}^{\pi} f(x)\,dx = \dfrac{1}{\pi}\int\limits_{-\pi}^{\pi}\left(x + x^2\right)dx$

$$= \dfrac{1}{\pi}\left[\dfrac{x^2}{2} + \dfrac{x^3}{3}\right]_{-\pi}^{\pi} = \dfrac{1}{\pi}\left\{\dfrac{\pi^2}{2} + \dfrac{\pi^3}{3} - \left(\dfrac{\pi^2}{2} - \dfrac{\pi^3}{3}\right)\right\}$$

$$= \dfrac{1}{\pi}\times\dfrac{2\pi^3}{3} = \dfrac{2\pi^2}{3}.$$

4. (*b*)

$$a_0 = \frac{1}{\pi}\int_{-\pi}^{\pi} f(x)\,dx = \frac{1}{\pi}\int_{-\pi}^{\pi}\left(x^2 + 4\sin x\cos x\right)dx$$

$$= \frac{1}{\pi}\int_{-\pi}^{\pi} x^2\,dx + \frac{2}{\pi}\int_{-\pi}^{\pi}\sin 2x\,dx$$

$$= \frac{1}{\pi}\left[\frac{x^3}{3}\right]_{-\pi}^{\pi} + \frac{2}{\pi}\times 0 \left(\because \int_{-\pi}^{\pi}\sin nx\,dx = 0\right)$$

$$= \frac{1}{\pi}\left\{\frac{\pi^3}{3} - \left(-\frac{\pi^3}{3}\right)\right\} = \frac{2\pi^2}{3}$$

$$a_n = \frac{1}{\pi}\int_{-\pi}^{\pi} f(x)\cos nx\,dx$$

$$= \frac{1}{\pi}\int_{-\pi}^{\pi}\left(x^2 + 4\sin x\cos x\right)\cos nx\,dx$$

$$= \frac{1}{\pi}\int_{-\pi}^{\pi} x^2\cos nx\,dx + \frac{2}{\pi}\int_{-\pi}^{\pi}\sin 2x\times\cos nx\,dx$$

$$= \frac{2}{\pi}\int_{-0}^{\pi} x^2\cos nx\,dx + \frac{1}{\pi}\int_{-\pi}^{\pi}(2\sin 2x\cos nx)\,dx$$

$(\because\quad x^2\cos nx$ is an even function of $x)$

$$= \frac{2}{\pi}\left\{\left[x^2\times\frac{\sin nx}{n}\right]_0^{\pi} - \int_0^{\pi}2x\times\frac{\sin nx}{n}\,dx\right\}$$

$$+ \frac{1}{\pi}\int_{-\pi}^{\pi}\left\{\sin[(n+2)x] - \sin[(n-2)x]\right\}dx$$

$$= \frac{2}{\pi}\left[0 - 0 - \frac{2}{n}\left\{\left[x\times\left(\frac{-\cos nx}{n}\right)\right]_0^{\pi} - \int_0^{\pi}1\times\left(\frac{-\cos nx}{n}\right)dx\right\}\right]$$

$$+ \frac{1}{\pi}\left[-\frac{\cos[(n+2)x]}{n+2} + \frac{\cos[(n-2)x]}{n-2}\right]_{-\pi}^{\pi}$$

$$= \frac{2}{\pi}\left\{\frac{2}{n^2}[x\cos nx]_0^{\pi} + \frac{1}{n^2}[\sin nx]_0^{\pi}\right\}$$

$$+ \frac{1}{\pi}\left\{-\frac{\cos(n+2)\pi}{n+2} + \frac{\cos((n-2)\pi)}{n-2}\right.$$

$$\left. + \frac{\cos((n+2)\pi)}{n+2} - \frac{\cos((n-2)\pi)}{n-2}\right\}$$

$$= \frac{2}{\pi}\left\{\frac{2}{n^2}\times\pi\cos n\pi + 0\right\} + \frac{1}{\pi}\times 0$$

$$= \frac{4}{n^2}\times(-1)^n$$

$$b_n = \frac{1}{\pi}\int_{-\pi}^{\pi}\left(x^2 + 4\sin x\cos x\right)\sin nx\,dx$$

$$= \frac{1}{\pi}\int_{-\pi}^{\pi} x^2\sin nx\,dx + \frac{2}{\pi}\int_{-\pi}^{\pi}\sin 2x\times\sin nx\,dx$$

$$= \frac{1}{\pi}\times 0 + \frac{2}{\pi}\times 0$$

$(\because\quad x^2\sin nx$ is an odd function)

$$= 0$$

$$\therefore \left|\sum_{n=0}^{\infty} a_n - \sum_{n=1}^{\infty} b_n\right| = \left|a_0 + \sum_{n=1}^{\infty} a_n - \sum_{n=1}^{\infty} 0\right|$$

$$= \left|\frac{2\pi^2}{3} + \sum_{n=1}^{\infty}\frac{4}{n^2}(-1)^n - 0\right|$$

$$= \left|\frac{2\pi^2}{3} - 4\left(\frac{1}{1^2} - \frac{1}{2^2} + \frac{1}{3^2}\cdots\cdots\infty\right)\right| = \frac{2\pi^2}{3} - 4\times\frac{\pi^2}{12} = \frac{\pi^2}{3}.$$

$$\left(\because\frac{1}{1^2} - \frac{1}{2^2} + \frac{1}{3^2} - \cdots\cdots\infty = \frac{\pi^2}{12}\right)$$

5. (*a*) We know that the Fourier series expansion of f(x) in (-c, c) is given by

$$f(x) = \frac{a_0}{2} + \sum_{n=1}^{\infty}\left[a_n\cos\left(\frac{n\pi x}{c}\right) + b_n\sin\left(\frac{n\pi x}{c}\right)\right]$$

where, $a_0 = \frac{1}{c}\int_{-c}^{c} f(x)\,dx,$

$$a_n = \frac{1}{c}\int_{-c}^{c} f(x)\cos\left(\frac{n\pi x}{c}\right)dx\ (n\geq 1)$$

and $\quad b_n = \frac{1}{c}\int_{-c}^{c} f(x)\sin\left(\frac{n\pi x}{c}\right)dx\ (n\geq 1)$

Here, $\quad c = 2$

$$\therefore \quad a_0 = \frac{1}{2}\int_{-2}^{2} f(x)\,dx = \frac{1}{2}\left[\int_{-2}^{0} 0\,dx + \int_0^2 1\,dx\right]$$

$$= \frac{1}{2}[0 + 2] = 1$$

$$a_n = \frac{1}{2}\int_{-2}^{2} f(x)\cos\left(\frac{n\pi x}{2}\right)dx = \frac{1}{2}\left[\int_{-2}^{0} 0\times\cos\left(\frac{n\pi x}{2}\right)dx\right.$$

$$\left. + \int_0^2 1\times\cos\left(\frac{n\pi x}{2}\right)dx\right]$$

$$= \frac{1}{2}\int_0^2\cos\left(\frac{n\pi x}{2}\right)dx = \frac{1}{2}\times\frac{2}{n\pi}\left[\sin\left(\frac{n\pi x}{2}\right)\right]_0^2$$

$$= \frac{1}{n\pi}\{\sin n\pi - \sin 0\} = 0$$

$$b_n = \frac{1}{2}\int_{-2}^2 f(x)\sin\left(\frac{n\pi x}{2}\right)dx$$

$$= \frac{1}{2}\left[\int_{-2}^{0} 0 \times \sin\left(\frac{n\pi x}{2}\right)dx + \int_{0}^{2} 1 \times \sin\left(\frac{n\pi x}{2}\right)dx\right]$$

$$= \frac{1}{2}\left\{0 - \frac{2}{n\pi}\left[\cos\left(\frac{n\pi x}{2}\right)\right]_{0}^{2}\right\}$$

$$= -\frac{1}{n\pi}\{\cos n\pi - \cos 0\}$$

$$= \frac{1}{n\pi}\{1 - (-1)^{n}\}$$

$$= \begin{cases} 0, & if \quad n = \text{even} \\ \dfrac{2}{n\pi}, & if \quad n = \text{odd} \end{cases}$$

$$\therefore \quad f(x) = \frac{1}{2} + \sum_{n=1}^{\infty}\left[0 + \frac{2}{n\pi} \times \sin\left(\frac{n\pi x}{2}\right)\right](n = odd)$$

$$= \frac{1}{2} + \frac{2}{\pi}\sum_{n=1}^{\infty}\frac{1}{n}\sin\left(\frac{n\pi x}{2}\right)(n = odd)$$

$$= \frac{1}{2} + \frac{2}{\pi}\left[\sin\left(\frac{\pi x}{2}\right) + \frac{1}{3}\sin\left(\frac{3\pi x}{2}\right) \right.$$
$$\left. + \frac{1}{5}\sin\left(\frac{5\pi x}{2}\right) + ...\infty\right]$$

6. (d)

$$a_0 = \frac{2}{4}\int_{0}^{4}f(x)dx \qquad (\because c = 4)$$

$$= \frac{1}{2}\left[\int_{0}^{2}2x\,dx + \int_{2}^{4}(8 - 2x)dx\right]$$

$$= \frac{1}{2}\left\{\left[x^2\right]_{0}^{2} + \left[8x - x^2\right]_{2}^{4}\right\}$$

$$= \frac{1}{2}\{4 + (32 - 16) - (16 - 4)\} = 4.$$

7. (a)

We know that the Fourier sine series of f(x) in (0, c) is given by

$$f(x) = \sum_{n=1}^{\infty}b_n\sin\left(\frac{n\pi x}{c}\right)$$

where, $b_n = \dfrac{2}{c}\displaystyle\int_{0}^{c}f(x)\sin\left(\dfrac{n\pi x}{c}\right)dx$

Here, $f(x) = 1$ and $c = \pi$

$$\therefore \quad b_n = \frac{2}{\pi}\int_{0}^{\pi}1 \times \sin nx\,dx = -\frac{2}{\pi}\times\left[\frac{\cos nx}{n}\right]_{0}^{\pi}$$

$$= -\frac{2}{\pi n}\{\cos n\pi - \cos 0\}$$

$$= -\frac{2}{\pi n}\{(-1)^{n} - 1\}$$

Then, $f(x) = -\displaystyle\sum_{n=1}^{\infty}\frac{2}{n\pi}\left\{(-1)^{n} - 1\right\}\sin nx.$

1. A function with period 2π is shown below:

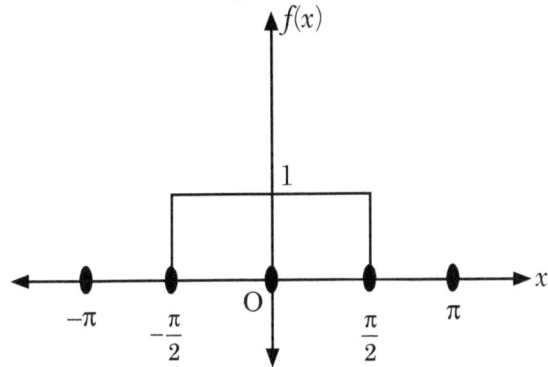

The Fourier series of the function is given by

$(a) \quad f(x) = \dfrac{1}{2} + \displaystyle\sum_{n=1}^{\infty}\frac{2}{n\pi}\sin\left(\frac{n\pi}{2}\right)\cos nx$

$(b) \quad f(x) = \displaystyle\sum_{n=1}^{\infty}\frac{2}{n\pi}\sin\left(\frac{n\pi}{2}\right)\cos nx$

$(c) \quad f(x) = \dfrac{1}{2} + \displaystyle\sum_{n=1}^{\infty}\frac{2}{n\pi}\sin\left(\frac{n\pi}{2}\right)\sin nx$

$(d) \quad f(x) = \displaystyle\sum_{n=1}^{\infty}\frac{2}{n\pi}\sin\left(\frac{n\pi}{2}\right)\sin nx$

(CE GATE 2000)

2. The Fourier series expansion of a symmetric and even function $f(x)$, where

$$f(x) = \begin{cases} 1 + \dfrac{2x}{\pi}, & -\pi \le x \le 0 \\ 1 - \dfrac{2x}{\pi}, & 0 \le x \le \pi \end{cases} \quad \text{is given by}$$

$(a) \quad \displaystyle\sum_{n=1}^{\infty}\frac{4}{\pi^2 n^2}(1 + \cos n\pi)$ $(b) \quad \displaystyle\sum_{n=1}^{\infty}\frac{4}{\pi^2 n^2}(1 - \cos n\pi)$

$(c) \quad \displaystyle\sum_{n=1}^{\infty}\frac{4}{\pi^2 n^2}(1 - \sin n\pi)$ $(d) \quad \displaystyle\sum_{n=1}^{\infty}\frac{4}{\pi^2 n^2}(1 + \sin n\pi)$

(CE GATE 2003)

3. The Fourier series of the function

$$f(x) = \begin{cases} 0, & -\pi < x \le 0 \\ \pi - x, & 0 < x < \pi \end{cases}$$

in the interval $[-\pi, \pi]$ is

$$f(x) = \frac{\pi}{4} + \frac{2}{\pi}\left[\frac{\cos x}{1^2} + \frac{\cos 3x}{3^2} + ...\right] +$$
$$\left[\frac{\sin x}{1} + \frac{\sin 2x}{2} + \frac{\sin 3x}{3}...\right]$$

The convergence of the above Fourier series at $x = 0$ gives

(a) $\displaystyle\sum_{n=1}^{\infty}\frac{1}{n^2}=\frac{\pi^2}{6}$ (b) $\displaystyle\sum_{n=1}^{\infty}\frac{(-1)^{n-1}}{n^2}=\frac{\pi^2}{12}$

(c) $\displaystyle\sum_{n=1}^{\infty}\frac{1}{(2n-1)^2}=\frac{\pi^2}{8}$ (d) $\displaystyle\sum_{n=1}^{\infty}\frac{(-1)^{n+1}}{2n-1}=\frac{\pi}{4}$

[CE GATE 2016]

4. A periodic signal $x(t)$ has a trigonometric Fourier series expansion

$$x(t) = a_0 + \sum_{n=1}^{\infty}\left[a_n\cos(n\omega_0 t)+b_n\sin(n\omega_0 t)\right]$$

$$x(t) = a_0 + \sum_{n=1}^{\infty}\left[a_n\cos(n\omega_0 t)b_n\sin(n\omega_0 t)\right]$$

If $x(t) = -x(-t) = -x\left(t-\dfrac{\pi}{\omega_0}\right)$, we can conclude that

(a) a_n are zero $\forall\, n$ and b_n are zero for n = even
(b) a_n are zero $\forall\, n$ and b_n are zero for n = odd
(c) a_n are zero for n = even and b_n are zero for n = odd
(d) a_n are zero for n = odd and b_n are zero for n = even   **[EC GATE 2017]**

5. The Fourier series expansion of the saw-toothed wave form $f(x) = x$ in $(-\pi, \pi)$ of period 2π gives the series $1-\dfrac{1}{3}+\dfrac{1}{5}-\dfrac{1}{4}+\ldots\infty = ?$

(a) $\dfrac{\pi}{2}$ (b) $\dfrac{\pi^2}{4}$

(c) $\dfrac{\pi^2}{16}$ (d) $\dfrac{\pi}{4}$

(EE, EC GATE 2017)

6. The Fourier cosine series for an even function f(x) is given by

$$f(x) = a_0 + \sum_{n=1}^{\infty}a_n\cos nx.$$

Then the value of the coefficient a_2 for the function $f(x) = \cos^2 x$ in $[0, \pi]$ is
(a) −0.5   (b) 0
(c) 0.5   (d) 1.0

(ME GATE 2018)

Answer Key				
1. (a)	2. (b)	3. (c)	4. (a)	5. (d)
6. (c)				

Explanation

1. (a) Here, $f(x) = f(x) = \begin{cases} 1, & -\dfrac{\pi}{2} \leq x \leq \dfrac{\pi}{2} \\ 0, & otherwise \end{cases}$

$$\therefore\quad a_0 = \frac{1}{\pi}\int_{-\pi}^{\pi}f(x)dx = \frac{1}{\pi}\int_{-\frac{\pi}{2}}^{\frac{\pi}{2}}1\,dx$$

$$= \frac{1}{\pi}[x]_{-\frac{\pi}{2}}^{\frac{\pi}{2}} = \frac{1}{\pi}\left\{\frac{\pi}{2}-\left(-\frac{\pi}{2}\right)\right\}$$

$$= 1$$

$$a_n = \frac{1}{\pi}\int_{-\pi}^{\pi}f(x)\cos nx\,dx$$

$$= \frac{1}{\pi}\int_{-\pi/2}^{\pi/2}1\times\cos nx\,dx$$

$$= \frac{1}{\pi}\left[\frac{\sin nx}{n}\right]_{-\frac{\pi}{2}}^{\pi/2} = \frac{1}{\pi n}\left\{\sin\left(\frac{n\pi}{2}\right)-\sin\left(-\frac{n\pi}{2}\right)\right\}$$

$$= \frac{1}{n\pi}\left\{\sin\left(\frac{n\pi}{2}\right)+\sin\left(\frac{n\pi}{2}\right)\right\} = \frac{2}{n\pi}\sin\left(\frac{n\pi}{2}\right)$$

$$b_n = \frac{1}{\pi}\int_{-\pi}^{\pi}f(x)\sin nx\,dx = \frac{1}{\pi}\int_{-\frac{\pi}{2}}^{\frac{\pi}{2}}1\times\sin nx\,dx$$

$$= 0\ (\because \sin nx \text{ is an odd function of } x)$$

Hence, the Fourier series for $f(x)$ is given by

$$f(x) = \frac{a_0}{2} + \sum_{n=1}^{\infty}(a_n\cos nx + b_n\sin nx)$$

$$= \frac{1}{2} + \sum_{n=1}^{\infty}\frac{2}{n\pi}\sin\left(\frac{n\pi}{2}\right)\cos nx.$$

2. (b)
$\because\ f(x)$ is even, so the Fourier series of $f(x)$ is given by

$$f(x) = \frac{a_0}{2} + \sum_{n=1}^{\infty}a_n\cos nx$$

Here, $a_0 = \dfrac{2}{\pi}\displaystyle\int_0^{\pi}f(x)dx = \dfrac{2}{\pi}\displaystyle\int_0^{\pi}\left(1-\dfrac{2x}{\pi}\right)dx$

$$= \frac{2}{\pi}\left[x-\frac{x^2}{\pi}\right]_0^{\pi} = \frac{2}{\pi}\left[\pi-\frac{\pi^2}{\pi}-0\right] = 0$$

$$a_n = \frac{2}{\pi}\int_0^{\pi}f(x)\cos nx\,dx = \frac{2}{\pi}\int_0^{\pi}\left(1-\frac{2x}{\pi}\right)\cos nx\,dx$$

$$= \frac{2}{\pi}\left\{\left[\left(1-\frac{2x}{\pi}\right)\frac{\sin nx}{n}\right]_0^{\pi}-\int_0^{\pi}\left(-\frac{2}{\pi}\right)\frac{\sin nx}{n}dx\right\}$$

$$= \frac{2}{\pi}\left\{0+\frac{2}{\pi n^2}\left[-\cos nx\right]_0^{\pi}\right\}$$

$(\because\ \sin 0 = 0,\ \sin n\pi = 0)$

$$= \frac{-4}{\pi^2 n^2}(\cos n\pi - \cos 0)$$

$$= \frac{4}{\pi^2 n^2}(1 - \cos n\pi)$$

Here, the required Fourier series is

$$f(x) = \sum_{n=1}^{\infty} \frac{4}{\pi^2 n^2}(1 - \cos n\pi).$$

3. (c) We know that a function $f(x)$ is discontinuous at $x = x_0$, then the Fourier series converges to

$$\frac{1}{2}\left[\lim_{x \to x_0^+} f(x) + \lim_{x \to x_0^-} f(x)\right]$$

Here, $f(x)$ is continuous at $x = 0$.
the Fourier series converges to

$$\frac{1}{2}\left[\lim_{x \to 0^+} f(x) + \lim_{x \to 0^-} f(x)\right]$$

i.e., $\frac{1}{2}\left[\lim_{x \to 0^+} (\pi - x) + \lim_{x \to 0^-} 0\right]$

i.e., $\frac{1}{2}[\pi - 0 + 0]$

i.e., $\frac{\pi}{2}$

Now putting $x = 0$ in the given Fourier series, we get,

$$f(0) = \frac{\pi}{4} + \frac{2}{\pi}\left[\frac{1}{1^2} + \frac{1}{3^2} + \frac{1}{5^2} + \ldots \infty\right]$$

$(\because \cos 0 = 1 \text{ and } \sin 0 = 0)$

or, $\frac{\pi}{2} = \frac{\pi}{4} + \frac{2}{\pi}\left[\frac{1}{1^2} + \frac{1}{3^2} + \frac{1}{5^2} + \ldots \infty\right]$

or, $\frac{1}{1^2} + \frac{1}{3^2} + \frac{1}{5^2} + \ldots \infty = \frac{\pi^2}{8}$

or, $\sum_{n=1}^{\infty} \frac{1}{(2n-1)^2} = \frac{\pi^2}{8}.$

4. (a)

$$x(t) = -x(t) \Rightarrow x(-t) = -x(t)$$

$\Rightarrow$ $x(t)$ is a odd function of t

$\Rightarrow$ a_n are zero $\forall n \in N$

Again, $x(t) = -x(-t) = -x\left(t - \frac{\pi}{\omega_0}\right)$

$\Rightarrow$ $x(t)$ contains only odd harmarics

$\Rightarrow$ $b_n = 0$ for $n =$ even.

5. (d)

$$f(x) = x \Rightarrow f(-x) = -x = -f(x)$$

$\Rightarrow$ $f(x)$ is an odd function of x

So its Fourier series is given by

$$f(x) = \sum_{n=1}^{\infty} b_n \sin nx,$$

where, $b_n = \frac{2}{\pi}\int_0^\pi f(x)\sin nx \, dx = \frac{2}{\pi}\int_0^\pi x \sin nx \, dx$

$$= \frac{2}{\pi}\left\{\left[x\left(-\frac{\cos nx}{n}\right)\right]_0^\pi - \int_0^\pi 1 \times \left(-\frac{\cos nx}{n}\right)dx\right\}$$

$$= \frac{2}{\pi}\left\{-\frac{\pi}{n}\cos n\pi - 0 + \frac{1}{n^2}[\sin nx]_0^\pi\right\}$$

$$= -\frac{2}{n}\cos n\pi \quad (\because \quad \sin 0 = 0, \sin n\pi = 0)$$

$$= -\frac{2}{n}(-1)^n$$

$$\therefore \quad f(x) = \sum_{n=1}^{\infty} b_n \sin nx = \sum_{n=1}^{\infty}\left(-\frac{2}{n}\right)(-1)^n \sin nx$$

$$\Rightarrow f\left(\frac{\pi}{2}\right) = \sum_{n=1}^{\infty}\left(-\frac{2}{n}\right)(-1)^n \sin\frac{n\pi}{2}$$

$$\Rightarrow \quad \frac{\pi}{2} = (-2)(-1)\sin\frac{\pi}{2} + \left(-\frac{2}{2}\right) \times 1 \times \sin\pi +$$

$$\left(-\frac{2}{3}\right)(-1)\sin\frac{3\pi}{2} + \left(-\frac{2}{4}\right) \times 1 \times \sin 2\pi$$

$$+ \left(-\frac{2}{5}\right)(-1)\sin\frac{5\pi}{2} + \cdots \infty$$

$$\Rightarrow \quad \frac{\pi}{2} = 2 + 0 + \frac{2}{3} \times (-1) + 0 + \frac{2}{5} \times (+1) + \cdots \infty$$

$$\Rightarrow 1 - \frac{1}{3} + \frac{1}{5} \cdots \infty = \frac{\pi}{4}.$$

6. (c) We know that for the Fourier cosine series in $[o, c]$

$$a_n = \frac{2}{c}\int_0^c f(x)\cos\left(\frac{n\pi x}{c}\right)dx$$

$$= \frac{2}{\pi}\int_0^\pi \cos^2 x \cos\left(\frac{n\pi x}{\pi}\right)dx (\because C = \pi)$$

$$= \frac{1}{\pi}\int_0^\pi (1 + \cos 2x)\cos nx \, dx$$

$$\therefore \quad a_2 = \frac{1}{\pi}\int_0^\pi (1 + \cos 2x)\cos 2x \, dx$$

$$= \frac{1}{2\pi}\int_0^\pi (2\cos 2x + 2\cos^2 2x) \, dx$$

$$= \frac{1}{2\pi}\int_0^\pi (2\cos 2x + 1 + \cos 4x) \, dx$$

$$= \frac{1}{2\pi}\left[2 \times \frac{\sin 2x}{2} + x + \frac{\sin 4x}{4}\right]_0^\pi = \frac{1}{2\pi} \times \pi = \frac{1}{2}.$$

$(\sin 2\pi = 0, \sin 4\pi = 0)$

Questions for Practice

1. Consider a function $f(x)$ defined by

$$f(x) = \begin{cases} x, & 0 < x < \pi \\ -x, & -\pi < x < 0 \end{cases}$$

Then the Fourier series expansion of $f(x)$ is given by

(a) $f(x) = \dfrac{\pi}{2} - \dfrac{4}{\pi} \displaystyle\sum_{n=1,3,5,7,...}^{\infty} \dfrac{1}{n^2}\cos nx$

(b) $f(x) = \dfrac{\pi^2}{2} + \dfrac{4}{\pi}\displaystyle\sum_{n=1}^{\infty} \dfrac{1}{n}\cos nx$

(c) $f(x) = -\dfrac{\pi}{2} - \dfrac{4}{\pi}\displaystyle\sum_{n=2,4,6,...}^{\infty} \dfrac{\sin nx}{n^2}$

(d) none of these

2. The Fourier series of the function $f(x)$, where
$f(x) = \begin{cases} 1-x, & -\pi < x < 0 \\ 1+x, & 0 \le x < \pi \end{cases}$ is given by

(a) $\dfrac{\pi^2}{2} + \dfrac{2}{\pi}\left(\cos x + \dfrac{1}{3}\cos 3x + \dfrac{1}{5}\cos 5x + ...\infty\right)$

(b) $\dfrac{\pi-2}{2} + \dfrac{4}{\pi}\left(\cos x + \dfrac{1}{9}\cos 3x + \dfrac{1}{25}\cos 5x + ...\infty\right)$

(c) $\dfrac{\pi+2}{2} + \dfrac{4}{\pi}\left(\cos x + \dfrac{1}{9}\cos 3x + \dfrac{1}{25}\cos 5x + ...\infty\right)$

(d) none of these

3. Consider the Fourier series of $f(x) = x - x^2$ in $-\pi < x < \pi$. Then which of the following is true?

(a) $\dfrac{\pi^2}{12} = \dfrac{1}{1^2} - \dfrac{1}{2^2} + \dfrac{1}{3^2} - \dfrac{1}{4^2} + ...\infty$

(b) $\dfrac{\pi^2}{12} = \dfrac{1}{1^2} + \dfrac{1}{2^2} + \dfrac{1}{3^2} + \dfrac{1}{4^2} + ...\infty$

(c) $\dfrac{\pi^2}{8} = 1 + \dfrac{1}{2^2} + \dfrac{1}{3^2} + ...\infty$

(d) none of these

4. Consider a function $f(x)$ given by
$f(x) = \begin{cases} -2, & -\pi < x < 0 \\ 2, & 0 < x < \pi \end{cases}$

Then what will be the value of b_n in the Fourier series expansion of $f(x)$?

(a) $b_n = \dfrac{2}{\pi}\left\{\dfrac{2}{n} + \dfrac{2(-1)^n}{n}\right\}$

(b) $b_n = 0$ if n is even
and $b_n = \dfrac{8}{nk}$ if n is odd.

(c) $b_n = \begin{cases} \dfrac{4}{nk} & \text{if } n \text{ is even} \\ 0, & \text{if } n \text{ is odd} \end{cases}$

(d) $b_n = \begin{cases} 0, & \text{if } n \text{ is even} \\ 2, & \text{if } n \text{ is odd} \end{cases}$

5. Consider the Fourier series expansion of $f(x) = |x|$ in $(-2, 2)$. Then which of the following is not correct?

(a) $a_0 = 2$

(b) $a_n = \dfrac{4}{\pi^2 n^2}\left\{(-1)^n - 1\right\}$

(c) $f(x)$ is an even function of x

(d) none of these

6. Which of the following is a half range cosine series for the function $f(x) = \begin{cases} 1, & \text{if } 0 < x < \dfrac{\pi}{2} \\ 0, & \text{if } \dfrac{\pi}{2} < x < 1 \end{cases}$

(a) $-\dfrac{1}{2} + \dfrac{4}{\pi}\left(\cos x - \dfrac{1}{3}\cos 3x + \dfrac{1}{5}\cos 5x - ...\infty\right)$

(b) $-\dfrac{1}{2} + \dfrac{2}{\pi}\left(\cos x - \dfrac{1}{3}\cos 3x + \dfrac{1}{5}\cos 5x - ...\infty\right)$

(c) $-\dfrac{1}{2} + \dfrac{2}{\pi}\left(\cos x + \dfrac{1}{3}\cos 3x + \dfrac{1}{5}\cos 5x + ...\infty\right)$

(d) none of these

7. Consider the Fourier series expansion of $f(x) = x\cos x$ in $-\pi < x < \pi$. Then $a_0 = ?$

(a) $-\dfrac{1}{3}\pi^2$

(b) 0

(c) 2

(d) $\dfrac{(-1)^2 2n}{n^2 - 1}$

7. (b)

Answer Key				
1. (a)	2. (c)	3. (a)	4. (b)	5. (d)
6. (b)	7. (b)			

Explanation

1. (a)
$f(x) = \begin{cases} -x, & -\pi < x < 0 \\ x, & 0 < x < \pi \end{cases}$
$= |x|, -\pi < x < x$

GRAPH THEORY

11.1 GRAPHS

11.1.1 Definition of a Graph

A graph G is an ordered pair (V, E) where $V (\neq \varnothing)$ (called the set of vertices or set of nodes) and E is called the set edges of the graph.

Every graph can be represented by a diagram in which each vertex is represented by a point and each edge is represented by a straight line (or a curve). For $e \in E$, if e is an edge joining the vertices u and v, then we write $e = (u, v)$.

Example:

Let $V = \{v_1, v_2, v_3, v_4\}$, $E = \{e_1, e_2, e_3, e_4\}$ and $e_1 = v_1$, v_2 $e_2 = (v_2, v_3)$, $e_3 = (v_1, v_4)$, $e_4 = (v_3, v_4)$.

Then the graph $G = (V, E)$ can be diagrammatically represented by

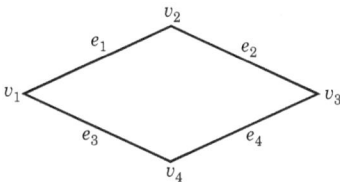

Remember:

(i) In any graph an edge should not pass through any vertices other than two end vertices of the edge.

(ii) In a graph $G = (V, E)$, the number of vertices is denoted by $|V|$ or $|V(G)|$; and the number of edges is denoted by $|E|$ or $|E(G)|$.

(iii) In a graph $G = (V, E)$ for $e \in E$ and $e = (u, v)$, we say that the vertices u and v

are adjacent. In other words, any pair of vertices which are connected by an edge, are called adjacent vertices.

Thus, in the previous example, the vertices v_1 and v_2 are adjacent, v_2 and v_3 are adjacent v_3 and v_4 are adjacent, v_1 and v_4 are adjacent.

(iv) A graph G is said to be a finite if the set of vertices and the set of edges are finite.

(v) A graph with "p" vertices and "q" edges is called a (p, q) graph. p is called the order of the graph and q is called the size of the graph.

11.1.2 Incidence

In a graph $G = (V, E)$, let $e \in E$ and $e = (u, v)$, *i.e.*, e is an edge joining the vertices u and v. Then e is said to be incident with the vertices u and v.

Example:

Consider the following graph:

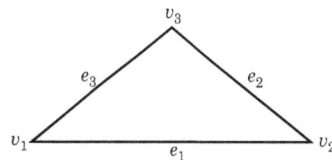

Here e_1 is incident with v_1 and v_2, e_2 is incident with v_2 and v_3, and e_3 is incident with v_1 and v_3.

11.1.3 Loops

An edge of a graph which joins a vertex to itself is called a loop. A loop is denoted by the circle symbol "O."

Example:

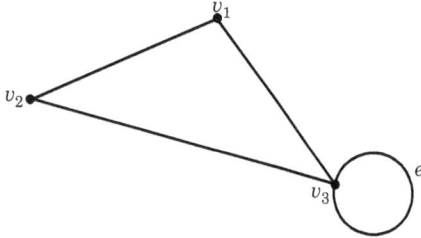

Here, the edge e joins the vertex v_3 to itself, *i.e.*, v_3. So "e" is a loop.

11.1.4 Parallel Edges

If in a graph, there exist a pair of vertices which are joined by more than one edges, then such edges are called parallel edges.

Example:

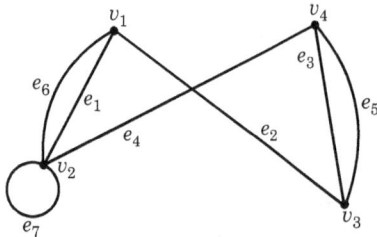

Here the vertices v_3 and v_4 are joined by the edges e_3 and e_5. So e_3 and e_5 are parallel edges.

Also the vertices v_1 and v_2 are joined by the edges e_1 and e_6. So e_1 and e_6 are also parallel edges.

11.1.5 Degree of a Vertex

Let $v \in V$ be a vertex in the graph $G = (V, E)$. Then degree of v is denoted by $\deg_G (v)$ or $d_G (v)$ and is defined by:

$\deg_G (v) = d_G(v)$ = number of edges of G which are incident with "v" (each loop is counted twice).

Example:

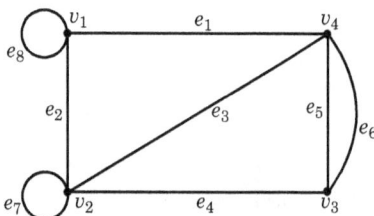

Vertex	Edges that the incident with the vertex	Degree of the vertex
v_1	e_8 (loop), e_1, e_2	$2 + 1 + 1 = 4$
v_2	e_7 (loop), e_2, e_4, e_3	$2 + 1 + 1 + 1 = 5$
v_3	e_4, e_5, e_6	$1 + 1 + 1 = 3$
v_4	e_1, e_3, e_5, e_6	$1 + 1 + 1 + 1 = 4$

Remember:

(*i*) If a graph $G = (V, E)$ contains "n" number of vertices $v_1, v_2, v_3, ..., v_n$

then $\displaystyle\sum_{i=1}^{n} \deg(v_i) = 2|E|$.

Thus, sum of degree of all vertices

= 2 × (total number of edges).

Explanation:

Consider the following graph:

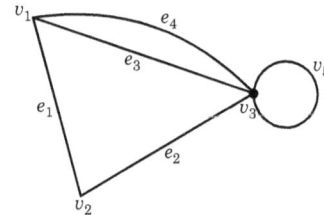

Here, $\deg_G (v_1) = 3$, $\deg_G(v_2) = 2$, $\deg_G(v_3) = 5$.

$\therefore \displaystyle\sum_{i=1}^{n} \deg(v_i) = 3 + 2 + 5 = 10 = 2 \times 5 = 2|E|$

(since E = set of edges = $\{e_1, e_2, e_3, e_4, e_5\}$, so $|E|$ = number of edges = 5).

(*ii*) In any graph, the number of vertices of odd degree is even.

Explanation:

Consider the following graph:

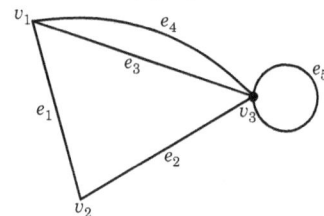

Here $\deg_G(v_1) = 3$, $\deg_G (v_2) = 2$, $\deg_G (v_3) = 5$. Thus, the vertices v_1 and v_3 have odd degree. So number of vertices of odd degree = 2, which is even.

(*iii*) A vertex of degree "0" is called an isolated vertex.

Example:

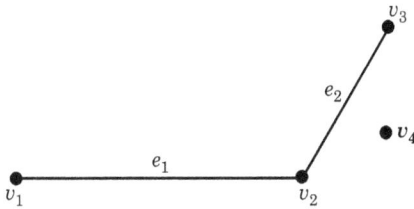

Here no edges are incident with the vertex v_4. So $\deg_G(v_4) = 0$. Here v_4 is an isolated vertex.

(*iv*) A vertex of degree "1" is called a "pendant vertex."

Example:

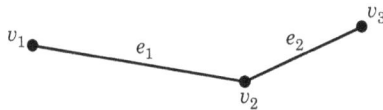

Here $\deg_G(v_1) = 1 = \deg_G(v_3)$. Thus, v_1 and v_3 are pendant vertices's.

11.1.6 Directed Graph

In a graph $G = (V, E)$, an edge $e \in E$ is called a directed edge if it is associated with an ordered pair of vertices, otherwise it is called an undirected edge. Thus, each directed edge contains a direction. If in a graph, every edge is directed then, the graph is called a directed graph.

Example:

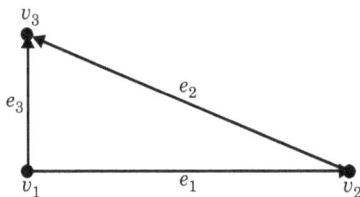

11.1.7 In Degree and Out Degree

If in a directed graph $G = (V, E)$, an edge $e \in E$, starts from the vertex u and ends in the vertex v, then the edge "e" is said to be incident out of the vertex u and incident into the vertex v. Then

(*i*) the number of edges incident out of the vertex u is called the "out degree of u" and is denoted by $\deg_G^-(u)$.

(*ii*) The number of edges incident into the vertex u is called the "in degree of v" and is denoted by $\deg_G^+(v)$.

Example:

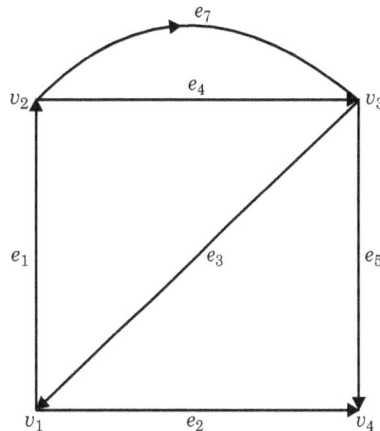

vertex	Out degree	In degree
v_1	2 (since the edges e_1 and e_2 are incident out from v_1)	1 (since the edge e_3 is incident into v_1)
v_2	2 (since the edges e_4 and e_7 are incident out from v_2)	1 (since the edge e_1 is incident into v_2)
v_3	2 (since the edges e_3 and e_5 are incident out from v_3.	2 (since the edges e_4 and e_7 are incident into v_3)
v_4	0 (since no edge is incident out from v_4.	2 (since the edges e_2 and e_5 are incident into v_4)

Here sum of out degrees = 6 = sum of in degrees = Total number of edges.

Remember:

If a directed graph $G = (V, E)$ contains "n" number of vertices namely v_1, v_2, v_3, ..., v_n then

$$\sum_{i=1}^{n} \deg_G^-(v_i) = \sum_{i=1}^{n} \deg_G^+(v_i) = |E|.$$

11.1.8 Minimum Degree and Maximum Degree

Let $G = (V, E)$ be a graph. Then

(*i*) the minimum of all degrees of the vertices of G is denoted by $\delta(G)$ and is

defined by $\delta(G) = \min\{\deg_G(v) : v \in V\}$.

(*ii*) the maximum of all degrees of the vertices of G is denoted by $\Delta(G)$ and is defined by $\Delta(G) = \max\{\deg_G(v) : v \in V\}$.

Example:

Here, $\deg_G(v_1) = 5$, $\deg_G(v_2) = 3$,

$\deg_G(v_3) = 4$.

Then

$$\delta(G) = \min\{\deg_G(v_1), \deg_G(v_2), \deg_G(v_3)\}$$
$$= \min\{5, 3, 4\} = 3$$

$$\Delta(G) = \max\{\deg_G(v_1), \deg_G(v_2), \deg_G(v_3)\}$$
$$= \max\{5, 3, 4\}$$
$$= 5$$

Remember:

For any graph $G = (V, E)$,

(i) $\delta(G) \le \Delta(G)$

(ii) $\delta(G) \le \dfrac{2|E|}{|V|} \le \Delta(G)$.

11.2 DIFFERENT TYPES OF GRAPHS

11.2.1 Mixed Graph

If in a graph, some edges are directed and some are undirected, then the graph is called a mixed graph.

Example:

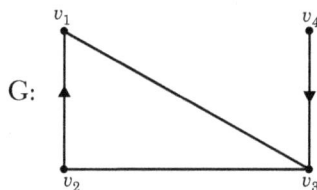

11.2.2 Multi Graph

If a graph contains some loops and parallel edges, then it is called a multigraph.

Example:

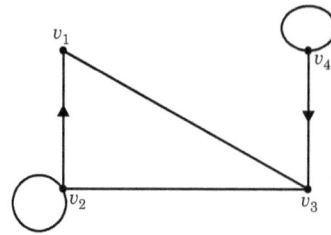

11.2.3 Simple Graph

A graph having no loops and parallel edges is called a simple graph. It both loops and parallel edges are there then the graph is called a pseudo graph.

Example:

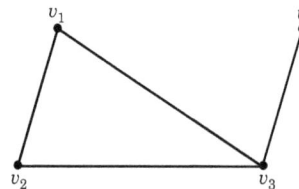

11.2.4 Trivial Graph

A graph which contains only one vertex but no edges is called a trivial graph.

Example:

G: • y

11.2.5 Null Graph

If a graph contains no edges, then it is called a null graph.

Example:

11.2.6 K-regular Graph

If in a simple graph, G each vertex has degree "K," then the graph is called K-regular graph (or regular graph of degree K).

Example:

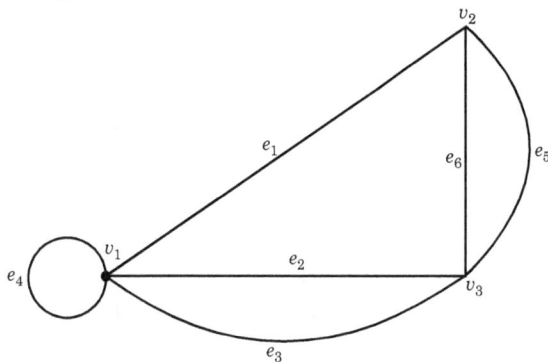

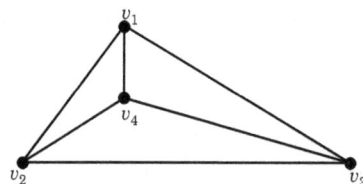

Here, $\deg_G(v_1) = \deg_G(v_2) = \deg_G(v_3)$

$= \deg_G(v_4) = 3$.

Therefore, G is a regular graph of degree 3.

Remember:

If G is a K-regular graph, then

$$\delta(G) = \frac{2|E|}{|V|} = \Delta(G).$$

11.2.7 Complete Graph

If G be a simple graph in which every pair of distinct vertices are adjacent, then G is called a complete graph. In other words, if in a graph, there is exactly one edge between each pair of vertices, then the graph is said to be complete. A complete graph with "n" vertices, is denoted by K_n.

Example:

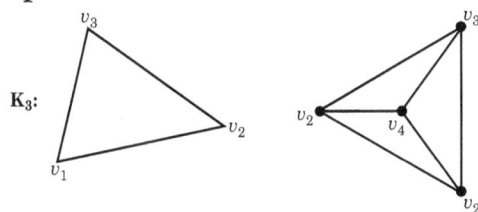

11.2.8 Bipartite Graph

Let $G = (V, E)$ be a graph and suppose that there exists two subsets X and Y of V such that

(*i*) $X \cap Y = \varnothing$

(*ii*) $X \cup Y = V$

(*iii*) every edge joins a vertex in X to a vertex in Y

(*iv*) No edge joins either two vertices in X or two vertices in Y.

Then the graph G is called a bipartite graph.

Example:

Let $G = (V, E)$ where $V = \{v_1, v_2, v_3, v_4\}$.

Let $X = \{v_1, v_3\}$, $Y = \{v_2, v_4\}$.

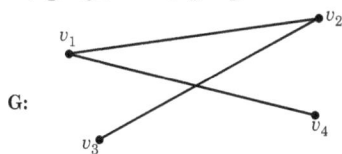

Here, G is a bipartite graph

11.2.9 Complete Bipartite Graph

Let $G = (V, E)$ be a graph and suppose there exists two subsets X and Y of V such that

(*i*) $X \cap Y = \varnothing$ (*ii*) $X \cup Y = V$

(*iii*) every vertex of X is adjacent to every vertex of Y.

Then the graph G is called a complete bipartite graph.

If $|X|$ = number of elements in $X = m$ and $|Y|$ = number of elements in $Y = n$, then complete bipartite graph is denoted by $K_{m,n}$.

Example:

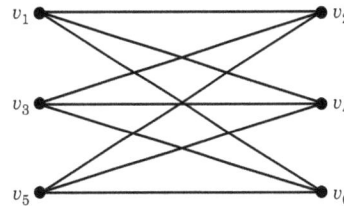

Here, $X = \{v_1, v_3, v_5\}$ and $Y = \{v_2, v_4, v_6\}$.

11.3 WALK AND PATH

11.3.1 Walk

Let $G = (V, E)$ be a graph which contains "n" number of vertices $v_1, v_2, v_3, ..., v_n$ and "m" number of edges $e_1, e_2, e_3, ..., e_m$. Then a walk in G is defined as a finite alternating sequence of vertices and edges starting and ending with verities of the form.

$$v_1, e_1, v_2, e_2, v_3, ..., v_{n-1}, e_m, v_n.$$

Example:

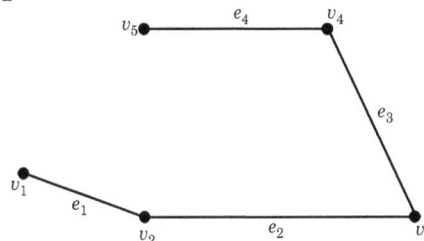

Here, $v_1, e_1, v_2, e_2, v_3, e_3, v_4, e_4, v_5$ is a walk.

Remember:

(*i*) If starting and ending vertices are not same, i.e., $v_1 \neq v_n$, then the walk is called open; otherwise, it is called a closed walk.

(*ii*) A walk is also known as "chain."

(*iii*) A walk, in which no edge is repeated is called a trial.

(*iv*) Length of the walk = number of edges.

11.3.2 Path and Circuit

In a nondirected graph G, a sequence P of zero or more edges of the form $(v_0, v_1), ..., (v_{n-1}, v_n)$ is called a path. A path is an open walk which can also be denoted by $v_0 - v_1 - v_2... -v_{n-1} - v_n$. If $v_o = v_n$, then the closed walk is called a circuit.

Example:

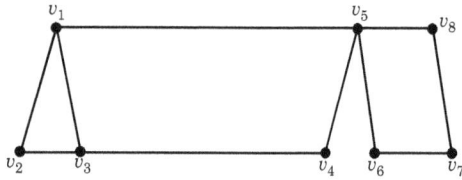

Here, $v_1 - v_2 - v_3 - v_1$ is a circuit.

$v_1 - v_3 - v_4 - v_5 - v_6 - v_7 - v_8$ is a path.

Remember:

(*i*) The total number of edges appearing in the path is called the *length of the path*.

(*ii*) A *trivial path* is a path having length zero.

(*iii*) A path is said to be a *simple* if all the edges and vertices in the path are distinct except possibly at end vertices.

(*iv*) Closed walk called a *circuit* if,

(*a*) length of the walk ≥ 1

(*b*) no edge is repeated in the walk.

(*c*) end vertices of the walk are same.

(*v*) A *cycle* is a circuit with no other repeated vertices except its end vertices.

Example:

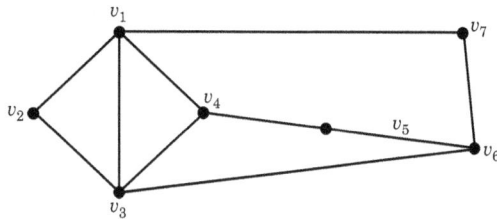

Path	Simple path	Closed path	Circuit	Cycle
$v_1 - v_2 - v_3 - v_4 - v_1$	Yes	Yes	Yes	Yes
$v_4 - v_5 - v_6 - v_3$	Yes	No	No	No
$v_2 - v_1 - v_7 - v_6 - v_3 - v_1 - v_2$	No	Yes	No	No
$v_3 - v_6$	Yes	No	No	No
$v_1 - v_3 - v_1$	No	Yes	No	No

11.4 Matrix Representation of Graphs

There are three ways of representing a graph by a matrix, namely adjacent matrix, incident matrix, and path matrix.

11.4.1 Adjacent Matrix

Let $G = (V, E)$ be a simple graph having n vertices $v_1, v_2, ..., v_n$. Then the adjacent matrix of G is denoted by $A(G)$ or A_G and is defined by $A_G = [a_{ij}]_{n \times n}$, where

$$a_{ij} = \begin{cases} 1, \text{if vertex } v_i \text{ is adjacent to vertex } v_j \\ 0, \text{otherwise} \end{cases}$$

Thus, A_G is a symmetric matrix, *i.e.*, $A_G^T = A_G$.

Example:

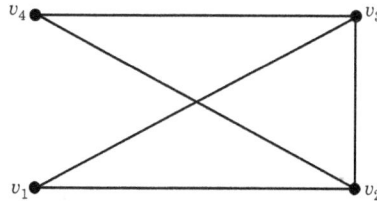

$$A_G = \begin{matrix} & \begin{matrix} v_1 & v_2 & v_3 & v_4 \end{matrix} \\ \begin{matrix} v_1 \\ v_2 \\ v_3 \\ v_4 \end{matrix} & \begin{pmatrix} 0 & 1 & 1 & 0 \\ 1 & 0 & 1 & 1 \\ 1 & 1 & 0 & 1 \\ 0 & 1 & 1 & 0 \end{pmatrix} \end{matrix}$$

Remember:

If G is a simple digraph, then

$$a_{ij} = \begin{cases} 1, \text{if} \left(v_i, v_j\right) \in E \\ 0, \text{otherwise} \end{cases}$$

Example:

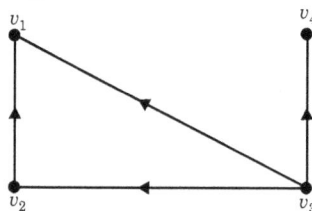

$$A_G = \begin{matrix} & \begin{matrix} v_1 & v_2 & v_3 & v_4 \end{matrix} \\ \begin{matrix} v_1 \\ v_2 \\ v_3 \\ v_4 \end{matrix} & \begin{pmatrix} 0 & 0 & 0 & 0 \\ 1 & 0 & 0 & 0 \\ 1 & 1 & 0 & 1 \\ 0 & 0 & 0 & 0 \end{pmatrix} \end{matrix}$$

11.4.2 Incidence Matrix

Let $G = (V, E)$ be a graph with n vertices $v_1, v_2, ...,$ v_n and m edges $e_1, e_2, ..., e_m$. Then the incidence matrix G is denoted by $I(G)$ or I_G and is defined by $I_G = [t_{ij}]_{n \times m}$, where

$$t_{ij} = \begin{cases} 1, \text{when vertex } v_i \text{ is incident with } e_j \\ 0, \text{otherwise} \end{cases}$$

Example:

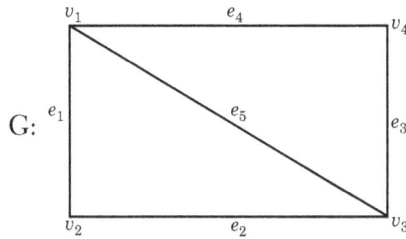

$$I_G = \begin{array}{c} \\ v_1 \\ v_2 \\ v_3 \\ v_4 \end{array} \begin{array}{ccccc} e_1 & e_2 & e_3 & e_4 & e_5 \\ \begin{pmatrix} 1 & 0 & 0 & 1 & 1 \\ 1 & 1 & 0 & 0 & 0 \\ 0 & 1 & 1 & 0 & 1 \\ 0 & 0 & 1 & 1 & 0 \end{pmatrix} \end{array}$$

11.4.3 Path Matrix

Let $G = (V, E)$ be a simple digraph with no parallel edges. Also, let G has n vertices $v_1, v_2, ..., v_n$. Then the path matrix G is denoted by $P(G)$ or P_G and is defined by $P_G = [p_{ij}]_{n \times n}$, where

$$p_{ij} = \begin{cases} 1, \text{if there is a path from } v_i \text{ to } v_j \\ 0, \text{ otherwise} \end{cases}$$

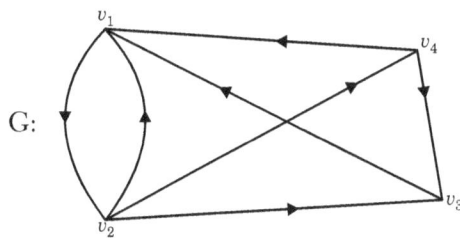

$$P_G = \begin{array}{c} \\ v_1 \\ v_2 \\ v_3 \\ v_4 \end{array} \begin{array}{cccc} v_1 & v_2 & v_3 & v_4 \\ \begin{pmatrix} 1 & 1 & 1 & 1 \\ 1 & 1 & 1 & 1 \\ 1 & 1 & 1 & 1 \\ 1 & 1 & 1 & 1 \end{pmatrix} \end{array}$$

11.5 Planar Graphs and Euler's Formula

11.5.1 Planar Graph

A graph G is said to be planar if it can be drawn in the plane without any crossover of edges.

Example:

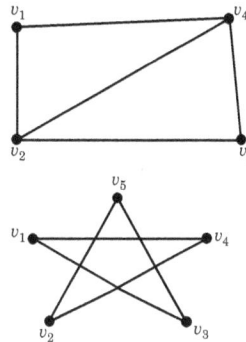

Remember:

A plane graph divides the whole plane into several region. A region is characterized by a cycle which form its boundary and the degree of a region is equal to the length of the cycle forming its boundary.

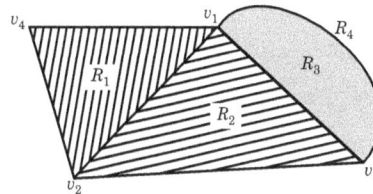

Here, the cycle $v_1 - v_2 - v_4 - v_1$ forms the boundary of the region R_1. Therefore, $\deg(R_1) = 3$.

The cycle $v_1 - v_2 - v_3 - v_1$ froms the boundary of the region R_2. Therefore, $\deg(R_2) = 3$.

The cycle $v_1 - v_3 - v_1$ forms the boundary of the region R_3. Therefore, $\deg(R_3) = 2$.

The cycle $v_1 - v_4 - v_2 - v_3 - v_1$ forms the boundary of the region R_4. Therefore, $\deg(R_4) = 4$. (R_4 is the exterior region and R_1, R_2, R_3 are interior regions).

Note:

If $G = (V, E)$ is a planar graph, then $\sum_i \deg(R_i) = 2|E|$, where R_i's are the region determined by G.

From the above example,

$$\sum_i \deg(R_i) = \deg(R_1) + \deg(R_2) + \deg(R_3) + \deg(R_4)$$

$$= 3 + 3 + 2 + 4 = 12 = 2 \times |E|$$

(since there are 6 edges in G)

11.5.2 Euler's Formula

If $G = (V, E)$ is a connected graph, then $|V| + |R| = |E| + 2$, where $|V|$ = no. of vertices, $|E|$ = no. of edges, $|R|$ = no. of regions.

Example:

Consider the example given below:

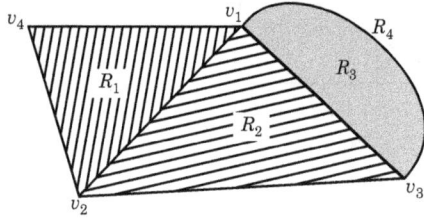

Here, $|V| = 4$, $|E| = 6$, $|R| = 4$.

$\therefore |V| + |R| = 8 = 6 + 2 = |E| + 2$.

Remember:

(*i*) If $G = (V, E)$ be a planar graph with $|E| > 1$, and every region is bounded by at least 3 edges then

(*a*) $|E| \le 3|V| - 6$

(*b*) $\exists$ a vertex such that $\deg_G(v) \le 5$

(*ii*) A complete graph K_n is planar if and any if $n \le 4$.

(*iii*) A complete graph $K_{m,n}$ is planar if and any if m $\le$ 2 or n $\le$ 2.

(*iv*) If $G = (V, E)$ be a connected planar graph with $|V| \ge 3$ and every region is bounded by at least 4 edges then $|R| \le 2|V| - 4$.

(*v*) A graph is a planar if and only if it contains subgraph homeomorphic to K_5 or $K_{3,3}$.

11.6 SUB GRAPHS AND ISOMORPHIC GRAPHS

11.6.1. Subgraphs

A graph H is called a sub graphs of a graph $G = (V, E)$ if

(*i*) the set of edges of H is a subset of E.

(*ii*) the set of vertices of H is a subset of V.

Example:

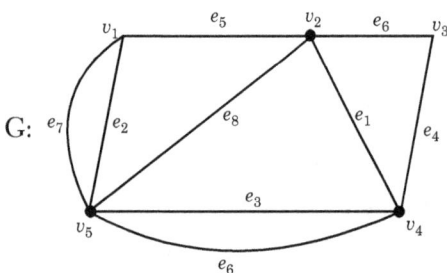

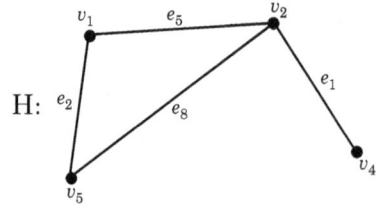

11.6.2 Isomorphic Graph

Two graphs $G_1 = (V_1, E_1)$ and $G_2 = (V_2, E_2)$ are said to be isomorphic if $\exists$ a function $f : V_1 \to V_2$ such that

(*i*) f is bijective, *i.e.*, one-one and onto

(*ii*) $e = (u, v)\ E_1 \Leftrightarrow (f(u),\ f(v)) \in E_2$; $e = (u,v) \in E_1 \Leftrightarrow (f(u), f(v)) \in E_2$; which means if there is an edge joining the vertices u and v in G_1, then there exist an edge joining the vertices $f(u)$ and $f(v)$ in G_2.

Example:

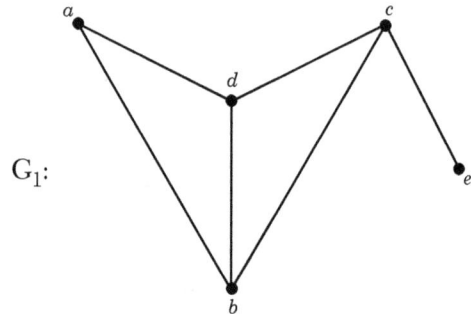

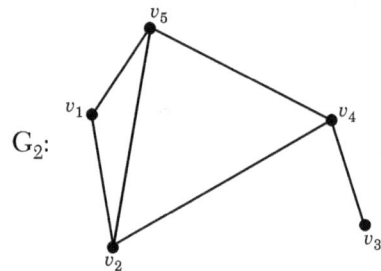

Here, $G_1 = (V_1, E_1)$ where $V_1 = \{a, b, c, d, e\}$ and $G_2 = (V_2, E_2)$ where $V_2 = \{v_1, v_2, v_3, v_4, v_5\}$.

Let $f : V_1 \to V_2$ be a mapping defined by $f(a) = v_1, f(b) = v_2, f(c) = v_4, f(d) = v_5, f(e) = v_3$.

Then clearly f is one-one onto.

Also there exist an edge between vertices a and b $\Leftrightarrow$ there exist an edge between vertices v_1 and v_2.

There exist an edge between vertices a and d $\Leftrightarrow$ there exist an edge between vertices v_1 and v_5.

There exist an edge between vertices b and d $\Leftrightarrow$ there exist an edge between vertices v_2 and v_5.

There exist an edge between vertices b and c $\Leftrightarrow$ there exist an edge between vertices v_2 and v_4.

There exist an edge between vertices d and c $\Leftrightarrow$ there exist an edge between vertices v_5 and v_4.

There exist an edge between vertices c and e $\Leftrightarrow$ there exist an edge between vertices v_3 and v_4.

Hence, the graphs G_1 and G_2 are isomorphic.

11.7 CONNECTEDNESS

11.7.1 Connected Graph

A graph is said to be connected if there is a path between any two of its vertex. Thus, in a graph, if starting from any vertex, we can reach to any other vertex through a path, then the graph is said to be connected.

Example:

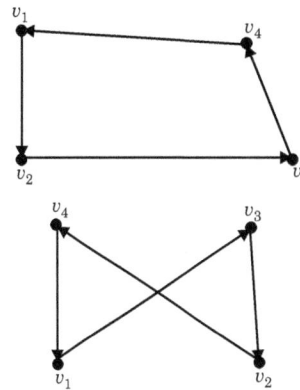

11.7.2 Strongly Connected Graph

If in a graph, for any pair of vertices, both the vertices of the pair are reachable from one to another, then the graph is called strongly connected.

Example:

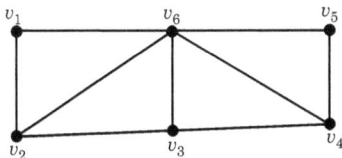

11.7.3 Weakly Connected Graph

A simple directed graph is said to be weakly connected if the corresponding undirected graph is connected.

Example:

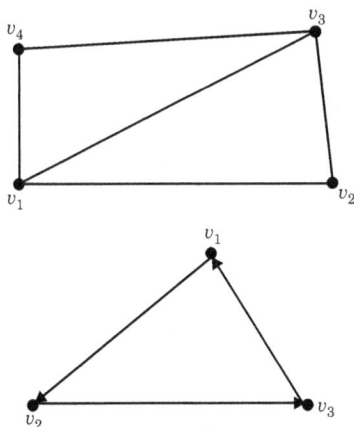

11.7.4 Component

A connected sub graph of a graph G which is not a proper subgraph of any other connected subgraph of G is called a component of G.

Remember:

(*i*) If a graph G has n vertices and $\delta(G) \geq \dfrac{n-1}{2}$ then G is connected

(*ii*) If G be a simple graph with n number of vertices and k number of components then G can have at most $\dfrac{1}{2}(n-k)(n-k+1)$ edges.

(*iii*) Every maximal connected sub graph of G is called a component.

(*iv*) If a connected planar simple graph is triangle free, then $|E| \leq 2|V| - 4$.

(*v*) If a planar graph has "p" components, then $|V| - |E| + |R| = p + 1$.

(*vi*) If G be a planar graph has "p" connected components, where each component contains at least three vertices, then $|E| \leq 3|V| - 6p$.

(*vii*) A graph is connected if and only if it has one and only one component.

11.7.5 Eulerian Graph

A path in a graph G is called an Euler path (or Eulerian trail) if it contains every edges of G exactly once. A graph G which has an Euler circuit is called an Eulerian graph. A closed walk in G is called an Euler circuit if it contains every edges of G exactly once.

Example:

Consider the graph G given below:

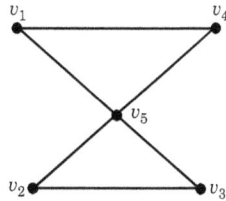

Clearly G is connected and all the vertices of G have even degree. So G has a Euler circuit. Consequently, G is a Eulerian graph.

Remember:

(i) A connected graph G is Eulerian if and only if G is Eulerian every vertex of G has even degree

(ii) A connected graph G has an Eulerian trail if and only if it has at most two odd vertices.

(iii) If a graph G has a vertex of odd degree, then no Euler circuit exists in G.

(iv) Fleury's algorithm can be used to find a Euler circuit.

11.7.6 Hamiltonian Graph

A path P: $v_0 - v_1 - ..... - v_{n-1} - v_n$ in the connected graph G is called a *Hamiltonian Path*, if $V_G = \{v_0, v_1, ..., v_{n-1}, v_n\}$ and $v_i \neq v_j$ for $0 \leq i < j \leq n$, *i.e.*, the Hamiltonian Path passes through each of the vertices in the graph exactly once.

A circuit (or cycle) C: $v_0 - v_1 - ..... - v_{n-1} - v_n - v_0$, $n > 1$, in the graph G is called a *Hamiltonian Circuit*, if P: $v_0 - v_1 - ..... - v_{n-1} - v_n - v_o$ is a Hamiltonian path. That is Hamiltonian circuit contains each vertex of G exactly once except the starting and ending vertex that appear twice and if we remove any one edge from the circuit we are left a Hamiltonian path.

A graph G is called a Hamiltonian graph if it contains a Hamiltonian circuit.

Example:

Consider the following graph G:

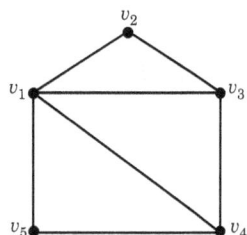

G has a Hamiltonian circuit C: $v_1 - v_5 - v_2 - v_3 - v_4 - v_1$. Hence, G is a Hamiltonian graph.

Remember:

(i) Let G be a simple connected graph with $n \geq 3$ vertices. If $\deg(v) \geq n/2$ for all vertices of G, then G is Hamiltonian.

(ii) A graph with n vertices and with no loops or parallel edges which has at least $\frac{1}{2}(n-1)(n-2)+2$ edges is Hamiltonian.

(iii) The complete graph K_n ($n \geq 3$) is Hamiltonian.

(iv) The complete bipartite graph $K_{m,n}$ is Hamiltonian if and only if $m = n$ and $n > 1$.

11.8 VERTEX AND EDGE CONNECTIVITY

11.8.1 Cut Vertex

Let v be a vertex of a connected graph G such that $G - v$ is not connected. Then the vertex v is called cut vertex.

In other words, if the removal of a vertex v from a connected graph G, makes the graph disconnected, then the vertex v is called a cut vertex.

Example:

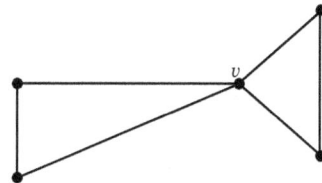

Here v is the cut vertex.

Remember:

(i) The removal of a cut vertex from a graph G, increase the number of components of G.

(ii) A vertices v in a connected graph $G = (V, E)$ is a cut vertex if and only if there exist vertices v_1 ($\neq v$) and v_2 ($\neq v$) such that every path connecting the vertices v_1 and v_2 contains the vertex v.

11.8.2 Cut Edge (Bridge)

Let "e" be an edge of a connected graph G such that $G - e$ is disconnected. Then the edge "e" is called cut edge. In other words, if the removal of an edge "e" from a connected graph G, makes the graph disconnected, then the edge "e" is called a cut edge.

Example:

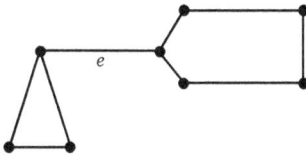

Here, e is a cut edge.

11.8.3 Cut Set

A cut set in a graph G is a set of edges whose removal from the graph; makes the graph disconnected, provided no proper subset of it disconnects the graph G.

Example:

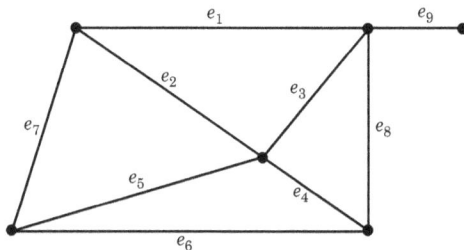

Here, some of cut set are $\{e_1, e_2, e_3, e_6\}$ and $\{e_1, e_2, e_7\}$

Note:

In the above example, $\{e_1, e_3, e_4, e_8\}$ is not a cut set, since one of its proper subset $\{e_1, e_2, e_8\}$ forms a cut set.

11.8.4 Edge Connectivity

The edge connectivity of a connected graph G is the minimum number of edges whose removal result in a disconnected or trivial graph. We denote the edge connectivity of a graph a by $\lambda(G)$.

Example:

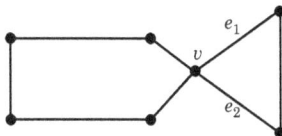

Here $\lambda(G) = 2$, since removal of the edges e_1 and e_2 from G results in a disconnected graph.

Remember:

(*i*) If G is a connected graph having a bridge, then $\lambda(G) = 1$

(*ii*) Edge connectivity of G = no. of edges in the smallest cut set of G.

11.8.5 Vertex Connectivity

The vertex connectivity of a connected graph G is the minimum number of vertices whose removal

result in a disconnected or trivial graph. we use the symbol $K(G)$ to denote the vertex connectivity of a graph G.

Example:

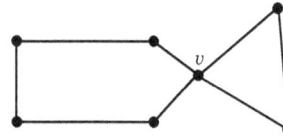

Here, $K(G) = 1$, since removal of the only one vertex "v" makes the graph G disconnected.

Remember:

(*i*) If a connected graph G has a cut edge (bridge), then $K(G) = 1$.

(*ii*) If $G = K_n$ (Complete graph with n vertices) then $K(G) = n–1$.

(*iii*) The vertex connectivity of a cyclic graph C_n is two.

(*iv*) For any connected graph G, $K(G) \leq \lambda(G) \leq \delta(G)$.

Remember:

(*i*) A cyclic graph of order "n" is a connected graph whose edges from a cycle of length "n." It is denoted by C_n.

Example:

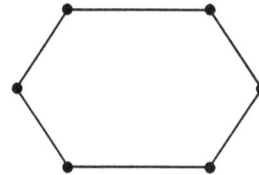

(*ii*) A path graph of order "n" is obtained by removing one edge from a cycle graph C_n. It is denoted by P_n.

Example:

11.9 GRAPH COLORING, MATCHING, AND COVERING

11.9.1 Vertex Coloring

A simple graph is said to be m-vertex colorable if it is possible to assign one color from a set of m-number of colors to the vertices such that no two adjacent vertices are assigned with the same color.

11.9.2 Chromatic Number

If a simple graph is k-vertex colorable but not $(k-1)$ vertex colorable, we say that the graph is k- chromatic or the chromatic number, denoted by $X(G)$ is "k."

Example:

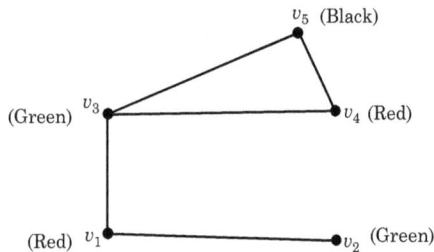

Let us assign red color to v_1 and v_4; green to v_2 and v_3; black to v_5.

Then, G is 3-vertex colorable and so $X(G) = 3$.

Remember:

(*i*) A k-chromatic graph needs at least k color for its coloring.

(*ii*) If a graph G has n-number of vertices, then $X(G) \leq n$.

(*iii*) If G is a null graph, then G is 1- chromatic graph.

(*iv*) A graph G is k-vertex colorable if and only if each block in it is k-vertex colorable.

(*v*) If the chromatic number of each block of a graph G is "k," then the chromatic number of the graph is "k."

(*vi*) The chromatic number of a complete graph with "n" vertices is "n."

(*vii*) Let G be a cycle with "n" vertices. Then,
$$\chi(G) = \begin{cases} 2, \text{ if } n \text{ is even} \\ 3, \text{ if } n \text{ is odd} \end{cases}$$

(*viii*) $X(K_{m,n}) = 2$, where $K_{m,n}$ is the bipartite graph.

(*ix*) The chromatic number of a graph with at least one edge is 2 if and only if the graph contains no cycles of odd length.

(*x*) If G is a planar graph, then $X(G) \leq 4$ (**Four color theorem**).

(*xi*) A planar graph can be colored with 5 colors (**Five color theorem**).

11.9.3 Matching

A matching in a graph G is a subset of edges in which no two edges are adjacent, *i.e*; no two edges in G have a common end vertex.

Example:

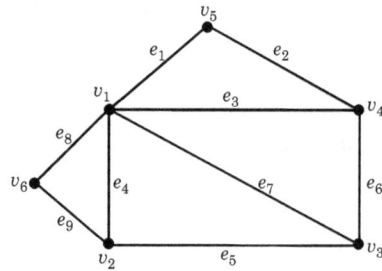

Here, $M = \{e_2, e_7, e_9\}$ is a matching.

Remember:

(*i*) If M is a matching in a graph G with edges set E, then M is called a maximal matching if $M \cup \{e\}$ is not a matching for arbitrary $e \in E - M$, *i.e.*, a maximal matching is a matching for which no edge in the graph can be added.

(*ii*) The maximal matching with the largest number of edges are called maximum matching.

(*iii*) A matching M in a graph G is called a perfect matching if every vertex of the graph G is the end of some edges in M.

11.9.4 Covering

A set of edges E of a graph G is said to cover the graph G if every vertex in G is incident with at least one edges in E. Here, the set E is called the edge covering of G.

Remember:

(*i*) The covering of an n-vertex graph will have at least $\left\lceil \dfrac{n}{2} \right\rceil$ edges.

(*ii*) A minimal covering of an n-vertex graph will contain not more than n-edges.

(*iii*) A covering exists for a graph if and only if the graph has no isolated vertex.

(*iv*) A covering E of a graph is minimal if and only if E contains no path of length three or more.

11.10 Tree

11.10.1 Definition

A tree is a connected graph without any circuits.

Example:

Remember:

(*i*) A tree is a connected and acyclic graph

(*ii*) A directed tree is a connected, acyclic, and directed graph.

(*iii*) A simple *un*directed graph G is a tree if and only if G is connected and has no cycles.

(*iv*) A trivial tree has exactly one vertex of degree 1.

(*v*) A tree with n vertices has exactly $n - 1$ edges.

(*vi*) Every non-trivial tree has at least 2 vertices of degree 1.

(*vii*) There is one and only one path between every pair of vertices in a tree T.

(*viii*) If in a graph there is one and only one path between every pair of vertices, then G is a tree.

(*ix*) Any connected acyclic graph with n vertices and n-1 edges is a tree.

(*x*) A graph is a tree if and only if it is minimally connected.

11.10.2 Spanning Tree

Let G be a connected graph. Then a sub graphs H of G is called a spanning tree of G if

(*i*) H is a tree

(*ii*) H contains all the vertices of G.

Example:

H is a spanning tree of G.

Remember:

(*i*) A non- directed graph G is connected if and only G contains a spanning tree.

(*ii*) The complete graph K_n has n^{n-2} different spanning trees.

Construction of Spanning Trees

(I) BFS (Breath First Search) Algorithm

Step-I: Chose a vertex arbitrarily from a connected graph G and designate it as a root.

Step-II: Add all the edges incident to this vertex so that no loop is formed. The new vertices added at this stage become the vertices at level 1 in the spanning tree.

Step-III: For each vertex at level 1, add each edge incident to this vertex as long as it does not produce any cycle. This produces the vertices at level 2 in the spanning tree.

Step-IV: Continue the same process until all the vertices of the given tree are added.

Example:

Consider the graph G.

G

We chose "a" as the root. Then add edges (a, c) and (a, b) as they are incident with "a." The vertices "b" and "c" are at level 1.

Next, we add the edge (c, d). The vertex "d" is at level 2.

Next, we consider the edges (d, b), (d, g), and (d, e) as they are incident with "d." Out of these only two edges (d, g), and (d, e) can be added [since the edge (d, b) forms a cycle]. Then the vertices "g" and "e" are at level 3.

Finally, we add the edge (e, f) at level 3.

The spanning tree T so obtained is given by

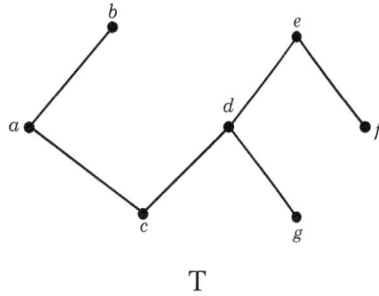

T

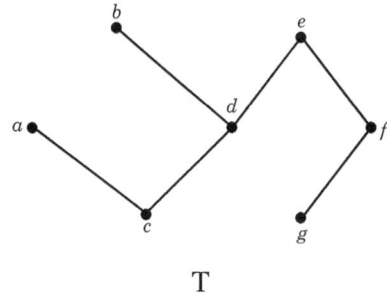

T

(II) DFS (Depth First Search) Algorithm

Step-I: Choose a vertex arbitrarily from the connected graph G and designate it as a root.

Step-II: Form a path starting from this root by successively adding edges as long as possible where each new edge is incident with the last vertex in the path and no cycle is formed.

Step-III: If the path goes through all vertices of the graph, the tree so obtained becomes a spanning tree. Otherwise, go to Step IV.

Step-IV: Move back to the next to last vertex in the path and if possible, form a new path starting from this vertex passing through the vertices that were not already visited. If this cannot be done, move back to another vertex in the path and repeat the process.

Step-V: Repeat the procedure until no more edges can be added.

Example:

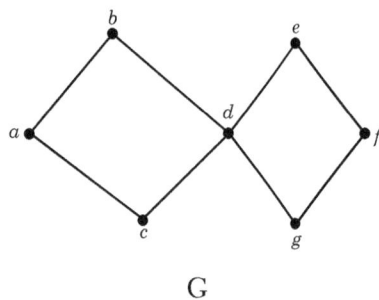

G

We chose "a" as the root. Then, successively adding edges we obtain a path a-c-d-e-f-g. Next, we move back to vertex "f." But there is no path beginning at "f" containing vertices not already visited. Next, we track back at the vertex "e." But no path begins at "e." So we move back to the vertex "d" and then form the path d-b. Thus, all the vertices are covered.

The spanning tree T so obtained is given by

11.10.3 Minimal Spanning Tree

A minimal spanning tree of a connected weighted graph G (weight is associated with every edge of G) is a spanning tree of G whose total weight is as small as possible.

(I) Prim's Algorithm

The stepwise Prim's algorithm to find a minimal spanning tree in a weighted graph is given below:

Let $G = (V, E)$ be a graph and $S = (V_S, E_S)$ be the spanning tree to be found from G.

Step-1: Select a vertex v_i of V and initialize $V_S = \{v_i\}$ and $E_S = \{ \}$.

Step-2: Select a nearest neighbor of v_i from V that is adjacent to some $v_j \in V_S$ and that edge (v_i, v_j) does not form a cycle with members edge of E_S. Set $V_S = V_S \cup \{v_i\}$ and $E_S = E_S \cup \{(v_i, v_j)\}$.

Step-3: Repeat step-2 until $|E_S| = |V| - 1$.

Example:

Consider the following graph:

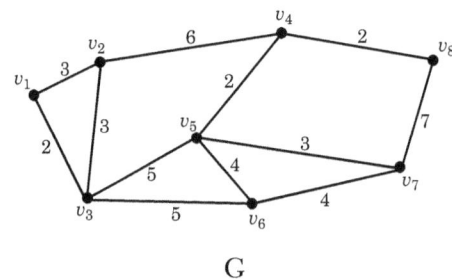

G

Let us begin with the node v_1 of the graph. Initialize $V_S = \{v_1\}$ and $E_S = \{ \}$. There are eight nodes so the spanning tree will have seven edges.

Step-1: Nodes v_2 and v_3 are neighbors, of v_1. Since v_3 is nearest to the node v_1, we select v_3. Thus, we have $V_S = \{v_1, v_3\}$ and $E_S = \{(v_1, v_3)\}$.

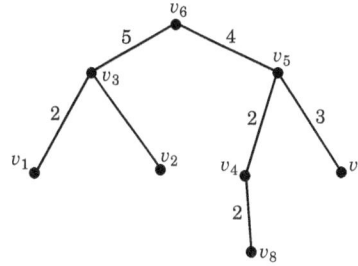

Step-2: Now node v_2 is neighbor of both v_1 and v_3. v_3 has nodes v_5 and v_6 as its neighbors. Since weight$(v_1, v_2) = 3$, weight$(v_3, v_2) = 3$, weight$(v_3, v_5) = 5$, and weight$(v_3, v_6) = 5$, so the nearest neighbor is v_2. We can select either the edge (v_1, v_2) or the edge (v_3, v_2). Let us select the edge (v_3, v_2). Then $V_S = \{v_1, v_3, v_2\}$ and $E_S = \{(v_1, v_3), (v_3, v_2)\}$.

Step-3: Now v_4, v_5, v_6 are neighbors of the nodes in V_S. The edge (v_1, v_2) cannot be considered as it forms cycle with the edges (v_1, v_3) and (v_3, v_2). Thus, we have to select from the edges (v_2, v_4), (v_3, v_5), (v_3, v_6). Since weight$(v_2, v_4) = 6$, weight$(v_3, v_5) = 5$, weight$(v_3, v_6) = 5$, so we may take either (v_3, v_5) or (v_3, v_6). Let us select (v_3, v_6). Then $V_S = \{v_1, v_3, v_2, v_6\}$ and $E_S = \{(v_1, v_3), (v_3, v_2), (v_3, v_6)\}$.

Step-4: Now, we have to select an edge from (v_2, v_4), (v_3, v_5), (v_6, v_5), (v_6, v_7). Since weight $(v_2, v_4) = 6$, weight$(v_3, v_5) = 5$, weight$(v_6, v_5) = 4$, and weight$(v_6, v_7) = 4$. We select the edge (v_6, v_5). Therefore, $V_S = \{v_1, v_3, v_2, v_6, v_5\}$ and $E_S = \{(v_1, v_3), (v_3, v_2), (v_3, v_6), (v_6, v_5)\}$.

Step-5: The selection of the edge (v_3, v_5) is ruled out as it forms a cycle with the edges (v_3, v_6) and (v_6, v_5). Thus, we have to select an edge from (v_2, v_4), (v_5, v_4), (v_6, v_7). Since weight$(v_2, v_4) = 6$, weight$(v_5, v_4) = 2$, weight$(v_6, v_7) = 4$, we select (v_5, v_4). Therefore, $V_S = \{v_1, v_3, v_2, v_6, v_5, v_4\}$ and $E_S = \{(v_1, v_3), (v_3, v_2), (v_3, v_6), (v_6, v_5), (v_5, v_4)\}$.

Step-6: Now, the edge (v_2, v_4) is ruled out as it forms cycle with the edges (v_3, v_2), (v_3, v_6), (v_6, v_5) and (v_5, v_4). Thus, we have to select an edge from (v_4, v_8), (v_5, v_7), (v_6, v_7). Since weight$(v_4, v_8) = 2$, weight$(v_5, v_7) = 3$, weight$(v_6, v_7) = 4$, we select the edge (v_4, v_8). Therefore, $V_S = \{v_1, v_3, v_2, v_6, v_5, v_4, v_8\}$ and $E_S = \{(v_1, v_3), (v_3, v_2), (v_3, v_6), (v_6, v_5), (v_5, v_4), (v_4, v_8)\}$.

Step-7: The edges left are (v_5, v_7), (v_8, v_7) and (v_6, v_7). Since weight$(v_5, v_7) = 3$, weight$(v_8, v_7) = 7$ and weight$(v_6, v_7) = 4$, we select the edge (v_5, v_7). Therefore, $V_S = \{v_1, v_3, v_2, v_6, v_5, v_4, v_8, v_7\}$ and $E_S = \{(v_1, v_3), (v_3, v_2), (v_3, v_6), (v_6, v_5), (v_5, v_4), (v_4, v_8), (v_5, v_7)\}$.

Since the number of edges in E_S is 7 (= total number of vertices - 1), the process terminates here. The spanning tree so obtained is given below:

(II) Kruskal's Algorithm

The stepwise Kruskal's algorithm to find a minimal spanning tree in a weighted graph is given below:

Let $G = (V, E)$ be a graph and $S = (V_S, E_S)$ be the spanning tree to be found from G.

Step-1: Arrange the edges of G in order of their increasing weights and chose the edge with minimum weight.

Step-2: Proceed sequentially, add each edge which does not result in a cycle until $n - 1$ edges are selected.

Example:

Consider the following graph:

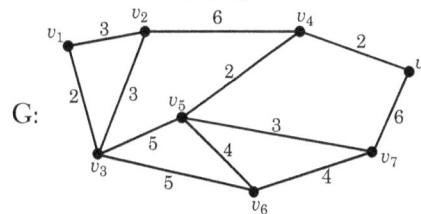

G:

Step-1: Since the edges (v_1, v_3), (v_5, v_4), and (v_4, v_8) have minimum weight 2 and they do not form a cycle, we select all of them and set $E_S = \{(v_1, v_3), (v_5, v_4), (v_4, v_8)\}$.

Step-2: Next edges with minimum weight 3 are (v_1, v_2), (v_2, v_3), and $(v_5, v_7)\}$. If we select both of (v_1, v_2) and (v_2, v_3), then a cycle will be formed. So we select only one of them, say (v_2, v_3). Also we can select (v_5, v_7). Therefore, $E_S = \{(v_1, v_3), (v_5, v_4), (v_4, v_8), (v_2, v_3), (v_5, v_7)\}$.

Step-3: Next edges with minimum weight 4 are (v_5, v_6) and $(v_6, v_7)\}$. If we select both of them, then a cycle will be formed. So we select only one of them, say (v_6, v_7). Therefore, $E_S = \{(v_1, v_3), (v_5, v_4), (v_4, v_8), (v_2, v_3), (v_5, v_7), (v_6, v_7)\}$.

Step-4: Next edges with minimum weight 5 are (v_3, v_5) and $(v_3, v_6)\}$. If we select both of them, then a cycle will be formed. So we select only one of them, say (v_3, v_5). Therefore, $E_S = \{(v_1, v_3), (v_5, v_4), (v_4, v_8), (v_2, v_3), (v_5, v_7), (v_6, v_7), (v_3, v_5)\}$.

Since the number of edges in E_S is 7 (= total number of vertices - 1), the process terminates here. The spanning tree so obtained is given below:

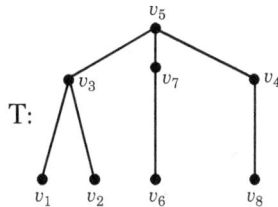

11.10.4 Binary Tree

A tree T is called a binary tree if

(i) There is exactly one vertex of degree two.

(ii) Each of the other vertices is of degree one or three.

Example:

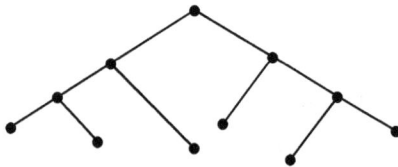

Remember:

(i) The number of vertices in a binary tree is odd.

(ii) The number of pendant vertices in a binary tree T with n vertices is $\dfrac{n+1}{2}$.

11.10.5 Rooted Tree

A rooted tree is a tree in which a particular vertex is distinguished from the others, which is called the root of the tree.

Remember:

(i) The level of a vertex is the number of edges along the unique path between it and the root. The level of the root is defined as "0."

(ii) The height of a rooted tree is the maximum level to any vertex of the tree.

(iii) The depth of a vertex "v" in a tree is the length of the path from the root to "v."

(iv) Given any internal vertex "v" of a rooted tree, the children of "v" are all those vertices that are adjacent to "v" and are one level farther away from the root than "v."

(v) If a vertex "v" has one or two children, then "v" is called an internal vertex.

(vi) The descendants of the vertex "v" is the set consisting of all the children of "v" together with the descents of those children.

Example:

Consider the rooted tree T given below:

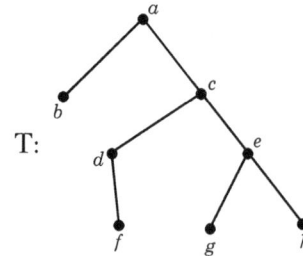

Here,

"a" is the root of the tree.

Height of the rooted tree is 3.

The root "a" is at level "0;" the vertices "b" and "c" are at level "1;" the vertices "d" and "e" are at level "2;" the vertices "f," "g," "h" are at level "3."

The internal vertices are c, d, e.

The children of "c" are d and e.

The descendants of "a" are b, c, d, e, f, g, h.

Remark:

(i) A rooted tree is called m-ary tree if every internal vertex has at most m children. A m-ary tree is called a full m-ary tree if every internal vertex has exactly m children. Thus, in particular, a full binary tree is a binary tree in which each internal vertex has exactly two children.

(ii) A full m-ary tree with "i" internal vertex has $n = mi + 1$ vertices.

(iii) The maximum number of vertices in a binary tree of depth "d" is $2^d - 1$.

11.10.6 Traversal of a Tree

Traversal of a tree is a process to traverse a tree in a systematic way so that each vertex is visited exactly once.

To implement this, a tree is considered to have three components: root, left subtree and right subtree. These three components can be arranged in six different ways: (left, root, right), (root, left, right), (left, right, root), (right, left, root), (right, root, left), and (root, right, left). The first three are used where as the last three combinations are of no use as it alternates the positions of a node in a positional tree.

In-order traversal: In this form of traversal, a tree is traversed in the sequence: left subtree, root, right subtree.

Pre-order traversal: In this form of traversal, a tree is traversed in the sequence: root, left subtree, right subtree.

Post-order traversal: In this form of travers-al, a tree is traversed in the sequence: left subtree, right subtree, root.

Consider the following tree:

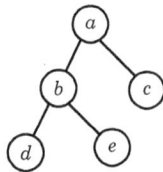

In the figure above, In-order traversal is $d\ b\ e$ $a\ c$, Preorder traversal is $a\ b\ d\ e\ c$ and Post-order traversal is $d\ e\ b\ c\ a$.

Fully Solved MCQs

1. Which of the following is a planar graph?

   (a) a complete graph with 4 vertices (K_4)
   (b) $K_{2,2}$
   (c) $K_{2,3}$
   (d) all of these

2. If f denotes the number of regions in a planar graph G, with n vertices, then

   (a) $n = f$ (b) $n \geq 2 + \dfrac{f}{2}$

   (c) $n \leq 2 + \dfrac{f}{2}$ (d) none of these

3. A nondirected graph has 10 edges. If the degree of each vertex is 2, then the number of vertex will be
   (a) 10   (b) 5   (c) 15   (d) 8

4. Let G be a nondirected graph with 14 edges. If G has 5 vertices each of degree 3 and rest have degree less than 3, then the minimum number of vertices of G is
   (a) 5   (b)   (c) 7   (d) 8

5. The incidence matrix of the following graph is

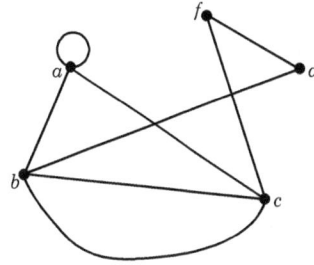

(a) $\begin{pmatrix} 1 & 1 & 0 & 0 & 0 & 0 & 0 & 0 \\ 0 & 0 & 1 & 1 & 1 & 0 & 0 & 0 \\ 1 & 0 & 0 & 0 & 0 & 1 & 0 & 0 \\ 0 & 0 & 0 & 0 & 1 & 1 & 1 & 1 \\ 0 & 1 & 0 & 1 & 0 & 0 & 1 & 1 \end{pmatrix}$

(b) $\begin{pmatrix} 0 & 1 & 0 & 0 & 0 & 0 & 1 & 0 \\ 0 & 0 & 1 & 1 & 1 & 0 & 0 & 0 \\ 1 & 0 & 0 & 0 & 0 & 1 & 0 & 0 \\ 0 & 0 & 0 & 0 & 1 & 0 & 1 & 1 \\ 1 & 1 & 0 & 1 & 0 & 0 & 1 & 1 \end{pmatrix}$

(c) $\begin{pmatrix} 1 & 1 & 0 & 0 & 0 & 0 & 0 & 1 \\ 0 & 1 & 1 & 1 & 1 & 0 & & 1 \\ 1 & 0 & 0 & 0 & 0 & 1 & 0 & 0 \\ 0 & 1 & 0 & 0 & 1 & 1 & 1 & 1 \\ 0 & 1 & 0 & 1 & 0 & 0 & 0 & 0 \end{pmatrix}$

(d) none of these

6. The adjacency matrix of the following graph is

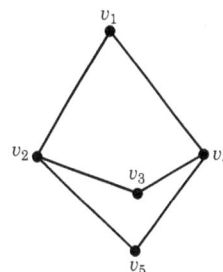

(a) $\begin{pmatrix} 0 & 1 & 0 & 1 & 0 \\ 1 & 0 & 1 & 0 & 1 \\ 0 & 1 & 0 & 1 & 0 \\ 1 & 0 & 1 & 0 & 1 \\ 0 & 1 & 0 & 1 & 0 \end{pmatrix}$

(b) $\begin{pmatrix} 1 & 1 & 0 & 1 & 0 \\ 1 & 1 & 1 & 1 & 1 \\ 0 & 1 & 0 & 1 & 0 \\ 1 & 0 & 1 & 0 & 1 \\ 0 & 1 & 1 & 1 & 0 \end{pmatrix}$

(c) $\begin{pmatrix} 0 & 1 & 0 & 1 & 0 \\ 1 & 0 & 1 & 0 & 0 \\ 0 & 1 & 0 & 1 & 0 \\ 1 & 0 & 1 & 1 & 1 \\ 0 & 0 & 0 & 0 & 0 \end{pmatrix}$

(d) none of these

7. The order and size of the following graph are, respectively

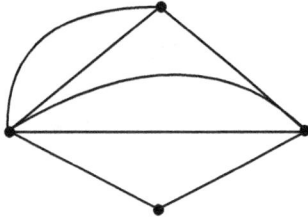

(a) 6, 6   (b) 6,7   (c) 4, 7   (d) 8,8

8. The chromatic number of the following graph is

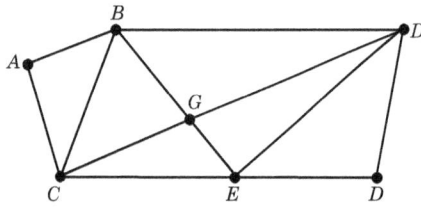

(a) 4   (b) 5   (c) 2   (d) 3

9. The chromatic number of the following graph is

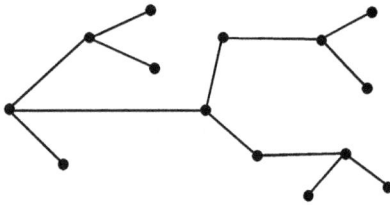

(a) 4   (b) 5   (c) 2   (d) 3

10. The chromatic number of the following graph is

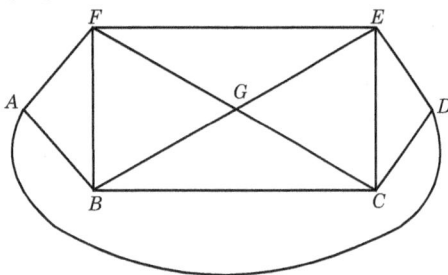

(a) 4   (b) 5   (c) 2   (d) 3

11. The complement of the following graph is

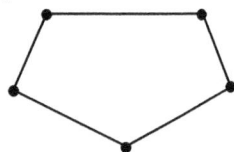

(a)

(b)

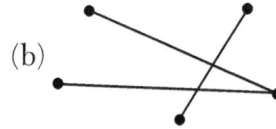

(c)

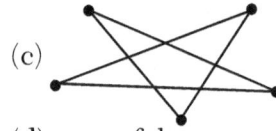

(d) none of these

12. Which of the following is not true?

(a) K_n has n vertices and $\frac{n(n-1)}{2}$ edges.

(b) $K_{m,n}$ has $m + n$ vertices and mn edges.

(c) $K_{m,n}$ has mn vertices and mn edges.

(d) K_2 has 2 vertices and 1 edge.

13. The size of a simple graph of order n cannot exceed

(a) $^nC_2$ (b) $^nC_3$ (c) $^nC_4$ (d) n

14. Which of the following is possible?

(a) a simple graph with 4 vertices and 8 edges

(b) a simple graph with 4 vertices and 6 edges

(c) a simple graph with 4 vertices and 7 edges

(d) a simple graph with 4 vertices and 9 edges

15. What is the size of a k-regular (n, q) graph

(a) $\frac{nk}{2}$ (b) $\frac{(n-1)k}{2}$

(c) $\frac{n(k-1)}{2}$ (d) nk

16. Which of the following is not possible?

(a) a 4-regular graph with 6 vertices

(b) a 3-regular graph with 14 vertices

(c) a 3-regular graph with 17 vertices

(d) a 5-regular graph with 8 vertices

17. If a graph G with "n" vertices is isomorphic to it is complement $\bar{G}$, then which of the following is true?

(a) n or n-1 must be a multiple of 2

(b) n or n-1 must be a multiple of 3

(c) n-1 must be a multiple of 4

(d) n or n-1 must be a multiple of 4

18. A simple graph with 6 vertices is connected if it has

(a) more than 10 edges

(b) more than 20 edges

(c) less than 10 edges

(d) less than 20 edges

19. The rank and nullity of the complete graph K_n are, respectively

(a) $n-1$ and $\frac{(n-1)(n-2)}{2}$

(b) n and $\dfrac{n-1}{2}$

(c) $n-1$ and $\dfrac{n-2}{2}$

(d) n and $\dfrac{n(n-1)}{2}$

20. The maximum degree of any vertex in a simple graph with 10 vertices is?

(a) 10　　(b) 5　　(c) 19　　(d) 9

21. The minimum and maximum number of edges of a simple graph with 8 vertices and 2 components are, respectively?

(a) 5,20　(b) 5, 21　(c) 6, 21　(d) 6, 19

22. The maximum number of vertices in a connected graph having 20 edges is____?

(a) 20　　(b) 21　　(c) 19　　(d) 10

23. How many subgraphs can be formed from a graph having 10 sides?

(a) 2^{10}　　(b) 2^{12}　　(c) 2^{9}　　(d) $2^{10}-1$

24. The minimum number of connected components of a graph with 14 vertices and 7 edges is___

(a) 7　　(b) 6　　(c) 5　　(d) 4

25. Maximum number of edges in a planar graph with n vertices is

(a) $3(n-2)$　(b) $3n+1$　(c) $2n+4$　(d) $2n$

26. Let G be a connected, undirected graph. A cut in G is a set of edges whose removal results in G being broken into two or more components, which are not connected with each other. The size of a cut is called its cardinality. A min-cut of G is a cut in G with minimum cardinality. Consider the following graph.

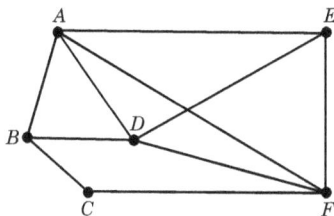

Which of the following sets of edges is a cut?

(a) $\{(A, B),(E, F),(B, D), (A, E), (A, D)\}$
(b) $\{(B, D),(C, F), (A, B)\}$
(c) $\{(A, B), (E, F)\}$
(d) none of these

27. Given T is a graph with "n" vertices. If T is connected and T has $(n-1)$ edges, then

(a) T is a tree
(b) T contains no cycles
(c) Every pair of vertices in T is connected by exactly one path
(d) All the above

28. Consider a rooted tree in which every node has at least 3 children. What will be the minimum number of nodes at level "i" ($i > 0$) of this tree? (assume that root is at level zero)

(a) 3^{i}　　(b) $3i$　　(c) 3　　(d) $3i + 1$

29. The maximum number of nodes in a binary tree of depth 5 is

(a) 28　　(b) 29　　(c) 30　　(d) 31

30. How many internal nodes are there in a full binary tree of height "k"?

(a) 2^{k}　(b) 2^{k-1}　(c) $2^{k} + 1$　(d) $2^{k}-1$

31. How many different trees are there with 5 nodes?

(a) 120　　(b) 125　　(c) 130　　(d) 135

32. Which of the following does not define a tree?

(a) A tree is a graph with no cycles
(b) A tree is a connected acyclic graph
(c) A tree is a connected graph with n-1 edges where n is the number of vertices
(d) A tree is an acyclic graph with $n-1$ edges where n is the number of vertices

33. Which two of the following statements are equivalent for an undirected graph G?

(i)　G is a tree

(ii)　There is at least one path between any two distinct vertices of G

(iii)　G contains no cycles and has $(n-1)$ edges.

(iv)　G has n edges

(a) (i) and (ii)　　　　(b) (i) and (iii)
(c) (i) and (iv)　　　　(d) (ii) and (iii)

34. Number of binary trees formed with 4 nodes are

(a) 8　　(b) 10　　(c) 12　　(d) 14

35. Consider the following statements:

(i)　A graph in which there is a unique path between every pair of vertices is a tree.

(ii)　A connected graph with $|e| = |v|-1$ edges is a tree.

(iii)　A graph with $|e| = |v|-1$ edges that has no circuit is a tree.

Which of the above statements is/are true?
(a) (*i*) and (*iii*) (b) (*ii*) and (*iii*)
(c) (*i*) and (*ii*) (d) all of these

36. The number of colors required to properly color the vertices of every planar graph is
(a) 2 (b) 3 (c) 4 (d) 5

37. To find the shortest path in a weighted graph, which of the following algorithms is not used?
(a) Warshall's algorithm
(b) Dijkstra's algorithm
(c) Kruskal's algorithm
(d) None of these

38. A graph is strongly connected if for all v_i, v_j ∈ G, both the (i, j) and (j, i)th cell in the path matrix are
(a) (0, 0) (b) (1, 0) (c) (0, 1) (d) (1, 1)

39. A full binary tree with *n* leaves contains
(a) *n* nodes (b) $\log_2 n$ nodes
(c) 2*n*–1 nodes (d) 2^n nodes

40. The number of edges which must be removed from a connected graph with "*n*" vertices and "*m*" edges to produce a spanning tree is
(a) *m* – *n* (b) *m* – *n* + 1
(c) *n* – *m* (d) *n* – *m* – 1

Answer Key				
1. (d)	**2.** (b)	**3.** (a)	**4.** (a)	**5.** (a)
6. (a)	**7.** (c)	**8.** (d)	**9.** (c)	**10.** (a)
11. (c)	**12.** (c)	**13.** (a)	**14.** (b)	**15.** (a)
16. (c)	**17.** (d)	**18.** (a)	**19.** (a)	**20.** (d)
21. (c)	**22.** (b)	**23.** (a)	**24.** (c)	**25.** (a)
26. (b)	**27.** (d)	**28.** (a)	**29.** (d)	**30.** (d)
31. (b)	**32.** (a)	**33.** (b)	**34.** (d)	**35.** (d)
36. (c)	**37.** (c)	**38.** (d)	**39.** (c)	**40.** (b)

Explanation

1. (*d*)

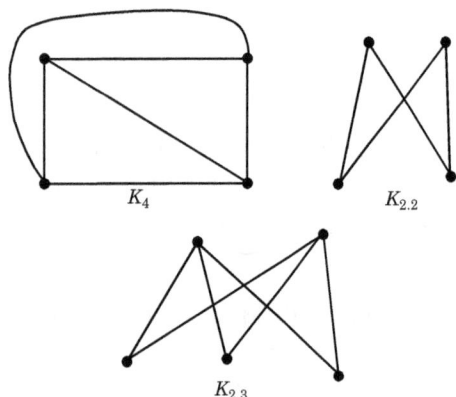

K_4 $K_{2,2}$

$K_{2,3}$

Each of these graphs can be drawn without crossover of edges. So, each of them is planar.

2. (*b*)

$$3f \geq 2|E| \Rightarrow |E| \geq \frac{3}{2}f$$

Now, by the Euler formula,

$$|V| = |E| - |R| + 2 \geq \frac{3}{2}f - f + 2$$

$$\Rightarrow n \geq \frac{f}{2} + 2$$

3. (*a*)

$$\sum_i \deg(V_i) = 2|E|$$

$$\Rightarrow 2|V| = 2 \times 10$$

$$\Rightarrow |V| = 10$$

4. (*a*)

$$\sum_i \deg(V_i) = 2|E| = 2 \times 14 = 28$$

Let *x* be the no. of vertices in the graph G having degree less than 3.

Again,

$$\sum_i \deg(V_i) < 5 \times 3 + 3x$$

$$\Rightarrow 28 < 15 + 3x \Rightarrow 3x > 13 \Rightarrow x > \frac{13}{3}.$$

∴ $x \geq 5$.

5. (*a*) Use the definition of incidence matrix.

6. (*a*) Use the definition of adjacent matrix.

7. (*c*) Order of the graph = no. of vertices = 4 and

size of the graph = no. of edges = 7.

8. (*d*)

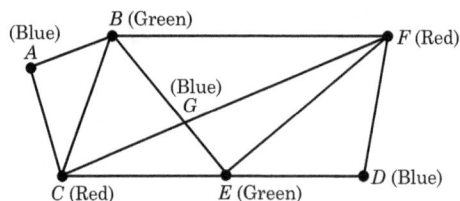

Here the graph is three vertex colorable. So the chromatic number of the graph is "3."

9. (*c*) Here the graph is a tree. So the chromatic number of the graph is "2."

10. (*a*)

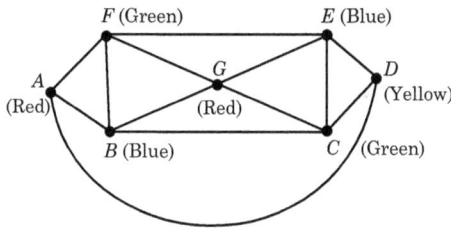

Here the graph is "4" vertex colorable. So the chromatic number of the graph is "4."

11. (*c*)

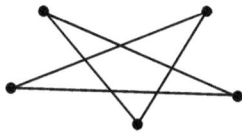

Remember:

The *complement* of a graph G is simple graph with the same vertex set as G and where two vertices "*u*" and "*v*" are adjacent only when they are not adjacent in G.

12. (*c*) K_n has n vertices and $\dfrac{n(n-1)}{2}$ edges.

$K_{m,n}$ has $m + n$ vertices and mn edges.

13. (*a*) Let $G = G(V, E)$ be a simple graph of order n. Then, V contains "n" elements as vertices and elements of E are distinct two elements subsets of V. But the number of ways we can choose two elements from V is $^nC_2$. Thus E contains maximum $^nC_2$ edges.

14. (*b*) We know that if a simple graph G has "n" vertices, then it has $\dfrac{n(n-1)}{2}$ edges.

Thus, a simple graph with 4 vertices has $\dfrac{4(4-1)}{2}$ *i.e*, 6 edges.

15. (*a*)

$$\sum_{i=1}^{n}\deg(v_i) = 2|E|$$

$$\Rightarrow \sum_{i=1}^{n} k = 2q$$

$$\left(\begin{array}{c}\because \text{a graph is } k\text{-regular} \\ \Rightarrow \text{every vertex has degree } 'k'\end{array}\right)$$

$$\Rightarrow nk = 2q \Rightarrow q = \frac{nk}{2}.$$

Hence, the size of the graph $= q = \dfrac{nk}{2}$.

16. (*c*) We know that a "k-regular" graph has $\dfrac{nk}{2}$ edges if "n" is the number of vertices.

Case-I: If $n = 6$, $k = 4$; then

the total number of edges $= \dfrac{6\times 4}{2} = 12$, which is a positive integer.

Hence, a 4-regular graph with 6 vertices exist.

Case-II: If $n = 14$, $k = 3$; then

the total number of edges $= \dfrac{14\times 3}{2} = 21$, which is a positive integer.

Hence, a 3-regular graph with 14 vertices exist.

Case-III: If $n = 17$, $k = 3$; then

total number of edges $= \dfrac{17\times 3}{2} = \dfrac{51}{2}$, which is not a positive integer.

Hence, a 3-regular graph with 17 vertices does not exist.

Case-IV: If $n = 8$, $k = 5$; then

the total number of edges $= \dfrac{8\times 5}{2} = 20$, which is a positive integer.

Hence, a 5-regular graph with 8 vertices exist.

17. (*d*) Since G is isomorphic to $\bar{G}$, so both of G and $\bar{G}$ have the same number of edges. Then we must have, total number of edges in G + total number of edges in $\bar{G}$ = total number of edges in K_n. This gives total number of edges in $G = \dfrac{n(n-1)}{4}$

$$\left[\because \text{total number of edges in } K_n \text{ is } \frac{n(n-1)}{2}\right].$$

Since $\dfrac{n(n-1)}{4}$ must be a positive integer, so either "n" or "n-1" must be a multiple of 4.

18. (*a*) We know that a simple graph with "n" vertices is connected if it has more than $\dfrac{(n-1)\times(n-2)}{2}$ edges

Therefore, a simple graph with 6 vertices is connected if it has more than $\dfrac{(6-1)\times(6-2)}{2}$ *i.e*; 10 edges.

19. (*a*) We know that if G is a graph with "n" vertices, "m" edges, and "k" components, then rank of $G = n$-k and nullity of $G = m$-$n + k$.

In the graph K_n, number of vertices $= n$, number of edges $= \dfrac{n(n-1)}{2}$ and number of components $= 1$.

Hence, the rank of K_n = n-1 and nullity of

$K_n = \dfrac{n(n-1)}{2} - n + 1 = \dfrac{(n-1)(n-2)}{2}$.

20. (d) We know that the maximum degree of any vertex in a simple graph with "n" vertices is n-1. Hence, the maximum degree of any vertex in a simple graph with 10 vertices is 10-1, *i.e*; 9.

21. (c) If G be a simple graph with "n" vertices and "k" components, then

$n - k \le$ no. of edges $\le \dfrac{(n-k)(n-k+1)}{2}$ ...(i)

Here, $n = 8$ and $k = 2$.

So (i) $\Rightarrow$

$8 - 2 \le$ no. of edges $\le \dfrac{(8-2)(8-2+1)}{2}$

or, $6 \le$ no. of edges ≤ 21.

22. (b)

We know that if G be a connected graph having "n" vertices and "m" edges, then $n - 1 \le m$. Here, $m = 20$. So $n - 1 \le 20$, *i.e*; $n \le 21$.

23. (a) Here, the number of subgraphs

$= {}^{10}C_0 + {}^{10}C_1 + {}^{10}C_2 + ... + {}^{10}C_{10} = 2^{10}$.

24. (c) Let "k" be the number of connected components, then we have

$14 - k \le 7 + 2$ *i.e*; $k \ge 5$.

25. (a) Let G be a planar graph with "n" vertices, "m" edges, and "k" regions.

Then, by the Euler's formula, $n - m + k = 2$ ...(i)

Again, $2m \ge 3k \Rightarrow k \le \dfrac{2m}{3}$

$\therefore$ (i) gives,

$2 = n - m + k \le n - m + \dfrac{2m}{3}$

$\Rightarrow n - \dfrac{m}{3} \ge 2 \Rightarrow m \le 3n - 6 = 3(n-2)$

26. (b) If we delete the edges BD, CF, and AB, then the graph becomes disconnected.

27. (d) We know that a connected graph with "n" vertices and n–1 edges edges is always a tree.

Hence, T is a tree.

Since a tree contains no cycles, so T has no cycles.

As T has no cycles, so only one path is possible between each pair of vertices in T.

28. (a) We know that the number of nodes at level "i" of a n–ary tree is n^i. Here, $n = 3$. Hence, the minimum number of nodes at level "i" ($i > 0$) of this tree = 3^i.

29. (d) The maximum number of nodes in a binary tree of depth 5 = $2^5 - 1 = 31$.

30. (d) A full binary tree with height k has $2^{k+1} - 1$ number of nodes and 2^k number of leaves.

Therefore, the number of internal nodes

$= (2^{k+1} - 1) - 2^k = 2^k - 1$.

31. (b) The number of different trees with "n" nodes = n^{n-2}. Here, $n = 5$ and so the number of different trees with "5" nodes is 125.

32. (a) Consider a graph G with three isolated vertices. Then G has no cycles and G is not a tree.

33. (b) We know that any tree with n number of vertices contains n–1 edges and no cycles.

34. (d) The number of binary trees formed with n nodes

$= \dfrac{{}^{2n}C_n}{n+1} = \dfrac{{}^8C_4}{4+1}$ (here $n = 4$)

$= \dfrac{8 \times 7 \times 6 \times 5}{5 \times 4 \times 3 \times 2 \times 1} = 14$.

35. (d) It follows from the definition of a tree.

36. (c) Result follows from the four color's theorem.

37. (c) Using Kruskal's algorithm, we can find out the minimum spanning tree which does not necessarily give the shortest path.

38. (d) In case of strongly connected graph, each vertex must be reachable from any other vertex.

39. (c) The total number of nodes

= total number of leaves + number of non-leaves

$= n + (n - 1) = 2n - 1$.

40. (b) The number of edges in spanning tree = $n - 1$.

Hence, the number of edges to be removed = $m - (n - 1) = m - n + 1$.

1. The minimum number of colors required to color the vertices of a cycle with "n" nodes (n = odd) in such a way that no two adjacent nodes have the same color is

   (a) 2   (b) 3
   (c) 4   (d) $n - 2[n/2] + 2$
   (CS GATE 2002)

2. Let G be an arbitrary graph with "n" nodes and "k" components. If a vertex is removed from G, the number of components in the resultant graph must necessary lie between

   (a) k and n (b) $k - 1$ and $k + 1$
   (c) $k - 1$ and $n - 1$ (d) $k + 1$ and $n - k$
   (CS GATE 2003)

3. How many perfect matching are there in a complete graph of six vertices?

   (a) 15   (b) 24   (c) 30   (d) 60
   (CS GATE 2003)

4. The minimum number of colors required to the color the following graph, such that no two adjacent vertices are assigned same color, is

   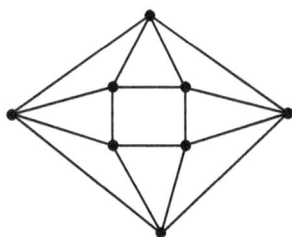

   (a) 2   (b) 3   (c) 4   (d) 5
   (CS GATE 2004)

5. How many graph on "n" labeled vertices exist which have at least $\dfrac{n^2 - 3n}{2}$ edges?

   (a) $^{(n^2-n)/2}C_{(n^2-3n)/2}$ (b) $\displaystyle\sum_{k=0}^{(n^2-3n)/2} {}^{(n^2-n)/2}C_k$

   (c) $^{(n^2-n)/2}C_n$ (d) $\displaystyle\sum_{k=0}^{n} {}^{(n^2-n)/2}C_k$

   (CS GATE 2004)

6. Let $G_1 = (V, E_1)$ and $G_2 = (V, E_2)$ be two connected graphs on the same vertex set V with more than two vertices. If $G_1 \cap G_2$ = $(V, E_1 \cap E_2)$ is not a connected graph, then the graph $G_1 \cup G_2 = (V, E_1 \cup E_2)$

   (a) Cannot have a cut vertex
   (b) Must have a cycle
   (c) Must have a cut-edge (bridge)
   (d) Has chromatic number strictly greater than those of G_1 and G_2.
   (CS GATE 2004)

7. What is the maximum number of edges in an acyclic undirected graph with "n" vertices?
   (a) $n - 1$ (b) n (c) $n + 1$ (d) $2n - 1$
   (IT GATE 2004)

8. What is the number of vertices in an undirected connected graph with 27 edges, 6 vertices of degree 2, 3 vertices of degree 4 and remaining of degree 3?
   (a) 10   (b) 11   (c) 18   (d) 19
   (IT GATE 2004)

9. An undirected graph G has "n" nodes. Its adjacency matrix is given by an $n \times n$ square matrix whose

   (*i*) diagonal elements are 0's and

   (*ii*) nondiagonal elements are 1's.

   Which one of the following is TRUE?

   (a) Graph G has no minimal spanning tree (MST)
   (b) Graph G has a unique MST of cost $n - 1$.
   (c) Graph G has multiple distinct MSTs, each of cost $n - 1$.
   (d) Graph G has multiple spanning trees of different costs.
   (CS GATE 2005)

10. Let G be a simple connected planar graph with 13 vertices and 19 edges. Then, the number of faces in the planar embedding of the graph is:

    (a) 6   (b) 8   (c) 9   (d) 13
    (CS GATE 2005)

11. Let G be a simple graph with 20 vertices and 100 edges. The size of the minimum vertex cover of G is 8. Then, the size of the maximum independent set of G is:

    (a) 12   (b) 8
    (c) Less than 8   (d) More than 12.
    (CS GATE 2005)

12. Which of the following graphs is not planar?

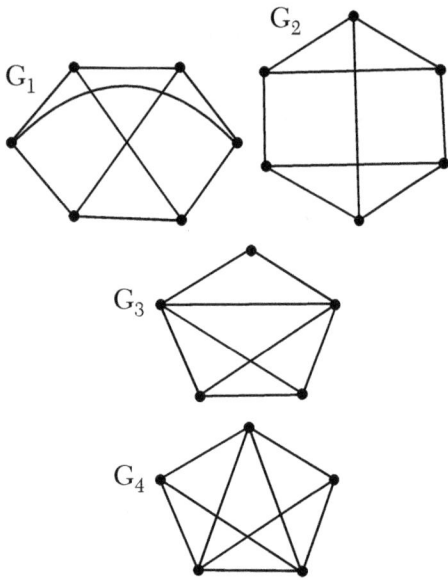

(a) G_1 (b) G_2 (c) G_3 (d) G_4

(CS GATE 2005)

13. The 2^n vertices of a graph G corresponds to all subsets of a set of size "n," for $n \geq 6$. Two vertices of G are adjacent if and only if the corresponding sets intersect in exactly two elements.

(i) The number of vertices of degree "0" in G is

(a) 1   (b) n (c) $n + 1$ (d) 2^n

(ii) The maximum degree of a vertex G is

(a) $^{n/2}C_2 \times 2^n$ (b) 2^{n-2}

(c) $2^{n-3} \times 3$ (d) 2^{n-1}

(iii) The number of connected components in G is

(a) n (b) $n + 2$ (c) $2^{n/2}$ (d) $2^n/n$

(CS GATE 2006)

14. Let T be a depth first search tree in an undirected graph G. Vertices "u" and "v" are leaves of this tree T. The degree of both "u" and "v" in G are at least 2. Then which of the following statements is true?

(a) There must be a vertex "w" adjacent to both "u" and "v" in G

(b) There must be a vertex "w" whose removal disconnects "u" and "v" in G

(c) There must exist a cycle in G containing "u" and "v"

(d) There must exist a cycle in G containing "u" and all its neighbors in G

(CS GATE 2006)

15. Which of the following graphs has an Eulerian circuit?

(a) Any k-regular graph where k is an even number.

(b) A complete graph on 90 vertices.

(c) The complement of a cycle on 25 vertices.

(d) None of these.

(CS GATE 2007)

16. Consider a weighted graph with positive edge weights and let (u, v) be an edge in the graph. It is known that the shortest path from the source vertex "s" to "u" has weight 53 and the shortest path from "s" to "v" has weight 65. Which one of the following statements is always true?

(a) weight of $(u, v) < 12$

(b) weight of $(u, v) = 12$

(c) weight of $(u, v) \geq 12$

(d) weight of $(u, v) > 12$

(IT GATE 2007)

17. What is the largest integer "m" such that every simple connected graph with "n" vertices and "n" edges contains at least "m" different spanning trees?

(a) 1   (b) 2   (c) 3   (d) n

(IT GATE 2007)

18. What is the chromatic number of the following graph?

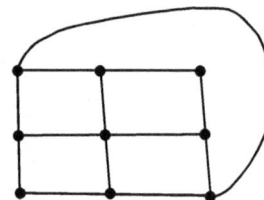

(a) 2   (b) 3   (c) 4   (d) 5

(IT GATE 2008)

19. Let G be a simple undirected graph. Some vertices of G are of an odd degree. Add a node "v" to G and make it adjacent to each odd degree vertex of G. The resultant graph will be

(a) regular   (b) complete

(c) Hamiltonian   (d) Euler

(IT GATE 2008)

20. Let G be a graph with "n" vertices and $2n-2$ edges. The edges of G can be partitioned into two edge-disjoint spanning trees. Which of the following is NOT true for *G*?

(a) For every subset of "*k*" vertices, the induced sub-graph has at most 2*k*-2 edges
(b) The minimum cut in G has at least two edges
(c) There are two edge-disjoint paths between every pair of vertices
(d) There are two vertex-disjoint paths between every pair of vertices

(CS GATE 2008)

21. What is the chromatic number of an n-vertex simple connected graph which does not contain any odd length cycle? (assume $n \geq 2$)

(a) 2   (b) 3   (c) $n-1$ (d) n

(CS GATE 2009)

22. Which one of the following is TRUE for any simple connected undirected graph with more than two vertices?

(a) No two vertices have the same degree
(b) At least two vertices have the same degree
(c) At least three vertices have the same degree
(d) All vertices have the same degree

(CS GATE 2009)

23. Let $G = (V, E)$ be a graph. Define $\xi(G) = \sum_d i_d \times d$, where i_d is the number of vertices of degree "*d*" in G. If S and T are two different trees with $\xi(S) = \xi(T)$, then

(a) $|S| = 2|T|$ (b) $|S| = |T| - 1$
(c) $|S| = |T|$ (d) $|S| = |T| + 1$

(CS GATE 2010)

24. K_4 and Q_3 are graphs with the following structures:

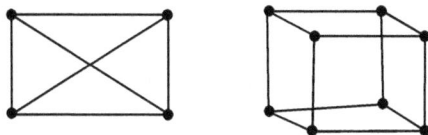

Which of the following statements is TRUE in relation to these graphs?

(a) K_4 is planar while Q_3 is not
(b) Both K_4 and Q_3 are planar
(c) Q_3 is planar while K_4 is not
(d) Neither Q_3 nor K_4 is planar

(CS GATE 2011)

25. An undirected graph G (V, E) contains "*n*" (*n* > 2) nodes namely $v_1, v_2, ..., v_n$. Two nodes v_i, v_j are connected if and only if $0 < |i - j| \leq 2$. Each edge (v_i, v_j) is assigned a weight $i + j$. A sample graph with $n = 4$ is shown below:

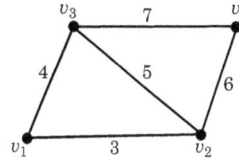

(*i*) What will be the cost of the minimum spanning tree (MST) of such a graph with "*n*" nodes?

(a) $\frac{1}{12}(11n^2 - 5n)$ (b) $n^2 - n + 1$
(c) $6n - 11$ (d) $2n + 1$

(CS GATE 2011)

(*ii*) The length of the path from v_5 to v_6 in the MST with $n = 10$ is

(a) 11   (b) 25   (c) 31   (d) 41

(CS GATE 2011)

26. Let G be a simple undirected planar graph on 10 vertices with 15 edges. If G is a connected graph, then the number of bounded faces in any embedding of G on the plane is equal to

(a) 3   (b) 4   (c) 5   (d) 6

(CS GATE 2012)

27. The maximum number of edges in a bipartite graph on 12 vertices is _____

(CS GATE 2014)

28. Let G be a graph with n vertices and m edges. What is the tightest upper bound on the running time Depth First Search on G, when G is represented as an adjacency matrix?

(a) $\Theta(n)$ (b) $\Theta(n + m)$
(c) $\Theta(n^2)$ (d) $\Theta(m^2)$

(CS GATE 2014)

29. Consider the tree arcs of a BFS traversal from a source node W in an unweighted, connected, undirected graph. The tree T formed by the three arcs is a data structure for computing

(a) the shortest path between every pair of vertices.

(b) the shortest path from W to every vertex in the graph.

(c) the shortest path from W to only those nodes that are levels of T.

(d) the longest path in the graph.

(CS GATE 2014)

30. Consider the directed graph given below.

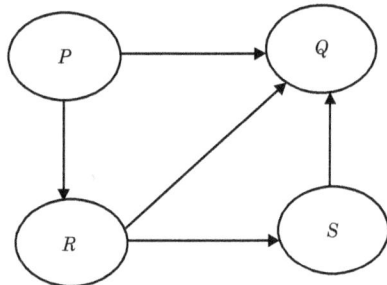

Which one of the following is TRUE?

(a) The graph does not have any topological ordering.

(b) Both $PQRS$ and $SRQP$ are topological orderings.

(c) Both $PSRQ$ and $SPRQ$ are topological orderings.

(d) $PSRQ$ is the only topological ordering.

(CS GATE 2014)

31. A cycle on n vertices is isomorphic to its complement. The value of n is _____?

(CS GATE 2014)

32. If G is a forest with n vertices and k connected components, how many edges does G have?

(a) $[n/k]$ (b) $[n/k] + 1$
(c) $n - k$ (d) $n - k + 1$

(CS GATE 2014)

33. Let δ denote the minimum degree of a vertex in a graph. For all planar graphs on n vertices with $\delta \geq 3$, which one of the following is TRUE?

(a) In any planar embedding, the number of faces is at least $\frac{n}{2} + 2$

(b) In any planar embedding, the number of faces is less than $\frac{n}{2} + 2$

(c) There is a planar embedding in which number of faces is less than $\frac{n}{2} + 2$

(d) There is a planar embedding in which number of faces is at most $\frac{n}{2} + 2$

(CS GATE 2014)

34. Let G be a connected planar graph with ten vertices. If the number of edges on each face is three, then the number of edges in G is _____?

(CS GATE 2015)

35. The minimum number of colors that is sufficient to vertex-color any planar graph is _____?

(CS GATE 2016)

36. Consider the weighted undirected graph with four vertices, where the weight of edge $\{i, j\}$ is given by the entry W_{ij} in the matrix W.

$$W = \begin{bmatrix} 0 & 2 & 8 & 5 \\ 2 & 0 & 5 & 8 \\ 8 & 5 & 0 & x \\ 5 & 8 & x & 0 \end{bmatrix}$$

The largest possible integer value of x, for which at least one shortest path between some pair of vertices will contain the edge with weight x is _____?

(CS GATE 2016)

37. Let G be a complete undirected graph on 4 vertices, having six edges with weights being 1,2,3,4,5, and 6. The maximum possible weight that a minimum weight spanning tree of G can have is _____?

(CS/IT GATE 2016)

38. Let $G = (V, E)$ is an undirected simple graph in which each edge has a distinct weight, and e is a particular edge of G. Which of the following statements about the minimum spanning trees (MSTs) of G is/ are TRUE?

I. If e is the lightest edge of some cycle in G, then every MST of G includes e.

II. If e is the heaviest edge of some cycle in G, then every MST of G excludes e.

(a) I only (b) II only
(c) Both I and II (d) Neither I nor II

(CS/IT GATE 2016)

39. Let *T* be a tree with 10 vertices. The sum of the degrees of all the vertices in *T* is ___

(CS (IT) GATE 2017)

40. *G* is undirected graph with *n* vertices and 25 edges such that each vertex of *G* has degree at least 3. Then the maximum possible value of *n* is _____

(CS (IT) GATE, 2017)

41. The chromatic number of the following graph is

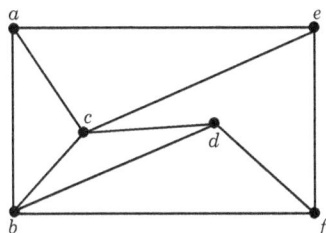

(CS (IT) GATE, 2018)

42. The post order traversal of a binary tree is 8, 9, 6, 7, 4, 5, 2, 3, 1. The in-order traversal of the same tree is 8, 6, 9, 4, 7, 2, 5, 1, 3. The height of a tree is the length of the longest path from the root to any leaf. The height of the binary tree above is_____

(CS (IT) GATE, 2018)

43. Consider the following undirected graph *G*:

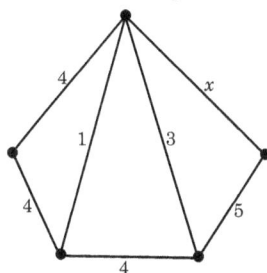

Choose a value for *x* that will maximize the number of minimum weight (MWSTs) of *G*. The number of MWSTs of *G* for this value of *x* is _____

(CS (IT) GATE, 2018)

Answer Key				
1. (d)	**2.** (a)	**3.** (a)	**4.** (c)	**5.** (d)
6. (b)	**7.** (a)	**8.** (d)	**9.** (c)	**10.** (b)
11. (a)	**12.** (a)	**13.** (i)-(c); (ii)-(c); (iii)-(b)			
14. (b)	**15.** (c)	**16.** (c)	**17.** (c)	**18.** (b)

19. (d)	**20.** (d)	**21.** (a)	**22.** (b)	**23.** (c)
24. (b)	**25.** (i)-(b); (ii)-(c)		**26.** (d)	**27.** 36
28. (c)	**29.** (b)	**30.** (c)	**31.** 5	**32.** (c)
33. (a)	**34.** 24	**35.** 4	**36.** 12	**37.** 7
38. (b)	**39.** 18	**40.** 16	**41.** 3	**42.** 4

Explanation

1. (*d*) Let $n = 3$. Then the minimum number of colors required to color the vertices of a cycle with "3" nodes is "3" which satisfies $n - 2[n/2] + 2$ (since $3 - 2[3/2] + 2 = 3 - 2 \times 1 + 2 = 3$).

Now let $n = 5$. Then the minimum number of colors required to color the vertices of a cycle with "5" nodes is "3" which satisfies $n - 2[n/2] + 2$ (since $5 - 2[5/2] + 2 = 5 - 2 \times 2 + 2 = 3$).

2. (*a*) There are two possibilities:

(*i*) If the removed vertex is an isolated vertex (it self a component), then the minimum number of components in the resultant graph will be "*k*."

(*ii*) If the removed vertex disconnects it is all components, then the maximum number of components in the resultant graph will be "*n*."

3. (*a*) The number of perfect matching in a graph with "*n*" vertices is $\frac{n(n-1)}{2}$. Here, $n = 6$.

So, the required number of perfect matching

$= \frac{6(6-1)}{2} = 15$.

4. (*c*)

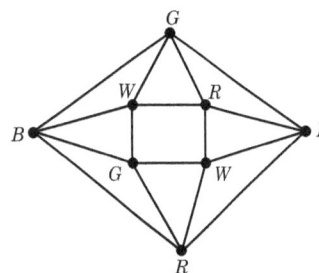

Here, "*G*" stands for the color "green," "*B*" stands for the color "black," "*W*" stands for the color "white" and "*R*" stands for the color "red."

Clearly "4" colors are required to the color the graph, such that no two adjacent vertices are assigned same color.

5. (*d*) The maximum number of edges in a simple graph

$$= \frac{n(n-1)}{2} = a \text{ (say)}.$$

Let $b = (n^2-3n)/2$.

Then, $a - b = n$ and so number of graphs with minimum b edges

$$= {}^aC_b + {}^aC_{b+1} + {}^aC_{b+2} + \dots\dots\dots + {}^aC_a$$

$$= {}^aC_{a-b} + {}^aC_{a-b-1} + {}^aC_{a-b-2} + \dots\dots\dots + {}^aC_{a-a}$$

$$= {}^aC_n + {}^aC_{n-1} + {}^aC_{n-2} + \dots\dots\dots + {}^aC_0$$

$$= {}^{\frac{n^2-n}{2}}C_n + {}^{\frac{n^2-n}{2}}C_{n-1} + {}^{\frac{n^2-n}{2}}C_{n-2} + \dots\dots + {}^{\frac{n^2-n}{2}}C_0$$

$$= \sum_{k=0}^{n} {}^{(n^2-n)/2}C_k$$

6. (*b*) Since G_1 and G_2 are two connected graphs on the same vertex set V with more than two vertices, so $G_1 \cup G_2$ must have a cycle.

7. (*a*) In a tree number of edges = no. of vertices–1 = n–1.

8. (*d*) Let "x" be the total number of vertices. Then, the sum of degrees of the vertices = 2 × total number of edges

$$\Rightarrow 6 \times 2 + 3 \times 4 + (x - 9) \times 3 = 2 \times 27$$

$$\Rightarrow 24 + 3x - 27 = 54 \Rightarrow x = 19.$$

Thus, the total number of vertices = 19.

9. (*c*) If all nondiagonal elements are 1, then every vertex is connected to every other vertex in the graph with an edge of weight 1. Such a graph has multiple distinct MSTs with cost n–1.

10. (*b*) For a planar embedding of a simple connected graph G with "n" vertices, "m" edges, and "r" regions (faces), we have by Euler's formula:

$n - m + r = 2$.

Here, $n = 13$ and $m = 19$. So $r = 8$.

11. (*a*) We know that if a graph has "n" vertices, "m" edges, and "x" be the minimum vertex cover of G, then the size of the maximum independent set of G is $n - x$.

Here, $n = 20$, $m = 100$, and $x = 8$. So the size of the maximum independent set of G is $20 - 8$, *i.e.*, 12.

12. (*a*) We know that a graph is planar if it can be drawn in a plane without any edge crossing. Among the given graphs, only G_1 can be redrawn in a plane without any crossing edges.

13. (*i*)-(*c*); (*ii*)-(*c*); (*iii*)-(*b*)
(*i*) No. of vertices with degree zero
= no. of subsets with size (≤ 1) = $n + 1$.

(*ii*) The maximum degree of a vertex G

$$\max_k \left({}^kC_2 \times 2^{n-k} \right) = {}^3C_2 \times 2^{n-3} = 3 \times 2^{n-k}$$

(*iii*) According to (*i*), $n + 1$ nodes of the graph are not connected to anyone while others are connected.

Hence, the total number of connected components = $n + 2$ ($n + 1$ connected components by each of the $n + 1$ vertices +1 connected component by remaining vertices).

14. (*b*) Since the vertices "u" and "v" are leaves of this depth first search tree T, so there must be a vertex "w" whose removal disconnects "u" and "v" in G.

15. (*c*) We know that a graph has an Eulerian circuit if following conditions are true.

(*i*) All vertices with nonzero degree are connected.

(*ii*) All vertices have even degree.

(a) Any k-regular graph (where k is an even number) is not Eulerian as a k regular graph may not be connected.

(b) A complete graph on 90 vertices is not Eulerian because all vertices have degree as 89.

(c) The complement of a cycle on 25 vertices is Eulerian. The reason is in a cycle of 25 vertices, all vertices have degree as 2 and in the complement graph, all vertices would have degree as 22 and so the graph would be connected.

16. (*c*) weight(s, u) + weight (u, v) $\geq$ weight(s, v)

$$\Rightarrow 53 + \text{weight} (u, v) \geq 65$$

$$\Rightarrow \text{weight} (u, v) \geq 12$$

17. (*c*) We know that a graph is connected if and only if all nodes can be traversed from each node.

For a simple connected graph with n nodes, there will be n-1 minimum number of edges.

Given that there are n edges which implies that a cycle is there in the graph. Now we can get a different spanning tree by removing one edge from the cycle, one at a time. Since the minimum cycle length can be 3, so, there must be at least 3 spanning trees in any such graph.

18. (b)

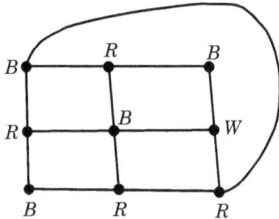

Here, "B" stands for the color "black," "W" stands for the color "white" and "R" stands for the color "red."

Clearly "3" colors are required to the color the graph, such that no two adjacent vertices are assigned same color. Hence, the chromatic number of the graph is "3."

19. (d) When we add a node "v" to the simple undirected graph G and make it adjacent to each odd degree vertex of G, the graph will contain vertices each of even degree. Hence, the resultant graph will become an Euler graph since we know that a connected graph G is Eulerian if and only if all vertices of G are of even degree of v is also even since number of odd degree vertices in a graph is even.

20. (d) Let us take two copies of K_4 (which is a complete graph on 4 vertices), say G_1 and G_2. Let $V(G_1) = \{a, b, c, d\}$ and $V(G_2) = \{m, n, p, q\}$.

Now let us construct a new graph G_3 by using these two graphs G_1 and G_2 by merging at a vertex, say (d, m). The resultant graph is two edge connected, and of minimum degree 2 but there exists a cut vertex, the merged vertex.

21. (a) We know that

(i) A simple graph with no odd cycles is a bipartite graph,

(ii) A bipartite graph can be colored with two colors.

Hence, the chromatic number of an n-vertex simple connected graph which does not contain any odd length cycle will be "2."

22. (b) The graph is simple implies neither loops nor parallel edges exist in the graph. Again the graph is connected implies that the degree of any vertex cannot be "0." As a result, the degree of any vertex will lie between 1 and n-1. Hence, at least two vertices will have the same degree.

23. (c) Here, $\xi(G)$ denotes the sum of all degrees in G. Since $\xi(S)$ and $\xi(T)$ are same for the two trees S and T, so both of them have same number of vertices.

24. (b) Both of K_4 and Q_3 can be drawn in a plane without any edge crossing.

25. (i)-(b); (ii)-(c)

To calculate MST, first we take the edges (v_1, v_2) and (v_1, v_3) as the two least cost edges. Next, we add the set of edges (v_2, v_4), (v_4, v_6), (v_6, v_8),... and (v_3, v_5), (v_5, v_7), (v_7, v_9), ...

Thus, we have two chains-one with odd labeled vertices and other with all even numbered vertices. These two chains merge at v_1.

Hence, the cost of the minimum spanning tree (MST) of such a graph with "n" nodes

= [weight (v_1, v_2) + weight (v_2, v_4) + weight (v_4, v_6) + weight (v_6, v_8) +.........] + [weight (v_1, v_3) + weight (v_3, v_5) + weight (v_5, v_7) + weight (v_7, v_9) + ...]

= [(1 + 2) + (2 + 4) + (4 + 6) + (6 + 8) +.......]+ [(1 + 3) + (3 + 5) + (5 + 7) + (7 + 9) +.........]

= 1 × 2 + [2 × 2 + 4 × 2 + 6 × 2 + 8 × 2 +.......]+ [3 × 2 + 5 × 2 + 7 × 2 + 9 × 2 +..........]

= n + (n–1) + 2 × [1 + 2 + 3 + 4 + ...+(n–2)]

= $n + (n-1) + 2 \times \dfrac{(n-2)}{2}\{1 + (n-2)\}$

= $n^2 - n + 1$.

(ii) Length of the path from $(v_5$ to $v_6)$ in the MST with $n = 10$

= [weight (v_1, v_2) + weight (v_2, v_4) + weight (v_4, v_6)] + [weight (v_1, v_3) + weight (v_3, v_5)]

= [(1 + 2) + (2 + 4) + (4 + 6)] + [(1 + 3) + (3 + 5)] = 31.

26. (d) We know that,

no. of vertices – no. of edges + no. of faces = 2.

Here no. of vertices = 10, number of edges = 15.

So the number of faces = 2 – 10 + 15 = 7.

Out of these seven faces, one is an unbounded face. Hence, total number of bounded faces = 6.

27. 36

We know that, the maximum number of edges in a bipartite graph = $\dfrac{n^2}{4}$.

Here, $n = 12$. So the maximum number of edges in a bipartite graph with 12 vertices is $\dfrac{12 \times 12}{4}$

$i.e; 36$ edges.

28. (c) In adjacency matrix representation, graph is represented as an $n \times n$ matrix. To do DFS, for every vertex, we traverse the row corresponding to that vertex to find all adjacent vertices. (In adjacency list representation we traverse only the adjacent vertices of the vertex). Therefore, time complexity becomes $\Theta(n^2.)$

29. (b) BFS is applied to find shortest path from the vertex u to the vertex v. According to the question, W is the source node. Therefore, the tree T formed by the three arcs is a data structure for computing the shortest path from W to every vertex in the graph.

30. (c) The graph does not contain any cycle, so there exist topological ordering.

Q depends on P, R, and S. R depends on P and S. Hence, both $PSRQ$ and $SPRQ$ are topological orderings.

31. 5

It can be easily verified that a cycle graph on $n = 1, 2, 3, 4$ vertices is not isomorphic to its complement. Now consider $n = 5$. Then, we have the following cycle:

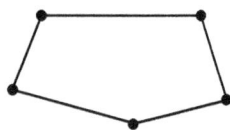

Its complement is given by

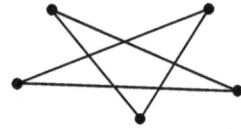

Clearly, these two graphs are isomorphic.

So, the value of n is 5.

32. (c) A tree has n vertices $\Rightarrow$ it has $n-1$ edges.

Therefore, to make it forest with k connected components, we have to delete $k-1$ edges from the tree.

Hence, G has total $(n-1) - (k-1)$ i.e; $(n-k)$ edges.

33. (a) $\delta \geq 3 \Rightarrow 3|V| \leq 2|E| \Rightarrow 3n \leq 2|E| \Rightarrow |E| \geq \dfrac{3n}{2}$.

Then, $|E| = |V| + |R| - 2$ (Euler's formula)

$\Rightarrow n + |R| - 2 \geq \dfrac{3n}{2} \Rightarrow |R| \geq \dfrac{n}{2} + 2$

34. 24

By Euler's formula, $|V| + |R| = |E| + 2.$

Given that, $|V| = 10$ and number of edges on each face = 3.

Thus, $3|R| = 2|E| \Rightarrow |R| = \dfrac{2}{3}|E|.$

Putting all the values in Euler's formula, we get,

$10 + \dfrac{2}{3}|E| = |E| + 2 \Rightarrow |E| = 24.$

35. 4

By the 4-color theorem, every planar graph is 4- colorable.

36. 12

Let $V = \{a, b, c, d\}$ be the set of vertices.

Then, the graph is given by

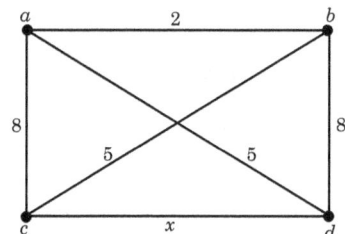

The paths between the vertices a and d are s follows:

c-d with weight "x,"

c-a-d with weight = 8 + 5 = 13,

c-b-d with weight = 5 + 8 = 13,

c-a-b-d with weight = 8 + 2 + 8 = 18.

Hence, the largest possible integer value of "x" will be 12.

37. 7

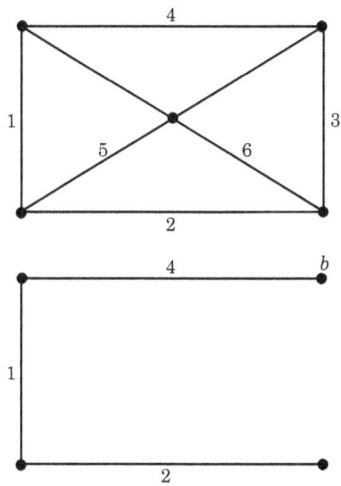

The maximum possible weight = 1 + 2 + 4 = 7.

38. (b) Since MST asked so if e any heaviest edge in cycle, it obviously will not be involved. Therefore, only statement (II) is true.

39. 18

The sum of degrees of all vertices

= 2 × Number of edges = 2 × 9 = 18.

$$\begin{bmatrix} \because \text{in a tree, total no. of edges} \\ = \text{total no. of vertices} - 1 = 10 - 1 = 9 \end{bmatrix}$$

40. 16

The sum of degree of n vertices =

2 × Number of edges

$\Rightarrow 2 \times 25 \geq 3 \times n$

(Since each vertex has degree at least 3)

$\Rightarrow n \leq \dfrac{50}{3} = 16.66 \Rightarrow n \leq 16.$

41. 3

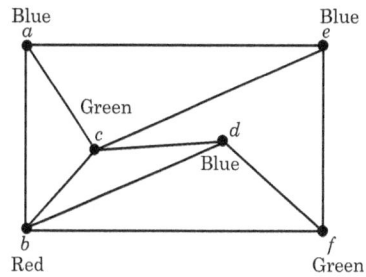

Thus, "3" colors (Red, Blue, and Green) are sufficient to the color the graph, such that no two adjacent vertices are assigned same color. Hence, the chromatic number of the graph is "3."

42. 4

The tree is given below:

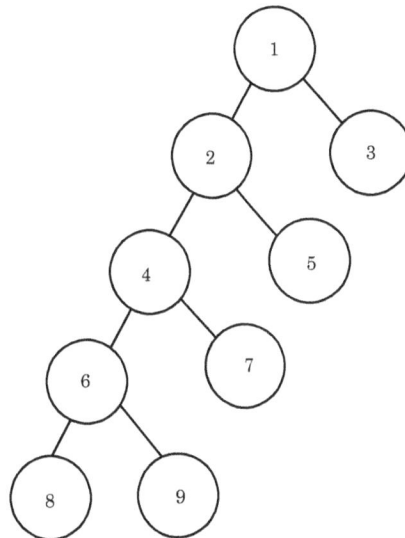

The height of the above tree = 1 + 1 + 1 + 1 = 4.

43. 4

For x = 1, 2, 3, 4 only 2 spanning trees are possible but for x = 5 total 4 spanning trees possible.

Questions for Practice

1. How many vertices does a regular graph of degree 4 with 10 edges have?

(a) 3   (b) 4   (c) 5   (d) 6

2. A simple graph with n vertices is connected if the graph has

(a) $\dfrac{(n-1)(n-2)}{2}$

(b) more than $\dfrac{(n-1)(n-2)}{2}$ edges

(c) less than $\frac{(n-1)(n-2)}{2}$ edges

(d) none of these

3. Which of the following graph is not regular?

   (a) K_n (b) C_n (c) W_n (d) Q_n

4. The number of distinct graph of order 5 and size 4 is

   (a) 6   (b) 5   (c) 4   (d) 3

5. The number of distinct graph of order n and size e is

   (a) n (b) e

   (c) min $\{n, e\}$ (d) none of these

6. An undirected graph possesses an Eulerian circuit if and only if it is connected and its vertices are

   (a) all of even degree
   (b) all of odd degree
   (c) of any degree
   (d) even in number

7. For the given graphs G_1, G_2, and G_3,

   G_1:

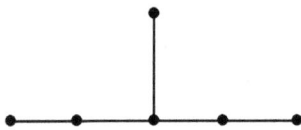

   G_2:

   G_3:

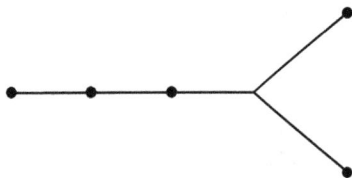

   (a) G_1 is isomorphic to G_2, and G_2 is isomorphic to G_3
   (b) G_1 is not isomorphic to G_2 and G_2 is not isomorphic to G_3
   (c) G_1 is isomorphic to G_2, but G_2 is not isomorphic to G_3
   (d) G_1 is not isomorphic to G_2, but G_2 is isomorphic to G_3

8. Which of the following statements is true regarding connected graphs?

   (a) No Eulerian graph is a Hamiltonian graph
   (b) Every Hamiltonian graph is an Euler graph
   (c) A graph can be neither Eulerian nor Hamiltonian
   (d) A Hamiltonian graph cannot be an Eulerian graph

9. Let G be a simple graph of size "e." Let $A(G)$, $C(G)$, and $K(G)$ be the incidence matrix, the circuit matrix, and cut-set matrix of G. Which of the following statement is false:

   (a) rank(A) + rank(C) = e
   (b) rank(C) + rank(K) = e
   (c) rank(A) = rank(C)
   (d) rank(A) + rank(C) + rank(K) = e

10. A graph with n vertices and $n-1$ edges that is not a tree is

   (a) Euler   (b) Circuit
   (c) Disconnected   (d) Connected

11. The length of a Hamiltonian path (if exists) in a connected graph of n vertices is

   (a) n (b) $n–1$ (c) $n + 1$ (d) $n/2$

12. A path in graph G, which contains every vertex of G once and only once is

   (a) Euler tour   (b) Hamiltonian path
   (c) Euler trail   (d) Hamiltonian tour

13. Every cut-set of a connected Euler graph has

   (a) an odd number of edges
   (b) an even number of edges
   (c) at least three edges
   (d) none of these

14. If G is a connected graph, then

   (a) G is unicursal if it has exactly one vertex of even degree
   (b) G is unicursal if it has exactly one vertex of odd degree
   (c) G is never unicursal
   (d) G is unicursal if it has no vertex of odd degree

15. Which of the following statement is true:

   (a) both K_5 and $K_{3,3}$ are planar
   (b) both K_5 and $K_{3,3}$ are nonplanar
   (c) K_5 is planar but $K_{3,3}$ is nonplanar
   (d) K_5 is nonplanar but $K_{3,3}$ is planar

16. Determine which of the following statements is not true

   (a) The Peterson graph is nonplanar
   (b) K_5 is nonplanar

(c) $K_{3,3}$ is nonplanar
(d) Dual graph of a planar graph is nonplanar

17. Which of the following statement is true
(a) K_4 is a self-dual
(b) if $G**$ is the dual of a dual of G, G, and $G**$ are always isomorphic
(c) $G**$ is always non-isomorphic to G
(d) if G_1 and G_2 are isomorphic, so are their duals

18. Determine which of the following statements is not true.
(a) The chromatic number of K_n is "n"
(b) The chromatic number of any cycle is 2
(c) The chromatic number of any nontrivial tree is 2
(d) Every planar graph is 4 colorable

19. The graph whose chromatic polynomial is $\lambda(\lambda-1)^5$ is
(a) a tree having 6 vertices
(b) K_6 (c) K_5
(d) A tree having 5 vertices

20. Let $\chi(G)$ be the (vertex) chromatic number of a graph G and δ, Δ, respectively, denote the minimum and maximum degree of a vertex of G, then
(a) $x > 1 + \Delta$ (b) $x > 1 + \delta$
(c) $x \le 1 + \Delta$ (d) $x \le 1 + \delta$

21. Let $X(G)$ be the (vertex) chromatic number of a graph G having maximum degree Δ, then
(a) if $X \le \Delta$, G is either a complete graph or has an odd cycle
(b) $X > \Delta$, G neither a complete graph nor has an odd cycle
(c) if G is either a complete graph or has an odd cycle, then $X > \Delta$
(d) if G neither a complete graph nor has an odd cycle, then $X \le \Delta$

22. Let $x(G)$ be the (vertex) chromatic number of a graph G of order n, then
(a) $\sqrt{n} < x(G) + x(\bar{G}) < n+1$
(b) $2\sqrt{n} < \chi(G) + \chi(\bar{G}) < n+1$
(c) $\sqrt{n} < \chi(G) + \chi(\bar{G}) < 2n$
(d) none of the above

23. Every complete graph is
(a) regular   (b) connected
(c) simple   (d) circuit

24. What is the minimum number of edges in a connected cyclic graph containing "n" vertices?
(a) n (b) $n + 1$ (c) $n-1$ (d) $2n$

25. Let G be a planar graph. Then G does not contain
(a) subgraph isomorphic to K_3 or $K_{3,3}$
(b) subgraph homomorphic to K_5 or $K_{3,3}$
(c) subgraph homomorphic to K_2 or $K_{2,2}$
(d) subgraph isomorphic to $K_{4,4}$

26. The adjacency matrix of the following graph is

(a) $\begin{bmatrix} 1 & 0 & 1 & 0 \\ 1 & 0 & 1 & 0 \\ 0 & 1 & 0 & 1 \\ 0 & 0 & 0 & 0 \end{bmatrix}$ (b) $\begin{bmatrix} 1 & 0 & 0 & 1 \\ 1 & 0 & 0 & 0 \\ 0 & 1 & 0 & 1 \\ 0 & 1 & 1 & 0 \end{bmatrix}$

(c) $\begin{bmatrix} 1 & 0 & 1 & 1 \\ 0 & 0 & 1 & 0 \\ 0 & 1 & 0 & 0 \\ 1 & 1 & 0 & 0 \end{bmatrix}$ (d) none of these

27. Consider a simple connected graph G with "n" vertices and "n"-edges ($n > 2$). Then, which of the following statement are true?
(a) G has no cycles.
(b) The graph obtained by removing any edge from G is not connected.
(c) G has at least two cycle.
(d) The graph obtained by removing any two edges from G is not connected.

28. Let G be a graph without any loop. If G has six vertices and ten edges, then the size of its incidence matrix is
(a) 10×6 (b) 6×10
(c) 10×10 (d) 6×6

29. If $A = (a_{ij})_{n \times n}$ be the adjacency matrix of a graph having "n" number of vertices. If $a_{33} = 1$, then which of the following is true?
(a) there exist a self loop at vertex v_3
(b) the vertex v_3 is connected with each vertex of the graph
(c) each edge is incident to v_3
(d) there exists at least one parallel edge joining v_2 and v_3

30. If $A = (a_{ij})_{5\times5}$ be the adjacency matrix of a graph. If $a_{24} = 0$, then which of the following is true?
(a) no edge connects the vertices v_2 and v_4
(b) v_2 and v_4 are adjacent vertices
(c) each of v_2 and v_4 is an isolated vertex
(d) none of the above

31. Which of the following graph is connected?

(a)   (b)

(c)   (d) none of these

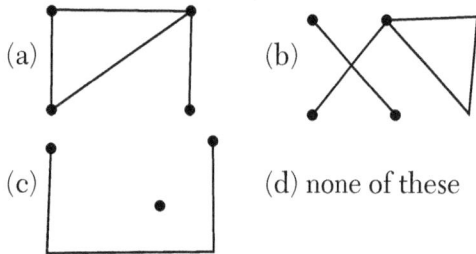

32. The number of distinct simple graphs with up to four nodes is:
(a) 2^5-1 (b) 2^4-1 (c) 2^7-1 (d) 2^4+1

33. Let G be a simple graph all of whose vertices have degree 3 and $|E| = 2|V|-3$. What can be said about G?
(a) G has 6 vertices and 9 edges
(b) G has 6 vertices and 12 edges
(c) G has 9 vertices and 6 edges
(d) None of these

34. The number of spanning trees in a complete graph G with "n" vertices is
(a) n^{n-2} (b) n^n (c) n^2 (d) n^{n-1}

35. A complete graph with "n" vertices is
(a) 2-chromatic (b) $n/2$ chromatic
(c) n-1 chromatic (d) n chromatic

Answer Key				
1. (c)	2. (b)	3. (a)	4. (a)	5. (d)
6. (a)	7. (b)	8. (c)	9. (d)	10. (c)
11. (b)	12. (b)	13. (b)	14. (d)	15. (b)
16. (d)	17. (a)	18. (b)	19. (b)	20. (c)
21. (d)	22. (b)	23. (b)	24. (a)	25. (b)
26. (a)	27. (d)	28. (b)	29. (a)	30. (a)
31. (a)	32. (b)	33. (a)	34. (a)	35. (d)

Hints

26. (a) Use the definition of adjacency matrix.

32. (b) The number of distinct simple graphs with up to "n" nodes is equal to 2^n-1.

GATE *2019* SOLVED PAPERS

TOPIC: LINEAR ALGEBRA

1. Consider the matrix $P = \begin{bmatrix} 1 & 1 & 0 \\ 0 & 1 & 1 \\ 0 & 0 & 1 \end{bmatrix}$

   The number of distinct eigenvalues of P is
   (a) 0   (b) 1
   (c) 3   (d) 2
   [ME GATE 2019]

2. The set of equations
   $x + y + z = 1$
   $ax - ay + 3z = 5$
   $5x - 3y + az = 6$

   has infinite solutions, if $a =$?
   (a) 4   (b) −4
   (c) −3   (d) 3
   [ME GATE 2019]

3. In matrix equation $AX = R$,

   $A = \begin{bmatrix} 4 & 8 & 4 \\ 8 & 16 & -4 \\ 4 & -4 & 15 \end{bmatrix}, X = \begin{Bmatrix} 2 \\ 1 \\ 4 \end{Bmatrix}$ and $R = \begin{Bmatrix} 32 \\ 16 \\ 64 \end{Bmatrix}$

   One of the eigenvalues of the matrix A is
   (a) 8   (b) 16
   (c) 15   (d) 4
   [ME GATE 2019]

4. Let X be a square matrix. Consider the following two statements on X.
   I.   X is invertible
   II. Determinant of X is non zero.
   Which one of the following is true?
   (a) I implies II; II does not imply I
   (b) I does not implies II; II does not imply I

(c) I and II are equivalent statements
(d) II implies I; I does not imply II
   [CS GATE 2019]

5. Consider the following matrix
   $R = \begin{bmatrix} 1 & 2 & 4 & 8 \\ 1 & 3 & 9 & 27 \\ 1 & 4 & 16 & 64 \\ 1 & 5 & 25 & 125 \end{bmatrix}$

   the absolute value of the product of eigenvalues of R is _____?
   [CS GATE 2019]

6. A 3×3 matrix has eignvalues 1, 2, and 5. The determinant of the matrix is _____?
   [CS GATE 2019]

7. The number of distinct eigenvalues of the matrix $A = \begin{vmatrix} 2 & 2 & 3 & 3 \\ 0 & 1 & 1 & 1 \\ 0 & 0 & 3 & 3 \\ 0 & 0 & 0 & 2 \end{vmatrix}$ is equals to _____
   [EC GATE 2019]

8. M is a 2×2 matrix with eigenvalues 4 and 9. The eigenvalues of M^2 are
   (a) 16 and 81   (b) 2 and 3
   (c) −2 and −3   (d) 4 and 9
   [EE GATE 2019]

9. The rank of the matrix, $M = \begin{bmatrix} 0 & 1 & 1 \\ 1 & 0 & 1 \\ 1 & 1 & 0 \end{bmatrix}$, is _____?
   [EE GATE 2019]

10. Consider a 2×2 matrix $M = \begin{bmatrix} v_1 & v_2 \end{bmatrix}$, where v_1 and v_2 are the column vectors. Suppose

$M^{-1} = \begin{bmatrix} u'_1 \\ u'_2 \end{bmatrix}$ where u'_1 and u'_2 are the row vectors. Consider the following statements:

Statement 1: $u'_1 v_1 = [1]$ *and* $u'_2 v_2 = [1]$

Statement 2: $u'_1 v_2 = [0]$ *and* $u'_2 v_1 = [0]$.

Which of the following options is correct?

(a) statement 2 is true and statement 1 is false

(b) statement 1 is true and statement 2 is false

(c) both statements are false

(d) both the statements are true

[EE GATE 2019]

Answer key				
1. (b)	**2.** (a)	**3.** (b)	**4.** (c)	**5.** 12
6. 10	**7.** 3.	**8.** (a)	**9.** 3	**10.** (d)

Explanations

1. (b) Since P is a upper triangular matrix, so its diagonal elements 1, 1, 1 are the eigenvalues. Thus the number of distinct eigenvalues of $P = 1$.

2. (a) Now the system has an infinite number of solutions

   if $\begin{vmatrix} 1 & 1 & 1 \\ a & -a & 3 \\ 5 & -3 & a \end{vmatrix} = 0$

   i.e., if $(-a^2 + 9) - (a^2 - 15) + 1(-3a + 5a) = 0$

   i.e., if $-a^2 + a + 12 = 0$

   i.e., if $(a - 4)(a + 3) = 0$

   i.e., if $a = 4, -3$

   Now for $a = 4$,

   $\Delta_1 = \begin{vmatrix} 1 & 1 & 1 \\ 5 & -a & 3 \\ 6 & -3 & a \end{vmatrix} = \begin{vmatrix} 1 & 1 & 1 \\ 5 & -4 & 3 \\ 6 & -3 & 4 \end{vmatrix}$

   $= (-16 + 9) - (20 - 18) + (-15 + 24) = 0,$

   $\Delta_2 = \begin{vmatrix} 1 & 1 & 1 \\ a & 5 & 3 \\ 5 & 6 & a \end{vmatrix} = \begin{vmatrix} 1 & 1 & 1 \\ 4 & 5 & 3 \\ 5 & 6 & 4 \end{vmatrix}$

   $= (20 - 18) - (16 - 15) + (24 - 25) = 0,$

   $\Delta_3 = \begin{vmatrix} 1 & 1 & 1 \\ a & -a & 5 \\ 5 & -3 & 6 \end{vmatrix} = \begin{vmatrix} 1 & 1 & 1 \\ 4 & -4 & 5 \\ 5 & -3 & 6 \end{vmatrix}$

   $= (-24 + 15) - (24 - 25) + (-12 + 20) = 0.$

Thus, for $a = 4$, $\Delta = \Delta_1 = \Delta_2 = \Delta_3 = 0$ and so by Cramer's rule, the system has an infinite number of solutions.

3. (b) $AX = R$

   $\Rightarrow \begin{bmatrix} 4 & 8 & 4 \\ 8 & 16 & -4 \\ 4 & -4 & 15 \end{bmatrix} \begin{Bmatrix} 2 \\ 1 \\ 4 \end{Bmatrix} = \begin{Bmatrix} 32 \\ 16 \\ 64 \end{Bmatrix} = 16 \begin{Bmatrix} 2 \\ 1 \\ 4 \end{Bmatrix},$

   which is of the form $AX = \lambda X$; where λ is an eigenvalue.

   ∴ One of the eigenvalues of the matrix must be 16

4. (c) $|A| \neq 0 \Leftrightarrow A$ is invertible matrix

   ∴ I and II are equivalent statements.

5. 12

   $\begin{vmatrix} 1 & 2 & 4 & 8 \\ 1 & 3 & 9 & 27 \\ 1 & 4 & 16 & 64 \\ 1 & 5 & 25 & 125 \end{vmatrix}$

   $= \begin{vmatrix} 1 & 2 & 4 & 8 \\ 0 & 1 & 5 & 19 \\ 0 & 2 & 12 & 56 \\ 0 & 3 & 21 & 117 \end{vmatrix} \begin{pmatrix} \text{by } R_2 \to R_2 - R_1, \\ R_3 \to R_3 - R_1, R_4 \to R_4 - R_1 \end{pmatrix}$

   $= \begin{vmatrix} 1 & 5 & 19 \\ 2 & 12 & 56 \\ 3 & 21 & 117 \end{vmatrix} = \begin{vmatrix} 1 & 5 & 19 \\ 0 & 2 & 18 \\ 0 & 6 & 60 \end{vmatrix} \begin{pmatrix} \text{by } R_2 \to R_2 - 2R_1, \\ R_3 \to R_3 - 3R_1 \end{pmatrix}$

   $= \begin{vmatrix} 2 & 18 \\ 6 & 60 \end{vmatrix} = 120 - 108 = 12.$

   ∴ The absolute value of the product of eigenvalues
   $= |\det(R)| = 12.$

6. 10

   Since determinant of a matrix is the product of its eigenvalues, so determinant of the given matrix
   $= 1 \times 2 \times 5 = 10.$

7. 3.

   Clearly the given matrix A is upper triangular. So its eigenvalues are its diagonal elements *i.e*; 2, 1, 3, 2.
   ∴ The number of distinct eigenvalues = 3.

8. (a) We know that if λ is an eigenvalue of a matrix M then λ^2 is an eigenvalues of the matrix M^2.

   Given, M is a 2×2 matrix with eigenvalues 4 and 9. So, the eigenvalues of M^2 are $4^2, 9^2$ i.e; 16, 81.

9. 3

The matrix $M = \begin{bmatrix} 0 & 1 & 1 \\ 1 & 0 & 1 \\ 1 & 1 & 0 \end{bmatrix}$ is a non-singular

square matrix of order 3. So, its rank = 3.

10. (d)

Let $M_{2\times 2} = \begin{bmatrix} a_{11} & a_{12} \\ a_{21} & a_{22} \end{bmatrix}$, where $v_1 = \begin{bmatrix} a_{11} \\ a_{21} \end{bmatrix}$,

$v_2 = \begin{bmatrix} a_{12} \\ a_{22} \end{bmatrix}$.

$\therefore M^{-1} = \dfrac{1}{a_{11}a_{22} - a_{21}a_{12}}\begin{bmatrix} a_{22} & -a_{12} \\ -a_{21} & a_{11} \end{bmatrix}$. Then

$u_1' = \dfrac{1}{a_{11}a_{22} - a_{21}a_{12}}\begin{bmatrix} a_{22} & -a_{12} \end{bmatrix}$,

$u_2' = \dfrac{1}{a_{11}a_{22} - a_{21}a_{12}}\begin{bmatrix} -a_{21} & a_{11} \end{bmatrix}$.

$\therefore u_1'v_1 = \begin{bmatrix} \dfrac{a_{22}}{a_{11}a_{22} - a_{21}a_{12}} & \dfrac{-a_{12}}{a_{11}a_{22} - a_{21}a_{12}} \end{bmatrix}\begin{bmatrix} a_{11} \\ a_{21} \end{bmatrix}$

$= \begin{bmatrix} a_{11}\left(\dfrac{a_{22}}{a_{11}a_{22} - a_{21}a_{12}}\right) + a_{21}\left(\dfrac{-a_{12}}{a_{11}a_{22} - a_{21}a_{12}}\right) \end{bmatrix}$

$= [1]$

Similarly, $u_2'v_2 = [1], u_1'v_2 = [0], u_2'v_1 = [0]$.
So, both statements are true.

TOPIC: CALCULUS

1. A parabola $x = y^2$ with $0 \le x \le 1$ is shown in the figure. The volume of the solid of rotation obtained by rotating the shaded area by $360°$ around the x axis is

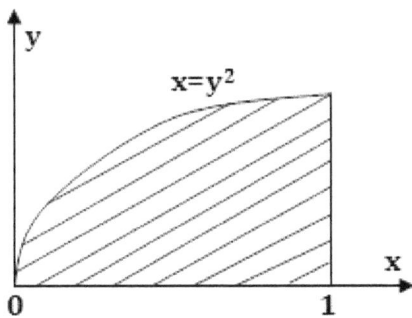

(a) $\dfrac{\pi}{4}$ (b) $\dfrac{\pi}{2}$

(c) 2π (d) π

[ME GATE 2019]

2. The value of the following definite integral $\int_1^e (x \ln x)dx$ is _____? (round off to three decimal places)

[ME GATE 2019]

3. Compute $\lim\limits_{x \to 3} \dfrac{x^4 - 81}{2x^2 - 5x - 3}$

(a) 108/7   (b) 1

(c) 53/12   (d) Limit does not exist

[CS GATE 2019]

4. The curve y = f(x) is such that the tangent to the curve at every point (x, y) has a y-axis intercept c, given by c = -y. Then f(x) is proportional to

(a) x^{-1} (b) x^2

(c) x^3 (d) x^4

[CS GATE 2019]

5. The value of the integral $\int_0^\pi \int_y^\pi \dfrac{\sin x}{x}dx\,dy$, is equal to_____?

[EC GATE 2019]

6. Consider a differentiable function f(x) on the set of real numbers such that f(–1) = 0 and $|f'(x)| \le 2$. Given these conditions, which one of the following inequalities is necessarily true for all $x \in [-2, 2]$?

(a) $f(x) \le 2|x+1|$ (b) $f(x) \le 2|x|$

(c) $f(x) \le \dfrac{1}{2}|x+1|$ (d) $f(x) \le \dfrac{1}{2}|x|$

[EC GATE 2019]

7. For a small value of h, the Taylor series expansion of f(x + h) is

(a) $f(x) - hf'(x) + \dfrac{h^2}{2}f''(x) - \dfrac{h^3}{3}f'''(x) + ...\infty$

(b) $f(x) - hf'(x) + \dfrac{h^2}{2!}f''(x) - \dfrac{h^3}{3!}f'''(x) + ...\infty$

(c) $f(x) + hf'(x) + \dfrac{h^2}{2}f''(x) + \dfrac{h^3}{3}f'''(x) + ...\infty$

(d) $f(x) + hf'(x) + \dfrac{h^2}{2!}f''(x) + \dfrac{h^3}{3!}f'''(x) + ...\infty$

[CE GATE 2019]

8. Which of the following is correct?

(a) $\lim\limits_{x \to 0} \dfrac{\sin 4x}{\sin 2x} = 1$ and $\lim\limits_{x \to 0} \dfrac{\tan x}{x} = 1$

(b) $\lim\limits_{x \to 0} \dfrac{\sin 4x}{\sin 2x} = \infty$ and $\lim\limits_{x \to 0} \dfrac{\tan x}{x} = 1$

(c) $\lim\limits_{x\to 0}\dfrac{\sin 4x}{\sin 2x}=2$ and $\lim\limits_{x\to 0}\dfrac{\tan x}{x}=\infty$

(d) $\lim\limits_{x\to 0}\dfrac{\sin 4x}{\sin 2x}=2$ and $\lim\limits_{x\to 0}\dfrac{\tan x}{x}=1$

[CE GATE 2019]

9. Which of the following is not a correct statement?

(a) The function $|x|$ has the global minima at $x=0$

(b) The function $\sqrt[x]{x}$ $(x>0)$ has the global minima at $x=e$

(c) The function $\sqrt[x]{x}$ has the global maxima at $x=e$

(d) The function x^3 has neither global minima nor global maxima

[CE GATE 2019]

10. The following inequality is true for all x close to zero.

$$2-\frac{x^2}{3}<\frac{x\sin x}{1-\cos x}<2$$

What is the value of $\lim\limits_{x\to 0}\dfrac{x\sin x}{1-\cos x}$?

(a) 2   (b) ½

(c) 0   (d) 1

[CE GATE 2019]

Answer key				
1. (b)	2. $\frac{1}{4}(e^2+1)$	3. (a)	4. (b)	5. 2
6. (a)	7. (d)	8. (d)	9. (b)	10. (a)

Explanation

1. (b) Required volume of the solid of rotation,

$$V=\int_{x=a}^{b}\pi y^2\,dx=\int_{x=0}^{1}\pi x\,dx=\pi\left[\frac{x^2}{2}\right]_0^1=\frac{\pi}{2}$$

2. $\frac{1}{4}(e^2+1)$

Let $I=\int_1^e(x\ln x)\,dx$.

Put $\ln x=t$. Then $x=e^t$ and so $dx=e^t dt$.

Also $x=1\Rightarrow t=\ln 1=0$ and $x=e\Rightarrow t=\ln e=1$.

$$\therefore I=\int_0^1 e^t\times t\,e^t\,dt=\int_0^1 te^{2t}\,dt$$

$$=\left[t\frac{e^{2t}}{2}\right]_0^1-\int_0^1\left[1\times\frac{e^{2t}}{2}\right]dt$$

$$=\left[\frac{e^2}{2}-0\right]-\left[\frac{e^{2t}}{4}\right]_0^1=\frac{e^2}{2}-\left[\frac{e^2}{4}-\frac{1}{4}\right]$$

$$=e^2\left[\frac{1}{2}-\frac{1}{4}\right]+\frac{1}{4}=\frac{1}{4}(e^2+1).$$

3. (a)

$$\lim_{x\to 3}\frac{x^4-81}{2x^2-5x-3}\qquad\left(\text{form }\frac{0}{0}\right)$$

$$=\lim_{x\to 3}\frac{4x^3}{4x-5}\qquad(\text{by the L'Hospital Rule})$$

$$=\frac{4\times 3^3}{4\times 3-5}=\frac{108}{7}.$$

4. (b) The equation of tangent is given by:

$$y=mx+c\Rightarrow y=mx-y\Rightarrow y=\frac{mx}{2}.$$

If the point of contact is (x,y), then

$$m=\frac{dy}{dx}.$$

So the equation of the tangent becomes,

$$y=\frac{x}{2}\frac{dy}{dx}\Rightarrow\frac{dy}{y}=2\frac{dx}{x}$$

Integrating, we get,

$$\log y=2\log x=\log x^2\text{ or, }y=x^2$$

Thus y i.e; $f(x)$ is proportional to x^2.

5. 2

Clearly, the region of integration is bounded by the lines $y=0,y=x,x=\pi$.

So by changing the order of integration, we have,

$$\int_0^{\pi}\int_y^{\pi}\frac{\sin x}{x}\,dx\,dy$$

$$=\int_{x=0}^{\pi}\left[\int_{y=0}^{x}\frac{\sin x}{x}\,dy\right]dx=\int_{x=0}^{\pi}\frac{\sin x}{x}[y]_0^x\,dx=\int_{x=0}^{\pi}\sin x\,dx$$

$$=[-\cos x]_0^{\pi}=-[\cos\pi-\cos 0]=-[-1-1]=2.$$

6. (a) Given, $|f'(x)|\le 2$, $f(-1)=0$.

Now $|f'(x)|\le 2\Rightarrow -2\le f'(x)\le 2$ where $x\in[-2,2]$

Using Lagrange's Mean Value Theorem over $[-1,2]$, we have,

$$f'(c) = \frac{f(2)-f(-1)}{2-(-1)}, \text{ for } c \in (-1,2)$$

$$\therefore -2 \le f'(x) \le 2$$

$$\Rightarrow -2 \le f'(c) \le 2 \Rightarrow -2 \le \frac{f(2)-f(-1)}{2-(-1)} \le 2$$

$$\Rightarrow -2 \le \frac{f(2)-0}{3} \le 2 \Rightarrow -6 \le f(2) \le 6,$$

which satisfies $f(x) \le 2|x+1|$.

7. (d)

8. (d)

$$\lim_{x\to 0}\frac{\sin 4x}{\sin 2x} = 2 \quad \left(\text{form }\frac{0}{0}\right)$$

$$= \lim_{x\to 0}\frac{4\cos 4x}{2\cos 2x} \quad (\text{by the L'Hospital Rule})$$

$$= \frac{4\cos 0}{2\cos 0} = \frac{4}{2} = 2.$$

By the fundamental formulas on limits,

$$\lim_{x\to 0}\frac{\tan x}{x} = 1.$$

9. (b)

$$y = x^{\frac{1}{x}}$$

$$\Rightarrow \log y = \log\left(x^{\frac{1}{x}}\right) = \frac{1}{x}\log x$$

$$\Rightarrow \frac{d}{dx}[\log y] = \frac{d}{dx}\left[\frac{1}{x}\log x\right]$$

$$\Rightarrow \frac{1}{y}\frac{dy}{dx} = \frac{x\times\frac{1}{x}-\log x\times 1}{x^2} = \frac{1-\log x}{x^2}$$

$$\Rightarrow \frac{dy}{dx} = y\times\frac{1-\log x}{x^2}$$

$$\therefore \frac{dy}{dx} = 0$$

$$\Rightarrow y\times\frac{1-\log x}{x^2} = 0 \Rightarrow 1-\log x = 0$$

$$\Rightarrow \log x = 1 = \log e \Rightarrow x = e$$

Now $\frac{d^2 y}{dx^2}$

$$= \frac{d}{dx}\left(\frac{dy}{dx}\right) = \frac{d}{dx}\left(y\times\frac{1-\log x}{x^2}\right)$$

$$= \frac{1-\log x}{x^2}\times\frac{dy}{dx} + y\left[\frac{-x^2\times\frac{1}{x}+(1-\log x)\times 2x}{x^4}\right]$$

$$= 0 + x^{\frac{1}{x}}\times\left[\frac{1-2\log x}{x^3}\right]\left(\because \frac{dy}{dx}=0\right)$$

$$\therefore \frac{d^2 y}{dx^2}\Bigg|_{x=e} = e^{\frac{1}{e}}\times\left[\frac{1-2\log e}{e^3}\right] = -\frac{e^{\frac{1}{e}}}{e^3} < 0.$$

Hence, the function $x^{\frac{1}{x}}$ has a local maximum value at $x = e$.

So option (b) cannot be correct.

10. (a)

$$\lim_{x\to 0}\frac{x\sin x}{1-\cos x} \quad \left(\text{form }\frac{0}{0}\right)$$

$$= \lim_{x\to 0}\frac{x\cos x+\sin x}{0+\sin x} \quad (\text{by the L'Hospital Rule})$$

$$= \lim_{x\to 0}\frac{x\cos x+\sin x}{\sin x} \quad \left(\text{form }\frac{0}{0}\right)$$

$$= \lim_{x\to 0}\frac{-x\sin x+\cos x+\cos x}{\cos x}$$

$$= \frac{0+\cos 0+\cos 0}{\cos 0} = \frac{1+1}{1} = 2.$$

TOPIC: ORDINARY DIFFERENTIAL EQUATIONS

1. For the equation $\frac{dy}{dx}+7x^2 y = 0$, if $y(0)=\frac{3}{7}$, then the value of $y(1)$ is

(a) $\frac{7}{3}e^{-7/3}$ (b) $\frac{3}{7}e^{-7/3}$

(c) $\frac{3}{7}e^{-3/7}$ (d) $\frac{7}{3}e^{-3/7}$

[ME GATE 2019]

2. The differential equation $\frac{dy}{dx}+4y=5$ is valid in the domain $0 \le x \le 1$ with $y(0) = 2.25$. The solution of the differential equation is

(a) $y = e^{-4x}+1.25$

(b) $y = e^{4x}+1.25$

(c) $y = e^{-4x}+5$

(d) $y = e^{4x}+5$

[ME GATE 2019]

3. A differential equation is given as

$$x^2\frac{d^2y}{dx^2}-2x\frac{dy}{dx}+2y=4$$

The solution of the differential equation in terms of arbitrary constants C_1 & C_2 is

(a) $y=C_1x^2+C_2x+4$

(b) $y=\dfrac{C_1}{x^2}+C_2x+4$

(c) $y=\dfrac{C_1}{x^2}+C_2x+2$

(d) $y=C_1x^2+C_2x+2$

[ME GATE 2019]

4. The families of curves represented by the solution of the equation $\dfrac{dy}{dx}=-\left(\dfrac{x}{y}\right)^n$ for $n=-1$ and $n=+1$, respectively, are

(a) Hyperbolas and circles
(b) Circles and hyperbolas
(c) Hyperbolas and parabolas
(d) Parabolas and circles.

[EC GATE 2019]

5. Consider the homogeneous ordinary differential equation: $x^2\dfrac{d^2y}{dx^2}-3x\dfrac{dy}{dx}+3y=0,x>0$.

With $y(x)$ as a general solution. Given that $y(1)=1$ & $y(2)=14$. Then the value of $y(1.5)$, (rounded off to two decimal places), is _____?

[EC GATE 2019]

6. An ordinary differential equation is given below:

$$\frac{dy}{dx}x\ln x=y$$

The solution for the above equation is (k denotes a constant)

(a) $y=k\ln x$ (b) $y=kx\ln x$
(c) $y=kxe^x$ (d) $y=kxe^{-x}$

[CE GATE 2019]

7. Consider the ordinary differential equation:

$$x^2\frac{d^2y}{dx^2}-2x\frac{dy}{dx}+2y=0.$$

Given that $y(1)=0$ & $y(2)=2$. Then the value of $y(3)$, (rounded off to one decimal place), is _____?

[CE GATE 2019]

Answer key				
1. (b)	**2.** (a)	**3.** (d)	**4.** (a)	**5.** 5.25
6. (a)	**7.** 6			

Explanation

1. (b) Given $\dfrac{dy}{dx}+7x^2y=0$ (i)

Comparing (1) with $\dfrac{dy}{dx}+Py=Q$ we get,

$P=7x^2,Q=0$.

So I.F $=e^{\int Pdx}=e^{\int 7x^2dx}=e^{\frac{7x^3}{3}}$

Hence, the solution of equation (1) is given by:

$$y\times I.F=\int Q\times I.F\,dx+C$$

or, $ye^{\frac{7x^3}{3}}=\int 0\times e^{\frac{7x^3}{3}}\,dx+C$

or, $ye^{\frac{7x^3}{3}}=C$

or, $y=Ce^{-\frac{7x^3}{3}}$ ....(2)

Given that $y=\dfrac{3}{7}$ at $x=0$.

$\therefore (2)\Rightarrow \dfrac{3}{7}=Ce^{-0}=C$

Hence, (2) gives, $y=\dfrac{3}{7}e^{-\frac{7x^3}{3}}$

$\therefore y(1)=\dfrac{3}{7}e^{-\frac{7}{3}}$.

2. (a) Comparing the given differential equation with $\dfrac{dy}{dx}+Py=Q$, we get $P=4,Q=5$.

Therefore, the integrating factor (I.F)

$$=e^{\int Pdx}=e^{\int 4dx}=e^{4x}.$$

Hence, the solution is given by:

$$y\times(I.F)=\int Q\times(I.F)dx+C$$

or, $ye^{4x}=\int 5e^{4x}dx+C=\dfrac{5}{4}e^{4x}+C$

or, $y=\dfrac{5}{4}+Ce^{-4x}$ ....(1)

Given that $y(0)=2.25$ i.e; at $x=0$, $y=2.25$. So (1) gives, $2.25=\dfrac{5}{4}+C\times e^0$ i.e; $C=2.25-1.25=1$.

Hence, the required solution is:

$y = \dfrac{5}{4} + e^{-4x}$ i.e; $y = 1.25 + e^{-4x}$.

3. (d) The given differential equation,

$x^2 \dfrac{d^2 y}{dx^2} - 2x \dfrac{dy}{dx} + 2y = 4$

$\Rightarrow \left[x^2 D^2 - 2xD + 2 \right] y = 4 \ldots\ldots\ldots(1)$

Let us put log$x$ = z.

Then, $\dfrac{dy}{dz} = x \dfrac{dy}{dx}$ and $x^2 \dfrac{d^2 y}{dx^2} = \dfrac{d^2 y}{dz^2} - \dfrac{dy}{dz}$.

So, (1) becomes,

$\dfrac{d^2 y}{dz^2} - \dfrac{dy}{dz} - 2 \dfrac{dy}{dz} + 2y = 4$

$\Rightarrow \dfrac{d^2 y}{dz^2} - 3 \dfrac{dy}{dz} + 2y = 4 \ldots\ldots\ldots(2)$

Consider $\dfrac{d^2 y}{dz^2} - 3 \dfrac{dy}{dz} + 2y = 0 \ldots\ldots\ldots(3)$

Let $y = e^{mz}$ be trial solution of (3).

Then $\dfrac{dy}{dz} = me^{mz}, \dfrac{d^2 y}{dz^2} = m^2 e^{mz}$.

Using these, we get from (3),

$m^2 - 3m + 2 = 0$ which gives, m = 1,2.

Since the values of m are real and distinct, so

$y_{CF} = C_1 e^{2z} + C_2 e^z = C_1 x^2 + C_2 x$.

Again from (2), we have, (taking $\theta \equiv \dfrac{d}{dz}$)

$y_{PI} = \dfrac{1}{\theta^2 - 3\theta + 2}(4) = \dfrac{1}{\theta^2 - 3\theta + 2}(4e^{0 \times z})$

$= \dfrac{1}{0^2 - 3 \times 0 + 2} \times 4e^{0 \times z} = 2$

So the complete solution is:

$y = y_{CF} + y_{PI}$

$= C_1 e^{2z} + C_2 e^z + 2$

$= C_1 x^2 + C_2 x + 2 \ \left[\because x = e^z \right]$

4. (a) Given,

$\dfrac{dy}{dx} = -\left(\dfrac{x}{y} \right)^n \Rightarrow \dfrac{dy}{dx} = -\dfrac{x^n}{y^n} \ldots\ldots\ldots(i)$

For $n = -1$; we have from (i) $\dfrac{dy}{dx} = -\dfrac{x^{-1}}{y^{-1}}$

$\Rightarrow \dfrac{dy}{dx} = -\dfrac{y}{x} \Rightarrow \dfrac{1}{y} dy = \dfrac{-1}{x} dx$

Integrating both side; we get

$\ln y = -\ln x + \ln c \Rightarrow \ln(xy) = \ln c \Rightarrow xy = c^2$,

which represents family of hyperbolas.

For n = 1; we have from (i), $\dfrac{dy}{dx} = -\dfrac{x}{y}$

$\Rightarrow ydy = -xdx$.

Integrating both sides; we get

$\dfrac{y^2}{2} = -\dfrac{x^2}{2} + \dfrac{k^2}{2} \Rightarrow x^2 + y^2 = k^2$, which repre-

sents a family of circles.

5. 5.25

The given differential equation,

$x^2 \dfrac{d^2 y}{dx^2} - 3x \dfrac{dy}{dx} + 3y = 0 \ldots\ldots(1)$

Let us put log$x$ = z.

Then, $\dfrac{dy}{dz} = x \dfrac{dy}{dx}$ and $x^2 \dfrac{d^2 y}{dx^2} = \dfrac{d^2 y}{dz^2} - \dfrac{dy}{dz}$.

So, (1) becomes,

$\dfrac{d^2 y}{dz^2} - \dfrac{dy}{dz} - 3 \dfrac{dy}{dz} + 3y = 0$

$\Rightarrow \dfrac{d^2 y}{dz^2} - 4 \dfrac{dy}{dz} + 3y = 0 \ldots\ldots\ldots(2)$

Let $y = e^{mz}$ be trial solution of (2).

Then, $\dfrac{dy}{dz} = me^{mz}, \dfrac{d^2 y}{dz^2} = m^2 e^{mz}$.

Using these, we get from (2),

$m^2 - 4m + 3 = 0$ which gives, m = 1,3.

Since the values of m are real and distinct, so solution of (2) is given by:

$y = C_1 e^{3z} + C_2 e^z = C_1 x^3 + C_2 x \ldots\ldots\ldots(3)$

Now, $y(1) = 1 \Rightarrow 1 = C_1 + C_2$ (using(3))$\ldots\ldots(4)$

& $y(2) = 14 \Rightarrow 14 = 2C_1 + 8C_2$ (using(3))

$\Rightarrow C_1 + 4C_2 = 7 \ldots\ldots(5)$

Solving (4) and (5), we get, $C_1 = -1, C_2 = 2$.

Thus, (3) becomes, $y = (-1)x + (2)x^3$.

$\therefore y(1.5) = (-1) \times (1.5) + 2 \times (1.5)^3 = 5.25$.

6. (a)

$\dfrac{dy}{dx} x \ln x = y$

$\Rightarrow \dfrac{dy}{y} = \dfrac{\frac{1}{x} dx}{\ln x} \Rightarrow \int \dfrac{dy}{y} = \int \dfrac{\frac{1}{x}}{\ln x} dx + \ln k$

$\Rightarrow \int \dfrac{dy}{y} = \int \dfrac{1}{t} dt + \ln k$

$\left[\text{putting } \ln x = t \text{ so that } \dfrac{1}{x} dx = dt \right]$

$\Rightarrow \ln y = \ln t + \ln k = \ln(tk) \Rightarrow y = tk \Rightarrow y = k \ln x$.

7. 6

The given differential equation,

$$x^2 \frac{d^2y}{dx^2} - 2x\frac{dy}{dx} + 2y = 0........(1)$$

Let us put $\log x = z$.

Then, $\frac{dy}{dz} = x\frac{dy}{dx}$ and $x^2\frac{d^2y}{dx^2} = \frac{d^2y}{dz^2} - \frac{dy}{dz}$.

So, (1) becomes,

$$\frac{d^2y}{dz^2} - \frac{dy}{dz} - 2\frac{dy}{dz} + 2y = 0$$

$$\Rightarrow \frac{d^2y}{dz^2} - 3\frac{dy}{dz} + 2y = 0.............(2)$$

Let $y = e^{mz}$ be trial solution of (2).

Then, $\frac{dy}{dz} = me^{mz}, \frac{d^2y}{dz^2} = m^2e^{mz}$.

Using these, we get from (2),

$m^2 - 3m + 2 = 0$ which gives, m = 1,2.

Since the values of m are real and distinct, so solution of (2) is given by:

$$y = C_1 e^{2z} + C_2 e^z = C_1 x^2 + C_2 x.............(3)$$

Now, $y(1) = 0 \Rightarrow 0 = C_1 + C_2$ (using(3)).........(4)

& $y(2) = 2 \Rightarrow 2 = 4C_1 + 2C_2$ (using(3))

$$\Rightarrow 2C_1 + C_2 = 1.......(5)$$

Solving (4) and (5), we get, $C_1 = 1, C_2 = -1$.

Thus, (3) becomes, $y = x^2 - x$.

$\therefore y(3) = 3^2 - 3 = 6$.

TOPIC: PARTIAL DIFFERENTIAL EQUATIONS

1. The partial differential equation

$$\frac{\partial^2 u}{\partial x^2} - C^2\left(\frac{\partial^2 u}{\partial x^2} + \frac{\partial^2 u}{\partial y^2}\right) = 0; \quad \text{where} \quad C \neq 0 \quad \text{is}$$

known as

(a) Wave equation   (b) Poisson's equation
(c) Laplace equation (d) Heat equation

[EE GATE 2019]

Answer key				
1. (a)				

TOPIC: COMPLEX ANALYSIS

1. A harmonic function is analytic if it satisfies the Laplace equation. If $u(x,y) = 2x^2 - 2y^2 + 4xy$ is a harmonic function, then its conjugate harmonic function $v(x,y)$ is

(a) $-4xy + 2y^2 - 2x^2 + $ constant

(b) $4xy - 2x^2 + 2y^2 +$ constant

(c) $2x^2 - 2y^2 + xy +$ constant

(d) $-4y^2 - 4xy +$ constant

[ME GATE 2019]

2. An analytic function $f(z)$ of complex variable $z = x + iy$ may be written as $f(z) = u(x,y) + iv(x,y)$. Then $u(x,y)$ and $v(x,y)$ must satisfy

(a) $\frac{\partial u}{\partial x} = \frac{\partial v}{\partial y}, \frac{\partial u}{\partial y} = \frac{\partial v}{\partial x}$

(b) $\frac{\partial u}{\partial x} = -\frac{\partial v}{\partial y}, \frac{\partial u}{\partial y} = \frac{\partial v}{\partial x}$

(c) $\frac{\partial u}{\partial x} = -\frac{\partial v}{\partial y}, \frac{\partial u}{\partial y} = -\frac{\partial v}{\partial x}$

(d) $\frac{\partial u}{\partial x} = \frac{\partial v}{\partial y}, \frac{\partial u}{\partial y} = -\frac{\partial v}{\partial x}$

[ME GATE 2019]

3. A complex function $f(z) = u(x,y) + iv(x,y)$ and its complex conjugate $\overline{f}(z) = u(x,y) - iv(x,y)$ are both analytic in the entire complex plane, where $z = x + iy$ and $i = \sqrt{-1}$. The function f is then given by

(a) $f(z) = x + iy$ (b) $f(z) = x^2 - y^2 + 2ixy$

(c) $f(z) = $ constant   (d) $f(z) = x^2 + y^2$

[IN GATE 2019]

4. Which one of the following functions is analytic over the entire complex plane?

(a) $\ln(z)$ (b) $\cos(z)$ (c) $e^{\frac{1}{z}}$ (d) $\frac{1}{1-z}$

[EC GATE 2019]

5. The value of contour integral $\frac{1}{2\pi j}\oint\left(z + \frac{1}{z}\right)^2 dz$ evaluated over the unit circle $|z| = 1$ is _____?

[EC GATE 2019]

6. Which one of the following functions is analytic in the region $|z| \leq 1$.

(a) $\dfrac{z^2 - 1}{z + j0.5}$ (b) $\dfrac{z^2 - 1}{z + 2}$

(c) $\dfrac{z^2 - 1}{z - 0.5}$ (d) $\dfrac{z^2 - 1}{z}$

[EE GATE 2019]

7. The closed loop integral $\displaystyle\oint_{|z|=5} \dfrac{z^3 + z^2 + 8}{z + 2} dz$

evaluated counterclockwise, is

(a) $4j\pi$ (b) $-4j\pi$

(c) $8j\pi$ (d) $-8j\pi$

[EE GATE 2019]

Answer key

1. (b)	2. (d)	3. (c)	4. (b)	5. 0
6. (b)	7. (c)			

Explanation

1. (b)

$u(x, y) = 2x^2 - 2y^2 + 4xy$

$\Rightarrow \dfrac{\partial u}{\partial x} = 4x + 4y, \dfrac{\partial u}{\partial y} = -4y + 4x.$

$v = v(x, y)$

$\Rightarrow dv = \dfrac{\partial v}{\partial x} dx + \dfrac{\partial v}{\partial y} dy$

$\Rightarrow dv = \left(-\dfrac{\partial u}{\partial y}\right) dx + \left(\dfrac{\partial u}{\partial x}\right) dy$

(by the CR equations)

$\Rightarrow dv = Mdx + Ndy$ (say)

$\Rightarrow \int dv = \int Mdx + \int Ndy + C$............(1)

Now $\int Mdx$

$= \int -(-4y + 4x) dx = \int (4y - 4x) dx$

$= 4yx - 4 \times \dfrac{x^2}{2} = 4yx - 2x^2,$

$\int Ndy$

$= \int (4x + 4y) dy = 4xy + 4 \times \dfrac{y^2}{2} = 4xy + 2y^2.$

$\therefore$ (1) gives, $v(x, y) = 4xy - 2x^2 + 2y^2 + C$

2. (d) By the Cauchy Riemann equations,

$\dfrac{\partial u}{\partial x} = \dfrac{\partial v}{\partial y}, \dfrac{\partial u}{\partial y} = -\dfrac{\partial v}{\partial x}.$

3. (c) Option (a) is false, since $f(z)$ is analytic but $\overline{f}(z)$ is not analytic.

Option (b) is false, since $f(z)$ is analytic, but $\overline{f}(z)$ is not analytic.

Option (d) is false, since $f(z)$ is not analytic.

Suppose $f(z)$ = constant and $f(z) = c_1 + ic_2$ where c_1 & c_2 are real constants.

Then, $u(x, y) = c_1, v(x, y) = c_2$.

$\therefore \dfrac{\partial u}{\partial x} = 0, \dfrac{\partial u}{\partial y} = 0, \dfrac{\partial v}{\partial x} = 0, \dfrac{\partial v}{\partial y} = 0.$

Thus, the CR equations are satisfied by $f(z)$.

$\therefore f(z) = c_1 + ic_2$ is analytic.

Similarly, conjugate of $f(z)$, i.e, $\overline{f}(z) = c_1 - ic_2$ is also analytic.

So option (c) is correct.

4. (b) $f(z) = \ln(z)$ is not analytic at z = 0 since

$f'(z) = \dfrac{1}{z}$ does not exist at $z = 0$.

$f(z) = e^{\frac{1}{z}}$ is not analytic at $z = 0$ as

$f'(z) = -\dfrac{1}{z^2} e^{\frac{1}{z}}$ does not exist at $z = 0$.

$f(z) = \dfrac{1}{1 - z}$ is not analytic at z = 1 since

$f'(z) = \dfrac{1}{(1 - z)^2}$ does not exist at $z = 1$.

Thus, options (a), (c), and (d) are not correct.

5. 0

$\dfrac{1}{2\pi j} \oint \left(z + \dfrac{1}{z}\right)^2 dz$

$= \dfrac{1}{2\pi j} \oint \dfrac{(z^2 + 1)^2}{z^2} dz$

$= \dfrac{1}{2\pi j} \oint \dfrac{f(z)}{(z - 0)^2} dz \left[\text{where } f(z) = (z^2 + 1)^2\right]$

$= \dfrac{1}{2\pi j} \times \dfrac{2\pi j}{1!} f'(z)\big|_{z=0}$

[by Cauchy's integral formula on higher order derivatives as the function as z = 0 is a singularity of $\dfrac{\left(z^2+1\right)^2}{z^2}$ which lies inside the

circle $|z|=1$]

$$= 4z\left(z^2+1\right)\Big|_{z=0} = 0.$$

6. (b) Consider the function $f(z)=\dfrac{z^2-1}{z+2}$.

Clearly $z=-2$ is a singular point of $f(z)$ and lies outside $|z|=1$

$\therefore \dfrac{z^2-1}{z+2}$ is analytic in the region $|z|\le 1$

7. (c) Clearly $\dfrac{z^3+z^2+8}{z+2}$ has a singularity at

$z=-2$ which lies inside the circle $C:|z|=5$. Therefore,

$$\oint_{|z|=5}\dfrac{z^3+z^2+8}{z+2}dz$$

$$= \oint_{|z|=5}\dfrac{f(z)}{z-(-2)}dz \text{ [where } f(z)=z^3+z^2+8]$$

$$= 2\pi j\times f(-2) \text{ [by Cauchy's integral formula]}$$

$$= 2\pi j\times\left((-2)^3+(-2)^2+8\right)=8\pi j.$$

TOPIC: NUMERICAL ANALYSIS

1. Evaluation of $\int_2^3 x^3 dx$ using a 2-equal-segment trapezoidal rule gives a value of _____?

[ME GATE 2019]

2. The derivative of $f(x)=\cos x$ can be estimated using the approximation given below:

$$f'(x)=\dfrac{f(x+h)-f(x-h)}{2h}.$$

The percentage error is calculated as given below:

$$\dfrac{\text{Exact value} - \text{Approximate value}}{\text{Exact value}}\times 100.$$

The percentage error in the derivative of $f(x)$

at $x=\dfrac{\pi}{6}$ radian, choosing $h=0.1$ radian, is

(a) >5%   (b) >0.1 % and <1%
(c) <0.1%   (d) >1% and <5%

[ME GATE 2019]

3. A series of perpendicular off sets taken from a curved boundary wall to a straight curve line at an interval of 6 m are 1.22, 1.67, 2.04, 2.34,

2.14, 1.87, and 1.15 m. Then area (in m2, round of to two decimal places) boundary by the survey line, curved boundary wall, the 1st and last offsets, determined using Simpson's 1/3rd rule is _____?

[CE GATE 2019]

Answer key				
1. 63	**2.** (b)	**3.** 68.50		

Explanation

1. 63

The Trapezoidal rule gives:

$$\int_a^b f(x)\,dx = \dfrac{h}{2}[(y_0+y_n)+2(y_1+y_2+.....+y_{n-1})]$$

where step size, $h=\dfrac{b-a}{n}$.

Here, $f(x)=x^3, a=2, b=4, n=2$.

$$\therefore h=\dfrac{4-2}{2}=1$$

x	2	3	4
$f(x)=x^3$	8 (y_0)	27 (y_1)	64 (y_2)

$$\therefore \int_2^4 x^3 dx$$

$$=\dfrac{h}{2}[(y_0+y_n)+2y_1]=\dfrac{1}{2}[(8+64)+2\times 27]=63.$$

2. (b)

$f(x)=\cos x$

$$\Rightarrow f'(x)=-\sin x \Rightarrow f'(x)\Big|_{x=\frac{\pi}{6}}=-\sin\dfrac{\pi}{6}=-0.5.$$

Thus, the exact value of the derivative = –0.5. The approximate value of the derivative

$$=f'(x)\approx \dfrac{f(x+h)-f(x-h)}{2h}$$

$$\Rightarrow f'(x)\approx \dfrac{\cos(x+h)-\cos(x-h)}{2h}$$

$$\Rightarrow f'\left(\dfrac{\pi}{6}\right)\approx \dfrac{\cos\left(\dfrac{\pi}{6}+0.1\right)-\cos\left(\dfrac{\pi}{6}-0.1\right)}{2\times 0.1} \text{ (}\because h=0.1)$$

$$=\dfrac{-2\sin\left(\dfrac{\pi}{6}\right)\sin(0.1)}{2\times 0.1}=\dfrac{-0.5\times\sin(0.1)}{0.1}\approx -0.4992$$

Thus, $f'\left(\dfrac{\pi}{6}\right)\approx -0.4992$ (Approximate value)

$\therefore$ Percentage error

$$= \left[\frac{-0.5-(0.4992)}{-0.5}\right]\times100 = 0.16\% \ .$$

3. 68.50

Simpson's 1/3rd rule gives:

$$\int_{x_0}^{x_n} f(x)\ dx = \frac{h}{3}[(y_0+y_n)+2(y_2+y_4+\ldots+y_{n-2})$$
$$+4(y_1+y_3+\ldots+y_{n-1})]$$

Here, $h=6, n=6$.

Therefore, the required area

$$= \frac{h}{3}[(y_0+y_6)+2(y_2+y_4)+4(y_1+y_3+y_5)]$$

$$= \frac{6}{3}[(1.22+1.15)+2(2.04+2.14)$$
$$+4(1.67+2.34+1.87)]$$

$$= 2\times[2.37+8.36+23.52] = 68.50\,\text{m}^2$$

TOPIC: VECTORS

1. The directional derivative of the function $f(x,y)=x^2+y^2$ along a line directed from $(0,0)$ to $(1,1)$ evaluated at the point $x=1, y=1$ is

(a) $4\sqrt{2}$ (b) $\sqrt{2}$

(c) $2\sqrt{2}$ (d) $\sqrt{2}$

[ME GATE 2019]

2. Given a vector $\vec{u} = \frac{1}{3}\left(-y^3\hat{i}+x^3\hat{j}+z^3\hat{k}\right)$ and $\hat{n}$ as the unit normal vector to the surface of the hemisphere S: $x^2+y^2+z^2=1, z\geq0$, the value of integral $\int(\vec{\nabla}\times\vec{u}).\hat{n}\,ds$ evaluated on the curved surface of the hemisphere S is

(a) $\frac{\pi}{2}$ (b) $\frac{\pi}{3}$

(c) π (d) $-\frac{\pi}{2}$

[ME GATE 2019]

3. $\vec{a},\vec{b},\vec{c}$ are orthogonal vectors. Given that $\vec{a}=\hat{i}+2\hat{j}+5\hat{k}$ and $\vec{b}=\hat{i}+2\hat{j}-\hat{k}$, the vector $\vec{c}$ is

(a) $\hat{i}+2\hat{j}+3\hat{k}$ (b) $2\hat{i}+\hat{j}$

(c) $2\hat{i}-\hat{j}$ (d) $4\hat{k}$

[CS GATE 2019]

4. The vector function $\vec{A}$ is given by $\vec{A}=\vec{\nabla}u$ where $u(x,y,z)$ is a scalar function. Then $\left|\vec{\nabla}\times\vec{A}\right|$ is

(a) -1 (b) 0

(c) 1 (d) ∞

[CS GATE 2019]

5. If $f=2x^3+3y^2+4z$, the value of line integral $\int_C grad\ f.d\vec{r}$ evaluated over contour C formed by the segment $(-3,-3,2)\rightarrow(2,-3,2)\rightarrow(2,6,2)\rightarrow(2,6,-1)$ is _____ ?

[EE GATE 2019]

6. If $\vec{A}=2x\hat{i}+3y\hat{j}+4z\hat{k}$ and $u=x^2+y^2+z^2$, then $div(u\vec{A})$ at $(1,1,1)$ is _____ ?

[EE GATE 2019]

7. What is the curl of the vector field $\vec{f}=2x^2y\hat{i}+5z^2\hat{j}-4yz\hat{k}$?

(a) $-14z\hat{i}-2x^2\hat{k}$ (b) $6z\hat{i}+4x\hat{j}-2x^2\hat{k}$

(c) $6z\hat{i}+8xy\hat{j}+2x^2y\hat{k}$

(d) $-14z\hat{i}+6y\hat{j}+2x^2\hat{k}$

[CE GATE 2019]

Answer key				
1. (c)	2. (a)	3. (c)	4. (b)	5. 139
6. 45	7. (a)			

Explanation

1. (c) Given, $f(x,y)=x^2+y^2$.

$$\therefore \nabla f = \hat{i}\frac{\partial}{\partial x}(x^2+y^2)+\hat{j}\frac{\partial}{\partial y}(x^2+y^2)$$
$$= 2x\hat{i}+2y\hat{j}$$

The line directed from $(0,0)$ to $(1,1)$ is $(1-0)\hat{i}+(1-0)\hat{j}$ i.e; $\hat{i}+\hat{j}$.

$\therefore$ The directional derivative of $f(x,y)$ at $(1,1)$ in the direction of $\hat{i}+\hat{j}$

$$= \left(\nabla f\right)_{(1,1)}\cdot\frac{(\hat{i}+\hat{j})}{\left|\hat{i}+\hat{j}\right|}$$

$$= \left(2\hat{i}+2\hat{j}\right)\cdot\frac{(\hat{i}+\hat{j})}{\left|\hat{i}+\hat{j}\right|} = \frac{2\times1+2\times1}{\sqrt{1^2+1^2}} = \frac{4}{\sqrt{2}} = 2\sqrt{2}.$$

2. (a)

$$\int_S (\vec{\nabla} \times \vec{u}).\hat{n}\,ds$$

$$= \int_C \vec{u}.d\vec{r} \quad \text{[by Stokes' theorem]}$$

$$= \int_C \left[-\frac{y^3}{3}dx + \frac{x^3}{3}dy + \frac{z^3}{3}dz \right]$$

$$= \int_C \left[-\frac{y^3}{3}dx + \frac{x^3}{3}dy \right] \quad [\because z = 0 \text{ on } C]$$

$$= \iint_R \left[\frac{\partial}{\partial x}\left(\frac{x^3}{3}\right) - \frac{\partial}{\partial y}\left(-\frac{y^3}{3}\right) \right] dx\,dy \quad \text{[by Green's theorem]}$$

$$= \iint_R [x^2 + y^2]\,dx\,dy = \int_{\theta=0}^{2\pi}\int_{r=0}^{1} r^2 \times r\,dr\,d\theta$$

[putting $x = r\cos\theta, y = r\sin\theta$ so that Jacobian $= r$]

$$= \int_{\theta=0}^{2\pi} d\theta \times \int_{r=0}^{1} r^3\,dr = [\theta]_0^{2\pi} \times \left[\frac{r^4}{4}\right]_0^1 = 2\pi \times \frac{1}{4} = \frac{\pi}{2}.$$

3. (c) $\vec{a}, \vec{b}, \vec{c}$ are orthogonal vectors

$$\Rightarrow \vec{a}.\vec{b} = 0, \vec{b}.\vec{c} = 0, \vec{a}.\vec{c} = 0.$$

Take $\vec{c} = 2\hat{i} - \hat{j}$. Then,

$$\vec{b}.\vec{c} = (\hat{i} + 2\hat{j} - \hat{k}).(2\hat{i} - \hat{j}) = 2 + (-2) + 0 = 0,$$

$$\vec{a}.\vec{c} = (\hat{i} + 2\hat{j} + 5\hat{k}).(2\hat{i} - \hat{j}) = 2 + (-2) + 0 = 0.$$

4. (b) $\vec{\nabla} \times \vec{A} = \vec{\nabla} \times \vec{\nabla}u = curl(grad\,u) = \vec{0}.$

$$\therefore \left|\vec{\nabla} \times \vec{A}\right| = |\vec{0}| = 0.$$

5. 139

$$f = 2x^3 + 3y^2 + 4z$$

$$\Rightarrow grad\,f = 6x^2\hat{i} + 6y\hat{j} + 4\hat{k}$$

$$\Rightarrow gard\,f.d\vec{r} = 6x^2dx + 6ydy + 4dz$$

$$= d\left(2x^3 + 3y^2 + 4z\right)$$

Now *curl grad f*

$$= \begin{vmatrix} \hat{i} & \hat{j} & \hat{k} \\ \frac{\partial}{\partial x} & \frac{\partial}{\partial y} & \frac{\partial}{\partial z} \\ 6x^2 & 6y & 4 \end{vmatrix}$$

$$= \hat{i}(0-0) - \hat{j}(0-0) + \hat{k}(0-0) = \hat{0}$$

Hence, *grad f* is irrotational and so the line integral does not dependent on the path of integration but only depends on the end points of integral.

$$\therefore \int_C gard\,f.d\vec{r}$$

$$= \int_C d\left(2x^3 + 3y^2 + 4z\right)$$

$$= \int_{(-3,-3,2)}^{(2,-3,2)} d\left(2x^3 + 3y^2 + 4z\right) + \int_{(2,-3,2)}^{(2,6,2)} d\left(2x^3 + 3y^2 + 4z\right)$$

$$+ \int_{(2,6,2)}^{(2,6,-1)} d\left(2x^3 + 3y^2 + 4z\right)$$

$$= \int_{(-3,-3,2)}^{(2,6,-1)} d\left(2x^3 + 3y^2 + 4z\right)$$

$$= \left[2x^3 + 3y^2 + 4z\right]_{(-3,-3,2)}^{(2,6,-1)}$$

$$= (2\times2^3 + 3\times6^2 - 4) - (-2\times3^3 + 3\times(-3)^2 + 4\times2)$$

$$= 120 - (-19) = 139.$$

6. 45

$$div(u\vec{A})$$

$$= div(2x(x^2 + y^2 + z^2)\hat{i} + 3y(x^2 + y^2 + z^2)\hat{j}$$

$$+ 4z(x^2 + y^2 + z^2)\hat{k}$$

$$= \frac{\partial}{\partial x}(2x(x^2 + y^2 + z^2)) + \frac{\partial}{\partial y}(3y(x^2 + y^2 + z^2))$$

$$+ \frac{\partial}{\partial z}(4z(x^2 + y^2 + z^2))$$

$$= 2(x^2 + y^2 + z^2) + 4x^2 + 3(x^2 + y^2 + z^2) + 6y^2$$

$$+ 4(x^2 + y^2 + z^2) + 8z^2.$$

$$\therefore div(u\vec{A})\Big|_{(1,1,1)}$$

$$= 2(1+1+1) + 4 + 3(1+1+1) + 6 + 4(1+1+1) + 8$$

$$= 45.$$

7. (a)

curl $\vec{f}$

$$= \begin{vmatrix} \hat{i} & \hat{j} & \hat{k} \\ \frac{\partial}{\partial x} & \frac{\partial}{\partial y} & \frac{\partial}{\partial z} \\ 2x^2y & 5z^2 & -4yz \end{vmatrix}$$

$$= \hat{i}(-4z - 10z) - \hat{j}(0-0) + \hat{k}(0 - 2x^2)$$

$$= -14z\hat{i} - 2x^2\hat{k}.$$

TOPIC: PROBABILITY & STATISTICS

1. The lengths of a large stock of titanium rods follow a normal distribution with mean (μ) of 440 mm and a standard deviation (σ)

of 1 mm. What is percentage of rods whose lengths lie between 438 mm and 441 mm?

(a) 86.64% (b) 68.4%
(c) 99.75% (d) 81.85%

[ME GATE 2019]

2. If x is the mean of data 3, x, 2 and 4, then the mode is _____ ?

[ME GATE 2019]

3. The probability that a part manufactured by a company will be defective is 0.05. If such parts are selected randomly and inspected, then the probability that at least two parts will be defective is _____ ? (round off to two decimal places).

[ME GATE 2019]

4. An array of 25 distinct elements is to be sorted using quick sort. Assume that the pivot element is chosen uniformly at random. The probability that the pivot element gets placed in the worst possible location in the first round of partitioning (rounded off to 2 decimal places) is _____ ?

[CS GATE 2019]

5. Suppose Y is distributed uniformly in the open interval (1, 6). The probability that the polynomial $3x^2 + 6xy + 3y + 6$ has only real roots is (rounded off to 1 decimal places) _____? **[CS GATE 2019]**

6. A box has eight red balls and eight green balls. Two balls are drawn randomly in succession from the box without replacement. The probability that the first ball drawn is red and the second ball drawn is green is

(a) 4/15 (b) 7/16
(c) 1/2 (d) 8/15

[CS GATE 2019]

7. The function p(x) is given $p(x) = \dfrac{A}{x^\mu}$ where A and μ are constants with $\mu > 1$ and $1 \le x < \infty$ and p(x) = 0 for $-\infty < x < 1$. For p(x) to be probability density function, the value of A should be equal to

(a) $\mu - 1$ (b) $\mu + 1$ (c) $\dfrac{1}{\mu - 1}$ (d) $\dfrac{1}{\mu + 1}$

[CS GATE 2019]

8. If X and Y are random variables such that $E(2X + Y) = 0$ and $E(X + 2Y) = 33$, then $E(X) + E(Y) = $ ___?

[EC GATE 2019]

9. Let Z be an exponential random variable with mean 1. That is, the cumulative distribution function of Z is given by

$$F_z(x) = \begin{cases} 1 - e^{-x} & \text{if } x \ge 0 \\ 0 & \text{if } x < 0 \end{cases}$$ then prob. (Z >2 | Z >1), rounded off two decimal places, is equal to _____?

[EC GATE 2019]

10. The probability of a resistor being defective is 0.02. There are 50 such resistors in a circuit. The probability of two or more defective resistor in the circuit (round off to two decimal places) is _____ ?

[EE GATE 2019]

11. The probability density function of a continuous random variable distributed uniformly between x and $y (y > x)$ is

(a) $\dfrac{1}{x - y}$ (b) $x - y$

(c) $y - x$ (d) $\dfrac{1}{y - x}$

[CE GATE 2019]

Answer key				
1. (d)	**2.** 3	**3.** 0.17	**4.** 0.08	**5.** 0.8
6. (a)	**7.** (a)	**8.** 11	**9.** $\dfrac{1}{e}$	**10.** 0.26
11. (d)				

Explanation

1. (d) Here mean (μ) = 440 mm and SD (σ) = 1mm

Let X be the random variable denoting the lengths of rods.

Here, we need to find $P[438 < X < 441]$.

Since the standard normal variable,

$Z = \dfrac{X - \mu}{\sigma}$ so

$X = 438 \Rightarrow Z = \dfrac{438 - 440}{1} = -2$ and

$X = 441 \Rightarrow Z = \dfrac{441 - 440}{1} = 1.$

$\therefore P[438 < X < 441]$

$= P[-2 < Z < 1] = P[-2 < Z < 0] + P[0 < Z < 1]$

$= \left(\dfrac{95.44}{2}\right)\% + \left(\dfrac{68.26}{2}\right)\% = 81.85\%$

2. 3

Mean $= \dfrac{3+x+2+4}{4} \Rightarrow x = \dfrac{9+x}{4} \Rightarrow x = 3.$

$\therefore$ The data values are 3, 3, 2 and 4 among which '3' is most frequently used. Therefore, Mode = 3.

3. 0.17

Let X be the random variable denoting the number of defective parts.

Then X follows a Binomial distribution with number of trials $(n) = 15$ and probability of success $(p) = 0.05$. Therefore, $q = 1-0.05 = 0.95$.

So, $P[X \geq 2] = 1 - P[X < 2]$

$= 1 - \{P(X=0) + P(X=1)\}$

$= 1 - \{^{15}C_0 p^{15} q^0 + {}^{15}C_1 p^{14} q\} = 1 - \{p^{15} + 15 p^{14} q\}$

$= 1 - \{(0.05)^{15} + 15(0.05)^{14} \times 0.05\} \approx 0.17$

4. 0.08

Clearly, either first place or the last place is the worst possible location for the pivot element.

Hence, the required probability $= \dfrac{2}{25} = 0.08$

5. 0.8

Since Y is distributed uniformly in the open interval (1, 6), so the p.m.f of Y is:

$f(y) = \begin{cases} \dfrac{1}{5}, & 1 \leq y \leq 6 \\ 0, & \text{otherwise} \end{cases}$

The polynomial $3x^2 + 6xy + 3y + 6$ has only real roots

if $3x^2 + 6xy + 3y + 6 = 0$ has a real root

i.e; if $(6y)^2 - 4 \times 3 \times (3y+6) \geq 0$

i.e; if $(y-2)(y+1) \geq 0$

i.e; if $(y-2) \geq 0, (y+1) \geq 0$

  or $(y-2) \leq 0, (y+1) \leq 0$

i.e; if $y \geq 2, y \geq -1$ or $y \leq 2, y \leq -1$

i.e; if $y \geq 2$ or $y \leq -1$

$\therefore$ Prob.$(3x^2 + 3xy + 3y + 6$ has only real roots)

$= P(Y \leq -1) + P(Y \geq 2)$

$= \int_{-\infty}^{-1} f(y)\,dy + \int_2^\infty f(y)\,dy = 0 + \int_2^6 \dfrac{1}{5}\,dy$

$= \dfrac{1}{5}[y]_2^6 = \dfrac{6-2}{5} = \dfrac{4}{5} = 0.8.$

6. (a) The total number of balls = 8 + 8 = 16.

Let A = the event of drawing a red ball in 1st withdraw & B = event of drawing a green ball in 2nd withdraw. Since the balls are drawn without replacement, so the required probability

$= P(A \cap B) = P(A) \times P(B/A)$

$= \dfrac{^8C_1}{^{16}C_1} \times \dfrac{^8C_1}{^{15}C_1} = \dfrac{1}{2} \times \dfrac{8}{15} = \dfrac{4}{15}.$

7. (a) $p(x)$ is a p.d.f

$\Rightarrow \int_{-\infty}^\infty p(x)dx = 1 \Rightarrow \int_{-\infty}^1 0\,dx + \int_1^\infty \dfrac{A}{x^\mu}dx = 1$

$\Rightarrow A\int_1^\infty x^{-\mu}dx = 1 \Rightarrow A\left[\dfrac{1}{(1-\mu)x^{\mu-1}}\right]_1^\infty = 1$

$\Rightarrow \dfrac{A}{1-\mu}[0-1] = 1 \Rightarrow A = \mu - 1.$

$\left(\because \mu > 1 \Rightarrow \dfrac{1}{x^{\mu-1}} \to 0 \text{ when } x \to \infty\right)$

8. 11

$E(2X+Y) = 0 \Rightarrow 2E(X) + E(Y) = 0 ....(i)$

$E(X+2Y) = 33 \Rightarrow E(X) + 2E(Y) = 33 ....(ii)$

Solving (i) and (ii) ; we have,

$E(X) = -11$ and $E(Y) = 22.$

$\therefore E(X) + E(Y) = -11 + 22 = 11.$

9. $\dfrac{1}{e}$

Given, $F_z(x) = \begin{cases} 1 - e^{-x} & \text{if } x \geq 0 \\ 0 & \text{if } x < 0 \end{cases}$

$\therefore F_z'(x) = f(x) = \begin{cases} e^x & \text{if } x \geq 0 \\ 0 & \text{if } x < 0 \end{cases}$

Now Prob$(z > 2 \mid z > 1)$

$$= \frac{\text{Prob}(z > 2 \cap z > 1)}{\text{Prob}(z > 1)}$$

$$= \frac{\text{Prob}(z > 2)}{\text{Prob}[z > 1]} = \frac{\int_{2}^{\infty} f(x)dx}{\int_{1}^{\infty} f(x)dx} = \frac{\int_{2}^{\infty} e^{-x}dx}{\int_{1}^{\infty} e^{-x}dx}$$

$$= \frac{\left[-e^{-x}\right]_{2}^{\infty}}{\left[-e^{-x}\right]_{1}^{\infty}} = \frac{0 + e^{-2}}{0 + e^{-1}} = \frac{e^{-2}}{e^{-1}} = \frac{1}{e}.$$

10. 0.26 The probability being defective, $p = 0.02$ and the number of resistor, $n = 50$.

$$\therefore \lambda = np = 50 \times 0.02 = 50 \times \frac{2}{100} = 1 \cdot$$

Let "X" be the random variable denoting the number of defective resistors. Clearly X follows the Poisson distribution with mean $\lambda = 1$.

So, the required probability

$$= P[x \geq 2] = 1 - P[x < 2]$$

$$= 1 - [P(x = 0) + P(x = 1)]$$

$$= 1 - \left[\frac{e^{-\lambda}\lambda^{0}}{0!} + \frac{e^{-\lambda}\lambda}{1!}\right]$$

$$= 1 - e^{-\lambda}[1 + \lambda] = 1 - e^{-1}[1 + 1] = 1 - \frac{2}{e} \approx 0.26$$

11. (d) The result follows from the definition of uniform probability distribution.

TOPICS: GRAPH THEORY

1. Let G be an undirected complete graph on n vertices, where n > 2. Then, the number of different Hamiltonian cycles in G is equal to

(a) $n!$

(b) $\dfrac{(n-1)!}{2}$

(c) 1

(d) $(n-1)!$

[CS GATE 2019]

2. Let T be a full binary tree with eight leaves (A full binary tree has every level full.). Suppose two leaves a and b of T are chosen uniformly and independently at random. The expected value of the distance between a and b in T (i.e., the number of edges in the unique path between a and b) is (rounded off to two decimal places)_____

[CS GATE 2019]

Explanation

1. (b), (c) In case of labeled nodes, for an undirected complete graph G, we have, number of Hamiltonian cycles $= \dfrac{(n-1)!}{2}$.

In case of unlabeled nodes, for an undirected complete graph G, every Hamiltonian cycles will be similar and so number of Hamiltonian cycles $= 1$.

2. 4.25

For a full binary tree T with eight leaves, the number of pairs in which two levels can be chosen independently$= 8 \times 8 = 64$.

Among them 8 pairs will at distance 0 unit from each others, 8 pairs will be at distance 2 unit from each other (Children of the same parent), 16 pairs will be distance 4 unit from each other and 32 pairs will be at distance 6 unit from each others.

Hence, the total distance $= 8 \times 0 + 8 \times 2 + 16 \times 4 + 32 \times 6 = 272$

So, the expected distance $= 272/64 = 4.25$.

TOPIC: LAPLACE TRANSFORMS

1. Let Y(s) be the unit-step response of a casual system having a transfer function $G(s) = \dfrac{3 - s}{(s+1)(s+3)}$ That is, $Y(s) = \dfrac{G(s)}{s}$.

The forced response of the system is

(a) $u(t) - 2e^{-t}u(t) + e^{-3t}u(t)$

(b) $2u(t)$

(c) $u(t)$

(d) $2u(t) - 2e^{-t}u(t) + e^{-3t}u(t)$

[EC GATE 2019]

2. The inverse Laplace transform of

$H(s) = \dfrac{s+3}{s^2 + 2s + 1}$ for $t \geq 0$ is

(a) $3te^{-t} + e^{-t}$

(b) $3e^{-t}$

(c) $4te^{-t} + e^{-t}$

(d) $2te^{-t} + e^{-t}$

[EE GATE 2019]

3. The output response of a system is denoted as y(t), and its Laplace transform is given by

$$Y(s) = \frac{10}{s\left(s^2 + s + 100\sqrt{2}\right)}.$$ The steady state value of y(t) is

(a) $\dfrac{1}{100\sqrt{2}}$

(b) $10\sqrt{2}$

(c) $\dfrac{1}{10\sqrt{2}}$

(d) $100\sqrt{2}$

[EE GATE 2019]

Answer key				
1. (a)	**2.** (d)	**3.** (c)		

Explanation

1. (a)

$$Y(s) = \frac{G(s)}{s} = \frac{3-s}{s(s+1)(s+3)}$$

$$= \frac{4-(1+s)}{s(s+1)(s+3)} = \frac{4}{s(s+1)(s+3)} - \frac{1}{s(s+3)}$$

$$= 4\frac{(s+1)-s}{s(s+1)(s+3)} - \frac{(s+3)-s}{3s(s+3)}$$

$$= \frac{4}{s(s+3)} - \frac{4}{(s+1)(s+3)} - \frac{1}{3s} + \frac{1}{3(s+3)}$$

$$= \frac{4}{3}\frac{(s+3)-s}{s(s+3)} - 2\frac{(s+3)-(s+1)}{(s+1)(s+3)} - \frac{1}{3s} + \frac{1}{3(s+3)}$$

$$= \frac{4}{3s} - \frac{4}{3(s+3)} - \frac{2}{s+1} + \frac{2}{s+3} - \frac{1}{3s} + \frac{1}{3(s+3)}$$

$$= \frac{1}{s} - \frac{2}{s+1} + \frac{1}{s+3}$$

$$\therefore y(t) = L^{-1}\{Y(s)\}$$

$$= L^{-1}\left[\frac{1}{s}\right] - 2L^{-1}\left[\frac{1}{s+1}\right] + L^{-1}\left[\frac{1}{s+3}\right]$$

$$= u(t) - 2e^{-t}u(t) + e^{-3t}u(t).$$

2. (d)

$$L^{-1}[H(s)]$$

$$= L^{-1}\left[\frac{s+3}{s^2+2s+1}\right] = L^{-1}\left[\frac{s+3}{(s+1)^2}\right]$$

$$= L^{-1}\left[\frac{s+1+2}{(s+1)^2}\right] = L^{-1}\left[\frac{1}{s+1} + \frac{2}{(s+1)^2}\right]$$

$$= L^{-1}\left[\frac{1}{s+1}\right] + 2L^{-1}\left[\frac{1}{(s+1)^2}\right]$$

$$= e^{-t} + 2te^{-t}.$$

3. (c) We need to find steady state values of $y(t)$ i.e, $y(\infty)$.

By the final value theorem,

$y(\infty)$

$$= \lim_{s\to0} sY(s) = \lim_{s\to0} s\frac{10}{s\left(s^2+s+100\sqrt{2}\right)}$$

$$= \lim_{s\to0}\frac{10}{\left(s^2+s+100\sqrt{2}\right)} = \frac{10}{0+0+100\sqrt{2}} = \frac{1}{10\sqrt{2}}.$$

GATE 2020 SOLVED PAPERS

TOPIC: LINEAR ALGEBRA

1. Consider the matrix $M = \begin{bmatrix} 1 & -1 & 0 \\ 1 & -2 & 1 \\ 0 & -1 & 1 \end{bmatrix}$. One of the eigenvectors of M is _____?

(a) $\begin{bmatrix} -1 \\ 1 \\ -1 \end{bmatrix}$ (b) $\begin{bmatrix} 1 \\ -1 \\ 1 \end{bmatrix}$

(c) $\begin{bmatrix} 1 \\ 1 \\ 1 \end{bmatrix}$ (d) $\begin{bmatrix} 1 \\ 1 \\ -1 \end{bmatrix}$

[IN GATE 2020]

2. Let A and B be two $n \times n$ matrices over real numbers. Let rank (M) and det (M) denote the rank and determinant of a matrix M, respectively. Consider the following statements:

I. $rank(AB) = rank(A) \times rank(B)$

II. $det(AB) = det(A) \times det(B)$

III. $rank(A + B) \le rank(A) + rank(B)$

IV. $det(A + B) \le det(A) + det(B)$

Which of the above statements are true?

(a) III and IV only   (b) II and III only

(c) I and II only   (d) I and IV only

[CSE/IT GATE 2020]

3. If $V_1, V_2, \ldots, V_6$ are six vectors in R^4, which of the following statements is false?

(a) If $\{V_1, V_3, V_5, V_6\}$ spans R^4, then it forms a basis for R^4

(b) Any four of these vectors form a basis for R^4

(c) These vectors are not linearly independent

(d) It is not necessary that these vectors span R^4

[EC GATE 2020]

4. Consider the following system of linear equations:

$$x_1 + 2x_2 = b_1, 2x_1 + 4x_2 = b_2, 3x_1 + 7x_2 = b_3, 3x_1 + 9x_2 = b_4$$

Which of the following conditions ensures that a solution exists for the above system?

(a) $b_3 = 2b_1, 3b_1 - 6b_3 + b_4 = 0$

(b) $b_3 = 2b_1, 6b_1 - 3b_3 + b_4 = 0$

(c) $b_2 = 2b_1, 3b_1 - 6b_3 + b_4 = 0$

(d) $b_2 = 2b_1, 6b_1 - 3b_3 + b_4 = 0$ **[EC GATE 2020]**

5. The number of purely real elements in a lower triangular representation of the given 3×3 matrix, obtained through the given decomposition is _____?

$$\begin{bmatrix} 2 & 3 & 3 \\ 3 & 2 & 1 \\ 3 & 1 & 7 \end{bmatrix} = \begin{bmatrix} a_{11} & 0 & 0 \\ a_{12} & a_{22} & 0 \\ a_{13} & a_{23} & a_{33} \end{bmatrix} \begin{bmatrix} a_{11} & a_{12} & a_{13} \\ 0 & a_{22} & a_{23} \\ 0 & 0 & a_{33} \end{bmatrix}$$

(a) 6   (b) 5   (c) 8   (d) 9

[EE GATE 2020]

6. A matrix P is given below:

$$P = \begin{bmatrix} 0 & 1 & 3 & 0 \\ -2 & 3 & 0 & 4 \\ 0 & 0 & 6 & 1 \\ 0 & 0 & 1 & 6 \end{bmatrix}$$ The eigenvalues of P are

(a) 1, 2, 3, 4   (b) 0, 3, 6, 6

(c) 3, 4, 5, 7   (d) 1, 2, 5, 7

[CE GATE 2020]

7. Consider the system of equations

The value of x_3 (round off to nearest integer), is __?

$$\begin{bmatrix} 1 & 3 & 2 \\ 2 & 2 & -3 \\ 4 & 4 & -6 \\ 2 & 5 & 2 \end{bmatrix}\begin{bmatrix} x_1 \\ x_2 \\ x_3 \end{bmatrix} = \begin{bmatrix} 1 \\ 1 \\ 2 \\ 1 \end{bmatrix}$$ **[CE GATE 2020]**

8. A matrix P is decomposed into its symmetric part S and skew-symmetric part V. If

$$S = \begin{bmatrix} -4 & 4 & 2 \\ 4 & 3 & \frac{7}{2} \\ 2 & \frac{7}{2} & 2 \end{bmatrix}, V = \begin{bmatrix} 0 & -2 & 3 \\ 2 & 0 & \frac{7}{2} \\ -3 & -\frac{7}{2} & 2 \end{bmatrix}, \text{ then } P \text{ is}$$

(a) $\begin{bmatrix} -4 & 2 & 5 \\ 6 & 3 & 7 \\ -1 & 0 & 2 \end{bmatrix}$ (b) $\begin{bmatrix} -2 & \frac{9}{2} & -1 \\ -1 & \frac{81}{4} & 11 \\ -2 & \frac{45}{2} & \frac{73}{4} \end{bmatrix}$

(c) $\begin{bmatrix} 4 & -6 & 1 \\ -2 & -3 & 0 \\ -5 & -7 & -2 \end{bmatrix}$ (d) $\begin{bmatrix} -4 & 6 & -1 \\ 2 & 3 & 0 \\ 5 & 7 & 2 \end{bmatrix}$

[ME GATE 2020]

9. Let I be a 100 dimensional identity matrix and E be the set of all its distinct real eigenvalues. The number of elements in E is _____ ?

[ME GATE 2020]

Answers key				
1. (c)	**2.** (b)	**3.** (b)	**4.** (d)	**5.** (c)
6. (d)	**7.** (3)	**8.** (a)	**9.** (1)	

Explanations

1. (c) $\det(M - \lambda I) = 0$

$$\Rightarrow \begin{vmatrix} 1-\lambda & -1 & 0 \\ 1 & -2-\lambda & 1 \\ 0 & -1 & 1-\lambda \end{vmatrix} = 0$$

$$\Rightarrow \begin{vmatrix} -\lambda & -1 & 0 \\ -\lambda & -2-\lambda & 1 \\ -\lambda & -1 & 1-\lambda \end{vmatrix} = 0 \text{ (by } C_1 \to C_1 + C_2 + C_3)$$

$$\Rightarrow -\lambda \begin{vmatrix} 1 & -1 & 0 \\ 1 & -2-\lambda & 1 \\ 1 & -1 & 1-\lambda \end{vmatrix} = 0$$

$\Rightarrow \lambda = 0$ is an eigenvalue

Now, $MX = \lambda X$

$$\Rightarrow \begin{bmatrix} 1 & -1 & 0 \\ 1 & -2 & 1 \\ 0 & -1 & 1 \end{bmatrix}\begin{bmatrix} x \\ y \\ z \end{bmatrix} = 0\begin{bmatrix} x \\ y \\ z \end{bmatrix}$$

$$\Rightarrow \begin{bmatrix} x-y \\ x-2y+z \\ -y+z \end{bmatrix} = \begin{bmatrix} 0 \\ 0 \\ 0 \end{bmatrix}$$

$\Rightarrow x-y=0, x-2y+z=0, -y+z=0$

$\Rightarrow x=y=z$

Therefore, the eigenvector corresponding to $\lambda = 0$ is

$$X = \begin{bmatrix} x \\ y \\ z \end{bmatrix} = \begin{bmatrix} x \\ x \\ x \end{bmatrix} = \begin{bmatrix} 1 \\ 1 \\ 1 \end{bmatrix} \text{ (for } x = 1)$$

2. (b) Let $A = \begin{pmatrix} 1 & 0 \\ 0 & 0 \end{pmatrix}$ and $B = \begin{pmatrix} 0 & 0 \\ 0 & 1 \end{pmatrix}$.

Then, $AB = \begin{pmatrix} 0 & 0 \\ 0 & 0 \end{pmatrix} = O$.

So $rank(AB) = 0$ but $rank(A) = 1 = rank(B)$.

Therefore, $rank(AB) \neq rank(A) \times rank(B)$.

Hence, the statement I is not correct.

Now, $A + B = \begin{pmatrix} 1 & 0 \\ 0 & 1 \end{pmatrix}$.

So $\det(A+B) = 1$ but $\det(A) = 0 = \det(B)$.

Therefore, $\det(A+B) > \det(A) + \det(B)$.

Hence, the statement IV is not correct.

3. (b) Let $V_1 = (1,0,1,0), V_2 = (-1,0,-1,0), V_3 = (0,0,1,1),$

$V_4 = (0,0,-1,-1)$.

Then, $V_1 + V_2 + V_3 + V_4 = (0,0,0,0) = O$, which shows that the set $\{V_1, V_2, V_3, V_4\}$ is not linearly independent and hence does not form a basis of R^4.

4. (d) The given system of equations can be expressed as

$$AX = B,$$

where $A = \begin{bmatrix} 1 & 2 \\ 2 & 4 \\ 3 & 7 \\ 3 & 9 \end{bmatrix}, X = \begin{bmatrix} x_1 \\ x_2 \end{bmatrix}, B = \begin{bmatrix} b_1 \\ b_2 \\ b_3 \\ b_4 \end{bmatrix}$

So, the augmented matrix

$$= [A:B] = \begin{bmatrix} 1 & 2 & b_1 \\ 2 & 4 & b_2 \\ 3 & 7 & b_3 \\ 3 & 9 & b_4 \end{bmatrix}$$

$$\sim \begin{bmatrix} 1 & 2 & b_1 \\ 0 & 0 & b_2 - 2b_1 \\ 0 & 1 & b_3 - 3b_1 \\ 0 & 3 & b_4 - 3b_1 \end{bmatrix} \begin{pmatrix} \text{by } R_2 \to R_2 - 2R_1, \\ R_3 \to R_3 - 3R_1, \\ R_4 \to R_4 - 3R_1 \end{pmatrix}$$

$$\sim \begin{bmatrix} 1 & 2 & b_1 \\ 0 & 3 & b_4 - 3b_1 \\ 0 & 1 & b_3 - 3b_1 \\ 0 & 0 & b_2 - 2b_1 \end{bmatrix} \left(\text{by } R_2 \leftrightarrow R_4 \right)$$

$$\sim \begin{bmatrix} 1 & 2 & b_1 \\ 0 & 0 & 6b_1 - 3b_3 + b_4 \\ 0 & 1 & b_3 - 3b_1 \\ 0 & 0 & b_2 - 2b_1 \end{bmatrix} \left(\text{by } R_2 \to R_2 - 3R_3 \right)$$

$$\sim \begin{bmatrix} 1 & 2 & b_1 \\ 0 & 0 & b_3 - 3b_1 \\ 0 & 1 & 6b_1 - 3b_3 + b_4 \\ 0 & 0 & b_2 - 2b_1 \end{bmatrix} \left(\text{by } R_2 \leftrightarrow R_3 \right)$$

Hence, rank$(a) = 2$.
Now, the given system has a solution
$\Leftrightarrow$ rank$([A:B]) = $ rank(A)
$\Leftrightarrow$ rank$([A:B]) = 2$
$\Leftrightarrow b_2 - 2b_1 = 0, 6b_1 - 3b_3 + b_4 = 0$
$\Leftrightarrow b_2 = 2b_1, 6b_1 - 3b_3 + b_4 = 0$

5. (c)

$$\begin{bmatrix} 2 & 3 & 3 \\ 3 & 2 & 1 \\ 3 & 1 & 7 \end{bmatrix} = \begin{bmatrix} a_{11} & 0 & 0 \\ a_{12} & a_{22} & 0 \\ a_{13} & a_{23} & a_{33} \end{bmatrix} \times \begin{bmatrix} a_{11} & a_{12} & a_{13} \\ 0 & a_{22} & a_{23} \\ 0 & 0 & a_{33} \end{bmatrix}$$

$$\Rightarrow \begin{bmatrix} a_{11}^2 & a_{11}a_{12} & a_{11}a_{13} \\ a_{21}a_{11} & a_{12}^2 + a_{22}^2 & a_{12}a_{13} + a_{22}a_{23} \\ a_{13}a_{11} & a_{13}a_{12} + a_{23}a_{22} & a_{13}^2 + a_{23}^2 + a_{33}^2 \end{bmatrix}$$

$$= \begin{bmatrix} 2 & 3 & 3 \\ 3 & 2 & 1 \\ 3 & 1 & 7 \end{bmatrix}$$

$\Rightarrow a_{11}^2 = 2, a_{11}a_{12} = 3, a_{11}a_{13} = 3, a_{21}a_{11} = 3,$
$\quad a_{12}^2 + a_{22}^2 = 2, a_{12}a_{13} + a_{22}a_{23} = 1,$
$\quad a_{13}^2 + a_{23}^2 + a_{33}^2 = 7$

$\Rightarrow a_{11} = \pm\sqrt{2}, a_{12} = \pm\dfrac{3}{\sqrt{2}}, a_{13} = \pm\dfrac{3}{\sqrt{2}}, a_{21} = \pm\dfrac{3}{\sqrt{2}},$

$a_{22}^2 = -\dfrac{5}{2}$ i.e; $a_{22} = \sqrt{\dfrac{5}{2}}i, a_{23} = -\dfrac{1}{a_{22}}(1 - a_{12}a_{13}),$

$a_{33}^2 = 7 - a_{13}^2 - a_{23}^2$

Clearly, each of a_{22}, a_{23}, a_{33} is a complex number.

Therefore, the number of purely real elements in the lower triangular representation of the given 3×3 matrix = 8.

6. (d)

$$|P| = \begin{vmatrix} 0 & 1 & 3 & 0 \\ -2 & 3 & 0 & 4 \\ 0 & 0 & 6 & 1 \\ 0 & 0 & 1 & 6 \end{vmatrix} = -(-2)\begin{vmatrix} 1 & 3 & 0 \\ 0 & 6 & 1 \\ 0 & 1 & 6 \end{vmatrix}$$

$$= 2[(36-1) - 3(0-0) + 0] = 70$$

$$= \text{Product of eigenvalues} = 1 \times 2 \times 5 \times 7$$

7. (3)

$$\begin{bmatrix} 1 & 3 & 2 \\ 2 & 2 & -3 \\ 4 & 4 & -6 \\ 2 & 5 & 2 \end{bmatrix} \begin{bmatrix} x_1 \\ x_2 \\ x_3 \end{bmatrix} = \begin{bmatrix} 1 \\ 1 \\ 2 \\ 1 \end{bmatrix} \Rightarrow \begin{bmatrix} x_1 + 3x_2 + 2x_3 \\ 2x_1 + 2x_2 - 3x_3 \\ 4x_1 + 4x_2 - 6x_3 \\ 2x_1 + 5x_2 + 2x_3 \end{bmatrix} = \begin{bmatrix} 1 \\ 1 \\ 2 \\ 1 \end{bmatrix}$$

$\Rightarrow x_1 + 3x_2 + 2x_3 = 1,$
$\quad 2x_1 + 2x_2 - 3x_3 = 1,$
$\quad 4x_1 + 4x_2 - 6x_3 = 2,$

$\quad 2x_1 + 5x_2 + 2x_3 = 1$
$\Rightarrow x_1 + 3x_2 + 2x_3 = 1 ....(1)$
$\quad 2x_1 + 2x_2 - 3x_3 = 1 ....(2)$
$\quad 2x_1 + 5x_2 + 2x_3 = 1 ....(3)$
$(3) - (2) \Rightarrow 3x_2 + 5x_3 = 0$

$$\Rightarrow x_3 = -\frac{3x_2}{5} ....(4)$$

Putting $x_3 = -\dfrac{3x_2}{5}$ we get from (2),

$$2x_1 = 1 - 2x_2 + 3x_3 = 1 - 2x_2 - \frac{9x_2}{5}$$

$$\Rightarrow x_1 = \frac{1}{2} - \frac{19x_2}{10} ....(5)$$

Using (4) and (5) we get from (1),

$$\frac{1}{2} - \frac{19x_2}{10} + 3x_2 - \frac{6x_2}{5} = 1$$

or, $\dfrac{-31x_2 + 30x_2}{10} = \dfrac{1}{2}$

or, $x_2 = -5$

$\therefore x_3 = -\dfrac{3}{5} \times (-5) = 3$

8. (a)

$$P = \frac{1}{2}(P + P^T) + \frac{1}{2}(P - P^T) = S + V \text{(given)}$$

$$\therefore \frac{1}{2}(P + P^T) = S = \begin{bmatrix} -4 & 4 & 2 \\ 4 & 3 & \frac{7}{2} \\ 2 & \frac{7}{2} & 2 \end{bmatrix} ....(i)$$

and

$$\frac{1}{2}(P - P^T) = V = \begin{bmatrix} 0 & -2 & 3 \\ 2 & 0 & \frac{7}{2} \\ -3 & -\frac{7}{2} & 2 \end{bmatrix} \quad \dots(ii)$$

$(i) + (ii) \Rightarrow P = S + V$

$$= \begin{bmatrix} -4 & 4 & 2 \\ 4 & 3 & \frac{7}{2} \\ 2 & \frac{7}{2} & 2 \end{bmatrix} + \begin{bmatrix} 0 & -2 & 3 \\ 2 & 0 & \frac{7}{2} \\ -3 & -\frac{7}{2} & 2 \end{bmatrix} = \begin{bmatrix} -4 & 2 & 5 \\ 6 & 3 & 7 \\ -1 & 0 & 2 \end{bmatrix}$$

9. (1) 1 is the only eigenvalue of an identity matrix of any order.

TOPIC: CALCULUS:

1. Consider the functions:

I. e^{-x} II. $x^2 - \sin x$

III. $\sqrt{x^3 + 1}$

Which one of the functions above is/are increasing everywhere in $[0, 1]$?

(a) I and III only   (b) III only

(c) II and III only   (d) II only

[CSE/IT GATE 2020]

2. For real numbers x and y, with $y = 3x^2 + 3x + 1$, the maximum and minimum value of y for $x \in [-2, 0]$ are, respectively _____ ?

(a) 1 and 1/4   (b) 7 and 1

(c) 7 and 1/4   (d) −2 and −1/2

[EE GATE 2020]

3. The partial derivative of the function

$f(x,y,z) = e^{1-x\cos y} + xze^{-\frac{1}{1+y^2}}$, w.r.t x at the point $(1, 0, e)$ is

(a) $\frac{1}{e}$ (b) 1   (c) −1   (d) 0

[EC GATE 2020]

4. Consider the function $f(x,y) = x^2 + y^2$. The minimum value of the function attains on the line $x + y = 1$ (rounded off to one decimal place) is _____ ?

[IN GATE 2020]

5. The value of $\lim\limits_{x \to \infty} \dfrac{\sqrt{9x^2 + 2020}}{x + 7}$ is

(a) $\dfrac{7}{9}$ (b) 1

(c) indeterminate   (d) 3

[CE GATE 2020]

6. The value of $\lim\limits_{x \to \infty} \dfrac{x^2 - 5x + 4}{4x^2 + 2x}$ is

(a) 1   (b) $\dfrac{1}{2}$ (c) $\dfrac{1}{4}$ (d) 0

[CE GATE 2020]

7. The value of $\lim\limits_{x \to 1} \left(\dfrac{1 - e^{-c(1-x)}}{1 - xe^{-c(1-x)}} \right)$ is

(a) $\dfrac{c+1}{c}$ (b) $c + 1$

(c) $\dfrac{c}{c+1}$ (d) c

[ME GATE 2020]

8. The area of an ellipse represented by an equation $\dfrac{x^2}{a^2} + \dfrac{y^2}{b^2} = 1$ is

(a) $\dfrac{\pi ab}{4}$ (b) $\dfrac{4\pi ab}{3}$

(c) $\dfrac{\pi ab}{2}$ (d) πab

[CE GATE 2020]

9. If C represents a line segment between $(0, 0, 0)$ and $(1, 1, 1)$ in the Cartesian coordinate system, the value (expressed as integer) of the line integral

$\int\limits_C \left[(y+z)dx + (x+z)dy + (x+y)dz \right]$ is

[CE GATE 2020]

10. Let $I = \int_{x=0}^{1} \int_{y=0}^{x^2} xy^2 \, dy \, dx$. Then, I may also be expressed as

(a) $\int_{y=0}^{1} \int_{x=\sqrt{y}}^{1} yx^2 \, dx \, dy$ (b) $\int_{y=0}^{1} \int_{x=\sqrt{y}}^{1} xy^2 \, dx \, dy$

(c) $\int_{y=0}^{1} \int_{x=0}^{\sqrt{y}} xy^2 \, dx \, dy$ (d) $\int_{y=0}^{1} \int_{x=0}^{\sqrt{y}} yx^2 \, dx \, dy$

[ME GATE 2020]

11. Define $[x]$ as the smallest integer less than or equal to x for each $x \in (-\infty, \infty)$. If $y = [x]$, then area under y for $x \in [1, 4]$ is _____ ?

(a) 6   (b) 3   (c) 4   (d) 1

[ME GATE 2020]

Answer key				
1. (b)	**2.** (c)	**3.** (d)	**4.** (0.5)	**5.** (d)
6. (c)	**7.** (c)	**8.** (d)	**9.** (3)	**10.** (a)
11. (a)				

Explanation

1. (b) Let $f(x) = \sqrt{x^3 + 1}$.

Then, $f'(x) = \dfrac{3x^2}{2\sqrt{x^3 + 1}} \geq 0 \quad \forall x \in [0, 1]$.

Hence, $\sqrt{x^3 + 1}$ is increasing everywhere

in $[0, 1]$

2. (c) $y = 3x^2 + 3x + 1$

$\Rightarrow \dfrac{dy}{dx} = 6x + 3$ and $\dfrac{d^2y}{dx^2} = 6$

Now, $\dfrac{dy}{dx} = 0 \Rightarrow 6x + 3 = 0 \Rightarrow x = -0.5$

At $x = -0.5$, $\dfrac{d^2y}{dx^2} = 6 > 0$.

Therefore, $x = -0.5$ is a point of local minima of y.

Hence, the maximum value of y in $[-2, 0]$

$= \max\{y(-2), y(0)\}$

$= \max\{3 \times (-2)^2 + 3 \times (-2) + 1, 0 + 0 + 1\} = 7$

Also, the minimum value of y in $[-2, 0]$

$= \min\{y(-2), y(0), y(-0.5)\}$

$= \min\{7, 1, 3 \times (-0.5)^2 + 3 \times (-0.5) + 1\}$

$= \min\left\{7, 1, \dfrac{1}{4}\right\} = \dfrac{1}{4}$.

3. (d)

$f(x, y, z) = e^{1 - x\cos y} + xze^{-\frac{1}{1+y^2}}$

$\Rightarrow \dfrac{\partial f}{\partial x} = e^{1 - x\cos y} \times (-\cos y) + ze^{-\frac{1}{1+y^2}}$

$\Rightarrow \dfrac{\partial f}{\partial x}\bigg|_{(1,0,e)} = -e^{1 - 1\cos 0} + e \times e^{-\frac{1}{1+0^2}} = 0$.

4. (0.5)

$x + y = 1 \Rightarrow y = 1 - x$.

$\therefore f(x, y) = x^2 + y^2 = x^2 + (1-x)^2 = g(x)$ (say)

Now $g(x) = x^2 + (1-x)^2$

$\Rightarrow g'(x) = 2x - 2(1-x) = 4x - 2, g''(x) = 4$

So $g'(x) = 0 \Rightarrow 4x - 2 = 0 \Rightarrow x = 0.5$

$\therefore g''(x)\big|_{x=0.5} = 4 > 0$

Thus, $x = 0.5$ is a point of local minima of $g(x)$.

At $x = 0.5, y = 1 - 0.5 = 0.5$.

Therefore, $(0.5, 0.5)$ is a point of minimum of $f(x, y)$.

Hence, the required minimum value of $f(x, y)$

$= f(0.5, 0.5) = 0.5^2 + 0.5^2 = 0.5$

5. (d) $\lim\limits_{x \to \infty} \dfrac{\sqrt{9x^2 + 2020}}{x + 7}$

$= \lim\limits_{x \to \infty} \dfrac{\sqrt{9 + \dfrac{2020}{x^2}}}{1 + \dfrac{7}{x}} = \dfrac{\sqrt{9 + 0}}{1} = 3$

6. (c) $\lim\limits_{x \to \infty} \dfrac{x^2 - 5x + 4}{4x^2 + 2x}$ $\left[\text{form } \dfrac{\infty}{\infty}\right]$

$= \lim\limits_{x \to \infty} \dfrac{\dfrac{d}{dx}(x^2 - 5x + 4)}{\dfrac{d}{dx}(4x^2 + 2x)}$ $\left[\text{by the L'Hospital rule}\right]$

$= \lim\limits_{x \to \infty} \dfrac{2x - 5}{8x + 2}$ $\left[\text{form } \dfrac{\infty}{\infty}\right]$

$= \lim\limits_{x \to \infty} \dfrac{\dfrac{d}{dx}(2x - 5)}{\dfrac{d}{dx}(8x + 2)}$ $\left[\text{by the L'Hospital rule}\right]$

$= \lim\limits_{x \to \infty} \dfrac{2}{8} = \dfrac{1}{4}$.

7. (c)

$\lim\limits_{x \to 1} \left(\dfrac{1 - e^{-c(1-x)}}{1 - xe^{-c(1-x)}} \right)$ $\left(\text{form } \dfrac{0}{0}\right)$

$= \lim\limits_{x \to 1} \dfrac{\dfrac{d}{dx}\left(1 - e^{-c(1-x)}\right)}{\dfrac{d}{dx}\left(1 - xe^{-c(1-x)}\right)}$ $\left(\text{by the L'Hospital rule}\right)$

$= \lim\limits_{x \to 1} \dfrac{-ce^{-c(1-x)}}{-\left\{e^{-c(1-x)} + cxe^{-c(1-x)}\right\}}$

$= \lim\limits_{x \to 1} \dfrac{-c}{-(1 + cx)} = \dfrac{c}{1 + c \times 1} = \dfrac{c}{c + 1}$

8. (d) $\dfrac{x^2}{a^2} + \dfrac{y^2}{b^2} = 1 \Rightarrow \dfrac{y^2}{b^2} = 1 - \dfrac{x^2}{a^2} \Rightarrow y = \dfrac{b}{a}\sqrt{a^2 - x^2}$

In the given ellipse, x varies from $-a$ to a.

Hence, the required area

$= 2 \times (\text{area of upper half of the ellipse})$

$= 2\int_{-a}^{a} \dfrac{b}{a}\sqrt{a^2 - x^2}\, dx$

$= \dfrac{4b}{a}\int_{0}^{a} \sqrt{a^2 - x^2}\, dx$

$\left(\because \sqrt{a^2 - x^2} \text{ is an even function}\right)$

$= \dfrac{4b}{a}\left[\dfrac{x}{2}\sqrt{a^2 - x^2} + \dfrac{a^2}{2}\sin^{-1}\dfrac{x}{a}\right]_{0}^{a}$

$= \dfrac{4b}{a}\left[\dfrac{a}{2}\sqrt{a^2 - a^2} + \dfrac{a^2}{2}\sin^{-1}\dfrac{a}{a} - (0 - 0)\right]$

$= \dfrac{4b}{a} \times \dfrac{a^2}{2}\sin^{-1} 1 = 2ab \times \dfrac{\pi}{2} = \pi ab$

9. (3) The given line integral

$= \int_C \left[(y + z)\, dx + (x + z)\, dy + (x + y)\, dz\right]$

$= \int_C \left[(y\, dx + x\, dy) + (y\, dz + z\, dy) + (x\, dz + z\, dx)\right]$

$= \int_{(0,0,0)}^{(1,1,1)} \left[d(xy) + d(yz) + d(xz)\right]$

$= \left[xy + yz + zx\right]_{(0,0,0)}^{(1,1,1)}$

$= (1 + 1 + 1) - (0 + 0 + 0)$

$= 3$

10. (*a*) The limits of integration are as follows:

$0 \le x \le 1, \ 0 \le y \le x^2$.

In other words, $0 \le y \le 1, \ \sqrt{y} \le x \le 1$.

Thus, changing the order of integration, we have,

$$I = \int_{x=0}^{1} \int_{y=0}^{x^2} xy^2 \, dy \, dx = \int_{y=0}^{1} \int_{x=\sqrt{y}}^{1} xy^2 \, dx \, dy$$

11. (*a*)

$$[x] = \begin{cases} 1, & 1 \le x < 2 \\ 2, & 2 \le x < 3 \\ 3, & 3 \le x < 4 \\ 4, & x = 4 \end{cases}$$

∴ Required area

$$= \int_1^4 y \, dx = \int_1^4 [x] \, dx = \int_1^2 1 \, dx + \int_2^3 2 \, dx + \int_3^4 3 \, dx$$

$$= [x]_1^2 + 2[x]_2^3 + 3[x]_3^4$$

$$= (2-1) + 2(3-2) + 3(4-3) = 6.$$

TOPIC: ORDINARY DIFFERENTIAL EQUATIONS

1. Consider the initial value problem below. The value of y at $x = ln2$ (rounded off to 3 decimal places) is _____

$$\frac{dy}{dx} = 2x - y, \ y(0) = 1 \qquad \textbf{[EE GATE 2020]}$$

2. The general solution of $\dfrac{d^2y}{dx^2} - 6\dfrac{dy}{dx} + 9y = 0$ is

(*a*) $y = c_1 e^{3x} + c_2 e^{-3x}$ (*b*) $y = (c_1 + c_2 x)e^{3x}$

(*c*) $y = (c_1 + c_2 x)e^{-3x}$ (*d*) $y = c_1 e^{3x}$

[EC GATE 2020]

3. Which of the following options contain two solutions of the differential equation $\dfrac{dy}{dx} = x(y-1)$?

(*a*) $\ln(y-1) = 2x^2 + C, \ y = 1$

(*b*) $\ln(y-1) = 2x^2 + C, \ y = -1$

(*c*) $\ln(y-1) = 0.5x^2 + C, \ y = 1$

(*d*) $\ln(y-1) = 0.5x^2 + C, \ y = -1$

[EC GATE 2020]

4. The ODE $\dfrac{d^2u}{dx^2} - 2x^2u + \sin x = 0$ is

(*a*) linear and homogeneous

(*b*) non-linear and homogeneous

(*c*) non-linear and non-homogeneous

(*d*) linear and non-homogeneous

[CE GATE 2020]

5. An ordinary differential equation is given below:

$$6\frac{d^2y}{dx^2} + \frac{dy}{dx} - y = 0 \text{ Its solution is}$$

(*a*) $y(x) = c_1 e^{\frac{x}{3}} + c_2 e^{-\frac{x}{2}}$ (*b*) $y(x) = c_1 x e^{\frac{-x}{3}} + c_2 e^{\frac{x}{2}}$

(*c*) $y(x) = c_1 e^{\frac{-x}{3}} + c_2 e^{\frac{x}{2}}$ (*d*) $y(x) = c_1 e^{\frac{-x}{3}} + c_2 x e^{\frac{x}{2}}$

[CE GATE 2020]

6. For the ordinary differential equation

$$\frac{d^2x}{dt^2} - 5\frac{dx}{dt} + 6x = 0 \text{ with initial conditions } x(0) =$$

0 and $\dfrac{dx}{dt}(0) = 10$, the solution is

(*a*) $-10e^{2t} + 10e^{3t}$ (*b*) $5e^{2t} + 6e^{3t}$

(*c*) $10e^{2t} + 10e^{3t}$ (*d*) $-5e^{2t} + 6e^{3t}$

[CE GATE 2020]

Answer key				
1. (0.886)	**2.** (*b*)	**3.** (*c*)	**4.** (*d*)	**5.** (*a*)
6. (*a*)				

Explanation

1. (0.886) The given equation is $\dfrac{dy}{dx} + y = 2x$ ...(i)

Integrating factor (*I.F.*) of (1) $= e^{\int dx} = e^x$

Therefore, the solution of the linear equation (1) is

$y^x = \int (I.F.) \times e^x \, dx + C$

or, $ye^x = \int 2xe^x \, dx + C$

or, $ye^x = 2xe^x - 2e^x + C$

or, $y = 2x - 2 + Ce^{-x}$

Now $y(0) = 1$ gives 0–2 + C = 1 *i.e*; C = 3. Therefore, the particular solution of (*i*) is given by

$y = 2x - 2 + 3e^{-x}$

∴ $y(\ln 2)$

$= 2\ln 2 - 2 + 3e^{-\ln 2} = 1.3862 - 2 + 3 \times \dfrac{1}{2} \approx 0.886$

2. (*b*) Let $y = e^{mx}$ be a trial solution of

$$\frac{d^2y}{dx^2} - 6\frac{dy}{dx} + 9y = 0 \qquad \qquad ...(i)$$

Then $\dfrac{dy}{dx} = me^{mx}$ and $\dfrac{d^2y}{dx^2} = m^2 e^{mx}$.

Therefore, (*i*) becomes, $m^2 e^{mx} - 6me^{mx} + 9 = 0$

$\Rightarrow m^2 - 6m + 9 = 0 \Rightarrow m = 3, 3$

Hence, the required general solution of (*i*) is

$y = (c_1 + c_2 x)e^{3x}$

3. (c) $\dfrac{dy}{dx} = x(y-1)$

$\Rightarrow \displaystyle\int \dfrac{dy}{y-1} = \int x\,dx + C$

$\Rightarrow \ln(y-1) = \dfrac{1}{2}x^2 + C = 0.5x^2 + C$

4. (d)

5. (a) Let $y = e^{mx}$ be a solution of

$6\dfrac{d^2y}{dx^2} + \dfrac{dy}{dx} - y = 0 \quad\dots\dots\dots(i)$

Then (1) gives,

$6m^2 e^{mx} + m e^{mx} - e^{mx} = 0$

$\Rightarrow 6m^2 + m - 1 = 0$

$\Rightarrow 6m^2 + 3m - 2m - 1 = 0$

$\Rightarrow 3m(2m+1) - (2m+1) = 0$

$\Rightarrow (2m+1)(3m-1) = 0$

$\Rightarrow m = -\dfrac{1}{2}, 3$

Hence, the solution of (i) is: $y(x) = c_1 e^{\frac{x}{3}} + c_2 e^{-\frac{x}{2}}$

6. (a) Suppose $x = e^{mt}$ be a trial solution of

$\dfrac{d^2x}{dt^2} - 5\dfrac{dx}{dt} + 6x = 0 \quad\dots\dots\dots\dots(1)$

Then, $\dfrac{d^2x}{dt^2} = m^2 e^{mt}$ and $\dfrac{dx}{dt} = m e^{mt}$

$\therefore (1)$ becomes,

$m^2 e^{mt} - 5m e^{mt} + 6 e^{mt} = 0$

$\Rightarrow m^2 - 5m + 6 = 0 \Rightarrow m = 2, 3$

Therefore, the general solution of (1) is given by

$x(t) = c_1 e^{2t} + c_2 e^{3t} \quad\dots\dots\dots(2)$

Now given that $x(0) = 0$

$\therefore (2)$ gives, $c_1 + c_2 = 0 \quad\dots\dots\dots(3)$

It is also given that $\left.\dfrac{dx}{dt}\right|_{t=0} = 10$

$\therefore (2)$ gives, $2c_1 + 3c_2 = 0 \quad\dots\dots\dots(4)$

Solving (3) and (4), we get $c_2 = 10$ and $c_1 = -10$

Hence, the required particular solution of (1) is

$x(t) = -10 e^{2t} + 10 e^{3t}$

TOPIC: PARTIAL DIFFERENTIAL EQUATIONS

1. The following partial differential equation is defined for $u = u(x,y)$: $\dfrac{\partial u}{\partial y} = \dfrac{\partial^2 u}{\partial x^2}, y \geq 0, x_1 \leq x \leq x_2$

The set of auxiliary conditions necessary to solve the equation uniquely, is

(a) One initial condition and two boundary conditions

(b) Three initial conditions

(c) Two initial conditions and one boundary condition

(d) Three boundary conditions

[CE GATE 2020]

2. In the following partial differential equation, θ is a function of t and z and D and K are functions of θ:

$D(\theta) = \dfrac{\partial^2 \theta}{\partial z^2} + \dfrac{\partial K(\theta)}{\partial z} - \dfrac{\partial \theta}{\partial t} = 0$

The above equation is

(a) a second order linear equation.

(b) a second order non-linear equation.

(c) a second degree non-linear equation.

(d) a second degree linear equation.

[CE GATE 2020]

Answer key				
1. (a)	**2.** (b)			

Explanation

1. (a) The given condition $\dfrac{\partial u}{\partial y} = \dfrac{\partial^2 u}{\partial x^2}$ is nothing but a one-dimensional heat equation. So, one initial condition and two boundary conditions are required to solve it.

2. (b)

TOPIC: COMPLEX ANALYSIS

1. The value of the following complex integral, with C representing the unit circle centered at origin in the counterclockwise sense is

$\displaystyle\int_C \dfrac{z^2 + 1}{z^2 - 2z}\,dz$

(a) πi (b) $-8\pi i$ (c) $-\pi i$ (d) $8\pi i$

[EE GATE 2020]

2. Let $f(z) = \dfrac{1}{z+a}$, $a > 0$. The value of the integral over a circle C with the center $(-a, 0)$ and radius $R > 0$ evaluated in the counterclockwise direction is _____.

(a) $2\pi i$ (b) 0

(c) $4\pi i$ (d) $-2\pi i$

[IN GATE 2020]

3. Which of the following functions $f(z)$, of the complex variable z, is not analytic at all the points of the complex plane?

(a) $f(z) = z^2$ (b) $f(z) = \sin z$
(c) $f(z) = \log z$ (d) $f(z) = e^z$

[ME GATE 2020]

4. An analytic function of a complex variable $z = x + iy \ (i = \sqrt{-1})$ is defined as $f(z) = x^2 - y^2 + i\psi(x,y)$ where $\psi(x,y)$ is a real function. The value of the imaginary part of $f(z)$ at $z = 1 + i$ is _____ ? (round off to 2 decimal places) **[ME GATE 2020]**

5. The function $f(z)$ of the complex variable $z = x + iy$, where $i = \sqrt{-1}$, is given as $f(z) = (x^3 - 3xy^2) + iv(x,y)$. For this function to be analytic, $v(x, y)$ should be

(a) $x^3 - 3x^2y + c$ (b) $3xy^2 - y^3 + c$
(c) $3x^2y - y^3 + c$ (d) $3x^2y^2 - y^3 + c$

[ME GATE 2020]

Answer key				
1. (c)	**2.** (a)	**3.** (c)	**4.** (2)	**5.** (c)

Explanation

1. (c) $z^2 - 2z \Rightarrow z(z-2) = 0 \Rightarrow z = 0, 2$

Hence, $\frac{z^2+1}{z^2-2z}$ has simple poles at $z = 0$ and $z = 2$.

The pole $z = 0$ lies inside the circle $C : |z| = 1$ where as the pole $z = 2$ lies outside the circle $C : |z| = 1$.

Now $\operatorname{Re} s(z = 0)$

$= \lim_{z\to 0}(z-0)\frac{z^2+1}{z^2-2z} = \lim_{z\to 0}\frac{z^2+1}{z-2}$

$= \frac{0^2+1}{0-2} = -\frac{1}{2}$.

Hence, by the Cauchy's residue theorem,

$\int_C \frac{z^2+1}{z^2-2z}dz$

$= 2\pi i \times$ the sum of the residues at the poles lying inside C

$= 2\pi i \times \left(-\frac{1}{2}\right) = -\pi i$

2. (a) $\int_C \frac{1}{z+a}dz$

$= \int_C \frac{g(z)}{z-(-a)}dz$ (where $g(z) = 1$)

$= 2\pi i \times g(-a)$ (by Cauchy's integral formula)

$= 2\pi i \times 1 = 2\pi i$

3. (c)

$f(z) = \log z \Rightarrow f'(z) = \frac{1}{z}$, which does not exist at z=0.

Hence, $f(z) = \log z$ is not analytic at $z = 0$.

4. (2) Suppose $\varphi(x,y) = x^2 - y^2$

Then $f(z) = \varphi(x,y) + i\psi(x,y)$ is analytic

$\Rightarrow \frac{\partial\psi}{\partial y} = \frac{\partial\varphi}{\partial x}$ and $\frac{\partial\psi}{\partial x} = -\frac{\partial\varphi}{\partial y}$

[By CR equations]

$\Rightarrow \frac{\partial\psi}{\partial y} = 2x$ and $\frac{\partial\psi}{\partial x} = 2y$

Now by total differential of Ψ, we have

$d\psi = \frac{\partial\psi}{\partial x}dx + \frac{\partial\psi}{\partial y}dy = 2y\,dx + 2x\,dy = 2d(xy)$

Integrating both sides, we get

$\int d\psi = 2\int d(xy) \Rightarrow \psi = 2xy$

$\therefore$ At $z = 1+i$, $\psi(x,y) = 2\times 1\times 1 = 2$.

5. (c) Comparing $f(z) = (x^3 - 3xy^2) + iv(x,y)$ with

$f(z) = u(x,y) + iv(x,y)$, we get,

$u(x,y) = x^3 - 3xy^2$

$\therefore \frac{\partial u}{\partial x} = 3x^2 - 3y^2$ and $\frac{\partial u}{\partial y} = -6xy$.

Now $f(z)$ is analytic $\Rightarrow$ the CR equations are satisfied

$\Rightarrow \frac{\partial u}{\partial x} = \frac{\partial v}{\partial y}$ and $\frac{\partial u}{\partial y} = -\frac{\partial v}{\partial x}$

$\Rightarrow \frac{\partial v}{\partial y} = 3x^2 - 3y^2$ and $\frac{\partial v}{\partial x} = 6xy$

Now, $dv = \frac{\partial v}{\partial x}dx + \frac{\partial v}{\partial y}dy$

$= 6xy\,dx + (3x^2 - 3y^2)dy$

$= M\,dx + N\,dy$ (say)

$\therefore \int dv = v = \int M\,dx + N\,dy$

Here $\int M\,dx = \int 6xy\,dx = 3x^2 y$

and $\int N\,dy = \int(3x^2 - 3y^2)dy = 3x^2 y - y^3$.

Hence, $v = 3x^2y - y^3 + c$ (constant)

TOPIC: NUMERICAL ANALYSIS

1. The integral $\int_0^1 (5x^3 + 4x^2 + 3x + 2)dx$ is estimated numerically using three alternative methods namely the rectangular, trapezoidal and Simpson's rule with a common step size. In this context, which one of the following statements is true?

(a) Simpson's rule as well as rectangular rule of estimation will give non–zero error

(b) Only Simpson's rule of estimation will give zero error

(c) Simpson's rule, rectangular rule as well as trapezoidal rule of estimation will give non–zero error

(d) Only rectangular rule of estimation will give zero error. **[CE GATE 2020]**

2. The evaluation of the definite integral $\int_{-1}^{1.4} x|x|\,dx$ by using Simpson's $1/3^{\text{rd}}$ rule with step size $h = 0.6$ yields

(a) 0.581   (b) 0.592

(c) 1.248   (d) 0.914

[ME GATE 2020]

3. For the integral $\int_{0}^{\frac{\pi}{2}} (8 + 4\cos x)\,dx$, the absolute percentage error in numerical evaluation with the Trapezoidal rule, only the end points, is _____ ? (round off to one decimal place) **[ME GATE 2020]**

Answer key				
1. (b)	**2.** (b)	**3.** (5.2)		

Explanation

1. (b) Simpson's rule gives the exact result for a polynomial of degree ≤ 3.

2. (b) Here, $f(x) = x|x|$ and $h = 0.6$

x	$-1 (= x_0)$	$-0.4 (= x_1)$	$0.2 (= x_2)$	$0.8 (= x_3)$	$1.4 (= x_4)$
$f(x)$	$-1 (= y_0)$	$-0.16 (= y_1)$	$0.04 (= y_2)$	$0.64 (= y_3)$	$1.96 (= y_4)$

By Simpson's $1/3^{\text{rd}}$ rule,

$$\int_{-1}^{1.4} x|x|\,dx$$

$$= \frac{h}{3}\left[(y_0 + y_4) + 2y_2 + 4(y_1 + y_3)\right]$$

$$= \frac{0.6}{3}\left[(-1 + 1.96) + 2 \times 0.04 + 4(-0.16 + 0.64)\right]$$

$$\approx 0.592$$

3. (5.2) $\int_{0}^{\frac{\pi}{2}} (8 + 4\cos x)\,dx$

$$= \left[8x + 4\sin x\right]_0^{\pi/2} = 8 \times \frac{\pi}{2} + 4 \times 1 = 4\pi + 4.$$

Then, exact value of the integral = $4\pi + 4$

Now, by Trapezoidal rule, approximate value of the integral

$$= \frac{h}{2}\left[f(0) + f\left(\frac{\pi}{2}\right)\right]$$

$$= \frac{\frac{\pi}{2} - 0}{2}\left[(8 + 4\cos 0) + \left(8 + 4\cos\frac{\pi}{2}\right)\right]$$

$$= 5\pi$$

$\therefore$ Absolute percentage error

$$= \frac{|\text{True value-approximate value}|}{\text{True value}} \times 100$$

$$= \frac{|4\pi + 4 - 5\pi|}{4\pi + 4} \times 100 = \frac{4 - \pi}{\pi + 1} \times 25 = 5.2\%$$

TOPIC: VECTORS

1. The vector function expressed by
$\vec{F} = \hat{a}_x(5y - k_1 z) + \hat{a}_y(3z + k_2 x) + \hat{a}_z(k_3 y - 4x)$
represents a conservative field, where $\hat{a}_x, \hat{a}_y, \hat{a}_z$ are unit vectors along x, y, and z directions, respectively. The values of constants k_1, k_2, k_3 are given by

(a) $k_1 = 3, k_2 = 8, k_3 = 5$

(b) $k_1 = 4, k_2 = 5, k_3 = 3$

(c) $k_1 = 3, k_2 = 3, k_3 = 7$

(d) $k_1 = 0, k_2 = 0, k_3 = 0$ **[EE GATE 2020]**

2. For a vector field $\vec{A}$, which one of the following is false?

(a) $\vec{\nabla} \times \vec{A}$ is another vector field

(b) $\vec{A}$ is irrotational if $\vec{\nabla}^2 \vec{A} = 0$

(c) $\vec{\nabla} \times (\vec{\nabla} \times \vec{A}) = \vec{\nabla}(\vec{\nabla}.\vec{A}) - \vec{\nabla}^2 \vec{A}$

(d) $\vec{A}$ is solenoidal if $\vec{\nabla}.\vec{A} = 0$

[EC GATE 2020]

3. The unit vectors along the mutually perpendicular x, y, and z axes are $\hat{i}, \hat{j}, \hat{k}$, respectively. Consider the plane $z = 0$ and two vectors $\vec{a}$ and $\vec{b}$ on that plane such that $\vec{a} \neq \alpha \vec{b}$ for any scalar α. A vector perpendicular to both $\vec{a}$ and $\vec{b}$ is _____

(a) $\hat{i} - \hat{j}$ (b) $\hat{i}$ (c) $-\hat{j}$ (d) $\hat{k}$

[IN GATE 2020]

4. For three vectors $\vec{A} = 2\hat{j} - 3\hat{k}, \vec{B} = -2\hat{i} + \hat{k}$ and $\vec{C} = 3\hat{i} - \hat{j}$, where $\hat{i}, \hat{j}, \hat{k}$ are unit vectors along the axes of a right handed rectangular/ Cartesian coordinate system, the value of $\vec{A}.(\vec{B} \times \vec{C}) + 6$ is _____ ?

[ME GATE 2020]

5. A vector field is defined as

$$\vec{f}(x,y,z)=\frac{x}{\left(x^2+y^2+z^2\right)^{\frac{3}{2}}}\hat{i}+\frac{y}{\left(x^2+y^2+z^2\right)^{\frac{3}{2}}}\hat{j}$$
$$+\frac{z}{\left(x^2+y^2+z^2\right)^{\frac{3}{2}}}\hat{k}$$

where $\hat{i},\hat{j},\hat{k}$ are unit vectors along the axes of a right handed rectangular/Cartesian coordinate system. The surface integral $\iint \vec{f}.d\vec{s}$ (where $d\vec{s}$ is an elementary surface area vector) evaluated over the inner and outer surface of a spherical shell formed by two concentric spheres with origin as the center, and internal and external radii of 1 and 2, respectively, is

(a) 4π (b) 0   (c) 2π (d) 8π

[ME GATE 2020]

6. The directional derivative of $f(x,y,z)=xyz$ at $(-1, 1, 3)$ in the direction of the vector $\hat{i}-2\hat{j}+2\hat{k}$ is

(a) $\dfrac{7}{3}$ (b) $3\hat{i}-3\hat{j}-\hat{k}$

(c) $-\dfrac{7}{3}$ (d) 7

[ME GATE 2020]

Answer key				
1. (b)	**2.** (b)	**3.** (d)	**4.** (6)	**5.** (b)
6. (a)				

Explanation

1. (b) $\vec{F}$ represents a conservative field

$\Rightarrow$ curl $\vec{F}=\vec{0}$

$$\Rightarrow \begin{vmatrix} \hat{a}_x & \hat{a}_y & \hat{a}_z \\ \dfrac{\partial}{\partial x} & \dfrac{\partial}{\partial y} & \dfrac{\partial}{\partial z} \\ 5y-k_1 z & 3z+k_2 x & k_3 y-4x \end{vmatrix}=\vec{0}$$

$$\Rightarrow (k_3-3)\hat{a}_x-(-4+k_1)\hat{a}_y+(k_2-5)\hat{a}_z$$
$$=0\hat{a}_x+0\hat{a}_y+0\hat{a}_z$$
$$\Rightarrow k_3-3=0, 4-k_1=0, k_2-5=0$$
$$k_3=3, k_1=4, k_2=5$$

2. (b)

3. (d) $\vec{a},\vec{b}$ are the two vectors on the plane $z=0$ such that $\vec{a}\ne\alpha\vec{b}$ for any scalar α.

$\Rightarrow \vec{a}$ and $\vec{b}$ are non-collinear on the plane $z=0$

$\Rightarrow \hat{k}$ is perpendicular to both $\vec{a}$ and $\vec{b}$

($\because$ each of $\hat{i}-\hat{j}, \hat{i},-\hat{j}$ lies on z=0 plane)

4. (6) $\vec{A}.\left(\vec{B}\times\vec{C}\right)$

$$=\left[\vec{A}\,\vec{B}\,\vec{C}\right]=\begin{vmatrix} 0 & 2 & -3 \\ -2 & 0 & 1 \\ 3 & 1 & 0 \end{vmatrix}$$
$$=0(0+1)-2(0-3)-3(2-0)=0$$
$$\therefore \vec{A}.\left(\vec{B}\times\vec{C}\right)+6=0+6=6$$

5. (b) Let $\vec{r}=x\hat{i}+y\hat{j}+z\hat{k}$

Then $r=|\vec{r}|=\sqrt{x^2+y^2+z^2}$

$$\therefore \vec{f}(x,y,z)=\frac{x\hat{i}+y\hat{j}+z\hat{k}}{\left(x^2+y^2+z^2\right)^{\frac{3}{2}}}=\frac{\vec{r}}{r^3}$$

So $div\ \vec{f}$

$$=\vec{\nabla}.\left(\frac{x\hat{i}+y\hat{j}+z\hat{k}}{r^3}\right)$$
$$=\frac{\partial}{\partial x}\left(\frac{x}{r^3}\right)+\frac{\partial}{\partial y}\left(\frac{y}{r^3}\right)+\frac{\partial}{\partial z}\left(\frac{z}{r^3}\right)$$
$$=\frac{1}{r^3}+\left(-\frac{3x}{r^4}\right)\frac{\partial r}{\partial x}+\frac{1}{r^3}+\left(-\frac{3y}{r^4}\right)\frac{\partial r}{\partial y}$$
$$+\frac{1}{r^3}+\left(-\frac{3z}{r^4}\right)\frac{\partial r}{\partial z}$$
$$=\frac{3}{r^3}-\frac{3}{r^4}\left(x\frac{\partial r}{\partial x}+y\frac{\partial r}{\partial y}+z\frac{\partial r}{\partial z}\right)$$
$$=\frac{3}{r^3}-\frac{3}{r^4}\times\left(x\times\frac{2x}{2\sqrt{x^2+y^2+z^2}}\right.$$
$$\left.+y\times\frac{2y}{2\sqrt{x^2+y^2+z^2}}+z\times\frac{2z}{2\sqrt{x^2+y^2+z^2}}\right)$$
$$=\frac{3}{r^3}-\frac{3}{r^4}\times\sqrt{x^2+y^2+z^2}=\frac{3}{r^3}-\frac{3}{r^4}\times r=0$$

Hence, by the Gauss-divergence theorem $\iint \vec{f}.d\vec{s}=\iiint div\ \vec{f}.d\vec{v}=0$

6. (a) Here, $\vec{\nabla} f=\hat{i}\dfrac{\partial}{\partial x}(xyz)+\hat{j}\dfrac{\partial}{\partial y}(xyz)+\hat{k}\dfrac{\partial}{\partial z}(xyz)$

$$=yz\hat{i}+xz\hat{j}+xy\hat{k}$$
$$\therefore \vec{\nabla} f\bigg|_{(-1,1,3)}=3\hat{i}-3\hat{j}-\hat{k}.$$

Also, $\hat{a}=\dfrac{\hat{i}-2\hat{j}+2\hat{k}}{\left|\hat{i}-2\hat{j}+2\hat{k}\right|}=\dfrac{\hat{i}-2\hat{j}+2\hat{k}}{\sqrt{1^2+(-2)^2+2^2}}$

$$=\frac{1}{3}\left(\hat{i}-2\hat{j}+2\hat{k}\right)$$

Hence, the required directional derivative

$$\vec{\nabla} f\bigg|_{(-1,1,3)}.\hat{a}=\frac{1}{3}\left(3\times 1+(-3)\times(-2)+(-1)\times 2\right)=\frac{7}{3}.$$

TOPIC: PROBABILITY & STATISTICS

1. For $n > 2$, let $a \in \{0,1\}^n$ be a nonzero vector. Suppose that x is chosen uniformly at random from $\{0,1\}^n$. Then the probability that $\sum_{i=1}^{n} a_i x_i$ is an odd number is _____ ?

[CSE/IT GATE 2020]

2. The two sides of a fair coin are labeled as 0 and 1. The coin is tossed two times independently. Let M and N denote the labels corresponding to the outcomes of those tosses. For a random variable X, defined as $X = \min(M, N)$, the expected value $E(X)$ (rounded off to two decimal places) is _____.

[EC GATE 2020]

3. X is a random variable with uniform probability density function in the interval $[-2, 10]$. For $Y = 2X - 6$, the conditional probability $P(Y \le 7 \mid X \ge 5)$ (rounded off to three decimal places) is _____. **[EC GATE 2020]**

4. A player throws a ball at a basket kept at a distance. The probability that the ball falls into the basket in a single attempt is 0.1. The player attempts to through the ball twice. Considering each attempt to be independent, the probability that this player puts the ball into the basket only in the second attempt (rounded off to two decimal places) is _____.

[IN GATE 2020]

5. A fair (unbiased) coin is tossed 15 times. The probability of gettings exactly 8 heads is _____? **[CE GATE 2020]**

6. The probability that a 50 years flood may not occur at all during 25 years life of a project is __? **[CE GATE 2020]**

7. Consider two exponentially distributed random variables X and Y, both having a mean of 0.50. Let $Z = X + Y$ and r be the correlation coefficient between X and Y. If the variance of Z equals zero, then the value of r is _____ (round off to 2 decimal places) **[ME GATE 2020]**

8. The sum of two normally distributed random variables X and Y is

(a) always normally distributed

(b) normally distributed only if X and Y are independent

(c) normally distributed only if X and Y have the same mean

(d) normally distributed only X and Y have the same standard deviation.

[ME GATE 2020]

9. A fair coin is tossed 20 times. The probability that heads will appear exactly 4 times in the first ten tosses and tails will appear exactly 4 times in the next ten tosses is ____ ? (round off to 3 decimal places). **[ME GATE 2020]**

Answer key			
1. (0.5)	**2.** (0.25)	**3.** (0.3)	**4.** (0.09)
5. (0.196)	**6.** (0.603)	**7.** (–1)	**8.** (a)
9. (0.042)			

Explanation

1. (0.5) Here $a_i = 0$ or 1 and $x_i = 0$ or 1. So $\sum_{i=1}^{n} a_i x_i$ takes values $0, 1, 2, 3, \ldots, n$. The numbers $0, 1, 2, 3, \ldots, n$ can be chosen in ${}^nC_0, {}^nC_1, {}^nC_2, \ldots, {}^nC_n$ ways, respectively.

Hence, prob. ($\sum_{i=1}^{n} a_i x_i$ is an odd number)

$= $ Prob. ($\sum_{i=1}^{n} a_i x_i = 1$ or 3 or 5 or ......)

$= $ Prob. ($\sum_{i=1}^{n} a_i x_i = 1$) + Prob. ($\sum_{i=1}^{n} a_i x_i = 3$)

$+ $ Prob. ($\sum_{i=1}^{n} a_i x_i = 5$) + ...

$$= \frac{{}^nC_1}{{}^nC_0 + {}^nC_1 + {}^nC_2 + \ldots + {}^nC_n}$$

$$+ \frac{{}^nC_3}{{}^nC_0 + {}^nC_1 + {}^nC_2 + \ldots + {}^nC_n}$$

$$+ \frac{{}^nC_5}{{}^nC_0 + {}^nC_1 + {}^nC_2 + \ldots + {}^nC_n} + \ldots$$

$$= \frac{{}^nC_1 + {}^nC_3 + {}^nC_5 + \ldots}{{}^nC_0 + {}^nC_1 + {}^nC_2 + \ldots + {}^nC_n} = \frac{2^{n-1}}{2^n} = 0.5$$

2. (0.25) Here, the sample space, $S = \{(0, 0), (0, 1), (1, 0), (1, 1)\}$.

Clearly, each of M and N takes the values 0 and 1.

Hence, X takes the values $\min(0, 0)$, $\min(1, 0)$, $\min(0, 1)$ and $(1,1)$ i.e; 0 and 1.

Now, $P(X = 0) = P(\{(0,0),(0,1),(1,0)\}) = \frac{3}{4}$ and

$$P(X=1) = P(\{(1,1)\}) = \frac{1}{4}.$$

Hence, $E(X)$

$$= 0 \times P(X=0) + 1 \times P(X=1)$$

$$= \frac{1}{4} = 0.25$$

3. (0.3) The p.m.f of X is given by

$$f(x) = \frac{1}{10-(-2)} = \frac{1}{12}, \quad -2 < x < 10$$

Since, $Y = 2X-6$, the p.m.f of Y is given by

$$f(y) = \frac{1}{14-(-10)} = \frac{1}{24}, \quad -10 < y < 14$$

$$(\because -2 < x < 10 \Rightarrow -4 < 2x < 20)$$

$$\Rightarrow -4-6 < 2x-6 < 20-6 \Rightarrow -10 < y < 14)$$

$$\therefore P(Y \le 7 \mid X \ge 5)$$

$$= \frac{P(Y \le 7 \cap X \ge 5)}{P(X \ge 5)} = \frac{P(Y \le 7 \cap Y \ge 4)}{P(X \ge 5)}$$

$$(\because X \ge 5 \Rightarrow 2X - 6 \ge 10 - 6 \Rightarrow Y \ge 4)$$

$$= \frac{P(4 \le Y \le 7)}{P(X \ge 5)} = \frac{\int_4^7 f(y)\,dy}{\int_5^{10} f(x)\,dx} = \frac{\int_4^7 \frac{1}{24}\,dy}{\int_5^{10} \frac{1}{12}\,dx}$$

$$= \frac{\frac{1}{24} \times (7-4)}{\frac{1}{12} \times (10-5)} = 0.3$$

4. (0.09) Required probability = Prob. (the player puts the ball into the basket only in the second attempt)

= Prob. (the player fails to put the ball into the basket in the first attempt) × Prob. (the player puts the ball into the basket in the second attempt)

= (1–0.1) × 0.1 = 0.09

5. (0.196) Here, the total number of trials $(n) = 15$. Prob. (success) = p = prob.(setting heads) = $\frac{1}{2}$

Let X be a random variable denoting the number of heads. Then X follows a binomial distribution.

∴ Required probability

$$= p(X=8) = 15_{C_8} \left(\frac{1}{2}\right)^8 \left(1-\frac{1}{2}\right)^{15-8}$$

$$= \frac{15 \times 14 \times 13 \times 12 \times 11 \times 10 \times 9 \times 8}{8 \times 7 \times 6 \times 5 \times 4 \times 3 \times 2 \times 1} \times \frac{1}{2^{15}}$$

$$= \frac{6435}{32768} \approx 0.196$$

6. (0.603) Required probability

$$= \left[1 - \text{prob.(flood does not occur)}\right]^{25}$$

$$= \left(1 - \frac{1}{50}\right)^{25} \approx 0.603$$

7. (–1) Let X and Y be two exponentially distributed random variables with parameters λ_1 and λ_2, respectively.

Then ATQ, $E(X) = E(Y) = 0.50$

$$\Rightarrow \frac{1}{\lambda_1} = \frac{1}{\lambda_2} = 0.50 \Rightarrow \lambda_1 + \lambda_2 = 2$$

Now $Z = X + Y$

$$\Rightarrow \text{Var}(Z) = \text{Var}(X+Y)$$

$$= \text{Var}(X) + \text{Var}(Y) + 2\,\text{cov}(X,Y)$$

$$\Rightarrow \text{Var}(X) + \text{Var}(Y) + 2\,\text{cov}(X,Y) = 0$$

$$[\because \text{Var}(Z) = 0 \,(\text{given})]$$

$$\Rightarrow \text{cov}(X,Y)$$

$$= -\frac{1}{2}\left[\text{Var}(X) + \text{Var}(Y)\right]$$

$$= -\frac{1}{2}\left[\frac{1}{\lambda_1^2} + \frac{1}{\lambda_2^2}\right] = -\frac{1}{2}\left[\frac{1}{4} + \frac{1}{4}\right] = -\frac{1}{4}$$

$$\therefore \rho(X,Y) = \frac{\text{cov}(X,Y)}{\sigma_X \sigma_Y} = \frac{-\frac{1}{4}}{\sqrt{\frac{1}{4}}\sqrt{\frac{1}{4}}} = -1.$$

8. (a)

9. (0.042) Let A = event that heads will appear exactly four times in the first ten tosses and

B = event that tails will appear exactly four times in the next ten tosses.

Then the required probability

$$= P(A \cap B) = P(A) \times P(B)$$

$$\left[\because A \text{ and } B \text{ are independent}\right]$$

$$= 10_{C_4}\left(\frac{1}{2}\right)^4 \left(1-\frac{1}{2}\right)^{10-4} \times 10_{C_4}\left(\frac{1}{2}\right)^4 \left(1-\frac{1}{2}\right)^{10-4}$$

$$\left[\text{using the binomial distribution}\right]$$

$$= \frac{10 \times 9 \times 8 \times 7}{4 \times 3 \times 2 \times 1} \times \frac{1}{2^{10}} \times \frac{10 \times 9 \times 8 \times 7}{4 \times 3 \times 2 \times 1} \times \frac{1}{2^{10}}$$

$$= \frac{44100}{2^{20}} \approx 0.042$$

TOPIC: LAPLACE TRANSFORMS

1. Consider a linear time invariant system whose input $r(t)$ and output $y(t)$ are related by the following differential equation:

$$\frac{d^2y(t)}{dt^2} + 4y(t) = 6r(t)$$

The poles of this system are at
(a) 2, –2   (b) 4, –4   (c) 2j, –2j   (d) 4j, –4j
[EE GATE 2020]

2. The solution of $\dfrac{d^2y}{dt^2} - y = 1$, which additionally

satisfies $y(t=0) = \dfrac{dy}{dt}(t=0)$ in the Laplace s-

domain is

(a) $\dfrac{1}{s(s+1)(s-1)}$ (b) $\dfrac{1}{s(s-1)}$

(c) $\dfrac{1}{s(s+1)}$ (d) $\dfrac{1}{(s-1)}$

[ME GATE 2020]

Answer key				
1. (c)	**2.** (a)			

Explanation

1. (c) $\dfrac{d^2y(t)}{dt^2} + 4y(t) = 6r(t)$

$\Rightarrow L\left(\dfrac{d^2y(t)}{dt^2}\right) + 4L(y(t)) = 6L(r(t))$

$\Rightarrow s^2Y(s) - sy(0) - y'(0) + 4Y(s) = 6R(s)$

$\left[\text{where } L(y(t)) = Y(s),\ L(r(t)) = R(s)\right]$

$\Rightarrow s^2Y(s) + 4Y(s) = 6R(s)$

(putting $y(0) = y'(0) = 0$)

$\Rightarrow \dfrac{Y(s)}{R(s)} = \dfrac{6}{s^2+4}$

Therefore, the poles are given by

$s^2 + 4 = 0$, *i.e*; $s = 2j, -2j$ (where $j = \sqrt{-1}$)

2. (a) $\dfrac{d^2y}{dt^2} - y = 1$

$\Rightarrow L\left(\dfrac{d^2y}{dt^2}\right) - L(y) = L(1)$

$\Rightarrow L\left(\dfrac{d^2y(t)}{dt^2}\right) - L\big(y(t)\big) = \dfrac{1}{s}$

$\Rightarrow s^2Y(s) - sy(0) - y'(0) - Y(s) = \dfrac{1}{s}$

$\Rightarrow Y(s) = \dfrac{1}{s(s^2-1)} = \dfrac{1}{s(s+1)(s-1)}.$

TOPIC: FOURIER SERIES

1. The Fourier series to represent $x - x^2$ for $-\pi \le x \le \pi$ is given by

$x - x^2 = a_0 + \sum_{n=1}^{\infty}(a_n \cos nx + b_n \sin nx)$. The value of

a_0 is   **[CE GATE 2020]**

Answer key				
1. (–6.58)				

Explanation

1. (–6.58)

$a_0 = \dfrac{1}{\pi}\int_{-\pi}^{\pi}\left(x - x^2\right)dx = \dfrac{1}{\pi}\left[\dfrac{x^2}{2} - \dfrac{x^3}{3}\right]_{-\pi}^{\pi}$

$= \dfrac{1}{\pi}\left[\left(\dfrac{\pi^2}{2} - \dfrac{\pi^3}{3}\right) - \left(\dfrac{\pi^2}{2} + \dfrac{\pi^3}{3}\right)\right]$

$= -\dfrac{2\pi^2}{3} = -6.58$

www.ingramcontent.com/pod-product-compliance
Lightning Source LLC
Chambersburg PA
CBHW080347220326
41598CB00030B/4633